Lecture Notes in Computer Science 8242

Commenced Publication in 1973
Founding and Former Series Editors:
Gerhard Goos, Juris Hartmanis, and Jan va

Stephen Wismath   Alexander Wolff (Eds.)

Graph Drawing

21st International Symposium, GD 2013
Bordeaux, France, September 23-25, 2013
Revised Selected Papers

 Springer

Volume Editors

Stephen Wismath
University of Lethbridge
Department of Mathematics
and Computer Science
4401 University Dr.
Lethbridge, AB T1K-3M4, Canada
E-mail: wismath@uleth.ca

Alexander Wolff
Universität Würzburg
Institut für Informatik
Am Hubland
97074 Würzburg, Germany

ISSN 0302-9743   e-ISSN 1611-3349
ISBN 978-3-319-03840-7   e-ISBN 978-3-319-03841-4
DOI 10.1007/978-3-319-03841-4
Springer Cham Heidelberg New York Dordrecht London

Library of Congress Control Number: 2013954558

CR Subject Classification (1998): G.2, F.2, I.3

LNCS Sublibrary: SL 1 – Theoretical Computer Science and General Issues

Typesetting: Camera-ready by author, data conversion by Scientific Publishing Services, Chennai, India

Printed on acid-free paper

Springer is part of Springer Science+Business Media (www.springer.com)

Preface

This volume contains the papers that were presented at the 21st International Symposium on Graph Drawing, which was held during September 23–25, 2013, in Bordeaux, France. The symposium was hosted by the University of Bordeaux I and was attended by 99 participants from 15 countries. We thank David Auber, the local arrangements chair, and his team for their warm hospitality and their effort to keep the conference fee low. Everyone enjoyed the nice weather, the beauty of Bordeaux, and the excellent cuisine.

As was the case for GD 2012, paper submissions were partitioned into two tracks and there was a separate poster track. Track 1 deals with combinatorial and algorithmic aspects; track 2 with visualization systems and interfaces. There was a record of 110 submissions: 94 papers—79 long (with 12 pages) and 15 short (with 6 pages)—and 16 posters (each with a two-page description). Each submission was reviewed by at least three Program Committee members. The committee decided to accept 42 papers and 12 posters. The acceptance rates were 36/72 in track 1, 6/22 in track 2, and 12/16 for posters. Only one short paper was accepted (in track 1). We thank the Program Committee, the sub-reviewers, the session chairs, and all authors for their hard work.

As a novelty, GD 2013 had *three* invited talks, one from applications, one from theory, and one from industry. Tamara Munzner from the University of British Columbia had a look at graph drawing through the lens of a framework for analyzing visualization methods. Emo Welzl from ETH Zurich showed us how to count crossing-free geometric graphs. And Joe Marks, co-founder and CTO of Upfront Analytics Ltd., analyzed the value of research in an industrial environment. We thank the speakers for their excellent talks, which complemented each other nicely and were very well received.

As another novelty, GD 2013 awarded prizes for the best presentations. The conference participants who stayed until the last talk voted for the winners. Michael Bannister from the University of California at Irvine received the largest number of votes and hence the first prize for his talk on the paper "Superpatterns and Universal Point Sets." Maarten Löffler from Utrecht University received the second prize for his talk "Colored Spanning Graphs for Set Visualization."

Following a well-established tradition, the 20th Annual Graph Drawing contest was held during the conference. It had two main categories: an off-line and an on-line challenge. This year's contest committee was chaired by Carsten Gutwenger (University of Dortmund). We thank the committee for preparing challenging problems and problem instances. A report of the contest is included in these proceedings.

We also wish to thank our sponsors; "diamond" sponsor Microsoft, "gold" sponsor Tom Sawyer Software, "silver" sponsor Vis4, "bronze" sponsor yWorks, and the local sponsors Région Aquitaine, CNRS, LaBRI, Université Bordeaux I, and LACUB. Without their support, the registration fees would have been roughly twice as high.

We thank Philipp Kindermann from the University of Würzburg for his technical support in producing these proceedings.

The 22nd International Symposium on Graph Drawing (GD 2014) will be held in Würzburg, Germany, September 24–26, 2014. Christian Duncan and Antonios Symvonis will be the program committee chairs; Alexander Wolff will be the organizing committee chair.

September 2013   Stephen Wismath
                                                                Alexander Wolff

Organization

Program Committee

Therese Biedl	University of Waterloo, Canada
Ulrik Brandes	Universität Konstanz, Germany
Emilio Di Giacomo	Università degli Studi di Perugia, Italy
Vida Dujmović	University of Ottawa, Canada
William S. Evans	University of British Columbia, Canada
Stefan Felsner	TU Berlin, Germany
Fabrizio Frati	University of Sydney, Australia
Yifan Hu	AT&T Labs Research, USA
Christophe Hurter	École Nationale de l'Aviation Civile, France
Andreas Kerren	Linnéuniversitetet, Sweden
William J. Lenhart	Williams College, USA
Martin Nöllenburg	KIT, Germany
János Pach	EPFL, Switzerland
Helen Purchase	University of Glasgow, UK
Falk Schreiber	Universität Halle, Germany
Andreas Spillner	Universität Greifswald, Germany
Antonios Symvonis	National Technical University of Athens, Greece
Alexandru Telea	Universiteit Groningen, The Netherlands
Csaba D. Tóth	CSUN and University of Calgary, Canada
Frank van Ham	IBM Software Group, The Netherlands
Sue Whitesides	University of Victoria, Canada
Steve Wismath (Co-chair)	University of Lethbridge, Canada
Alexander Wolff (Co-chair)	Universität Würzburg, Germany

Organizing Committee

David Auber (Chair)	LaBRI, Université de Bordeaux, France
Romain Bourqui	LaBRI, Université de Bordeaux, France
Guy Melanon	LaBRI, Université de Bordeaux, France
Bruno Pinaud	LaBRI, Université de Bordeaux, France

Graph Drawing Contest Committee

Christian Duncan	Quinnipiac University, USA
Carsten Gutwenger (Chair)	TU Dortmund, Germany
Lev Nachmanson	Microsoft Research, USA
Georg Sander	IBM Frankfurt, Germany

Additional Reviewers

Aerts, Nieke
Angelini, Patrizio
Asinowski, Andrei
Bachmaier, Christian
Bekos, Michael
Binucci, Carla
Bläsius, Thomas
Bose, Prosenjit
Cornelsen, Sabine
Czauderna, Tobias
Da Lozzo, Giordano
Di Battista, Giuseppe
Didimo, Walter
Dragicevic, Pierre
Dumitrescu, Adrian
Dwyer, Tim
Erten, Cesim
Fulek, Radoslav
Gansner, Emden
Gemsa, Andreas
Grilli, Luca
Hartmann, Anja
Hoffmann, Udo
Hong, Seok-Hee
Huron, Samuel
Joret, Gwenaël
Keszegh, Balázs
Klein, Karsten
Kleist, Linda

Langerman, Stefan
Lazard, Sylvain
Maheshwari, Anil
Mchedlidze, Tamara
Montecchiani, Fabrizio
Morin, Pat
Mustaţă, Irina-Mihaela
Niedermann, Benjamin
Nocaj, Arlind
Ortmann, Mark
Pampel, Barbara
Patrignani, Maurizio
Poon, Sheung-Hung
Pálvölgyi, Dömötör
Radoičić, Radoš
Roselli, Vincenzo
Rutter, Ignaz
Saeedi, Noushin
Schaefer, Marcus
Scheepens, Roeland
Schulz, Hans-Joerg
Sheffer, Adam
Tóth, Géza
Ueckerdt, Torsten
Verbeek, Kevin
Walny, Jagoda
Wood, David R.
Wybrow, Michael
Zimmer, Björn

Sponsors

Diamond Sponsor	Gold Sponsor

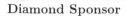

Silver Sponsor	Bronze Sponsor

Other Sponsors

Invited Talks

Graph Drawing through the Lens of a Framework for Analyzing Visualization Methods

Tamara Munzner

University of British Columbia, Department of Computer Science
Vancouver BC, Canada
tmm@cs.ubc.ca, http://www.cs.ubc.ca/~tmm

Abstract. The visualization community has drawn heavily on the algorithmic and systems-building work that has appeared with the graph drawing literature, and in turn has been a fertile source of applications. In the spirit of further promoting the effective transfer of ideas between our two communities, I will discuss a framework for analyzing the design of visualization systems. I will then analyze a range of graph drawing techniques through this lens. In the early stages of a project, this sort of analysis may benefit algorithm developers who seek to identify open problems to attack. In later project stages, it could guide algorithm developers in characterizing how newly developed layout methods connect with the tasks and goals of target users in different application domains.

The Counting of Crossing-Free Geometric Graphs — Algorithms and Combinatorics

Emo Welzl[*]

Institute for Theoretical Computer Science
ETH Zurich, CH-8092 Zurich, Switzerland
emo@inf.ethz.ch

Abstract. We are interested in the understanding of crossing-free geometric graphs–these are graphs with an embedding on a given planar point set where the edges are drawn as straight line segments without crossings. Often we are restricted to certain types of graphs, most prominently triangulations, but also spanning cycles, spanning trees, or (perfect) matchings (and crossing-free partitions), among others. A primary goal is to enumerate, to count, or to sample graphs of a certain type for a given point set–so these are algorithmic questions–, or to give estimates for the maximum and minimum number of such graphs on any set of n points–these are problems in extremal combinatorial geometry.

In this talk we will encounter some of the recent developments (since my GD'06 talk). Among others, I will show some of the new ideas for providing extremal estimates, e.g. for the number of crossing-free spanning cycles: the support-refined estimate for cycles versus triangulations, the use of pseudo-simultaneously flippable edges in triangulations, and the employment of Kasteleyn's beautiful linear algebra method for counting perfect matchings in planar graphs–here, interestingly, in a weighted version. Moreover, Raimund Seidel's recent 2^n-algorithm for counting triangulations is discussed, with its extensions by Manuel Wettstein to other types of graphs (e.g. crossing-free perfect matchings).

Keywords: computational geometry, geometric graphs, counting, sampling, enumeration.

---

[*] Supported by EuroCores/EuroGiga/ComPoSe SNF grant 20GG21_134318/1.

The Value of Research

Joe Marks

Upfront Analytics Ltd.
[1]Marine Terrace, Dun Laoghaire, Co. Dublin, Ireland

Abstract. Which outcome would you prefer for your current research project: a paper with 500 citations, or a business worth $5M? Or maybe it doesn't matter to you as long as you're doing interesting work and having fun. But even if academic and industrial researchers tend to value research differently, deciding what to work on is the first and fundamental step in any inquiry, and so it is worth thinking about the process and criteria that researchers use for project selection.

To get you thinking about this issue, I will present a selection of recent industrial projects with which I was involved as a researcher or manager. In the tradition of reality TV, you the audience will be invited to predict a 10-year citation count, estimate a dollar value, and assess general coolness for each of the projects presented. Active participation is encouraged! Although the focus is on research from the media and entertainment industry, hopefully you will come away with thoughts about how to estimate the value of your own research from both an academic and a commercial perspective. I will end with some speculative potential graph-drawing projects to which we can apply our just-practiced judgment on project selection.

Table of Contents

Practical Graph Drawing

Subgraphs

Crossings

Geometric Graphs and Geographic Networks

Angular Restrictions

Grids

Curves and Routes

Graph Drawing Contest

Posters

On the Upward Planarity of Mixed Plane Graphs[*]

Fabrizio Frati[1], Michael Kaufmann[2], János Pach[3],
Csaba D. Tóth[4], and David R. Wood

[1] School of Information Technologies, The University of Sydney, Australia
brillo@it.usyd.edu.au
[2] Wilhelm-Schickard-Institut für Informatik, Universität Tübingen, Germany
mk@informatik.uni.tuebingen.de
[3] EPFL, Lausanne, Switzerland and Rényi Institute, Budapest, Hungary
pach@cims.nyu.edu
[4] California State University Northridge, USA and University of Calgary, Canada
cdtoth@acm.org
[5] School of Mathematical Sciences, Monash University, Melbourne, Australia
david.wood@monash.edu

Abstract. A *mixed plane graph* is a plane graph whose edge set is partitioned into a set of directed edges and a set of undirected edges. An *orientation* of a mixed plane graph G is an assignment of directions to the undirected edges of G resulting in a directed plane graph $\mathbf{G}$. In this paper, we study the computational complexity of testing whether a given mixed plane graph G is *upward planar*, i.e., whether it admits an orientation resulting in a directed plane graph $\mathbf{G}$ such that $\mathbf{G}$ admits a planar drawing in which each edge is represented by a curve monotonically increasing in the y-direction according to its orientation.

Our contribution is threefold. First, we show that the upward planarity testing problem is solvable in cubic time for *mixed outerplane graphs*. Second, we show that the problem of testing the upward planarity of mixed plane graphs reduces in quadratic time to the problem of testing the upward planarity of *mixed plane triangulations*. Third, we exhibit linear-time testing algorithms for two classes of mixed plane triangulations, namely *mixed plane 3-trees* and mixed plane triangulations in which *the undirected edges induce a forest*.

1 Introduction

Upward planarity is the natural extension of planarity to directed graphs. When visualizing a directed graph, one usually requires an *upward drawing*, that is, a drawing in which the directed edges flow monotonically in the y-direction. A drawing is *upward planar* if it is planar and upward. Testing whether a directed graph G admits an upward planar drawing is NP-hard [9], however, it is polynomial-time solvable if G has a *fixed planar embedding* [2], if it has a *single-source* [3,13], if it is *outerplanar* [15], or if it is

---

[*] Pach was supported by Hungarian Science Foundation EuroGIGA Grant OTKA NN 102029, by Swiss National Science Foundation Grants 200020-144531 and 200021-137574, and by NSF Grant CCF-08-30272. Tóth was supported in part by NSERC (RGPIN 35586) and NSF (CCF-0830734).

S. Wismath and A. Wolff (Eds.): GD 2013, LNCS 8242, pp. 1–12, 2013.

a *series-parallel graph* [7]. *Exponential-time algorithms* [1] and *FPT algorithms* [12] for upward planarity testing are known.

In this paper we deal with *mixed graphs*. A mixed graph is a graph whose edge set is partitioned into a set of directed edges and a set of undirected edges. Mixed graphs unify the expressive power of directed and undirected graphs, as they allow one to simultaneously represent hierarchical and non-hierarchical relationships. A number of problems on mixed graphs have been studied, e.g., *coloring mixed graphs* [11,17] and *orienting mixed graphs to satisfy connectivity requirements* [5,6].

Upward planarity generalizes to mixed graphs as follows. A drawing of a mixed graph is *upward planar* if it is planar, every undirected edge is a y-monotone curve, and every directed edge is an arc with monotonically increasing y-coordinates. Hence, testing the upward planarity of a mixed graph is equivalent to testing whether its undirected edges can be oriented to produce an upward planar directed graph. Since the upward planarity testing problem is NP-hard for directed graphs [9], it is NP-hard for mixed graphs as well. Binucci and Didimo [4] studied the problem of testing the upward planarity of *mixed plane graphs*, that is, of mixed graphs with a given plane embedding. They describe an ILP formulation for the problem and present experiments showing the efficiency of their solution. Different graph drawing questions on mixed graphs (related to crossing and bend minimization) have been studied in [8,10].

We show the following results.

In Section 3 we show that the upward planarity testing problem can be solved in $O(n^3)$ time for n-vertex *mixed outerplane graphs*. Our dynamic programming algorithm uses a characterization for the upward planarity of directed plane graphs due to Bertolazzi et al. [2], and it tests the upward planarity of a mixed outerplane graph G based on the upward planarity of two subgraphs of G.

In Section 4 we show that, for every n-vertex mixed plane graph G, there exists an $O(n^2)$-vertex *mixed plane triangulation G'* such that G is upward planar if and only if G' is upward planar. As a consequence, the problem of testing the upward planarity of mixed plane graphs is polynomial-time solvable (NP-hard) if and only if the problem of testing the upward planarity of mixed plane triangulations is polynomial-time solvable (resp., NP-hard).

In Section 5, motivated by the previous result, we present linear-time algorithms to test the upward planarity of two classes of mixed plane triangulations, namely *mixed plane 3-trees* and mixed plane triangulations in which *the undirected edges induce a forest*. The former algorithm uses dynamic programming, while the latter algorithm uses induction on the number of undirected edges in the mixed plane triangulation.

Because of space limitations, some proofs are omitted or sketched in this extended abstract. Complete proofs are available in the full version of the paper.

2   Preliminaries

A planar drawing of a graph determines a circular ordering of the edges incident to each vertex. Two planar drawings of the same graph are *equivalent* if they determine the same circular orderings around each vertex. A *planar embedding* is an equivalence class of planar drawings. A planar drawing partitions the plane into topologically connected regions, called *faces*. The unbounded face is the *outer face* and the bounded faces are the *internal faces*. An edge of G incident to the outer face (not incident to the outer face) is called *external* (resp., *internal*). Two planar drawings with the same planar embedding have the same faces. However, they could still differ in their outer faces. A *plane embedding* is a planar embedding together with a choice for the outer face. A *plane graph* is a graph with a given plane embedding. An *outerplane graph* is a plane graph whose vertices are all incident to the outer face. A *plane triangulation* is a plane graph whose faces are delimited by 3-cycles. An *outerplane triangulation* is an outerplane graph whose internal faces are delimited by 3-cycles.

A *block* of a graph $G(V, E)$ is a maximal (both in terms of vertices and in terms of edges) 2-connected subgraph of G; in particular, an edge of G whose removal disconnects G is considered as a block of G. In this paper, when talking about the connectivity of mixed graphs or directed graphs, we always refer to the connectivity of their underlying undirected graphs.

A vertex v in a directed graph is a *sink* (*source*) if every edge incident to v is incoming at v (resp., outgoing at v). A vertex v in a directed plane graph is *bimodal* if the incoming edges at v are consecutive in the cyclic ordering of edges incident to v (which implies that the outgoing edges at v are also consecutive). A directed plane graph is *bimodal* if every vertex is bimodal. A vertex v in a 2-connected directed outerplane graph is a *sink-switch* (*source-switch*) if the two external edges incident to v are both incoming (resp., outgoing) at v.

Bertolazzi et al. [2] characterized the directed plane graphs that are upward planar. In this paper, we will use such a characterization when dealing with two specific classes of directed plane graphs, namely directed outerplane triangulations and directed plane triangulations. Thus, we state such a characterization directly for such graph classes.

Theorem 1 ([2]). *A directed outerplane triangulation G is upward planar if and only if it is acyclic, it is bimodal, and the number of sources plus the number of sinks in G equals the number of sink-switches (or source-switches) plus one.*

Theorem 2 ([2]). *A directed plane triangulation G is upward planar if and only if it is acyclic, it is bimodal, and G has exactly one source and one sink that are incident to the outer face of G.*

A mixed plane graph is upward planar if and only if each of its connected components is upward planar. Thus, without loss of generality, we only consider connected mixed plane graphs. In the following lemma, we show that a stronger condition can in fact be assumed for each considered plane graph G, namely that G is *2-connected*.

Lemma 1. *Every n-vertex mixed plane graph G can be augmented with new edges and vertices to a 2-connected mixed plane graph G' with $O(n)$ vertices such that G is upward planar if and only if G' is. If G is outerplane, than G' is also outerplane. Moreover, G' can be constructed from G in $O(n)$ time.*

Proof Sketch: While G has a cutvertex c that is incident to a face f (if G is outerplane, then f is its outer face), we consider two edges (v_1, c) and (u_2, c) that are consecutively incident to c in G and that belong to different blocks of G. We add a vertex w inside f and connect it to v_1 and u_2. The repetition of such an augmentation leads to a 2-connected mixed plane graph G' satisfying the conditions of the lemma. □

3   Upward Planarity Testing for Mixed Outerplane Graphs

This section is devoted to the proof of the following theorem.

Theorem 3. *The upward planarity of an n-vertex mixed outerplane graph can be tested in $O(n^3)$ time.*

Let G be any n-vertex mixed outerplane graph. By Lemma 1, an $O(n)$-vertex 2-connected mixed outerplane graph G^* can be constructed in $O(n)$ time such that G is upward planar if and only if G^* is.

We introduce some notation and terminology. Let u and v be distinct vertices of G^*. We denote by $G^* + (u, v)$ the graph obtained from G^* by adding edge (u, v) if it is not already in G^*, and by $G^* - u$ the graph obtained from G^* by deleting u and its incident edges. Consider an orientation $\boldsymbol{G}^*$ of G^*. A vertex is *sinky* (*sourcey*) in $\boldsymbol{G}^*$ if it is a sink-switch but not a sink (if it is a source-switch but not a source, resp.). A vertex that is neither a sink, a source, sinky, nor sourcey is *ordinary*; that is, v is ordinary if the two external edges incident to v are one incoming at v and one outgoing at v in $\boldsymbol{G}^*$. We say the *status* of a vertex of G^* in $\boldsymbol{G}^*$ is sink, source, sinky, sourcey, or ordinary.

First note that G^* is upward planar if and only if there is an upward planar directed outerplane triangulation T of G^*, that is, if and only if G^* can be augmented to a mixed outerplane triangulation, and the undirected edges of such a triangulation can be oriented in such a way that the resulting directed outerplane triangulation T is upward planar. The approach of our algorithm is to determine if there is such a T using recursion. The algorithm can be easily modified to produce T if it exists.

We observe that a directed outerplane triangulation T is acyclic if and only if every 3-cycle in T is acyclic. One direction is trivial. Conversely, suppose that T contains a directed cycle. Let C be a shortest directed cycle of T. If C is a 3-cycle, then we are done. Otherwise, an edge $(x, y) \notin C$ exists in T between two vertices x and y both in C. Thus, $C + (x, y)$ contains two shorter cycles, one of which is a directed cycle, contradicting the choice of C. Hence, to ensure the acyclicity of a directed outerplane triangulation, it suffices to ensure that its internal faces are acyclic.

A *potential edge* of G^* is a pair of distinct vertices x and y in G^* such that $G^* + (x, y)$ is outerplane, which is equivalent to saying that x and y are incident to a common internal face of G^* (notice that an edge of G^* is a potential edge of G^*). Fix some external edge r of G^*, called the *root edge*. Let $e = \{x, y\}$ be an internal potential edge of G^*. Then $\{x, y\}$ separates G^*, that is, G^* contains two subgraphs G_1^* and G_2^*, such that $G^* = G_1^* \cup G_2^*$ and $V(G_1^* \cap G_2^*) = \{x, y\}$. (Thus, there is no edge between $G_1^* - x - y$ and $G_2^* - x - y$.) W.l.o.g., $r \in E(G_1^*)$. Let $G_e^* := G_2^* + (x, y)$. Observe that G_e^* is a 2-connected mixed outerplane graph with e incident to the outer face. Also, let $e = \{x, y\} \neq r$ be an external potential edge of G^*. Then, we define G_e^* to be the 2-vertex graph containing the single edge (x, y). Further, let $G_r^* := G^*$. For any (internal or external) potential edge $e = \{x, y\}$ of G^* and for an orientation $\overrightarrow{xy}$ of e, let $G_{\overrightarrow{xy}}^*$ be G_e^* with e oriented $\overrightarrow{xy}$. Define a partial order $\prec$ on the potential edges of G^* as follows. For distinct potential edges e and f of G^*, say $e \prec f$ if both end-vertices of f are in G_e^*. Loosely speaking, $e \prec f$ if $G^* + e + f$ is outerplane and e is "between" r and f.

A *potential arc* of G^* is a potential edge that is assigned an orientation preserving its orientation in G^*. So if e is an undirected edge of G^* or a potential edge not in

G^*, then there are two potential arcs associated with e, while if e is a directed edge of G^*, then there is one potential arc associated with e. If a potential arc $\overrightarrow{xy}$ is part of a triangulation T of G^*, then x is a source, sourcey, or ordinary, and y is a sink, sinky, or ordinary in $G^*_{\overrightarrow{xy}}$. We define the *status* of $\overrightarrow{xy}$ in $G^*_{\overrightarrow{xy}}$ as an ordered pair S of $S(x) \in \{\text{source, sourcey, ordinary}\}$ and $S(y) \in \{\text{sink, sinky, ordinary}\}$.

We now define a function $\text{UP}(\overrightarrow{xy}, S)$, that takes as an input a potential arc $\overrightarrow{xy}$ and a status S of $\overrightarrow{xy}$, and has value "true" if and only if there is an upward planar directed outerplane triangulation $T_{\overrightarrow{xy}}$ of $G^*_{\overrightarrow{xy}}$ that respects $S(x)$ and $S(y)$; notice that, if $\overrightarrow{xy}$ is external and does not correspond to r, then $T_{\overrightarrow{xy}}$ is a single edge.

First, the values of $\text{UP}(\overrightarrow{xy}, S)$ can be computed in total $O(n)$ time for all the external potential arcs $\overrightarrow{xy}$ of G^* not corresponding to r and for all statuses of $\overrightarrow{xy}$. Indeed, $\text{UP}(\overrightarrow{xy}, S)$ is true if and only if $S(x) = \text{source}$ and $S(y) = \text{sink}$.

We show below that, for each potential arc $\overrightarrow{xy}$ in G^* that is internal or that is external and corresponds to r, and for each status S of $\overrightarrow{xy}$, the value of $\text{UP}(\overrightarrow{xy}, S)$ can be computed in $O(n)$ time from values associated to potential arcs corresponding to potential edges e with $\{x, y\} \prec e$. Since there are at most $n(n + 1)$ potential arcs and nine statuses for each potential arc, all the values of $\text{UP}(\overrightarrow{xy}, S)$ can be computed in $O(n^3)$ time by dynamic programming in reverse order to a linear extension of $\prec$. Then, there is an upward planar directed outerplane triangulation of G^* if and only if $\text{UP}(\overrightarrow{xy}, S)$ is true for some orientation $\overrightarrow{xy}$ of r and some status S of $\overrightarrow{xy}$.

Let $\overrightarrow{xy}$ be a potential arc that is internal to G^* or that corresponds to r. Let S be a status of $\overrightarrow{xy}$. Suppose that $\text{UP}(\overrightarrow{xy}, S)$ is true. Then, there is an upward planar directed outerplane triangulation $T_{\overrightarrow{xy}}$ of $G^*_{\overrightarrow{xy}}$ that respects $S(x)$ and $S(y)$. Such a triangulation contains a vertex $z \in V(G^*_{xy}) - x - y$ such that (x, y, z) is an internal face of $T_{\overrightarrow{xy}}$. Since $T_{\overrightarrow{xy}}$ has edge (x, y) oriented from x to y, then edges (x, z) and (y, z) cannot be simultaneously incoming at x and outgoing at y, respectively, as otherwise $T_{\overrightarrow{xy}}$ would contain a directed cycle, which is not possible by Theorem 1. Hence, edges (x, z) and (y, z) in $T_{\overrightarrow{xy}}$ are either outgoing at x and incoming at y, or outgoing at x and outgoing at y, or incoming at x and incoming at y, respectively.

Now, for any status S of $\overrightarrow{xy}$ and for a particular vertex $z \in V(G^*_{xy}) - x - y$, we characterize the conditions for which an upward planar directed outerplane triangulation $T_{\overrightarrow{xy}}$ exists that respects $S(x)$ and $S(y)$ and that contains edges (x, z) and (y, z) oriented according to each of the three orientations described above.

Lemma 2. *There is an upward planar directed outerplane triangulation $T_{\overrightarrow{xy}}$ that respects $S(x)$ and $S(y)$, that contains edge (x, z) outgoing at x, and that contains edge (z, y) incoming at y, if and only if $\overrightarrow{xz}$ and $\overrightarrow{zy}$ are potential arcs of G^* and there are statuses S_1 of $\overrightarrow{xz}$ and S_2 of $\overrightarrow{zy}$ such that the following conditions hold: (a) $S_1(x) = S(x) \in \{\text{source, sourcey, ordinary}\}$, (b) $S_2(y) = S(y) \in \{\text{sink, sinky, ordinary}\}$, (c) $S_1(z) \in \{\text{sink, ordinary}\}$, (d) $S_2(z) \in \{\text{source, ordinary}\}$, (e) $S_1(z) = \text{sink or } S_2(z) = \text{source}$, and (f) both $\text{UP}(\overrightarrow{xz}, S_1)$ and $\text{UP}(\overrightarrow{zy}, S_2)$ are true.*

Proof: ($\Longrightarrow$) Let $T_{\overrightarrow{xy}}$ be an upward planar directed outerplane triangulation of $G^*_{\overrightarrow{xy}}$ that respects $S(x)$ and $S(y)$, that contains edge (x, z) outgoing at x, and that contains edge (z, y) incoming at y. Then, $\overrightarrow{xz}$ and $\overrightarrow{zy}$ are potential arcs of G^*. Further, $T_{\overrightarrow{xy}}$ determines upward planar directed outerplane triangulations $T_{\overrightarrow{xz}}$ and $T_{\overrightarrow{zy}}$

respectively of $G^*_{\overrightarrow{xz}}$ and $G^*_{\overrightarrow{zy}}$ (where $T_{\overrightarrow{xz}}$ and $T_{\overrightarrow{zy}}$ are single edges if $\overrightarrow{xz}$ and $\overrightarrow{zy}$ are external, respectively), as well as statuses S_1 and S_2 of $\overrightarrow{xz}$ and $\overrightarrow{zy}$, respectively, such that (f) both $\mathrm{UP}(\overrightarrow{xz}, S_1)$ and $\mathrm{UP}(\overrightarrow{zy}, S_2)$ are true. Since $\overrightarrow{xy}$ and $\overrightarrow{xz}$ are consecutive outgoing arcs at x, we have (a) $S_1(x) = S(x) \in \{$source, sourcey, ordinary$\}$. Similarly, (b) $S_2(y) = S(y) \in \{$sink, sinky, ordinary$\}$. Since $\overrightarrow{xz}$ is incoming at z, we have $S_1(z) \in \{$sink, ordinary, sinky$\}$. However, if $S_1(z) = $ sinky, then z is not bimodal in $T_{\overrightarrow{xy}}$. Thus (c) $S_1(z) \in \{$sink, ordinary$\}$. Similarly, (d) $S_2(z) \in \{$source, ordinary$\}$. Finally, if z is ordinary in both $T_{\overrightarrow{xz}}$ and $T_{\overrightarrow{zy}}$, then z is not bimodal in $T_{\overrightarrow{xy}}$. Thus (e) $S_1(z) = $ sink or $S_2(z) = $ source.

($\Longleftarrow$) Let $T_{\overrightarrow{xz}}$ be an upward planar directed outerplane triangulation of $G^*_{\overrightarrow{xz}}$ respecting S_1 ($T_{\overrightarrow{xz}}$ is a single edge if $\overrightarrow{xz}$ is external). Let $T_{\overrightarrow{zy}}$ be an upward planar directed outerplane triangulation of $G^*_{\overrightarrow{zy}}$ respecting S_2 ($T_{\overrightarrow{zy}}$ is a single edge if $\overrightarrow{zy}$ is external). Such triangulations exist because $\mathrm{UP}(\overrightarrow{xz}, S_1)$ and $\mathrm{UP}(\overrightarrow{zy}, S_2)$ are true. Let $T_{\overrightarrow{xy}}$ be the triangulation of $G^*_{\overrightarrow{xy}}$ determined from $T_{\overrightarrow{xz}}$ and $T_{\overrightarrow{zy}}$ by adding the arc $\overrightarrow{xy}$. Since $T_{\overrightarrow{xz}}$, $T_{\overrightarrow{zy}}$, and (x, y, z) are acyclic, $T_{\overrightarrow{xy}}$ is acyclic. Since x is bimodal in $T_{\overrightarrow{xz}}$, it is bimodal in $T_{\overrightarrow{xy}}$. Similarly, y is bimodal in $T_{\overrightarrow{xy}}$. As described above, the conditions on $S_1(z)$ and $S_2(z)$ imply that z is bimodal in $T_{\overrightarrow{xy}}$. Every other vertex is bimodal in $T_{\overrightarrow{xy}}$ because it is bimodal in $T_{\overrightarrow{xz}}$ or in $T_{\overrightarrow{zy}}$. Hence, $T_{\overrightarrow{xy}}$ is bimodal.

Let s_1, t_1 and w_1 (s_2, t_2 and w_2; s, t and w) be the number of sources, sinks, and sink-switches in $T_{\overrightarrow{xz}}$ (resp., in $T_{\overrightarrow{zy}}$; resp., in $T_{\overrightarrow{xy}}$), respectively. By Theorem 1, $s_i + t_i = w_i + 1$, for $i \in \{1, 2\}$. If z is a sink in $T_{\overrightarrow{xz}}$ and ordinary in $T_{\overrightarrow{zy}}$, then $s = s_1 + s_2$, $t = t_1 + t_2 - 1$ (for z), and $w = w_1 + w_2$. If z is a source in $T_{\overrightarrow{zy}}$ and ordinary in $T_{\overrightarrow{xz}}$, then $s = s_1 + s_2 - 1$ (for z), $t = t_1 + t_2$, and $w = w_1 + w_2$. If z is a sink in $T_{\overrightarrow{xz}}$ and a source in $T_{\overrightarrow{zy}}$, then $s = s_1 + s_2 - 1$ (for z) and $t = t_1 + t_2 - 1$ (for z) and $w = w_1 + w_2 - 1$ (for z). In all three cases, it follows that $s + t = w + 1$.

By Theorem 1, $T_{\overrightarrow{xy}}$ is upward planar. By construction, $T_{\overrightarrow{xy}}$ respects $S(x)$ and $S(y)$ and contains edge (x, z) outgoing at x and edge (z, y) incoming at y. □

Lemma 3. *There is an upward planar directed outerplane triangulation $T_{\overrightarrow{xy}}$ that respects $S(x)$ and $S(y)$ and that contains edges (x, z) and (y, z) incoming at z if and only if $\overrightarrow{xz}$ and $\overrightarrow{yz}$ are potential arcs of G^* and there are statuses S_1 of $\overrightarrow{xz}$ and S_2 of $\overrightarrow{yz}$ such that the following conditions hold: (a) $S_1(x) = S(x) \in \{$source, sourcey, ordinary$\}$, (b) $S(y) \in \{$sinky, ordinary$\}$, (c) $S_2(y) \in \{$source, ordinary$\}$, (d) $S(y) = $ ordinary if and only if $S_2(y) = $ source, (e) $S(y) = $ sinky if and only if $S_2(y) = $ ordinary, (f) $S_1(z) \in \{$sink, sinky, ordinary$\}$, (g) $S_2(z) \in \{$sink, sinky, ordinary$\}$, (h) $S_1(z) \in \{$sink, ordinary$\}$ or $S_2(z) = $ sink, (i) $S_2(z) \in \{$sink, ordinary$\}$ or $S_1(z) = $ sink, and (j) both $\mathrm{UP}(\overrightarrow{xz}, S_1)$ and $\mathrm{UP}(\overrightarrow{yz}, S_2)$ are true.*

Lemma 4. *There is an upward planar directed outerplane triangulation $T_{\overrightarrow{xy}}$ that respects $S(x)$ and $S(y)$ and that contains edges (z, x) and (z, y) outgoing at z if and only if $\overrightarrow{zx}$ and $\overrightarrow{zy}$ are potential arcs of G^* and there are statuses S_1 of $\overrightarrow{zx}$ and S_2 of $\overrightarrow{zy}$ such that the following conditions hold: (a) $S_2(y) = S(y) \in \{$sink, sinky, ordinary$\}$, (b) $S(x) \in \{$sourcey, ordinary$\}$, (c) $S_1(x) \in \{$sink, ordinary$\}$, (d) $S(x) = $ ordinary if and only if $S_1(x) = $ sink, (e) $S(x) = $ sourcey if and only if $S_1(x) = $ ordinary, (f) $S_1(z) \in \{$source, sourcey, ordinary$\}$, (g) $S_2(z) \in \{$source, sourcey, ordinary$\}$, (h)*

$S_1(z) \in \{source, ordinary\}$ or $S_2(z) = source$, (i) $S_2(z) \in \{source, ordinary\}$ or $S_1(z)=source$, and (j) both $UP(\overrightarrow{zx}, S_1)$ and $UP(\overrightarrow{zy}, S_2)$ are true.

For any status S of $\overrightarrow{xy}$ and for a particular vertex $z \in V(G^*_{xy}) - x - y$, it can be checked in $O(1)$ time whether an upward planar directed outerplane triangulation $T_{\overrightarrow{xy}}$ exists that respects $S(x)$ and $S(y)$ and that contains edges (x, z) and (y, z) by checking whether the conditions in at least one of Lemmata 2-4 are satisfied. Further, $UP(\overrightarrow{xy}, S)$ is true if and only if there exists a vertex $z \in V(G^*_{xy}) - x - y$ such that an upward planar directed outerplane triangulation $T_{\overrightarrow{xy}}$ exists that respects $S(x)$ and $S(y)$ and that contains edges (x, z) and (y, z). Thus, we can determine $UP(\overrightarrow{xy}, S)$ in $O(n)$ time since there are less than n possible choices for z.

This completes the proof of Theorem 3. The time complexity analysis can be strengthened as follows. Suppose that every internal face of G^* has at most t vertices. Then each vertex v is incident to less than $t \cdot \deg_{G^*}(v)$ potential edges and the total number of potential arcs is less than $2 \sum_v t \cdot \deg_{G^*}(v) \leq 8tn$. Since each potential arc has nine statuses, and since there are less than t choices for z, the time complexity is $O(t^2 n)$. In particular, if G^* is an outerplane triangulation, then the time complexity is $O(n)$.

4   Reducing Mixed Plane Graphs to Mixed Plane Triangulations

This section is devoted to the proof of the following theorem.

Theorem 4. *Let G be an n-vertex mixed plane graph. There exists an $O(n^2)$-vertex mixed plane triangulation G' such that G is upward planar if and only if G' is. Moreover, G' can be constructed from G in $O(n^2)$ time.*

Proof: By Lemma 1, an $O(n)$-vertex 2-connected mixed plane graph G^* can be constructed in $O(n)$ time such that G is upward planar if and only if G^* is.

We show how to construct a graph G' satisfying the statement of the theorem. In order to construct G', we augment G^* in several steps. At each step, vertices and edges are inserted inside a face f of G^* delimited by a cycle C_f with $n_f \geq 4$ vertices. Such an insertion is done in such a way that one of the faces that is created by the insertion of vertices and edges into f has $n_f - 1$ vertices, while all the other such faces have 3 vertices. The repetition of such an augmentation yields the desired graph G'.

We now describe how to augment G^*. Consider any face f of G^* delimited by a cycle C_f with $n_f \geq 4$ vertices. Let $(u_1, u_2, \ldots, u_{n_f})$ be the clockwise order of the vertices along C_f starting at any vertex. Insert a cycle C'_f inside f with $n_f - 1$ vertices $v_1, v_2, \ldots, v_{n_f - 1}$ in this clockwise order along C'_f. For any $1 \leq i \leq n_f - 1$, insert edges (v_i, u_i) and (v_i, u_{i+1}) inside C_f and outside C'_f; also, insert edge (v_1, u_{n_f}) inside cycle $(u_{n_f}, u_1, v_1, v_{n_f - 1})$. All the edges inserted in f are undirected. See Fig. 1. Denote by G'_f the graph consisting of cycle C_f together with the vertices and edges inserted in f. Observe that the face of G'_f delimited by C'_f has $n_f - 1$ vertices, while all the other faces into which f is split by the insertion of x_f and of its incident edges have 3 vertices.

We show that G^* before the augmentation is upward planar if and only if G^* after the augmentation is upward planar. One implication is trivial, given that G^* before the augmentation is a subgraph of G^* after the augmentation. For the other implication,

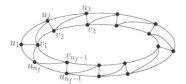

Fig. 1. Augmentation of a face f

it suffices to prove that, for any upward planar orientation C_f of C_f, there exists an upward planar orientation G'_f of G'_f that coincides with C_f when restricted to C_f.

Consider an upward planar drawing Γ_f of C_f with orientation C_f. We shall place the vertices of C'_f inside f in Γ_f, thus obtaining a drawing Γ'_f of G'_f.

Pach and Tóth [14] proved that any planar drawing of a graph G in which all the edges are y-monotone can be triangulated by the insertion of y-monotone edges inside the faces of G (the result in [14] states that the addition of a vertex might be needed to triangulate the outer face of G, which however is not the case if the outer face is bounded by a simple cycle, as in our case). Hence, there exists an index j, with $1 \leq j \leq n_f$, such that a y-monotone curve can be drawn connecting u_{j-1} and u_{j+1} inside f.

If $j < n_f$, then for $1 \leq i \leq j-1$, we place v_i inside f close to u_i, with $y(v_i) \neq y(u_i)$, so that y-monotone curves can be drawn inside f connecting v_i with u_{i-1}, with u_i, and with u_{i+1} (we draw y-monotone curves corresponding to edges of G'_f). Then, we place v_j inside f close to u_{j+1}, with $y(v_j) \neq y(u_{j+1})$, so that y-monotone curves can be drawn inside f connecting v_j with u_{j-1}, with u_j, with u_{j+1}, and with u_{j+2} (we in fact draw y-monotone curves corresponding to edges of G'_f). This is possible, since a y-monotone curve can be drawn inside f connecting v_j and u_j, by construction, and since a y-monotone curve can be drawn inside f connecting u_{j-1} and u_{j+1}, by assumption, hence a y-monotone curve can be drawn inside f connecting v_j and u_{j-1}. Then, for $j+1 \leq i \leq n_f-1$, we place v_i inside f close to u_{i+1}, with $y(v_i) \neq y(u_{i+1})$, so that y-monotone curves can be drawn inside f connecting v_i with u_i, with u_{i+1}, and with u_{i+2} (we in fact draw y-monotone curves corresponding to edges of G'_f). For any $1 \leq i \leq n_f-1$, since y-monotone curves can be drawn inside f connecting v_i with the vertices of C_f to which v_{i-1} and v_{i+1} are close, y-monotone curves can be drawn inside f representing the edges of C'_f (we in fact draw such curves). If $j = n_f$, the drawing is constructed analogously by placing v_i inside f close to u_i, for any $1 \leq j \leq n_f-1$.

The number of vertices of the mixed plane triangulation G' resulting from the augmentation is $O(n^2)$. Namely, the number of vertices inserted inside a face f of G^* with n_f vertices is $(n_f - 1) + (n_f - 2) + \cdots + 3$, hence the number of vertices of G' is $\sum_f (n_f(n_f - 1)/2 - 3) = O(n^2)$, given that $\sum_f n_f \in O(n)$ (where the sums are over all the faces of G^*). Finally, the augmentation of G^* to G' can be easily performed in a time that is linear in the size of G', hence quadratic in the size of the input graph. $\qquad\square$

Corollary 1. *The problem of testing the upward planarity of mixed plane graphs is polynomial-time equivalent to the problem of testing the upward planarity of mixed plane triangulations.*

5   Upward Planarity Testing of Mixed Plane Triangulations

In this section we show how to test in linear time the upward planarity of two classes of mixed plane triangulations.

A *plane 3-tree* is a plane triangulation that can be constructed as follows. Denote by H_{abc} a plane 3-tree whose outer face is delimited by a cycle (a, b, c), with vertices a, b, and c in this clockwise order along the cycle. A cycle (a, b, c) is the only plane 3-tree H_{abc} with three vertices. Any plane 3-tree H_{abc} with $n > 3$ vertices can be constructed from three plane 3-trees H_{abd}, H_{bcd}, and H_{cad} by identifying the vertices incident to their outer faces with the same label. See Fig. 2(a).

Theorem 5. *The upward planarity of an n-vertex mixed plane 3-tree can be tested in $O(n)$ time.*

Consider an n-vertex mixed plane 3-tree H_{uvw}. We define a function $\mathrm{UP}(xy, H_{abc})$ as follows. For each graph H_{abc} in the construction of H_{uvw} and for any distinct $x, y \in \{a, b, c\}$ we have that $\mathrm{UP}(xy, H_{abc})$ is true if and only if there exists an upward planar orientation of H_{abc} in which cycle (a, b, c) has x has a source and y as a sink.

Observe that H_{uvw} is upward planar if and only if $\mathrm{UP}(xy, H_{uvw})$ is true for some $x, y \in \{u, v, w\}$ with $x \neq y$. The necessity comes from the fact that, in any upward planar orientation of H_{uvw}, the cycle delimiting the outer face of H_{uvw} has exactly one source x and one sink y, by Theorem 2. The sufficiency is trivial.

We show how to compute the value of $\mathrm{UP}(xy, H_{abc})$, for each graph H_{abc} in the construction of H_{uvw}.

If $|H_{abc}| = 3$, then let $x, y, z \in \{a, b, c\}$ with $x \neq y$, $x \neq z$, and $y \neq z$. Then, $\mathrm{UP}(xy, H_{abc})$ is true if and only if edges (x, y), (x, z), and (z, y) are not prescribed to be outgoing at y, outgoing at z, and outgoing at y, respectively. Hence, if $|H_{abc}| = 3$ the value of $\mathrm{UP}(xy, H_{abc})$ can be computed in $O(1)$ time.

Second, if $|H_{abc}| > 3$, denote by H_{abd}, H_{bcd}, and H_{cad} the three graphs that compose H. We have the following:

Lemma 5. *For any distinct $x, y, z \in \{a, b, c\}$, $\mathrm{UP}(xy, H_{abc})$ is true if and only if:*
(1) $\mathrm{UP}(xy, H_{xyd})$, $\mathrm{UP}(xd, H_{zxd})$, and $\mathrm{UP}(zy, H_{yzd})$ are all true; or
(2) $\mathrm{UP}(xy, H_{xyd})$, $\mathrm{UP}(xz, H_{zxd})$, and $\mathrm{UP}(dy, H_{yzd})$ are all true.

Proof Sketch: ($\Longrightarrow$) Assume that H_{abc} has an upward planar orientation $\boldsymbol{H}_{abc}$ with x and y as a source and sink in $\{a, b, c\}$, respectively, (let $z \in \{a, b, c\}$ with $z \neq x, y$). Edge (z, d) might be outgoing or incoming at z, as in Figs. 2(b) and 2(c), respectively. In the first case, $\mathrm{UP}(xy, H_{xyd})$, $\mathrm{UP}(zy, H_{yzd})$, and $\mathrm{UP}(xd, H_{zxd})$ are all true, while in the second case $\mathrm{UP}(xy, H_{xyd})$, $\mathrm{UP}(dy, H_{yzd})$, and $\mathrm{UP}(xz, H_{zxd})$ are all true.

($\Longleftarrow$) Consider the case in which $\mathrm{UP}(xy, H_{xyd})$, $\mathrm{UP}(xd, H_{zxd})$, and $\mathrm{UP}(zy, H_{yzd})$ are all true, the other case is analogous. Then, there exist upward planar orientations $\boldsymbol{H}_{xyd}$, $\boldsymbol{H}_{zxd}$, and $\boldsymbol{H}_{yzd}$ of H_{xyd}, H_{zxd}, and H_{yzd} with x and y, with x and d, and with z and y as a source and sink, respectively. Orientations $\boldsymbol{H}_{xyd}$, $\boldsymbol{H}_{zxd}$, and $\boldsymbol{H}_{yzd}$ together yield an orientation $\mathrm{UP}(xy, H_{xyz})$ of $\boldsymbol{H}_{xyz}$, which is upward planar by Theorem 2. □

For each graph H_{abc} in the construction of H_{uvw} and for any distinct $x, y \in \{a, b, c\}$, the conditions in Lemma 5 can be computed in $O(1)$ time by dynamic programming. Thus, the running time of the algorithm is $O(n)$. This concludes the proof of Theorem 5.

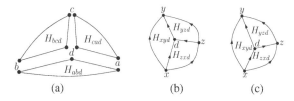

Fig. 2. (a) Construction of a plane 3-tree H_{abc} with $n > 3$ vertices. (b)-(c) Distinct orientations of edge (z, d) in two upward planar orientations of H_{abc}.

We now deal with mixed plane triangulations with no cycle of undirected edges.

Theorem 6. *The upward planarity of an n-vertex mixed plane triangulation in which the undirected edges induce a forest can be tested in $O(n)$ time.*

Proof: Let G be an n-vertex mixed plane triangulation. Let F be the set of undirected edges of G. We assume that F contains no external edge of G. Indeed, F contains at most two external edges: We can guess the orientation of all the external edges in F, and for each of the four possibilities, independently, test the upward planarity for the mixed graph G in which only the internal edges of F are undirected.

We prove the statement by induction on the size of F.

If $|F| = 0$, then G is a directed plane triangulation and its upward planarity can be tested in linear time by checking whether G satisfies the conditions in Theorem 2.

If $|F| > 0$, consider a leaf v in the forest whose edge set is F. Denote by (v, w) the only undirected edge of G incident to v. By the assumptions, (v, w) is an internal edge of G. Let (v, w, x_1) and (v, w, x_2) be the internal faces of G incident to edge (v, w).

Suppose that both edges (x_1, v) and (x_2, v) are incoming at v. If v has an outgoing incident edge, then by the bimodality condition in Theorem 2, edge (v, w) is incoming at v in every upward planar orientation of G. Suppose that v has no outgoing incident edge. If v is the sink of G (recall that the edges incident to the outer face of G are directed), then edge (v, w) is incoming at v in every upward planar orientation of G, by the single sink condition in Theorem 2. Otherwise, edge (v, w) is outgoing at v in every upward planar orientation of G, again by the single sink condition in Theorem 2.

Analogously, if both (x_1, v) and (x_2, v) are outgoing at v, the orientation of edge (v, w) can be decided without loss of generality.

Assume that (x_1, v) and (x_2, v) are incoming and outgoing at v, respectively, the case in which they are outgoing and incoming at v is analogous. We have two cases.

Case 1: (x_1, x_2) is an edge of G. By the acyclicity condition in Theorem 2, edge (x_1, x_2) is outgoing at x_1 in every upward planar orientation of G.

If $\deg(v) = 3$, then remove v and its incident edges from G, obtaining a mixed plane triangulation G' with one fewer undirected edge than G. Inductively test whether G' admits an upward planar orientation. If not, then G does not admit any upward planar orientation, either. If G' admits an upward planar orientation $\mathbf{G'}$, then construct an upward drawing Γ' of $\mathbf{G'}$; insert v in Γ' inside cycle (w, x_1, x_2), so that $y(v) > y(x_1)$, $y(v) < y(x_2)$, and $y(v) \neq y(w)$. Draw y-monotone curves connecting v with each of w, x_1, and x_2. The resulting drawing Γ of G is an upward planar orientation $\mathbf{G}$ of G, provided that it coincides with G' when restricted to G', the edges (x_1, v) and (x_2, v) are drawn as y-monotone curves according to their orientations, and the edge (v, w) is drawn as a y-monotone curve.

If $\deg(v) > 3$, then the cycle (w, x_1, x_2) does not delimit a face of G, and it contains non-empty sets V' and V'' of vertices in its interior and its exterior, respectively. Then, two upward planarity tests can be performed, namely one for the subgraph G' of G induced by $V' \cup \{w, x_1, x_2\}$, and one for the subgraph G'' of G induced by $V'' \cup \{w, x_1, x_2\}$. If one of the tests fails, then G admits no upward planar orientation. Otherwise, upward planar orientations $\boldsymbol{G'}$ of G' and $\boldsymbol{G''}$ of G'' together form an upward planar orientation $\boldsymbol{G}$ of G, provided that each edge of (w, x_1, x_2) has the same orientation in $\boldsymbol{G'}$ and in $\boldsymbol{G''}$.

Case 2: (x_1, x_2) is not an edge of G. Remove (v, w) from G and insert a directed edge (x_1, x_2) outgoing at x_1 inside face (x_1, v, x_2, w). This results in a graph G' with one fewer undirected edge than G. We show that G is upward planar iff G' is.

Suppose that G admits an upward planar orientation $\boldsymbol{G}$. Let Γ be an upward planar drawing of $\boldsymbol{G}$. Remove edge (v, w) from G in Γ. Draw edge (x_1, x_2) inside cycle $C_f = (x_1, v, x_2, w)$, thus ensuring the planarity of the resulting drawing Γ' of G', following closely the drawing of path (x_1, v, x_2), thus ensuring the upwardness of Γ'.

Suppose that G' admits an upward planar orientation $\boldsymbol{G'}$. Let Γ' be an upward planar drawing of $\boldsymbol{G'}$. Remove (x_1, x_2) from Γ'. Since $\boldsymbol{G'}$ is acyclic, C_f has three possible orientations in $\boldsymbol{G'}$. In Orientation 1, w is its source and x_2 its sink; in Orientation 2, x_1 is its source and w its sink; finally, in Orientation 3, x_1 is its source and x_2 its sink. If C_f is oriented in $\boldsymbol{G'}$ as in Orientation 1 (as in Orientation 2), then draw edge (v, w) inside C_f in Γ', thus ensuring the planarity of the resulting drawing Γ of G, following closely the drawing of path (w, x_1, v) (resp., of path (v, x_2, w)), thus ensuring the upwardness of Γ. If C_f is oriented in $\boldsymbol{G'}$ as in Orientation 3, slightly perturb the position of the vertices in Γ' so that $y(v) \neq y(w)$. Draw edge (v, w) in Γ' as follows. Suppose that $y(v) < y(w)$, the other case being analogous. Draw a line segment inside C_f starting at v and slightly increasing in the y-direction, until reaching path (x_1, w, x_2). Then, follow such a path to reach w. This results in an upward drawing of edge (v, w) inside C_f, hence in an upward planar drawing of G.

Finally, the running time of the described algorithm is clearly $O(n)$.   □

6 Conclusions

We considered the problem of testing the upward planarity of mixed plane graphs. We proved that the upward planarity testing problem is $O(n^3)$-time solvable for mixed outerplane graphs. It would be interesting to investigate whether our techniques can be strengthened to deal with larger classes of mixed plane graphs, e.g. series-parallel plane graphs. Also, since testing upward planarity is a polynomial-time solvable problem for directed outerplanar graphs [15], it might be polynomial-time solvable for mixed outerplanar graphs without a prescribed plane embedding as well.

We proved that the upward planarity testing problem for mixed plane graphs is polynomial-time equivalent to the upward planarity testing problem for mixed plane triangulations (and showed two classes of mixed plane triangulations for which the problem can be solved efficiently). This, together with the characterization of the upward planarity of directed plane triangulations in terms of acyclicity, bimodality, and uniqueness of the sources and sinks (see [2] and Theorem 2), might indicate that a polynomial-time algorithm for testing the upward planarity of mixed plane triangulations should be pursued. On the other hand, Patrignani [16] proved that testing the existence of an acyclic and bimodal orientation for a mixed plane graph is NP-hard.

Acknowledgments. Thanks to Hooman Reisi Dehkordi, Peter Eades, Graham Farr, Seok-Hee Hong, and Brendan McKay for useful discussions on the problems considered in this paper.

References

1. Bertolazzi, P., Di Battista, G., Didimo, W.: Quasi-upward planarity. Algorithmica 32(3), 474–506 (2002)
2. Bertolazzi, P., Di Battista, G., Liotta, G., Mannino, C.: Upward drawings of triconnected digraphs. Algorithmica 12(6), 476–497 (1994)
3. Bertolazzi, P., Di Battista, G., Mannino, C., Tamassia, R.: Optimal upward planarity testing of single-source digraphs. SIAM J. Comput. 27(1), 132–169 (1998)
4. Binucci, C., Didimo, W.: Upward planarity testing of embedded mixed graphs. In: Speckmann, B. (ed.) GD 2011. LNCS, vol. 7034, pp. 427–432. Springer, Heidelberg (2011)
5. Boesch, F., Tindell, R.: Robbins's theorem for mixed multigraphs. Amer. Math. Monthly 87(9), 716–719 (1980)
6. Chung, F., Garey, M., Tarjan, R.: Strongly connected orientations of mixed multigraphs. Networks 15(4), 477–484 (1985)
7. Didimo, W., Giordano, F., Liotta, G.: Upward spirality and upward planarity testing. SIAM J. Discrete Math. 23(4), 1842–1899 (2009)
8. Eiglsperger, M., Kaufmann, M., Eppinger, F.: An approach for mixed upward planarization. J. Graph Algorithms Appl. 7(2), 203–220 (2003)
9. Garg, A., Tamassia, R.: On the computational complexity of upward and rectilinear planarity testing. SIAM J. Comput. 31(2), 601–625 (2001)
10. Gutwenger, C., Jünger, M., Klein, K., Kupke, J., Leipert, S., Mutzel, P.: A new approach for visualizing uml class diagrams. In: Diehl, S., Stasko, J., Spencer, S. (eds.) SOFTVIS, pp. 179–188. ACM (2003)
11. Hansen, P., Kuplinsky, J., de Werra, D.: Mixed graph colorings. Math. Meth. Oper. Res. 45(1), 145–160 (1997)
12. Healy, P., Lynch, K.: Two fixed-parameter tractable algorithms for testing upward planarity. Int. J. Found. Comput. Sci. 17(5), 1095–1114 (2006)
13. Hutton, M.D., Lubiw, A.: Upward planarity testing of single-source acyclic digraphs. SIAM J. Comput. 25(2), 291–311 (1996)
14. Pach, J., Tóth, G.: Monotone drawings of planar graphs. J. Graph Theory 46(1), 39–47 (2004)
15. Papakostas, A.: Upward planarity testing of outerplanar dags. In: Tamassia, R., Tollis, I.G. (eds.) GD 1994. LNCS, vol. 894, pp. 298–306. Springer, Heidelberg (1995)
16. Patrignani, M.: Finding bimodal and acyclic orientations of mixed planar graphs is NP-complete. Tech. Rep. RT-DIA-188-2011, Dept. Comp. Sci. Autom., Roma Tre Univ. (2011)
17. Sotskov, J.N., Tanaev, V.S.: Chromatic polynomial of a mixed graph. Vestsi Akademii Navuk BSSR. Ser. Fiz. Mat. Navuk 6, 20–23 (1976)

Upward Planarity Testing:
A Computational Study

Markus Chimani[1] and Robert Zeranski[2]

[1] Theoretical Computer Science, Osnabrück University, Germany
[2] Algorithm Engineering, Friedrich-Schiller-University Jena, Germany
markus.chimani@uni-osnabrueck.de, robert.zeranski@uni-jena.de

Abstract. A directed acyclic graph (DAG) is *upward planar* if it can be drawn without any crossings while all edges—when following them in their direction—are drawn with strictly monotonously increasing y-coordinates. Testing whether a graph allows such a drawing is known to be NP-complete, but there is a substantial collection of different algorithmic approaches known in literature.

In this paper, we give an overview of the known algorithms, ranging from combinatorial FPT and branch-and-bound algorithms to ILP and SAT formulations. Most approaches of the first class have only been considered from the theoretical point of view, but have never been implemented. For the first time, we give an extensive experimental comparison between virtually all known approaches to the problem.

Furthermore, we present a new SAT formulation based on a recent theoretical result by Fulek et al. [8], which turns out to perform best among all known algorithms.

1 Introduction

When drawing directed graphs, in particular DAGs, one often wants to make the edges' orientations clearly recognizable by having all edges pointing in the same general direction, w.l.o.g. upward. A *y-monotone* drawing is thus one, where the curves representing the edges have strictly monotonously increasing y-coordinates when traversing them from their source to the target vertices. More formally, a y-monotone edge intersects any horizontal line at most once, while its source vertex is drawn below its target vertex.

A second central concept in graph drawing is planarity, i.e., we want to avoid crossing edges if possible. The question of *upward planarity* of a DAG G is hence the question whether there exists a crossing-free y-monotone drawing of G. While (undirected) planarity is linear time solvable, upward planarity turns out to be NP-complete to decide [9]. Nonetheless, due to the problem's centrality, several exponential-time algorithms have been developed.

A core result is that the problem becomes polynomial time solvable if the graph's embedding (i.e., the order of the edges around their vertices) is fixed [1]. Based thereon, the historically first algorithms are fixed-parameter tractable (FPT) algorithms where the parameters are essentially bounding the (in general

S. Wismath and A. Wolff (Eds.): GD 2013, LNCS 8242, pp. 13–24, 2013.

exponential) number of possible embeddings; the algorithms are testing upward planarity for each possible embedding [10]. The process can be sped up using a polynomial time algorithm to solve the important special case of series-parallel graphs [7]. The special case of single-source DAGs is also polynomial time solvable [1,5,11,13] but not the focus of this paper.

A different approach is based on relaxing the upward requirement (*quasi-upward planarity* [2], see below) and considering the optimization problem to minimize the violating edges. There, the embedding enumeration is coupled with a sophisticated method to obtain upper and lower bounds for partially embedded graphs, allowing for a branch-and-bound algorithm. Finally, the most recent approach [3] is to formulate the problem as an integer linear program (ILP) or boolean satisfiability problem (SAT), to be solved with a corresponding solver.

In Section 2, we summarize the core ideas of these algorithms. We also present a *new* SAT formulation based on a recent theoretical result by Fulek et al. [8].

Before this paper, the only reported implementations were for the branch-and-bound and the ILP/SAT approach. The former implementation is in fact considering the more general optimization problem (instead of the decision version), and both implementations are based on two different underlying libraries. This made a direct comparison worrisome. In this paper (Section 3), we report on our consistent implementations of all the discussed algorithms. They share as much code as was feasibly possible, to maximize the fairness of the comparison. Hence, we are for the first time able to make substantiated claims about the algorithms' respective applicability in practice.

2   Algorithms

We always consider a DAG $G = (V, E)$ to test for upward planarity. A *combinatorial* embedding of G is specified by cyclically ordering the edges around their incident vertices. A *planar* embedding additionally chooses an external face.

2.1   FPT Algorithms

A fixed-parameter tractable (FPT) algorithm, with respect to some parameter k, is an algorithm with running time $\mathcal{O}(f(k) \cdot poly(n))$ where $poly(n)$ is a polynomial function in the size of the input (here, $n := |G|$), and $f(k)$ is any computable function (typically an exponential function) only dependent on k. A central ingredient of all known combinatorial FPT algorithms is the following result [1]:

Theorem 1 (Bertalozzi, Di Battista, Liotta, Mannino). *Let $G = (V, E)$ be an* embedded *DAG. There is an algorithm testing whether this embedding of G allows an upward planar drawing in $\mathcal{O}(n^2)$ time.*

Let G be planar and biconnected. We can decompose the underlying undirected graph into its triconnectivity components in linear time. These can be efficiently organized as an *SPQR-tree* [6]. For notational simplicity, we may talk about an *SPR-tree* (as Q-nodes, representing single edges, are not necessary):

The SPR-tree $\mathcal{T}(G)$ is a tree with three kinds of *nodes*: S- and P-nodes represent serial and parallel components, respectively; R-nodes represent planar triconnected components. We call these components the *skeletons* associated to the nodes. An edge in a skeleton S may be real or *virtual*; in the latter case, it represents a subgraph, described by the subtree attached to S's node in $\mathcal{T}(G)$.

Embedding Enumeration (EE). The natural approach to test upward planarity of an unembedded graph is to test every possible embedding of G, using the algorithm of Theorem 1. As the number of embeddings is, in general, exponential in the size of the graph, one has to seek for a meaningful parameter to bound the number of embeddings. The SPR-tree can be used to efficiently enumerate all possible embeddings of a graph. We can bound the number of embeddings by $\mathcal{O}(t! \cdot 2^t)$, where t is the number of nodes in our SPR-tree, which leads to an overall running time of $\mathcal{O}(t! \cdot 2^t \cdot n^2)$ to test upward planarity [10].

In the same publication, a kernelization algorithm for sparse (not necessarily biconnected) graphs is presented. Using the following preprocessing steps (until none is applicable anymore), leaves a graph with at most $30k^2 + 2k$ vertices and at most $(2k + 1)!$ embeddings; thereby, $k = |E| - |V|$ is the number of edges (minus 1) the (preprocessed) graph has more than a tree. Let a *chain* be a path in G where all inner vertices have degree 2, The preprocessing steps are: (PP1) remove vertices of degree 1; (PP2) replace chains where each inner vertex v has $indeg(v) = outdeg(v) = 1$ by single correspondingly oriented edges; (PP3) remove chains where both end vertices coincide; (PP4) for each set of *parallel* chains, remove all but one chain (parallel chains are those that have a common start vertex, a common end vertex, and an identical sequence of edge orientations along the chain). This preprocessing requires $\mathcal{O}(n^2)$ time.

After the preprocessing steps, again, all embeddings are tested in overall $\mathcal{O}(n^2 + k^4 \cdot (2k + 1)!)$ time. Observe that (PP1)–(PP4) are valid in general (although one has to specifically consider the case of biconnectivity-breaking PP4). Hence, when testing all embeddings after preprocessing, we in fact obtain an algorithm—denoted by EE in the following—with running time $\mathcal{O}(\min(t! \cdot 2^t \cdot n^2, n^2 + k^4 \cdot (2k + 1)!))$ for biconnected DAGs.

Upward Spirality (SPIR). Consider the SPR-tree rooted at some arbitrary node. Informally, *upward spirality* is a measure of how much a skeleton is "rolled up" around its poles (the end nodes of the virtual edge representing the node's parent). Furthermore, one has to distinguish several *pole categories*, i.e., local properties of the embedding around the pole vertices. For details of the definitions and the following algorithms cf. [7].

For series-parallel graphs, upward spirality allows to develop a polynomial dynamic programming algorithm that traverses the SPR-tree bottom up—recall that for such graphs it only has S- and P-nodes. At each node μ, we store a set of feasible (spirality/pole category)-pairs to upward embed the graph encoded by the SPR-tree rooted at μ. This information can then be used to obtain a corresponding set for the parent node, etc. We denote this algorithm by SPIR-sp. Its running time is $\mathcal{O}(n^4)$, but there are large constants hidden in the $\mathcal{O}$-notation, cf. Section 3.1.

Using this algorithm as a building block, one can establish an FPT algorithm for general DAGs, where the different configurations of the R-nodes w.r.t. each other have to be enumerated. This leads to a running time of $\mathcal{O}(d^r n^3 + dr^2 n + d^2 n^2)$ where d is the largest diameter of any skeleton and r is the number of R-nodes. We call this algorithm SPIR.

2.2 Branch-and-Bound via Quasi-Upward Planarity (BB)

In a *quasi-upward planar* drawing, we relax upwardness such that each edge only has to be drawn y-monotonously within an arbitrary small neighborhood of its incident vertices. In [2], a branch-and-bound algorithm is established which produces a quasi-upward drawing maximizing the number of fully y-monotone edges: For a given embedding, the minimum number of non-y-monotone edges can be computed in $\mathcal{O}(n^2 \log n)$ time using minimum cost-flow techniques.

Now, we can consider all possible embeddings of the graph via the SPR-tree. At any moment we have a *partial* embedding—several embeddings are fixed while the others are free. The algorithm in [2] is able to compute upper and lower bounds for the number of non-y-monotone edges in this case. If the current lower bound is worse than the global upper bound, we can avoid testing all embeddings further down in the search tree, which have the same fixed embeddings in common.

Observe that a DAG is upward planar iff there is a quasi-upward planar drawing where all edges are y-monotone. Hence, when testing upward planarity, we can prune a partial embedding in the search tree whenever we obtain a lower bound strictly greater than 0. We denote this algorithm by BB. Formally, this algorithms could also be considered an FPT algorithm with the same worst-case running time as EE.

2.3 SAT Formulations

A SAT formulation of a decision problem instance $\mathcal{I}$ is a propositional logic formula that is satisfiable if and only if $\mathcal{I}$ has the answer *true*. The formula is typically given in conjunctive normal form, i.e., as a set of clauses, each of which has to be satisfied. Each clause is a disjunction of (possibly negated) variables. For the sake of readability, we will provide *rules* as propositional formulae; it is straight-forward to transform them into their corresponding clauses.

Ordered Embeddings (OE). An edge e *dominates* an edge f if there is a directed path (possibly of length 0) from e's target to f's source vertex in G. Clearly, f has to be drawn above e in any upward drawing. A pair of edges is *non-dominating* if neither edge dominates the other. Let $\mathcal{N}$ denote the set of all non-dominating edge pairs of G.

In [3], a SAT formulation based on *ordered embeddings* has been proposed. We consider a strict total (vertical) order of the vertices together with a strict partial (horizontal) order of the edges; edges are comparable w.r.t. this order iff they are non-dominating each other. We model the vertical order by introducing boolean variables $\tau(v, w)$ for each proper pair of vertices v and w. Intuitively, $\tau(v, w) = \texttt{true}$ means that v is drawn *below* w. Since the vertex order is to be strict, $\tau(v, w) = \texttt{false}$ means that v is *above* w. We may use the shorthand $\tau(w, v) := \neg \tau(v, w)$ for notational simplicity. To establish a strict total order, it then suffices to require transitivity via (R_τ^t). The *upward rules* (R^u) ensure that all edges are drawn upward.

$$\tau(u, v) \wedge \tau(v, w) \to \tau(u, w) \qquad \forall \text{ pairwise distinct } u, v, w \in V. \qquad (R_\tau^t)$$

$$\tau(v, w) = \texttt{true} \qquad \forall \, (v, w) \in E. \qquad (R^u)$$

Similarly, we can establish the horizontal order of the edges by introducing variables $\sigma(e, f)$ for each pair $\{e, f\} \in \mathcal{N}$. Thereby (if both edges are vertically overlapping), $\sigma(e, f) = \texttt{true}$ implies that e is to the *left* of f, and the satisfied shorthand $\sigma(f, e) := \neg \sigma(e, f)$ implies that e is to the *right* of f. Again, we simply require transitivity:

$$\sigma(e, f) \wedge \sigma(f, g) \to \sigma(e, g) \qquad \forall \, \{e, f\}, \{f, g\}, \{e, g\} \in \mathcal{N}. \qquad (R_\sigma^t)$$

Based on this (upward) order system, we can establish planarity using surprisingly simple *planarity rules*: We only have to ensure that two adjacent edges e and f are on the same side of g (non-incident to the common vertex of e and f) if they both vertically overlap with g. Let $e \cap f$ denote the common vertex of two edges e, f, and $\mathcal{P} := \{(e, f, g) \mid \{e, g\}, \{f, g\} \in \mathcal{N} \wedge e \cap f \notin g\}$ the set of edge-triplets as described above. We have:

$$\Big(\tau(x, e \cap f) \wedge \tau(e \cap f, y) \Big) \to \Big(\sigma(e, g) \leftrightarrow \sigma(f, g) \Big) \quad \forall \, (e, f, g = (x, y)) \in \mathcal{P}. \quad (R^p)$$

The collection of the above rules allows a satisfying truth assignment to τ and σ if and only if G is upward planar [3]. Given such an assignment, it is trivial to construct the embedding in linear time. We denote this formulation by OE.

Hanani-Tutte type characterization (FPSS). The classical Hanani-Tutte theorem shows that a graph drawn such that all pairs of non-adjacent edges cross an even number of times is planar. A similar result has been established by Pach and Tóth [12], and was only recently improved upon by Fulek et al. [8]:

Theorem 2 (Pach, Tóth; Fulek, Pelsmajer, Schaefer, Štefankovič). *Let $G = (V, E)$ be a DAG. If G has a y-monotone[1] drawing such that every pair of non-adjacent edges crosses an even number of times, then there is a y-monotone planar embedding of G with the same location of vertices.*

To prove this theorem, [8] gives a quadratic time algorithm testing whether G allows a y-monotone drawing with prespecified vertex positions. This is essentially done by solving an equation system over (e, v)-*moves*. Such a move redraws the edge e by deforming it, close to the y-coordinate of v, into a horizontal "spike" that passes around v. There is a y-monotone embedding iff there

---

[1] In the original publications, these theorems were stated in terms of x-monotonicity.

is a set of (e, v)-moves turning a y-monotone drawing into a drawing in which every non-adjacent pair of edges crosses an even number of times. A simple set of only two equation classes suffices to describe all possible selections of (e, v)-moves that may lead to an even-crossing solution. These equations, in fact, resemble a 2SAT formulation (each clause has at most 2 literals), with the only prerequisite to know the vertical relationship between the DAG's vertices. This allows us to cast this powerful theoretical tool into a new SAT formulation not unlike OE:

We reuse the above boolean variables τ and rules $(R_\tau^t) \cup (R^u)$ to guarantee an upward strict total order on V. We can then use these variables as indicators to activate or deactivate the above 2SAT-clauses to link move-variables. For every edge e and node v, we introduce a boolean variable $\varrho(e, v)$ that indicates whether we perform an (e, v)-move. Let s_e and t_e denote the start and target vertex of an edge e, respectively. We obtain

$$\Big(\tau(s_e, s_f) \wedge \tau(s_f, t_e) \wedge \tau(t_e, t_f)\Big) \rightarrow \Big(\varrho(e, s_f) \leftrightarrow \neg\varrho(f, t_e)\Big) \quad \forall\, e, f \in E \quad (R_0^m)$$

$$\Big(\tau(s_e, s_f) \wedge \tau(t_f, t_e)\Big) \rightarrow \Big(\varrho(e, s_f) \leftrightarrow \varrho(e, t_f)\Big) \quad \forall\, e, f \in E \quad (R_1^m)$$

where the subformulae after the implication are the clauses from the 2SAT suggested by [8], for the scenario described by the rules' left-hand sides.

Corollary 1. *Let $G = (V, E)$ be a DAG. G is upward planar if and only if the formula composed of rules $(R_\tau^t) \cup (R^u) \cup (R_0^m) \cup (R_1^m)$ is satisfiable.*

Hybrid formulations. Constructing an upward planar embedding (in polynomial time) using only a feasible truth assignment for FPSS is non-trivial, and in fact an open problem.[2] In order to obtain an embedding from the FPSS formulation, we consider two variants of hybridizing FPSS and OE, as extracting an embedding is trivial for the latter. Recall that both formulations use the same τ variables to establish a vertical order of the vertices. Hence we can simply put all rules together in one large formulation, denoted by HF.

Alternatively (denoted by HL), we can first compute a satisfying assignment to FPSS. If it exists, we can "learn" the subsolution for the τ variables to fix the τ variables in OE and solve the so-restricted OE to obtain a matching σ assignment.

3   Experiments

3.1   Considered Algorithms and Their Implementations

Overall, we consider the practical performances of the following algorithms:

FPT	b&b	SAT	ILP (see note below)
EE, SPIR-sp, (SPIR)	BB	OE, FPSS, HF, HL	iOE, iFPSS, iHF, iHL

---

[2] An algorithm achieving this is currently under development [personal communication with Marcus Schaefer]

All experiments were performed on an Intel Xeon E5520, 2.27 GHz, 8 GB RAM running Debian 6. We implemented the algorithms as part of the Open Graph Drawing Framework (www.ogdf.net), using *minisat* as our SAT solver and CPLEX 12 as our ILP solver. For both solvers, we used their default settings. For all considered instances (available at www.cs.uos.de/theoinf), we first performed the preprocessing steps (PP1)–(PP4), as described in Section 2.1.

A note on the ILPs. Given the SAT formulations, it is straight-forward to construct integer linear programs (using only binary variables) along the same lines, for which to test feasibility. This has been done for OE in [3], leading to a vastly weaker practical performance than the SAT approach. We note that FPSS (and the hybridizations) also allow such ILPs. We performed all the below experiments also for the ILP variants. However, also for iFPSS and the hybridizations, the pure feasibility testing functionality of ILP solvers—lacking sophisticated propagation and backtracking—is clearly too weak to allow compatible performance. We will hence disregard the ILP approaches in the following discussions.

A note on BB. The branch-and-bound algorithm has only been devised for biconnected graphs. Our implementation hence inherits this restriction. As described above, we strengthened BB for the special case of upward planarity testing by pruning any subproblems with a lower bound larger than 0.

In [2], the practical performance of an implementation (within GDToolkit) has been reported on a set of random instances, denoted by *BDD* in the following. Out of the originally 300 considered instances, only 200 are still available[3]. Of those, we can discard 139, as they are not DAGs. The largest remaining instances have 100 vertices and take 0.03 seconds on average (cf. Table 2). In [2]—on a clearly slower PC and not restricted to the pure upward planarity test!—the graphs with 100 vertices require roughly 100 seconds on average. This gives us a hint that our implementation is not vastly inefficient.

A note on EE. As BB is already restricted to biconnected graphs, we also restricted our implementation of EE to this case, to be able to efficiently generate all embeddings purely via the SPR-tree. Our implementation is able to enumerate ≈100,000 (bimodal) embeddings per second (disregarding any testing time).

A note on SPIR-sp and SPIR. The spirality-based algorithms are very theoretically demanding algorithms. We provide the seemingly first implementation of the polynomial case for series-parallel graphs (SPIR-sp). However, the algorithm's theoretical beauty is unfortunately not matched with practicability. E.g., when combining tables in the bottom-up dynamic programming, there is a very large number of possible combinations of choices to check. Although this number is bounded by a constant, we often need to check close to the theoretical worst-case of 2^{49} ($\approx 5.6 \times 10^{14}$) combinations. This constitutes the main bottleneck of the algorithm; further theoretical research may be able to bring down this vast number. In our implementation, we parallelized this checking via OpenMP to mitigate the effect. However, it remains to slow down the algorithm dramatically.—We will see experimental evidence in Figure 3 below.

---

[3] Personal communication with Walter Didimo.

Table 1. # of solved instances within given time frames in seconds (*Rome* and *North*)

time (s)	Rome					North			
	[0, 0.01]	(0.01, 0.1]	(0.1, 1]	(1, 10]	> 10	[0, 0.01]	(0.01, 0.1]	(0.1, 1]	> 1
FPSS	1922	993	46	4	2	710	120	6	0
OE	1474	1223	257	11	2	625	172	39	0
HL	1384	1324	253	4	2	614	185	37	0
HF	1393	1311	254	6	3	609	187	40	0

The series-parallel algorithm SPIR-sp, as a base case for SPIR, can essentially be used as a lower bound for the running time of the latter, when applying it on the SPR-subtrees induced by only S- and P-nodes. This resembles the situation that SPIR could decide the R-nodes and all their embedding combinations (i.e., the reason for the exponential running time) in no time. Since obtaining these bounds is already not competitive enough, we refrained from a full implementation of the even more demanding enumeration procedures within SPIR.

3.2 Evaluation

In the course of our investigation, we will observe the following central findings throughout all benchmark instances, summarized here:

F1 FPSS gives the best solution time over virtually all scenarios. When extraction of the embedding is required, HL dominates OE; both dominate HF.
F2 FPSS (and to some extent OE) are rather independent of the number of embeddings and on whether we consider a yes- or a no-instance.
F3 Generally, all SAT approaches are preferable over their competitors.
F4 SPIR (and SPIR-sp) and the ILP variants are not competitive.
F5 EE and BB work well when the number of embeddings is small. For trivial instances, and for large instances with few embeddings, EE and BB can dominate the SAT approaches.
F6 EE is usually far weaker for no-instances than for yes-instances (as all embeddings have to be checked), while the effective pruning of BB allows it to typically solve the former faster than the latter.

Instances from Literature. There are mainly three instance sets that have been used in the context of upward planarity:

The *Rome* graphs [14] are (originally undirected) graphs with up to 100 vertices. They can be directed canonically to obtain DAGs. The *North* DAGs [4] were originally collected by AT&T and Stephen North. After filtering for bimodal planar graphs, 2967 and 836 remain, respectively. As some of our algorithms are restricted to biconnected graphs, we also consider sets *Rome2*, *North2* that are generated from the above by planar biconnectivity augmentation, obtaining (after filtering for bimodal planar graphs) 2671 and 780 instances, respectively. Furthermore, we consider the 61 *BDD* instances [2], described above in the context of BB's implementation.

We observe in Table 1 that all SAT formulations solve the *North* instances very fast: each is solved within 1 second, most of them (≈80%) in less than 0.01

Table 2. Number of solved instances (yes- and no-instances) within given time frames in seconds (*Rome2*, *North2*, and *BDD*). In brackets, are the number of no-instances.

time (s)	Rome2			North2				BDD			
	[0,0.1]	(0.1,1]	> 1	[0,0.1]	(0.1,1]	(1,10]	> 10	[0,0.1]	(0.1,1]	(1,10]	> 10
FPSS	2474 (26)	195 (6)	2 (2)	711 (33)	68 (7)	1 (0)	0 (0)	27 (3)	23 (1)	11 (2)	1 (1)
OE	1743 (12)	915 (17)	13 (5)	600 (24)	149 (13)	31 (3)	0 (0)	24 (3)	25 (1)	12 (2)	1 (1)
HL	1844 (26)	820 (6)	7 (2)	610 (33)	143 (7)	27 (0)	0 (0)	24 (3)	26 (1)	11 (2)	1 (1)
HF	1743 (11)	923 (19)	14 (5)	605 (28)	143 (9)	32 (3)	0 (0)	21 (3)	27 (1)	13 (2)	1 (1)
EE	2665 (29)	5 (4)	1 (1)	753 (27)	5 (1)	6 (4)	16 (8)	62 (7)	0 (0)	0 (0)	0 (0)
BB	2635 (22)	22 (9)	4 (3)	667 (14)	30 (11)	13 (1)	70 (11)	58 (4)	4 (3)	0 (0)	0 (0)

seconds. Generally, FPSS is the fastest SAT, solving 99% of *Rome* in under 0.1 seconds, whereas OE achieves a ratio of 90% ($\rightarrow$**F1**).

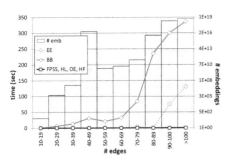

Fig. 1. *North2*; avg. runtime vs. # of edges. The different SATs are visually indistinguishable.

On the biconnected benchmark sets, we can compare the SAT approaches to the other implementations (Table 2): For *Rome2* and *BDD*, both EE and BB seem faster than any SAT formulation. This is mainly due to the triviality of many instances and the necessary overhead of SAT formulations and solvers. Already when considering ≤ 1 second computation time, all approaches solved nearly the same number of instances. We observe that instances of both these sets have only very few embeddings (*Rome2*: max. 36,864, avg. 372), *BDD*: max. 11,528, avg. 512) ($\rightarrow$**F5**). As an interesting side note, the lone instance in the last *BDD* column requires roughly 200 seconds for all SAT approaches; it is solved trivially by BB and EE as it has only two planar bimodal embeddings to check. The *North2* instances, however, have many embeddings (avg: 10^{17}), cf. Fig. 1. There, all SAT formulations dominate EE and BB for the non-trivial instances ($\rightarrow$**F1,F5**).

Constructed Instances. The above instances are very simple, small, and are solved too quickly to deduce general findings. Therefore, we consider a set *Rand* of generated biconnected instances with $n = 50, 100, 150, 200$ nodes and density $|E|/|V| = 1.2, 1.4, \ldots, 2.4$. As suggested in [2], we start with a triangle graph and iteratively perform random edge-subdivisions and face-splits (adding an edge within a face). Now, we orient this embedded graph to be upward planar, and invert $i = 1\%, 2\%, 3\%, 4\%$ of the edges (retaining the DAG property). We generate 1120 instances, 10 per possible parameter setting, which have up to 2.5×10^{15} embeddings (3.7×10^{12} on average). See Fig. 2 for a detailed graphical analysis. We use a timeout of 600 seconds, using this value for unsolved instances when averaging. Overall (Fig. 2, top-left), we can observe FPSS < HL < OE < HF < BB < EE for the running times, FPSS being the clear winner ($\rightarrow$**F1,F3,F5,F6**). The

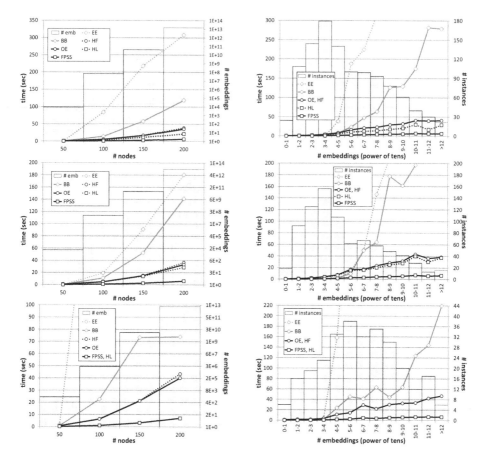

Fig. 2. *Rand* instances; avg. runtime vs. # of nodes (left) or embeddings (right). The latter is given as powers of 10, i.e., $> 10^{12}$ embeddings are considered. We group SAT formulations if they are visually indistinguishable. The first row considers all instances; the second and third row considers only the yes- and no-instances, respectively.

SATs, in particular FPSS, seem nearly oblivious to the number of embeddings in the graph (Fig. 2, top-right, →**F2**). It is instructive to consider the yes- and no-instances independently: We see that EE performs somewhat reasonable for yes-instances, but fails for no-instances (where it has to check all bimodal embeddings). In contrast to this, BB becomes even faster for the latter instances, due to its efficient pruning of large classes of "hopeless" subembeddings (→**F6**). The SAT approaches behave very similar for both kinds of instances (→**F2**).

Now, we consider biconnected series-parallel graphs SP to evaluate SPIR-sp. We generated a test set of 4500 instances (10 instances per parameter setting) with $m = 20, 40, \ldots, 300$ edges. They are constructed bottom up with probability $p = 0.1, 0.3, \ldots, 0.9$ to perform a serialization instead of a parallelization. We embed the graph, choose an upward planar orientation for the edges, and invert

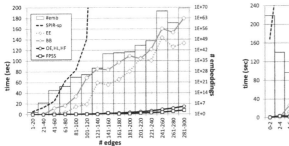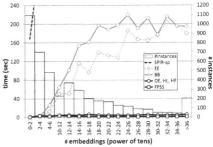

Fig. 3. *SP* instances; avg. runtime vs. # of edges (left) and embeddings (right)

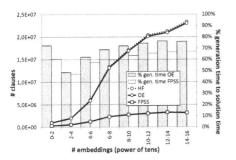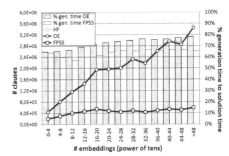

Fig. 4. *Rand* (left) and *SP* (right); the line plots show the number of generated clauses, relative to the number of embeddings (as a power of 10). The bars show the percentage of the running time spent to generate the formula (in contrast to solving it).

$i = 0, 10, 20, 30, 40, 50\%$ of the edges (retaining the DAG property). Again, we use a timeout of 600 seconds, using this value for unsolved instances when averaging. SPIR-sp—although formally the only polynomial time algorithm in this comparison—offers the clearly weakest performance, solving no instance with over 100 vertices in under 5 minutes and running into the time limit for all graphs with more than 120 edges (Fig. 3, left). The picture is analogous (Fig. 3, right) when looking at the runtime depending on the number of embeddings (→**F4**). On *SP*, EE performs better than BB, but this is since nearly 90% of the instances happen to be upward planar. Considering yes- and no-instances independently, we can observe the same pattern as for *Rand* (→**F6**).

Details on SAT. Although FPSS and OE have the same number of clauses in $\mathcal{O}$-notation—dominated by (R_τ^t) over the common τ variables—the former has considerably fewer additional clauses. In fact, this seems to be one of the main reasons of FPSS's superior performance. To back-up this assumption, consider the line diagrams in Fig. 4. They show that FPSS is rather independent of the number of embeddings (→**F2**). Impressively, the (minor) difference between OE and HF (= "OE ∪FPSS") shows that the number of clauses FPSS has to consider additionally to (R_τ^t) is negligible.

The bar diagrams in Fig. 4 show that the SATs spend a large portion of their time ($\approx 70\%$) with (trivially) *constructing* the formula. This explains the overhead for trivial instances where EE and BB can be faster than SAT ($\rightarrow$**F5**).

Acknowledgements. We thank Marcus Schaefer for pointing us to [8] and its potential applicability within our SAT approach, Walter Didimo for helpful discussions on BB, and Fabrice Stellmacher and Kerstin Gössner for implementation support with BB and SPIR, respectively.

References

1. Bertolazzi, P., Di Battista, G., Liotta, G., Mannino, C.: Upward drawings of tri-connected digraphs. Algorithmica 12(6), 476–497 (1994)
2. Bertolazzi, P., Di Battista, G., Didimo, W.: Quasi-upward planarity. Algorithmica 32(3), 474–506 (2002)
3. Chimani, M., Zeranski, R.: Upward planarity testing via SAT. In: Didimo, W., Patrignani, M. (eds.) GD 2012. LNCS, vol. 7704, pp. 248–259. Springer, Heidelberg (2013)
4. Di Battista, G., Garg, A., Liotta, G., Parise, A., Tamassia, R., Tassinari, E., Vargiu, F., Vismara, L.: Drawing directed acyclic graphs: An experimental study. Int. J. of Computational Geometry and Appl. 10(6), 623–648 (2000)
5. Di Battista, G., Liu, W.P., Rival, I.: Bipartite graphs, upward drawings, and planarity. Information Processing Letters 36(6), 317–322 (1990)
6. Di Battista, G., Tamassia, R.: On-line planarity testing. SIAM J. on Computing 25, 956–997 (1996)
7. Didimo, W., Giordano, F., Liotta, G.: Upward spirality and upward planarity testing. SIAM J. on Discrete Mathematics 23(4), 1842–1899 (2009)
8. Fulek, R., Pelsmajer, M.J., Schaefer, M., Štefankovič, D.: Hanani–Tutte, monotone drawings, and level-planarity. Thirty Essays on Geom. Graph Th., 263–287 (2013)
9. Garg, A., Tamassia, R.: On the computational complexity of upward and rectilinear planarity testing. SIAM J. on Computing 31(2), 601–625 (2002)
10. Healy, P., Lynch, K.: Fixed-parameter tractable algorithms for testing upward planarity. In: Vojtáš, P., Bieliková, M., Charron-Bost, B., Sýkora, O. (eds.) SOFSEM 2005. LNCS, vol. 3381, pp. 199–208. Springer, Heidelberg (2005)
11. Hutton, M.D., Lubiw, A.: Upward planar drawing of single source acyclic digraphs. In: Proc. SODA 1991, pp. 203–211 (1991)
12. Pach, J., Toth, G.: Monotone drawings of planar graphs. J. of Graph Theory 46(1), 39–47 (2004)
13. Papakostas, A.: Upward planarity testing of outerplanar dags. In: Tamassia, R., Tollis, I.G. (eds.) GD 1994. LNCS, vol. 894, pp. 298–306. Springer, Heidelberg (1995)
14. Welzl, E., Di Battista, G., Garg, A., Liotta, G., Tamassia, R., Tassinari, E., Vargiu, F.: An experimental comparison of four graph drawing algorithms. Computational Geometry 7, 303–325 (1997)

Characterizing Planarity by the Splittable Deque

Christopher Auer, Franz J. Brandenburg,
Andreas Gleißner, and Kathrin Hanauer

University of Passau, Germany
{auerc,brandenb,gleissner,hanauer}@fim.uni-passau.de

Abstract. A *graph layout* describes the processing of a graph G by a data structure $\mathcal{D}$, and the graph is called a $\mathcal{D}$-graph. The vertices of G are totally ordered in a *linear layout* and the edges are stored and organized in $\mathcal{D}$. At each vertex, all edges to predecessors in the linear layout are removed and all edges to successors are inserted. There are intriguing relationships between well-known data structures and classes of planar graphs: The stack graphs are the outerplanar graphs [4], the queue graphs are the arched leveled-planar graphs [12], the 2-stack graphs are the subgraphs of planar graphs with a Hamilton cycle [4], and the deque graphs are the subgraphs of planar graphs with a Hamilton path [2]. All of these are proper subclasses of the planar graphs, even for maximal planar graphs.

We introduce *splittable deques* as a data structure to capture planarity. A splittable deque is a deque which can be split into sub-deques. The splittable deque provides a new insight into planarity testing by a game on switching trains. Here, we use it for a linear-time planarity test of a given rotation system.

1 Introduction

In a graph layout, the vertices are processed according to a total order, which is called *linear layout*. The edges correspond to data items that are inserted to and removed from a data structure: Each edge is inserted at the end vertex that occurs first according to the linear layout and is removed at its other end vertex. These operations obey the principles of the underlying data structure, such as "last-in, first-out" for a stack or "first-in, first-out" for a queue. Stack layouts (also known as book embeddings) and queue layouts have been studied extensively, e.g., in [4,5,7,8,10–12,16,18], and are used for 3D drawings of graphs [16], in VLSI design [5] and in other application scenarios [12]. Moreover, Gauss codes and permutation networks of two parallel stacks are characterized by two-stack graphs [14].

Graph layouts are a powerful tool to study planar graphs. A graph G is a $\mathcal{D}$-graph if it has a layout in $\mathcal{D}$. The stack graphs are the outerplanar graphs, and the 2-stack graphs are the subgraphs of planar graphs with a Hamiltonian cycle [4]. Heath et al. [8,12] have characterized queue graphs as the arched leveled-planar graphs. Such graphs have a planar drawing with vertices placed

S. Wismath and A. Wolff (Eds.): GD 2013, LNCS 8242, pp. 25–36, 2013.

on levels. Inter-level edges connect vertices between two adjacent levels and intra-level edges (the *arches*) connect the left-most vertex to accessible vertices on the right side. In [2], we have characterized the proper leveled-planar graphs, i.e., the arched leveled-planar graphs without arches, as the bipartite queue graphs. Graph layouts can be extended to subdivisions where edges of the graph are replaced by paths. A graph is planar if and only if it has a subdivision that has a layout in two stacks [8].

In [1, 2], we have studied double-ended queue (*deque*) layouts: A deque has two ends, a head and a tail, and items can be inserted and removed at both sides. It can emulate two stacks and additionally allows for queue edges, i.e., edges inserted and removed at opposite sides. In [2], we have shown that the surplus power of a deque in comparison to two stacks captures the difference between Hamiltonian paths and cycles: A graph is a deque (2-stack) graph if and only if it is the subgraph of a planar graph with a Hamiltonian path (cycle). In fact, a planar embedding of a graph with a Hamiltonian path reflects the way the edges are processed in the deque: Fig. 1(c) shows an embedded graph with the Hamiltonian path $1, 2, 3, 4, 11, 12, 13$, which is the linear layout. An edge to the right of the path, e.g., edges e_6 and e_9, is inserted and removed at the tail of the deque whereas the queue edges change sides, e.g., e_3 and e_4 are inserted at the tail and removed at the head. Although more powerful than two stacks, not all planar graphs are deque graphs since there are maximal planar graphs with no Hamiltonian path and the respective decision problem is $\mathcal{NP}$-hard [2]. Yannakakis has shown that four stacks are sufficient and necessary for all planar graphs [17]. However, there are non-planar graphs that have a layout in four and even three stacks. This raises the following question: What is the additional operation a deque must perform to layout exactly the planar graphs? It turns out that the ability to split the deque into pieces is the adequate operation. We show that a graph is planar if and only if it is a *splittable deque (SD)* graph.

Our proof takes an algorithmic viewpoint: We give a linear-time algorithm to test whether or not a rotation system is planar, which uses the SD to process all edges. A rotation system defines the counterclockwise order of edges around each vertex and it is planar if it admits a plane drawing of the graph. In a nutshell, the algorithm is a depth-first search (DFS) which tries to process all edges in the SD according to the rotation system. Planarity follows if this is possible. Otherwise, an edge that cannot be processed, e.g., removed from the deque, causes a crossing. The algorithm is a means to an end for our characterization of planarity and there are other algorithms especially designed for solving the same problem [6]. Nevertheless, our algorithm has the benefit that it operates on an elementary data structure, i.e., a deque, which is very simple in comparison to other ones used for general planarity tests [15]. Note that any two-cell embedding on a surface of genus k can be defined by a rotation system. In particular, it is not sufficient to only test locally at each face whether the rotation system causes crossings. Another challenge are crossings between edges incident to the same vertex, which are ignored in the general case. As the SD exploits the structure of a graph's DFS tree, our characterization is related to the characterization of planarity by de Fraysseix et al. [9].

The SD also provides a playful characterization of planarity: At GD 2012, we presented the poster "Testing Planarity by Switching Trains" [3]. There, the edges are modeled as cars which have to be appended at the head and tail of a train which models the deque. The vertices are train stations which are the sources and destinations of the cars and, at junctions, the train can be split. We also implemented a Java game[1], which uses the time-reversed variant of the SD, i. e., the mergeable deque. The player is asked to switch the cars such that all can be removed at their destination station. The graph underlying a game level is obtained from a GraphML file. If it is possible to bring all cars to their destination without an error, then the underlying graph is planar.

2 Preliminaries

We consider simple, undirected, and connected graphs $G = (V, E)$ with vertices V and edges E such that $|V| \geq 2$. A graph $G = (V, E)$ is planar if it has a plane drawing which maps the vertices to distinct points in the plane and edges $\{u, v\}$ to Jordan arcs from u to v such that Jordan arcs do not cross except at common end points. A *rotation system* $\mathcal{R}_v$ defines a cyclic order of edges around each vertex v. From a plane drawing, we obtain a *planar rotation system* which is the counterclockwise ordering of the edges around each vertex in the drawing. Given a rotation system $\mathcal{R}_v$, each edge e has a successor edge $\mathrm{Succ}_v(e)$ and a predecessor edge $\mathrm{Pre}_v(e)$ at vertex v.

A *DFS tree* $\mathcal{T} = (V, E_\mathcal{T})$ is a rooted, directed spanning tree of G obtained from a DFS traversal starting at a root vertex r. We assume that the *tree edges* $E_\mathcal{T}$ are directed from the parent to its children. We denote by $u \to v$ that $(u, v) \in E_\mathcal{T}$. By $u \overset{+}{\to} v$, we denote a path of tree edges (at least one) from u to v. Vertex u is an *ancestor* of v and v is a *descendant* of u. By $u \overset{*}{\to} v$, we denote $u = v$ or $u \overset{+}{\to} v$. $\mathcal{T}$ partitions E into tree edges $E_\mathcal{T}$ and *forward edges* F. For each forward edge $\{u, v\} \in F$, there is a path $u \overset{+}{\to} v$ where u is an ancestor of v.

A *linear layout* if a total ordering $\prec$ of the vertices. If $u \prec v$, then u is called *predecessor* of v and v *successor* of u. In a graph layout, a vertex v can be seen as a processing unit that receives as *input* a data structure from which v's edges to predecessors are removed and edges to successors are inserted. Insertions and removals obey the modus operandi of the data structure. The resulting data structure is the *output* of the vertex and the input to the immediate successor. The input to the first and output of the last vertex is empty.

A *deque* is a doubly linked list whose content is denoted by $\mathcal{C} = (e_1, \ldots, e_k)$, where e_1 is at the *head* $\mathbf{h}$ and e_k is at the *tail* $\mathbf{t}$. The empty deque is denoted by (). We denote by $e_i \in \mathcal{C}$ that e_i is in $\mathcal{C}$ and by $e_i \ll_\mathcal{C} e_j$ that $i < j$ in $\mathcal{C}$. We omit the subscript in $\ll_\mathcal{C}$ if it is clear from the context which configuration is meant.

---

[1] http://www.infosun.fim.uni-passau.de/br/games/derail.jar

An edge $e \in \mathcal{C}$ can be removed if it is situated at the head or the tail of $\mathcal{C}$. In the following, we denote by $\mathcal{C}_v$ the input of vertex v.

Let $G = (V, E)$ be a graph endowed with a rotation system and assume that G contains Hamiltonian path $p = (v_1, \ldots, v_n)$. Path p is a (degenerate) DFS tree of G and it induces a linear layout with $v_i \prec v_j$ if and only if $i < j$ for all $1 \leq i, j \leq n$. The edges on p are directed from each vertex to its immediate successor. A rotation system of G defines the order in which the edges are inserted to and removed from a deque and Algorithm 1 shows how: It takes as input a vertex v and its rotation system $\mathcal{R}_v$ along with a deque $\mathcal{C}_v$. e_p and e_s are the edges to the immediate predecessor and successor of v on p, respectively. The value $\perp$ indicates that v is the first or last vertex. Let v_i with $1 < i < n$ be an inner vertex of p. Its rotation system is sketched in Fig. 1(a). In Algorithm 1, all edges $e_1^h, \ldots, e_k^h$ between e_p and e_s are inserted and removed at the head in counterclockwise order of the rotation system (lines 1–4). An edge is removed if it points to a predecessor and it is inserted if it points to a successor. We say that these edges are to the *left* of p at v. Then, at v, all edges $e_1^t, \ldots, e_l^t$ to the *right* of p between e_p and e_s are processed at the deque's tail in reversed, i.e., clockwise, order of the rotation system (lines 5–10). Whereas an edge can always be inserted, removing might not be possible if the edge is not accessible at the corresponding side of the deque. In this case, Algorithm 1 aborts and returns $\perp$. At the endpoints v_1 and v_n of p, the rotation system can be divided at an arbitrary position. In this case, Algorithm 1 processes all edges at the head. Note that the edges of the Hamiltonian path p are not processed in the deque. The reason is that these edges can always be processed canonically without interfering with any other edges: First of all, the edge to the immediate predecessor is removed at the head and, last of all, the edge to the immediate successor is inserted at the head. Hence, we can safely ignore all edges on p. We say that a rotation system *admits* a deque layout if subsequently calling ProcessDeque for all vertices $v_1, \ldots, v_n$ in order never returns $\perp$ and the input to v_1 is the empty deque and so is the output of v_n. From [2], we obtain the following proposition.

Proposition 1. *The rotation system of a graph with a Hamiltonian path is planar if and only if it admits a deque layout.*

For an example, consider Fig. 1(c). The input of vertex 3 is the deque with content (e_4, e_3) and edge e_6 is inserted to the tail of the deque, which results in (e_4, e_3, e_6). Consider edge e' which crosses e_6. This is also reflected in the deque layout: At vertex 4, e_4 is removed at the deque's head and e' inserted to the tail resulting in (e_3, e_6, e'). At vertex 11, e_6 must be removed at the tail which is not possible since e' is in its way.

3   The Splittable Deque

The SD enhances the deque by allowing it to be split. This is the dual to the mergeable deque of Kosaraju [13]. In order to define SD layouts, we generalize

---

Algorithm 1. ProcessDeque

Input: vertex v with rotation system $\mathcal{R}_v = (e_1, \ldots, e_k)$, deque $\mathcal{C}_v$, edges e_p (e_s)
to immediate predecessor (successor); or $\perp$ if v is first (last)

Output: new content of the deque or $\perp$ if deque layout is not possible

1 $e \leftarrow \mathrm{Succ}_v(e_p)$ if $e_p \neq \perp$, else $\mathrm{Succ}_v(e_s)$
2 **while** $e \neq e_s \wedge e \neq e_p$ **do**
3   **if** e *is edge to successor* **then** $\mathcal{C}_v$.insertAtHead(e)
4   **else** $\mathcal{C}_v$.removeAtHead(e) or return $\perp$ if not possible $e \leftarrow \mathrm{Succ}_v(e)$

5 **if** $e_s \neq \perp \wedge e_p \neq \perp$ **then** v is an inner vertex
6      $e \leftarrow \mathrm{Pre}_v(e_p)$
7   **while** $e \neq e_s$ **do**
8   **if** e *is edge to successor* **then** $\mathcal{C}_v$.insertAtTail(e)
9   **else** $\mathcal{C}_v$.removeAtTail(e) or return $\perp$ if not possible
10         $e \leftarrow \mathrm{Pre}_v(e)$

11 **return** $\mathcal{C}_v$

---

linear layouts to tree layouts. A *tree layout* is an ordered DFS tree $\mathcal{T} = (V, E_\mathcal{T})$ of a graph G, i.e., a DFS tree in which the children of each inner vertex are totally ordered from left to right. Remember that the linear layout of a deque layout describes the processing pipeline, i.e., if u is the immediate predecessor of v, then the output of u is the input of v. With a tree layout $\mathcal{T}$, a vertex v can have multiple immediate successors, namely, its children $w_1, w_2, \ldots, w_l$ in order. For a graph G, let $\mathcal{T} = (V, E_\mathcal{T})$ be a tree layout with root r. The input $\mathcal{C}_r = ()$ of root r is empty. Let $\mathcal{C}_v = (e_1, \ldots, e_k)$ be the input of vertex $v \in V$ and let $w_1, \ldots, w_l$ be the children of v in $\mathcal{T}$ in order. Assume for now that v has at least one child. At first, $\mathcal{C}_v$ is split into $l \leq 1$ consecutive and disjoint pieces $c_{w_1}, \ldots, c_{w_l}$, where $\mathcal{C}_v$ is the concatenation of $c_{w_1}, \ldots, c_{w_l}$. These l pieces constitute l new SDs. Each forward edge from an ancestor of v in $\mathcal{T}$ has to be removed and each forward edge to a descendant has to be inserted at one of these SDs. The SD obtained from c_{w_i} after all removals and insertions is the input $\mathcal{C}_{w_i}$ of child w_i. If v is a leaf, its output must be empty. If $\mathcal{T}$ is a path, the SD is never split and behaves like a deque. A graph is an *SD graph* if it has a tree layout such that all edges can be processed in the SD.

For an example, consider the graph in Fig. 1(b) whose rotation system can be obtained from the drawing. The vertices are numbered according to a DFS run starting at root 1. The red, dashed edges are ignored for the moment. All tree edges are directed from parent to children and drawn bold. In Fig. 1(d), the tree layout as defined by the DFS tree is displayed where the children are ordered from left to right according to the rotation system. In fact, the rotation system as shown in Fig. 1(d) is equal to the one obtained from Fig. 1(b). In the example, the input SD of vertex 4 is $\mathcal{C}_4 = (e_5, e_2, e_1, e_4, e_3, e_6)$. At vertex 4, the SD is split into three pieces $c_5 = (e_5, e_2)$, $c_{10} = (e_1)$, and $c_{11} = (e_4, e_3, e_6)$. Afterwards, forward edge e_4 is removed at the head of c_{11}. We obtain $\mathcal{C}_5 = (e_5, e_2)$, $\mathcal{C}_{10} = (e_1)$, and $\mathcal{C}_{11} = (e_3, e_6)$ as inputs to vertices 5, 10, and 11, respectively. In principle, $\mathcal{C}_4$ can

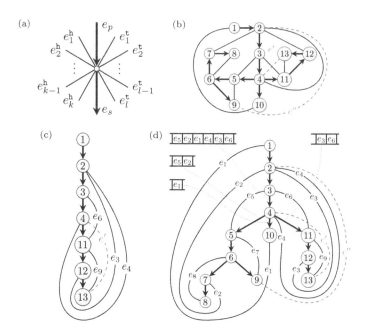

Fig. 1. (a) shows how a path splits the rotation system of a vertex and at which side of the deque the edges are processed. A planar graph with its DFS tree (directed and bold drawn edges) is shown in (b) and its tree layout in (d) along with the input SD for some vertices. (c) shows a path from the root to a leaf in the DFS tree, which corresponds to a deque layout.

also be split such that $c_{10} = (e_1, e_4)$ and $c_{11} = (e_3, e_6)$ and then e_4 is removed at the tail of c_{10}. This ambiguity occurs at vertices with more than one child and does not influence the property of a graph of being an SD graph.

4   Testing Planarity of a Rotation System by the SD

In this section, we prove our main result.

Theorem 1. *A graph G is an SD graph if and only if G is planar.*

To prove this theorem, we use Algorithm 2 which uses the SD to test whether a given rotation system is planar. By showing its correctness, Theorem 1 follows. Algorithm 2 defines the recursive routine `PlanarRS` which takes as input a vertex v endowed with a rotation system, a tree layout $\mathcal{T}$, the input SD $\mathcal{C}_v$ for v, and the tree edge e_a from its parent p in $\mathcal{T}$ or $\perp$ if v is the root. The tree layout $\mathcal{T}$ is obtained from a prior DFS run, where the children of each vertex v are ordered from left to right as given by the rotation system. In a nutshell, `PlanarRS` first splits $\mathcal{C}_v$ into pieces $c_{w_1}, \ldots, c_{w_l}$, one for each of v's children $w_1, \ldots, w_l$, and then removes and inserts all forward edges according to the rotation system of

---

Algorithm 2. `PlanarRS`

Input: vertex v with rotation system $\mathcal{R}_v = (e_1, \ldots, e_k)$, tree layout $\mathcal{T} = (V, E_\mathcal{T})$, SD $\mathcal{C}_v$, tree edge e_p from parent p or $\bot$ if v is the root

Output: **true** $\mathcal{R}$ is planar; **false** if one of $e_1, \ldots, e_k$ causes a crossing

1   **if** v *is leaf* **then return** *whether* $\texttt{ProcessDeque}(v, \mathcal{R}_v, \mathcal{C}_v, e_p, \bot) = ()$

2   $w_1, \ldots, w_l \leftarrow$ children of v in $E_\mathcal{T}$

3   Split $\mathcal{C}_v$ into pieces $c_{w_1}, \ldots, c_{w_l}$ such that $\forall e \in c_{w_i} : e$ is removed at v, w_i or one of w_i's descendents or return **false** if this is not possible

4   **foreach** $w_i \in \{w_1, \ldots, w_l\}$ **do** $\tilde{\mathcal{R}}_i \leftarrow []$; $removed[w_i] \leftarrow$ **false** $\mathcal{S} \leftarrow$ empty stack

5   $e_1, \ldots, e_k \leftarrow$ rotation system of v with $e_1 = e_p$ if v is not root; else $e_1 = (v, w_1)$

    foreach $e = e_1, \ldots, e_k$ **do**

6      $w_i \leftarrow$ child w_i with $e \in c_{w_i}$, e is removed at descendant of w_i, or $e = (v, w_i)$

7   **if** e *is forward edge* **then** $\tilde{\mathcal{R}}_i.\text{append}(e)$ **if** $\mathcal{S}.\text{top}() = w_i$ **then continue** with next edge in line 5 **if** $w_i \in \mathcal{S}$ **then**

8   **while** $\mathcal{S}.\text{top}() \neq w_i$ **do**

9          $w_{i'} \leftarrow \mathcal{S}.\text{pop}()$; $removed[w_{i'}] \leftarrow$ **true**

10          $\mathcal{C}_{w_{i'}} \leftarrow \texttt{ProcessDeque}(v, \tilde{\mathcal{R}}_{i'}, c_{w_{i'}}, e_p, (v, w_{i'}))$

11   **if** $\mathcal{C}_{w_{i'}} = \bot$ **then return false else if** $\neg\texttt{PlanarRS}(w_{i'}, \mathcal{R}_{w_{i'}}, \mathcal{T}, \mathcal{C}_{w_{i'}}, (v, w_{i'}))$ **then return false**

12   **else if** $\neg removed[w_i]$ **then** $\mathcal{S}.\text{push}(w_i)$ **else return false**

13 **while** $\neg\mathcal{S}.\text{isEmpty}()$ **do**

14     $w_i \leftarrow \mathcal{S}.\text{pop}()$

15     $\mathcal{C}_{w_i} \leftarrow \texttt{ProcessDeque}(v, \tilde{\mathcal{R}}_i, c_{w_i}, e_p, (v, w_i))$

16   **if** $\mathcal{C}_{w_i} = \bot$ **then return false else if** $\neg\texttt{PlanarRS}(w_i, \mathcal{R}_{w_i}, \mathcal{T}, \mathcal{C}_{w_i}, (v, w_i))$ **then return false**

17 **return true**

---

v. Afterwards, it recursively calls `PlanarRS` for all its children. If at some point, the SD cannot be split adequately or an edge cannot be removed, **false** is returned and propagated back to the initial caller of `PlanarRS`. Otherwise, **true** is returned. In the following, we assume that in a drawing of G which respects the rotation system no pair of edges crosses more than once and no edge crosses any of the tree edges: As the tree layout $\mathcal{T}$ itself contains no cycles, it is always planar regardless of the rotation system. Also, all forward edges can be drawn such that they cause no crossing with any tree edge. This is also reflected in the SD where all tree edges can be processed canonically without interfering with any forward edge: After splitting the SD, the tree edge from the parent can be removed at the head of the first SD and, as the last step, each tree edge to child w_i can be inserted at the head of the SD for child w_i. As with the deque, we only consider forward edges in the SD layout.

To actually insert and remove forward edges to an SD, `PlanarRS` uses the routine `ProcessDeque`. The observation behind is that a deque is a special case of the SD, namely, whenever the tree layout is a path: Let G be a graph endowed with a rotation system and $\mathcal{T}$ be a tree layout of G. Further, let p be a path from the root to a leaf of $\mathcal{T}$ and denote by G_p the subgraph of G induced by p which inherits G's rotation system. For instance, the subgraph G_p for the root-to-leaf path $p = 1 \overset{+}{\to} 13$ in Fig. 1(d) is shown in Fig. 1(c). G_p's rotation system is planar if and only if it admits a deque layout by Proposition 1. Hence, if G's rotation system is planar, then the rotation system on every root-to-leaf path admits a deque layout. In `PlanarRS`, all edges that lie on a common root-to-leaf path p are processed in the SD just like in a deque. If this is possible, then the rotation system of each root-to-leaf path is planar. This is already reflected in line 1: If v is a leaf, then the SD is not split and must be emptied by `ProcessDeque` and `true` is returned if and only if `ProcessDeque` returns an empty SD.

In line 3, the SD $\mathcal{C}_v$ is split into pieces $c_{w_1}, \ldots, c_{w_l}$ such that for each edge $e \in c_{w_i}$ edge e is removed at v, w_i or one of w_i's descendants. If this is not possible, `false` is returned. Consider edge e'' in Figs. 1(b) and (d), which crosses edge e_1. Edge e_1 is inserted at the head at vertex 1 and e'' at the tail at vertex 2, i.e., $e_1 \ll e''$. At vertex 4, the deque has to be split such that $e_1 \in c_{10}$ and $e'' \in c_5$. However, as $e_1 \ll e''$ and vertex 5 is left of vertex 10, this is not possible. In general, we obtain the following lemma:

Lemma 1. *Let e and e' be two forward edges in $\mathcal{C}_v$ such that e (e') is removed at w_i ($w_{i'}$) or one of its descendants. It is possible to split $\mathcal{C}_v$ such that $e \in c_{w_i}$ and $e' \in c_{w_{i'}}$ if and only if e and e' do not cross.*

Proof. We assume that e and e' are inserted at u and u' and removed at x and x', respectively. Since e and e' are forward edges, there are paths $p = u \overset{+}{\to} v \to w_i \overset{*}{\to} x$ and $p' = u' \overset{+}{\to} v \to w_{i'} \overset{*}{\to} x'$.

$\Leftarrow$: We prove the contrapositive and assume that $\mathcal{C}_v$ cannot be split such that $e \in c_{w_i}$ and $e' \in c_{w_{i'}}$. Either e and e' are inserted to the same side of the deque or to different sides. For the first case, assume that e and e' are both inserted at the head and, w. l. o. g., w_i is left of $w_{i'}$ in the total order of children at v. This implies that $e' \ll_{\mathcal{C}_v} e$. This situation is depicted in Fig. 2(a). Remember that `PlanarRS` inserts and removes edges to the SD as with a deque. Hence, both edges e and e' are to the left of p and p' at u and u', respectively. There is a cycle formed by path p and forward edge e. In a drawing of G which respects the rotation system, this circle encloses a region R (dark shaded in Fig. 2(a)) such that R does not contain $w_{i'}$. As $e' \ll_{\mathcal{C}_v} e$, e is inserted before e' and, thus, either u is an ancestor of u', or $u = u'$ and the rotation system at u is $\mathcal{R}_u = (\ldots, e, \ldots, e', \ldots, e_d, \ldots)$, where e_d is the tree edge from u to u's child on p. In either case, the edge curve of e' starts within region R and must end outside R which inevitably causes a crossing with e. The reasoning if e and e' are inserted at the tail is analogous.

In the second case, e and e' are inserted at different sides where, w. l. o. g., e is inserted at the tail and e' at the head (cf. Fig. 2(b)). Hence, $e' \ll_{\mathcal{C}_v} e$. Since $\mathcal{C}_v$ cannot be split adequately, this implies that w_i must be left of $w_{i'}$ at v. Again, let R be the region enclosed by p and e (dark shaded in Fig. 2(b)) such that R contains $w_{i'}$. As e is to the right of p at u and e' to the left of p' at u', the

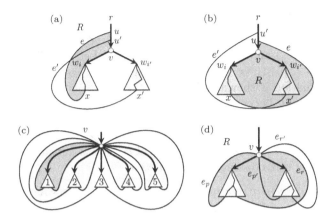

Fig. 2. The two cases where the SD cannot be split appropriately (Figs. (a) and (b)). Fig. (c) sketches nesting sectors and Fig. (d) clashing sectors.

edge curve of e' starts outside of R and must reach x' within R which leads to a crossing with e.

$\Rightarrow$: Again, we prove the contrapositive. We have two cases: Either e and e' lie on the same side of p and p' at u and u', respectively, or on different sides. For the first case, we assume w. l. o. g. that w_i is left of $w_{i'}$ at v. If a crossing between e and e' is unavoidable, then u is either an ancestor of u', or $u = u'$ and rotation system of u is $\mathcal{R}_u = (\ldots, e, \ldots, e', \ldots, e_d, \ldots)$, where e_d is the tree edge from u to u's child on p (Fig. 2(a)). In either case, e is inserted at the head before e' and $e' \ll_{\mathcal{C}_v} e$. Thus, $\mathcal{C}_v$ cannot be split such that $e \in c_{w_i}$ and $e' \in c_{w_{i'}}$. For the second case, e lies to the right of p and e' to the left of p' (w. l. o. g.). Thus, e is inserted at the tail and e' at the head and $e' \ll_{\mathcal{C}_v} e$. If a crossing between e and e' is unavoidable, then w_i is to the left of $w_{i'}$ at v (cf. Fig. 2(b)) and, again, $\mathcal{C}_v$ cannot be split such that $e \in c_{w_i}$ and $e' \in c_{w_{i'}}$. $\square$

Let $\mathcal{R}_v = (e_1, \ldots, e_k)$ be the rotation system of v such that e_1 is the tree edge from v's parent if v is an inner vertex of $\mathcal{T}$. If v is the root, then e_1 is the tree edge to v's first child w_1. After $\mathcal{C}_v$ is successfully split into pieces $c_{w_1}, \ldots, c_{w_l}$, **PlanarRS** finds for each piece a subsequence of $\mathcal{R}_v$ which contains all forward edges that must be removed from and inserted to c_{w_i}. A *sector* $\tilde{\mathcal{R}}_i$ of child w_i is a subsequence of v's rotation system such that all edges in $\tilde{\mathcal{R}}_i$ are forward edges incident to an ancestor of v or a descendant of w_i. Further, for each forward edge e there is exactly one sector $\tilde{\mathcal{R}}_i$ with $e \in \tilde{\mathcal{R}}_i$. If the rotation system is planar, then all sectors must properly nest: Consider Fig. 2(c) in which v has five children corresponding to five subtrees. In the rotation system of v, the sector that corresponds to subtree 2 encloses the sector corresponding to subtree 1, and the sector belonging to 3 encloses both. Let $\tilde{\mathcal{R}}_i = (e_p, \ldots, e_q)$ and $\tilde{\mathcal{R}}_{i'} = (e_{p'}, \ldots, e_{q'})$ be two sectors of $\mathcal{R}_v = (e_1, \ldots, e_k)$. We say that $\tilde{\mathcal{R}}_i$ and $\tilde{\mathcal{R}}_{i'}$ *clash* if there exist edges $e_r \in \tilde{\mathcal{R}}_i$ and $e_{r'} \in \tilde{\mathcal{R}}_{i'}$ with $p < p' < r < r'$ (see Fig. 2(d)) or $r < r' < q < q'$.

Lemma 2. *In a planar rotation system, no pair of sectors clash.*

Proof. Assume for contradiction that G's rotation system is planar but the sectors $\tilde{\mathcal{R}}_i = (e_p, \ldots, e_q)$ and $\tilde{\mathcal{R}}_{i'} = (e_{p'}, \ldots, e_{q'})$ clash. Let $e_r \in \tilde{\mathcal{R}}_i$ and $e_{r'} \in \tilde{\mathcal{R}}_{i'}$ be edges with $p < p' < r < r'$. The reasoning for the case $r < r' < q < q'$ is similar. The situation is shown in Fig. 2(d). There is a circle formed by v, e_p, e_r and a simple path between the endpoints of e_p and e_r distinct from v. In a plane drawing of G respecting the rotation system, this circle encloses a region R (shaded in Fig. 2(d)) such that it contains the endpoints of $e_{p'}$ and $e_{r'}$ other than v. As $p < p' < r < r'$, the edge curve of $e_{r'}$ starts outside of R and ends inside which leads to a crossing; a contradiction. □

Corollary 1. *In a planar rotation system, all pairs of sectors $\tilde{\mathcal{R}}_i = (e_p, \ldots, e_q)$, $\tilde{\mathcal{R}}_{i'} = (e_{p'}, \ldots, e_{q'})$ with $p < p'$, are either disjoint, i.e., $p \le q < p' \le q'$, or nesting, i.e., $p < p' < q' < q$.*

If all sectors are disjoint or nesting, we can construct a plane drawing at vertex v given plane drawings of all subgraphs that belong to the subtrees of v's children. To test if the sectors are nesting, PlanarRS uses a stack in which v's children $w_1, \ldots, w_l$ are inserted and removed. Further, for each child w_i, it maintains the boolean variable removed$[w_i]$ which stores if w_i has been removed from the stack. PlanarRS subsequently processes all edges e of v in order of the rotation system (line 5). In line 6, the child w_i "responsible" for e, is determined, i.e., either e is the tree edge from v to w_i, or e is a forward edge and must be removed from c_{w_i} or e must be inserted at v and is removed at a descendant of w_i. If e is a forward edge, it is appended to sector $\tilde{\mathcal{R}}_i$. If w_i is currently on top of the stack, no further action is needed (line 7). If $w_i \in \mathcal{S}$, all children $w_{i'}$ on the stack are removed until w_i is on top. For all removed children $w_{i'}$, removed$[w_{i'}]$ is set to **true**. If $w_i \notin \mathcal{S}$ and removed$[w_i] = $ **false**, w_i is pushed onto the stack (line 12).

If w_i is not in the stack and has previously been removed, **false** is returned (line 12) as the rotation system is not planar for the following reasons: Child w_i has been removed in a previous iteration when another child $w_{i'}$ further below in the stack needed to be on top. Let $\tilde{\mathcal{R}}_i = (e_p, \ldots, e_q)$ be the sector of w_i and $\tilde{\mathcal{R}}_{i'} = (e_{p'}, \ldots, e_{q'})$ the sector of $w_{i'}$. $w_{i'}$ has been inserted to $\mathcal{S}$ before w_i and, thus, $p' < p$. In the iteration when w_i is removed from $\mathcal{S}$, the edge of the iteration is $e_{r'} \in \mathcal{R}_{w_{i'}}$. Further, there is an edge $e_r \in \mathcal{R}_{w_i}$ when w_i would be reinserted with $r' < r$. Altogether we get $p' < p < r' < r$ and, hence, $\tilde{\mathcal{R}}_i$ and $\tilde{\mathcal{R}}_{i'}$ clash. Thus, the rotation system of G is not planar by Lemma 2.

Whenever a child $w_{i'}$ is removed from $\mathcal{S}$, it must never be inserted again and, hence, $\tilde{\mathcal{R}}_{i'}$ must contain all edges to be processed in $c_{w_{i'}}$. In line 10, ProcessDeque is called with sector $\tilde{\mathcal{R}}_{i'}$ and $c_{w_{i'}}$ as parameters. Remember that ProcessDeque needs two edges on a path which divide the rotation system into a left and right half. Here, these two edges are the tree edge from the parent of v, if existent, and the tree edge from v to $w_{i'}$. If the return value $\mathcal{C}_{w_{i'}}$ of ProcessDeque is $\bot$, not all edges could be processed in the deque and the rotation system is not planar by Proposition 1. Thus, **false** is returned in line 11.

Otherwise, `PlanarRS` is called recursively on $w_{i'}$ with input deque $\mathcal{C}_{w_{i'}}$ (line 11). After all forward edges of the rotation system are processed, all children remaining in $\mathcal{S}$ are removed (lines 13–16) and `ProcessDeque` and `PlanarRS` are called. If all calls of `PlanarRS` return `true`, the rotation system of each root-to-leaf path is planar and so are the rotation systems at each vertex with more than one child.

Lemma 3. `PlanarRS` *in Algorithm 2 returns* `true` *if and only if the rotation system of its input graph is planar.*

Since `PlanarRS` obeys the SD's modus operandi, Theorem 1 follows. Each edge is inserted and removed exactly once. Further, in the loop from lines 5–12 each forward edge is processed at most twice during the algorithm and, by using the DFS numbers, it is possible to decide in time $\mathcal{O}(1)$ in which subtree the end vertex of a forward edge lies. Also, each vertex is inserted at most twice to the stack for each edge. The operation that needs more arguing is splitting the deque (line 3) for which also a linear running time can be achieved.

Whenever `PlanarRS` returns `false`, the edges that cause a crossing can be determined: If one of the calls of `ProcessDeque` returns $\bot$, then an edge could not be removed from the SD since at least one other edge is blocking its way. Hence, these edges must cross. If the SD cannot be split adequately (line 3), then we obtain one of the situations as in Figs. 2(a) or (b) and the edges which prevent the SD from being split adequately are those that cause a crossing. Last, if some of v's children w_i were reinserted into the stack (lines 12 and 16), then the corresponding sectors would be clashing and, hence, there exist two edges that cross according to the proof of Lemma 2 (see Fig. 2(d)).

5   Conclusion

We characterized planarity by graph layouts in the splittable deque (SD): Although a stack, two stacks, or the deque characterize large classes of planar graphs, they do not capture all. We enhanced the deque by a split-operation and showed that it characterizes planarity. For our proof, we devised a linear-time algorithm operating on the SD to test the planarity of a rotation system. If it is not planar, the operations on the SD indicate crossing edges. Our test also works for graphs with multi-edges. Given a rotation system, it defines the order of order of insertions and removals to the SD. Conversely, given the order of insertions and removals to the SD, we can find a (planar) rotation system. So a planarity testing algorithm can use the SD to find an embedding of a graph.

References

1. Auer, C., Bachmaier, C., Brandenburg, F.J., Brunner, W., Gleißner, A.: Plane drawings of queue and deque graphs. In: Brandes, U., Cornelsen, S. (eds.) GD 2010. LNCS, vol. 6502, pp. 68–79. Springer, Heidelberg (2011)

2. Auer, C., Gleißner, A.: Characterizations of deque and queue graphs. In: Kolman, P., Kratochvíl, J. (eds.) WG 2011. LNCS, vol. 6986, pp. 35–46. Springer, Heidelberg (2011)

3. Auer, C., Gleißner, A., Hanauer, K., Vetter, S.: Testing planarity by switching trains. In: Didimo, W., Patrignani, M. (eds.) GD 2012. LNCS, vol. 7704, pp. 557–558. Springer, Heidelberg (2013)

4. Bernhart, F., Kainen, P.: The book thickness of a graph. J. Combin. Theory, Ser. B 27(3), 320–331 (1979)

5. Chung, F.R.K., Leighton, F.T., Rosenberg, A.L.: Embedding graphs in books: A layout problem with applications to VLSI design. SIAM J. Algebra. Discr. Meth. 8(1), 33–58 (1987)

6. Donafee, A., Maple, C.: Planarity testing for graphs represented by a rotation scheme. In: Banissi, E., Börner, K., Chen, C., Clapworthy, G., Maple, C., Lobben, A., Moore, C.J., Roberts, J.C., Ursyn, A., Zhang, J. (eds.) Proc. Seventh International Conference on Information Visualization, IV 2003, pp. 491–497. IEEE Computer Society, Washington, DC (2003)

7. Dujmović, V., Wood, D.R.: On linear layouts of graphs. Discrete Math. Theor. Comput. Sci. 6(2), 339–358 (2004)

8. Dujmović, V., Wood, D.R.: Stacks, queues and tracks: Layouts of graph subdivisions. Discrete Math. Theor. Comput. Sci. 7(1), 155–202 (2005)

9. de Fraysseix, H., Rosenstiehl, P.: A depth-first-search characterization of planarity. In: Graph Theory, Cambridge (1981); Ann. Discrete Math., vol. 13, pp. 75–80. North-Holland, Amsterdam (1982)

10. Heath, L.S., Leighton, F.T., Rosenberg, A.L.: Comparing queues and stacks as mechanisms for laying out graphs. SIAM J. Discret. Math. 5(3), 398–412 (1992)

11. Heath, L.S., Pemmaraju, S.V., Trenk, A.N.: Stack and queue layouts of directed acyclic graphs: Part I. SIAM J. Comput. 28(4), 1510–1539 (1999)

12. Heath, L.S., Rosenberg, A.L.: Laying out graphs using queues. SIAM J. Comput. 21(5), 927–958 (1992)

13. Kosaraju, S.R.: Real-time simulation of concatenable double-ended queues by double-ended queues (preliminary version). In: Proc. 11th Annual ACM Symposium on Theory of Computing, STOC 1979, pp. 346–351. ACM, New York (1979)

14. Rosenstiehl, P., Tarjan, R.E.: Gauss codes, planar hamiltonian graphs, and stack-sortable permutations. J. of Algorithms 5, 375–390 (1984)

15. Shih, W.K., Hsu, W.L.: A new planarity test. Theor. Comput. Sci. 223(1-2), 179–191 (1999)

16. Wood, D.R.: Queue layouts, tree-width, and three-dimensional graph drawing. In: Agrawal, M., Seth, A.K. (eds.) FSTTCS 2002. LNCS, vol. 2556, pp. 348–359. Springer, Heidelberg (2002)

17. Yannakakis, M.: Four pages are necessary and sufficient for planar graphs. In: Proc. of the 18th Annual ACM Symposium on Theory of Computing, STOC 1986, pp. 104–108. ACM, New York (1986)

18. Yannakakis, M.: Embedding planar graphs in four pages. J. Comput. Syst. Sci. 38(1), 36–67 (1989)

Strip Planarity Testing

Patrizio Angelini[1], Giordano Da Lozzo[1], Giuseppe Di Battista[1], and Fabrizio Frati[2]

[1] Dipartimento di Ingegneria, Roma Tre University, Italy
{angelini,dalozzo,gdb}@dia.uniroma3.it
[2] School of Information Technologies, The University of Sydney, Australia
brillo@it.usyd.edu.au

Abstract. In this paper we introduce and study the *strip planarity testing* problem, which takes as an input a planar graph $G(V, E)$ and a function $\gamma : V \to \{1, 2, \ldots, k\}$ and asks whether a planar drawing of G exists such that each edge is monotone in the y-direction and, for any $u, v \in V$ with $\gamma(u) < \gamma(v)$, it holds $y(u) < y(v)$. The problem has strong relationships with some of the most deeply studied variants of the planarity testing problem, such as *clustered planarity*, *upward planarity*, and *level planarity*. We show that the problem is polynomial-time solvable if G has a fixed planar embedding.

1 Introduction

Testing the planarity of a given graph is one of the oldest and most deeply investigated problems in algorithmic graph theory. A celebrated result of Hopcroft and Tarjan [20] states that the planarity testing problem is solvable in linear time.

A number of interesting variants of the planarity testing problem have been considered in the literature [25]. Such variants mainly focus on testing, for a given planar graph G, the existence of a planar drawing of G satisfying certain *constraints*. For example the *partial embedding planarity* problem [1,22] asks whether a plane drawing $\mathcal{G}$ of a given planar graph G exists in which the drawing of a subgraph H of G in $\mathcal{G}$ coincides with a given drawing $\mathcal{H}$ of H. *Clustered planarity testing* [10,23], *upward planarity testing* [5,16,21], *level planarity testing* [24], *embedding constraints planarity testing* [17], *radial level planarity testing* [4], and *clustered level planarity testing* [14] are further examples of problems falling in this category.

In this paper we introduce and study the *strip planarity testing* problem, which is defined as follows. The input of the problem consists of a planar graph $G(V, E)$ and of a function $\gamma : V \to \{1, 2, \ldots, k\}$. The problem asks whether a *strip planar* drawing of (G, γ) exists, i.e. a planar drawing of G such that each edge is monotone in the y-direction and, for any $u, v \in V$ with $\gamma(u) < \gamma(v)$, it holds $y(u) < y(v)$. The name "strip" planarity comes from the fact that, if a strip planar drawing Γ of (G, γ) exists, then k disjoint horizontal strips $\gamma_1, \gamma_2, \ldots, \gamma_k$ can be drawn in Γ so that γ_i lies below γ_{i+1}, for $1 \leq i \leq k - 1$, and so that γ_i contains a vertex x of G if and only if $\gamma(x) = i$, for $1 \leq i \leq k$. It is not difficult to argue that strips $\gamma_1, \gamma_2, \ldots, \gamma_k$ can be given as part of the input, and the problem is to decide whether G can be planarly drawn so that each edge is monotone in the y-direction and each vertex x of G with $\gamma(x) = i$ lies in the strip γ_i. That is, arbitrarily predetermining the placement of the strips does not alter the possibility of constructing a strip planar drawing of (G, γ).

S. Wismath and A. Wolff (Eds.): GD 2013, LNCS 8242, pp. 37–48, 2013.

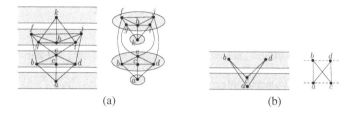

(a)   (b)

Fig. 1. (a) A negative instance (G, γ) of the strip planarity testing problem whose associated clustered graph $C(G, T)$ is c-planar. (b) A positive instance (G, γ) of the strip planarity testing problem that is not level planar.

Before presenting our result, we discuss the strong relationships of the strip planarity testing problem with three famous graph drawing problems.

Strip Planarity and Clustered Planarity. The *c-planarity testing* problem takes as an input a *clustered graph* $C(G, T)$, that is a planar graph G together with a rooted tree T, whose leaves are the vertices of G. Each internal node μ of T is called *cluster* and is associated with the set V_μ of vertices of G in the subtree of T rooted at μ. The problem asks whether a *c-planar drawing* exists, that is a planar drawing of G together with a drawing of each cluster $\mu \in T$ as a simple closed region R_μ so that: (i) if $v \in V_\mu$, then $v \in R_\mu$; (ii) if $V_\nu \subset V_\mu$, then $R_\nu \subset R_\mu$; (iii) if $V_\nu \cap V_\mu = \emptyset$, then $R_\nu \cap R_\mu = \emptyset$; and (iv) each edge of G intersects the border of R_μ at most once. Determining the time complexity of testing the c-planarity of a given clustered graph is a long-standing open problem. See [10,23] for two recent papers on the topic. An instance (G, γ) of the strip planarity testing problem naturally defines a clustered graph $C(G, T)$, where T consists of a root having k children $\mu_1, \dots, \mu_k$ and, for every $1 \leq j \leq k$, cluster μ_j contains every vertex x of G such that $\gamma(x) = j$. The c-planarity of $C(G, T)$ is a necessary condition for the strip planarity of (G, γ), since suitably bounding the strips in a strip planar drawing of (G, γ) provides a c-planar drawing of $C(G, T)$. However, the c-planarity of $C(G, T)$ is not sufficient for the strip planarity of (G, γ) (see Fig. 1(a)). It turns out that strip planarity testing *coincides* with a special case of a problem opened by Cortese et al. [8,9] and related to c-planarity testing. The problem asks whether a graph G can be planarly embedded "inside" an host graph H, which can be thought as having "fat" vertices and edges, with each vertex and edge of G drawn inside a prescribed vertex and a prescribed edge of H, respectively. It is easy to see that the strip planarity testing problem coincides with this problem in the case in which H is a path.

Strip Planarity and Level Planarity. The *level planarity testing* problem takes as an input a planar graph $G(V, E)$ and a function $\gamma : V \to \{1, 2, \dots, k\}$ and asks whether a planar drawing of G exists such that each edge is monotone in the y-direction and each vertex $u \in V$ is drawn on the horizontal line $y = \gamma(u)$. The level planarity testing (and embedding) problem is known to be solvable in linear time [24], although a sequence of incomplete characterizations by forbidden subgraphs [15,18] (see also [13]) has revealed that the problem is not yet fully understood. The similarity of the level planarity testing problem with the strip planarity testing problem is evident: They have the same input, they both require planar drawings with y-monotone edges, and they both

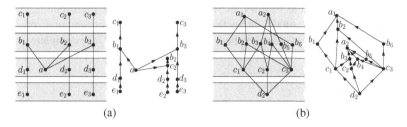

Fig. 2. Two negative instances (G_1, γ_1) (a) and (G_2, γ_2) (b) whose associated directed graphs are upward planar, where G_1 is a tree and G_2 is a subdivision of a triconnected plane graph

constrain the vertices to lie in specific regions of the plane; they only differ for the fact that such regions are horizontal lines in one case, and horizontal strips in the other one. Clearly the level planarity of an instance (G, γ) is a sufficient condition for the strip planarity of (G, γ), as a level planar drawing is also a strip planar drawing. However, it is easy to construct instances (G, γ) that are strip planar and yet not level planar, even if we require that the instances are *strict*, i.e., no edge (u, v) is such that $\gamma(u) = \gamma(v)$. See Fig. 1(b). Also, the approach of [24] seems to be not applicable to test the strip planarity of a graph. Namely, Jünger et al. [24] visit the instance (G, γ) one level at a time, representing with a PQ-tree [7] the possible orderings of the vertices in level i that are consistent with a level planar embedding of the subgraph of G induced by levels $\{1, 2, \ldots, i\}$. However, when visiting an instance (G, γ) of the strip planarity testing problem one strip at a time, PQ-trees seem to be not powerful enough to represent the possible orderings of the vertices in strip i that are consistent with a strip planar embedding of the subgraph of G induced by strips $\{1, 2, \ldots, i\}$.

Strip Planarity and Upward Planarity. The *upward planarity testing* problem asks whether a given directed graph $\overrightarrow{G}$ admits an *upward planar drawing*, i.e., a drawing which is planar and such that each edge is represented by a curve monotonically increasing in the y-direction, according to its orientation. Testing the upward planarity of a directed graph $\overrightarrow{G}$ is an $\mathcal{NP}$-hard problem [16], however it is polynomial-time solvable, e.g., if $\overrightarrow{G}$ has a fixed embedding [5], or if it has a single-source [21]. A strict instance (G, γ) of the strip planarity testing problem naturally defines a directed graph $\overrightarrow{G}$, by directing an edge (u, v) of G from u to v if $\gamma(u) < \gamma(v)$. It is easy to argue that the upward planarity of $\overrightarrow{G}$ is a necessary and not sufficient condition for the strip planarity of (G, γ) (see Figs. 2(a) and 2(b)). Roughly speaking, in an upward planar drawing different parts of the graph are free to "nest" one into the other, while in a strip planar drawing, such a nesting is only allowed if coherent with the strip assignment.

In this paper, we show that the strip planarity testing problem is polynomial-time solvable for planar graphs with a fixed planar embedding. Our approach consists of performing a sequence of modifications to the input instance (G, γ) (such modifications consist mainly of insertions of graphs inside the faces of G) that ensure that the instance satisfies progressively stronger constraints while not altering its strip planarity. Eventually, the strip planarity of (G, γ) becomes equivalent to the upward planarity of its associated directed graph, which can be tested in polynomial time.

The paper is organized as follows. In Section 2 we give some preliminaries; in Section 3 we prove our result; finally, in Section 4 we conclude and present open problems. For space limitations, proofs are sketched or omitted; refer to [3] for complete proofs.

2   Preliminaries

A planar drawing of a graph determines a circular ordering of the edges incident to each vertex. Two drawings of the same graph are *equivalent* if they determine the same circular orderings around each vertex. A *planar embedding* (or *combinatorial embedding*) is an equivalence class of planar drawings. A planar drawing partitions the plane into topologically connected regions, called *faces*. The unbounded face is the *outer face*. Two planar drawings with the same combinatorial embedding have the same faces. However, such drawings could still differ for their outer faces. A *plane embedding* of a graph G is a planar embedding of G together with a choice for its outer face. In this paper, we will assume all the considered graphs to have a prescribed plane embedding.

For the sake of simplicity of description, in the following we assume that the considered plane graphs are 2-*connected*, unless otherwise specified. We will sketch in the conclusions how to extend our results to simply-connected and even non-connected plane graphs. We now define some concepts related to strip planarity.

An instance (G, γ) of the strip planarity testing problem is *strict* if it contains no intra-strip edge, where an edge (u, v) is *intra-strip* f $\gamma(u) = \gamma(v)$. An instance (G, γ) of strip planarity is *proper* if, for every edge (u, v) of G, it holds $\gamma(v) - 1 \leq \gamma(u) \leq \gamma(v) + 1$. Given any non-proper instance of strip planarity, one can replace every edge (u, v) such that $\gamma(u) = \gamma(v) + j$, for some $j \geq 2$, with a path $(v = u_1, u_2, \ldots, u_{j+1} = u)$ such that $\gamma(u_{i+1}) = \gamma(u_i) + 1$, for every $1 \leq i \leq j$, thus obtaining a proper instance (G', γ') of the strip planarity testing problem. It is easy to argue that (G, γ) is strip planar if and only if (G', γ') is strip planar. In the following, we will assume all the considered instances of the strip planarity testing problem to be proper.

Let (G, γ) be an instance of the strip planarity testing problem. A path $(u_1, \ldots, u_j)$ in G is *monotone* if $\gamma(u_i) = \gamma(u_{i-1}) + 1$, for every $2 \leq i \leq j$. For any face f in G, we denote by C_f the simple cycle delimiting the border of f. Let f be a face of G, let u be a vertex incident to f, and let v and z be the two neighbors of u on C_f. We say that u is a *local minimum* for f if $\gamma(v) = \gamma(z) = \gamma(u) + 1$, and it is a *local maximum* for f if $\gamma(v) = \gamma(z) = \gamma(u) - 1$. Also, we say that u is a *global minimum* for f (a *global maximum* for f) if $\gamma(w) \geq \gamma(u)$ (resp. $\gamma(w) \leq \gamma(u)$), for every vertex w incident to f. A global minimum u_m and a global maximum u_M for a face f are *consecutive* in f if no global minimum and no global maximum exists in one of the two paths connecting u_m and u_M in C_f. A local minimum u_m and a local maximum u_M for a face f are *visible* if one of the paths P connecting u_m and u_M in C_f is such that, for every vertex u of P, it holds $\gamma(u_m) < \gamma(u) < \gamma(u_M)$.

Definition 1. *An instance* (G, γ) *of the strip planarity problem is* quasi-jagged *if it is strict and if, for every face* f *of* G *and for any two visible local minimum* u_m *and local maximum* u_M *for* f, *one of the two paths connecting* u_m *and* u_M *in* C_f *is monotone.*

Definition 2. *An instance (G, γ) of the strip planarity problem is* jagged *if it is strict and if, for every face f of G, any local minimum for f is a global minimum for f, and every local maximum for f is a global maximum for f.*

3 How to Test Strip Planarity

In this section we show an algorithm to test strip planarity.

3.1 From a General Instance to a Strict Instance

In this section we show how to reduce a general instance of the strip planarity testing problem to an equivalent strict instance.

Lemma 1. *Let (G, γ) be an instance of the strip planarity testing problem. Then, there exists a polynomial-time algorithm that either constructs an equivalent strict instance (G^*, γ^*) or decides that (G, γ) is not strip planar.*

Consider any intra-strip edge (u, v) in G, if it exists. We distinguish two cases.

In *Case 1*, (u, v) is an edge of a 3-cycle (u, v, z) that contains vertices in its interior in G. Observe that, $\gamma(u) - 1 \leq \gamma(z) \leq \gamma(u) + 1$. Denote by G' the plane subgraph of G induced by the vertices lying outside cycle (u, v, z) together with u, v, and z (this graph might coincide with cycle (u, v, z) if such a cycle delimits the outer face of G); also, denote by G'' the plane subgraph of G induced by the vertices lying inside cycle (u, v, z) together with u, v, and z. Also, let $\gamma'(x) = \gamma(x)$, for every vertex x in G', and let $\gamma''(x) = \gamma(x)$, for every vertex x in G''. We have the following:

Claim 1. *(G, γ) is strip planar if and only if (G', γ') and (G'', γ'') are both strip planar.*

The strip planarity of (G'', γ'') can be tested in linear time as follows.

If $\gamma''(z) = \gamma''(u)$, then (G'', γ'') is strip planar if and only if $\gamma''(x) = \gamma''(u)$ for every vertex x of G'' (such a condition can clearly be tested in linear time). For the necessity, 3-cycle (u, v, z) is entirely drawn in $\gamma''(u)$, hence all the internal vertices of G'' have to be drawn inside $\gamma''(u)$ as well. For the sufficiency, G'' has a plane embedding by assumption, hence any planar y-monotone drawing (e.g. a straight-line drawing where no two vertices have the same y-coordinate) respecting such an embedding and contained in $\gamma''(u)$ is a strip planar drawing of (G'', γ'').

If $\gamma''(z) = \gamma''(u) - 1$ (the case in which $\gamma''(z) = \gamma''(u) + 1$ is analogous), then we argue as follows: First, a clustered graph $C(G'', T)$ can be defined such that T consists of two clusters μ and ν, respectively containing every vertex x of G'' such that $\gamma''(x) = \gamma''(u) - 1$, and every vertex x of G'' such that $\gamma''(x) = \gamma''(u)$. We show that (G'', γ'') is strip planar if and only if $C(G'', T)$ is c-planar. For the necessity, it suffices to observe that a strip planar drawing of (G'', γ'') is also a c-planar drawing of $C(G'', T)$. For the sufficiency, if $C(G'', T)$ admits a c-planar drawing, then it also admits a c-planar *straight-line* drawing $\Gamma(C)$ in which the regions $R(\mu)$ and $R(\nu)$ representing μ and ν, respectively, are *convex* [2,12]. Assuming w.l.o.g. up to a rotation of $\Gamma(C)$ that $R(\mu)$ and $R(\nu)$ can be separated by a horizontal line, we have that disjoint horizontal strips can be drawn containing $R(\mu)$ and $R(\nu)$. Slightly perturbing the positions of the vertices so that no two of them have the same y-coordinate ensures that the the edges are y-monotone, thus resulting in a strip planar drawing of (G'', γ''). Finally, the c-planarity of a clustered graph containing two clusters can be decided in linear time, as independently proved by Biedl et al. [6] and by Hong and Nagamochi [19].

In *Case 2*, a 3-cycle (u, v, z) exists that contains no vertices in its interior in G. Then, *contract* (u, v), that is, identify u and v to be the same vertex w, whose incident edges are all the edges incident to u and v, except for (u, v); the clockwise order of the edges incident to w is: All the edges that used to be incident to u in the same clockwise order starting at (u, v), and then all the edges that used to be incident to v in the same clockwise order starting at (v, u). Denote by G' the resulting graph. Since G is plane, G' is plane; since G contains no 3-cycle (u, v, z) that contains vertices in its interior, G' is simple. Let $\gamma'(x) = \gamma(x)$, for every vertex $x \neq u, v$ in G, and let $\gamma'(w) = \gamma(u)$. We have the following.

Claim 2. (G', γ') *is strip planar if and only if* (G, γ) *is strip planar.*

Claims 1 and 2 imply Lemma 1. Namely, if (G, γ) has no intra-strip edge, there is nothing to prove. Otherwise, (G, γ) has an intra-strip edge (u, v), hence either Case 1 or Case 2 applies. If Case 2 applies to (G, γ), then an instance (G', γ') is obtained in linear time containing one less vertex than (G, γ). By Claim 2, (G', γ') is equivalent to (G, γ). Otherwise, Case 1 applies to (G, γ). Then, either the non-strip planarity of (G, γ) is deduced (if (G'', γ'') is not strip planar), or an instance (G', γ') is obtained containing at least one less vertex than (G, γ) (if (G'', γ'') is strip planar). By Claim 1, (G', γ') is equivalent to (G, γ). The repetition of such an argument either leads to conclude in polynomial time that (G, γ) is not strip planar, or leads to construct in polynomial time a strict instance (G^*, γ^*) of strip planarity equivalent to (G, γ).

3.2   From a Strict Instance to a Quasi-Jagged Instance

In this section we show how to reduce a strict instance of the strip planarity testing problem to an equivalent quasi-jagged instance. Again, for the sake of simplicity of description, we assume that every considered instance (G, γ) is 2-connected.

Lemma 2. *Let* (G, γ) *be a strict instance of the strip planarity testing problem. Then, there exists a polynomial-time algorithm that constructs an equivalent quasi-jagged instance* (G^*, γ^*) *of the strip planarity testing problem.*

Consider any face f of G containing two visible local minimum and maximum u_m and u_M, respectively, such that no path connecting u_m and u_M in C_f is monotone. Insert a monotone path connecting u_m and u_M inside f. Denote by (G^+, γ^+) the resulting instance of the strip planarity testing problem. We have the following claim:

Claim 3. (G^+, γ^+) *is strip planar if and only if* (G, γ) *is strip planar.*

Proof Sketch: The necessity is trivial. For the sufficiency, consider any strip planar drawing Γ of (G, γ). Denote by P the path connecting u_m and u_M along C_f and such that $\gamma(u_m) < \gamma(v) < \gamma(u_M)$ holds for every internal vertex v of P. Because of the existence of some parts of the graph that "intermingle" with P, it might not be possible to draw a y-monotone curve inside f connecting u_m and u_M in Γ. Thus, a part of Γ has to be horizontally shrunk, so that it moves "far away" from P, thus allowing for the monotone path connecting u_m and u_M to be drawn as a y-monotone curve inside f. This results in a strip planar drawing of (G^+, γ^+).   □

Claim 3 implies Lemma 2, as proved in the following.

First, the repetition of the above described augmentation leads to a quasi-jagged instance (G^*, γ^*). In fact, whenever the augmentation is performed, the number of triples (v_m, v_M, g) such that vertices v_m and v_M are visible local minimum and maximum for face g, respectively, and such that both paths connecting v_m and v_M along C_f are not monotone decreases by 1, thus eventually the number of such triples is zero, and the instance is quasi-jagged.

Second, (G^*, γ^*) can be constructed from (G, γ) in polynomial time. Namely, the number of pairs of visible local minima and maxima for a face g of G is polynomial in the number of vertices of g. Hence, the number of triples (v_m, v_M, g) such that vertices v_m and v_M are visible local minimum and maximum for face g, over all faces of G, is polynomial in n. Since a linear number of vertices are introduced in G whenever the augmentation described above is performed, it follows that the the construction of (G^*, γ^*) from (G, γ) can be accomplished in polynomial time.

Third, (G^*, γ^*) is an instance of the strip planarity testing problem that is equivalent to (G, γ). This directly comes from repeated applications of Claim 3.

3.3   From a Quasi-Jagged Instance to a Jagged Instance

In this section we show how to reduce a quasi-jagged instance of the strip planarity testing problem to an equivalent jagged instance. Again, for the sake of simplicity of description, we assume that every considered instance (G, γ) is 2-connected.

Lemma 3. *Let (G, γ) be a quasi-jagged instance of the strip planarity testing problem. Then, there exists a polynomial-time algorithm that constructs an equivalent jagged instance (G^*, γ^*) of the strip planarity testing problem.*

Consider any face f of G that contains some local minimum or maximum which is not a global minimum or maximum for f, respectively. Assume that f contains a local minimum v which is not a global minimum for f. The case in which f contains a local maximum which is not a global maximum for f can be discussed analogously. Denote by u (denote by z) the first global minimum or maximum for f that is encountered when walking along C_f starting at v while keeping f to the left (resp. to the right).

We distinguish two cases, namely the case in which u is a global minimum for f and z is a global maximum for f (Case 1), and the case in which u and z are both global maxima for f (Case 2). The case in which u is a global maximum for f and z is a global minimum for f, and the case in which u and z are both global minima for f can be discussed symmetrically.

In *Case 1*, denote by Q the path connecting u and z in C_f and containing v. Consider the internal vertex v' of Q that is a local minimum for f and that is such that

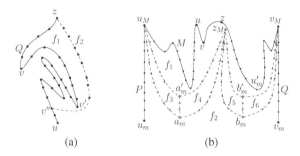

Fig. 3. Augmentation of (G, γ) inside a face f in: (a) Case 1 and (b) Case 2

$\gamma(v') = \min_{u'} \gamma(u')$ among all the internal vertices u' of Q that are local minima for f. Traverse Q starting from u, until a vertex v'' is found with $\gamma(v'') = \gamma(v')$. Notice that, the subpath of Q between u and v'' is monotone. Insert a monotone path connecting v'' and z inside f. See Fig. 3(a). Denote by (G^+, γ^+) the resulting instance of the strip planarity testing problem. We have the following claim:

Claim 4. *Suppose that Case 1 is applied to a quasi-jagged instance (G, γ) to construct an instance (G^+, γ^+). Then, (G^+, γ^+) is strip planar if and only if (G, γ) is strip planar. Also, (G^+, γ^+) is quasi-jagged.*

Proof Sketch: The necessity is trivial. For the sufficiency, consider any strip planar drawing Γ of (G, γ). First, Γ is modified so that v'' has y-coordinate smaller than every local minimum of Q different from u. Then, a y-monotone curve can be drawn inside f connecting v'' and z, thus resulting in a strip planar drawing of (G^+, γ^+). $\square$

In *Case 2*, denote by M a *maximal* path that is part of C_f, whose end-vertices are two global maxima u_M and v_M for f, that contains v in its interior, and that does not contain any global minimum in its interior. By the assumptions of Case 2, such a path exists. Assume, w.l.o.g., that face f is to the right of M when walking along M starting at u_M towards v_M. Possibly $u_M = u$ and/or $v_M = z$. Let u_m (v_m) be the global minimum for f such that u_m and u_M (resp. v_m and v_M) are consecutive global minimum and maximum for f. Possibly, $u_m = v_m$. Denote by P the path connecting u_m and u_M along C_f and not containing v. Also, denote by Q the path connecting v_m and v_M along C_f and not containing v. Since M contains a local minimum among its internal vertices, and since (G, γ) is quasi-jagged, it follows that P and Q are monotone.

Insert the plane graph $A(u_M, v_M, f)$ depicted by white circles and dashed lines in Fig. 3(b) inside f. Consider a local minimum $u'_m \in M$ for f such that $\gamma(u'_m) = \min_{v'_m} \gamma(v'_m)$ among the local minima v'_m for f in M. Set $\gamma(z_M) = \gamma(u_M)$, set $\gamma(a_m) = \gamma(b_m) = \gamma(u_m)$, and set $\gamma(a'_m) = \gamma(b'_m) = \gamma(u'_m)$. The dashed lines connecting a_m and u_M, connecting a'_m and u_M, connecting a_m and z_M, connecting a'_m and z_M, connecting b_m and z_M, connecting b'_m and z_M, connecting b_m and v_M, connecting b'_m and v_M, connecting a_m and a'_m, and connecting b_m and b'_m represent monotone paths. Denote by (G^+, γ^+) the resulting instance of the strip planarity testing problem. We have the following claim:

Claim 5. *Suppose that Case 2 is applied to a quasi-jagged instance (G, γ) to construct an instance (G^+, γ^+). Then, (G^+, γ^+) is strip planar if and only if (G, γ) is strip planar. Also, (G^+, γ^+) is quasi-jagged.*

Proof Sketch: The necessity is trivial. For the sufficiency, consider any strip planar drawing Γ of (G, γ). If P is to the left of Q, then a region R is defined as the region delimited by P, by M, by Q, and by the horizontal line delimiting $\gamma(u_m)$ from above. Then, the part of Γ that lies inside R is redrawn so that it lies inside a region $R_Q \subset R$ arbitrarily close to Q. Such a redrawing "frees" space for the drawing of $A(u_M, v_M, f)$ inside f, which results in a strip planar drawing of (G^+, γ^+). If P is to the right of Q, then M might "wiggle" to the right of P and to the left of Q. Thus, we first horizontally shrink a part of Γ that "intermingles" with P and Q, and we then draw $A(u_M, v_M, f)$ using its four monotone paths connecting global minima with global maxima in order to "circumvent" M. This results in a strip planar drawing of (G^+, γ^+).   □

Claims 4–5 imply Lemma 3, as proved in the following.

First, we prove that the repetition of the above described augmentation leads to a jagged instance (G^*, γ^*) of the strip planarity testing problem. For an instance (G, γ) and for a face g of G, denote by $n(g)$ the number of vertices that are local minima for g but not global minima for g, plus the number of vertices that are local maxima for g but not global maxima for g. Also, let $n(G) = \sum_g n(g)$, where the sum is over all faces g of G. We claim that, when one of the augmentations of Cases 1 and 2 is performed and instance (G, γ) is transformed into an instance (G^+, γ^+), we have $n(G^+) \leq n(G) - 1$. The claim implies that eventually $n(G^*) = 0$, hence (G^*, γ^*) is jagged.

We prove the claim. When a face f of G is augmented as in Case 1 or in Case 2, for each face $g \neq f$ and for each vertex u incident to g, vertex u is a local minimum, a local maximum, a global minimum, or a global maximum for g in (G^+, γ^+) if and only if it is a local minimum, a local maximum, a global minimum, or a global maximum for g in (G, γ), respectively. Hence, it suffices to prove that $\sum n(f_i) \leq n(f) - 1$, where the sum is over all the faces f_i that are created from the augmentation inside f.

Suppose that Case 1 is applied to insert a monotone path between vertices v'' and z inside f. Such an insertion splits f into two faces, which we denote by f_1 and f_2, as in Fig. 3(a). Face f_2 is delimited by two monotone paths, hence $n(f_2) = 0$. Every vertex inserted into f is neither a local maximum nor a local minimum for f_1. As a consequence, no vertex x exists such that x contributes to $n(f_1)$ and x does not contribute to $n(f)$. Further, vertex v' is a global minimum for f_1, by construction, and it is a local minimum but not a global minimum for f. Hence, v' contributes to $n(f)$ and does not contribute to $n(f_1)$. It follows that $n(f_1) + n(f_2) \leq n(f) - 1$.

Suppose that Case 2 is applied to insert plane graph $A(u_M, v_M, f)$ inside face f. Such an insertion splits f into six faces, which are denoted by $f_1, \ldots, f_6$, as in Fig. 3(b). Every vertex of $A(u_M, v_M, f)$ incident to a face f_i, for some $1 \leq i \leq 6$, is either a global maximum for f_i, or a global minimum for f_i, or it is neither a local maximum nor a local minimum for f_i. As a consequence, no vertex x exists such that x contributes to some $n(f_i)$ and x does not contribute to $n(f)$. Further, for each vertex x that contributes to $n(f)$, there exists at most one face f_i such that x contributes to $n(f_i)$. Finally, vertex u'_m of M is a global minimum for f_1, by construction, and it is a local minimum but not a global minimum for f. Hence, u'_m contributes to $n(f)$ and does not contribute to $n(f_i)$, for any $1 \leq i \leq 6$. It follows that $\sum_{i=1}^{6} n(f_i) \leq n(f) - 1$.

Second, (G^*, γ^*) can be constructed from (G, γ) in polynomial time. Namely, the number of local minima (maxima) for a face f that are not global minima (maxima) for f is at most the number of vertices of f. Hence, the number of such minima and maxima over all the faces of G, which is equal to $n(G)$, is linear in n. Since a linear number of vertices are introduced in G whenever the augmentation described above is performed, and since the augmentation is performed at most $n(G)$ times, it follows that the construction of (G^*, γ^*) can be accomplished in polynomial time.

Third, (G^*, γ^*) is an instance of the strip planarity testing problem that is equivalent to (G, γ). This directly comes from repeated applications of Claims 4 and 5.

3.4   Testing Strip Planarity for Jagged Instances

In this section we show how to test in polynomial time whether a jagged instance (G, γ) of the strip planarity testing problem is strip planar. Recall that the associated directed graph of (G, γ) is the directed plane graph $\overrightarrow{G}$ obtained from (G, γ) by orienting each edge (u, v) in G from u to v if and only if $\gamma(v) = \gamma(u) + 1$. We have the following:

Lemma 4. *A jagged instance (G, γ) of the strip planarity testing problem is strip planar if and only if the associated directed graph $\overrightarrow{G}$ of (G, γ) is upward planar.*

Proof Sketch: The necessity is trivial. For the sufficiency, we first insert dummy edges in $\overrightarrow{G}$ to augment it to a *plane st-digraph* $\overrightarrow{G}_{st}$, which is an upward planar directed graph with exactly one source s and one sink t incident to its outer face [11]. Each face f of $\overrightarrow{G}_{st}$ consists of two monotone paths, called *left path* and *right path*, where the left path has f to the right when traversing it from its source to its sink. The inserted dummy edges only connect two sources or two sinks of each face of $\overrightarrow{G}$. Since (G, γ) is jagged, the end-vertices of each dummy edge are in the same strip.

We divide the plane into k horizontal strips. We compute an upward planar drawing of $\overrightarrow{G}_{st}$ starting from a y-monotone drawing of the leftmost path of $\overrightarrow{G}_{st}$ and adding to the drawing one face at a time, in an order corresponding to any linear extension of the partial order of the faces induced by the directed dual graph of $\overrightarrow{G}_{st}$ [11]. When a face is added to the drawing, its left path is already drawn as a y-monotone curve. We draw the right path of f as a y-monotone curve in which each vertex u lies inside strip $\gamma(u)$, hence the rightmost path of the graph in the current drawing is always represented by a y-monotone curve. A strip planar drawing of (G, γ) can be obtained from the drawing of $\overrightarrow{G}_{st}$ by removing the dummy edges.   □

We thus obtain the following:

Theorem 1. *The strip planarity testing problem can be solved in polynomial time for instances (G, γ) such that G is a plane graph.*

Proof: By Lemmata 1–3, it is possible to reduce in polynomial time any instance of the strip planarity testing problem to an equivalent jagged instance (G, γ). By Lemma 4, (G, γ) is strip planar if and only if the associated directed plane graph $\overrightarrow{G}$ of (G, γ) is upward planar. Finally, by the results of Bertolazzi et al. [5], the upward planarity of $\overrightarrow{G}$ can be tested in polynomial time.   □

4 Conclusions

In this paper, we introduced the strip planarity testing problem and showed how to solve it in polynomial time if the input graph is 2-connected and has a prescribed plane embedding. We now sketch how to remove the 2-connectivity requirement.

Suppose that the input graph (G, γ) is simply-connected (possibly not 2-connected). The algorithmic steps are the same. The transformation of a general instance into a strict instance is exactly the same. The transformation of a strict instance into a quasi-jagged instance has some differences with respect to the 2-connected case. In fact, the visibility between local minima and maxima for a face f of G is redefined with respect to *occurrences* of such minima and maxima along f. Thus, the goal of such a transformation is to create an instance in which, for every face f and for every pair of visible occurrences $\sigma_i(u_m)$ and $\sigma_j(u_M)$ of a local minimum u_m and a local maximum u_M for f, respectively, there is a monotone path between $\sigma_i(u_m)$ and $\sigma_j(u_M)$ in C_f. This is done with the same techniques as in Claim 3. The transformation of a quasi-jagged instance into a jagged instance is almost the same as in the 2-connected case, except that the 2-connected components of G inside a face f have to be suitably squeezed along the monotone paths of f to allow for a drawing of a monotone path between v'' and z or for a drawing of plane graph $A(u_M, v_M, f)$ This is done with the same techniques as in Claims 4 and 5. Finally, the proof of the equivalence between the strip planarity of a jagged instance and the upward planarity of its associated directed graph holds as it is.

Suppose now that the input graph (G, γ) is not connected. Test individually the strip planarity of each connected component of (G, γ). If one of the tests fails, then (G, γ) is not strip planar. Otherwise, construct a strip planar drawing of each connected component of (G, γ). Place the drawings of the connected components containing edges incident to the outer face of G side by side. Repeatedly insert connected components in the internal faces of the currently drawn graph (G', γ) as follows. If a connected component (G_i, γ) of (G, γ) has to be placed inside an internal face f of (G', γ), check whether $\gamma(u_M) \leq \gamma(u_M^f)$ and whether $\gamma(u_m) \geq \gamma(u_m^f)$, where u_M (u_m) is a vertex of (G_i, γ) such that $\gamma(u_M)$ is maximum (resp. $\gamma(u_m)$ is minimum) among the vertices of G_i, and where u_M^f (u_m^f) is a vertex of C_f such that $\gamma(u_M^f)$ is maximum (resp. $\gamma(u_m^f)$ is minimum) among the vertices of C_f. If the test fails, then (G, γ) is not strip planar. Otherwise, using a technique analogous to the one of Claim 3, a strip planar drawing of (G', γ) can be modified so that two consecutive global minimum and maximum for f can be connected by a y-monotone curve $\mathcal{C}$ inside f. Suitably squeezing a strip planar drawing of (G_i, γ) and placing it arbitrarily close to $\mathcal{C}$ provides a strip planar drawing of $(G' \cup G_i, \gamma)$. Repeating such an argument leads either to conclude that (G, γ) is not strip planar, or to construct a strip planar drawing of (G, γ).

The main question raised by this paper is whether the strip planarity testing problem can be solved in polynomial time or is rather $\mathcal{NP}$-hard for graphs without a prescribed plane embedding. The problem is intriguing even if the input graph is a tree.

References

1. Angelini, P., Di Battista, G., Frati, F., Jelínek, V., Kratochvíl, J., Patrignani, M., Rutter, I.: Testing planarity of partially embedded graphs. In: Charikar, M. (ed.) SODA 2010, pp. 202–221. ACM (2010)

2. Angelini, P., Frati, F., Kaufmann, M.: Straight-line rectangular drawings of clustered graphs. Discrete & Computational Geometry 45(1), 88–140 (2011)
3. Angelini, P., Da Lozzo, G., Di Battista, G., Frati, F.: Strip planarity testing of embedded planar graphs. ArXiv e-prints 1309.0683 (September 2013)
4. Bachmaier, C., Brandenburg, F.J., Forster, M.: Radial level planarity testing and embedding in linear time. JGAA 9(1), 53–97 (2005)
5. Bertolazzi, P., Di Battista, G., Liotta, G., Mannino, C.: Upward drawings of triconnected digraphs. Algorithmica 12(6), 476–497 (1994)
6. Biedl, T.C., Kaufmann, M., Mutzel, P.: Drawing planar partitions II: HH-drawings. In: Hromkovič, J., Sýkora, O. (eds.) WG 1998. LNCS, vol. 1517, pp. 124–136. Springer, Heidelberg (1998)
7. Booth, K.S., Lueker, G.S.: Testing for the consecutive ones property, interval graphs, and graph planarity using PQ-tree algorithms. J. Comput. Syst. Sci. 13(3), 335–379 (1976)
8. Cortese, P.F., Di Battista, G., Patrignani, M., Pizzonia, M.: Clustering cycles into cycles of clusters. JGAA 9(3), 391–413 (2005)
9. Cortese, P.F., Di Battista, G., Patrignani, M., Pizzonia, M.: On embedding a cycle in a plane graph. Discrete Mathematics 309(7), 1856–1869 (2009)
10. Di Battista, G., Frati, F.: Efficient c-planarity testing for embedded flat clustered graphs with small faces. JGAA 13(3), 349–378 (2009)
11. Di Battista, G., Tamassia, R.: Algorithms for plane representations of acyclic digraphs. Theor. Comput. Sci. 61, 175–198 (1988)
12. Eades, P., Feng, Q., Lin, X., Nagamochi, H.: Straight-line drawing algorithms for hierarchical graphs and clustered graphs. Algorithmica 44(1), 1–32 (2006)
13. Estrella-Balderrama, A., Fowler, J.J., Kobourov, S.G.: On the characterization of level planar trees by minimal patterns. In: Eppstein, D., Gansner, E.R. (eds.) GD 2009. LNCS, vol. 5849, pp. 69–80. Springer, Heidelberg (2010)
14. Forster, M., Bachmaier, C.: Clustered level planarity. In: Van Emde Boas, P., Pokorný, J., Bieliková, M., Štuller, J. (eds.) SOFSEM 2004. LNCS, vol. 2932, pp. 218–228. Springer, Heidelberg (2004)
15. Fowler, J.J., Kobourov, S.G.: Minimum level nonplanar patterns for trees. In: Hong, S.-H., Nishizeki, T., Quan, W. (eds.) GD 2007. LNCS, vol. 4875, pp. 69–75. Springer, Heidelberg (2008)
16. Garg, A., Tamassia, R.: On the computational complexity of upward and rectilinear planarity testing. SIAM J. Comput. 31(2), 601–625 (2001)
17. Gutwenger, C., Klein, K., Mutzel, P.: Planarity testing and optimal edge insertion with embedding constraints. JGAA 12(1), 73–95 (2008)
18. Healy, P., Kuusik, A., Leipert, S.: A characterization of level planar graphs. Discrete Mathematics 280(1-3), 51–63 (2004)
19. Hong, S.H., Nagamochi, H.: Two-page book embedding and clustered graph planarity. Tech. Report 2009-004, Dept. of Applied Mathematics & Physics, Kyoto University (2009)
20. Hopcroft, J.E., Tarjan, R.E.: Efficient planarity testing. J. ACM 21(4), 549–568 (1974)
21. Hutton, M.D., Lubiw, A.: Upward planarity testing of single-source acyclic digraphs. SIAM J. Comput. 25(2), 291–311 (1996)
22. Jelínek, V., Kratochvíl, J., Rutter, I.: A kuratowski-type theorem for planarity of partially embedded graphs. Comput. Geom. Theory Appl. 46(4), 466–492 (2013)
23. Jelínková, E., Kára, J., Kratochvíl, J., Pergel, M., Suchý, O., Vyskocil, T.: Clustered planarity: Small clusters in cycles and Eulerian graphs. JGAA 13(3), 379–422 (2009)
24. Jünger, M., Leipert, S., Mutzel, P.: Level planarity testing in linear time. In: Whitesides, S.H. (ed.) GD 1998. LNCS, vol. 1547, pp. 224–237. Springer, Heidelberg (1999)
25. Schaefer, M.: Toward a theory of planarity: Hanani-tutte and planarity variants. In: Didimo, W., Patrignani, M. (eds.) GD 2012. LNCS, vol. 7704, pp. 162–173. Springer, Heidelberg (2013)

Morphing Planar Graph Drawings Efficiently[*]

Patrizio Angelini[1], Fabrizio Frati[2], Maurizio Patrignani[1], and Vincenzo Roselli[1]

[1] Engineering Department, Roma Tre University, Italy
{angelini,patrignani,roselli}@dia.uniroma3.it
[2] School of Information Technologies, The University of Sydney, Australia
brillo@it.usyd.edu.au

Abstract. A morph between two straight-line planar drawings of the same graph is a continuous transformation from the first to the second drawing such that planarity is preserved at all times. Each step of the morph moves each vertex at constant speed along a straight line. Although the existence of a morph between any two drawings was established several decades ago, only recently it has been proved that a polynomial number of steps suffices to morph any two planar straight-line drawings. Namely, at SODA 2013, Alamdari *et al.* [1] proved that any two planar straight-line drawings of a planar graph can be morphed in $O(n^4)$ steps, while $O(n^2)$ steps suffice if we restrict to maximal planar graphs.

In this paper, we improve upon such results, by showing an algorithm to morph any two planar straight-line drawings of a planar graph in $O(n^2)$ steps; further, we show that a morph with $O(n)$ steps exists between any two planar straight-line drawings of a series-parallel graph.

1 Introduction

A *planar morph* between two planar drawings of the same plane graph is a continuous transformation from the first drawing to the second one such that planarity is preserved at all times. The problem of deciding whether a planar morph exists for any two drawings of any graph dates back to 1944, when Cairns [7] proved that any two straight-line drawings of a maximal planar graph can be morphed one into the other while maintaining planarity. In 1981, Grünbaum and Shephard [10] introduced the concept of *linear morph*, that is a continuous transformation in which each vertex moves at uniform speed along a straight-line trajectory. With this further requirement, however, planarity cannot always be maintained for any pair of drawings. Hence, the problem has been subsequently studied in terms of the existence of a sequence of linear morphs, also called *morphing steps*, transforming a drawing into another while maintaining planarity. The first result in this direction is the one of Thomassen [13], who proved that a sequence of morphing steps always exists between any two straight-line drawings of the same plane graph. Further, if the two input drawings are convex, this property is maintained throughout the morph, as well. However, the number of morphing steps used by the algorithm of Thomassen might be exponential in the number of vertices.

---

[*] Part of the research was conducted in the framework of ESF project 10-EuroGIGA-OP-003 GraDR "Graph Drawings and Representations".

S. Wismath and A. Wolff (Eds.): GD 2013, LNCS 8242, pp. 49–60, 2013.

Recently, the problem of computing planar morphs gained increasing research attention. The case in which edges are not required to be straight-line segments has been addressed in [11], while morphs between orthogonal graph drawings preserving planarity and orthogonality have been explored in [12]. Morphs preserving more general edge directions have been considered in [6]. Also, the problem of "topological morphing", in which the planar embedding is allowed to change, has been addressed in [2].

In a paper appeared at SODA 2013, Alamdari *et al.* [1] tackled again the original setting in which edges are straight-line segments and linear morphing steps are required. Alamdari *et al.* presented the first morphing algorithms with a polynomial number of steps in this setting. Namely, they presented an algorithm to morph straight-line planar drawings of maximal plane graphs with $O(n^2)$ steps and of general plane graphs with $O(n^4)$ steps, where n is the number of vertices of the graph.

In this paper we improve upon the result of Alamdari *et al.* [1], providing a more efficient algorithm to morph general plane graphs. Namely, our algorithms uses $O(n^2)$ linear morphing steps. Further, we provide a morphing algorithm with a linear number of steps for a non-trivial class of planar graphs, namely series-parallel graphs. These two main results are summarized in the following theorems.

Theorem 1. *Let Γ_a and Γ_b be two drawings of the same plane series-parallel graph G. There exists a morph $\langle \Gamma_a, \ldots, \Gamma_b \rangle$ with $O(n)$ steps transforming Γ_a into Γ_b.*

Theorem 2. *Let Γ_s and Γ_t be two drawings of the same plane graph G. There exists a morph $\langle \Gamma_s, \ldots, \Gamma_t \rangle$ with $O(n^2)$ steps transforming Γ_s into Γ_t.*

The rest of the paper is organized as follows. Section 2 contains preliminaries and basic terminology. Section 3 describes an algorithm to morph series-parallel graphs. Section 4 describes an algorithm to morph plane graphs. Section 5 provides geometric details for the morphs described in Sections 3 and 4. Finally, Section 6 contains conclusions and open problems. Because of space limitations, some proofs are omitted or sketched. Full proofs can be found in the extended version of the paper [4].

2   Preliminaries

A *straight-line planar drawing* Γ (in the following simply *drawing*) of a graph $G(V, E)$ maps vertices in V to distinct points of the plane and edges in E to non-intersecting open straight-line segments between their end-vertices. Given a vertex v of a graph G, we denote by $\deg(v)$ the *degree* of v in G, that is, the number of vertices adjacent to v. A planar drawing Γ partitions the plane into connected regions called *faces*. The unbounded face is the *external face*. Also, Γ determines a clockwise order of the edges incident to each vertex. Two planar drawings are *equivalent* if they determine the same clockwise ordering of the incident edges around each vertex and if they have the same external face. A *planar embedding* is an equivalence class of planar drawings. A *plane* graph is a planar graph with a given planar embedding.

A *series-parallel graph* G is a planar graph that does not contain the complete graph on four vertices as a minor. A *plane* series-parallel graph is a graph together with a planar embedding. Let G be a plane biconnected series-parallel graph and let e be an edge incident to its outer face. Graph G has a unique *decomposition tree* T_e rooted at e having nodes of three types: Q-, S-, and P-nodes. A Q-node represents a single edge, while an S-node (a P-node) μ represents a series (a parallel, respectively) composition of the sugraphs associated to the subtrees of T_e rooted at the children of μ. An embedding of G naturally induces an ordering for the children of each node of T_e.

A *(linear) morphing step* $\langle \Gamma_1, \Gamma_2 \rangle$, also referred to as *linear morph*, of two straight-line planar drawings Γ_1 and Γ_2 of a plane graph G is a continuous transformation of Γ_1 into Γ_2 such that all the vertices simultaneously start moving from their positions in Γ_1 and, moving along a straight-line trajectory, simultaneously stop at their positions in Γ_2 so that no crossing occurs between any two edges during the transformation. A *morph* $\langle \Gamma_s, \ldots, \Gamma_t \rangle$ of two straight-line planar drawings Γ_s into Γ_t of a plane graph G is a finite sequence of morphing steps that transforms Γ_s into Γ_t. Let u and w be two vertices of G such that edge (u, w) belongs to G and let Γ be a straight-line planar drawing of G. The *contraction* of u onto w results in (i) a graph $G' = G/(u, w)$ not containing u and such that each edge (u, x) of G is replaced by an edge (w, x) in G', and (ii) a straight-line drawing Γ' of G' such that each vertex different from v is mapped to the same point as in Γ. In the following, the contraction of an edge (u, w) will be only applied if the obtained drawing Γ' is planar. The *uncontraction* of u from w in Γ' yields a straight-line planar drawing Γ'' of G. A morph in which contractions are performed, possibly together with other morphing steps, is a *pseudo-morph*. Let v be a vertex of G and let G' be the graph obtained by removing v and its incident edges from G. Let Γ' be a planar straight-line drawing of G'. The *kernel* of v in Γ' is the set P of points such that straight-line segments can be drawn in Γ' connecting each point $p \in P$ to each neighbor of v in G without intersecting any edge in Γ'.

3   Morphing Series-Parallel Graph Drawings in $O(n)$ Steps

In this section we show an algorithm to compute a pseudo-morph between any two drawings of the same plane series-parallel graph G. In Section 3.1 we assume that G is biconnected, and in Section 3.2 we show how to remove this assumption, thus proving the following theorem.

Theorem 3. *Let Γ_a and Γ_b be two drawings of the same plane series-parallel graph G. There exists a pseudo-morph $\langle \Gamma_a, \ldots, \Gamma_b \rangle$ with $O(n)$ steps transforming Γ_a into Γ_b .*

3.1   Biconnected Series-Parallel Graphs

Our approach consists of morphing any drawing Γ of a biconnected plane series-parallel graph G into a "canonical drawing" Γ^* of G in a linear number of steps. As a consequence, any two drawings Γ_1 and Γ_2 of G can be transformed one into the other in a linear number of steps, by morphing Γ_1 to Γ^* and Γ^* to Γ_2.

A *canonical drawing* Γ^* of a biconnected plane series-parallel graph G is defined as follows. The decomposition tree T_e of G is traversed top-down and a suitable geometric region of the plane is assigned to each node μ of T_e; such a region will contain the drawing of the series-parallel graph associated with μ. The regions assigned to the nodes

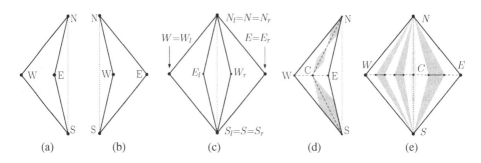

Fig. 1. (a) A left boomerang. (b) A right boomerang. (c) A diamond. (d) Diamonds inside a boomerang. (e) Boomerangs (and a diamond) inside a diamond.

of T_e are similar to those used in [5,3] to construct monotone drawings. Namely, we define three types of regions: Left boomerangs, right boomerangs, and diamonds. A *left boomerang* is a quadrilateral with vertices N, E, S, and W such that E is inside triangle $\triangle(N, S, W)$, where $|\overline{NE}| = |\overline{SE}|$ and $|\overline{NW}| = |\overline{SW}|$ (see Fig. 1(a)). A *right boomerang* is defined symmetrically, with E playing the role of W, and vice versa (see Fig. 1(b)). A *diamond* is a convex quadrilateral with vertices N, E, S, and W, where $|\overline{NW}| = |\overline{NE}| = |\overline{SW}| = |\overline{SE}|$. Observe that a diamond contains a left boomerang N_l, E_l, S_l, W_l and a right boomerang N_r, E_r, S_r, W_r, where $S = S_l = S_r$, $N = N_l = N_r$, $W = W_l$, and $E = E_r$ (see Fig. 1(c)).

We assign boomerangs (either left or right, depending on the embedding of G) to S-nodes and diamonds to P- and Q-nodes, as follows.

First, consider the Q-node ρ corresponding to the root edge e of G. Draw edge e as a segment between points $(0, 1)$ and $(0, -1)$. Also, if ρ is adjacent to an S-node μ, then assign to μ the left boomerang $N = (0, 1), E = (-1, 0), S = (0, -1), W = (-2, 0)$ or the right boomerang $N = (0, 1), E = (2, 0), S = (0, -1), W = (1, 0)$, depending on the embedding of G; if ρ is adjacent to a P-node μ, then associate to μ the diamond $N = (0, 1), E = (+2, 0), S = (0, -1), W = (-2, 0)$.

Then, consider each node μ of $T_e(G)$ according to a top-down traversal.

If μ is an S-node (see Fig. 1(d)), let N, E, S, W be the boomerang associated with it and let α be the angle $\widehat{WNE}$. We associate diamonds to the children $\mu_1, \mu_2, \ldots, \mu_k$ of μ as follows. Consider the midpoint C of segment $\overline{WE}$. Subdivide $\overline{NC}$ into $\lceil \frac{k}{2} \rceil$ segments with the same length and $\overline{CS}$ into $\lfloor \frac{k}{2} \rfloor$ segments with the same length. Enclose each of such segments $\overline{N_i S_i}$, for $i = 1, \ldots, k$, into a diamond N_i, E_i, S_i, W_i, with $\widehat{W_i N_i E_i} = \alpha$, and associate it with child μ_i of μ.

If μ is a P-node (see Fig. 1(e)), let N, E, S, W be the diamond associated with it. Associate boomerangs and diamonds to the children $\mu_1, \mu_2, \ldots, \mu_k$ of μ as follows. If a child μ_l of μ is a Q-node, then left boomerangs are associated to $\mu_1, \ldots, \mu_{l-1}$, right boomerangs are associated to $\mu_{l+1}, \ldots, \mu_k$, and a diamond is associated to μ_l. Otherwise, right boomerangs are associated to all of $\mu_1, \mu_2, \ldots, \mu_k$. We assume that a child μ_l of μ that is a Q-node exists, the description for the case in which no child of μ is a Q-node being similar and simpler. We describe how to associate left boomerangs to the children $\mu_1, \mu_2, \ldots, \mu_{l-1}$ of μ. Consider the midpoint C of segment $\overline{WE}$ and consider $2l$ equidistant points $W = p_1, \ldots, p_{2l} = C$ on segment $\overline{WC}$. Associate each child μ_i, with $i = 1, \ldots, l-1$, to the quadrilateral $N_i = N, E_i = p_{2i}, S_i = S, W_i = p_{2i+1}$. Right boomerangs are associated to $\mu_{l+1}, \mu_{l+2}, \ldots, \mu_k$ in a symmetric way. Finally, associate μ_l to any diamond such that $N_l = N, S_l = S, W_l$ is any point between C and E_{l-1}, and E_l is any point between C and W_{l+1}.

If μ is a Q-node, let N, E, S, W be the diamond associated with it. Draw the edge corresponding to μ as a straight-line segment between N and S.

Observe that the above described algorithm constructs a drawing of G, that we call the *canonical drawing* of G. We now argue that no two edges e_1 and e_2 intersect in the canonical drawing of G. Consider the lowest common ancestor ν of the Q-nodes τ_1 and τ_2 of T_e representing e_1 and e_2, respectively. Also, consider the children ν_1 and ν_2 of ν such that the subtree of T_e rooted at ν_i contains τ_i, for $i = 1, 2$. Such children are associated with internally-disjoint regions of the plane. Since the subgraphs G_1 and G_2 of G corresponding to ν_1 and ν_2, respectively, are entirely drawn inside such regions, it follows that e_1 and e_2 do not intersect except, possibly, at common endpoints.

In order to construct a pseudo-morph of a straight-line planar drawing $\Gamma(G)$ of G into its canonical drawing $\Gamma^*(G)$, we do the following: (i) We perform a contraction of a vertex v of G into a neighbor of v, hence obtaining a drawing $\Gamma(G')$ of a graph G' with $n - 1$ vertices; (ii) we inductively construct a pseudo-morph from $\Gamma(G')$ to the canonical drawing $\Gamma^*(G')$ of G'; and (iii) we uncontract v and perform a sequence of morphing steps to transform $\Gamma^*(G')$ into the canonical drawing $\Gamma^*(G)$ of G.

We describe the three steps in more detail.

Let $T_e(G)$ be the decomposition tree of G rooted at some edge e incident to the outer face of G. Consider a P-node ν such that the subtree of $T_e(G)$ rooted at ν does not contain any other P-node. This implies that all the children of ν, with the exception of at most one Q-node, are S-nodes whose children are Q-nodes. Hence, the series-parallel graph $G(\nu)$ associated to ν is composed of a set of paths connecting its poles s and t. Let p_1 and p_2 be two paths joining s and t and such that their union is a cycle $\mathcal{C}$ not containing other vertices in its interior (see Fig. 2(a)). Such paths exist given that the "rest of the graph" with respect to ν is in the outer face of $G(\nu)$, since the root e of $T_e(G)$ is incident to the outer face of G. Internally triangulate $\mathcal{C}$ by adding dummy edges (dashed edges of Fig. 2). Cycle $\mathcal{C}$ and the added dummy edges yield a drawing of a biconnected outerplane graph O which, hence, has at least two vertices of degree two.

If there exists a vertex v with $\deg(v) = 2$ and $v \neq s, t$ (*Case 1*), then apply the following contraction. Assume that v belongs to p_2. Since O is internally triangulated, both the neighbors v_1 and v_2 of v belong to p_2, and they are joined by a dummy edge. We obtain $\Gamma(G')$ from $\Gamma(G)$ by contracting v onto one of its neighbors, while preserving planarity (see Figs. 2(a) and 2(b)). If p_2 contains more than two edges (*Case 1.1*), then p_2 is replaced in G' with a path p_2' that contains edge (v_1, v_2) and does not contain vertex v. Otherwise, p_2 contains exactly two edges (v, v_1) and (v, v_2). If there exists edge (v_1, v_2) in G (*Case 1.2*), then $G' = G \setminus \{v\}$. Finally, if edge (v_1, v_2) does not exist in G (*Case 1.3*), then p_2 is replaced in G' with edge (v_1, v_2). Otherwise, the only two vertices of degree 2 in O are s and t (*Case 2*). In this case, one of the two vertices u_1 and u_2 of O adjacent to s has degree 3, say u_2 (since removing s and its incident

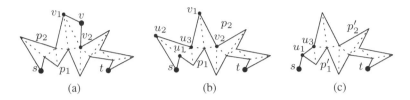

Fig. 2. The internally triangulated cycle C formed by paths p_1 and p_2. Dummy edges are drawn as dashed lines. (a–b) Vertex v of degree 2 can be contracted onto v_1. (b–c) Vertex u_2 of degree 3 can be contracted onto u_1.

edges from O yields another biconnected outerplane graph with two vertices of degree 2, namely t and one of u_1 and u_2). We obtain $\Gamma(G')$ from $\Gamma(G)$ by contracting u_2 onto u_1. Let u_3 be the neighbor of u_1 and u_2 different from s. Since the edges incident to u_2 are contained into triangles $\triangle_{s,u_1,u_2}$ and $\triangle_{u_1,u_2,u_3}$ during the contraction, planarity is preserved (see Figs. 2(b) and 2(c)). Let p_2' be the path composed of edge (u_1, u_3) and of the subpath of p_2 between u_3 and t, and let p_1' be the subpath of p_1 between u_1 and t. Note that G' contains edge (u_1, u_3) and does not contain vertex u_2. In both *Case 1* and *Case 2*, the decomposition tree $T_e(G')$ of G' differs from the decomposition tree $T_e(G)$ of G only "locally" to v. A precise description of the differences between $T_e(G)$ and $T_e(G')$ can be found in the extended version of the paper [4].

Let $\Gamma(G')$ be the drawing of the graph $G' = G \setminus \{v\}$ obtained after the contraction performed in *Case 1* or *Case 2*. Inductively construct a pseudo-morphing from $\Gamma(G')$ to the canonical drawing $\Gamma^*(G')$ of G' in $c \cdot (n-1)$ steps, where c is a constant. Drawing $\Gamma^*(G)$ can be obtained from $\Gamma^*(G')$ by uncontracting v and by performing a constant number of morphing steps, as described in the following.

Here we only describe how to obtain $\Gamma^*(G)$ from $\Gamma^*(G')$ if *Case 1.1* was applied to contract v into one of its neighbors in p_2. The other cases can be handled in a similar way (a full description can be found in the extended version of the paper [4]).

Drawings $\Gamma^*(G')$ and $\Gamma^*(G)$ coincide except for the fact that path p_2 in $\Gamma^*(G)$ contains v, while path p_2' in $\Gamma^*(G')$ does not contain v. Paths p_2' and p_2 are drawn inside two equal boomerangs in $\Gamma^*(G')$ and in $\Gamma^*(G)$, respectively, however v and some of the vertices of p_2' need to be moved in order to obtain the drawing of p_2 as in $\Gamma^*(G')$. Namely, the drawing $\Gamma^*(p_2')$ of p_2' inside the boomerang N, E, S, W associated to τ_2 in $\Gamma^*(G')$ is composed of edges lying on two straight-line segments $\overline{NC}$ and $\overline{SC}$, where C is the midpoint of segment $\overline{EW}$ (see Fig. 3(a)). The drawing $\Gamma^*(p_2)$ of p_2 in $\Gamma^*(G)$ also lies inside N, E, S, W and is composed of edges lying on $\overline{NC}$ and $\overline{SC}$, but vertices lie on different points (see Fig. 3(e)).

With one morphing step, uncontract v from the vertex it had been contracted onto and place it on any point of segment $\overline{v_1v_2}$ (note that edge (v_1, v_2) exists in G' and not in G; see Fig. 3(b)). Then, in order to redistribute the vertices of p_2 on $\overline{NC}$ and $\overline{SC}$, perform the following operation. Assume w.l.o.g. that s is on point N and t is on point S in $\Gamma^*(G')$ and in $\Gamma^*(G)$. Consider the vertices $w \in p_2$ and $w' \in p_2'$ that are placed on point C in $\Gamma^*(G)$ and $\Gamma^*(G')$, respectively. Note that either $w = w'$ or $(w, w') \in p_2$. If $w = w'$, either the subpath $p_2(s, w)$ of p_2 between s and w or the subpath $p_2(w, t)$ of p_2 between w and t has the same drawing in $\Gamma^*(G)$ and $\Gamma^*(G')$, say $p_2(w, t)$ has

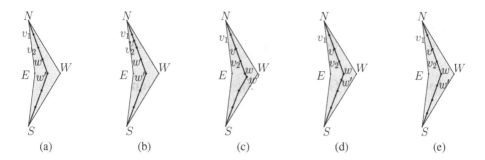

Fig. 3. Construction of $\Gamma^*(G)$ from $\Gamma^*(G')$ when *Case* 1.1 applied. (a) $\Gamma^*(p_2')$. The boomerang associated to τ_2 is light-grey. (b) Vertex v is uncontracted and placed on segment $\overline{v_1 v_2}$. (c) Vertices on the path between s and w are placed in their final position, and vertex w' is placed arbitrarily close to C on the elongation of $\overline{NC}$. (d) Vertex w' is placed on $\overline{CS}$. (e) Vertices on the path between w and t are placed in their final position, hence obtaining $\Gamma^*(G)$.

such a property. With one morphing step move the vertices of $p_2(s, w)$ on segment $\overline{NC}$ till reaching their positions in $\Gamma^*(G)$. If $w \neq w'$, assume without loss of generality that $w \in p_2(s, w')$. With one morphing step, move the vertices of $p_2(s, w)$ and vertex w' along the line through N and C, so that the vertices of $p_2(s, w)$ reach their positions in $\Gamma^*(G)$ and w' is placed arbitrarily close to C on the elongation of $\overline{NC}$ (see Fig. 3(c)). With a second morphing step, move w' to any point of $\overline{SC}$ between w and its other neighbor in p_2 (see Fig. 3(d)). Finally, with a third morphing step, move the vertices of $p_2(w, t)$ on segment $\overline{SC}$ till reaching their positions in $\Gamma^*(G)$ (see Fig. 3(e)).

3.2 Simply-Connected Series-Parallel Graphs

In this section we show how, by preprocessing the input drawings Γ_a and Γ_b of any series-parallel graph G, the algorithm presented in Section 3.1 can be used to compute a pseudo-morph $M = \langle \Gamma_a, \ldots, \Gamma_b \rangle$. The idea is to augment both Γ_a and Γ_b to two drawings Γ_a' and Γ_b' of a biconnected series-parallel graph G', compute the morph $M' = \langle \Gamma_a', \ldots, \Gamma_b' \rangle$, and obtain M by restricting M' to the vertices and edges of G.

This augmentation is performed on G by repeatedly applying the following lemma.

Lemma 1. *Let v be a cut-vertex of a plane series-parallel graph G with n_b blocks. Let $e_1 = (u, v)$ and $e_2 = (w, v)$ be two consecutive edges in the circular order around v such that e_1 belongs to block b_1 of G and e_2 belongs to block $b_2 \neq b_1$ of G. The graph G^* obtained from G by adding a vertex z and edges (u, z) and (w, z) is a plane series-parallel graph with $n_b - 1$ blocks.*

Observe that, when augmenting G to G^*, both Γ_a and Γ_b can be augmented to two planar straight-line drawings Γ_a^* and Γ_b^* of G^* by placing vertex z suitably close to v and with direct visibility to vertices u and w, as in the proof of Fáry's Theorem [9]. By repeatedly applying such an augmentation we obtain a biconnected series-parallel graph G' and its drawings Γ_a' and Γ_b', whose number of vertices and edges is linear in the size of G. Hence, the algorithm described in Section 3.1 can be applied to obtain a pseudo-morph $\langle \Gamma_a, \ldots, \Gamma_b \rangle$, thus proving Theorem 3. We will show in Section 5 how to obtain a morph starting from the pseudo-morph computed in this section.

4   Morphing Plane Graph Drawings in $O(n^2)$ Steps

In this section we prove the following theorem.

Theorem 4. *Let Γ_s and Γ_t be two drawings of the same plane graph G. There exists a pseudo-morph $\langle \Gamma_s, \ldots, \Gamma_t \rangle$ with $O(n^2)$ steps transforming Γ_s into Γ_t.*

Preliminary Definitions. Let Γ be a planar straight-line drawing of a plane graph G. A face f of G is *empty* in Γ if it is delimited by a simple cycle. Consider a vertex v of G and let v_1 and v_2 be two of its neighbors. Vertices v_1 and v_2 are *consecutive* neighbors of v if no edge appears between edges (v, v_1) and (v, v_2) in the circular order of the edges around v in Γ. Let v be a vertex with $\deg(v) \leq 5$ such that each face containing v on its boundary is empty. We say that v is *contractible* [1] if, for each two neighbors u_1 and u_2 of v, edge (u_1, u_2) exists in G *if and only if* u_1 and u_2 are consecutive neighbors of v. We say that v is *quasi-contractible* if, for each two neighbors u_1 and u_2 of v, edge (u_1, u_2) exists in G *only if* u_1 and u_2 are consecutive neighbors of v. In other words, no edge exists between non-consecutive neighbors of a contractible or quasi-contractible vertex; also, each face incident to a contractible vertex v is delimited by a 3-cycle, while a face incident to a quasi-contractible vertex might have more than three incident vertices. We have the following.

Lemma 2. *Every planar graph contains a quasi-contractible vertex.*

Further, given a neighbor x of v, we say that v is *x-contractible* onto x in Γ if: (i) v is quasi-contractible, and (ii) the contraction of v onto x in Γ results in a straight-line planar drawing Γ' of $G' = G/(v, x)$.

The Algorithm. We describe the main steps of our algorithm to pseudo-morph a drawing Γ_s of a plane graph G into another drawing Γ_t of G.

First, we consider a quasi-contractible vertex v of G, that exists by Lemma 2. Second, we compute a pseudo-morph with $O(n)$ steps of Γ_s into a drawing Γ_s^x of G and a pseudo-morph with $O(n)$ steps of Γ_t into a drawing Γ_t^x of G, such that v is x-contractible onto the same neighbor x both in Γ_s^x and in Γ_t^x. We will describe later how to perform these pseudo-morphs. Third, we contract v onto x both in Γ_s^x and in Γ_t^x, hence obtaining two drawings Γ_s' and Γ_t' of a graph $G' = G/(v, x)$ with $n - 1$ vertices. Fourth, we recursively compute a pseudo-morph transforming Γ_s' into Γ_t'. This completes the description of the algorithm for constructing a pseudo-morphing transforming Γ_s into Γ_t. Observe that the algorithm has $p(n) \in O(n^2)$ steps, thus proving Theorem 4. Namely, as it will be described later, $O(n)$ steps suffice to construct pseudo-morphings of Γ_s and Γ_t into drawings Γ_s^x and Γ_t^x of G, respectively, such that v is x-contractible onto the same neighbor x both in Γ_s^x and in Γ_t^x. Further, two steps are sufficient to contract v onto x in both Γ_s^x and Γ_t^x, obtaining drawings Γ_s'

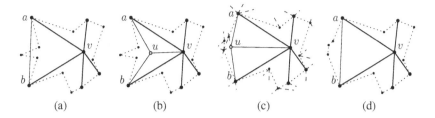

Fig. 4. Vertex v and its neighbors. (a) Vertices a and b do not have direct visibility and the triangle $\langle a, b, v \rangle$ is not empty. (b) A vertex u is added suitably close to v and connected to v, a, and b. (c) The output of CONVEXIFIER on the quadrilateral $\langle a, b, v, u \rangle$. (d) Vertex u and its incident edges can be removed in order to insert edge (a, b).

and Γ_t', respectively. Finally, the recursion on Γ_s' and Γ_t' takes $p(n-1)$ steps. Thus, $p(n) = p(n-1) + O(n) \in O(n^2)$. We will show in Section 5 how to obtain a morph starting from the pseudo-morph computed in this section.

We remark that our approach is similar to the one proposed by Alamdari *et al.* [1]. In [1] Γ_s and Γ_t are augmented to drawings of the same maximal planar graph with $m \in O(n^2)$ vertices, and a morph with $O(m^2)$ steps is constructed between two drawings of the same m-vertex maximal planar graph. This results in a morphing between Γ_s and Γ_t with $O(n^4)$ steps. Here, we also augment Γ_s and Γ_t to drawings of maximal planar graphs. However, we only require that the two maximal planar graphs coincide in the subgraph induced by the neighbors of v. Since this can be achieved by adding a constant number of vertices to Γ_s and Γ_t, namely one for each of the at most five faces v is incident to, our morphing algorithm has $O(n^2)$ steps.

Making v x-contractible. Let v be a quasi-contractible vertex of G. We show an algorithm to construct a pseudo-morph with $O(n)$ steps transforming any straight-line planar drawing Γ of G into a straight-line planar drawing Γ' of G such that v is x-contractible onto any neighbor x. If v has degree 1, then it is contractible into its unique neighbor in Γ, and there is nothing to prove.

In order to transform Γ into Γ', we use a support graph S and its drawing Σ, initially set equal to G and Γ, respectively. The goal is to augment S and Σ so that v becomes a contractible vertex of S. In order to do this, we have to add to S an edge between every two consecutive neighbors of v. However, the insertion of these edges might not be possible in Σ, as it might lead to a crossing or to enclose some vertex inside a cycle delimited by v and by two consecutive neighbors of v (see Fig. 4(a)).

Let a and b be two consecutive neighbors of v. If the closed triangle $\langle a, b, v \rangle$ does not contain any vertex other than a, b, and v, then add edge (a, b) to S and to Σ as a straight-line segment. Otherwise, proceed as follows.

Let Σ_u be the drawing of a plane graph S_u obtained by adding a vertex u and the edges (u, v), (u, a), and (u, b) to Σ and to S, in such a way that the resulting drawing is straight-line planar and each face containing u on its boundary is empty. As in the proof of Fáry's Theorem [9], a position for u with such properties can be found in Σ, suitably close to v. See Fig. 4(b). Augment Σ_u to the drawing Θ of a maximal plane graph T

by first adding three vertices p, q, and r to Σ_u, so that triangle $\langle p, q, r \rangle$ encloses the rest of the drawing, and then adding dummy edges [8]. If edge (a, b) has been added in this augmentation (this can happen if a and b share a face not having v on its boundary), subdivide (a, b) in Θ (namely, replace (a, b) with edges (a, w) and (w, b), placing w along the straight-line segment connecting a and b) and triangulate the two faces vertex w is incident to. Next, apply the algorithm described in [1], that we call CONVEXIFIER, to construct a morph of Θ into a drawing Θ' of T in which polygon $\langle a, v, b, u \rangle$ is convex. The input of algorithm CONVEXIFIER consists of a planar straight-line drawing Γ^* of a plane graph G^* and of a set of at most five vertices of G^* inducing a biconnected outerplane graph not containing any other vertex in its interior in Γ^*. The output of algorithm CONVEXIFIER is a sequence of $O(n)$ linear morphing steps transforming Γ^* into a drawing of G^* in which the at most five input vertices bound a convex polygon. Since, by construction, vertices a, v, b, u satisfy all such requirements, we can apply algorithm CONVEXIFIER to Θ and to a, v, b, u, hence obtaining a morph with $O(n)$ steps transforming Θ into the desired drawing Θ' (see Fig. 4(c)). Let Σ'_u be the drawing of S_u obtained by restricting Θ' to vertices and edges of S_u. Since $\langle a, v, b, u \rangle$ is a convex polygon containing no vertex of S_u in its interior, edge (u, v) can be removed from Σ'_u and an edge (a, b) can be introduced in Σ'_u, so that the resulting drawing Σ' is planar and cycle (a, b, v) does not contain any vertex in its interior (see Fig. 4(d)).

Once edge (a, b) has been added to S (either in Σ or after the described procedure transforming Σ into Σ'), if $\deg(v) = 2$ then v is both a-contractible and b-contractible. Otherwise, consider a new pair of consecutive vertices of v not creating an empty triangular face with v, if any, and apply the same operations described before.

Once every pair of consecutive vertices has been handled, vertex v is contractible in S. Let Σ_v be the current drawing of S. Augment Σ_v to the drawing Θ_v of a triangulation T_v (by adding three vertices and a set of edges), contract v onto a neighbor w such that v is w-contractible (one of such neighbors always exists, given that v is contractible), and apply CONVEXIFIER to the resulting drawing Θ'_v and to the neighbors of v to construct a morphing Θ'_v to a drawing Σ'_v in which the polygon defined by such vertices is convex. Drawing Γ' of G in which v is x-contractible for any neighbor x of v is obtained by restricting Σ'_v to the vertices and the edges of G. We can now contract v onto x in Γ' and recur on the obtained graph (with $n - 1$ vertices) and drawing.

It remains to observe that, given a quasi-contractible vertex v, the procedure to construct a pseudo-morph of Γ into Γ' consists of at most $\deg(v) + 1$ executions of CONVEXIFIER, each requiring a linear number of steps [1]. As $\deg(v) \leq 5$, the procedure to pseudo-morph Γ into Γ' has $O(n)$ steps. This concludes the proof of Theorem 4.

5   Transforming a Pseudo-Morph into a Morph

In this section we show how to obtain an actual morph M from a given pseudo-morph $\mathcal{M}$, by describing how to compute the placement and the motion of any vertex v that has been contracted during $\mathcal{M}$. By applying this procedure to Theorems 3 and 4, we obtain a proof of Theorems 1 and 2.

Let Γ be a drawing of a graph G and let $\mathcal{M} = \langle \Gamma, \ldots, \Gamma^* \rangle$ be a pseudo-morph that consists of the contraction of a vertex v of G onto one of its neighbors x, followed by a pseudo-morph $\mathcal{M}'$ of the graph $G' = G/(v, x)$, and then of the uncontraction of v.

The idea of how to compute M from $\mathcal{M}$ is the same as in [1]: Namely, morph M is obtained by (i) recursively converting $\mathcal{M}'$ into a morph M'; (ii) modifying M' to a morph M'_v obtained by adding vertex v (and its incident edges) to each drawing of M', in a suitable position; (iii) replacing the contraction of v onto x, performed in $\mathcal{M}$, with a linear morph that moves v from its initial position in Γ to its position in the first drawing of M'_v; and (iv) replacing the uncontraction of v, performed in $\mathcal{M}$, with a linear morph that moves v from its position in the last drawing of M'_v to its final position in Γ^*. Note that, in order to guarantee the planarity of M when adding v to any drawing of M' in order to obtain M'_v, vertex v must lie inside its kernel. Since vertex x lies in the kernel of v (as x is adjacent to all the neighbors of v in G'), we achieve this property by placing v suitably close to x, as follows.

At any time instant t during M', there exists an $\epsilon_t > 0$ such that the disk D centered at x with radius ϵ_t does not contain any vertex other than x. Let ϵ be the minimum among the ϵ_t during M'. We place vertex v at a suitable point of a sector S of D according to the following cases. **Case (a): v has degree 1 in G.** Sector S is defined as the intersection of D with the face containing v in G. See Fig. 5(a). **Case (b): v has degree 2 in G.** Sector S is defined as the intersection of D with the face containing v in G and with the halfplane defined by the straight-line passing through x and r, and containing v in Γ. See Fig. 5(b). Otherwise, $\deg(v) \geq 3$ in G'. Let (r, v) and (l, v) be the two edges such that (r, v), (x, v), and (l, v) are clockwise consecutive around v in G. Observe that edges (r, x) and (l, x) exist in G'. Assume that x, r, and l are not collinear in any drawing of M', as otherwise we can slightly perturb such a drawing without compromising the planarity of M'. Let α_i be the angle $\widehat{lxr}$ in any intermediate drawing of M'. **Case (c): $\alpha_i < \pi$.** Sector S is defined as the intersection of D with the wedge delimited by edges (x, r) and (x, l). See Fig. 5(c). **Case (d): $\alpha_i > \pi$.** Sector S is defined as the intersection of D with the wedge delimited by the elongations of (x, r) and (x, l) emanating from x. See Fig. 5(d). By exploiting the techniques shown in [1], the motion of v can be computed according to the evolution of S over M', thus obtaining a planar morph M'_v.

Observe that, in the algorithm described in Section 4, the vertex x onto which v has been contracted might be not adjacent to v in G. However, since a contraction has been performed, x is adjacent to v in one of the graphs obtained when augmenting G during the algorithm. Hence, a morph of G can be obtained by applying the above procedure to the pseudo-morph computed on this augmented graph and by restricting it to the vertices and edges of G.

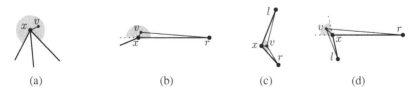

(a)   (b)   (c)   (d)

Fig. 5. Sector S (in grey) when: (a) $\deg(v) = 1$, (b) $\deg(v) = 2$, and (c)-(d) $\deg(v) \geq 3$.

6 Conclusions and Open Problems

In this paper we studied the problem of designing efficient algorithms for morphing two planar straight-line drawings of the same graph. We proved that any two planar straight-line drawings of a series-parallel graph can be morphed with $O(n)$ linear morphing steps, and that a planar morph with $O(n^2)$ linear morphing steps exists between any two planar straight-line drawings of any planar graph.

It is a natural open question whether the bounds we presented are optimal or not. We suspect that planar straight-line drawings exist requiring a linear number of steps to be morphed one into the other. However, no super-constant lower bound for the number of morphing steps required to morph planar straight-line drawings is known. It would be interesting to understand whether our techniques can be extended to compute morphs between any two drawings of a *partial planar* 3-*tree* with a linear number of steps. We recall that, as observed in [1], a linear number of morphing steps suffices to morph any two drawings of a *maximal planar* 3-*tree*.

References

1. Alamdari, S., Angelini, P., Chan, T.M., Di Battista, G., Frati, F., Lubiw, A., Patrignani, M., Roselli, V., Singla, S., Wilkinson, B.T.: Morphing planar graph drawings with a polynomial number of steps. In: SODA 2013, pp. 1656–1667 (2013)
2. Angelini, P., Cortese, P.F., Di Battista, G., Patrignani, M.: Topological morphing of planar graphs. In: Tollis, I.G., Patrignani, M. (eds.) GD 2008. LNCS, vol. 5417, pp. 145–156. Springer, Heidelberg (2009)
3. Angelini, P., Didimo, W., Kobourov, S., Mchedlidze, T., Roselli, V., Symvonis, A., Wismath, S.: Monotone drawings of graphs with fixed embedding. Algorithmica, 1–25 (2013)
4. Angelini, P., Frati, F., Patrignani, M., Roselli, V.: Morphing planar graph drawings efficiently. CoRR cs.CG (2013), http://arxiv.org/abs/1308.4291
5. Angelini, P., Colasante, E., Di Battista, G., Frati, F., Patrignani, M.: Monotone drawings of graphs. J. of Graph Algorithms and Appl. 16(1), 5–35 (2012); In: Brandes, U., Cornelsen, S. (eds.) GD 2010. LNCS, vol. 6502, pp. 13–24. Springer, Heidelberg (2011)
6. Biedl, T.C., Lubiw, A., Spriggs, M.J.: Morphing planar graphs while preserving edge directions. In: Healy, P., Nikolov, N.S. (eds.) GD 2005. LNCS, vol. 3843, pp. 13–24. Springer, Heidelberg (2006)
7. Cairns, S.S.: Deformations of plane rectilinear complexes. American Math. Monthly 51, 247–252 (1944)
8. Chazelle, B.: Triangulating a simple polygon in linear time. Discrete & Computational Geometry 6(5), 485–524 (1991)
9. Fáry, I.: On straight line representation of planar graphs. Acta Univ. Szeged. Sect. Sci. Math. 11, 229–233 (1948)
10. Grunbaum, B., Shephard, G.: The geometry of planar graphs. Cambridge University Press (1981), http://dx.doi.org/10.1017/CBO9780511662157.008
11. Lubiw, A., Petrick, M.: Morphing planar graph drawings with bent edges. Electronic Notes in Discrete Mathematics 31, 45–48 (2008)
12. Lubiw, A., Petrick, M., Spriggs, M.: Morphing orthogonal planar graph drawings. In: SODA 2006, pp. 222–230. ACM (2006)
13. Thomassen, C.: Deformations of plane graphs. J. Comb. Th., Series B 34, 244–257 (1983)

Graph Drawing through the Lens of a Framework for Analyzing Visualization Methods (Invited Talk, Extended Abstract)

Tamara Munzner

University of British Columbia, Department of Computer Science
Vancouver BC, Canada
tmm@cs.ubc.ca
http://www.cs.ubc.ca/~tmm

Abstract. The visualization community has drawn heavily on the algorithmic and systems-building work that has appeared with the graph drawing literature, and in turn has been a fertile source of applications. In the spirit of further promoting the effective transfer of ideas between our two communities, I will discuss a framework for analyzing the design of visualization systems. I will then analyze a range of graph drawing techniques through this lens. In the early stages of a project, this sort of analysis may benefit algorithm developers who seek to identify open problems to attack. In later project stages, it could guide algorithm developers in characterizing how newly developed layout methods connect with the tasks and goals of target users in different application domains.

1 Introduction

Visualization researchers and practitioners have long drawn on the algorithmic work conducted by the graph drawing community, and in turn have helped establish connections between that community and end users for specific application areas. Moreover, the network data that is the focus of the graph drawing community's efforts can be considered as a special case of the broader spectrum of data that is of interest in visualization, and thus its general principles are relevant.

I propose analysis of visualization techniques through *methods*; that is, an enumeration of the design space of techniques in terms of specific sets of choices. This kind of analysis supports thinking systematically about the space of possibilities. It may help a designer in the early stages of developing a new technique to identify gaps in the previous work to address. It can also be used to characterize existing work, in service of matching up which algorithms and techniques are suitable for which real-world problems. Further reading about this analysis framework can be found in an existing book chapter [13] and a forthcoming book [14]. These sources include many more references to the extensive related work that underlies this framework, which I do not directly include here.

In this talk, I begin with a distinction between four levels of visualization design, and continue with a brief discussion of abstraction for data. I introduce the

S. Wismath and A. Wolff (Eds.): GD 2013, LNCS 8242, pp. 61–70, 2013.

principles of marks and channels, and discuss the use of space in a visualization context. I continue with further examples of analysis drawn from graph drawing, and then conclude.

2 Levels of Visualization Design

In recent work, I proposed separating the design concerns of visualization into four levels: domain problem, data and task abstraction, visual encoding and interaction technique, and algorithm, as shown in Figure 1 [10]. In that paper, I also discuss the problem of how to validate designs at each of these levels. In this talk I will emphasize techniques, which are at a level just above the algorithm level that is the focus of much of the research from the graph drawing community. I also discuss the abstraction level briefly, to provide background context.

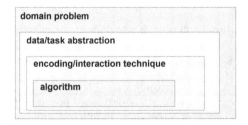

Fig. 1. Four levels of visualization design concerns [10]

My characterization here focuses on one major issue: how is space used? This question is an explicit consideration in visualization, but the motivation is not quite so obvious when considered purely from the perspective of problems that arise in graph drawing. I conjecture that the reason for this difference is that the very common cases in graph drawing, such as force-directed placement with node-link representations, or compound graphs that combine an underlying network with a hierarchy on top of it, are not trivial to analyze. My goal is to encourage more upwards characterization to map from algorithms up to techniques; that is, where algorithms are characterized in terms of the visual encoding and interaction techniques that they support.

When considering the four levels of design, another obvious route of attack is downwards from the top level of a domain problem; that is, to design a visualization system intended to solve some specific problem for a set of target users who have real data and real tasks. This sort of problem-driven work, often called *design studies* in the visualization literature has rich and interesting challenges, many of which are quite different than those that arise from technique-driven work. A detailed discussion of these issues appears elsewhere, in a recent paper on the methodology of design studies [17], and is beyond the scope of this talk.

3 Abstraction for Data

For the purposes of this framework, I will define only two basic types of data abstractions. At the dataset level, there are two major dataset types: *tables* and *networks*. In a simple table, I will call the rows *items* and the columns *attributes*. In a network, I distinguish between two kinds of items: *nodes*, and the *links* between them; either can have attributes. Obviously, the network dataset type is the focus of interest in graph drawing, but I will also present some analyses of table data as part of building up the framework. Attributes also have types. *Categorical* attributes have no implicit ordering, in contrast to *ordered* attributes; these are split into *quantitative* attributes that support full arithmetic operations such as addition or subtraction, in contrast to *ordinal*. For example, type of fruit is a categorical attribute; weight is quantitative; T-shirt size is ordinal.

The common case in visualization of complex, real-world data is that the designer will need to derive additional data beyond the original dataset. This derived data might be new attributes, or even a transformation from one dataset type to another, as with transforming a network into a table or vice versa. One example of a derived quantitative attribute computed from an original network is the Strahler number, a node-based centrality metric. Auber proposed exploiting it for fast interactive rendering of large graphs: by drawing nodes in priority order according to this attribute, a comprehensible skeleton of the network results from drawing only a small fraction of the nodes, in contrast to the poor results from drawing a random sampling [2].

4 Principles of Marks and Channels

I will introduce the idea of breaking down a visual encoding in terms of marks and channels by first considering some easy cases from statistical graphics that show tabular data: bar charts and scatterplots. These plots are straightforward to break down into *marks*, namely geometric primitives that represent items, and visual *channels* that control the appearance of marks. Marks are classified by their dimensionality: points, lines, areas, or volumes. Visual channels include spatial position, color, shape, size, orientation/tilt, and many others. A simple bar chart uses line marks, and encodes one attribute according to vertical spatial position channel. A scatterplot uses point marks, and encodes two attributes: one with the vertical spatial position channel, and one with horizontal position. A third attribute can be added to a scatterplot by encoding with the color channel, and a fourth by encoding with the size channel. The principles of marks and channels can be used to analyze more complex visual encoding techniques beyond these simple statistical graphics.

In addition to marks that represent items or nodes, marks may represent links. Link marks should implicitly convey the idea of relationships between items at a perceptual level. There are two particularly perceptually appropriate ways to do so: containment and connection. Containment uses an area mark to enclose a set of other marks within it; connection uses a line mark to directly connect two other marks together. I use the terms *connection* and *containment* for link marks, in contrast to *line* and *area* for item marks, to underscore that they communicate relationships between multiple items. A third perceptual way to indicate relationship is *proximity*, where items that are close to each other are implicitly perceived as being more related than those that are far apart. It is not possible to directly use proximity as a mark type, but in the next section I will discuss where it fits within the analysis of space use in visualization.

A crucial aspect of visual channels is that they also have implicit perceptual types, and these can and should be matched with attribute types. Some channels intrinsically convey *how much* in a way that maps well to ordered attributes, such as the spatial position along a common scale, or the length of a line mark, or the size of a point mark. Other channels convey *what* in a way that maps well to categorical attributes, such as what spatial region a mark is within, or what color a mark is, or what shape a mark is. The channels associated with the use of space have the strongest perceptual impact, leading to my choice to emphasize the spatial channels in this talk. The other channels can also be roughly ranked in terms of perceptual impact. In another talk, I discuss the underlying reasons for these rankings and a number of visualization principles that arise from them [11].

5   Using Space

I now discuss in more detail the ways to use space in the design of visual encoding and interaction techniques, emphasizing the different possible uses of spatial channels to control the appearance of marks.

I distinguish between five ways to use space: *use* given data; *express* values; and *separate*, *order*, and *align* regions. In the first case, the spatial layout is given, whereas in the other four the use of space is chosen. Using the data as given is the common case with geographic data. Although there are still many nuances in design considerations, as discussed in the cartographic literature, the fundamental use of space is constrained by this choice. This approach is also the common case when dealing with scalar, vector, or tensor spatial fields, where data is sampled at many points in the field, as in volume graphics and flow visualization. Of course, the existence of spatial data does not dictate its use as the fundamental use of space in a visual encoding; a designer may still choose to derive additional data and use space differently, as discussed below. I will not discuss this case further in this talk, where I focus on choosing the use of space. Although sometimes networks are drawn using given geographic data for node positions, the common case in graph drawing is on making choices about the use of space.

The case of expressing values spatially closely follows the discussion of marks and channels in the previous section: a quantitative attribute is encoded using the spatial position of a mark. Scatterplots are the quintessential example of expressing values in this way. In contrast, the other three uses of space pertain to establishing *regions*. Separating space into distinct regions, where each region shows something different, has major implications for how we perceive the structure of the dataset. Spatial proximity strongly implies grouping at a perceptual level, and so items in different regions are perceived as being in different categories. The regions themselves can be ordered with respect to each other, for example in a data-driven way according to an ordered attribute. Finally, regions can also be aligned to a shared baseline. Lengths and positions can be compared with higher precision between aligned regions than with unaligned regions, again for fundamental perceptual reasons. A 1D alignment is a list, while aligning in 2D yields a matrix, and in 3D a volumetric grid. In any of these cases, recursive subdivision is possible to accommodate hierarchical attribute structure.

The most extreme form of separation between regions is to divide the display into multiple separate views. There are three major approaches of combining views: showing multiple views side by side, superimposing multiple views on top of each other, and having a single view that changes over time. When superimposing multiple views as layers, they must all have a shared spatial layout. A single changing view is the common case for interactive navigation. Using multiple views side by side is a particularly powerful method because of the principle that "eyes beat memory" [11]. It is easy to compare by moving one's eyes between side by side views, where the views act as external cognitive supports. It is harder to compare a visible item to the memory of what one saw before, because of the limits of internal working memory.

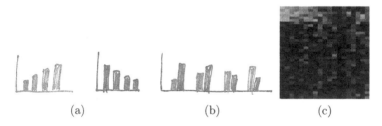

(a)   (b)   (c)

Fig. 2. Three ways to show a multidimensional table. a) Separate bar charts. b) Single interleaved bar chart. c) Heatmap. Figure credit: `http://commons.wikimedia.org/wiki/File:Heatmap.png`

Another seemingly simple example from statistical graphics is nevertheless a good example of the more complex uses of space: a multidimensional table of data with three attributes, one quantitative and two categorical. One categorical attribute is the type of export and there are two possible values: wine or cheese. The other categorical attribute is the name of the city, and there are four possible values. The quantitative attribute is the value in euro of the city's exports for a type over a year. We might encode this table as two separate bar charts, one for wine and one for cheese, with simple line marks in each, as in Figure 2a. An alternative is one interleaved bar chart where each item of data is depicted with two marks side by side, as in Figure 2b. We now have the vocabulary to analyze this choice in more detail in terms of channels and attributes: with separate charts, we have first separated into two large regions based on one categorical attribute, export type, then separated each of those into four smaller regions based on the other categorical attribute, city name. These subregions are aligned, and within each a single line mark expresses a value. The subregions are also ordered, for example either alphabetically by the name of the city, or in a data-driven way by the value of quantitative attribute, the exports. With interleaved charts, there is only one level of separation into regions: there are four regions, one for each item. Each region shows information about two attributes, as two side-by-side marks. The value of description at this level of detail is that we can reason about what kinds of information can be easily perceived by the viewer based on the partition into regions. In the first case, with two separate simple charts, the partition into one region for each type of export allows the easy perception of trends for that type. In the second case, having the two marks side by side allows the easy comparison of the export mix for a particular city.

We can now consider a quite different visual encoding of a heatmap display, which shows exactly the same data abstraction: a table with one quantitative attribute indexed by two categorical attributes. In a heatmap, regions are separated and aligned into a 2D matrix, and within each cell of the matrix an area mark is used in conjunction with the color channel to encode a quantitative attribute. The scale of the data is different: there are dozens or hundreds of values for each categorical attributes. In Figure 2c, these are genes and experimental conditions, and the quantitative attribute is the expression level of a gene in a specific condition.

I now turn from table to network datasets. A matrix view of a network is essentially the same as a heatmap, in terms of both data abstraction choice and use of space. The transformation at the data abstraction level is to transform the original network into a table, where a list of the nodes in the graph is used as both of the categorical attributes, and the weighted edge between a pair of nodes is the quantitative attribute. Thus, a cell in the matrix shows the presence or absence of an edge.

In contrast, the most common case in graph drawing is to use link marks to explicitly show links. As the name suggests, all node-link diagrams are an instance of using link connection marks. These diagrams best support tasks that pertain to the topological structure of networks [7], for example path tracing. In tree drawing, containment is also used; for example, all treemap variants are an instance of using link containment marks. These diagrams typically use the size channel to encode attribute values, either just for leaf nodes or recursively for interior nodes as well, and thus support tasks that pertain to understanding those attribute values.

In addition to the five choices for the use of spatial channels, it is useful to consider the orientation of the spatial axes within the layout. The two most common cases in graph drawing are *rectilinear* and *radial*. In 2D, the rectilinear choice yields Cartesian coordinates and the radial choice is polar coordinates. A third choice is to orient all of the axes *parallel* to each other. The limitations of these choices have been investigated in the visualization literature. Rectilinear layouts have the obvious scalability issue that the number of axes is highly constrained. A 2D rectilinear layout allows very high-precision perception of information encoded with spatial position. While 3D rectilinear layouts are possible, there are many perceptual problems in using three dimensions of space to encode nonspatial data [11]. Four or more rectilinear axes cannot be directly encoded. There has also been empirical work to investigate the strengths and weaknesses of radial layouts, in light of the known limitation that we perceive angles with less accuracy than lengths [5].

The work of McGuffin and Robert is a good example of analyzing many different tree drawing methods according to the efficiency of their use of space [8]. The *information density* of a diagram is as a measure of the amount of information encoded vs. the amount of unused space; there is a tradeoff between encoding as much information as possible, and the potential for visual clutter or other legibility problems. The examples that they analyze can be considered in terms of the methods that I have covered: whether connection or containment link marks are used, whether the layout is rectilinear or radial, how the spatial position channels are used to encode information. The information encoded in these diagrams is the link relationships, the depth in the tree of a node, and the order of siblings. Other analysis considerations are whether some information is encoded redundantly, for example through both spatial position and explicit link marks, and whether any arbitrary information is expressed through the use of spatial ordering. For example, in some trees, sibling order is not specifically defined, yet there is a visible spatial order.

Force-directed placement is a widely adopted approach for visually encoding network datasets. The visual encoding is in some sense straightforward because it is a node-link diagram: point marks represent nodes, and connection marks represent links as lines. However, considering the meaning of spatial position is somewhat tricky, because no meaning is directly encoded; instead it is left free to minimize crossings. Thus, the semantics of proximity are mixed: sometimes it is meaningful, for example when a cluster of nodes placed near to each other truly reflects strong interconnections of links between them. Sometimes it is an arbitrary artifact of the layout algorithm, and two nodes that happen to be nearby in one layout may be quite far in another. There is also an interesting tension between proximity cues and edge length: long edges are more visually salient than short ones that connect nodes close to each other.

A great deal of work has been devoted to developing better algorithms for force-directed placement through multilevel methods; the sfdp algorithm is one example [6]. The data abstraction is more complex, namely a compound graph: in addition to the original network dataset, the additional derived data of a cluster hierarchy atop the original network is computed. Although this hierarchy is used within the algorithm, it is not shown explicitly within the drawing. Thus, the fundamental use of space is the same as with simpler versions of force-directed placement; this multilevel approaches is an example of a better algorithm for the same visual encoding technique.

The GrouseFlocks system for the interactive analysis of compound graphs [1] has some instructive similarities and differences from sfdp. The data abstraction is the same, a compound graph. However, the visual encoding is different, as shown in Figure 3c. In addition to connection marks for the network links

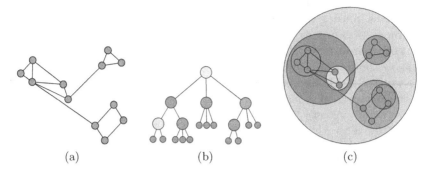

Fig. 3. Compound graph representation in GrouseFlocks [1]. a) Original network. b) Cluster hierarchy. c) Visual encoding in tool, fully expanded.

and point marks for the nodes, the hierarchy links are shown with containment marks. The system features dynamic interaction in order to support large datasets, where typically only an interactively-selected subset of the full compound graph is shown. The user can expand and contract individual metanodes in the hierarchy, which also shows the associated nodes from the base network.

6   Further Analysis Examples

In the talk, I will analyze three more systems within this framework: Cerebral [3, 4], Constellation [12, 15], and Noack's LinLog energy model [16]. All three are examples of design motivated by explicit prior analysis of the use of space.

The Cerebral system [3, 4] features both multiple side by side views, and superimposed layers within each view. The network layout is designed to mimic the semantics of hand-drawn diagrams of biological networks. I will analyze its visual encoding and design choices pertaining to the use of space in detail.

The Constellation system [12, 15] features a complex multi-level linguistic network that is laid out with spatial position reflecting specific attributes, where edge crossings are resolved using perceptual layers rather than through algorithmically reducing the number of instances.

Noack's LinLog energy model is designed to reveal clusters in data, by requiring that edges between clusters are longer than those within [16]. Noack specifically indicates that his approach uses the same minimization algorithms as previous work, and frames the energy model in terms of its visual results. I thus consider it a contribution at the visual encoding technique level, even though that exact vocabulary out of the visualization community does not appear in the paper. I note that it was published at a previous Graph Drawing conference, in contrast to the many other examples in this talk that come out of the visualization literature. I encourage more papers like this that can act as bridges between the communities!

7   Conclusions

I have discussed a general framework for systematically analyzing visualization techniques in terms of the methods of using space, and applied it to graph drawing examples in particular. It relies on breaking down visual encodings into marks that are geometric primitives representing either nodes or links, and channels that control their appearance to encode attributes. In this talk I focus on the channels related to the use of space. The simple case is using spatial position to express a quantitative attribute value, but space can also be separated into regions that partition according to a categorical attribute to indicate groups via proximity. These regions can also be ordered and aligned. This framework is a mix of ideas of that are widespread in the visualization literature and those that are new; for a more detailed discussion of the previous work, see the existing chapter [13] and the forthcoming full book [14]. These sources also describe principles in detail, and also the more complete analysis framework from which the subset discussed here was drawn. The full framework includes the nonspatial channels in addition to the spatial ones and a much more detailed discussion of methods for combining multiple views. It covers interactive techniques in addition to visual encoding techniques, particularly in terms of methods for the reduction of the amount of data shown.

This kind of analysis can guide the development of new techniques, or be used to characterize existing ones. While sometimes it is easy to map from a specific algorithm to the visual encoding technique that it supports, for example when the mapping is explicitly discussed in a paper or in work that it directly cites, sometimes this mapping is difficult to reverse-engineer. Algorithm descriptions may not facilitate analysis of the resulting visual encoding, either for the use of space or for other channels. In these cases, the line between technique and algorithm can be blurry: does a new algorithm support an existing technique, or does it result in a new one? Carrying out more such characterization may facilitate the transfer of algorithms from the graph drawing community to the visualization community. It is also important to characterize mappings between the other levels of visualization design [9], but this important question is beyond the scope of this talk.

Of course, characterization according to this sort of framework is only one of many possible ways to analyze graph drawing and visualization approaches. Benchmarks and complexity analysis are a different way to compare approaches, as are user studies in the form of controlled experiments or more qualitative investigation of how people use visualization systems [10].

References

[1] Archambault, D., Munzner, T., Auber, D.: GrouseFlocks: Steerable exploration of graph hierarchy space. IEEE Trans. on Visualization and Computer Graphics 14(4), 900–913 (2008)

[2] Auber, D.: Using Strahler numbers for real time visual exploration of huge graphs. In: Intl. Conf. Computer Vision and Graphics, pp. 56–69 (2002)

[3] Barsky, A., Gardy, J.L., Hancock, R.E., Munzner, T.: Cerebral: A Cytoscape plugin for layout of and interaction with biological networks using subcellular localization annotation. Bioinformatics 23(8), 1040–1042 (2007)

[4] Barsky, A., Munzner, T., Gardy, J., Kincaid, R.: Cerebral: Visualizing multiple experimental conditions on a graph with biological context. IEEE Trans. Visualization and Computer Graphics (Proc. InfoVis 2008) 14(6), 1253–1260 (2008)

[5] Diehl, S., Beck, F., Burch, M.: Uncovering strengths and weaknesses of radial visualizations - an empirical approach. IEEE Trans. Visualization and Computer Graphics (Proc. InfoVis) 16(6), 935–942 (2010)

[6] Hu, Y.: Efficient and high quality force-directed graph drawing. The Mathematica Journal 10, 37–71 (2005)

[7] Lee, B., Plaisant, C., Parr, C.S., Fekete, J.D., Henry, N.: Task taxonomy for graph visualization. In: Proc. AVI Workshop on Beyond Time and Errors: Novel Evaluation Methods for Information Visualization (BELIV). ACM Press (2006)

[8] McGuffin, M.J., Robert, J.M.: Quantifying the space-efficiency of 2D graphical representations of trees. Information Visualization 9(2), 115–140 (2010)

[9] Meyer, M., Sedlmair, M., Munzner, T.: The four-level nested model revisited: Blocks and guidelines. In: Proc. Workshop on Beyond Time and Errors: Novel Evaluation Methods for Information Visualization (BELIV) (2012)

[10] Munzner, T.: A nested model for visualization design and validation. IEEE Trans. Visualization and Computer Graphics (TVCG) 15(6), 921–928 (2009)

[11] Munzner, T.: Visualization Principles. VizBi 2011 Keynote (2011), `http://www.cs.ubc.ca/~tmm/talks.html#vizbi11`

[12] Munzner, T.: Constellation: Linguistic semantic networks. In: Interactive Visualization of Large Graphs and Networks (PhD thesis), ch. 5, pp. 105–122. Stanford University Department of Computer Science (2000)

[13] Munzner, T.: Visualization. In: Shirley, P., Marschner, S. (eds.) Fundamentals of Graphics, ch. 27, 3rd edn., vol. ch. 27, pp. 675–707. AK Peters (2009b)

[14] Munzner, T.: Visualization Analysis and Design: Principles, Abstractions, and Methods. AK Peters (to appear, 2014)

[15] Munzner, T., Guimbretière, F., Robertson, G.: Constellation: A visualization tool for linguistic queries from MindNet. In: Proc. IEEE Symposium on Information Visualization (InfoVis), pp. 132–135 (1999)

[16] Noack, A.: An energy model for visual graph clustering. In: Liotta, G. (ed.) GD 2003. LNCS, vol. 2912, pp. 425–436. Springer, Heidelberg (2004)

[17] Sedlmair, M., Meyer, M., Munzner, T.: Design study methodology: Reflections from the trenches and the stacks. IEEE Trans. Visualization and Computer Graphics (TVCG) 18(12), 2431–2440 (2012)

A Linear-Time Algorithm for Testing Outer-1-Planarity[*][**]

Seok-Hee Hong[1], Peter Eades[1], Naoki Katoh[2], Giuseppe Liotta[3],
Pascal Schweitzer[4], and Yusuke Suzuki[5]

[1] University of Sydney, Australia
{seokhee.hong,peter.eades}@sydney.edu.au
[2] Kyoto University, Japan
naoki@archi.kyoto-u.ac.jp
[3] University of Perugia, Italy
liotta@diei.unipg.it
[4] ETH, Switzerland
pascal@mpi-inf.mpg.de
[5] Niigata University, Japan
y-suzuki@math.sc.niigata-u.ac.jp

Abstract. A graph is *1-planar* if it can be embedded in the plane with at most one crossing per edge. A graph is *outer-1-planar* if it has an embedding in which every vertex is on the outer face and each edge has at most one crossing. We present a linear time algorithm to test whether a graph is outer-1-planar. The algorithm can be used to produce an outer-1-planar embedding in linear time if it exists.

1 Introduction

A recent research topic in topological graph theory is the study of graphs that are *almost planar* in some sense. Examples of such almost planar graphs are *1-planar* graphs, which can be embedded in a plane with at most one crossing per edge.

Ringel [3] introduced 1-planar graphs in the context of simultaneously coloring vertices and faces of planar graphs. Subsequently, various aspects of 1-planar graphs have been investigated. Borodin [4] gives colouring methods for 1-planar graphs. Pach and Toth [5] prove that a 1-planar graph with n vertices has at most $4n - 8$ edges, which is a tight upper bound. There are a number of structural results on 1-planar graphs [6], and *maximal* 1-planar embeddings [7] (a 1-planar embedding of a graph G is *maximal*, if no edge can be added without violating the 1-planarity of G).

The class of 1-planar graphs is not closed under edge contraction; accordingly, computational problems seem difficult. Korzhik and Mohar proved that testing 1-planarity

---

[*] This paper is an extended abstract. For omitted proofs, see the full version of this paper [1]. The problem studied in this paper was initiated at the Port Douglas Workshop on Geometric Graph Theory, June, 2011, held in Australia, organized by Peter Eades and Seok-Hee Hong, supported by IPDF funding from the University of Sydney.

[**] Independently, another linear time algorithm is reported in [2].

S. Wismath and A. Wolff (Eds.): GD 2013, LNCS 8242, pp. 71–82, 2013.
© Springer International Publishing Switzerland 2013

of a graph is NP-complete [8]. On the positive side, it has been shown that the problem of testing maximal 1-planarity of a graph G can be solved in linear time if a *rotation system* (i.e., the circular ordering of edges for each vertex) is given by Eades et al [9].

The existence of a 1-planar embedding does not guarantee the existence of a straight-line 1-planar drawing, as shown by Eggleton [10] and Thomassen [11]. However, recently Hong et al. [12] give a linear time testing algorithm, and a linear time drawing algorithm to construct such a drawing if it exists. Very recently, the more general problem on straight-line drawability of embedded graphs is studied by Nagamochi [13].

Eggleton [10] introduced the investigation of *outer-1-planar graphs*: a graph is outer-1-planar if it has a 1-planar drawing in which every vertex is on the outer face. Examples of outer-1-planar graph drawings are shown in Fig. 1(a), (b), (c) and (d); in Fig. 1(e) a graph that has no outer-1-planar drawing is illustrated. Eggleton describes a number of geometric, topological, and combinatorial properties of outer-1-planar graphs.

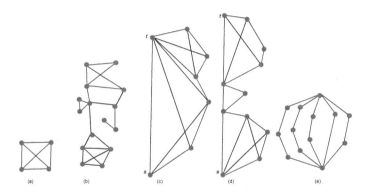

Fig. 1. *(a), (b), (c) and (d) are examples of outer-1-planar graph drawings. (a) illustrates the only triconnected outer-1-planar graph. (c) and (d) are examples of graphs that are one-sided outer-1-planar (OSO1P) with respect to (s, t). (d) illustrates a graph with no outer-1-planar drawing.*

In this paper, we investigate algorithmics of outer-1-planar graphs. More specifically, we describe a linear time algorithm to test outer-1-planarity of a given graph G.

Theorem 1. *There is a linear time algorithm to test whether a graph is outer-1-planar. The algorithm produces an outer-1-planar embedding if it exists.*

To prove Theorem 1, we define a sub-class of outer-1-planar graphs as follows. Suppose that G is a graph with vertices s and t. Let $G_{+(s,t)}$ denote the graph obtained by adding the edge (s, t), if this edge is not already in G. If $G_{+(s,t)}$ has an outer-1-planar embedding in which the edge (s, t) is completely on the outer face, then we say that G is *one-sided-outer-1-planar (OSO1P)* with respect to (s, t). Examples of such graphs are shown in Fig. 1(c) and (d). For these graphs, we prove the following result.

Theorem 2. *There is a linear time algorithm to test whether a biconnected graph G is one-sided outer-1-planar with respect to a given edge (s, t) of G. The algorithm produces a one-sided outer-1-planar embedding if it exists.*

Section 4 describes an algorithm to test whether a graph is one-sided outer-1-planar, and Section 5 shows how to use one-sided outer-1-planarity to test outer-1-planarity. The adaption of the algorithms of Sections 4 and 5 to construct an embedding is straightforward, and described in Section 6. In conclusion, Section 7 cites drawing algorithms and discusses future work.

2   Terminology

In this Section we define the terminology used throughout the paper.

A *topological graph* or *embedding* $G = (V, E)$ is a representation of a simple graph in the plane where each vertex is a point and each edge is a Jordan arc between the points representing its endpoints.

Two edges *cross* if they have a point in common, other than their endpoints. The point in common is a *crossing*. To avoid some pathological cases, some standard non-degeneracy conditions apply: (1) two edges intersect in at most one point; (2) an edge does not contain a vertex other than its endpoints; (3) no edge crosses itself; (4) edges must not meet tangentially; (5) no three edges share a crossing point; and (6) no two edges that share an endpoint cross.

A topological graph is *1-planar* if no edge has more than one crossing. A graph is *1-planar* if it has a 1-planar embedding. A graph is *outer-1-planar* if it has a 1-planar embedding in which every vertex is on the outer face. The aim of this paper is to give an algorithm to test whether a graph is outer-1-planar, and to provide an outer-1-planar embedding if it exists.

Our algorithm uses an *SPQR tree* to represent the decomposition of a biconnected graph into triconnected components. We recall some basic terminology of SPQR trees; for details, see [14]. Each node ν in the SPQR tree is associated with a graph called the *skeleton* of ν, denoted by $\sigma(\nu)$. There are four types of nodes ν in the SPQR tree: (1) S-nodes, where $\sigma(\nu)$ is a simple cycle with at least 3 vertices; (2) P-nodes, where $\sigma(\nu)$ consists of two vertices connected by at least 3 edges; (3) Q-nodes, where $\sigma(\nu)$ consists of two vertices connected by a real edges and a virtual edge; and (4) R-nodes, where $\sigma(\nu)$ is a simple triconnected graph. We treat the SPQR tree as a rooted tree by choosing an arbitrary node as its root. Note that every leaf is a Q-node and that the root is not a Q-node.

Let ρ be the parent of an internal node ν. The graph $\sigma(\rho)$ has exactly one *virtual edge* e in common with $\sigma(\nu)$; this is the *parent virtual edge* of $\sigma(\nu)$, and a *child virtual edge* in $\sigma(\rho)$. We denote the graph formed by the union of $\sigma(\nu)$ over all descendants ν of ρ by G_ρ.

If G is an outer-1-planar graph, then $\sigma(\nu)$ and G_ν are outer-1-planar graphs, using the embedding induced from G. If G_ν is a one-sided outer-1-planar (OSO1P) graph with respect to the parent virtual edge (s, t) of ν then we say that ν is a one-sided outer-1-planar (OSO1P) node with respect to (s, t).

For this paper, we need to define a specific type of S-node. Suppose that μ is an S-node with parent separation pair (u, v). A *tail at u* for μ is a Q-node child (that is, a real edge) with parent virtual edge (u, x) for some vertex x.

Further, we need to define a specific type of P-node. A P-node ν is *almost one-sided outer-1-planar (AOSO1P) with respect to (the directed edge)* (s, t) if G_ν consists of a parallel composition of an OSO1P graph with respect to (s, t) and an S-node μ such that μ has a tail at t and μ is OSO1P with respect to (s, t). See Fig. 2 for examples.

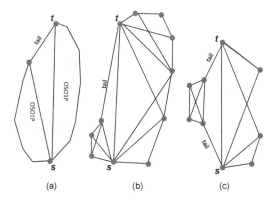

Fig. 2. *An AOSO1P graph consists of a parallel composition of an OSO1P graph and an OSO1P S-node with a tail. (a) The general shape of a graph that is AOSO1P with respect to (s, t). (b) A graph that is AOSO1P with respect to (s, t). (c) A graph that is AOSO1P with respect to both (s, t) and (t, s).*

3   Structural Results

In this Section we present structural results that support the algorithms defined in the subsequent sections.

First we note that the only triconnected outer-1-planar graph is K_4, embedded as depicted in Fig. 1(a).

Lemma 1. *If G is outer-1-planar and triconnected, then G is isomorphic to K_4 and every outer-1-planar drawing of G has exactly one crossing.*

Proof. Suppose that G is an outer-1-planar embedding of a triconnected graph; we can assume that G is maximal in the sense that no edge can be added without destroying the property of outer-1-planarity. Eggleton [10] shows that the outer face is a simple cycle γ.

Suppose that (a, b) and (c, d) are a pair of edges, neither on γ, that cross at point p. Suppose that a precedes c in clockwise order around γ. Suppose that there is at least one vertex v that lies between a and c on γ, as shown in Fig. 3.

All edges incident with the vertices between a and c on γ must have both endpoints in the region r bounded by γ, the curve ap and the curve cp. Removing a and c separates v from the remainder of the graph; this contradicts the triconnectivity of G, and we can deduce that there is no such vertex v.

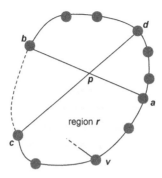

Fig. 3. *Here the pair a, c must be a separation pair, since all edges incident with vertices between a and c on the outer face γ must have both endpoints in the region r.*

Using the same argument, we can show that the only vertices on γ are a, b, c and d; thus G is K_4. □

Secondly, we note that we can restrict our attention to the biconnected case.

Lemma 2. *A graph is outer-1-planar if and only if its biconnected components are outer-1-planar.*

Next we present a simple fact about how an edge can cross a cycle in an outer-1-planar embedding.

Lemma 3. *Suppose that γ is a cycle in an outer-1-planar graph G and G' is an outer-1-planar embedding of G. Suppose that an edge (u, v) is not on γ but crosses an edge of γ in G'. Then either u or v is on γ.*

The next result is a relatively technical but fundamental Lemma about embeddings of paths which share endpoints, illustrated in Fig. 4. A path from a vertex s to a vertex t is *non-trivial* if it contains more than two vertices. If an edge from a path p_1 crosses an edge from a path p_2 then we say that p_1 *crosses* p_2.

Lemma 4. *Suppose that P is a set of paths between two vertices s and t. Let G be the union of the paths in P, and let G' be an outer-1-planar embedding of G. Then $|P| \leq 5$, and:*

(a) *If $|P| \geq 3$ and an edge from one non-trivial path $p_1 \in P$ crosses an edge from another non-trivial path $p_2 \in P$ then this crossing occurs between an edge incident with s and an edge incident with t.*

(b) *If $|P| = 3$ and all paths in P are non-trivial, then there are two paths p_1 and p_2 in P such that there is exactly one crossing between edges of p_1 and edges of p_2; furthermore, every edge in the third path is on the outer face.*

(c) *If $|P| = 3$ and one path in P is trivial and is on the outer face, then there are two paths p_1 and p_2 in P such that there is exactly one crossing between edges of p_1 and edges of p_2.*

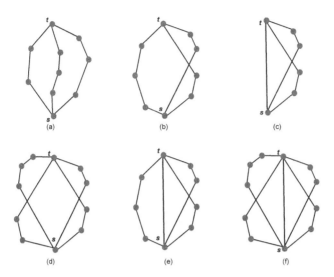

Fig. 4. *Embeddings of paths that share endpoints. (a) A planar embedding. (b) An outer-1-planar embedding of 3 non-trivial paths. (c) An outer-1-planar embedding of 3 paths, where one path is trivial. (d) An outer-1-planar embedding of 4 non-trivial paths. (e) An outer-1-planar embedding of 4 paths, where one path is trivial. (f) An outer-1-planar embedding of 5 paths.*

(d) *If P contains 4 non-trivial paths, then we can divide P into two pairs of paths $\{p_1, p_2\}$ and $\{p_3, p_4\}$ such that there is exactly one crossing between edges of p_1 and edges of p_2 and exactly one crossing between edges of p_3 and edges of p_4, and there are no other crossings.*

(e) *If $|P| = 4$ and P contains a trivial path, then there are two paths p_1 and p_2 in P such that there is exactly one crossing between edges of p_1 and edges of p_2, and one non-trivial path in P is on the outer face.*

(f) *If $|P| = 5$ then one path in P is trivial, and we can divide the other paths into two pairs of paths $\{p_1, p_2\}$ and $\{p_3, p_4\}$ such that there is exactly one crossing between edges of p_1 and edges of p_2 and exactly one crossing between edges of p_3 and edges of p_4, and there are no other crossings.*

4   Testing OSO1P and AOSO1P

In this Section we describe a linear time algorithm that takes a graph G and vertices s and t of G as input and tests whether G has an OSO1P or AOSO1P embedding with respect to (s, t).

From Lemma 2, we only need to consider biconnected graphs. Our algorithm computes the SPQR tree T and then works from the leaves of T upward toward the root, computing boolean labels $OSO1P(\nu, s, t)$ and $AOSO1P(\nu, s, t)$ that indicate whether

ν is OSO1P or AOSO1P with respect to (s, t). The label $OSO1P(\nu, s, t)$ is computed for each node ν of T, and the label $AOSO1P(\nu, s, t)$ is computed for every P-node ν.

Algorithm test-One-Sided-Outer-1-Planar

1. Construct the SPQR tree T of G.
2. Traverse T bottom up, and for each node ν of T with parent virtual edge (s, t):

   (a) `if` ν is a Q-node `then return true`.
   (b) `elseif` ν is an R-node `then return` $OSO1P(\nu, s, t)$ as described in Section 4.1, using the values $OSO1P(\nu', s', t')$ for each child ν' of ν with child virtual edge (s', t').
   (c) `elseif` ν is a P-node `then return` $OSO1P(\nu, s, t)$ and $AOSO1P(\nu, s, t)$, as described in Section 4.2, using the values $OSO1P(\nu', s', t')$ for each child ν' of ν with child virtual edge (s', t').
   (d) `else` /* ν is an S-node */ `then return` $OSO1P(\nu, s, t)$, as described in Section 4.3, using the values $OSO1P(\nu', s', t')$ and $AOSO1P(\nu, s', t')$ for each child ν' of ν with child virtual edge (s', t').

The time complexity of Step 1 is linear [14,15]. Step 2(a) is trivial and takes constant time for each Q-node. We show below that Steps 2(b), 2(c), and 2(d) each take time proportional to the number of children of the node ν. Summing over all nodes of the SPQR tree T results in linear time for the whole algorithm.

The cases for R-nodes and P-nodes are quite straightforward, and we deal with them first in Sections 4.1 and 4.2. The extension for the computation of the AOSO1P property for P-nodes is again straightforward and described in Section 4.2. The case for S-nodes is a little more involved, and we deal with this in Section 4.3.

4.1   R-nodes

Lemma 1 can be generalised to R-nodes as in the following Lemma.

Lemma 5. *Suppose that ν is an R-node of the SPQR tree of a biconnected graph G; suppose that (u, v) is its parent virtual edge. Then G_ν is OSO1P with respect to (u, v) if and only if:*

1. *$\sigma(\nu)$ is isomorphic to K_4; and*
2. *an edge (u, a) of $\sigma(\nu)$ with $a \neq v$ incident with u represents a child Q-node of ν, an edge (v, b) of $\sigma(\nu)$ with $b \neq u$ represents a child Q-node of ν, and (u, a) crosses (v, b); and*
3. *for every child ν' of ν, ν' is OSO1P with respect to (c, d), where the parent virtual edge of ν' is (c, d).*

An algorithm to test whether an R-node ν is OSO1P can be derived directly from Lemma 5. It is clear that, given the boolean labels $OSO1P(\nu_i, s, t)$ for each child ν_i of ν, the algorithm runs in time proportional to the number of children of ν.

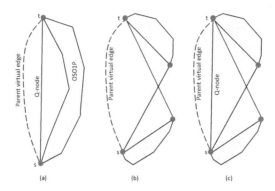

Fig. 5. *Possibilities for an OSO1P P-node*

4.2 P-nodes

Here we use Lemma 4 to show that a P-node can have at most three children, as depicted in Fig. 5. More specifically, we have the following Lemma:

Lemma 6. *Suppose that ν is a P-node of the SPQR tree of a biconnected graph G; suppose that (s,t) is its parent virtual edge. Then G_ν is OSO1P with respect to (s,t) if and only if either:*

(a) ν has two children, of which one is a Q-node (s,t), and the other is OSO1P with respect to (s,t); or

(b) ν has two children, of which one is an S-node with tail at s which is OSO1P with respect to (s,t), and the other is an S-node with tail at t which is OSO1P with respect to (s,t); or

(c) ν has three children, of which one is a Q-node (s,t), one is an S-node with tail at s which is OSO1P with respect to (s,t), and the other is an OSO1P S-node with tail at t which is OSO1P with respect to (s,t).

It is straightforward to extend Lemma 6 to test whether a node ν is AOSO1P, using the definition of AOSO1P together with Lemma 6.

Algorithms to test whether a P-node ν is OSO1P or AOSO1P can be derived directly from Lemma 6. It is clear that, given the boolean labels $OSO1P(\nu', s, t)$ for each child ν' of ν, the algorithm runs in time proportional to the size of the skeleton $\sigma(\nu)$ of ν.

4.3 S-nodes

Suppose that ν is an S-node with children $\nu_1, \nu_2, \ldots, \nu_k$, where the parent virtual edge of ν_i is (s_{i-1}, s_i), as shown in Fig. 6(a).

If every child ν_i is OSO1P with respect to (s_{i-1}, s_i), then clearly ν is OSO1P with respect to (s_0, s_k); however, the converse is false. Consider the example shown in Fig. 6(b). Here ν is OSO1P with respect to (s_0, s_k). However, the child ν_2 is not OSO1P

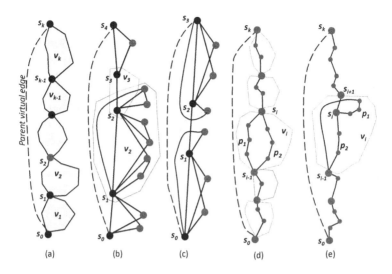

Fig. 6. *(a) An S-node. (b) An OSO1P S-node with a child (ν_2) that is not OSO1P. (c) An S-node that satisfies the conditions of Lemma 7 but is not OSO1P. (d) Two paths p_1 and p_2 in the graph G_{ν_i}. (e) The path p_1 crosses the edge (s_i, s_{i+1}).*

with respect to (s_1, s_2) (by Lemma 6). Note that ν_3 is a Q-node, and an edge from the skeleton of ν_2 crosses this edge. In fact the example shown in Fig. 6(b) illustrates the necessary conditions stated in the next lemma.

Lemma 7. *Suppose that ν is an S-node with children $\nu_1, \nu_2, \ldots, \nu_k$, where the parent virtual edge of ν_i is (s_{i-1}, s_i). Suppose that G_ν is OSO1P with respect to (s_0, s_k). Then for $1 \leq i \leq k$, either:*

(a) ν_i is OSO1P with respect to (s_{i-1}, s_i); or
(b) $i < k$, ν_i is AOSO1P with respect to (s_i, s_{i-1}), and ν_{i+1} is a Q-node; or
(c) $i > 1$, ν_i is AOSO1P with respect to (s_{i-1}, s_i), and ν_{i-1} is a Q-node.

Lemma 7 gives necessary conditions for an S-node to be OSO1P. However the conditions are not sufficient. Consider, for example, the graph shown in Fig. 6(c). This satisfies the necessary conditions as in Lemma 7, but it is not OSO1P. The problem is that the Q-node represented by the edge (s_1, s_2) has two crossings, one with an edge of the AOSO1P graph at the top and one with an edge of the AOSO1P graph at the bottom. Nevertheless, this situation does not occur when $k = 2$, and we shall show that in this case the conditions of Lemma 7 are sufficient. One can express sufficient conditions for an S-node to be OSO1P in a recursive way, as in the following Lemma. If ν is an S-node with children $\nu_1, \nu_2, \ldots, \nu_k$ then we denote the series combination of graphs $G_{\nu_1}, G_{\nu_2}, \ldots, G_{\nu_k}$ by $G(\nu_1, \nu_2, \ldots, \nu_k)$.

Lemma 8. *Suppose that ν is an S-node with children $\nu_1, \nu_2, \ldots, \nu_k$, where the parent virtual edge of ν_i is (s_{i-1}, s_i). Then G_ν is OSO1P with respect to (s_0, s_k) if and only if either:*

1. G_{ν_1} is OSO1P with respect to (s_0, s_1) and $G(\nu_2, \nu_3, \ldots, \nu_k)$ is OSO1P with respect to (s_1, s_k); or
2. ν_1 is a Q-node, G_{ν_2} is AOSO1P with respect to (s_1, s_2), and $G(\nu_3, \nu_4, \ldots, \nu_k)$ is OSO1P with respect to (s_2, s_k); or
3. G_{ν_1} is AOSO1P with respect to (s_1, s_0), ν_2 is a Q-node, and $G(\nu_3, \nu_4, \ldots, \nu_k)$ is OSO1P with respect to (s_2, s_k).

Lemma 8 leads to the recursive algorithm for S-nodes; see [1]. The algorithm runs in time proportional to the number of children of ν.

This completes the proof of Theorem 2.

5 Testing Outer-1-Planarity

Once we compute the labels $OSO1P(\nu, s, t)$ and $AOSO1P(\nu, s, t)$ for all internal nodes ν of the SPQR tree, we can test whether the whole graph (that is, the root ρ) is outer-1-planar. This requires separate tests depending on the type of the root node. See Fig. 7.

We can require the root node to be an R-node or a P-node, since if the SPQR tree contains no R-node and no P-node, then the graph is a cycle and thus outerplanar. Both tests for an R-node and a P-node are detailed below.

For R-nodes, we have the following Lemma.

Lemma 9. *Suppose that ρ is an R-node at the root of the SPQR tree. Then G is outer-1-planar if and only if*

1. *$\sigma(\rho)$ is isomorphic to K_4, and*
2. *at least two children of ρ are Q-nodes, and*
3. *for every child node ν' of $\sigma(\rho)$ with parent virtual edge (a, b), $G_{\nu'}$ is OSO1P with respect to (a, b).*

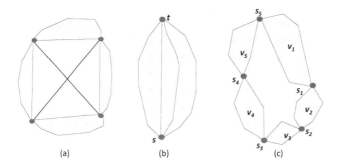

Fig. 7. *(a) An R-node at the root. (b) P-node at the root. (c) S-node at the root.*

It is clear that one can test the conditions of Lemma 9 in constant time, as long as the labels $OSO1P(\nu', a, b)$ for all children ν' of ρ have already been computed.

For P-nodes, the following result applies.

Lemma 10. *Suppose that ρ is a P-node at the root of the SPQR tree. Then G is outer-1-planar if and only if it is a parallel composition of two OSO1P graphs.*

Using Lemma 4, one can show that the number of children of a P-node ρ at the root is bounded (in fact, at most 5). It follows that the number of ways to partition the children is bounded by a constant. Thus we can define a constant time algorithm to implement Lemma 10 (given that the labels $OSO1P(\nu', a, b)$ for all children ν' of ρ have already been computed).

This completes the proof of Theorem 1.

6 Outer-1-Planar Embedding

One can construct a one-sided outer-1-planar embedding of an input graph G using an extension of the methods in Section 4. The methods for R-nodes and S-nodes described in Lemmas 5 and 6 define crossings; treating these crossings as dummy vertices gives a planar graph G^*. A one-sided outer-1-planar embedding of G is a specific planar embedding of G^*.

Every planar embedding of G^* is defined by an orientation and an ordering for nodes ν in the SPQR tree with respect to the parent separation pair of ν. For P-nodes, R-nodes, and S-nodes, it is possible to "flip" the orientation of ν around its parent separation pair. For P nodes, a left-right order for the children can be chosen. To produce an outer-1-planar embedding we use the same bottom-up strategy as in **Algorithm test-One-Sided-Outer-1-Planar** in Section 4. Throughout the algorithm we maintain an embedding; in particular we keep track of the outside face. At each node ν, we "flip" ν so that all vertices of G_ν lie on the outside face. Also, at each P-node ν, we order the children of ν so that all vertices of G_ν lie on the outside face. This requires linear time manipulation of the SPQR tree, using methods outlined in [14].

7 Conclusion

The algorithm presented in this paper takes a graph G as input and determines whether it has an outer-1-planar embedding. We show that if such an embedding does exist, then we can compute it in linear time.

Given the topological embedding computed by our algorithm, the methods of Eggleton [10] can be used to construct a straight-line drawing. In fact, Eggleton gives conditions that determine whether a given set of points support a straight-line outer-1-planar drawing. Dekhordi et al. [16] show further that every outer-1-planar topological embedding has a straight-line RAC (right-angle crossing) drawing, at the cost of exponential area.

Many algorithms for drawing outerplanar graphs exist, with a number of properties (see [17,18]). It would be interesting to see if these results extend to outer-1-planar graphs.

References

1. Hong, S.H., Eades, P., Katoh, N., Liotta, G., Schweitzer, P., Suzuki, Y.: A linear-time algorithm for testing outer-1-planarity. TR-IT-IVG-2013-01. Technical Report, School of IT, University of Sydney (June 2013)
2. Auer, C., Bachmaier, C., Brandenburg, F.J., Gleißner, A., Hanauer, K., Neuwirth, D., Reislhuber, J.: Recognizing outer 1-planar graphs in linear time. In: Wismath, S., Wolff, A. (eds.) GD 2013. LNCS, vol. 8242, pp. 107–118. Springer, Heidelberg (2013)
3. Ringel, G.: Ein Sechsfarbenproblem auf der Kugel. Abh. Math. Sem. Univ. Hamburg 29, 107–117 (1965)
4. Borodin, O.V.: Solution of the Ringel problem on vertex-face coloring of planar graphs and coloring of 1-planar graphs. Metody Diskret. Analiz. 41, 12–26 (1984)
5. Pach, J., Tóth, G.: Graphs drawn with few crossings per edge. Combinatorica 17, 427–439 (1997)
6. Fabrici, I., Madaras, T.: The structure of 1-planar graphs. Discrete Mathematics 307, 854–865 (2007)
7. Suzuki, Y.: Re-embeddings of maximum 1-planar graphs. SIAM J. Discrete Math. 24, 1527–1540 (2010)
8. Korzhik, V.P., Mohar, B.: Minimal obstructions for 1-immersions and hardness of 1-planarity testing. Journal of Graph Theory 72, 30–71 (2013)
9. Eades, P., Hong, S.-H., Katoh, N., Liotta, G., Schweitzer, P., Suzuki, Y.: Testing maximal 1-planarity of graphs with a rotation system in linear time - (extended abstract). In: Didimo, W., Patrignani, M. (eds.) GD 2012. LNCS, vol. 7704, pp. 339–345. Springer, Heidelberg (2013)
10. Eggleton, R.: Rectilinear drawings of graphs. Utilitas Mathematica 29, 149–172 (1986)
11. Thomassen, C.: Rectilinear drawings of graphs. Journal of Graph Theory 12, 335–341 (1988)
12. Hong, S.H., Eades, P., Liotta, G., Poon, S.H.: Fáry's theorem for 1-planar graphs. In: [19], pp. 335–346
13. Nagamochi, H.: Straight-line drawability of embedded graphs. Technical Report 2013-005, Department of Applied Mathematics and Physics, Kyoto University, Japan (2013)
14. Battista, G.D., Tamassia, R.: On-line maintenance of triconnected components with spqr-trees. Algorithmica 15, 302–318 (1996)
15. Gutwenger, C., Mutzel, P.: A Linear Time Implementation of SPQR-Trees. In: Marks, J. (ed.) GD 2000. LNCS, vol. 1984, pp. 77–90. Springer, Heidelberg (2001)
16. Dehkordi, H.R., Eades, P.: Every outer-1-plane graph has a right angle crossing drawing. Int. J. Comput. Geometry Appl. 22, 543–558 (2012)
17. Frati, F.: Straight-line drawings of outerplanar graphs in o(dn log n) area. Comput. Geom. 45, 524–533 (2012)
18. Knauer, K.B., Micek, P., Walczak, B.: Outerplanar graph drawings with few slopes. In: [19], pp. 323–334
19. Gudmundsson, J., Mestre, J., Viglas, T. (eds.): COCOON 2012. LNCS, vol. 7434. Springer, Heidelberg (2012)

Straight-Line Grid Drawings of 3-Connected 1-Planar Graphs

Md. Jawaherul Alam[1,*], Franz J. Brandenburg[2,**], and Stephen G. Kobourov[1,*]

[1] Department of Computer Science, University of Arizona, USA
{mjalam,kobourov}@cs.arizona.edu
[2] University of Passau, 94030 Passau, Germany
brandenb@informatik.uni-passau.de

Abstract. A graph is 1-planar if it can be drawn in the plane such that each edge is crossed at most once. In general, 1-planar graphs do not admit straight-line drawings. We show that every 3-connected 1-planar graph has a straight-line drawing on an integer grid of quadratic size, with the exception of a single edge on the outer face that has one bend. The drawing can be computed in linear time from any given 1-planar embedding of the graph.

1 Introduction

Since Euler's Königsberg bridge problem dating back to 1736, planar graphs have provided interesting problems in theory and in practice. Using the elaborate techniques of a canonical ordering and Schnyder realizers, every planar graph can be drawn on a grid of quadratic size, and such drawings can be computed in linear time [15, 21]. The area bound is asymptotically optimal, since the nested triangle graphs are planar graphs and require $\Omega(n^2)$ area [10]. The drawing algorithms were refined to improve the area requirement or to admit convex representations, i.e., where each inner face is convex [5,8] or strictly convex [1].

However, most graphs are nonplanar and recently, there have been many attempts to study larger classes of graphs. Of particular interest are 1-planar graphs, which in a sense are one step beyond planar graphs. They were introduced by Ringel [20] in an attempt to color a planar graph and its dual. Although it is known that a 3-connected planar graph and its dual have a straight-line 1-planar drawing [24] and even on a quadratic grid [13], little is known about general 1-planar graphs. It is NP-hard to recognize 1-planar graphs [16, 18] in general, although there is a linear-time testing algorithm [11] for maximal 1-planar graphs (i.e., where no additional edge can be added without violating 1-planarity) given the the circular ordering of incident edges around each vertex. A 1-planar graph with n vertices has at most $4n - 8$ edges [4, 14, 19] and this upper bound is tight. On the other hand straight-line drawings of 1-planar graphs may have at most $4n - 9$ edges and this bound is tight [9]. Hence not all 1-planar graphs admit straight-line drawings. Unlike planar graphs, maximal 1-planar graphs can be much sparser with only $2.64n$ edges [6].

---

* Research supported in part by NSF grants CCF-1115971 and DEB 1053573.
** Research supported by the Deutsche Forschungsgemeinschaft, DFG, grant Br 835/18-1.

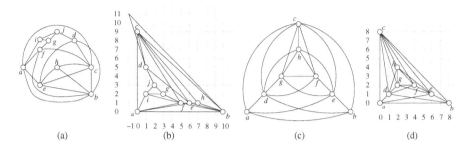

Fig. 1. (a)–(b) A 3-connected 1-planar graph and its straight-line grid drawing (with one bend in one edge), (c)–(d) another 3-connected 1-planar graph and its straight-line grid drawing

Thomassen [23] refers to 1-planar graphs as graphs with *cross index 1* and proved that an embedded 1-planar graph can be turned into a straight-line drawing if and only if it excludes B- and W-configurations; see Fig. 2. These forbidden configurations were first discovered by Eggleton [12] and used by Hong *et al.* [17], who show that the configurations can be detected in linear time if the embedding is given. They also proved that there is a linear time algorithm to convert a 1-planar embedding without B- and W-configurations into a straight-line drawing, but without bounds for the drawing area.

In this paper we settle the straight-line grid drawing problem for 3-connected 1-planar graphs. First we compute a *normal form* for an embedded 1-planar graph with no B-configuration and at most one W-configuration on the outer face. Then, after augmenting the graph with as many planar edges as possible and then deleting the crossing edges, we find a 3-connected planar graph, which is drawn with strictly convex faces using an extension of the algorithm of Chrobak and Kant [8]. Finally the pairs of crossing edges are reinserted into the convex faces. This gives a straight-line drawing on a grid of quadratic size with the exception of a single edge on the outer face, which may need one bend (and this exception is unavoidable); see Fig. 1. In addition, the drawing is obtained in linear time from a given 1-planar embedding.

2   Preliminaries

A *drawing* of a graph G is a mapping of G into the plane such that the vertices are mapped to distinct points and each edge is a Jordan arc between its endpoints. A drawing is *planar* if the Jordan arcs of the edges do not cross and it is *1-planar* if each edge is crossed at most once. Note that crossings between edges incident to the same vertex are not allowed. For example, K_5 and K_6 are 1-planar graphs. An *embedding* of a graph is planar (resp. 1-planar) if it admits a planar (resp. 1-planar) drawing. An embedding specifies the *faces*, which are topologically connected regions. The unbounded face is the *outer face*. A face in a planar graph is specified by a cyclic sequence of edges on its boundary (or equivalently by the cyclic sequence of the endpoints of the edges).

Accordingly, a *1-planar embedding* $\mathcal{E}(G)$ specifies the faces in a 1-planar drawing of G including the outer face. A 1-planar embedding is a witness for 1-planarity. In particular, $\mathcal{E}(G)$ describes the pairs of crossing edges and the faces where the edges

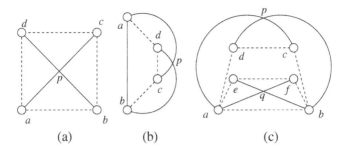

Fig. 2. (a) An augmented X-configuration, (b) an augmented B-configuration, (c) an augmented W-configuration. The graphs induced by the solid edges are called an X-configuration (a), a B-configuration (b), and a W-configuration (c).

cross and has linear size. Each pair of *crossing edges* (a, c) and (b, c) induces a *crossing point* p. Call the segment of an edge between the vertex and the crossing point a *half-edge*. Each half-edge is *impermeable*, analogous to the edges in planar drawings, in the sense that no edge can cross such a half-edge without violating the 1-planarity of the embedding. The non-crossed edges are called *planar*. A *planarization* $G^\times$ is obtained from $\mathcal{E}(G)$ by using the crossing points as regular vertices and replacing each crossing edge by its two half-edges. A 1-planar embedding $\mathcal{E}(G)$ and its planarization share equivalent embeddings, and each face is given by a list of edges and half-edges defining it, or equivalently, by a list of vertices and crossing points of the edges and half edges.

Eggleton [12] raised the problem of recognizing 1-planar graphs with rectilinear drawings. He solved this problem for outer-1-planar graphs (1-planar graphs with all vertices on the outer-cycle) and proposed three forbidden configurations. Thomassen [23] solved Eggleton's problem and characterized the rectilinear 1-planar embeddings by the exclusion of B- and W-*configurations*; see Fig. 2. Hong *et al.* [17], obtain a similar characterization where the B- and W-configurations are called the "Bulgari" and "Gucci" graphs. They also show that all occurrences of these configurations can be computed in linear time from a given 1-planar embedding.

Definition 1. *Consider a 1-planar embedding $\mathcal{E}(G)$:*

A B-configuration consists of an edge (a, b) and two edges (a, c) and (b, d) which cross in some point p such that c and d lie in the interior of the triangle (a, b, p). Here (a, b) is called the base *of the configuration.*

An X-configuration consists of a pair (a, c) and (b, d) of crossing edges which does not form a B-configuration.

A W-configuration consists of two pairs of edges (a, c), (b, d) and (a, f), (b, e) which cross in points p and q, such that c, d, e, f lie in the interior of the quadrangle a, p, b, q. Here again the edge (a, b), if present is the base.

Observe that for all these configurations the base edges may be crossed by another edge, whereas the crossing edges are impermeable; see Fig 2.

Thomassen [23] and Hong *et al.* [17] proved that for a 1-planar embedding to admit straight-line drawing, B- and W-configurations must be excluded:

Proposition 1. *A 1-planar embedding $\mathcal{E}(G)$ admits a straight-line drawing with a topologically equivalent embedding if and only if it does not contain a B- or a W-configuration.*

Augment a given 1-planar embedding $\mathcal{E}(G)$ by adding as many edges to $\mathcal{E}(G)$ as possible so that G remains a simple graph and the newly added edges are planar in $\mathcal{E}(G)$. We call such an embedding a *planar-maximal* embedding of G and the operation *planar-maximal augmentation*. (Note that Hong *et al.* [17] color the planar edges of a 1-planar embedding red and call a planar-maximal augmentation a *red augmentation*.) The *planar skeleton* $\mathcal{P}(\mathcal{E}(G))$ consists of the planar edges of a planar-maximal augmentation. It is a planar embedded graph, since all pairs of crossing edges are omitted. Note that the planar augmentation and the planar skeleton are defined for an embedding, not for a graph. A graph may have different embeddings which give rise to different configurations and augmentations. The notion of planar-maximal embedding is different from the notions of maximal 1-planar embeddings and maximal 1-planar graphs, which are such that the addition of any edge violates 1-planarity (or simplicity) [6].

The following claim, proven in many earlier papers [6, 14, 17, 22, 23], shows that a crossing pair of edges induces a K_4 in planar-maximal embedding, since missing edges of a K_4 can be added without inducing new crossings.

Lemma 1. *Let $\mathcal{E}(G)$ be a planar-maximal 1-planar embedding of a graph G and let (a, c) and (b, d) be two crossing edges. Then the four vertices $\{a, b, c, d\}$ induce a K_4.*

By Lemma 1, for a planar-maximal embedding each X-, B-, and W-configuration is augmented by additional edges. Here we define these augmented configurations.

Definition 2. *Let $\mathcal{E}(G)$ be a planar-maximal 1-planar embedding of a graph G. An augmented X-configuration consists of a K_4 with vertices (a, b, c, d) such that the edges (a, c) and (b, d) cross inside the quadrangle abcd. An augmented B-configuration consists of a K_4 with vertices (a, b, c, d) such that the edges (a, c) and (b, d) cross beyond the boundary of the quadrangle abcd. An augmented W-configuration consists of two K_4's (a, b, c, d) and (a, b, e, f) one of which is in an augmented X-configuration and the other in an augmented B-configuration.*

For an augmented X- or augmented B-configuration, the edges not inducing a crossing with other edges in the configuration define a cycle, we call it the skeleton. *In each configuration, the edges on the outer-boundary of the embedded configuration and not inducing a crossing with other edges in the configuration are the* base edges.

Using the results of Thomassen [23] and Hong *et al.* [17], we can now characterize when a planar-maximal 1-planar embedding of a graph admits a straight-line drawing:

Lemma 2. *Let $\mathcal{E}(G)$ be a planar-maximal 1-planar embedding of a graph G. Then there is a straight-line 1-planar drawing of G with a topologically equivalent embedding as $\mathcal{E}(G)$ if and only if $\mathcal{E}(G)$ does not contain an augmented B-configuration.*

Proof. Assume $\mathcal{E}(G)$ contains an augmented B-configuration. Then it contains a B-configuration and has no straight-line 1-planar drawing by Proposition 1. Conversely, if $\mathcal{E}(G)$ has no straight-line 1-planar drawing then by Proposition 1 it contains at least one B- or W-configuration. Since Γ is a planar-maximal embedding, by Lemma 1 each crossing edge pair in $\mathcal{E}(G)$ induces a K_4. Thus the dotted edges in Fig. 2(b)–(c) must be present in any B- or W- configuration, inducing an augmented B-configuration.   □

The *normal form* for an embedded 1-planar graph $\mathcal{E}(G)$ is obtained by first adding the four planar edges to form a K_4 for each pair of crossing edges while routing them closely to the crossing edges and then removing old duplicate edges if necessary. Such an embedding of a 1-planar graph is a normal embedding of it. A *normal planar-maximal augmentation* for an embedded 1-planar graph is obtained by first finding a normal form of the embedding and then by a planar-maximal augmentation.

Lemma 3. *Given a 1-planar embedding $\mathcal{E}(G)$, the normal planar-maximal augmentation of $\mathcal{E}(G)$ can be computed in linear time.*

Proof. First augment each crossing of two edges (a, c) and (b, d) to a K_4, such that the edges $(a, b), (b, c), (c, d), (d, a)$ are added and in case of a duplicate the former edge is removed. Then all augmented X-configurations are empty and contain no vertices inside their skeletons. Next triangulate all faces which do not contain a half-edge, a crossing edge, or a crossing point. Each step can be done in linear time.   □

3   Characterization of 3-Connected 1-Planar Graphs

Here we characterize 3-connected 1-planar graphs by a normal embedding, where the crossings are augmented to K_4's such that the resulting augmented X-configurations have vertex-empty skeletons and there is no augmented B-configuration except for at most one augmented W-configuration with a pair of crossing edges in the outer face.

Let $\mathcal{E}(G)$ be a 1-planar embedding of a graph G. Each pair of crossing edges induces a crossing point and the crossing edges and their half-edges are *impermeable* as they cannot be crossed by other edges without violating 1-planarity. An *impermeable path* in $\mathcal{E}(G)$ is an internally-disjoint sequence $P = v_1, p_1, v_2, p_2, \ldots, v_n, p_n, v_{n+1}$, where $v_1, v_2, \ldots, v_{n+1}$ are (regular) vertices of G, $p_1, p_2, \ldots, p_n$ are crossing points in $\mathcal{E}(G)$ and $(v_i, p_i), (p_i, v_{i+1})$ for each $i \in \{1, 2, \ldots, n\}$ are half edges. If $v_{n+1} = v_0$, then P is an *impermeable cycle*. An impermeable cycle is *separating* when it has vertices both inside and outside of it, since deleting its vertices disconnects G.

Lemma 4. *Let $G = (V, E)$ be a 3-connected 1-planar graph with a planar-maximal 1-planar embedding $\mathcal{E}(G)$. Then the following conditions hold.*

A.   (i) *Two augmented B-configurations or two augmented X-configurations cannot be on the same side of a common base edge.*

      (ii) *Suppose an augmented B-configuration B and an augmented X-configuration X are on the same side of a common base edge (a, b). Let p and q be the crossing points for X and B, respectively and let $R(X)$ and $R(B)$ be the regions inside the skeletons of X and B. Then all vertices of $V \setminus \{a, b\}$ are inside the impermeable cycle $apbq$ if $R(X) \subset R(B)$; otherwise all vertices of $V \setminus \{a, b\}$ are outside the impermeable cycle $apbq$.*

B.   (i) *If two augmented B-configurations are on opposite sides of a common base edge (a, b), with crossing points p and q, respectively, then all the vertices of $V \setminus \{a, b\}$ are inside the impermeable cycle $apbq$.*

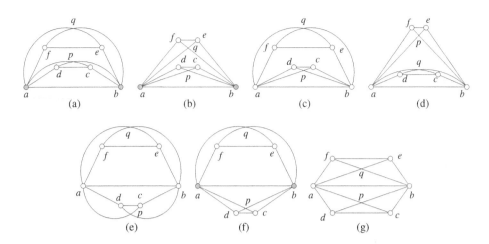

Fig. 3. Illustration for the proof of Lemma 4

   (ii) If two augmented X-configurations are on opposite sides of a common base edge (a, b), with crossing points p and q, respectively, then all the vertices of $V \setminus \{a, b\}$ are outside the impermeable cycle $apbq$.

   (iii) An augmented B-configuration and an augmented X-configuration cannot share a common base edge from opposite sides.

Proof. Condition A.(i) and B.(iii) hold because each of these configurations induces a separating impermeable $apbq$ cycle in $\mathcal{E}(G)$ with only two (regular) vertices from G, a contradiction with the 3-connectivity of G; see Fig. 3(a)–(b) and (f). Similarly, if any of the Conditions A.(ii) and B.(i)–(ii) is not satisfied, then the impermeable cycle $apbq$ becomes separating and hence the pair $\{a, b\}$ becomes separation pair of G, again a contradiction with the 3-connectivity of G; see Fig. 3(c)–(d), (e) and (g).   □

Corollary 1. *Let G be a 3-connected 1-planar graph with a planar-maximal 1-planar embedding $\mathcal{E}(G)$. Then no three crossing edge-pairs in $\mathcal{E}(G)$ share the same base edge.*

Proof. Each crossing edge pair induces either an augmented B- or an augmented X-configuration. This fact along with Lemma 4[A.(i), B.(iii)] yields the corollary.   □

Lemma 5. *Let G be a 3-connected 1-planar graph. Then there is a planar-maximal 1-planar embedding $\mathcal{E}(G^*)$ of a supergraph G^* of G so that $\mathcal{E}(G^*)$ contains at most one augmented W-configuration in the outer face and no other augmented B-configuration, and each augmented X-configuration in $\mathcal{E}(G^*)$ contains no vertex inside its skeleton.*

Proof. Let $\mathcal{E}(G)$ be a 1-planar embedding of G. We claim that by a normal planar-maximal augmentation of $\mathcal{E}(G)$ we get the desired embedding of a supergraph of G. Note that due to the edge-rerouting this operation converts any B-configuration whose base is not shared with another configuration into an X-configuration; see Fig. 4(a). If a base edge is shared by two B-configurations, they are converted into one W-configuration and by Lemma 4 this W-configuration is on the outer face; see Fig. 4(b).

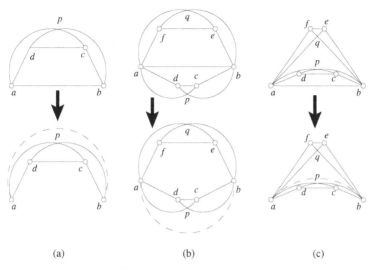

Fig. 4. Illustration for the proof of Lemma 5

By Corollary 1, a base edge cannot be shared by more than two B-configurations. Furthermore this operation does not create any new B-configuration. It also makes the skeleton of any augmented X-configuration vertex-empty; by Lemma 4 a base edge can be shared by at most two augmented X-configurations from opposite sides and if it is shared by two augmented X-configurations, the interior of the induced impermeable cycle is empty; see Fig. 4(c).   □

Lemma 5 together with Proposition 1 implies the following:

Theorem 1. *A 3-connected 1-planar graph admits a straight-line 1-planar drawing except for at most one edge in the outer face.*

4   Grid Drawings

In the previous section we showed that a 3-connected 1-planar graph has a straight-line 1-planar drawing, with the exception of a single edge in the outer face. We now strengthen this result and show that there is straight-line grid drawing with $O(n^2)$ area, which can be constructed in linear time from a given 1-planar embedding.

The algorithm takes an embedding $\mathcal{E}(G)$ and computes a normal planar-maximal augmentation. Consider the planar skeleton $\mathcal{P}(\mathcal{E}(G))$ for the embedding. If there is an augmented W-configuration and a crossing in the outer face, one crossing edge on the outer face is kept and the other crossing edge is treated separately. Thus the outer face of $\mathcal{P}(\mathcal{E}((G))$ is a triangle and the inner faces are triangles or quadrangles. Each quadrangle comes from an augmented X-configuration. It must be drawn strictly convex, such that the crossing edges can be re-inserted. This is achieved by an extension of the convex grid drawing algorithm of Chrobak and Kant [8], which itself is an extension of the shifting method of de Fraysseix, Pach and Pollack [15]. Since the faces are at most quadrangles, we can avoid three collinear vertices and the degeneration to a triangle by an extra unit shift. Note that our drawing algorithm achieves an area of $(2n-2) \times (2n-3)$, while the general algorithms for strictly convex grid drawings [1, 7] require larger area, since strictly convex drawings of n-gons need $\Omega(n^3)$ area [2]. Barany and Rote give a strictly convex grid drawing of a planar graph on a $14n \times 14n$ grid if the faces are at most 4-gons, and on a $2n \times 2n$ grid if, in addition, the outer face is a triangle. However, their approach is quite complex and does not immediately yield these bounds. It is also not clear how to use this approach for planar graphs in our 1-planar graph setting, in particular when we have an unavoidable W-configuration in the outer face.

The algorithm of Chrobak and Kant and in particular the computation of a canonical decomposition presumes a 3-connected planar graph. Thus the planar skeleton of a 3-connected 1-planar graph must be 3-connected, which holds except for the K_4, when it is embedded as an augmented X-configuration. This results parallels the fact that the planarization of a 3-connected 1-planar graph is 3-connected [14].

Lemma 6. *Let G be a graph with a planar-maximal 1-planar embedding $\mathcal{E}(G)$ such that it has no augmented B-configuration and each augmented X-configuration in $\mathcal{E}(G)$ has no vertex inside its skeleton. Then the planar skeleton $\mathcal{P}(\mathcal{E}(G))$ is 3-connected.*

We will prove Lemma 6 by showing that there is no separation pair in $\mathcal{P}(\mathcal{E}(G))$. First we obtain a planar graph H from G as follows. Let (a, c) and (b, d) be a pair of crossing edges that form an augmented X-configuration X in Γ. We then delete the two edges (a, c), (b, d); add a vertex u and the edges (a, u), (b, u), (c, u), (d, u) to triangulate the face $abcd$. Call v a *cross-vertex* and call this operation *cross-vertex insertion* on X. We then obtain H from G by cross-vertex insertion on each augmented X-configuration. Call H a *planarization* of G and denote the set of all the cross-vertices by U. Then $\mathcal{P}(\mathcal{E}(G)) = H \setminus U$. Before proving Lemma 6 we consider several properties of H, the planarization of the 1-planar graph.

Lemma 7. *Let $G = (V, E)$ be a graph with a planar-maximal 1-planar embedding $\mathcal{E}(G)$ such that $\mathcal{E}(G)$ contains no augmented B-configuration and each augmented X-configuration in $\mathcal{E}(G)$ contains no vertex inside its skeleton. Let H be a planarization of G, where U is the set of cross-vertices. Then the following conditions hold.*

(a) H is a maximal planar graph (except if H is the K_4 in an X-configuration)
(b) Each vertex of U has degree 4.
(c) U is an independent set of H.
(d) There is no separating triangle of H containing any vertex from U.
(e) There is no separating 4-cycle of H containing two vertices from U.

Proof. For convenience, we call each vertex in $V - U$ a *regular vertex*.

(a) Since H is a planar graph, we only show that each face of H is a triangle. Each crossing edge pair in Γ induces an augmented X-configuration whose skeleton has no vertex in its interior. Hence each face of H containing a crossing vertex is a triangle. Again, Hong *et al.* [17] showed that in a planar-maximal 1-planar embedding a face with no crossing vertices is a triangle. Thus H is a maximal planar graph.

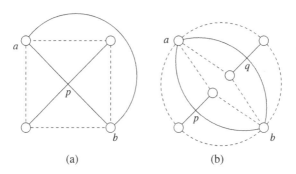

Fig. 5. Illustration for the proof of Lemma 6

(b)–(c) These two conditions follow from the fact that the neighborhood of each crossing vertex consists of exactly four regular vertices that form the skeleton of the corresponding augmented X-configuration.

(d) For a contradiction suppose a vertex $u \in U$ participates in a separating triangle T of H. Since the neighborhood of u forms the skeleton of the corresponding augmented X-configuration X, the other two vertices, say a and b, in T are regular vertices. The edge (a, b) cannot form a base edge for X, since if it did, then the interior of the separating triangle T would be contained in the interior of the skeleton for X and hence would be empty. Assume therefore that a and b are not consecutive on the skeleton of X. In this case the edge (a, b) is a crossing edge in G and hence has been deleted when constructing H; see Fig. 5(a).

(e) Suppose two vertices $u, v \in U$ participate in a separating 4-cycle T of H. Due to Condition (c), assume that $T = aubv$, where a, b are regular vertices. If the two vertices a, b are adjacent in H, assume without loss of generality that the edge (a, b) is drawn inside the interior of T. Then the interior of at least one of the two triangles abu and abv is non-empty, inducing a separating triangle in H, a contradiction with Condition (d). We thus assume that the two vertices a and b are not adjacent in H. Then for both the augmented X-configurations X and Y, corresponding to the two crossing vertices u and v, the two vertices u and v are not consecutive on their skeleton. This implies that the crossing edge (a, b) participates in two different augmented X-configurations in Γ, again a contradiction; see Fig. 5(b). $\qquad\square$

We are now ready to prove Lemma 6.

Proof (Lemma 6). Assume for a contradiction that $\mathcal{P}(\mathcal{E}(G))$ is not 3-connected. Then there exists some separation pair $\{a, b\}$ in $\mathcal{P}(\mathcal{E}(G))$. Let H be the planarization of G, where U is the set of cross-vertices. Then $S = U \cup \{a, b\}$ is a separating set for H. Take a minimal separating set $S' \subset S$ such that no proper subset of S' is a separating set. Since H is a maximal planar graph (from Lemma 7(a)), S' forms a separating cycle [3]. The 3-connectivity of the maximal planar graph H implies $|S'| \geq 3$. Again since S' contains at most two regular vertices a, b and no two cross-vertices can be adjacent in H (Lemma 7(c)), $|S'| < 5$. Hence S' is a separating triangle or a separating 4-cycle with at most two regular vertices; we get a contradiction with Lemma 7(d)–(e). $\qquad\square$

Finally, we describe our algorithm for straight-line grid drawings. This drawing algorithm is based on an extension of the algorithm of Chrobak and Kant [8] for computing a convex drawing of a planar 3-connected graph. For convenience we refer to this algorithm as the CK-algorithm and we begin with a brief overview. Let $G = (V, E)$ be an embedded 3-connected graph and let (u, v) be an edge on the outer-cycle of G. The CK-algorithm starts by computing a *canonical decomposition* of G, which is an ordered partition $V_1, V_2, \ldots, V_t$ of V such that the following conditions hold:

(i) For each $k \in \{1, 2, \ldots, t\}$, the graph G_k induced by the vertices $V_1 \cup \ldots \cup V_k$ is 2-connected and its outer-cycle C_k contains the edge (u, v).
(ii) G_1 is a cycle, V_t is a singleton $\{z\}$, where $z \notin \{u, v\}$ is on the outer-cycle of G.
(iii) For each $k \in \{2, \ldots, t - 1\}$ the following conditions hold:
  – If V_k is a singleton $\{z\}$, then z is on the outer face of G_{k-1} and has at least one neighbor in $G - G_k$.
  – If V_k is a chain $\{z_1, \ldots, z_l\}$, each z_i has at least one neighbor in $G - G_k$, z_1, z_l have one neighbor each on C_{k-1} and no other z_i has neighbors on G_{k-1}.

For each $k \in \{1, 2, \ldots, t\}$, we say that the vertices that belong to V_k have *rank k*. We call a vertex of G_k *saturated* if it has no neighbor in $G - G_k$. The CK-algorithm starts by drawing the edge (u, v) with a horizontal line-segment of unit length. Then for $k = 1, 2, \ldots, t$, it incrementally completes the drawing of G_k. Let $C_{k-1} = \{(u = w_1, \ldots, w_p, \ldots, w_q, \ldots, w_r = v)\}$ with $1 \leq p < q \leq r$ where w_p and w_q are the leftmost and the rightmost neighbor of vertices in V_k. Then the vertices of V_k are placed above the vertices $w_p, \ldots, w_q$. Assume that $V_k = \{z_1, \ldots, z_l\}$. Then z_1 is placed on the vertical line containing w_p if w_p is saturated in G_k; otherwise it is placed on the vertical line one unit to the right of w_p. On the other hand, z_l is placed on the negative diagonal line (i.e., with $-45°$ slope) containing w_q. If v_k is a singleton then $z = z_1 = z_l$ is placed at the intersection of these two lines. Otherwise (after necessary shifting of w_q and other vertices), the vertices $z_1, \ldots z_l$ are placed on consecutive vertical lines one unit apart from each other. In order to make sure that this shifting operation does not disturb planarity or convexity, each vertex v is associated with an "under-set" $U(v)$ and whenever v is shifted, all vertices in $U(v)$ are also shifted along with v. Thus the edges between vertices of any $U(v)$ are in a sense *rigid*.

Theorem 2. *Given a 1-planar embedding $\mathcal{E}(G)$ of a 3-connected graph G, a straight-line drawing on the $(2n - 2) \times (2n - 3)$ grid can be computed in linear time. Only one edge on the outer face may require one bend.*

Proof. Assume that $\mathcal{E}(G)$ is a normal planar-maximal embedding; otherwise we compute one by a normal planar-maximal augmentation in linear time by Lemma 3. Consider the planar skeleton $\mathcal{P}(\mathcal{E}(G))$. If there is no unavoidable W-configuration on the outer face of the maximal planar augmentation, then the outer-cycle of $\mathcal{P}(\mathcal{E}(G))$ is a triangle. Otherwise we add one of the crossing edges in the outer face to $\mathcal{P}(\mathcal{E}(G))$ to make the outer-cycle a triangle. The other crossing edge is treated separately. By Lemma 6, $\mathcal{P}(\mathcal{E}(G))$ is 3-connected, its outer face is a triangle (a, b, c) and the inner faces are triangles or quadrangles, where the latter result from augmented X-configurations and are in one-to-one correspondence to pairs of crossing edges.

We wish to obtain a planar straight-line grid drawing of $\mathcal{P}(\mathcal{E}(G))$ such that all quadrangles are strictly convex. Although the CK-algorithm draws any 3-connected planar graph of n vertices on a grid of size $(n - 1) \times (n - 1)$ with convex faces, the faces are not necessarily strictly convex [8]. Hence we must modify the algorithm so that all quadrangles are strictly convex. Note that by the assignment of the under-sets, the CK-algorithm guarantees that once a face is drawn strictly convex, it would remain strictly convex after any subsequent shifting of vertices.

For $\mathcal{P}(\mathcal{E}(G))$ each V_k is either a single vertex or a pair with an edge, since the faces are at most quadrangles. If V_k is an edge (z_1, z_2) then, by the definition of the canonical decomposition, exactly one quadrangle face $w_p z_1 z_2 w_q$ is formed and by construction this face is drawn convex. We thus assume that V_k contains a single vertex, say v. Let $C_{k-1} = \{(u = w_1, \ldots, w_p, \ldots, w_q, \ldots, w_r = v)\}$ with $1 \leq p < q \leq r$ where w_p and w_q are the leftmost and the rightmost neighbors of vertices in V_k. Then the new faces created by the insertion of v are all drawn strictly convex unless there is some quadrangle $v w_{p'-1} w_{p'} w_{p'+1}$ where $p < p' < q$ and $w_{p'-1}, w_{p'}, w_{p'+1}$ are collinear in the drawing of G_{k-1}. In this case $w_{p'}$ must be saturated in G_{k-1} and this occurs in the CK-algorithm only when the line containing $w_{p'-1}, w_{p'}, w_{p'+1}$ is a vertical line or a negative diagonal (with $-45°$ slope). In the former case, w_{p-1} should have also been saturated in G_{k-1}, which is not possible since v is its neighbor. It is thus sufficient to ensure that no saturated vertex of G_k is in the negative diagonal of both its left and right neighbors on C_k. We do this by the following extension of the CK-algorithm.

Suppose v is placed above w_q with slope -45, w_q was placed above its rightmost lower neighbor $w'_{q'}$ with slope -45, and there is a quadrangle $(v, w_q, w'_{q'}, u)$ for some vertex u with higher rank to be placed later. Then shift $w'_{q'}$ by one extra unit to the right when v or u is placed. This implies a bend at w_q and a strictly convex angle above w_q.

The CK-algorithm starts by placing the first two vertices one unit away and it requires a unit shift to the right for each following vertex. On the other hand, a 1-planar graph has at most $n - 2$ pairs of crossing edges. Hence, there are $g \leq n - 3$ augmented X-configurations, each of which induces a quadrangle in the planar skeleton. Thus the width and height are $n - 1 + g$, which is bounded by $2n - 4$. The vertices a, b, c of the outer triangle are placed at the grid points $(0, 0), (0, n - 1 + g), (n - 1 + g, 0)$.

If the graph had an unavoidable W-configuration in the outer face, we need a post-processing phase to draw the extra edge (b, d), which induces a crossing with the edge (a, c). Since a is the leftmost lower neighbor of d when d is placed and d is not saturated, d is placed at $(1, j)$ for some $j < n - 2 + g$. Shift b one unit to the right, insert a bend at $(-1, n + g)$, one diagonal unit left above c and route (b, d) via the bend point. □

5   Conclusion and Future Work

We showed that 3-connected 1-planar graphs can be embedded on $O(n) \times O(n)$ integer grid, so that edges are drawn as straight-line segments (except for at most one edge on the outerface that requires a bend). Moreover, the algorithm is simple and runs in linear time given a 1-planar embedding. Note that even a path may require exponential area for a given fixed 1-planar embedding, e.g., [17]. Recognition of 1-planar graphs is NP-hard [18]. How hard is the recognition of planar-maximal 1-planar graphs?

References

1. Bárány, I., Rote, G.: Strictly convex drawings of planar graphs. Documenta Mathematica 11, 369–391 (2006)
2. Bárány, I., Tokushige, N.: The minimum area of convex lattice n-gons. Combinatorica 24(2), 171–185 (2004)
3. Baybars, I.: On k-path hamiltonian maximal planar graphs. Discrete Mathematics 40(1), 119–121 (1982)
4. Bodendiek, R., Schumacher, H., Wagner, K.: Bemerkungen zu einem Sechsfarbenproblem von G. Ringel. Abh. aus dem Math. Seminar der Univ. Hamburg 53, 41–52 (1983)
5. Bonichon, N., Felsner, S., Mosbah, M.: Convex drawings of 3-connected plane graphs. Algorithmica 47(4), 399–420 (2007)
6. Brandenburg, F.J., Eppstein, D., Gleißner, A., Goodrich, M.T., Hanauer, K., Reislhuber, J.: On the density of maximal 1-planar graphs. In: Didimo, W., Patrignani, M. (eds.) GD 2012. LNCS, vol. 7704, pp. 327–338. Springer, Heidelberg (2013)
7. Chrobak, M., Goodrich, M.T., Tamassia, R.: Convex drawings of graphs in two and three dimensions. In: Symposium on Computational Geometry, pp. 319–328 (1996)
8. Chrobak, M., Kant, G.: Convex grid drawings of 3-connected planar graphs. International Journal of Computational Geometry and Applications 7(3), 211–223 (1997)
9. Didimo, W.: Density of straight-line 1-planar graph drawings. Information Processing Letter 113(7), 236–240 (2013)
10. Dolev, D., Leighton, T., Trickey, H.: Planar embedding of planar graphs. In: Advances in Computing Research, pp. 147–161 (1984)
11. Eades, P., Hong, S.H., Katoh, N., Liotta, G., Schweitzer, P., Suzuki, Y.: Testing maximal 1-planarity of graphs with a rotation system in linear time. In: Didimo, W., Patrignani, M. (eds.) GD 2012. LNCS, vol. 7704, pp. 339–345. Springer, Heidelberg (2013)
12. Eggleton, R.B.: Rectilinear drawings of graphs. Utilitas Math. 29, 149–172 (1986)
13. Erten, C., Kobourov, S.G.: Simultaneous embedding of a planar graph and its dual on the grid. Theory of Computing Systems 38(3), 313–327 (2005)
14. Fabrici, I., Madaras, T.: The structure of 1-planar graphs. Discrete Mathematics 307(7-8), 854–865 (2007)
15. de Fraysseix, H., Pach, J., Pollack, R.: How to draw a planar graph on a grid. Combinatorica 10(1), 41–51 (1990)
16. Grigoriev, A., Bodlaender, H.L.: Algorithms for graphs embeddable with few crossings per edge. Algorithmica 49(1), 1–11 (2007)
17. Hong, S.-H., Eades, P., Liotta, G., Poon, S.-H.: Fáry's theorem for 1-planar graphs. In: Gudmundsson, J., Mestre, J., Viglas, T. (eds.) COCOON 2012. LNCS, vol. 7434, pp. 335–346. Springer, Heidelberg (2012)
18. Korzhik, V.P., Mohar, B.: Minimal obstructions for 1-immersions and hardness of 1-planarity testing. Journal of Graph Theory 72(1), 30–71 (2013)
19. Pach, J., Tóth, G.: Graphs drawn with a few crossings per edge. Combinatorica 17, 427–439 (1997)
20. Ringel, G.: Ein Sechsfarbenproblem auf der Kugel. Abh. aus dem Math. Seminar der Univ. Hamburg 29, 107–117 (1965)
21. Schnyder, W.: Embedding planar graphs on the grid. In: Johnson, D.S. (ed.) Symposium on Discrete Algorithms, pp. 138–148 (1990)
22. Suzuki, Y.: Re-embeddings of maximum 1-planar graphs. SIAM Journal of Discrete Mathematics 24(4), 1527–1540 (2010)
23. Thomassen, C.: Rectilinear drawings of graphs. Journal of Graph Theory 12(3), 335–341 (1988)
24. Tutte, W.T.: How to draw a graph. Proc. London Math. Society 13(52), 743–768 (1963)

New Bounds on the Maximum Number of Edges in k-Quasi-Planar Graphs

Andrew Suk[1],[*] and Bartosz Walczak[2],[3],[**]

[1] Massachusetts Institute of Technology, Cambridge, MA
asuk@math.mit.edu
[2] École Polytechnique Fédérale de Lausanne, Switzerland
bartosz.walczak@epfl.ch
[3] Theoretical Computer Science Department, Faculty of Mathematics
and Computer Science, Jagiellonian University, Kraków, Poland
walczak@tcs.uj.edu.pl

Abstract. A topological graph is k-*quasi-planar* if it does not contain k pairwise crossing edges. An old conjecture states that for every fixed k, the maximum number of edges in a k-quasi-planar graph on n vertices is $O(n)$. Fox and Pach showed that every k-quasi-planar graph with n vertices and no pair of edges intersecting in more than $O(1)$ points has at most $n(\log n)^{O(\log k)}$ edges. We improve this upper bound to $2^{\alpha(n)^c} n \log n$, where $\alpha(n)$ denotes the inverse Ackermann function, and c depends only on k. We also show that every k-quasi-planar graph with n vertices and every two edges have at most one point in common has at most $O(n \log n)$ edges. This improves the previously known upper bound of $2^{\alpha(n)^c} n \log n$ obtained by Fox, Pach, and Suk.

1 Introduction

A *topological graph* is a graph drawn in the plane so that its vertices are represented by points and its edges are represented by curves connecting the corresponding points. The curves are always *simple*, that is, they do not have self-intersections. The curves are allowed to intersect each other, but they cannot pass through vertices except for their endpoints. Furthermore, the edges are not allowed to have tangencies, that is, if two edges share an interior point, then they must properly cross at that point. We only consider graphs without parallel edges or loops. Two edges of a topological graph *cross* if their interiors share a point. A topological graph is *simple* if any two of its edges have at most one point in common, which can be either a common endpoint or a crossing.

It follows from Euler's polyhedral formula that every topological graph on n vertices and with no two crossing edges has at most $3n - 6$ edges. A graph

---

[*] Supported by an NSF Postdoctoral Fellowship and by Swiss National Science Foundation Grant 200021-125287/1.

[**] Supported by Swiss National Science Foundation Grant 200020-144531 and by Polish MNiSW Grant 884/N-ESF-EuroGIGA/10/2011/0 within the ESF EuroGIGA project GraDR.

S. Wismath and A. Wolff (Eds.): GD 2013, LNCS 8242, pp. 95–106, 2013.

is called *k-quasi-planar* if it can be drawn as a topological graph with no k pairwise crossing edges. Hence, a graph is 2-quasi-planar if and only if it is planar. According to an old conjecture (see Problem 1 in Section 9.6 of [4]), for any fixed $k \geq 2$ there exists a constant c_k such that every k-quasi-planar graph on n vertices has at most $c_k n$ edges. Agarwal, Aronov, Pach, Pollack, and Sharir [2] were the first to prove this conjecture for *simple* 3-quasi-planar graphs. Later, Pach, Radoičić, and Tóth [14] generalized the result to *all* 3-quasi-planar graphs. Ackerman [1] proved the conjecture for $k = 4$.

For larger values of k, several authors have proved upper bounds on the maximum number of edges in k-quasi-planar graphs under various conditions on how the edges are drawn. These include but are not limited to [5,7,8,15,19]. In 2008, Fox and Pach [7] showed that every k-quasi-planar graph with n vertices and no pair of edges intersecting in more than t points has at most $n(\log n)^{c \log k}$ edges, where c depends only on t. In this paper, we improve the exponent of the polylogarithmic factor from $O(\log k)$ to $1 + o(1)$ for fixed t.

Theorem 1. *Every k-quasi-planar graph with n vertices and no pair of edges intersecting in more than t points has at most $2^{\alpha(n)^c} n \log n$ edges, where $\alpha(n)$ denotes the inverse of the Ackermann function, and c depends only on k and t.*

Recall that the *Ackermann function* $A(n)$ is defined as follows. Let $A_1(n) = 2n$, and $A_k(n) = A_{k-1}(A_k(n-1))$ for $k \geq 2$. In particular, we have $A_2(n) = 2^n$, and $A_3(n)$ is an exponential tower of n *two*'s. Now let $A(n) = A_n(n)$, and let $\alpha(n)$ be defined as $\alpha(n) = \min\{k \geq 1 \colon A(k) \geq n\}$. This function grows much slower than the inverse of any primitive recursive function.

For *simple* topological graphs, Fox, Pach, and Suk [8] showed that every k-quasi-planar simple topological graph on n vertices has at most $2^{\alpha(n)^c} n \log n$ edges, where c depends only on k. We establish the following improvement.

Theorem 2. *Every k-quasi-planar simple topological graph on n vertices has at most $c_k n \log n$ edges, where c_k depends only on k.*

We start the proofs of both theorems with a reduction to the case of topological graphs containing an edge that intersects every other edge. This reduction introduces the $O(\log n)$ factor for the bound on the number of edges. Then, the proof of Theorem 1 follows the approaches of Valtr [19] and Fox, Pach, and Suk [8], using a result on generalized Davenport-Schinzel sequences, which we recall in Section 3. Although the proofs in [19] and [8] heavily depend on the assumption that any two edges have at most one point in common, we are able to remove this condition by establishing some technical lemmas in Section 4. In Section 5, we finish the proof of Theorem 1. The proof of Theorem 2, which relies on a recent coloring result due to Lasoń, Micek, Pawlik, and Walczak [10], is given in Section 6.

2   Initial Reduction

We call a collection C of curves in the plane *decomposable* if there is a partition $C = C_1 \cup \cdots \cup C_w$ such that each C_i contains a curve intersecting all other curves in C_i, and for $i \neq j$, no curve in C_i crosses nor shares an endpoint with a curve in C_j.

Lemma 1 (Fox, Pach, Suk [8]). *There is an absolute constant $c > 0$ such that every collection C of $m \geq 2$ curves such that any two of them intersect in at most t points has a decomposable subcollection of size at least $\frac{cm}{t \log m}$.*

In the proofs of both Theorem 1 and Theorem 2, we establish a (near) linear upper bound on the number of edges under the additional assumption that the graph has an edge intersecting every other edge. Once this is achieved, we use the following lemma to infer an upper bound for the general case.

Lemma 2 (implicit in [8]). *Let G be a topological graph on n vertices such that no two edges have more than t points in common. Suppose that for some constant β, every subgraph G' of G containing an edge that intersects every other edge of G' has at most $\beta|V(G')|$ edges. Then G has at most $c_t \beta n \log n$ edges, where c_t depends only on t.*

Proof. By Lemma 1, there is a decomposable subset $E' \subset E(G)$ such that $|E'| \geq c'_t |E(G)| / \log |E(G)|$, where c'_t depends only on t. Hence there is a partition $E' = E_1 \cup \cdots \cup E_w$, such that each E_i has an edge e_i that intersects every other edge in E_i, and for $i \neq j$, the edges in E_i are disjoint from the edges in E_j. Let V_i denote the set of vertices that are the endpoints of the edges in E_i, and let $n_i = |V_i|$. By the assumption, we have $|E_i| \leq \beta n_i$ for $1 \leq i \leq w$. Hence

$$\frac{c'_t |E(G)|}{\log |E(G)|} \leq |E'| \leq \sum_{i=1}^{w} \beta n_i \leq \beta n.$$

Since $|E(G)| \leq n^2$, we obtain $|E(G)| \leq 2(c'_t)^{-1} \beta n \log n$.   □

3   Generalized Davenport-Schinzel Sequences

A sequence $S = (s_1, \ldots, s_m)$ is called l-*regular* if any l consecutive terms of S are pairwise different. For integers $l, m \geq 2$, the sequence $S = (s_1, \ldots, s_{lm})$ is said to be of *type* up(l, m) if the first l terms are pairwise different and $s_i = s_{i+l} = \cdots = s_{i+(m-1)l}$ for $1 \leq i \leq l$. In particular, every sequence of type up(l, m) is l-regular. For convenience, we will index the elements of an up(l, m) sequence as

$$S = (s_{1,1}, \ldots, s_{l,1}, \ s_{1,2}, \ldots, s_{l,2}, \ \ldots, \ s_{1,m}, \ldots, s_{l,m}),$$

where $s_{1,1}, \ldots, s_{l,1}$ are pairwise different and $s_{i,1} = \cdots = s_{i,m}$ for $1 \leq i \leq l$.

Theorem 3 (Klazar [9]). *For $l \geq 2$ and $m \geq 3$, every l-regular sequence over an n-element alphabet that does not contain a subsequence of type up(l, m) has length at most*

$$n \cdot l \cdot 2^{(lm-3)} \cdot (10l)^{10\alpha(n)^{lm}}.$$

For more results on generalized Davenport-Schinzel sequences, see [13,16,17].

4   Intersection Pattern of Curves

In this section, we will prove several technical lemmas on the intersection pattern of curves in the plane. We will always assume that no two curves are tangent, and that if two curves share an interior point, then they must properly cross at that point.

Lemma 3. *Let λ_1 and λ_2 be disjoint simple closed curves. Let C be a collection of m curves with one endpoint on λ_1, the other endpoint on λ_2, and no other common points with λ_1 or λ_2. If no k members of C pairwise cross, then C contains $\lceil m/(k-1)^2 \rceil$ pairwise disjoint members.*

Proof. Let G be the intersection graph of C. Since G does not contain a clique of size k, by Turán's theorem, $|E(G)| \leq (1 - 1/(k-1))m^2/2$. Hence there is a curve $a \in C$ and a subset $S \subset C$, such that $|S| \geq m/(k-1) - 1$ and a is disjoint from every curve in S. We order the elements in $S \cup \{a\}$ as $a_0, a_1, \ldots, a_{|S|}$ in clockwise order as their endpoints appear on λ_1, starting with $a_0 = a$. Now we define the partial order $\prec$ on the pairs in S so that $a_i \prec a_j$ if $i < j$ and a_i is disjoint from a_j. A simple geometric observation shows that $\prec$ is indeed a partial order. Since S does not contain k pairwise crossing members, by Dilworth's theorem [6], $S \cup \{a\}$ contains $\lceil m/(k-1)^2 \rceil$ pairwise disjoint members.   □

A collection of curves with a common endpoint v is called a *fan* with *apex* v. Let $C = \{a_1, \ldots, a_m\}$ be a fan with apex v, and $\gamma = \gamma_1 \cup \cdots \cup \gamma_m$ be a curve with endpoints p and q partitioned into m subcurves $\gamma_1, \ldots, \gamma_m$ that appear in order along γ from p to q. We say that C is *grounded* by $\gamma_1 \cup \cdots \cup \gamma_m$, if

(i) γ does not contain v,
(ii) each a_i has its other endpoint on γ_i.

We say that C is *well-grounded* by $\gamma_1 \cup \cdots \cup \gamma_m$ if C is grounded by $\gamma_1 \cup \cdots \cup \gamma_m$ and each a_i intersects γ only within γ_i. Note that both notions depend on a particular partition $\gamma = \gamma_1 \cup \cdots \cup \gamma_m$.

Lemma 4. *Let $C = \{a_1, \ldots, a_m\}$ be a fan grounded by a curve $\gamma = \gamma_1 \cup \cdots \cup \gamma_m$. If each a_i intersects γ in at most t points, then there is a subfan $C' = \{a_{i_1}, \ldots, a_{i_r}\} \subset C$ with $i_1 < \cdots < i_r$ and $r = \lfloor \log_{t+1} m \rfloor$ that is grounded by a subcurve $\gamma' = \gamma'_1 \cup \cdots \cup \gamma'_r \subset \gamma$. Moreover,*

(i) $\gamma'_j \supset \gamma_{i_j}$ *for $1 \leq j \leq r$,*
(ii) a_{i_j} *intersects γ' only within $\gamma'_1 \cup \cdots \cup \gamma'_j$ for $1 \leq j \leq r$.*

Proof. We proceed by induction on m. The base case $m \leq t$ is trivial. Now assume that $m \geq t + 1$ and the statement holds up to $m - 1$. Since a_1 intersects γ in at most t points, there exists an integer j such that a_1 is disjoint from $\gamma_j \cup \gamma_{j+1} \cup \cdots \cup \gamma_{j+\lfloor m/(t+1) \rfloor - 1}$. By the induction hypothesis applied to $\{a_j, a_{j+1}, \ldots, a_{j+\lfloor m/(t+1) \rfloor - 1}\}$ and the curve $\gamma_j \cup \gamma_{j+1} \cup \cdots \cup \gamma_{j+\lfloor m/(t+1) \rfloor - 1}$, we obtain a subfan $C^* = \{a_{i_2}, \ldots, a_{i_r}\}$ of $r - 1 = \lfloor \log_{t+1} \lfloor m/(t+1) \rfloor \rfloor = \lfloor \log_{t+1} m \rfloor - 1$ curves, and a subcurve $\gamma^* = \gamma'_2 \cup \cdots \cup \gamma'_r \subset \gamma_j \cup \gamma_{j+1} \cup \cdots \cup \gamma_{j+\lfloor m/(t+1) \rfloor - 1}$ with the desired properties. Let γ_1' be the subcurve of γ obtained by extending the endpoint of γ_1 to the endpoint of γ^* along γ so that $\gamma_1' \supset \gamma_1$. Set $\gamma' = \gamma_1' \cup \gamma^*$. Hence the collection of curves $C' = \{a_1\} \cup C^*$ and γ' have the desired properties. □

Lemma 5. *Let $C = \{a_1, \ldots, a_m\}$ be a fan grounded by a curve $\gamma = \gamma_1 \cup \cdots \cup \gamma_m$. If each a_i intersects γ in at most t points, then there is a subfan $C' = \{a_{i_1}, \ldots, a_{i_r}\} \subset C$ with $i_1 < \cdots < i_r$ and $r = \lfloor \log_{t+1} \log_{t+1} m \rfloor$ that is well-grounded by a subcurve $\gamma' = \gamma_1' \cup \cdots \cup \gamma_r' \subset \gamma$. Moreover, $\gamma_j' \supset \gamma_{i_j}$ for $1 \le j \le r$.*

Proof. We apply Lemma 4 to C and $\gamma = \gamma_1 \cup \cdots \cup \gamma_m$ to obtain a subcollection $C^* = \{a_{j_1}, a_{j_2}, \ldots, a_{j_{m^*}}\}$ of $m^* = \lfloor \log_{t+1} m \rfloor$ curves, and a subcurve $\gamma^* = \gamma_1^* \cup \cdots \cup \gamma_{m^*}^* \subset \gamma$ with the properties listed in Lemma 4. Then we apply Lemma 4 again to C^* and γ^* with the elements in C^* in reverse order. By the second property of Lemma 4, the resulting subcollection $C' = \{a_{i_1}, \ldots, a_{i_r}\}$ of $r = \lfloor \log_{t+1} \log_{t+1} m \rfloor$ curves is well-grounded by a subcurve $\gamma' = \gamma_1' \cup \cdots \cup \gamma_r' \subset \gamma$, and by the first property we have $\gamma_j' \supset \gamma_{i_j}$ for $1 \le j \le r$. □

We say that fans $C_1, \ldots, C_l$ are *simultaneously grounded* (*simultaneously well-grounded*) by a curve $\gamma = \gamma_1 \cup \cdots \cup \gamma_m$ to emphasize that they are grounded (well-grounded) by γ with *the same* partition $\gamma = \gamma_1 \cup \cdots \cup \gamma_m$.

Lemma 6. *Let $C_1, \ldots, C_l$ be l fans with $C_i = \{a_{i,1}, \ldots, a_{i,m}\}$, simultaneously grounded by a curve $\gamma = \gamma_1 \cup \cdots \cup \gamma_m$. If each $a_{i,j}$ intersects γ in at most t points, then there are indices $j_1 < \cdots < j_r$ with $r = \lfloor \log_{t+1}^{(2l)} m \rfloor$ (2l-times iterated logarithm of m) and a subcurve $\gamma' = \gamma_1' \cup \cdots \cup \gamma_r' \subset \gamma$ such that*

(i) *the subfans $C_i' = \{a_{i,j_1}, \ldots, a_{i,j_r}\} \subset C_i$ for $1 \le i \le l$ are simultaneously well-grounded by $\gamma_1' \cup \cdots \cup \gamma_r'$,*

(ii) *$\gamma_s' \supset \gamma_{j_s}$ for $1 \le s \le r$.*

Proof. We proceed by induction on l. The base case $l = 1$ follows from Lemma 5. Now assume the statement holds up to $l - 1$. We apply Lemma 5 to the fan $C_1 = \{a_{1,1}, \ldots, a_{1,m}\}$ and the curve $\gamma = \gamma_1 \cup \cdots \cup \gamma_m$, to obtain a subfan $C_1^* = \{a_{1,w_1}, \ldots, a_{1,w_s}\} \subset C_1$ with $w_1 < \cdots < w_s$ and $s = \lfloor \log_{t+1} \log_{t+1} m \rfloor$ that is well-grounded by a subcurve $\gamma^* = \gamma_1^* \cup \cdots \cup \gamma_s^* \subset \gamma$ and satisfies $\gamma_i^* \supset \gamma_{w_i}$ for $1 \le i \le s$. For $2 \le i \le l$, let $C_i^* = \{a_{i,w_1}, \ldots, a_{i,w_s}\} \subset C_i$. Now we apply the induction hypothesis on the collection of $l - 1$ fans $C_2^*, \ldots, C_l^*$ that are simultaneously grounded by the curve $\gamma^* = \gamma_1^* \cup \cdots \cup \gamma_s^*$. Hence we obtain indices $j_1 < \cdots < j_r$ with $r = \lfloor \log_{t+1}^{(2l-2)} s \rfloor = \lfloor \log_{t+1}^{(2l)} m \rfloor$ and a subcurve $\gamma' = \gamma_1' \cup \cdots \cup \gamma_r' \subset \gamma^*$ such that each subfan $C_i' = \{a_{i,j_1}, \ldots, a_{i,j_r}\} \subset C_i$ with $2 \le i \le l$ is well-grounded by $\gamma_1' \cup \cdots \cup \gamma_r'$, and moreover $\gamma_z' \supset \gamma_z^* \supset \gamma_z$ for $1 \le z \le r$. By setting $C_1' = \{a_{1,j_1}, \ldots, a_{1,j_r}\} \subset C_1^*$, the collection $C_1', \ldots, C_l'$ is simultaneously well-grounded by the subcurve $\gamma' = \gamma_1' \cup \cdots \cup \gamma_r' \subset \gamma$. □

Let $C = \{a_1, \ldots, a_m\}$ be a fan with apex v grounded by a curve $\gamma = \gamma_1 \cup \cdots \cup \gamma_m$ with endpoints p and q. We say that a_i is *left-sided* (*right-sided*) if moving along a_i from v until we reach γ for the first time, and then turning left (right) onto the curve γ, we reach the endpoint q (p). We say that C_i is *one-sided*, if the curves in C_i are either all left-sided or all right-sided.

Lemma 7. *Let* $C_1, \ldots, C_l$ *be* l *fans with* $C_i = \{a_{i,1}, \ldots, a_{i,m}\}$, *simultaneously grounded by a curve* γ. *Then there are indices* $j_1 < \cdots < j_r$ *with* $r = \lceil m/2^l \rceil$ *such that the subfans* $C_i' = \{a_{i,j_1}, \ldots, a_{i,j_r}\} \subset C_i$ *for* $1 \leq i \leq l$ *are one-sided.*

Proof. We proceed by induction on l. The base case $l = 1$ is trivial since at least half of the curves in $C_1 = \{a_{1,1}, \ldots, a_{1,m}\}$ form a one-sided subset. For the inductive step, assume that the statement holds up to $l - 1$. Let $C_1^* = \{a_{1,w_1}, \ldots, a_{1,w_{\lceil m/2 \rceil}}\}$ with $w_1 < \cdots < w_{\lceil m/2 \rceil}$ be a subset of $\lceil m/2 \rceil$ curves that is one-sided. For $i \geq 2$, set $C_i^* = \{a_{i,w_1}, \ldots, a_{i,w_{\lceil m/2 \rceil}}\}$. Then apply the induction hypothesis on the $l - 1$ fans $C_2^*, \ldots, C_l^*$, to obtain indices $j_1 < \cdots < j_r$ with $r = \lceil \lceil m/2 \rceil / 2^{l-1} \rceil = \lceil m/2^l \rceil$ such that the subfans $C_i' = \{a_{i,j_1}, \ldots, a_{i,j_r}\} \subset C_i^*$ for $2 \leq i \leq l$ are one-sided. By setting $C_1' = \{a_{1,j_1}, \ldots, a_{1,j_r}\} \subset C_1^*$, the collection $C_1', \ldots, C_l'$ have the desired properties. □

Since at least half of the fans obtained from Lemma 7 are either left-sided or right-sided, we have the following corollary.

Corollary 1. *Let* $C_1, \ldots, C_{2l}$ *be* $2l$ *fans with* $C_i = \{a_{i,1}, \ldots, a_{i,m}\}$, *simultaneously grounded by a curve* γ. *Then there are indices* $i_1 < \cdots < i_l$ *and* $j_1 < \cdots < j_r$ *with* $r = \lceil m/2^{2l} \rceil$ *such that the subfans* $C_{i_w}' = \{a_{i_w,j_1}, \ldots, a_{i_w,j_r}\} \subset C_{i_w}$ *for* $1 \leq w \leq l$ *are all left-sided or all right-sided.*

By combining Lemma 6 and Corollary 1, we easily obtain the following lemma which will be used in Section 5.

Lemma 8. *Let* $C_1, \ldots, C_{2l}$ *be* $2l$ *fans with* $C_i = \{a_{i,1}, \ldots, a_{i,m}\}$, *simultaneously grounded by a curve* $\gamma = \gamma_1 \cup \cdots \cup \gamma_m$. *Then there are indices* $i_1 < \cdots < i_l$ *and* $j_1 < \cdots < j_r$ *with* $r = \lceil \lfloor \log_{t+1}^{(4l)} m \rfloor / 2^{2l} \rceil$ *and a subcurve* $\gamma^* = \gamma_1^* \cup \cdots \cup \gamma_r^* \subset \gamma$ *such that*

(i) *the subfans* $C_{i_w}' = \{a_{i_w,j_1}, \ldots, a_{i_w,j_r}\} \subset C_{i_w}$ *for* $1 \leq w \leq l$ *are simultaneously well-grounded by* $\gamma_1^* \cup \cdots \cup \gamma_r^*$,
(ii) $\gamma_s^* \supset \gamma_{j_s}$ *for* $1 \leq s \leq r$,
(iii) *the subfans* $C_{i_1}', \ldots, C_{i_l}'$ *are all left-sided or all right-sided.*

5   Proof of Theorem 1

By Lemma 2 and the fact that the function $\alpha(n)$ is non-decreasing, it is enough to prove that every k-quasi-planar topological graph on n vertices such that no two edges have more than t points in common and there is an edge that intersects every other edge has at most $2^{\alpha(n)^c} n$ edges, where c depends only on k and t.

Let G be a k-quasi-planar graph on n vertices with no two edges intersecting in more than t points. Let $e_0 = pq$ be an edge that intersects every other edge of G. Let $V_0 = V(G) \setminus \{p, q\}$ and E_0 be the set of edges with both endpoints in

V_0. Hence we have $|E_0| > |E(G)| - 2n$. Assume without loss of generality that no two elements of E_0 cross e_0 at the same point.

By a well-known fact (see e.g. Theorem 2.2.1 in [3]), there is a bipartition $V_0 = V_1 \cup V_2$ such that at least half of the edges in E_0 connect a vertex in V_1 to a vertex in V_2. Let E_1 be the set of these edges. For each vertex $v_i \in V_1$, consider the graph G_i whose each vertex corresponds to the subcurve γ of an edge $e \in E_1$ such that

(i) e is incident to v_i,
(ii) the endpoints of $\gamma \subset e$ are v_i and the first intersection point in $e \cap e_0$ as moving from v_i along e.

Two vertices are adjacent in G_i if the corresponding subcurves cross. Each graph G_i is isomorphic to the intersection graph of a collection of curves with one endpoint on a simple closed curve λ_1 and the other endpoint on a simple closed curve λ_2 and with no other points in common with λ_1 or λ_2. To see this, enlarge the point v_i and the curve e_0 a little, making them simple closed curves λ_1 and λ_2, and shorten the curves γ appropriately, so as to preserve all crossings between them. Since no k of these curves pairwise intersect, by Lemma 3, G_i contains an independent set of size $\lceil |V(G_i)|/(k-1)^2 \rceil$. We keep all edges corresponding to the elements of this independent set, and discard all other edges incident to v_i. After repeating this process for all vertices in V_1, we are left with at least $\lceil |E_1|/(k-1)^2 \rceil$ edges, forming a set E_2. We continue this process on the vertices in V_2 and the edges in E_2. After repeating this process for all vertices in V_2, we are left with at least $\lceil |E_2|/(k-1)^2 \rceil$ edges, forming a set E'. Thus $|E(G)| < 2(k-1)^4 |E'| + 2n$. Now, for any two edges $e_1, e_2 \in E'$ that share an endpoint, the subcurves $\gamma_1 \subset e_1$ and $\gamma_2 \subset e_2$ described above must be disjoint.

For each edge $e \in E'$, fix an arbitrary intersection point $s \in e \cap e_0$ to be the *main intersection point* of e and e_0. Let $e_1, \ldots, e_{|E'|}$ denote the edges in E' listed in the order their main intersection points appear on e_0 from p to q, and let $s_1, \ldots, s_{|E'|}$ denote these points respectively. We label the endpoints of each e_i as p_i and q_i, as follows. As we move along e_0 from p to q until we arrive at s_i, then we turn left and move along e_i, we finally reach p_i, while as we turn right at s_i and move along e_i, we finally reach q_i. We define sequences $S_1 = (p_1, \ldots, p_{|E'|})$ and $S_2 = (q_1, \ldots, q_{|E'|})$. They are sequences of length $|E'|$ over the $(n-2)$-element alphabet V_0.

Lemma 9 (Valtr [19]). *For $2l \geq 1$, at least one of the sequences S_1, S_2 defined above contains a $2l$-regular subsequence of length at least $\lceil |E'|/(8l) \rceil$.*

The proof of Lemma 9 can be copied almost verbatim from the proof of Lemma 5 in [19]. Indeed, the only fact about the sequences S_1 and S_2 it uses is that the edges $e_{j_1}, e_{j_1+1}, \ldots, e_{j_2}$ are spanned by the vertices $p_{j_1}, \ldots, p_{j_2}$ and $q_{j_1}, \ldots, q_{j_2}$, for any $j_1 < j_2$.

For the rest of this section, we set $l = 2^{k^2+2k}$ and m to be such that $(\log_{t+1}^{(4l)} m)/2^{2l} = 3 \cdot 2^k - 4$.

Lemma 10. *Neither of the sequences S_1 and S_2 has a subsequence of type* $\mathrm{up}(2l, m)$.

Proof. By symmetry, it suffices to show that S_1 does not contain a subsequence of type up$(2l, m)$. We will prove that the existence of such a subsequence would imply that G has k pairwise crossing edges. Let

$$S = (s_{1,1}, \ldots, s_{2l,1}, \ s_{1,2}, \ldots, s_{2l,2}, \ \ldots, \ s_{1,m}, \ldots, s_{2l,m})$$

be a subsequence of S_1 of type up$(2l, m)$ such that the first $2l$ terms are pairwise distinct and $s_{i,1} = \cdots = s_{i,m} = v_i$ for $1 \leq i \leq 2l$. For $1 \leq j \leq m$, let $a_{i,j}$ be the subcurve of the edge corresponding to the entry $s_{i,j}$ in S_1 between the vertex v_i and the main intersection point with e_0. Let $C_i = \{a_{i,1}, \ldots, a_{i,m}\}$ for $1 \leq i \leq 2l$. Hence $C_1, \ldots, C_{2l}$ are $2l$ fans with apices $v_1, \ldots, v_{2l}$ respectively. Clearly, there is a partition $e_0 = \gamma_1 \cup \cdots \cup \gamma_m$ such that $C_1, \ldots, C_{2l}$ are simultaneously grounded by $\gamma_1 \cup \cdots \cup \gamma_m$.

We apply Lemma 8 to the fans $C_1, \ldots, C_{2l}$ that are simultaneously grounded by $\gamma_1 \cup \cdots \cup \gamma_m$ to obtain indices $i_1 < \cdots < i_l$ and $j_1 < \cdots < j_r$ with $r = (\log_{l+1}^{(4l)} m)/2^{2l} = 3 \cdot 2^k - 4$ and a subcurve $\gamma^* = \gamma_1^* \cup \cdots \cup \gamma_r^* \subset e_0$ such that

(i) the subfans $C'_{i_w} = \{a_{i_w,j_1}, \ldots, a_{i_w,j_r}\} \subset C_{i_w}$ for $1 \leq w \leq l$ are simultaneously well-grounded by $\gamma_1^* \cup \cdots \cup \gamma_r^*$,

(ii) $\gamma_z^* \supset \gamma_{j_z}$ for $1 \leq z \leq r$,

(iii) the subfans $C'_{i_1}, \ldots, C'_{i_l}$ are all left-sided or all right-sided.

We will only consider the case that $C'_{i_1}, \ldots, C'_{i_l}$ are left-sided, the other case being symmetric.

Now for $1 \leq w \leq l$ and $1 \leq z \leq r$, we define the subcurve $a^*_{w,z} \subset a_{i_w,j_z}$ whose endpoints are v_{i_w} and the first point from $a_{i_w,j_z} \cap \gamma^*$ as moving from v_{i_w} along a_{i_w,j_z}. Hence the interior of $a^*_{w,z}$ is disjoint from γ^*. Let $A^*_w = \{a^*_{w,1}, \ldots, a^*_{w,r}\}$ for $1 \leq w \leq l$. Note that any two curves in A^*_w do not cross by construction, and all curves in A^*_w enter γ^* from the same side. For simplicity, we will call this the *left side* of γ^* and we will relabel the apices of the fans $A^*_1, \ldots, A^*_l$ from $v_{i_1}, \ldots, v_{i_l}$ to $v_1, \ldots, v_l$. To finally reach a contradiction, we prove the following.

Claim 1. *For $l = 2^{k^2+2k}$ and $r = 3 \cdot 2^k - 4$, among the l fans $A^*_1, \ldots, A^*_l$ with the properties above, there are k pairwise crossing curves.*

The proof follows the argument of Lemma 4.3 in [8]. We proceed by induction on k. The base case $k = 1$ is trivial. For the inductive step, assume the statement holds up to $k - 1$. For simplicity, we let $a^*_{i,j} = a^*_{i,j'}$ for all $j \in \mathbb{Z}$, where $j' \in \{1, \ldots, r\}$ is such that $j \equiv j' \pmod{r}$. Consider the fan A^*_1, which is of size r. By construction of A^*_1, the arrangement $A^*_1 \cup \{\gamma^*\}$ partitions the plane into r regions. By the pigeonhole principle, there is a subset $V' \subset \{v_1, \ldots, v_l\}$ of size

$$|V'| = \frac{l-1}{r} = \frac{2^{k^2+2k} - 1}{3 \cdot 2^k - 4},$$

such that all the vertices in V' lie in the same region. Let $j_0 \in \{1, \ldots, r\}$ be an integer such that V' lies in the region bounded by $a^*_{1,j_0}, a^*_{1,j_0+1}$, and γ^*.

Let $v_i \in V'$ and $1 < j_1 < r$, and consider the curve a^*_{i,j_0+j_1}. Recall that a^*_{i,j_0+j_1} is disjoint from $\gamma^*_{j_0} \cup \gamma^*_{j_0+1}$ and thus intersects $a^*_{1,j_0} \cup a^*_{1,j_0+1}$. Let $a \subset a^*_{i,j_0+j_1}$

be the maximal subcurve with an endpoint on γ^* whose interior is disjoint from $a^*_{1,j_0} \cup a^*_{1,j_0+1}$. If a intersects a^*_{1,j_0+1} (i.e. the second endpoint of a lies on a^*_{1,j_0+1}), then v_i and the left side of $\gamma^*_{j_0+2} \cup \cdots \cup \gamma^*_{j_0+j_1-1}$ lie in different connected components of $\mathbb{R}^2 \setminus (a^*_{1,j_0+1} \cup \gamma^* \cup a)$. Likewise, if a intersects a^*_{1,j_0}, then v_i and the left-side of $\gamma^*_{j_0+j_1+1} \cup \cdots \cup \gamma^*_{j_0+r-1}$ lie in different connected components of $\mathbb{R}^2 \setminus (a^*_{1,j_0} \cup \gamma^* \cup a)$.

If a intersects a^*_{1,j_0+1}, then all curves $a^*_{i,j_0+2}, \ldots, a^*_{i,j_0+j_1-1}$ must also cross a^*_{1,j_0+1}. Indeed, they connect v_i with the left-side of $\gamma^*_{j_0+2} \cup \cdots \cup \gamma^*_{j_0+j_1-1}$, but their interiors are disjoint from γ^* and a^*_{i,j_0+j_1}. Likewise, if a intersects a^*_{1,j_0}, then all curves $a^*_{i,j_0+j_1+1}, \ldots, a^*_{i,j_0+r-1}$ must also cross a^*_{1,j_0}. Therefore, we have the following.

Claim 2. *For half of the vertices $v_i \in V'$, the curves emanating from v_i satisfy one of the following:*

  (i) $a^*_{i,j_0+2}, a^*_{i,j_0+3}, \ldots, a^*_{i,j_0+r/2}$ *all cross* a^*_{1,j_0+1},
  (ii) $a^*_{i,j_0+r/2+1}, a^*_{i,j_0+r/2+2}, \ldots, a^*_{i,j_0+r-1}$ *all cross* a^*_{1,j_0}.

We keep all curves satisfying Claim 2, and discard all other curves. Since $r/2-2 = 3 \cdot 2^{k-1} - 4$ and

$$\frac{|V'|}{2} \geq \frac{l-1}{2r} = \frac{2^{k^2+2k}-1}{6 \cdot 2^k - 8} \geq 2^{(k-1)^2+2(k-1)},$$

by Claim 2, we can apply the induction hypothesis on these remaining curves which all cross a^*_{1,j_0+1} or a^*_{1,j_0}. Hence we have found k pairwise crossing edges, and this completes the proof of Claim 1 and thus Lemma 10. $\qquad\square$

We are now ready to prove Theorem 1.

Proof (Theorem 1). By Lemma 9 we know that, say, S_1 contains a $2l$-regular subsequence of length $\lceil |E'|/(8l) \rceil$. By Theorem 3 and Lemma 10, this subsequence has length at most

$$n \cdot 2l \cdot 2^{(2lm-3)} \cdot (20l)^{10\alpha(n)^{2lm}}.$$

Therefore, we have

$$\left\lceil \frac{|E'|}{8l} \right\rceil \leq n \cdot 2l \cdot 2^{(2lm-3)} \cdot (20l)^{10\alpha(n)^{2lm}},$$

which implies

$$|E'| \leq 8n \cdot 2l^2 \cdot 2^{(2lm-3)} \cdot (20l)^{10\alpha(n)^{2lm}}.$$

Since $l = 2^{k^2+2k}$ and m depends only on k and t, for sufficiently large c (depending only on k and t) and $\alpha(n) \geq 2$, we have

$$|E(G)| < 2(k-1)^4|E'| + 2n \leq 2^{\alpha(n)^c} n,$$

which completes the proof of Theorem 1. $\qquad\square$

6   Proof of Theorem 2

A family of curves in the plane is *simple* if any two of them share at most one
point. A family C of curves is K_k-*free* if the intersection graph of C is K_k-free,
that is, no k curves in C pairwise intersect. By $\chi(C)$ we denote the chromatic
number of the intersection graph of C, that is, the minimum number of colors
that suffice to color the curves in C so that no two intersecting curves receive
the same color.

Let ℓ be a horizontal line in the plane. Our proof of Theorem 2 is based on
the following result, proved in [10] in a more general setting, for simple K_k-free
families of compact arc-connected sets in the plane whose intersections with a
line ℓ are non-empty segments.

Theorem 4 (Lasoń, Micek, Pawlik, Walczak [10]). *Every simple K_k-free
family of curves C all intersecting ℓ at exactly one point satisfies $\chi(C) \leq a_k$,
where a_k depends only on k.*

Special cases of Theorem 4 have been proved by McGuinness [11] for $k = 3$ and
by Suk [18] for y-monotone curves and any k. We will also use the following
graph-theoretical result.

Lemma 11 (McGuinness [12]). *Let G be a graph, $\prec$ be a total ordering of
$V(G)$, and $a, b \geq 0$. For $u, v \in V(G)$, let $G(u, v)$ denote the subgraph of G
induced by the vertices strictly between u and v in $\prec$. If $\chi(G) > 2^{a+b+1}$, then
there is an induced subgraph H of G such that $\chi(H) > 2^a$ and $\chi(G(u, v)) \geq 2^b$
for any $uv \in E(H)$.*

Let β be a segment in ℓ. We will consider curves crossing β at exactly one
point, always assuming that this intersection point is distinct from the endpoints
of β. Any such curve γ is partitioned by β into two subcurves: γ^+ that enters
β from above and γ^- that enters β from below, both including the intersection
point of β and γ.

Lemma 12. *Let C be a simple K_k-free family of curves all crossing β at exactly
one point. If $\gamma_1^+ \cap \gamma_2^+ = \emptyset$ and $\gamma_1^- \cap \gamma_2^- = \emptyset$ for any $\gamma_1, \gamma_2 \in C$, then $\chi(C) \leq
2^{3k-6}$.*

Proof. We proceed by induction on k. The base case $k = 2$ is trivial. For the
induction step, assume $k \geq 3$ and the statement holds up to $k-1$. Assume for the
sake of contradiction that $\chi(C) > 2^{3k-6}$. Let $\prec$ be the ordering of C according
to the left-to-right order of the intersection points with β. Apply Lemma 11 with
$a = 0$ and $b = 3k - 7$. It follows that there are two intersecting curves $\delta_1, \delta_2 \in C$
such that $\chi(C(\delta_1, \delta_2)) \geq 2^{3k-7}$, where $C(\delta_1, \delta_2) = \{\gamma \in C : \delta_1 \prec \gamma \prec \delta_2\}$. The
curves β, δ_1 and δ_2 together partition the plane into two regions R^+ and R^- so
that for $\gamma \in C(\delta_1, \delta_2)$, γ^+ enters β from the side of R^+, while γ^- enters β from
the side of R^-. Take any $\gamma_1, \gamma_2 \in C(\delta_1, \delta_2)$ that intersect at a point p. It follows
from the assumptions of the lemma that $p \in \gamma_1^+ \cap \gamma_2^-$ or $p \in \gamma_1^- \cap \gamma_2^+$. If $p \in R^+$,
then one of γ_1^-, γ_2^- (whichever contains p) must intersect δ_1 or δ_2. Similarly,

if $p \in R^-$, then one of γ_1^+, γ_2^+ must intersect δ_1 or δ_2. In both cases, one of γ_1, γ_2 intersects δ_1 or δ_2. Let C_1 and C_2 consist of those members of $C(\delta_1, \delta_2)$ that intersect δ_1 and δ_2, respectively. Clearly, both C_1 and C_2 are K_{k-1}-free, and thus the induction hypothesis yields $\chi(C_1) \leq 2^{3k-9}$ and $\chi(C_2) \leq 2^{3k-9}$. Moreover, $\chi\big(C(\delta_1, \delta_2) \setminus (C_1 \cup C_2)\big) \leq 1$ as $C(\delta_1, \delta_2) \setminus (C_1 \cup C_2)$ is independent by the assumption $\gamma_1^+ \cap \gamma_2^+ = \emptyset$ and $\gamma_1^- \cap \gamma_2^- = \emptyset$ for any $\gamma_1, \gamma_2 \in C$. To conclude, $\chi(C(\delta_1, \delta_2)) \leq 2 \cdot 2^{3k-9} + 1 < 2^{3k-7}$, which is a contradiction. $\qquad\square$

Now we prove the following theorem, which can also be generalized to simple K_k-free families of compact arc-connected sets in the plane whose intersections with a segment β are non-empty subsegments.

Theorem 5. *Every simple K_k-free family of curves C all crossing β at exactly one point satisfies $\chi(C) \leq b_k$, where b_k depends only on k.*

Proof. Assume without loss of generality that no curve in C passes through the endpoints of β. One can transform the family $C^+ = \{\gamma^+ : \gamma \in C\}$ into a family $\tilde{C}^+ = \{\tilde{\gamma}^+ : \gamma \in C\}$ so that

- $\tilde{C}^+$ is simple,
- each $\tilde{\gamma}^+$ is entirely contained in the upper half-plane delimited by ℓ,
- $\tilde{\gamma}_1^+$ and $\tilde{\gamma}_2^+$ intersect if and only if γ_1^+ and γ_2^+ intersect.

Similarly, one can transform the family $C^- = \{\gamma^- : \gamma \in C\}$ into a family $\tilde{C}^- = \{\tilde{\gamma}^- : \gamma \in C\}$ so that

- $\tilde{C}^-$ is simple,
- each $\tilde{\gamma}^-$ is entirely contained in the lower half-plane delimited by ℓ,
- $\tilde{\gamma}_1^-$ and $\tilde{\gamma}_2^-$ intersect if and only if γ_1^- and γ_2^- intersect.

The curves $\tilde{\gamma}^+$ and $\tilde{\gamma}^-$ are respectively the upper and lower parts of the curve $\tilde{\gamma} = \tilde{\gamma}^+ \cup \tilde{\gamma}^-$ intersecting ℓ at exactly one point. The family $\tilde{C} = \{\tilde{\gamma} : \gamma \in C\}$ is clearly simple and K_k-free. Therefore, by Theorem 4, $\chi(\tilde{C}) \leq a_k$. Fix a proper a_k-coloring ϕ of $\tilde{C}$ and consider the set C_i consisting of those $\gamma \in C$ for which $\phi(\tilde{\gamma}) = i$. It follows that $\gamma_1^+ \cap \gamma_2^+ = \emptyset$ and $\gamma_1^- \cap \gamma_2^- = \emptyset$ for any $\gamma_1, \gamma_2 \in C_i$. Therefore, by Lemma 12, $\chi(C_i) \leq 2^{3k-6}$. Summing up over all colors used by ϕ we obtain $\chi(C) \leq 2^{3k-6} a_k$. $\qquad\square$

Proof (Theorem 2). By Lemma 2, it is enough to prove that every k-quasi-planar simple topological graph on n vertices that contains an edge intersecting every other edge has at most $c_k n$ edges, where c_k depends only on k.

Let G be a k-quasi-planar simple topological graph on n vertices, and let pq be an edge that intersects every other edge. Remove all edges with an endpoint at p or q except the edge pq. Shorten each curve representing a remaining edge by a tiny bit at both endpoints, so that curves sharing an endpoint become disjoint, while all crossings are preserved. The resulting set of curves C is simple and K_k-free and contains a curve γ crossing every other curve in C. Therefore, $C \setminus \{\gamma\}$ is K_{k-1}-free and $|C \setminus \{\gamma\}| > |E(G)| - 2n$. Since C can be transformed into an equivalent set of curves so that γ becomes the horizontal segment β, Theorem 5 yields $\chi(C \setminus \{\gamma\}) \le b_{k-1}$. Consequently, $C \setminus \{\gamma\}$ contains an independent set S of size

$$|S| \ge \frac{|C \setminus \{\gamma\}|}{b_{k-1}} > \frac{|E(G)| - 2n}{b_{k-1}}.$$

The edges of G corresponding to the curves in S form a planar subgraph of G, which implies $|S| < 3n$. The two inequalities give $|E(G)| < (3b_{k-1} + 2)n$.   □

References

1. Ackerman, E.: On the maximum number of edges in topological graphs with no four pairwise crossing edges. Discrete Comput. Geom. 41, 365–375 (2009)
2. Agarwal, P.K., Aronov, B., Pach, J., Pollack, R., Sharir, M.: Quasi-planar graphs have a linear number of edges. Combinatorica 17, 1–9 (1997)
3. Alon, N., Spencer, J.: The Probabilistic Method, 3rd edn. Wiley Interscience (2008)
4. Brass, P., Moser, W., Pach, J.: Research Problems in Discrete Geometry. Springer (2005)
5. Capoyleas, V., Pach, J.: A Turán-type theorem on chords of a convex polygon. J. Combin. Theory Ser. B 56, 9–15 (1992)
6. Dilworth, R.P.: A decomposition theorem for partially ordered sets. Ann. Math. 51, 161–166 (1950)
7. Fox, J., Pach, J.: Coloring K_k-free intersection graphs of geometric objects in the plane. European J. Combin. 33, 853–866 (2012)
8. Fox, J., Pach, J., Suk, A.: The number of edges in k-quasi-planar graphs. SIAM J. Discrete Math. 27, 550–561 (2013)
9. Klazar, M.: A general upper bound in extremal theory of sequences, Comment. Math. Univ. Carolin. 33, 737–746 (1992)
10. Lasoń, M., Micek, P., Pawlik, A., Walczak, B.: Coloring intersection graphs of arc-connected sets in the plane (manuscript)
11. McGuinness, S.: Colouring arcwise connected sets in the plane I. Graphs Combin. 16, 429–439 (2000)
12. McGuinness, S.: On bounding the chromatic number of L-graphs. Discrete Math. 154, 179–187 (1996)
13. Nivasch, G.: Improved bounds and new techniques for Davenport-Schinzel sequences and their generalizations. J. Assoc. Comput. Machin. 57, 1–44 (2010)
14. Pach, J., Radoičić, R., Tóth, G.: Relaxing planarity for topological graphs. In: Győri, E., Katona, G.O.H., Lovász, L., Fleiner, T. (eds.) More Sets, Graphs and Numbers, Bolyai Soc. Math. Stud., vol. 15, pp. 285–300. Springer (2006)
15. Pach, J., Shahrokhi, F., Szegedy, M.: Applications of the crossing number. J. Graph Theory 22, 239–243 (1996)
16. Pettie, S.: Generalized Davenport-Schinzel sequences and their 0-1 matrix counterparts. J. Combin. Theory Ser. A 118, 1863–1895 (2011)
17. Pettie, S.: On the structure and composition of forbidden sequences, with geometric applications. In: 27th ACM Symposium on Computational Geometry, pp. 370–379. ACM (2011)
18. Suk, A.: Coloring intersection graphs of x-monotone curves in the plane. Combinatorica (to appear)
19. Valtr, P.: On geometric graphs with no k pairwise parallel edges. Discrete Comput. Geom. 19, 461–469 (1998)

Recognizing Outer 1-Planar Graphs in Linear Time[*],[**]

Christopher Auer, Christian Bachmaier, Franz J. Brandenburg,
Andreas Gleißner, Kathrin Hanauer, Daniel Neuwirth, and Josef Reislhuber

University of Passau, 94030 Passau, Germany
{auerc,bachmaier,brandenb,gleissner,hanauer,
neuwirth,reislhuber}@fim.uni-passau.de

Abstract. A graph is outer 1-planar (*o1p*) if it can be drawn in the plane such that all vertices are on the outer face and each edge is crossed at most once. *o1p* graphs generalize outerplanar graphs, which can be recognized in linear time and specialize 1-planar graphs, whose recognition is $\mathcal{NP}$-hard.

Our main result is a linear-time algorithm that first tests whether a graph G is *o1p*, and then computes an embedding. Moreover, the algorithm can augment G to a maximal *o1p* graph. If G is not *o1p*, then it includes one of six minors (see Fig. 3), which are also detected by the recognition algorithm. Hence, the algorithm returns a positive or negative witness for *o1p*.

1 Introduction

Planar graphs are one of the most studied areas in graph theory and an important class in graph drawing. Outerplanar graphs are in turn an important subfamily of planar graphs. Here, all vertices are on the outer face and edges do not cross. Every outerplanar graph has at least two vertices of degree two, which is used for a recognition in linear time [16].

There were several attempts to generalize planarity to graphs that are "almost" planar in some sense. Such attempts are important as many graphs are not planar. One generalization is 1-planar graphs, which were introduced by Ringel [17] in an approach to color a planar graph and its dual. A graph is 1-planar if it can be drawn in the plane such that each edge is crossed at most once and incident edges do not cross. 1-planar graphs are a hot topic in graph drawing, see also [1, 4–6, 8, 9, 13, 15].

The combination of 1-planarity and outerplanarity leads to *o1p* graphs, which are graphs with an embedding in the plane with all vertices on the outer face and at most one crossing per edge. They were introduced by Eggleton [10] who called

---

[*] This work was supported in part by the Deutsche Forschungsgemeinschaft (DFG) grant Br835/18-1.

[**] A linear-time algorithm for testing outer 1-planarity was independently obtained by Hong et al. and appears in these proceedings [14].

S. Wismath and A. Wolff (Eds.): GD 2013, LNCS 8242, pp. 107–118, 2013.

them outerplanar graphs with edge crossing number one. He showed that edges of maximal *o1p* graphs do not cross in the outer face and each face is incident to at most one crossing, from which he concluded that every *o1p* graph has an *o1p* drawing with straight-line edges and convex (inner) faces. Thomassen [18] generalized Eggleton's result and characterized the class of 1-planar graphs which admit straight-line drawings by the exclusion of so-called B- and W-configurations in embeddings. These configurations were rediscovered by Hong et al. [13], who also provide a linear-time drawing algorithm that starts from a given embedding.

From the algorithmic perspective there is a big step from zero to some crossings. It is well-known that planar graphs can be recognized in linear time, and there are linear-time algorithms to construct an embedding and drawings, e. g., straight-line drawings and visibility representations in quadratic area. On the contrary, dealing with crossings generally leads to $\mathcal{NP}$-hard problems. It is $\mathcal{NP}$-hard to recognize 1-planar graphs [15], even if the graph is given with a rotation system, which determines the cyclic ordering of the edges at each vertex [2]. 1-planarity remains $\mathcal{NP}$-hard even if the treewidth is bounded [3]. There also is no efficient algorithm to compute the crossing number of a graph [12] and to compute the number of crossings induced by the insertion of an edge into a planar graph [5]. However, there is a linear-time recognition algorithm of Eades et al. [8] for maximal 1-planar graphs, which needs a given rotation system.

In this paper we study *o1p* graphs. Our main result is a linear-time recognition algorithm for *o1p* graphs. This is the first efficient algorithm for a test of 1-planarity that takes solely a graph as input. Our recognition algorithm is based on SPQR-trees. It analyzes its nodes and then either computes an *o1p* embedding or detects one of six minors. If the graph is *o1p*, it can be augmented to a maximal *o1p* graph. In a maximal *o1p* graph, adding a new edge violates its defining property. From the structure of a maximal *o1p* graph we derive that every *o1p* graph is planar. Thus, they are subgraphs of planar graphs with a Hamiltonian cycle, and the SPQR-tree reveals a treewidth of at most three.

2   Preliminaries

We consider simple, undirected graphs $G = (V, E)$ with n vertices and m edges. The graphs are biconnected, otherwise, the components are treated separately. A *drawing* of a graph is a mapping of G into the plane such that the vertices are mapped to distinct points and each edge is a Jordan arc between its endpoints. A drawing is *planar* if the Jordan arcs of the edges do not cross and it is *1-planar* if each edge is crossed at most once. Accordingly, a graph is planar (1-planar) if it has a planar (1-planar) drawing. Crossings of edges with the same endpoint, i. e., *incident* edges, are excluded. A planar drawing of a graph partitions the plane into *faces*. A face is specified by a cyclic sequence of edges that forms its boundary. The set of all faces forms the *embedding* of the graph. In 1-planar drawings, every crossing divides an edge into two *edge segments*. An uncrossed edge consists of one segment. Therefore, a face of a *1-planar embedding* is specified by a cyclic list of edge segments.

A graph G is *outerplanar* if it has a planar drawing with all vertices on one distinguished face. This face is referred to as the *outer face* and corresponds to the unbounded, external face in a drawing on the plane. G is *maximal* outerplanar if no further edge can be added without violating outerplanarity. Then, the edges on the outer face form a Hamiltonian cycle. A graph G is *outer 1-planar*, *o1p* for short, if it has a drawing with all vertices on the outer face and such that each edge is crossed at most once. G is *maximal o1p* if the addition of any edge violates outer 1-planarity.

In an *o1p* embedding, an edge is either *crossing* or *plane* (*non-crossing*). We say that it is *inner*, if none of its segments is part of the boundary of the outer face. Analogously, an edge is *outer*, if it is entirely part of this boundary. Observe that a crossed edge cannot be outer. If the embedding is maximal, we can classify every edge as *outer* or *inner*.

Maximal outerplanar graphs have a unique embedding. This does no longer hold for maximal *o1p* graphs. Consider a graph with 6 vertices and 11 edges consisting of two K_4s. If the left K_4 is fixed, the right can be flipped. In order to gain more insight into the structure of an *o1p* graph G, we consider its *SPQR-tree* $\mathcal{T}$. SPQR-trees were first introduced by Di Battista and Tamassia [7] and provide a description of how the graph is composed. Fig. 2(a) depicts an example graph along with its SPQR-tree in Fig. 2(b). In the definition we adopt here, the SPQR-tree is unrooted. The nodes of $\mathcal{T}$ either represent a series composition (S), a parallel composition (P), a single edge (Q), or a triconnected component (R). Associated with each node μ of $\mathcal{T}$ is a graph that is homeomorphic to a subgraph of G and called the *skeleton* skel(μ) of μ. In its original definition, every edge $e = \{u, v\}$ of a skeleton, except for one of each Q-node, is a *virtual edge*, i.e., an edge that represents the subgraph of G which connects u and v. This subgraph is also referred to as the *expansion graph* expg(e) of e. For every virtual edge e in the skeleton of a node μ, there is another node ν that refines the structure of expg(e). We say that ν is the *refining* node refn(e) of e. This link is represented by an edge between μ and ν in $\mathcal{T}$ and we say that μ and ν are *adjacent* in $\mathcal{T}$. Therefore, every leaf of $\mathcal{T}$ is a Q-node. For simplification, we represent edges of the graph directly in the skeleton of an S-, P-, or R-node, so that we can neglect Q-nodes. We also call these edges *non-virtual*. Observe that all nodes are always as large as possible, so neither two S-nodes nor two P-nodes may be adjacent. For a more detailed introduction to SPQR-trees, the reader is referred to [7].

3   Recognition

There are linear-time algorithms for the recognition of (maximal) outerplanar graphs, that use the fact that there are at least two vertices of degree two. A single K_4 implies that this property no longer holds for *o1p* graphs. In contrast, the recognition of 1-planar graphs is $\mathcal{NP}$-hard [15], even if the graphs are given with a rotation system [2].

Algorithm 1. *o1p* Recognition

1: **procedure** TEST OUTER 1 PLANARITY(G)
2:     **if** G is not planar **then return** $\perp$
3:     $\mathcal{T} \leftarrow$ SPQR-tree of G
4:     **for all** R- and P-nodes $\mu \in \mathcal{T}$ **do**
5:         **if** μ is R-node **then**
6:             **if** skel$(\mu) \neq K_4$ **or** contains vertex incident to > 2 virtual edges **then**
7:                 **return** $\perp$ ▷ Lemma 1, Corollary 1
8:             **for all** neighbors ν of μ **do**
9:                 **if** ν is S-node or R-node **then** insert plane edge   ▷ Proposition 1
10:         **else if** μ is P-node **then**
11:             **if** skel(μ) contains > 4 virtual edges **then return** $\perp$ ▷ Corollary 1
12:             **else if** μ has only virtual edges **then** insert plane edge   ▷ Lemma 4
13:     compute mapping $\mathcal{C}$
14:     $\mathbb{P}_F \leftarrow$ {fixable P-nodes} ; $\mathbb{P}_N \leftarrow$ {P-nodes with crossings, but none fixable}
15:     **while** $\mathbb{P}_F \cup \mathbb{P}_N \neq \emptyset$ **do**
16:         **while** $\mathbb{P}_F \neq \emptyset$ **do**
17:             remove next P-node π from $\mathbb{P}_F$ with fixable S-nodes σ_1, σ_2
18:             $z \leftarrow$ FIX CROSSING AT PNODE$(G, \mathcal{T}, \pi, \sigma_1, \sigma_2)$
19:             **if** $z = \perp$ **then return** $\perp$
20:             **for all** $\pi' \in z$ **do** update $\mathcal{C}$
21:                 **if** π' is fixable **then** move π' from $\mathbb{P}_N$ to $\mathbb{P}_F$
22:         **if** $\mathbb{P}_N \neq \emptyset$ **then**   ▷ Lemma 5
23:             choose any element π of $\mathbb{P}_N$ with S-nodes σ_1, σ_2 conformant to $\mathcal{C}$
24:             $z \leftarrow$ FIX CROSSING AT PNODE$(G, \mathcal{T}, \pi, \sigma_1, \sigma_2)$
25:             **for all** $\pi' \in z$ **do** update $\mathcal{C}$
26:                 **if** π' is fixable **then** move π' from $\mathbb{P}_N$ to $\mathbb{P}_F$
27:     **for all** S-/P-/R-nodes $\mu \in \mathcal{T}$ **do** fix embedding
28:     **return** 2-clique-sum of skeleton embeddings

Theorem 1. *There is a linear-time algorithm to test whether a biconnected graph G is o1p and, if so, returns an embedding.*

We prove this theorem by first establishing a number of necessary conditions for a graph to have an *o1p* embedding. At the same time, we implement a linear-time algorithm (Algorithm 1) that checks these conditions and, if positive, constructs an *o1p* embedding of the input graph. The algorithm starts by ensuring that the input graph is planar (cf. Corollary 4) and computes its SPQR-tree. Both subroutines take $\mathcal{O}(n)$ time [11]. Observe that, although the graph will be augmented during the next steps, it remains *o1p* and therefore also planar. Consequently, the number of nodes in $\mathcal{T}$ always is in $\mathcal{O}(n)$. In a second step, we show that the conditions are not only necessary, but also sufficient.

We start with two observations regarding *o1p* embeddings. For maximal 1-planar embeddings, a well-known fact is that every crossing induces a K_4. This holds in an even tightened form for *o1p* embeddings:

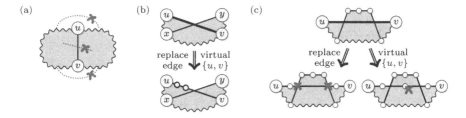

Fig. 1. (a): Proposition 2, (b) and (c): Proposition 3

Proposition 1 ([6]). *Let $\{a, b\}$ and $\{c, d\}$ be a pair of crossing edges in an o1p embedding of a maximal o1p graph. Then the vertices a, b, c, and d form a K_4 and the edges $\{a, b\}$, $\{b, c\}$, $\{c, d\}$, and $\{d, a\}$ are plane.*

Consider a plane, inner edge $\{u, v\}$ in an *o1p* embedding of a graph G. Then, $\{u, v\}$ "partitions" the embedding and the deletion of u and v disconnects G (cf. Fig. 1(a)).

Proposition 2. *Every plane, inner edge in an o1p embedding connects a separation pair.*

Let $\mathcal{T}$ be the SPQR-tree of an *o1p* graph G.

Lemma 1. *The skeleton of every R-node is a K_4.*

Proof. Recall that outerplanar graphs are series-parallel. Hence, the SPQR-tree of an outerplanar graph has no R-nodes. Let μ be an R-node in $\mathcal{T}$. Then, $\text{skel}(\mu)$ must be embedded such that at least two edges cross, e.g., edges $\{a, b\}$ and $\{c, d\}$. By Proposition 1, the vertices a, b, c, and d form a K_4.

There must be an embedding of $\text{skel}(\mu)$ such that all vertices are on the boundary of the same face. Suppose $\text{skel}(\mu)$ has more than four vertices. Then, at least one of $\{a, b\}$, $\{b, c\}$, $\{c, d\}$, and $\{d, a\}$ is an inner edge. By Proposition 1, all of them are plane. As an inner edge cannot be virtual, by Proposition 2, $\text{skel}(\mu)$ has a separation pair, so $\text{skel}(\mu)$ is not triconnected, a contradiction.   □

Instead of considering the possible embeddings of G on the whole, we study those of the skeletons of the nodes in $\mathcal{T}$. As G is *o1p*, there must be an embedding of every skeleton of $\mathcal{T}$ such that the 2-clique-sums over all skeletons result in an *o1p* embedding of G. In short, a *2-clique-sum* combines two graphs by selecting an edge (2-clique) in each one and glueing them together at those edges. The selected edges are removed from the new graph. If the input graphs were embedded, the embedding is inherited for the 2-clique-sum such that in each case the other graph takes the position of the removed edge.

Consequently, we need to derive properties of *o1p* embeddings of skeletons. Like in usual *o1p* embeddings, there must be a face (the outer face) such that all vertices lie on its boundary. However, as virtual edges represent entire subgraphs, they demand special attention. Observe that the expansion graph of every virtual edge contains, besides the separation pair, at least one more vertex. Consider

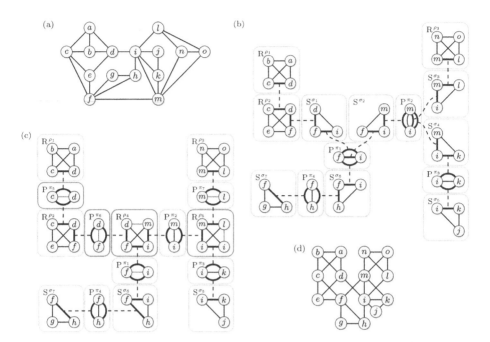

Fig. 2. Input graph (a), its SPQR-tree (b), the SPQR-tree after the algorithm (c) (new edges and nodes colored), and the found *o1p* embedding (d).

the virtual edge $\{u, v\}$ in Fig. 1(b). The crossing edge $\{x, y\}$ partitions $\{u, v\}$ into two segments, hence, $\mathrm{expg}(\{u, v\})$ must be embedded such that it replaces the edge segment of $\{u, v\}$ that lies on the outer face. Suppose a virtual edge e is embedded such that it has at least two crossings. Then there is either an edge in $\mathrm{expg}(e)$ that is crossed more than once or at least one vertex is enclosed between two crossings and hence does not lie on the outer face (cf. Fig. 1(c)).

Proposition 3. *Every virtual edge may consist of at most two edge segments and the embedding must be such that at least one segment is part of the boundary of the outer face.*

Observe that in contrast to the *o1p* embedding of the whole graph, we must allow the crossing of two virtual edges with a common end vertex in the embedding of a skeleton. We qualify the virtual edges that must always be embedded plane.

Lemma 2. *Let μ be a node of $\mathcal{T}$ and let $e = \{u, v\}$ be a virtual edge in $\mathrm{skel}(\mu)$. If both u and v have degree > 1 in $\mathrm{expg}(e)$, then e must be embedded plane.*

Proof. Suppose e is embedded such that it crosses another edge e', which can be virtual or not. In either case, e' may be crossed at most once. As $\mathrm{skel}(\mu)$ is biconnected and $\mathrm{expg}(e)$ contains at least one additional vertex, in $\mathrm{expg}(e)$, either all edges incident to u or all edges incident to v must be crossed in order to have all vertices lie on the outer face. If both u and v have degree > 1 in $\text{expg}(e)$, e' has at least two crossings. □

Note that unlike planar embeddings, neither the skeleton of an S-node nor that of an R-node has a unique $o1p$ embedding. However, Lemma 2 limits the number of possible $o1p$ embeddings for skeletons considerably:

Lemma 3. *Let μ be a node in $\mathcal{T}$. Then for every virtual edge $e = \{u, v\}$ in $\text{skel}(\mu)$ holds:*
If $\text{refn}(e)$ is a P- or an R-node, then e must be embedded plane in $\text{skel}(\mu)$.
If $\text{refn}(e)$ is an S-node whose skeleton is the cycle graph $(u, c_1, c_2, \ldots, c_k, v, u)$, then e must be embedded such that the segment incident to u (v) lies on the outer face if the edge $\{u, c_1\}$ ($\{c_k, v\}$) is virtual.

Proof. If $\text{refn}(e)$ is a P- or an R-node, both u and v have degree > 1 in $\text{expg}(e)$, hence by Lemma 2, e must be embedded plane. Suppose $\text{refn}(e)$ is an S-node whose skeleton is the cycle graph $(u, c_1, c_2, \ldots, c_k, v, u)$. As the embedding must be such that all vertices lie on the outer face, only the edges $\{u, c_1\}$ or $\{c_k, v\}$ may be crossed. Recall that by the structure of an SPQR-tree, if $\{u, c_1\}$ ($\{c_k, v\}$) is virtual, then $\text{refn}(\{u, c_1\})$ ($\text{refn}(\{c_k, v\})$) is either a P- or an R-node, so $\{u, c_1\}$ ($\{c_k, v\}$) must be embedded plane. □

Lemma 1, Proposition 3, and Lemma 3 allow us to draw the following conclusion:

Corollary 1. *Every virtual edge in an S-node must be embedded plane.*
The skeleton of every R-node contains at most four virtual edges, which must be embedded plane, and no vertex may be incident to more than two virtual edges. The skeleton of a P-node may have at most 4 virtual edges.

The conditions for R-nodes are easily checked by Algorithm 1 in time $\mathcal{O}(1)$ per R-node. Additionally, if an R-node is adjacent to another R-node or an S-node, then one of the edges of the K_4 is not present. For an example, see the R-nodes ρ_1 and ρ_2 in Fig. 2(b). By Proposition 1, however, the edge may be inserted and is always plane. Observe that this introduces a new P-node π_5 in Fig. 2(c). As an R-node may have at most four neighbors and as the SPQR-tree can be updated in $\mathcal{O}(1)$ time, this modification takes constant time, too.

The following lemma allows us to insert a non-virtual edge in every P-node, if none is present. In Fig. 2(b), this would apply, e. g., to π_1.

Lemma 4. *Let u, v be the vertices in the skeleton of a P-node without non-virtual edges. Then, the insertion of $\{u, v\}$ does not violate outer 1-planarity and $\{u, v\}$ is plane for every o1p embedding of G.*

Proof. Let π be a P-node with separation pair u, v, that is connected by virtual edges only. According to the definition of SPQR-trees, every skeleton of a P-node has at least three edges. Hence, π is adjacent to at least three other nodes. Subsequently, at least two virtual edges must be refined by S-nodes and are embedded with a crossing. This results in a crossing of two non-virtual edges in G that are, by Lemma 3, incident to u and v, respectively. By Proposition 1, the edge $\{u, v\}$ can always be inserted and is plane. □

Again, Algorithm 1 can check these two conditions and augment the graph for a P-node in time $\mathcal{O}(1)$, which results in a running time of $\mathcal{O}(n)$ for ll. $4 - 12$.

Consider a P-node π with vertices u, v. If skel(π) has at most two virtual edges, they can be embedded without a crossing and such that both completely lie on the outer face. Suppose skel(π) has at least three virtual edges. In consequence of Proposition 3, two of them must cross each other. In Fig. 2(b), this holds for π_1 and π_2. We say that a P-node π *claims* a non-virtual edge e, and express this by defining the mapping $\mathcal{C}(e) = \pi$, if e is crossed in every embedding of skel(π) that conforms with Lemma 3. Observe that $\mathcal{C}$ is uniquely defined, since G is *o1p* and thus, no edge may be crossed more than once. We say that an embedding of the skeleton of a P-node is *admissible* if it conforms with Lemma 3 and does not imply the crossing of non-virtual edges claimed by other P-nodes. In Fig. 2(b), e. g., π_1 has two admissible embeddings, but both imply crossing the edge $\{f, m\}$, either by $\{d, i\}$ or by $\{h, i\}$. Hence, π_1 claims $\{f, m\}$. Computing $\mathcal{C}$ involves checking the embeddings of all P-nodes. As every P-node has at most four virtual edges, there are at most $\binom{4}{2} \cdot 2 = 12$ embeddings. Hence, the total time needed for this step is in $\mathcal{O}(n)$.

If every admissible embedding of skel(π) yields the same set of edges that are crossed, then π is called *fixable*. Let e, e' be two virtual edges that are embedded crossing each other. Observe that in this case, two S-nodes, namely refn(e) and refn(e'), are "crossing". By Proposition 1, the crossing can be augmented to a K_4. The insertion of these additional edges transforms the crossing S-nodes into an R-node that represents the K_4. In Fig. 2(b), this happens to π_1, σ_1, and σ_2. If the skeleton of an S-node previously had exactly three vertices, it is now completely contained in the K_4. Otherwise, its number of vertices is reduced by exactly 1, i. e., the vertex u or v, respectively. Note that completing the K_4 may affect the number of admissible embeddings and hence the fixability of other P-nodes if there was an admissible embedding of their skeletons that implied crossing one of e or e'. Algorithm 2 checks whether the virtual edges may cross each other and fixes the embedding of π. The next lemma enables us to also proceed when there is no fixable P-node.

Lemma 5. *Let π be a non-fixable P-node. If $\mathcal{T}$ has no fixable P-nodes, then every admissible embedding of* skel(π) *maintains at least one admissible embedding for every other P-node.*

Proof. Consider the fixing procedure of an embedding for a P-node π and S-nodes σ' and σ''. Let e' and e'' be the non-virtual edges that are crossed thereby. This affects the number of admissible embeddings for the skeletons of at most two other P-nodes π' and π'', that are adjacent to σ' and σ'', respectively. Observe that $\pi' \neq \pi''$, as $\mathcal{T}$ is a tree, and that every non-virtual edge is represented in the skeleton of exactly one node of $\mathcal{T}$.

Consider π'. W. l. o. g., let e' be the non-virtual edge in skel(σ') that is crossed after the fixing. Then, the number of admissible embeddings of skel(π') is reduced by exactly those that implied crossing e', too. However, π' did not claim e', so there is at least one other admissible embedding of skel(π'). The same argument holds for π'' and e''. $\square$

Algorithm 2. Fix Embedding of P-node with two crossing S-nodes

---

1: **procedure** FIXCROSSINGATPNODE(G, $\mathcal{T}$, P-Node π, S-Node σ_1, S-Node σ_2)
2:     Let u, v be the separation pair of π,
3:     Let $(u, c_1, \ldots, c_k, v, u)$ be the cycle in skel(σ_1).
4:     Let $(u, d_1, \ldots, d_l, v, u)$ be the cycle in skel(σ_2).
5:     **if** $\{c_k, v\}$ virtual **or** $\{u, d_1\}$ virtual **then**
6:         **if** $\{u, c_1\}$ virtual **or** $\{d_l, v\}$ virtual **then return** $\bot$
7:         **else** swap roles of σ_1, σ_2
8:     $\mathbb{P}_d \leftarrow \emptyset$ ▷ possibly affected P-nodes
9:     **if** $k > 1$ **then** insert edge $\{u, c_k\}$ in G, update $\mathcal{T}$
10:         **if** $\{c_{k-1}, c_k\}$ virtual **then** add its associated P-node to $\mathbb{P}_d$
11:     **else if** $\{u, c_k\}$ virtual **then** add its associated P-node to $\mathbb{P}_d$
12:     **if** $l > 1$ **then** insert edge $\{v, d_1\}$ in G, update $\mathcal{T}$
13:         **if** $\{d_1, d_2\}$ virtual **then** add its associated P-node to $\mathbb{P}_d$
14:     **else if** $\{v, d_1\}$ virtual **then** add its associated P-node to $\mathbb{P}_d$
15:     insert edge $\{c_k, d_1\}$, update $\mathcal{T}$
16:     **if** π has two (other) virtual edges **then** add π to $\mathbb{P}_d$
17:     **return** $\mathbb{P}_d$

Hence, by applying Lemma 5, we can step by step fix all embeddings of the skeletons of P-nodes with at least three virtual edges. Afterwards, every P-node has exactly two virtual edges and one non-virtual (cf. Fig. 2(c)). In Algorithm 1, this corresponds to ll. 15 – 26. FIXCROSSINGATPNODE takes $\mathcal{O}(1)$ time per call and there are embeddings of at most $\mathcal{O}(n)$ P-nodes to fix. Hence, the time for this part is $\mathcal{O}(n)$. The algorithm concludes by selecting an admissible embedding for all P- and R-nodes. All remaining S-nodes are embedded as plane cycles. The embedding for G is obtained via the 2-clique-sums of all skeleton embeddings (cf. Fig. 2(d)). Consequently, Algorithm 1 has a running time of $\mathcal{O}(n)$.

It remains to show that all conditions presented so far are also sufficient for a graph to be *o1p*. Every skeleton is, taken by itself, embedded *o1p*. Consider the 2-clique-sum of two skeleton embeddings. This operation glues both graphs together at two virtual edges. After the augmentation of Algorithm 1, every virtual edge is embedded such that it lies on the outer face. Hence, in the resulting embedding, every vertex still lies on the outer face and every edge is crossed at most once. With this, the outer 1-planarity of the whole embedding follows by structural induction.

Lemma 6. *A graph G is o1p if and only if it is a subgraph of a graph H with SPQR-tree $\mathcal{T}$ such that R-nodes and S-nodes are adjacent to P-nodes only, every skeleton of an R-node is a K_4, and every skeleton of a P-node has exactly one non-virtual and two virtual edges.*

This concludes the proof of Theorem 1. Additionally, if a graph is *o1p*, Algorithm 1 also provides an *o1p* embedding. With some extra effort, we can augment G to maximality. Consider the supergraph H constructed from G by Algorithm 1 and its SPQR-tree. It may have S-nodes with four or more vertices. As all remaining S-nodes are embedded plane, we can insert a plane edge between two non-adjacent vertices, which splits the S-node into two smaller S-nodes with an intermediate P-node. This procedure can be repeated until all S-nodes are triangles. Next, consider a P-node that is adjacent to exactly two S-nodes, e. g., π_4 in Fig. 2(c). Then, we can insert a crossing edge ($\{g, i\}$ in the example) that augments the subgraph to a K_4. As a result, the nodes π_4, σ_6, and σ_7 are replaced by a new R-node. We denote by H^+ this supergraph of H. Its SPQR-tree consists of R-nodes, of which each corresponds to a K_4 and S-nodes, of which each corresponds to a triangle. R- and S-nodes are only connected via P-nodes, which in turn have exactly two virtual edges and one non-virtual. Consider an embedding of H^+. It has a tree-like structure that consists of K_4s and triangles (K_3s) that share an edge if and only if their corresponding R- and S-nodes are connected via a P-node. As no P-node is adjacent to two S-nodes, triangles can only share an edge with K_4s. Suppose H^+ was not maximal. If we were able to insert an inner, plane edge, this would correspond to inserting a P-node into the SPQR-tree of H^+. However, no two P-nodes may be adjacent. Inserting an inner, crossed edge is equal to augmenting two triangles to a K_4, which is impossible, too, as no P-node is adjacent to two S-nodes. Finally, consider adding an edge to the outer face. As every crossing has been augmented to a K_4, the boundary of the outer face consists of a plane Hamiltonian cycle. Hence, every additional edge would shield at least one vertex from the outer face. Consequently, we can easily extend Algorithm 1 such that it maximizes the input graph. Additionally, we receive another characterization:

Lemma 7. *A graph G is maximal o1p if and only if the conditions for H in Lemma 6 hold and in its SPQR-tree, no P-node is adjacent to more than one S-node and the skeleton of every S-node is a cycle of length three.*

The argument above also implies that every embedded maximal *o1p* graph is maximal for all *o1p* embeddings.

Corollary 2. *A graph G is maximal o1p if it has a maximal o1p embedding.*

Note that the embedding of a maximal *o1p* graph is fixed if and only if that of the skeleton of every R-node is. This, in turn, is the case iff it contains at least two incident virtual edges.

Corollary 3. *The embedding of a maximal o1p graph is unique up to inversion if and only if the skeleton of every R-node of its SPQR-tree contains a vertex that is incident to exactly two virtual edges.*

Another consequence of Lemma 7 is, that every maximal *o1p* graph is composed of triangles and K_4s. Changing the embedding of the K_4s, we obtain:

Corollary 4. *Every o1p graph is planar and has treewidth at most three.*

Observe that if the step that augments a P-node with two adjacent triangle S-nodes to a K_4 is omitted, we obtain a plane-maximal *o1p* graph, i. e., every additional edge either violates outer 1-planarity or introduces a new crossing. Equivalently, we can also adjust Algorithm 1 to test (plane) *o1p* maximality.

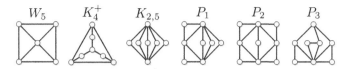

Fig. 3. Set M of minors of non-*o1p* graphs

Corollary 5. *There is a linear-time algorithm to test whether a graph is maximal (plane-maximal) o1p or to augment an o1p graph to a maximal (plane-maximal) o1p graph.*

From the recognition algorithm, we can immediately derive minors of non-*o1p* graphs: If the algorithm returns $\perp$, the graph at hand contains at least one of the *o1p* minors M as depicted in Fig. 3.

Theorem 2. *If a graph is not o1p, it contains at least one graph in M as a minor. Further, M is minimal and every graph in M is not o1p while removing or contracting an edge makes it o1p.*

Note that a graph might still be *o1p* even if it contains a graph in M as a minor, as outer 1-planar graphs are not closed under taking minors. The first minor W_5 is the wheel with five vertices, which is the smallest triconnected graph that is not *o1p* (Lemma 1). W_5 occurs in ll. 2 and 6 of Algorithm 1. If $\perp$ is returned in l. 2, then the graph contains a K_5 or $K_{3,3}$ as minor and W_5 is a minor of both. In l. 6, the first of the two checks implies W_5: If the R-node contains more than four vertices, $\perp$ is returned and the whole graph contains a planar triconnected component with at least four vertices, which always contains a W_5 as a minor. If the second condition in l. 6 is true, then the R-node at hand is a K_4 that contains a vertex v incident to three virtual edges. As the expansion graph of a virtual edge has a path with two edges as minor, we obtain K_4^+ in Fig. 3, where vertex v is in the center. If a P-node has at least five virtual edges (l. 11), then the $K_{2,5}$ is the minor. The remaining minors P_1, P_2, and P_3 can occur when fixing the embedding of a P-node with two crossing S-nodes. Consider l. 6 in Algorithm 2. If $\{u, d_1\}$ and $\{u, c_1\}$ are virtual, then u is incident to virtual edges in both S-nodes σ_1 and σ_2. If u is incident to at least one other virtual edge in π in whose expansion graph, u has at least degree two, then π has no admissible embedding and we obtain P_3 as minor. By a complete case differentiation, P_1 and P_2 can also be identified as minors.

4   Conclusion

We have designed a linear-time recognition algorithm for *o1p* that uses the SPQR-tree and returns a witness in terms of an *o1p* embedding or detects one of six minors, respectively.

Are there other classes of 1-planar graphs which can be recognized efficiently?

References

1. Alam, M.J., Brandenburg, F.J., Kobourov, S.G.: Straight-line drawings of 3-connected 1-planar graphs. In: Wismath, S., Wolff, A. (eds.) GD 2013. LNCS, vol. 8242, pp. 83–94. Springer, Heidelberg (2013)
2. Auer, C., Brandenburg, F.J., Gleißner, A., Reislhuber, J.: On 1-planar graphs with rotation systems. Tech. Rep. MIP 1207, University of Passau (2012)
3. Bannister, M.J., Cabello, S., Eppstein, D.: Parameterized complexity of 1-planarity. In: Dehne, F., Solis-Oba, R., Sack, J.-R. (eds.) WADS 2013. LNCS, vol. 8037, pp. 97–108. Springer, Heidelberg (2013)
4. Brandenburg, F.J., Eppstein, D., Gleißner, A., Goodrich, M.T., Hanauer, K., Reislhuber, J.: On the density of maximal 1-planar graphs. In: Didimo, W., Patrignani, M. (eds.) GD 2012. LNCS, vol. 7704, pp. 327–338. Springer, Heidelberg (2013)
5. Cabello, S., Mohar, B.: Adding one edge to planar graphs makes crossing number and 1-planarity hard. Tech. Rep. arXiv:1203.5944 (cs.CG), Computing Research Repository (CoRR) (March 2012)
6. Dehkordi, H.R., Eades, P.: Every outer-1-plane graph has a right angle crossing drawing. Internat. J. Comput. Geom. Appl. 22(6), 543–558 (2012)
7. Di Battista, G., Tamassia, R.: On-line planarity testing. SIAM J. Comput. 25(5), 956–997 (1996)
8. Eades, P., Hong, S.H., Katoh, N., Liotta, G., Schweitzer, P., Suzuki, Y.: Testing maximal 1-planarity of graphs with a rotation system in linear time. In: Didimo, W., Patrignani, M. (eds.) GD 2012. LNCS, vol. 7704, pp. 339–345. Springer, Heidelberg (2013)
9. Eades, P., Liotta, G.: Right angle crossing graphs and 1-planarity. In: Speckmann, B. (ed.) GD 2011. LNCS, vol. 7034, pp. 148–153. Springer, Heidelberg (2011)
10. Eggleton, R.B.: Rectilinear drawings of graphs. Utilitas Math. 29, 149–172 (1986)
11. Gutwenger, C., Mutzel, P.: A linear time implementation of SPQR-trees. In: Marks, J. (ed.) GD 2000. LNCS, vol. 1984, pp. 77–90. Springer, Heidelberg (2001)
12. Hliněný, P.: Crossing number is hard for cubic graphs. J. Combin. Theory, Ser. B 96(4), 455–471 (2006)
13. Hong, S.-H., Eades, P., Liotta, G., Poon, S.-H.: Fáry's theorem for 1-planar graphs. In: Gudmundsson, J., Mestre, J., Viglas, T. (eds.) COCOON 2012. LNCS, vol. 7434, pp. 335–346. Springer, Heidelberg (2012)
14. Hong, S.H., Eades, P., Naoki, K., Liotta, G., Schweitzer, P., Suzuki, Y.: A linear-time algorithm for testing outer-1-planarity. In: Wismath, S., Wolff, A. (eds.) GD 2013. LNCS, vol. 8242, pp. 71–82. Springer, Heidelberg (2013)
15. Korzhik, V.P., Mohar, B.: Minimal obstructions for 1-immersion and hardness of 1-planarity testing. J. Graph Theor. 72, 30–71 (2013)
16. Mitchell, S.L.: Linear algorithms to recognize outerplanar and maximal outerplanar graphs. Inform. Process. Lett. 9(5), 229–232 (1979)
17. Ringel, G.: Ein Sechsfarbenproblem auf der Kugel. Abh. aus dem Math. Seminar der Univ. Hamburg 29, 107–117 (1965)
18. Thomassen, C.: Rectilinear drawings of graphs. J. Graph Theor. 12(3), 335–341 (1988)

Straight Line Triangle Representations

Nieke Aerts and Stefan Felsner

Technische Universität Berlin,
Institut für Mathematik, Berlin, Germany
{aerts,felsner}@math.tu-berlin.de

Abstract. A straight line triangle representation (SLTR) of a planar
graph is a straight line drawing such that all the faces including the
outer face have triangular shape. Such a drawing can be viewed as a
tiling of a triangle using triangles with the input graph as skeletal structure. In this paper we present a characterization of graphs that have an
SLTR that is based on flat angle assignments, i.e., selections of angles
of the graph that have size π in the representation. We also provide a
second characterization in terms of contact systems of pseudosegments.
With the aid of discrete harmonic functions we show that contact systems of pseudosegments that respect certain conditions are stretchable.
The stretching procedure is then used to get straight line triangle representations. Since the discrete harmonic function approach is quite flexible
it allows further applications, we mention some of them.

The drawback of the characterization of SLTRs is that we are not able
to effectively check whether a given graph admits a flat angle assignment
that fulfills the conditions. Hence it is still open to decide whether the
recognition of graphs that admit straight line triangle representation is
polynomially tractable.

1 Introduction

In this paper we study a representation of planar graphs in the classical setting,
i.e., vertices are represented by points in the Euclidean plane and edges by non-
crossing continuous curves connecting the points. We aim at classifying the class
of planar graphs that admit a straight line representation in which all faces are
triangles. Haas et al. present a necessary and sufficient condition for a graph to
be a pseudo-triangulation [8], however this condition is not sufficient for a graph
to have a straight line triangle representation (e.g. see Fig. 2 and [1]). There
have been investigations of the problem in the dual setting, i.e., in the setting
of side contact representations of planar graphs with triangles. Gansner, Hu and
Kobourov show that outerplanar graphs, grid graphs and hexagonal grid graphs
are Touching Triangle Graphs (TTGs). They give a linear time algorithm to
find the TTG [7]. Alam, Fowler and Kobourov [2] consider proper TTGs, i.e.,
the union of all triangles of the TTG is a triangle and there are no holes. They
give a necessary and a stronger sufficient condition for biconnected outerplanar
graphs to be TTG, a characterization, however, is missing. Kobourov, Mondal
and Nishat present construction algorithms for proper TTGs of 3-connected

S. Wismath and A. Wolff (Eds.): GD 2013, LNCS 8242, pp. 119–130, 2013.

cubic graphs and some grid graphs. They also present a decision algorithm for testing whether a 3-connected planar graph is proper TTG [10].

Here is the formal introduction of the main character for this paper.

Definition 1. *A plane drawing of a graph such that*
- *all the edges are straight line segments and*
- *all the faces, including the outer face, bound a non-degenerate triangle is called a* Straight Line Triangle Representation *(SLTR).*

    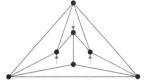

Fig. 1. A graph and one of its SLTRs   **Fig. 2.** A Flat Angle Assignment (given by the arrows) that is not an SLTR

Clearly every straight line drawing of a triangulation is an SLTR. So the class of planar graphs admitting an SLTR is rich. On the other hand, graphs admitting an SLTR cannot have a cut vertex. Indeed, as shown below (Prop. 1), graphs admitting an SLTR are well connected. Being well connected, however, is not sufficient as shown e.g. by the cube graph.

To simplify the discussion we assume that the input graph is given with a plane embedding and a selection of three vertices of the outer face that are designated as corner vertices for the outer face. These three vertices are called *suspension vertices*. If needed, an algorithm may try all triples of vertices as suspensions.

If a degree two vertex has an angle of size π in one of its incident faces, then it also has an angle of size π in the face on the other side. Hence, this vertex and its two incident edges can be replaced by a single edge connecting the two neighbors of the vertex. Such an operation is called a *vertex reduction*. The only angles of an SLTR whose size exceeds π are the outer angles at the outer triangle. Therefore, we can use vertex reductions to eliminate all the degree two vertices, except for degree two vertices that are suspensions.

A plane graph G with suspensions s_1, s_2, s_3 is said to be *internally 3-connected* when the addition of a new vertex v_∞ in the outer face, that is made adjacent to the three suspension vertices, yields a 3-connected graph.

Proposition 1. *If a graph G admits an SLTR with s_1, s_2, s_3 as corners of the outer triangle and no vertex reduction is possible, then G is internally 3-connected.*

Proof. Consider an SLTR of G. Suppose there is a separating set U of size 2. It is enough to show that each component of $G \setminus U$ contains a suspension vertex, so that $G + v_\infty$ is not disconnected by U. Since G admits no vertex reduction every degree two vertex is a suspension. Hence, if C is a component and $C \cup U$ induces a path, then there is a suspension in C. Otherwise consider the convex hull of $C \cup U$ in the SLTR. The convex corners of this hull are vertices that expose an angle of size at least π. Two of these large angles may be at vertices of U but there is at least one additional large angle. This large angle must be the outer angle at a vertex that is an outer corner of the SLTR, i.e., a suspension. □

From Prop. 1 it follows that any graph that is not internally 3-connected but does admit an SLTR, is a subdivision of an internally 3-connected graph. Therefore we may assume that the graphs we consider are internally 3-connected.

In Section 2 we present necessary conditions for the existence of an SLTR in terms of what we call a flat angle assignment. A flat angle assignment that fulfills the conditions is shown to induce a partition of the set of edges into a set of pseudosegments. Finally, with the aid of discrete harmonic functions we show that in our case the set of pseudosegments is stretchable. Hence, the necessary conditions are also sufficient. The drawback of the characterization is that we are not aware of an effective way of checking whether a given graph admits a flat angle assignment that fulfills the conditions.

In Section 3 we consider further applications of the stretching approach. First we look at flat angle assignments that yield faces with more than three corners. Then we proceed to prove a more general result about stretchable systems of pseudosegments with our technique. The result is not new, de Fraysseix and Ossona de Mendez have investigated stretchability conditions for systems of pseudosegments in [3,4,5]. The counterpart to Theorem 2 can be found in [5, Theorem 38]. The proof there is based on a long and complicated inductive construction.

2   Necessary and Sufficient Conditions

Consider a plane, internally 3-connected graph $G = (V, E)$ with suspensions given. Suppose that G admits an SLTR. This representation induces a set of *flat angles*, i.e., incident pairs (v, f) such that vertex v has an angle of size π in the face f.

Since G is internally 3-connected every vertex has at most one flat angle. Therefore, the flat angles can be viewed as a partial mapping of vertices to faces. Since the outer angle of suspension vertices exceeds π, suspensions have no flat angle. Since each face f (including the outer face) is a triangle, each face has precisely three angles that are not flat. In other words every face f has $|f| - 3$ incident vertices that are assigned to f. This motivates the definition:

Definition 2. *A flat angle assignment (FAA) is a mapping from a subset U of the non-suspension vertices to faces such that*
$[C_v]$ *Every vertex of U is assigned to at most one face,*
$[C_f]$ *For every face f, precisely $|f| - 3$ vertices are assigned to f.*

Not every FAA induces an SLTR. An example is given in Fig. 2. Hence, we have to identify another condition. To state this we need a definition. Let H be a connected subgraph of the plane graph G. The *outline cycle* $\gamma(H)$ *of H is* the closed walk corresponding to the outer face of H. An *outline cycle* of G is a closed walk that can be obtained as outer cycle of some connected subgraph of G. Outline cycles may have repeated edges and vertices, see Fig. 3. The interior $\text{int}(\gamma)$ of an outline cycle $\gamma = \gamma(H)$ consists of H together with all vertices, edges and faces of G that are contained in the area enclosed by γ.

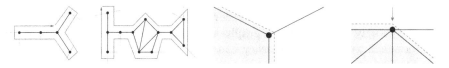

Fig. 3. Examples of outline cycles   **Fig. 4.** Combinatorially Convex Corners

Proposition 2. *An SLTR obeys the following condition C_o:*

[C_o] *Every outline cycle that is not the outline cycle of a path, has at least three geometrically convex corners.*

Proofs of Propositions 2 and 3 have been moved to the appendix.

Condition C_o has the disadvantage that it depends on a given SLTR, hence, it is useless for deciding whether a planar graph G admits an SLTR. The following definition allows to replace C_o by a combinatorial condition on an FAA.

Definition 3. *Given an FAA ψ. A vertex v of an outline cycle γ is a combinatorial convex corner for γ with respect to ψ if*

- *v is a suspension vertex, or*
- *v is not assigned and there is an edge e incident to v with $e \notin \text{int}(\gamma)$, or*
- *v is assigned to a face f, $f \notin \text{int}(\gamma)$ and there exists an edge e incident to v with $e \notin \text{int}(\gamma)$.*

In Fig. 4 an unassigned and an assigned combinatorially convex corner are shown. The grey area represents the interior of some outline cycle and the arrow represents the assignment of the vertex to the face in which the arrow is drawn.

Proposition 3. *Let G admit an SLTR Γ, that induces the FAA ψ and let H be a connected subgraph of G. If v is a geometrically convex corner of the outline cycle $\gamma(H)$ in Γ, then v is a combinatorially convex corner of $\gamma(H)$ with respect to ψ.*

The proposition enables us to replace the condition on geometrically convex corners w.r.t. an SLTR by a condition on combinatorially convex corners w.r.t. an FAA.

[C_o^*] Every outline cycle that is not the outline cycle of a path, has at least three combinatorially convex corners.

From Prop. 2 and Prop. 3 it follows that this condition is necessary for an FAA that induces an SLTR. In Thm. 1 we prove that if an FAA obeys C_o^* then it induces an SLTR. The proof is constructive. In anticipation of this result we say that an FAA obeying C_o^* is a *good flat angle assignment* and abbreviate it as a *GFAA*.

Next we show that a GFAA induces a contact family of pseudosegments. This family of pseudosegments is later shown to be stretchable, i.e., it is shown to be homeomorphic to a contact system of straight line segments.

Definition 4. *A* contact family of pseudosegments *is a family* $\{c_i\}_i$ *of simple curves* $c_i : [0,1] \to \mathbb{R}^2$, *with different endpoints, i.e.,* $c_i(0) \neq c_i(1)$, *such that any two curves* c_j *and* c_k *(*$j \neq k$*) have at most one point in common. If so, then this point is an endpoint of (at least) one of them.*

A GFAA ψ on a graph G gives rise to a relation ρ on the edges: Two edges, both incident to v and f are in relation ρ if and only if v is assigned to f. The transitive closure of ρ is an equivalence relation on the edges of G.

Proposition 4. *The equivalence classes of edges of* G *defined by* ρ *form a contact family of pseudosegments.*

Proof. Let the equivalence classes of ρ be called arcs.

Condition C_v ensures that every vertex is interior to at most one arc. Hence, the arcs are simple curves and no two arcs cross.

Every arc has two distinct endpoints, otherwise it would be a cycle and its outline cycle has only one combinatorially convex corner. If an arc touched itself, the outline cycle of this equivalence class would have at most one combinatorially convex corner. This again contradicts C_o^*.

If two arcs share two points, the outline cycle has at most two combinatorially convex corners. This again contradicts C_o^*.

We conclude that the family of arcs satisfies the properties of a contact family of pseudosegments.   □

Definition 5. *Let* Σ *be a family of pseudosegments and let* S *be a subset of* Σ. *A point* p *of a pseudosegment from* S *is a* free point *for* S *if*
  1. *p is an endpoint of a pseudosegment in S, and*
  2. *p is not interior to a pseudosegment in S, and*
  3. *p is incident to the unbounded region of S, and*
  4. *p is incident to the unbounded region of Σ or*
     p is incident to a pseudosegment that is not in S.

With Lem. 1 we prove that the family of pseudosegments Σ that arises from a GFAA has the following property
[C_P] Every subset S of Σ with $|S| \geq 2$ has at least three free points.

Lemma 1. *Let* ψ *a GFAA on a plane, internally 3-connected graph* G. *For every subset* S *of the family of pseudosegments associated with* ψ, *it holds that, if* $|S| \geq 2$ *then* S *has at least 3 free points.*

Proof. Let S be a subset of the contact family of pseudosegments defined by the GFAA (Prop. 4).

Each pseudosegment of S corresponds to a path in G. Let H be the subgraph of G obtained as union of the paths of pseudosegments in S. We assume that H

is connected and leave the discussion of the cases where it is not to the reader. If H itself is not a path, then by C_o^* the outline cycle $\gamma(H)$ must have at least three combinatorially convex corners. Every combinatorially convex corner of $\gamma(H)$ is a free point of S.

If S induces a path, then the two endpoints of this path are free points for S. Moreover, there exists at least one vertex v in this path which is an endpoint for two pseudosegments and not an interior point for any. Now there must be an edge e incident to v, such that $e \notin S$, therefore v is a free point for S.   □

Given an internally 3-connected, plane graph G with a GFAA. To find a corresponding SLTR we aim at representing each of the pseudosegments induced by the FAA as a straight line segment. If this can be done, every assigned vertex will be between its two neighbors that are part of the same pseudosegment. This property can be modeled by requiring that the coordinates $p_v = (x_v, y_v)$ of the vertices of G satisfy a harmonic equation at each assigned vertex.

Fig. 5. A stretched representation of a contact family of pseudosegments that arises from a GFAA in the graph of Fig 2

Indeed if uv and vw are edges belonging to a pseudosegment s, then the coordinates satisfy

$$x_v = \lambda_v x_u + (1 - \lambda_v)x_w \qquad \text{and} \qquad y_v = \lambda_v y_u + (1 - \lambda_v)y_w \qquad (1)$$

where the parameter λ_v can be chosen arbitrarily from $(0, 1)$. These are the harmonic equations for v.

In the SLTR every unassigned vertex v is placed in a weighted barycenter of its neighbors. In terms of coordinates this can be written as

$$x_v = \sum_{u \in N(v)} \lambda_{vu} x_u, \qquad y_v = \sum_{u \in N(v)} \lambda_{vu} y_u. \qquad (2)$$

These are the harmonic equations for an unassigned vertex v. The λ_{vu} can be chosen arbitrarily in the range set by the convexity conditions: $\sum_{u \in N(v)} \lambda_{vu} = 1$ and $\lambda_{vu} > 0$.

Vertices whose coordinates are not restricted by harmonic equations are called *poles*. In our case the suspension vertices are the three poles of the harmonic functions for the x and y-coordinates. The coordinates for the suspension vertices are fixed as the corners of some non-degenerate triangle, this adds six equations to the linear system.

The theory of harmonic functions and applications to (plane) graphs are nicely explained by Lovász [11]. The following proposition is taken from Chapter 3 of [11].

Proposition 5. *For every choice of the parameters λ_v and λ_{vu} complying with the conditions, the system has a unique solution.*

Now we state our main result, it shows that the necessary conditions are also sufficient.

Theorem 1. *Let G be an internally 3-connected, plane graph and Σ a family of pseudosegments associated to an FAA, such that each subset $S \subseteq \Sigma$ has three free points or cardinality at most one. The unique solution of the system of equations that arises from Σ is an SLTR.*

Proof. The proof consists of 7 arguments, which together yield that the drawing induced from the GFAA is a non-degenerate, plane drawing. The proof has been inspired by the proof of Colin de Verdière [6] for convex straight line drawings of plane graphs via spring embeddings.

1. *Pseudosegments become Segments.* Let $(v_1, v_2), (v_2, v_3), \ldots, (v_{k-1}, v_k)$ be the set of edges of a pseudosegment defined by ψ. The harmonic conditions for the coordinates force that v_i is placed between v_{i-1} and v_{i+1} for $i = 2, .., k-1$. Hence all the vertices of the pseudosegment are placed on the segment with endpoints v_1 and v_k.

2. *Convex Outer Face.* The outer face is bounded by three pseudosegments and the suspensions are the endpoints for these three pseudosegments. The coordinates of the suspensions (the poles of the harmonic functions) have been chosen as corners of a non-degenerate triangle and the pseudosegments are straight line segments, therefore the outer face is a triangle and in particular convex.

3. *No Concave Angles.* Every vertex, not a pole, is forced either to be on the line segment between two of its neighbors (if assigned) or in a weighted barycenter of all its neighbors (otherwise). Therefore every non-pole vertex is in the convex hull of its neighbors. This implies that there are no concave angles at non-poles.

4. *No Degenerate Vertex.* A vertex is degenerate if it is placed on a line, together with at least three of its neighbors. Suppose there exists a vertex v, such that v and at least three of its neighbors are placed on a line ℓ. Let S be the connected component of pseudosegments that are aligned with ℓ, such that S contains v. The set S contains at least two pseudosegments. Therefore S must have at least three free points, v_1, v_2, v_3.

By property 4 in the definition of free points, each of the free points is incident to a segment that is not aligned with ℓ. Suppose the free points are not suspension vertices. If v_i is interior to $s_i \in S$, then s_i has an endpoint on each side of ℓ. If v_i is not assigned by the GFAA it is in the strict convex hull of its neighbors, hence, v_i is an endpoint of a segment reaching into each of the two half-planes defined by ℓ.

Now suppose v_1 and v_2 are suspension vertices[1] and consider the third free point, v_3. If v_3 is interior to a pseudosegment not on ℓ, then one endpoint of this pseudosegment lies outside the convex hull of the three suspensions, which is a contradiction. Hence it is not interior to any pseudosegment and at least one of its neighbors does not lie on ℓ, but then v_3 should be in a weighted barycenter of its neighbors, hence again we would find a vertex outside the convex hull of

---

[1] Not all three suspension vertices lie on one line, hence at least one of the three free points is not a suspension.

the suspension vertices. Therefore at most one of the free points is a suspension and ℓ is incident to at most one of the suspension vertices.

In any of the above cases each of v_1, v_2, v_3 has a neighbor on either side of ℓ.

Let n^+ and $n^- = -n^+$ be two normals for line ℓ and let p^+ and p^- be the two poles, that maximize the inner product with n^+ resp. n^-. Starting from the neighbors of the v_i in the positive halfplane of ℓ we can always move to a neighbor with larger[2] inner product with n^+ until we reach p^+. Hence v_1, v_2, v_3 have paths to p^+ in the upper halfplane of ℓ and paths to p^- in the lower halfplane. Since v_1, v_2, v_3 also have a path to v we can contract all vertices of the upper and lower halfplane of ℓ to p^+ resp. p^- and all inner vertices of these paths to v to produce a $K_{3,3}$ minor of G. This is in contradiction to the planarity of G. Therefore, there is no degenerate vertex.

5. *Preservation of Rotation System.* Let $\theta(v) = \sum_f \theta(v, f)$ denote the sum of the angles around an interior vertex. Here f is a face incident to v and $\theta(v, f)$ is the (smaller!) angle between the two edges incident to v and f in the drawing obtained by solving the harmonic system. If the incident faces are oriented consistently around v, then the angles sum up to 2π, otherwise $\theta(v) > 2\pi$ (see Fig. 6). We do not consider the outer face in the sums so that the b vertices incident to the outer face contribute a total angle of $(b-2)\pi$ to the inner faces.

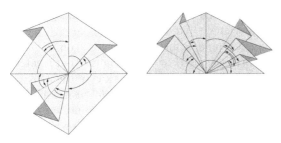

Now consider the sum $\theta(f) = \sum_v \theta(v, f)$ of the angles of a face f. At each vertex incident to f the contribution $\theta(v, f)$ is at most of size π. A closed polygonal chain with k corners, selfintersecting or not, has a sum of inner angles equal to $(k-2)\pi$. Therefore $\theta(f) \leq (|f| - 2)\pi$. The sum over all vertices $\sum_v \theta(v)$ and the sum over all faces $\sum_f \theta(f)$ must be equal since they count the same angles in two different ways.

Fig. 6. Vertices with their surrounding faces not oriented consistently

$$(|V| - b)2\pi + (b-2)\pi \leq \sum_v \theta(v) = \sum_f \theta(f) \leq ((2|E| - b) - 2(|F| - 1))\pi \quad (3)$$

This yields $|V| - |E| + |F| \leq 2$. Since G is planar Euler's formula implies equality. Therefore $\theta(v) = 2\pi$ for every interior vertex v and the faces must be oriented consistently around every vertex, i.e. the rotation system is preserved. Note that the rotation system could have been flipped, between clockwise and counterclockwise but then it is flipped at every vertex.

6. *No Crossings.* Suppose two edges cross. On either side of both of the edges there is a face, therefore there must be a point p in the plane which is covered by at least two faces. Outside of the drawing there is only the unbounded face.

---

[2] If n^+ is perpendicular to another segment this may not be possible. In this case we can use a slightly perturbed vector n_ε^+ to break ties.

Move along a ray, that does not pass through a vertex of the graph, from p to infinity. A change of the cover number, i.e. the number of faces by which the point is covered, can only occur when crossing an edge. But if the cover number changes then the rotation system at a vertex of that edge must be wrong. This would contradict the previous item. Therefore a crossing cannot exist.

7. *No Degeneracy.* Suppose there is an edge of length zero. Since every vertex has a path to each of the three suspensions there has to be a vertex a that is incident to an edge of length zero and an edge ab of non-zero length. Following the direction of forces we can even find such a vertex-edge pair with b contributing to the harmonic equation for the coordinates of a. We now distinguish two cases.

If a is assigned, it is on the segment between b and some b', together with the neighbor of the zero length edge this makes three neighbors of a on a line. Hence, a is a degenerate vertex. A contradiction.

If a is unassigned it is in the convex hull of its neighbors. However, starting from a and using only zero-length edges we eventually reach some vertex a' that is incident to an edge $a'b'$ of non-zero length, such that b' is contributing to the harmonic equation for the coordinates of a'. Vertex a' has the same position as a and is also in the convex hull of its neighbors. This makes a crossing of edges unavoidable. A contradiction. Hence, there are no edges of length zero.

Suppose there is an angle of size zero. Since every vertex is in the convex hull of its neighbors there are no angles of size larger than π. Moreover there are no crossings, hence the face with the angle of size zero is stretching along a line segment with two angles of size zero. Since there are no edges of length zero and all vertices are in the convex hull of their neighbors, all but two vertices of the face must be assigned to this face. Therefore, there are two pseudosegments bounding this face, which have at least two points in common, this contradicts that Σ is a family of pseudosegments. We conclude that there is no degeneracy.

From items 1–7 we conclude that the drawing is plane and thus an SLTR.   □

3   Further Applications of the Proof Technique

We have shown that a graph G has an SLTR exactly if it admits an FAA satisfying C_v, C_f and C_o^*. Conditions C_v and C_o^* are necessary for the proof that the system of pseudosegments corresponding to the FAA is stretchable. Condition C_f, however, is only needed to make all the faces triangles. Modifying condition C_f allows for further applications of the stretching technique. Of course we still need at least three corners for every face. Also we have to make sure that all the non-suspension vertices of the outer face are assigned to the outer face. Together this makes the modified face condition:

[C_f^*]   For every face f, at most $|f| - 3$ vertices are assigned to f and all non-suspension vertices of the outer face f^o are assigned to f^o.

If we use the empty flat angle assignment, i.e., if the harmonic equations of all non-suspensions are of type (2), then we obtain a drawing such that all non-suspension vertices are in the barycenter of their neighbors. This is the Tutte drawing [12] with asymmetric elastic forces given by the parameters λ_{uv}, see also [11]. Note that in this case the existence of at least three combinatorially convex corners at an outline cycle (condition C_o^*) follows from the internally 3-connectedness of the graph.

The construction of Section 2 also applies when

- the assignment has $|f| - i$ vertices assigned to every inner face f, for $i = 4, 5$ (drawing with only convex 4-gon or only convex 5-gon faces.)
- the assignment has some number c_f of corners at inner face f (drawing with convex faces of prescribed complexity).

The drawback is that again in these cases we do not know how to find an FAA that fulfills C_o^*.

In [9] Kenyon and Sheffield study T-graphs in the context of dimer configurations (weighted perfect matchings). In our terminology T-graphs correspond to straight line representations such that each non-suspension is assigned. In [9] the straight line representations of T-graphs are obtained by analyzing random walks. Cf. [11] for further connections between discrete harmonic functions and Markov chains.

Stretchability of Systems of Pseudosegments. A contact system of pseudosegments is *stretchable* if it is homeomorphic to a contact system of straight line segments. De Fraysseix and Ossona de Mendez characterized stretchable systems of pseudosegments [3,4,5]. They use the notion of an extremal point.

Definition 6. *Let Σ be a family of pseudosegments and let S be a subset of Σ. A point p is an* extremal point *for S if*

1. *p is an endpoint of a pseudosegment in S, and*
2. *p is not interior to a pseudosegment in S, and*
3. *p is incident to the unbounded region of S.*

Theorem 2 (De Fraysseix & Ossona de Mendez [5, Theorem 38]).
A contact family Σ of pseudosegments is stretchable if and only if each subset $S \subseteq \Sigma$ of pseudosegments with $|S| \geq 2$, has at least 3 extremal points.

Our notion of a free point (Def. 5) is more restrictive than the notion of an extremal point. In the following we show that there is no big difference. First in Prop. 6 we show that in the case of families of pseudosegments that live on a plane graph via an FAA, the two notions coincide. Then we continue by reproving Thm. 2 as a corollary of Thm. 1.

Proposition 6. *Let G be an internally 3-connected, plane graph and Σ a family of pseudosegments associated to an FAA, such that each subset $S \subseteq \Sigma$ has three extremal points or cardinality at most one. The unique solution of the system of equations corresponding to Σ, is an SLTR.*

Proof. Note that in the proof of Thm. 1 the notion of free points is only used to show that there is no degenerate vertex. We show how to modify this part of the argument for the case of extremal points:

Consider again the set S of pseudosegments aligned with ℓ. We will show that all extremal points are also free points. Let p an extremal point of S. Assuming that p is not free we can negate item 4. from Def. 5, i.e., all the pseudosegments for which p is an endpoint are in S. By 3-connectivity p is incident to at least three pseudosegments, all of which lie on the line ℓ. Since all regions are bounded by three pseudosegments and p is not interior to a segment of S, all the regions incident to p must lie on ℓ. But then p is not incident to the unbounded region of S, hence p is not an extremal point. Therefore all extremal points of S are also free points of S. Prop. 6 now follows from Thm. 1.   □

Proof (of Thm. 2). Let Σ a contact family of pseudosegments which is stretchable. Consider a set $S \subseteq \Sigma$ of cardinality at least two in the stretching, i.e., in the segment representation. Endpoints (of segments) on the boundary of the convex hull of S are extremal points. There are at least three of them unless S lies on a line ℓ. In the latter case, there is a point q on ℓ that is the endpoint of two colinear segments. This is a third extremal point.

Conversely, assume that each subset $S \subseteq \Sigma$ of pseudosegments, with $|S| \geq 2$, has at least 3 extremal points. We aim at applying Prop 6. To this end we construct an extended system Σ^+ of pseudosegments in which every region is bounded by precisely three pseudosegments.

First we take a set Δ of three pseudosegments that intersect like the three sides of a triangle so that Σ is in the interior. The corners of Δ are chosen as suspensions and the sides of Δ are deformed such that they contain all extremal points of the family Σ. Let the new family be Σ'.

Next we add new *protection points*, these points ensure that the pseudosegments of Σ' will be mapped to straight lines. For each inner region R in Σ', for each pseudosegment s in R, we add a protection point for each visible side of s. The protection point is connected to the endpoints of s, with respect to R from the visible side of s.

Now the inner part of R is bounded by an alternating sequence of endpoints of Σ' and protection points. We connect two protection points if they share a neighbor in this sequence. Last we add a *triangulation point* in R and connect it to all protection points of R.

This construction yields a family Σ^+ of pseudosegments such that every region is bounded by precisely three pseudosegments and every subset $S \subseteq \Sigma^+$ has at least 3 extremal points, unless it has cardinality one.

Let V be the set of points of Σ^+ and E the set of edges induced by Σ^+. It follows from the construction that $G = (V, E)$ is internally 3-connected.

Fig. 7.   Protection points in red and the triangulation point in cyan for two faces of some Σ'

By Prop. 6 the graph $G = (V, E)$ together with Σ^+ is stretchable to an SLTR. Removing the protection points, triangulation points and their incident edges yields a contact system of straight line segments homeomorphic to Σ.   □

4 Conclusion and Open Problems

We have given necessary and sufficient conditions for a 3-connected planar graph to have an SLT Representation. Given an FAA and a set of rational parameters $\{\lambda_i\}_i$, the solution of the harmonic system can be computed in polynomial time. Checking whether a solution is degenerate can also be done in polynomial time. Hence, we can decide in polynomial time whether a given FAA corresponds to an SLTR. In other words, checking whether a given FAA is a GFAA can be done in polynomial time. However, most graphs admit different FAAs of which only some are good. We are not aware of an effective way of finding a GFAA. Therefore we have to leave this problem open: Is the recognition of graphs that have an SLTR (GFAA) in P?

Given a 3-connected planar graph and a GFAA, interesting optimization problems arise, e.g. find the set of parameters $\{\lambda_i\}_i$ such that the smallest angle in the graph is maximized, or the set of parameters such that the length of the shortest edge is maximized.

References

1. Aerts, N., Felsner, S.: Henneberg steps for Triangle Representations, http://page.math.tu-berlin.de/~aerts/pubs/ptsltr.pdf
2. Alam, M.J., Fowler, J., Kobourov, S.G.: Outerplanar graphs with proper touching triangle representations (unpublished manuscript)
3. de Fraysseix, H., de Mendez, P.O.: Stretching of jordan arc contact systems. In: Liotta, G. (ed.) GD 2003. LNCS, vol. 2912, pp. 71–85. Springer, Heidelberg (2004)
4. de Fraysseix, H., Ossona de Mendez, P.: Contact and intersection representations. In: Pach, J. (ed.) GD 2004. LNCS, vol. 3383, pp. 217–227. Springer, Heidelberg (2005)
5. de Fraysseix, H., de Mendez, P.O.: Barycentric systems and stretchability. Discrete Applied Mathematics 155, 1079–1095 (2007)
6. de Verdière, Y.C.: Comment rendre géodésique une triangulation d'une surface? L'Enseignement Mathématique 37, 201–212 (1991)
7. Gansner, E.R., Hu, Y., Kobourov, S.G.: On Touching Triangle Graphs. In: Brandes, U., Cornelsen, S. (eds.) GD 2010. LNCS, vol. 6502, pp. 250–261. Springer, Heidelberg (2011)
8. Haas, R., Orden, D., Rote, G., Santos, F., Servatius, B., Servatius, H., Souvaine, D.L., Streinu, I., Whiteley, W.: Planar minimally rigid graphs and pseudo-triangulations. Comput. Geom. 31, 31–61 (2005)
9. Kenyon, R., Sheffield, S.: Dimers, tilings and trees. J. Comb. Theory, Ser. B 92, 295–317 (2004)
10. Kobourov, S.G., Mondal, D., Nishat, R.I.: Touching triangle representations for 3-connected planar graphs. In: Didimo, W., Patrignani, M. (eds.) GD 2012. LNCS, vol. 7704, pp. 199–210. Springer, Heidelberg (2013)
11. Lovász, L.: Geometric representations of graphs, Draft version (December 11, 2009), http://www.cs.elte.hu/~lovasz/geomrep.pdf
12. Tutte, W.T.: How to draw a graph. Proc. of the London Math. Society 13, 743–767 (1963)

Extending Partial Representations
of Circle Graphs[*]

Steven Chaplick[1], Radoslav Fulek[1], and Pavel Klavík[2,**]

[1] Department of Applied Mathematics, Faculty of Mathematics and Physics,
Charles University, Malostranské náměstí 25, 118 00 Prague, Czech Republic
chaplick@kam.mff.cuni.cz, radoslav.fulek@gmail.com
[2] Computer Science Institute, Faculty of Mathematics and Physics,
Charles University, Malostranské náměstí 25, 118 00 Prague, Czech Republic
klavik@iuuk.mff.cuni.cz

Abstract. *The partial representation extension problem* is a recently introduced generalization of the recognition problem. A *circle graph* is an intersection graph of chords of a circle. We study the partial representation extension problem for circle graphs, where the input consists of a graph G and a partial representation $\mathcal{R}'$ giving some pre-drawn chords that represent an induced subgraph of G. The question is whether one can extend $\mathcal{R}'$ to a representation $\mathcal{R}$ of the entire G, i.e., whether one can draw the remaining chords into a partially pre-drawn representation.

Our main result is a polynomial-time algorithm for partial representation extension of circle graphs. To show this, we describe the structure of all representation a circle graph based on split decomposition. This can be of an independent interest.

1 Introduction

Graph drawings and visualizations are important topics of graph theory and computer science. A frequently studied type of representations are so-called *intersection representations*. An intersection representation of a graph represents its vertices by some objects and encodes its edges by intersections of these objects, i.e., two vertices are adjacent if and only if the corresponding objects intersect. Classes of intersection graphs are obtained by restricting these objects; e.g., *interval graphs* are intersection graphs of intervals of the real line, *string graphs* are intersection graphs of strings in plane, and so on. These representations are well-studied; see e.g. [25].

For a fixed class $\mathcal{C}$ of intersection-defined graphs, a very natural computational problem is *recognition*. It asks whether an input graph G belongs to $\mathcal{C}$. In this paper, we study a recently introduced generalization of this problem called *partial representation extension* [19]. Its input gives with G a part of the representation and the problem asks whether this partial representation can be extended to a representation of the entire G; see Fig. 1 for an illustration. We show that this problem can be solved in polynomial time for the class of *circle graphs*.

---

[*] Supported by ESF Eurogiga project GraDR as GAČR GIG/11/E023.
[**] Supported by Charles University as GAUK 196213.

S. Wismath and A. Wolff (Eds.): GD 2013, LNCS 8242, pp. 131–142, 2013.

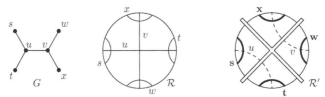

Fig. 1. On the left, a circle graph G with a representation $\mathcal{R}$ is given. A partial representation $\mathcal{R}'$ given on the right with the pre-drawn chords **s**, **t**, **w**, and **x** is not extendible. The chords are depicted as arcs to make the figure more readable.

Circle Graphs. Circle graphs are intersection graphs of chords of a circle. They were first considered by Even and Itai [12] in the early 1970s in study of stack sorting techniques. Other motivations are due to their relations to Gauss words [11] (see Fig. 2) and matroid representations [10,5]. Circle graphs are also important regarding rank-width [22].

Let $\chi(G)$ denote chromatic number of G, and let $\omega(G)$ denote the clique-number of G. Trivially we have $\omega(G) \leq \chi(G)$ and the graphs for which every induced subgraph satisfies equality are the well-known *perfect graphs* [6]. In general, the difference between these two numbers can be arbitrarily high, e.g, there is a triangle-free graph with arbitrary high chromatic number. Circle graphs are known to be *almost perfect* which means that $\chi(G) \leq f(\omega(G))$ for some function f. The best known result for circle graphs [20] states that $f(k)$ is $\Omega(k \log k)$ and $\mathcal{O}(2^k)$.

The complexity of recognition of circle graphs was a long standing open problem; see [25] for an overview. The first results, e.g., [12], gave existential characterizations which did not give polynomial-time algorithms. The mystery whether circle graphs can be recognized in polynomial time frustrated mathematicians for some years. It was resolved in the mid-1980s and several polynomial-time algorithms were discovered [4,13,21] (in time $\mathcal{O}(n^7)$ and similar). Later, a more efficient algorithm [24] based on *split decomposition* was given, and the current state-of-the-art recognition algorithm [14] runs in a quasi-linear time in the number of vertices and the number of edges of the graph.

The Partial Representation Extension Problem. It is quite surprising that this very natural generalization of the recognition problem was considered only recently. It is currently an active area of research which is inspiring a deeper investigation of many classical graph classes. For instance, a recent result of Angelini et al. [1] states that the problem is decidable in linear time for planar

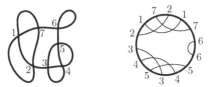

Fig. 2. A self-intersecting closed curve with n intersections numbered $1, \ldots, n$ corresponds to a representation of circle graph with the vertices $1, \ldots, n$ where the endpoints of the chords are placed according to the order of the intersections along the curve.

graphs. On the other hand, Fáry's Theorem claims that every planar graph has a straight-line embedding, but extension of such an embedding is NP-hard [23].

In the context of intersection-defined classes, this problem was first considered in [19] for interval graphs. Currently, the best known results are linear-time algorithms for interval graphs [3] and proper interval graphs [17], a quadratic-time algorithm for unit interval graphs [17], and polynomial-time algorithms for permutation and function graphs [16]. For chordal graphs (as subtree-in-a-tree graphs) several versions of the problems were considered [18] and all of them are NP-complete.

The Structure of Representations. To solve the recognition problem for G, one just needs to build a single representation. However, to solve the partial representation extension problem, the structure of all representations of G must be well understood. A general approach used in the above papers is the following. We first derive necessary and sufficient constraints from the partial representation $\mathcal{R}'$. Then we efficiently test whether some representation $\mathcal{R}$ satisfies these constraints. If none satisfies them, then $\mathcal{R}'$ is not extendible. And if some $\mathcal{R}$ satisfies them, then it extends $\mathcal{R}'$.

It is well-known that the split decomposition [8, Theorem 3] captures the structure of all representations of circle graphs. The standard recognition algorithms produce a special type of representations using split decomposition as follows. We find a *split* in G, construct two smaller graphs, build their representation recursively, and then join these two representations to produce $\mathcal{R}$. In Section 3, we give a simple recursive descriptions of all possible representations based on splits. Our result can be interpreted as "describing a structure like PQ-trees for circle graphs." It is possible that the proof techniques from other papers on circle graphs such as [7,14] would give a similar description. However, these techniques are more involved than our approach which turned out to be quite elementary and simple.

Restricted Representations. The partial representation extension problem belongs to a larger group of problems dealing with *restricted representations of graphs*. These problems ask whether there is some representation of an input graph G satisfying some additional constraints. We describe two examples of these problems.

An input of the *simultaneous representations problem*, shortly SIM, consists of graphs $G_1, \ldots, G_k$ with some vertices common for all the graphs. The problem asks whether there exist representations $\mathcal{R}_1, \ldots, \mathcal{R}_k$ representing the common vertices the same. This problem is polynomially solvable for permutation and comparability graphs [15]. They additionally show that for chordal graphs it is NP-complete when k is part of the input and polynomially solvable for $k = 2$. For interval graphs, a linear-time algorithm is known for $k = 2$ [3] and the complexity is open in general. For some classes, these problems are closely related to the partial representation extension problems. For example, there is an FPT algorithm for interval graphs with the number of common vertices as the parameter [19], and partial representations of interval graphs can be extended in linear time by reducing it to a corresponding simultaneous representations problem [3].

The *bounded representation problem* [17] prescribes bounds for each vertex of the input graph and asks whether there is some representation satisfying these bounds. For circle graphs, the input specifies a pair of arcs (A_v, A'_v) of the circle for each chord v and a solution is required to have one endpoint of v in A_v and the other one in A'_v. This problem is clearly a generalization of partial representation extension since one can describe a partial representation using singleton arcs. It is known to be polynomially solvable for interval and proper interval graphs [2], and surprisingly it is NP-complete for unit interval graphs [17]. The complexity for other classes of graphs is not known.

Our Results. We study the following problem (see Section 2 for definitions):

Problem:	Partial Representation Extension – REPEXT(CIRCLE)
Input:	A circle graph G and a partial representation $\mathcal{R}'$.
Output:	Is there a representation $\mathcal{R}$ of G extending $\mathcal{R}'$?

In Section 3, we describe a simple structure of all representations. This is used in Section 4 to obtain our main algorithmic result:

Theorem 1. *The problem* REPEXT(CIRCLE) *can be solved in polynomial time.*

To spice up our results, we show in the full version of the paper the following.

Proposition 2. *For k part of the input, the problem* SIM(CIRCLE) *is* NP*-complete.*

Corollary 3. *The problem* SIM(CIRCLE) *is* FPT *in the size of the common subgraph.*

2   Definitions and Preliminaries

Circle Representations. A *circle graph representation* $\mathcal{R}$ is a collection of chords a circle $\{C_u \mid u \in V(G)\}$ such that C_u intersects C_v if and only if $uv \in E(G)$. A graph is a *circle graph* if it has a circle representation, and we denote the class of circle graphs by CIRCLE.

Notice that the representation of a circle graph is completely determined by the circular order of the endpoints of the chords in the representation, and two chords C_u and C_v cross if and only if their endpoints alternate in this order. For convenience we label both endpoints of the chord representing a vertex by the same label as the vertex.

A *partial representation* $\mathcal{R}'$ is a representation of an induced subgraph G'. The vertices of G' are *pre-drawn* vertices and the chords of $\mathcal{R}'$ are *pre-drawn chords*. A representation $\mathcal{R}$ *extends* $\mathcal{R}'$ if $C_u = C'_u$ for every $u \in V(G')$.

Word Representations. A sequence τ over an alphabet of symbols Σ is a *word*. A *circular word* represents a set of words which are cyclical shifts of one another. In the sequel, we represent a circular word by a word from its corresponding set of words. We denote words and circular words by small Greek letters.

For a word τ and a symbol u we write $u \in \tau$, if u appears at least once in τ. Thus, τ is also used to denote the set of symbols occurring in τ. A word τ is a *subword* of σ, if τ appears consecutively in σ. A word τ is a *subsequence* of σ, if the word τ can be obtained from σ by deleting some symbols. We say that u *alternates* with v in τ, if $uvuv$ or $vuvu$ is a subsequence of τ. The corresponding definitions also apply to circular words. If σ and τ are two words, we denote their concatenation by $\sigma\tau$.

From now on each representation $\mathcal{R}$ of G corresponds to the unique circular word τ over V. The word τ is obtained by the circular order of the endpoints of the chords in $\mathcal{R}$ as they appear along the circle when traversed clockwise. The occurrences of u and v alternate in τ if and only if $uv \in E(G)$. For example $\mathcal{R}$ in Fig. 1 corresponds to the circular word $\tau = susxvxtutwvw$.

Let G be a circle graph, and let $\mathcal{R}$ be its representation with the corresponding circular word τ. If G' is an induced subgraph of G, then the subsequence of τ consisting of the vertices in G' is a circular word σ. This σ corresponds to a representation $\mathcal{R}'$ of G' which is extended by $\mathcal{R}$.

3   Structure of Representations of Splits

Let G be a connected graph. A *split* of G is a partition of the vertices of G into four parts A, B, $\mathfrak{s}(A)$ and $\mathfrak{s}(B)$, such that:

- For every $a \in A$ and $b \in B$, we have $ab \in E(G)$.
- There is no edge between $\mathfrak{s}(A)$ and $B \cup \mathfrak{s}(B)$, and between $\mathfrak{s}(B)$ and $A \cup \mathfrak{s}(A)$.
- Both sides of the split have at least two vertices: $|A \cup \mathfrak{s}(A)| \geq 2$ and $|B \cup \mathfrak{s}(B)| \geq 2$.

Fig. 3 shows two possible representations of a split. Notice that a split is uniquely determined just by the sets A and B, since $\mathfrak{s}(A)$ consists of connected components of $G \setminus (A \cup B)$ attached to A, and similarly for $\mathfrak{s}(B)$ and B. We refer to this split as a split *between* A and B.

In this section, we examine the recursive structure of every possible representation of G based on splits.

3.1   Split Structure of a Representation

Let $\mathcal{R}$ be a representation of a graph G with a split between A and B. The representation $\mathcal{R}$ corresponds to a unique circular word τ and we consider the

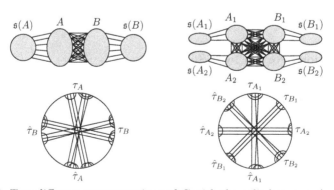

Fig. 3. Two different representations of G with the split between A and B.

circular subsequence γ induced by $A \cup B$. The maximal subwords of γ consisting of A alternate with the maximal subwords of γ consisting of B. We denote all these maximal subwords $\gamma_1, \ldots, \gamma_{2k}$ according to their circular order; so $\gamma = \gamma_1 \gamma_2 \cdots \gamma_{2k}$. Without loss of generality, we assume that γ_1 consists of symbols from A. We call γ_i an A-*word* when i is odd, and a B-*word* when i is even.

We first investigate for each γ_i which symbols it contains.

Lemma 4. *For the subwords $\gamma_1, \ldots, \gamma_k$ the following holds:*
(a) Each γ_i contains each symbol at most once.
(b) The opposite words γ_i and γ_{i+k} contains the same symbols.
(c) Let $i \neq j$. If $x \in \gamma_i$ and $y \in \gamma_j$, then $xy \in E(G)$.

Proof (Sketch). These three properties are easily forced by the structure of the split; see the full version.   □

If $A, B \subset V(G)$ give rise to a split in G, we call the vertices of A and B the *long vertices* with respect to the split between A and B. Similarly the vertices $\mathfrak{s}(A)$ and $\mathfrak{s}(B)$ are called *short vertices* with respect to the split between A and B. In the sequel, if the split is clear from the context, we will just call some vertices long and some vertices short.

Consider a connected component C of $\mathfrak{s}(A)$ (for a component of $\mathfrak{s}(B)$ the same argument applies) and consider the subsequence of τ induced by $A \cup B \cup C$. By Lemma 4(a)-(b) and the fact that no vertex of $\mathfrak{s}(A)$ is adjacent to B, this subsequence almost equals γ. The only difference is that one subword γ_i is replaced by a subword which additionally contains all occurrences of the vertices of C. By accordingly adding the vertices of all components of $\mathfrak{s}(A)$ and $\mathfrak{s}(B)$ to γ, we get τ. Thus, τ consists of the circular subwords $\tau_1, \ldots, \tau_{2k}$ concatenated in this order, where τ_i is obtained from γ_i by adding the components of $\mathfrak{s}(A)$ or $\mathfrak{s}(B)$ attached to it. In particular, we also have the following:

Lemma 5. *If two long vertices $x, y \in A$ are connected by a path having the internal vertices in $\mathfrak{s}(A)$, then x and y belong to the same pair γ_i and γ_{i+k} in any representation.*

Proof. If x and y belong to different subwords γ_i and γ_j, where $i < j$ and $j \neq i, i + k$, of γ, by Lemma 4(a)-(b) any path connecting x and y has an internal vertex adjacent to a vertex of B. However, no vertex in $\mathfrak{s}(A)$ is adjacent to a vertex of B.   □

3.2   Conditions Forced by a Split

Now, we want to investigate the opposite relation. Namely, what can one say about a representation from the structure of a split? Suppose that x and y are two long vertices. We want to know the properties of x and y which force every representation $\mathcal{R}$ to have a subword γ_i of γ containing both x and y.

We define a relation $\sim$ on $A \cup B$ where $x \sim y$ means that x, y has to be placed in the same subword γ_i of γ. This relation is given by two conditions:
(C1) Lemma 4(c) states that if $xy \notin E(G)$, then $x \sim y$, i.e., if x and y are placed in different subwords, then C_x intersects C_y.

(C2) Lemma 5 gives $x \sim y$ when x and y are connected by a non-trivial path with all the inner vertices in $\mathfrak{s}(A) \cup \mathfrak{s}(B)$.

Let us take the transitive closure of $\sim$, which we denote by $\sim$ thereby slightly abusing the notation. Thus, we obtain an equivalence relation $\sim$ on $A \cup B$. Notice that every equivalence class of $\sim$ is either fully contained in A or in B. For the graph in Fig. 3, the relation $\sim$ has four equivalence classes A_1, A_2, B_1 and B_2.

Now, let Φ be an equivalence class of $\sim$. We denote by $\mathfrak{s}(\Phi)$ the set consisting of all the vertices in the connected components of $G \setminus (A \cup B)$ which have a vertex adjacent to a vertex of Φ. Since $\sim$ satisfies (C2), we know that the sets $\mathfrak{s}(\Phi)$ of the equivalent classes of $\sim$ define a partition of $\mathfrak{s}(A) \cup \mathfrak{s}(B)$.

Recognition Algorithms Based on Splits. The splits are used in the current state-of-the-art algorithms for recognizing circle graphs. If a circle graph contains no split, it is called a *prime graph*. The representation of a prime graph is uniquely determined (up to the orientation of the circle) and can be constructed efficiently. There is an algorithm which finds a split between two sets A and B in linear time [9]. In fact, the entire *split decomposition tree* (i.e., the recursive decomposition tree obtained via splits) can be found in linear time. Usually the representation $\mathcal{R}$ is constructed as follows.

We define two graphs G_A and G_B where G_A is a subgraph of G induced by the vertices corresponding to $A \cup \mathfrak{s}(A) \cup \{v_A\}$ where the vertex v_A is adjacent to all the vertices in A and non-adjacent to all the vertices in $\mathfrak{s}(A)$, and G_B is defined similarly for B, $\mathfrak{s}(B)$, and v_B. Then we apply the algorithm recursively on G_A and G_B and construct their representations $\mathcal{R}_A$ and $\mathcal{R}_B$; see Fig. 4. It remains to join the representations $\mathcal{R}_A$ and $\mathcal{R}_B$ in order to construct $\mathcal{R}$.

To this end we take $\mathcal{R}_A$ and replace C_{v_A} by the representation of $B \cup \mathfrak{s}(B)$ in $\mathcal{R}_B$. More precisely, let the circular ordering of the endpoints of chords defined by $\mathcal{R}_A$ be $v_A \tau_A v_A \hat{\tau}_A$ and let the circular ordering defined by $\mathcal{R}_B$ be $v_B \tau_B v_B \hat{\tau}_B$. The constructed $\mathcal{R}$ has the corresponding circular ordering $\tau_A \tau_B \hat{\tau}_A \hat{\tau}_B$. It is easy to see that $\mathcal{R}$ is a correct circle representation of G.

Structure of All Representations. The above algorithm constructs a very specific representation $\mathcal{R}$ of G, and a representation like the one in Fig. 3 on the right cannot be constructed by the algorithm. In what follows we describe a structure of all the representations of G, based on different circular orderings of the classes of $\sim$. First, we show that every representation obtained in this way is a correct representation of G. Second, we prove that every representation $\mathcal{R}$ of G can be constructed like this.

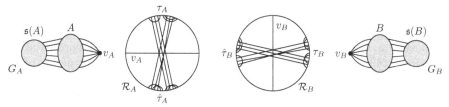

Fig. 4. The graphs G_A and G_B together with some constructed representations $\mathcal{R}_A$ and $\mathcal{R}_B$. By joining these representations, we get the left representation of Fig. 3.

We choose an arbitrary circular ordering $\Phi_1, \ldots, \Phi_\ell$ of the classes of $\sim$. Let G_i be a graph constructed from G by contracting the vertices of $V(G) \setminus (\Phi_i \cup \mathfrak{s}(\Phi_i))$ into one vertex v_i; i.e., G_i is defined similarly to G_A and G_B above. Let $\mathcal{R}_1, \ldots, \mathcal{R}_\ell$ be arbitrary representations of $G_1, \ldots, G_\ell$. We join these representations as follows. Let $v_i \tau_i v_i \hat{\tau}_i$ be the circular ordering of $\mathcal{R}_i$. We construct $\mathcal{R}$ as the circular ordering

$$\tau_1 \tau_2 \ldots \tau_{k-1} \tau_k \hat{\tau}_1 \hat{\tau}_2 \ldots \hat{\tau}_{k-1} \hat{\tau}_k. \tag{1}$$

In Fig. 3, we obtain the representation on the left by the circular ordering $A_1 A_2 B_1 B_2$ of the classes of $\sim$ and the representation on the right by $A_1 B_1 A_2 B_2$.

Lemma 6. *Every circular ordering (1) constructed as above defines a circle representation of G.*

Proof. Every long vertex $u \in \Phi_i$ alternates with v_i in $\mathcal{R}_i$ and every short vertex $v \in \mathfrak{s}(\Phi_i)$ has both occurrences either in τ_i, or in $\hat{\tau}_i$, since it is not adjacent to v_i. Thus, we get a correct representation $\mathcal{R}$ of G.   □

We now show that this approach can construct every representation of G.

Lemma 7. *Let τ be the circular word corresponding to a representation $\mathcal{R}$ of G. Then the symbols of $\Phi_i \cup \mathfrak{s}(\Phi_i)$ form exactly two subwords of τ.*

Proof (Sketch). The conditions (C1) and (C2) imply consecutivity for some pairs x and y. The statement is then proved by applying induction on the classes of $\sim$; see the full version for details.   □

Now, we are ready to prove the main structural proposition, which is inspired by Section IV.4 of the thesis of Naji [21].

Proposition 8. *Let $\sim$ be the equivalence relation defined by (C1) and (C2) on $A \cup B$. Then every representation $\mathcal{R}$ corresponds to some circular ordering $\Phi_1, \ldots, \Phi_\ell$ and to some representations $\mathcal{R}_1, \ldots, \mathcal{R}_\ell$ of $G_1, \ldots, G_\ell$. More precisely, $\mathcal{R}$ can be constructed by arranging $\mathcal{R}_1, \ldots, \mathcal{R}_\ell$ as in (1): $\tau_1 \ldots \tau_k \hat{\tau}_1 \ldots \hat{\tau}_k$.*

Proof. Let $\mathcal{R}$ be any representation with the corresponding circular word τ. According to Lemma 7, we know $\Phi_i \cup \mathfrak{s}(\Phi_i)$ forms two subwords τ_i and $\hat{\tau}_i$ of τ. For $i \neq j$, the edges between Φ_i and Φ_j form a complete bipartite graph. The subwords τ_i, $\hat{\tau}_i$, τ_j and $\hat{\tau}_j$ alternate, i.e., appear as $\tau_i \tau_j \hat{\tau}_i \hat{\tau}_j$ or $\tau_j \tau_i \hat{\tau}_j \hat{\tau}_i$ in τ. Thus, if we start from some point along the circle, the order of τ_i's gives a circular ordering $\Phi_1, \ldots, \Phi_\ell$ of the classes. The representation $\mathcal{R}_i$ is given by the circular ordering $v_i \tau_i v_i \hat{\tau}_i$.   □

4   Algorithm

In this section, we give a polynomial-time algorithm for the partial representation extension problem of circle graphs. Our algorithm is based on the structure of all representations described in Section 3. We assume that the graph is connected and we deal with disconnected graphs in the full version of this paper.

Overview. Let τ' be the circular word corresponding to the given partial rep. $\mathcal{R}'$. We want to extend τ' to a circular word τ corresponding to a rep. $\mathcal{R}$ of G.

Our algorithm proceeds recursively via split decomposition.

1. If G is prime, we have two possible representations (one is reversal of the other) and we test whether one of them is compatible with $\mathcal{R}'$.
2. Otherwise, we find a split and compute the $\sim$ relation.
3. We test whether some ordering $\Phi_1, \ldots, \Phi_\ell$ of these classes along the circle is compatible with the partial representation $\mathcal{R}'$. This order is partially prescribed by short chords and long chords of $\mathcal{R}'$.
4. If no ordering is compatible, we stop and output "no". If there is an ordering which is compatible with $\mathcal{R}'$, we recurse on the graphs $G_1, \ldots, G_\ell$ constructed according to the equivalence classes of $\sim$.

Now we describe everything in detail.

Splits. Now we assume that the graph is not prime, otherwise the problem is easy to solve (details will be given in the full version). A split between A and B is called *trivial* if for one side, let us say A, we have $|A| = 1$ and $|\mathfrak{s}(A)| = 1$. If G contains only trivial splits, then we call it *trivial*. For technical purposes, we assume that the split is non-trivial, again the full version of this paper contains the details.

So we have a non-trivial split between A and B which can be constructed in polynomial time [9]. We compute the equivalence relation $\sim$ and we want to find an ordering of its equivalence classes. For a class Φ of $\sim$, we define the *extended class* Ψ of $\sim$ as $\Phi \cup \mathfrak{s}(\Phi)$. We can assume that each extended class has a vertex pre-drawn in the partial representation, otherwise any representation of it is good. So $\sim$ has ℓ equivalence classes, and all of them appear in τ'.

Now, τ' is composed of k *maximal subwords*, each containing only symbols of one extended class Ψ. We denote these maximal subwords as $\tau'_1, \ldots, \tau'_k$ according to their circular order in τ', so $\tau' = \tau'_1 \cdots \tau'_k$. According to Proposition 8, each extended class Ψ corresponds to at most two different maximal subwords. Also, if two extended classes Ψ and $\hat{\Psi}$ correspond to two different maximal subwords, then occurrences of these subwords in τ' alternate. Otherwise we reject the input.

Case 1: An extended class corresponds to two maximal subwords.
We denote this class by Ψ_1 and put this class as first in the ordering. By renumbering, we may assume that Ψ_1 corresponds to τ'_1 and τ'_t. Then one circular order of the classes can be determined as follows. We have $\Psi_1 < \Psi$ for any other class Ψ. Let Ψ_i and Ψ_j be two distinct classes. If Ψ_i corresponds to τ'_a and Ψ_j corresponds to τ'_b such that either $a < b < t$ or $t < a < b$, we put $\Psi_i < \Psi_j$. We obtain the ordering of the classes as any linear extension of $<$. One can observe that $<$ is acyclic, otherwise the maximal subwords would not alternate correctly.

Now, we have ordered the extended classes $\Psi_1, \ldots, \Psi_\ell$ and the corresponding classes $\Phi_1, \ldots, \Phi_\ell$. We construct each G_i with the vertices $\Psi_i \cup \{v_i\}$ as in Section 3.2, so v_i is adjacent to Φ_i and non-adjacent to $\mathfrak{s}(\Phi_i)$. As the partial representation $\mathcal{R}'_i$ of G_i, we put the word $v_i \tau'_i v_i \tau'_j$ where Ψ_i corresponds to τ_i and τ_j (possibly one of them is empty). We test recursively, whether each representation $\mathcal{R}'_i$ of G_i is extendible to a representation of $\mathcal{R}_i$. If yes, we join $\mathcal{R}_1, \ldots, \mathcal{R}_\ell$ as in Proposition 8. Otherwise, the algorithm outputs "no".

Lemma 9. *For Case 1, the representation $\mathcal{R}'$ is extendible if and only if the representations $\mathcal{R}'_1, \ldots, \mathcal{R}'_\ell$ of the graphs $G_1, \ldots, G_\ell$ are extendible.*

Proof. Suppose that $\mathcal{R}$ extends $\mathcal{R}'$. According to Proposition 8, the representations of $\Psi_1, \ldots, \Psi_\ell$ are somehow ordered along the circle, and so we obtain representations $\mathcal{R}_1, \ldots, \mathcal{R}_\ell$ extending $\mathcal{R}'_1, \ldots, \mathcal{R}'_\ell$.

For the other implications, we just take $\mathcal{R}_1, \ldots, \mathcal{R}_\ell$ and put them in $\mathcal{R}$ together as in (1). The ordering $<$ was constructed exactly in such a way that $\mathcal{R}$ extends $\mathcal{R}'$.   □

Case 2: No extended class corresponds to two maximal subwords

In this case, we have the ordering of the classes according to their appearance in τ', so Ψ_i corresponds to the subword τ'_i. According to Proposition 8, we know that in any representation $\mathcal{R}$ of G the class Ψ_i corresponds to two subwords τ_i and $\hat{\tau}_i$. The difficulty here arises from the potential for τ'_i to be a subsequence of only one of τ_i and $\hat{\tau}_i$.

We solve this as follows. Instead of constructing just one graph G_i with one partial representation $\mathcal{R}'_i$, we construct an additional graph $\widetilde{G}_i$ with a partial representation $\widetilde{\mathcal{R}}'_i$ as follows. The graph $\widetilde{G}_i$ is G_i with an additional leaf w_i attached to v_i. The partial representation $\mathcal{R}'_i$ corresponds to the word $\tau'_i v_i v_i$ and the partial representation $\widetilde{\mathcal{R}}'_i$ corresponds to $\tau'_i w_i w_i$. The difference is that $\widetilde{\mathcal{R}}'_i$ is less restrictive and only one endpoint of v_i is prescribed (i.e., the location of the "other" end of v_i is not restricted). We can easily observe that if $\mathcal{R}'_i$ is extendible, then $\widetilde{\mathcal{R}}'_i$ is also extendible.

The following lemma is fundamental for the algorithm, and it states that at most one class can be forced to use $\widetilde{G}_i$ with $\widetilde{\mathcal{R}}'_i$, if τ' is extendible:

Lemma 10. *The representation $\mathcal{R}'$ is extendible if and only if $\widetilde{\mathcal{R}}'_i$ is extendible for some i and $\mathcal{R}'_j$ is extendible for all $j \neq i$.*

Proof. Suppose that $\mathcal{R}_j$ corresponding to a word $v_j \tau_j v_j \hat{\tau}_j$ is an extension of $\mathcal{R}'_j$ for $j \neq i$. And let $\mathcal{R}_i$ corresponding to a word $w_i v_i w_i \tau_i v_i \hat{\tau}_i$ be an extension of $\widetilde{\mathcal{R}}'_i$. Then the representation $\mathcal{R}$ (after removing w_i) constructed as in (1) extends $\mathcal{R}'$.

For the other implication, suppose that $\mathcal{R}$ extends $\mathcal{R}'$. For contradiction, suppose that two distinct partial representations $\mathcal{R}'_i$ and $\mathcal{R}'_j$ are not extendible. According to Proposition 8, the representation $\mathcal{R}$ gives a representation $\mathcal{R}_i$ corresponding to $v_i \tau_i v_i \hat{\tau}_i$ of G_i and $\mathcal{R}_j$ corresponding to $v_j \tau_j v_j \hat{\tau}_j$ of G_j. But since both Ψ_i and Ψ_j correspond to single maximal words of τ', we have that τ'_i is a subsequence of τ_i or $\hat{\tau}_i$, or τ'_j is a subsequence of τ_j or $\hat{\tau}_j$, and so $\mathcal{R}'_i$ or $\mathcal{R}'_j$ is extendible. Contradiction.   □

For the algorithm, we can efficiently test which of $\mathcal{R}'_i$ and $\widetilde{\mathcal{R}}'_i$ are extendible with the pseudocode of Algorithm 1.

Analysis of the Algorithm. By using the established results, we show that the partial representation extension problem of circle graphs can be solved in polynomial time.

Lemma 11. *The described algorithm correctly decides whether the partial representation $\mathcal{R}'$ of G' is extendible.*

Algorithm 1. Subroutine for Case 2.

---
1. Let Ψ_1 be the largest class (i.e., $|\Psi_i| \leq n/2$ for $i > 1$).
2. **If** $\mathcal{R}'_2, \ldots, \mathcal{R}'_\ell$ are extendible **then**
3.   **If** $\widetilde{\mathcal{R}}'_1$ is extendible **then** ACCEPT **else** REJECT.
4. **Else if** only $\mathcal{R}'_i$ is not extendible **then**
5.   **If** $\widetilde{\mathcal{R}}'_i$ and $\mathcal{R}'_1$ are extendible **then** ACCEPT **else** REJECT.
6. **Else** REJECT.

Proof (Sketch). We just put together the lemmas which are already proved; see the full version for details. □

The next lemma states that the algorithm runs in polynomial time. A precise time analysis depends on algorithm used for split decomposition, and on the order in which we choose splits for recursion. We avoid this technical analysis and just note that the degree of the polynomial is reasonable small. Certainly, it would be easy to show the complexity of order $\mathcal{O}(nm)$.

Lemma 12. *The running time of the algorithm is polynomial.* □

Proof (Theorem 1). The result is implied by Lemma 11 and Lemma 12. □

5   Conclusions

The structural results described in Section 3, namely Proposition 8, are the main new tools developed in this paper. Using it, one can easily work with the structure of all representations which is a key component of the algorithm of Section 4 that solves the partial representation extension problem for circle graphs. The algorithm works with the recursive structure of all representations and matches the partial representation on it. Proposition 8 also seems to be useful in attacking the following open problems:

Question 13. What is the complexity of SIM(CIRCLE) for a fixed number k of graphs? In particular, what is it for $k = 2$?

Recall that in the bounded representation problem, we give for some chords two circular arcs and we want to construct a representation which places endpoints into these circular arcs.

Question 14. What is the complexity of the bounded representation problem for circle graphs?

References

1. Angelini, P., Battista, G.D., Frati, F., Jelínek, V., Kratochvíl, J., Patrignani, M., Rutter, I.: Testing planarity of partially embedded graphs. In: SODA 2010, pp. 202–221 (2010)
2. Balko, M., Klavík, P., Otachi, Y.: Bounded representations of interval and proper interval graphs. In: ISAAC (to appear, 2013)

3. Bläsius, T., Rutter, I.: Simultaneous PQ-ordering with applications to constrained embedding problems. In: SODA 2013, pp. 1030–1043 (2013)
4. Bouchet, A.: Reducing prime graphs and recognizing circle graphs. Combinatorica 7(3), 243–254 (1987)
5. Bouchet, A.: Unimodularity and circle graphs. Discrete Mathematics 66(1-2), 203–208 (1987)
6. Chudnovsky, M., Robertson, N., Seymour, P., Thomas, R.: The strong perfect graph theorem. Annals of Mathematics 164, 51–229 (2006)
7. Courcelle, B.: Circle graphs and monadic second-order logic. J. Applied Logic 6(3), 416–442 (2008)
8. Cunningham, W.: Decomposition of directed graphs. SIAM J. Alg. and Disc. Methods 3, 214–228 (1982)
9. Dahlhaus, E.: Parallel algorithms for hierarchical clustering and applications to split decomposition and parity graph recognition. Journal of Algorithms 36(2), 205–240 (1998)
10. de Fraysseix, H.: Local complementation and interlacement graphs. Discrete Mathematics 33(1), 29–35 (1981)
11. de Fraysseix, H., de Mendez, P.O.: On a characterization of gauss codes. Discrete & Computational Geometry 22(2), 287–295 (1999)
12. Even, S., Itai, A.: Queues, stacks, and graphs. In: Kohavi, Z., Paz, A. (eds.) Theory of Machines and Computation, pp. 71–76 (1971)
13. Gabor, C.P., Supowit, K.J., Hsu, W.: Recognizing circle graphs in polynomial time. J. ACM 36(3), 435–473 (1989)
14. Gioan, E., Paul, C., Tedder, M., Corneil, D.: Practical and efficient circle graph recognition. Algorithmica, 1–30 (2013)
15. Jampani, K.R., Lubiw, A.: The simultaneous representation problem for chordal, comparability and permutation graphs. Journal of Graph Algortihms and Applications 16(2), 283–315 (2012)
16. Klavík, P., Kratochvíl, J., Krawczyk, T., Walczak, B.: Extending partial representations of function graphs and permutation graphs. In: Epstein, L., Ferragina, P. (eds.) ESA 2012. LNCS, vol. 7501, pp. 671–682. Springer, Heidelberg (2012)
17. Klavík, P., Kratochvíl, J., Otachi, Y., Rutter, I., Saitoh, T., Saumell, M., Vyskočil, T.: Extending partial representations of proper and unit interval graphs (in preparation, 2013)
18. Klavík, P., Kratochvíl, J., Otachi, Y., Saitoh, T.: Extending partial representations of subclasses of chordal graphs. In: Chao, K.-M., Hsu, T.-S., Lee, D.-T. (eds.) ISAAC 2012. LNCS, vol. 7676, pp. 444–454. Springer, Heidelberg (2012)
19. Klavík, P., Kratochvíl, J., Vyskočil, T.: Extending partial representations of interval graphs. In: Ogihara, M., Tarui, J. (eds.) TAMC 2011. LNCS, vol. 6648, pp. 276–285. Springer, Heidelberg (2011)
20. Kostochka, A., Kratochvíl, J.: Covering and coloring polygon-circle graphs. Discrete Mathematics 163(1-3), 299–305 (1997)
21. Naji, W.: Graphes de Cordes: Une Caracterisation et ses Applications. PhD thesis, l'Université Scientifique et Médicale de Grenoble (1985)
22. Oum, S.: Rank-width and vertex-minors. J. Comb. Theory, Ser. B 95(1), 79–100 (2005)
23. Patrignani, M.: On extending a partial straight-line drawing. In: Healy, P., Nikolov, N.S. (eds.) GD 2005. LNCS, vol. 3843, pp. 380–385. Springer, Heidelberg (2006)
24. Spinrad, J.P.: Recognition of circle graphs. J. of Algorithms 16(2), 264–282 (1994)
25. Spinrad, J.P.: Efficient Graph Representations. Field Institute Monographs (2003)

On Balanced +-Contact Representations

Stephane Durocher* and Debajyoti Mondal**

Department of Computer Science, University of Manitoba, Canada
{durocher,jyoti}@cs.umanitoba.ca

Abstract. In a +-contact representation of a planar graph G, each vertex is represented as an axis-aligned plus shape consisting of two intersecting line segments (or equivalently, four axis-aligned line segments that share a common endpoint), and two plus shapes touch if and only if their corresponding vertices are adjacent in G. Let the four line segments of a plus shape be its arms. In a c-balanced representation, $c \leq 1$, every arm can touch at most $\lceil c\Delta \rceil$ other arms, where Δ is the maximum degree of G. The widely studied T- and L-contact representations are c-balanced representations, where c could be as large as 1. In contrast, the goal in a c-balanced representation is to minimize c. Let c_k, where $k \in \{2,3\}$, be the smallest c such that every planar k-tree has a c-balanced representation. In this paper we show that $1/4 \leq c_2 \leq 1/3 (= b_2)$ and $1/3 < c_3 \leq 1/2 (= b_3)$. Our result has several consequences. Firstly, planar k-trees admit 1-bend box-orthogonal drawings with boxes of size $\lceil b_k \Delta \rceil \times \lceil b_k \Delta \rceil$, which generalizes a result of Tayu, Nomura, and Ueno. Secondly, they admit 1-bend polyline drawings with $2\lceil b_k \Delta \rceil$ slopes, which is significantly smaller than the 2Δ upper bound established by Keszegh, Pach, and Pálvölgyi for arbitrary planar graphs.

1 Introduction

In a contact representation of a planar graph G, the vertices of G are represented using different non-overlapping geometric shapes (e.g., lines, triangles, or circles) and the adjacencies are represented by the contacts of the corresponding objects. Contact representations arise in many applied fields, such as cartography, VLSI floor-planning, and data visualization, which has motivated extensive research over the past several decades. In this paper we examine +-*contact representations* of planar graphs, i.e., each vertex in such a representation Γ corresponds to an axis-aligned plus shape, two plus shapes never cross, but touch if and only if their corresponding vertices are adjacent in the input planar graph. Let the four orthogonal parts associated with a plus symbol be its *left, right, up* and *down arms*. We call Γ a c-balanced representation, where $c \leq 1$, if every arm in Γ touches at most $\lceil c\Delta \rceil$ other arms, where Δ is the maximum degree of the

* Work of the author is supported in part by the Natural Sciences and Engineering Research Council of Canada (NSERC).
** Work of the author is supported in part by a University of Manitoba Graduate Fellowship.

S. Wismath and A. Wolff (Eds.): GD 2013, LNCS 8242, pp. 143–154, 2013.

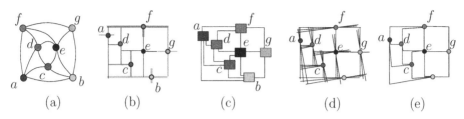

Fig. 1. (a) A graph G with $\Delta = 5$. (b) A (1/2)-balanced $\maltese$-contact representation of G. (c) A box-orthogonal drawing of G. (d)–(e) A transformation into a polyline drawing.

underlying graph. The horizontal (or vertical) segments of two touching plus shapes in Γ may be collinear, e.g., the shapes representing the vertices e and g in Figures 1(a)–(b).

In 1994, de Fraysseix et al. [1] gave an algorithm to construct contact representations of planar graphs with axis-aligned T shapes. Many studies followed to characterize classes of planar graphs that admit contact representations with shapes simpler than T, such as axis-aligned segments [2]and L shapes [3]. L- and T-contact representations can be viewed as c-balanced $\maltese$-contact representations, however, c may be required to be as large as 1. On the other hand, in a c-balanced representation, our goal is to minimize c.

Box-Orthogonal Drawings with Small Boxes of Constant Aspect Ratio. Balanced $\maltese$-contact representations are useful in the study of box-orthogonal drawings in $\mathbb{R}^2$. A *k-bend box-orthogonal drawing* of a planar graph G is a planar drawing of G, where each vertex is represented as an axis-aligned box and each edge is drawn as an orthogonal polygonal chain with at most k bends. Every $\maltese$-contact representation can be transformed into a box-orthogonal drawing [4], as shown in Figure 1(c). Some important aesthetics of a box-orthogonal drawing are the number of bends per edge, and the aspect ratio and size of the boxes. Biedl and Kaufmann [4] showed that every planar graph admits a 1-bend box-orthogonal drawing on an integer grid, but the width or height of a box in such a drawing could be as large as Δ. A c-balanced $\maltese$-contact representation implies a 1-bend box-orthogonal drawing with boxes of size $\lceil c\Delta \rceil \times \lceil c\Delta \rceil$.

Orthogonal drawings are box-orthogonal drawings with boxes of degenerate shapes, i.e., points. The graphs that admit orthogonal drawings are of maximum degree four. Hence a 0- and 1-bend orthogonal drawing gives a (1/4)-balanced $\maltese$-contact representation. There have been several attempts in the literature to characterize the graphs that admit 0- and 1-bend orthogonal drawings [5,6]. Recently, Tayu, Nomura, and Ueno [7] showed that every 2-tree with maximum degree four admits a 1-bend orthogonal drawing. In this paper we show that 2-trees and planar 3-trees admit (1/3)- and (1/2)-balanced $\maltese$-contact representations, respectively, and thus admit 1-bend box-orthogonal drawings with boxes of size $\lceil \Delta/3 \rceil \times \lceil \Delta/3 \rceil$ and $\lceil \Delta/2 \rceil \times \lceil \Delta/2 \rceil$, respectively.

Planar Slope Number with One Bend Per Edge. A *k-bend polyline drawing* of a planar graph G is a planar drawing Γ of G, where each vertex is represented as a point and each edge is drawn as a polygonal chain with at most k bends. Γ is a t-*slope drawing* of G if the number of distinct slopes used by the line segments in Γ is at most t. The *planar slope number* of G is the smallest number t such that G admits a t-slope 0-bend drawing. A rich body of literature examines planar slope number of different subclasses of planar graphs [8,9,10]. Keszegh et al. [11] proved a q^Δ upper bound on the planar slope number, where q is a constant. They also showed that every planar graph G admits a 1-bend polyline drawing with at most 2Δ slopes, by a transformation from T-contact representations into 1-bend polyline drawings, as follows. Replace each vertical (respectively, horizontal) arm with Δ closely spaced nearly vertical (respectively, horizontal) slopes, e.g., see Figure 1(d). Finally, choose the bend points from the intersection points of these slopes such that the resulting drawing remains planar, e.g., see Figure 1(e). In this paper we show that 2-trees and planar 3-trees admit (1/3)- and (1/2)-balanced ✚-contact representations, respectively, and thus admit 1-bend polyline drawings with at most $2\lceil \Delta/3 \rceil$ slopes, and $2\lceil \Delta/2 \rceil$ slopes, respectively.

2   Definitions and Preliminary Approach

In this section we introduce some definitions and construct (1/2)-balanced ✚-contact representations for 2-trees.

A 2-*tree*, or *series-parallel graph* (SP graph) G is a two-terminal directed simple graph with $n \geq 2$ vertices, which is defined recursively as follows.

(a) If $n = 2$, then G has a single edge (u, v), where either u or v is the source and the other vertex is the sink.
(b) If $n > 2$, then G can be constructed from two SP graphs G_1 and G_2 from one of the following two operations, e.g., see Figure 2(a).
    - *Series Composition*: Identify the sink of G_1 with the source of G_2.
    - *Parallel Composition*: Identify the source and sink of G_1 with the source and sink of G_2, respectively. Finally, identify any parallel edges.

A c-balanced representation of a given SP graph G can be constructed as follows. Construct a rectangle R and place the source and sink of G at the top-left and bottom-right corners, respectively. Initially each edge of R can have $\lceil c\Delta \rceil$ contact points. If G is formed by a series composition of two SP graphs G_1 and G_2, then we split R into four rectangles, e.g., see Figure 2(b), and draw G_1 and G_2 into the top-left and bottom-right rectangles, respectively. If G is formed by a parallel composition of G_1 and G_2, then we take two copies R_i, $i \in \{1, 2\}$, of R and draw G_i inside R_i (later on we merge these two drawings inside R). In both series and parallel cases, we distribute the available contact points among the subproblems, i.e., we compute the recursive drawings with bounded number of contact points on the edges of their bounding rectangles. In our algorithms, we specify the distribution of contact points so that we can merge the recursively computed drawings maintaining planarity.

Let h be an arm of some vertex while constructing a ✚-contact representation. By the *number of free points* of h we refer to the number of other arms that can touch h, which we denote by $f(h)$. If $f(h) = 0$, then we say h is *saturated*, otherwise h is *unsaturated*. The *center* of a vertex is the point, at which all four of its arms meet. For a center m, we denote by m_l, m_r, m_u, m_d the left, right, up and down arms of m, respectively. *Distributing* an integer z among the arms of m in some order $\sigma = (m_d, m_r, m_u, m_l)$ is an operation that finds the first arm h such that $z \leq \sum_{h' \leq_\sigma h} f(h')$, then sets $f(h) = z - \sum_{h' <_\sigma h} f(h')$, and finally, for all arms h'' subsequent to h, sets $f(h'') = 0$. Such an operation is defined only when $z \leq \sum_h f(h)$. By $d_i(v, G)$ and $d_o(v, G)$ we denote the in-degree and out-degree of vertex v in G. We omit the term G if it is clear from the context. We now present a construction of $(1/2)$-balanced representations of SP graphs.

Lemma 1. *Let G be a SP graph with source s and sink t, and let G' be the graph obtained from G by deleting the edge (s, t), if such an edge exists. Let $R = abcd$ be an axis-parallel rectangle such that s and t lie on the opposite corners a and c, respectively. Assume that $f(a_d), f(a_r), f(c_l), f(c_u)$ are prespecified. If $d_o(s, G') \leq f(a_d) + f(a_r)$ and $d_i(t, G') \leq f(c_l) + f(c_u)$, then G' admits a $(1/2)$-balanced $\textbf{+}$-contact representation Γ in R satisfying the following property.*

($\star$) *The number of contact points at each arm incident to s and t in Γ is at most the number of free points specified for that arm as input.*

Proof. We employ an induction on the number of vertices n of G. If $n = 2$, then G' consists of two vertices of degree zero, i.e., s and t, that lie on the two opposite corners a and c of R, respectively. It is now straightforward to verify Property ($\star$). Hence assume that $n > 2$, and the lemma holds for every G that has fewer than n vertices. We now consider the case when G has n vertices.

Since G is a SP graph and $n > 2$, G' must be a SP graph, i.e., G' is obtained either by a series combination or a parallel composition of some SP graphs G_1 and G_2. Let s_i and t_i be the source and sink of G_i, respectively, where $i \in \{1, 2\}$. We now consider two cases depending on the composition of G_1 and G_2 in G'.

Case 1 (Series Composition): In this case $s = s_1$, $t_1 = s_2$ and $t_2 = t$. We first define two rectangles R_1 and R_2 inside R, where G_1 and G_2 will be drawn, respectively. To construct R_1 and R_2 we first add a vertex r inside R, which corresponds to the center of vertex $t_1(= s_2)$. We then draw four orthogonal line segments re, rm, rg, rh such that $e \in ab, m \in bc, g \in cd, h \in ad$. Then $R_1 = aerh$ and $R_2 = rmcg$, as in Figure 2(b). We set $f(r_l) = f(r_r) = f(r_u) = f(r_d) = \lceil \Delta/2 \rceil$, and then assign the free points of s and t to s_1 and t_2, respectively.

If the edge (s_1, t_1) exists, then we draw (s_1, t_1) either along the polygonal chain ahr or aer, depending on whether $f(a_d) = 0$ or not. If the edge (s_2, t_2) exists, then we draw (s_2, t_2) either along the polygonal chain rgc or rmc, depending on whether $f(c_l) = 0$ or not. Here we consider the case when both (s_1, t_1) and (s_2, t_2) exist (the other cases can be treated similarly). Figure 2(c) shows such an example, where $f(a_d) \neq 0$ and $f(c_l) = 0$. Observe that while drawing (s_1, t_1) and (s_2, t_2), we use some free points of s_1 and t_2. Therefore, we decrease the free points by one for each arm that helps routing (s_1, t_1) and (s_2, t_2), e.g., see Figures 2(d)–(e). Since $d_o(s, G') \leq f(a_d) + f(a_r)$ in R, we have $d_o(s_1, G_1 \setminus (s_1, t_1)) \leq f(a'_d) + f(a'_r)$, where a' represents a in R_1. Since $d_i(t_1, G_1 \setminus (s_1, t_1)) \leq \Delta - 1$, we have $d_i(t_1, G_1 \setminus (s_1, t_1)) \leq f(r_u) + f(r_l)$ in R_1. Therefore, we can

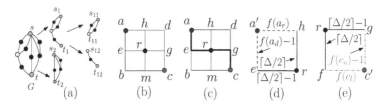

Fig. 2. (a) A few steps of series-parallel decomposition for some graph G, according to the definition. (b) Computation of R_1 and R_2. (c) Drawing (s_1, t_1) and (s_2, t_2). (d)–(e) Illustration for free points.

inductively draw G_i inside R_i, $i \in \{1, 2\}$. It is straightforward to merge these drawings by appropriate scaling. Since the drawings inside R_i maintain Property $(\star)$, the merged drawing also satisfies that property.

Case 2 (Parallel Composition): In this case $s = s_1 = s_2$ and $t = t_1 = t_2$. We first create two copies R_1 and R_2 of R, i.e., $R_1 = a'b'c'd'$ and $R_2 = a''b''c''d''$, where G_1 and G_2 will be drawn, respectively. Figure 3 illustrates an example.

We now define the free points of the arms of s_1, t_1 and s_2, t_2 that are inside R_1 and R_2, respectively. We distribute $d_o(s_1)$ among a'_d and a'_r in this order, i.e., we set $f(a'_d) = \min\{f(a_d), d_o(s_1)\}$, and $f(a'_r) = \max\{0, d_o(s_1) - f(a'_d)\}$. Similarly, distribute $d_i(t_1)$ among c'_l and c'_u in this order, i.e., set $f(c'_l) = \min\{f(c_l), d_i(t_1)\}$, and $f(c'_u) = \max\{0, d_i(t_1) - f(c'_l)\}$, e.g., Figure 3(b). The number of free points of s_2 and t_2 is the number of free points of s and t that remains after assigning free points to s_1 and t_1, as shown in Figure 3(c).

Since $d_o(s) \leq f(a_d) + f(a_r)$ and $d_o(s) = d_o(s_1) + d_o(s_2)$, according to our assignment of free points, $d_o(s_1) \leq f(a'_d) + f(a'_r)$. Similarly, since $d_i(t) \leq f(c_l) + f(c_u)$ and $d_i(t) = d_i(t_1) + d_i(t_2)$, we obtain $d_i(t_1) \leq f(c'_l) + f(c'_u)$. It is now straightforward to observe that $d_o(s_2) \leq f(a''_d) + f(a''_r)$ and $d_i(t_2) \leq f(c''_l) + f(c''_u)$. Therefore, by induction, can draw G_1 and G_2 inside R_1 and R_2, respectively.

The drawing of G_1 takes consecutive free points from the arms of s (respectively, t) in anticlockwise (respectively, clockwise) order. The drawing of G_2 takes the remaining consecutive free points in the same order. Therefore, one can merge the two drawings inside R_1 and R_2 avoiding edge crossings inside R. The details are omitted due to space constraints. □

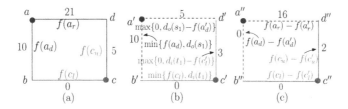

Fig. 3. (a) R. (b)–(c) Computation of R_1 and R_2. The numbers exterior to the three rectangles illustrate a concrete example, where $d_o(s_1) = 15$ and $d_i(t_1) = 3$.

Theorem 1. *Every SP graph G has a $(1/2)$-balanced ✚-contact representation.*

Proof. If the source s and sink t of G are not adjacent, then by Lemma 1, G admits the required representation. Otherwise, let G' be the graph $G \setminus (s, t)$. By Lemma 1, G' admits a $(1/2)$-balanced ✚-contact representation inside a rectangle $R = abcd$, where s and t lie on the opposite corners a and c, respectively, and the free points $f(a_d), f(a_r), f(c_l), f(c_u)$ are prespecified such that $d_o(s, G') \leq f(a_d) + f(a_r)$ and $d_i(t, G') \leq f(c_l) + f(c_u)$.

We define $f(a_d) = \lceil \Delta/2 \rceil - 1, f(a_r) = \lceil \Delta/2 \rceil, f(c_l) = \lceil \Delta/2 \rceil - 1$ and $f(c_u) = \lceil \Delta/2 \rceil$. Since $d_o(s, G') \leq \Delta - 1$ and $d_i(t, G') \leq \Delta - 1$, the conditions $d_o(s, G') \leq f(a_d) + f(a_r)$ and $d_i(t, G') \leq f(c_l) + f(c_u)$ hold. Hence by Lemma 1, we can compute a $(1/2)$-balanced ✚-contact representation Γ' of G' inside R. Finally, we draw the edge (s, t) along the polygonal chain abc. □

3   Balanced Representations for 2-Trees $(c = 1/3)$

The idea of the algorithm for computing $(1/3)$-balanced ✚-contact representations is similar to that of Section 2, however, here the construction is more involved. Let $\overline{uv}$ denote the line segment from u to v. We first prove the following lemma, which is similar to Lemma 1.

Lemma 2. *Let G be a SP graph with source s and sink t, and let G' be the graph obtained from G by deleting the edge (s, t), if such an edge exists. Let $R = k_1 k_2 k_3 k_4$ be an axis-parallel rectangle such that s and t are centered at $a \in \overline{k_1 k_2}$ and $c \in \overline{k_2 k_3}$, respectively, but not at k_2. Assume that the free points of the arms of s and t that lie on R are prespecified. Let x (respectively, y) be the total number of free points of all arms of s (respectively, t) that lie inside R. If $d_o(s, G') \leq x$ and $d_i(t, G') \leq y$, then G' admits a $(1/3)$-balanced ✚-contact representation Γ in R satisfying the following property.*

(✭) *The number of contact points at each arm incident to s and t in Γ is at most the number of free points specified for that arm as input.*

Proof. We employ an induction on the number of vertices n of G. The case when $n = 2$ is straightforward, hence we now assume that $n > 2$, and the lemma holds for every G that has fewer than n vertices. We now consider the case when G has n vertices. Since G is a SP graph and $n > 2$, G' must be a SP graph, i.e., G' is obtained either by a series or a parallel composition of some SP graphs G_1 and G_2. Let s_j and t_j be the source and sink of G_j, respectively, where $j \in \{1, 2\}$. We consider two cases depending on the composition of G_1 and G_2 in G'.

Case 1 (Series Composition): We first construct two rectangular regions R_1 and R_2 inside R, where G_1 and G_2 will be drawn, respectively, and then define the free points. In the following we construct R_1 and R_2 assuming that

$d_i(t_1) \geq 2\lceil \Delta/3 \rceil$. Therefore, we ensure that three of the arms of t_1 lie in R_1 and one of the arms of s_2 lies in R_2. The case when $d_i(t_1) < \lceil \Delta/3 \rceil$ (i.e., $d_o(s_2) \geq 2\lceil \Delta/3 \rceil$) is symmetric. By slightly modifying the construction[1] we can deal with the case when $\lceil \Delta/3 \rceil \leq d_i(t_1) < 2\lceil \Delta/3 \rceil$. We omit the details due to space constraints.

A. Determine the leftmost arm h in the sequence a_d, a_r, a_u that is not saturated.
B. Determine the leftmost arm h' in the sequence c_l, c_u, c_r that is not saturated.
C. If h and h' lie on the boundary of R, then we compute R_1 and R_2 according to the cases (C_1)–(C_3). Figure 4 shows that the case analysis is exhaustive by examining all possible positions of a and c in R. In Figure 4, the point r corresponds to the center of $t_1 (= s_2)$.

(C_1) If h is parallel to h' and $h = a_r$ (i.e., Column 3 of Row 1 in Figure 4), then we draw a straight line pq such that p, q are two points on h, h', respectively. Let r and r' be two distinct points on pq such that $\text{dist}(p, r) < \text{dist}(p, r')$. We then draw a line segment $r'z \perp pq$, such that $z \in c_u$. R_1 and R_2 are the rectangles that contain the unsaturated arms, i.e., in this case R_1 (respectively, R_2) is the rectangle with diagonal $r'k_4$ (respectively, $r'c$). Sometimes R_i, $i \in \{1, 2\}$, may not contain the center of the corresponding source and sink. In such a case, we add a dummy copy of the source or sink, e.g., see the gray diamond shapes in Figure 4. Note that while computing the drawing of G_1 inside R_1 inductively, we rotate R_1 by $90°$ anticlockwise such that the preconditions of the induction hold. Furthermore, we define $f(s_1z) = 0$ such that no unnecessary adjacencies are created in the inductive drawing. Since the addition of dummy copy of a source or sink is straightforward, we do not explicitly describe them in the subsequent cases.

(C_2) If h is parallel to h' and $h \in \{a_d, a_u\}$ (i.e., Column 2 of Row 1 and Columns 1–2 of Row 2 in Figure 4), then we draw a straight line pq such that p, q are two points on h, h', respectively. Let r be a point on pq. We then draw a line segment $rz \perp pq$, such that $z \in c_l$.

(C_3) Otherwise, $h \perp h'$ (i.e., Columns 1 and 4 of Row 1, Columns 3–4 of Row 2, and Row 3 in Figure 4). Here we draw a polygonal chain p, r, q such that p, q are two points on h, h', respectively, $pr \perp rq$. We then draw a line segment $rz \perp rq$, such that either $z \in k_2k_3$ (when rq is horizontal), or $z \in k_3k_4$ (when rq is vertical).

An interesting case is shown in Column 2 of Row 3 in Figure 4, where the dummy vertex is placed in the proper interior of the segment qk_3 instead of placing it on q. The reason is to respect the precondition of the induction that s_2 and t_2 should not lie on q. Here we set the free points of the left and up arms of t_2 to 0 to avoid any unnecessary adjacencies in the recursive construction.

---

[1] Here we ensure that at least two arms of t_1 (respectively, s_2) lie in R_1 (respectively, R_2). At most one arm of t_1 on the boundary of R_1 may coincide with an arm of s_2 on the boundary of R_2, where we assign the free points depending on $d_i(t_1)$.

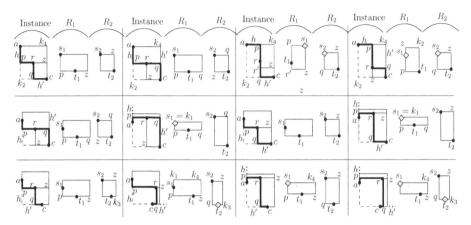

Fig. 4. Computation of R_1 and R_2 when h and h' lie on the boundary of R, and $d_i(t_1) \geq \lceil \Delta/3 \rceil$. **(Case $a = k_1$, $c = k_3$):** Row 1. **(Case $a \neq k_1$, $c = k_3$):** Row 2. **(Case $a = k_1$, $c \neq k_3$):** Symmetric to Row 2. **(Case $a \neq k_1$, $c \neq k_3$):** Row 3.

D. Otherwise, at least one of h, h' is in the proper interior of R. In this scenario we consider the following cases depending on the positions of a and c in R.
   (D_1) If $a \neq k_1$ and $c = k_3$ (i.e., Row 1 of Figure 5), then we follow (C_2) or (C_3), depending on whether $h \| h'$ or $h \perp h'$, setting $p = r$.
   (D_2) If $a = k_1$ and $c \neq k_3$, then the computation is symmetric to (D_1).
   (D_3) Otherwise, both h and h' may lie in the proper interior of R. In this case, if h' lies on the boundary of R, then the computation of R_1 and R_2 is shown in Row 2 of Figure 5. Otherwise, h' lies in the proper interior of R, and the computation of R_1 and R_2 depends on whether $f(c_r) \neq 0$ (i.e., see Row 3 of Figure 5) and $f(c_r) = 0$ (i.e., Row 4 in Figure 5). The details are omitted due to space constraints.

Computation of free points: If R_2 contains an arm of r that does not lie on the boundary of R_1, then we set $f(r_l) = f(r_r) = f(r_u) = f(r_d) = \lceil \Delta/3 \rceil$. Otherwise, R_2 contains only one arm of r and it is shared with R_1, i.e., Columns 3–4 of Row 1 in Figure 4, and Row 4 of Figure 5. In such a case, we assign $\lceil \Delta/3 \rceil$ free points to the arms of r that are not shared, and for the shared arm we assign $d_o(s_2, G_2)$ free points in R_2 and $\lceil \Delta/3 \rceil - d_o(s_2, G_2)$ free points in R_1.

We now assign the free points of s and t to s_1 and t_2, respectively, and place $t_1(= s_2)$ on r. If the edge (s_i, t_i) exists, $i \in \{1, 2\}$, then we draw (s_i, t_i) along h and h', as shown in bold in Figures 4 and 5. Observe that while drawing (s_i, t_i), we use some free points of s_1 and t_2. Therefore, we decrease the free points by one for each arm that helps routing (s_i, t_i). Since R_1 includes all unsaturated arms of s that lie in R, the number of free points of a in R_1 is at least $d_o(s_1, G_1 \setminus (s_1, t_1))$. According to our assignment of free points, the number of free points of r in R_1 is at least $d_i(t_1, G_1 \setminus (s_1, t_1))$. Therefore, we can inductively draw G_1 inside R_1, and similarly G_2 inside R_2.

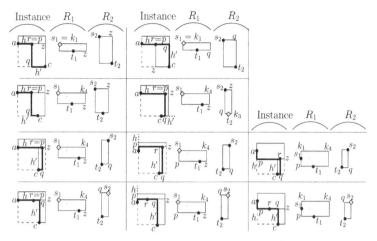

Fig. 5. Computation of R_1 and R_2 when at least one of h, h' lie in the proper interior of R, and $d_i(t_1) \geq \lceil \Delta/3 \rceil$. (**Case** $a \neq k_1$ **and** $c = k_3$): Row 1. (**Case** $a = k_1$ **and** $c \neq k_3$): Symmetric to Row 1. (**Case** $a \neq k_1$ **and** $c \neq k_3$): Rows 2–4.

While computing the drawings of G_1 and G_2 inductively, sometimes we rotated R_1 and R_2 anticlockwise. Therefore, before merging such a drawing, we rotate it clockwise by the same amount. Furthermore, while computing the drawings of G_1 and G_2, sometimes we added some dummy source and sink. For any arm h of the dummy vertex, that is not a part of the arm of its real copy (e.g., see the illustration in Case (C_1)), we set $f(h) = 0$. Therefore, the merged drawing correctly realizes all adjacencies. Since the drawings inside R_1 and R_2 maintains Property (✦), the merged drawing also satisfies that property.

Case 2 (Parallel Composition): In this case $s = s_1 = s_2$ and $t = t_1 = t_2$. We first create two copies R_1 and R_2 of R, i.e., $R_1 = a'b'c'd'$ and $R_2 = a''b''c''d''$, where G_1 and G_2 will be drawn, respectively. We now define the free points. Recall that in Case 2 of Lemma 1, we distributed $d_o(s_1)$ among $f(a'_d), f(a'_r)$, and $d_i(t_1)$ among $f(c'_l), f(c'_u)$. Since here we may have at most three arms of s and t inside R, we distribute $d_o(s_1)$ among a'_d, a'_r and a'_u in this order, and similarly, distribute $d_i(t_1)$ among c'_l, c'_u and c'_r in this order. The number of free points in the arms of s_2 and t_2 is determined by the free points of s and t that remains after assigning free points to s_1 and t_1.

According to our assignment of free points, $d_o(s_1) = f(a'_d) + f(a'_r) + f(a'_u)$. Since $d_o(s) \leq f(a_d) + f(a_r) + f(a_u)$ and $d_o(s) = d_o(s_1) + d_o(s_2)$, the inequality $d_o(s_2) \leq f(a''_d) + f(a''_r) + f(a''_u)$ holds. Similarly, $d_i(t_1) = f(c'_l) + f(c'_u) + f(c'_r)$ and $d_i(t_2) \leq f(c''_l) + f(c''_u) + f(c''_r)$. Therefore, by induction, we can draw G_1 and G_2 inside R_1 and R_2, respectively.

The idea of merging the drawings of G_1 and G_2 into R is similar to the Case 2 of Lemma 1. Observe that the drawing of G_1 takes consecutive free points from the arms of s (respectively, t) in anticlockwise (respectively, clockwise) order. On the other hand, the drawing of G_2 takes the remaining consecutive free points from the arms of s (respectively, t) in anticlockwise (respectively, clockwise) order. Therefore, one can merge the two drawings inside R_1 and R_2 avoiding edge crossings inside R. Since the drawings inside R_1 and R_2 maintains Property ($\star$), the combined drawing also satisfy that property.   □

Theorem 2. *Every SP graph G has a $(1/3)$-balanced $\bf{+}$-contact representation, but not necessarily a $(1/4 - \epsilon)$-balanced representation, for any $\epsilon > 0$.*

Proof. The proof for the upper bound is analogous to the proof of Theorem 1. The only difference is that here we use Lemma 2 instead of Lemma 1. The proof for the lower bound is implied by SP graphs with $\Delta \geq 4$ and $\Delta \bmod 4 = 0$.   □

4   Balanced Representations of Planar 3-Trees ($c = 1/2$)

In this section we show that planar 3-trees admit $(1/2)$-balanced representations. A *planar 3-tree* G with $n \geq 3$ vertices is a triangulated planar graph such that if $n > 3$, then G contains a vertex whose deletion yields a planar 3-tree with $n - 1$ vertices. Let x, y, z be a cycle in G. By G_{xyz} we denote the subgraph induced by x, y, z and the vertices that lie interior to the cycle. Every planar 3-tree G with $n > 3$ vertices contains a vertex that is the common neighbor of all three outer vertices of G. We call this vertex the *representative vertex* of G. Let p be the representative vertex of G and let a, b, c be the three outer vertices of G, as in Figure 6(a). The subgraphs G_{abp}, G_{bcp} and G_{cap} are planar 3-trees. Let G'_{abp}, G'_{bcp} and G'_{cap} be the subgraphs obtained by deleting the outer edges of G_{abp}, G_{bcp} and G_{cap}, respectively. These subgraphs the three *nested components* of G. By $d(u, G)$, we denote the degree of vertex u in G. Given a planar 3-tree G and a rectangle R, we recursively divide R into three sub-rectangles where the nested components of G will be drawn. We first prove the following lemma.

Lemma 3. *Let G be a planar 3-tree with outer vertices a, b, c and representative vertex p, and let G' be the graph obtained from G by deleting the outer edges of G. Let $R = k_1 k_2 k_3 k_4$ be an axis-parallel rectangle such that a, b, c lie on $k_1 k_2$, $k_2 k_3$ and k_4, respectively. Assume that the number of free points of each arm of a, b, c is prespecified. If the inequalities $d(a, G') \leq f(a_d) + f(a_u)$, $d(b, G') \leq f(b_l) + f(b_r)$, and $d(c, G') \leq f(c_l) + f(c_d)$ hold, then G' admits a $(1/2)$-balanced $\bf{+}$-contact representation Γ in R satisfying the following property.*

($\maltese$) *The number of contact points at each arm incident to a, b and c in Γ is at most the number of free points specified for that arm as input.*

Proof. We employ an induction on the number of vertices of G. If $n = 3$, then G' consists of only three isolated vertices a, b and c that lie on $k_1 k_2$, $k_2 k_3$ and k_4, respectively. It is now straightforward to verify Property ($\maltese$). Hence we assume that $n \geq 4$ and the lemma holds for all G with smaller than n vertices. We now consider the case when G has n vertices.

   We first compute three sub-rectangles R_1, R_2 and R_3, where G'_{abp}, G'_{bcp} and G'_{cap} will be drawn, respectively. Define h to be either a_u or a_d depending on whether $d(a, G'_{abp}) \geq f(a_d)$ or not. Similarly, define h' to be either b_r or b_l

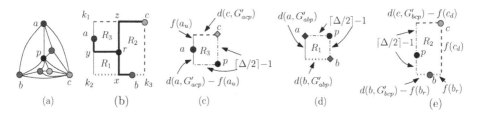

Fig. 6. (a) A plane 3-tree. (b) Computation of R_1, R_2 and R_3, where $d(a, G'_{abp}) < f(a_d)$, $d(b, G'_{abp}) < f(b_l)$, and $d(c, G'_{bcp}) \geq f(c_d)$. (c)–(e) Assignment of free points.

depending on whether $d(b, G'_{abp}) \geq f(b_l)$ or not. Since the inequalities $f(a_d) + f(a_u) \geq d(a, G') \geq 1$ and $f(b_l) + f(b_r) \geq d(b, G') \geq 1$ hold, h and h' must be unsaturated. Let x and y be two points on h' and h, respectively, as shown in Figure 6(b). Draw two line segments $xr \perp h'$ and $yr \perp h$ such that they meet at point r. Define h'' to be the arm c_l or c_d depending on whether $d(c, G'_{bcp}) \geq f(c_d)$ or not. Since $f(c_l) + f(c_d) \geq d(c, G') \geq 1$, h'' must be unsaturated. We draw an orthogonal line segment rz such that $z \in h''$. Observe that rx, ry and rz divides R into three sub-rectangles R_1, R_2 and R_3, i.e., the sub-rectangles that contain corners k_2, k_3 and k_1, respectively.

We place the vertex p on r, draw the edges $(a, p), (b, p)$ and (c, p) along ry, rx and rz, respectively, and then assign $\lceil \Delta/2 \rceil - 1$ free points at each arm of r. To define the free points of the other arms of R_i, we distribute the free points of a, b and c as follows. We distribute $d(a, G'_{abp})$ among a_d and a_u (in R_1), and $d(a, G'_{acp})$ among a_u and a_d (in R_3). We then distribute $d(b, G'_{abp})$ among b_l and b_r (in R_1), and $d(b, G'_{bcp})$ among b_r and b_l (in R_2). Finally, we distribute $d(c, G'_{bcp})$ among c_d and c_l (in R_2), and $d(c, G'_{acp})$ among c_l and c_d (in R_3).

Let G'_i be the nested component of G that corresponds to R_i, $i \in \{1, 2, 3\}$. Observe that some outer vertices of G'_i may not lie on R_i. Hence we cannot directly apply the induction hypothesis. Hence for each vertex a, b or c that does not lie on the boundary of R_i but belongs to G'_i, we add a dummy copy of that vertex at x, y or z, respectively. Furthermore, for each arm h of the dummy copy that is not a part of any arm of its real copy, we set $f(h) = 0$, e.g., $f(c_d) = 0$ in Figure 6(c). Consequently, the recursively computed drawings do not create any unnecessary adjacencies. Observe that each R_i now meets the preconditions of the induction, as shown in Figures 6(c)–(e), and hence we inductively draw G'_i inside R_i. To apply the induction, we need to be careful of the vertex that play the role of k_4, i.e., the corner having exactly two arms inside the rectangle that are perpendicular to each other, e.g., the position of p in Figures 6(c)–(d), and the position of c in Figure 6(e). Each R_i contains exactly one of r and c at one of its four corners, which plays the role of k_4 in R_i. Since the smaller drawings satisfy Property (✱), the final drawing satisfies Property (✱). □

Theorem 3. *Every planar 3-tree G admits a $(1/2)$-balanced +-contact representation, but not necessarily a $(1/3)$-balanced representation.*

Proof. Let a, b, c be the outer vertices of G, and let $R = k_1 k_2 k_3 k_4$ be an axis-parallel rectangle. Place a, b, c on $k_1 k_2$, $k_2 k_3$ and k_4, respectively, draw the outer edges of G along the boundary of R, and finally, assign $\lceil \Delta/2 \rceil - 1$ free points at each arm of a, b and c. Let G' be the graph obtained by removing the outer edges of G. By Lemma 3, G' has a $(1/2)$-balanced ✛-contact representation in R. The lower bound that $c > 1/3$ is implied by the graph K_4.   □

5   Conclusion

We have proved that 2-trees (respectively, planar 3-trees) admit c-balanced ✛-contact representations, where $1/4 \le c \le 1/3$ (respectively, $1/3 < c \le 1/2$). A natural open question is to find tight bounds on c. Although our representations for planar 3-trees preserve input embedding, the representations for 2-trees do not have this property. Thus it would be interesting to examine whether there exist algorithms for $(1/3)$-balanced representations of 2-trees that preserve input embedding. Another intriguing open question is to characterize planar graphs that admit c-balanced ✛-contact representations, for small fixed values c.

References

1. de Frayseix, H., de Mendez, P.O., Rosenstiehl, P.: On triangle contact graphs. Combinatorics, Probability and Computing 3(2), 233–246 (1994)
2. Czyzowicz, J., Kranakis, E., Urrutia, J.: A simple proof of the representation of bipartite planar graphs as the contact graphs of orthogonal straight line segments. Information Processing Letters 66(3), 125–126 (1998)
3. Kobourov, S.G., Ueckerdt, T., Verbeek, K.: Combinatorial and geometric properties of planar Laman graphs. In: Khanna, S. (ed.) Proceedings of the 24th Annual ACM-SIAM Symposium on Discrete Algorithms (SODA), pp. 1668–1678. SIAM (2013)
4. Biedl, T.C., Kaufmann, M.: Area-efficient static and incremental graph drawings. In: Burkard, R.E., Woeginger, G.J. (eds.) ESA 1997. LNCS, vol. 1284, pp. 37–52. Springer, Heidelberg (1997)
5. Kant, G.: Drawing planar graphs using the canonical ordering. Algorithmica 16(1), 4–32 (1996)
6. Zhou, X., Nishizeki, T.: Orthogonal drawings of series-parallel graphs with minimum bends. SIAM Journal on Discrete Mathematics 22(4), 1570–1604 (2008)
7. Tayu, S., Nomura, K., Ueno, S.: On the two-dimensional orthogonal drawing of series-parallel graphs. Discrete Applied Mathematics 157(8), 1885–1895 (2009)
8. Dujmović, V., Suderman, M., Wood, D.R.: Graph drawings with few slopes. Computational Geometry 38(3), 181–193 (2007)
9. Lenhart, W., Liotta, G., Mondal, D., Nishat, R.I.: Planar and plane slope number of partial 2-trees. In: Wismath, S., Wolff, A. (eds.) GD 2013. LNCS, vol. 8242, pp. 412–423. Springer, Heidelberg (2013)
10. Jelínek, V., Jelínková, E., Kratochvíl, J., Lidický, B., Tesar, M., Vyskocil, T.: The planar slope number of planar partial 3-trees of bounded degree. Graphs and Combinatorics 29(4), 981–1005 (2013)
11. Keszegh, B., Pach, J., Pálvölgyi, D.: Drawing planar graphs of bounded degree with few slopes. SIAM Journal on Discrete Mathematics 27(2), 1171–1183 (2013)

Strongly-Connected Outerplanar Graphs with Proper Touching Triangle Representations

J. Joseph Fowler

Department of Computer Science, University of Arizona, Tucson, AZ, USA
fowler@email.arizona.edu

Abstract. A *proper touching triangle representation* $\mathcal{R}$ of an n-vertex planar graph consists of a triangle divided into n non-overlapping triangles. A pair of triangles are considered to be adjacent if they share a partial side of positive length. Each triangle in $\mathcal{R}$ represents a vertex, while each pair of adjacent triangles represents an edge in the planar graph. We consider the problem of determining when a proper touching triangle representation exists for a *strongly-connected outerplanar graph*, which is biconnected and after the removal of all degree-2 vertices and outeredges, the resulting *connected* subgraph only has chord edges (w.r.t. the original graph). We show that such a graph has a proper representation if and only if the graph has at most two *internal faces* (*i.e.,* faces with no outeredges).

1 Introduction

Although the node-link model has been the traditional form of drawing a planar graph $G(V, E)$, many application areas demand alternate models of representing graphs, such as polygon *edge-contact representations*. Here vertices are represented by simple polygons and edges are represented by adjacent polygons that have at least a partial side in common. As pointed out by de Fraysseix *et al.* [2], one can easily find such a representation of any planar graph with non-convex polygons with complexity as high as $|V| - 1$, where much area is unused leading to many gaps and holes within the representation. Recently, convex hexagons have been shown to always be sufficient in producing hole-free representations [1], although, 6-sided polygons are sometimes necessary. The problem thus arises in determining which classes of planar graphs can be represented by polygons with fewer than six sides.

In this context, we focus on the case of minimal polygonal complexity, where all the representing polygons are triangles. Specifically, an n-vertex *touching triangle graph* (TTG) has a representation $\mathcal{R}$ where each vertex is represented by one of n non-overlapping triangles and each edge is represented by a pair of adjacent triangles in $\mathcal{R}$. Again, triangles are only considered to be adjacent if they share at least a partial side of positive length. Thus, pairs of triangles having only point contacts are not considered to be adjacent, and hence, do not represent edges.

The most natural looking edge-contact representations have triangular boundaries where their interiors contains no gaps or holes. If this is the case, then $\mathcal{R}$ is a *proper* TTG representation, and the graph is a proper TTG. Visually, $\mathcal{R}$ can be thought of as a triangle that has been subdivided into n non-overlapping triangles, where adjacent

S. Wismath and A. Wolff (Eds.): GD 2013, LNCS 8242, pp. 155–160, 2013.

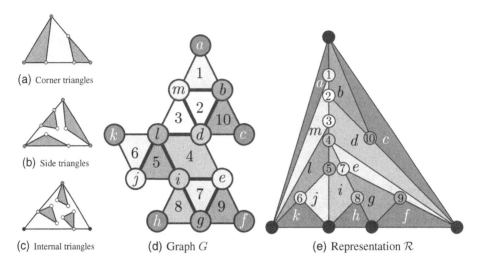

(a) Corner triangles

(b) Side triangles

(c) Internal triangles

(d) Graph G

(e) Representation $\mathcal{R}$

Fig. 1. (a–c) Three types of representing triangles; (d) a strongly-connected outerplanar graph G with two internal faces: 2 and 7; and (e) a proper TTG representation $\mathcal{R}$ of G

triangles do not necessarily share entire sides as they do in a triangulation. Not all planar graphs are TTGs [3], let alone proper. While it was also shown in [3] that all biconnected outerplanar graphs have hole-free TTG representations, their boundaries are not necessarily triangular, and hence, are not necessarily proper. This raises the question as to which outerplanar graphs have proper TTG representations.

A proper outerplanar TTG has several restrictions. Degree-1 vertices can only be represented by *corner triangles of $\mathcal{R}$* with two edges along the boundary T of $\mathcal{R}$, while degree-2 vertices can also be represented by *side triangles of $\mathcal{R}$* with one side along T. All other vertices are represented by *internal triangles of $\mathcal{R}$*; *cf.* Figs. 1(a)–1(c).

A biconnected outerplanar graph G is *strongly-connected* if after the removal of all degree-2 vertices and outeredges, the resulting *connected* subgraph only has chord edges (w.r.t. G) as in Fig. 1(d). Such graphs are not necessarily maximal. We characterize this graph class in terms of *internal faces* (*i.e.* faces with no outeredges) as follows:

(1) First, we construct proper TTG representations for strongly-connected outerplanar graphs, as in Fig. 1(e), using a *chord-to-endpoint assignment* that pairs each chord (except for one) with a distinct vertex that is also an endpoint of the chord.

(2) Second, we show that having at most two internal faces is sufficient when the graph is strongly-connected, since a chord-to-endpoint assignment exists in this case.

(3) Third, we finish our characterization by proving that having at most two internal faces is also necessary in order for G to have a proper TTG representation.

To the best of our knowledge, the only other results specifically for proper TTGs are in [4], where a fixed-parameter tractable decision algorithm for 3-connected planar max-degree-Δ graphs is described, and where it it shown that planar 3-connected cubic graphs are proper TTGs.

2   Proper Strongly-Connected Outerplanar TTGs

Let G be a strongly-connected outerplanar graph. A *strong reversed peeling order* σ is an ordering of the inner faces $F_1, \ldots, F_q$ of G such that the subgraph $G_i = F_1 \cup \cdots \cup F_i$ is also strongly-connected for each $i \in [1 .. q]$. We index the chords $\mathcal{C} = \{c_2, \ldots, c_q\}$ of G by σ such that c_i is the chord of face F_i that becomes an outeredge of G_{i-1} when F_i is "peeled" from G_i. Thus, c_i is common to $F_{i'}$ and F_i for some $i' < i$. Finally, a *chord-to-endpoint assignment* $\tau : \mathcal{C}' \rightarrow V'$ of a strong reversed peeling order σ assigns the chords $\mathcal{C}' = \{c_3, \ldots, c_q\} = \mathcal{C} \setminus \{c_2\}$ to a subset of their endpoints $V' = \{v_3, \ldots, v_q\}$ such that $\tau(c_i) = v_i$ is an endpoint of c_i for each $i \in [3 .. q]$ where $v_i \neq v_j$ if $i \neq j$.

Claim 1. *If G is a strongly-connected outerplanar graph, then G has a strong reversed peeling order σ.*

Proof. Faces F_1 and F_2 of σ can be any pair of adjacent faces. For $i \in [2 .. q-1]$, assume that subgraph $G_i = F_1 \cup \cdots \cup F_i$, has the connected *chord subgraph* $H_i = c_2 \cup \cdots \cup c_i$, where c_i is the chord of face F_i that was added to G_{i-1}. At least one outeredge of G_i is a chord in G. Otherwise, the outerface of G_i, which is a separating cycle C in G, would have a cut-vertex in G—all the cut edges in G are chords, none of which can be in C—violating the biconnectivity of G. Since G is strongly-connected, every chord c in the outerface of G_i must be incident to some chord of H_i. Thus, the next face F_{i+1} in σ can be any remaining face whose chord c_{i+1} is an outeredge of G_i. $\square$

Lemma 2. *If G is a strongly-connected outerplanar graph for which there exists a chord-to-endpoint assignment τ, then G has a proper touching triangle representation.*

Proof. Let σ be the strong reversed peeling order of the faces $F_1, \ldots, F_q$ of G given by the order of chords as assigned by τ. We apply induction on k and assume that we have a proper TTG representation $\mathcal{R}_k$ for G_k, and show how to modify $\mathcal{R}_k$ to obtain $\mathcal{R}_{k+1}$ for $G_{k+1} = G_k \cup F_{k+1}$. When G_1 has a single face F_1 with the edge (u, v) connected by the chain of vertices $x_1, \ldots, x_i$, Fig. 2(a) gives a proper TTG representation $\mathcal{R}_1$ for G_1. Next, when G_2 has two faces with the common chord (u, v) (where $y_1, \ldots, y_j$ forms a chain in F_2), Fig. 2(b) gives a proper TTG representation $\mathcal{R}_2$ for G_2 where the triangle $\triangle u$ representing u in $\mathcal{R}_1$ was subdivided into a total of $j + 1$ triangles.

For $k \in [2 .. q-1]$, we also assume by induction on k that triangle $\triangle v_k$ representing $\tau(c_k) = v_k$ is a side triangle in $\mathcal{R}_k$. This holds in the base case of $k = 2$, since $\mathcal{R}_2$ in

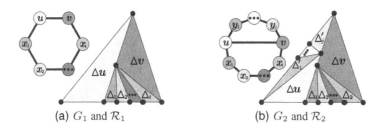

(a) G_1 and $\mathcal{R}_1$ (b) G_2 and $\mathcal{R}_2$

Fig. 2. Proper TTG representations of strongly-connected outerplanar graphs with 1 or 2 faces

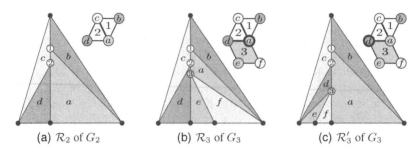

(a) $\mathcal{R}_2$ of G_2 (b) $\mathcal{R}_3$ of G_3 (c) $\mathcal{R}'_3$ of G_3

Fig. 3. For chord $c_3 = (a, d)$, either triangle $\triangle a$ or $\triangle d$ in $\mathcal{R}_2$ in (a) is divided into triangles $\triangle \tau(c_3), \triangle e, \triangle f$ to form $\mathcal{R}_3$ if $\tau(c_3) = a$ as in (b) or $\mathcal{R}'_3$ if $\tau(c_3) = d$ as in (c)

Fig. 2(b) only has side triangles. This allows us to subdivide $\triangle v_k$ in $\mathcal{R}_k$ a total of j_k times to obtain $\mathcal{R}_{k+1}$ in which an internal triangle now represents v_k and j_k new side triangles represent the j_k degree-2 vertices in F_k that were added to G_k to form G_{k+1}. Figure 3 illustrates this process. The chord $c_3 = (a, d)$ in G_3 either has $\tau(c_3) = a$ or $\tau(c_3) = d$. This results in $\mathcal{R}_3$ in Fig. 3(b) (or in $\mathcal{R}'_3$ in Fig. 3(c)) after dividing the side triangle $\triangle a$ (or $\triangle d$) in $\mathcal{R}_2$ of G_2 in Fig. 3(a) into the internal triangle $\triangle a$ (or $\triangle d$) and the two side triangles $\triangle e$ and $\triangle f$. As a consequence, each chord endpoint v_k has its representing triangle $\triangle v_k$ subdivided once, at which point $\triangle v_k$ is an internal triangle in $\mathcal{R}_{k'}$ for $k' > k$. However, this is not a problem in maintaining our inductive hypothesis for $k + 1$ since τ assigns each chord to a distinct vertex in G. $\square$

Lemma 3. *If G is a strongly-connected outerplanar graph with at most two internal faces, then G is a proper touching triangle graph.*

Proof. We construct a chord-to-endpoint assignment τ, where Lemma 2 then implies that G is a proper TTG. If G has no internal faces, its connected chord subgraph H is acyclic, and hence, a tree. Let σ be a strong reversed peeling order of faces $F_1, \ldots, F_q$ for G given by Claim 1. Each face F_i was picked so that its chord c_i is a leaf edge in the subtree of chords $H_i = c_2 \cup \cdots \cup c_i$ of G_i. Thus, we can assign $\tau(c_i) = u_i$, where u_i is the endpoint of c_i that is a leaf node in H_i, for $i \in [3 \,..\, q]$. Both endpoints of the first chord c_2 are left unassigned by τ.

If G has one internal face F, we apply Claim 1 and assign chords as before with the following exceptions: We set $F_1 = F$ and faces $F_2, \ldots, F_j$ as the adjacent faces of F_1, where face F_{i+1} is incident to face F_i for $i \in \{2 \,..\, j - 1\}$. Since chord c_j forms the cycle C (edges of F_1) when added to H_j, we can set $\tau(c_j)$ to be the common endpoint of the chords c_j and c_2, which leaves one endpoint of c_2 unassigned.

Lastly, when G has two internal faces, we apply Claim 1 as follows: We set F_1 and F_k in σ to be the two internal faces in G such that k is minimal. The chords of G_k contain a path p connecting F_1 to F_k. For τ, we assign each chord of p to the endpoint first encountered along p so that c_2 and c_k each have exactly one assigned endpoint. To σ we add each remaining face F_i adjacent first to F_1 (and then to F_k) starting from the endpoints of p in cyclic order along each respective face. For τ, we assign chord c_i of each cycle to its newly added endpoint in H_i—except for the last chord, call it c', in F_1

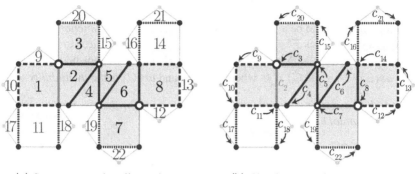

(a) Strong reversed peeling order σ (b) Chord-to-endpoint assignment τ

Fig. 4. Example of determining σ and τ for strongly-connected outerplanar graph G with two internal faces F_1 and F_8. Subgraph G_8 (dark gray faces) is the minimal strongly-connected subgraph containing both F_1 and F_8, whose chord subgraph H_8 (solid chords) is a caterpillar. Subgraph G_{14} (light/dark gray faces) has faces $F_9, ..., F_{11}$ and $F_{12}, ..., F_{14}$ added in cyclic order along F_1 and F_8, resp., each starting from an endpoint of path p (white vertices). The chords of path p are assigned the endpoints first encountered along p from F_1 to F_8. The dashed chords of F_1 and F_8 are assigned endpoints next along each cycle (starting from endpoints of p). Chord c_i for $i \in \{4, 6, 15, ..., 22\}$ is assigned its endpoint that was not in the chord subgraph H_{i-1}.

(or in F_k). Given the greedy assignment of chords along p, c' has an available endpoint in common with c_2 in F_1 (or c_k in F_k). Each remaining face can be added to σ and have its chord assigned by τ as before. Figure 4 illustrates this procedure. □

For the next lemma, we consider the *representing dual graph* $\mathcal{G}_{\mathcal{R}}$ (the graph formed by the representation $\mathcal{R}$) of a proper TTG G. With respect to the triangular boundary T of a proper TTG representation $\mathcal{R}$, each vertex or edge of $\mathcal{G}_{\mathcal{R}}$ (common to one or more representing triangles) is either *external* if along the boundary T or *internal* if inside T. Likewise, we term the faces of $\mathcal{G}_{\mathcal{R}}$ as being either *external faces* if they correspond to corner or side triangles or *internal faces*, otherwise. We have the following relationships between G and $\mathcal{G}_{\mathcal{R}}$: (1) each vertex v in G corresponds to a bounded triangular face F_v in $\mathcal{G}_{\mathcal{R}}$ and (2) each bounded face f in G corresponds to an internal vertex v_F in $\mathcal{G}_{\mathcal{R}}$.

Clearly, the angle of a vertex v_F in $\mathcal{G}_{\mathcal{R}}$ along the face F_v in $\mathcal{G}_{\mathcal{R}}$ can be at most $180°$. If the angle is less than $180°$, then v_F is a corner of the triangular face F_v. Each internal vertex v_F in $\mathcal{G}_{\mathcal{R}}$ has at most one $180°$ angle since $deg(v_F) > 2$ in $\mathcal{G}_{\mathcal{R}}$. For example, the angle of vertex v_1 in $\mathcal{R}$ in Fig. 1(e) is $180°$ for face F_b, but not for faces F_a and F_m. Face F_v in $\mathcal{G}_{\mathcal{R}}$ representing v has exactly three vertices (each with an angle less than $180°$) if either (i) F_v is an external face where $deg(v) = 2$ or (ii) F_v is an internal face where $deg(v) = 3$. Otherwise, F_v has at least $ch(v) = max\{deg(v) - 3, 0\}$ vertices in $\mathcal{G}_{\mathcal{R}}$ whose internal angles are $180°$ along the boundary of F_v. These correspond to $ch(v)$ faces in G that are incident to v.

We denote $ch(v)$ as the *charge* of v since each incident face F in G can *dissipate* at most one charge from v. This is done by having the internal angle for v_F be $180°$ along the face F_v representing v in $\mathcal{G}_{\mathcal{R}}$. However, face F can dissipate at most one charge from an incident vertex, since v_F has at most one $180°$ angle. For example, vertex d in

Fig. 1(d) has charge $ch(d) = 2$ dissipated by faces F_3 and F_{10}, where the corresponding internal vertices v_3 and v_{10} have $180°$ angles for face F_d in Fig. 1(e).

Thus, in order for G to be a TTG, there must exist a *discharge function*, $\pi : \mathcal{F}' \to V$ that assigns a subset faces $\mathcal{F}' \subseteq \mathcal{F}$ (the faces of G) to incident vertices to fully dissipate the *total charge* $ch(G) = \sum_{v \in V(G)} ch(v)$ of G. Hence, G cannot be a proper TTG graph if $av(G) = q - ch(G) < 0$, where $q = |\mathcal{F}|$ and $av(G)$ is the *availability* of G.

Lemma 4. *If G is a strongly-connected outerplanar graph with more than two internal faces, then G cannot be a proper touching triangle graph.*

Proof. We apply Claim 1 to get a strong reversed peeling order σ of the faces $F_1, \dots, F_q$ of G, where the subgraph $G_i = F_1 \cup \cdots \cup F_i$ is strongly-connected. If G_i does not contain any internal faces, then H_i is a tree. When peeling face F_i from G_i to obtain G_{i-1}, chord c_i cannot have both of its endpoints u_i and v_i in H_{i-1}. Otherwise, c_i would form a cycle in H_i. Thus, one endpoint $v_i \notin H_{i-1}$ of c_i has $deg(v_i) = 2$ in G_{i-1}. While both degrees of u_i and v_i increased by 1 in going from G_{i-1} to G_i, only $deg(u_i) > 3$ in G_i, while $deg(v_i) = 3$ in G_i. Thus, the total charge $ch(G_i) = ch(G_{i-1}) + 1$ increases by one, so that the availability $av(G_i) = av(G_{i-1})$ remains constant.

If G_i contains a new internal face C that G_{i-1} does not, then both endpoints u_i and v_i of c_i must be in H_{i-1} in order for c_i to form the new cycle C in H_i. Hence, both $deg(u_i) > 3$ and $deg(v_i) > 3$ in G_i, so that $ch(G_i) = ch(G_{i-1}) + 2$, and as a result, $av(G_i) = av(G_{i-1}) - 1$ where the availability decreases by one.

Initially, the availability is at most 2, where G_2 has maximum degree 3 with the two faces F_1 and F_2 so that $av(G_2) = 2$. Consequently, G can have at most two internal faces, before the availability drops below 0, preventing it from being a TTG.   □

We conclude by combining Lemmas 3 and 4 to give the main theorem of the paper.

Theorem 5. *A strongly-connected outerplanar graph G has a proper touching triangle representation if and only if G has at most two internal faces.*

Acknowledgments. We thank our colleagues for helpful insights: Jawaherul Alam, Michael Kaufman, Stephen Kobourov, Martin Nöllenburg, Ignaz Rutter, Alexander Wolff, and many others.

References

1. Duncan, C.A., Gansner, E.R., Hu, Y.F., Kaufmann, M., Kobourov, S.G.: Optimal polygonal representation of planar graphs. Algorithmica 63(3), 672–691 (2012)
2. de Fraysseix, H., de Mendez, P.O., Rosenstiehl, P.: On triangle contact graphs. Combinatorics, Probability and Computing 3, 233–246 (1994)
3. Gansner, E.R., Hu, Y., Kobourov, S.G.: On touching triangle graphs. In: Brandes, U., Cornelsen, S. (eds.) GD 2010. LNCS, vol. 6502, pp. 250–261. Springer, Heidelberg (2011)
4. Kobourov, S.G., Mondal, D., Nishat, R.I.: Touching triangle representations for 3-connected planar graphs. In: Didimo, W., Patrignani, M. (eds.) GD 2012. LNCS, vol. 7704, pp. 199–210. Springer, Heidelberg (2013)

Achieving Good Angular Resolution in 3D Arc Diagrams

Michael T. Goodrich and Paweł Pszona

Dept. of Computer Science
University of California, Irvine

Abstract. We study a three-dimensional analogue to the well-known graph visualization approach known as *arc diagrams*. We provide several algorithms that achieve good angular resolution for 3D arc diagrams, even for cases when the arcs must project to a given 2D straight-line drawing of the input graph. Our methods make use of various graph coloring algorithms, including an algorithm for a new coloring problem, which we call *localized edge coloring*.

1 Introduction

An *arc diagram* is a two-dimensional graph drawing where the vertices of a graph, G, are placed on a one-dimensional curve (typically a straight line) and the edges of G are drawn as circular arcs that may go outside that curve (e.g., see [1,2,6,8,19,20,23]). By way of analogy, we define a *three-dimensional arc diagram* to be a drawing where the vertices of a graph, G, are placed on a two-dimensional surface (such as a sphere or plane) and the edges of G are drawn as circular arcs that may go outside that surface. (See Fig. 1.) This 3D drawing paradigm is used, for example, to draw geographic networks or flight networks (e.g., see [3]).

In this paper, we are interested in the angular resolution of 3D arc diagrams, that is, the smallest angle determined by the tangents at a vertex, v, to two

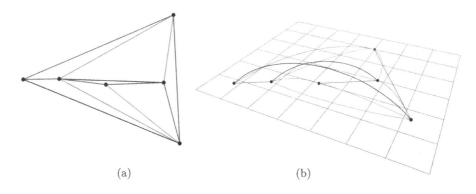

(a)	(b)

Fig. 1. A graph rendered (a) as a straight-line drawing and (b) as a 3D arc diagram

S. Wismath and A. Wolff (Eds.): GD 2013, LNCS 8242, pp. 161–172, 2013.

arcs incident to v in such a drawing. Specifically, we provide algorithms for achieving good angular resolution in 3D arc diagrams where the (*base*) surface that contains the vertices for the graph, G, is a sphere or a plane. Moreover, for the 3D arc diagrams that we consider in this paper, we assume that all the edges of G are drawn to protrude out of only one side of the base surface.

1.1 Previous Related Results

The term "*arc diagram*" was defined in 2002 by Wattenberg [23], but the drawing paradigm actually can be traced back to the 1960's, including work by Saaty [20] and Nicholson [19]. Also, earlier work by Brandes [2] explores symmetry in arc diagrams, earlier work by Cimikowski and Shope [6] explores heuristics for minimizing the number of arc crossings, and earlier work by Djidjev and Vrt'o [8] explores lower bounds for the crossing numbers of such drawings. Most recently, Angelini *et al.* [1] show that there is a universal set of $O(n)$ points on a parabola that allows any planar graph to be drawn as a planar arc diagram.

In terms of previous work on arc diagrams for optimizing the angular resolution of such drawings, Duncan *et al.* [11] give a complete characterization of which regular graphs can be drawn as arc diagrams with vertices placed on a circle and perfect angular resolution, using a drawing style inspired by the artist, Mark Lombardi, where edges are drawn using circular arcs so as to achieve good angular resolution. With respect to a lower bound for this drawing style, Cheng *et al.* [5] give a planar graph with bounded degree, d, that requires exponential area if it is drawn as a plane graph with circular-arc edges and angular resolution $\Omega(1/d)$. Even so, it is possible to draw any planar graph as a plane graph with poly-line or poly-circular edges to achieve polynomial area and $\Omega(1/d)$ angular resolution, based on results by a number of authors (e.g., see Brandes *et al.* [4], Cheng *et al.* [5], Duncan *et al.* [9,11], Garg and Tamassia [15], Goodrich and Wagner [17], and Gutwenger and Mutzel [18]).

In addition, several researchers have investigated how to achieve good angular resolution for various straight-line drawings of graphs. Duncan *et al.* [10] show that one can draw an ordered tree of degree d as a straight-line planar drawing with angular resolution $\Omega(1/d)$. Formann *et al.* [14] show that any graph of degree d has a straight-line drawing with polynomial area and angular resolution $\Omega(1/d^2)$, and this can be improved to be $\Omega(1/d)$ for planar graphs, albeit with a drawing that may not be planar.

We are not familiar with any previous work on achieving good angular resolution for 3D arc diagrams, but there is previous related work on other types of 3D drawings [7]. For instance, Brandes *et al.* [3] show that one can achieve $\Omega(1/d)$ angular resolution for 3D geometric network drawings, but their edges are curvilinear splines, rather than simple circular arcs. Garg *et al.* [16] study 3D straight-line drawings so as to satisfy various resolution criteria, but they do not constrain vertices to belong to a 2D surface. In addition, Eppstein *et al.* [12] provide an algorithm for achieving optimal angular resolution in 3D drawings of low-degree graphs using poly-line edges.

1.2   Our Results

In this paper, we give several algorithms for achieving good angular resolution for 3D arc diagrams. In particular, we show the following for a graph, G, with maximum degree, d:

- We can draw G as a 3D arc diagram with an angular resolution of $\Omega(1/d)$ ($\Omega(1/d^{1/2})$ if G is planar) using straight-line segments and vertices placed on a sphere.
- We can draw G as a 3D arc diagram with an angular resolution of $\Omega(1/d)$ using circular arcs that project perpendicularly to a given straight-line drawing for G in a base plane, no matter how poor the angular resolution of that projected drawing.
- If a straight-line 2D drawing of G already has an angular resolution of $\Omega(1/d)$ in a base plane, $\mathcal{P}$, then we can draw G as a 3D arc diagram with an angular resolution of $\Omega(1/d^{1/2})$ using circular arcs that project perpendicularly to the given drawing of G in $\mathcal{P}$.
- Given any 2D straight-line drawing of G in a base plane, $\mathcal{P}$, we can draw G as a 3D arc diagram with an angular resolution of $\Omega(1/d^{1/2})$ using circular arcs that project to the edges of the drawing of G in $\mathcal{P}$, with each arc possibly using a different projection direction.

Our algorithms make use of various graph coloring methods, including an algorithm for a new coloring problem, which we call *localized edge coloring*.

Note that $O(1/d^{1/2})$ is an upper bound on the resolution of a 3D arc drawing of G, as maximizing the smallest angle between two edges around a vertex, v, is equivalent to maximizing smallest distance between intersections of a unit sphere centered at v, and lines tangent to edges incident to v, which is known as the *Tammes problem* [21]. The $O(1/d^{1/2})$ upper bound is due to Fejes Tóth [13].

2   Preliminaries

In this section, we provide formal definitions of two notions of 3D arc diagrams.

We extend the notion of arc diagrams and define *3D arc diagram drawings* of a graph, G, to be 3D drawings that meet the following criteria:

(1) nodes (vertices) are placed on a single (*base*) sphere or plane
(2) each edge, e, is drawn as a *circular arc*, i.e., a contiguous subset of a circle
(3) all edges lie entirely on one side of the base sphere or plane.

In addition, if the base surface is a plane, $\mathcal{P}_1$, then each circular edge, e, which belongs to a plane, $\mathcal{P}_2$, forms the same angle, $\alpha_e \leq \pi/2$, in $\mathcal{P}_2$, at its two endpoints. Moreover, in this case, each edge projects (perpendicularly) to a straight line segment in $\mathcal{P}_1$. An example of such an arc is shown in Fig. 2a.

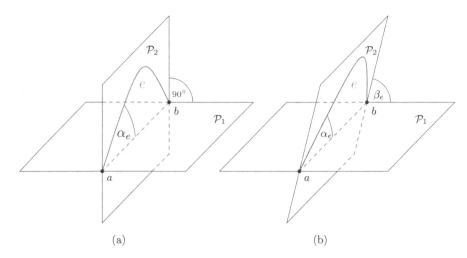

Fig. 2. Edge $e = (a, b)$ drawn as (a) circular arc with angle α_e; (b) slanted circular arc with angles (α_e, β_e).

For 3D arc diagrams restricted to use a base plane, $\mathcal{P}_1$ (rather than a sphere), by modifying the second condition, we obtain a definition of *slanted 3D arc diagram drawings*.

(2′) each edge e is a *circular arc* that lies on a plane, $\mathcal{P}_2$, that contains both endpoints of e and forms an angle, $\beta_e < \pi/2$, with the base plane, $\mathcal{P}_1$; the edge, e, forms the same angle, $\alpha_e \leq \pi/2$, in $\mathcal{P}_2$, at its two endpoints.

Note that in this case each circular edge, e, joining vertices a and b, in a slanted 3D arc diagram, projects to a straight line segment, $L = ab$, in the base plane, $\mathcal{P}_1$, using a direction perpendicular to L in $\mathcal{P}_2$. Still, a perpendicular projection of the drawing onto the base plane, $\mathcal{P}_1$, is not necessarily a straight-line drawing of G and may not even be planar. For an example, see Fig. 2b.

3   Localized Edge Coloring

Recall that a *vertex coloring* of a graph is an assignment of colors to vertices so that every vertex is given a color different from those of its adjacent vertices, and an *edge coloring* is an assignment of colors to a graph's edges so that every edge is given a color different from its incident edges. A well-known greedy algorithm can color any graph with maximum degree, d, using $d + 1$ colors, and Vizing's theorem [22] states that edges of an undirected graph G can similarly be colored with $d + 1$ colors, as well.

Assuming we are given an undirected graph G together with its combinatorial embedding on a plane (i.e., the order of edges around each vertex, which is also known as a *rotation system*), we introduce a localized notion of an edge coloring, which will be useful for some of our results regarding 3D arc diagrams. Given an even integer parameter, L, we define an *L-localized edge coloring* to be an edge coloring that satisfies the following condition:

> Suppose an edge $e = (u, v)$ has color c, and let $(l_1, l_2, \ldots, l_i = e, \ldots l_k)$ be a clockwise ordering of edges incident to u. Then none of the edges $l_{i-L/2}, l_{i-L/2+1}, \ldots, l_{i-1}, l_{i+1}, \ldots, l_{i+L/2}$, that is, the $L/2$ edges before e and $L/2$ edges after e in the ordering, has color c. (Note that, by symmetry, the same goes for edges around v.)

Thus, a valid d-localized edge coloring is also a valid classical edge coloring. We call the set, $\{l_{i-L/2}, l_{i-L/2+1}, \ldots, l_{i-1}, l_{i+1}, \ldots, l_{i+L/2}\}$, the *L-neighborhood of e around u.*

As with the greedy approach to vertex coloring, an L-localized edge coloring can be found by a simple greedy algorithm that incrementally assigns colors to edges, one at a time. Each edge $e = (u, v)$ is colored with color c that does not appear in both L-neighborhoods of e (around u and around v). Using reasonable data structures, this greedy algorithm can be implemented to run in $O(mL)$ time, for a graph with m edges, and combining it with Vizing's theorem [22], allows us to find an edge coloring that uses at most $\min\{d, 2L\} + 1$ colors.

4   Improving Resolution via Edge Coloring

As mentioned above, we define the angle between two incident arcs in the 3D arc diagram to be the angle between lines tangent to the arcs at their common endpoint. In order to reason about angles in 3D, the following lemma will prove useful.

Lemma 1. *Consider two segments l_1, l_2 that share a common endpoint that lies on a plane $\mathcal{P}$ (see Fig. 3). If both l_1 and l_2 form angle $\beta \leq \pi/4$ with their projections onto $\mathcal{P}$, and projections of l_1 and l_2 onto $\mathcal{P}$ form angle α, then δ, the angle between l_1 and l_2, is at least $\alpha/2$.*

Proof. Assume w.l.o.g. that $|l_1| = |l_2| = 1$. The distance d between endpoints of l_1 and l_2 is the same as the distance between endpoints of projections of l_1 and l_2 onto $\mathcal{P}$ (because both l_1 and l_2 form angle β with $\mathcal{P}$). Lengths of the projections are $\cos \beta$, and by the law of cosines,

$$d^2 = \cos^2 \beta + \cos^2 \beta - 2 \cos \beta \cos \beta \cos \alpha = 2 \cos^2 \beta (1 - \cos \alpha).$$

On the other hand, again by the law of cosines,

$$d^2 = |l_1|^2 + |l_2|^2 - 2|l_1||l_2| \cos \delta = 2(1 - \cos \delta).$$

Comparing the two yields

$$2 \cos^2 \beta (1 - \cos \alpha) = 2(1 - \cos \delta),$$

which leads to

$$\cos \delta = 1 - \cos^2 \beta (1 - \cos \alpha).$$

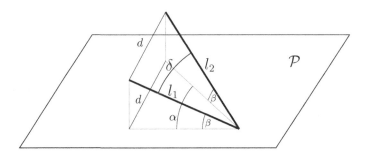

Fig. 3. Illustration of Lemma 1

For $\beta \leq \pi/4$,

$$\cos \delta \leq \cos \frac{\alpha}{2},$$

which means that

$$\delta \geq \frac{\alpha}{2}.$$

$\square$

In addition, the following lemma will also be useful in our results.

Lemma 2. *Consider two segments, l_1 and l_2, that share a common endpoint, with l_1 lying on a plane $\mathcal{P}$ (see Fig. 4). If l_2 forms angle $\beta < \pi/4$ with its projection onto $\mathcal{P}$, then δ, the angle between l_1 and l_2, is at least β.*

Proof. Assume w.l.o.g. that $|l_1| = |l_2| = 1$. Length of a, the projection of l_2 onto $\mathcal{P}$, is $\cos \beta$, and h, the distance of l_2's endpoint from $\mathcal{P}$ is $\sin \beta$. Let α be the angle between l_1 and a, and let b be the segment connecting their endpoints. By the law of cosines,

$$|b|^2 = |a|^2 + |l_1|^2 - 2|a||l_1|\cos \alpha = \cos^2 \beta + 1 - 2\cos \beta \cos \alpha.$$

Then,

$$|d|^2 = |h|^2 + |b|^2 = \sin^2 \beta + \cos^2 \beta + 1 - 2\cos \alpha \cos \beta = 2(1 - \cos \alpha \cos \beta).$$

Again, by the law of cosines,

$$|d|^2 = |l_1|^2 + |l_2|^2 - 2|l_1||l_2|\cos \delta = 2(1 - \cos \delta).$$

Comparing the two yields

$$\cos \delta = \cos \alpha \cos \beta.$$

Since $\cos \alpha \leq 1$, we get

$$\cos \delta \leq \cos \beta,$$

and it follows that $\delta \geq \beta$. $\square$

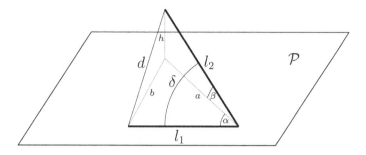

Fig. 4. Illustration of Lemma 2

4.1 Vertices on a Sphere

In this subsection, we consider the angular resolution obtained in a 3D arc diagram using straight-line edges drawn between vertices placed on a sphere. The two algorithms we present here are inspired by a two-dimensional drawing algorithm by Formann *et al.* [14]. Our main result is the following.

Theorem 1. *Let $G = (V, E)$ be a graph of degree d. There is a 3D straight-line drawing of G with an angular resolution of $\Omega(1/d)$, with the vertices of G placed on the surface on a sphere.*

Proof. Let $G^2 = (V, E^2)$ be the square of G, that is the graph with the same set of vertices as G, and an edge between vertices (u, v) if there is a path of length ≤ 2 between u and v in G. Since G has degree d, G^2 has degree $\leq d(d-1) < d^2$. Therefore, we can color the vertices of G^2 with at most d^2 colors, with the requirement that adjacent vertices have different colors.

We place the vertices on a unit sphere $\mathcal{S}$. We define d^2 *cluster positions* as follows. First, we cut the circle with $d + 1$ uniformly spaced parallel planes (see Fig. 5), such that the maximum distance between the center of $\mathcal{S}$ and a plane is h (thus, the distance between two neighboring planes is $2h/d$). Then, we uniformly place d points on each resulting circle. These are the *cluster positions*.

Since a coloring $\mathcal{C}$ of G^2 uses $\leq d^2$ colors, we can assign distinct cluster positions to colors in $\mathcal{C}$. To obtain a drawing of G, we place all vertices of the same color in $\mathcal{C}$ on the sphere, $\mathcal{S}$, within a small distance, ϵ, around this color's cluster position, and draw edges in E as straight lines. We can remove any intersections by perturbing the vertices slightly.

The claim is that the resulting drawing has resolution $\Omega(1/d)$. Indeed, by setting $h = \pi/(\sqrt{1 + \pi^2})$, we get $\Omega(1/d)$ minimal distance between any two planes, and $\Omega(1/d)$ minimal distance between any two cluster positions on the same plane. So, the distance between any two cluster positions is at least $\Omega(1/d)$.

Now let us consider any angle $\sphericalangle abc$ formed by edges (a, b) and (b, c). The edges forming $\sphericalangle abc$ define a plane, $\mathcal{P}$, whose intersection with $\mathcal{S}$ is a circle, C. Angle $\sphericalangle abc$ is inscribed in C, and based on the arc $\widehat{ac}$. Therefore, any other angle inscribed in C and based on $\widehat{ac}$ has the same size, in particular the one formed

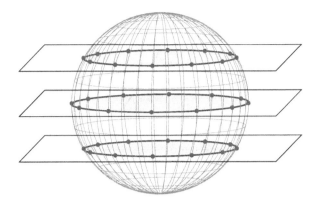

Fig. 5. Sphere cut with equidistant planes. Red points are the *cluster positions.*

by an isosceles triangle $\triangle adc$. Since $|ad| = |cd| \leq 2$ ($\mathcal{S}$ has radius 1), and $|ac|$ is at least $\Omega(1/d)$, then $|\sphericalangle abc| = |\sphericalangle adc|$ and is at least $\Omega(1/d)$.   □

In addition, we also have the following.

Corollary 1. *Let $G = (V, E)$ be a planar graph of degree d. There is a 3D straight-line drawing of G with an angular resolution of $\Omega(1/d^{1/2})$, with the vertices of G placed on the surface of a sphere.*

Proof. The proof is a direct consequence of applying the algorithm from the proof of Theorem 1 and the fact that the degree of G^2, the square of a planar graph, G, has degree $O(d)$ [14].   □

Thus, we can produce 3D arc diagram drawings of planar graphs that achieve an angular resolution that is within a constant factor of optimal. Admittedly, this type of drawing is probably not going to be very pretty when rendered, say, as a video fly-through on a 2D screen, as this type of drawing is unlikely to project to a planar drawing in any direction.

4.2   Stationary Vertices

In this subsection, we show how to overcome the drawback of the above method, in that we show how to start with any existing 2D straight-line drawing and dramatically improve the angular resolution for that drawing using a 3D arc diagram rendering that projects perpendicularly to the 2D drawing.

Theorem 2. *Let $D(G)$ be a straight-line drawing of a graph, G, with arbitrary, but distinct, placements for its vertices in the base plane. There is a 3D arc diagram drawing of G with the same vertex placements as $D(G)$ and with an angular resolution at least $\Omega(1/d)$, where d is the degree of G, regardless of the angular resolution of $D(G)$.*

Proof. Since we are not allowed to move vertices, and edges have to lie on planes perpendicular to the *base plane*, we are restricted to selecting angles α_e for edges e of G. We do it by utilizing classical edge coloring, observing that the "entry" and "exit" angles for each vertex need to match.

First, we compute an edge coloring $\mathcal{C}$ of G with c colors ($c \leq d + 1$). Then, for each edge e, if its color in $\mathcal{C}$ is i ($i = 0, 1, \ldots, c - 1$), we set its angle to be $\alpha_e = i \cdot \pi/4(c - 1)$. For any two edges e_1, e_2, the difference between their angles α_{e_1} and α_{e_2} is at least $\pi/4(c - 1)$ (let $\alpha_{e_1} < \alpha_{e_2}$; consider the plane, $\mathcal{P}$, determined by both tangent lines having angle α_{e_1}; the angle between e_2 and the plane $\mathcal{P}$, on which tangent of e_1 lies, is $\alpha_{e_2} - \alpha_{e_1}$). Therefore, by Lemma 2, the angle between e_1 and e_2 in the arc diagram is also at least $\pi/4(c - 1) = \Omega(1/d)$.

It is unlikely that any pairs of the arcs touch each other in 3D, but if any pair of them do touch, we can perturb one of them slightly to eliminate the crossing, while still keeping the angular separation for every pair of incident edges to be $\Omega(1/d)$. $\qquad\square$

In addition, through the use of a slanted 3D arc diagram rendering, we can produce a drawing with angular resolution that is within a constant factor of optimal, with each arc projecting to its corresponding straight-line edge in some direction.

Theorem 3. *Let $D(G)$ be a straight-line drawing of a graph, G, with arbitrary, but distinct, placements for its vertices in the base plane. There is a slanted 3D arc diagram drawing of G with the same vertex placements as $D(G)$ and with an angular resolution at least $\Omega(1/d^{1/2})$, where d is the degree of G, regardless of the angular resolution of $D(G)$.*

Proof. Let C be a set of $\lceil d^{1/2} \rceil + 1$ uniformly distributed angles from 0 to $\pi/4$. Define a set of $d + 1$ "colors" as distinct pairs, (α, β), where α and β are each in C. Compute an edge coloring of G using these colors. Now let e be an edge in G, which is colored with (α, β). Draw the edge, e, using a circular arc that lies in a plane, $\mathcal{P}$, that makes an angle of α with the *base plane* and which has a tangent in P that forms an angle of β at each endpoint of e. (For instance, in Fig. 1b, we give a slanted 3D arc diagram based on the edge coloring of the graph in Fig. 1a, corresponding to the following (α_e, β_e) "colors:" $(0°, 0°)$, $(22.5°, 0°)$, $(45°, 0°)$, $(22.5°, 22.5°)$, $(45°, 45°)$.)

The claim is that every pair of incident edges is separated by an angle of size at least $\Omega(1/d^{1/2})$. So suppose e and f are two edges incident on the same vertex, v. Let (α_e, β_e) be the color of e and let (α_f, β_f) be the color of f. Since e and f are incident and we computed a valid coloring for G, $\alpha_e \neq \alpha_f$ or $\beta_e \neq \beta_f$. In either case, this implies that e and f are separated by an angle of size at least $\Omega(1/d^{1/2})$ (by Lemma 1 if $\beta_e = \beta_f$, by Lemma 2 otherwise), which establishes the claim. As previously, we can perturb the arcs to eliminate crossings in 3D. $\qquad\square$

Thus, we can achieve optimal angular resolution in a 3D arc diagram for any graph, G, to within a constant factor, for any arbitrary placement of vertices of G

in the plane. Note, however, that even if $D(G)$ is planar, the 3D arc diagram this algorithm produces, when projected to the *base plane*, may create edge crossings in the projected drawing. It would be nice, therefore, to have 3D arc diagrams that could have good angular resolution and also have planar perpendicular projections in the *base plane*.

4.3 Free Vertices

In this section, we show how to take any 2D straight-line drawing with good angular resolution and convert it to a 3D arc diagram with angular resolution that is within a constant factor of optimal. Moreover, this is the result that makes use of a localized edge coloring.

Theorem 4. *Let $D(G)$ be a straight-line drawing of a graph, G, with arbitrary, but distinct, placement for its vertices in the base plane, and $\Omega(1/d)$ angular resolution. There is a 3D arc diagram drawing of G with the same vertex placements as $D(G)$ and with angular resolution at least $\Omega(1/d^{1/2})$, where d is the degree of G, such that all arcs project perpendicularly as straight lines onto the base plane.*

Proof. The algorithm is similar to the one from the proof of Theorem 2. This time, however, we first compute an L-localized edge coloring, $\mathcal{C}$, of G utilizing c colors ($c \le 2L+1$). Then, as previously, we assign angle $\alpha_e = i \cdot \pi/4(c-1)$ to an edge e of color i in $\mathcal{C}$ ($i = 0, 1, \ldots, c-1$).

Let us consider two arcs, e and f, incident on a vertex, v. If $\alpha_e \ne \alpha_f$, then the angle between e and f is at least $\pi/(4c) = \Omega(1/L)$, by Lemma 2. Otherwise, $\alpha_e = \alpha_f$, and e and f have the same color in $\mathcal{C}$. By the definition of L-localized edge coloring, e and f are separated by at least $L/2$ edges around v. Because $D(G)$ has resolution $\Omega(1/d)$, the angle between e and f in $D(G)$ is $\Omega(L/d)$. Thus, by Lemma 1, the angle between e and f is also $\Omega(L/d)$. Therefore, the angle between e and f is $\Omega(\min\{1/L, L/d\})$. We achieve the advertised angular resolution by setting $L = d^{1/2}$.   □

Theorem 4 shows that we can achieve $\Omega(1/d^{1/2})$ angular resolution in a 3D arc diagram drawing of a graph, G, with arcs projecting perpendicularly onto the base plane as straight-line segments, if there is a straight-line drawing of G on a plane with an angular resolution of $\Omega(1/d)$. The following is an immediate consequence.

Corollary 2. *There is a 3D arc diagram drawing of any planar graph, G, with straight-line projection onto the base plane, and an angular resolution of $\Omega(1/d^{1/2})$.*

Proof. By [14], we can draw G in a straight-line manner on a plane with an angular resolution of $\Omega(1/d)$.   □

Admittedly, the 2D projection of this graph is not necessarily planar. We can nevertheless also achieve the following.

Corollary 3. *There is a 3D arc diagram drawing of any ordered tree, T, with straight-line projection onto the base plane, and an angular resolution of $\Omega(1/d^{1/2})$.*

Proof. By Duncan *et al.* [10], we can draw T in a straight-line manner on a plane with an angular resolution of $\Omega(1/d)$. □

In addition, the area of the projection of the drawings produced by the previous two corollaries is polynomial.

5   Conclusion

We have given efficient algorithms for drawing 3D arc diagrams that achieve polynomial area in the base plane or sphere that contains all the vertices while also achieving good angular resolution. Since our algorithms deal with arc intersections via arc perturbation, the results may not be satisfactory, as the perturbed edges will still be very close. Therefore, one direction for future work is a related resolution question of what volumes are achievable if, in addition to angular resolution, we also insist that every circular arc always be at least unit distance from every other non-incident arc edge.

Acknowledgements. We thank Joe Simons, Michael Bannister, Lowell Trott, Will Devanny, and Roberto Tamassia for helpful discussions regarding angular resolution in 3D drawings.

References

1. Angelini, P., Eppstein, D., Frati, F., Kaufmann, M., Lazard, S., Mchedlidze, T., Teillaud, M., Wolff, A.: Universal point sets for planar graph drawings with circular arcs (2013) (manuscript)
2. Brandes, U.: Hunting down Graph B. In: Kratochvíl, J. (ed.) GD 1999. LNCS, vol. 1731, pp. 410–415. Springer, Heidelberg (1999)
3. Brandes, U., Shubina, G., Tamassia, R.: Improving angular resolution in visualizations of geographic networks. In: Leeuw, W., Liere, R. (eds.) Data Visualization, Eurographics, pp. 23–32. Springer (2000)
4. Brandes, U., Shubina, G., Tamassia, R.: Improving angular resolution in visualizations of geographic networks. In: Proc. Joint Eurographics — IEEE TCVG Symposium on Visualization (VisSym 2000) (2000)
5. Cheng, C., Duncan, C., Goodrich, M., Kobourov, S.: Drawing planar graphs with circular arcs. Discrete & Computational Geometry 25(3), 405–418 (2001)
6. Cimikowski, A., Shope, P.: A neural-network algorithm for a graph layout problem. IEEE Trans. on Neural Networks 7(2), 341–345 (1996)
7. Cohen, R., Eades, P., Lin, T., Ruskey, F.: Three-dimensional graph drawing. Algorithmica 17(2), 199–208 (1997)
8. Djidjev, H.N., Vrt'o, I.: An improved lower bound for crossing numbers. In: Mutzel, P., Jünger, M., Leipert, S. (eds.) GD 2001. LNCS, vol. 2265, pp. 96–101. Springer, Heidelberg (2002)

9. Duncan, C.A., Eppstein, D., Goodrich, M.T., Kobourov, S.G., Löffler, M.: Planar and poly-arc Lombardi drawings. In: Speckmann, B. (ed.) GD 2011. LNCS, vol. 7034, pp. 308–319. Springer, Heidelberg (2011)

10. Duncan, C.A., Eppstein, D., Goodrich, M.T., Kobourov, S.G., Nöllenburg, M.: Drawing trees with perfect angular resolution and polynomial area. In: Brandes, U., Cornelsen, S. (eds.) GD 2010. LNCS, vol. 6502, pp. 183–194. Springer, Heidelberg (2011)

11. Duncan, C.A., Eppstein, D., Goodrich, M.T., Kobourov, S.G., Nöllenburg, M.: Lombardi drawings of graphs. In: Brandes, U., Cornelsen, S. (eds.) GD 2010. LNCS, vol. 6502, pp. 195–207. Springer, Heidelberg (2011)

12. Eppstein, D., Löffler, M., Mumford, E., Nöllenburg, M.: Optimal 3d angular resolution for low-degree graphs. In: Brandes, U., Cornelsen, S. (eds.) GD 2010. LNCS, vol. 6502, pp. 208–219. Springer, Heidelberg (2011)

13. Fejes Tóth, L.: Über die Abschätzung des kürzesten Abstandes zweier Punkte eneis auf einer Kugelfläche liegenden Punktsystems. Jbf. Deutsch. Math. Verein 53, 66–68 (1943)

14. Formann, M., Hagerup, T., Haralambides, J., Kaufmann, M., Leighton, F.T., Symvonis, A., Welzl, E., Woeginger, G.J.: Drawing graphs in the plane with high resolution. In: FOCS, pp. 86–95. IEEE Computer Society (1990)

15. Garg, A., Tamassia, R.: Planar drawings and angular resolution: Algorithms and bounds. In: van Leeuwen, J. (ed.) ESA 1994. LNCS, vol. 855, pp. 12–23. Springer, Heidelberg (1994)

16. Garg, A., Tamassia, R., Vocca, P.: Drawing with colors. In: Díaz, J. (ed.) ESA 1996. LNCS, vol. 1136, pp. 12–26. Springer, Heidelberg (1996)

17. Goodrich, M.T., Wagner, C.G.: A framework for drawing planar graphs with curves and polylines. Journal of Algorithms 37(2), 399–421 (2000)

18. Gutwenger, C., Mutzel, P.: Planar polyline drawings with good angular resolution. In: Whitesides, S.H. (ed.) GD 1998. LNCS, vol. 1547, pp. 167–182. Springer, Heidelberg (1999)

19. Nicholson, T.: Permutation procedure for minimising the number of crossings in a network. Proc. of the Inst. of Electrical Engineers 115(1), 21–26 (1968)

20. Saaty, T.L.: The minimum number of intersections in complete graphs. Proceedings of the National Academy of Sciences of the United States of America 52(3), 688–690 (1964)

21. Tammes, P.: On the origin of number and arrangements of the places of exit on the surface of pollen grains. Rec. Trav. Bot. Neerl. 27, 1–81 (1930)

22. Vizing, V.G.: On an estimate of the chromatic class of a p-graph. Diskret. Analiz. 3, 25–30 (1964)

23. Wattenberg, M.: Arc diagrams: Visualizing structure in strings. In: IEEE Symp. on Information Visualization (InfoVis), pp. 110–116 (2002)

A Duality Transform for Constructing Small Grid Embeddings of 3D Polytopes[*]

Alexander Igamberdiev and André Schulz

Institut für Mathematische Logik und Grundlagenforschung,
Universität Münster, Germany
{alex.igamberdiev,andre.schulz}@uni-muenster.de

Abstract. We study the problem of how to obtain an integer realization of a 3d polytope when an integer realization of its dual polytope is given. We focus on grid embeddings with small coordinates and develop novel techniques based on Colin de Verdière matrices and the Maxwell–Cremona lifting method.

As our main result we show that every truncated 3d polytope with n vertices can be realized on a grid of size polynomial in n. Moreover, for a class $\mathcal{C}$ of simplicial 3d polytopes with bounded vertex degree, at least one vertex of degree 3, and polynomial size grid embedding, the dual polytopes of $\mathcal{C}$ can be realized on a polynomial size grid as well.

1 Introduction

By Steinitz's theorem the graphs of convex 3d polytopes[1] are exactly the planar 3-connected graphs [16]. Several methods are known for realizing a planar 3-connected graph G as a polytope with graph G on the grid [4,7,11,12,13,15]. It is challenging to find algorithms that produce polytopes with small integer coordinates. Having a realization with small grid size is a desirable feature, since then the polytope can be stored and processed efficiently. Moreover, grid embeddings imply good vertex and edge resolution. Hence, they produce "readable" drawings.

In 2d, it is well known that planar 3-connected graphs with n vertices can be drawn on a $O(n) \times O(n)$ grid without crossings [5], and a drawing with convex faces can be realized on a $O(n^{3/2} \times n^{3/2})$ grid [2]. For the realization as a polytope the best algorithm guarantees an integer embedding with coordinates at most $O(147.7^n)$ [3,11]. The current best lower bound is $\Omega(n^{3/2})$ [1]. Closing this large gap is probably one of the most interesting open problems in lower dimensional polytope theory. Recently, progress has been made for a special class of 3d polytopes, the so-called *stacked polytopes*. A *stacking operation* replaces a triangular face of a polytope with a tetrahedron, while maintaining the convexity of the embedding. A polytope that can be constructed from a tetrahedron and a

---

[*] This work was funded by the German Research Foundation (DFG) under grant SCHU 2458/2-1.
[1] In our terminology polytopes are always considered *convex*.

S. Wismath and A. Wolff (Eds.): GD 2013, LNCS 8242, pp. 173–184, 2013.

sequence of stacking operation is called stacked polytope. The graphs of stacked polytopes are planar 3-trees. Stacked polytopes can be embedded on a grid that is polynomial in n [6]. This is, however, the only nontrivial polytope class for which such an algorithm is known.

In this paper we introduce a duality transform that maintains a polynomial grid size. In other words, we provide a technique that takes a grid embedding of a polytope with graph G and generates a grid embedding of a polytope whose skeleton is G^*, the dual graph of G. We call a 3d polytope with graph G^* a dual polytope. If the original polytope has integer coordinates bounded by a polynomial in n, then the dual polytope obtained with our techniques has also integer coordinates bounded by a (different) polynomial in n. Our methods can only be applied to special polytopes. Namely, we require that the graph of the polytope is a triangulation (the polytope is simplicial), that it contains a K_4, and that the maximum vertex degree is bounded.

For the class of stacked polytopes (although their maximum vertex degree is not bounded) we can also apply our approach to show that all graphs dual to planar 3-trees can be embedded as polytopes on a polynomial size grid. These polytopes are known as truncated polytopes. Truncated polytopes are simple polytope, that can be generated from a tetrahedron and a sequence of *truncations*. A truncation is the dual operation to stacking. This means that a degree-3 vertex of the polytope is cut off by adding a new bounding hyperplane that separates this vertex from the remaining vertices of the polytope. We show that all truncated polytopes can be realized with integer coordinates in $O(n^{44})$. The approach for this class is more direct, since stronger results for realizations of stacked polytopes on the grid are known [6].

Duality. There exist several natural approaches how to construct for a given polytope a dual. The most prominent construction is polarity. Let P be some polytope that contains the origin. Then $P^* = \{y \in \mathbb{R}^d : x^T y \leq 1 \text{ for all } x \in P\}$ is a realization of a polytope dual to P, called its *polar*. The vertices of P^* are intersection points of planes with integral normal vectors, and hence not necessarily integer points. In order to scale to integrality one has to multiply P^* with the product of all denominators of its vertex coordinates, which may cause an exponential increase of the grid size.

A second approach uses the classic Maxwell–Cremona correspondence technique (also known as lifting approach) [10], which is applied in many embedding algorithms for 3d polytope realization. The idea here is to first draw the graph of the polytope as a convex 2d embedding with an additional equilibrium condition. The equilibrium condition guarantees that the 2d drawing is a projection of a convex 3d polytope, furthermore the polytope can be reconstructed from its projection in a canonical way (called lifting) in linear time. There is a classical transformation that constructs for a 2d drawing in equilibrium a 2d drawing of its dual graph, also in equilibrium. This drawing is called the *reciprocal diagram*. The induced lifting realizes the dual polytope, but it does not provide small integer coordinates for two reasons. First, the weights that define the equilibrium of the reciprocal diagram are the reciprocals of the weights in the original graph.

Second, the lifting realizes the dual polytope in projective space with one point "over the horizon". The second property can be "fixed" with a projective transformation. This, however, makes a large scaling factor for an integer embedding unavoidable. Also the reciprocal weights are difficult to handle without scaling by a large factor.

Structure and notation. As a novelty we work with Colin de Verdière matrices to construct *small* grid embeddings. In order to make these techniques (as introduced by Lovász) applicable we extend this framework slightly; see Sect. 2. In Sect. 3 we then present the main idea, combining the classical lifting approach with the methods of Sect. 2, which finds applications in the following sections, where the results on truncated polytopes and triangulations are presented.

Throughout the paper we denote by G the graph of the original polytope, and by G^* its dual graph. For any graph H we write $V(H)$ for its vertex set, $E(H)$ for its edge set and $N(v, H)$ for the set of neighbors of a vertex v in H. Since we consider 3-connected planar graphs, the facial structure of the graph is predetermined up to a global reflection [17]. The set of faces is therefore predetermined, and we name it $F(H)$. For convenience we denote an edge (v_i, v_j) as (i, j). A face spanned by vertices v_i, v_j, and v_k is denoted as $(v_i v_j v_k)$. A graph obtained from H by stacking a vertex v_1 on a face $(v_2 v_3 v_4)$, is denoted as $\mathrm{Stack}(H; v_1; v_2 v_3 v_4)$. For convenience we use $|p|$ for the Euclidean norm of the vector p. We denote the maximum vertex degree of a graph G as Δ_G. Finally, we write $G[X]$ for the induced subgraph of a vertex set $X \subseteq V(G)$.

2   3d Representations with CDV Matrices

In this section we review some of the methods Lovász introduced in his paper on Steinitz representations [9]. In our constructions throughout the paper every face of any graph is realized such that all its vertices lie on a common plane. From this perspective drawings of graphs in $\mathbb{R}^3$ and the realizations of their corresponding polyhedra are the same objects.

Definition 1. *We call a straight-line embedding* $(u_1, \dots, u_n) \in (\mathbb{R}^3)^n$ *of a planar 3-connected graph G in $\mathbb{R}^3$ a* cone-convex embedding *iff the cones over its faces,* $C_f = \{\lambda x \mid x \in f, \lambda > 0\}$, $f \in F(G)$ *are convex and have disjoint interiors.*

In other words, an embedding is a cone-convex embedding if its projection to the sphere $S = \{|x| = 1\}$ is a convex drawing of G with edges drawn as geodesic arcs. We remark that the vertices of a cone-convex embedding are not supposed to form a convex polytope.

Definition 2. *Let* $(u_1, \dots, u_n)$ *be an embedding of a graph G into $\mathbb{R}^d$. We call a symmetric matrix* $M = [M_{ij}]_{1 \leq i,j \leq n}$ *a* CDV matrix *of the embedding if*

1. $M_{ij} = 0$ for $i \neq j, (i,j) \notin E(G)$, and
2. $\sum_{1 \leq j \leq n} M_{ij} u_j = 0$ for $1 \leq i \leq n$.

We call a CDV Matrix positive *if* $M_{ij} > 0$ *for all* $(i, j) \in E(G)$.

We call the second condition in the above definition the *CDV equilibrium condition*.

The CDV equilibrium condition can also be expressed in a slightly different, more geometric form as

$$\sum_{j \in N(i,G)} M_{ij} u_j = -M_{ii} u_i \quad \text{for } 1 \le i \le n. \tag{1}$$

Hence, a positive CDV Matrix witnesses that every vertex of the embedding can be written as a convex combination of its neighbors using symmetric weights. The following lemma appears in [9], we include the proof since it illustrates how to construct a realization out of a CDV matrix.

Lemma 1 (Lemma 4, [9]). *Let $(u_1, \ldots, u_n)$ be a cone-convex embedding of a graph G with a positive CDV matrix $[M_{ij}]$. Then every face f in G can be assigned a vector ϕ_f s.t. for each adjacent face g and separating edge (i, j)*

$$\phi_f - \phi_g = M_{ij}(u_i \times u_j), \tag{2}$$

where f lies to the left and g lies to the right from $\overrightarrow{u_i u_j}$. The set of vectors $\{\phi_f\}$ is uniquely defined up to translations.

Proof. To construct the family of vectors $\{\phi_f\}$, we start by assigning an arbitrary value to ϕ_{f_0} (for an arbitrary face f_0); then we proceed iteratively. To prove the consistency of the construction, we show that the vectors $(\phi_f - \phi_g)$ sum to zero over every cycle in G^*. Since G as well as G^* is planar and 3-connected, it suffices to check this condition for all elementary cycles of G^*, which are the faces of G^*. Let $\tau(i)$ denote the set of counterclockwise oriented edges of the face in G^* dual to $v_i \in V(G)$. Then, combining 1 and 2 yields

$$\sum_{(f,g) \in \tau(i)} (\phi_f - \phi_g) = \sum_{j \in N(i,G)} M_{ij}(u_i \times u_j) = u_i \times \left(\sum_{j \in N(i,G)} M_{ij} u_j \right) = u_i \times (-M_{ii} u_i) = 0.$$

The vectors $\{\phi_f\}$ are unique up to the initial choice of ϕ_{f_0}. □

Note that there is a canonical way to derive a CDV matrix from a 3d polytope [9]. Every 3d embedding of a graph G as a polytope (u_i) possesses a positive CDV matrix defined by the vertices (ϕ_i) of its polar and equation (2). We refer to this matrix as the *canonical CDV matrix*.

The following theorem, which is a variation of Lemma 5 in [9], is the main tool in our construction.

Theorem 1 (based on Lovász [9]). *Let $(u_1, \ldots, u_n)$ be a cone-convex embedding of a graph G and M a positive CDV matrix for this embedding. Then for any set of vectors $\{\phi_f\}_{f \in F(G)}$ fulfilling (2), the convex hull $\mathrm{Conv}(\{\phi_f\}_{f \in F(G)})$ is a convex polytope with graph G^*; and the isomorphism between G^* and the skeleton of $\mathrm{Conv}(\{\phi_f\}_{f \in F(G)})$ is given by $f \to \phi_f$.*

The proof of the theorem is included in the full version of the paper. It relies on a projection of the cone-convex embedding onto the sphere and an appropriate "scaling" of the CDV matrix.

3 Construction of Cone-Convex Embeddings

In this section we describe how to go from a convex 2d embedding with a positive equilibrium stress to a cone-convex 3d embedding with a positive CDV matrix.

Definition 3. *We call a set of reals $\{\omega_{ij}\}_{(i,j)\in E(G)}$ an* equilibrium stress *for an embedding $(u_1, \ldots, u_n)$ of a graph G into $\mathbb{R}^d$ if for each $i \in V(G)$*

$$\sum_{j \in N(i,G)} \omega_{ij}(u_j - u_i) = 0.$$

We call an equilibrium stress of a 2d embedding with a distinguished boundary face f_0 positive *if it is positive on every edge that does not belong to f_0.*

The concept of equilibrium stresses plays a central role in the classical Maxwell–Cremona lifting approach and it is also a crucial concept in our embedding algorithm. The equilibrium stress on a realization of a complete graph arises as a "building block" in later constructions. The complete graph K_n, embedded in $\mathbb{R}^{n-2}$, has a unique equilibrium stress up to multiplication with a scalar. This stress has an easy expression in terms of volumes related to the embedding. We use the *square bracket notation*[2]

$$[q_i q_j q_k q_l] := \det \begin{pmatrix} x_i & x_j & x_k & x_l \\ y_i & y_j & y_k & y_l \\ z_i & z_j & z_k & z_l \\ 1 & 1 & 1 & 1 \end{pmatrix}, \quad \text{where } q = \begin{pmatrix} x \\ y \\ z \end{pmatrix},$$

to obtain a formulation for the equilibrium stress on the K_5 embedding.

Lemma 2 (Rote, Santos, and Streinu [14]). *Let $(u_0, u_1, \ldots, u_4)$ be an integer embedding of the complete graph K_5 onto $\mathbb{R}^3$. Then the set of real numbers:*

$$\omega_{ij} := [u_{i-2} u_{i-1} u_{i+1} u_{i+2}][u_{j-2} u_{j-1} u_{j+1} u_{j+2}]$$

(indices in cyclic notation) defines an integer equilibrium stress on this embedding.

Theorem 2. *Let $(p_2, \ldots, p_n)$ be a convex 2d drawing of a planar 3-connected graph $G_\uparrow$ with positive equilibrium stress $\{\omega_{ij}\}$ and designated triangular face $f_0 = (p_2 p_3 p_4)$ embedded as the boundary face. Then we can define a cone convex embedding (q_i) of the graph $G = \text{Stack}(G_\uparrow; v_1; v_2 v_3 v_4)$ into $\mathbb{R}^3$ equipped with a positive CDV matrix $[M_{ij}]$, such that*

$$M_{ij} = \omega_{ij} \quad \text{for each internal edge } (i, j) \text{ of the 2d drawing of } G_\uparrow$$

and each entry of M is bounded by $O(n \max_{ij} |\omega_{ij}| \cdot \max_i |p_i|^6)$.

---

[2] For 2d vectors $[p_i p_j p_k]$ is defined similarly.

Proof. We can assume that $(0,0)^T$ lies inside the embedding of f_0. Let $(q_1, \ldots, q_n)$ be the embedding of the graph G, defined as follows: The embedding of $G_\uparrow$ is realized in the plane $\{z = 1\}$ and the stacked vertex is placed at $(0, 0, -1)^T$. The embedding is cone-convex since it describes a tetrahedron containing the origin with one face that is refined with a plane convex subdivision.

Following the structure of $G = \text{Stack}(G_\uparrow; v_1; v_2 v_3 v_4)$, we decompose G into two subgraphs: $G_\uparrow = G[\{v_2, \ldots, v_n\}]$ and $G_\downarrow := G[\{v_1, v_2, v_3, v_4\}]$.

We first compute a CDV matrix $[M'_{ij}]_{2 \leq i,j \leq n}$ for the embedding $(q_2, \ldots, q_n)$ of $G_\uparrow$. The plane embedding (p_i) of $G_\uparrow$ has the equilibrium stress $\{\omega_{ij}\}_{2 \leq i,j \leq n}$. Since $\{q_2, \ldots, q_n\}$ is just a translation of $\{p_2, \ldots, p_n\}$, clearly, $\{\omega_{ij}\}_{2 \leq i,j \leq n}$ is as well an equilibrium stress for the embedding $(q_2, \ldots, q_n)$ and we can assign:

$$M'_{ij} := \begin{cases} -\sum_{k \in N(i,G_\uparrow)} \omega_{ik} & i = j, \\ \omega_{ij} & (i,j) \in E(G_\uparrow), \\ 0 & \text{else.} \end{cases}$$

Now we check the CDV equilibrium condition: for every $2 \leq i \leq n$

$$\sum_{2 \leq j \leq n} M'_{ij} q_j = \sum_{j \in N(i,G_\uparrow)} M'_{ij} q_j + M'_{ii} q_i = \sum_{j \in N(i,G_\uparrow)} M'_{ij}(q_j - q_i) + (M'_{ii} + \sum_{j \in N(i,G_\uparrow)} M'_{ij}) q_i$$

$$= \sum_{j \in N(i,G_\uparrow)} \omega_{ij}(q_j - q_i) + (M'_{ii} + \sum_{j \in N(i,G_\uparrow)} M'_{ij}) q_i = 0.$$

The last transition holds since both summands equal 0. Hence, $[M'_{ij}]$ is a valid CDV matrix for the embedding $(q_i)_{2 \leq i \leq n}$ of $G_\uparrow$.

As a second step we compute a CDV matrix $[M''_{ij}]_{1 \leq i,j \leq 4}$ for the embedding of the tetrahedron $G_\downarrow$. We apply Lemma 2 for the embedding of the K_5 formed by $\{q_0 = (0,0,0)^T, q_1, q_2, q_3, q_4\}$ and receive an equilibrium stress $\{\omega''_{ij}\}_{0 \leq i,j \leq 4}$. We can now derive a CDV matrix $[M''_{ij}]_{1 \leq i,j \leq 4}$ for the tetrahedron $\{q_1, q_2, q_3, q_4\}$ based on the equilibrium stresses $\{\omega''_{ij}\}_{0 \leq i,j \leq 4}$ as follows: We set

$$M''_{ij} := \begin{cases} -\sum_{0 \leq j \leq 4, j \neq i} \omega''_{ij}, & i = j, \\ \omega''_{ij}, & \text{otherwise,} \end{cases}$$

and see that the CDV equilibrium condition holds, by noting

$$\forall i \quad \sum_{1 \leq j \leq 4} M''_{ij} q_j = \sum_{1 \leq j \leq 4, j \neq i} M''_{ij} q_j + M''_{ii} q_i = \sum_{1 \leq j \leq 4, j \neq i} \omega''_{ij} q_j + M''_{ii} q_i$$

$$= \sum_{0 \leq j \leq 4, j \neq i} \omega''_{ij}(q_j - q_i) - \omega''_{i0} q_0 + (\sum_{0 \leq j \leq 4, j \neq i} \omega''_{ij} + M''_{ii}) q_i = 0.$$

The last transition holds since $\sum_{0 \leq j \leq 4, j \neq i} \omega''_{ij}(q_j - q_i) = 0$ by the definition of $\{\omega_{ij}\}$, $q_0 = 0$, and $\sum_{0 \leq j \leq 4, j \neq i} \omega''_{ij} + M''_{ii} = 0$ due to the choice of M''_{ii}. One can easily check that as soon as the origin lies inside the tetrahedron $\{q_1, q_2, q_3, q_4\}$ all entries M''_{ij} have the same sign. We can assume that $[M''_{ij}]$ is positive, otherwise we reorder the vertices $\{v_2, v_3, v_4\}$.

In the final step we extend the two CDV matrices M' and M'' to G and combine them. Clearly, a CDV matrix padded with zeros remains a CDV matrix. Furthermore, any linear combination of CDV matrices is again a CDV matrix. Thus, we form a CDV matrix for the whole embedding $(q_1, \ldots, q_n)$ of G by setting:

$$M := M' + \lambda M'',$$

where λ is a positive integer chosen such that M is a positive CDV matrix. This can be done as follows.

Recall that $\{\omega_{ij}\}$ is a positive stress and $[M''_{ij}]$ is a positive CDV matrix. Hence, the only six entries in $[M_{ij}]$ that may be negative are: M_{23}, M_{34} and M_{42} (and their symmetric entries), for which $M_{ij} := M'_{ij} + \lambda M''_{ij}$ with $M'_{ij} < 0$ and $M''_{ij} > 0$. Thus, we choose λ such that M is positive at these entries. To satisfy this condition we pick

$$\lambda = \left\lfloor \max_{(i,j)\in\{(2,3),(3,4),(4,2)\}} (|M'_{ij}|/|M''_{ij}|) \right\rfloor + 1.$$

To bound M_{ij} we notice that the entries of M''_{ij} are strictly positive integers, so $\lambda = O(\max |M'_{ij}|)$, while $|M'_{ij}| = O(n \cdot \max |\omega_{ij}|)$ and $|M''_{ij}| = O(\max |\omega''_{ij}|) = O(\max |p_i|^6)$. The bound $|M_{ij}| = O(n \cdot \max_{ij} |\omega_{ij}| \cdot \max_i |p_i|^6)$ follows. $\square$

4 Realizations of Truncated Polytopes

In this section we sum up previous results in Theorem 3 and present an embedding algorithm for truncated 3d polytopes in Theorem 4. We will apply Theorem 3 also in the more general setup of Sect. 5.

Theorem 3. *Let $G = \mathrm{Stack}(G_\uparrow; v_1; v_2v_3v_4)$ and $(p_2, \ldots, p_n)$ be an integer planar embedding of $G_\uparrow$ with boundary face $(v_2v_3v_4)$ and with positive integer equilibrium stress $\{\omega_{ij}\}$. Then one can construct a grid embedding (ϕ_f) of a convex polytope with graph G^* such that*

$$|\phi_f| = O(n^2 \cdot \max |\omega_{ij}| \cdot \max |p_i|^8).$$

Proof. We first apply Theorem 2 to obtain a cone-convex embedding $(q_1, \ldots, q_n)$ of G with a positive CDV matrix $[M_{ij}]_{1\leq i,j\leq n}$. We then apply Lemma 1 and obtain a family of vectors $\{\phi_f\}_{f\in F(G^*)}$ fulfilling

$$\phi_f - \phi_g = M_{ij}(q_i \times q_j), \quad \forall(f,g) \text{ dual to } (i,j) - \text{edges of } G^* \text{ and } G.$$

Due to Theorem 1 the vectors $\{\phi_f\}$ form a realization of G^* as a polytope.

To finish the proof we estimate how large the coordinates of the embedding (ϕ_f) are. To do so, let us again follow the construction of (ϕ_f) as outlined in the proof of Lemma 1. We pick one face as $f_0 \in F(G)$, and assign $\phi_{f_0} = (0,0,0)^T$. Let us now evaluate ϕ_{f_k} for some face $f_k \in F(G)$. The following algebraic expression holds for all values $\{\phi_{f_i}\}$:

$$\phi_{f_k} = \phi_{f_0} + (\phi_{f_1} - \phi_{f_0}) + \ldots + (\phi_{f_{k-1}} - \phi_{f_{k-2}}) + (\phi_{f_k} - \phi_{f_{k-1}}).$$

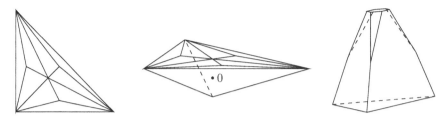

Fig. 1. A 2d embedding of $G_\uparrow$ (left), the cone-convex embedding of G (center), and the resulting embedding of the dual (right).

Let us now consider the shortest path $f_0, f_1, \ldots, f_k$ in G^* connecting the faces f_0 and f_k. Clearly, k is less than $2n - 3$, and hence

$$|\phi_{f_k}| \leq 2n \cdot \max_{(f_a, f_b) \in E(G^*)} |\phi_{f_a} - \phi_{f_b}| = 2n \cdot \max_{v_i, v_j \in V(G)} |M_{ij}(q_i \times q_j)|$$

$$= O(n \cdot (n \cdot \max |\omega_{ij}| \cdot \max |p_i|^6) \cdot \max |q_i|^2) = O(n^2 \cdot \max |\omega_{ij}| \cdot \max |p_i|^8).$$

The bound for the entries of M is due to Theorem 2.   □

Next we apply Theorem 3 to construct an integer polynomial size grid embedding for truncated polytopes. To construct small integer 2d embeddings with small integer equilibrium stresses we use a Lemma by Demaine and Schulz [6], which states that the graph of a stacked polytope with n vertices and any distinguished face f_0 can be embedded on a $10n^4 \times 10n^4$ grid with boundary face f_0 and with integral positive equilibrium stress $\{\omega_{ij}\}$ such that, for every edge (i, j), we have $|\omega_{ij}| = O(n^{10})$.

Theorem 4. *Any truncated 3d polytope with n vertices can be realized with integer coordinates of size $O(n^{44})$.*

Proof. Let G^* be the graph of the truncated polytope and $G := (G^*)^*$ its dual. Clearly, G is the graph of a stacked polytope with $(n + 4)/2$ vertices. We denote the last stacking operation (for some sequence of stacking operations producing G) as the stacking of the vertex v_1 onto the face $(v_2 v_3 v_4)$ of the graph $G_\uparrow := G[V \setminus \{v_1\}]$. The graph $G_\uparrow$ is again a stacked graph, and hence, by the Lemma of Demaine and Schulz, there exists an embedding $(p_i)_{2 \leq i \leq n}$ of $G_\uparrow$ into $\mathbb{Z}^2$ with an equilibrium stress $\{\omega_{ij}\}$ satisfying the properties of Theorem 3. We apply the theorem and obtain a polytope embedding (ϕ_f) of G^* with bound $|\phi_f| = O(n^2 \cdot \max |\omega_{ij}| \cdot \max |p_i|^8) = O(n^{44})$.   □

Figure 1 shows an example of our method. The computations for this example are included in the full paper.

5   A Dual Transform for Simplicial Polytopes

As we have seen a small grid embedding of a 3d polytope can be computed when a small integer (though, not necessarily convex) embedding of its dual polytope with a small integral positive CDV matrix is known. However, if one wants to build a dual for an already embedded polytope, one usually does not possess such a matrix. The canonical CDV matrix associated with any embedding of a 3d polytope is not helpful, since its entries, when scaled to integers, might become exponentially large. We show in this section how one can tackle this problem for a special class of polytopes. In particular, we require that the original polytope is simplicial, it contains a vertex of degree 3, and its maximum vertex degree is bounded.

Before proceeding, let us review how the canonical stress associated with an orthogonal projection of a 3d polytope in the $\{z = 0\}$ plane can be described. The assignment of heights to the interior vertices of a 2d embedding resulting in a polyhedral surface is called a (polyhedral) *lifting*. By the Maxwell-Cremona correspondence the equilibrium stresses of a 2d embedding of a planar 3-connected graph and its liftings are in 1-1 correspondence. Moreover, the bijection between liftings and stresses can be defined as follows. Let (p_i) be a 2d drawing of a triangulation and let (q_i) be the 3d embedding induced by some lifting. We map this lifting to the equilibrium stress $\{\omega_{ij}\}$ by assigning to every edge (i, j) separating the faces $(v_i v_j v_k)$ (on the left) and $(v_i v_j v_l)$ (on the right)

$$\omega_{ij} := \frac{[q_i q_j q_k q_l]}{[p_i p_j p_k][p_l p_j p_i]}. \tag{3}$$

This mapping is a bijection between the space of liftings and the space of equilibrium stresses. The expression (3) is a slight reformulation of the form presented in Hopcroft and Kahn [8, Equation 11].

We continue by studying the spaces of equilibrium stresses for triangulations. A graph formed by a cycle $v_1, \ldots, v_n$ with an additional vertex v_0, called *center*, that is adjacent to every other vertex, is called a *wheel*; we denote it as $W(v_0; v_1 \ldots v_n)$. A wheel that is a subgraph of a triangulation G with $v_i \in V(G)$ as center is denoted by W_i. Every triangulation can be "covered" with a set of wheels $\{W_i\}_{v_i \in V(G)}$, such that every edge is covered four times.

Lemma 3. *Let $(p_0, \ldots, p_n)$ be an embedding of a wheel $W(v_0; v_1 \ldots v_n)$ in $\mathbb{R}^2$. Then the following expression defines an equilibrium stress:*

$$\omega_{ij} = \begin{cases} -1/[p_i p_{i+1} p_0] & j = i+1, 1 \le i \le n, \\ [p_{i-1} p_i p_{i+1}]/([p_{i-1} p_i p_0][p_i p_{i+1} p_0]) & j = 0, 1 \le i \le n. \end{cases}$$

The equilibrium stress for the embedding (p_i) is unique up to a renormalization.

Proof. This stress coincides with (3) from the lifting of W with $z_0 = 1$ and $z_i = 0$ for $1 \le i \le n$ and so is an equilibrium stress. The space of the stresses is 1-dimensional, since the space of the polyhedral liftings is 1-dimensional. $\square$

Definition 4. *1. For a wheel W embedded in the plane we refer to the equilibrium stress defined in Lemma 3 as its* small atomic stress *and denote it as $\{\omega_{ij}^a(W)\}$.*

2. *We call the stress* $\{\Omega_{ij}^a(W)\}$ *that is obtained by the renormalization of* $\{\omega_{ij}^a(W)\}$ *by the factor* $\prod_{1\leq j\leq n}[p_j p_{j+1} p_0]$, *the large atomic stress of* W.

We point out that the large atomic stresses are products of $\deg(v_0) - 1$ triangle areas multiplied by 2, and so, $\{\Omega_{ij}^a(W)\}$ is a set of integers if W is realized with integer coordinates.

Theorem 5 (Wheel-decomposition). *Let G be a triangulation. Every equilibrium stress $\{\omega_{ij}\}$ of an embedding $(p_1,\ldots,p_n)$ of G can be expressed as a linear combination of the small atomic stresses on the wheels $\{W_i\}$:*

$$\omega = \sum_{i\leq n} \alpha_i \omega^a(W_i),$$

where the coefficients α_i are the heights (i.e., z-coordinates) of the corresponding vertices v_i in the Maxwell–Cremona lifting of $(p_1,\ldots,p_n)$ induced by $\{\omega_{ij}\}$.

Proof. Let $(q_1,\ldots,q_n)$ be the Maxwell-Cremona lifting of (p_i) by means of the stress $\{\omega_{ij}\}$. We rewrite this stress (given by Equation 3) using

$$[q_1 q_2 q_3 q_4] = \sum_{1\leq i\leq 4} (-1)^{i+1} z_i [p_{i+1} p_{i+2} p_{i+3}],$$

(with cyclic notation for indices) and obtain

$$\omega_{ij} = z_i \frac{[p_j p_k p_l]}{[p_i p_j p_k][p_l p_j p_i]} + z_j \frac{[p_l p_k p_i]}{[p_i p_j p_k][p_l p_j p_i]} - z_k \frac{1}{[p_i p_j p_k]} - z_l \frac{1}{[p_l p_j p_i]},$$

which is exactly the decomposition of ω_{ij} into small atomic wheel stresses.   □

Theorem 6. *Let $(q_1,\ldots,q_n)$ be an embedding of a triangulation G into $\mathbb{Z}^3$, whose projection $(p_1,\ldots,p_n)$ to the plane $\{z = 0\}$ is a noncrossing embedding of G with boundary face $(v_1 v_2 v_3)$. Then one can construct a positive integer equilibrium stress $\{\omega_{ij}\}$ for the embedding $(p_1,\ldots,p_n)$ such that*

$$|\omega_{ij}| < (\max_{i\leq n} |q_i|)^{2\Delta_G + 5}.$$

Proof. We start with the equilibrium stress $\{\widetilde{\omega}_{ij}\}$ as specified by (3) for the embedding (p_i). Since all the coordinates are integers, all stresses are bounded by

$$\frac{1}{L^4} \leq \frac{1}{|[p_i p_j p_k]||[p_l p_j p_i]|} \leq |\widetilde{\omega}_{ij}| \leq |[q_i q_j q_k q_l]| \leq L^3,$$

for $L = 2\max_{i\leq n} |q_i|$.

We are left with making these stresses integral while preserving a polynomial bound. The stress $\{\widetilde{\omega}_{ij}\}$ can be written as a linear combination of *large* atomic stresses of the wheels W_i by means of the Wheel-decomposition Theorem,

$$\widetilde{\omega}_{ij} = \alpha_i \Omega_{ij}^a(W_i) + \alpha_j \Omega_{ij}^a(W_j) + \alpha_k \Omega_{ij}^a(W_k) + \alpha_l \Omega_{ij}^a(W_l).$$

Since all points p_i have integer coordinates, the large atomic stresses are integers as well. Moreover, each of them, as a product of $\deg(v_k) - 1$ triangle areas, is bounded by $|\Omega_{ij}^a(W_k)| \leq L^{2(\Delta_G - 1)}$.

To make the $\widetilde{w}_{ij}$s integral we round the coefficients α_i down. To guarantee that the rounding does not alter the signs of the stress, we scale the atomic stresses (before rounding) with the factor

$$C = 4 \max_{i,j,k} |\Omega_{ij}^a(W_k)| / \min_{i,j} |\widetilde{w}_{ij}|$$

and define as the new stress:

$$\omega_{ij} := \lfloor C\alpha_i \rfloor \Omega_{ij}^a(W_i) + \lfloor C\alpha_j \rfloor \Omega_{ij}^a(W_j) + \lfloor C\alpha_k \rfloor \Omega_{ij}^a(W_k) + \lfloor C\alpha_l \rfloor \Omega_{ij}^a(W_l).$$

Clearly,

$$|\omega_{ij} - C\widetilde{w}_{ij}| = \left| \sum_{\tau=i,j,k,l} (\lfloor C\alpha_\tau \rfloor - C\alpha_\tau)\Omega_{ij}^a(W_\tau) \right|$$

$$\leq \sum_{\tau=i,j,k,l} |\Omega_{ij}^a(W_\tau)| \leq 4 \max_{i,j,k} |\Omega_{ij}^a(W_k)| = C \min_{i,j} |\widetilde{w}_{ij}| \leq C|\widetilde{w}_{ij}|$$

and so $\operatorname{sign}(\omega_{ij}) = \operatorname{sign}(C\widetilde{w}_{ij}) = \operatorname{sign}(\widetilde{w}_{ij})$.

Therefore, the constructed equilibrium stress $\{\omega_{ij}\}$ is integral and positive. We conclude the proof with an upper bound on its size. Since $C < 4\,L^{2(\Delta_G - 1)}\,L^4$,

$$|\omega_{ij}| \leq \left| \sum_{\tau=i,j,k,l} (C\alpha_\tau \pm 1)\Omega_{ij}^a(W_\tau) \right| \leq C|\widetilde{w}_{ij}| + \sum_{\tau=i,j,k,l} |\Omega_{ij}^a(W_\tau)|$$

$$\leq C \max |\widetilde{w}_{ij}| + 4 \max |\Omega_{ij}^a(W_k)| \leq 4\,L^{2\Delta_G + 2} \cdot L^3 + 4\,L^{2\Delta_G - 2} = O(L^{2\Delta_G + 5}).$$

$\square$

Combining Theorem 6 and Theorem 3 leads to the following result:

Theorem 7. *Let $(q_2, \ldots, q_n)$ be an integer embedding of a simplicial 3d polytope with graph $G_\uparrow$, such that the orthogonal projection into the plane $\{z = 0\}$ gives a planar 2d embedding $(p_2, \ldots, p_n)$ with boundary face $(v_2 v_3 v_4)$. Then we can construct an embedding $(\phi_f)_{f \in F(G)}$ of a graph dual to $G = \mathrm{Stack}(G_\uparrow; v_1; v_2 v_3 v_4)$ with integer coordinates bounded by*

$$|\phi_f| = O(n^2 \max |q_i|^{2\Delta_G + 13}).$$

We remark that the algorithms following the lifting approach generate embeddings that fulfill the conditions of the above theorem. Using a more technical analysis we can even show that the following stronger version of Theorem 7 holds. The proof of the theorem can be found in the full version of the paper.

Theorem 8. *Let G be a triangulation with at least one vertex of degree 3, and let (q_i) be an integer realization of G as a convex polytope. Then there is a realization $(\phi_f)_{f \in F(G)}$ of the dual graph G^* as a convex polytope with integer coordinates bounded by*

$$|\phi_f| < \max |q_i|^{O(\Delta_G)}.$$

References

1. Andrews, G.E.: A lower bound for the volume of strictly convex bodies with many boundary lattice points. Trans. Amer. Math. Soc. 99, 272–277 (1961)
2. Bárány, I., Rote, G.: Strictly convex drawings of planar graphs. Documenta Math. 11, 369–391 (2006)
3. Buchin, K., Schulz, A.: On the number of spanning trees a planar graph can have. In: de Berg, M., Meyer, U. (eds.) ESA 2010, Part I. LNCS, vol. 6346, pp. 110–121. Springer, Heidelberg (2010)
4. Das, G., Goodrich, M.T.: On the complexity of optimization problems for 3-dimensional convex polyhedra and decision trees. Computational Geometry: Theory and Applications 8(3), 123–137 (1997)
5. de Fraysseix, H., Pach, J., Pollack, R.: How to draw a planar graph on a grid. Combinatorica 10(1), 41–51 (1990)
6. Demaine, E.D., Schulz, A.: Embedding stacked polytopes on a polynomial-size grid. In: Proc. 22nd ACM-SIAM Symposium on Discrete Algorithms (SODA), pp. 1177–1187. ACM Press (2011)
7. Eades, P., Garvan, P.: Drawing stressed planar graphs in three dimensions. In: Brandenburg, F.J. (ed.) GD 1995. LNCS, vol. 1027, pp. 212–223. Springer, Heidelberg (1996)
8. Hopcroft, J.E., Kahn, P.J.: A paradigm for robust geometric algorithms. Algorithmica 7(4), 339–380 (1992)
9. Lovász, L.: Steinitz representations of polyhedra and the Colin de Verdière number. J. Comb. Theory, Ser. B 82(2), 223–236 (2001)
10. Maxwell, J.C.: On reciprocal figures and diagrams of forces. Phil. Mag. Ser. 27, 250–261 (1864)
11. Mor, A.R., Rote, G., Schulz, A.: Small grid embeddings of 3-polytopes. Discrete & Computational Geometry 45(1), 65–87 (2011)
12. Onn, S., Sturmfels, B.: A quantitative Steinitz' theorem. Beiträge zur Algebra und Geometrie 35, 125–129 (1994)
13. Richter-Gebert, J.: Realization Spaces of Polytopes. Lecture Notes in Mathematics, vol. 1643. Springer (1996)
14. Rote, G., Santos, F., Streinu, I.: Expansive motions and the polytope of pointed pseudo-triangulations. Discrete and Computational Geometry–The Goodman-Pollack Festschrift 25, 699–736 (2003)
15. Schulz, A.: Drawing 3-polytopes with good vertex resolution. Journal of Graph Algorithms and Applications 15(1), 33–52 (2011)
16. Steinitz, E.: Polyeder und Raumeinteilungen. In: Encyclopädie der mathematischen Wissenschaften, vol. 3-1-2 (Geometrie), ch. 12, pp. 1–139. B.G. Teubner, Leipzig (1916)
17. Whitney, H.: Congruent graphs and the connectivity of graphs. Amer. J. Math. 54, 150–168 (1932)

Block Additivity of $\mathbb{Z}_2$-Embeddings

Marcus Schaefer[1] and Daniel Štefankovič[2]

[1] School of Computing, DePaul University, Chicago, IL 60604
mschaefer@cs.depaul.edu
[2] Computer Science Department, University of Rochester, Rochester, NY 14627-0226
stefanko@cs.rochester.edu

Abstract. We study embeddings of graphs in surfaces up to $\mathbb{Z}_2$-homology. We introduce a notion of genus mod 2 and show that some basic results, most noteworthy block additivity, hold for $\mathbb{Z}_2$-genus. This has consequences for (potential) Hanani-Tutte theorems on arbitrary surfaces.

1 Introduction

A graph G *embeds* in a surface $\mathcal{S}$ if it can be drawn in $\mathcal{S}$ so that no pair of edges cross. In this paper we want to relax the embedding condition using $\mathbb{Z}_2$-homology, that is, we are only interested in the parity of the number of crossings between independent edges; in terms of algebraic topology we are studying the "mod 2 homology of the deleted product of the graph" [1]. We say a graph $\mathbb{Z}_2$-*embeds* in $\mathcal{S}$ if it can be drawn in $\mathcal{S}$ so that every pair of independent edges crosses evenly. This approach is inspired by two Hanani-Tutte theorems which, for the plane [2] and the projective plane [3], show that embeddability is equivalent to $\mathbb{Z}_2$-embeddability. For other surfaces, it is only known that $\mathbb{Z}_2$-embeddability is a necessary condition for embeddability. In this paper we want to lay the foundations for a study of $\mathbb{Z}_2$-embeddings of graphs in surfaces which may, at some point, lead to a proof of the Hanani-Tutte theorem for arbitrary surfaces. Our main result is that if we define the notion of $\mathbb{Z}_2$-genus as a homological invariant of $\mathbb{Z}_2$-embeddings, then block additivity holds for $\mathbb{Z}_2$-genus just as it does for the standard notion of genus (as proved by Battle, Harary, Kodama and Youngs for the orientable case, and by Stahl and Beineke in the non-orientable case, see [4, Section 4.4]). This implies that a counterexample to the Hanani-Tutte theorem on an arbitrary surface can be assumed to be 2-connected (Corollary 1).

2 $\mathbb{Z}_2$-Embeddings

2.1 Definition

In the introduction we defined a $\mathbb{Z}_2$-embedding in a surface $\mathcal{S}$, as a drawing of a graph G in which every pair of independent edges crosses evenly. In this section, we want to develop a more algebraic version of this definition, which separates the topology of the surface from the drawing. We start with the plane, and then add crosscaps and handles.

S. Wismath and A. Wolff (Eds.): GD 2013, LNCS 8242, pp. 185–195, 2013.

Pick an initial drawing D of $G = (V, E)$ in the plane. For edges $e, f \in E$ let $i_D(e, f)$ be the number of crossings of e and f in the drawing D. We want to extend the drawing to a surface $\mathcal{S}$ with c crosscaps. Since we only plan to keep track of the parity of the number of crossings of independent edges and hence we use $\mathbb{Z}_2$-homology. For each edge we have a vector $y_e \in \mathbb{Z}_2^c$ where $(y_e)_i = 1$ if e is pulled through the i-th crosscap an odd number of times (in a drawing, we can deform e so it passes through the i-th crosscap; this changes the crossing parity of e with any edge that passes through the i-th crosscap an odd number of times). We also allow changing the planar part of the drawing—for each edge we have a vector $x_e \in \mathbb{Z}_2^V$ where $(x_e)_v = 1$ indicates that we made an (e, v)-move, that is, we pull the edge e over v (this changes the crossing parity between e and any edge incident to v). We say that the initial drawing together with $\{x_e\}_{e \in E}$ and $\{y_e\}_{e \in E}$ is a $\mathbb{Z}_2$-embedding of G in $\mathcal{S}$ if for each pair of independent edges $e = \{u, v\}, f = \{s, t\}$ we have

$$i_D(e, f) + (x_e)_s + (x_e)_t + (x_f)_u + (x_f)_v + y_e^T y_f \equiv 0 \pmod 2. \tag{1}$$

All congruences in this paper are modulo 2, so to simplify notation, we drop (mod 2) from now on. See Figure 1 for a $\mathbb{Z}_2$-embedding of K_5 in the projective plane, illustrating the effect of crosscap- and (e, v)-moves. This definition is equivalent to the more intuitive definition given in the introduction (see, for example, Levow [5, Theorem 3]). We say that the drawing is *orientable* if $y_e^T y_e \equiv 0$ for every $e \in E$ (that is, every y_e has an even number of ones).

Handles can be dealt with in the following way: For each handle we have three coordinates in y_e and the possible settings for these coordinates are 000, 110, 101, and 011. We extend the definitions of $\mathbb{Z}_2$-embeddings and orientability given earlier to y_e containing handles. Note that each of the four vectors modeling an edge passing through handle has an even number of ones, so if the surface contains only handles, then any drawing on it is orientable by the earlier definition.

A handle and a crosscap are equivalent to three crosscaps (Dyck, see [6, Section 1.2.4]): in the $\mathbb{Z}_2$-homology this corresponds to the following bijection (we replace the $3 + 1$ coordinates in y_e by 3 coordinates in y_e):

$$\begin{aligned} &000\,0 \leftrightarrow 000, 000\,1 \leftrightarrow 111, 011\,0 \leftrightarrow 011, 011\,1 \leftrightarrow 100, \\ &110\,0 \leftrightarrow 110, 110\,1 \leftrightarrow 001, 101\,0 \leftrightarrow 101, 101\,1 \leftrightarrow 010, \end{aligned} \tag{2}$$

where the first three coordinates on the left hand sides correspond to the handle. Note that (2) preserves the parity of the number of ones (and thus the orientability of the drawing). Also note that (2) is linear (add vector 111 times the last coordinate to the vector of the first three coordinates) and hence preserves the dimension of the space generated by the $\{y_e\}_{e \in E}$.

Remark 1 ($\mathbb{Z}_2$-drawings). Call a drawing D of a graph G in the plane together with $\{x_e\}_{e \in E}$ and $\{y_e\}_{e \in E}$ a $\mathbb{Z}_2$-*drawing*, and define $i_{D,x,y}(e, f) := i_D(e, f) + (x_e)_s + (x_e)_t + (x_f)_u + (x_f)_v + y_e^T y_f$. With this notion of $\mathbb{Z}_2$-drawing, we can model drawings of graphs in a surface up to the $\mathbb{Z}_2$-homology we are interested in: If D is

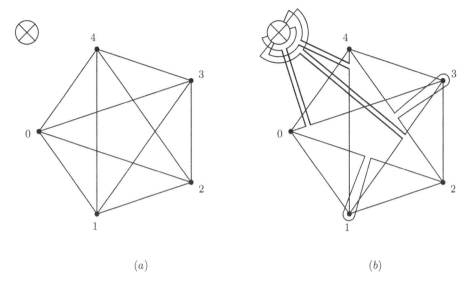

(a)   (b)

Fig. 1. (a) shows the initial drawing of $G = K_5$ in the projective plane. (b) shows a $\mathbb{Z}_2$-embedding of G in the projective plane with $(x_{02})_1 = (x_{24})_3 = 1$, $y_{03} = y_{14} = y_{24} = 1$ (dropping the subscript for the single crosscap) and all other values being zero.

a drawing of a graph G in some surface $\mathcal{S}$, then there is a $\mathbb{Z}_2$-drawing (D', x, y) of G in $\mathcal{S}$ so that $i_D(e, f) \equiv i_{D',x,y}(e, f)$ for every pair (e, f) of independent edges. As mentioned earlier, a result like this (with a slightly different model) was stated by Levow [5]. In the plane, algebraic topologists would phrase this as saying that any two drawings differ by a coboundary, or that they define the same cohomology class in the second symmetric cohomology, see, for example, [7, Section 4.6].

By the observations in Remark 1, any embedding in a surface $\mathcal{S}$ can be considered a $\mathbb{Z}_2$-embedding, so that having a $\mathbb{Z}_2$-embedding is a necessary condition for embeddability in a surface. Hanani-Tutte theorems state that this condition is also sufficient. As we mentioned earlier, this is only known for the plane [2] and the projective plane [3].

Remark 2 (Crosscaps versus Handles in $\mathbb{Z}_2$-Embeddings). Suppose that a surface $\mathcal{S}$ contains c crosscaps and h handles. By the classification theorem for surfaces, each handle is equivalent to two crosscaps, as long as $c > 0$. The same is true for $\mathbb{Z}_2$-embeddability: If $c > 0$, then $\mathbb{Z}_2$-embeddability in $\mathcal{S}$ is equivalent to $\mathbb{Z}_2$-embeddability in a surface with $c + 2h$ crosscaps—we apply (2) and convert each handle into 2 crosscaps; the transformation is possible because $c > 0$. If $c = 0$, then $\mathbb{Z}_2$-embeddability in $\mathcal{S}$ is equivalent—again using (2)—to $\mathbb{Z}_2$-embeddability in a surface with $2h + 1$ crosscaps where drawings are restricted to be orientable. The orientability ensures that when applying (2) to convert crosscaps into handles, the single crosscap left at the end is not used by any edge and hence can be discarded.

With this terminology, we can now define $\mathbb{Z}_2$-homological variants of the genus and the Euler genus of a graph. We write $\mathbf{g}(G)$ and $\mathbf{eg}(G)$ for the traditional genus and Euler genus of G (following [4]).

Definition 1 ($\mathbb{Z}_2$-genus and $\mathbb{Z}_2$-Euler genus). *If a graph G has a $\mathbb{Z}_2$-embedding in an orientable surface with h handles, but not in any surface with fewer handles, we write $\mathbf{g}_0(G) = h$ and call h the $\mathbb{Z}_2$-genus of G. If G has a $\mathbb{Z}_2$-embedding in a surface S with c crosscaps and h handles, but not in any surface with a smaller value of $2h + c$, we write $\mathbf{eg}_0(G) = 2h + c$ and call $2h + c$ the $\mathbb{Z}_2$ Euler genus of G.*

By definition, we have $\mathbf{g}_0(G) \leq \mathbf{g}(G)$ and $\mathbf{eg}_0(G) \leq \mathbf{eg}(G)$, where $\mathbf{g}(G)$ is the genus of G and $\mathbf{eg}(G)$ is the Euler genus of G.

2.2   Basic Properties

We derive some basic properties of $\mathbb{Z}_2$-embeddings. We call two graphs G and H *disjoint* if $V(G) \cap V(H) = \emptyset$.

Lemma 1. *Let G be a graph $\mathbb{Z}_2$-embedded in a (possibly non-orientable) surface S. Let C_1 and C_2 be two disjoint cycles in G. Then*

$$\sum_{e \in C_1, f \in C_2} y_e^T y_f \equiv 0. \tag{3}$$

Let $e_1 \in C_1$ and $e_2 \in C_2$. Suppose that all edges e in $C_1 \setminus \{e_1\}$ have $y_e = 0$ and all edges $f \in C_2 \setminus \{e_2\}$ have $y_f = 0$. Then

$$y_{e_1}^T y_{e_2} \equiv 0. \tag{4}$$

Proof. We have

$$\sum_{e \in C_1, f \in C_2} \left(i_D(e, f) + (x_e)_s + (x_e)_t + (x_f)_u + (x_f)_v + y_e^T y_f \right) \equiv 0, \tag{5}$$

where s, t are endpoints of f and u, v are endpoints of e (note that s, t, u and v vary over the terms in the sum). The equality in (5) follows from the disjointness of C_1 and C_2 (any $e \in C_1$ and $f \in C_2$ are independent and hence (1) has to be satisfied).

We have

$$\sum_{e \in C_1, f \in C_2} \left(i_D(e, f) + (x_e)_s + (x_e)_t + (x_f)_u + (x_f)_v \right) \equiv 0, \tag{6}$$

since (6) corresponds to a drawing in the plane (and two transversally intersecting cycles in the plane intersect evenly; the cycles have to intersect transversally since they are disjoint).

Combining (5) with (6) we obtain (3). Equation (4) is an immediate corollary (since terms in (3) other than $y_{e_1}^T y_{e_2}$ are zero). $\qquad \square$

Lemma 1 suggests that orthogonality plays a role in understanding $\mathbb{Z}_2$-embeddings. We will need the following fact about vector spaces over finite fields (see [8, Section 2.3]).

Lemma 2. *Let A be a subspace of $\mathbb{Z}_2^t$. Then for $A^\perp := \{x \in \mathbb{Z}_2^t \mid (\forall y \in A)\ x^T y \equiv 0\}$ we have*

$$\dim A + \dim A^\perp = t.$$

Let A, B be subspaces of $\mathbb{Z}_2^t$ with $A \subseteq B$. Then for $A^{\perp B} := \{x \in B \mid (\forall y \in A)\ x^T y \equiv 0\} = A^\perp \cap B$ we have

$$\dim A + \dim A^{\perp B} = \dim B + \dim \operatorname{rad} B,$$

where $\operatorname{rad} B := B^{\perp B}$ is the radical *of B.*

The dimension of the $\mathbb{Z}_2$-embedding (the dimension of the space spanned by $\{y_e\}_{e \in E}$) is closely related to its $\mathbb{Z}_2$-genus. In Lemma 3 we extend this result to $\mathbb{Z}_2$-Euler genus.

Lemma 3. *Let G be a graph $\mathbb{Z}_2$-embedded in a (possibly non-orientable) surface S. Assume that the drawing is orientable. Let d be the dimension (over $\mathbb{Z}_2$) of the vector space generated by the edge vectors $\{y_e\}_{e \in E}$. Then G can be $\mathbb{Z}_2$-embedded in an orientable surface of genus $\lfloor d/2 \rfloor$.*

Proof. In the light of Remark 2 we can assume that S has t crosscaps (and no handles). We are going to remove the crosscaps from S one by one. Let S be the space generated by the edge vectors $\{y_e\}_{e \in E}$. Let $d = \dim S$. Let $T = S^\perp$ be the space of vectors that are perpendicular to S. Assume that T contains a vector z such that $z \not\equiv 0$ and $z^T z \equiv 0$ (the computations are in $\mathbb{Z}_2$). Rearranging coordinates, if necessary, we can assume that $z_1 = 1$. To each y_e for which $(y_e)_1 = 1$ we add z. This transformation has the following properties:

- it is linear ($y \mapsto y + y_1 z$) and hence the dimension of S cannot increase,
- it preserves orientability (since $(y + z)^T (y + z) \equiv y^T y$),
- it preserves the parity of the number of crossings for every independent pair of edges e, f (since $(y_e + z)^T (y_f + z) \equiv y_e^T y_f$ and $(y_e + z)^T y_f \equiv y_e^T y_f$; here we use the fact that $T = S^\perp$).

After the transformation, the first crosscap is not used by any edge and hence we can remove it thus decreasing t. We repeat the crosscap removal process as long as such a z exists. We distinguish three cases depending on whether the process stops with $d \le t - 2$, $d = t - 1$ or $d = t$.

If $d \le t - 2$, then T always contains $z \ne 0$ with $z^T z \equiv 0$ (since by a dimension argument there are two distinct vectors z_1, z_2 in $T \setminus \{0\}$ and then one of $z_1, z_2, z_1 + z_2$ satisfies $z^T z \equiv 0$) and hence we can always remove a crosscap in this case. Therefore, the process ends up with either $d = t$ or $d = t - 1$, $T = \langle z \rangle$, and $z^T z \equiv 1$. If $d = t$, then we convert the crosscaps back into $\lfloor d/2 \rfloor$ handles (if t is odd we end up with $(t - 1)/2$ handles, if t is even we add an extra crosscap and end up with $t/2$ handles, in both cases use Remark 2 on being able to drop a crosscap in an orientable embedding).

The final case to handle is $d = t - 1$. Let k be the number of ones in z. W.l.o.g. the first k coordinates of z are 1 and the rest are 0. Note that k is odd and that $z^T y_e \equiv 0$ for each e (that is, if we restrict our attention to the first k crosscaps, the drawing is orientable). Hence we can convert the first k crosscaps into $(k-1)/2$ handles. Then—as in the $d = t$ case—we convert the remaining $t - k$ crosscaps into $\lfloor (t-k)/2 \rfloor$ handles. In total, we have

$$(k-1)/2 + \lfloor (t-k)/2 \rfloor = \lfloor (t-1)/2 \rfloor = \lfloor d/2 \rfloor$$

handles.   □

For non-orientable surfaces we have the following analogue of Lemma 3, replacing the notion of genus by Euler genus.

Lemma 4. *Let G be a graph $\mathbb{Z}_2$-embedded in a (possibly non-orientable) surface $\mathcal{S}$. Let d be the dimension (over $\mathbb{Z}_2$) of the vector space generated by the edge vectors $\{y_e\}_{e \in E}$. Then G can be $\mathbb{Z}_2$-embedded in a (possibly non-orientable) surface of Euler genus d.*

Proof. The proof is almost the same as the proof of Lemma 3. We first convert handles to crosscaps and work on a surface with crosscaps only. We again remove crosscaps one by one until we end up with $d = t$ or with $d = t - 1$, $T = \langle z \rangle$, and $z^T z \equiv 1$. In the case that $d = t$ we are done.

In the case $d = t - 1$ we assume, as in the proof of Lemma 3, that the first k coordinates of z are 1 and the rest are 0 and convert the first k crosscaps into $(k-1)/2$ handles. We leave the remaining crosscaps as they are. The Euler genus of the resulting surface is $2((k-1)/2) + t - k = t - 1 = d$.   □

We end this section with a more complex move that allows us to zero out the labels of all edges in a spanning forest.

Lemma 5. *Suppose G is $\mathbb{Z}_2$-embedded on a surface $\mathcal{S}$, and F is a spanning forest of G. Then there is a $\mathbb{Z}_2$-embedding of G on $\mathcal{S}$ in which all edges of F are labeled with zero vectors.*

Proof. By Remark 2 we can assume that $\mathcal{S}$ is a surface with $c > 0$ crosscaps. Choose $z \in \mathbb{Z}_2^c$ and $v \in V$. Consider the following collection of moves: 1) add z to y_e for all e that are adjacent to v, and 2) for every f not adjacent to v and so that $y_f^T z \equiv 1$ perform an (f, v)-move. This collection of moves preserves the parity of the number of crossings between any pair of independent edges. Moreover, if z contains an even number of ones, the parity of the number of ones in no y-label is changed. Pick a root for each component of F, orient the edges of F away from the root, and process the edges in each component in a breadth-first traversal; for each edge e in this traversal, we turn its label into the zero vector, by performing the collection of moves above with $z = y_e$ and v the head of e. Note that this changes the label of e into the zero vector, without affecting the labels of any edges that have already been processed (since F is a forest, and $y_f^T z \equiv 0$ for edges f already processed, because $y_f = 0$ for those edges). If the $\mathbb{Z}_2$-embedding was orientable to begin with, it remains so, since z is chosen from the set of existing labels, all of which contain an even number of ones originally and throughout the relabeling. □

3 Block Additivity mod $\mathbb{Z}_2$

As a warm-up we show the additivity of genus over connected components (a result that is nearly obvious for embeddings).

Lemma 6. *The $\mathbb{Z}_2$-genus of a graph is the sum of the $\mathbb{Z}_2$-genera of its connected components.*

Proof. Let G be a graph. Let $g := \mathbf{g}_0(G)$ be the $\mathbb{Z}_2$-genus of G. By Remark 2 we have an orientable drawing of G on the surface with $t := 2g + 1$ crosscaps. Assume that G is the disjoint union of G_1 and G_2. Let F_1 be a maximum spanning forest of G_1 and F_2 be a maximum spanning forest of G_2. We can assume (see Lemma 5) that the y_e-labels for edges e in F_1 and F_2 are zero.

Let e_1 be an edge in $G_1 - F_1$ and let e_2 be an edge in $G_2 - F_2$. Let C_1 be the unique cycle in $F_1 + e_1$ and let C_2 be the unique cycle in $F_2 + e_2$. Note that C_1 and C_2 are disjoint (since G_1 and G_2 are disjoint) and hence by Lemma 1 we have

$$y_{e_1}^T y_{e_2} \equiv 0; \tag{7}$$

that is, the vectors y_{e_1} and y_{e_2} are perpendicular. Let S_1 be the vector space generated by the y_e-labels on the edges in G_1 and let S_2 be the vector space generated by the y_e-labels on the edges in G_2. Then $S_1 \perp S_2$ and hence

$$\dim S_1 + \dim S_2 \leq t = 2g + 1. \tag{8}$$

By Lemma 3, we can $\mathbb{Z}_2$-embed G_i in an orientable surface with $\lfloor (\dim S_i)/2 \rfloor$ handles. Note

$$\lfloor (\dim S_1)/2 \rfloor + \lfloor (\dim S_2)/2 \rfloor \leq g$$

and hence $\mathbf{g}_0(G_1) + \mathbf{g}_0(G_2) \leq \mathbf{g}_0(G)$. □

Again, one also has the analogue of Lemma 6 for non-orientable surfaces.

Lemma 7. *The $\mathbb{Z}_2$-Euler genus of a graph is the sum of the $\mathbb{Z}_2$-Euler genera of its connected components.*

Proof. The proof is the same as the proof of Lemma 6 except the final part. We have

$$\dim S_1 + \dim S_2 \leq t, \tag{9}$$

where $t := \mathbf{eg}_0(G)$ is the $\mathbb{Z}_2$-Euler genus of G. By Lemma 4 we can draw G_i in a surface with $\dim S_i$ crosscaps. Hence $\mathbf{eg}_0(G_1) + \mathbf{eg}_0(G_2) \leq \mathbf{eg}_0(G)$. □

We are ready now to establish additivity of $\mathbb{Z}_2$-genus and $\mathbb{Z}_2$-Euler genus over 2-connected components (blocks).

Theorem 1. *The $\mathbb{Z}_2$-genus of a graph is the sum of the $\mathbb{Z}_2$-genera of its blocks. The $\mathbb{Z}_2$-Euler genus of a graph is the sum of the $\mathbb{Z}_2$-Euler genera of its blocks.*

Proof. There is a large shared part in the arguments for $\mathbb{Z}_2$-genus and $\mathbb{Z}_2$-Euler genus (only the initial setup and the final drawing step are different).

The initial setup for the $\mathbb{Z}_2$-genus case is the following. Let $G = (V, E)$ be a connected graph (we can assume this by Lemma 6) and let $g := \mathbf{g}_0(G)$ be the $\mathbb{Z}_2$-genus of G. Thus we have an orientable $\mathbb{Z}_2$-embedding of G on the surface with $t := 2g + 1$ crosscaps. Let B be the subspace of $\mathbb{Z}_2^t$ consisting of vectors with an even number of ones (we will keep our drawing orientable, that is, all the edge labels will be from B). Note that

$$\operatorname{rad} B = \{0\}, \tag{10}$$

since each vector in $B \setminus \{0\}$ has a zero and a one. Let $\hat{t} := \dim B = t - 1$.

The initial setup for the $\mathbb{Z}_2$-Euler genus case is the following. Let $G = (V, E)$ be a connected graph (Lemma 7), and let $g := \mathbf{eg}_0(G)$ be the $\mathbb{Z}_2$-Euler genus of G. Thus we have a $\mathbb{Z}_2$-embedding of G on the surface with t crosscaps. In this case we let $B := \mathbb{Z}_2^t$. (And we trivially have (10).) Let $\hat{t} := \dim B = t$.

Let v be a cut vertex of G. Let $G_1 = (V_1, E_1)$ be a block of G containing v and let $G_2 = (V_2, E_2)$ be the union of the remaining blocks (note that $V_1 \cap V_2 = \{v\}$). Let $T = (V, F)$ be a depth-first search (DFS) spanning tree of G with the exploration starting at v. We can assume $y_e = 0$ for the edges in F (see Lemma 5). The reason for taking a DFS spanning tree is that we will need the following property: if $e \in E \setminus F$ is not adjacent to v then the unique cycle in $F + e$ does not contain v.

Let S_i be the vector space generated by the y_e-labels of $e \in E_i$ that are not adjacent to v. Let Z_i be the vector space generated by the y_e-labels of $e \in E_i$ that are adjacent to v. See Figure 2. Our plan is to modify the y_e-labels of the edges adjacent to v (changing Z_1, Z_2) so that: 1) no new odd crossings between independent edges are introduced, and 2) after the modification $\dim(S_1 + Z_1) + \dim(S_2 + Z_2) \le t$.

We can modify the y_e of an edge e adjacent to v by adding any vector in $O := (S_1 + S_2)^{\perp_B}$. This does not change the parity of the number of crossings between independent pairs of edges (for f that are not adjacent to v we have $y_f^T(y_e + z) \equiv y_f^T y_e$; and f which are adjacent to v are not independent of e). Note that the modification in the $\mathbb{Z}_2$-genus case also preserves orientability (by choice of B).

We are going to modify the y_e-labels of edges adjacent to v (by adding vectors in O) as follows. For $i \in \{1, 2\}$ we do the following. Let $a := \dim(Z_i \cap S_i)$, $b := \dim(Z_i \cap (S_i + O))$, and $c := \dim(Z_i)$. Let $z_1, \ldots, z_c$ be a basis of Z_i such that 1) $z_1, \ldots, z_a$ is a basis of $Z_i \cap S_i$, 2) $z_1, \ldots, z_b$ is a basis of $Z_i \cap (S_i + O)$, and 3) $z_{a+1}, \ldots, z_b \in Z_i \cap O$. Such a basis can be constructed as follows: first apply the Steinitz exchange lemma on a basis of $Z_i \cap S_i$ and a basis of $Z_i \cap O$ (the

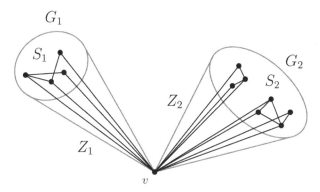

Fig. 2. Graph G with cutvertex v, block G_1 and union of remaining blocks G_2. Vector space Z_i (S_i) is generated by labels of edges in G_i (not) incident to v.

basis of $Z_i \cap S_i$ will be extended by vectors in the basis of $Z_i \cap O$ to a basis of $Z_i \cap (O + S_i)$); then apply the Steinitz exchange lemma on the resulting basis and a basis of Z_i. For each edge in G_i adjacent to v we relabel $y_e = \alpha_1 z_1 + \cdots + \alpha_c z_c$ by setting $\alpha_{a+1} = \alpha_{a+2} = \cdots = \alpha_b = 0$ (note that this corresponds to adding an element of O to y_e). After the modification (which also changed Z_i) we have that $z_1, \ldots, z_a$ is a basis of $Z_i \cap (S_i + O)$ and also a basis of $Z_i \cap S_i$. Thus we have

$$Z_i \cap (S_i + O) = Z_i \cap S_i. \tag{11}$$

Let $e_1 \in E_1 \setminus F$ and let $e_2 \in E_2 \setminus F$ be such that e_2 is not adjacent to v. Let C_1 be the unique cycle in $F + e_1$ and let C_2 be the unique cycle in $F + e_2$. Note that C_2 does not contain v (since F is a DFS spanning tree and e_2 is not adjacent to v). Thus C_1 and C_2 are disjoint (C_1 is in G_1, C_2 is in G_2 and $V_1 \cap V_2 = \{v\}$) and hence, by Lemma 1, we have

$$y_{e_1}^T y_{e_2} \equiv 0.$$

Thus $(Z_1 + S_1) \subseteq S_2^{\perp B}$ and since $O \subseteq S_2^{\perp B}$ (by the definition of O) we also have $(Z_1 + S_1 + O) \subseteq S_2^{\perp B}$. By symmetry we also have $(Z_2 + S_2 + O) \subseteq S_1^{\perp B}$ and hence, by Lemma 2 and (10), we obtain

$$\dim(Z_i + S_i + O) + \dim(S_{3-i}) \leq \hat{t}. \tag{12}$$

Thus we have (using $\dim(A + B) = \dim(A) + \dim(B) - \dim(A \cap B)$)

$$\dim(Z_i) + \dim(S_i + O) - \dim(Z_i \cap (S_i + O)) + \dim(S_{3-i}) \leq \hat{t},$$

and (11) yields

$$\dim(Z_i) + \dim(S_i + O) - \dim(Z_i \cap S_i) + \dim(S_{3-i}) \leq \hat{t}. \tag{13}$$

Adding (13) for $i = 1, 2$ and simplifying (again using $\dim(A + B) = \dim(A) + \dim(B) - \dim(A \cap B)$) we obtain

$$\dim(Z_1 + S_1) + \dim(Z_2 + S_2) + \dim(S_1 + O) + \dim(S_2 + O) \leq 2\hat{t}. \tag{14}$$

We have

$$\dim(S_1 + O) + \dim(S_2 + O)$$
$$= \dim(S_1) + \dim(O) - \dim(S_1 \cap O) + \dim(S_2) + \dim(O) - \dim(S_2 \cap O)$$
$$= \dim(S_1 + S_2) + \dim(S_1 \cap S_2) + 2\dim(O) - \dim(S_1 \cap O) - \dim(S_2 \cap O)$$
$$= \hat{t} + \dim(S_1 \cap S_2) + \dim(O) - \dim(S_1 \cap O) - \dim(S_2 \cap O)$$
$$\geq \hat{t} + \dim(S_1 \cap S_2 \cap O) + \dim(O) - \dim(S_1 \cap O) - \dim(S_2 \cap O)$$
$$= \hat{t} + \dim(O) - \dim((S_1 + S_2) \cap O) \geq \hat{t},$$

where in the first, second, and fourth equality we used $\dim(A+B)+\dim(A\cap B) = \dim(A)+\dim(B)$; in the third equality we used $\dim(S_1+S_2)+\dim(O) = \hat{t}$ (which follows from the definition of O and Lemma 2); in the first and the last inequality we used the monotonicity of dimension,

Plugging $\dim(S_1 + O) + \dim(S_2 + O) \geq \hat{t}$ into (14) we obtain

$$\dim(Z_1 + S_1) + \dim(Z_2 + S_2) \leq \hat{t} \leq t. \tag{15}$$

The $\mathbb{Z}_2$-genus case of the lemma now follows from Lemma 3, using the argument from the proof of Lemma 6 (the final part after equation (8)) giving us $\mathbb{Z}_2$-embeddings of G_1 and G_2 on two surfaces which have g handles total. In the $\mathbb{Z}_2$-Euler genus case we apply Lemma 4, using the argument from the proof of Lemma 7 (the final part after equation (9)).   □

The Hanani-Tutte theorem for surface $\mathcal{S}$ would—if true—state that if a graph has a $\mathbb{Z}_2$-embedding on surface $\mathcal{S}$, then it can be embedded on $\mathcal{S}$. Theorem 1 implies that in a search for counterexamples we can concentrate on 2-connected graphs. For the projective plane this result was obtained using much simpler means in [3, Lemma 2.4].

Corollary 1. *A minimal counterexample to the Hanani-Tutte theorem on any surface $\mathcal{S}$ is 2-connected.*

Proof. Suppose G is a minimal counterexample to Hanani-Tutte on some surface $\mathcal{S}$. If G is not connected, let $G_1, \ldots, G_k$, $k \geq 2$ be its connected components. If $\mathcal{S}$ is orientable, let g be the genus of $\mathcal{S}$. Then $g \geq \mathbf{g}_0(G) = \sum_{i=1}^{k} \mathbf{g}_0(G_i) = \sum_{i=1}^{k} \mathbf{g}(G_i) = \mathbf{g}(G)$, where the first equality is true by Lemma 6, and the second equality because G is minimal (and the third is a standard property of the genus of a graph). Therefore, $\mathbf{g}(G) \leq g$, so G can be embedded in $\mathcal{S}$, meaning it cannot be a counterexample. If $\mathcal{S}$ is non-orientable we can make essentially the same argument with Lemma 7 replacing Lemma 6, and Euler genus replacing genus. If G is connected, but not 2-connected, we repeat the same argument with Theorem 1 replacing the two lemmas, and using block additivity of (Euler) genus to conclude that $\sum_{i=1}^{k} \mathbf{g}(G_i) = \mathbf{g}(G)$ where the G_i are the blocks of G having cutvertex v in common.   □

4   Questions

The (Euler) genus of a graph is an obvious upper bound on the $\mathbb{Z}_2$-(Euler) genus of a graph, but are they always the same?

Conjecture 1. The $\mathbb{Z}_2$-(Euler) genus of a graph equals its (Euler) genus.

The truth of this conjecture would imply the Hanani-Tutte theorem for arbitrary surfaces, so we have to leave the question open. The block additivity result from Section 3 implies that a minimal counterexample to the conjecture (if it exists) and, thereby, to the Hanani-Tutte theorem on an arbitrary surface, can be assumed to be 2-connected (since it cannot have a cut-vertex).

A much more modest goal than Conjecture 1 would be to bound the standard (Euler) genus in the $\mathbb{Z}_2$-(Euler) genus: Are there functions f and g so that $\mathbf{g}(G) \leq f(\mathbf{g}_0(G))$ and $\mathbf{eg}(G) \leq g(\mathbf{eg}_0(G))$?

In the absence of a proof of Conjecture 1, we can ask what other results for (Euler) genus also hold for $\mathbb{Z}_2$-(Euler) genus. For example, is the computation of the $\mathbb{Z}_2$-(Euler) genus **NP**-hard (as it is for (Euler) genus [9])? And is $\mathbb{Z}_2$-embeddability decidable in polynomial time for a fixed surface $\mathcal{S}$ (as it is for embeddability [10])? One could also try to extend the block additivity result: if G_1 and G_2 are two edge-disjoint graphs with $|V(G_1) \cap V(G_2)| = 2$, is it true that $|\mathbf{g}_0(G) - (\mathbf{g}_0(G_1) + \mathbf{g}_0(G_2))| \leq 1$? (This inequality is known to be true for the standard genus, a result by Decker, Glover, Huneke, and Stahl, see [4, Section 4.4]).

References

1. Sarkaria, K.S.: A one-dimensional Whitney trick and Kuratowski's graph planarity criterion. Israel J. Math. 73(1), 79–89 (1991)
2. Chojnacki (Haim Hanani), C.: Über wesentlich unplättbare Kurven im dreidimensionalen Raume. Fundamenta Mathematicae 23, 135–142 (1934)
3. Pelsmajer, M.J., Schaefer, M., Stasi, D.: Strong Hanani–Tutte on the projective plane. SIAM Journal on Discrete Mathematics 23(3), 1317–1323 (2009)
4. Mohar, B., Thomassen, C.: Graphs on surfaces. Johns Hopkins Studies in the Mathematical Sciences. Johns Hopkins University Press, Baltimore (2001)
5. Levow, R.B.: On Tutte's algebraic approach to the theory of crossing numbers. In: Hoffman, F., Levow, R.B., Thomas, R.S.D. (eds.) Proceedings of the Third Southeastern Conference on Combinatorics, Graph Theory and Computing, Boca Raton, Fla., Florida Atlantic University, pp. 315–314 (1972)
6. Stillwell, J.: Classical topology and combinatorial group theory. Second edn, 2nd edn. Graduate Texts in Mathematics, vol. 72. Springer, New York (1993)
7. de Longueville, M.: A course in topological combinatorics. In: Universitext, Springer, New York (2013)
8. Babai, L., Frankl, P.: Linear algebra methods in combinatorics with applications to geometry and computer science. Department of Computer Science, The University of Chicago (1992)
9. Thomassen, C.: The graph genus problem is NP-complete. J. Algorithms 10(4), 568–576 (1989)
10. Mohar, B.: A linear time algorithm for embedding graphs in an arbitrary surface. SIAM J. Discrete Math. 12(1), 6–26 (1999)

Exploiting Air-Pressure to Map Floorplans on Point Sets

Stefan Felsner[*]

Institut für Mathematik, Technische Universität Berlin, Germany
felsner@math.tu-berlin.de

Abstract. We prove a conjecture of Ackerman, Barequet and Pinter. Every floorplan with n segments can be embedded on every set of n points in generic position. The construction makes use of area universal floorplans also known as area universal rectangular layouts.

The notion of area used in our context depends on a nonuniform density function. We, therefore, have to generalize the theory of area universal floorplans to this situation. The method is then used to prove a result about accommodating points in floorplans that is slightly more general than the conjecture of Ackerman et al.

1 Introduction

In our context a *floorplan* is a partition of a rectangle into a finite set of interior-disjoint rectangles. A floorplan is *generic* if it has no cross, i.e., no point where four rectangles of the partition meet. A *segment* of a floorplan is a maximal nondegenerate interval that belongs to the union of the boundaries of the rectangles. Segments are either horizontal or vertical. The segments of a generic floorplan are internally disjoint. Two floorplans F and F' are *weakly equivalent* if there exist bijections $\phi : S_H(F) \to S_H(F')$ and $\phi : S_V(F) \to S_V(F')$ between their horizontal and vertical segments such that segment s has an endpoint on segment t in F iff $\phi(s)$ has an endpoint on $\phi(t)$. A set P of points in $\mathbb{R}^2$ is *generic* if no two points from P have the same x or y coordinate. Section 2 provides a more comprehensive overview of definitions and notions related to floorplans.

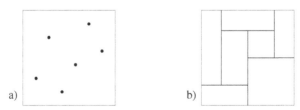

Fig. 1. A generic set of six points and a generic floorplan with six segments

Let P be a set of n points in a rectangle R and let F be a generic floorplan with n internal segments. A *cover map* from F to P is a floorplan F' weakly equivalent to F with outer rectangle R such that every internal segment of F' contains exactly one point from P. Figure 2 shows an example.

---

[*] Partially supported by ESF EuroGIGA projects Compose and GraDR.

S. Wismath and A. Wolff (Eds.): GD 2013, LNCS 8242, pp. 196–207, 2013.

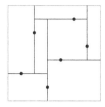 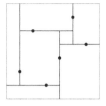

Fig. 2. Two cover maps from the floorplan of Fig. 1.b to the point set of Fig. 1.a

In this paper we answer a question of Ackerman et al. [1] by proving Theorem 1. The proof of the theorem and some variants and generalizations is the subject of Section 4.

Theorem 1. *If P is a generic set of n points and F is a generic floorplan with n internal segments, then there is a cover map from F to P.*

The proof is based on results about area representations of floorplans. The following theorem is known, it has been proven with quite different methods, see [14], [11], [5].

Theorem 2. *Let F be a floorplan with rectangles $R_1, \ldots, R_{n+1}$, let A be a rectangle and let $w : \{1, \ldots, n+1\} \to \mathbb{R}_+$ be a weight function with $\sum_i w(i) = \mathsf{area}(A)$. There exists a unique floorplan F' contained in A that is weakly equivalent to F such that the area of the rectangle $R'_i = \phi(R_i)$ is exactly $w(i)$.*

In Section 3 we prove the generalization of Theorem 2 that will be needed for the proof of Theorem 1. In the generalized theorem (Theorem 3) the weight of a rectangle is measured as integral over some density function instead of the area.

2 Floorplans and Graphs

A *floorplan* is a dissection of a rectangle into a finite set of interiorly disjoint rectangles. From a given floorplan F we can obtain several graphs and additional structure. We introduce two of these and use them to define notions of equivalence for floorplans.

The Rectangular Dual. Let $R(F)$ be the set of rectangles of a floorplan F. It is convenient to include the enclosing rectangle in the set $R(F)$. The *rectangular dual* of F is the graph $G^*(F)$ with vertex set $R(F)$ and edges joining pairs of rectangles that share a boundary segment. Usually the notion of a rectangular dual is used in the other direction, i.e., it is assumed that a planar graph G is given and the quest is for a floorplan F such that $G = G^*(F)$. It is convenient to extend a floorplan F with four rectangles that frame F as shown in Figure 3. In the dual of an extended floorplan F_+ we omit the vertex that corresponds to the enclosing rectangle. With this twist in the definition of the dual we get: The dual $G^*_+(F)$ of the extended floorplan of a generic F is a 4-connected inner triangulation of a 4-gon. Indeed this is the characterization of the duals of extended generic floorplans. Buchsbaum et al. [3] and Felsner [6] provide many pointers to the literature related to floorplans and rectangular duals.

The Transversal Structure. The transversal structure (also known as regular edge labeling) associated to a floorplan F is an orientation and coloring of the edges of the extended dual $G^*_+(F)$.

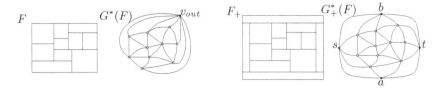

Fig. 3. A floorplan F, the extended floorplan F_+, and the duals $G^*(F)$ and $G^*_+(F)$

Let G be an inner triangulation of a 4-gon with outer vertices s, a, t, b in counterclockwise order. A *transversal structure* for G is an orientation and 2-coloring of the inner edges of G such that two local conditions hold: 1) All edges incident to s, a, t and b are blue outgoing, red outgoing, blue ingoing, and red ingoing, respectively. 2) The edges incident to an inner vertex v come in clockwise order in four nonempty blocks consisting solely of red ingoing, blue ingoing, red outgoing, and blue outgoing edges, respectively.

Transversal structures have been studied in [9], [10], and in [12]. A proof of the following proposition can e.g. be found in [6].

Proposition 1. *Every transversal structure of an inner triangulation G of a 4-gon with outer vertices s, a, t, b is induced by a floorplan F with $G = G^*_+(F)$ and each floorplan F induces a transversal structure on $G^*_+(F)$.*

Figure 4 indicates the correspondence between floorplans and transversal structures.

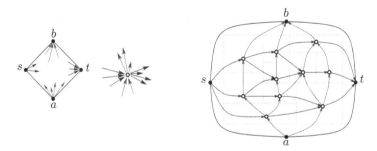

Fig. 4. The two local conditions and an example of a transversal structure together with a corresponding floorplan

The Segment Contact Graph. Recall that we call a floorplan *generic* if it has no cross, i.e., no point where four rectangles meet. A *segment* of a floorplan is a maximal nondegenerate interval that belongs to the union of the boundaries of the rectangles. In the generic case intersections between segments only occur between horizontal and vertical segments and they involve an endpoint of one of the segments, i.e., they are contacts. If a floorplan has a cross at point p we can break one of the two segments that contain p into two to get a system of interiorly disjoint segments.

The *segment contact graph* $G_{\text{seg}}(F)$ of a floorplan F is the bipartite planar graph whose vertices are the segments of F and edges correspond to contacts between segments. From Figure 5 we see that $G_{\text{seg}}(F)$ is indeed planar and that the faces of $G_{\text{seg}}(F)$ are in bijection with the rectangles of F and are uniformly of degree 4. Therefore $G_{\text{seg}}(F)$ is a maximal bipartite planar graph, i.e., a quadrangulation.

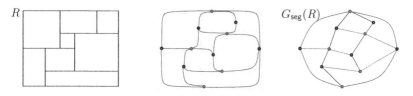

Fig. 5. A floorplan F and two drawings of its segment contact graph $G_{\text{seg}}(F)$

The Separating Decomposition. The separating decomposition associated to a floorplan is an orientation and coloring of the edges of the segment contact graph.

Let Q be a quadrangulation, we call the color classes of the bipartition white and black and name the two black vertices on the outer face s and t. A *separating decomposition* of Q is an orientation and coloring of the edges of Q with colors red and blue such that two conditions hold: 1) All edges incident to s are ingoing red and all edges incident to t are ingoing blue. 2) Every vertex $v \neq s, t$ is incident to a nonempty interval of red edges and a nonempty interval of blue edges. If v is white, then, in clockwise order, the first edge in the interval of a color is outgoing and all the other edges of the interval are incoming. If v is black, the outgoing edge is the last one in its color in clockwise order (see the left part of Figure 6).

Separating decompositions have been studied in [4], [8], and [7]. To us they are of interest because of the following proposition proved e.g. in [6].

Proposition 2. *A floorplan F induces a separating decomposition on its segment contact graph $G_{\text{seg}}(F)$ and every separating decomposition of a planar quadrangulation Q corresponds to a floorplan F with $Q = G_{\text{seg}}(F)$.*

The right part of Figure 6 indicates the correspondence between floorplans and separating-decompositions.

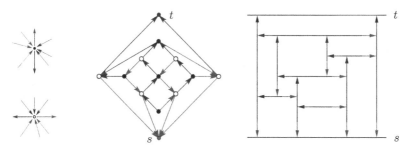

Fig. 6. The rule for black and white vertices. A separating decomposition S and a floorplan F corresponding to S.

2.1 Notions of Equivalence for Floorplans

Definition 1. *Two floorplans are* weakly equivalent *if they induce the same separating decomposition.*

Definition 2. *Two floorplans are* strongly equivalent *if they induce the same transversal structure.*

In the introduction we said that two floorplans F and F' are weakly equivalent if there exist bijections $\phi : S_H(F) \to S_H(F')$ and $\phi : S_V(F) \to S_V(F')$ between their horizontal and vertical segments such that segment s has an endpoint on segment t in F iff $\phi(s)$ has an endpoint on $\phi(t)$. We claim that this yields the same equivalence classes as Definition 1. Clearly, if F and F' induce the same separating decomposition then they are weakly equivalent in the above sense. For the converse first observe that the segment contact graphs of F and F' are isomorphic, i.e., $G_{\text{seg}}(F) = G_{\text{seg}}(F')$. Now define an orientation Q_F on $Q = G_{\text{seg}}(F)$ by orienting s to t iff segment s has an endpoint on segment t. Let $Q_{F'}$ be the orientation defined on Q using the segments of F'. Observe that $Q_F = Q_{F'}$. Since a separating decomposition is uniquely determined by the underlying 2-orientation (see, [4] or [7]) we conclude that F and F' induce the same separating decomposition, i.e., $SD_F = SD_{F'}$. This implies that F and F' are weakly equivalent in the sense of Definition 1.

Eppstein et al. [5] use the term *layout* instead of floorplan. Their equivalent layouts correspond to strongly equivalent floorplans and order-equivalent layouts to weakly equivalent floorplans. Asinowski et al. [2] study independent notions of R-equivalence and S-equivalence for floorplans.

3   Realizing Weighted Floorplans via Air-Pressure

In this section we prove a generalization of Theorem 2 to situations where the "area" of a rectangle is replaced by the mass defined through a density distribution.

Let $\mu : [0,1]^2 \to \mathbb{R}_+$ be a density function on the unit square whose total mass is 1, i.e., $\int_0^1 \int_0^1 \mu(x,y)dxdy = 1$. We assume that μ can be integrated over axis aligned rectangles and all fibers μ_x and μ_y can be integrated over intervals. Moreover, we require that integrating μ over a nondegenerate rectangle and fibers over nondegenerate intervals always yields nonzero values. The *mass* of an axis aligned rectangle $R \subseteq [0,1]^2$ is defined as $m(R) = \int\int_R \mu(x,y)dxdy$.

Theorem 3. *Let $\mu : [0,1]^2 \to \mathbb{R}_+$ be a density function of total mass 1. If F is a floorplan with rectangles $R_1, \ldots, R_{n+1}$ and $w : \{1, \ldots, n+1\} \to \mathbb{R}_+$ a positive weight function with $\sum_1^{n+1} w(i) = 1$ then there exists a unique floorplan F' in the unit square that is weakly equivalent to F such that $m(R_i) = w(i)$ for each rectangle R_i.*

Our proof follows the air-pressure paradigm as proposed by Izumi, Takahashi and Kajitani [11]. We first describe the idea: Consider a realization of F in the unit square and compare the mass $m(R_i)$ to the intended mass $w(i)$. The quotient of these two values can be interpreted as the pressure inside the rectangle. Integrating this pressure along a side of the rectangle yields the force by which R_i is pushing against the segment that contains the side. The difference of pushing forces from both sides of a segment yields the effective force acting on the segment. The intuition is that shifting a segment in direction of the effective force yields a better balance of pressure in the rectangles. We will show that iterating such improvement steps drives the realization of F towards a situation with $m(R_i) = w(i)$ for all i, i.e., the procedure converges towards the floorplan F' whose existence we want to show.

In [11] the air-pressure paradigm was used for situations where the mass of a rectangle is its area. The authors observed fast convergence experimentally but they had no proof of convergence. Here we provide such a proof for the more general case of weights given by integrals over a density function.

A proof of Theorem 3 can also be given along the lines of the proof of Theorem 2 that has been given by Eppstein et al. in [5]. This approach has been taken by Schrenzenmaier [13]. The resulting proof is quite compact, however, it has the disadvantage of being purely existential. Schrenzenmaier also has a java implementation of the air-pressure approach that solves moderate size instances quickly.

Let $R_i = [x_l, x_r] \times [y_b, y_t]$ be a rectangle of F. Recall that the mass of R_i is $m(R_i) = \int_{x_l}^{x_r} \int_{y_b}^{y_t} \mu(x, y) dy dx$. The pressure $p(i)$ in R_i is the fraction of the intended mass $w(i)$ and the actual mass $m(R_i)$, i.e., $p(i) = \frac{w(i)}{m(R_i)}$. Let s be a segment of F and let R_i be one of the rectangles with a side in s. Let s be vertical with x-coordinate x_s and let $s \cap R_i$ span the interval $[y_b(i), y_t(i)]$. The (undirected) *force imposed on s by R_i* is the pressure $p(i)$ of R_i times the density dependent length of the intersection.

$$f(s, i) = \frac{w(i)}{m(R_i)} \int_{y_b(i)}^{y_t(i)} \mu(x_s, y) dy = p(i) \int_{y_b(i)}^{y_t(i)} \mu_{x_s}(y) dy.$$

The *force acting on s* is obtained as a sum of the directed forces imposed on s by incident rectangles.

$$f(s) = \sum_{R_i \text{ left of } s} f(s, i) \quad - \sum_{R_i \text{ right of } s} f(s, i).$$

Symmetric definitions apply to horizontal segments.

Balance for Rectangles and Segments

A segment s is in *balance* if $f(s) = 0$. A rectangle R_i is in *balance* if $p(i) = 1$, i.e., if $w(i) = m(R_i)$.

Lemma 1. *If all rectangles R_i of F are in balance, then all segments are in balance.*

Proof. Since all rectangles are in balance we can eliminate the pressures from the definition of the $f(s, i)$. With this simplification we get for a vertical segment s

$$f(s) = \sum_{R_i \text{ left of } s} \int_{y_b(i)}^{y_t(i)} \mu_{x_s}(y) dy - \sum_{R_j \text{ right of } s} \int_{y_b(j)}^{y_t(j)} \mu_{x_s}(y) dy.$$

Hence $f(s) = M_s - M_s = 0$, where M_s is the integral of the fiber density μ_{x_s} along s. The symmetric argument applies to horizontal segments. $\qquad \square$

Interestingly, the converse of the lemma also holds.

Proposition 3. *If all segments of F are in balance, then all rectangles are in balance.*

Proof. Suppose that F balances all segments but not all rectangles. Choose some τ with $\min_i p(i) < \tau \le \max_i p(i)$. Let T_τ be the union of all rectangles R_i whose pressure exceeds τ and let Γ_τ be the boundary of T_τ.

Claim. The boundary Γ_τ of T_τ contains no complete segment.

Suppose Γ_τ contains the vertical segment s such that T_τ is left of s. Let I be a nontrivial interval on s that is defined as intersection of a rectangle R_i that has its right edge on s and a rectangle R_j that has its left edge on s. The force acting on s along I is $p(i) \int_I \mu_{x_s}(y)dy - p(j) \int_I \mu_{x_s}(y)dy$. Since $\int_I \mu_{x_s}(y)dy > 0$ by our assumption on μ and $p(i) > p(j)$ by definition of T_τ the force is positive. This holds for every interval I on s, hence, the overall force $f(s)$ acting on s is also positive. This contradicts the assumption that s is in balance and completes the proof of the claim. $\triangle$

Let s_0 be any segment which contributes to Γ_τ. From the lemma we know that at some interior point of segment s_0 the boundary leaves s_0 and continues along another segment s_1. Again, the boundary has to leave s_1 at some interior point to continue on s_2. Because this procedure always follows the boundary of T which is a region defined by a union of rectangles in F the sequence of segments has to get back to segment s_0, i.e., there is an k such that $s_k = s_0$.

From the definition of the separating decomposition SD_F corresponding to F we find that $s_0 \leftarrow s_1 \leftarrow s_2 \leftarrow \ldots \leftarrow s_{k-1} \leftarrow s_0$ is a directed cycle in SD_F. The four segments of the enclosing square of F do not contribute to the boundary of T_τ, simply because they cannot belong to a directed cycle of SD_F.

Recall the assumption that F balances all segments but not all rectangles. Let s be the vertical segment with maximal x-coordinate among all vertical segments that contribute to a boundary Γ_τ for some τ. From the choice of s it is clear that T_τ is to the left of s. Consider the segment $s' = s_{k-1}$ following $s = s_0$ in the cycle $s_0 \leftarrow \ldots \leftarrow s_{k-1} \leftarrow s_0$ in SD_F corresponding to Γ_τ. Left from the contact point p of s and s' the segment s' is part of the boundary Γ_τ of T_τ. From the choice of τ and s it follows that to the right of p the rectangles on both sides of s' have the same pressure $p(i)$. Otherwise the right part of s' would belong to some boundary $\Gamma_{\tau'}$ and the vertical segment following s' on $\Gamma_{\tau'}$ is in contradiction to the choice of s.

Now consider $f(s')$ and split the contributions to this force at p. Left of p the pressure on the side of T_τ exceeds the pressure from the other side. Right of p the rectangles on both sides of p have the same pressure. This shows that in contradiction to the assumption $f(s') \neq 0$. This completes the proof of the proposition. $\square$

Balancing Segments and Optimizing the Entropy

Proposition 4. *If a segment s of F is unbalanced, then we can keep all the other segments at their position and shift s parallel to a position where it is in balance. The resulting floorplan F' is weakly equivalent to F.*

Proof. We consider the case of a vertical segment, the horizontal case is symmetric. Let x_s be the x-coordinate of s. With S_- and S_+ we denote the sets of rectangles in F that touch s from the left respectively from the right. Let R_l be the rectangle with a left boundary of maximal x-coordinate x_l in S_- and let R_r be the rectangle with a right boundary of minimal x-coordinate x_r in S_+. Note that if t satisfies $x_l < t < x_r$ then segment s can be shifted parallel to the position $x_s = t$ and the resulting floorplan is weakly equivalent to F.

For $t \in (x_l, x_r)$ we define $h(t)$ as the force acting on s when the segment it is shifted to $x_s = t$. We observe:

- The pressure $p(i)$ depends continuously on t for all rectangles $R_i \in S_- \cup S_+$.
- The value of $\int_I \mu_t(y) dy$ is a continuous function of t for all intervals I.

Hence, $h(t)$ is a continuous function. With t approaching x_l from the right the area of R_l tends to zero. Hence, the mass $m(R_l)$ also tends to zero and the pressure $p(l)$ tends to infinity. Since $\int_{y_b(l)}^{y_t(l)} \mu_t(y) dy > 0$ we conclude that $h(t) \to +\infty$ with $t \to x_l$. A similar reasoning involving R_r shows that $h(t) \to -\infty$ with $t \to x_r$. It follows that there is some $t_0 \in (x_l, x_r)$ with $h(t_0) = 0$. Hence, if we shift s to the position $x_s = t_0$ the force acting on s vanishes and s is in balance. $\qquad\square$

Definition 3. *The* entropy *of a rectangle R_i of F is defined as $-w(i) \log p(i)$. The* entropy *of the floorplan F is*

$$E = \sum -w(i) \log p(i)$$

The proof of Theorem 3 will be completed after showing the following

(1)   The entropy E is always nonpositive.
(2) $E = 0$ if and only if all rectangles R_i of F are in balance.
(3)   Shifting an unbalanced segment s into its balance position increases the entropy.
(4)   The process of repeatedly shifting unbalanced segments into their balance position makes F converge to a floorplan F' such that the entropy of F' is zero.
(5)   The solution is unique.

The first two of these statements are shown in the next lemma.

Lemma 2. *The entropy E is always nonpositive and $E = 0$ if and only if all rectangles R_i of F are in balance.*

Proof. We use that $p(i) > 0$ and hence $\log p(i) \geq (1 - \frac{1}{p(i)}) = (1 - \frac{m(R_i)}{w(i)})$. For the entropy of R_i we get $-w(i) \log p(i) \leq -w(i)(1 - \frac{m(R_i)}{w(i)}) = m(R_i) - w(i)$. This yields

$$E = \sum_i -w(i) \log p(i) \leq \sum_i m(R_i) - \sum_i w(i) = 1 - 1 = 0$$

The equality $E = 0$ is equivalent to equality for each summand. Hence $0 = \log p(i) = (1 - \frac{m(R_i)}{w(i)})$ and $m(R_i) = w(i)$ for all i. $\qquad\square$

Lemma 3. *Shifting an unbalanced segment s into its balance position increases the entropy.*

Proof. We consider a vertical segment s as in the proof of Proposition 4 and assume $f(s) > 0$. Let t_0 be the first zero of $h(t)$ right of x_s. For all $t \in [x_s, t_0)$ the force $h(t)$ acting on s is positive, i.e., pushing s to the right.

Let $E(t)$ be the entropy of the floorplan when s is shifted towards the position $x_s = t_0$. We consider $E(t)$ as a function of t.

Claim. $\dfrac{d}{dt} E(t) = h(t)$.

Only rectangles touching s change their contribution to $E(t)$. Let $R_i = [x_l, t] \times [y_1, y_2]$ be a rectangle on the left of s, i.e., $R_i \in S_-$, and t is the x-coordinate of the right side of R_i. Hence

$$\frac{d}{dt}(-w(i)\log p(i)) = -w(i)\frac{1}{p(i)}p'(i) = -w(i)\frac{m(R_i)}{w(i)}\frac{d}{dt}\frac{w(i)}{m(R_i)} =$$

$$-w(i)\frac{m(R_i)}{w(i)}w(i)\frac{-m'(R_i)}{m^2(R_i)} = \frac{w(i)}{m(R_i)}m'(R_i) = \frac{w(i)}{m(R_i)}\frac{d}{dt}\int_{x_l}^t\int_{y_1}^{y_2}\mu(x,y)dydx =$$

$$\frac{w(i)}{m(R_i)}\int_{y_1}^{y_2}\mu_t(y)dy = p(i)\int_{y_1}^{y_2}\mu_t(y)dy.$$

When $R_i \in S_+$ the mass $m(R_i)$ is decreasing with t so that $m'(R_i)$ is negative and $\frac{d}{dt}(-w(i)\log p(i)) = -p(i)\int_{y_1}^{y_2}\mu_t(y)dy$. Summing this over all rectangles incident to s we obtain that $\frac{d}{dt}E(t) = h(t)$. This is the claim. $\triangle$

While shifting s from the initial position x_s to t_0 we have $h(t) > 0$. The claim implies that the derivative of the entropy is positive and, hence, the entropy is increasing. $\square$

We continue with item (4) from our program. To prove this, however, we have to add a condition to the process of balancing segments. The iteration has to be performed such that no unbalanced segment can be ignored. A rule is called *nonignoring* if it complies with this condition. Here are two examples of nonignoring selection rules: 1) Choose the segment for balancing uniformly at random from the set of unbalanced segments. 2) Always choose the segment so that the increase of the entropy is as large as possible.

Proposition 5. *Let $F_0, F_1, F_2, \ldots$ be a sequence of floorplans where F_{i+1} is obtained from F_i by balancing an unbalanced segment from F_i. If the selection of segments is nonignoring, then there is a subsequence $G_0, G_1, \ldots$ of floorplans that has a limit $G = \lim G_i$ and the entropy of the floorplan G is zero.*

Proof. Let $s_1, s_2, \ldots, s_n$ be the inner segments of F. Floorplans that are weakly equivalent to F can be encoded by the coordinate vector of the segments. This vector z in $\mathbb{R}^n$ has the value $z(i) = x_s$ if s_i is a vertical segment and $z(i) = y_s$ if s_i is horizontal. A sequence of floorplans is converging if the corresponding coordinate vectors converge.

Consider the sequence of coordinate vectors $z_0, z_1, \ldots$ of the given sequence of floorplans. Since each of the coordinates of these vectors is from the interval $(0, 1)$, there is a convergent subsequence. Let $G_0, G_1, \ldots$ be the corresponding convergent sequence of floorplans and let e_i be the entropy of G_i. From Lemma 3 we know that the e_i are an increasing sequence of negative numbers. Assume that the sequence e_i converges to $-a \neq 0$. Consider the limit $G = \lim G_i$. Since the entropy of G is $-a < 0$ there is an unbalanced rectangle R_j in G (Lemma 2) and, hence, there is an unbalanced segment s in G (Proposition 3). Let Δ be the increase of the entropy that comes from balancing s in G. Now, for all i greater than a sufficiently large N the floorplan G_i is so close to G that balancing s in G_i implies an increase of entropy of at least $\Delta/2$. For all i greater than a sufficiently large M we also have $e_i > -a - \Delta/2$. It follows that the unbalanced segment s was not used for balancing in any G_i with $i \geq \max(M, N)$. This is in contradiction to the assumption that the process is nonignoring. $\square$

Actually, a stronger statement is true. The full sequence $F_0, F_1, F_2, \ldots$ is also converging. To prove this we need the uniqueness from Proposition 6. In fact if G is the unique floorplan that is weakly equivalent to F and has has $m(R_i) = w(i)$ for all i, then it follows from the continuity of the entropy that there is an $\varepsilon > 0$ such that all floorplans whose entropy is larger than $-\varepsilon$ have a coordinate vector that is δ_ε close to the coordinate vector of G. This implies that $\lim F_i = G$.

Proposition 6. *For every floorplan F with $n + 1$ rectangles and every positive weight function $w : \{1, \ldots, n+1\} \to \mathbb{R}_+$ with $\sum_i w(i) = 1$ there is a unique floorplan F' in the unit square that is weakly equivalent to F and has $m(R_i) = w(i)$ for all i.*

A proof of the proposition can be found in the paper by Eppstein et al. [5].

4 Accommodating Floorplans on Point Sets

Let P be a generic set of n points in a rectangle $\mathcal{R}$. Let F be a generic floorplan and S be a subset of the segments of F of size n. A *cover map* from (F, S) to P is a floorplan F' with outer rectangle $\mathcal{R}$ that is weakly equivalent to F such that every segment from $S' = \phi(S)$ contains a point from P. The following is a generalization of Theorem 1.

Theorem 4. *If P is a generic set of n points in a rectangle $\mathcal{R}$ and F is a generic floorplan with a prescribed subset S of the segments of size n, then there is a cover map F' from (F, S) to P.*

Proof. We use Theorem 3 as a tool for the proof. First we transform the point set P into a suitable density distribution $\mu = \mu_P$ inside $\mathcal{R}$. This density is defined as the sum of a uniform distribution μ_1 with $\mu_1(q) = 1/\text{area}(\mathcal{R})$ for all $q \in \mathcal{R}$ and a distribution μ_2 that represents the points of P. Choose some $\Delta > 0$ such that $\|p - p'\| > 3\Delta$ for all $p, p' \in P$. Define $\mu_2 = \sum_{p \in P} \mu_p$ where $\mu_p(q)$ takes the value $(\Delta^2\pi)^{-1}$ on the disk $D_\Delta(p)$ of radius Δ around p and value 0 for q outside of this disk.

For a density ν defined on $\mathcal{R}$ and a rectangle $R \subseteq \mathcal{R}$ let $\nu(R)$ be the integral of the density ν over R. Since $\mu_1(\mathcal{R}) = 1$ and $\mu_p(\mathcal{R}) = 1$ for all $p \in P$ the total mass of $\mathcal{R}$ is $\mu(\mathcal{R}) = 1 + n$.

Transform the floorplan F into a floorplan F_S depending on the set S of segments that have to cover points of P: inflate every segment in S to form a thin rectangle. This description is not very formal but with a look at Figure 7 should make clear how to produce F_S from F and S. Let $\mathcal{S}$ be the set of new rectangles obtained by inflating segments from S.

Define weights for the rectangles of F_S as follows. If F_S has r rectangles we define $w(R) = 1 + 1/r$ if $R \in \mathcal{S}$ and $w(R) = 1/r$ for all the rectangles of F_S that came from rectangles of F. The total weight, $\sum_R w(R) = 1 + n$ equals the total mass $\mu(\mathcal{R})$.

The data $\mathcal{R}$ with μ and F_S with w constitute, up to scaling of $\mathcal{R}$ and w, a set of inputs for Theorem 3. From the conclusion of the theorem we obtain a floorplan F_S' weakly equivalent to F_S such that $m(R) = \int\int_R \mu(x, y)dxdy = w(R)$ for all rectangles.

The definition of the weight function w and the density μ is so that F_S' should be close to a cover map from (F, S) to P: Only the rectangles $R \in \mathcal{S}$ that have been constructed by inflating segments may contain a disk $D_\Delta(p)$ and each of these rectangles

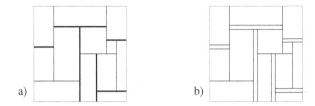

Fig. 7. Floorplans F with as prescribed subset S of segments (bold) and the floorplan F_S obtained by doubling the segments of S

may contain at most one of the disks. This suggests a correspondence $S \leftrightarrow P$. However, a rectangle $R \in S$ may use parts of several discs to accumulate mass. To find a correspondence between S and P we define a bipartite graph G whose vertices are the points in P and the rectangles in S:

- A pair (p, R) is an edge of G iff $R \cap D_\Delta(p) \neq \emptyset$ in F'_S.

The proof of the theorem will be completed by proving two claims: 1) G admits a perfect matching. 2) From F'_S and a perfect matching M in G we can produce a floorplan F' that realizes the cover map from (F, S) to P.

For the first of the claims we check Hall's matching condition: Consider a subset A of S. Since F_S is realizing the prescribed weights we have $m(A) = \mu(A) = \sum_{R \in A} \mu(R) = \sum_{R \in A} w(R) = |A|(1 + 1/r)$. Since $\mu_1(A) < 1$ and $\mu_p(A) \leq 1$ for all $p \in P$ there must be at least $|A|$ points $p \in P$ with $\mu_p(A) > 0$, these are the points that have an edge to a rectangle from A in G. We have thus shown that every set A of inflated segments is incident to at least $|A|$ points in G, hence, there is an injective mapping $\alpha : S \rightarrow P$ such that $R \cap D_\Delta(\alpha(R)) \neq \emptyset$ in F'_S for all $R \in S$.

To prove the second claim we have to construct a floorplan F' that realizes the cover map from (F, S) to P: Let s be a segment in S and let R_s be the rectangle in F'_S that corresponds to s. If s is horizontal we define s' to be the unique maximal horizontal segment in R_s whose y-coordinate is as close to the y-coordinate of the point $\alpha(R_s)$ as possible. For vertical segments we focus on the x-coordinate. For segments s of F that do not belong to S set $s' = s$. The collection $\{s' : s$ segment in $F\}$ of segments may fail to be a floorplan, see e.g. Figure 8.b. However, if s_1 and s_2 are segments of F such that s_1 has one of its endpoints on s_2 and $s_2 \in S$ then we can extend s'_1 into R_{s_2} to recover the contact with s'_2. By doing this for all qualifying pairs s'_1, s'_2 we get a floorplan, see Figure 8.c. This floorplan is weakly equivalent to F but there may still be segments of S that do not quite cover the assigned point. By construction the distance from a segment to its assigned point is at most Δ. Since Δ is small compared to the distances of points in P we can shift all segments orthogonally to make them cover their assigned points. Again this may spoil the floorplan property, see e.g. Figure 8.d. However, enlarging or shortening of segments by an amount of at most Δ at the ends finally yields the floorplan F' that realizes the cover from (F, S) to P.   □

The topic of [1] was the study of the number $Z(P)$ of rectangulations of a generic point set P. In our terminology this is the total number of cover maps from floorplans with n inner segments to a generic point set P with n points. Theorem 4 implies that this number is at least as large as the number of weak equivalence classes of floorplans.

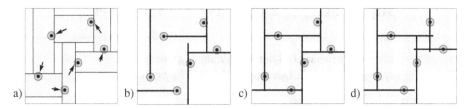

Fig. 8. a) A solution F'_S for the instance from Fig. 1. The arrows indicate a matching α. b) Segments $s \in S$ shifted to their optimal position in R_s. c) Enlarged segments recover the contacts. d) Some segments s are moved outside R_s to cover the corresponding points $\alpha(R_s)$. Small final adjustments (clipping and enlarging) yield F'.

This is the Baxter number B_{n+1} which is known to be of order $\Theta(8^{n+1}/(n+1)^4)$. In [1] an upper bound for $Z(P)$ of order $O(20^n/n^4)$ is shown. To improve this bound remains an intriguing problem.

References

1. Ackerman, E., Barequet, G., Pinter, R.Y.: On the number of rectangulations of a planar point set. J. Combin. Theory Ser. A 113, 1072–1091 (2006)
2. Asinowski, A., Barequet, G., Bousquet-Mélou, M., Mansour, T., Pinter, R.: Orders induced by segments in floorplan partitions and (2-14-3,3-41-2)-avoiding permutations (2010) (submitted), arXiv:1011.1889
3. Buchsbaum, A.L., Gansner, E.R., Procopiuc, C.M., Venkatasubramanian, S.: Rectangular layouts and contact graphs. ACM Trans. Alg. 4, 28 pages (2008)
4. de Fraysseix, H., Ossona de Mendez, P.: On topological aspects of orientation. Discr. Math. 229, 57–72 (2001)
5. Eppstein, D., Mumford, E., Speckmann, B., Verbeek, K.: Area-universal and constrained rectangular layouts. SIAM J. Computing 41, 537–564 (2012)
6. Felsner, S.: Rectangle and square representations of planar graphs. In: Pach, J. (ed.) Thirty Essays on Geometric Graph Theory, pp. 213–248. Springer, New-York (2012)
7. Felsner, S., Fusy, É., Noy, M., Orden, D.: Bijections for Baxter families and related objects. Journal of Comb. Theory A 18, 993–1020 (2011)
8. Felsner, S., Huemer, C., Kappes, S., Orden, D.: Binary labelings for plane quadrangulations and their relatives. Discr. Math. & Theor. Comp. Sci. 12, 115–138 (2010)
9. Fusy, É.: Combinatoire des cartes planaires et applications algorithmiques. PhD thesis, LIX Ecole Polytechnique (2007), http://www.lix.polytechnique.fr/~fusy/Articles/these_eric_fusy.pdf
10. Fusy, É.: Transversal structures on triangulations: A combinatorial study and straight-line drawings. Discr. Math. 309, 1870–1894 (2009)
11. Izumi, T., Takahashi, A., Kajitani, Y.: Air-pressure model and fast algorithms for zero-wasted-area layout of general floorplan. IEICE Trans. Fundam. Electron., Commun. and Comp. Sci. E81-A, 857–865 (1998)
12. Kant, G., He, X.: Regular edge labeling of 4-connected plane graphs and its applications in graph drawing problems. Theor. Comput. Sci. 172, 175–193 (1997)
13. Schrenzenmaier, H.: Ein Luftdruckparadigma zur Optimierung von Zerlegungen. Bachelor's Thesis, TU Berlin (2013)
14. Wimer, S., Koren, I., Cederbaum, I.: Floorplans, planar graphs, and layouts. IEEE Transactions on Circuits and Systems 35, 267–278 (1988)

Superpatterns and Universal Point Sets

Michael J. Bannister, Zhanpeng Cheng,
William E. Devanny, and David Eppstein *

Department of Computer Science, University of California, Irvine, USA

Abstract. An old open problem in graph drawing asks for the size of a *universal point set*, a set of points that can be used as vertices for straight-line drawings of all n-vertex planar graphs. We connect this problem to the theory of permutation patterns, where another open problem concerns the size of *superpatterns*, permutations that contain all patterns of a given size. We generalize superpatterns to classes of permutations determined by forbidden patterns, and we construct superpatterns of size $n^2/4 + \Theta(n)$ for the 213-avoiding permutations, half the size of known superpatterns for unconstrained permutations. We use our superpatterns to construct universal point sets of size $n^2/4 - \Theta(n)$, smaller than the previous bound by a 9/16 factor. We prove that every proper subclass of the 213-avoiding permutations has superpatterns of size $O(n \log^{O(1)} n)$, which we use to prove that the planar graphs of bounded pathwidth have near-linear universal point sets.

1 Introduction

Fary's theorem tells us that every planar graph can be drawn with its edges as non-crossing straight line segments. As usually stated, this theorem allows the vertex coordinates of the drawing to be drawn from an uncountable and unbounded set, the set of all points in the plane. It is natural to ask how tightly we can constrain the set of possible vertices. In this direction, the *universal point set problem* asks for a sequence of point sets $U_n \subseteq \mathbf{R}^2$ such that every graph with n vertices can be straight-line embedded with vertices in U_n and such that the size of U_n is as small as possible.

So far the best known upper bounds for this problem have considered sets U_n of a special form: the intersection of the integer lattice with a convex polygon. In 1988 de Fraysseix, Pach and Pollack showed that a triangular set of lattice points within a rectangular grid of $(2n-3) \times (n-1)$ points forms a universal set of size $n^2 - O(n)$ [1,2], and in 1990 Schnyder found more compact grid drawings within the lower left triangle of an $(n-1) \times (n-1)$ grid [3], a set of size $n^2/2 - O(n)$. Using the method of de Fraysseix et al., Brandenburg found that a triangular subset of a of a $\frac{4}{3}n \times \frac{2}{3}n$ grid, of size $\frac{4}{9}n^2 + O(n)$, is universal [4]. Until now his bound has remained the best known.

On the other side, Dolev, Leighton, and Trickey [5] used the *nested triangles graph* to show that rectangular grids that are universal must have size at least $n/3 \times n/3$, or with a fixed choice of planar embedding and outer face $2n/3 \times 2n/3$. Thus, if we wish to find subquadratic universal point sets we must consider sets not forming a grid. However,

---

* This research was supported in part by the National Science Foundation under grants 0830403 and 1217322, and by the Office of Naval Research under MURI grant N00014-08-1-1015.

S. Wismath and A. Wolff (Eds.): GD 2013, LNCS 8242, pp. 208–219, 2013.

the known lower bounds that do not make this grid assumption are considerably weaker. In 1988 de Fraysseix, Pach and Pollack proved the first nontrivial lower bounds of $n + \Omega(\sqrt{n})$ for a general universal point set [1]. This was later improved to $1.098n - o(n)$ by Chrobak and Payne [2]. Finally, Kurowski improved the lower bound to $1.235n$ [6], which is still the best known lower bound.[1]

With such a large gap between these lower bounds and Brandenburg's upper bound, obtaining tighter bounds remains an important open problem in graph drawing [8].

Universal point sets have also been considered for subclasses of planar graphs. For instance, every set of n points in general position (no three collinear) is universal for the n-vertex outerplanar graphs [9]. Universal point sets of size $O\left(n(\log n / \log \log n)^2\right)$ exist for simply-nested planar graphs (graphs that can be decomposed into nested induced cycles) [10], and planar 3-trees have universal point sets of size $O(n^{5/3})$ [11]. Based in part on the results in this paper, the graphs of line and pseudoline arrangements have been shown to have universal point sets of size $O(n \log n)$ [12].

In this paper we provide a new upper bound on universal point sets for general planar graphs, and improved bounds for certain restricted classes of planar graphs. We approach these problems via a novel connection to a different field of study than graph drawing, the study of patterns in permutations.[2] A permutation σ is said to *contain* the pattern π (also a permutation) if σ has a (not necessarily contiguous) subsequence whose elements are in the same relative order with respect to each other as the elements of π. The permutations that do not contain any pattern in a given set F of forbidden patterns are said to be *F-avoiding*; we define $S_n(F)$ to be the length-n permutations avoiding F. Researchers in permutation patterns have defined a *superpattern* to be a permutation that contains all length-n permutations among its patterns, and have studied bounds on the lengths of these patterns [14, 15], culminating in a proof by Miller that there exist superpatterns of length $n^2/2 + \Theta(n)$ [16]. We generalize this concept to an $S_n(F)$-superpattern, a permutation that contains all possible patterns in $S_n(F)$; we prove that for certain sets F, the $S_n(F)$-superpatterns are much shorter than Miller's bound.

As we show, the existence of small $S_n(213)$-superpatterns leads directly to small universal point sets for arbitrary planar graphs, and the existence of small $S_n(F)$-superpatterns for F containing 213 leads to small universal point sets for subclasses of the planar graphs. Our method constructs a universal set U that has one point for each each element of the superpattern σ. It uses two different traversals of a depth-first-search tree of a canonically oriented planar graph G to derive a permutation $\mathrm{cperm}(G)$ from G, and it uses the universality of σ to find $\mathrm{cperm}(G)$ as a pattern in σ. Then, the positions of the elements of this pattern in σ determine the assignment of the corresponding vertices of G to points in U, and we prove that this assignment gives a planar embedding of G. A similar but simpler reduction uses S_n-superpatterns to construct universal point sets for *dominance drawings* of transitively reduced *st*-planar graphs.

---

[1] The validity of this result was originally questioned by Mondal [7], but later confirmed.

[2] A different connection between permutation patterns and graph drawing is being pursued independently by Bereg, Holroyd, Nachmanson, and Pupyrev, in connection with bend minimization in bundles of edges that realize specified permutations [13].

Specifically our contributions include proving the existence of:

- superpatterns for 213-avoiding permutations of size $n^2/4 + \Theta(n)$;
- universal point sets for planar graphs of size $n^2/4 - \Theta(n)$;
- universal point sets for dominance drawings of size $n^2/2 + \Theta(n)$;
- superpatterns for every proper subclass of the 213-avoiding permutations of size $O(n \log^{O(1)} n)$;
- universal point sets for graphs of bounded pathwidth of size $O(n \log^{O(1)} n)$; and
- universal point sets for simply-nested planar graphs of size $O(n \log n)$.

In addition, we prove that every superpattern for $\{213, 132\}$-avoiding permutations has length $\Omega(n \log n)$, which in turn implies that every superpattern for 213-avoiding permutations has length $\Omega(n \log n)$. It was known that S_n-superpatterns must have quadratic length—otherwise they would not have enough length-n subsequences to cover all $n!$ permutations [14]—but such counting arguments cannot provide nonlinear bounds for $S_n(F)$-superpatterns due to the now-proven Stanley–Wilf conjecture that $S_n(F)$ grows singly exponentially [17]. Instead, our proof finds an explicit set of $\{213, 132\}$-avoiding permutations whose copies within a superpattern cannot share many elements.

2 Preliminaries

Let S_n denote the set of all *permutations* of the numbers from 1 to n. We will normally specify a permutation as a sequence of numbers: for instance, the six permutations in S_3 are 123, 132, 213, 231, 312, and 321. If π is a permutation, then we write π_i for the element in the ith position of π, and $|\pi|$ for the number of elements in π.

We say that a permutation π is a *subpattern* of a permutation σ of length n if there exists a sequence of integers $1 \le \ell_1 < \ell_2 < \cdots < \ell_{|\pi|} \le n$ such that $\pi_i < \pi_j$ if and only if $\sigma_{\ell_i} < \sigma_{\ell_j}$. In other words, π is a subpattern of σ if π is order-isomorphic to a subsequence of σ. We say that a permutation σ *avoids* a permutation ϕ if σ does not contain ϕ as a subpattern. A *permutation class* is a set of permutations with the property that all patterns of a permutation in the class also belong to the class; every permutation class may be defined by a (not necessarily finite) set of *forbidden permutations*, the minimal patterns that do not belong to the class. Define $S_n(\phi_1, \ldots, \phi_k)$ to be the set of all length-n permutations that avoid all of the (forbidden) patterns $\phi_1, \ldots, \phi_k$. Given a set of permutations $P \subseteq S_n$, we define a *P-superpattern* to be a permutation σ with the property that every $\pi \in P$ is a subpattern of σ.

One of the most important permutation classes in the study of permutation patterns is the class of *stack-sortable permutations* [18], the permutations that avoid the pattern 231. Knuth's discovery that these are exactly the permutations that can be sorted using a single stack [19] kicked off the study of permutation patterns. The 213-avoiding permutations that form the focus of our research are related to the 231-avoiding permutations by a simple transformation, the replacement of each value i in a permutation by the value $n + 1 - i$, that does not affect the existence or size of superpatterns.

Given a permutation π we define a *column* of π to be a maximal ascending run of π, and we define a *row* of π to be a maximal ascending run in π^{-1}, where a *run* is a contiguous monotone subsequence of the permutation. We define a *block* of π to be a set of consecutive integers that appear contiguously (but not necessarily in order) in π. For instance, $\{3,4,5\}$ forms a block in 14352. (Our definition of rows and columns differs from that of Miller [16]: for our definition, the intersection of a row and column is a block that could contain more than one element, whereas in Miller's definition a row and column intersect in at most one element.)

We define the *chessboard representation* of a permutation π to be an $r \times c$ matrix $M = \text{chessboard}(\pi)$, where r is number of rows in π and c is the number of columns in π, such that $M(i,j)$ is the number of points in the intersection of the ith column and the jth row of π. An example of a chessboard representation can be seen in Figure 1. To recover a permutation from its chessboard representation, start with the lowest row and work upwards assigning an ascending subsequence of values to the squares of each row in left to right order within each row. If a square has label i, allocate i values for it. Then, after this assignment has been made, traverse each column in left-to-right order, within each column listing in ascending order the values assigned to each square of the column. The sequence of values listed by this column traversal is the desired permutation.

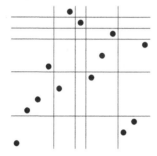
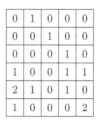

Fig. 1. The permutation $\pi = 1\ 4\ 5\ 8\ 6\ 13\ 12\ 7\ 9\ 11\ 2\ 3\ 10$ represented by its scatterplot (the points (i, π_i)) with lines separating its rows and columns (left), and by chessboard(π) (right)

3   From Superpatterns to Universal Point Sets

In this section, we show how 213-avoiding superpatterns can be turned into universal point sets for planar graphs. Let G be a planar graph. We assume G is *maximal planar*, meaning that no additional edges can be added to G without breaking its planarity; this is without loss of generality, because a point set that is universal for maximal planar graphs is universal for all planar graphs. Additionally, we assume that G has a fixed plane embedding; for maximal planar graphs, such an embedding is determined by the choice of which of the triangles of G is to be the outer face.

3.1   Canonical Representation

As in the grid drawing method of de Fraysseix, Pach and Pollack [1], we use *canonical representations* of planar graphs; these are sequences $v_1, v_2, \ldots v_n$ of the vertices of the given maximal plane graph G with the following properties:

- $v_1v_2v_n$ is the outer triangle of the embedding, in clockwise order v_1, v_n, v_2.
- Each vertex v_i with $i \geq 3$ has two or more earlier neighbors in the sequence, and these neighbors form a contiguous subset of the cyclic ordering of neighbors around v_i in the embedding of G.

Given a canonical representation, let G_k be the induced subgraph of G with vertex set $\{v_1, v_2, \ldots v_k\}$. Then G_k is necessarily 2-connected; its induced embedding has as its exterior face a simple cycle C_k containing v_k and the edge v_1v_2, and the neighbors of v_k in G_k form an induced path in C_{k-1}.

As de Fraysseix, Pach and Pollack proved, every embedded maximal planar graph has at least one canonical representation. For the rest of this section, we will assume that the vertices v_i of the given maximal planar graph G are numbered according to such a representation. The definition of a canonical representation implies that the outer face of G is the triangle $v_1v_2v_n$; we will assume that this triangle is oriented so that v_1, v_n, v_2 are in clockwise order.

For each vertex v_i with $i > 2$, let parent(v_i) be the most clockwise smaller-numbered neighbor of v_i. By following a path of edges from vertices to their parents, each vertex can reach v_1, so these edges form a tree ctree(G) having v_1 as its root; this same tree may also be obtained by orienting each edge of G from lower to higher numbered vertices, and then performing a depth-first search of the resulting oriented graph that visits the children of each vertex in clockwise order, starting from v_1.[3] For each vertex v_i of G, let pre(v_i) be the position of v_i in a pre-order traversal of ctree(G) that visits the children of each node in clockwise order, and let post(v_i) be the position of v in a sequence of the nodes of ctree(G) formed by reversing a post-order clockwise traversal. These two numbers may be used to determine the ancestor-descendant relationships in ctree(G): a node v_i is an ancestor of a node v_j if and only if both pre$(v_i) <$ pre(v_j) and post$(v_i) <$ post(v_j) [20].

Lemma 1. *Let G be a canonically-represented maximal plane graph, and renumber the vertices of G in order by their values of* post(v_i). *Then the result is again a canonical representation of the same embedding of G, and for each induced subgraph G_k of this new canonical representation, the clockwise ordering of the vertices along the exterior cycle C_k is in sorted order by the values of* pre(v_i).

Proof. The fact that post(v_i) gives a canonical representation comes from the fact that it is a reverse postorder traversal of a depth-first search tree. Reverse postorder traversal gives a topological ordering of every directed acyclic graph, from which it follows that every vertex in G has the same set of earlier neighbors when ordered by post(v_i) as it did in the original ordering.

The statement on the ordering of the vertices of C_k follows by induction, from the fact that v_k has a larger value of pre(v_i) than its earliest incoming neighbor (its parent in ctree(G)) and a smaller value than all of its other incoming neighbors.   □

---

[3] Although we do not use this fact, ctree(G) is also part of a Schnyder decomposition of G, together with a second tree rooted at v_2 connecting each vertex to its most counterclockwise earlier neighbor and a third tree rooted at v_n connecting each vertex to the later vertex whose addition removes it from C_k.

Let cperm(G) be the permutation in which, for each vertex v_i, the permutation value in position pre(v_i) is post(v_i). That is, cperm(G) is the permutation given by traversing ctree(G) in preorder and listing for each vertex of the traversal the number post(v_i).

Lemma 2. *For every canonically-represented maximal planar graph G, the permutation* $\pi = $ cperm(G) *is 213-avoiding.*

Proof. Let $i < j < k$ be an arbitrary triple of indexes in the range from 1 to n, corresponding to the vertices u_i, u_j and u_k. If π_j is not the smallest of these three values, then π_i, π_j, and π_k certainly do not form a 213 permutation pattern. If π_j is the smallest of these three values, then, since pre(u_i) < pre(u_j) but post(u_i) > post(u_j), u_i is not an ancestor or descendant of u_j, and u_j is an ancestor of u_k. Therefore u_i is also not an ancestor or descendant of u_k, from which it follows that $\pi_i > \pi_k$ and the pattern formed by π_i, π_j, and π_k is 312 rather than 213. Since the choice of indices was arbitrary, no three indices can form a 213 pattern and π is 213-avoiding. □

We observe that cperm(G) has some additional structure, as well: its first element is 1, its second element is n, and its last element is 2.

3.2 Stretching a Permutation

It is natural to represent a permutation σ by the points with Cartesian coordinates (i, σ_i), but for our purposes we need to stretch this representation in the vertical direction; we use a transformation closely related to one used by Bukh, Matoušek, and Nivasch [21] for weak epsilon-nets, and by Fulek and Tóth [11] for universal point sets for plane 3-trees. Letting $q = |\sigma|$, we define

$$\text{stretch}(\sigma) = \left\{ (i, q^{\sigma_i}) \mid 1 \le i \le q \right\}.$$

Lemma 3. *Let σ be an arbitrary permutation with $|\sigma| = q$, let p_i denote the point in* stretch(σ) *corresponding to position i in σ, let i and j be two indices with $\sigma_i < \sigma_j$, and let m be the absolute value of the slope of line segment $p_i p_j$. Then $q^{\sigma_j - 1} \le m < q^{\sigma_j}$.*

Proof. The minimum value of m is obtained when $|i - j| = q - 1$ and $\sigma_i = \sigma_j - 1$, for which $q^{\sigma_j - 1} = m$. The maximum value of m is obtained when $|i - j| = 1$ and $\sigma_i = 1$, for which $m = q^{\sigma_j} - q < q^{\sigma_j}$. □

Lemma 4. *Let σ be an arbitrary permutation with $|\sigma| = q$, let p_i denote the point in* stretch(σ) *corresponding to position i in σ, and let i, j, and k be three indices with* $\max\{\sigma_i, \sigma_j\} < \sigma_k$ *and $i < j$. Then the clockwise ordering of the three points p_i, p_j, and p_k is p_i, p_k, p_j.*

Proof. The result follows by using Lemma 3 to compare the slopes of the two line segments $p_i p_j$ and $p_i p_k$. □

Lemma 5. *Let σ be an arbitrary permutation with $|\sigma| = q$, let p_i denote the point in* stretch(σ) *corresponding to position i in σ, and let h, i, j, and k be four indices with* $\max\{\sigma_h, \sigma_i, \sigma_j\} < \sigma_k$ *and $h < j$. Then line segments $p_h p_j$ and $p_i p_k$ cross if and only if both $h < i < j$ and $\max\{\sigma_h, \sigma_j\} > \sigma_i$.*

Proof. A crossing occurs between two line segments if and only if the endpoints of each segment are on opposite sides of the line through the other segment. The endpoints of $p_i p_k$ are on opposite sides of line $p_h p_j$ if and only if the two triangles $p_h p_i p_j$ and $p_h p_k p_j$ have opposite orientations; with the assumption that σ_k is the largest of the three values, this is equivalent by Lemma 4 to the condition that σ_i is not the second-largest. The endpoints of $p_h p_j$ are on opposite sides of line $p_i p_k$ if and only if the two triangles $p_i p_h p_k$ and $p_i p_j p_k$ have opposite orientations; this is equivalent by Lemma 4 to the condition that $h < i < j$.   □

3.3   Universal Point Sets

If σ is any permutation, we define augment(σ) to be a permutation of length $|\sigma| + 3$, in which the first element is 1, the second element is $|\sigma| + 3$, the last element is 2, and the remaining elements form a pattern of type σ. It follows from Lemma 2 that, if σ is an $S_{n-3}(213)$-superpattern and if G is an arbitrary n-vertex maximal plane graph, then cperm(G) is a pattern in augment(σ).

Theorem 1. *Let σ be an $S_{n-3}(213)$-superpattern, and let $U_n =$ stretch(augment(σ)). Then U_n is a universal point set for planar graphs on n vertices.*

Proof. Let G be an arbitrary maximal plane graph, let $v_1, v_2, \ldots v_n$ be a canonical representation of G, and let x_i denote a sequence of positions in augment(σ) that form a pattern of type cperm(G), with position x_i in augment(σ) corresponding to position pre(v_i) in cperm(G). Let $q = |\text{augment}(\sigma)|$, and for each i, let $y_i = q^j$ where j is the value of augment(σ) at position x_i. Embed G by placing vertex v_i at the point $(x_i, y_i) \in U_n$.

Let $v_h, v_i, v_j,$ and v_k be four vertices in G such that $v_h v_j$ and $v_i v_k$ are edges in G. We may choose these indices in such a way that post(v_k) is larger than the post values of the other three vertices. If these two edges crossed in the given embedding of G, then by Lemma 5 we would necessarily have pre(v_h) < pre(v_i) < pre(v_j), and post(v_i) < max{post(v_h), post(v_j)}. By Lemma 1, v_i would not be on the outside face of the graph induced by the vertices with post values at most max{post(v_h), post(v_j)}, and could not be a neighbor of v_k in the canonical representation given by the post values. This contradiction shows that no crossing is possible, so the embedding is planar.   □

4 $S_n(213)$-Superpatterns

In this section we construct a $S_n(213)$-superpattern of size $n^2/4 + n + ((-1)^n - 1)/8$. An exhaustive computer search has shown that this size is minimal for $n \le 6$. We start with a lemma about $S_n(213)$-superpatterns with n rows and n columns (the minimal amount of each), demonstrating their recursive structure.

Lemma 6. *If σ is a $S_n(213)$-superpattern and has n rows and n columns, then the permutation described by the intersection of columns $n - j + 1$ to $n - i + 1$ and rows i to j of σ is a $S_{j-i+1}(213)$-superpattern.*

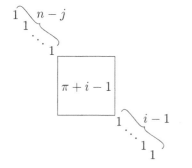

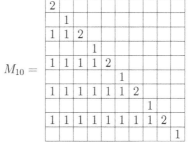

Fig. 2. The permutation τ constructed from π in the proof of Lemma 6

Fig. 3. The base case permutations for constructing μ_n for $n > 2$ and their chessboard representations

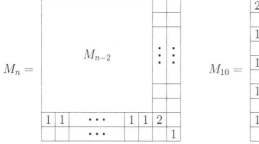

Fig. 4. The inductive construction of chessboard(μ_n) from chessboard(μ_{n-2}). Cells of the matrix containing zero are shown as blank

Proof. Let π be an arbitrary 213-avoiding permutation of length $j - i + 1$ and consider the n-element 213-avoiding permutation

$$\tau = n(n-1)\ldots(j+1)(\pi_1 + i - 1)(\pi_2 + i - 1)\ldots(\pi_{j-i+1} + i - 1)(i-1)(i-2)\ldots321.$$

(See Fig. 2.) By the assumption that σ is a superpattern, τ has an embedding into σ. Because there are $n - j$ descents in τ before the first element of the form $\pi_i + i - 1$, this embedding cannot place any element $\pi_i + i - 1$ into the first $n - j$ columns of σ. Similarly because there are i descents in τ after the last element of the form $\pi_i + i - 1$, this embedding cannot place any element $\pi_i + i - 1$ into the last $i - 1$ columns of σ. By a symmetric argument, the elements of the form $\pi_i + i - 1$ cannot be embedded into the $i - 1$ lowest rows nor the $n - j$ highest rows of σ. Therefore these elements, which form a pattern of type π, must be embeddable into σ inclusively between column $n - j + 1$, column $n - i + 1$, row i, and row j. Since π was arbitrary, this part of σ must be universal for permutations of length $j - i + 1$, as claimed. □

We define a permutation μ_n, which will be shown to be a $S_n(213)$-superpattern, by describing chessboard$(\mu_n) = M_n$. In our construction M_n and μ_n have n columns and

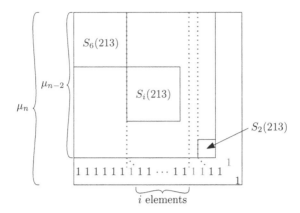

Fig. 5. A partial embedding of the red elements, showing where the remaining blocks can be fit into the columns of μ_{n-2}

n rows. The bottom two rows of M_n contain the values $M_n(n,1) = M_n(i,2) = 1$ for $1 \le i \le n-2$, and $M_n(n-1,2) = 2$, with all other values in these rows zero. The values in the top $n-2$ rows are given recursively by $M_n(1:n-2,3:n) = M_{n-2}$, again with all values outside this submatrix zero. The base cases of μ_1 and μ_2 are shown in Figure 3 and the inductive definition and an example are shown in Figure 4.

Theorem 2. *The permutation μ_n is a $S_n(213)$-superpattern. Thus there exists a $S_n(213)$-superpattern whose size is $n^2/4 + n + ((-1)^n - 1)/8$.*

Proof. It can be easily verified that μ_i is a $S_i(213)$-superpattern when $1 \le i \le 2$. Let π be an arbitrary 213-avoiding permutation of length n. We will show that π can be embedded into μ_n.

Case 1: $\pi_n = 1$

　　Let $\pi_{i_1} \ldots \pi_{i_k}$ be the second lowest row of π. Because π is 213-avoiding, $i_k = n-1$ and for all j, $\pi_{i_j} = j + 1$. We embed this bottom row by mapping π_n to the bottom right element of μ_n and π_{i_j} to the i_j-th position of the second lowest row of μ_n.

Case 2: $\pi_n \ne 1$

　　Let $\pi_{i_1} \ldots \pi_{i_k}$ be the lowest row of π. Similarly to Case 1, because π is 213-avoiding, $i_k = n$ and for all j, $\pi_{i_j} = j$. We embed this bottom row by mapping π_{i_j} to the i_j-th position of the second lowest row of μ_n.

　　These partial embeddings maintain the ordering of the lowest and possibly second lowest row of π. To finish the embedding, the remaining elements need to be fit into the copy of μ_{n-2}. Recall that a block of a permutation is a contiguous subsequence of consecutive values. Because π is 213-avoiding, the remaining elements of π form disjoint blocks that fit between the elements embedded so far. If one block is to the right of another in π, it has smaller values. Let π_{i_1} be the leftmost element that has been embedded on the second row of μ_n. Then there are $i_1 - 1$ elements before it in π and $i_1 - 1$ columns before where it was embedded in μ_n. By Lemma 6, these elements can fit into the last $i_1 - 1$ rows of these columns. Let π_{i_j} and $\pi_{i_{j+1}}$ be two adjacent elements embedded in the second lowest row of μ_n. Between these two there are at least $i_{j+1} - i_j - 1$ columns of μ_{n-2} available: the column above π_{i_s} to the column before $\pi_{i_{j+1}}$. So again by Lemma 6, the $i_{j+1} - i_j - 1$ elements between π_{i_j} and $\pi_{i_{j+1}}$ can be fit into the last $i_{j+1} - i_j - 1$ rows of those columns. (See Fig. 5) Because $i_k = n - 1$ in Case 1 and n in Case 2, there is no block after π_{i_k}. Therefore π can be embedded into μ_n and μ_n is a $S_n(213)$-superpattern. □

Combining Theorem 2 with Theorem 1, the following is immediate:

Theorem 3. *The n-vertex planar graphs have universal point sets of size $n^2/4 - \Theta(n)$.*

5 Dominance Drawing

A *dominance drawing* of a directed acyclic graph [22] is a drawing of the graph in the plane such that each edge is directed upwards and to the right, such that the axis-aligned bounding box of every edge contains no vertices other than its endpoints, and such that no edge can be added to the drawing preserving these properties. The graphs with planar dominance drawings are exactly the transitively reduced *st*-planar graphs, i.e. the planar directed acyclic graphs in which there is one source and one sink, both on the outer face, and in which each edge forms the only directed path connecting its two endpoints.

If a graph has a dominance drawing D, then it has a drawing in which the points are in general position, and the points in this case can be thought of as representing a permutation π_D, where the positions of the elements in the permutation are the positions of the points in the sorted order by their x coordinates and the values of these elements are the positions in the sorted order by the y coordinates. Any two point sets with the same two sorted orders may be used as the basis for a dominance drawing combinatorially equivalent to D. In particular, if π_D appears as a pattern in another permutation σ, then the subset of the points (i, σ_i) corresponding to elements of π_D may be used to draw the same graph. This gives us the following result:

Theorem 4. *If σ is a superpattern for the length-n permutations, then the set of points (i, σ_i) is universal for dominance drawings of n-vertex transitively reduced st-planar graphs.*

Combining this result with Miller's bound on superpatterns [16] shows that dominance drawings have universal point sets of size $n^2/2 + \Theta(n)$, half the size of the point sets given by previous methods based on $n \times n$ grids.

Not every permutation is of the form π_D for a planar dominance drawing D; for instance the permutation 2143 corresponds to a drawing that has a crossing. However, every permutation π forms a pattern in a larger permutation σ that does define a planar dominance drawing, constructed from the Dedekind–MacNeille completion of a partially ordered set associated to π [23]. For this reason, the permutations that define planar dominance drawings have no forbidden patterns. However, in later research, we have shown that the dominance drawings of some other classes of graphs have forbidden patterns, leading to smaller universal sets for these drawings [24].

6 Additional Results

In the full version of the paper, we provide the following results.

- We prove that, for every 213-avoiding permutation ϕ, the $\{213, \phi\}$-avoiding permutations have superpatterns of near-linear size. In particular, the $\{213, 312\}$-avoiding permutations and the $\{213, 3412\}$-avoiding permutations have superpatterns of linear size, and the minimum size of a superpattern for the $\{213, 132\}$-avoiding permutations is $\Theta(n \log n)$ (despite these permutations being equinumerous with the $\{213, 312\}$-avoiding permutations). We define the *Strahler number* of any 213-avoiding permutation from a forest derived from its chessboard representation, and we show that if a permutation ϕ has Strahler number s then the $\{213, \phi\}$-avoiding permutations have superpatterns of size $O(n \log^{s-1} n)$.
- We prove that, for every integer w, there exists a pattern ϕ such that the planar graphs of pathwidth at most w correspond to a $\{213, \phi\}$-avoiding permutation (using the same correspondence between graphs and permutations as in Section 3). As a consequence, the planar graphs of bounded pathwidth have universal point sets of size $O(n \log^{O(1)} n)$.
- We improve the bound of Angelini et al. [10] on universal point sets for simply-nested planar graphs from $O\left(n(\log n / \log \log n)^2\right)$ to $O(n \log n)$.

For space reasons we defer the proofs of these results to the full version of the paper.

7 Conclusion

In this paper we have constructed universal point sets for planar graphs of size $n^2/4 - \Theta(n)$, and of subquadratic size for graphs of bounded pathwidth. In the process of building these constructions we have provided a new connection between universal point sets and permutation superpatterns. We have also, for the the first time, provided nontrivial upper bounds and lower bounds on the size of superpatterns for restricted classes of permutations. We leave the following problems open for future research:

- Which natural subclasses of planar graphs (beyond the bounded-pathwdith graphs) can be represented by permutations in a proper subclass of $S_n(213)$?
- Can we reduce the gap between our $O(n^2)$ upper bound and $\Omega(n \log n)$ lower bound for $S_n(213)$-superpatterns?
- Our construction uses area exponential in n^2; how does constraining the area to a smaller bound affect the number of points in a universal point set?

References

1. de Fraysseix, H., Pach, J., Pollack, R.: How to draw a planar graph on a grid. Combinatorica 10(1), 41–51 (1990); Originally presented at STOC 1988
2. Chrobak, M., Payne, T.: A linear-time algorithm for drawing a planar graph on a grid. Inform. Proc. Lett. 54(4), 241–246 (1995)

3. Schnyder, W.: Embedding planar graphs on the grid. In: Proc. 1st ACM–SIAM Symp. on Discrete Algorithms, pp. 138–148 (1990)

4. Brandenburg, F.J.: Drawing planar graphs on $\frac{8}{9}n^2$ area. In: Proc. Int. Conf. Topological and Geometric Graph Theory. Electronic Notes in Discrete Mathematics, vol. 31, pp. 37–40. Elsevier (2008)

5. Dolev, D., Leighton, T., Trickey, H.: Planar embedding of planar graphs. Advances in Computing Research 2, 147–161 (1984)

6. Kurowski, M.: A 1.235 lower bound on the number of points needed to draw all n-vertex planar graphs. Inform. Proc. Lett. 92(2), 95–98 (2004)

7. Mondal, D.: Embedding a Planar Graph on a Given Point Set. Master's thesis, Dept. of Computer Science, U. of Manitoba (2012)

8. Demaine, E., O'Rourke, J.: Smallest Universal Set of Points for Planar Graphs (Problem 45). In: Demaine, E., Mitchell, J.S.B., O'Rourke, J. (eds.) The Open Problems Project (2002-2012)

9. Gritzmann, P., Mohar, B., Pach, J., Pollack, R.: Embedding a planar triangulation with vertices at specified positions. Amer. Math. Monthly 98(2), 165–166 (1991)

10. Angelini, P., Di Battista, G., Kaufmann, M., Mchedlidze, T., Roselli, V., Squarcella, C.: Small point sets for simply-nested planar graphs. In: van Kreveld, M., Speckmann, B. (eds.) GD 2011. LNCS, vol. 7034, pp. 75–85. Springer, Heidelberg (2011)

11. Fulek, R., Tóth, C.D.: Universal point sets for planar three-trees. In: Dehne, F., Solis-Oba, R., Sack, J.-R. (eds.) WADS 2013. LNCS, vol. 8037, pp. 341–352. Springer, Heidelberg (2013)

12. Eppstein, D.: Drawing arrangement graphs in small grids, or how to play Planarity. In: Wismath, S., Wolff, A. (eds.) GD 2013. LNCS, vol. 8242, pp. 436–447. Springer, Heidelberg (2013)

13. Bereg, S., Holroyd, A.E., Nachmanson, L., Pupyrev, S.: Drawing permutations with few corners. In: Wismath, S., Wolff, A. (eds.) GD 2013. LNCS, vol. 8242, pp. 488–499. Springer, Heidelberg (2013)

14. Arratia, R.: On the Stanley-Wilf conjecture for the number of permutations avoiding a given pattern. Elect. J. Comb. 6, N1 (1999)

15. Eriksson, H., Eriksson, K., Linusson, S., Wästlund, J.: Dense packing of patterns in a permutation. Ann. Comb. 11(3-4), 459–470 (2007)

16. Miller, A.: Asymptotic bounds for permutations containing many different patterns. J. Comb. Theory A 116(1), 92–108 (2009)

17. Marcus, A., Tardos, G.: Excluded permutation matrices and the Stanley–Wilf conjecture. J. Comb. Theory A 107(1), 153–160 (2004)

18. Rotem, D.: Stack sortable permutations. Discrete Math. 33(2), 185–196 (1981)

19. Knuth, D.: Vol. 1: Fundamental Algorithms. In: The Art of Computer Programming. Addison-Wesley, Reading (1968)

20. Dietz, P.F.: Maintaining order in a linked list. In: Proc. 15th ACM Symp. on Theory of Computing, pp. 122–127 (1982)

21. Bukh, B., Matoušek, J., Nivasch, G.: Lower bounds for weak epsilon-nets and stair-convexity. Israel J. Math. 182, 199–208 (2011)

22. Di Battista, G., Eades, P., Tamassia, R., Tollis, I.G.: 4.7 Dominance drawings. In: Graph Drawing: Algorithms for the Visualization of Graphs, pp. 112–127. Prentice-Hall (1999)

23. Eppstein, D., Simons, J.A.: Confluent Hasse diagrams. In: van Kreveld, M., Speckmann, B. (eds.) GD 2011. LNCS, vol. 7034, pp. 2–13. Springer, Heidelberg (2011)

24. Bannister, M.J., Devanny, W.E., Eppstein, D.: Small superpatterns for dominance drawing (manuscript 2013)

Simultaneous Embedding:
Edge Orderings, Relative Positions, Cutvertices[*]

Thomas Bläsius, Annette Karrer, and Ignaz Rutter[**]

Karlsruhe Institute of Technology (KIT)
firstname.lastname@kit.edu

Abstract. A simultaneous embedding of two graphs $G^①$ and $G^②$ with common graph $G = G^① \cap G^②$ is a pair of planar drawings of $G^①$ and $G^②$ that coincide on G. It is an open question whether there is a polynomial-time algorithm that decides whether two graphs admit a simultaneous embedding (problem SEFE).

In this paper, we present two results. First, a set of three linear-time preprocessing algorithms that remove certain substructures from a given SEFE instance, producing a set of equivalent SEFE instances without such substructures. The structures we can remove are (1) cutvertices of the *union graph* $G^① \cup G^②$, (2) cutvertices that are simultaneously a cutvertex in $G^①$ and $G^②$ and that have degree at most 3 in G, and (3) connected components of G that are biconnected but not a cycle.

Second, we give an $O(n^2)$-time algorithm for SEFE where, for each pole u of a P-node μ (of a block) of the input graphs, at most three virtual edges of μ contain common edges incident to u. All algorithms extend to the sunflower case.

1 Introduction

A simultaneous embedding of two graphs $G^①$ and $G^②$ with common graph $G = G^① \cap G^②$ is a pair of planar drawings of $G^①$ and $G^②$, that coincide on G. The problem to decide whether a simultaneous embedding exists is called SEFE (simultaneous embedding with fixed edges). This definition extends to more than two graphs. For three graphs SEFE is NP-complete [7]. In the *sunflower case* it is required that every pair of input graphs has the same intersection. See [2] for a survey on SEFE and related problems.

There are two fundamental approaches to solving SEFE in the literature. The first approach is based on the characterization of Jünger and Schulz [10] stating that finding a simultaneous embedding of two graphs $G^①$ and $G^②$ with common graph G is equivalent to finding planar embeddings of $G^①$ and $G^②$ that induce the same embedding on G. The second very recent approach by Schaefer [11] is based on Hanani-Tutte-style redrawing results. One tries to characterize the existence of a SEFE via the existence of drawings where no two independent edges of the same graph cross an odd number of times. The existence of such drawings can be expressed using a linear system of boolean equations.

When following the first approach, we need two things to describe the planar embedding of the common graph G. First, for each vertex v, a cyclic order of incident edges

---

[*] Partly done within GRADR – EUROGIGA project no. 10-EuroGIGA-OP-003.
[**] Work was supported by a fellowship within the Postdoc-Program of the German Academic Exchange Service (DAAD).

S. Wismath and A. Wolff (Eds.): GD 2013, LNCS 8242, pp. 220–231, 2013.

around v. Second, for every pair of connected components H and H' of G, the face f of H containing H'. We call this relationship the *relative position* of H' with respect to H. To find a simultaneous embedding, one needs to find a pair of planar embeddings that induce the same cyclic edge orderings (*consistent edge orderings*) and the same relative positions (*consistent relative positions*) on the common graph G.

Most previous results use the first approach but none of them considers both consistent edge orderings and relative positions. Most of them assume the common graph to be connected or to contain no cycles. The strongest results of this type are the two linear-time algorithms for the case that G is biconnected by Haeupler et al. [9] and by Angelini et al. [1] and a quadratic-time algorithm for the case where $G^①$ and $G^②$ are biconnected and G is connected [4]. In the latter result, SEFE is modeled as an instance of the problem SIMULTANEOUS PQ-ORDERING. On the other hand, there is a linear-time algorithm for SEFE if the common graph consists of disjoint cycles [3], which requires to ensure consistent relative positions but makes edge orderings trivially consistent.

The advantage of the second approach is that it implicitly handles both, consistent edge orderings and consistent relative positions, at the same time. Thus, the results by Schaefer [11] are the first that handle SEFE instances where the common graph consists of several, non-trivial connected components. He gives a polynomial-time algorithm for the cases where each connected component of the common graph is biconnected or has maximum degree 3. Although this approach is conceptionally simple, very elegant, and combines several notions of planarity within a common framework, it has two disadvantages. The running time of the algorithms are quite high and the high level of abstraction makes it difficult to generalize the results, e.g., to the sunflower case.

Contribution & Outline. In this paper, we follow the first approach and show how to enforce consistent edge orderings and consistent relative positions at the same time, by combining different recent approaches, namely the algorithm by Angelini et al. [1] and result on SIMULTANEOUS PQ-ORDERING [4] for consistent edge orderings and the result on disjoint cycles [3] for consistent relative positions. To handle relative positions of connected components to each other without knowing their embedding, we show that these relative positions can be expressed in terms of relative positions with respect to a cycle basis. In addition to that, we are able to handle certain cutvertices of $G^①$ and $G^②$.

More precisely, we classify a vertex v to be a *union cutvertex*, a *simultaneous cutvertex*, and an *exclusive cutvertex* if v is a cutvertex of $G^① \cup G^②$, of $G^①$ and $G^②$ but not of $G^① \cup G^②$, and of $G^①$ but not $G^②$ or the other way around, respectively. We say that v has *common-degree k* if it is a common vertex with degree k in G. We present three preprocessing algorithms that simplify given instances of SEFE; see Section 3. They remove union cutvertices and simultaneous cutvertices with common-degree 3 (note that simultaneous cutvertices with common degree less than 3 cannot exist), and replace connected components of G that are biconnected by cycles. They run in linear time and can be applied independently. The latter algorithm together with the linear-time algorithm for disjoint cycles [3] improves the result by Schaefer [11] for instances where every connected component of G is biconnected to linear time and the sunflower case.

In Section 4 we show how to solve instances that have *common P-node degree 3* and contain neither union nor simultaneous cutvertices in quadratic time. An instance has common P-node degree k if, for each pole u of a P-node μ (of a block) of the input graphs, at most k virtual edges of μ contain common edges incident to u. Together with the preprocessing steps, this includes the case where every connected component of G is biconnected, has maximum degree 3, or is outerplanar with maximum degree 3 cutvertices. As before, this approach also applies to the sunflower case.

2   Preliminaries

Connectivity & SPQR-trees. A graph is *connected* if there exists a path between any pair of vertices. A *separating k-set* is a set of k vertices whose removal disconnects the graph. Separating 1-sets and 2-sets are *cutvertices* and *separation pairs*, respectively. A connected graph is *biconnected* if it has no cut vertex and *triconnected* if it has no separation pair. The maximal biconnected components of a graph are called *blocks*. The *split components* with respect to a separating k-set are the maximal subgraphs that are not disconnected by removing the separating k-set.

The SPQR-tree $\mathcal{T}$ of a biconnected graph G represents the decomposition of G along its *split pairs*, where a split pair is either a separating pair or a pair of adjacent vertices [6]. We consider the SPQR-tree to be unrooted, representing embeddings on the sphere, i.e., planar embeddings without a designated outer face.

Let $\{s, t\}$ be a split pair and let H_1 and H_2 be two subgraphs of G such that $H_1 \cup H_2 = G$ and $H_1 \cap H_2 = \{s, t\}$. Consider the tree consisting of two nodes μ_1 and μ_2 associated with the graphs $H_1 + \{s, t\}$ and $H_2 + \{s, t\}$, respectively. These graphs are called *skeletons* of the nodes μ_i, denoted by $\mathrm{skel}(\mu_i)$, and the special edge $\{s, t\}$ is a *virtual edge*. The edge connecting the nodes μ_1 and μ_2 associates the virtual edges in $\mathrm{skel}(\mu_1)$ and $\mathrm{skel}(\mu_2)$ with each other. The *expansion graph* of a virtual edge $\{s, t\}$ is the subgraph of G it represents, that is in $\mathrm{skel}(\mu_1)$ the expansion graph of $\{s, t\}$ is H_2 and the expansion graph of $\{s, t\}$ in $\mathrm{skel}(\mu_2)$ is H_1. A combinatorial embedding of G uniquely induces a combinatorial embedding of $\mathrm{skel}(\mu_1)$ and $\mathrm{skel}(\mu_2)$ and vice versa.

Applying this kind of decomposition systematically yields the SPQR-tree $\mathcal{T}$. The skeletons of the internal nodes of $\mathcal{T}$ are either a cycle (S-node), a bunch of parallel edges (P-node) or a triconnected planar graph (R-node). The leaves are Q-nodes, and their skeleton consists of two vertices connected by a virtual and a normal edge. Thus, the only possible embedding choices are flipping skeletons of R-nodes and ordering the edges in skeletons of P-nodes. The SPQR-tree can be computed in linear time [8].

Let $\mathcal{T}^\text{①}$ by the SPQR-tree of a block of $G^\text{①}$ in an instance of SEFE and let G be the common graph. Let further μ be a P-node of $\mathcal{T}^\text{①}$. We say that μ has *common P-node degree k* if both vertices in $\mathrm{skel}(\mu)$ are incident to common edges in the expansion graphs of at most k virtual edges (note that these can be different edges for the two vertices). We say that $G^\text{①}$ has common P-node degree k if each P-node in the SPQR-tree of each block of $G^\text{①}$ has common P-node degree k. If this is the case for $G^\text{①}$ and $G^\text{②}$, we say that the instance of SEFE has common P-node degree k.

PQ-trees. A PQ-tree, originally introduced by Booth and Lueker [5], is a tree, whose inner nodes are either P-nodes or Q-nodes (note that these P-nodes have nothing to do with the P-nodes of the SPQR-tree). The order of edges around a P-node can be ordered arbitrarily, the edges around a Q-node are fixed up to a flip. In this way, a PQ-tree represents a set of orders on its leaves. A rooted PQ-tree represents linear orders, an unrooted

PQ-tree represents cyclic orders (in most cases we consider unrooted PQ-trees). Given a PQ-tree T and a subset S of its leaves, there exists another PQ-tree T' representing exactly the orders represented by T where the elements in S are consecutive. The tree T' is the *reduction* of T with respect to S. The *projection* of T to S is a PQ-tree with leaves S representing exactly the orders on S that are represented by T.

The problem SIMULTANEOUS PQ-ORDERING has several PQ-trees sharing some leaves as input, that are related by identifying some of their leaves [4]. More precisely, every instance is a directed acyclic graph, where each node is a PQ-tree, and each arc (T, T') has the property that there is an injective map from the leaves of the *child* T' to the leaves of the *parent* T. For each PQ-tree in such an instance, one wants to find an order of its leaves such that for every arc (T, T') the order chosen for the parent T is an extension of the order chosen for the child T' (with respect to the injective map). We will later use instances of SIMULTANEOUS PQ-ORDERING to express relations between orderings of edges around vertices.

3   Preprocessing Algorithms

In this section, we present several algorithms that can be used as a preprocessing of a given SEFE instance. The result is usually a set of SEFE instances that admit a solution if and only if the original instance admits one. The running time of the preprocessing algorithms is linear, and so is the total size of the equivalent set of SEFE instances. Each of the preprocessing algorithms removes certain types of structures form the instance, in particular from the common graph. Namely, we show that we can eliminate union cutvertices, simultaneous cutvertices with common-degree 3, and connected components of G that are biconnected but not a cycle. None of these algorithms introduces new cutvertices in G or increases the degree of a vertex. Thus, the preprocessing algorithms can be successively applied to a given instance, removing all the claimed structures.

Let $(G^{①}, G^{②})$ be a SEFE instance with common graph $G = G^{①} \cap G^{②}$. We can equivalently encode such an instance in terms of its *union graph* $G_{\cup} = G^{①} \cup G^{②}$, whose edges are labeled $\{1\}$, $\{2\}$, or $\{1, 2\}$, depending on whether they are contained exclusively in $G^{①}$, exclusively in $G^{②}$, or in G, respectively. Any graph with such an edge coloring can be considered as a SEFE instance. Since sometimes the coloring version is more convenient, we use these notions interchangeably throughout this section.

3.1   Union Cutvertices and Simultaneous Cutvertices

It is not hard to see that union cutvertices of a SEFE instance can be used to split it into independent instances. A simultaneous cutvertex with common-degree 3 can be modified as in Fig. 1, yielding an equivalent instance. Exhaustively applying these ideas, yields the following results; proofs are omitted due to space constraints.

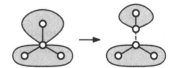

Fig. 1. A simultaneous cutvertex with common-degree 3. The gray regions are the split components of $G^{①}$, the new dashed edge belongs to $G^{②}$.

Theorem 1. *There is a linear-time algorithm that decomposes a SEFE instance into an equivalent set of SEFE instances that do not contain union cutvertices.*

Theorem 2. *Let* $(G^{①}, G^{②})$ *be an instance of* SEFE *such that every simultaneous cutvertex has common-degree 3. An equivalent instance without simultaneous cutvertices can be computed in linear time.*

3.2   Connected Components that are Biconnected

Let $(G^{①}, G^{②})$ be a SEFE instance. Throughout this section, we assume without loss of generality that $G^{①}$ and $G^{②}$ are connected [3] and that the common graph G is an induced subgraph of $G^{①}$ and $G^{②}$. The latter can be achieved by subdividing each exclusive edge once, which clearly does not alter the existence of a SEFE.

Let C be a connected component of G that is a cycle. A *union bridge* of $G^{①}$ and $G^{②}$ with respect to C is a connected component of $G_{\cup} - C$ together with all its attachment vertices on C. Similarly, there are ①-*bridges* and ②-*bridges*, which are connected components of $G^{①} - C$ and $G^{②} - C$ together with their attachment vertices on C, respectively. We say that two bridges B_1 and B_2 *alternate* if there are attachments a_1, b_1 of B_1 and attachments a_2, b_2 of B_2, such that the order along C is $a_1 a_2 b_1 b_2$. We have the following lemma.

Lemma 1. *Let* $G^{①}$ *and* $G^{②}$ *be two planar graphs and let* C *be a connected component of the common graph that is a cycle. Then the graphs* $G^{①}$ *and* $G^{②}$ *admit a* SEFE *where* C *is the boundary of the outer face if and only if (i) each union bridge admits a* SEFE *together with* C *and (ii) no two* ①-*bridges of* C *alternate for* $i = 1, 2$.

Proof. Clearly the conditions are necessary; we prove sufficiency. Let $B_1, \ldots, B_k$ be the union bridges with respect to C, and let $(\mathcal{E}_1^{①}, \mathcal{E}_1^{②}), \ldots, (\mathcal{E}_k^{①}, \mathcal{E}_k^{②})$ be the corresponding simultaneous embeddings of B_i together with C, which exist by condition (i). Note that each union bridge is connected, and hence all its edges and vertices are embedded on the same side of C. After possibly flipping some of the embeddings, we may assume that each of them has C with the same clockwise orientation as the outer face.

We now glue $\mathcal{E}_1^{①}, \ldots, \mathcal{E}_k^{①}$ to an embedding $\mathcal{E}^{①}$ of $G^{①}$, which is possible by condition (ii). In the same way, we find an embedding $\mathcal{E}^{②}$ of $G^{②}$ from $\mathcal{E}_1^{②}, \ldots, \mathcal{E}_k^{②}$. We claim that $(\mathcal{E}_1, \mathcal{E}_2)$ is a SEFE of $G^{①}$ and $G^{②}$. For the consistent edge orderings, observe that any common vertex v with common-degree at least 3 is contained, together with all neighbors, in some union bridge B_i. The compatibility of the edge ordering follows since $(\mathcal{E}_i^{①}, \mathcal{E}_i^{②})$ is a SEFE. Concerning the relative position of a vertex v and some common cycle C', we note that the relative positions clearly coincide in $\mathcal{E}^{①}$ and $\mathcal{E}^{②}$ for $C = C'$. Otherwise C' is contained in some union bridge. If v is embedded in the interior of C' in one of the two embeddings, then it is contained in the same union bridge as C', and the compatibility follows. If this case does not apply, it is embedded outside of C' in both embeddings, which is compatible as well.   □

Now consider a connected component C of the common graph G of a SEFE instance such that C is biconnected. Such a component is called *2-component*. If C is a cycle, it is a *trivial* 2-component. We define the union bridges, and the ①- and ②- bridges of $G^{①}$ and $G^{②}$ with respect to C as above. We call an embedding $\mathcal{E}$ of C together with an assignment of the union bridges to its faces *admissible* if and only if, (i) for each union bridge, all attachments are incident to the face to which it is assigned, and (ii) no two ①- or ②-bridges that are assigned to the same face alternate. For a union bridge B, let C_B denote the cycle consisting of the attachments of B in the ordering of an arbitrary cycle of G containing all the attachments. It can be shown that this cycle is uniquely determined. Let G_B denote the graph consisting of B and C_B. We call these graphs the *union bridge graphs*.

Lemma 2. *Let $G^①$ and $G^②$ be two connected planar graphs and let C be 2-component of the common graph G. Then the graphs $G^①$ and $G^②$ admit a* SEFE *if and only if (i) C admits an admissible embedding, and (ii) each union bridge graph admits a* SEFE. *If a* SEFE *exists, the embedding of C can be chosen as an arbitrary admissible embedding.*

Proof. Clearly, a SEFE of $G^①$ and $G^②$ defines an embedding of C and a bridge assignment that is admissible. Moreover, it induces a SEFE of each union bridge graph.

Conversely, assume that C admits an admissible embedding and each union bridge graph admits a SEFE. We obtain a SEFE of $G^①$ and $G^②$ as follows. Embed C with the admissible embedding and consider a face f of this embedding with facial cycle C_f. Let $B_1, \ldots, B_k$ denote the union bridges that are assigned to this face, and let $(\mathcal{E}_1^①, \mathcal{E}_1^②), \ldots, (\mathcal{E}_k^①, \mathcal{E}_k^②)$ be simultaneous embeddings of the bridge graphs G_B. By subdividing the cycle C_B, in each of the embeddings, we may assume that the outer face of each B_i in the embedding $(\mathcal{E}_i^①, \mathcal{E}_i^②)$ is the facial cycle C_f with the same orientation in each of them. By Lemma 1, we can hence combine them to a single SEFE of all union bridges whose outer face is the cycle C_f. We embed this SEFE into the face f of C. Since we can treat the different faces of C independently, applying this step for each face yields a SEFE of $G^①$ and $G^②$ with the claimed embedding of C.   □

Lemma 2 suggests a simple strategy for reducing SEFE instances containing non-trivial 2-components. Namely, take such a component, construct the corresponding union bridge graphs, where C occurs only as a cycle, and find an admissible embedding of C. Finding an admissible embedding for C can be done as follows. To enforce the non-overlapping attachment property, replace each ①-bridge of C by a *dummy ①-bridge* that consists of a single vertex that is connected to the attachments of that bridge via edges in $E^①$. Similarly, we replace ②-bridges, which are connected to attachments via exclusive edges in $E^②$. We seek a SEFE of the resulting instance (where the common graph is biconnected), additionally requiring that dummy bridges belonging to the same union bridge are embedded in the same face. We also refer to such an instance as SEFE *with union bridge constraints*. A slight modification of the algorithm by Angelini et al. [1] can decide the existence of such an embedding in polynomial time. It then remains to treat the bridge graphs. Exhaustively applying Lemma 2 results in a set of SEFE instances where each 2-component is trivial.

Linear-Time Decomposition. We now show that the set of instances resulting from exhaustively applying Lemma 2 can be computed in linear time.

Theorem 3. *Given a* SEFE *instance, an equivalent set of instances of total linear size such that each 2-component of these instances is trivial can be computed in linear time.*

Let G be a planar graph and let $C_1, \ldots, C_k$ be connected components of G. We are interested in simultaneously determining for each component C_i the number of incident bridges, and for each such bridge its attachment vertices. For this, we introduce the notion of *subbridges*. A subbridge of G with respect to $C_1, \ldots, C_k$ is a maximal connected subgraph of G that does not become disconnected by removing all vertices of one component C_i. It is readily seen that each C_i-bridge B contains a unique subbridge S incident to C_i and that the attachments of S at C_i are exactly the attachments of B at C_i. We will thus rather work with the subbridges than the actual bridges as they represent the same information but in a more compact way. Our reduction now works in three phases.

1. Compute for each 2-component of G the number of ①-, ②-bridges, for each such bridge its attachments, and the grouping of these bridges into union bridges.
2. Find for each 2-component an admissible embedding with respect to its bridges.
3. Compute for each subbridge of G with respect to its 2-components a corresponding instance where each 2-component has been replaced by a suitable cycle.

The correctness of this approach descends from Lemma 2, since the set of instances computed by the procedure is exactly the one that can be obtained by exhaustively applying this lemma. The details of the implementation of this procedure are deferred to the full version of this paper. Here we only sketch the main ideas. For step 1, we exploit the fact that, after contracting each connected component of G that is biconnected to a single vertex, (almost) every such component is a cutvertex, and the union subbridges are essentially the blocks of the resulting graph. We can then traverse for each cutvertex its incident edges and label them by the block (subbridge) containing them. This allows us to construct the dummy-bridges and union bridges that are solved in step 2. For step 2, we modify the algorithm due to Angelini et al.[1]. Augmenting it such that it computes admissible embedding in polynomial time is straightforward. Achieving linear running time is quite technical and, like the linear version of the original algorithm, requires some intricate data structures. Step 3 is finally implemented by taking the admissible embeddings from step 2. We then traverse each such face exactly one, and construct, during this traversal, the corresponding cycles in all incident subbridges that are embedded in this face.

4 Instances with Common P-Node Degree 3

We consider instances of SEFE that have common P-node degree 3. Recall that a simultaneous embedding must induce consistent edge orderings and consistent relative positions on the common graph. We show how to address both requirements separately, by formulating necessary and sufficient constraints using linear equations over $\mathbb{F}_2$. Both resulting systems of equations share all variables representing embedding choices. Satisfying both sets of linear equations at the same time then solves SEFE.

Before we can follow this strategy, we need to address one problem. The relative position of a component H' of G with respect to another connected component H, denoted by $\mathrm{pos}_H(H')$, is the face of H containing H'. However, the set of faces of H depends on the embedding of H. To be able to handle relative positions independently from edge orderings, we need to express the relative positions independently from faces.

4.1 Relative Positions with Respect to a Cycle Basis

A *generalized cycle* C in a graph H is a subset of its edges such that every vertex of H has an even number of incident edges in C. The *sum* $C + C'$ of two generalized cycles is the symmetric difference between the edge sets, i.e., an edge e is contained in $C + C'$ if and only if it is contained in C or in C' but not in both. The resulting edge set $C + C'$ is again a generalized cycle. The set of all generalized cycles in H is a vector space over $\mathbb{F}_2$. A basis of this vector space is called *cycle basis* of H.

Instead of considering the relative position $\mathrm{pos}_H(H')$ of a connected component H' with respect to another component H, we choose a cycle basis $\mathcal{C}$ of H and show that the relative positions of H' with respect to the cycles in $\mathcal{C}$ suffice to uniquely define $\mathrm{pos}_H(H')$, independent from the embedding of H. We assume H to be biconnected. All results can be extended to connected graphs by using a cycle basis for each block.

Let $C_0, \ldots, C_k$ be the set of facial cycles with respect to an arbitrary planar embedding of H. The set $\mathcal{C} = \{C_1, \ldots, C_k\}$ obtained by removing one of the facial cycles is a cycle basis of G. A cycle basis that can be obtained in this way is called *planar cycle basis*. In the following we assume all cycle bases to be planar cycle bases. Moreover, we consider all cycles to have an arbitrary but fixed orientation, which has the effect, that $\mathrm{pos}_C(p)$ for any cycle C and any point p can have either the value LEFT or RIGHT.

Theorem 4. *Let H be a planar graph embedded on the sphere, let p be a point on the sphere, and let $\mathcal{C} = \{C_1, \ldots, C_k\}$ be an arbitrary planar cycle basis of H. Then the face containing p is determined by the relative positions $\mathrm{pos}_{C_i}(p)$ for $1 \leq i \leq k$.*

Proof (sketch). Clearly, the point p and the face f containing p have to lie on the same side of each of the cycles in $\mathcal{C}$. It remains to show that the face with this property is unique. Let C be the facial cycle of f and let $C = C_1 + \cdots + C_\ell$ be the linear combination of basis cycles of C. The *position vector* of a point p with respect to the facial cycle C is $\mathrm{pos}(p) = (\mathrm{pos}_{C_1}(p), \ldots, \mathrm{pos}_{C_\ell}(p))$. It can be seen that inside f, the vector $\mathrm{pos}(p)$ has a different parity of values LEFT than outside, which shows, that no other face can have the same relative positions with respect to all cycles in $\mathcal{C}$. $\square$

4.2 Consistent Edge Orderings

We first assume that the graphs $G^{①}$ and $G^{②}$ are biconnected. There exists an instance of SIMULTANEOUS PQ-ORDERING that has a solution if and only if $G^{①}$ and $G^{②}$ admit embeddings with consistent edge ordering [4]. This solution is based on the *PQ-embedding representation*, an instance of SIMULTANEOUS PQ-ORDERING representing all embeddings of a biconnected planar graph. We describe this embedding representation and show how to simplify it for instances that have common P-node degree 3.

For each vertex $v^{①}$ of $G^{①}$, the PQ-embedding representation, denoted by $D(G^{①})$, contains the *embedding tree* $T(v^{①})$ having a leaf for each edge incident to $v^{①}$, representing all possible orders of edges around $v^{①}$. For every P-node $\mu^{①}$ in the SPQR-tree $\mathcal{T}^{①}$ of $G^{①}$ that contains $v^{①}$ in $\mathrm{skel}(\mu^{①})$ there is a P-node in $T(v^{①})$ representing the choice to reorder the virtual edges in $\mathrm{skel}(\mu^{①})$. Similarly, for every R-node $\mu^{①}$ in $\mathcal{T}^{①}$ containing $v^{①}$ there is a Q-node in $T(v^{①})$ whose flip corresponds to the flip of $\mathrm{skel}(\mu^{①})$. As the orders of edges around different vertices of $G^{①}$ cannot be chosen independently

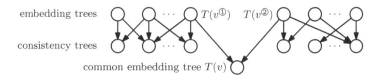

Fig. 2. The Q-embedding representations of $G^{①}$ and $G^{②}$ and one common embedding tree

from each other, so called *consistency trees* are added as common children to enforce Q-nodes stemming from the same R-node in $\mathcal{T}^{①}$ to have the same flip and P-nodes stemming from the same P-node to have consistent (i.e., opposite) orders. Every solution of the resulting instance corresponds to a planar embedding of $G^{①}$ and vice versa [4].

As we are only interested in the order of common edges, we modify $D(G^{①})$ by projecting each PQ-tree to the leaves representing common edges. As $G^{①}$ and $G^{②}$ have common P-node degree 3, all P-nodes of the resulting PQ-trees have degree 3 and can be assumed to be Q-nodes representing a binary decision. We call the resulting instance *Q-embedding representation* and denote it by $D(G^{①})$. Let $\mu^{①}$ be an R-node of the SPQR-tree $\mathcal{T}^{①}$ whose embedding has influence on the ordering of common edges around a vertex. It is not hard to see that the Q-embedding representation contains a consistency tree consisting of a single Q-node representing the flip of $\mathrm{skel}(\mu^{①})$. We associate the binary variable $\mathrm{ord}(\mu^{①})$ with this decision. For a P-node $\mu^{①}$ we get a similar result. Let $u^{①}$ and $v^{①}$ be the nodes in $\mathrm{skel}(\mu^{①})$. If the consistency tree enforcing a consistent decision in the embedding trees $T(u^{①})$ and $T(v^{①})$ has degree 3, its flip represents the embedding decision for $\mathrm{skel}(\mu^{①})$ and we again get a binary variable $\mathrm{ord}(\mu^{①})$. Otherwise, this consistency tree contains two or less leaves and can be ignored. Then the choices for the Q-nodes corresponding to $\mu^{①}$ in $T(u^{①})$ and $T(v^{①})$ are independent and we get one binary variable for each of these Q-nodes. We denote these variables by $\mathrm{ord}(\mu_u^{①})$ and $\mathrm{ord}(\mu_v^{①})$. We call these variables *PR-ordering variables*.

For a common vertex v occurring as $v^{①}$ and $v^{②}$ in $G^{①}$ and $G^{②}$, respectively, we can ensure a consistent edge ordering by adding a so called *common embedding tree* $T(v)$ as child of the embedding trees $T(v^{①})$ and $T(v^{②})$ in the Q-embedding representations of $G^{①}$ and $G^{②}$; see Fig. 2. We get the following lemma.

Lemma 3. *Let $G^{①}$ and $G^{②}$ be two biconnected graphs with common P-node degree 3. Requiring the common edges to be ordered consistently is equivalent to satisfying a system of linear equations M_{ord} over $\mathbb{F}_2$ with the following properties.*

(i) All equations in M_{ord} are of the type $x + y = c$ for $c \in \mathbb{F}_2$.

(ii) M_{ord} contains all PR-ordering variables.

(iii) M_{ord} has linear size and can be computed in linear time.

In the following, we extend this result to the case where we allow exclusive cutvertices. Let $B_1^{①}, \ldots, B_k^{①}$ be the blocks of $G^{①}$ and let $B_1^{②}, \ldots, B_\ell^{②}$ be the blocks of $G^{②}$. We say that embeddings of these blocks have *blockwise consistent edge orderings* if for every pair of blocks $B_i^{①}$ and $B_j^{②}$ sharing a vertex v the edges incident to v they share are ordered consistently. To have consistent edge orderings, it is obviously necessary to have blockwise consistent edge orderings.

When composing the embeddings of two blocks that share a cutvertex, the edges of each of the two blocks have to appear consecutively (note that this is no longer true for three or more blocks), which leads to another necessary condition. Let v be an exclusive cutvertex of $G^{①}$. Then v is contained in a single block of $G^{②}$ whose embedding induces an order $O^{②}$ on all common edges incident to v. For every pair of blocks $B_i^{①}$ and $B_j^{②}$ containing v, the edges in $B_i^{①}$ must appear consecutively in the order of the edges incident to v in $B_i^{①}$ and $B_j^{②}$ that is induced by $O^{②}$. If this is true for every exclusive cutvertex, we say that the embeddings have *pairwise consecutive blocks*.

Lemma 4. *Two graphs without simultaneous cutvertices admit embeddings with consistent edge orderings if and only if their blocks admit embeddings that have blockwise consistent edge orderings and pairwise consecutive blocks.*

To extend Lemma 3 to the case where we allow exclusive cutvertices, we enforce blockwise consistent edge orderings and pairwise consecutive blocks by adding additional PQ-trees to the above instance of SIMULTANEOUS PQ-ORDERING. As before, we get direct access to the embedding chosen for each block, via the PR-ordering variables. We want to get access to the ordering of common edges around a cutvertex v of $G^{①}$ in a similar way. Let $B^{②}$ be a block that contains the common edge e incident to v and let e_1 and e_2 be two common edges incident to v that are contained in a different block. We use the *cutvertex-ordering variable* $\mathrm{ord}(e_1, e_2, B^{②})$ to denote the order of e_1, e_2, and e. Note that this is independent from the choice of the edge e of $B^{②}$. To decrease the number of variables, we only consider those variables that are *required by a cycle basis* $\mathcal{C}$, where $\mathrm{ord}(e_1, e_2, B^{②})$ is required by $\mathcal{C}$ if e_1 and e_2 share a cycle in $\mathcal{C}$.

Lemma 5. *Given two graphs without union or simultaneous cutvertices with common P-node degree 3, requiring the common edges to be ordered consistently is equivalent to satisfying a system of linear equations M_{ord} with the following properties.*
  (i) All equations in M_{ord} are of the type $x + y = 0$ or $x + y = 1$.
  (ii) M_{ord} contains all PR-ordering variables and all cutvertex-ordering variables required by a cycle basis of the common graph.
  (iii) M_{ord} has size $O(\min\{n^2, n\Delta^2\})$ (where Δ is the maximum degree in the common graph) and can be computed in linear time in its size.

4.3 Consistent Relative Positions

In this section, we present a system of linear equations M_{pos} containing the PR-ordering and cutvertex-ordering variables such that satisfying M_{pos} is equivalent to requiring consistent relative positions for an instance of SEFE. Let H and H' be two connected components of the common graph G. To represent the relative position $\mathrm{pos}_{H'}(H)$ of H with respect to H', we use the relative positions $\mathrm{pos}_C(H)$ of H with respect to cycles C in the cycle basis of H' (Theorem 4). To get binary variables, we use $\mathrm{pos}_C(H) = 0$ if H lies to the right of C and $\mathrm{pos}_C(H) = 1$ if H lies to the left of C. When we consider the graph $G^{①}$ containing G, it is known that the value of $\mathrm{pos}_C(H)$ is determined by a single, very local embedding decision of $G^{①}$ [3]. In the following we consider the three possible cases that $\mathrm{pos}_C(H)$ is determined by an R-node, by a P-node or by a cutvertex.

R-Node. If $\mathrm{pos}_C(H)$ is determined by an R-node μ, then C induces a cycle κ in $\mathrm{skel}(\mu)$ and parts of H are contained in a virtual edge ε not contained in κ. The relative position of H with respect to C is the same as the position of ε with respect to κ [3]. As the value of $\mathrm{pos}_\kappa(\varepsilon)$ changes, when the embedding of $\mathrm{skel}(\mu)$ changes, we can simply set $\mathrm{pos}_C(H) + \mathrm{ord}(\mu) = c$ (where $c \in \mathbb{F}_2$ depends on the reference embedding of $\mathrm{skel}(\mu)$ and the orientation of C). Note that this implicitly ensures the consistency of all relative positions that are determined by thee embedding of $\mathrm{skel}(\mu)$.

P-Node. If $\mathrm{pos}_C(H)$ is determined by the embedding of $\mathrm{skel}(\mu)$ of a P-node μ, then C induces a cycle κ (of length 2) in $\mathrm{skel}(\mu)$ and H is completely contained in a single edge ε of $\mathrm{skel}(\mu)$ not belonging to κ. Again $\mathrm{pos}_C(H)$ in G is the same as $\mathrm{pos}_\kappa(\varepsilon)$ in $\mathrm{skel}(\mu)$. However, this time the embedding choices of $\mathrm{skel}(\mu)$ are more complicated than to flip or not to flip. Thus, we have to consider all relative positions decided by the embedding of $\mathrm{skel}(\mu)$ at once, to get the dependencies between them.

We only consider the case where the common graph induces paths between the vertices of $\mathrm{skel}(\mu)$ in the expansion graphs of three edges ε_1, ε_2, and ε_3 (all other cases are simpler as μ has common P-node degree 3). Cycles in the cycle basis $\mathcal{C}$ can induce three different cycles, namely $\kappa_{1,2}$, $\kappa_{2,3}$, and $\kappa_{1,3}$ consisting of the virtual edges $(\varepsilon_1, \varepsilon_2)$, $(\varepsilon_2, \varepsilon_3)$, and $(\varepsilon_1, \varepsilon_3)$, respectively. For every virtual edge $\varepsilon \neq \varepsilon_i$, we get the three variables $\mathrm{pos}_{\kappa_{1,2}}(\varepsilon)$, $\mathrm{pos}_{\kappa_{2,3}}(\varepsilon)$, and $\mathrm{pos}_{\kappa_{1,3}}(\varepsilon)$ determining the position of ε with respect to these three cycles. Recall that the variable $\mathrm{ord}(\mu)$ determines the ordering of the three edges ε_1, ε_2, and ε_3. Consider the case that $\mathrm{ord}(\mu) = 0$ and assume that the reference order of ε_1, ε_2, and ε_3 as well as the orientation of the cycles $\kappa_{1,2}$, $\kappa_{2,3}$, and $\kappa_{1,3}$ is as shown in Fig. 3. Then either all three relative positions have the value 0 (which corre-

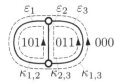

Fig. 3. A P-node with the three cycles $\kappa_{1,2}$, $\kappa_{1,2}$, and $\kappa_{1,2}$ (dashed)

sponds to RIGHT), or exactly two relative position has the value 1. Thus, a combination of values for the positions $\mathrm{pos}_{\kappa_{1,2}}(\varepsilon)$, $\mathrm{pos}_{\kappa_{2,3}}(\varepsilon)$, and $\mathrm{pos}_{\kappa_{1,3}}(\varepsilon)$ is possible if and only if there is an even number of 1s. When setting $\mathrm{ord}(\mu) = 1$, there need to be an odd number of 1s. This yields the constraint $\mathrm{ord}(\mu) + \mathrm{pos}_{\kappa_{1,2}}(\varepsilon) + \mathrm{pos}_{\kappa_{2,3}}(\varepsilon) + \mathrm{pos}_{\kappa_{1,3}}(\varepsilon) = 0$ (a different reference embedding or different orientations of the cycles may lead to a 1 on the right-hand side of the equation). As for R-nodes we set $\mathrm{pos}_C(H) + \mathrm{pos}_{\kappa_{i,j}}(\varepsilon) = c$ ($c \in \mathbb{F}_2$), if C induces $\kappa_{i,j}$ in $\mathrm{skel}(\mu)$ and H is contained in the expansion graph of ε.

Cutvertex. The relative position $\mathrm{pos}_C(H)$ is determined by a cutvertex v of $G^\oplus$ if C contains v and H lies in a split component S (with respect to v) different from the split component containing C. We can either embed the whole split component S to the left or to the right of C. For this decision, we introduce the variable $\mathrm{pos}_C(S)$. Clearly, we get $\mathrm{pos}_C(H) = \mathrm{pos}_C(S)$ for every connected component H of G in S. Moreover, there are no further constraints on the relative positions determined by the embedding at the cutvertex v [4]. In case the split component S contains a common edge incident to v, fixing $\mathrm{pos}_C(S)$ is equivalent fixing the cutvertex-ordering variable $\mathrm{ord}(e_1, e_2, B)$, where e_1 and e_2 are the edges in C incident to v and B is the block in S containing v. Thus, we get $\mathrm{ord}(e_1, e_2, B) + \mathrm{pos}_C(S) = c$ ($c \in \mathbb{F}_2$). Together with the constraints for the relative positions determined by the P- and R-nodes, we get the following lemma.

Lemma 6. *Let $G^①$ and $G^②$ be two graphs without union or simultaneous cutvertices with common P-node degree 3. Requiring the relative positions to be consistent is equivalent to satisfying a system of linear equations M_{pos} with the following properties.*

 (i) All equations in M_{pos} are of the type $x + y = c$ (with $c \in \mathbb{F}_2$) except for a linear number of equations of size 4.
 (ii) M_{pos} contains all PR-ordering variables and all cutvertex-ordering variables required by a cycle basis of G.
 (iii) M_{pos} has quadratic size and can be computed in quadratic time.

Lemma 5 and Lemma 6 yield the following theorem. We obtain the quadratic running time by first eliminating all equations of size 1, and then solving the remaining system of $O(n)$ linear equations of size 4 with the algorithm by Wiedemann [12].

Theorem 5. SEFE *can be solved in quadratic time for two graphs without union or simultaneous cutvertex with common P-node degree 3.*

References

1. Angelini, P., Di Battista, G., Frati, F., Patrignani, M., Rutter, I.: Testing the simultaneous embeddability of two graphs whose intersection is a biconnected or a connected graph. Journal of Discrete Algorithms 14, 150–172 (2012)
2. Bläsius, T., Kobourov, S.G., Rutter, I.: Simultaneous Embedding of Planar Graphs. In: Handbook of Graph Drawing and Visualization, pp. 349–381. Chapman and Hall/CRC (2013)
3. Bläsius, T., Rutter, I.: Disconnectivity and relative positions in simultaneous embeddings. In: Didimo, W., Patrignani, M. (eds.) GD 2012. LNCS, vol. 7704, pp. 31–42. Springer, Heidelberg (2013)
4. Bläsius, T., Rutter, I.: Simultaneous PQ-ordering with applications to constrained embedding problems. In: Proc. 24th ACM-SIAM Sympos. Discrete Algorithm, SODA 2013, pp. 1030–1043. ACM (2013)
5. Booth, K.S., Lueker, G.S.: Testing for the consecutive ones property, interval graphs, and graph planarity using PQ-tree algorithms. Journal of Computer and System Sciences 13, 335–379 (1976)
6. Di Battista, G., Tamassia, R.: On-line maintenance of triconnected components with SPQR-trees. Algorithmica 15(4), 302–318 (1996)
7. Gassner, E., Jünger, M., Percan, M., Schaefer, M., Schulz, M.: Simultaneous graph embeddings with fixed edges. In: Fomin, F.V. (ed.) WG 2006. LNCS, vol. 4271, pp. 325–335. Springer, Heidelberg (2006)
8. Gutwenger, C., Mutzel, P.: A linear time implementation of SPQR-trees. In: Marks, J. (ed.) GD 2000. LNCS, vol. 1984, pp. 77–90. Springer, Heidelberg (2001)
9. Haeupler, B., Jampani, K., Lubiw, A.: Testing simultaneous planarity when the common graph is 2-connected. J. Graph Algorithms Appl. 17(3), 147–171 (2013)
10. Jünger, M., Schulz, M.: Intersection graphs in simultaneous embedding with fixed edges. Journal of Graph Algorithms and Applications 13(2), 205–218 (2009)
11. Schaefer, M.: Toward a theory of planarity: Hanani-tutte and planarity variants. In: Didimo, W., Patrignani, M. (eds.) GD 2012. LNCS, vol. 7704, pp. 162–173. Springer, Heidelberg (2013)
12. Wiedemann, D.: Solving sparse linear equations over finite fields. IEEE Transactions on Information Theory 32(1), 54–62 (1986)

Sketched Graph Drawing:
A Lesson in Empirical Studies

Helen C. Purchase

School of Computing Science, University of Glasgow, UK

Abstract. This paper reports on a series of three similar graph drawing empirical studies, and describes the results of investigating subtle variations on the experimental method. Its purpose is two-fold: to report the results of the experiments, as well as to illustrate how easy it is to inadvertently make conclusions that may not stand up to scrutiny. While the results of the initial experiment were validated, instances of speculative conclusions and inherent bias were identified. This research highlights the importance of stating the limitations of any experiment, being clear about conclusions that are speculative, and not assuming that (even minor) experimental decisions will not affect the results.

Keywords: graph sketching, empirical studies, replication, limitations.

1 Introduction

This paper reports on a series of three similar experiments with a common aim: to determine the graph drawing layout principles favored when participants draw graphs. The first experiment of the series was peer-reviewed and published in a reputable journal [1].

Typically in information visualization or HCI research, researchers run an experiment, form conclusions, publish, and then move on to the next interesting question. The second and third experiments reported here arose from a reflective critique of the first experiment. In this case, the experimenter (the author of this paper) did not "publish and move on", but explored subtle aspects of the experimental design, attempting to replicate and confirm the initial results.

The focus of this paper is therefore two-fold: to present the experiments and their results, but also to describe a process whereby revisiting empirical work has highlighted interesting facts about the process of conducting empirical studies.

This motivation for this paper relates to the idea of reflective critique, an idea borrowed from the practice of action learning projects. Action learning projects [2] do not aim to address specified research questions, and their approach is formative rather than summative, resulting in a cycle of continuous improvement. This approach is unlike the typical experimental research project, where a research question is clearly defined, a methodology is designed to address it, data is collected, and final conclusions made and published. In contrast, an action learning methodology encourages honest reflection on outcomes, and these reflections are fed into another cycle of the process.

This paper, therefore, reports on the results of two experiments conducted 'after the fact', i.e. after conclusions from the first experiment had been disseminated, with the latter two experiments being designed as a result of reflection and critique.

S. Wismath and A. Wolff (Eds.): GD 2013, LNCS 8242, pp. 232–243, 2013.

2   Background

The design of automatic graph layout algorithms tend to be based on common 'aesthetic principles', for example, the elimination of edge crossings or the minimisation of adjacent edge angles. Early experimental research investigating how graphs may best be laid out tended to use a task-based performance approach (e.g. [3,4,5]), and established key findings such as the fact that a high occurrence of crossed edges reduced performance and prominent depiction of symmetry increased performance.

More recent empirical research has instead focussed on the manner in which participants create their own visual layout of relational information as a graph drawing. Van Ham and Rogowitz [6] (HR08) asked participants to adjust manually the layout of existing graph drawings. They used four graphs of 16 nodes, each with two clusters separated by one, two, three and four edges respectively. The graphs were presented in a circular and a spring layout [7], giving a total of eight starting diagrams, shown in random order. They collected 73 unique drawings, and found that most participants separated the two clusters, that the human drawings contained 60% fewer edge crossings than the automatically produced drawings, and that humans did not value uniform edge length as much as the spring algorithm did.

Dwyer et. al [8] (D+09) performed a similar hands-on experiment, asking participants to lay out two social networks, each with a circular initial arrangement. Participants were encouraged to lay the graphs out in a way that would best support the identification of cliques, chains, cut nodes and leaf nodes. With a focus on the process of layout rather than on the product, the only observation that they make about the graphs produced is that users removed edge crossings.

The first experiment in the series reported here (**Experiment 1**, [1]) was designed to address a similar research goal as HR08 and D+09, using a different methodology. The research question is *Which graph drawing layout principles do people favour when creating their own drawings of graphs?*

There are five main design features of Experiment 1 that differentiate it from HR08 and D+09, the *differentiating design features* (DDFs):

DDF-1. The participants had to both draw the graph, as well as lay it out, a more complex task than both HR08 and D+09;

DDF-2. The participants drew the graphs from scratch, so were not biased by any initial layout (both HR08 and D+09 used an initial configuration);

DDF-3. A sketching tool was used, so the physical drawing process was unhindered by an intermediate editing interface;

DDF-4. Video data was collected, so both the process and product of creation were able to be analysed (this was done by D+09, but not HR08);

DDF-5. Layout preferences were discussed with the participants in a post-experiment interview (this was done by D+09, but not HR08).

Four graphs were used, two practise graphs and two experimental graphs. Data on product, process and preferences were collected[1].

---

[1] Note that the publication arising from Experiment 1 also included graphs drawn in a point-and-click mode, but only the sketched graph experiment is considered here.

The several conclusions of **Experiment 1** as published [1] included:

CONC-1 The layout principles that participants favoured during the process of laying out their drawings were often not evident in the final product.

CONC-2 The principle of fixing nodes and edges to an underlying unit grid was prominent.

2.1 Reflective Critique: Issues Arising

After peer-review, publication, presentation, and independent citation of this first experiment and its results, audience members at two seminars suggested some subtle variations on the experimental method: not new research questions, simply small amendments. As is typical in such situations, this author responded that such variations could be addressed as part of 'future work'.

It was not necessary to investigate these issues (the paper had already been published after all), but they led the experimenter to reflect on the research, and to question to what extent results might be biased by a method.

Three issues arose as a result of this reflection:

- If participants compromise their layout design during creation of the drawing (CONC-1), does this mean that they are not happy with their final product? **Experiment 2** addressed the issue of whether participants were satisfied with their final drawing, or whether they expressed disappointment that they were unable to conform to their desired layout.
- If participants favour a grid-based layout (CONC-2), would they prefer a drawing laid out using an algorithm that aligns nodes and edges to an underlying grid to their own? **Experiment 2** investigated whether the participants preferred their own sketched graph drawing to a similar one that conformed to a grid layout.
- Was the tendency to favour a grid layout and straight lines (CONC-2) a consequence of the way in which the graph information stimuli were presented as a list of edges? **Experiment 3** investigated whether representing the graph structure in an alternative text format also produces results that favour grid layout.

3 The Experiments

3.1 Experimental Process for all Three Experiments

The primary research question in all three of these graph sketching experiments is: *Which graph drawing layout principles do people favour when creating their own drawings of graphs?* Participants were asked to draw graphs and their drawings were analysed for evidence of common graph drawing layout principles.

Equipment: A graph-drawing sketch tool, SketchNode [9] was used, allowing nodes, edges and node labels to be sketched, edited and moved with a stylus on a tablet screen, laid flat.

Task: Participants were given a textual description of a graph and asked to draw it in SketchNode, with the instruction to *Please draw this graph as best as you can so to make it "easy to understand"*. They were deliberately not given any further instruction as to what "easy to understand" means, nor primed with any information about common graph layout principles (e.g. minimising edge crossings, use of straight edges, etc,). They had as long as they liked to draw and adjust the layout of the graphs.

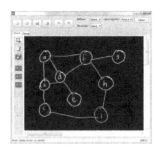

Graphs: Two experimental graphs were used in all three experiments: graph A (10 nodes, 14 edges) and graph B (10 nodes, 18 edges).

Experimental method: Participants completed required ethical procedures and provided demographic information. All the relevant features of SketchNode were demonstrated, and participants were given the opportunity to practice and ask questions. Two practice graphs ensured participants were comfortable with the task and system.

The two experimental graphs A and B, were then presented to the participants, with the graph edges presented in different random order for each participant, and counterbalanced between participants. At the end of the experiment, the participants were asked "Why did you arrange the graphs in the way you did?" in a recorded interview.

The participants: Participants in all three experiments were of a similar profile: a mixture of students and non-students, of both sexes, with around half the student participants in each experiment studying some form of computer science.

3.2 Differences between the Experiments

In **Experiment 1**, the most important differentiating feature in comparison with prior research was the way the graphs were presented (DDF-2). HR08 and D+09 presented their graphs as graph drawings which already had some layout properties (circular and spring in HR08 and circular in D+09). So as not to bias the participants towards any layout principles, the graphs in Experiment 1 were presented in textual form, as a list of edges (Table 1, column 1).

Experiments 2 and 3 addressed four issues arising from Experiment 1.

Compromised layout: CONC-1 suggests participants may not have been entirely happy with their final drawing, as they had been obliged to compromise their favoured layout as the graph became more complex. **Experiment 2** investigated the extent of this compromise, and whether participants acknowledged it.

The first research question for this experiment was "*Do participants like the layout of their final product?*" We speculated they would express dissatisfaction with their final product. Once they had drawn their graph, we asked them to indicate on a scale of 1–5 how "happy" they were with their drawing.

Preference for a grid: CONC-2: Experiment 1 suggested a grid layout was favoured; we anticipated participants would prefer a grid layout to their own.

The second research question was "*Do participants prefer their sketched drawing to be laid out in a grid format to their own layout?*"

Experiment 2 used the two automatic graph layout algorithms in SketchNode: spring (based on Fruchterman and Reingold [10]) and grid (placing nodes and edges on the lines and vertices of an underlying unit grid). After layout, the visual sketched appearance of the nodes and labels remains the same, so the resultant diagram can be directly compared with original sketched drawing.

At the end of the sketching stage of Experiment 2, the participant's own drawing was laid out using these two algorithms. Participants were asked to rank the three drawings according to their preference.

In an attempt to eliminate any personal bias or recency effects, a willing subset of the participants chose between hand-drawn and the two automatically laid out drawings two weeks after the experiment.

Validation: As both changes to the experimental method for Experiment 2 were post-experiment activities, Experiment 2 also served as a means to validate the results of Experiment 1.

Effect of graph format: One of the main differentiating features between Experiment 1 and prior research was that the graph information was presented in abstract form, rather than as a graph drawing (DDF-2), and the participants drew the graphs from scratch. Here we investigated whether even this abstract form had produced a bias.

The research question for this experiment was "*Does the format in which the graph structure is represented affect the layout of the sketched graph drawings produced?*" For **Experiment 3**, we presented the graphs as an adjacency list (Table 1, column 3), and followed exactly the same process as Experiment 1. This format is visually quite different from the simple list of edges, as each edge is not clearly specified as an individual pair, and it is more obvious which nodes have a higher degree. We wished to investigate whether a format that does not focus on the individual node pairs (as in Experiment 1) would still result in user-sketched drawings that conform to a grid structure.

Table 1 shows the differences between the experiments, as well as those factors that remained the same. Figure 1 shows example sketches from all three experiments.

4   Data Analysis

Compromised layout (Experiment 2): The participants in Experiment 2 indicated how happy they were with their own sketches (five-point scale, 5=perfectly happy). Graph A's mean: 4.14, graph Bs' mean: 3.5; both graphs together: 3.91.

Participants were asked what they didn't like about their drawings, and how they would improve them. None mentioned that they would have liked to conform to a grid layout; most comments related to local issues like the size and shape of the nodes, and connections between the nodes and edges. The few comments that referred to overall layout of the drawing were concerned with spreading the nodes out, symmetry and circular layout. There were also several comments about the need to plan in advance.

Table 1. Summary of the differences between the three experiments

	Experiment 1	Experiment 2	Experiment 3
experimental graphs	A $(n = 10, e = 14)$ B $(n = 10, e = 18)$	as Experiment 1	as Experiment 1
experimental task	After introduction and training activities, participants sketched the two graphs	as Experiment 1	as Experiment 1
equipment	SketchNode	as Experiment 1	as Experiment 1
number of participants	17	22	26
form of graph presented to participants	(A, D) (A, C) (B, D) (C, D) (B, C) (B, E) (C, E) (E, J) (F, G) (J, F) (F, I) (G, I) (J, H) (I, H)	as Experiment 1	S | U R Q | S V | W Z Y | X V R | U W | Y U | Z T | R Q S X | Z
post-experiment discussion	none	participants indicated how happy they were with their drawing	as Experiment 1
post-experiment ranking	none	participants ranked sketched drawings against associated spring and grid drawings	as Experiment 1

Preference for a grid (Experiment 2): Once participants had sketched their graph, the two graph layout algorithms (Section 3.2 above) were applied to their drawing.

A three-way-set (TWS) is a set of three drawings: a participant's sketched drawing, and two versions of this sketch produced by the algorithms. Each participant has a TWS-GA and TWS-GB. Figure 2 shows a TWS for one of the participant's graphs.

Participants ranked the drawings in their own TWS-GA and TWS-GB at the end of Experiment 2. After two weeks, we contacted all participants for a follow-up ranking experiment; fourteen took part. They ranked their own TWS-GA and TWS-GB (as before), as well as the TWS-GA and TWS-GB for two other participants. They were not told their own drawings were included in these sets (Table 2).

The data were analysed with Friedman tests with adjusted pairwise comparisons. The only significant results related to graph B (highlighted in Table 2):

---

[2] The notational convention is: graph A drawn by participant 3 is 3A; graph B drawn by participant number 8 is 8B. The suffix -1, -2 or -3 indicates experiment 1, 2 or 3.

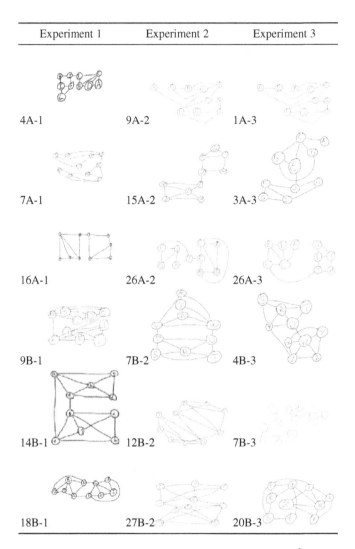

	Experiment 1	Experiment 2	Experiment 3

4A-1   9A-2   1A-3

7A-1   15A-2   3A-3

16A-1   26A-2   26A-3

9B-1   7B-2   4B-3

14B-1   12B-2   7B-3

18B-1   27B-2   20B-3

Fig. 1. Example drawings from all three experiments[2]

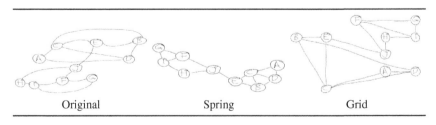

Original   Spring   Grid

Fig. 2. The TWS-GA for participant 17. Note that the hand-drawn nature and form of the nodes and their labels is retained.

Table 2. Mean ranking of TWSs (3=most preferred). Significant results in italics.

	number of participants : TWSs	mean rank: sketch	mean rank: spring	mean rank: grid	% times sketch preferred	% times grid preferred
After drawing the graph (own TWS)	22:22	GA 2.45	1.77	1.77	50%	23%
		GB *2.54*	*1.41*	2.05	68%	23%
Two weeks later (own TWS)	14:14	GA 2.07	1.86	2.07	36%	29%
		GB 1.79	1.79	2.43	29%	43%
Two weeks later (own + two other TWSs)	14:42	GA 2.00	1.79	2.21	36%	36%
		GB *1.90*	*1.60*	*2.50*	26%	55%

- immediately after the experiment, the participants' ranking for their own sketched graph was higher than that for the spring layout ($p < 0.001$);
- two weeks after the experiment, the participants' ranking for the grid layout was higher than the spring layout (p ¡ 0.001), and higher than their own drawing (p = 0.019) in the set of three sketched graphs that included their own.

Validation (Experiment 2): Experiment 1 found that participants appeared to favour a grid layout and horizontal and vertical edges in their sketched graph drawings. We analysed the 44 graph drawings from Experiment 2 for the following key layout features (Table 3):

- *Number of edge crossings:* points outside of the node boundaries where one or more edges cross.
- *Number of straight lines.* A visual assessment as to whether an edge was intended to be straight was agreed by two independent coders.
- *Number of vertical or horizontal edges, and grid structure.* A visual assessment was agreed by two independent coders as to whether edges were intended to be horizontal or vertical, and whether drawings had been drawn with a grid in mind.

Independent samples two-tailed t-tests were conducted using the 34 drawings from Experiment 1 and the 44 drawings from Experiment 2 (Table 3).

Effect of graph format (Experiment 3): We analysed the 52 graph drawings from Experiment 3 using the same layout features as for Experiments 1 and 2.

Independent samples two-tailed t-tests were conducted using the 34 drawings from Experiment 1 and the 52 drawings from Experiment 3 were used (Table 3).

In the interviews, as before, no participants spoke directly of a grid layout; there were comments about local features (size of the nodes, the need for straight lines), and layout (crossings, symmetry, distance between nodes). When asked why they drew the graph as they did, many said the adjacency list itself suggested the structure of the drawing: they 'worked from top to bottom.'

Table 3. Comparing key layout features in the graphs produced in the experiments.

	E1 $n = 34$	E2 $n = 44$	E3 $n = 52$	validation: E1 v. E2		effect of format: E1 vs E3	
				t	p	t	p
Number of crosses/drawing	2.21	4.05	2.38	1.410	0.163	0.216	0.829
Number of crosses/drawing (excl. outliers)	1.24	2.05	1.62	1.400	0.166	0.776	0.440
% HV edges	33.6%	28.0%	23.3%	1.232	0.222	2.842	0.006^a
% straight edges	88.1%	85.8%	90.3%	0.472	0.638	0.541	0.590
Drawn with grid in mind	10(29%)	7(16%)	4(8%)	1.433	0.156	2.478	0.007^b

[a] More HV edges in Experiment 1. [b] More grids in Experiment 1

5 Results

Compromised layout: We wished to see whether participants expressed any dissatisfaction with their drawings as a result of having to compromise the layout while creating the drawing: a stated conclusion of Experiment 1.

In general, the participants were satisfied with the layout of their own drawings: their satisfaction ratings were high; they expressed little dissatisfaction, and did not mention that they would have liked to have been able to conform to a grid.

This is an instance of *stating a conclusion based on insufficient data*. In Experiment 1, observation of the videos (by two independent coders) suggested that participants conformed to a grid in the initial stages of drawing, but that this layout feature was abandoned later in the drawing process. This conclusion was, however, simply suggestive, and there was no qualitative interview data to back it up. It appears that, even if layout compromises had been made, participants were not aware of having done so.

Preference for a grid: We wished to see whether participants would prefer a grid-based layout to their own layout.

While initially they preferred their own drawings in Experiment 2, when the recency effect of having just drawn the graph was eliminated, the grid layout was ranked as substantially better than the sketches for the more complex graph. However, none of the participants mentioned a grid formation or horizontal or vertical edges in their comments.

It appears that participants know what they like when they see it (and when it is not in competition with a drawing that they know is their own), but cannot independently articulate the layout features that contribute to what they like.

This is an instance of *stating an incorrect conclusion based on qualitative data*. Despite the Experiment 1 drawings being analyzed by two independent coders who formed the same conclusion (that participants preferred a grid), this inferred conclusion does not hold when more direct data is collected.

Validation: The fact that there are no significant differences in the values for the key metrics for Experiment 2 suggests validation of the results of Experiment 1: when the graph is presented as a list of edges, participants tend to favour a grid layout and straight lines.

This is an instance of *validating data*. Using identical methodologies for experiments 1 and 2, we would expect similar values for the dependent data variables.

Effect of graph format: We wished to see whether the form in which the graph stimuli are presented would affect the results.

The drawings produced in Experiment 3 from adjacency list stimuli did not favour a grid layout; there is a significant difference on the key metrics of horizontal and vertical edges, and grid formation, between Experiments 1 and 3.

The format in which the relational graph information is presented to the participants thus affects the final layout of their drawing, a result echoing experimental results on visual metaphors [11]. The participants spoke of 'following the table from top to bottom' in Experiment 3; it is likely that the participants in Experiment 1 followed the edge list from left to right.

There is an irony here: Experiment 1 presented the graph as a list of edges so to address possible layout biases in HR08 and D+09, who presented graphs as drawings. It seems even using a simple list of edges can introduce a bias.

This is an instance of *unintended bias*. Even a simple (and seemingly innocuous) decision in Experiment 1 introduced a factor that biased the results.

Table 4. Summary and comparison of the findings for the three experiments

	Experiment 1	Experiment 2	Experiment 3
compromised layout	assumed, from video data	*not proven*	not investigated
preferring a grid over own drawing	assumed, from sketch graph drawing data	*some support after elimination of recency effects*	not investigated
validation	not applicable	*validation of key graph drawing metrics*	not appropriate
effect of graph format	not considered	not investigated	*effect found*

6   Discussion

6.1   Implications for this Research

Our speculation that participants prefer the result of a grid-based algorithm over their own drawing was partially confirmed, but only when the 'personal pride' factor had diminished over time, and for the more complex graph. It is still clear, however, that while participants might prefer a grid layout in both creation and recognition, they are less able to articulate this preference.

The most surprising result is that how the graph is presented to the participants has a significant effect on the form of their drawing — this was something that had not been considered originally. It suggests that the only way bias may be eliminated entirely would be by asking the participant to draw a graph based on their internal cognitive structure, and not on an externally visible form. This may mean describing a scenario to the participant while attempting to avoid verbal bias (for example, a social network) and then asking the participant to draw the graph representing the relational information.

Of course, this story could not simply end here, as there are several outstanding issues to address about all three of these experiments. Do these results extend to larger graphs? What would happen if the participants were all novice computer users? Or if a digital whiteboard were used? Of if participants were told that the graphs related to a domain (e.g. a transport network or a circuit diagram)? Or if they were explicitly advised to plan in advance? Further experiments would no doubt shed more light on these initial studies (and would probably reveal some unexpected results).

6.2 Implications for Experimental Research

The results of our investigation of subtle experimental variations suggest that:

- There will always be a bias relating to the manner in which information is presented to an experimental participant;
- Even if the results of an experiment are validated by repeating it, these results may still be compromised by bias;
- Firm experimental conclusions need validation through replication and multiple sources of data.

We did not set out to investigate the effect of experimental subtleties: our original aim was not to run a series of comparative experiments. If it had been, then we might have conducted a broader experiments-within-an-experiment study, preferably using the same participants throughout, and followed a systematic process of enquiry. No: we initially set out to investigate what happens when participants draw graphs from scratch — and we published a peer-reviewed paper in a reputable journal presenting the results of this study, as is common practice: researchers run an experiment, collect data, form conclusions, and publish. And then typically move on to their next experiment.

The contribution of this paper is therefore broader than the simple 'run an experiment and report' model: by reflecting on and critiquing our own experimental work, and investigating issues arising from the critique, we have demonstrated the limitations of this common practice.

It is rare that researchers repeat an experiment with subtle variations — doing so has revealed that there is still much to learn about the nature of user-sketched graphs, that even a carefully-conducted experiment may have flaws, that there is value in not simply moving on to the next 'big' question, and that repeating an experiment so as to investigate subtleties may produce surprising results. All experiments have limitations — no experiment can ever be perfect. This paper demonstrates the importance of acknowledging these limitations, of validating results where possible, and of reading published experimental results with healthy critical attitude.

Acknowledgments. SketchNode is the result of significant efforts of Beryl Plimmer and Hong Yul, of the University of Auckland. The experiments were conducted by

Christopher Pilcher, Rosemary Baker, Anastasia Giachanou, and Gareth Renaud. John Hamer assisted with data analysis, coding and editing. Ethical approval was given by the Universities of Auckland and Glasgow.

References

1. Purchase, H.C., Pilcher, C., Plimmer, B.: Graph drawing aesthetics — created by users not algorithms. IEEE Trans. Vis. and Computer Graphics 18(1), 81–92 (2012)
2. Kemmis, S., McTaggart, R.: The Action Research Planner, 3rd edn. Deakin University Press (1988)
3. Huang, W.: Using eye-tracking to investigate graph layout effects. In: Hong, S.H., Ma, K.L. (eds.) Proc. Asia Pacific Symp. on Visualisation, pp. 97–100. IEEE (2007)
4. Ware, C., et al.: Cognitive measurements of graph aesthetics. Information Visualization 1(2), 103–110 (2002)
5. Purchase, H.C., Cohen, R.F., James, M.: An experimental study of the basis for graph drawing algorithms. ACM J. Experimental Algorithmics 2(4), 1–17 (1997)
6. van Ham, F., Rogowitz, B.E.: Perceptual organisation in user-generated graph layouts. IEEE Trans. Vis. and Computer Graphics 14(6), 1333–1339 (2008)
7. Gansner, E.R., Koren, Y., North, S.C.: Graph Drawing by Stress Majorization. In: Pach, J. (ed.) GD 2004. LNCS, vol. 3383, pp. 239–250. Springer, Heidelberg (2005)
8. Dwyer, T., et al.: A comparison of user-generated and automatic graph layouts. IEEE Trans. on Vis. and Computer Graphics 15(6), 961–968 (2009)
9. Plimmer, B., Purchase, H.C., Yang, H.Y.: SketchNode: Intelligent sketching support and formal diagramming. In: Brereton, M., Viller, S., Kraal, B. (eds.) OzChi Conference, Brisbane, Australia, pp. 136–143 (2010)
10. Fruchterman, T.M.J., Reingold, E.M.: Graph drawing by force-directed placement. Software — Practice and Experience 21(11), 1129–1164 (1991)
11. Ziemkiewicz, C., Kosara, R.: The shaping of information by visual metaphors. IEEE Trans. on Vis. and Computer Graphics 14(6), 1269–1276 (2008)

Many-to-One Boundary Labeling
with Backbones

Michael A. Bekos[1], Sabine Cornelsen[2], Martin Fink[3],
Seok-Hee Hong[4], Michael Kaufmann[1], Martin Nöllenburg[5],
Ignaz Rutter[5], and Antonios Symvonis[6]

[1] Institute for Informatics, University of Tübingen, Germany
{bekos,mk}@informatik.uni-tuebingen.de
[2] Department of Computer and Information Science, University of Konstanz
sabine.cornelsen@uni-konstanz.de
[3] Lehrstuhl für Informatik I, Universität Würzburg, Germany
martin.a.fink@uni-wuerzburg.de
[4] School of Information Technologies, University of Sydney
shhong@it.usyd.edu.au
[5] Institute of Theoretical Informatics, Karlsruhe Institute of Technology, Germany
{noellenburg,rutter}@kit.edu
[6] School of Applied Mathematics and Physical Sciences, NTUA, Greece
symvonis@math.ntua.gr

Abstract. In this paper we study *many-to-one boundary labeling with backbone leaders*. In this model, a horizontal backbone reaches out of each label into the feature-enclosing rectangle. Feature points associated with this label are linked via vertical line segments to the backbone. We present algorithms for label number and leader-length minimization. If crossings are allowed, we aim to minimize their number. This can be achieved efficiently in the case of fixed label order. We show that the corresponding problem in the case of flexible label order is NP-hard.

1 Introduction

Boundary labeling was developed by Bekos et al. [2] as a framework and an algorithmic response to the poor quality (feature occlusion, label overlap) of specific labeling applications. In boundary labeling, labels are placed at the boundary of a rectangle and are connected to their associated features via arcs referred to as *leaders*. Leaders attach to labels at *label ports*. A survey by Kaufmann [4] presents different boundary labeling models that have been studied so far.

In *many-to-one boundary labeling* each label is associated to more than one feature point. This model was formally introduced by Lin et al. [7], who assumed that each label has one port for each connecting feature point (see Fig. 1a) and showed that several crossing minimization problems are NP-complete and, subsequently, developed approximation and heuristic algorithms. In a variant of this model, referred to as *boundary labeling with hyperleaders*, Lin [6] resolved the multiple port issue by joining together all leaders attached to a common label with a vertical line segment in the track-routing area (see Fig. 1b). At the cost of label duplications, leader crossings could be eliminated.

S. Wismath and A. Wolff (Eds.): GD 2013, LNCS 8242, pp. 244–255, 2013.
© Springer International Publishing Switzerland 2013

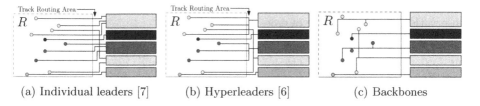

(a) Individual leaders [7]   (b) Hyperleaders [6]   (c) Backbones

Fig. 1. Different types of many-to-one labelings

We study *many-to-one boundary labeling with backbone leaders* (for short, *backbone labeling*). In this model, a horizontal backbone reaches out of each label into the feature-enclosing rectangle. Feature points that need to be connected to a label are linked via vertical line segments to the label's backbone (*backbone leaders*; see Fig. 1c). Formally, we are given a set $P = \{p_1, \ldots, p_n\}$ of n points in an axis-aligned rectangle R, where each point $p \in P$ is assigned a color $c(p)$ from a color set C. We also assume that the points are in general position and sorted in decreasing order of y-coordinates, with p_1 being the topmost. Our goal is to place colored labels to the left or right side of R and assign each point $p \in P$ to a label $l(p)$ of color $c(p)$. A *backbone labeling* for a set of colored points P is a set $\mathcal{L}$ of colored labels and a mapping of each point $p \in P$ to some $c(p)$-colored label in $\mathcal{L}$, so that (i) each point is connected to a label of the same color, and (ii) there are no backbone leader overlaps. A *crossing-free* backbone labeling is one without leader crossings.

The number of labels of a specific color may be unlimited or bounded by $K \geq |C|$. If $K = |C|$, all points of the same color are associated with a common label. One may restrict the maximum number of allowed labels for each color in C separately by specifying a *color vector* $\boldsymbol{k} = (k_1, \ldots, k_{|C|})$. A backbone labeling that satisfies all of the restrictions on the number of labels is called *feasible*.

Our goal is to find feasible backbone labelings that optimize different quality criteria. We study three different quality criteria, *label number minimization* (Section 2), *total leader length minimization* (Section 3), and *crossing minimization* (Section 4). The first two require crossing-free leaders. We consider both *finite backbones* and *infinite backbones*. Finite backbones extend horizontally from the label to the furthest point connected to the backbone, whereas infinite backbones span the whole width of the rectangle (thus one could use duplicate labels on both sides). Our algorithms also vary depending on whether the order of the labels is fixed or flexible and whether more than one label per color class can be used. Note that, due to space constraints some of our proofs are only sketched. Detailed proofs can be found in the technical report [1].

2   Minimizing the Total Number of Labels

In this section we minimize the total number of labels in a crossing-free solution, i.e., we set $K = n$ so that there is effectively no upper bound on the number of labels. We first consider the case of infinite backbones and present an important observation on the structure of crossing-free labelings.

Lemma 1. *Let p_i, p_{i+1} be two vertically consecutive points. Let p_j $(j < i)$ be the first point above p_i with $c(p_j) \neq c(p_i)$, and let $p_{j'}$ $(j' > i + 1)$ be the first point below p_{i+1} with $c(p_{j'}) \neq c(p_{i+1})$, if such points exist. In any crossing-free backbone labeling with infinite backbones, p_i and p_{i+1} are vertically separated by at most 2 backbones and any separating backbone has color $c(p_i), c(p_{i+1}), c(p_j)$, or $c(p_{j'})$.*

Sketch of Proof. In a crossing-free solution any infinite backbone splits the drawing into two independent subinstances above and below the backbone. Clearly, a backbone traversing a point has to be of the same color. On the other hand, we can check that a backbone lying between two points p_i and p_{i+1} can only have color $c(p_i), c(p_{i+1})$, or the color of the next point of distinct color above p_i or of the one below p_{i+1}.   □

Clearly, if all points have the same color, one label always suffices. Even in an instance with two colors, one label per color is enough. However, if a third color is involved, then many labels may be required. We sketch how to find an optimum solution in $O(n)$ time. First, we replace any maximal set of identically colored consecutive points by the topmost point in the set. One can show that an optimum solution of the original instance can be easily obtained from an optimum solution of the reduced instance, in which no two consecutive points have the same color. We solve the reduced instance using dynamic programming.

Theorem 1. *Let $P = \{p_1, p_2, \ldots, p_n\}$ be an input point set consisting of n points sorted from top to bottom. Then, a crossing-free labeling of P with the minimum number of infinite backbones can be computed in $O(n)$ time.*

Sketch of Proof. We store a table nl of values $nl(i, \text{cur}, c_{\text{bak}}, c_{\text{free}})$ representing the minimum number of backbones needed above or at point p_i such that the lowest backbone is c_{bak}-colored, the lowest backbone goes through p_i if the flag $\text{cur} = \text{true}$ and lies above p_i otherwise, and the (single) point between p_i and the lowest backbone (if $\text{cur} = \text{false}$) has color c_{free}. By careful case analysis, we can see that any entry of the table can be computed in constant time.   □

We now consider finite backbones. First, note that we can slightly shift the backbones in a given solution so that backbones are placed only in gaps between points. We number the gaps from 0 to n where gap 0 is above and gap n is below all points. Suppose a point p_l lies between a backbone of color c in gap g and a backbone of color c' in gap g' with $0 \leq g < l \leq g' \leq n$ such that both backbones horizontally extend to at least the x-coordinate of p_l. Let $R(g, g', l)$ be the rectangle bounded by these two backbones, the vertical line through p_l and the right side of R. Suppose all points except the ones in $R(g, g', l)$ are already labeled. An optimum solution for connecting the points in R cannot reuse any backbone except for the two backbones in gaps g and g'; hence, it is independent of the rest of the solution. We use this observation for solving the problem by dynamic programming.

Theorem 2. *Given a set P of n colored points and a color set C, we can compute a feasible labeling of P with the minimum number of finite backbones in $O(n^4|C|^2)$ time.*

Sketch of Proof. For $0 \le g \le g' \le n$, $l \in \{g, \ldots, g'\} \cup \{\emptyset\}$, and two colors c and c' let $T[g, c, g', c', l]$ be the minimum number of additional labels that are needed for labeling all points in the rectangle $R(g, g', l)$ under the assumption that there is a backbone of color c in gap g, a backbone of color c' in gap g', between these two backbones there is no backbone placed yet, and they both extend to the left of p_l. Note that for $l = \emptyset$ the rectangle is empty and $T[g, c, g', c', \emptyset] = 0$. Finally, let $\bar{c} \notin C$ be a dummy color, and let $p_{\bar{l}}$ be the leftmost point. Then, the value $T[0, \bar{c}, n, \bar{c}, \bar{l}]$ is the minimum number of labels needed for labeling all points. By careful case analysis, we can compute each of the $(n+1) \times |C| \times (n+1) \times |C| \times (n+1)$ entries of table T in $O(n)$ time.   □

3   Length Minimization

In this section we minimize the total length of all leaders in a crossing-free solution, either including or excluding the horizontal lengths of the backbones. We distinguish between a global bound K on the number of labels or a color vector $\boldsymbol{k}$ of individual bounds per color. We first consider the case of infinite backbones and use a parameter λ to distinguish the two minimization goals, i.e., we set $\lambda = 0$, if we want to minimize only the sum of the length of all vertical segments and we set λ to be the width of the rectangle R if we also take the length of the backbones into account. We further assume that $p_1 > \cdots > p_n$ are the y-coordinates of the input points.

Single Color. If all points have the same color, we seek for a set of at most K y-coordinates where we draw the backbones and connect each point to its nearest one, i.e., we must solve the following problem: Given n points with y-coordinates $p_1 > \ldots > p_n$, find a set S of at most K y-coordinates that minimizes

$$\lambda \cdot |S| + \sum_{i=1}^{n} \min_{y \in S} |y - p_i|. \tag{1}$$

Note that we can optimize the value in Eq. (1) by choosing $S \subseteq \{p_1, \ldots, p_n\}$. Hence, the problem can be solved in $O(Kn)$ time if the points are sorted according to their y-coordinates using the algorithm of Hassin and Tamir [3]. Note that the problem corresponds to the K-median problem if $\lambda = 0$.

Multiple Colors. If the input points have different colors, we can no longer assume that all backbones go through one of the given n points. However, by Lemma 1, it suffices to add between any pair of vertically consecutive points two additional candidates for backbone positions, plus one additional candidate above all points and one below all points. Hence, we have a set of $3n$ candidate lines at y-coordinates

$$p_1^- > p_1 > p_1^+ > p_2^- > p_2 > p_2^+ > \cdots > p_n^- > p_n > p_n^+ \qquad (2)$$

where for each i the values p_i^- and p_i^+ are as close to p_i as the label heights allow. Clearly, a backbone through p_i can only be connected to points with color $c(p_i)$. If we use a backbone through p_i^- (or p_i^+, respectively), it will have the same color as the first point below p_i (or above p_i, respectively) that has a different color than p_i. Hence, the colors of all candidates are fixed or the candidate will never be used as a backbone. For an easier notation, we denote the ith point in Eq. (2) by y_i and its color by $c(y_i)$. We solve the problem using dynamic programming.

Theorem 3. *A minimum length backbone labeling with infinite backbones for n points with $|C|$ colors can be computed in $O(n^2 \cdot \prod_{i=1}^{|C|} k_i)$ time if at most k_i labels are allowed for color i, $i = 1, \ldots, |C|$ and in $O(n^2 \cdot K)$ time if in total at most K labels are allowed.*

Sketch of Proof. For each $i = 1, \ldots, 3n$, and for each vector $\boldsymbol{k'} = (k_1', \ldots, k_{|C|}')$ with $k_1' \leq k_1, \ldots, k_{|C|}' \leq k_{|C|}$, let $L(i, \boldsymbol{k'})$ denote the minimum length of a feasible backbone labeling of $p_1, \ldots, p_{\lfloor \frac{i+1}{3} \rfloor}$ using k_c' infinite backbones of color c for $c = 1, \ldots, |C|$ such that the bottommost backbone is at position y_i, if such a labeling exists. Otherwise $L(i, \boldsymbol{k'}) = \infty$. One can show that the values $L(i, \boldsymbol{k'})$ can be computed recursively in $\mathcal{O}(n^2 \prod_{i=1}^{|C|} k_i)$ time in total. Let S be the set of candidates y_i such that all points below y_i have the same color as y_i. Then, we can compute the minimum total length of a backbone labeling of $p_1, \ldots, p_n$ with at most k_c, $c = 1, \ldots, |C|$ labels per color c by the following formula:

$$\min_{y_i \in S \cup \{p_n^+\}, k_1' \leq k_1, \ldots, k_{|C|}' \leq k_{|C|}} \left(L(i, k_1', \ldots, k_{|C|}') + \sum_{\frac{i+2}{3} \leq x \leq n} (y_i - p_x) \right).$$

If we bound the total number of labels by K, we obtain a similar dynamic program with the corresponding values $L(i, k)$, $i = 1, \ldots, 3n$, $k < K$. $\square$

We now turn our attention to the case of finite backbones and sketch how to modify the dynamic program for minimizing the total number of labels (see Theorem 1) to minimize the total leader length.

Theorem 4. *Given a set P of n colored points, a color set C, and a label bound K (or color vector $\boldsymbol{k}$), we can compute a feasible labeling of P with finite backbones that minimizes the total leader length in time $O(n^7 |C|^2 K^2)$ (or $O(n^7 |C|^2 (\prod_{c \in C} k_c)^2))$.*

Sketch of Proof. We change the meaning of an entry in the table T to denote the additional length of segments and backbones needed for labeling the points of the subinstance. Moreover, the precise positions of backbones matter for length

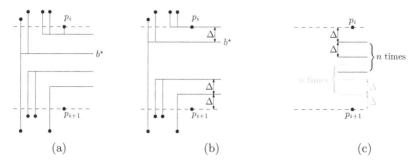

Fig. 2. (a) Longest backbone $b^\star$ splitting the backbones between p_i and p_{i+1}. (b) Backbones placed with the minimum leader length. (c) Candidate positions for backbones inside the gap

minimization. A clear candidate set is the set of the y-coordinates of input points which may be used by a backbone of the same color. We can also identify candidates for backbones inside a gap between points p_i and p_{i+1}. We observe that the longest backbone $b^\star$ inside the gap splits all other backbones lying between p_i and p_{i+1}; see Fig. 2a. The backbones above $b^\star$ connect only to points above and, hence, must be placed as close to p_i as possible for length minimization. Symmetrically, the backbones below $b^\star$ connect only to the bottom and must be placed as close to p_{i+1} as possible.

For avoiding overlaps and to accommodate labels with fixed heights, we enforce a minimum distance $\Delta > 0$ between pairs of backbones, as well as backbones and differently colored points. Then, for the labels close to p_i, we get a sequence of consecutive candidate positions with distance Δ below p_i; see Fig. 2b and 2c. Symmetrically, there is such a sequence above p_{i+1}. Any such sequence contains up to n points (less if the gap is too small). Note that the two sequences might overlap; we can, however, easily ensure that no two backbones with distance less than Δ are used in the dynamic program. To address entries in T we use the $O(n^2)$ candidate positions (input points and positions in gaps) instead of the gaps; no position can be used twice.

As a final step, we integrate the global value K or the color vector $\boldsymbol{k}$ as a bound on the allowed numbers of labels. To this end, we add additional dimensions for K or for $k_c, c \in C$ to the table that specify the remaining available numbers of labels in the subinstance. □

4 Crossing Minimization

In this section we allow crossings between backbone leaders, which generally allows us to use fewer labels. We concentrate on minimizing the number of crossings for the case $K = |C|$, i.e., one label per color, and distinguish fixed and flexible label orders.

4.1   Fixed y-Order of Labels

In this part, we assume that the color set C is ordered and we require that for each pair of colors $i < j$, the i-colored label is above the j-colored label.

Infinite Backbones. Observe that it is possible to slightly shift the backbones of a solution without increasing the number of crossings so that no backbone contains a point. So, the backbones can be assumed to be in the gaps between vertically consecutive points; we number the gaps from 0 to n, as in Section 2.

Theorem 5. *Given a set P of n colored points and an ordered color set C, a backbone labeling with one label per color, labels in the given color order, infinite backbones, and minimum number of crossings can be computed in $O(n|C|)$ time.*

Proof. Suppose that we fix the position of the i-th backbone to gap g. For $1 \leq i \leq |C|$ and $0 \leq g \leq n$, let $\mathrm{cross}(i, g)$ be the number of crossings of the vertical segments of the non-i-colored points when the color-i backbone is placed at gap g. Note that this number depends only on the y-ordering of the backbones, which is fixed, and not on their actual positions. So, we can precompute the table cross, using dynamic programming, as follows. All table entries of the form $\mathrm{cross}(\cdot, 0)$ can be clearly computed in $O(n)$ time. Then, $\mathrm{cross}(i, g) = \mathrm{cross}(i, g-1) + 1$, if the point between gaps $g - 1$ and g has color j and $j > i$. In the case where the point between gaps $g - 1$ and g has color j and $j < i$, $\mathrm{cross}(i, g) = \mathrm{cross}(i, g - 1) - 1$. If it has color i, then $\mathrm{cross}(i, g) = \mathrm{cross}(i, g - 1)$. From the above, it follows that the computation of table cross takes $O(n|C|)$ time.

Now, we use another dynamic program to compute the minimum number of crossings. Let $T[i, g]$ denote the minimum number of crossings on the backbones $1, \ldots, i$ in any solution subject to the condition that the backbones are placed in the given ordering and backbone i is positioned in gap g. Clearly $T[0, g] = 0$ for $g = 0, \ldots, n$. Moreover, we have $T[i, g] = \min_{g' \leq g} T[i - 1, g'] + \mathrm{cross}(i, g)$. Having pre-computed table cross and assuming that for each entry $T[i, g]$, we also store the smallest entry of row $T[i, \cdot]$ to the left of g, each entry of table T can by computed in constant time. Hence, table T can be filled in time $O(n|C|)$. Then, the minimum crossing number is $\min_g T[|C|, g]$. A corresponding solution can be found by backtracking in the dynamic program.   □

Finite Backbones. We can easily modify the approach used for infinite backbones to minimize the number of crossings for finite backbones, if the y-order of labels is fixed, as the following theorem shows.

Theorem 6. *Given a set P of n colored points and an ordered color set C, a backbone labeling with one label per color, labels in the given order, finite backbones, and minimum number of crossings can be computed in $O(n|C|)$ time.*

Proof. We present a dynamic program similar to the one presented in the proof of Theorem 5. Recall that all points of the same color are routed to the same label and the order of the labels is fixed, i.e., the label of the i-colored points is above the label of the j-colored points, when $i < j$. Here, the computation of the number of crossings when fixing a backbone at a certain position should take into consideration that the backbones are not of infinite length. Recall that the dynamic program could precompute these crossings, by maintaining an $n \times |C|$ table cross, in which each entry $\text{cross}(i, g)$ corresponds to the number of crossings of the non-i-colored points when the color-i-backbone is placed at gap g, for $1 \le i \le |C|$ and $0 \le g \le n$. In our case, $\text{cross}(i, g) = \text{cross}(i, g - 1) + 1$, if the point between gaps $g - 1$ and g is right of the leftmost of the i-colored points and has color j s.t. $j > i$. In the case, where the point between gaps $g - 1$ and g is right of the leftmost of the i-colored points and has color j and $j < i$, $\text{cross}(i, g) = \text{cross}(i, g - 1) - 1$. Otherwise, $\text{cross}(i, g) = \text{cross}(i, g - 1)$. Again, all table entries of the form $\text{cross}(\cdot, 0)$ can be clearly computed in $O(n)$ time. $\square$

4.2   Flexible y-Order of Labels

In this part the order of labels is no longer given and we need to minimize the number of crossings over all label orders. While there is an efficient algorithm for infinite backbones, the problem is NP-complete for finite backbones.

Infinite Backbones. We give an efficient algorithm for the case that there are $K = |C|$ fixed label positions $y_1, \ldots, y_K$ on the right boundary of R, e.g., uniformly distributed.

Theorem 7. *Given a set P of n colored points, a color set C, and a set of $|C|$ fixed label positions, we can compute in $O(n + |C|^3)$ time a feasible backbone labeling with infinite backbones that minimizes the number of crossings.*

Proof. First observe that if the backbone of color $k, 1 \le k \le |C|$ is placed at position $y_i, 1 \le i \le |C|$, then the number of crossings created by the vertical segments leading to this backbone is fixed, since all label positions will be occupied by an infinite backbone. This crossing number $\text{cr}(k, i)$ can be determined in $O(n_k + |C|)$ time, where n_k is the number of points of color k. In fact, by a sweep from top to bottom, we can even determine all crossing numbers $\text{cr}(k, \cdot)$ for backbone $k, 1 \le k \le |C|$ in time $O(n_k + |C|)$. Now, we construct an instance of a weighted bipartite matching problem, where for each position $y_i, 1 \le k \le |C|$ and each backbone $k, 1 \le k \le |C|$, we establish an edge (k, i) of weight $\text{cr}(k, i)$. In total, this takes $O(n + |C|^2)$ time. The minimum-cost weighted bipartite matching problem can be solved in time $O(|C|^3)$ with the Hungarian method [5] and yields a backbone labeling with the minimal number of crossings. $\square$

Finite Backbones. Next, we consider the variant with finite backbones and prove that it is NP-hard to minimize the number of crossings. For simplicity, we

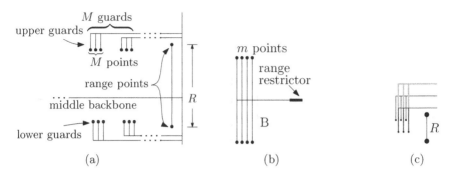

Fig. 3. (a) The range restrictor gadget, (b) a blocker gadget, (c) crossings caused by a pair of an upper and a lower guard that are positioned on the same side outside range R.

allow points that share the same x- or y-coordinates. This can be remedied by a slight perturbation. Our arguments do not make use of this special situation, and hence carry over to the perturbed constructions. We first introduce a number of gadgets that are required for our proof and sketch their properties.

The first one is the *range restrictor gadget*. Its construction consists of the middle backbone, whose position will be restricted to a given range R, and an upper and a lower *guard gadget* that ensure that positioning the middle backbone outside range R creates many crossings. We assume that the middle backbone is connected to at least one point further to the left such that it extends beyond all points of the guard gadgets. The middle backbone is connected to two *range points* whose y-coordinates are the upper and lower boundary of the range R. Their x-coordinates are such that they are on the right of the points of the guard gadgets. A *guard* consists of a backbone that connects to a set of M points, where $M > 1$ is an arbitrary number. The M points of a guard lie left of the range points. The upper guard points are horizontally aligned and lie slightly below the upper bound of range R. The lower guard points are placed such that they are slightly above the lower bound of range R. We place M upper and M lower guards such that the guards form pairs for which the guard points overlap horizontally. The upper (resp. lower) guard gadget is formed by the set of upper (resp. lower) guards. We call M the *size* of the guard gadgets. The next lemma shows properties of the range restrictor.

Lemma 2. *The backbones of the range restrictor can be positioned such that there are no crossings. If the middle backbone is positioned outside the range R, there are at least $M − 1$ crossings.*

Proof. The first statement is illustrated in Fig. 3a. To prove the second statement, assume to the contrary that the middle backbone is positioned outside range R, say w.l.o.g. below range R, and that there are fewer than $M − 1$ crossings. Observe that all guards must be positioned above the middle backbone, as a guard below the middle backbone would create M crossings, namely between the middle backbone and the segments connecting the points of the guard to its backbone. So, the middle backbone is the lowest. Now observe that any guard that is positioned below the upper range point crosses the segment that connects this range point to the middle backbone. To avoid having $M - 1$ crossings, at least $M + 1$ guards (both upper and lower) must be positioned above range R. Hence, there is at least one pair consisting of an upper and a lower guard that are both positioned above range R. This, independent of their ordering, creates at least $M - 1$ crossings, a contradiction; see Fig. 3c, where the two alternatives for the lower guard are drawn in black and bold gray, respectively.   □

Let B be an axis-aligned rectangular box and R a small interval that is contained in the range of y-coordinates spanned by B. A *blocker gadget* of *width* m consists of a backbone that connects to $2m$ points, half of which are on the top and bottom side of B, respectively. A range restrictor gadget is used to restrict the backbone of the blocker to the range R; see Fig. 3b. Note that, due to the range restrictor, this drawing is essentially fixed. We say that a backbone *crosses* the blocker gadget if its backbone crosses box B. It is easy to see that any backbone that crosses a blocker gadget creates m crossings, where m is the width of the blocker. We are now ready to present the NP-hardness reduction.

Theorem 8. *Given a set of n input points in k different colors and an integer Y it is NP-complete to decide whether a backbone labeling with one label per color and flexible y-order of the labels that has at most Y leader crossings exists.*

Proof. The proof is by reduction from the NP-complete Fixed Linear Crossing Number problem [8]: Given a graph $G = (V, E)$, a bijective function $f \colon V \to \{1, \ldots, |V|\}$, and an integer Z, is there a drawing of G with the vertices placed on a horizontal line (*spine*) in the order specified by f and the edges drawn as semi-circles above or below the spine so that there are at most Z crossings? Masuda et al. [8] showed that the problem is NP-complete, even if G is a matching.

Let G be a matching. Then, the number of vertices is even and we can assume that the vertices $V = \{v_1, \ldots, v_{2n}\}$ are indexed in the order specified by f, i.e., $f(v_i) = i$ for all i. We also direct each edge $\{v_i, v_j\}$ with $i < j$ from v_i to v_j. Let $\{u_1, \ldots, u_n\}$ be the ordered source vertices and let $\{w_1, \ldots, w_n\}$ be the ordered sink vertices; see Fig. 4a. In our reduction we will create an edge gadget for every edge of G. The gadget consists of five blocker gadgets and one *side selector gadget*. Each of the six sub-gadgets uses its own color and thus defines one backbone. The edge gadgets are ordered from left to right according to the sequence of source vertices $(u_1, \ldots, u_n)$; see Fig. 4b.

The edge gadgets are placed symmetrically with respect to the x-axis. We create $2n + 1$ special rows above the x-axis and $2n + 1$ special rows below, indexed by $-(2n+1), -2n, \ldots, 0, \ldots, 2n, 2n+1$. The gadget for an edge (v_i, v_j) uses five blocker gadgets (denoted as *central, upper, lower, upper gap*, and *lower gap* blockers) in two different columns to create two small gaps in rows j and $-j$, see the hatched blocks in the same color in Fig. 4b. The upper and lower blockers extend vertically to rows $2n + 1$ and $-2n - 1$. The gaps are intended to create two alternatives for routing the backbone of the side selector. Every backbone that starts left of the two gap blockers is forced to cross at least one of

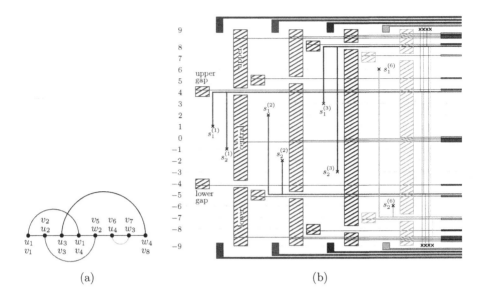

(a)   (b)

Fig. 4. (a) An input instance with four edges, (b) Sketch of the reduction for the graph of Fig. 4a. Hatched rectangles correspond to blockers, thick segments to side selectors, and filled shapes to guard gadgets or range restrictor gadgets.

these five blocker gadgets as long as it is vertically placed between rows $2n + 1$ and $-2n - 1$. The blockers have width $m = 8n^2$. Their backbones are fixed to lie between rows 0 and -1 for the central blocker, between rows $2n$ and $2n + 1$ ($-2n$ and $-2n - 1$) for the upper (resp. lower) blocker, and between rows j and $j + 1$ ($-j$ and $-j - 1$) for the upper (resp. lower) gap blocker.

The *side selector* consists of two horizontally spaced *selector points* $s_1^{(i)}$ and $s_2^{(i)}$ in rows i and $-i$ located between the left and right blocker columns. They have the same color and thus define one joint backbone that is supposed to pass through one of the two gaps in an optimal solution. The n edge gadgets are placed from left to right in the order of their source vertices; see Fig. 4b. The backbone of every selector gadget is vertically restricted to the range between rows $2n + 1$ and $-2n - 1$ in any optimal solution by augmenting each selector gadget with a range restrictor gadget. So, we add two more points for each selector to the right of all edge gadgets, one in row $2n + 1$ and one in row $-2n - 1$. They are connected to the selector backbone. In combination with a corresponding upper and lower guard gadget of size $M = \Omega(n^4)$ between the two selector points $s_1^{(i)}$ and $s_2^{(i)}$ this achieves the range restriction according to Lemma 2.

We now claim that in a crossing-minimal labeling the backbone of the selector gadget for every edge (v_i, v_j) passes through one of its two gaps in rows j or $-j$. The proof of this claim is based on three different options for placing a selector backbone: (a) outside its range restriction, i.e., above row $2n + 1$ or below row $-2n - 1$, (b) between rows $2n + 1$ and $-2n - 1$, but outside one of the two gaps, and (c) in rows j or $-j$, i.e., inside one of the gaps. By this claim and the fact that violating any range restriction immediately causes M crossings, we can assume that every backbone adheres to the rules, i.e., stays within its range as defined by the range restriction gadgets or passes through one of its two gaps.

One can show that an optimal solution of the backbone labeling instance I_G created for a matching G with n edges has $X + 2Z$ crossings, where X is a constant depending on G, and Z is the minimum number of crossings of G in the Fixed Linear Crossing Number problem. The detailed proof is based on carefully counting crossings in four different cases, depending on which types of backbones and vertical segments intersect. It turns out that almost all crossings are fixed (yielding the number X), except for those of selector backbones with vertical selector segments for which the two underlying edges (v_i, v_j) and (v_k, v_l) with $i < k$ are *interlaced*, i.e., $i < k < j < l$ holds (yielding the number $2Z$). Note that we can guess an order of the backbones and apply Theorem 6 to compute the minimum crossing number, which concludes the NP-completeness proof. □

Acknowledgements. This work was started at the Bertinoro Workshop on Graph Drawing 2013. M. Nöllenburg received financial support by the Concept for the Future of KIT. The work of M. A. Bekos and A. Symvonis is implemented within the framework of the Action "Supporting Postdoctoral Researchers" of the Operational Program "Education and Lifelong Learning" (Action's Beneficiary: General Secretariat for Research and Technology), and is co-financed by the European Social Fund (ESF) and the Greek State. We also acknowledge partial support by GRADR – EUROGIGA project no. 10-EuroGIGA-OP-003.

References

1. Bekos, M.A., Cornelsen, S., Fink, M., Hong, S., Kaufmann, M., Nöllenburg, M., Rutter, I., Symvonis, A.: Many-to-one boundary labeling with backbones. CoRR (2013), arXiv:1308.6801
2. Bekos, M.A., Kaufmann, M., Symvonis, A., Wolff, A.: Boundary labeling: Models and efficient algorithms for rectangular maps. Computational Geometry 36(3), 215–236 (2007)
3. Hassin, R., Tamir, A.: Improved complexity bounds for location problems on the real line. Operations Research Letters 10(7), 395–402 (1991)
4. Kaufmann, M.: On map labeling with leaders. In: Albers, S., Alt, H., Näher, S. (eds.) Efficient Algorithms. LNCS, vol. 5760, pp. 290–304. Springer, Heidelberg (2009)
5. Kuhn, H.W.: The Hungarian method for the assignment problem. Naval Research Logistics Quarterly 2(1-2), 83–97 (1955)
6. Lin, C.C.: Crossing-free many-to-one boundary labeling with hyperleaders. In: Proc. IEEE Pacific Visualization Symp. (PacificVis 2010), pp. 185–192. IEEE (2010)
7. Lin, C.C., Kao, H.J., Yen, H.C.: Many-to-one boundary labeling. Journal of Graph Algorithms and Applications 12(3), 319–356 (2008)
8. Masuda, S., Nakajima, K., Kashiwabara, T., Fujisawa, T.: Crossing minimization in linear embeddings of graphs. IEEE Trans. Computers 39(1), 124–127 (1990)

Streamed Graph Drawing
and the File Maintenance Problem

Michael T. Goodrich and Paweł Pszona

Dept. of Computer Science, University of California, Irvine

Abstract. In *streamed graph drawing*, a planar graph, G, is given incrementally as a data stream and a straight-line drawing of G must be updated after each new edge is released. To preserve the mental map, changes to the drawing should be minimized after each update, and Binucci *et al.* show that exponential area is necessary for a number of streamed graph drawings for trees if edges are not allowed to move at all. We show that a number of streamed graph drawings can, in fact, be done with polynomial area, including planar streamed graph drawings of trees, tree-maps, and outerplanar graphs, if we allow for a small number of coordinate movements after each update. Our algorithms involve an interesting connection to a classic algorithmic problem—the *file maintenance problem*—and we also give new algorithms for this problem in a framework where bulk memory moves are allowed.

1 Introduction

In the *streamed graph drawing* framework, which was introduced by Binucci *et al.* [4,3], a graph, G, is incrementally presented as a data stream of its vertices and edges, and a drawing of G must be updated after each new edge is released. So as to preserve the *mental map* [6,9] of the drawing, this framework also requires that changes to the drawing of G should be minimized after each update. Indeed, to achieve this goal, Binucci *et al.* took the extreme position of requiring that once an edge is drawn no changes can be made to that edge. They showed that, under this restriction, exponential area is necessary and sufficient for planar drawings of trees under various orderings for how the vertices and edges of the trees are presented.

In light of recent results regarding the mental map [1], however, we now know that moving a small number of vertices or edges in a drawing of a graph does not significantly affect readability in a negative way. Therefore, in this paper, we choose to relax the requirement that there are no changes to the drawing of the graph after an update and instead allow a small number of coordinate movements after each such update. In this paper, we study planar streamed graph drawing schemes for trees, tree-maps, and outerplanar graphs, showing that polynomial area is achievable for such streamed graph drawings if small changes to the drawings are allowed after each update. Our results are based primarily on an interesting connection between streamed graph drawing and a classic algorithmic problem, the *file maintenance problem*.

S. Wismath and A. Wolff (Eds.): GD 2013, LNCS 8242, pp. 256–267, 2013.

In the *file maintenance problem* [12], we wish to maintain an ordered set, S, of n elements, such that each element, x in S, is assigned a unique integer label, $L(x)$, in the range $[0, N]$, where x comes before y if and only if $L(x) < L(y)$. In the classic version of this problem, N is restricted to be $O(n)$, with the motivation that the integer labels are addresses or pseudo-addresses for memory locations where the elements of S are stored[1]. If N is only restricted to be polynomial in n, then this is known as the *online list labeling problem* [2,5]. In either case, the set, S, can be updated by issuing a command, insertAfter(x, y), where y is to be inserted to be immediately after $x \in S$ in the ordering, or insertBefore(x, y), where y is to be inserted to be immediately before $x \in S$ in the ordering. The goal is to minimize the number of elements in S needing to be relabeled as a result of such an update.

Previous Related Results. For the file maintenance problem, Willard [12] gave a rather complicated solution that achieves $O(\log^2 n)$ relabelings in the worst case after each insertion, and this result was later simplified by Bender *et al.* [2]. For the online list labeling problem, Dietz and Sleator [5] give an algorithm that achieves $O(\log n)$ amortized relabelings per insertion, and $O(\log^2 n)$ in the worst-case, using an $O(n^2)$ tag range. Their solution was simplified by Bender *et al.* [2] with the same bounds. Recently, Kopelowitz [8] has given an algorithm that achieves $O(\log n)$ worst-case relabelings after each insertion, using a polynomial bound for N.

For streamed tree drawings, as we mentioned above, Binucci *et al.* [4,3] show that exponential area is required for planar drawings of trees, depending on the order in which vertices and edges are introduced (e.g, BFS, DFS, etc.).

Our Results. For the context of this paper, we focus on planar drawings of graphs, so we consider a drawing to consist essentially of a set of non-crossing line segments. For traditional drawings of trees and outerplanar graphs, the endpoints of the segments correspond to vertices and the segments represent edges. In tree-map drawings, each vertex v of a tree T is represented by a rectangle, R, such that the children of v are represented by rectangles inside R that share portions of at least two sides of R. Thus, in a tree-map drawing, the line segments correspond to the sides of rectangles.

We present new streamed graph drawing algorithms for general trees, tree-maps, and outerplanar graphs that keep the area of the drawing to be of polynomial size and allow new edges to arrive in any order, provided the graph is connected at all times. After each update to a graph is given, we allow a small number of, say, a polylogarithmic number of the endpoints of the segments in the drawing to move to accommodate the representation of the new edge. We alternatively consider these to be movements of either individual endpoints or sets of at most B endpoints, for a parameter B, provided that each set of such

---

[1] For instance, in the EDT text editor developed for the DEC PDP-11 series of minicomputers, each line was assigned a pseudo line number, 1.0000, 2.0000, and so on, and if a new line was to be introduced between two existing lines, x and y, it was given as a default label the average of the labels of x and y as its label.

endpoints is contained in a given convex region, R, and all the endpoints in this region are translated by the same vector. We call such operations the *bulk moves*.

All of our methods are based on our showing interesting connections between the streamed graph drawing problems we study and the file maintenance problem. In addition to utilizing existing algorithms for the file maintenance problem in our graph drawing schemes, we also give a new algorithm for this classic problem in a framework where bulk memory moves are allowed, and we show how this solution can also be applied to streamed graph drawing.

2   Building Blocks

The Ordered Streaming Model. We start with the description of the model under which we operate. At each time $t \geq 1$, a new edge, e, of a graph, G, arrives and has to be incorporated immediately into a drawing of G, using line segments whose endpoints are placed at grid points with integer coordinates. Since we are focused on planar drawings in this paper, together with the edge, e, we also get the information of its relative position among the neighbors of e's endpoints (i.e., for every vertex we know the clockwise order of its neighbors and e's placement in this order). At all times, the current graph, G, is connected, and the edges never disappear (infinite persistence).

Incidentally, the streaming model of Binucci *et al.* [4] is slightly different, in that edges arrive without the order information in their model. Under that model, they have given an $\Omega(2^{\frac{n}{8}})$ area lower bound for binary tree drawing and an $\Omega(n(d-1)^n)$ lower bound for drawing trees with maximum degree $d > 2$. These bounds stand when the algorithm is not allowed to move any vertex. However, they only apply to a very restricted class of algorithms, namely *predefined-location algorithms*, which are non-adaptive algorithms whose behavior does not depend on the order in which the edges arrive or the previously drawn edges. Also, as noted above, Binucci *et al.* do not allow for vertices to move once they are placed. As we show in the following theorem, even with the added information regarding the relative placement of an edge among its neighbors incident on the same vertex, if we don't allow for vertex moves, we must allow for exponential area. In the full version of this paper [7] we prove the following.

Theorem 1. *Under the ordered streaming model without vertex moves, any tree drawing algorithm requires $\Omega(2^{n/2})$ area in the worst case.*

File Maintenance with Bulk Moves. Here we consider the file maintenance problem and the online list labeling problem variants where we allow for bulk moves[2] of an interval of B labels, for some parameter B. We have already mentioned the known results for the file maintenance problem, where the only type of relabelings that are allowed are for individual elements, in which $O(\log^2 n)$

---

[2] Note that bulk moves are also motivated for the original file maintenance problem if we define the complexity of a solution in terms of the number of commands that are sent to a DMA controller for bulk memory-to-memory moves.

worst-case relabelings suffice for each update when N is $O(n)$ [2,12] and $O(\log n)$ suffice in the worst-case when N is a polynomial in n [8].

Bulk moves allow us to improve on these bounds. We have achieved several tradeoffs between the operation count and the size of B. Of course, if B is n, then achieving constant number of operations is easy, since we can maintain the n elements to have the indices 1 to n, and with each insertion, at some rank i, we simply move the elements currently from i to n up by one, as a single bulk move. Theorem 2 summarizes the rest of our results.

Theorem 2. *We can achieve the following bounds:*

1. *$O(1)$ worst-case relabeling bulk moves suffice for the file maintenance problem if $B = n^{1/2}$.*
2. *$O(1)$ worst-case relabeling bulk moves suffice for the online list maintenance problem if $B = \log n$.*
3. *$O(\log n)$ worst-case relabeling bulk moves suffice for the file maintenance problem if $B = \log n$.*

Proof.

1. This is accomplished in an amortized way by partitioning the array into $n^{1/2}$ chunks of size at most $2n^{1/2}$. Whenever a chunk, i, grows to have size $n^{1/2}$, we move all the chunks to the right of i by one chunk (using $O(n^{1/2})$ bulk moves). Then we split the chunk i in two, keeping half the items in chunk i and moving half to chunk $i + 1$. These moves are charged to the previous $n^{1/2}/2$ insertions in chunk i. Turning this bound into a worst-case bound is then done using standard de-amortization techniques.
2. This is accomplished by modifying a two-level structure of Kopelowitz [8]. Kopelowitz used the top level of this structure to maintain order of sublists of size $O(\log n)$ each. Order within each sublist was maintained using standard file maintenance problem solutions. Our modification is that each sublist is now represented as a small subarray of size $O(\log n)$ and operations on the top level of Kopelowitz's structure are simulated using bulk moves on a big array containing all concatenated subarrays.
3. This is accomplished by using the method of Bender *et al.* [2] and noting that each insertion in their scheme uses a process where each substep involves moving a contiguous subarray of size $O(\log n)$ using $O(\log n)$ individual moves. Each such move can alternatively be done using $O(1)$ bulk moves of subarrays of size $\log n$. □

3   Streamed Graph Drawing of Trees

In this section, we present several algorithms for upward grid drawings of trees in the ordered streaming model. The algorithms are designed to accommodate different types of vertex moves allowed. For example, by a *bulk move*, we mean a move that translates all segment endpoints that belong to a given convex region,

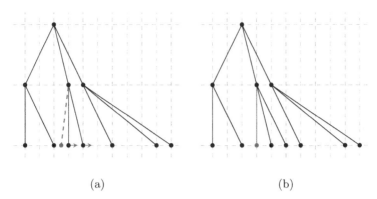

(a)   (b)

Fig. 1. Illustrating an insertion for our tree-drawing scheme: (a) determining relative position for the new (dashed, red) edge; (b) tree after edge insert and related vertex moves

R, by the same vector. This corresponds to the observation [13] that moving a small number of elements in the same direction is easy to follow and does not interfere with the ability to understand the structure of the drawing (as long as there are no intersections).

Let G be a tree that is revealed one edge at a time, keeping the graph connected. Algorithm 1 selects one endpoint of the first edge, r, puts it at position $(0,0)$, and produces an upward straight-line grid drawing of G, level-by-level, with each edge from parent to child pointed downwards. (If a new edge is ever revealed for the current root, we simply recalibrate what we are calling position $(0,0)$ without changing the position of the vertices already drawn.) For the kth level, L_k, with n_k nodes, we place nodes in positions $(0,-k)$ through $(N,-k)$ in the order of their parents (to avoid intersections), where $N \geq n_k$. When a new edge is added, we locate the position (row and position in the row) of the new node and insert the new node after its predecessor (or before its successor), shifting other nodes on this level as needed to make room for the new node. (See Fig. 1.) The details are as shown in Algorithm 1.

Input: $e = (a,b)$, the edge to be added; b is the new vertex
1: $k \leftarrow b$'s distance from r
2: determine c, b's predecessor (or successor) in level L_k
3: perform L_k.insertAfter(c,b) (or L_k.insertBefore(c,b)), giving b integer label $L(b)$
4: move vertices whose labels have changed in the previous step
5: place b at $(L(b),-k)$ and draw e

Algorithm 1. Generic insertion algorithm for upward straight-line grid streamed tree drawing

Drawing the tree in this fashion ensures there are no intersections (edges connect only vertices in two neighboring levels), even as the vertices are shifted (relative order of vertices stays the same). In addition, there are at most $O(n)$ levels, and at most $O(n)$ nodes per level.

Theorem 3. *Depending on the implementation for the insertBefore and insertAfter methods, Algorithm 1 maintains a straight-line upward grid drawing of a tree in the ordered streaming model to have the following possible performance bounds:*

1. *$O(n^2)$ area and $O(1)$ vertex moves per insertion if bulk moves of size $n^{1/2}$ are allowed.*
2. *$O(n^2)$ area and $O(\log n)$ vertex moves per insertion if bulk moves of size $\log n$ are allowed.*
3. *$O(n^2)$ area and $O(\log^2 n)$ vertex moves per insertion if bulk moves are not allowed.*
4. *polynomial area and $O(1)$ vertex moves per insertion if bulk moves of size $\log n$ are allowed.*
5. *polynomial area and $O(\log n)$ vertex moves per insertion if bulk moves are not allowed.*

Proof. The claimed bounds follow immediately from Theorem 2.   □

Note that $\Omega(n^2)$ area is necessary in the worst case for an upward straight-line grid drawing of a tree if siblings are always placed on the same level.

4   Streamed Graph Drawing of Tree-Maps

A tree-map, M, is a visualization technique introduced by Shneiderman [10], which draws a rooted tree, T, as a collection of nested rectangles. The root, r, of T is associated with a rectangle, R, and if r has k children, then R is partitioned into k sub-rectangles using line segments parallel to one of the coordinate axes (say, the x-axis), with each such rectangle associated with one of the children of r. Then, each child rectangle is recursively partitioned using line segments parallel to the other coordinate axis (say, the y-axis). (See Fig. 2.)

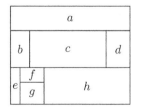 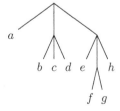

Fig. 2. A tree-map and its associated tree

We assume in this case that a tree, G, is released one edge at a time, as in the previous section. We assume inductively that we have a tree-map drawn for G, with a global set, X, of all x-coordinates maintained for the rectangle boundaries and a global set, Y, of all y-coordinates maintained for the rectangle boundaries. When an edge, e, of a rectangle has to be moved, the largest segment containing e is moved accordingly. Our insertion method is shown in Algorithm 2 (for brevity, the case when a vertex has no predecessors among its siblings is omitted).

Input: $e = (a, b)$, the edge to be added; b is a new child vertex
1: Let R be the rectangle for a, and let z be the primary axis for R (w.l.o.g., $z = x$)
2: **if** b has no siblings **then**
3:    $R_b \leftarrow R$ (and give it primary axis orthogonal to z)
4: **else if then**
5: **else**
6:    determine c, b's predecessor sibling (w.l.o.g.), and let R_c be c's rectangle
7:    perform X.insertAfter($R_c.x_{\max}, b$), giving b integer label $L(b)$
8:    move segment endpoints whose labels have changed in the previous step
9:    $R_b \leftarrow$ the rectangle in R with left boundary $R_c.x_{\max}$ and right boundary $L(b)$
10: **end if**

Algorithm 2. Generic insertion algorithm for streamed tree-map drawing

Theorem 4. *Depending on the implementation for the insertBefore and insertAfter methods, Algorithm 2 maintains a tree-map drawing of a tree in the ordered streaming model to have the following possible performance bounds:*

1. *$O(n^2)$ area and $O(1)$ x- and y-coordinate moves per insertion if bulk moves of size $n^{1/2}$ are allowed.*
2. *$O(n^2)$ area and $O(\log n)$ x- and y-coordinate moves per insertion if bulk moves of size $\log n$ are allowed.*
3. *$O(n^2)$ area and $O(\log^2 n)$ x- and y-coordinate moves per insertion if bulk moves are not allowed.*
4. *polynomial area and $O(1)$ x- and y-coordinate moves per insertion if bulk moves of size $\log n$ are allowed.*
5. *polynomial area and $O(\log n)$ x- and y-coordinate moves per insertion if bulk moves are not allowed.*

Proof. The claimed bounds follow immediately from Theorem 2.   □

5   Streamed Graph Drawing of Outerplanar Graphs

Our algorithm for drawing outerplanar graphs in the streaming model is based on a well-known fact about outerplanar graphs, namely that any outerplanar graph may be drawn with straight-line edges and without intersections in such a way that the vertices are placed on a circle [11].

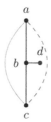

Fig. 3. Situation where information about order of edges around vertex is insufficient. Initially, there are vertices a, b, c and edges (a, b), (b, c). When a new edge (a, c) is added, it can be drawn in two ways (solid green or dashed red) – ordering of edges does not specify which one to choose. When edge (b, d) arrives (with edge order (a, d, c) around b), if the dashed red edge location was selected, there is no way to move the vertices without intersections to produce an outerplanar drawing.

As previously, we assume that each new edge comes with the information about its relative placement among its endpoints' incident edges. In other words, for each vertex, we know the clockwise order of its incident edges. Fig. 3 shows a situation when this information alone is not enough, however.

Nevertheless, this type of problem can only happen when the new edge connects two vertices of degree 1 as shown below.

Lemma 1. *If at least one of the newly connected vertices has degree > 1, the information about relative order of incident edges suffice.*

Proof. Consider the situation shown in Fig. 4. (p, q) is the new edge. The graph is connected, and the path between p and q is shown. The initial direction of the edge (bold part) is determined by the ordering of edges around p. Then the edge can either go around r (shown in dashed red) or not (solid green). Obviously, the dashed red edge location is invalid, as it would surround r with a face, violating the requirement that each vertex belongs to the outer plane. Therefore, there is only one possibility for correctly drawing the edge. □

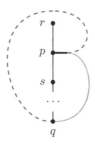

Fig. 4. Of the two possibilities of drawing new edge (p, q), only solid green is valid

---

Invariant: vertices are placed on the circle in the same order as they appear on the
  outer face of the drawing

Input: outerplanar drawing of graph G on a circle; $e = (p, q)$ – edge to be added

1: **if** q is a new vertex **then**
2:    place q on the circle according to ordering that includes e
3: **else**
4:    add a *virtual* arc e' connecting p and q according to the specification of e s.t.
      e' does not intersect any existing edge
5:    calculate order O of vertices on the outer plane (taking e' into account)
6:    move vertices into place on the circle according to O
7: **end if**
8: draw e

Algorithm 3. Adding new edge to outerplanar drawing of graph G

It follows that when the new edge connects two vertices of degree 1, additional
information (such as relative order of the vertices on the outer face) is necessary.

We present our streamed drawing algorithm for an outerplanar graph, G, in
terms of placing vertices of G on a circle. We will later derive an algorithm for
drawing G using grid points. We show in Algorithm 3 how to handle adding a
new edge to the graph.

As mentioned previously, maintaining the invariant guarantees planarity of
the drawing. We now show that vertices can move into place without causing
any intersections in the process.

Lemma 2. *Moving a vertex v inside the circle along a trajectory that does not
intersect any edge non-incident to v does not introduce any intersections and
maintains the order of edges around v.*

Proof. Consider the drawing with edges incident to v removed (marked with
dashed lines in Fig. 5). The face to which v belongs (limited by edges and circle
boundary) is the area where v can move. Because it is convex, as v moves, its
incident edges never intersect the boundaries of the area (other than at their
incident vertices), and the relative order of the edges stays the same.   □

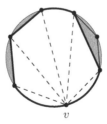

Fig. 5. Vertex v can move in the (convex) white area without causing intersections or
edge order changes

Lemma 3. *A vertex, v, that moves into its new position can do so without crossing any edges.*

Proof. Consider the face (in the sense of Lemma 2 and Fig. 5) of the drawing that v belongs to. The drawing is still outerplanar after adding the *virtual* arc (which is not necessarily straight-line), and therefore at least part of the circle, C', that forms this face's border still belongs to the outer face of the drawing (see vertex c in Fig. 6). By Lemma 2, v can move to C' without crossing any edges. When there are more vertices whose destination is the same part of the circle, C'', (vertices d, e and f in Fig. 6), they must form a path with inner vertices all having degree 2. After adding the new edge, their order on the circle (and hence on C'') is the reverse of their current order, so their (straight-line) trajectories do not cross. Since they form a path, edges between them will not intersect as they move into place.   □

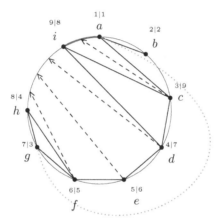

Fig. 6. Adding edge (a, g). *Virtual* arc is drawn in green dots. Part of the circle that lies in the outer plane and is reachable from c is shown in bold red. Numbers above vertices denote the order of vertex before and after the edge is added. Dashed lines are the trajectories of the vertices that need to move to maintain the invariant.

Corollary 1. *Algorithm 3 maintains an outerplanar drawing of a graph G as new edges are added to it.*

Extending Algorithm 3 to placing nodes on a grid is straightforward. Instead of a circle, we operate on a set of grid points in convex position that are approximately circular. We apply one of the algorithms for the file maintenance problem or the online list labeling problem for maintaining order of vertices in this set. When such an algorithm would move vertex v, we first check if there is an unused grid point between new neighbors of v on the circle. If so, we simply move the vertex to that point. Otherwise, we "reserve" the destination for v by inserting a stub vertex in the correct place (between new neighbors of v) on the circle. The list labeling algorithm will move vertices around the circle (without changing order on the circle, so it will not cause any intersections) to make room for this stub. Afterwards, we move v into its reserved position.

Lemma 4. *Vertex v is moved in line 6 of Algorithm 3 at most $deg(v) - 1$ times.*

Proof. A vertex v is moved when the new edge forms a *shortcut* that bypasses v on the outer face. v can appear at most $deg(v)$ times on the outer face, so after $deg(v) - 1$ moves, there will be only one valid position for v on the outer face, so it cannot be bypassed anymore. (See Fig. 7.) □

Fig. 7. Vertex v has degree 3. After two shortcuts (solid green lines) around v have been added, adding a third (dashed red line) would completely surround v, violating outerplanarity. Arrows show direction of edges on the outer plane.

Theorem 5. *The grid-based version of Algorithm 3 maintains an outerplanar drawing of a graph G and has the following update performances: uses $O(\log n)$ amortized moves per vertex, and*

1. *$O(n^3)$ area and $O(\log^2 n)$ vertex moves per edge insertion.*
2. *polynomial area and $O(\log n)$ vertex moves per edge insertion.*

In addition, each of the above complexity bounds applies in an amortized sense per vertex in the drawing.

Proof. By Corollary 1, we know that the algorithm maintains an outerplanar drawing of G. For the area bounds, the file maintenance algorithms requires $O(n)$ available integer tags (in this case, points in convex position) to handle n elements. Since m grid points in (strict) convex position require $O(m^3)$ area, the streamed drawing algorithm therefore uses $O(n^3)$ area in such cases. Likewise, it uses polynomial area when using a solution to the online list labeling problem.

By Lemma 4, each vertex v is moved by Algorithm 3 at most $deg(v) - 1$ times. Each such move requires at most one insertion into the list for the file maintenance or list maintenance algorithm, so there are at most $O(1)$ such insertions per vertex in the amortized sense (sum of degrees of all vertices in an outerplanar graph is $O(n)$). □

Note that we cannot immediately apply our results for bulk moves, unless we restrict our attention to possible vertex points that are uniformly distributed on a circle and moves that involve rotations of intervals of points around this circle.

6   Conclusion

We provided a revised approach to streamed graph drawing, utilizing solutions to the file maintenance problem, either on a level-by-level basis (for level drawings of trees), a cross-product basis (for tree-maps), or a circular/convex-position basis (for outerplanar graphs). For future work, it would be interesting to find other applications of this problem in streamed or dynamic graph drawing applications.

Acknowledgements. We would like to thank Alex Nicolau and Alex Veidenbaum for helpful discussions regarding the file maintenance problem. This work was supported in part by the NSF, under grants 1011840 and 1228639.

References

1. Archambault, D., Purchase, H.C.: Mental map preservation helps user orientation in dynamic graphs. In: Didimo, W., Patrignani, M. (eds.) GD 2012. LNCS, vol. 7704, pp. 475–486. Springer, Heidelberg (2013)
2. Bender, M.A., Cole, R., Demaine, E.D., Farach-Colton, M., Zito, J.: Two simplified algorithms for maintaining order in a list. In: Möhring, R.H., Raman, R. (eds.) ESA 2002. LNCS, vol. 2461, pp. 152–164. Springer, Heidelberg (2002)
3. Binucci, C., Brandes, U., Battista, G.D., Didimo, W., Gaertler, M., Palladino, P., Patrignani, M., Symvonis, A., Zweig, K.: Drawing trees in a streaming model. Information Processing Letters 112(11), 418–422 (2012)
4. Binucci, C., Brandes, U., Di Battista, G., Didimo, W., Gaertler, M., Palladino, P., Patrignani, M., Symvonis, A., Zweig, K.: Drawing trees in a streaming model. In: Eppstein, D., Gansner, E.R. (eds.) GD 2009. LNCS, vol. 5849, pp. 292–303. Springer, Heidelberg (2010)
5. Dietz, P., Sleator, D.: Two algorithms for maintaining order in a list. In: 19th ACM Symp. on Theory of Computing (STOC), pp. 365–372 (1987)
6. Eades, P., Lai, W., Misue, K., Sugiyama, K.: Preserving the mental map of a diagram. In: Proceedings of Compugraphics, pp. 24–33 (1991)
7. Goodrich, M.T., Pszona, P.: Cole's parametric search technique made practical. CoRR (arXiv ePrint), abs/1308.6711 (2013)
8. Kopelowitz, T.: On-line indexing for general alphabets via predecessor queries on subsets of an ordered list. In: FOCS, pp. 283–292. IEEE Computer Society (2012)
9. Misue, K., Eades, P., Lai, W., Sugiyama, K.: Layout adjustment and the mental map. Journal of Visual Languages & Computing 6(2), 183–210 (1995)
10. Shneiderman, B.: Tree visualization with tree-maps: 2-d space-filling approach. ACM Trans. Graph. 11(1), 92–99 (1992)
11. Tamassia, R.: Handbook of Graph Drawing and Visualization. Discrete Mathematics and Its Applications. Chapman & Hall/CRC (2007)
12. Willard, D.E.: Good worst-case algorithms for inserting and deleting records in dense sequential files. In: ACM SIGMOD, pp. 251–260 (1986)
13. Yantis, S.: Multielement visual tracking: Attention and perceptual organization. Cognitive Psychology 24(3), 295–340 (1992)

COAST: A Convex Optimization Approach to Stress-Based Embedding

Emden R. Gansner, Yifan Hu, and Shankar Krishnan

AT&T Labs - Research, Florham Park, NJ

Abstract. Visualizing graphs using virtual physical models is probably the most heavily used technique for drawing graphs in practice. There are many algorithms that are efficient and produce high-quality layouts. If one requires that the layout also respect a given set of non-uniform edge lengths, however, force-based approaches become problematic while energy-based layouts become intractable. In this paper, we propose a reformulation of the stress function into a two-part convex objective function to which we can apply semi-definite programming (SDP). We avoid the high computational cost associated with SDP by a novel, compact re-parameterization of the objective function using the eigenvectors of the graph Laplacian. This sparse representation makes our approach scalable. We provide experimental results to show that this method scales well and produces reasonable layouts while dealing with the edge length constraints.

1 Introduction

For visualizing general undirected graphs, algorithms based on virtual physical models are some of the most frequently used drawing methods. Among these, the spring-electrical model [7,8] treats edges as springs that pull nodes together, and nodes as electrically-charged entities that repel each other. Efficient and effective implementations [13,14,26] usually utilize a multilevel approach and fast force approximation with a suitable spatial data structure, and can scale to millions of vertices and edges while still producing high-quality layouts.

In certain instances, the graph may assign non-uniform lengths to its edges, and the layout problem will have the additional constraint of trying to match these lengths. A suitable formulation of the spring-electrical model that works well when edges have predefined target lengths is still an open problem.

In contrast, the (full) stress model assumes that there are springs connecting all vertex pairs of the graph. Assuming we have a graph $G = (V, E)$, with V the set of vertices and E the set of edges, the energy of this spring system is

$$\sum_{i,j \in V} w_{ij} \left(\left\| x_i - x_j \right\| - d_{ij} \right)^2, \tag{1}$$

where d_{ij} is the ideal distance between vertices i and j, and w_{ij} is a weight factor. The weight factor can modify the impact of an error. Weights can be arbitrary but are frequently taken as a negative power of d_{ij}, thus lessening the error for larger ideal distances. A layout that minimizes this stress energy is taken as an optimal layout of the

S. Wismath and A. Wolff (Eds.): GD 2013, LNCS 8242, pp. 268–279, 2013.

graph. The justification for this is clear: in most cases, it is not possible to find a drawing that respects all of the edge lengths, while expression (1) is basically the weighted mean square error of a drawing. (See also the work of Brandes and Pich [2].)

The stress model has its roots in multidimensional scaling (MDS) [19] which was eventually applied to graph drawing [16,20]. Note that typically we are given only the ideal distance between vertices that share an edge, which is taken to be unit length for graphs without predefined edge lengths. For other vertex pairs, a common practice is to define d_{ij} as the length of a shortest path between vertex i and j. Such a treatment, however, means that an all-pairs shortest path problem must be solved. Johnson's algorithm [15] takes $O(|V|^2 \log |V| + |V||E|)$ time, and $O(|V|^2)$ memory. (A slightly faster, but still quadratic, algorithm is also known [23].) For large graphs, such complexities make solving the full stress model infeasible.

A number of techniques have been proposed for circumventing this problem, typically focused on approximate solutions, using only a few computed distances, or approximating the shortest path calculations. Gansner et al. [12] proposed another approach for solving the "stress model" efficiently, by redefining the problem. The key idea was to note that only the edge distances are given, while using shortest path lengths for the remainder is somewhat arbitrary, and could be replaced with some other constraint that is faster to compute but still works in terms of layout quality. This led them to propose a two-part modified stress function

$$\sum_{\{i,j\} \in E} w_{ij} \left(\|x_i - x_j\| - d_{ij} \right)^2 - \alpha H(x), \tag{2}$$

where the first term encodes the stress associated with the given distances, and the second handles the remaining pairs.

In this paper, we also consider minimizing a two-part modified stress function. However, our formulation is such that the objective function is convex. More specifically, it is quartic in the positions of the nodes, and can be expressed as a quadratic function of auxiliary variables, where each of the auxiliary variables is a product of positions. We solve the problem by projecting the positions into a subspace spanned by the eigenvectors of the Laplacian, and transform the minimization problem into one of convex programming. We call our technique COAST (Convex Optimization Approach to STress-based emdedding).

The rest of the paper is organized as follows. In Section 2, we discuss related work. Section 3 gives the COAST model, and discusses a way to solve the model by semidefinite programming. Section 4 evaluates our algorithm experimentally by comparing it with some of the existing fast approximate stress models. Section 5 presents a summary and topics for further study.

2   Related Work

Most of the earlier approaches [24,10,1,17,12] for efficiently handling graph drawings with edge lengths relied on approximately minimizing the stress model, typically using some sparse model [10]. One notable effort is that of PivotMDS of Brandes and Pich

[1]. This is an approximation algorithm which only requires distance calculations from all nodes to a few chosen nodes.

While PivotMDS is very efficient and works well for some graphs, for graphs such as trees, multiple nodes can share the same position. Khoury et al. [17] approximate the solution of the linear system in a stress majorization procedure [10] by a low-rank singular value decomposition (SVD). They used a result of Drineas et al. [6] which states that for a matrix with well-distributed SVD values, the SVD values and left SVD vectors of the submatrix consisting of randomly sampled columns of the original matrix are a good approximation to the corresponding SVD values and vectors of the original matrix. With this result, they were able to calculate only the shortest paths from a selected number of nodes, as in PivotMDS. The method avoided having nodes in a tree-like graph being embedded into the same position by using the exact (instead of approximate) right-hand-side of the stress majorization procedure, using an observation that this right-hand-side can be calculated efficiently for the special case of $w_{ij} = 1/d_{ij}$.

The work most akin to that presented here is the maxent-stress model [12]. That approach borrows from the principle of maximal entropy, which says that items should be placed uniformly in the absence of constraints. The model tries to minimize the local stresses, while selecting a layout that maximizes the dispersion of nodes. This leads to the function shown in expression (2), where typically $H(x) = ln_{\{i,j\} \notin E} \|x_i - x_j\|$. The authors introduce an algorithm, called force-augmented stress majorization, to minimize this objective function.

Although it essentially ignores edge lengths, the binary stress model of Koren and Çivril [18] is stylistically related, in that the first term attempts to specify edge lengths (as uniformly 0) and the second term has the effect of uniformly spacing the nodes. Specifically, there is a distance of 0 between nodes sharing an edge, and a distance of 1 otherwise, giving the model

$$\sum_{\{i,j\} \in E} \|x_i - x_j\|^2 + \alpha \sum_{\{i,j\} \notin E} (\|x_i - x_j\| - 1)^2.$$

Similarly, Noack [21,22] has proposed the LinLog model and, more generally, the r-PolyLog model,

$$\sum_{\{i,j\} \in E} \|x_i - x_j\|^r - \sum_{i,j \in V} ln\|x_i - x_j\|,$$

where, in particular, the second term is suggestive of the use of entropy in the maxent-stress model.

The most significant attempt to use a force-directed approach for encoding edge distances was the GRIP algorithm [9]. The multilevel coarsening uses maximal independent set based filtration, with the length of an edge at a coarse level computed from lengths of its composite edges. On coarse levels, the algorithm uses a version of the Kamada-Kawai algorithm [16] applied to each node within a local neighborhood of the original graph, thus handling the relevant edge lengths. On the finest level, however, a localized Fruchterman-Reingold algorithm [8] is used, with no modeling of edge lengths.

In the area of data clustering, Chen and Buja [3] present LMDS, a model based on localized versions of MDS. Algebraically, this reduces to

$$\sum_{\{i,j\}\in S} \left(\|x_i - x_j\| - d_{ij} \right)^2 - t \sum_{(i,j)\notin S} \|x_i - x_j\|,$$

where S contains $\{i,j\}$ if node j is among the k nearest neighbors if i. It is difficult to determine how scalable this approach is but some tests indicate it is not appropriate for graph drawing.

3   The COAST Algorithm

Let $G = (V,E)$ denote an undirected graph, with the node set (vertices) V and edge set E. We use $n = |V|$ for the number of vertices in G. We assume that each edge (i,j) has a desired length d_{ij} with weight w_{ij}. Typically, one sets $w_{ij} = 1/d_{ij}^2$, but our analysis does not require that assumption. We wish to embed G into d-dimensional Euclidean space. Let x_i represent the coordinates of vertex i in $\mathbb{R}^d$, and let P be the $n \times d$ matrix whose rows are the x_i. We define the Gram matrix $X = (x_{ij})$ where $x_{ij} = x_i \cdot x_j$, the matrix of inner products. It is well known that X is a positive semi-definite matrix.

We consider minimizing a two-part modified stress function:

$$T(P) = \sum_{\{i,j\}\in E} (w_{ij}\|x_i - x_j\|^2 - w_{ij}d_{ij}^2)^2 - t\lambda \sum_{(i,j)\notin E} \|x_i - x_j\|^2, \tag{3}$$

where the first term attempts to assign edges their ideal edge lengths, and the second term separates unrelated nodes as much as possible. The parameter t can be used to balance the two terms, emphasizing either conformity to the specified edge lengths (small t) or uniform placement (large t). Without loss of generality, we can assume a zero mean for the x_i, i.e., $\sum_i x_i = 0$. We set $\lambda = |E|/ \left(\binom{n}{2} - |E| + 1 \right)$ to balance the relative size of the two terms, as suggested by Chen and Buja [3]. To minimize $T(P)$, let T_1 and T_2 be the first and second terms of T, respectively, so that $T = T_1 - T_2$, and consider the first term. We have the following derivation:

$$T_1 = \sum_{\{i,j\}\in E} \left\{ w_{ij}(x_{ii} - x_{ij} - x_{ji} + x_{jj}) - w_{ij}d_{ij}^2 \right\}^2$$

$$= \sum_{\{i,j\}\in E} \left\{ w_{ij}Tr(E_{ij}X) - w_{ij}d_{ij}^2 \right\}^2. \tag{4}$$

where $Tr()$ is the trace function and $E_{ij} = (e_{kl})$ is the $n \times n$ matrix with

$$e_{kl} = \begin{cases} 1, & \text{if } k = l = i \text{ or } k = l = j \\ -1, & \text{if } k = i \text{ and } l = j \\ -1, & \text{if } k = j \text{ and } l = i \\ 0, & \text{otherwise} \end{cases}$$

Using standard properties of the trace, the expression (4) can be rewritten as

$$\sum_{\{i,j\}\in E} w_{ij}^2 \{vec(E_{ij})^T \mathscr{X} - d_{ij}^2\}^2, \tag{5}$$

where $\mathscr{X} = vec(X)$ and $vec()$ is the matrix vectorization operator.

Functions defined on nodes of a graph can be well approximated by the eigenvectors of the graph Laplacian [4], and the smoother the function is, fewer eigenvectors are required to approximate it well. It is reasonable to assume that the function that embeds the vertices in $\mathbb{R}^d$ is smooth over the graph. Therefore, the bottom k eigenvectors of the graph's Laplacian provide a good sparse basis for the position vectors. Typical values of k range from 10-30 depending on the size of the graph. Let $Q \in \mathbb{R}^{n \times k}$ be the matrix composed of the eigenvectors of the Laplacian corresponding to the k smallest eigenvalues, ignoring the eigenvalue 0. It is well known that the eigenvector corresponding to eigenvalue 0 accounts for the center of mass of the function. Removing it from consideration automatically places the embedding at the origin. We can then find k vectors y_l in $\mathbb{R}^k$ so that we can write each x_i as $\sum_l q_{il} y_l$ where $q_i = (q_{i1}, q_{i2}, \ldots, q_{ik})$ is the ith row of Q. If we then define the $k \times k$ positive semi-definite matrix $Y = (y_{ij})$ where $y_{ij} = y_i \cdot y_j$, we have

$$X = PP^T = QYQ^T.$$

Using $\mathscr{X} = vec(X)$ and letting $\mathscr{Y} = vec(Y)$, we can rewrite the above as

$$\mathscr{X} = (Q \otimes Q)\mathscr{Y},$$

where $\otimes$ is the Kronecker product. Using this in expression (5), we have

$$T_1 = \sum_{\{i,j\}\in E} w_{ij}^2 \{vec(E_{ij})^T (Q \otimes Q)\mathscr{Y} - d_{ij}^2\}^2. \tag{6}$$

Since $x_i - x_j = \sum_l (q_{il} - q_{jl})y_l$, it is fairly straightforward to see that the following holds:

$$vec(E_{ij})^T (Q \otimes Q) = (q_i - q_j) \otimes (q_i - q_j).$$

Applying this to equation (6), we have

$$T_1 = \sum_{\{i,j\}\in E} w_{ij}^2 \{(q_i - q_j) \otimes (q_i - q_j)\mathscr{Y} - d_{ij}^2\}^2$$

$$= \sum_{\{i,j\}\in E} w_{ij}^2 \mathscr{Y}^T [((q_i - q_j) \otimes (q_i - q_j))^T ((q_i - q_j) \otimes (q_j - q_i))]\mathscr{Y} -$$

$$2 \sum_{\{i,j\}\in E} w_{ij}^2 d_{ij}^2 ((q_i - q_j) \otimes (q_i - q_j))\mathscr{Y} + \sum_{\{i,j\}\in E} w_{ij}^2 d_{ij}^4.$$

Now, turning to the second term of $T(P)$, we have

$$T_2 = t\lambda \sum_{(i,j)\notin E} \|x_i - x_j\|^2$$

$$= t\lambda \left\{ \sum_{i>j} \|x_i - x_j\|^2 - \sum_{\{i,j\}\in E} \|x_i - x_j\|^2 \right\}. \tag{7}$$

Lemma 1. $\sum_{i>j} \|x_i - x_j\|^2 = nTr(Y)$ and $\sum_{\{i,j\}\in E} \|x_i - x_j\|^2 = ((q_i - q_j) \otimes (q_i - q_j))\mathcal{Y}$.

Proof. Because the x_i have zero mean, the first summation is equal to $n\sum_i \|x_i\|^2 = nTr(X) = nTr(Y)$. $\qquad\qquad\qquad\qquad\qquad\qquad\qquad\qquad\qquad\qquad\qquad\qquad\qquad\square$

Using lemma 1, we can rewrite equation (7) as

$$T_2 = t\lambda\{nTr(Y) - ((q_i - q_j) \otimes (q_i - q_j))\mathcal{Y}\}$$
$$= t\lambda\{nvec(I)^T - \sum_{\{i,j\}\in E} ((q_i - q_j) \otimes (q_i - q_j))\}\mathcal{Y}.$$

Combining our recastings of the two terms of equation (3), we have:

$$T(P) = T_1 - T_2$$
$$= \mathcal{Y}^T \left[\sum_{\{i,j\}\in E} w_{ij}^2 \{((q_i - q_j) \otimes (q_i - q_j))^T ((q_i - q_j) \otimes (q_i - q_j))\} \right] \mathcal{Y} -$$
$$\left[\sum_{\{i,j\}\in E} (2w_{ij}^2 d_{ij}^2 - t\lambda)((q_i - q_j) \otimes (q_i - q_j)) - nt\lambda vec(I)^T \right] \mathcal{Y} +$$
$$\sum_{\{i,j\}\in E} w_{ij}^2 d_{ij}^4.$$

To simplify the exposition, we can write $T(P)$ as $\mathcal{Y}^T A\mathcal{Y} + \mathbf{b}^T \mathcal{Y} + \text{constant}$. Since A and Y are symmetric positive semi-definite matrices, this is a convex function inside the semi-definite cone. It can be solved easily by any off-the-shelf semi-definite program (SDP). SDP is usually inefficient, taking cubic time in the size of the variables and constraints. A key novelty in our approach is the use of the approximation using the graph Laplacian. Instead of minimizing with n^2 variables, our re-parameterization with Y reduces the number of variables to k^2. This is usually constant for most graphs and hence makes our approach scalable. Because of the special structure of our problem, we can further improve the running time by converting our quadratically-constrained SDP to a Semidefinite Quadratic Linear Program (SQLP) and use a specialized solver like SDPT3 [25]. Details of this conversion are given in the report [11].

4 Experimental Results

We implemented the COAST algorithm in a combination of Python, Matlab and C code. The main parts consist of forming the matrix A and vector b, calculating the eigenvectors of the Laplacian, and solving the optimization problem. Time for the last part is dependent only on the number of eigenvectors k, hence is constant for a fixed number of eigenvectors. For graphs of size up to 100,000, the minimization using SQLP takes less than 10 seconds inside Matlab.

We tested the COAST algorithm for solving the quartic stress model on a range of graphs. For comparison, we also tested PivotMDS; PivotMDS(1), which uses Pivot-MDS, followed by a sparse stress majorization; the maxent-stress model Maxent; and

Table 1. Algorithms tested

Algorithm	Model	Fits distances?
COAST	quartic stress model	Yes. Edges only
PivotMDS	approx. strain model	Yes/No
PivotMDS(1)	PivotMDS + sparse stress	Yes.
Maxent	PivotMDS + maxent-stress	Yes.
FSM	full stress model	Yes. All-pairs

the full stress model, using stress majorization. We summarize all the tested algorithms in Table 1.

With the exception of graph gd, which is an author collaboration graph of the International Symposium on Graph Drawing between 1994-2007, the graphs used are from the University of Florida Sparse Matrix Collection [5]. Our selection is exactly the same as that used by Gansner et al. [12]. Two of the graphs (commanche and luxembourg) have associated pre-defined non-unit edge lengths. In our study, a rectangular matrix, or one with an asymmetric pattern, is treated as a bipartite graph. Test graph sizes are given in Table 2.

Table 2. Test graphs. Graphs marked * have pre-specified non-unit edge lengths. Otherwise, unit edge length is assumed.

Graph	$|V|$	$|E|$	description
gd	464	1311	Collaboration graph
btree	1023	1022	Binary tree
1138_bus	1138	1358	Power system
qh882	1764	3354	Quebec hydro power
lp_ship041	2526	6380	Linear programming
USpowerGrid	4941	6594	US power grid
commanche*	7920	11880	Helicopter
bcsstk31	35586	572913	Automobile component
luxembourg*	114599	119666	Luxembourg street map

Tables 3 and 4 present the outcomes for two graphs. (The drawings for all of the graphs tested, with additional color detail, can be found in the companion report [11].) Following Brandes and Pich [2], each drawing has an associated error chart. In an error chart, the x-axis gives the graph distance bins, the y-axis is the difference between the actual geometric distance in the layout and the graph distance. The chart shows the median (black line), the 25 and 75 percentiles (gray band) and the min/max errors (gray lines) that fall within each bin. For ease of understanding, we plot graph distance against distance error, instead of graph distance vs. actual distance as suggested by Brandes and Pich [2]. Because generating the error chart requires an all-pairs shortest paths calculation, we provide this chart only for graphs with less than 10,000 nodes.

Table 3. Drawings and error charts of the tested algorithms for `btree`

PivotMDS	PivotMDS(1)	Maxent	COAST	FSM

With the error chart, we also include a graph distance distribution curve (line with dots), representing the number of vertex pairs in each graph distance bin. This distribution depends on the graph, and is independent of the drawing. In making the error charts, the layout is scaled to minimize the full stress model (1), with $w_{ij} = 1/d_{ij}^2$.

As an example, the error chart for PivotMDS on `btree` (Table 3, column 1, bottom) shows that, on average, the median line is under the x-axis for small graph distances. This means that the PivotMDS layout under-represents the graph distance between vertex pairs that are a few hops away. This is because it collapses branches of tree-like structures. The leaves of such structures tend to be a few hops away, but are now positioned very near to each other. To some extent the same under-representation of graph distance for vertex pairs that are a few hops away is seen for PivotMDS and PivotMDS(1) on other non-rigid graphs, including `1138_bus`, `btree`, `lp_ship041` and `USpowerGrid`. Compared with PivotMDS and PivotMDS(1), the median line for Maxent (column 3) does not undershoot the x-axes as much.

Comparing the COAST layouts with the others, we note that it appears to track the x-axis more tightly and uniformly than the others, except for large lengths where, in certain cases, it dives significantly. In general, COAST has a more consistent bias for under-representation than the other layouts. The others tend to under-represent short lengths and over-represent long lengths. Visually, most of the COAST layouts are satisfactory, certainly avoiding the limitations of PivotMDS. For example, although it does not capture the symmetry of `btree` as well as Maxent, it does a better job of handling the details.

While visually comparing drawings made by different algorithms is informative, and may give an overall impression of the characteristics of each algorithm, such inspection is subjective. Ideally we would prefer to rely on a quantitative measure of performance. However such a measure is not easy to devise. For example, if we use sparse stress as our measure, PivotMDS, which minimizes sparse stress, is likely to come out best,

Table 4. Drawings and error charts of the tested algorithms for lp_ship041

PivotMDS	PivotMDS(1)	Maxent	COAST	FSM

despite its shortcomings. As a compromise, we propose to measure full stress, as defined by (1), with $w_{ij} = 1/d_{ij}^2$. Bear in mind that this measure naturally favors the full stress model.

Table 5 gives the full stress measure achieved by each algorithm, as well as the corresponding timings. Because it is expensive to calculate all-pairs shortest paths, we restrict experimental measurement to graphs with less than 10,000 nodes. From the table we can see that, as expected, FSM is the best, because it tries to optimize this measure. We note that COAST is mostly competitive with the other non-FSM layouts.

As for timings, COAST, although a hybrid implementation, is comparable with Maxent, and appears to work well on large graphs.

4.1   Measuring Precision of Neighborhood Preservation

Sometimes, in embedding high dimensional data into a lower dimension, one is interested in preserving the neighborhood structure. In such a situation, exact replication of distances between objects becomes a secondary concern.

For example, imagine a graph where each node is a movie. Based on some recommender algorithm, an edge is added between two movies if the algorithm predicts that a user who likes one movie would also like the other, with the length of the edge defined as the distance (dissimilarity) between the two movies. The graph is sparse because only movies that are strongly similar are connected by an edge. For a visualization of this data to be helpful, we need to embed this graph in 2D in such a way that, for each node (movie), nodes in its neighborhood in the layout are very likely to be similar to this node. This would allow the user to explore movies that are more likely of interest to her by examining, in the visualization, the neighborhoods of the movies she knew and liked.

Table 5. Full stress measure ($\times 1000$) and CPU time (in seconds) for PivotMDS, PivotMDS(1), Maxent, COAST and FSM. Smaller is better. We limit the measurements to graphs with less than 10,000 nodes and 10 hours of CPU time. A "-" is used to denote these missing data points.

Graph	PivotMDS		PivotMDS(1)		Maxent		COAST		FSM	
gd	19	0.3	15	0.3	12	0.8	13	4.6	10	2.3
btree	130	1.1	110	1.1	64	2.7	89	0.4	60	10.0
1138_bus	78	0.1	67	0.2	45	2.1	58	3.4	40	16.0
qh882	147	0.1	120	0.3	103	2.2	184	2.7	84	39.0
lp_ship04l	667	0.1	769	0.1	363	2.2	368	5.0	251	58.0
USpowerGrid	1124	0.1	932	0.9	1018	6.5	1073	5.3	702	272.0
commanche	2305	0.2	1547	0.9	1545	9.0	2853	8.8	654	1025.0
bcsstk31	-	2.4	-	21.6	-	102.0	-	226.7	-	-
luxembourg	-	2.4	-	630.0	-	209.0	-	128.9	-	-

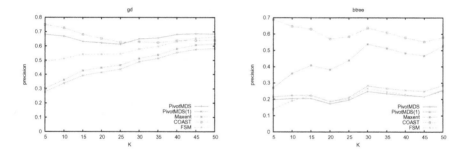

Fig. 1. Precision of neighborhood preservation of the algorithms, as a function of K. The higher the precision, the better.

Following Gansner et al. [12], we look at the *precision of neighborhood preservation*. We are interested in answering the question: if we see vertices nearby in the embedding, how many of these are actually also neighbors in the graph space? We define the precision of neighborhood preservation as follows. For each vertex i, K neighboring vertices of i in the layout are chosen. These K vertices are then checked to see if their graph distance is less than a threshold $d(K)$, where $d(K)$ is the distance of the K-th closest vertex to i in the graph space. The percentage of the K vertices that are within the threshold, averaged over all vertices i, is taken as the precision. Note that precision (the fraction of retrieved instances that are relevant) is a well-known concept in information retrieval. Chen and Buja [3] use a similar concept called *LC meta-criteria*.

Figure 1 gives the precision as a function of K for two representative graphs. (Figures for the remaining are available in the report [11].) From the figure, it is seen that, in general, COAST has the highest, or nearly the highest, precision. PivotMDS(1) tends to have low precision. The precision of other algorithms, including Maxent, tends to be between these two extremes.

Overall, precision of neighborhood preservation is a way to look at one aspect of embedding not well-captured by the full stress objective function, but is important to applications such as recommendations. COAST performs well in this respect.

5   Conclusion and Future Work

In this paper, we described a new technique for graph layout that attempts to satisfy edge length constraints. This technique uses a modified two-part stress function, one part for the edge lengths, the other to guide the relative placement of other node pairs. The stress is quartic in the positions of the nodes, and can be transformed to a form that is suitable for solution using convex programming. The results produced are good and the algorithm is scalable to large graphs.

Although the performance of the COAST algorithm is already competitive, we rely on an *ad hoc* implementation using a combination of Python, Matlab and C code. It would be very desirable to re-implement the algorithm entirely in C.

Our technique follows the general strategy of doing length-sensitive drawings for large graphs by reformulating the energy function, keeping the core length constraints, and then applying some appropriate mathematical machinery. Variations of this technique have been successfully used by others [17,12]. It would be interesting to explore additional adaptations of this approach.

Acknowledgements. We would like to thank the reviewers for their helpful comments.

References

1. Brandes, U., Pich, C.: Eigensolver methods for progressive multidimensional scaling of large data. In: Kaufmann, M., Wagner, D. (eds.) GD 2006. LNCS, vol. 4372, pp. 42–53. Springer, Heidelberg (2007)
2. Brandes, U., Pich, C.: An experimental study on distance-based graph drawing. In: Tollis, I.G., Patrignani, M. (eds.) GD 2008. LNCS, vol. 5417, pp. 218–229. Springer, Heidelberg (2009)
3. Chen, L., Buja, A.: Local multidimensional scaling for nonlinear dimension reduction, graph drawing, and proximity analysis. J. Amer. Statistical Assoc. 104, 209–219 (2009)
4. Chung, F.R.K.: Spectral Graph Theory (CBMS Regional Conference Series in Mathematics, No. 92). American Mathematical Society, Providene (1996)
5. Davis, T.A., Hu, Y.: U. of Florida Sparse Matrix Collection. ACM Transaction on Mathematical Software 38, 1–18 (2011),
   http://www.cise.ufl.edu/research/sparse/matrices/
6. Drineas, P., Frieze, A.M., Kannan, R., Vempala, S., Vinay, V.: Clustering large graphs via the singular value decomposition. Machine Learning 56, 9–33 (2004)
7. Eades, P.: A heuristic for graph drawing. Congressus Numerantium 42, 149–160 (1984)
8. Fruchterman, T.M.J., Reingold, E.M.: Graph drawing by force directed placement. Software - Practice and Experience 21, 1129–1164 (1991)
9. Gajer, P., Goodrich, M.T., Kobourov, S.G.: A multi-dimensional approach to force-directed layouts of large graphs. In: Marks, J. (ed.) GD 2000. LNCS, vol. 1984, pp. 211–221. Springer, Heidelberg (2001)

10. Gansner, E.R., Koren, Y., North, S.C.: Graph drawing by stress majorization. In: Pach, J. (ed.) GD 2004. LNCS, vol. 3383, pp. 239–250. Springer, Heidelberg (2005)
11. Gansner, E.R., Hu, Y., Krishnan, S.: COAST: A convex optimization approach to stress-based embedding (2013), http://arxiv.org/abs/1308.5218
12. Gansner, E.R., Hu, Y., North, S.C.: A maxent-stress model for graph layout. IEEE Trans. Vis. Comput. Graph. 19(6), 927–940 (2013)
13. Hachul, S., Jünger, M.: Drawing large graphs with a potential-field-based multilevel algorithm. In: Pach, J. (ed.) GD 2004. LNCS, vol. 3383, pp. 285–295. Springer, Heidelberg (2005)
14. Hu, Y.: Efficient and high quality force-directed graph drawing. Mathematica Journal 10, 37–71 (2005)
15. Johnson, D.B.: Efficient algorithms for shortest paths in sparse networks. J. ACM 24(1), 1–13 (1977)
16. Kamada, T., Kawai, S.: An algorithm for drawing general undirected graphs. Information Processing Letters 31, 7–15 (1989)
17. Khoury, M., Hu, Y., Krishnan, S., Scheidegger, C.: Drawing large graphs by low-rank stress majorization. Computer Graphics Forum 31(3), 975–984 (2012)
18. Koren, Y., Çivril, A.: The binary stress model for graph drawing. In: Tollis, I.G., Patrignani, M. (eds.) GD 2008. LNCS, vol. 5417, pp. 193–205. Springer, Heidelberg (2009)
19. Kruskal, J.B.: Multidimensioal scaling by optimizing goodness of fit to a nonmetric hypothesis. Psychometrika 29, 1–27 (1964)
20. Kruskal, J.B., Seery, J.B.: Designing network diagrams. In: Proc. First General Conference on Social Graphics, pp. 22–50. U. S. Department of the Census, Washington, D.C. (July 1980), Bell Laboratories Technical Report No. 49
21. Noack, A.: Energy models for graph clustering. J. Graph Algorithms and Applications 11(2), 453–480 (2007)
22. Noack, A.: Modularity clustering is force-directed layout. Physical Review E 79, 026102 (2009)
23. Pettie, S.: A new approach to all-pairs shortest paths on real-weighted graphs. Theoretical Computer Science 312(1), 47–74 (2004)
24. de Silva, V., Tenenbaum, J.B.: Global versus local methods in nonlinear dimensionality reduction. In: Advances in Neural Information Processing Systems 15, pp. 721–728. MIT Press, Cambridge (2003)
25. Tütüncü, R.H., Toh, K.C., Todd, M.J.: Solving semidefinite-quadratic-linear programs using SDPT3. Mathematical Programming 95(2), 189–217 (2003)
26. Walshaw, C.: A multilevel algorithm for force-directed graph drawing. J. Graph Algorithms and Applications 7, 253–285 (2003)

Colored Spanning Graphs for Set Visualization*

Ferran Hurtado[1], Matias Korman[1], Marc van Kreveld[2], Maarten Löffler[2],
Vera Sacristán[1], Rodrigo I. Silveira[4,1], and Bettina Speckmann[3]

[1] Dept. de Matemàtica Aplicada II, Universitat Politècnica de Catalunya, Spain
{ferran.hurtado,matias.korman,vera.sacristan}@upc.edu
[2] Dept. of Computing and Information Sciences, Utrecht University, the Netherlands
{m.loffler,m.j.vankreveld}@uu.nl
[3] Dept. of Mathematics and Computer Science,
Technical University Eindhoven, The Netherlands
speckman@win.tue.nl
[4] Dept. de Matemática, Universidade de Aveiro, Portugal
rodrigo.silveira@ua.pt

Abstract. We study an algorithmic problem that is motivated by ink minimization for sparse set visualizations. Our input is a set of points in the plane which
are either blue, red, or purple. Blue points belong exclusively to the blue set, red
points belong exclusively to the red set, and purple points belong to both sets.
A *red-blue-purple spanning graph* (RBP spanning graph) is a set of edges connecting the points such that the subgraph induced by the red and purple points is
connected, and the subgraph induced by the blue and purple points is connected.

We study the geometric properties of minimum RBP spanning graphs and the
algorithmic problems associated with computing them. Specifically, we show that
the general problem is NP-hard. Hence we give an $(\frac{1}{2}\rho+1)$-approximation, where
ρ is the Steiner ratio. We also present efficient exact solutions if the points are
located on a line or a circle. Finally we consider extensions to more than two sets.

1 Introduction

Visualizing sets and their elements is a recurring theme in information visualization.
Sets arise in many application areas, as varied as social network analysis (grouping individuals into communities), linguistics (related words), or geography (related places).
Among the oldest representations for sets are Venn diagrams [11] and their generalization, Euler diagrams. These representations are natural and effective for a small number
of elements and sets. However, for larger numbers of sets and more complicated intersection patterns more intricate solutions are necessary. The last years have seen a steady

* M.L. was supported by the Netherlands Organisation for Scientific Research (NWO) under
grant 639.021.123. F. H., M. K., V. S. and R.I. S. were partially supported by ESF EURO-
CORES programme EuroGIGA, CRP ComPoSe: grant EUI-EURC-2011-4306, and by project
MINECO MTM2012-30951. F. H., V. S. and R.I. S. were supported by project Gen. Cat. DGR
2009SGR1040. M. K. was supported by the Secretary for Universities and Research of the Ministry of Economy and Knowledge of the Government of Catalonia and the European Union.
R. I. S. was funded by the FP7 Marie Curie Actions Individual Fellowship PIEF-GA-2009-
251235 and by FCT through grant SFRH/BPD/88455/2012.

S. Wismath and A. Wolff (Eds.): GD 2013, LNCS 8242, pp. 280–291, 2013.
© Springer International Publishing Switzerland 2013

stream of developments in this direction, both for the situation where the location of set elements can be freely chosen and for the important special case that elements have to be drawn at particular fixed positions (for example, restaurant locations on a city map).

Our paper is motivated by some recent approaches that use very sparse enclosing shapes when depicting sets. LineSets [3] are the most minimal of all, reducing the geometry to a single continuous line per set which connects all elements. Both Kelp Diagrams [10] and its successor KelpFusion [15] are based on sparse spanning graphs, essentially variations of minimal spanning trees for different distance measures. These methods attempt to reduce visual clutter by reducing the amount of "ink" (see Tufte's rule [23]) necessary to connect all elements of a set. However, although the results are visually pleasing, neither method does use the optimal amount of ink. In this paper we explore the algorithmic questions that arise when computing spanning graphs for set visualization which are optimal with respect to ink usage.

Problem Statement. Our input is a set of n points in the plane. Each point is a member of one or more sets. We mostly consider the case where there are exactly two sets, namely a red and a blue set. A point is red if it is part only of the red set and it is blue if it is part only of the blue set. A point that is part of both the red and the blue set is purple.

A *red-blue-purple spanning graph* (RBP spanning graph) for a set of points that are red, blue and purple is a set of edges connecting the points such that the subgraph induced by the red and purple points is connected, and the subgraph induced by the blue and purple points is connected. A *minimum RBP spanning graph* for a set of points that are red, blue and purple is a red-blue-purple spanning graph that has minimum weight (total edge length) (see Figure 1). In this paper we consider the algorithmic problems associated with computing minimum RBP spanning graphs.

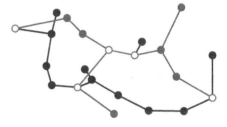

Fig. 1. A minimum RBP spanning graph

Results and Organization. We first review related work. In Section 2 we describe and prove various useful properties of (minimum) RBP spanning graphs. Then, in Section 3, we consider the two special cases where the points are located on a line or on a circle. This setting is meaningful if the elements of the sets are not associated with a specific location (for example, social networks or software systems). Here visualizations frequently choose to arrange elements in simple configurations such as lines or circles. We give an $O(n)$ time algorithm for points on a line, assuming that the input is already sorted. For points on a circle we exploit a structural result which allows us to use dynamic programming in $O(k^3 + n)$ time, where k is the number of purple points. In Section 4 we prove that computing a minimum RBP spanning graph is NP-hard in general. Hence, in Section 5 we turn to approximations. We describe an $O(n \log n)$ algorithm that computes a $(\frac{1}{2}\rho + 1)$-approximation of the minimum RBP spanning graph, where ρ is the Steiner ratio. Finally, in Section 6 we discuss various extensions for situations with more than two sets. Due to space constraints some proofs have been deferred to the full version.

Related Work. In recent years a number of papers explored the problem of automatically drawing Euler diagrams, for example, Simonetto and Auber [20], Stapleton *et al.* [21], and Henry Riche and Dwyer [13]. These methods assume that the locations of the set elements can be freely chosen. An important variation is the case that elements have to be drawn at fixed positions. Collins *et al.* [9] presented *Bubble Sets*, a method based on isocontours. A similar approach was suggested by Byelas and Telea [7]. *LineSets* by Alper *et al.* [3] attempt to improve the overall readability by the minimalist approach of drawing a single line per set. Dinkla *et al.* [10] introduced *Kelp Diagrams* which use a sparse spanning graph, essentially a minimum spanning tree with some additional edges. Kelp Diagrams were extended by Meulemans *et al.* [15] to a hybrid technique named *KelpFusion* which uses a mix of hulls and lines to generate fitted boundaries.

Sets defined over points in the plane can be interpreted as an embedding of a *hypergraph* where the points are vertices and each set is a hyperedge connecting an arbitrary number of vertices. Drawings of hypergraphs have been discussed in several papers (e.g., see Brandes *et al.* [5] and references therein).

Also in the area of discrete and computational geometry many problems on colored point sets have been studied. Possibly the most famous example is the *Ham-Sandwich Theorem*: given a set of $2n$ red points and $2m$ blue points in general position in the plane, there is always a line ℓ such that each open halfplane bounded by ℓ contains exactly n red points and m blue points. There have been many variations on this theorem and also many other results on finding configurations or geometric graphs with constraints depending on colors. We refer the interested reader to the survey [14] and to Chapter 8 in the collection of research problems [6].

From an algorithmic point of view, many problems have been considered, here we mention only a few of them. The *bichromatic closest pair* (e.g., see Preparata and Shamos [19] Section 5.7), the *chromatic nearest neighbor search* (see Mount et al. [18]), the problems on finding *smallest color-spanning objects* (see Abellanas *et al.* [1]), the *colored range searching* problems (see Agarwal *et al.* [2]), and the *group Steiner tree* problem where, for a graph with colored vertices, the objective is to find a minimum weight subtree that covers all colors (see Mitchell [16], Section 7.1). Finally, Tokunaga [22] considers a set of red or blue points in the plane and computes two geometric spanning trees of the blue and the red points such that they intersect in as few points as possible.

2   Properties of RBP Spanning Graphs

We call an edge of a RBP spanning graph *red* if it connects two red points or a red and a purple point. We call an edge *blue* if it connects two blue points or a blue and a purple point. We call an edge *purple* if it connects two purple points. A minimum RBP spanning graph does not contain edges between a red and a blue point. The subgraph induced by the red and purple points in a minimum RBP spanning graph is a spanning tree (and the analogous statement holds for the blue and purple points). Figure 2 (a) illustrates the trade-off between red, blue, and purple edges in a minimum RBP spanning graph.

Every red edge in a minimum RBP spanning graph also occurs in a minimum spanning tree of only the red and purple points. The corresponding statement is true also

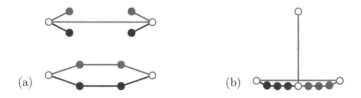

Fig. 2. (a) Two examples of minimum RBP spanning graphs of similar configurations of points. (b) A minimum RBP spanning graph with a purple edge crossing.

for the blue edges, but not for the purple edges. That is, there can be purple edges in a minimum RBP spanning graph which do not occur in a minimum spanning tree of only the purple points.

It is easy to see that a minimum RBP spanning graph is not necessarily planar. Red and blue edges are mostly independent and they can easily cross. Moreover, a red and a purple edge can cross, a blue and a purple edge can cross, and even two purple edges can cross in a minimum RBP spanning graph, as shown in Figure 2 (b). In fact, a single purple edge can cross arbitrarily many purple edges. The intricate construction and the additional observations necessary to prove this are relegated to the full version.

Lemma 1. *A purple edge in an optimal RBP spanning graph can cross $\Theta(n)$ other purple edges.*

Just as with standard minimum spanning trees the degree of the points in a minimum RBP spanning graph is bounded.

Lemma 2. *The maximum degree of a point in a minimum RBP spanning graph is at most 18.*

This bound can be attained by a purple point p: Let p be the center of a regular hexagon with radius 3 and two more regular hexagons with radius 1, one slightly rotated. Place a purple point on each corner of the large hexagon, place red points on the corners of one of the smaller hexagons, and blue points on the corners of the other one. Then the star graph with p at the center is a minimum RBP spanning graph. A similar construction shows that there is a point set such that the unique minimum RBP spanning graph requires a point of degree 15; a higher degree is never necessary. A red or blue point can have degree at most 6, degree 5 is the highest degree that can be enforced.

3   Points on a Line or on a Circle

Here we describe efficient algorithms to find a minimum RBP spanning graph if the points lie on a line or on a circle. In the circle case, we first present additional geometric observations which allow us to use dynamic programming.

3.1   Points on a Line

Given a problem instance S, we number the purple points $p_1, \ldots, p_k$ from left to right. For any $1 \leq i \leq k - 1$, let S_i be the set of points between p_i and p_{i+1} (including both

Fig. 3. A minimum RBP spanning graph of points on a line, and its partition into sets S_i (some edges are depicted by curved arcs for clarity).

p_i and p_{i+1}). We also define S_0 to contain p_1 and all red/blue points to its left, and S_k to contain p_k and all red/blue points to its right (see Figure 3). First, we show that each subset can be treated independently.

Lemma 3. *Let S be a set of red, blue and purple points located on a line, and let G^* be a minimum RBP spanning graph of S. Then for any edge $qq' \in G^*$, the points q and q' are contained in S_j, for some j.*

Using this lemma it is straightforward to obtain an efficient algorithm.

Theorem 1. *Let S be a set of n red, blue and purple points located on a line. We can compute a minimum RBP spanning graph of S in $O(n)$ time, provided that the points are sorted along the line.*

3.2   Points on a Circle

We proved in Lemma 1 that for points in general position a purple edge can cross many other purple edges. Even if the points lie in convex position, purple edges can cross each other (see Figure 2 (right)). But below we prove that for points on a circle the situation is structurally different and purple edges cannot cross any other edges.

Lemma 4. *Let S be a set of red, blue and purple points located on a circle. A minimum RBP spanning graph of S cannot have a purple edge crossing any other edge.*

Proof. Let G be a minimum RBP spanning graph in which two edges $e_1 = vv'$ and $e_2 = ww'$ cross. We will perform a local transformation that will reduce the weight of G, contradicting the minimality of G.

First assume that both e_1 and e_2 are purple, and consider the four red paths in G that start at either v or v' and end at w or w'. Without loss of generality, we can assume that the minimum-link path among the four (i.e., the path with smallest number of edges) connects v and w. Let π_R be such path; note that π_R cannot use edge vv' nor edge ww'. Likewise, let π_B be the minimum-link blue path among those that connect v or v' with w or w'. We now distinguish three cases depending on the number of shared endpoints between π_R and π_B (see Figure 4):

π_R **and** π_B **share both endpoints.** We replace edges vv' and ww' with edges vw' and $v'w$. The resulting graph G' clearly has smaller weight than G. We now prove that G' is indeed spanning. First consider the red tree in G': the removal of edge ww' created two components. Moreover, points v and w' must belong to different components (otherwise, the edge ww' would create a cycle in G). Thus, by adding

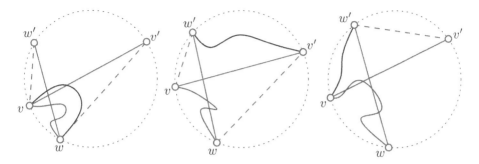

Fig. 4. The three cases in the proof of Lemma 4, local transformations are shown by dashed purple edges. No assumptions are made on the number of crossings between π_R, π_B, vv', and ww'.

the edge vw' we reconnect the two components again. Likewise, the removal of edge vv' creates two red components that are reconnected with the edge ww'. That is, graph G' also spans red. We repeat the same reasoning for blue and obtain that G' is a RBP spanning graph with smaller weight than G, a contradiction.

π_R **and** π_B **share no endpoints.** We proceed as in the previous case, replacing edges vv' and ww' by vw' and $v'w$. The argumentation is identical to the previous case.

π_R **and** π_B **share one endpoint.** We can assume that v is the shared endpoint, and that the other endpoint of π_B is w' (see Figure 4, right). In this case, both red and blue paths from v' to both w and w' in G use the edge vv'. Then we can replace this edge by either $v'w$ or $v'w'$ and maintain the spanning property. Using the fact that the four vertices are on the boundary of a circle, it is easy to see that either $||v'w|| < ||v'v||$ or $||v'w'|| < ||v'v||$ must hold, thus one of the two resulting graphs will have smaller weight.

If one of the edges is not purple the situation is easier, since we need to consider only one color. We assume that the edge e_2 is red, and that the path from v to w does not use e_1 nor e_2. Then we can replace the edge ww' by either vw' or $v'w'$ to obtain a graph of smaller weight. Note that, since we are changing a red edge, the spanning property of blue cannot be altered, and the lemma is shown. $\qquad\square$

Next we present another crossing property that will be useful for our algorithm.

Corollary 1. *Let S be a set of red, blue and purple points located on the boundary of a circle. In a minimum RBP spanning graph G of S, no red or blue edge of G can cross a segment between two purple points.*

Proof. Let p, p' be two purple points, and assume that a red edge $rr' \in G$ crosses the segment pp'. As in the proof of Lemma 4, we can assume that the red path from p to r does not pass through neither p' nor r'. Then, we replace the edge rr' by either $r'p$ or $r'p'$ to obtain a RBP spanning graph of smaller weight. $\qquad\square$

Using these geometric observations and a few others proved in the full version we can now compute a minimum RBP spanning graph with dynamic programming.

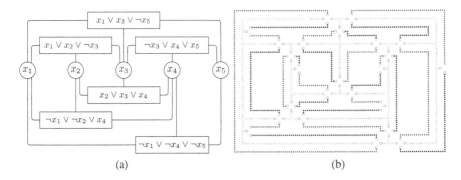

(a)   (b)

Fig. 5. (a) A planar 3-SAT formula. (b) The corresponding set of red, blue, and purple points.

Theorem 2. *Let S be a set of n red, blue and purple points located on a circle. We can compute a minimum RBP spanning graph of S in $O(k^3 + n)$ time, where k is the number of purple points.*

4   NP-hardness

Computing a minimum RBP spanning graph is NP-hard. We prove this by a reduction from planar 3-SAT. Figure 5 illustrates the global construction: an embedded 3-SAT formula, and the corresponding set of red, blue, and purple points that we construct. There is a value W such that a solution of weight less than W exists if and only if the 3-SAT formula is satisfiable.

The construction consists of two parts. First, we have a *variable gadget* for each variable x_i in the 3-SAT formula. Such a gadget consists of a purple *skeleton* and a red/blue *loop* around the skeleton. The loop consists of alternating densely-sampled blue and red chains, separated by a pair of purple points at distance $\sqrt{2}$ from each other. For each such pair of purple points there is also another purple point at distance 1 from both points, which is connected to the skeleton. We call such a group of 3 purple points a *switch*. Figure 6 (a) shows an example.

First, we observe that if the red, blue and purple chains are sampled sufficiently densely, then they will be connected in any optimal solution; hence, we ignore the costs of these connections from now on. There are exactly two ways to optimally connect the

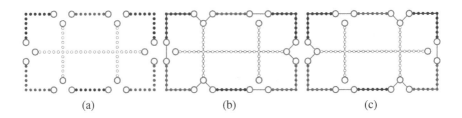

(a)   (b)   (c)

Fig. 6. (a) The input for a variable gadget. (b) One possible optimal solution of this construction, which represents the value *true*. (c) The other optimal solution represents the value *false*.

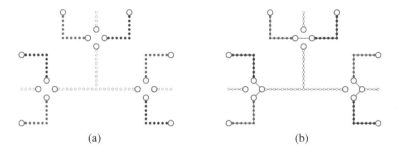

Fig. 7. (a) The input for a clause gadget. (b) One possible solution where the clause is satisfied. The left and top literals are in their *true* states; the right literal is in its *false* state.

remaining components: we alternatingly connect the purple points on opposite sides of a switch to each other, or both to the skeleton. Figures 6(b) and 6(c) illustrate the two solutions. These will correspond to the values *true* and *false* of the variable.

Next, we have *clause gadgets* for all clauses of the 3-SAT formula. These occur at places where the variable loops of the three involved variables get close to each other, and we make sure that there is a switch in each of them (from red to blue if the variable occurs positively in the clause, from blue to red otherwise). For these switches, we place a fourth point, also at distance 1 from both ends of the switch, and we connect these three extra points by a clause skeleton. Figure 7(a) shows an example.

Lemma 5. *There is a value W such that an RBP spanning graph of length less than W exists if and only if the 3-SAT formula is satisfiable.*

Proof. Suppose the planar 3-SAT formula has n variables and m clauses, and let k_i be the number of clauses that variable x_i appears in. There are $2k_i$ switches per variable. The total number of switches is $2\sum_{i=1}^{n} k_i = 6m$. We ignore the red, blue and purple chains; we only argue about the total length of purple edges within the switches.

Within variable x_i, there is one skeleton, k_i blue pieces, and k_i red pieces, which means that there are $3k_i + 1$ components in either the red or the blue tree that need to be connected. For this, we need to add $3k_i$ edges. Within each switch, the purple points that are connected to the red or blue paths must certainly get a purple edge. Since we add only $3k_i$ edges, half of the switches get only one edge; these edges must then connect the blue-path purple point to the red-path purple point and have length $\sqrt{2}$. The other half of the switches are connected via the skeleton with two edges of length 1. So, the total length within a variable is at least $k_i\sqrt{2} + 2k_i$. This is also possible to achieve, as seen in Figure 6(b) and 6(c).

Now, the clause skeletons have to be connected to at least one of the variables. For each of them, we need one more edge in one of the switches. The cheapest possible way of doing this is by using the groups that had expensive edges (of length $\sqrt{2}$) and using two normal edges (total length 2) instead. So, globally, of the $6m$ switches, at least $4m$ need two edges and the ones with a single edge must have length at least $\sqrt{2}$. Thus, the total length must be at least $W = (8 + \sqrt{2})m$. We claim that a solution of this length exists if and only if the formula is satisfiable. ☐

Theorem 3. *Computing a minimum RBP spanning graph is NP-hard.*

5   Approximation

A simple approximation algorithm determines the red-purple minimum spanning tree and the blue-purple minimum spanning tree, and takes the union of their edges. It is easy to see that this is a 2-approximation algorithm that requires $O(n \log n)$ time.

Another approximation algorithm, A, starts by computing the minimum spanning tree of the purple edges, and then adds the red and blue points in an optimal manner in the style of Kruskal's algorithm for minimum spanning trees. Algorithm A can also be implemented to run in $O(n \log n)$ time by computing the Delaunay triangulation of the red and purple points and of the blue and purple points. It is easy to argue that A also is a 2-approximation algorithm but interestingly, we can prove a better bound (close to 1.6) by expressing the approximation factor in the Steiner ratio ρ. Gilbert and Pollak [12] conjectured that $\rho = \frac{2}{\sqrt{3}} \approx 1.15$, but this conjecture has not been proved yet.[1] Chung and Graham [8] showed a bound of ≈ 1.21, which is the best-known upper bound on ρ.

Theorem 4. *Approximation algorithm A is a $(\frac{1}{2}\rho + 1)$-approximation of the minimum RBP spanning graph, where ρ is the Steiner ratio. The approximation is not a c-approximation for any constant $c < 1 + \frac{1}{\sqrt{3}}$.*

Proof. Algorithm A is not a c-approximation for any $c < 1 + \frac{1}{\sqrt{3}}$ (see Figure 8). Hence our approximation analysis is tight if the Gilbert-Pollak conjecture is true.

Next we prove our claim on the approximation factor. Let R, B, and P be sets of red, blue, and purple points. Let G^* be their minimum RBP spanning graph. Let R^* be the red edges, B^* the blue edges, and P^* the purple edges in G^*. Let A be the algorithm that computes a spanning graph by taking the minimum spanning tree of the purple points, and then adding the red and blue points optimally. We denote the resulting graph on $R \cup B \cup P$ by G', and its red, blue and purple edges by R', B', and P'.

Suppose first that G^* has no purple edges. Then algorithm A gives extra length in terms of purple edges equal to the MST of the purple points, denoted $||P'||$. The optimal

Fig. 8. A minimum RBP spanning graph and the RBP spanning graph on the same points obtained by approximation algorithm A

---

[1] A proof of the conjecture by Du and Hwang, "A proof of Gilbert-Pollak Conjecture on the Steiner ratio", *Algorithmica* 7:121–135 (1992), turned out to be incorrect.

graph G^* must connect all purple points simultaneously through a red spanning tree and through a blue spanning tree whose lengths are $||R^*||$ and $||B^*||$. Algorithm A has a total length of red edges of $||R'|| \leq ||R^*||$ and a total length of blue edges of $||B'|| \leq ||B^*||$. Hence the approximation ratio of A in case of absence of purple edges in the optimal solution is

$$\frac{||P'|| + ||R'|| + ||B'||}{||R^*|| + ||B^*||} \leq \frac{||P'|| + ||R^*|| + ||B^*||}{||R^*|| + ||B^*||}.$$

This ratio is maximized when R lies very densely on the Steiner Minimum Tree of P, and the same is true for B. Due to the density, algorithm A will choose nearly the full length of the Steiner Minimum Tree of P as well, once for red and once for blue. The approximation ratio is then smaller than but arbitrarily close to

$$\frac{MST(P) + 2 \cdot SMT(P)}{2 \cdot SMT(P)} \leq \frac{\rho \cdot SMT(P) + 2 \cdot SMT(P)}{2 \cdot SMT(P)} = \frac{\rho + 2}{2} = \frac{1}{2}\rho + 1,$$

where $SMT(P)$ is the Steiner Minimum Tree of P (or its length) and $MST(P)$ is the Minimum Spanning tree of P (or its length).

Next, suppose that G^* has a set P^* of purple edges, and assume them fixed. We will reason about sets of red, blue and purple points for which the algorithm A performs as poorly as possible in terms of approximation ratio.

If G^* has any red point r that has a single red edge incident to it in R^*, then this edge will connect r to the closest red or purple point, otherwise G^* is not optimal. Algorithm A will choose exactly the same edge in its solution. Hence, the approximation ratio of A for the points $R \setminus \{r\}$, B, and P is higher than for the points R, B, and P. The same is true for a blue or purple point that has a single incident edge in G^*. So we can restrict ourselves to analyzing point sets whose optimal solution does not have any leafs in G^*.

Let $B@R$ be a set of blue points infinitesimally close to the locations of the red points, and let $R@B$ be a set of red points infinitesimally close to the locations of the blue points. Now we can compare the approximation ratio of A (i) on P, R, and B, (ii) on P, R, and $B@R$, and (iii) on P, $R@B$, and B, and notice that at least one of (ii) and (iii) gives an approximation ratio at least as high as for (i). Hence, we can restrict ourselves to analyzing point sets where the red and blue points lie at basically the same positions (but they are not purple points).

The edges of P^* partition the purple points of P into a number of purple components which are connected by a red spanning forest and a blue spanning forest. We have

$$||P'|| \leq ||P^*|| + \rho||R^*||,$$

because in G^* the red (blue) connections between the purple components cannot be shorter than the Steiner Minimum Forest of the purple components. By the observations above we can assume that all red and blue points are used in the red and blue spanning forests that connect the purple components. Hence the approximation ratio is

$$\frac{||P'|| + ||R'|| + ||B'||}{||P^*|| + ||R^*|| + ||B^*||} \leq \frac{||P'|| + 2||R'||}{||P^*|| + 2||R^*||} \leq \frac{||P^*|| + \rho||R^*|| + 2||R^*||}{||P^*|| + 2||R^*||}.$$

This ratio is maximized when $||P^*|| = 0$, in which case we get exactly the same ratio as above, when no purple edges are present. $\square$

It is possible that a PTAS exists for our problem, but it is not clear whether the techniques of Arora [4] or Mitchell [17] for the Euclidean traveling salesperson problem can be applied since RBP spanning graphs are not planar, and the number of crossings of a single edge can be large.

6   Extensions and Future Work

Beyond Purple. So far we considered the case where there are exactly two sets, the red set and the blue set, leading to an input with red, blue, and purple points. In general we might have k different sets, all denoted by primary colors. For instance, for $k = 3$ we could have red, blue and yellow sets, which leads to three secondary colors (purple, orange, green) and one tertiary color (black). The objective is again to minimize the total length of a multi-colored spanning graph which has the property that the subgraphs induced by the red, blue, and yellow sets are connected (see Figure 9 (a)).

This problem is clearly still NP-hard. The 2-approximation immediately generalizes to a 3-approximation (or a k-approximation for k primary colors). We can improve on this by incorporating our $(1 + \frac{1}{2}\rho)$-approximation algorithm to obtain a $(2 + \frac{1}{2}\rho)$-approximation for three sets, or more generally a $(\lceil \frac{1}{2}k \rceil + \lfloor \frac{1}{2}k \rfloor \frac{1}{2}\rho)$-approximation for k sets. Interestingly, our algorithms for points on a line or on a circle are not straightforward to generalize; these problems remain open.

Line Drawings. Another extension is motivated by LineSets [3]. Returning for the moment to the setting with two sets (and red, blue, and purple points), we now wish to compute a minimum RBP spanning graph such that the subgraphs induced by the red and blue sets are paths. That is, the red and purple edges form a path connecting all red and purple points, and the blue and purple edges form a path connecting all blue and purple points (see Figure 9 (b)).

This problem is NP-hard since TSP is hard. Nonetheless, we can obtain a $(2 + \varepsilon)$-approximation by independently computing an approximate TSP for the blue and purple points and for the red and purple points, and simply taking the union. An approach similar to the spanning tree case seems to fail and hence a better solution remains an open problem. The question also remains open for points on a line or on a circle.

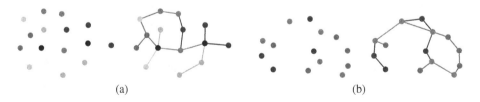

(a)   (b)

Fig. 9. (a) A set of multicolored points representing red, blue, and yellow sets and a corresponding spanning graph. (b) A set of red, blue and purple points, and a graph that connects all red points in a path and all blue points in a path.

References

1. Abellanas, M., Hurtado, F., Icking, C., Klein, R., Langetepe, E., Ma, L., Palop, B., Sacristán, V.: Smallest color-spanning objects. In: Meyer auf der Heide, F. (ed.) ESA 2001. LNCS, vol. 2161, pp. 278–289. Springer, Heidelberg (2001)
2. Agarwal, P.K., Govindarajan, S., Muthukrishnan, S.M.: Range searching in categorical data: Colored range searching on grid. In: Möhring, R.H., Raman, R. (eds.) ESA 2002. LNCS, vol. 2461, pp. 17–28. Springer, Heidelberg (2002)
3. Alper, B., Riche, N., Ramos, G., Czerwinski, M.: Design study of LineSets, a novel set visualization technique. IEEE TVCG 17(12), 2259–2267 (2011)
4. Arora, S.: Polynomial time approximation schemes for euclidean traveling salesman and other geometric problems. J. ACM 45(5), 753–782 (1998)
5. Brandes, U., Cornelsen, S., Pampel, B., Sallaberry, A.: Path-based supports for hypergraphs. J. Discrete Algorithms 14, 248–261 (2012)
6. Brass, P., Moser, W.O.J., Pach, J.: Research Problems in Discrete Geometry. Springer, New York (2005)
7. Byelas, H., Telea, A.: Towards realism in drawing areas of interest on architecture diagrams. Visual Languages and Computing 20(2), 110–128 (2009)
8. Chung, F., Graham, R.: A new bound for Euclidean Steiner minimum trees. Ann. N.Y. Acad. Sci. 440, 328–346 (1986)
9. Collins, C., Penn, G., Carpendale, S.: Bubble Sets: Revealing set relations with isocontours over existing visualizations. IEEE TVCG 15(6), 1009–1016 (2009)
10. Dinkla, K., van Kreveld, M., Speckmann, B., Westenberg, M.A.: Kelp Diagrams: Point set membership visualization. Computer Graphics Forum 31(3), 875–884 (2012)
11. Edwards, A.W.F.: Cogwheels of the mind. John Hopkins University Press (2004)
12. Gilbert, E., Pollak, H.: Steiner minimal trees. SIAM J. Appl. Math. 16, 1–29 (1968)
13. Henry Riche, N., Dwyer, T.: Untangling Euler diagrams. IEEE TVCG 16(6), 1090–1099 (2010)
14. Kaneko, A., Kano, M.: Discrete geometry on red and blue points in the plane – a survey. In: Discrete and Comp. Geometry, The Goodman-Pollack Festschrift, pp. 551–570 (2003)
15. Meulemans, W., Henry Riche, N., Speckmann, B., Alper, B., Dwyer, T.: KelpFusion: a hybrid set visualization technique. In: IEEE TVCG (to appear, 2013)
16. Mitchell, J.S.B.: Geometric shortest paths and network optimization. In: Handbook of Computational Geometry, pp. 633–701 (1998)
17. Mitchell, J.S.B.: Guillotine subdivisions approximate polygonal subdivisions: A simple polynomial-time approximation scheme for geometric TSP, k-MST, and related problems. SIAM J. Comput. 28(4), 1298–1309 (1999)
18. Mount, D.M., Netanyahu, N.S., Silverman, R., Wu, A.Y.: Chromatic nearest neighbor searching: A query sensitive approach. Computational Geometry: Theory and Applications 17(3-4), 97–119 (2000)
19. Preparata, F.P., Shamos, M.I.: Computational Geometry: An Introduction. Springer, New York (1985)
20. Simonetto, P., Auber, D.: Visualise undrawable Euler diagrams. In: Proc. 12th Conf. on Information Visualisation, pp. 594–599 (2008)
21. Stapleton, G., Rodgers, P., Howse, J., Zhang, L.: Inductively generating Euler diagrams. IEEE TVCG 17(1), 88–100 (2011)
22. Tokunaga, S.: Intersection number of two connected geometric graphs. Information Processing Letters 59(6), 331–333 (1996)
23. Tufte, E.R.: The Visual Display of Quantitative Information. Graphics Press (1983)

Drawing Non-Planar Graphs
with Crossing-Free Subgraphs[*]

Patrizio Angelini[1], Carla Binucci[2], Giordano Da Lozzo[1], Walter Didimo[2],
Luca Grilli[2], Fabrizio Montecchiani[2], Maurizio Patrignani[1], and Ioannis G. Tollis[3]

[1] Università Roma Tre, Italy
{angelini,dalozzo,patrigna}@dia.uniroma3.it
[2] Università degli Studi di Perugia, Italy
{binucci,didimo,grilli,montecchiani}@diei.unipg.it
[3] Univ. of Crete and Institute of Computer Science-FORTH, Greece
tollis@ics.forth.gr

Abstract. We initiate the study of the following problem: *Given a non-planar graph G and a planar subgraph S of G, does there exist a straight-line drawing Γ of G in the plane such that the edges of S are not crossed in Γ?* We give positive and negative results for different kinds of spanning subgraphs S of G. Moreover, in order to enlarge the subset of instances that admit a solution, we consider the possibility of bending the edges of $G \setminus S$; in this setting different trade-offs between number of bends and drawing area are given.

1 Introduction

Lots of papers in graph drawing address the problem of computing drawings of non-planar graphs with the goal of mitigating the negative effect that edge crossings have on the drawing readability. Many of these papers describe crossing minimization methods, which are effective and computationally feasible for relatively small and sparse graphs (see [8] for a survey). Other papers study which non-planar graphs can be drawn such that the "crossing complexity" of the drawing is somewhat controlled, either in the number or in the type of crossings. They include the study of *k-planar drawings*, in which each edge is crossed at most k times (see, e.g., [7,11,12,15,16,20,24]), of *k-quasi planar drawings*, in which no k pairwise crossing edges exist (see, e.g., [1,2,10,23,26,28]), and of *large angle crossing drawings*, in which any two crossing edges form a sufficiently large angle (see [14] for a survey). Most of these drawings exist only for sparse graphs.

In this paper we initiate the study of a new graph drawing problem concerned with the drawing of non-planar graphs. Namely: *Given a non-planar graph G and a planar subgraph S of G, decide whether G admits a drawing Γ such that the edges of S are not crossed in Γ, and compute Γ if it exists.*

Besides its intrinsic theoretical interest, this problem is also of practical relevance in many application domains. Indeed, distinct groups of edges in a graph may have different semantics, and a group can be more important than another for some applications;

---

[*] Work on these results began at the 8th Bertinoro Workshop on Graph drawing. Discussion with other participants is gratefully acknowledged. Part of the research was conducted in the framework of ESF project 10-EuroGIGA-OP-003 GraDR "Graph Drawings and Representations".

S. Wismath and A. Wolff (Eds.): GD 2013, LNCS 8242, pp. 292–303, 2013.
© Springer International Publishing Switzerland 2013

in this case a visual interface might attempt to display more important edges in a planar way. Again, the user could benefit from a layout in which a spanning connected subgraph is drawn crossing free, since it would support the user to quickly recognize paths between any two vertices, while keeping the other edges of the graph visible.

We remark that the problem of recognizing specific types of subgraphs that are not self-crossing (or that have few crossings) in a given drawing Γ, has been previously studied (see, e.g., [17,19,22,25]). This problem, which turns out to be NP-hard for most different kinds of instances, is also very different from our problem. Indeed, in our setting the drawing is not the input, but the output of the problem. Also, we require that the given subgraph S is not crossed by any edge of the graph, not only by its own edges.

In this paper we concentrate on the case in which S is a spanning subgraph of G and consider both straight-line and polyline drawings of G. Namely:

(i) In the straight-line drawing setting we prove that if S is any given spanning spider or caterpillar, then a drawing of G where S is crossing free always exists; such a drawing can be computed in linear time and requires polynomial area (Section 3.1). We also show that this positive result cannot be extended to any spanning tree, but we describe a large family of spanning trees that always admit a solution, and we show that any graph G contains such a spanning tree; unfortunately, our drawing technique for trees may require exponential area. Finally, we characterize the instances $\langle G, S \rangle$ that admit a solution when S is a spanning triconnected subgraph, and we provide a polynomial-time testing and drawing algorithm, whose layouts have polynomial area (Section 3.2).

(ii) We investigate polyline drawings where only the edges of $G \setminus S$ are allowed to bend. In this setting, we show that all spanning trees can be realized without crossings in a drawing of G of polynomial area, and we describe efficient algorithms that provide different trade-offs between number of bends per edge and drawing area (Section 4). Also, in Section 5 we briefly discuss a characterization of the instances $\langle G, S \rangle$ that admit a drawing when S is any given biconnected spanning subgraph.

Due to space restrictions, some proofs are omitted or only sketched in the text; full proofs for all results can be found in the [4].

2   Preliminaries and Definitions

We assume familiarity with basic concepts of graph drawing and planarity (see, e.g., [9]). Let $G(V, E)$ be a graph and let Γ be a drawing of G in the plane. If all vertices and edge bends of Γ have integer coordinates, then Γ is an *integer grid drawing* of G, and the *area* of Γ is the area of the minimum bounding box of Γ. Otherwise, suppose that Γ is not an integer grid drawing and let d_{min} be the minimum distance between two points of Γ on which either vertices or bends are drawn. In this case, the *area* of Γ is defined as the area of the minimum bounding box of a drawing obtained by scaling Γ by a constant c such that $c \times d_{min} = 1$; this corresponds to establish a certain resolution rule between vertices and bends of Γ, which is comparable to that of an integer grid drawing.

Let $G(V, E)$ be a graph and let $S(V, W)$, $W \subseteq E$, be a spanning subgraph of G. A straight-line drawing Γ of G such that S is crossing-free in Γ (i.e., such that crossings occur only between edges of $E \setminus W$) is called a *straight-line compatible drawing* of $\langle G, S \rangle$. If each edge of $E \setminus W$ has at most k bends in Γ (but still the subdrawing of S is straight-line and crossing-free), Γ is called a *k-bend compatible drawing* of $\langle G, S \rangle$.

If S is a rooted spanning tree of G such that every edge of $G \setminus S$ connects either vertices at the same level of S or vertices that are on consecutive levels, then we say that S is a *BFS-tree* of G.

A *star* is a tree $T(V, E)$ such that all its vertices but one have degree one, that is, $V = \{u, v_1, v_2, \ldots, v_k\}$ and $E = \{(u, v_1), (u, v_2), \ldots, (u, v_k)\}$; any subdivision of T (including T), is a *spider*: vertex u is the *center* of the spider and each path from u to v_i is a *leg* of the spider. A *caterpillar* is a tree such that removing all its leaves (and their incident edges) results in a path, which is called the *spine* of the caterpillar. The one-degree vertices attached to a spine vertex v are called the *leaves* of v.

In the remainder of the paper we implicitly assume that G is always a connected graph (if the graph is not connected, our results apply for any connected component).

3   Straight-Line Drawings

We start studying straight-line compatible drawings of pairs $\langle G, S \rangle$: Section 3.1 concentrates on the case in which S is a spanning tree, while Section 3.2 investigates the case in which S is a spanning triconnected graph.

3.1   Spanning Trees

The simplest case is when S is a given Hamiltonian path of G; in this case Γ can be easily computed by drawing all vertices of S in convex position, according to the ordering they occur in the path. In the following we prove that in fact a straight-line compatible drawing Γ of $\langle G, S \rangle$ can be always constructed in the more general cases in which S is a spanning spider (Theorem 1), or a spanning caterpillar (Theorem 2), or a BFS-tree (Theorem 3); our construction techniques guarantee polynomial-area drawings for spiders and caterpillars, while require exponential area for BFS-trees. On the negative side, we show that if S is an arbitrary spanning tree, a straight-line compatible drawing of $\langle G, S \rangle$ may not exist (Lemmas 1 and 2).

Theorem 1. *Let G be a graph with n vertices and m edges, and let S be a spanning spider of G. There exists an integer grid straight-line compatible drawing Γ of $\langle G, S \rangle$. Drawing Γ can be computed in $O(n + m)$ time and has $O(n^3)$ area.*

Proof. Let u be the center of S and let $\pi_1, \pi_2, \ldots, \pi_k$ be the legs of S. Also, denote by v_i the vertex of degree one of leg π_i $(1 \leq i \leq k)$. Order the vertices of S distinct from u such that: (i) the vertices of each π_i are ordered in the same way they appear in the simple path of S from u to v_i; (ii) the vertices of π_i precede those of π_{i+1} $(1 \leq i \leq k-1)$. If v is the vertex at position j $(0 \leq j \leq n-2)$ in the ordering defined above, draw v at coordinates (j^2, j). Finally, draw u at coordinates $(0, n-2)$. With this strategy, all vertices of S are in convex position, and they are all visible from u in such a way that no edge incident to u can cross other edges of Γ. Hence, the edges of S do not cross other edges in Γ. The area of Γ is $(n-2)^2 \times (n-2) = O(n^3)$ and Γ is constructed in linear time.   □

The next algorithm computes a straight-line compatible drawing of $\langle G, S \rangle$ when S is a spanning caterpillar. Theorem 2 proves its correctness, time and area requirements. Although the drawing area is still polynomial, the layout is not an integer grid drawing.

Algorithm. STRAIGHT-LINE-CATERPILLAR. Denote by $u_1, u_2, \ldots, u_k$ the vertices of the spine of S. Also, for each spine vertex u_i $(1 \leq i \leq k)$, let $v_{i1}, \ldots, v_{in_i}$ be its leaves in S (refer to the bottom image in Fig. 1(a)). The algorithm temporarily adds to S and G some dummy vertices, which will be removed in the final drawing. Namely, for each u_i, it attaches to u_i two dummy leaves, s_i and t_i. Also, it adds a dummy spine vertex u_{k+1} attached to u_k and a dummy leaf s_{k+1} of u_{k+1} (see the top image in Fig. 1(a)). Call G' and S' the new graph and the new caterpillar obtained by augmenting G and S with these dummy vertices.

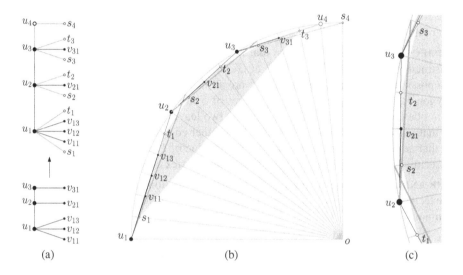

(a)   (b)   (c)

Fig. 1. Illustration of Algorithm STRAIGHT-LINE-CATERPILLAR: (a) a caterpillar S and its augmented version S'; (b) a drawing of S'; edges of the graph connecting leaves of S are drawn in the gray (convex) region; (c) enlarged detail of the picture (b)

The construction of a drawing Γ' of G' is illustrated in Fig. 1(b). Consider a quarter of circumference C with center o and radius r. Let N be the total number of vertices of G'. Let $\{p_1, p_2, \ldots, p_N\}$ be N equally spaced points along C in clockwise order, where $\overline{op_1}$ and $\overline{op_N}$ are a horizontal and a vertical segment, respectively. For each $1 \leq i \leq k$, consider the ordered list of vertices $L_i = \{u_i, s_i, v_{i1}, \ldots v_{in_i}, t_i\}$, and let L be the concatenation of all L_i. Also, append to L the vertices u_{k+1} and s_{k+1}, in this order. Clearly the number of vertices in L equals N. For a vertex $v \in L$, denote by $j(v)$ the position of v in L. Vertex u_i is drawn at point $p_{j(u_i)}$ $(1 \leq i \leq k)$; also, vertices u_{k+1} and s_{k+1} are drawn at points p_{N-1} and p_N, respectively. Each leaf v of S' will be suitably drawn along radius $\overline{op_{j(v)}}$ of C. More precisely, for any $i \in \{1, \ldots, k\}$, let a_i be the intersection point between segments $\overline{p_{j(u_i)}p_{j(s_{i+1})}}$ and $\overline{op_{j(s_i)}}$, and let b_i be the intersection point between segments $\overline{p_{j(u_i)}p_{j(u_{i+1})}}$ and $\overline{op_{j(t_i)}}$. Vertices s_i and

t_i are drawn at points a_i and b_i, respectively. Also, let A_i be the circular arc that is tangent to $\overline{p_{j(u_i)}p_{j(u_{i+1})}}$ at point b_i, and that passes through a_i; vertex v_{ih} is drawn at the intersection point between A_i and $\overline{op_{j(v_{ih})}}$ ($1 \le h \le n_i$).

Once all vertices of G' are drawn, each edge of G' is drawn in Γ' as a straight-line segment between its end-vertices. Drawing Γ is obtained from Γ' by deleting all dummy vertices and their incident edges.

Theorem 2. *Let G be graph with n vertices and m edges, and let S be a spanning caterpillar of G. There exists a straight-line compatible drawing Γ of $\langle G, S \rangle$. Drawing Γ can be computed in $O(n + m)$ time in the real RAM model[1] and has $O(n^2)$ area.*

Proof sketch: Let Γ be the output of Algorithm STRAIGHT-LINE-CATERPILLAR. We first prove that Γ is a straight-line compatible drawing of $\langle G, S \rangle$, and then we analyze time complexity and area requirement. We adopt the same notation used in the description of the algorithm.

CORRECTNESS. We have to prove that in Γ the edges of S are never crossed. Our construction places all spine vertices of S' (and hence of S) in convex position. It is also possible to see that the leaves of S' are all in convex position and form a convex polygon P. Since by construction the edges of S are all outside P in Γ, these edges cannot be crossed by edges of G connecting two leaves of S. Also, it is immediate to see that an edge of S cannot be crossed by another edge of S and it is not difficult to see that an edge of S cannot be crossed by an edge of G connecting either two non-consecutive spine vertices or a leaf of S to a spine vertex of S.

TIME AND AREA REQUIREMENT. Clearly, the construction of Γ' (and then of Γ) can be executed in linear time, in the real RAM model. About the area, let $d_{\min}$ be the minimum distance between any two vertices of Γ. It can be proved that if we require $d_{\min} \ge 1$ then $r < \frac{\sqrt{2}}{\beta}$, for $\beta = \theta(\frac{1}{N})$. Thus, the area of Γ is $O(N^2) = O(n^2)$. □

The next lemmas show that, unfortunately, Theorem 1 and Theorem 2 cannot be extended to any spanning tree S, that is, there are pairs $\langle G, S \rangle$ that do not admit a straight-line compatible drawing, even if S is a ternary or a binary tree.

Lemma 1. *Let G be the complete graph on 13 vertices and let S be a complete rooted ternary tree that spans G. There is no straight-line compatible drawing of $\langle G, S \rangle$.*

Proof sketch: Suppose, for a contradiction, that a straight-line compatible drawing Γ of $\langle G, S \rangle$ exists. Let r be the root of S (see Fig. 2(a)). Note that r is the only vertex of S with degree 3. Let u, v, w be the three neighbors of r in S. Two are the cases: either one of u, v, w (say u) lies inside triangle $\triangle(r, v, w)$ (Case 1, see Fig. 2(b)); or r lies inside triangle $\triangle(u, v, w)$ (Case 2).

In Case 1, consider a child u_1 of u. Vertex u_1 is placed in Γ in such a way that u lies inside either triangle $\triangle(u_1, r, w)$ or triangle $\triangle(u_1, r, v)$; assume the former (see Fig. 2(c)). Then, consider another child u_2 of u; in order for edge (u, u_2) not to cross any edge, also u_2 has to lie inside $\triangle(u_1, r, w)$, in such a way that both u and u_1 lie inside triangle $\triangle(u_2, r, v)$. This implies that u lies inside $\triangle(u_1, r, u_2)$ (see Fig. 2(d)),

---

[1] We also assume that basic trigonometric functions are executed in constant time.

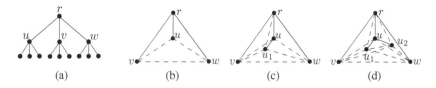

Fig. 2. Illustration for Lemma 1: (a) A complete rooted ternary tree with 13 vertices. (b) Case 1 in the proof; u lies inside $\triangle(r, v, w)$. (c) Placement of u_1. (d) Placement of u_2.

together with its last child u_3. However, u_3 cannot be placed in any of the three triangles in which $\triangle(u_1, r, u_2)$ is partitioned by the edges (of S) connecting u to u_1, to r, and to u_2, respectively, without introducing any crossing involving edges of S, a contradiction. Case 2 can be analyzed with analogous considerations. □

The proof strategy of Lemma 2 is similar to that of Lemma 1.

Lemma 2. *Let G be the complete graph on 22 vertices and let S be a complete unrooted binary tree that spans G. There is no straight-line compatible drawing of $\langle G, S \rangle$.*

In the light of Lemmas 1 and 2, it is natural to ask whether there are specific subfamilies of spanning trees S (other than paths, spiders, and caterpillars) such that a straight-line compatible drawing of $\langle G, S \rangle$ always exists. The next algorithm gives a positive answer to this question: it computes a straight-line compatible drawing when S is a BFS-tree of G. Theorem 3 proves the algorithm correctness, its time complexity, and its area requirement.

Algorithm. STRAIGHT-LINE-BFS-TREE. Let u be the root of S (which is at level 0) and let $u_{l1}, \ldots, u_{lk_l}$ be the vertices at level $l \in \{1, \ldots, d\}$, where d is the depth of S. The algorithm temporarily adds to S and G some dummy vertices, which will be removed in the final drawing. Namely, for each u_{li}, $1 \leq l \leq d - 1$ and $1 \leq i \leq k_l$, it attaches to u_{li} one more (leftmost) child s_{li}. Also, it attaches to root u a dummy (rightmost) child t. Denote by G' and S' the new graph and the new tree, respectively. Notice that S' is still a BFS-tree of G'. The algorithm iteratively computes a drawing Γ' of G'. For $l = 1, \ldots, d$, the algorithm defines a circumference C_l with center $o = (0, 0)$ and radius $r_l < r_{l-1}$ ($C_1, \ldots, C_d$ are concentric). The vertices of level l are drawn on the quarter of C_l going from point $(-r_l, 0)$ to point $(0, r_l)$ clockwise.

   Let $\{u_{11}, \ldots, u_{1k_1}, t\}$ be the ordered list of the children of root u and let $\{p_{11}, \ldots, p_{1k_1}, p_t\}$ be $k_1 + 1$ equally spaced points along C_1 in clockwise order, where $\overline{op_{11}}$ and $\overline{op_t}$ are a horizontal and a vertical segment, respectively. Vertex u_{1j} is drawn on p_{1j} ($1 \leq j \leq k_1$) and vertex t is drawn on p_t. Also, u is drawn on point $(-r_1, r_1)$.

   Assume now that all vertices $u_{l1}, \ldots, u_{lk_l}$ of level l have been drawn ($1 \leq l \leq d-1$) in this order on the sequence of points $\{q_1, \ldots, q_{k_l}\}$, along C_l. The algorithm draws the vertices of level $l + 1$ as follows. Let $\overline{q_i q_{i+1}}$ be the chords of C_l, for $1 \leq i \leq k_l - 1$, and let c_l be the shortest of these chords. The radius r_{l+1} of C_{l+1} is chosen arbitrarily in such a way that C_{l+1} intersects c_l in two points and $r_{l+1} < r_l$. This implies that C_{l+1} also intersects every chord $\overline{q_i q_{i+1}}$ in two points. For $1 \leq i \leq k_l$, denote by $L(u_{li}) = \{v_1, \ldots, v_{n_{li}}\}$ the ordered list of children of u_{li} in G'. Also, let a_i be the intersection point between $\overline{q_i q_{i+1}}$ and C_{l+1} that is closest to q_i, and let ℓ_i be the line through q_i tangent to C_{l+1}; denote by b_i the tangent point between ℓ_i and C_{l+1}. Let A_{l+1} be the arc of C_{l+1} between a_i and b_i, and let $\{p_0, p_1, \ldots, p_{n_{li}}\}$ be $n_{li} + 1$ equally spaced points along A_{l+1} in clockwise order. For $v \in L(u_{li})$, denote by $j(v)$ the position of v in $L(u_{li})$. Vertex v_j is drawn on $p_{j(v_j)}$ $(1 \leq j \leq n_{li})$ and vertex s_{li} is drawn on p_0.

Once all vertices of G' are drawn each edge of G' is drawn in Γ' as a straight-line segment between its end-vertices. Drawing Γ is obtained from Γ' by deleting all dummy vertices and their incident edges.

Theorem 3. *Let G be a graph with n vertices and m edges, and let S be a BFS-tree of G. There exists a straight-line compatible drawing Γ of $\langle G, S \rangle$. Drawing Γ can be computed in $O(n + m)$ time in the real RAM model.*

It is worth observing that any graph G admits a BFS-tree rooted at an arbitrarily chosen vertex r of G. Thus, each graph admits a straight-line drawing Γ such that one of its spanning trees S is never crossed in Γ. Unfortunately, the compatible drawing computed by Algorithm STRAIGHT-LINE-BFS-TREE may require area $\Omega(2^n)$.

3.2 Spanning Triconnected Subgraphs

Here we focus on triconnected spanning subgraph S of G. Clearly, since every tree can be augmented with edges to become a triconnected graph, Lemmas 1 and 2 imply that, if S is a triconnected graph, a straight-line compatible drawing of $\langle G, S \rangle$ may not exist. The next theorem characterizes those instances for which such a drawing exists.

Theorem 4. *Let $G(V, E)$ be a graph, $S(V, W)$ be a spanning planar triconnected subgraph of G, and $\mathcal{E}$ be the unique planar (combinatorial) embedding of S (up to a flip). A straight-line compatible drawing Γ of $\langle G, S \rangle$ exists if and only if: (1) Each edge $e \in E \setminus W$ connects two vertices belonging to the same face of $\mathcal{E}$. (2) There exists a face f of $\mathcal{E}$ containing three vertices such that any pair u, v of them does not separate in the circular order of f the end-vertices $x, y \in f$ of any other edge in $E \setminus W$.*

Proof sketch: Since $\mathcal{E}$ is unique the necessity of Condition 1 is trivial. Its sufficiency would be also trivial if S admitted a convex drawing where the external face is a triangle. Otherwise, suppose that v_1, v_2, and v_3 are three vertices of a face f satisfying Condition 2. Dummy vertices can be added to S among v_1, v_2, and v_3 in order to have a triangular face to be used as external face when computing the convex drawing (for example, using the algorithm in [27]). The necessity of Condition 2 follows from considering any three vertices on the convex hull of a compatible drawing of $\langle G, S \rangle$.   □

The next algorithm exploits Theorem 4 in order to decide in polynomial time whether $\langle G, S \rangle$ admits a straight-line compatible drawing.

Algorithm. STRAIGHT-LINE-TRICONNECTED. Let $\mathcal{E}$ be the unique planar embedding of S (up to a flip). The algorithm verifies that each edge of $E \setminus W$ satisfies Condition 1 of Theorem 4 and that there exists a face f of $\mathcal{E}$ containing three vertices v_1, v_2, and v_3, that satisfy Condition 2 of Theorem 4. If both conditions hold, then v_1, v_2, and v_3 can be used to find a straight-line compatible drawing Γ of $\langle G, S \rangle$ as described in the proof of Theorem 4.

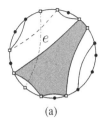
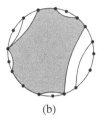

(a)   (b)

Fig. 3. Two consecutive steps of Algorithm STRAIGHT-LINE-TRICONNECTED. (a) The outer-plane graph G_f; the shaded face is `full` (the others are `empty`); the dash-dot edge e is the next edge of E_f to be considered; edges in E_χ are drawn as dashed lines; white squares are vertices of V_χ. (b) Graph G_f after the update due to edge e.

Condition 1 is verified as follows. Construct an auxiliary graph S' from S by subdividing each edge e of W with a dummy vertex v_e. Also, for each face f of $\mathcal{E}$ add to S' a vertex v_f and connect v_f to all non-dummy vertices of f. We have that two vertices of V belong to the same face of $\mathcal{E}$ if and only if their distance in S' is two.

To test Condition 2 of Theorem 4 we perform the following procedure on each face f of $\mathcal{E}$, restricting our attention to the set E_f of edges in $E \setminus W$ whose end-vertices belong to f. We maintain an auxiliary outerplane graph G_f whose vertices are the vertices V_f of f. Each internal face of G_f is either marked as `full` or as `empty`. Faces marked `full` are not adjacent to each other. Intuitively, we have that any three vertices of an `empty` face satisfy Condition 2, while all triples of vertices of a `full` face do not. We initialize G_f with the cycle composed of the vertices and the edges of f and mark its unique internal face as `empty`. At each step an edge e of E_f is considered and G_f is updated. If adding e to G_f splits a single face marked `empty`, we update G_f by splitting such a face into two `empty` faces. If the end-vertices of e belong to a single face marked `full`, we ignore e. Otherwise, adding e to G_f would cross several edges and faces (see Fig. 3(a)). Consider the set E_χ of internal edges of G_f crossed by e. Define a set of vertices V_χ of G_f with the end-vertices of e, the end-vertices of edges of E_χ that are incident to two `empty` faces, the vertices of the `full` faces traversed by e. Remove all edges in E_χ from G_f. Mark the face f' obtained by such a removal as `empty`. Form a new face f_χ inside f' with all vertices in V_χ by connecting them as they appear in the circular order of f, and mark f_χ as `full` (see Fig. 3(b)).

When all the edges of E_f have been considered, if G_f has an internal face marked as `empty`, any three vertices of this face satisfy Condition 2. Else, G_f has a single internal face marked `full` and all triples of vertices of f do not satisfy Condition 2.

Theorem 5. *Let $G(V, E)$ be a graph and let $S(V, W)$ be a spanning triconnected planar subgraph of G. There exists an $O(|V| \times |E \setminus W|)$-time algorithm that decides whether $\langle G, S \rangle$ admits a straight-line compatible drawing Γ and, in the positive case, computes it on an $O(|V|^2) \times O(|V|^2)$ grid.*

Proof sketch: Algorithm STRAIGHT-LINE-TRICONNECTED constructs Γ. Its correctness trivially descends from Theorem 4. Regarding the time complexity, the unique planar embedding $\mathcal{E}$ of S can be computed in $O(|V|)$ time. The auxiliary graph S' for

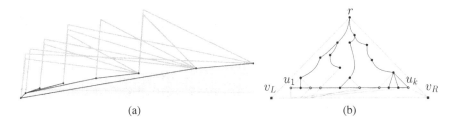

Fig. 4. Illustration of: (a) Algorithm ONE-BEND TREE and (b) Algorithm THREE-BEND TREE; a graph G with a given spanning tree S (black edges)

testing Condition 1 can be constructed in time linear in the size of S. Since S' is a planar graph, deciding if two vertices have distance two can be done in constant time [21]. Thus, testing Condition 1 for all edges in $E \setminus W$ can be done in $O(|V| + |E \setminus W|)$ time. While verifying Condition 1, E_f can be computed, for each face f of $\mathcal{E}$, in $O(|V| + |E \setminus W|)$ time. Since for each face f of $\mathcal{E}$, the size of G_f is $O(|V_f|)$, adding edges in E_f has time complexity $O(|E_f| \times |V_f|)$. Overall, we have that the time complexity of testing Condition 2 is $O(|E \setminus W| \times |V|)$, which gives the time complexity of the whole algorithm. Regarding the area, the algorithm in [5] can be used to obtain in linear time a straight-line grid drawing of S on an $O(|V|^2) \times O(|V|^2)$ grid; this drawing is strictly convex.   □

4 Polyline Drawings

We now prove that, using bends along the edges of $G \setminus S$ allows us to compute compatible drawings of pairs $\langle G, S \rangle$ for every spanning tree S of G; such drawings are on a polynomial-area grid. In particular, since edge bends are negatively correlated to the drawing readability, we want to compute k-bend compatible drawings for small values of k. We provide algorithms that offer different trade-offs between number of bends and drawing area. In Section 5 we briefly discuss some preliminary results about 1-bend compatible drawings of $\langle G, S \rangle$ when S is a biconnected spanning subgraph.

Let $G(V, E)$ be a graph with n vertices and m edges, and let $S(V, W)$ be any spanning tree of G. We denote by $x(v)$ and $y(v)$ the x- and the y-coordinate of a vertex v, respectively. The next algorithm computes a 1-bend compatible drawing of $\langle G, S \rangle$.

Algorithm. ONE-BEND TREE. The algorithm works in two steps (refer to Fig. 4(a)).

STEP 1: Consider a point set of size n such that for each point p_i, the x- and y-coordinates of p_i are i^2 and i, respectively. Construct a straight-line drawing of S by placing the vertices on points p_i, $1 \leq i \leq n$, according to a DFS traversal.

STEP 2: Let v_i be the vertex placed on point p_i. For each $i \in \{1, \ldots, n\}$, draw each edge $(v_i, v_j) \in E \setminus W$ such that $j > i$ as a polyline connecting p_i and p_j, and bending at point $(i^2 + 1, n + c)$ where c is a progressive counter, initially set to one.

Theorem 6. *Let $G(V, E)$ be a graph with n vertices and m edges, and let $S(V, W)$ be any spanning tree of G. There exists a 1-bend compatible drawing Γ of $\langle G, S \rangle$. Drawing Γ can be computed in $O(n + m)$ time and has $O(n^2(n + m))$ area.*

Proof sketch: The algorithm that computes Γ is Algorithm ONE-BEND TREE. Note that the drawing of S contained in Γ is planar, and that the edges in $E \setminus W$ are drawn outside the convex region containing the drawing of S. About area requirements, the width of Γ is $O(n^2)$, by construction, while the height of Γ is given by the y-coordinate of the topmost bend point, that is $n + m$. $\square$

Next, we describe an algorithm that constructs 3-bend compatible drawings of pairs $\langle G, S \rangle$ with better area bounds than Algorithm ONE-BEND TREE for sparse graphs.

Algorithm. THREE-BEND TREE. The algorithm works in four steps (see Fig. 4(b)).

STEP 1: Let G' be the graph obtained from G by subdividing each edge $(v_i, v_j) \in E \setminus W$ with two dummy vertices $d_{i,j}$ and $d_{j,i}$. Let S' be the spanning tree of G', rooted at any non-dummy vertex r, obtained by deleting all edges connecting two dummy vertices. Clearly, every dummy vertex is a leaf of S'.

STEP 2: For each vertex of S', order its children arbitrarily, thus inducing a left-to-right order of the leaves of S'. Rename the leaves of S' as $u_1, \ldots, u_k$ following this order. For each $i \in \{1, \ldots, k-1\}$, add an edge (u_i, u_{i+1}) to S'. Also, add to S' two dummy vertices v_L and v_R, and edges $(v_L, r), (v_R, r), (v_L, u_1), (u_k, v_R), (v_L, v_R)$.

STEP 3: Construct a straight-line grid drawing Γ' of S', as described in [18], in which edge (v_L, v_R) is drawn as a horizontal segment on the outer face, vertices $u_1, \ldots, u_k$ all lie on points having the same y-coordinate Y, and the rest of S' is drawn above such points. Remove from Γ' the vertices and edges added in STEP 2.

STEP 4: Compute a drawing Γ of G such that each edge in W is drawn as in Γ', while each edge $(v_i, v_j) \in E \setminus W$ is drawn as a polyline connecting v_i and v_j, bending at $d_{i,j}$, at $d^{j,i}$, and at a point $(c, Y-1)$ where c is a progressive counter, initially set to $x(u_1)$.

Theorem 7. *Let $G(V, E)$ be a graph with n vertices and m edges, and let $S(V, W)$ be any spanning tree of G. There exists a 3-bend compatible drawing Γ of $\langle G, S \rangle$. Drawing Γ can be computed in $O(n + m)$ time and has $O((n + m)^2)$ area.*

Proof sketch: The algorithm that computes Γ is Algorithm THREE-BEND TREE. The drawing of S contained in Γ is planar ([18]) and lies above the horizontal line $y = Y$. The area bounds descend from the construction and from the area bounds of [18]. $\square$

We finally remark that there exists a drawing algorithm that computes 4-bend compatible drawings that are more readable than those computed by Algorithm THREE-BEND TREE. Although the area of these drawings is still $O((n + m)^2)$, they have optimal crossing angular resolution, i.e., edges cross only at right angles. Drawings of this type are called *RAC drawings* and are widely studied in the literature [13,14].

5 Discussion

We initiated the study of a new problem in graph drawing, i.e., computing a drawing Γ of a non-planar graph G such that a desired subgraph $S \subseteq G$ is crossing-free in Γ. In the setting where edges are straight-line segments and S is a spanning tree of G, we showed that Γ does not always exist; also, we provided existential and algorithmic results for meaningful subfamilies of spanning trees and we described a linear-time testing and drawing algorithm when S is a spanning triconnected subgraph. One of the main problems still open in this setting is the following: *Given a graph G and a spanning tree S of G, what is the complexity of deciding whether $\langle G, S \rangle$ admits a straight-line compatible drawing?* This problem can be also studied when S is a biconnected spanning subgraph, trying to extend the characterization of Theorem 4. Another interesting problem is to extend the results of Lemmas 1 and 2 in order to give a characterization of what spanning trees S of a complete graph can be always realized.

Allowing bends on the edges of $G \setminus S$, a drawing Γ exists for any given spanning tree S; we described several efficient algorithms that offer different compromises between drawing area and number of bends. Also, in this setting we have a characterization of which pairs $\langle G, S \rangle$ admit a 1-bend compatible drawing when S is a biconnected spanning subgraph. Namely, a necessary and sufficient condition is that S has a planar embedding such that for each edge e of $G \setminus S$ the end-vertices of e belong to the same face of S (as for Condition 1 of Theorem 4). Given such an embedding one can: (i) add a dummy vertex inside each face of S and connect it to all the vertices of the face; (ii) construct a planar straight-line drawing of the resulting graph, and (iii) construct a 1-bend compatible drawing where each edge (u, v) of $G \setminus S$ has a bend-point coinciding with the dummy vertex of the face containing u and v. A small perturbation of the bend-points will avoid that two of them coincide. An algorithm that tests the condition above can be derived as a simplification of the algorithm in [3], used to test the existence of a Simultaneous Embedding with Fixed Edges (SEFE) of two graphs [6]. Finally, we remark that Algorithm ONE-BEND TREE can be adapted to find a 1-bend compatible drawing when S is an outerplanar graph with the same bounds stated by Theorem 6. Many problems for k-compatible drawings are still open. Among them: trying to reduce the area bounds when S is a tree and devising algorithms for computing grid 1-bend compatible drawings of feasible $\langle G, S \rangle$ when S is biconnected.

References

1. Ackerman, E.: On the maximum number of edges in topological graphs with no four pairwise crossing edges. Discrete & Computational Geometry 41(3), 365–375 (2009)
2. Ackerman, E., Tardos, G.: On the maximum number of edges in quasi-planar graphs. Journal of Combinatorial Theory, Ser. A 114(3), 563–571 (2007)
3. Angelini, P., Di Battista, G., Frati, F., Patrignani, M., Rutter, I.: Testing the simultaneous embeddability of two graphs whose intersection is a biconnected or a connected graph. Journal of Discrete Algorithms 14, 150–172 (2012)
4. Angelini, P., Binucci, C., Da Lozzo, G., Didimo, W., Grilli, L., Montecchiani, F., Patrignani, M., Tollis, I.G.: Drawings of non-planar graphs with crossing-free subgraphs. ArXiv e-prints 1308.6706 (September 2013)
5. Bárány, I., Rote, G.: Strictly convex drawings of planar graphs. Documenta. Math. 11, 369–391 (2006)
6. Blasiüs, T., Kobourov, S.G., Rutter, I.: Simultaneous embedding of planar graphs. In: Tamassia, R. (ed.) Handbook of Graph Drawing and Visualization. CRC Press (2013)
7. Brandenburg, F.J., Eppstein, D., Gleißner, A., Goodrich, M.T., Hanauer, K., Reislhuber, J.: On the density of maximal 1-planar graphs. In: Didimo, W., Patrignani, M. (eds.) GD 2012. LNCS, vol. 7704, pp. 327–338. Springer, Heidelberg (2013)

8. Buchheim, C., Chimani, M., Gutwenger, C., Jünger, M., Mutzel, P.: Crossings and planariza-tion. In: Tamassia, R. (ed.) Handbook of Graph Drawing and Visualization. CRC Press (2013)
9. Di Battista, G., Eades, P., Tamassia, R., Tollis, I.G.: Graph Drawing. Prentice Hall, Upper Saddle River (1999)
10. Di Giacomo, E., Didimo, W., Liotta, G., Montecchiani, F.: h-quasi planar drawings of bounded treewidth graphs in linear area. In: Golumbic, M.C., Stern, M., Levy, A., Morgenstern, G. (eds.) WG 2012. LNCS, vol. 7551, pp. 91–102. Springer, Heidelberg (2012)
11. Di Giacomo, E., Didimo, W., Liotta, G., Montecchiani, F.: Area requirement of graph draw-ings with few crossings per edge. Computational Geometry 46(8), 909–916 (2013)
12. Didimo, W.: Density of straight-line 1-planar graph drawings. Information Processing Let-ters 113(7), 236–240 (2013)
13. Didimo, W., Eades, P., Liotta, G.: Drawing graphs with right angle crossings. Theoretical Computer Science 412(39), 5156–5166 (2011)
14. Didimo, W., Liotta, G.: The crossing angle resolution in graph drawing. In: Pach, J. (ed.) Thirty Essays on Geometric Graph Theory. Springer (2013)
15. Eades, P., Hong, S.H., Katoh, N., Liotta, G., Schweitzer, P., Suzuki, Y.: Testing maximal 1-planarity of graphs with a rotation system in linear time - (extended abstract). In: Didimo, W., Patrignani, M. (eds.) GD 2012. LNCS, vol. 7704, pp. 339–345. Springer, Heidelberg (2013)
16. Hong, S.-H., Eades, P., Liotta, G., Poon, S.-H.: Fáry's theorem for 1-planar graphs. In: Gud-mundsson, J., Mestre, J., Viglas, T. (eds.) COCOON 2012. LNCS, vol. 7434, pp. 335–346. Springer, Heidelberg (2012)
17. Jansen, K., Woeginger, G.J.: The complexity of detecting crossingfree configurations in the plane. BIT Numerical Mathematics 33(4), 580–595 (1993)
18. Kant, G.: Drawing planar graphs using the canonical ordering. Algorithmica 16(1), 4–32 (1996)
19. Knauer, C., Schramm, É., Spillner, A., Wolff, A.: Configurations with few crossings in topo-logical graphs. Computational Geometry 37(2), 104–114 (2007)
20. Korzhik, V.P., Mohar, B.: Minimal obstructions for 1-immersions and hardness of 1-planarity testing. Journal of Graph Theory 72(1), 30–71 (2013)
21. Kowalik, L., Kurowski, M.: Short path queries in planar graphs in constant time. In: Larmore, L.L., Goemans, M.X. (eds.) STOC 2003, pp. 143–148. ACM (2003)
22. Kratochvíl, J., Lubiv, A., Nešetřil, J.: Noncrossing subgraphs in topological layouts. SIAM Journal on Discrete Mathematics 4(2), 223–244 (1991)
23. Pach, J., Shahrokhi, F., Szegedy, M.: Applications of the crossing number. Algorith-mica 16(1), 111–117 (1996)
24. Pach, J., Tóth, G.: Graphs drawn with few crossings per edge. Combinatorica 17(3), 427–439 (1997)
25. Rivera-Campo, E., Urrutia-Galicia, V.: A sufficient condition for the existence of plane span-ning trees on geometric graphs. Computational Geometry 46(1), 1–6 (2013)
26. Suk, A.: k-quasi-planar graphs. In: Speckmann, B. (ed.) GD 2011. LNCS, vol. 7034, pp. 266–277. Springer, Heidelberg (2011)
27. Tutte, W.T.: How to draw a graph. Proceedings of the London Mathematical Society s3-13(1), 743–767 (1963)
28. Valtr, P.: On geometric graphs with no k pairwise parallel edges. Discrete & Computational Geometry 19(3), 461–469 (1998)

Exploring Complex Drawings via Edge Stratification

Emilio Di Giacomo[1], Walter Didimo[1], Giuseppe Liotta[1],
Fabrizio Montecchiani[1], and Ioannis G. Tollis[2]

[1] Dip. di Ingegneria Elettronica e dell'Informazione, Università degli Studi di Perugia
{digiacomo,didimo,liotta,montecchiani}@diei.unipg.it
[2] Univ. of Crete and Institute of Computer Science-FORTH, Greece
tollis@ics.forth.gr

Abstract. We propose an approach that allows a user to explore a layout produced by any graph drawing algorithm, in order to reduce the visual complexity and clarify its presentation. Our approach is based on *stratifying* the drawing into *layers* with desired properties; layers can be explored and combined by the user to gradually acquire details. We present stratification heuristics, a user study, and an experimental analysis that evaluates how our stratification heuristics behave on the drawings computed by a variety of popular force-directed algorithms.

1 Introduction

Graph drawing algorithms are used in many applications to visualize networked information. Among them, force-directed algorithms are the most popular and are widely adopted to compute drawings in which vertices are represented as small circles and edges are drawn as straight-line segments. Of course, the chosen algorithm is of great importance in creating a readable visualization. However, when the graph is complex (large or locally dense) a high number of edge crossings is typically unavoidable; this is the case, for example, of most small world and scale-free graphs (see, e.g., [13,30]). It is well known that a high number of edge crossings seriously affects the drawing readability [26,27], and makes it hard to perform detailed tasks based on visual inspection. These tasks include finding the shortest path between two given vertices, finding the vertices that are adjacent to both, or even determining the degree of a vertex.

In this paper we propose a new approach to support the user in the visual inspection of complex drawings. Namely, given a drawing Γ of a graph $G(V, E)$, we aim at partitioning the set of edges E into subsets $E_1, E_2, \ldots, E_h$, such that the subdrawing $\Gamma_i \subseteq \Gamma$ of each subgraph $G_i(V, E_i)$ guarantees some desired readability property (in each subdrawing, the vertices remain fixed in their original positions as determined in Γ). For example, a user could prefer to see Γ_i without any edge crossing, i.e., as planar, or that any two crossing edges form a sufficiently large angle. We say that Γ is *stratified* into a set of *layers* Γ_i, each containing all the vertices of Γ (in their original positions) but only a portion of the edge set. The user can then interact with this edge stratification, by exploring one layer at a time, or by arbitrarily combining multiple layers into a single view. The edges of each layer are assigned the same color and different colors are used for the edges of different layers. The main advantage of this approach is that

S. Wismath and A. Wolff (Eds.): GD 2013, LNCS 8242, pp. 304–315, 2013.

users can get multiple readable views of different portions of the drawing, with the possibility of simplifying the total amount of information, thus allowing them to gradually acquire details by exploring or combining layers. On the negative side, from the cognitive point of view, the user has to face the difficulty of making sense of a distributed information. To deal with this difficulty in practical terms, it is crucial to minimize the number of layers required to stratify the drawing so that the desired readability property is guaranteed for each layer. The main contribution of this paper is as follows:

(i) We define an edge stratification model and the related optimization problems. Then, we give a general framework to solve these problems for several desired readability properties of the layers, and we describe heuristics within this framework (Section 3).

(ii) We present the results of a user study aimed at understanding the effectiveness of the proposed approach for executing tasks based on visual inspection (Section 4). These results highlight the usefulness of edge stratification, especially for some of these tasks and for some specific readability properties of the layers.

(iii) We present an experimental analysis that compares the number of layers required to stratify drawings computed by a variety of popular force-directed algorithms, using our heuristics (Section 5). On one side, these experiments suggest that for some of the computed drawings the number of layers required by some edge stratification is a more reliable measure of the drawing visual complexity with respect to the number of edge crossings. On the other side, the results show that most of the force-directed algorithms that we have considered guarantee a strong correlation between number of crossings and number of layers in the stratifications of their drawings. We interpret this behavior as a positive feature of the drawing algorithms, which witnesses a quite uniform distribution of the crossings in the drawing.

In Fig. 1 we give an example of how the number of layers produced by the stratification heuristics in this paper can be used to measure the readability of different drawings of a same graph. Fig. 1(a) shows a drawing Γ_1 of a graph with 50 vertices and 150 edges, computed by Fruchterman-Reingold's algorithm [17] and containing $1,395$ edge crossings; Fig. 1(b) shows a drawing Γ_2 of the same graph, computed by Kamada-Kawai's algorithm [21] and having $1,437$ edge crossings. Applying our stratification heuristic to compute planar layers, Γ_1 requires 8 layers while drawing Γ_2 requires 7 layers; even if we require crossing angles of at least $\frac{\pi}{4}$ in each layer, our stratification heuristic generates 4 layers for Γ_1 (Fig. 1(c)) and only 3 layers for Γ_2 (Fig. 1(d)). Indeed, despite its higher number of crossings, drawing Γ_2 appears more readable, due to a more uniform distribution of the crossings and a better area. Namely, in the figure the two drawings are scaled to fit in the same bounding box; if we scale the drawings such that they satisfy the same resolution rule, drawing Γ_1 has twice the area of drawing Γ_2.

2 Related Work

To reduce the negative impact of edge crossings, different constraints on the type of crossings have been studied. Some of them require that edges cross only at large angles, or that each edge is crossed at most a limited number of times, or even that only few pairwise crossing edges are allowed. A very limited list of papers includes [10,11,12]. Unfortunately, only restricted sub-families of sparse graphs admit drawings that respect

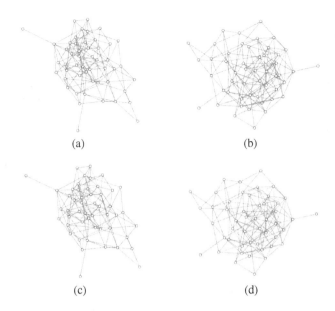

(a)   (b)

(c)   (d)

Fig. 1. (a-b) Drawings of the same graph computed by two different force-directed algorithms. (c-d) The same drawings in (a-b) stratified with layers having crossing angles of at least $\frac{\pi}{4}$; layers are conveyed with different edge colors.

these constraints. Also, an impressive set of crossing minimization methods are proposed in graph drawing (see [8] for a survey). However, these methods become computationally expensive (or even unfeasible) for large and dense graphs.

Many visualization techniques that compute a *hierarchical clustering* of the vertices and that allow users to interactively explore it are also known (see , e.g., [1,3,30]). The levels of a cluster hierarchy are a sort of vertex stratification, which allows users to control the amount of information displayed in the same view. Other well-studied approaches that facilitate in the visual exploration of networked data are based on node or edge *filtering*, *grouping*, and *motif simplification* (see, e.g., [1,29]).

Our edge stratification approach does not aim at computing drawings of graphs with controlled visual complexity, but rather it starts from a drawing of a graph and aims at supporting its exploration and analysis by distributing the whole drawing information into a set of logical layers with desired edge crossings properties. This idea is somewhat related to the notion of *geometric thickness* of a graph G [14,15], which is the minimum number of colors that can be assigned to the edges of G, such that there exists a straight-line drawing of G where no two edges of the same color cross. Similar to thickness, our stratification defines an edge coloring; the difference is that stratification is executed on a specific drawing, which cannot be changed. Hence, if the geometric thickness can be used as a measure of the graph complexity, the minimum size of a stratification of a straight-line drawing can be used as a measure of the drawing visual complexity in terms of number, types, and distribution of its edge crossings. Clearly, the size of a stratification into planar layers of a drawing Γ of G cannot be smaller than the geometric thickness of G.

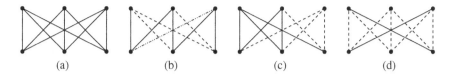

Fig. 2. (a) A straight-line drawing Γ of $K_{3,3}$. (b) A stratification $S(\Gamma, \text{PLANARITY})$. (c) A stratification $S(\Gamma, \text{LAC}(\frac{\pi}{3}))$. (d) A stratification $S(\Gamma, 1\text{-PLANARITY})$. In each stratification different dash styles for the edges represent different layers.

Other approaches have been proposed for supporting the analysis of a given a drawing. One of the most popular is *edge bundling*, which deforms and groups together edges that are similar according to some metric (see [32] for a survey). Another approach, called *geometric graph generalization*, reduces vertex and/or edge clutter by collapsing groups of vertices that are geometrically close to one another into a single point [7]. Differently from our stratification, edge bundling and geometric graph generalization modify the input drawing, emphasizing its skeletal structure at the expenses of loss of details. We finally mention another recent technique, which aims to visually simplify or remove edge crossings in a given drawing by displaying only portions of the crossing edges; the modified drawings are called *partial edge drawings* [6].

3 Stratification: Model and Algorithms

Here we formally describe our edge stratification model and related algorithms.

Model. Let $G(V, E)$ be a graph and let Γ be a straight-line drawing of G. Given a subset $E' \subseteq E$, $G[E']$ denotes the subgraph $G'(V, E')$ of G, and $\Gamma[E']$ is the subdrawing of $G[E']$ in Γ. Also, let $\mathcal{P}$ denote a desired geometric property of a drawing of a graph. An *edge stratification* (or simply a *stratification*) of Γ with respect to $\mathcal{P}$, also denoted as $S(\Gamma, \mathcal{P})$, is a partition of the edges of G into h subsets $E_1, E_2, \ldots, E_h$ such that, for every $i \in \{1, \ldots, h\}$, property $\mathcal{P}$ holds for $\Gamma[E_i]$. Each subdrawing $\Gamma[E_i]$ is called a *layer* of $S(\Gamma, \mathcal{P})$, and the *size* of $S(\Gamma, \mathcal{P})$ is the number h of its layers. We study the following general optimization problem.

Problem 1. – MINGENERALSTRATIFICATION: Given a straight-line drawing Γ and a geometric property $\mathcal{P}$, find a stratification $S(\Gamma, \mathcal{P})$ of minimum size.

In particular, we focus on the following geometric properties $\mathcal{P}$: (i) PLANARITY: the drawing is crossing free; (ii) LAC(α): any two crossing edges of the drawing form an angle of at least α radians (LAC stands for *large angle crossing*); (iii) k-PLANARITY: each edge of the drawing is crossed at most by k edges ($k \geq 1$). Each property gives rise to a specialized version of Problem 1 (see Fig. 2 for an example of the different types of stratification):

Problem 2. – MINPLANARSTRATIFICATION: Given a straight-line drawing Γ, find a stratification $S(\Gamma, \text{PLANARITY})$ of minimum size.

Problem 3. – MINLACSTRATIFICATION: Given a straight-line drawing Γ and a constant $\alpha \in (0, \frac{\pi}{2}]$, find a stratification $S(\Gamma, \text{LAC}(\alpha))$ of minimum size.

Problem 4. – MINk-PLANARSTRATIFICATION: Given a straight-line drawing Γ and a constant $k > 0$, find a stratification $S(\Gamma, k\text{-PLANARITY})$ of minimum size.

It is natural to ask what is the complexity of the stratification problems defined above. We prove that they are difficult at least as the well-known problem called *classification*, restricted to planar graphs with maximum vertex degree 4 or 5; this problem is conjectured to be NP-hard [9]. Namely, an *edge coloring* of a graph $G(V, E)$ is an assignment of edge colors such that adjacent edges have different colors. The minimum number of colors of an edge coloring of G is called the *chromatic index* of G. It is known that the chromatic index of a graph is either $\Delta(G)$ or $\Delta(G) + 1$, where $\Delta(G)$ is the maximum vertex degree of G [31]. The classification problem is the problem of deciding whether a graph G has chromatic index $\Delta(G)$ or $\Delta(G) + 1$, and it is NP-complete in general [20]. Restricting the input graph to be planar, the classification problem can be reduced to our stratification problems (we omit details due to space limitations).

Algorithms. Let Γ be a drawing of $G(V, E)$. To solve our different stratification problems on Γ we provide heuristics based on a common unified framework. It exploits an enhanced version of the *crossing graph* of Γ, which is a graph $\chi_\Gamma(V_\chi, E_\chi)$ having a vertex for each edge of Γ, i.e., $V_\chi = E$, and an edge for each pair of crossing edges of Γ, i.e., $E_\chi = \{(e_1, e_2) | e_1, e_2 \in E \text{ and } e_1 \text{ and } e_2 \text{ cross in } \Gamma\}$. In our enhanced version of χ_Γ, we add a weight to each edge $(e_1, e_2) \in E_\chi$, equal to the minimum angle formed by e_1 and e_2 at their crossing point in Γ. Given the one-to-one correspondence between the edges of Γ and the vertices of χ_Γ, an edge stratification $S(\Gamma, \mathcal{P}) = \{E_1, \ldots, E_h\}$ corresponds to coloring the vertices of χ_Γ such that the subgraph induced by all vertices with the same color satisfies a property $\mathcal{P}'$ that is the "translation" of $\mathcal{P}$ to the crossing graph. Namely, if $V_\chi^i \subseteq V_\chi$ is the color class associated with E_i $(1 \le i \le h)$, we have that: (i) $\mathcal{P} =$PLANARITY translates into $\mathcal{P}' =$INDEPENDENTSET: the vertices of V_χ^i form an independent set in χ_Γ; (ii) $\mathcal{P} = k$-PLANARITY translates into $\mathcal{P}' =$MAXDEGREE-k: the subgraph of χ_Γ induced by V_χ^i has vertex degree at most k; (iii) $\mathcal{P} =$LAC(α) translates into $\mathcal{P}' =$EDGEWEIGHT(α): the subgraph of χ_Γ induced by V_χ^i has no edge weight less than α.

Hence, computing a stratification $S(\Gamma, \mathcal{P}) = \{E_1 \ldots, E_h\}$ is equivalent to computing a coloring $C(\chi_\Gamma, \mathcal{P}') = \{V_\chi^1, \ldots, V_\chi^h\}$ of the vertices of χ, such that the subgraph induced by each V_χ^i satisfies property $\mathcal{P}'$. In particular, Problem MINPLANARSTRATIFICATION equals to the classical *minimum vertex coloring* problem on χ_Γ, which consists of coloring the vertices of χ_Γ with the minimum number of colors, such that no vertices with the same color are adjacent. Problems MINLACSTRATIFICATION can be reduced to a minimum vertex coloring problem on χ_Γ by applying a pre-processing step that removes from χ_Γ all the edges whose weight is at least α. Problem MINk-PLANARSTRATIFICATION corresponds to a generalization of the minimum vertex coloring on χ_Γ, which allows each vertex to have at most k adjacent vertices of its same color. Given this strong correlation among all problems, we solve them with a unified framework that is an adaptation of a heuristic for the minimum vertex coloring problem, called *sequential coloring* [5]. It has been shown to be more effective with respect to other heuristics for the minimum vertex coloring, and can be easily adapted to all our variants of this problem. Our unified heuristic framework works as follows.

Let $\mathcal{P}'$ be the desired property for the subgraph induced by each color class. The vertices $\{v_1, \ldots, v_{|E|}\}$ of χ_Γ are processed one per time; the first vertex is assigned to color class V_χ^1. If vertices $v_1, v_2, \ldots, v_{i-1}$ have been assigned to the color classes $V_\chi^1, V_\chi^2, \ldots, V_\chi^k$, the next vertex v_i is assigned to the color class V_χ^j, where j is the minimum value for which $V_\chi^j \cup \{v_i\}$ satisfies property $\mathcal{P}'$; if no such j exists, v_i is assigned to a new color class V_χ^{k+1}. Several different criteria can be used to choose the next vertex v_i to be processed. We choose vertex v_i with the highest *degree of saturation*, which is the number of different colors assigned to the neighbors of v_i. This strategy has been experimentally proven to give good performance in terms of number of colors used [5]. The time complexity of our heuristic can be evaluated as follows. By using a brute-force approach the crossing graph can be computed in $O(m^2)$, where m is the number of edges of Γ. Using the degree of saturation as the criterium for the vertex selection, the time complexity of the sequential coloring heuristic is $O(N^2 \log N)$, where N is the number of vertices of the graph to be colored. Since in our case $N = m$, the overall time complexity of our heuristic is $O(m^2 \log n)$.

4   User Study

To evaluate the usefulness of our approach, we performed a user study where different interfaces based on edge stratification are compared with an interface where drawings are not stratified. The stratifications were computed with the heuristic framework described in Section 3 and the geometric properties considered were PLANARITY and LAC($\frac{\pi}{4}$). In particular, we chose $\frac{\pi}{4}$ as minimum crossing angular resolution, because we observed that this value gives rise to limited number of layers without affecting too much the readability of each layer (see also [12]). Also, in the experiment we decided not to evaluate stratifications obtained for k-PLANARITY for two reasons: (i) comparing too many interfaces would have taken to the users a very long time to complete their test; (ii) there were not significant differences between the sizes of the stratifications $S(\Gamma, \text{PLANARITY})$ and $S(\Gamma, k\text{-PLANARITY})$ (considering small values of k) for the drawings Γ of our benchmark, thus there was no clear advantage in using k-PLANARITY with respect to PLANARITY from the practical point of view. Clearly, this last observation motivates the study of more effective heuristics to compute a stratification $S(G, k\text{-PLANARITY})$, when $k > 0$. Alternatively, we could consider large values of k, however, this would strongly reduce the readability of the layers.

The goal of our study was to address the following two research questions: **(Q1).** Given a straight-line drawing of a graph, does stratification assist in the reading of the relational information represented by the graph? **(Q2).** If the first question is settled in the affirmative, is one of the two considered geometric properties (PLANARITY and LAC($\frac{\pi}{4}$)) more effective in assisting the reading of such relational information?

We performed a within-subjects experiment involving 40 participants. We used 5 different drawings; for each drawing, the participants had to solve 3 different tasks, using 3 different user interfaces. Thus, a trial was represented by the triple $\langle$drawing, task, interface$\rangle$ and the number of trials for each participant was $5 \times 3 \times 3 = 45$.

Table 1. The graphs of the user study

graph	vertices	edges	density	stratification layers PLANARITY	stratification layers LAC($\frac{\pi}{4}$)
lesmis	77	2,148	27.9	4	3
football	115	613	5.33	7	4
gd01	249	635	2.55	6	4
organization	165	726	4.4	7	4
scalefree	204	803	3.94	9	7

Drawings. We chose 5 different complex graphs modeling both real and artificial networks of different type, and we drew them using the OGDF [1] implementation of the multi-level force-directed algorithm FMMM [19]. The chosen graphs are: lesmis, a coappearance graph of characters in the novel "Les Miserables" [22]; football, a graph of American football games [18]; gd01, a graph drawing self-reference network used in the GD 2001 contest [25]; organization, a social network modeling the relationships among employees in a private company [24]; scalefree, a scale-free network generated using the Barabási-Albert model [2]. For each graph, Table 1 reports the number of vertices, the number of edges, the density (ratio between number of edges and number of vertices), and the number of stratification layers for properties PLANARITY and LAC($\frac{\pi}{4}$).

Tasks. We chose 3 tasks as representative of the possible tasks involving graph reading. We aimed at having tasks significantly different one to another, easy to understand by non-expert users, and requiring both local and global explorations of the drawing. We considered: the *Shortest Path (SP)* task, which asks "How long is the shortest path between the two highlighted vertices?"; the *Common Adjacent (CO)* task, which asks "What is the number of adjacent vertices shared by the two highlighted vertices?"; and the *Degree (DE)* task, which asks "What is the degree of the highlighted vertex?". Similar tasks are used in other experiments on graph reading (see, e.g., [28]). The vertices highlighted for each task were chosen without looking at the stratification layers.

User interfaces. The three user interfaces differ from one another only by the geometric property used to compute a stratification. The *Planar Stratification (PS)* interface uses PLANARITY; the *LAC Stratification (LS)* interface uses LAC($\frac{\pi}{4}$); the *Overview (OV)* interface does not use any stratification (there is only one layer). To limit the number of layers in PS and LS, we halted the edge partitioning process when 80% of the edges of the drawing were placed in some layer. The remaining edges were all put in an additional layer, for which no geometric property is guaranteed. In every interface the edges of each layer were displayed with the same color and we used a different color for the edges of each layer. We chose the colors so to maximize the readability over a black background and so that different colors can be easily distinguished. Every interface allowed the user to select any combination of the available layers, so that only the edges in the selected layers were displayed on the screen; the edges in the non-selected

---

[1] http://www.ogdf.net

layers were sketched as transparent light gray segments (alpha=0.2). In this way, the user could focus on the selected layers only, still keeping in mind that the remaining edges were part of the drawing and should possibly be considered to properly solve the task. Thus, the user's strategy to explore the drawing varied from looking at one single layer per time to looking at the drawing as a whole. Also, the vertices were always drawn as white circles except the highlighted vertices of each trial, which were drawn red and slightly larger. Such a consistent environment did not require to the user any cognitive shift to move from one interface to another. Finally, the participants were not aware of the criteria used to stratify the drawings.

Experimental Procedure. Before starting the experiments, the participants received a brief tutorial introducing the basic concepts on graphs. Also, an explanation of the tasks and of the user interfaces was given with practical examples. The 45 trials were preceded by 4 training trials whose results were not taken into account, although the participants were not aware of it. Regardless of the task, participants had to answer each question by entering a number in a text box, with no time limit. However, they were asked to spend at most 3 minutes for each question. In order to counter the learning effect, in each experiment the 45 stimuli appeared in a randomized order, and the system randomly flipped and rotated each drawing. Between a question and the next one, participants could take a short break, without the possibility of exchanging information.

Results. 40 volunteering students (with age from 19 to 25) in Computer Engineering took part in the experiments. We recorded their answers and the time spent for each question. We compared the performance of OV, PS, and LS in terms of absolute error rate (the absolute value of the difference between the user answer and the correct answer) and response time. First of all, we performed a Shapiro-Wilk test (α=0.05) to determine whether the data was normally distributed or not. We found that none of the considered populations was normally distributed. Thus, we performed a non-parametric analysis exploiting repeated measures Friedman tests (α=0.05), with post-hoc pairwise comparisons. We applied Bonferroni corrections on the pairwise comparisons setting $\alpha < 0.017$. We obtained these results (see Table 2). (i) Considering all the tasks, both PS and LS significantly outperformed OV in terms of absolute error rate. This improvement in the accuracy comes together with a slower response time: PS and LS show slower performance than OV, and LS is faster than PS. (ii) For task SP, the mean absolute error rate of LS is smaller than the one of OV and PS, although such a difference turns out to be not statistically significant. In terms of response time, again, OV is faster than PS and LS, while LS is faster than PS. (iii) For task CA, the three interfaces led to comparable performance in terms of absolute error rate. In terms of response time, the situation is similar to the previous cases, OV is faster than PS and LS, while LS is faster than PS. (iv) For task DE, both PS and LS outperformed OV in terms of absolute error rate, and the difference turns out to be statistically significant. Also, PS led to more accurate results than LS, still with statistical significance. About the response time, OV and LS behave similarly and slightly faster than PS.

Discussion. The results of the user study show an improvement in terms of accuracy in the reading of the displayed graphs when using stratification. The PS interface outperformed both LS and OV for most of the considered tasks. We conclude that stratifying

Table 2. Results of the user study. The mean values and the pairwise significance between each user interface are shown for absolute error rate and time

		Overall	SP	CA	DE			Overall	SP	CA	DE
Abs. Error Rate	mean OV	2.74	1.84	1.93	4.43	**Time (sec.)**	mean OV	54.36	60.47	61.23	41.36
	mean PS	1.87	1.79	1.96	1.86		mean PS	80.01	91.78	98.26	50.01
	mean LS	1.92	1.57	1.98	2.22		mean LS	68.73	77.46	85.71	43.00
	OV vs PS	< .001	n.s.	n.s.	< .001		OV vs PS	< .001	< .001	< .001	.001
	OV vs LS	< .001	n.s.	n.s.	< .001		OV vs LS	< .001	< .001	< .001	n.s.
	PS vs LS	n.s.	n.s.	n.s.	.010		PS vs LS	< .001	.002	.001	< .001

the drawing into planar layers gave a significant help to the participants. The task that received more significant advantage from both PS and LS is task DE. Indeed, counting the degree of a vertex can be quite hard when the drawing is cluttered around the highlighted vertex; on the other hand, by selecting a layer per time, one can effectively cope with such clutter and the partial degree of the vertex can be easily counted, at the expenses of a negligible increase of the response time (less than 10 seconds in the average). Moreover, according to this strategy, having planar layers guarantees that the region of the drawing around the highlighted vertex is crossing free, and hence clearer.

5   Comparison of Graph Drawing Algorithms

We describe a second experiment, which compares several force-directed algorithms to answer the following research questions: **(Q3).** What is the force-directed drawing algorithm for which the computed layouts require the smallest number of layers with properties PLANARITY and LAC($\frac{\pi}{4}$), when stratified with our heuristics? **(Q4).** How are the number of crossings and the number of layers correlated?

We tested 5 different force-directed algorithms on a benchmark of 105 graphs, thus collecting 525 drawings; for each drawing Γ, we measured its number of crossings, the size of a stratification $S(\Gamma, \text{PLANARITY})$, and the size of a stratification $S(\Gamma, \text{LAC}(\frac{\pi}{4}))$ computed by our heuristics. Observe that, since the size of the crossing graph can be $\Omega(n^4)$, coloring this graph exactly is often prohibitive even for small graphs.

Algorithms. We tested the whole set of force-directed algorithms available in the OGDF library: Kamada-Kawai (KK) [21], Fruchterman-Reingold(FR) [17], GEM (GEM) [16], FM^3 (FMMM) [19], and Stress majorization (SM) [4]. Since tuning is a critical issue for force-directed techniques, it is worth remarking that we initialized the algorithms by using the default parameters set by their implementations in the library.

Benchmark. We ran the drawing algorithms on a benchmark of 105 complex graphs, organized in three groups: UniformRandGraphs, containing 40 random graphs generated with a uniform probability distribution; for each $n \in \{100, 200, \ldots, 400\}$ we generated 10 graphs with density in the interval $[2, 6]$. ScaleFreeRandGraphs, containing 60 small-world and scale-free graphs generated with the LFR algorithm [23], already used to generate graphs in previous extensive experimental works [13]; for each

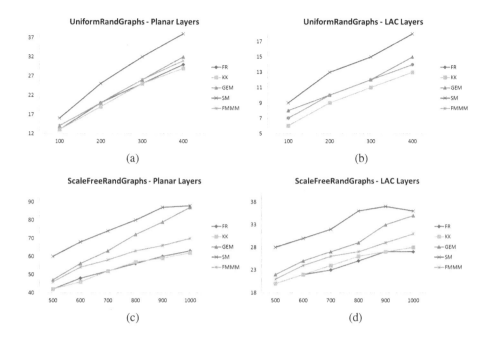

Fig. 3. Mean number of planar and LAC($\frac{\pi}{4}$) layers on UniformRandGraphs (a-b) and ScaleFreeRandGraphs (c-d). The x-axis reports the number of vertices.

$n \in \{500, 600, \dots, 1000\}$ we generated 10 graphs with density in the interval $[4, 8]$. UserStudyGraphs, which contains the 5 graphs used in the user study.

Results. (i) On UniformRandGraphs, the layouts of FR, GEM, and FMMM require a comparable number of layers regardless the geometric property. KK outperforms the other algorithms for LAC($\frac{\pi}{4}$) layers, while SM always led to the highest number of layers. See Figs. 3(a) and 3(b). (ii) On ScaleFreeRandGraphs, the layouts of FR and KK most frequently require the smallest number of layers, regardless the geometric property; FMMM led to slightly bigger numbers, while again SM led to the largest number of layers. Finally, GEM behaves similarly to FMMM for small values of n, and approaches SM as n grows. See Figs. 3(c) and 3(d). (iii) On UserStudyGraphs, the layouts computed by FR are those that more often require the smallest number of layers. KK and FMMM behave similarly, GEM is slightly worse, and SM most frequently led to the largest number of layers. We omit the charts due to space limitations.

To evaluate the correlation between layers and crossings, we executed a Kendall's τ test ($\alpha < 0.01$). The test showed a strong correlation ($0.9 < r < 0.96$) between number of crossings and number of layers (both planar and LAC($\frac{\pi}{4}$)), for all the algorithms.

Discussion. Concerning question Q3, the results show that FR and KK have the best performance in terms of number of layers required to stratify their layouts (both planar and LAC($\frac{\pi}{4}$)), while SM is always the worst for the graphs in our benchmark. About question Q4, we interpret the strong correlation between the number of crossings and the number of layers as a positive feature of the drawing algorithms, which witnesses a quite uniform distribution of the crossings in the drawing. We also observe that, for some instances, there are pairs of drawings $\langle \Gamma_1, \Gamma_2 \rangle$ computed by different algorithms such that Γ_1 has more crossings but less layers than Γ_2. By visually inspecting these instances, the crossings in the drawings with fewer layers appear to be more evenly distributed, resulting in a more readable drawing in spite of the greater number of crossings. This seems to confirm our intuition that the number of layers is also related to the distribution of the crossings in the drawing area, and suggests that the number of layers could be a more reliable measure of the drawing visual complexity with respect to the number of crossings. Note that just measuring the crossing spatial distribution in a drawing does not necessarily yield to similar conclusions; indeed, a drawing with low average crossing spatial distribution may still contain very cluttered spots surrounded by crossing-free regions, which might cause a high number of layers.

6 Conclusions and Future Research Directions

Our framework, based on the use of the crossing graph, is not suited for very large graphs. The number of crossings can be $\Omega(n^4)$ and hence, even for relatively small complex drawings, the crossing graph can be so large that the performance of our heuristic degrades both in terms of space and time. It would be useful to devise more efficient heuristics that do not make use of crossing graphs. Also, in our user interface we presented the different layers using colors and allowing users to select any subset of layers. It would be interesting to design new visualization paradigms to effectively present stratified drawings. A 2.5D visualization could be an interesting option to explore.

References

1. Auber, D., Chiricota, Y., Jourdan, F., Melançon, G.: Multiscale visualization of small world networks. In: InfoVis 2003, pp. 75–81. IEEE (2003)
2. Barabasi, A.-L., Albert, R.: Emergence of Scaling in Random Networks. Science 286(5439), 509–512 (1999)
3. Batagelj, V., Brandenburg, F., Didimo, W., Liotta, G., Palladino, P., Patrignani, M.: Visual analysis of large graphs using (X,Y)-clustering and hybrid visualizations. IEEE TVCG 17(11), 1587–1598 (2011)
4. Brandes, U., Pich, C.: More flexible radial layout. In: Eppstein, D., Gansner, E.R. (eds.) GD 2009. LNCS, vol. 5849, pp. 107–118. Springer, Heidelberg (2010)
5. Brélaz, D.: New methods to color the vertices of a graph. Comm. ACM 22, 251–256 (1979)
6. Bruckdorfer, T., Cornelsen, S., Gutwenger, C., Kaufmann, M., Montecchiani, F., Nöllenburg, M., Wolff, A.: Progress on partial edge drawings. In: Didimo, W., Patrignani, M. (eds.) GD 2012. LNCS, vol. 7704, pp. 67–78. Springer, Heidelberg (2013)
7. Brunel, E., Gemsa, A., Krug, M., Rutter, I., Wagner, D.: Generalizing geometric graphs. In: Speckmann, B. (ed.) GD 2011. LNCS, vol. 7034, pp. 179–190. Springer, Heidelberg (2011)
8. Buchheim, C., Chimani, M., Gutwenger, C., Jünger, M., Mutzel, P.: Crossings and planarization. In: Tamassia, R. (ed.) Handbook of Graph Drawing and Visualization. CRC Press (2013)
9. Chrobak, M., Nishizeki, T.: Improved edge-coloring algorithms for planar graphs. J. Algo. 11(1), 102–116 (1990)

10. Di Giacomo, E., Didimo, W., Liotta, G., Montecchiani, F.: *h*-quasi planar drawings of bounded treewidth graphs in linear area. In: Golumbic, M.C., Stern, M., Levy, A., Morgenstern, G. (eds.) WG 2012. LNCS, vol. 7551, pp. 91–102. Springer, Heidelberg (2012)
11. Di Giacomo, E., Didimo, W., Liotta, G., Montecchiani, F.: Area requirement of graph drawings with few crossings per edge. Comp. Geom. 46(8), 909–916 (2013)
12. Didimo, W., Liotta, G.: The crossing angle resolution in graph drawing. In: Pach, J. (ed.) Thirty Essays on Geometric Graph Theory. Springer (2012)
13. Didimo, W., Montecchiani, F.: Fast layout computation of hierarchically clustered networks: Algorithmic advances and experimental analysis. In: IV 2012, pp. 18–23 (2012)
14. Dillencourt, M.B., Eppstein, D., Hirschberg, D.S.: Geometric thickness of complete graphs. Jour. Graph. Alg. and Appl. 4(3), 5–17 (2000)
15. Duncan, C.A., Eppstein, D., Kobourov, S.G.: The geometric thickness of low degree graphs. In: SoCG 2004, pp. 340–346. ACM (2004)
16. Frick, A., Ludwig, A., Mehldau, H.: A fast adaptive layout algorithm for undirected graphs. In: Tamassia, R., Tollis, I.G. (eds.) GD 1994. LNCS, vol. 894, pp. 388–403. Springer, Heidelberg (1995)
17. Fruchterman, T.M.J., Reingold, E.M.: Graph drawing by force-directed placement. Softw. Pract. Exper. 21(11), 1129–1164 (1991)
18. Girvan, M., Newman, M.E.J.: Community structure in social and biological networks. PNAS 99(12), 7821–7826 (2002)
19. Hachul, S., Jünger, M.: Drawing large graphs with a potential-field-based multilevel algorithm. In: Pach, J. (ed.) GD 2004. LNCS, vol. 3383, pp. 285–295. Springer, Heidelberg (2005)
20. Holyer, I.: The NP-completeness of edge-coloring. SIAM J. Comput. 10(4), 718–720 (1981)
21. Kamada, T., Kawai, S.: An algorithm for drawing general undirected graphs. Inf. Process. Lett. 31(1), 7–15 (1989)
22. Knuth, D.E.: The Stanford Graphbase: A Platform for Combinatorial Computing. Addison-Wesley Professional (1993)
23. Lancichinetti, A., Fortunato, S., Radicchi, F.: Benchmark graphs for testing community detection algorithms. Phys. Rev. E 78(4) (2008)
24. Michael Fire, Y.E., Puzis, R.: Organization mining using online social networks (2012), http://proj.ise.bgu.ac.il/sns
25. Mutzel, P., Jünger, M., Leipert, S. (eds.): GD 2001. LNCS, vol. 2265. Springer, Heidelberg (2002)
26. Purchase, H.C.: Effective information visualisation: a study of graph drawing aesthetics and algorithms. Interact. Comput. 13(2), 147–162 (2000)
27. Purchase, H.C., Carrington, D.A., Allder, J.-A.: Empirical evaluation of aesthetics-based graph layout. Empir. Softw. Eng. 7(3), 233–255 (2002)
28. Purchase, H.C., Hamer, J., Nöllenburg, M., Kobourov, S.G.: On the usability of lombardi graph drawings. In: Didimo, W., Patrignani, M. (eds.) GD 2012. LNCS, vol. 7704, pp. 451–462. Springer, Heidelberg (2013)
29. Shneiderman, B., Dunne, C.: Interactive network exploration to derive insights: Filtering, clustering, grouping, and simplification. In: Didimo, W., Patrignani, M. (eds.) GD 2012. LNCS, vol. 7704, pp. 2–18. Springer, Heidelberg (2013)
30. van Ham, F., van Wijk, J.J.: Interactive visualization of small world graphs. In: InfoVis 2004, pp. 199–206. IEEE (2004)
31. Vizing, V.G.: On an estimate of the chromatic class of a *p*-graph. Diskret. Analiz No. 3, 25–30 (1964)
32. Zhou, H., Xu, P., Yuan, X., Qu, H.: Edge bundling in information visualization. Tsinghua Science and Technology 18(2), 145–156 (2013)

Drawing Planar Graphs with a Prescribed Inner Face

Tamara Mchedlidze, Martin Nöllenburg, and Ignaz Rutter

Institute of Theoretical Informatics, Karlsruhe Institute of Technology (KIT), Germany

Abstract. Given a plane graph G (i.e., a planar graph with a fixed planar embedding) and a simple cycle C in G whose vertices are mapped to a convex polygon, we consider the question whether this drawing can be extended to a planar straight-line drawing of G. We characterize when this is possible in terms of simple necessary conditions, which we prove to be sufficient. This also leads to a linear-time testing algorithm. If a drawing extension exists, it can be computed in the same running time.

1 Introduction

The problem of extending a partial drawing of a graph to a complete one is a fundamental problem in graph drawing that has many applications, e.g., in dynamic graph drawing and graph interaction. This problem has been studied most in the planar setting and often occurs as a subproblem in the construction of planar drawings.

The earliest example of such a drawing extension result are so-called Tutte embeddings. In his seminal paper "How to Draw a Graph" [10], Tutte showed that any triconnected planar graph admits a convex drawing with its outer vertices fixed to an arbitrary convex polygon. The strong impact of this fundamental result is illustrated by its, to date, more than 850 citations and the fact that it received the "Best Fundamental Paper Award" from GD'12. The work of Tutte has been extended in several ways. In particular, it has been strengthened to only require polynomial area [4], even in the presence of collinear points [3]. Hong and Nagamochi extended the result to show that triconnected graphs admit convex drawings when their outer vertices are fixed to a star-shaped polygon [5]. For general subdrawings, the decision problem of whether a planar straight-line drawing extension exists is NP-hard [9]. Pach and Wenger [8] showed that every subdrawing of a planar graph that fixes only the vertex positions can be extended to a planar drawing with $O(n)$ bends per edge and that this bound is tight. The drawing extension problem has also been studied in a topological setting, where edges are represented by non-crossing curves. In contrast to the straight-line variant, it can be tested in linear time whether a drawing extension of a given subdrawing exists [1]. Moreover, there is a characterization via forbidden substructures [6].

In this paper, we study the problem of finding a planar straight-line drawing extension of a plane graph for which an arbitrary cycle has been fixed to a convex polygon. It is easy to see that a drawing extension does not always exist in this case; see Fig. 1(a). Let G be a plane graph and let C be a simple cycle of G represented by a convex polygon Γ_C in the plane. The following two simple conditions are clearly necessary for the existence of a drawing extension: (i) C has no chords that must be embedded outside of C and (ii) for every vertex v with neighbors on C that must be embedded outside of

S. Wismath and A. Wolff (Eds.): GD 2013, LNCS 8242, pp. 316–327, 2013.

C there exists a placement of v outside Γ_C such that the drawing of the graph induced by C and v is plane and bounded by the same cycle as in G. We show in this paper that these two conditions are in fact sufficient. Both conditions can be tested in linear time, and if they are satisfied, a corresponding drawing extension can be constructed within the same time bound.

Our paper starts with some necessary definitions (Section 2) and useful combinatorial properties (Section 3). The idea of our main result has two steps. We first show in Section 4 that the conditions are sufficient if Γ_C is one-sided (i.e., it has an edge whose incident inner angles are both less than 90°). Afterward, we show in Section 5 that, for an arbitrary convex polygon Γ_C, we can place the neighborhood of C in such a way that the drawing is planar, and such that the boundary C' of its outer face is a one-sided polygon $\Gamma_{C'}$. Moreover, our construction ensures that the remaining graph satisfies the conditions for extendability of $\Gamma_{C'}$. The general result then follows directly from the one-sided case.

2 Definitions and a Necessary Condition

Plane Graphs and Subgraphs. A graph $G = (V, E)$ is *planar* if it has a drawing Γ in the plane $\mathbb{R}^2$ without edge crossings. Drawing Γ subdivides the plane into connected regions called *faces*; the unbounded region is the *outer* and the other regions are the *inner* faces. The boundary of a face is called *facial cycle*, and *outer cycle* for the outer face. The cyclic ordering of edges around each vertex of Γ together with the description of the external face of G is called an *embedding* of G. A graph G with a planar embedding is called *plane graph*. A *plane subgraph H* of G is a subgraph of G together with a planar embedding that is the restriction of the embedding of G to H.

Let G be a plane graph and let C be a simple cycle of G. Cycle C is called *strictly internal*, if it does not contain any vertex of the outer face of G. A chord of C is called *outer* if it lies outside C in G. A cycle without outer chords is called *outerchordless*. The *subgraph of G inside C* is the plane subgraph of G that is constituted by vertices and edges of C and all vertices and edges of G that lie inside C.

Connectivity. A graph G is *k-connected* if removal of any set of $k - 1$ vertices of G does not disconnect the graph. For $k = 2, 3$ a k-connected graph is also called *biconnected* and *triconnected*, respectively. An internally triangulated plane graph is triconnected if and only if there is no edge connecting two non-consecutive vertices of its outer cycle (see, for example, [2]).

Star-shaped and One-sided Polygons. Let Π be a polygon in the plane. Two points inside or on the boundary of Π are mutually *visible*, if the straight-line segment connecting them belongs to the interior of Π. The *kernel* $K(\Pi)$ of polygon Π is the set of all the points inside Π from which all vertices of Π are visible. We say that Π is *star-shaped* if $K(\Pi) \neq \emptyset$. We observe that the given definition of a star-shape ensures that its kernel has a positive area.

A convex polygon Π with k vertices is called *one-sided*, if there exists an edge e (i.e., a line segment) of Π such that the orthogonal projection to the line supporting e

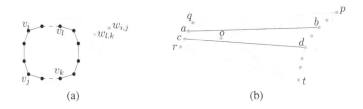

Fig. 1. Convex polygon of cycle C is denoted by black. Vertex $w_{i,j}$ cannot be placed on the plane without changing the embedding or intersecting C. Vertices $w_{i,j}$ and $w_{l,k}$ are petals of C, where $w_{l,k} \prec w_{i,j}$. Petal $w_{l,k}$ is realizable, while petal $w_{i,j}$ is not. (b) Illustration of Fact 1.

maps all polygon vertices actually onto segment e. Then e is called the *base edge* of Π. Without loss of generality let $e = (v_1, v_k)$ and $v_1, \ldots, v_k$ be the clockwise ordered sequence of vertices of Π.

Extension of a Drawing. Let G be a plane graph and let H be a plane subgraph of G. Let Γ_H be a planar straight-line drawing of H. We say that Γ_H is *extendable* if drawing Γ_H can be completed to a planar straight-line drawing Γ_G of the plane graph G. Then Γ_G is called an *extension* of Γ_H. A planar straight-line drawing of G is called *convex*, if every face of G (including the outer face) is represented as a convex polygon.

The following theorem by Hong and Nagamochi [5] shows the extendability of a prescribed star-shaped outer face of a plane graph.

Theorem 1 (Hong, Nagamochi [5]). *Every drawing of the outer face f of a 3-connected graph G as a star-shaped polygon can be extended to a planar drawing of G, where each internal face is represented by a convex polygon. Such a drawing can be computed in linear time.*

Petals and Stamens. Let G be a plane graph, and let P_{uv} be a path in G between vertices u and v. Its subpath from vertex a to vertex b is denoted by $P_{uv}[a, b]$. Let C be a simple cycle of G, and let $v_1, \ldots, v_k$ be the vertices of C in clockwise order. Given two vertices v_i and v_j of C, we denote by $C[v_i, v_j]$ the subpath of C encountered when we traverse C clockwise from v_i to v_j. Assume that C is represented by a convex polygon Γ_C in the plane. We say that a vertex v_i, $1 \le i \le k$ of Γ_C is *flat*, if $\angle v_{i-1} v_i v_{i+1} = \pi$. Throughout this paper, we assume that convex polygons do not have flat vertices.

A vertex $w \in V(G) \setminus V(C)$ adjacent to at least two vertices of C and lying outside C in G, is called a *petal* of C (see Figure 1(a)). Consider the plane subgraph G' of G induced by the vertices $V(C) \cup \{w\}$. Vertex w appears on the boundary of G' between two vertices of C, i.e. after some $v_i \in V(C)$ and before some $v_j \in V(C)$ in clockwise order. To indicate this fact, we will denote petal w by $w_{i,j}$. Edges $(w_{i,j}, v_i)$ and $(w_{i,j}, v_j)$ are called the *outer* edges of petal $w_{i,j}$. The subpath $C[v_i, v_j]$ of C is called *base* of the petal $w_{i,j}$. A vertex v_f is called *internal*, if it appears on C after v_i and before v_j in clockwise order. A petal $w_{i,i+1}$ is called *trivial*. A vertex of $V(G) \setminus V(C)$ adjacent to exactly one vertex of C is called a *stamen* of C.

Let v be a petal of C and let u be either a petal or a stamen of C, we say that u is *nested* in v, and denote this fact by $u \prec v$, if u lies in the cycle delimited by the base and the outer edges of petal v. For two stamens u and v, neither $u \prec v$ nor $v \prec u$. So for each pair of stamens or petals u and v we have either $u \prec v$, or $v \prec u$, or none of these. This relation $\prec$ is a partial order. A petal or a stamen u of C is called *outermost* if it is maximal with respect to $\prec$.

Necessary Petal Condition. Let again G be a plane graph and let C be an outerchordless cycle of G represented by a convex polygon Γ_C in the plane. Let $w_{i,j}$ be a petal of C. Let G' be the plane subgraph of G, induced by the vertices $V(C) \cup \{w_{i,j}\}$. We say that $w_{i,j}$ is *realizable* if there exists a planar drawing of G' which is an extension of Γ_C. This gives us the necessary condition that Γ_C is *extendable only if each petal of C is realizable*. In the rest of the paper we prove that this condition is sufficient.

3   Combinatorial Properties of Graphs and Petals

In this section, we derive several properties of petals in graphs, which we use throughout the construction of the drawing extension in the remaining parts of this paper. Due to space constraints the proofs can be found in the full version of this paper [7]. Proposition 1 allows us to restrict our attention to maximal plane graphs for which the given cycle C is strictly internal. The remaining lemmas are concerned with the structure of the (outermost) petals of C in such a graph.

Proposition 1. *Let G be a plane graph on n vertices and let C be a simple outerchordless cycle of G. There exists a plane supergraph G' of G with $O(n)$ vertices such that (i) G' is maximal, (ii) there are no outer chords of C in G', (iii) each petal of G' with respect to C is either trivial or has the same neighbors on C as in G, and (iv) C is strictly internal to G.*

In the following we assume that our given plane graph is maximal, and the given cycle is strictly internal, otherwise Proposition 1 is applied.

Lemma 1. *Let G be a triangulated planar graph with a strictly internal outerchordless cycle C. Then the following statements hold. (i) Each vertex of C that is not internal to an outermost petal is adjacent to two outermost petals. (ii) There is a simple cycle C' whose interior contains C and that contains exactly the outermost stamens and petals of C.*

Lemma 2. *Let G be a maximal planar graph with a strictly internal outerchordless cycle C. Let u and v be two adjacent vertices on C that are not internal to the same petal. Then there exists a third vertex w of C such that there exist three chordless disjoint paths from u, v and w to the vertices of the outer face of G such that none of them contains other vertices of C.*

4   Extension of a One-sided Polygon

Let G be a plane graph, and let C be a simple outerchordless cycle, represented by a one-sided polygon Γ_C. In this section, we show that if each petal of C is realizable, then Γ_C is extendable to a straight-line drawing of G. This result serves as a tool for the general case, which is shown in Section 5.

The drawing extension we produce preserves the outer face, i.e., if the extension exists, then it has the same outer face as G. It is worth mentioning that, if we are allowed to change the outer face, the proof becomes rather simple, as the following claim shows.

Claim 1. *Let G be a maximal plane graph and let C be an outerchordless cycle of G, represented in the plane by a one-sided polygon Γ_C. Then drawing Γ_C is extendable.*

Proof. Let (v_1, v_k) be the base edge of Γ_C. Edge (v_1, v_k) is incident to two faces of G, to a face f_{in} inside C and to a face f_{out} outside C. We select f_{out} as the outer face of G. With this choice, edge (v_1, v_k) is on the outer face of G. Let v be the third vertex of this face. We place the vertex v far enough from Γ_C, so that all vertices of Γ_C are visible from v. Thus, we obtain a planar straight-line drawing of the subgraph G_v induced by the vertices $V(C) \cup \{v\}$, such that each face is represented by a star-shaped polygon. Each subgraph of G inside a face of G_v is triconnected, and therefore, we can complete the existing drawing to a straight-line planar drawing of G, by Theorem 1. □

In the rest of the section we show that extendability of Γ_C can be efficiently tested, even if the outer face of G has to be preserved. The following simple geometric fact will be used in the proof of the result of this section (see Figure 1(b) for the illustration).

Fact 1. *Let $pqrt$ be a convex quadrilateral and let o be the intersection of its diagonals. Let S_{pt} be a one-sided convex polygon with base $\overline{pt}$, that lies inside triangle $\triangle opt$. Let $\overline{ab}$ and $\overline{cd}$ be such that $b, d \in S_{pt}$, ordered clockwise as t, d, b, p and $a, c \in \overline{qr}$, ordered as q, a, c, r. Then, neither $\overline{ab}$ and $\overline{cd}$ intersect each other, nor do they intersect a segment between two consecutive points of S_{pt}.*

We are now ready to prove the main result of this section.

Theorem 2. *Let G be a maximal plane graph and C be a strictly internal simple outerchordless cycle of G, represented in the plane by a one-sided polygon Γ_C. If every petal of C is realizable, then Γ_C is extendable.*

Proof. Let $v_1, \ldots, v_k$ be the clockwise ordering of the vertices of C, so that (v_1, v_k) is the base of Γ_C. We rotate Γ_C so that (v_1, v_k) is horizontal. Let a, b, c be the vertices of the external face of G, in clockwise order, see Fig. 2. By Lemma 2, there exists a vertex $v_j, 1 < j < k$, such that there exist chordless disjoint paths between v_1, v_j, v_k, and the vertices a, b, c, respectively. Without loss of generality assume they are P_{v_1a}, P_{v_jb} and P_{v_kc}. Some vertices of P_{v_1a} and P_{v_kc} are possibly adjacent to each other, as well as to the boundary of C. Depending on these adjacencies, we show how to draw the paths P_{v_1a}, P_{v_kc} and how to place vertex b, so that the graph induced by these vertices and cycle C is drawn with star-shaped faces. Then, the drawing of G can be completed by applying Theorem 1. Let v_i be the topmost vertex of Γ_C. It can happen that there are two adjacent topmost vertices v_i and v_{i+1}. However, v_{i-1} and v_{i+2} are lower, since Γ_C does not contain flat vertices. In the following, we assume that v_i and v_{i+1} have the same y-coordinate. The case when v_i is unique can be seen as a special case where $v_i = v_{i+1}$. Without loss of generality assume that $i + 1 \leq j \leq k - 1$, the case where $2 \leq j \leq i$

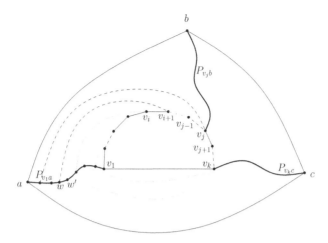

Fig. 2. Illustration for the proof of Theorem 2. Edges between $C[v_1, v_i]$ and $P_{v_1a}[v_1, w'] \cup \{v_k\}$ are gray. Edges between $C[v_{i+1}, v_j]$ and $P_{v_1a}[w, a]$ are dashed.

is treated symmetrically. Notice that the presence of the path P_{v_jb} ensures that edges between vertices of P_{v_1a} and P_{v_kc} can only lie in the interior of the cycle delimited by these paths and edges (v_1, v_k) and (a, c) (refer to Figure 2). Consider a vertex of P_{v_1a} which is a petal of C. The base of such a petal cannot contain edge (v_{k-1}, v_k), since this would cause a crossing with P_{v_kc}. Moreover, if the base of this petal contains edge (v_1, v_k), then it cannot contain any edge (v_f, v_{f+1}) for $i \leq f < j$, since otherwise this petal would not be realizable. Thus a vertex of P_{v_1a} is either adjacent to v_k or to a vertex v_f, where $i + 1 \leq f \leq j$, but not both. It is worth mentioning that a vertex of P_{v_1a} cannot be adjacent to any v_f, $j + 1 \leq f \leq k - 1$, since such an adjacency would cause a crossing either with P_{v_jb} or with P_{v_kc}.

Let ℓ_a, ℓ and ℓ_c be three distinct lines through v_j that lie clockwise between the slopes of edges (v_{j-1}, v_j) and (v_j, v_{j+1}) (see Figure 3). Such lines exist since Γ_C does not contain flat vertices. Let ℓ_i be the line through v_i with the slope of (v_{i-1}, v_i). Let ℓ_a^1 be the half-line originating at an internal point of (v_1, v_k) towards $-\infty$, slightly rotated counterclockwise from the horizontal position, so that it crosses ℓ_i. Let q denote the intersection point of ℓ_a^1 and ℓ_i. Let p be any point on ℓ_a^1 further away from v_1 than q. Let ℓ_a^2 be the line through p with the slope of ℓ. By construction of lines ℓ_a, ℓ and ℓ_c, line ℓ_a^2 crosses ℓ_a above the polygon Γ_C at point p_a and line ℓ_c below this polygon at point p_c.

Let G' be the plane subgraph of G induced by the vertices of C, P_{v_1a}, and P_{v_kc}. The outer cycle of G' consists of edge (a, c) and a path P_{ac} between vertices a and c.

Claim 2. *The vertices of P_{v_1a} and P_{v_kc} can be placed on lines ℓ_a^1, ℓ_a^2 and ℓ_c such that in the resulting straight-line drawing of G', path P_{ac} is represented by an x-monotone polygonal chain, and the inner faces of G' are star-shaped polygons.*

The vertices of P_{v_1a} will be placed on line ℓ_a^1 between points p and q and on line ℓ_a^2 above point p_a. The vertices of P_{v_kc} will be placed on ℓ_c below p_c. In order to place

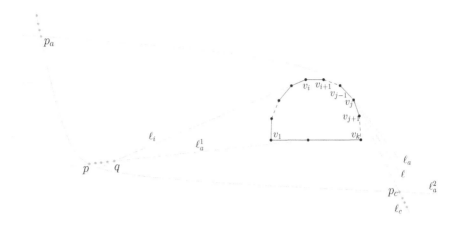

Fig. 3. Illustration for the proof of Theorem 2. For space reasons lines were shown by curves.

the vertices, we need to understand how the vertices of $P_{v_1 a}$ are adjacent to vertices of C. As we travel on $P_{v_1 a}$ from v_1 to a, we first meet all vertices adjacent to $v_1, \ldots, v_i$ and then all vertices adjacent to $v_{i+1}, \ldots, v_j$, since G is a planar graph. Let w be the first vertex of $P_{v_1 a}$ adjacent to v_f, $i + 1 \leq f \leq j$, and let w' be the vertex preceding w on $P_{v_1 a}$. We place vertices of $P_{v_1 a}[v_1, w']$, in the order they appear in the path, on line ℓ_a^1, between q and p, in increasing distance from v_1. We place all vertices of $P_{v_1 a}[w, a]$ on ℓ_a^2 above p_a in increasing distance from p. We draw the edges between the vertices of C and $P_{v_1 a}$. Notice that vertex w might not exist, since it might happen that none of the vertices of $P_{v_1 a}$ is adjacent to v_f, $i + 1 \leq f \leq k$. In this case all vertices of $P_{v_1 a}$ are placed on line ℓ_a^1, between q and p. In the following, we show that the constructed drawing is planar. Notice that the quadrilateral formed by the points w, a, v_j, v_{i+1} is convex, by the choice of line ℓ_a^2 and the positions of vertices w and a on it. Also, notice that the points of vertices $v_{i+1}, \ldots, v_j$ form a one-sided polygon with base segment $\overline{v_{i+1} v_j}$, which lies in the triangle $\triangle o v_j v_{i+1}$, where o is the intersection of $\overline{v_{i+1} a}$ and $\overline{v_j w}$. Thus, by Fact 1, the edges connecting $C[v_{i+1}, v_j]$ and $P_{v_1 a}[w, a]$ do not cross each other. By applying Fact 1, we can also prove that edges connecting $P_{v_1 a}[v_1, w']$ with $C[v_1, v_i]$, cross neither each other, nor Γ_C. Recalling that vertices of $P_{v_1 a}[v_1, w']$ can be also adjacent to v_k, we notice that these edges also do not cross Γ_C, by the choice of line ℓ_a^1. Finally, path $P_{v_1 a}$ is chordless, and therefore the current drawing is planar. Notice that the subpath of $P_{a,c}$ that has already been drawn is represented by an x-monotone chain. We next draw the vertices of $P_{v_k c}$. We observe that in the already constructed drawing path $P_{v_1 a}$ taken together with edge (v_1, v_k) is represented by an x-monotone chain, each point of which is visible from any point below the line ℓ_a^2. This means that any point below line ℓ_a^2, can be connected by a straight-line segment to the vertices $V(P_{v_1 a}) \cup \{v_k\}$ without creating any crossing either with $P_{v_1 a}$ or with (v_1, v_k). We also notice that any of the vertices $v_j, \ldots, v_k$ can be connected to a point of ℓ_c, without intersecting Γ_C. Recall that p_c denotes the intersection point of ℓ_c and ℓ_a^2. Thus we place the vertices of $P_{v_k c}$ on the line ℓ_c, below ℓ_a^2, in increasing distance from point p_c. Applying Fact 1 we can prove that the edges induced by $\{v_j, \ldots, v_k\} \cup V(P_{v_k c})$ are

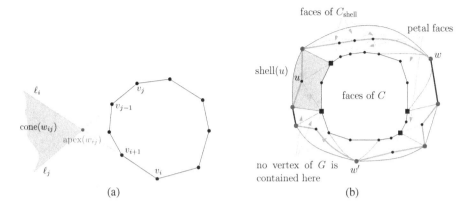

Fig. 4. (a) Vertex $w_{i,j}$ is the petal of C with base $C[v_i, v_j]$. Point apex$(w_{i,j})$ is red, region cone$(w_{i,j})$ is gray. (b) Graph G_{shell}. Polygon Γ_C is black. Cycle C_{shell} is bold gray. Cycle C'_{shell} is blue. Graph G'_{shell} is comprised by blue, red and black edges. Vertices of B are squares.

drawn without crossings. Edges between $P_{v_k c}$ and $P_{v_1 a}$ cross neither $P_{v_1 a}$, nor (v_1, v_k) by the choice of lines ℓ_c and ℓ_a^2.

We have constructed a planar straight-line drawing of G'. We notice that path P_{ac} is drawn as an x-monotone polygonal chain. We also notice that the faces of G', created when placing vertices of $P_{v_1 a}$ (resp. $P_{v_k c}$) are star-shaped and have their kernels arbitrarily close to the vertices of $P_{v_1 a}$ (resp. $P_{v_k c}$).

Notice that vertex b is possibly adjacent to some of the vertices of P_{ac}. Thus, placing b at an appropriate distance above P_{ac}, the edges between b and P_{ac} can be drawn straight-line without intersecting P_{ac} and therefore no other edge of G'. The faces created when placing b are star-shaped and have their kernels arbitrarily close to b. We finally apply Theorem 1. □

5   Main Theorem

Let G be a maximal plane graph and C be a strictly internal simple outerchordless cycle of G, represented by an arbitrary convex polygon Γ_C in the plane. In Theorem 3 we prove that it is still true that if each petal of C is realizable, then Γ_C is extendable. Before stating and proving Theorem 3, we introduce notation that will be used through this section.

Recall that $v_1, \ldots, v_k$ denote the vertices of C. Let $w_{i,j}$ be an outermost petal of C in G. Let ℓ_i (resp. ℓ_j) be a half-line with the slope of edge (v_i, v_{i+1}) (resp. (v_{j-1}, v_j)) originating at v_i (resp. v_j) (see Figure 4(a)). Since $w_{i,j}$ is realizable, lines ℓ_i and ℓ_j intersect. Denote by apex$(w_{i,j})$ their intersection point and by cone$(w_{i,j})$ the subset of $\mathbb{R}^2$ that is obtained by the intersection of the half-planes defined by ℓ_i and ℓ_j, not containing Γ_C. It is clear that any internal point of cone$(w_{i,j})$ is appropriate to draw $w_{i,j}$ so that the plane subgraph of G induced by $V(C) \cup \{w_{i,j}\}$ is crossing-free. For consistency, we also define cone(w) and apex(w) of an outer stamen w of C as follows.

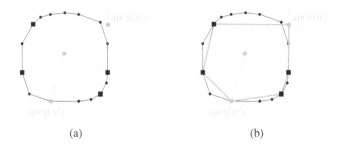

<table>
<tr><td>(a)</td><td>(b)</td></tr>
</table>

Fig. 5. Construction of drawing of graph G_{shell} shown in Figure 4(b). (a) Apex points are gray, points of B are black squares. (b) Polygon Π is gray, lines $\{\ell(w) \mid w \in S \cap C'_{\text{shell}}\}$ are dashed.

Assume that w is adjacent to $v_i \in C$. Then $\text{cone}(w) \subset \mathbb{R}^2$ is the union of the half-planes defined by lines of edges (v_{i-1}, v_i) and (v_i, v_{i+1}), that do not contain Γ_C. We set $\text{apex}(w) = v_i$.

Let P (resp. S) denote the set of outermost petals (resp. stamens) of C in G. By Lemma 1, there exists a cycle C_{shell} in G that contains exactly $P \cup S$. Let G_{shell} denote the plane subgraph of G induced by the vertices of C and C_{shell}. (Figure 4(b)). Let C'_{shell} denote the outer cycle of G_{shell}. We denote the graph consisting of C, C'_{shell} and edges between them by G'_{shell}. Each petal or stamen of C, say w, that belongs to C_{shell} but not to C'_{shell}, belongs to a face of G'_{shell}. We denote this face by $\text{shell}(w)$. We categorize the faces of G_{shell} as follows. The faces that lie inside cycle C are called *faces of C*. The faces that are bounded only by C_{shell} and its chords, are called *faces of C_{shell}*. Notice that each face of C_{shell} is a triangle. Notice that a face of G_{shell} that is comprised by two consecutive edges adjacent to the same vertex of C (not belonging to C), is a triangle, and contains no vertex of $G \setminus G_{\text{shell}}$, since both facts would imply that the taken edges are not consecutive. Finally, faces bounded by a subpath of C and two edges adjacent to the same petal, are called *petal faces*. The plane subgraph of G inside a petal face is triangulated and does not have a chord connecting two vertices of its outer face, and therefore is triconnected. Thus we have the following

Observation 1. *Each vertex of $G \setminus G_{\text{shell}}$ either lies in a face of C, or in a face that is a triangle, or in a petal face, or outside C'_{shell}. Each subgraph of G inside a petal face is triconnected.*

Theorem 3. *Let G be a maximal plane graph and let C be a strictly internal simple outerchordless cycle of G, represented by a convex polygon Γ_C in the plane. Γ_C is extendable to a straight-line drawing of G if and only if each petal of C is realizable.*

Proof. The condition that each petal of C is realizable is clearly necessary. Next we show that it is also sufficient.

We first show how to draw the graph G'_{shell}. Afterward we complete it to a drawing of G_{shell}. Our target is to represent C'_{shell} as a one-sided polygon, so that Theorem 2 can be applied for the rest of G that lies outside C'_{shell}. We first decide which edge of C'_{shell} to "stretch", i.e., which edge will serve as base edge of the one-sided polygon

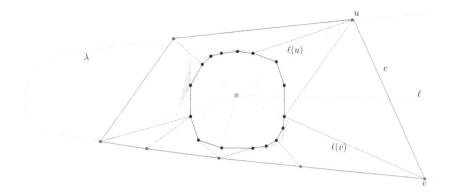

Fig. 6. Construction of Case 1. Corresponding G_{shell} is shown in Figure 4(b).

for representing C'_{shell}. In order to be able to apply Theorem 2, this one-sided polygon should be such that each petal of C'_{shell} is realizable. Thus we choose the base edge e of C'_{shell} as follows. If C'_{shell} contains an edge on the outer face of G, we choose e to be this edge. Otherwise, we choose an edge e, such that at least one of the end vertices of e is adjacent to an outermost petal of C'_{shell} in G. Such a choice of e ensures that each petal of C'_{shell} is realizable.

Claim 3. *Polygon Γ_C can be extended to a straight-line drawing of graph G'_{shell}, such that its outer face C'_{shell} is represented by a one-sided polygon with base edge e. Moreover, C'_{shell} contains in its interior all points of $\{\text{apex}(w) \mid w \in C_{\text{shell}}\}$.*

Recall that P (resp. S) denotes the set of outermost petals (resp. stamens) of C in G. Let B denote the set of vertices of C, to which stamens $S \cap C'_{\text{shell}}$ are adjacent (refer to Figure 4(b)). By construction of the apex points, the set $\{\text{apex}(w) \mid w \in P \cap C'_{\text{shell}}\} \cup B$ is in convex position, and we denote by Π its convex hull. Polygon Π may be degenerate, and may contain only a single vertex or a single edge. We treat these cases separately to complete the construction of the drawing of the graph G'_{shell}. Next, we explain the construction in the non-degenerate case; the degenerate cases are covered in the full version of this paper [7].

Let p be a point inside Π. Let $\ell(w)$ denote a half-line from p through w, where w is a vertex of Π. If we order the constructed half-lines around p, any two consecutive lines have between them an angle less than π. If $w \in B$, we substitute $\ell(w)$ by the same number of slightly rotated lines as the number of stamens of C'_{shell} adjacent to w, without destroying the order (refer to Figure 5(b)). Thus, for each $w \in C'_{\text{shell}}$, a line $\ell(w)$ is defined. Notice that, for any $w \in P \cap C'_{\text{shell}}$, line $\ell(w)$ passes through $\text{apex}(w)$, and the infinite part of $\ell(w)$ lies in $\text{cone}(w)$. Thus, for any position of w on a point of $\ell(w) \cap \text{cone}(w)$, edges between C and w do not cross Γ_C. For a stamen $w \in S \cap C'_{\text{shell}}$, line $\ell(w)$ crosses $\text{cone}(w)$ very close to $\text{apex}(w)$, and its infinite part lies in $\text{cone}(w)$. Thus, similarly, for any position of w on a point of $\ell(w) \cap \text{cone}(w)$, edges between C and w do not cross Γ_C.

Recall that $e = (u, v)$ is the edge of C'_{shell} that we have decided to "stretch". Recall also that $\ell(u)$ and $\ell(v)$ are consecutive in the sequence of half-lines we have constructed. Let κ be a circle around Γ_C that contains in the interior the polygon Π and the set of points $\{\text{apex}(w) | w \in C_{\text{shell}}\}$. Let ℓ be a half-line bisecting the angle between $\ell(u)$ and $\ell(v)$ (refer to Figure 6). Let λ be a parabola with ℓ as axis of symmetry and the center of κ as focus. We position and parametrize λ such that it does not cross ℓ and κ.

With this placement of λ, each half-line $\ell(w)$, $w \in \Pi$, crosses λ, moreover, intersections with $\ell(u)$ and $\ell(v)$ are on different branches of λ and appear last on them as we walk on λ from its origin to infinity. Let Π' be the convex polygon comprised by the intersection points of lines $\{\ell(w) : w \in V(C'_{\text{shell}})\}$ with λ. We make λ large enough, so that the polygon Π' still contains the circle κ in the interior. As a results, for each $w \in C$, $\text{cone}(w) \cap \Pi'_{\text{in}} \neq \emptyset$, where Π'_{in} denotes the interior of Π'. This concludes the proof of the claim in the non-degenerate case.

Let Γ'_{shell} be the constructed drawing of G'_{shell}. Recall that each petal or stamen w of C, that does not belong to C'_{shell}, lies in a face of G'_{shell}, denoted by $\text{shell}(w)$. Let $\Gamma_{\text{shell}(w)}$ denote the polygon representing face $\text{shell}(w)$ in Γ'_{shell}. By construction, $\text{cone}(w) \cap \Gamma_{\text{shell}(w)} \neq \emptyset$. We next explain how to extend the drawing of G'_{shell} to the drawing of G_{shell}. For each edge (u, v) of C'_{shell}, we add a convex curve, lying close enough to this edge inside Γ'_{shell}. Let μ be the union of these curves for all edges of C'_{shell}. We notice that we can place them so close to C'_{shell} that all the points of $\{\text{apex}(w) \mid w \in C\}$ are still in the interior of μ. Thus μ is intersected by all the sets $\text{cone}(w)$, for each $w \in C$. We place each vertex w of $C_{\text{shell}} \setminus C'_{\text{shell}}$ on $\mu \cap \text{cone}(w)$ in the order they appear in C_{shell}. Since all edges induced by C_{shell} lie outside of C_{shell}, and both end points of such an edge are placed on a single convex curve, they can be drawn straight without intersecting each other, or other edges of G_{shell}. Thus, we have constructed a planar extension of Γ_C to a drawing of G_{shell}, call it Γ_{shell}.

Recall the definitions of faces of C, faces of C_{shell} and petal faces from the beginning of this section. The faces of C appear in Γ_{shell} as convex polygons. The faces of C_{shell} are triangles, and the petal faces of G_{shell} are star-shapes whose kernel is close to the corresponding petal. By Observation 1, each vertex of $G \setminus G_{\text{shell}}$ either lies in a face of C, or in a face that is a triangle, or in a petal face, or outside C'_{shell}. Moreover a subgraph of G inside a petal face is triconnected. Thus, by multiple applications of Theorem 1, we can extend the drawing of G_{shell} to a straight-line planar drawing of the subgraph of G inside C'_{shell}.

Finally, notice that in the constructed drawing of G_{shell} each petal of its outer cycle, i.e. C'_{shell}, is realizable. This is by the choice of edge e. Moreover, by construction of G_{shell}, C'_{shell} has no outer chords. In case C'_{shell} is not strictly internal, we apply Proposition 1, to construct a maximal plane graph G', such that G is a plane subgraph of G', C'_{shell} is a strictly internal outerchordless cycle of G' and each petal of C'_{shell} is realizable. Then, we apply Theorem 2, to complete the drawing of G', lying outside C'_{shell}. Finally, we remove the edges of G' that do not belong to G.   □

We conclude with the following general statement, that follows from Proposition 1, Theorem 3 and one of the known algorithms that constructs drawing of a planar graph with a prescribed outer face (e.g. [4,10] or Theorem 1).

Corollary 1. *Let G be a plane graph and H be a biconnected plane subgraph of G. Let Γ_H be a straight-line convex drawing of Γ_H. Γ_H is extendable to a planar straight-line* *drawing of G if and only if the outer cycle of H is outerchordless and each petal of the outer cycle of H is realizable.*

6   Conclusions

In this paper, we have studied the problem of extending a given convex drawing of a cycle of a plane graph G to a planar straight-line drawing of G. We characterized the cases when this is possible in terms of two simple necessary conditions, which we proved to also be sufficient. We note that it is easy to test whether the necessary conditions are satisfied in linear time. It is readily seen that our proof of existence of the extension is constructive and can be carried out in linear time. As an extension of our research it would be interesting to investigate whether more envolved necessary conditions are sufficient for more general shape of a cycle, for instance a star-shaped polygon.

Acknowledgments. M.N. received financial support by the Concept for the Future of KIT. I.R. was supported by a fellowship within the Postdoc-Program of the German Academic Exchange Service (DAAD). Part of this work was done within GRADR – EUROGIGA project no. 10-EuroGIGA-OP-003.

References

1. Angelini, P., Di Battista, G., Frati, F., Jelínek, V., Kratochvíl, J., Patrignani, M., Rutter, I.: Testing planarity of partially embedded graphs. In: 21st Annual ACM-SIAM Symposium on Discrete Algorithms (SODA 2010), pp. 202–221. SIAM (2010)
2. Avis, D.: Generating rooted triangulations without repetitions. Algorithmica 16, 618–632 (1996)
3. Chambers, E.W., Eppstein, D., Goodrich, M.T., Löffler, M.: Drawing graphs in the plane with a prescribed outer face and polynomial area. Journal of Graph Algorithms and Applications 16(2), 243–259 (2012)
4. Duncan, C.A., Goodrich, M.T., Kobourov, S.G.: Planar drawings of higher-genus graphs. Journal of Graph Algorithms and Applications 15(1), 7–32 (2011)
5. Hong, S.-H., Nagamochi, H.: Convex drawings of graphs with non-convex boundary constraints. Discrete Applied Mathematics 156(12), 2368–2380 (2008)
6. Jelínek, V., Kratochvíl, J., Rutter, I.: A Kuratowski-type theorem for planarity of partially embedded graphs. Computational Geometry Theory & Applications 46(4), 466–492 (2013)
7. Mchedlidze, T., Nöllenburg, M., Rutter, I.: Drawing planar graphs with a prescribed inner face. CoRR, abs/1308.3370 (2013)
8. Pach, J., Wenger, R.: Embedding planar graphs at fixed vertex locations. In: Whitesides, S.H. (ed.) GD 1998. LNCS, vol. 1547, pp. 263–274. Springer, Heidelberg (1999)
9. Patrignani, M.: On extending a partial straight-line drawing. In: Healy, P., Nikolov, N.S. (eds.) GD 2005. LNCS, vol. 3843, pp. 380–385. Springer, Heidelberg (2006)
10. Tutte, W.T.: How to draw a graph. Proc. London Math. Soc. 13(3), 743–768 (1963)

Metro-Line Crossing Minimization:
Hardness, Approximations, and Tractable Cases

Martin Fink[1] and Sergey Pupyrev[2,3,*]

[1] Lehrstuhl für Informatik I, Universität Würzburg, Germany
[2] Department of Computer Science, University of Arizona, USA
[3] Institute of Mathematics and Computer Science, Ural Federal University, Russia

Abstract. Crossing minimization is one of the central problems in graph drawing. Recently, there has been an increased interest in the problem of minimizing crossings between paths in drawings of graphs. This is the *metro-line crossing minimization* problem (MLCM): Given an embedded graph and a set L of simple paths, called *lines*, order the lines on each edge so that the total number of crossings is minimized. So far, the complexity of MLCM has been an open problem. In contrast, the problem variant in which line ends must be placed in outermost position on their edges (MLCM-P) is known to be NP-hard.

Our main results answer two open questions: (i) We show that MLCM is NP-hard. (ii) We give an $O(\sqrt{\log |L|})$-approximation algorithm for MLCM-P.

1 Introduction

In metro maps and transportation networks, some edges, that is, railway tracks or road segments, are used by several lines. Usually, lines that share an edge are drawn individually along the edge in distinct colors; see Fig. 1. Often, some lines must cross, and one normally wants to have as few crossings of metro lines as possible. In the *metro-line crossing minimization* problem (MLCM), the goal is to order different metro-lines along each edge of the underlying network so that the total number of crossings is minimized. Although the problem has been studied, many questions remain open.

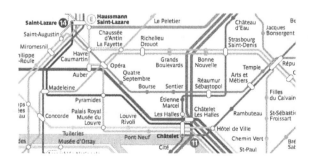

Fig. 1. A part of the official metro map of Paris

* Research supported in part by NSF grant DEB 1053573.

S. Wismath and A. Wolff (Eds.): GD 2013, LNCS 8242, pp. 328–339, 2013.

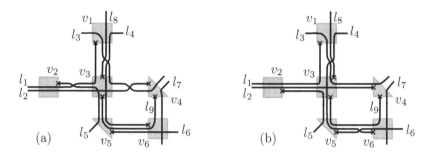

Fig. 2. Nine lines on a portion of an underlying network with 6 vertices and 8 edges. (a) $\pi_{v_3 v_4} = (l_3, l_2)$ and $\pi_{v_3 v_1} = (l_1, l_8, l_4, l_3)$. The lines l_3 and l_4 have an unavoidable edge crossing on $\{v_1, v_3\}$. In contrast, the crossing of l_2 and l_3 on $\{v_3, v_4\}$ is avoidable. In v_3 there is an unavoidable vertex crossing of the lines l_2 and l_8. As the vertex crossing of l_2 and l_5 in v_3 is avoidable the solution is not feasible. (b) A feasible solution satisfying the periphery condition.

Apart from the visualization of metro maps, the problem has various applications including the visual representation of biochemical pathways. In very-large-scale integration (VLSI) design, there is the closely related problem of minimizing intersections between nets (physical wires) [8,10]. Net patterns with fewer crossings have better electrical characteristics and require less area. In graph drawing, the number of edge crossings is one of the most important aesthetic criteria. In *edge bundling*, groups of edges are drawn close together—like metro lines—emphasizing the structure of the graph; minimizing crossings between parallel edges arises as a subproblem [14].

Problem Definitions. The input is a planarly embedded graph $G = (V, E)$ and a set L of simple paths in G. We call G the *underlying network*, the vertices *stations*, and the paths *lines*. The endpoints v_0, v_k of a line $(v_0, \ldots, v_k) \in L$ are *terminals*, and the vertices $v_1, \ldots, v_{k-1}$ are *intermediate stations*. For each edge $e = (u, v) \in E$, let $L_e = L_{uv}$ be the set of lines passing through e.

Following previous work [2,12], we use the *k-side* model; each station v is represented by a polygon with k sides, where k is the degree of v in G; see Fig. 2. For $k \leq 2$ a rectangle is used. Each side of the polygon is called a *port* of v and corresponds to an incident edge $(v, u) \in E$. A line $(v_0, \ldots, v_k)$ is represented by a polyline starting at a port of v_0 (on the boundary of the polygon), passing through two ports of v_i for $1 \leq i < k$, and ending at a port of v_k. For each port of $u \in V$ corresponding to $(u, v) \in E$, we define the *line order* $\pi_{uv} = (l_1 \ldots l_{|L_{uv}|})$ as an ordered sequence of the lines in L_{uv}, which specifies the clockwise order at which the lines L_{uv} are connected to the port of u with respect to the center of the polygon. Note that there are two different line orders π_{uv} and π_{vu} on any edge (u, v) of the network. A *solution*, or a *line layout*, specifies line orders π_{uv} and π_{vu} for each edge $(u, v) \in E$.

A line crossing is a crossing between polylines corresponding to a pair of lines. We distinguish two types of crossings; see Fig. 2(a). An *edge crossing* between lines l and l' occurs whenever $\pi_{uv} = (\ldots l \ldots l' \ldots)$ and $\pi_{vu} = (\ldots l \ldots l' \ldots)$ for some edge $(u, v) \in E$. We now consider the concatenated cyclic sequence π_u of the orders

$\pi_{uv_1}, \ldots, \pi_{uv_k}$, where $(u, v_1), \ldots, (u, v_k)$ are the edges incident to u in clockwise order. A *vertex crossing* between l and l' occurs in u if $\pi_u = (\ldots l \ldots l' \ldots l \ldots l' \ldots)$. Intuitively, the lines change their relative order inside u. A crossing is called *unavoidable* if the lines cross in each line layout; otherwise it is *avoidable*. A crossing is unavoidable if neither l nor l' have a terminal on their common subpath and the lines split on both ends of this subpath in such a way that their relative order has to change; see Fig. 2. By checking all pairs of lines, we can determine all unavoidable crossings in $O(|L|^2|E|)$ time. Following previous work, we insist that (i) *avoidable vertex crossings are not allowed* in a solution, that is, these crossings are not hidden below a station symbol, and (ii) *unavoidable vertex crossings are not counted* since they occur in any solution.

A pair of lines may share several common subpaths, and the lines may cross multiple times on the subpaths. For simplicity of presentation, we assume that there is at most one common subpath of two lines. Our results do, however, also hold for the general case as every common subpath can be considered individually.

Problem variants. Several variants of the problem have been considered in the literature. The original metro-line crossing minimization problem is formulated as follows.

Problem 1 (MLCM). For a given instance (G, L), find a line layout with the minimum number of crossings.

In practice, it is desirable to avoid gaps between adjacent lines; to this end, every line is drawn so that it starts and terminates at the topmost or bottommost end of a port; see Fig. 2(b). In fact, many manually created maps follow this *periphery condition* introduced by Bekos et al. [4]. Formally, we say that a line order π_{uv} at the port of u satisfies the periphery condition if $\pi_{uv} = (l_1 \ldots l_p \ldots l_q \ldots l_{|L_{uv}|})$, where u is a terminal for the lines $l_1, \ldots, l_p, l_q, \ldots, l_{|L_{uv}|}$ and u is an intermediate station for the lines $l_{p+1}, \ldots, l_{q-1}$. The problem is known as MLCM *with periphery condition*.

Problem 2 (MLCM-P). For a given instance (G, L), find a line layout, subject to the periphery condition on every port, with the minimum number of crossings.

In the special case of MLCM-P with *side assignment* (MLCM-PA), the input additionally specifies for each line end on which side of its port it terminates; Nöllenburg [12] showed that MLCM-PA is computationally equivalent to the version of MLCM in which all lines terminate at vertices of degree one.

As MLCM and MLCM-P are NP-hard even for very simple networks, we introduce the additional constraint that no line is a subpath of another line. Indeed, this is often the case for bus and metro transportation networks; if, however, there is a line that is a subpath of a longer line then one can also visualize it as a part of the longer line. We call the problems with this new restriction PROPER-MLCM and PROPER-MLCM-P.

Previous Work. Benkert et al. [5] described a quadratic-time algorithm for MLCM when the underlying network consists of a single edge with attached leaves, leaving open the complexity status of MLCM.

Bekos et al. [4] studied MLCM-P and proved that the variant is NP-hard on paths. Motivated by the hardness, they introduced the variant MLCM-PA and studied the

Table 1. Overview of results for the metro-line crossing minimization problem

problem	graph class	result	reference
MLCM	caterpillar	NP-hard	Thm. 1
MLCM	single edge	$O(|L|^2)$-time algorithm	[5]
MLCM	general graph	crossing-free test	Thm. 2
MLCM-P	path	NP-hard	[2]
MLCM-P	general graph	ILP	[3]
MLCM-P	general graph	$O(\sqrt{\log|L|})$-approximation	Thm. 5
MLCM-P	general graph	crossing-free test	Thm. 3
PROPER-MLCM-P	general graph with consistent lines	$O(|L|^3)$-time algorithm	Thm. 7
MLCM-PA	general graph	$O(|V| + |E| + |V||L|)$-time	[14]
MLCM-PA	general graph	crossing-free test	[12]

problem on simple networks. Later, polynomial-time algorithms for MLCM-PA were found with gradually improving running time by Asquith et al. [3], Argyriou et al. [2], and Nöllenburg [12], until Pupyrev et al. [14] presented a linear-time algorithm. Asquith et al. [3] formulated MLCM-P as an integer linear program that finds an optimal solution for the problem on general graphs. Note that in the worst case this approach requires exponential time. Fink and Pupyrev studied a variant of MLCM in which a crossing between two blocks of lines is counted as a single crossing [7]. Okamoto et al. [13] presented exact and fixed-parameter tractable algorithms for MLCM-P on paths.

In the circuit design community (VLSI), Groeneveld [8] considered the problem of adjusting the routing so as to minimize crossings between the pairs of nets, which is equivalent to MLCM-PA, and suggested an algorithm for general graphs. Marek-Sadowska et al. [10] considered a related problem of distributing the line crossings among edges of the underlying graph in order to simplify the net routing.

Our Results. Table 1 summarizes our contributions and previous results. We first prove that the unconstrained variant MLCM is NP-hard even on caterpillars (paths with attached leaves), thus, answering an open question of Benkert et al. [5] and Nöllenburg [11]. As crossing minimization is hard, it is natural to ask whether there exists a *crossing-free* solution. We show that there is a crossing-free solution if and only if there is no pair of lines forming an unavoidable crossing.

We then study MLCM-P. Argyriou et al. [2] and Nöllenburg [11] asked for an approximation algorithm. To this end, we develop a 2SAT model for the problem. Using the model, we get an $O(\sqrt{\log|L|})$-approximation algorithm for MLCM-P. This is the first approximation algorithm in the context of metro-line crossing minimization. We also show how to find a crossing-free solution (if it exists) in polynomial time. Moreover, we prove that MLCM-P is fixed-parameter tractable with respect to the maximum number of allowed crossings by using the fixed-parameter tractability of 2SAT.

We then study the new variant PROPER-MLCM-P and show how to solve it on caterpillars, *left-to-right trees* (considered in [4,2]), and other instances described in Section 4. An optimal solution can be found by applying a maximum flow algorithm on a certain graph. This is the first polynomial-time exact algorithm for the variant in which avoidable crossings may be present in an optimal solution.

2 The MLCM Problem

We begin with MLCM, the most general formulation of the problem, and show that it is hard to decide whether there is a solution with at most $k > 0$ crossings, even if the underlying network is a caterpillar. In contrast, we give a polynomial-time algorithm for deciding whether there exists a crossing-free solution.

Theorem 1. MLCM *is NP-hard on caterpillars.*

Proof. We prove hardness by reduction from MLCM-P which is known to be NP-hard on paths [2]. Suppose we have an instance of MLCM-P consisting of a path $G = (V, E)$ and lines L on the path. We want to decide whether it is possible to order the lines with periphery condition and at most k crossings.

We create a new underlying network $G' = (V', E')$ by adding vertices and edges to G. We assume that G is embedded along a horizontal line and specify positions relative to this line. For each edge $e = (u, v) \in E$, we add vertices $u_1, u_2, v_1,$ and v_2 and edges $(u, u_1), (u, u_2), (v, v_1),$ and (v, v_2) such that v_1 and u_1 are above the path and v_2 and u_2 are below the path. Next, we add $\ell = |L|^2$ lines from u_1 to v_2, and ℓ lines from u_2 to v_1 to $L' \supseteq L$; see Fig. 3. We call the added structure the *red cross* of e and the added lines *red lines*. We claim that there is a number K such that there is a solution of MLCM-P on (G, L) with at most k crossings if and only if there is solution of MLCM on (G', L') with at most $k + K$ crossings.

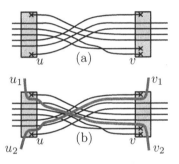

Fig. 3. (a) MLCM-P-solution on (u, v). (b) Insertion of a red cross

Let $e = (u, v) \in E$ be an edge of the path, and let $l \in L_e$. If l has its terminals on u and v, that is, completely lies on e, it never has to cross in G or G'; hence, we assume such lines do not exist. Assume l has none of its terminals on u or v. It has to cross all 2ℓ lines of the red cross of e. Finally, suppose l has just one terminal at a vertex of e, say on u. If the line end of l at u is above the edge (u, u_1) in the order π_{uv}, then it has to cross all red lines from u_2 to v_1 but can avoid the red lines from u_1 to v_2, that is, ℓ crossings with red lines are necessary. Symmetrically, if the line end is below (u, u_2) then only the ℓ crossings with the red lines from u_1 to v_2 are necessary. If the terminal is between the edges (u, u_1) and (u, u_2) then all 2ℓ red edges must be crossed. There are, of course, always ℓ^2 unavoidable internal crossings of the red cross of e.

Let $\ell_e = \ell_e^t + \ell_e^m$ be the number of lines on e, where ℓ_e^t and ℓ_e^m are the numbers of lines on e that do or do not have a terminal at u or v, respectively. In any solution there are at least $\ell_e^t \cdot \ell + 2 \cdot \ell_e^m \cdot \ell + \ell^2$ crossings on e in which at least one red line is involved. It is easy to see that placing a terminal between red lines leaving towards a leaf never brings an advantage. On the other hand, if just a single line has an avoidable crossing with a block of red lines, the number of crossings increases by $\ell = |L|^2$, which is more than the number of crossings in any solution for (G, L) without double crossings. Hence, any optimal solution of the lines in G' has no avoidable crossings with red blocks and,

Fig. 4. A separator l of lines l_1 and l_2 **Fig. 5.** Unavoidable crossing of 2 separators of l_1 and l_3

therefore, satisfies the periphery condition; thus, after deleting the added edges and red lines, we get a feasible solution for MLCM-P on G.

Let $K := |E| \cdot \ell^2 + \sum_{e \in E} (\ell_e^t + 2\ell_e^m) \cdot \ell$ be the minimum number of crossings with red lines involved on G'. Suppose we have an MLCM-solution on G' with at most $K+k$ crossings. Then, after deleting the red lines, we get a feasible solution for MLCM-P on G with at most k crossings. On the other hand, if we have an MLCM-P-solution on G with k crossings, then we can insert the red lines with just K new crossings: Suppose we want to insert the block of red lines from u_1 to v_2 on an edge $e = (u, v) \in E$. We start by putting them immediately below the lines with a terminal on the top of u. Then we cross all lines below until we see the first line that ends on the bottom of v and, hence, must not be crossed by this red block. We go to the right and just keep always directly above the block of lines that end at the bottom side of v; see Fig. 3. Finally, we reach v and have not created any avoidable crossing. Once we have inserted all blocks of red lines, we get a solution for the lines on G with exactly $K + k$ crossings.   □

Crossing-Free Instances. Given an instance of MLCM, we want to check whether there exists a solution without any crossings. If there exists such a crossing-free solution then there cannot be a pair of lines with an unavoidable crossing. We show that this condition is already sufficient. We sketch the proofs; see full version for the complete proof [6].

We assume that no line is a subpath of another line as a subpath can be reinserted parallel to the longer line in a crossing-free solution. Consider a pair of lines l_1, l_2 whose common subpath P starts in u and ends in v. If u (similarly, v) is not a terminal for both l_1 and l_2 then there is a unique relative order of the lines along P in any crossing-free solution. Hence, we assume u is a terminal for l_1, v is a terminal for l_2, and we call such a pair *overlapping*. Suppose there is a *separator* for l_1 and l_2, that is, a line l on the subpath of l_1 and l_2 that has to be below l_1 and above l_2 (or the other way round) as shown in Fig. 4. Then, l_1 has to be above l_2 in a crossing-free solution. The only remaining case is a pair of lines l_1, l_2 without a separator. Suppose l_1, l_2 is chosen such that the number of edges of the common subpath is minimum. If there exists a crossing-free solution then there is also a crossing-free solution in which l_1 and l_2 are immediate neighbors in the orders on their common subpath; see full version [6].

Theorem 2. *Any instance of* MLCM *without unavoidable crossings has a crossing-free solution.*

Proof (sketch). We can merge a pair of overlapping lines without a separator into a new line. This merging step does not introduce an unavoidable crossing. We iteratively perform merging steps until any overlapping pair has a separator. There might be multiple separators for a pair, but all of them separate the pair in the same relative order;

otherwise, there would be a pair of separators with an unavoidable crossing; see Fig. 5. After the merging steps, we get a relative order for every pair of lines sharing an edge. One can show that the relative orders are acyclic. We build a crossing-free solution by putting all lines in the (only) right order on the edge. As the relative order of any pair of lines is the same on any edge, there cannot be a crossing.   □

The proof yields an $O(|L|^2|E|)$-time algorithm for finding crossing-free solutions.

3   The MLCM-P Problem

Let $(G = (V, E), L)$ be an instance of MLCM-P. Our goal is to decide for each line end on which side of its terminal port it should lie. We arbitrarily choose one side of each port and call it "top", the opposite side is called "bottom". For each line l starting at vertex u and ending at vertex v, we create binary variables l_u and l_v, which are true if and only if l terminates at the top side of the respective port. We formulate the problem of finding a truth assignment that minimizes the number of crossings as a 2SAT instance. Note that Asquith et al. [3] already used 2SAT clauses as a tool for developing their ILP for MLCM, where the variables represent above/below relations between line ends. In contrast, in our model a variable directly represents the position of a line on the top or bottom side of a port. We first prove a simple property of lines.

Lemma 1. *Let l, l' be a pair of lines sharing a terminal. We can transform any solution in which l and l' cross to a solution with fewer crossings in which the lines do not cross.*

Proof. Assume l and l' cross in a solution. We switch the positions of line ends at the common terminal v between l and l' and reroute the two lines between the crossing's position and v. By reusing the route of l for l' and vice versa, the number of crossings does not increase. On the other hand, the crossing between l and l' is eliminated.   □

Let l, l' be two lines whose common subpath P starts at vertex u and ends at v. Terminals of l and l' that lie on P can only be at u or v. If neither l nor l' has a terminal on P, then a crossing of the lines does not depend on the positions of the terminals; hence, we assume that there is a terminal at u or v. A possible crossing between l and l' is modeled by a 2SAT formula, the *crossing formula*, consisting of at most two clauses. This formula evaluates to true if and only if l and l' do not cross. For simplicity, we assume that the top sides of the terminal ports of u and v are located on the same side of P; see Fig. 6. If it is not the case, a variable l_u should be substituted with its inverse $\neg l_u$ in the formula. Note that generating all crossing formulas needs $O(|E||L|^2)$ time. We consider several cases; see also the illustrations in the full version [6].

(f1) Suppose u and v are terminals for l and intermediate stations for l', that is, l is a subpath of l'. Then, l does not cross l' if and only if both terminals of l lie on the same side of P. This is expressed by the crossing formula $(l_u \wedge l_v) \vee (\neg l_u \wedge \neg l_v) \equiv (\neg l_u \vee l_v) \wedge (l_u \vee \neg l_v)$, which may occur multiple times, caused by a different l'.

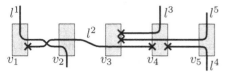

(a) An instance (G, L) of PROPER-MLCM-P

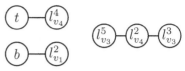

(b) Graph G_{ab} for the instance (G, L)

Fig. 6. A small instance of MLCM-P. The generated 2SAT formulas are: $(l^2_{v_1})$ for the crossing of l^1 and l^2; $(\neg l^4_{v_4})$ for the crossing of l^5 and l^4; $(l^2_{v_4} \vee l^3_{v_3}) \wedge (\neg l^2_{v_4} \vee \neg l^3_{v_3})$ for the crossing of l^2 and l^3; $(l^2_{v_4} \vee l^5_{v_3}) \wedge (\neg l^2_{v_4} \vee \neg l^5_{v_3})$ for the crossing of l^2 and l^5.

(f2) Suppose u is a terminal for l and intermediate for l', and v is a terminal for l' and intermediate for l. Then there is no crossing if and only if both terminals lie on opposite sides of P. This is described by the formula $(l_u \wedge \neg l'_v) \vee (\neg l_u \wedge l'_v) \equiv (l_u \vee l'_v) \wedge (\neg l_u \vee \neg l'_v)$.

(f3) Suppose both l and l' terminate at the same vertex u or v. By Lemma 1, a solution of MLCM-P with a crossing of l and l' can be transformed into a solution in which l and l' do not cross. Hence, we do not introduce formulas in this case.

(f4) In the remaining case, there is only one terminal of l and l' on P. Without loss of generality, let l terminate at u. A crossing is triggered by a single variable. Depending on the fixed terminals or leaving edges at v and u, we get the single clause (l_u) or $(\neg l_u)$. The same clause can occur multiple times, caused by different lines l'.

Crossing-free solutions. Note that the 2SAT formulation of the problem yields an algorithm for deciding whether there exists a crossing-free solution of an MLCM-P instance. First, we check for unavoidable crossings by analyzing every pair of lines individually. Second, the 2SAT model is satisfiable if and only if there is a solution of the MLCM-P instance without avoidable crossing. Since 2SAT can be solved in linear time and there are at most $|L|^2$ crossing formulas, we conclude as follows.

Theorem 3. *Deciding whether there exists a crossing-free solution for* MLCM-P *can be accomplished in* $O(|E||L|^2)$ *time.*

For MLCM the existence of a crossing-free solution is equivalent to the absence of unavoidable crossings. In contrast, there is no such simple criterion for MLCM-P.

Fixed-parameter tractability. We can use the 2SAT model for obtaining a fixed-parameter tractable algorithm on the number k of allowed crossings. We must show that we can check in $f(k) \cdot \mathrm{poly}(\mathcal{I})$ time whether there is a solution with at most k avoidable crossings, where f must be a computable function and $\mathcal{I}$ is the input size.

First, note that minimizing the number of crossings is the same as maximizing the number of satisfied clauses in the corresponding 2SAT instance. Maximizing the number of satisfied clauses, or solving the MAX-2SAT problem, is NP-hard.

However, the problem of deciding whether it is possible to remove a given number k of m 2SAT clauses so that the formula becomes satisfiable is fixed-parameter tractable with respect to the parameter k [15]. This yields the following theorem.

Theorem 4. MLCM-P *is fixed-parameter tractable with respect to the maximum al-lowed number of avoidable crossings.*

Proof. We show that the SAT formula can be made satisfiable by removing at most k clauses if and only if there is a solution with at most k crossings.

First, suppose it is possible to remove at most k clauses from the 2SAT model so that there is a truth assignment satisfying all remaining clauses. Fix such a truth assignment, and consider the corresponding assignment of sides to the terminals. Any crossing leads to an unsatisfied clause in the SAT formula, and no two crossings share an unsatisfied clause. Hence, we have a side assignment that causes at most k crossings.

Now, we assume that there is an assignment of sides for all terminals that causes at most k crossings. We know that in the corresponding truth assignment for all pairs of clauses of types (f1)–(f2) of the SAT model at most one is unsatisfied. Hence, there are at most k unsatisfied clauses since any crossing just leads to a single unsatisfied clause. The removal of these clauses creates a new, satisfiable formula.   □

Using the $O(15^k km^3)$-time algorithm for 2SAT [15] our algorithm has a running time of $O(15^k \cdot k \cdot |L|^6 + |L|^2|E|)$.

Approximating MLCM-P. The proof of Theorem 4 yields that the number of crossings in a crossing-minimal solution of MLCM-P equals the minimum number of clauses that we need to remove from the 2SAT formula in order to make it satisfiable. Fur-thermore, a set of k clauses, whose removal makes the 2SAT formula satisfiable, cor-responds to an MLCM-P solution with at most k crossings. Hence, an approximation algorithm for the problem of making a 2SAT formula satisfiable by removing the min-imum number of clauses (also called MIN 2CNF DELETION) yields an approximation for MLCM-P of the same quality. As there is an $O(\sqrt{\log m})$-approximation algorithm for MIN 2CNF DELETION [1], we have the following result.

Theorem 5. *There is an $O(\sqrt{\log |L|})$-approximation algorithm for* MLCM-P.

4   The PROPER-MLCM-P Problem

In this section we consider the PROPER-MLCM-P problem, where no line in L is a subpath of another line. First we focus on graphs whose underlying network is a caterpillar. There, the top and bottom sides of ports are given naturally; see Fig. 6.

Based on the 2SAT model described in the previous section, we construct a graph G_{ab}, which has a vertex l_u for each variable of the model and two additional vertices b and t. Since no line is a subpath of another line, our 2SAT model has only the two types of crossing formulas (f2) and (f4); compare Section 3. For case (f2), we create an edge (l_u, l'_v). The edge models a possible crossing between lines l and l'; that is, the lines cross if and only if l terminates on top (bottom) of u and l' terminates on top (bottom) of v. For a crossing formula of type (l_u) (case (f4)), we add an edge (b, l_u) to G_{ab}; similarly, we add an edge (t, l_u) for a formula $(\neg l_u)$; see Fig. 6(b) for an example.

Any truth assignment to the variables is equivalent to a b-t cut in G_{ab}, that is, a cut separating b and t. Indeed, any edge in the graph models the fact that two lines should

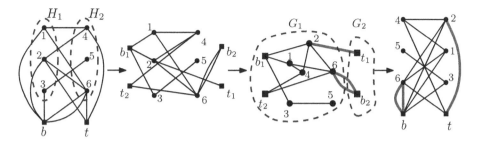

Fig. 7. Solving MIN-UNCUT on an almost bipartite graph. The maximum flow (minimum cut) with value 3 results in vertex partitions $V_b^1 = \{b_1, 4, 5, 6\}$, $V_t^1 = \{t_2, 1, 2, 3\}$, $V_b^2 = \{b_2\}$, and $V_t^2 = \{t_1\}$. The optimal partition $S_b = \{b, 4, 5, 6\}$, $S_t = \{t, 1, 2, 3\}$ induces 3 uncut edges $(b, 6), (b, 6), (t, 2)$.

not be assigned to the same side as they would cause a crossing otherwise. Hence, any line crossing corresponds to an *uncut* edge. Therefore, for minimizing the number of crossings, we need to solve the known MIN-UNCUT problem, which asks for a partitioning of the vertices of a graph into sets S_t, S_b so that the number of uncut edges $((v, u)$ with $v, u \in S_t$ or $v, u \in S_b)$ is minimized. Although MIN-UNCUT is NP-hard, it turns out that the graph G_{ab} has a special structure, which we call *almost bipartite*.

Definition 1. *A graph $G = (V, E)$ is called* almost bipartite *if it is a union of a bipartite graph $H = (V_H, E_H)$ and two additional vertices b, t whose edges may be incident to vertices of both partitions of H, that is, $V = V_H \cup \{b\} \cup \{t\}$ and $E = E_H \cup E'$, where $E' \subseteq \{(b, v) \mid v \in V\} \cup \{(t, v) \mid v \in V\}$.*

The bipartition is given by the fact that "left" (similarly, "right") terminals of two lines can never be connected by an edge in G_{ab}. We show that MIN-UNCUT can be solved optimally for almost bipartite graphs.

Theorem 6. MIN-UNCUT *can be solved in $O(|V|^3)$ on almost bipartite graphs.*

Proof. Almost bipartite graphs are a subclass of *weakly bipartite graphs*. It is known that MIN-UNCUT can be solved in polynomial time on weakly bipartite graphs using the ellipsoid method [9]. We present a simple and efficient combinatorial algorithm.

The special vertices b and t have to belong to different partitions of G_{ab}. We create a new graph G' from G_{ab}. We split vertex b into b_1, b_2 and t into t_1, t_2 such that b_1 and t_1 are connected to the vertices of the first partition H_1 of H, and b_2 and t_2 are connected to the second partition H_2. Formally, for each edge $(b, v) \in E, v \in H_1$, we create an edge (b_1, v); for each edge $(b, v) \in E, v \in H_2$, we create an edge (v, b_2). Similarly, edges (v, t_1) are created for all $(t, v) \in E, v \in H_1$, and edges (t_2, v) are created for all $(t, v) \in E, v \in H_2$. The construction is illustrated in Fig. 7.

Now, for each edge (u, v) of G' we assign capacity 1, and compute a maximum flow between the pair of sources b_1, t_2 to the pair of sinks b_2, t_1. This can be done in $O(|V|^3)$ time using a maximum flow algorithm with a supersource (connected to b_1 and t_2) and a supersink (connected to b_2 and t_1). Indeed, there is an integral maximum flow in G'.

A maximum flow corresponds to a maximum set of edge-disjoint paths starting at b_1 or t_2 and ending at b_2 or t_1. Such a path corresponds to one of the following structures in the original graph G: (i) an odd cycle containing vertex b (a cycle with an odd number of edges); (ii) an odd cycle containing vertex t; (iii) an even path between b and t.

Note that if a graph has an odd cycle then at least one of the edges of the cycle belongs to the same partition in any solution of MIN-UNCUT. The same holds for an even path connecting b and t in G since b and t have to belong to different partitions. Since the maximum flow corresponds to the edge-disjoint odd cycles and even paths in G, the value of the flow is a lower bound for a solution of MIN-UNCUT.

Let us prove that the value of the maximum flow in G' is also an upper bound. By Menger's theorem, this value is the cardinality of a minimum edge cut separating sources and sinks. Let E^* be the minimum edge cut and let G_1 and G_2 be the correspondent disconnected subgraphs of G'; see Fig. 7. Note that G_1 is bipartite since $H \cap G_1$ is bipartite; vertex b_1 is only connected to vertices of H_1 and vertex t_2 is only connected to vertices of H_2. Therefore, there is a 2-partition of vertices of G_1 such that b_1 and t_2 belong to different partitions; let us denote the partitions V_b^1 and V_t^1. Similarly, there is a 2-partition of G_2 into V_b^2 and V_t^2 with $b_2 \in V_b^2$ and $t_1 \in V_t^2$. We combine these partitions so that $S_b = \{b\} \cup \left(V_b^1 \cup V_b^2\right) \setminus \{b_1, b_2\}$ and $S_t = \{t\} \cup \left(V_t^1 \cup V_t^2\right) \setminus \{t_1, t_2\}$, which yield the required partition of vertices of G for MIN-UNCUT. The set of uncut edges is E^*, which completes the proof of the theorem. $\qquad\square$

As a direct corollary, we get a $O(|L|^3)$-time algorithm for PROPER-MLCM-P on caterpillars. I can be applied for some other underlying networks. Let $(G = (V, E), L)$ be an instance of PROPER-MLCM-P. The lines L have *consistent* directions on G if the lines can be directed so that for each edge $e \in E$ all lines L_e have the same direction. If the underlying graph is a path then we can consistently direct the lines from left to right. Similarly, consistent line directions exist for "left-to-right" [4,2] and "upward" [7] trees, that is, trees for which there is an embedding with all lines being monotone in some direction. It is easy to test whether there are consistent line directions by giving an arbitrary direction to some first line, and then applying the same direction on all lines sharing edges with the first line until all lines have directions or an inconsistency is found. Hence, we get the following result; see full version for the proof [6].

Theorem 7. PROPER-MLCM-P *can be solved in* $O(|L|^3)$ *time for instances* (G, L) *admitting consistent line directions.*

5   Conclusion and Open Problems

We proved that MLCM is NP-hard and presented an $O(\sqrt{\log |L|})$-approximation algorithm for MLCM-P, as well as an exact $O(|L|^3)$-time algorithm for PROPER-MLCM-P on instances with consistent line directions. We also suggested polynomial-time algorithms for crossing-free solutions for MLCM and MLCM-P. From a theoretical point of view, there are still interesting open problems: 1. Is there an approximation algorithm for MLCM? 2. Is there a constant-factor approximation algorithm for MLCM-P? 3. What is the complexity status of PROPER-MLCM/PROPER-MLCM-P in general?

On the practical side, the visualization of the computed line crossings is a possible future direction. So far, the focus has been on the number of crossings, although two line orders with the same crossing number may look quite differently [7]. For example, a metro line is easy to follow if it has few bends. Hence, an interesting question is how to visualize metro lines using the minimum total number of bends.

Acknowledgments. We thank Martin Nöllenburg, Jan-Henrik Haunert, Joachim Spoerhase, Lukas Barth, Stephen Kobourov, and Sankar Veeramoni for discussions about problem variants. We are especially grateful to Alexander Wolff for help with the paper.

References

1. Agarwal, A., Charikar, M., Makarychev, K., Makarychev, Y.: $O(\sqrt{\log n})$ approximation algorithms for min UnCut, min 2CNF deletion, and directed cut problems. In: STOC 2005, pp. 573–581. ACM, New York (2005)
2. Argyriou, E.N., Bekos, M.A., Kaufmann, M., Symvonis, A.: On metro-line crossing minimization. Journal of Graph Algorithms and Applications 14(1), 75–96 (2010)
3. Asquith, M., Gudmundsson, J., Merrick, D.: An ILP for the metro-line crossing problem. In: Harland, J., Manyem, P. (eds.) CATS 2008. CRPIT, vol. 77, pp. 49–56. Australian Computer Society (2008)
4. Bekos, M.A., Kaufmann, M., Potika, K., Symvonis, A.: Line crossing minimization on metro maps. In: Hong, S.-H., Nishizeki, T., Quan, W. (eds.) GD 2007. LNCS, vol. 4875, pp. 231–242. Springer, Heidelberg (2008)
5. Benkert, M., Nöllenburg, M., Uno, T., Wolff, A.: Minimizing intra-edge crossings in wiring diagrams and public transportation maps. In: Kaufmann, M., Wagner, D. (eds.) GD 2006. LNCS, vol. 4372, pp. 270–281. Springer, Heidelberg (2007)
6. Fink, M., Pupyrev, S.: Metro-line crossing minimization: Hardness, approximations, and tractable cases. ArXiv e-print abs/1306.2079 (2013), http://arxiv.org/abs/1306.2079
7. Fink, M., Pupyrev, S.: Ordering metro lines by block crossings. In: Chatterjee, K., Sgall, J. (eds.) MFCS 2013. LNCS, vol. 8087, pp. 397–408. Springer, Heidelberg (2013)
8. Groeneveld, P.: Wire ordering for detailed routing. IEEE Des. Test 6(6), 6–17 (1989)
9. Grötschel, M., Pulleyblank, W.: Weakly bipartite graphs and the Max-Cut problem. Operations Research Letters 1(1), 23–27 (1981)
10. Marek-Sadowska, M., Sarrafzadeh, M.: The crossing distribution problem. IEEE Transactions on CAD of Integrated Circuits and Systems 14(4), 423–433 (1995)
11. Nöllenburg, M.: Network Visualization: Algorithms, Applications, and Complexity. Ph.D. thesis, Fakultät für Informatik, Universität Karlsruhe (TH) (2009)
12. Nöllenburg, M.: An improved algorithm for the metro-line crossing minimization problem. In: Eppstein, D., Gansner, E.R. (eds.) GD 2009. LNCS, vol. 5849, pp. 381–392. Springer, Heidelberg (2010)
13. Okamoto, Y., Tatsu, Y., Uno, Y.: Exact and fixed-parameter algorithms for metro-line crossing minimization problems. ArXiv e-print abs/1306.3538 (2013)
14. Pupyrev, S., Nachmanson, L., Bereg, S., Holroyd, A.E.: Edge routing with ordered bundles. In: van Kreveld, M.J., Speckmann, B. (eds.) GD 2011. LNCS, vol. 7034, pp. 136–147. Springer, Heidelberg (2012)
15. Razgon, I., O'Sullivan, B.: Almost 2-SAT is fixed-parameter tractable. Journal of Computer and System Sciences 75(8), 435–450 (2009)

Fixed Parameter Tractability of Crossing Minimization of Almost-Trees

Michael J. Bannister, David Eppstein, and Joseph A. Simons

Department of Computer Science, University of California, Irvine

Abstract. We investigate exact crossing minimization for graphs that differ from trees by a small number of additional edges, for several variants of the crossing minimization problem. In particular, we provide fixed parameter tractable algorithms for the 1-page book crossing number, the 2-page book crossing number, and the minimum number of crossed edges in 1-page and 2-page book drawings.

1 Introduction

Graphs that differ from a tree by the inclusion of a small number of edges arise in many applications; such graphs are called *almost-trees*. Almost-trees can be found in the areas of biology, medicine, operations research, sociology, genealogy, distributed systems, and telecommunications, and in each of these applications it is important to find effective visualizations[1]. One of the most important criteria for the aesthetics and readability of a graph drawing is its number of its crossings. Although crossing minimization problems tend to be NP-complete, we may hope that the graphs arising in applications are not hard instances for these problems, allowing us to find optimal drawings for them efficiently. In this paper we prove that almost-trees are indeed not hard instances by designing algorithms for crossing minimization of almost-trees that are fixed-parameter tractable when parameterized by the number of extra non-tree edges in these graphs.

Many different variants of the crossing number have been studied, depending on what types of drawing are allowed and what we count as a crossing [1]. The most frequently studied is the topological crossing number, $cr(G)$, which counts the number of crossings in a unrestricted placement of vertices and edges in the plane. In this paper we consider also the 1-page and 2-page crossing numbers, denoted $cr_1(G)$ and $cr_2(G)$ respectively. The 1-page crossing number counts the minimal number of crossings in a drawing where all the vertices of G are placed on a straight line, and all edges must be placed to one side of the line. The 2-page crossing number is defined similarly: all vertices of G are placed on a straight line and edges may be assigned to either side of the line, but are not allowed to cross the line. In both 1-page and 2-page drawings it is not uncommon to place the vertices on a circle instead of a straight line; this does not change the crossing structure of the drawing. In addition to the number of crossings we consider the number of crossed edges for these drawing styles, denoted $cre_1(G)$ and $cre_2(G)$.

---

[1] We provide more details about these applications in Section 2.

S. Wismath and A. Wolff (Eds.): GD 2013, LNCS 8242, pp. 340–351, 2013.

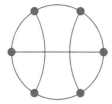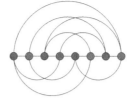

Fig. 1. Left: 1-page circular embedding with two crossings and three crossed edges. Right: 2-page linear embedding of $K_{4,4}$ with four crossings and eight crossed edges.

Following Gurevich, Stockmeyer and Vishkin [2] we define a *k-almost-tree* to be a graph such that every biconnected component of the graph has cyclomatic number at most k, where the *cyclomatic number* is the difference between the number of edges in a graph and in one of its maximal spanning forests. The k-almost-tree parameter has been used in past fixed-parameter algorithms [3–9], and will play the same role in our algorithms for crossing minimization.

Grohe and later Kawarabayashi and Reed showed the topological crossing number to be fixed parameter tractable for its natural parameter [10, 11]; the same is true for odd crossing number [12]. Because the topological crossing number is at most quadratic in the k-almost-tree parameter, $cr(G)$ is also fixed parameter tractable for k-almost-trees. However, to our knowledge no fixed parameter tractable algorithms were known for computing 1-page or 2-page crossing numbers. Indeed, for the 2-page problem, determining whether a graph can be drawn with zero crossings is already NP-complete [13], so to achieve fixed parameter tractability we must use some other parameter such as the k-almost-tree parameter rather than using the crossing number itself as a parameter.

Our main results are that $cr_1(G)$, $cr_2(G)$, $cre_1(G)$, and $cre_2(G)$ are all fixed-parameter tractable for almost-trees. As with previous work on parameterized algorithms for crossing numbers [10–12], our algorithms have a high dependence on their parameters. Making our algorithms more practical by reducing this dependence remains an open problem.

2   Application Domains

Examples of k-almost-trees can be found in biological gene expression networks, where vertices represent genes and edges represent correlations between pairs of genes. The k-almost-tree structure of such graphs has been exploited in parameterized algorithms for finding sequences of valid labelings of genes as active or inactive [5]. The parameter k has also been used in algorithms for continuous facility location where weighted edges represent a road network on which to efficiently place facilities serving clients [2]. Intraprogram communication networks whose vertices represent modules of a distributed system and whose edges represent communicating pairs of modules also have an almost-tree structure that has been exploited for parameterized algorithms [3].

A typical example of almost-trees arises when studying the spread of sexually transmitted infections, where sexual networks are constructed by voluntary survey. In these graphs vertices represent people who have received treatment, and edges represent their reported sexual parters. Analysis of these networks allows researchers to identify the growth and decline phases of an outbreak, and the general spread of the disease [14–16].

Another type of social network represents the business dealings of individuals and business entities. Examples of these networks can be found in the art of Mark Lombardi, an artist famous for his drawings of networks connecting the key players of conspiracy theories [17]. Many of Lombardi's networks show an almost-tree structure; the Lippo Group Shipping network listed below is one.

The directed acyclic graphs originating from genealogical data where edges represent parental relationships on the vertices are another example of k-almost-trees, when viewed as undirected graphs, since in modern societies marriage between close relatives is rare. Similar types of graphs also come from animal pedigrees, academic family trees, and organizational lineages [18].

Utility networks such as telecommunication networks and power grids also form an almost-tree structure, where additional edges beyond those of a spanning tree provide load balancing and redundancy. Since such links are expensive they are placed in the network sparingly.

In order to visually distinguish the tree-like parts of these graphs from the parts with nontrivial connectivity, we may use a *sunburst* style (Figure 2) in which the 2-*core* of the graph (the part of a graph which is left after repeatedly removing all degree one vertices [19]) is drawn with a one-page circular layout and the rest of the graph extends outwards using a radial layout on concentric circles. In this style, crossings occur only within the inner one-page layout, motivating our interest in crossing minimization for one-page drawings of almost-trees.

We collect statistics for several real world graphs in Table 1. The table shows vertex and edge counts (n and m), the cyclomatic number $a = m - n + 1$, the

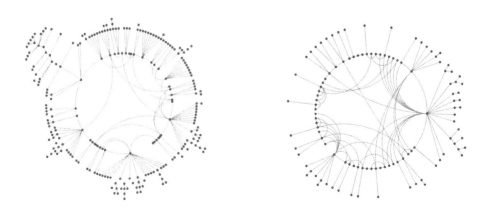

Fig. 2. Two sunburst drawings. Left: An HIV infection graph. Right: Lombardi's World Finance Miami graph.

k-almost-tree parameter k, and the vertex and edge counts for the 2-core (n_2 and m_2). For most of these graphs the parameters a and k are low.

Table 1. Statistics for real-world almost-trees

Name	n	m	a	k	n_2	m_2
Gonorrhoea sexual network 1 [15]	38	39	2	2	9	10
Gonorrhoea sexual network 2 [15]	84	90	7	4	22	28
Lippo Group Shipping [17]	96	112	17	16	45	61
Global International Airways and Indian Springs State Bank [17]	82	99	18	15	33	50
Gondola Genealogy [20]	242	255	14	14	50	63
HIV [14]	243	257	15	12	39	53
Power Grid [21]	4941	6594	1654	1516	3353	5006

3   The Kernel

Our fixed-parameter algorithms use the *kernelization method*. In this method we find a polynomial time transformation from an arbitrary input instance to a *kernel*, an instance whose size is bounded by a fixed function $f(k)$ of the parameter value, and then apply a non-polynomial algorithm to the kernel. In this section we outline the general method for kernelization that we use in our fixed parameter algorithms, based on a similar kernelization by Bannister, Cabello and Eppstein [22] for a different problem, 1-planarity testing.

We first describe our kernelization for cyclomatic number, which starts by reducing the graph to its 2-core. The 2-core of a graph can be found in linear time by initializing a queue of degree one vertices, repeatedly finding and removing vertices from the queue and the graph, and updating the degree and queue membership of the neighbor of each removed vertex.The 2-core consists of vertices of degree at least three connected to each other by paths of degree two vertices. The following lemma bounds the numbers of high degree vertices and maximal paths of degree two vertices.

Lemma 1. *If G is a graph with cyclomatic number k and minimum degree three then G has at most $2k - 2$ vertices and at most $3k - 3$ edges. Furthermore, this bound is tight. As a consequence, the 2-core of a graph with cyclomatic number k has at most $2k - 2$ vertices of degree at least three, and at most $3k - 3$ maximal paths of degree two vertices.*

Proof. Double counting yields $2(n - 1 + k) \geq 3n$, simplifying to $n \leq 2k - 2$. A spanning tree of G has at most $2k - 3$ edges, and there are k edges outside the tree, from which the bound on edges follows. For a graph realizing the upper bound consider any cubic graph with $2k - 2$ vertices. □

The final step in this kernelization is to reduce the length of the maximal degree two paths. Depending on the specific problem, we will determine a maximal allowed path length $\ell(k)$, and if any paths exceed this length we will shorten them to length exactly $\ell(k)$. After this step the kernel will have $O(k\ell(k))$ edges and vertices, bounded by a function of k.

To change the parameter of our algorithms from the cyclomatic number to the k-almost-tree parameter, we first decompose the graph into its biconnected components. These components have a tree structure and in most drawing styles they can be embedded separately without introducing crossings. We then kernelize and optimally embed each biconnected component individually.

4 1-page Crossing Minimization

Minimizing crossings in 1-page drawings is important for several drawing styles, but is NP-hard [23], leading Baur and Brandes to develop fast practical heuristics for reducing but not optimizing the number of crossings [24]. As we now show, crossing minimization and crossed edge minimization in 1-page drawings of k-almost-trees is fixed-parameter tractable in the parameter k. We use the kernelization of Section 3, keeping one vertex per maximal degree two path.

Lemma 2. *Let G have cyclomatic number k and let K be the kernel constructed from G with $\ell(k) = 2$. Then*

1. *K has at most $5k$ vertices and $6k$ edges;*
2. *$\mathrm{cr}_1(G) = \mathrm{cr}_1(K)$;*
3. *$\mathrm{cre}_1(G) = \mathrm{cre}_1(K)$.*

Proof. (1) After reducing a graph with cyclomatic number k to its 2-core and reducing all maximal degree two paths to single edges we have a graph with $2k - 2$ vertices and $3k - 3$ edges, by Lemma 1. Since we are then adding one vertex back to every path that was not a single edge in the original graph, K has at most $5k - 5 \leq 5k$ vertices and $6k - 6 \leq 6k$ edges.

(2) First we show that $\mathrm{cr}_1(G) \leq \mathrm{cr}_1(K)$. Suppose that K has been embedded in one page with the minimum number of crossings. Every degree two vertex v in K corresponds to a path of degree two vertices in G. We can expand this path in a small neighborhood of v without introducing any new crossings. After expanding all degree two paths we have an embedding of the 2-core of G. Now each of the remaining vertices corresponds to a tree in G. Since trees can be embedded in one page without crossings, we can expand each tree in a small neighborhood of its corresponding vertex without introducing further crossings.

Now we show that $\mathrm{cr}_1(K) \leq \mathrm{cr}_1(G)$. Suppose that G is embedded on one page with minimum crossings. Reduce G to its 2-core; this does not increase crossings. Let u and v be two adjacent degree-two vertices of G, let e be the edge between u and v and let f be the edge from v not equal to e. Now, change the embedding of G by keeping u fixed and moving v next to u, rerouting f along the former path used by both e and f. This change moves all crossings from e to f but does not create new crossings, so it produces another minimum-crossing embedding. After this change, edge e may be contracted, again without changing the crossing number. Repeatedly moving one of two adjacent degree-two vertices and then contracting their connecting edge eventually produces an embedding of K whose crossing number equals that of G.

(3) Follows from the proof of (2) with minor modification. □

Lemma 3. *If G is a graph with n vertices and m edges, then $\mathrm{cr}_1(G)$ and $\mathrm{cre}_1(G)$ can be computed in $O(n!)$ time.*

Proof. We place the vertices in an arbitrary order on a circle, and compute the number of crossings or crossed edges for this layout. Then we use the Steinhaus–Johnson–Trotter algorithm [25] to list the $(n-1)!$ permutations of all but one vertex efficiently, with consecutive permutations differing by a transposition. When a transposition swaps u and v, the number of crossings (or crossed edges) in the new layout can be updated from its previous value in $O(\deg(u)+\deg(v)) = O(n)$ time, as in [24]. This yields a total run time of $O(n!)$. □

Combining the above lemmas, we apply the non-polynomial time algorithm only on the kernel of the graph to achieve the following fixed parameter result.

Theorem 1. *If G is a graph with cyclomatic number k, then $\mathrm{cr}_1(G)$ and $\mathrm{cre}_1(G)$ can be computed in $O((5k)! + n)$ time. If G is a k-almost-tree, then $\mathrm{cr}_1(G)$ and $\mathrm{cre}_1(G)$ can be computed in $O((5k)!n)$ time.*

In Section 6 we show how to improve the base of the factorial in this bound by applying fast matrix multiplication algorithms.

5   2-page Crossing Minimization

In this section we consider the problem of 2-page crossing minimization. I.e., we seek a circular arrangement of the vertices of a graph G, and an assignment of the edges to either the interior or exterior of the circle, such that the total number of crossings is minimized. As in the 1-page case, we consider minimizing both the number of crossings $\mathrm{cr}_2(G)$ and the number of crossed edges $\mathrm{cre}_2(G)$.

There are two sources of combinatorial complexity for this problem, the vertex ordering and the edge assignment. However, even when the vertex ordering is fixed, choosing an edge assignment to minimize crossings is NP-hard [26]. The hard instances of this problem can be chosen to be perfect matchings (with k-almost-tree parameter zero), so unless P = NP there can be no FPT algorithm for the version of the problem with a fixed vertex ordering. Paradoxically, we show that requiring the algorithm to choose the ordering as well as the edge assignment makes the problem easier. A straightforward exact algorithm considers all $2^m(n-1)!$ possible configurations and chooses the one minimizing the total number of crossings, running in $O(2^m n!)$ time. We will combine this fact with our kernelization to produce an FPT algorithm.

We will give a sequence of reduction rules that transforms any drawing of G into a drawing with the same number of crossings and crossed edges, in which the lengths of all paths are bounded by a function $f(k)$ of the parameter k. These reductions will justify the correctness of our kernelization using the same function. Our reductions are based on the observation that, if uv is an uncrossed edge, and u and v are consecutive vertices on the spine, then edge uv can be contracted without changing the crossing number or number of uncrossed edges. A given layout may not have any uncrossed edges connecting consecutive vertices, but we will show that, for a graph with a long degree two path, the layout can be modified to produce edges of this type without changing its crossings.

Lemma 4. *Let G be a graph with cyclomatic number k. Then there exists a 2-page drawing with at most k crossed edges, and at most $\binom{k}{2}$ crossings.*

Proof. Remove k edges from G to produce a forest, F. Draw F without crossings on one page, and draw the remaining k edges on the other page. Only the k edges in the second page may participate in a crossing. □

We classify the possible configurations of pairs of consecutive edges of a degree two path, up to horizontal and vertical symmetries, into four possible types: *m, s, rainbow,* and *spiral*, as depicted in Figure 3.

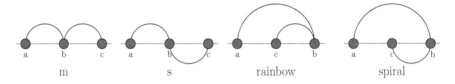

Fig. 3. Up to horizontal and vertical symmetry, the only possible arrangements of two consecutive edges are m, s, rainbow, and spiral

Lemma 5. *If a layout contains a pair of edges ab and bc of m or rainbow type with edge bc uncrossed and with b and c both having degree two, then it can be reduced without changing its crossings by a rearrangement followed by a contraction of the edge bc.*

Proof. In either configuration we move vertex b adjacent to vertex c, on the opposite side of c from its other neighbor, as demonstrated in Figure 4. Since the edge bc is uncrossed this transformation does not change the crossing structure of the drawing. Now that b and c are placed next to each other the edge bc may be contracted. □

Lemma 6. *If a layout contains a pair of uncrossed edges ab and bc of s or spiral type, with a, b, and c all having degree two, then the layout can be reduced without changing its crossings by a rearrangement and contraction.*

Proof. We assume by symmetry that a is the leftmost of the three vertices, edge ab is in the upper page, and edge bc is in the lower page. Let x be the neighbor of

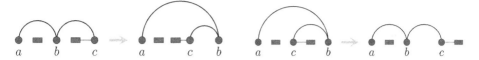

Fig. 4. The m reduction (left) and the rainbow reduction (right) shown with an edge into the β region

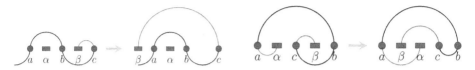

Fig. 5. The s reduction (left) shown with an edge into β and the spiral reduction (right) shown with edges into α and β

a that is not b and let y be the neighbor of c that is not b. We may assume that edge xa is in the lower page, for if it were in the upper page then edge ab would be part of an m or rainbow configuration and could be reduced by Lemma 5. By the same reasoning we may assume that cy is in the upper page.

First we consider s configurations. Let α be the set of vertices between a and b, and let β be the set of vertices between b and c. Then β can have no incoming edges in the lower page, because cy is upper and bc blocks all other edges. Therefore, we may move β directly to the left of a, as in Figure 5. Since edges ab and bc are uncrossed this transformation does not change the crossing structure of the drawing. We can then contract edge ab.

For the spiral, assume by symmetry that c is between a and b. Let α be the set of vertices between a and c, and let β be the set of vertices between c and b. Because cy is assumed to be in the upper page, and bc blocks all other lower edges, β can have no incoming lower edges; however, it might have edges in the upper page connecting it to α, so we must be careful to avoid twisting those connections and introducing new crossings. In this case, we move β between a and α and contract edge bc. □

As shown above, if any degree two path has at least four edges and two consecutive uncrossed edges, then we can apply one of the reduction rules and reduce the number of edges. For this reason we define the kernel K for computing $cr_2(G)$ using the method in Section 3, with the bound $\ell(k) = 2k^2$ on the length of the maximal degree two paths. Similarly, we define the the kernel L for computing $cre_2(G)$ by setting $\ell(k) = 2k$.

Lemma 7. *Let G be a graph with cyclomatic number k. Then,*

1. *K has at most $6k^3$ vertices and $6k^3$ edges;*
2. *$cr_2(G) = cr_2(K)$;*
3. *L has at most $6k^2$ vertices and $6k^2$ edges;*
4. *$cre_2(G) = cre_2(L)$*

Proof. (1) Since we have at most $2k^2$ vertices per maximal degree two path, the total number of vertices is at most $2k^2(3k-3) + 2k - 2 \leq 6k^3$. The number of edges is at most $2k^2(3k-3) + (2k-2) + (k-1) \leq 6k^3$.

(2) The proof that $\mathrm{cr}_2(G) \leq \mathrm{cr}_2(K)$ is that same as in Lemma 2. To see that $\mathrm{cr}_2(K) \leq \mathrm{cr}_2(G)$ we suppose that G has been given an embedding that minimizes $\mathrm{cr}_2(G)$. The total number of crossings in such an embedding is bounded above by $\binom{k}{2} < k^2/2$, and in turn the number of crossed edges is less than k^2. Thus any maximal degree two path in G with length greater than $2k^2$ can be shortened.

(3) and (4) The proof follows by the same argument as in (1) and (2), noting that there always exists a drawing with at most k crossed edges.   □

We apply the straightforward exact algorithm to the kernel of the graph to achieve the following result:

Theorem 2. *If G is a graph with cyclomatic number k, then $\mathrm{cr}_2(G)$ can be computed in $O(2^{6k^3}(6k^3)!+n)$ time, and $\mathrm{cre}_2(G)$ can be computed in $O(2^{6k^2}(6k^2)!+n)$ time. If G is a k-almost-tree, then $\mathrm{cr}_2(G)$ and $\mathrm{cre}_2(G)$ can be computed in $O(2^{6k^3}(6k^3)!n)$ time and $O(2^{6k^2}(6k^2)!n)$ time respectively.*

6   Matrix Multiplication Improvement

The asymptotic run time for processing each bicon-nected component in both the one page and two page cases can be further improved using matrix multiplication to find the minimum weight triangle in a graph [27].

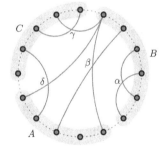

Fig. 6. Types of crossings

We begin with the 1-page case, in which we improve the run time to $O(k^{O(1)}(5k)!^{\omega/3})$ where $\omega < 2.3727$ is the exponent for matrix multiplica-tion [28]. Let $N \leq 5k$ be the number of vertices in the kernel K, and for simplicity of exposition, as-sume that N is a multiple of 3. We construct a new graph G' as follows. For each subset $S \subset K$ of $N/3$ vertices in the original kernel K, and for each ordering of S, we create one vertex in G'. Thus, the number of vertices in G' is $(N/3)! \cdot \binom{N}{N/3} = O\left((N!)^{1/3}\right)$. We add edges in G' between pairs of vertices that represent disjoint subsets. G' has a triangle for every triple of subsets that form a proper partition of V in G. Thus, each triangle corresponds to an assignment of the vertices to three uni-formly sized regions and a distinct ordering of the vertices in each region, which together form a complete layout of G.

We assign a weight to each edge in G' based on the number of edge crossings in G between the vertices in the corresponding regions. There are four possible types of crossing, represented by α, β, γ and δ in Figure 6. For a crossing of type α, in which all endpoints of a pair of crossing edges in G are contained in the same region B, we add $1/2$ to the weights of edges AB and BC in G'. For β, in which a pair of crossing edges in G both start in a region A and end in another region B, we add 1 to the weight of edge AB in G'. For γ, in which three endpoints of a pair of edges lie in the same region C, and the fourth lies in a different region B, we add 1 to the weight of edge BC in G'. Finally, for δ, in which a pair of crossed edges both have an endpoint in one region A, but their other endpoint in two different regions B and C, we add 1/2 to the weight of edge AC and 1/2 to the weight of edge AB in G'. With these weights, the total weight of a triangle in G' equals the number of edge crossings in the corresponding layout. The edge weights for G' can be computed in $O(k^{O(1)}(5k)!^{2/3})$ time.

To find the minimum weight triangle we construct the weighted adjacency matrix A, where $A_{i,j}$ is given the weight of the edge from i to j or infinity if no such edge exists. We then compute the min-plus matrix product of A with itself, which is defined by $[A \star A]_{i,j} = \min_k A_{i,k} + A_{k,j}$. The weight of a minimum weight triangle in A then corresponds to the minimum entry in $A + A \star A$. From the minimum weight and corresponding i and j the triangle can be found in linear time. Thus, the runtime is dominated by computing $A \star A$, which can be done in $O(k^{O(1)}(5k)!^{\omega/3})$ time using fast matrix multiplication [29, 30].

For the 2-page case we consider each of the 2^M edge page assignments separately, computing the minimum crossing drawing for this assignment using matrix multiplication. As before we construct a graph G' with weighted edges between compatible vertices, such that a minimum weight triangle in G' corresponds to a minimum weight drawing. Matrix multiplication is then used to find this minimal weight triangle for each page assignment, yielding a running time of $O(2^M (N!)^{\omega/3})$, where N is the number of vertices and M is the number of edges in the kernel. Thus, we have the following result:

Theorem 3. *If G is a graph with cyclomatic number k, then we can compute:*

- $\mathrm{cr}_1(G)$ *and* $\mathrm{cre}_1(G)$ *in* $O(k^{O(1)}(5k)!^{\omega/3} + n)$ *time;*
- $\mathrm{cr}_2(G)$ *in* $O(2^{6k^3}(6k^3)!^{\omega/3} + n)$ *time;*
- $\mathrm{cre}_2(G)$ *in* $O(2^{6k^2}(6k^2)!^{\omega/3} + n)$ *time.*

7   Conclusion

We have given new fixed parameter algorithms for computing the minimum number of edge crossings and minimum number of crossed edges in 1-page and 2-page embeddings of k-almost trees. To our knowledge, these are the only parameterized exact algorithms for these drawing styles.

We leave the following questions open to future research:

- For 2-page embeddings, the hardness of finding uncrossed drawings [13] shows that crossing minimization cannot be FPT in its natural parameter, the number of crossings. What about 1-page embeddings?
- Can the dependence on k be reduced to singly exponential?
- What other NP-hard problems in graph drawing are FPT with respect to k?

Acknowledgements. We thank Emma S. Spiro whose work with social networks led us to consider drawing almost-trees, and an anonymous reviewer for helpful suggestions in our 2-page kernelization. This research was supported in part by the National Science Foundation under grant 0830403 and 1217322, and by the Office of Naval Research under MURI grant N00014-08-1-1015.

References

[1] Pach, J., Tóth, G.: Which crossing number is it anyway? J. Combin. Theory Ser. B 80(2), 225–246 (2000)

[2] Gurevich, Y., Stockmeyer, L., Vishkin, U.: Solving NP-hard problems on graphs that are almost trees and an application to facility location problems. J. ACM 31(3), 459–473 (1984)

[3] Fernández-Baca, D.: Allocating modules to processors in a distributed system. IEEE Transactions on Software Engineering 15(11), 1427–1436 (1989)

[4] Kloks, T., Bodlaender, H., Müller, H., Kratsch, D.: Computing treewidth and minimum fill-in: All you need are the minimal separators. In: Lengauer, T. (ed.) ESA 1993. LNCS, vol. 726, pp. 260–271. Springer, Heidelberg (1993)

[5] Akutsu, T., Hayashida, M., Ching, W.K., Ng, M.K.: Control of Boolean networks: Hardness results and algorithms for tree structured networks. J. Theor. Bio. 244(4), 670–679 (2007)

[6] Fiala, J., Kloks, T., Kratochvíl, J.: Fixed-parameter complexity of λ-labelings. Discrete Appl. Math. 113(1), 59–72 (2001)

[7] Coppersmith, D., Vishkin, U.: Solving NP-hard problems in 'almost trees': Vertex cover. Discrete Appl. Math. 10(1), 27–45 (1985)

[8] Bodlaender, H.: Dynamic algorithms for graphs with treewidth 2. In: van Leeuwen, J. (ed.) WG 1993. LNCS, vol. 790, pp. 112–124. Springer, Heidelberg (1994)

[9] Fürer, M.: Counting perfect matchings in graphs of degree 3. In: Kranakis, E., Krizanc, D., Luccio, F. (eds.) FUN 2012. LNCS, vol. 7288, pp. 189–197. Springer, Heidelberg (2012)

[10] Grohe, M.: Computing crossing numbers in quadratic time. In: ACM Symp. Theory of Computing (STOC 2001), pp. 231–236 (2001)

[11] Kawarabayashi, K.I., Reed, B.: Computing crossing number in linear time. In: ACM Symp. Theory of Computing (STOC 2007), pp. 382–390 (2007)

[12] Pelsmajer, M.J., Schaefer, M., Štefankovič, D.: Crossing number of graphs with rotation systems. In: Hong, S.-H., Nishizeki, T., Quan, W. (eds.) GD 2007. LNCS, vol. 4875, pp. 3–12. Springer, Heidelberg (2008)

[13] Chung, F.R.K., Leighton, F.T., Rosenberg, A.L.: Embedding graphs in books: A layout problem with applications to VLSI design. SIAM J. Alg. Disc. Meth. 8(1), 33–58 (1987)

[14] Potterat, J.J., Phillips-Plummer, L., Muth, S.Q., Rothenberg, R.B., Woodhouse, D.E., Maldonado-Long, T.S., Zimmerman, H.P., Muth, J.B.: Risk network structure in the early epidemic phase of HIV transmission in Colorado Springs. Sexually Transmitted Infections 78(suppl. 1), i159–i163 (2002)

[15] De, P., Singh, A.E., Wong, T., Yacoub, W., Jolly, A.M.: Sexual network analysis of a gonorrhoea outbreak. Sexually Transmitted Infections 80(4), 280–285 (2004)

[16] Potterat, J.J., Muth, S.Q., Rothenberg, R.B., Zimmerman-Rogers, H., Green, D.L., Taylor, J.E., Bonney, M.S., White, H.A.: Sexual network structure as an indicator of epidemic phase. Sexually Transmitted Infections 78(suppl. 1), i152–i158 (2002)

[17] Lombardi, M., Hobbs, R.: Mark Lombardi: Global Networks. Independent Curators (2003)
[18] Butts, C.T., Acton, R.M., Marcum, C.S.: Interorganizational collaboratio in the Hurricane Katrina response. J. Social Structure 13(1) (2012)
[19] Seidman, S.B.: Network structure and minimum degree. Social Networks 5(3), 269–287 (1983)
[20] de Nooy, W., Mrvar, A., Batagelj, V.: Exploratory Social Network Analysis with Pajek. Cambridge University Press (2012)
[21] Watts, D., Strogatz, S.: Collective dynamics of 'small-world' networks. Nature 393, 440–442 (1998)
[22] Bannister, M.J., Cabello, S., Eppstein, D.: Parameterized complexity of 1-planarity. In: Dehne, F., Solis-Oba, R., Sack, J.-R. (eds.) WADS 2013. LNCS, vol. 8037, pp. 97–108. Springer, Heidelberg (2013)
[23] Masuda, S., Kashiwabara, T., Nakajima, K., Fujisawa, T.: On the NP-completeness of a computer network layout problem. In: 20th IEEE Symp. Circuits and Systems, pp. 292–295 (1987)
[24] Baur, M., Brandes, U.: Crossing reduction in circular layouts. In: Hromkovič, J., Nagl, M., Westfechtel, B. (eds.) WG 2004. LNCS, vol. 3353, pp. 332–343. Springer, Heidelberg (2004)
[25] Sedgewick, R.: Permutation generation methods. ACM Comput. Surv. 9(2), 137–164 (1977)
[26] Masuda, S., Nakajima, K., Kashiwabara, T., Fujisawa, T.: Crossing minimization in linear embeddings of graphs. IEEE Trans. Computers 39(1), 124–127 (1990)
[27] Williams, R.: A new algorithm for optimal 2-constraint satisfaction and its implications. Theor. Comput. Sci. 348(2-3), 357–365 (2005)
[28] Williams, V.V.: Multiplying matrices faster than Coppersmith–Winograd. In: ACM Symp. Theory of Computing (STOC 2012), pp. 887–898 (2012)
[29] Alon, N., Galil, Z., Margalit, O.: On the exponent of the all pairs shortest path problem. J. Comput. Sys. Sci. 54(2), 255–262 (1997)
[30] Yuval, G.: An algorithm for finding all shortest paths using $N^{2.81}$ infinite-precision multiplications. Inf. Proc. Lett. 4(6), 155–156 (1976)

Strict Confluent Drawing

David Eppstein[1], Danny Holten[2], Maarten Löffler[3],
Martin Nöllenburg[4], Bettina Speckmann[5], and Kevin Verbeek[6]

[1] Computer Science Department, University of California, Irvine, USA
eppstein@uci.edu
[2] Synerscope BV, Eindhoven, the Netherlands
danny.holten@synerscope.com
[3] Department of Computing and Information Sciences, Utrecht University, The Netherlands
m.loffler@uu.nl
[4] Institute of Theoretical Informatics, Karlsruhe Institute of Technology, Germany
noellenburg@kit.edu
[5] Department of Mathematics and Computer Science,
Technical University Eindhoven, The Netherlands
speckman@win.tue.nl
[6] Department of Computer Science, University of California, Santa Barbara, USA
kverbeek@cs.ucsb.edu

Abstract. We define *strict confluent drawing*, a form of confluent drawing in which the existence of an edge is indicated by the presence of a smooth path through a system of arcs and junctions (without crossings), and in which such a path, if it exists, must be unique. We prove that it is NP-complete to determine whether a given graph has a strict confluent drawing but polynomial to determine whether it has an *outerplanar* strict confluent drawing with a fixed vertex ordering (a drawing within a disk, with the vertices placed in a given order on the boundary).

1 Introduction

Confluent drawing is a style of graph drawing in which edges are not drawn explicitly; instead vertex adjacency is indicated by the existence of a smooth path through a system of arcs and junctions that resemble train tracks. These types of drawings allow even very dense graphs, such as complete graphs and complete bipartite graphs, to be drawn in a planar way [4]. Since its introduction, there has been much subsequent work on confluent drawing [7,6,9,10,13,17], but the complexity of confluent drawing has remained unclear: how difficult is it to determine whether a given graph has a confluent drawing? Confluent drawings have a certain visual similarity to a graph drawing technique called *edge bundling* [3,5,11,12,14], in which "similar" edges are routed together in "bundles", but we note that these drawings should be interpreted differently. In particular, sets of edges bundled together form visual junctions, however, interpreting them as confluent junctions can create false adjacencies.

Formally, a confluent drawing may be defined as a collection of *vertices*, *junctions* and *arcs* in the plane, such that all arcs are smooth and start and end at either a junction or a vertex, such that arcs intersect only at their endpoints, and such that all arcs that meet at a junction share the same tangent line there. A confluent drawing D represents a graph G defined as follows: the vertices of G are the vertices of D, and there is an

S. Wismath and A. Wolff (Eds.): GD 2013, LNCS 8242, pp. 352–363, 2013.

edge between two vertices u and v if and only if there exists a smooth path in D from u to v that does not pass any other vertex. (In some variants of confluent drawing an additional restriction is made that the smooth path may not intersect itself [13]; however, this constraint is not relevant for our work.)

Contribution. In this paper we introduce a subclass of confluent drawings, which we call *strict* confluent drawings. Strict confluent drawings are confluent drawings with the additional restrictions that between any pair of vertices there can be *at most one* smooth path, and there cannot be any paths from a vertex to itself. Figure 1 illustrates the forbidden configurations. To avoid irrelevant components in the drawing, we also require all arcs of the drawing to be part of at least one smooth path representing an edge. We

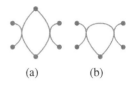

Fig. 1. (a) A drawing with a duplicate path. (b) A drawing with a self-loop.

believe that these restrictions may make strict drawings easier to read, by reducing the ambiguity caused by the existence of multiple paths between vertices. In addition, as we show, the assumption of strictness allows us to completely characterize their complexity, the first such characterization for any form of confluence on arbitrary undirected graphs.

We prove the following:

- It is NP-complete to determine whether a given graph has a strict confluent drawing.
- For a given graph, with a given cyclic ordering of its vertices, there is a polynomial time algorithm to find an *outerplanar* strict confluent drawing, if it exists: this is a drawing in a disk, with the vertices in the given order on the boundary of the disk
- When a graph has an outerplanar strict confluent drawing, an algorithm based on circle packing can construct a layout of the drawing in which every arc is drawn using at most two circular arcs.

See Fig. 2(a) for an example of an outerplanar strict confluent drawing. Previous work on *tree-confluent* [13] and *delta-confluent drawings* [6] characterized special cases of outerplanar strict confluent drawings as being the chordal bipartite graphs and distance-hereditary graphs respectively, so these graphs as well as the outerplanar graphs are all outerplanar strict confluent. The six-vertex wheel graph in Fig. 2(b) provides an example

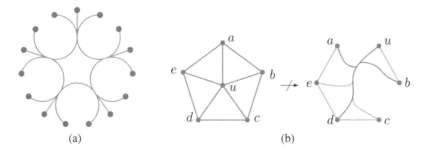

(a)   (b)

Fig. 2. (a) Outerplanar strict confluent drawing of the GD2011 contest graph. (b) A graph with no outerplanar strict confluent drawing.

of a graph that does not have an outerplanar strict confluent drawing. (The central vertex u needs to be placed between two of the outer vertices, say, a and b. The smooth path from u to the opposite vertex d separates a and b, so there must be a junction shared by the u–d and a–b paths, creating a wrong adjacency with d.)

2   Preliminaries

Let $G = (V, E)$ be a graph. We call an edge e in a drawing D *direct* if it consists only of a single arc (that does not pass through junctions). We call the angle between two consecutive arcs at a junction or vertex *sharp* if the two arcs do not form a smooth path; each junction has exactly two angles that are not sharp, and every angle at a vertex is sharp (so the number of sharp angles equals the degree of the vertex).

Lemma 1. *Let G be a graph, and let $E' \subseteq E$ be the edges of E that are incident to at least one vertex of degree 2. If G has a strict confluent drawing D, then it also has a strict confluent drawing D' in which all edges in E' are direct.*

Proof. Let v be a degree-2 vertex in G with two incident edges e and f. We consider the representation of e and f in D and modify D so that e and f are single arcs. There are two cases. If e and f leave v on two disjoint paths, then these paths have only merge junctions from v's perspective. We can simply separate these junctions from e and f as shown in Fig. 3(a). If, on the other hand, e and f share the same path leaving v, then their paths split at some point. We need to reroute the merge junctions prior to the split and separate the merge junctions after the split as shown in Fig. 3(b). This is always possible since v has no other incident edges. Because D was strict and these changes do not affect strictness, D' is still a strict confluent drawing and edges e and f are direct.   □

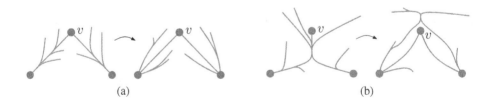

(a)   (b)

Fig. 3. The two cases of creating single arcs for edges incident to a degree-2 vertex

Lemma 2. *Let G be a graph. If G has no $K_{2,2}$ as a subgraph, whose vertices have degrees ≥ 3 in G, then G has a strict confluent drawing if and only if G is planar.*

Proof. Since every planar drawing is also a strict confluent drawing, that implication is obvious. So let D be a strict confluent drawing for a graph G without a $K_{2,2}$ subgraph, whose vertices have degrees ≥ 3 in G. Since larger junctions, where more than three arcs meet, can easily be transformed into an equivalent sequence of binary junctions, we can assume that every junction in D is binary, i.e., two arcs merge into one (or, from a different perspective, one arc splits into two). By Lemma 1 we can further transform D so that all edges incident to degree-2 vertices are direct. Now for any vertex u in D none of its outgoing paths to some neighbor v can visit a merge junction before visiting a split junction as this would imply either a non-strict drawing or a $K_{2,2}$ subgraph with vertex degrees ≥ 3. So the sequence of junctions on any u-v path consists of a number of split junctions followed by a number of merge junctions. But any such path can be unbundled from its junctions to the left and right and turned into a direct edge without creating arc intersections as illustrated in Fig. 4. This shows that D can be transformed into a standard planar drawing of G. □

Fig. 4. Any strict confluent drawing of a graph without a $K_{2,2}$ subgraph can be transformed into a standard planar drawing

Lemma 3 characterizes the combinatorial complexity of strict confluent drawings. Its proof is found in the full paper [8] and uses Euler's formula and double counting.

Lemma 3. *The combinatorial complexity of any strict confluent drawing D of a graph G, i.e., the number of arcs, junctions, and faces in D, is linear in the number of vertices of G.*

Lemma 3 is in contrast to previous methods for confluently drawing interval graphs [4] and for drawing confluent Hasse diagrams [9], both of which may produce (non-strict) drawings with quadratically many features.

3   Computational Complexity

We will show by a reduction from planar 3-SAT [15] that it is NP-complete to decide whether a graph G has a strict confluent drawing in which all edges incident to degree-2 vertices are direct. By Lemma 1, this is enough to show that it is also NP-complete to decide if G has any strict confluent drawing.

Consider the subdivided grid graph (a grid with one extra vertex on each edge). In this graph, all edges are adjacent to a degree 2 vertex. Since a grid graph more than one square wide has only one fixed planar embedding (up to choice of the outer face), the subdivided grid graph has only one confluent embedding in which all edges are direct. We will base our construction on a number of such grids.

Let S be a planar 3-SAT formula. Globally speaking, we will create a grid graph for each variable of S, of size depending on the number of clauses that the variable appears in. The external edges of this grid graph are alternatingly colored green and red. We connect the variable graphs by identifying certain vertices: for each of the three variables that appear in a clause, we select one subdivided edge (that is, three vertices connected by two edges) on the outer face, and identify the endpoints of these edges

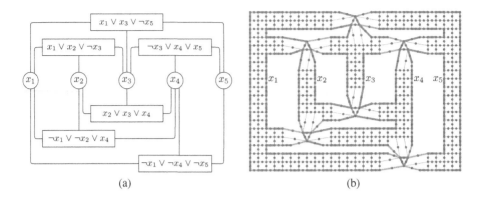

(a)   (b)

Fig. 5. (a) A planar 3-SAT formula. (b) The corresponding global frame of the construction: one grid graph per variable, with some vertices identified at each clause. Green boundary edges correspond to positive literals, red edges to negated literals. For easier readability the grids in this figure are larger than strictly necessary.

Fig. 6. K_4 and its two strict confluent drawings, without moving the vertices and keeping all arcs inside the convex hull of the vertices

into a triangle of subdivided edges (that is, a 6-cycle). We choose a green edge for a positive occurrence of the variable and a red edge for a negated occurrence. This will become clear below. We call the resulting graph F the *frame* of the construction; all edges of F are adjacent to a degree-2 vertex and F has only one planar embedding (up to choice of the outer face). Figure 5 shows an example.

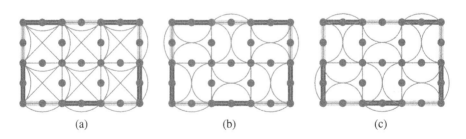

(a)   (b)   (c)

Fig. 7. (a) A variable gadget consists of a grid of K_4's. Green (light) edges of the frame highlight normal literals, red (dark) edges negated ones. (b) One of the two possible strict confluent drawings, corresponding to the value *true*. (c) The other strict confluent drawing, corresponding to *false*.

The main idea of the construction is based on the fact that K_4, when drawn with all four vertices on the outer face, has exactly two strict confluent drawings: we need to create a junction that merges the diagonal edges with one pair of opposite edges, and we can choose the pair. Figure 6 illustrates this. We will add a copy of K_4 to every cell of the frame graph F. Recall that every cell, except for the triangular clause faces, is a subdivided square (that is, an 8-cycle). We add K_4 on the four grid vertices (not the subdivision vertices). The edges that connect external grid vertices are called *literal edges*. Figure 7(a) shows this for a small grid. Since neighboring grid cells share a (subdivided) edge, the K_4's are not edge-independent. This implies that in a strict confluent drawing, we cannot "use" such a common edge in both cells. Therefore, we need to orient the K_4-

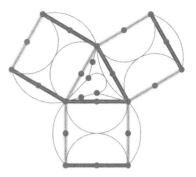

Fig. 8. Three variables attached to a clause gadget. The top left variable occurs in the clause as a positive literal, the others as negative literals. The clause can be satisfied because the top right variable is set to *false*.

junctions alternatingly, as illustrated in Figures 7(b) and 7(c). If the grid is sufficiently large (every cell is part of a larger at least size-(2×2) grid) these choices are completely propagated through the entire grid, so there are two structurally different possible embeddings, which we use to represent the values *true* and *false* of the corresponding variable. For every green edge of the frame in the *true* state and every red edge in the *false* state there is one remaining literal edge in the outer face, which can still be drawn either inside or outside their grid cells. In the opposite states these literal edges are needed inside the grid cells to create the K_4 junctions. The availability of at least one literal edge (corresponding to a *true* literal) is important for satisfying the clause gadgets, which we describe next.

Inside each triangular clause face, we add the graph depicted in Figure 9(a). This graph has several strict confluent drawings; however, in every drawing at least one of the three outer edges needs to be drawn inside the subdivided triangle.

Lemma 4. *There is no strict confluent drawing of the clause graph in which all three long edges are drawn outside. Moreover, there is a strict confluent drawing of the clause graph with two of these edges outside, for every pair.*

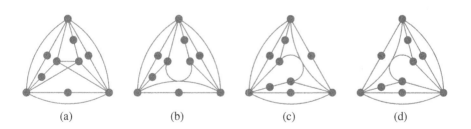

(a)   (b)   (c)   (d)

Fig. 9. (a) The input graph of the clause. (b, c, d) Three different strict confluent drawings

Proof. Recall that by Lemma 1 the subdivided triangle must be embedded as a 6-cycle of direct arcs. To prove the first part of the lemma, assume that the triangle edges are all drawn outside this cycle. The remainder of the graph has no 4-cycles without subdivision vertices (that is, no $K_{2,2}$ with higher-degree vertices), so by Lemma 2 it can only have a strict confluent drawing if it is planar. However, it is a subdivided K_5, which is not planar. To prove the second part of the lemma, we refer to Figures 9(b), 9(c) and 9(d).   □

This describes the reduction from a planar 3-SAT instance to a graph consisting of variable and clause gadgets. Next we show that this graph has a strict confluent drawing if and only if the planar 3-SAT formula is satisfiable. For a given satisfying assignment we choose the corresponding embeddings of all variable gadgets. The assignment has at least one *true* literal per clause, and correspondingly in each clause gadget one of the three literal edges can be drawn inside the clause triangle, allowing a strict confluent drawing by Lemma 4. Conversely, in any strict confluent drawing, each clause must be drawn with at least one literal edge inside the clause triangle by Lemma 4, so translating the state of each variable gadget into its truth value yields a satisfying assignment.

To show that testing strict confluence is in NP, recall that by Lemma 3 the combinatorial complexity of the drawing is linear in the number of vertices. Thus the existence of a drawing can be verified by guessing its combinatorial structure and verifying that it is planar and a drawing of the correct graph.

Theorem 1. *Deciding whether a graph has a strict confluent drawing is NP-complete.*

4   Outerplanar Strict Confluent Drawings

For a graph G with a fixed cyclic ordering of its vertices, we can test in polynomial time whether an outerplanar strict confluent drawing with this vertex ordering exists, and, if so, construct one. This algorithm uses the closely related notion of a canonical diagram of G, which is unique and exists if and only if an outerplanar strict confluent drawing exists. From the canonical diagram a confluent drawing can be constructed. We further show that the drawing can be constructed such that every arc consists of at most two circular arcs.

4.1   Canonical Diagrams

We define a *canonical diagram* to be a collection of junctions and arcs connecting the vertices in the given order on the outer face (as in a confluent drawing), but with some of the faces of the diagram *marked*, satisfying additional constraints enumerated below. Figure 10 shows a canonical diagram and an outerplanar strict confluent drawing of the same graph. In such a diagram, a *trail* is a smooth curve from one vertex to another that follows the arcs (as in a confluent drawing) but is allowed to cross the interior of marked faces from one of its sharp corners to another. The constraints are:

- Every arc is part of at least one trail.
- No two trails between the same two vertices can follow different sequences of arcs and faces.

Fig. 10. Three views of the same graph as a node-link diagram (left), canonical diagram (center), and outerplanar strict confluent drawing (right)

- Each marked face must have at least four angles, all of which are sharp.
- Each arc must have either sharp angles or vertices at both of its ends.
- For each junction j with exactly two arcs in each direction, let f and f' be the two faces with sharp angles at j. Then it is not allowed for f and f' to both be either marked or to be a triangle (a face with three angles, all sharp).

Let j be a junction of a canonical diagram D. Then define the *funnel* of j to be the 4-tuple of vertices a, b, c, d where a is the vertex reached by a path that leaves j in one direction and continues as far clockwise as possible, b is the most counterclockwise vertex reachable in the same direction from j, c is the most clockwise vertex reachable in the other direction, and d is the most counterclockwise vertex reachable in the other direction. Note that none of the paths from j to a, b, c, and d can intersect each other without contradicting the uniqueness of trails. We call the circular intervals of vertices $[a, b]$ and $[c, d]$ (in the counterclockwise direction) the *funnel intervals* of the respective funnel. We say a circular interval $[a, b]$ is *separated* if either a and b are not adjacent in G, or there exists a junction in the canonical diagram with funnel intervals $[a, e]$ and $[f, b]$, where $e, f \in [a, b]$.

A canonical diagram represents a graph G in which the edges in G correspond to trails in the diagram. As we show in the full paper [8], a graph G has a canonical diagram if and only if it has an outerplanar strict confluent drawing, and if a canonical diagram exists then it is unique.

4.2 Algorithm

By using the properties of canonical diagrams (see the full paper [8]), we may obtain an algorithm that constructs a canonical diagram and strict confluent drawing of a given cyclically-ordered graph G, or reports that no drawing exists, in time and space $O(n^2)$. This bound is optimal in the worst case, as it matches the input size of a graph that may have quadratically many edges.

Steps 1–3 of the algorithm, detailed below, build some simple data structures that speed up the subsequent computations. Step 4 discovers all of the funnels in the input, from which it constructs a list of all of the junctions of the canonical diagram. Step 5 connects these junctions into a planar drawing, a subset of the canonical diagram. Step 6 builds a graph for each face of this drawing that will be used to complete it into the entire canonical diagram, and step 7 uses these graphs to find the remaining arcs of the diagram and to determine which faces of the diagram are marked. Step 8 checks that the diagram constructed by the previous steps correctly represents the input graph, and step 9 splits the marked faces, converting the diagram into a strict confluent drawing.

1. Number the vertices clockwise around the boundary cycle from 0 to $n - 1$.
2. Build a table, containing for each pair i, j, the number of ordered pairs (i', j') with $i' \leq i, j' \leq j$, and vertices i' and j' adjacent in G. By performing a constant number of lookups in this table we may determine in constant time how many edges exist between any two disjoint intervals of the boundary cycle.
3. Build a table that lists, for each ordered pair u, v of vertices, the neighbor w of u that is closest in clockwise order to v. That is, w is adjacent to u, and the interval from v clockwise to w contains no other neighbors of u. The table entries for u can be found in linear time by a single counterclockwise scan. Repeat the same construction in the opposite orientation.
4. For each separated interval $[a, b]$, let c be the next neighbor of a that is counterclockwise of b, and let d be the next neighbor of b that is clockwise of a. If (i) c is a neighbor of b, (ii) d is a neighbor of a, (iii) a is the next neighbor of c that is counterclockwise of d, and (iv) b is the next neighbor of d that is clockwise of c, then (if a confluent diagram exists) a, b, c, d must form the funnel of a junction, and all funnels have this form. We check all circular intervals in increasing order of their cardinalities. For each discovered funnel, we mark the intervals that are separated by the corresponding junction. This way we can check in $O(1)$ time whether a circular interval is separated. If the number of funnels exceeds the linear bound of Lemma 3 on the number of junctions in a confluent drawing, abort the algorithm.
5. Create a junction for each of the funnels found in step 4. For each vertex v, make a set J_v of the junctions whose funnel includes that vertex; if they are to be drawn as part of a canonical diagram, the junctions of J_v need to be connected to v by a confluent tree. For any two junctions in J_v, it is possible to determine in constant time whether one is an ancestor of another in this tree, or if not whether one is clockwise of the other, by examining the cyclic ordering of vertices in their funnels. Construct the trees of junctions and their planar embedding in this way. The result of this stage of the algorithm should be a planar embedding of part of the canonical diagram consisting of all vertices and junctions, and the subset of the arcs that are part of a path from a junction to one of its funnel vertices. Check that the embedding is planar by computing its Euler characteristic, and abort the algorithm if it is not.
6. For each face f of the drawing created in step 5, and each pair j, j' of junctions belonging to f, use the data structure from step 2 to test whether there is an edge whose trail passes through both j and j'. This results in a graph H_f in which the vertices represent the vertices or junctions on the boundary of f and the edges represent pairs of vertices or junctions that must be connected, either by an arc or by shared membership in a marked face. The remaining arcs to be drawn in f will be exactly the edges of H_f that are not crossed by other edges of H_f; the marked faces in f will be exactly the faces that contain pairs of crossing edges of H_f.
7. Within each face f of the drawing so far, build a table using the same construction as in step 2 that can be used to determine the existence of a crossing edge for an

edge in H_f in constant time. Use this data structure to identify the crossed edges, and draw an arc in f for each uncrossed edge. For each face g of the resulting subdivision of f, if g has four or more vertices or junctions, find two pairs that would cross and test whether both pairs correspond to edges in H_f; if so, mark g.

8. Construct a directed graph that has a vertex for each vertex of G, two vertices for each junction of the diagram (one in each direction), two directed edges for each arc, and a directed edge for each ordered pair of sharp angles that are non-consecutive in a marked face. By performing a depth-first search in this graph, determine whether there exist multiple smooth paths in the resulting drawing from any vertex of G to any other point in the drawing, and abort the algorithm if any such pair of paths is found. Determine the set of vertices of G reachable from v and verify that it is the same set of vertices that are reachable in the original graph. Additionally, verify that the diagram satisfies the requirements in the definition of a canonical diagram. Abort the algorithm if any inconsistency is found in this step.

9. Convert the canonical diagram into a confluent drawing and return it.

Theorem 2. *For a given n-vertex graph G, and a given circular ordering of its vertices, it is possible to determine whether G has an outerplanar strict confluent drawing with the given vertex ordering, and if so to construct one, in time $O(n^2)$.*

4.3 Drawings with Low Curve Complexity

Suppose that we are given a topological description of an outerplanar strict confluent drawing D of a connected graph G, describing the tangency pattern and ordering of the arcs at each junction. It still remains to draw D (or possibly an equivalent but combinatorially different outerplanar strict confluent drawing) in the plane using concrete curves for its arcs. If we ignore the tangency requirements at its junctions, the arcs and junctions of D form a planar graph, but applying standard planar graph drawing methods will generate arcs that may not be smooth and that are not tangent to each other at the junctions. So how are we to draw D? Here we use a circle packing method to draw D with a small number of circular arcs for each arc of D. Thus, these drawings have low *curve complexity* in the sense of Bekos et al. [1], but with this complexity measured along arcs of the confluent diagram rather than edges of another type of graph drawing.

Given such a drawing D, let D' be a modified version of D in which every junction is incident to exactly three arcs, formed from D by suppressing two-arc junctions and splitting junctions with more than three arcs. Assume also (again by adding more junctions if necessary) that each vertex in D' has only a single arc incident to it.

Given the topological diagram D', we form a planar graph H that has a vertex for each vertex or junction of D', and an edge for each arc of D'. Additionally, we create an edge in H for each two vertices that are consecutive in the cyclic ordering of the vertices around the disk containing the drawing.

Lemma 5. *H is planar, 3-regular, and 3-vertex-connected.*

Proof. Planarity and 3-regularity follow immediately from the construction of H. Every two vertices of G are connected by three vertex-disjoint paths in H: at least one (not necessarily a smooth path) through D, using the assumption that G is connected, and two more around the boundary of the disk. Therefore, if H were not 3-vertex-connected, only one of its 3-connected components could contain vertices of G. The other components would either contain components of D that are not part of any smooth path between vertices of G (forbidden in a strict confluent drawing) or would contain more than one smooth path between the same sets of vertices (also forbidden).   □

Theorem 3. *Let D be an outerplanar strict confluent drawing of a graph G, given topologically but not geometrically. Then we can construct an outerplanar strict confluent drawing of G in which each arc of the drawing is represented by a smooth curve that is either a circular arc or the union of two circular arcs.*

Proof. By the Koebe–Thurston–Andreev circle packing theorem, there exists a system C of circles representing the faces of H, such that two circles are adjacent exactly when the corresponding faces share an edge. We may assume (by performing a Möbius transformation if necessary) that the outer circle of this circle packing corresponds to the outer face of H. C may be found efficiently (although not in strongly polynomial time) by a numerical iteration that quickly converges to the system of radii of the circles, from which their centers can also be computed easily [2,16].

Each vertex of G corresponds in C to one of the triangular gaps between the outer circle and two other circles, and may be placed at the point of tangency of the two non-outer circles (one of the vertices of this triangle); see Fig. 11. The junctions in D' lie at the meeting point of three faces of H, and correspond in C to the remaining triangular gaps between three circles. A confluent drawing of G may be formed by removing the outer circle, removing all circular arcs bounding the triangular gaps incident to the outer circle, and in each remaining triangular gap removing the arc that is on the other side of the sharp angle. The resulting drawing contracts some edges of D' to form junctions with four incident arcs, but this does not affect the correctness of the drawing. In the resulting drawing, arcs of the diagram that have merge points or vertices at both of their endpoints are drawn as two circular arcs (possibly both from the same circle); other arcs of the diagram are drawn as a single circular arc.   □

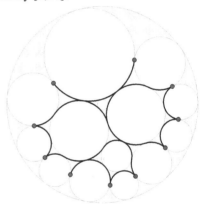

Fig. 11. Constructing an outerplanar strict confluent drawing from a circle packing. The vertices of the drawing correspond to triangular gaps adjacent to the outer circle, and the junctions to the remaining triangular gaps.

5   Conclusions

We have shown that, in confluent drawing, restricting attention to the strict drawings allows us to completely characterize their complexity, and we have also shown that outerplanar strict confluent drawings with a fixed vertex ordering may be constructed in polynomial time. The most pressing problem left open by this research is to recognize the graphs that have outerplanar strict confluent drawings, without imposing a fixed vertex order. Can we recognize these graphs in polynomial time?

Acknowledgements. This work originated at Dagstuhl seminar 13151, *Drawing Graphs and Maps with Curves.* D.E. was supported in part by the National Science Foundation under grants 0830403 and 1217322, and by the Office of Naval Research under MURI grant N00014-08-1-1015. M.L. was supported by the Netherlands Organisation for Scientific Research (NWO) under grant 639.021.123. M.N. received financial support by the 'Concept for the Future' of KIT under grant YIG 10-209.

References

1. Bekos, M.A., Kaufmann, M., Kobourov, S.G., Symvonis, A.: Smooth orthogonal layouts. In: Didimo, W., Patrignani, M. (eds.) GD 2012. LNCS, vol. 7704, pp. 150–161. Springer, Heidelberg (2013)
2. Collins, C.R., Stephenson, K.: A circle packing algorithm. Comput. Geom. Theory Appl. 25(3), 233–256 (2003)
3. Cui, W., Zhou, H., Qu, H., Wong, P.C., Li, X.: Geometry-based edge clustering for graph visualization. IEEE TVCG 14(6), 1277–1284 (2008)
4. Dickerson, M., Eppstein, D., Goodrich, M.T., Meng, J.Y.: Confluent drawings: Visualizing non-planar diagrams in a planar way. J. Graph. Algorithms Appl. 9(1), 31–52 (2005)
5. Dwyer, T., Marriott, K., Wybrow, M.: Integrating edge routing into force-directed layout. In: Kaufmann, M., Wagner, D. (eds.) GD 2006. LNCS, vol. 4372, pp. 8–19. Springer, Heidelberg (2007)
6. Eppstein, D., Goodrich, M.T., Meng, J.Y.: Delta-confluent drawings. In: Healy, P., Nikolov, N.S. (eds.) GD 2005. LNCS, vol. 3843, pp. 165–176. Springer, Heidelberg (2006)
7. Eppstein, D., Goodrich, M.T., Meng, J.Y.: Confluent layered drawings. Algorithmica 47(4), 439–452 (2007)
8. Eppstein, D., Holten, D., Löffler, M., Nöllenburg, M., Speckmann, B., Verbeek, K.: Strict confluent drawing. CoRR, abs/1308.6824 (2013)
9. Eppstein, D., Simons, J.A.: Confluent hasse diagrams. In: Speckmann, B. (ed.) GD 2011. LNCS, vol. 7034, pp. 2–13. Springer, Heidelberg (2011)
10. Hirsch, M., Meijer, H., Rappaport, D.: Biclique edge cover graphs and confluent drawings. In: Kaufmann, M., Wagner, D. (eds.) GD 2006. LNCS, vol. 4372, pp. 405–416. Springer, Heidelberg (2007)
11. Holten, D.: Hierarchical edge bundles: visualization of adjacency relations in hierarchical data. IEEE TVCG 12(5), 741–748 (2006)
12. Holten, D., van Wijk, J.J.: Force-Directed Edge Bundling for Graph Visualization. Computer Graphics Forum 28(3), 983–990 (2009)
13. Hui, P., Pelsmajer, M.J., Schaefer, M., Štefankovič, D.: Train tracks and confluent drawings. Algorithmica 47(4), 465–479 (2007)
14. Hurter, C., Ersoy, O., Telea, A.: Graph Bundling by Kernel Density Estimation. Computer Graphics Forum 31(3pt. 1), 865–874 (2012)
15. Lichtenstein, D.: Planar formulae and their uses. SIAM J. Comput. 11(2), 329–343 (1982)
16. Mohar, B.: A polynomial time circle packing algorithm. Discrete Math. 117(1-3), 257–263 (1993)
17. Quercini, G., Ancona, M.: Confluent drawing algorithms using rectangular dualization. In: Brandes, U., Cornelsen, S. (eds.) GD 2010. LNCS, vol. 6502, pp. 341–352. Springer, Heidelberg (2011)

A Ramsey-Type Result for Geometric ℓ-Hypergraphs

Dhruv Mubayi[1,*] and Andrew Suk[2,**]

[1] University of Illinois at Chicago, Chicago, IL
mubayi@math.uic.edu
[2] Massachusetts Institute of Technology, Cambridge, MA
asuk@math.mit.edu

Abstract. Let $n \geq \ell \geq 2$ and $q \geq 2$. We consider the minimum N such that whenever we have N points in the plane in general position and the ℓ-subsets of these points are colored with q colors, there is a subset S of n points all of whose ℓ-subsets have the same color and furthermore S is in convex position. This combines two classical areas of intense study over the last 75 years: the Ramsey problem for hypergraphs and the Erdős-Szekeres theorem on convex configurations in the plane. For the special case $\ell = 2$, we establish a single exponential bound on the minimum N such that every complete N-vertex geometric graph whose edges are colored with q colors, yields a monochromatic convex geometric graph on n vertices.

For fixed $\ell \geq 2$ and $q \geq 4$, our results determine the correct exponential tower growth rate for N as a function of n, similar to the usual hypergraph Ramsey problem, even though we require our monochromatic set to be in convex position. Our results also apply to the case of $\ell = 3$ and $q = 2$ by using a geometric variation of the Stepping-up lemma of Erdős and Hajnal. This is in contrast to the fact that the upper and lower bounds for the usual 3-uniform hypergraph Ramsey problem for two colors differ by one exponential in the tower.

1 Introduction

The classic 1935 paper of Erdős and Szekeres [12] entitled *A Combinatorial Problem in Geometry* was the starting point of a very rich discipline within combinatorics: Ramsey theory (see, e.g., [15]). The term *Ramsey theory* refers to a large body of deep results in mathematics which have a common theme: "Every large system contains a large well-organized subsystem." Motivated by the observation that any five points in the plane in general position[1] must contain four members in convex position, Esther Klein asked the following.

Problem 1. For every integer $n \geq 2$, determine the minimum $f(n)$ such that any set of $f(n)$ points in the plane in general position contains n members in convex position.

---

* Research partially supported by NSF grant DMS-0969092 and DMS-1300138.
** Supported by an NSF Postdoctoral Fellowship and by Swiss National Science Foundation Grant 200021-125287/1.

[1] A planar point set P is in *general position* if no three members are collinear.

S. Wismath and A. Wolff (Eds.): GD 2013, LNCS 8242, pp. 364–375, 2013.

Celebrated results of Erdős and Szekeres [12,13] imply that

$$2^{n-1} + 1 \leq f(n) \leq \binom{2n-4}{n-2} = 2^{2n(1+o(1))}. \tag{1}$$

They conjectured that $f(n) = 2^{n-1} + 1$, and Erdős offered a \$500 reward for a proof. Despite much attention over the last 75 years, the constant factors in the exponents have not been improved.

In the same paper, Erdős and Szekeres [12] gave another proof of a classic result due to Ramsey [23] on hypergraphs. An ℓ-uniform hypergraph H is a pair (V, E), where V is the vertex set and $E \subset \binom{V}{\ell}$ is the set of edges. We denote $K_n^\ell = (V, E)$ to be the complete ℓ-uniform hypergraph on an n-element set V, where $E = \binom{V}{\ell}$. When $\ell = 2$, we write $K_n^2 = K_n$. Motivated by obtaining good quantitative bounds on $f(n)$, Erdős and Szekeres looked at the following problem.

Problem 2. For every integer $n \geq 2$, determine the minimum integer $r(K_n, K_n)$ such that any two-coloring on the edges of a complete graph G on $r(K_n, K_n)$ vertices yields a monochromatic copy of K_n.

Erdős and Szekeres [12] showed that $r(K_n, K_n) \leq 2^{2n}$. Later, Erdős [9] gave a construction showing that $r(K_n, K_n) > 2^{n/2}$. Despite much attention over the last 65 years, the constant factors in the exponents have not been improved.

Generalizing Problem 2 to q-color and ℓ-uniform hypergraphs has also be studied extensively. Let $r(K_n^\ell; q)$ be the least integer N such that any q-coloring on the edges of a complete N-vertex ℓ-uniform hypergraph H yields a monochromatic copy of K_n^ℓ. We will also write

$$r(K_n^\ell; q) = r(\underbrace{K_n^\ell, K_n^\ell, \ldots, K_n^\ell}_{q \text{ times}}).$$

Erdős et al. [10,11] showed that there are positive constants c and c' such that

$$2^{cn^2} < r(K_n^3, K_n^3) < 2^{2^{c'n}}. \tag{2}$$

They also conjectured that $r(K_n^3, K_n^3) > 2^{2^{cn}}$ for some constant $c > 0$, and Erdős offered a \$500 reward for a proof. For $\ell \geq 4$, there is also a difference of one exponential between the known upper and lower bounds for $r(K_n^\ell, K_n^\ell)$, namely,

$$\mathrm{twr}_{\ell-1}(cn^2) \leq r(K_n^\ell, K_n^\ell) \leq \mathrm{twr}_\ell(c'n), \tag{3}$$

where c and c' depend only on ℓ, and the tower function $\mathrm{twr}_\ell(x)$ is defined by $\mathrm{twr}_1(x) = x$ and $\mathrm{twr}_{i+1} = 2^{\mathrm{twr}_i(x)}$. As Erdős and Rado [11] have shown, the upper bound in equation (3) easily generalizes to q colors, implying that $r(K_n^\ell; q) \leq \mathrm{twr}_\ell(c'n)$, where $c' = c'(\ell, q)$. On the other hand, for $q \geq 4$ colors, Erdős and Hajnal (see [15]) showed that $r(K_n^\ell; q)$ does indeed grow as a ℓ-fold exponential tower in n, proving that $r(K_n^\ell; q) = \mathrm{twr}_\ell(\Theta(n))$. For $q = 3$ colors, Conlon et al. [6] modified the construction of Erdős and Hajnal to show that $r(K_n^\ell, K_n^\ell, K_n^\ell,) > \mathrm{twr}_\ell(c \log^2 n)$.

Interestingly, both Problems 1 and 2 can be asked simultaneously for geometric graphs, and a similar-type problem can be asked for geometric ℓ-hypergraphs. A *geometric ℓ-hypergraph H in the plane* is a pair (V, E), where V is a set of points in the plane in general position, and $E \subset \binom{V}{\ell}$ is a collection of ℓ-tuples from V. When $\ell = 2$ ($\ell = 3$), edges are represented by straight line segments (triangles) induced by the corresponding vertices. The sets V and E are called the *vertex set* and *edge set* of H, respectively. A geometric hypergraph H is *convex*, if its vertices are in convex position.

Geometric graphs ($\ell = 2$) have been studied extensively, due to their wide range of applications in combinatorial and computational geometry (see [17,18,22]). Complete convex geometric graphs are very well understood, and are some of the most *well-organized* geometric graphs (if not the most). Many long-standing problems on complete geometric graphs, such as its crossing number [2], number of halving-edges [26], and size of crossing families [3], become trivial when its vertices are in convex position. There has also been a lot of research on geometric 3-hypergraphs in the plane, due to its connection to the *k-set problem* in $\mathbb{R}^3$ (see [7,21,24]). In this paper, we study the following problem which combines Problems 1 and 2.

Problem 3. Determine the minimum integer $g(K_n^\ell; q)$ such that any q-coloring of the edges of a complete geometric ℓ-hypergraph H on $g(K_n^\ell; q)$ vertices yields a complete monochromatic convex ℓ-hypergraph on n vertices.

We will also write

$$g(K_n^\ell; q) = g(\underbrace{K_n^\ell, \ldots, K_n^\ell}_{q \text{ times}}).$$

Clearly we have $g(K_n^\ell; q) \geq \max\{r(K_n^\ell; q), f(n)\}$. An easy observation shows that by combining equations (1) and (3), we also have

$$g(K_n^\ell; q) \leq f(r(K_n^\ell; q)) \leq \mathrm{twr}_{\ell+1}(cn),$$

where $c = c(\ell, q)$. Our main results are the following two exponential improvements on the upper bound of $g(K_n^\ell; q)$.

Theorem 1. *For geometric graphs, we have*

$$2^{q(n-1)} < g(K_n; q) \leq 2^{8qn^2 \log(qn)}.$$

The argument used in the proof of Theorem 1 above extends easily to hypergraphs, and for each fixed $\ell \geq 3$ it gives the bound $g(K_n^\ell; q) < \mathrm{twr}_\ell(O(n^2))$. David Conlon pointed out to us that one can improve this slightly as follows.

Theorem 2. *For geometric ℓ-hypergraphs, when $\ell \geq 3$ and fixed, we have*

$$g(K_n^\ell; q) \leq \mathrm{twr}_\ell(cn),$$

where $c = O(q \log q)$.

By combining Theorems 1, 2, and the fact that $g(K_n^\ell; q) \geq r(K_n^\ell; q)$, we have the following corollary.

Corollary 1. *For fixed ℓ and $q \geq 4$, we have $g(K_n^\ell; q) = \mathrm{twr}_\ell(\Theta(n))$.*

As mentioned above, there is an exponential difference between the known upper and lower bounds for $r(K_n^3, K_n^3)$. Hence, for two-colorings on geometric 3-hypergraphs in the plane, equation (2) implies

$$g(K_n^3, K_n^3) \geq r(K_n^3, K_n^3) \geq 2^{cn^2}.$$

Our next result establishes an exponential improvement in the lower bound of $g(K_n^3, K_n^3)$, showing that $g(K_n^3, K_n^3)$ does indeed grow as a 3-fold exponential tower in a power of n. One noteworthy aspect of this lower bound is that the construction is a geometric version of the famous Stepping-up lemma of Erdős and Hajnal [10] for sets. While it is a major open problem to apply this method to $r(K_n^3, K_n^3)$ and improve the lower bound in equation (2), we are able to achieve this in the geometric setting as shown below.

Theorem 3. *For geometric 3-hypergraphs in the plane, we have*

$$g(K_n^3, K_n^3) \geq 2^{2^{cn}},$$

where c is an absolute constant. In particular, $g(K_n^3, K_n^3) = \mathrm{twr}_3(\Theta(n))$.

2 Proof of Theorems 1 and 2

Before proving Theorems 1 and 2, we will first define some notation. We let $V = \{p_1, \ldots, p_N\}$ be a set of N points in the plane in general position ordered from left to right according to x-coordinate, that is, for $p_i = (x_i, y_i) \in \mathbb{R}^2$, we have $x_i < x_{i+1}$ for all i. For $i_1 < \cdots < i_t$, we say that $X = (p_{i_1}, \ldots, p_{i_t})$ forms an t-*cup* (t-*cap*) if X is in convex position and its convex hull is bounded above (below) by a single edge. See Figure 1. When $t = 3$, we will just say X is a cup or a cap.

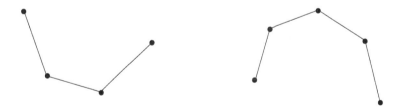

Fig. 1. A 4-cup and a 5-cap

Proof of Theorem 1. We first prove the upper bound. Let $G = (V, E)$ be a complete geometric graph on $N = 2^{8qn^2 \lceil \log(qn) \rceil}$ vertices such that the vertices $V = \{v_1, \ldots, v_N\}$ are ordered from left to right according to x-coordinate. Let χ be a q-coloring on the edge set E. We will recursively construct a sequence of vertices $p_1, \ldots, p_t$ from V and a subset $S_t \subset V$, where $t = 0, 1, \ldots, qn^2$ (when $t = 0$ the sequence is empty), such that the following holds.

1. for any vertex p_i, $i = 1, \ldots, qn^2$, all pairs (p_i, p) where $p \in \{p_j : j > i\} \cup S_t$ have the same color, which we denote by $\chi'(p_i)$,
2. for every pair of vertices p_i and p_j, where $i < j$, either (p_i, p_j, p) is a cap for all $p \in \{p_k : k > j\} \cup S_t$, or (p_i, p_j, p) is a cup for all $p \in \{p_k : k > j\} \cup S_t$,
3. the set of points S_t lies to the right of the point p_t, and
4. $|S_t| \geq \frac{N}{q^t t!} - t$.

We start with no vertices in the sequence ($t = 0$), and set $S_0 = V$. After obtaining vertices $\{p_1, \ldots, p_t\}$ and S_t, we define p_{t+1} and S_{t+1} as follows. Let $p_{t+1} = (x_{t+1}, y_{t+1}) \in \mathbb{R}^2$ be the smallest indexed element in S_t (the left-most point), and let H be the right half-plane $x > x_{t+1}$. We define t lines $l_1, \ldots, l_t$ such that l_i is the line going through points p_i, p_{t+1}. Note that the arrangement $\bigcup_{i=1}^{t} l_i$ partitions the right half-plane H into $t + 1$ cells. See Figure 2. Since V is in general position, by the pigeonhole principle, there exists a cell $\Delta \subset H$ that contains at least $(|S_t| - 1)/(t + 1)$ points of S_t.

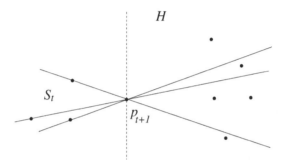

Fig. 2. Lines $l_1, \ldots, l_t$ partitioning the half-plane H

Let us call two elements $v_1', v_2' \in \Delta \cap S_t$ *equivalent* if $\chi(p_{t+1}, v_1') = \chi(p_{t+1}, v_2')$. Hence, there are at most q equivalence classes. By setting S_{t+1} to be the largest of those classes, we have the recursive formula

$$|S_{t+1}| \geq \frac{|S_t| - 1}{(t + 1)q}.$$

Substituting in the lower bound on $|S_t|$, we obtain the desired bound

$$|S_{t+1}| \geq \frac{N}{(t + 1)! q^{t+1}} - (t + 1).$$

This shows that we can construct the sequence $p_1, \ldots, p_{t+1}$ and the set S_{t+1} with the desired properties. For $N = 2^{8qn^2 \lceil \log(qn) \rceil}$, we have

$$|S_{qn^2}| \geq \frac{2^{8qn^2 \log(qn)}}{(qn^2)! q^{qn^2}} - qn^2 \geq 1. \tag{4}$$

Hence, $P_1 = \{p_1, \ldots, p_{qn^2}\}$ is well defined. Since χ' is a q-coloring on P_1, by the pigeonhole principle, there exists a subset $P_2 \subset P_1$ such that $|P_2| = n^2$, and every vertex has the same color. By construction of P_2, every pair in P_2 has the same color. Hence these vertices induce a monochromatic geometric graph.

Now let $P_2 = \{p'_1, \ldots, p'_{n^2}\}$. We define partial orders $\prec_1$, $\prec_2$ on P_2, where $p'_i \prec_1 p'_j$ ($p'_i \prec_2 p'_j$) if and only if $i < j$ and the set of points $P_2 \setminus \{p'_1, \ldots, p'_j\}$ lies above (below) the line going through points p'_i and p'_j. See Figure 3. By construction of P_2, $\prec_1$, $\prec_2$ are indeed partial orders and every two elements in P_2 are comparable by either $\prec_1$ or $\prec_2$. By Dilworth's theorem [8] (see also Theorem 1.1 in [14]), there exists a chain $p^*_1, \ldots, p^*_n$ of length n with respect to one of the partial orders. Hence $(p^*_1, \ldots, p^*_n)$ forms either an n-cap or an n-cup. Therefore, these vertices induce a complete monochromatic convex geometric graph.

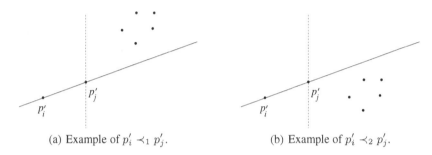

(a) Example of $p'_i \prec_1 p'_j$.   (b) Example of $p'_i \prec_2 p'_j$.

Fig. 3. Partial orders $\prec_1$, $\prec_2$

For the lower bound, we proceed by induction on q. The base case $q = 1$ follows by taking the complete geometric graph on 2^{n-1} vertices, whose vertex set does not have n members in convex position. This is possible by the construction of Erdős and Szekeres [13]. Let G_0 denote this geometric graph. For $q > 1$, we inductively construct a complete geometric graph $G = (V, E)$ on $2^{(q-1)(n-1)}$ vertices, and a coloring $\chi :$ $E \to \{1, 2, \ldots, q-1\}$ on the edges of G such that G does not contain a monochromatic convex geometric graph on n vertices. Now we replace each vertex $v_i \in G$ with a very small copy[2] of G_0, which we will denote as G_i, where all edges in G_i are colored with the color q, and all edges between G_i and G_j have color $\chi(v_i v_j)$. Then we have a complete geometric graph G' on

$$2^{(q-1)(n-1)}2^{n-1} = 2^{q(n-1)}$$

vertices, such that G' does not contain a monochromatic convex graph on n vertices. □

By following the proof above, one can show that $g(K_n^\ell; q) \le \mathrm{twr}_\ell(O(n^2))$. However, the following short argument due to David Conlon gives a better bound. The proof uses an old idea of Tarsi (see [21] Chapter 3) that yields an upper bound on $f(n)$.

Lemma 1. *For geometric 3-hypergraphs, we have* $g(K_n^3; q) \le r(K_n^3; 2q) \le 2^{2^{cn}}$, *where* $c = O(q \log q)$.

---

[2] Obtained by an affine transformation.

Proof. Let $H = (V, E)$ be a complete geometric 3-hypergraph on $N = r(K_n^3; 2q)$ vertices, and let χ be a q coloring on the edges of H. By fixing an ordering on the vertices $V = \{v_1, \ldots, v_N\}$, we say that a triple $(v_{i_1}, v_{i_2}, v_{i_3})$, $i_1 < i_2 < i_3$, has a *clockwise (counterclockwise) orientation*, if $v_{i_1}, v_{i_2}, v_{i_3}$ appear in clockwise (counterclockwise) order along the boundary of $conv(v_{i_1} \cup v_{i_2} \cup v_{i_3})$. Hence by Ramsey's theorem, there are n points from V for which every triple has the same color and the same orientation. As observed by Tarsi (see Theorem 3.8 in [25]), these vertices must be in convex position. □

Lemma 2. *For $\ell \geq 4$ and $n \geq 4^\ell$, we have $g(K_n^\ell; q) \leq r(K_n^\ell; q + 1) \leq \mathrm{twr}_\ell(cn)$, where $c = O(q \log q)$.*

Proof. Let $H = (V, E)$ be a complete geometric ℓ-hypergraph on $N = r(K_n^\ell; q + 1)$ vertices, and let χ be a q coloring on the ℓ-tuples of V with colors $1, 2, \ldots, q$. Now if an ℓ-tuple from V is not in convex position, we change its color to the new color $q + 1$. By Ramsey's theorem, there is a set $S \subset V$ of n points for which every ℓ-tuple has the same color. Since $n \geq 4^\ell$, by the Erdős-Szekeres Theorem, S contains ℓ members in convex position. Hence, every ℓ-tuple in S is in convex position, and has the same color which is not the new color $q + 1$. Therefore S induces a monochromatic convex geometric ℓ-hypergraph. □

Theorem 2 now follows by combining Lemmas 1 and 2.

3   A Lower Bound Construction for Geometric 3-hypergraphs

In this section, we will prove Theorem 3, which follows immediately from the following lemma.

Lemma 3. *For sufficiently large n, there exists a complete geometric 3-hypergraph $H = (V, E)$ in the plane with $2^{2^{n/2}}$ vertices, and a two-coloring χ' on the edge set E, such that H does not contain a monochromatic convex 3-hypergraph on $2n$ vertices.*

Proof. Let G be the complete graph on $2^{n/2}$ vertices, where $V(G) = \{1, \ldots, 2^{n/2}\}$, and let χ be a red-blue coloring on the edges of G such that G does not contain a monochromatic complete subgraph on n vertices. Such a graph does indeed exist by a result of Erdős [9], who showed that $r(K_n, K_n) > 2^{n/2}$. We will use G and χ to construct a complete geometric 3-hypergraph H on $2^{2^{n/2}}$ vertices, and a coloring χ' on the edges of H, with the desired properties.

Set $M = 2^{n/2}$. We will recursively construct a point set P_t of 2^t points in the plane as follows. Let P_1 be a set of two points in the plane with distinct x-coordinates. After obtaining the point set P_t, we define P_{t+1} as follows. We inductively construct two copies of P_t, $L = P_t$ and $R = P_t$, and place L to the left of R such that all lines determined by pairs of points in L go below R and all lines determined by pairs of points of R go above L. Then we set $P_{t+1} = L \cup R$. See Figure 4.

Let $P_M = \{p_1, \ldots, p_{2^M}\}$ be the set of 2^M points in the plane, ordered by increasing x-coordinate, from our construction. Note that P_M contains 2^{M-t} disjoint copies of P_t. For $i < j$, we define

$$\delta(p_i, p_j) = \max\{t : p_i, p_j \text{ lies inside a copy of } P_t = L \cup R, \text{ and } p_i \in L, p_j \in R\}.$$

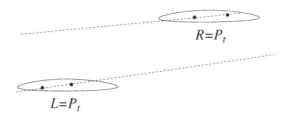

Fig. 4. Constructing P_{t+1} from P_t

Note that

Property A: $\delta(p_i, p_j) \neq \delta(p_j, p_k)$ for every triple $i < j < k$,
Property B: for $i_1 < \cdots < i_n$, $\delta(p_{i_1}, p_{i_n}) = \max_{1 \leq j \leq n-1} \delta(p_{i_j}, p_{i_j+1})$.

Now we define a red-blue coloring χ' on the triples of P_M as follows. For $i < j < k$,

$$\chi'(p_i, p_j, p_k) = \chi(\delta(p_i, p_j), \delta(p_j, p_k)).$$

Now we claim that the geometric 3-hypergraph $H = (P_M, E)$ does not contain a monochromatic convex 3-hypergraph on $2n$ vertices. For sake of contradiction, let $S = \{q_1, \ldots, q_{2n}\}$ be a set of $2n$ points from P_M, ordered by increasing x-coordinate, that induces a red convex 3-hypergraph. Set $\delta_i = \delta(q_i, q_{i+1})$.

Case 1. Suppose that there exists a j such that $\delta_j, \delta_{j+1}, \ldots, \delta_{j+n-1}$ forms a monotone sequence. First assume that

$$\delta_j > \delta_{j+1} > \cdots > \delta_{j+n-1}.$$

Since G does not contain a red complete subgraph on n vertices, there exists a pair $j \leq i_1 < i_2 \leq j + n - 1$ such that $(\delta_{i_1}, \delta_{i_2})$ is blue. But then the triple $(q_{i_1}, q_{i_2}, q_{i_2+1})$ is blue. Indeed, by Property B,

$$\delta(q_{i_1}, q_{i_2}) = \delta(q_{i_1}, q_{i_1+1}) = \delta_{i_1}.$$

Therefore, since $\delta_{i_1} > \delta_{i_2}$ and $(\delta_{i_1}, \delta_{i_2})$ is blue, the triple $(q_{i_1}, q_{i_2}, q_{i_2+1})$ must also be blue which is a contradiction. A similar argument holds if $\delta_j < \delta_{j+1} < \cdots < \delta_{j+n-1}$.

Case 2. Suppose we are not in Case 1. For $2 \leq i \leq 2n$, we say that i is a *local minimum* if $\delta_{i-1} > \delta_i < \delta_{i+1}$, a *local maximum* if $\delta_{i-1} < \delta_i > \delta_{i+1}$, and a *local extremum* if it is either a local minimum or a local maximum. This is well defined by Property A.

Observation 4. *For $2 \leq i \leq 2n$, i is never a local minimum.*

Proof. Suppose $\delta_{i-1} > \delta_i < \delta_{i+1}$ for some i, and suppose that $\delta_{i-1} \geq \delta_{i+1}$. We claim that $q_{i+1} \in \text{conv}(q_{i-1}, q_i, q_{i+2})$. Indeed, since $\delta_{i-1} \geq \delta_{i+1} > \delta_i$, this implies that $q_{i-1}, q_i, q_{i+1}, q_{i+2}$ lies inside a copy of $P_{\delta_{i-1}} = L \cup R$, where $q_{i-1} \in L$ and $q_i, q_{i+1}, q_{i+2} \in R$. Since $\delta_{i+1} > \delta_i$, this implies that q_i, q_{i+1}, q_{i+2} lie inside a copy $P_{\delta_{i+1}} = L' \cup R' \subset R$, where $q_i, q_{i+1} \in L'$ and $q_{i+2} \in R'$.

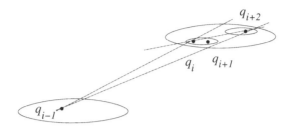

Fig. 5. Point $q_{i+1} \in \mathrm{conv}(q_{i-1}, q_i, q_{i+2})$

Notice that all lines determined by q_i, q_{i+1}, q_{i+2} go above the point q_{i-1}. Therefore q_{i+1} must lie above the line that goes through the points q_{i-1}, q_{i+2}, and furthermore, q_{i+1} must lie below the line that goes through the points q_{i-1}, q_i. Since $\delta_{i+1} > \delta_i$, the line through q_i, q_{i+1} must go below the point q_{i+2}, and therefore $q_{i+1} \in \mathrm{conv}(q_{i-1}, q_i, q_{i+2})$. See Figure 5. If $\delta_{i-1} < \delta_{i+1}$, then a similar argument shows that $q_i \in \mathrm{conv}(q_{i-1}, q_{i+1}, q_{i+2})$.

$\square$

Since $\delta_1, \ldots, \delta_{2n}$ does not have a monotone subsequence of length n, it must have at least two local extrema. Since between any two local maximums there must be a local minimum, we have a contradiction by Observation 4. This completes the proof.

$\square$

4   Concluding Remarks

For $q \geq 4$ colors and $\ell \geq 2$, we showed that $g(K_n^\ell; q) = \mathrm{twr}_\ell(\Theta(n))$. Our bounds on $g(K_n^\ell; q)$ for $q \leq 3$ can be summarized in the following table.

	$q = 2$	$q = 3$
$\ell = 2$	$2^{\Omega(n)} < g(K_n, K_n) \leq 2^{O(n^2 \log n)}$	$2^{\Omega(n)} < g(K_n; 3) \leq 2^{O(n^2 \log n)}$
$\ell = 3$	$g(K_n^3, K_n^3) = 2^{2^{\Theta(n)}}$	$g(K_n^3; 3) = 2^{2^{\Theta(n)}}$
$\ell \geq 4$	$g(K_n^\ell, K_n^\ell) \geq \mathrm{twr}_{\ell-1}(\Omega(n^2))$	$g(K_n^\ell; 3) \geq \mathrm{twr}_\ell(\Omega(\log^2 n))$
	$g(K_n^\ell, K_n^\ell) \leq \mathrm{twr}_\ell(O(n))$	$g(K_n^\ell; 3) \leq \mathrm{twr}_\ell(O(n))$

Off-diagonal. The Ramsey number $r(K_s, K_n)$ is the minimum integer N such that every red-blue coloring on the edges of a complete N-vertex graph G, contains either a red clique of size s, or a blue clique of size n. The off-diagonal Ramsey numbers, $r(K_s, K_n)$ with s fixed and n tending to infinity, have been intensively studied. For example, it is known [1,4,5,20] that $R_2(3, n) = \Theta(n^2/\log n)$ and, for fixed $s > 3$,

$$c_1(\log n)^{1/(s-2)} \left(\frac{n}{\log n}\right)^{(s+1)/2} \leq r(K_s, K_n) \leq c_2 \frac{n^{s-1}}{\log^{s-2} n}. \tag{5}$$

Another interesting variant of Problem 3 is the following off-diagonal version.

Problem 4. Determine the minimum integer $g(K_s, K_n)$, such that any red-blue coloring on the edges of a complete geometric graph G on $g(K_s, K_n)$ vertices, yields either a red convex geometric graph on s vertices, or a blue convex geometric graph on n vertices.

For fixed s, one can show that $g(K_s, K_n)$ grows single exponentially in n. In particular

$$2^{n-1} + 1 \leq g(K_s, K_n) \leq 4^{4^s n}.$$

The lower bound follows from the fact that $g(K_s, K_n) \geq f(n)$. The upper bound follows from the inequalities

$$g(K_s, K_n) \leq r(K_{4^s}, K_{4^n}) \leq (4^n)^{4^s}.$$

Indeed if G contains a red-clique of size 4^s, then by the Erdos-Szkeres Theorem there must be a red convex geometric graph on s vertices. Likewise, If G contains a blue clique of size 4^n, then there must be a blue convex geometric graph on n vertices.

Higher Dimensions. Generalizing Problem 1 to higher dimensions has also been studied. Let $f_d(n)$ be the smallest integer such that any set of at least $f_d(n)$ points in $\mathbb{R}^d$ in general position[3] contains n members in convex position. The following upper and lower bounds were obtained by Károlyi [16] and Károlyi and Valtr [19] respectively,

$$2^{cn^{1/(d-1)}} \leq f_d(n) \leq \binom{2n - 2d - 1}{n - d} + d = 2^{2n(1+o(1))}.$$

A *geometric ℓ-hypergraph H in d-space* is a pair (V, E), where V is a set of points in general position in $\mathbb{R}^d$, and $E \subset \binom{V}{\ell}$ is a collection of ℓ-tuples from V. When $\ell \leq d+1$, ℓ-tuples are represented by $(\ell - 1)$-dimensional simplices induced by the corresponding vertices.

Problem 5. Determine the minimum integer $g_d(K_n^\ell; q)$, such that any q-coloring on the edges of a complete geometric ℓ-hypergraph H in d-space on $g_d(K_n^\ell; q)$ vertices, yields a monochromatic complete convex ℓ-hypergraph on n vertices.

When $d = 2$, we write $g_2(K_n^\ell; q) = g(K_n^\ell; q)$. Clearly

$$g_d(K_n^\ell; q) \geq \max\{f_d(n), R(K_n^\ell; q)\}.$$

---

[3] A point set P in $\mathbb{R}^d$ is in general position, if no $d + 1$ members lie on a common hyperplane.

One can also show that $g_d(K_n^\ell; q) \leq g(K_n^\ell; q)$. Indeed, for any complete geometric ℓ-hypergraph $H = (V, E)$ in d-space with a q-coloring χ on $E(H)$, one can obtain a complete geometric ℓ-hypergraph in the plane $H' = (V', E')$, by projecting H onto a 2-dimensional subspace $L \subset \mathbb{R}^d$ such that V' is in general position in L. Thus we have

$$g_d(K_n; q) \leq g(K_n; q) \leq 2^{cn^2 \log n},$$

where $c = O(q \log q)$, and for $\ell \geq 3$

$$g_d(K_n^\ell; q) \leq g(K_n^\ell; q) \leq \text{twr}_\ell(c'n^2),$$

where $c' = c'(q, \ell)$.

Acknowledgment. We thank David Conlon for showing us an improved version of Theorem 2, shown in Section 2.

References

1. Ajtai, M., Komlós, J., Szemerédi, E.: A note on Ramsey numbers. J. Combin. Theory Ser. A 29, 354–360 (1980)
2. Ábrego, B.M., Fernández-Merchant, S., Salazar, G.: The rectilinear crossing number of K_n: Closing in (or are we?), Thirty Essays on Geometric Graph Theory. In: Pach, J. (ed.) Algorithms and Combinatorics, vol. 29, pp. 5–18. Springer (2012)
3. Aronov, B., Erdős, P., Goddard, W., Kleitman, D.J., Klugerman, M., Pach, J., Schulman, L.J.: Crossing families. Combinatorica 14, 127–134 (1994)
4. Bohman, T.: The triangle-free process. Adv. Math. 221, 1653–1677 (2009)
5. Bohman, T., Keevash, P.: The early evolution of the H-free process. Invent. Math. 181, 291–336 (2010)
6. Conlon, D., Fox, J., Sudakov, B.: Journal of the American Mathematical Society 23, 247–266 (2010)
7. Dey, T., Pach, J.: Extremal problems for geometric hypergraphs. Discrete Comput. Geom. 19, 473–484 (1998)
8. Dilworth, R.P.: A decomposition theorem for partially ordered sets. Ann. of Math. 51, 161–166 (1950)
9. Erdős, P.: Some remarks on the theory of graphs. Bull. Amer. Math. Soc. 53, 292–294 (1947)
10. Erdős, P., Hajnal, A., Rado, R.: Partition relations for cardinal numbers. Acta Math. Acad. Sci. Hungar. 16, 93–196 (1965)
11. Erdős, P., Rado, R.: Combinatorial theorems on classifications of subsets of a given set. Proc. London Math. Soc. 3, 417–439 (1952)
12. Erdős, P., Szekeres, G.: A combinatorial problem in geometry. Compos. Math. 2, 463–470 (1935)
13. Erdős, P., Szekeres, G.: On some extremum problems in elementary geometry. Ann. Univ. Sci. Budapest. Eötvös Sect. Math. 3(4), 53–62 (1960-1961)
14. Fox, J., Pach, J., Sudakov, B., Suk, A.: Erdős-Szekeres-type theorems for monotone paths and convex bodies. Proceedings of the London Mathematical Society 105, 953–982 (2012)
15. Graham, R.L., Rothschild, B.L., Spencer, J.H.: Ramsey Theory, 2nd edn. Wiley, New York (1990)
16. Károlyi, G.: Ramsey-remainder for convex sets and the Erdős-Szekeres theorem. Dsicrete Appl. Math. 109, 163–175 (2001)

17. Károlyi, G., Pach, J., Tóth, G.: Ramsey-type results for geometric graphs, I. Disc. Comp. Geom. 18, 247–255 (1997)
18. Károlyi, G., Pach, J., Tóth, G., Valtr, P.: Ramsey-type results for geometric graphs, II. Disc. Comp. Geom. 20, 375–388 (1998)
19. Károlyi, G., Valtr, P.: Point configurations in d-space without large subsets in convex position. Disc. Comp. Geom. 30, 277–286 (2003)
20. Kim, J.H.: The Ramsey number $R(3, t)$ has order of magnitude $t^2 / \log t$. Random Structures Algorithms 7, 173–207 (1995)
21. Matoušek, J.: Lectures on Discrete Geometry. Springer-Verlag New York, Inc. (2002)
22. Pach, J., Agarwal, P.: Combinatorial geometry. Wiley-Interscience, New York (1995)
23. Ramsey, F.P.: On a problem in formal logic. Proc. London Math. Soc. 30, 264–286 (1930)
24. Suk, A.: A note on geometric 3-hypergraphs, Thirty Essays on Geometric Graph Theory. In: Pach, J. (ed.) Algorithms and Combinatorics, vol. 29, pp. 489–498. Springer (2012)
25. Van Lint, J.H., Wilson, R.M.: A Course in Combinatorics. Cambridge University Press (2001)
26. Wagner, U.: k-sets and k-facets, Discrete and Computational Geometry - 20 Years Later. In: Goodman, E., Pach, J., Pollack, R. (eds.) Contemporary Mathematics, vol. 453, pp. 443–514. American Mathematical Society (2008)

Minimum Length Embedding of Planar Graphs at Fixed Vertex Locations

Timothy M. Chan, Hella-Franziska Hoffmann,
Stephen Kiazyk, and Anna Lubiw

David R. Cheriton School of Computer Science,
University of Waterloo, Waterloo, ON, Canada
{tmchan,hrhoffmann,skiazyk,alubiw}@uwaterloo.ca

Abstract. We consider the problem of finding a planar embedding of a graph at fixed vertex locations that minimizes the total edge length. The problem is known to be NP-hard. We give polynomial time algorithms achieving an $O(\sqrt{n} \log n)$ approximation for paths and matchings, and an $O(n)$ approximation for general graphs.

1 Introduction

Suppose we want to draw a planar graph and the vertex locations are specified. Such a planar drawing always exists, although not necessarily with straight line edges. Pach and Wenger [1] showed how to construct a drawing using $O(n)$ bends on each edge, where n is the number of vertices. We consider an equally natural optimization criterion—to minimize the total edge length.

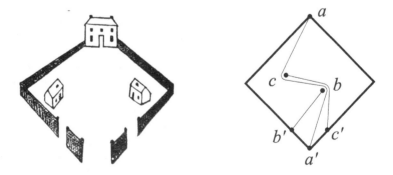

Fig. 1. A puzzle from Loyd [2]—connect each house to the opposite gate with non-crossing paths. On the left is the minimum length solution to an asymmetric version.

For example, Figure 1 shows a puzzle disseminated by Sam Loyd [2, p. 27]. The goal is to connect each house with the gate opposite its door via non-crossing paths. There are two distinct solutions but if the points are shifted as shown on the right in Figure 1, there is a unique shortest solution.

S. Wismath and A. Wolff (Eds.): GD 2013, LNCS 8242, pp. 376–387, 2013.

In this example the fixed outer wall plays a significant role, so the example demonstrates a more general problem—to extend a planar drawing of a subgraph to a planar drawing of the whole graph minimizing the total length. Fixed edges in the drawing act as *obstacles*. We call this *Minimum Length Planar Drawing of Partially Embedded Graphs*.

Angelini et al. [3] gave a linear time algorithm to decide if a planar drawing of a subgraph can be extended to a planar drawing of the whole graph. Our problem is the optimization version, to minimize the total edge length.

We will restrict attention to the case where all vertex positions are fixed. Furthermore, most of our results are for the case when none of the edges are fixed. We call this *Minimum Length Planar Drawing [or Embedding] at Fixed Vertex Locations*. This problem is very interesting even for special graphs such as matchings and paths.

When the edges to be added form a matching, the problem is to join specified pairs of points via non-crossing paths of minimum total length. The case when there are no obstacles was considered by Liebling et al. [4] in 1995. They gave some heuristics based on finding a short non-crossing tour of the points and then "wrapping" the matching edges around the tour. We use this same technique for our approximation results. They also proved that for points in the unit square a shortest non-crossing matching has length $O(n\sqrt{n})$ and there are examples realizing this bound. The lower bound (due to Peter Shor) relies on the existence of expander graphs with large crossing number. In 1996 Bastert and Fekete [5] proved that the problem is NP-hard, even with no obstacles. There is also substantial work on the case where there are obstacles (i.e. when some edges are fixed)—specifically when the points lie on the boundary of a polygon [6], or multiple polygons [7] (in which case the run time is exponential in the number of polygons). We give more details below in Section 1.1, and also discuss related work on finding "thick" paths that are separated from each other.

The problem of Minimum Length Planar Embedding at Fixed Vertex Locations is also interesting when the graph we want to embed is a path. This version of the problem was formulated by Polishchuk and Mitchell [8]. Their main goal was to find a minimum length tour that visits a given sequence of convex bodies in $\mathbb{R}^d$ (see [9] for the planar case), without regard to whether the path is self-intersecting, but in their conclusion section they ask about finding a minimum length non-crossing tour for a sequence of points.

Our Contributions. We give polynomial time approximation algorithms for Minimum Length Planar Embedding at Fixed Vertex Locations. In the case of general planar graphs we achieve an approximation ratio of $O(n)$. In the case of a matching or a path we achieve an approximation ratio of $O(\sqrt{n}\log n)$. Our main technique is to route graph edges around a carefully chosen path or tree defined on the input points in the plane.

1.1   Related Work

Bastert and Fekete [5] prove that Minimum Length Planar Drawing at Fixed Vertex Locations is NP-hard when the graph is a matching.

Patrignani [10] proved that it is NP-hard to decide if a planar drawing of a subgraph can be extended to a planar straight-line drawing of the whole graph.

Papadopoulou [6] gave an efficient algorithm for finding minimum length non-crossing paths joining pairs of points on the boundary of a polygon. In this case each path is a shortest path, but Papadopoulou finds them more efficiently than the obvious approach. Erickson and Nayyeri [7] extended this to points on the boundaries of h polygonal obstacles. Their algorithm has a running time that is linear for fixed h, but grows exponentially in h.

The difficulty with multiple polygons is deciding which homotopy of the paths gives minimum length. If the homotopy is specified the problem is easy [11,12].

In the aforementioned results on shortest non-crossing paths, one issue is that paths will overlap in general even though crossings are forbidden (see Figure 1 for an example). In practical applications we often need paths that are disjoint and maintain some minimum separation from each other. This issue is addressed in papers about drawing graphs with "thick" edges. Duncan et al. [13] show how to find thick shortest homotopic paths. Polishcuk and Mitchell [14] show how to find shortest thick disjoint paths joining endpoints on the boundaries of polygonal obstacles (with exponential dependence on the number of obstacles). They also show hardness results, including hardness of approximation.

In our problem the correspondence between the vertices and the fixed points in the plane is given. There is a substantial body of work where the correspondence of vertices to points is not fixed. Cabello [15] showed that it is NP-hard to decide if there is a correspondence that allows a straight-line planar drawing. Many special cases have been classified as polynomial time or NP-complete. A related problem is to find small *universal point sets* on which all planar graphs can be straight-line embedded (see [16]).

A problem related to minimum length planar embedding at fixed vertex locations is to draw planar graphs so that each edge is a monotone path of axis-parallel line segments. Any such path is a shortest path in the L_1 or Manhattan metric, and these drawings are called Manhattan-geodesic embeddings. This concept was introduced by Katz et al. [17]. They considered the case where the graph is a matching and showed that the problem is NP-hard when the drawing is restricted to a grid, but solvable in polynomial time otherwise.

We restricted the general problem of extending a partial planar embedding to the case where all vertex positions are fixed. The case where some vertices are free to move is also very interesting and is related to work on Steiner trees with fixed tree topology [18] and Steiner trees with obstacles [19]. Finally, one may consider drawing graphs at fixed vertex locations but allowing edges to cross. This is interesting and non-trivial when crossings must have large angles [20].

For other geometric graph augmentation problems see the survey by Hurtado and Tóth [21].

1.2   Definitions and Basic Observations

We consider the following problem called *Minimum Length Planar Embedding at Fixed Vertex Locations:* Given a planar graph $G = (\{v_1, \ldots, v_n\}, E)$ and a set of points $P = \{p_1, \ldots, p_n\}$, find a planar embedding of G in the plane that places vertex v_i at point p_i and minimizes the total edge length.

Edges of the embedding are allowed to overlap, but they must be non-crossing (i.e. infinitesimally deformable into disjoint paths). In the following we will refer to the vertices and their respective fixed locations interchangeably. The Euclidean distance between two points $p, q \in \mathbb{R}^2$ is denoted $d(p, q)$.

Observation 1. $L = \sum_{(v_i, v_j) \in E} d(p_i, p_j)$ *is a trivial lower bound for the total length of any planar embedding of graph G at fixed vertex locations P.*

One approach for finding short embeddings at fixed vertex locations is to draw each edge (v_i, v_j) as a curve whose length is within a constant factor of distance $d(p_i, p_j)$. Unfortunately such a planar drawing does not always exist; see Fig. 2.

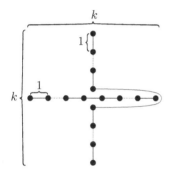

Fig. 2. Example for which any solution contains at least one edge of length greater than $k = n/2 - 1$

Fig. 3. Example for which any optimal solution contains no straight line edges

The example in Fig. 3 shows an instance where no straight line edge is included in any optimal solution, which means that obvious greedy algorithms for the problem fail. This was first observed by Liebling et al. [4].

Note that any edge of an optimal embedding bends only at vertex locations. Thus any optimal embedding lives in some underlying triangulation of the point set. Given the triangulation, the problem becomes that of finding short non-crossing paths in a planar graph. This problem was first proposed by Takahashi et al. [22] who considered the case of terminal points on two faces. Erickson and Nayyeri [7] say that the general problem is NP-hard, citing Bastert and Fekete [5]. Unfortunately, we cannot find this result in the version of the report that we have.

Instead of fixing the underlying triangulation, we use a carefully chosen path or tree as the layout for our embeddings.

2   Embedding a Path or Matching

In this section we give polynomial time approximation algorithms for the case where G is a path or a matching. Our starting point is a 1-dimensional version of the problem that will be the basis for all further results. We give a polynomial time (exact) algorithm for the case where G is a path, and the points lie on a line. We note that Liebling et al. [4] apparently knew the analogous result for the case when G is a matching, since they say that the edges of a matching can be "wrapped around" any non-self-intersecting tour of the points. They give no details of how to do the wrapping, and we consider the details worth explaining.

Fig. 4. A minimum length embedding of a path on fixed points that lie on a line. Edges are drawn with gaps between them for clarity only.

Lemma 1. *There is a polynomial time algorithm to find a minimum length embedding of a path on fixed vertex positions that lie on a line.*

Proof. Without loss of generality, assume that all the points lie on a horizontal line. See Fig. 4. This allows us to speak of "above" and "below". We draw the edges in order along the path. Draw edge (v_i, v_{i+1}) as a curve from point p_i to point p_{i+1} that stays below all edges drawn so far, but stays above all later points $p_j, j > i + 1$. This ensures that later edges of the path can reach their endpoints without crossing earlier edges. □

As noted by Liebling et al., this idea of "weaving" the edges through the points can be extended to the the case where the points lie on a simple (i.e. non-self-intersecting) curve in the plane. (In fact, the idea even extends to a tree, as we shall see in Section 3.) If one can find a simple curve C passing through every point in P, then "weaving" the edges of the path along the curve creates a path of length at most

$$\sum_{i=1}^{n-1} d_C(p_i, p_{i+1}), \qquad (1)$$

where $d_C(p_i, p_{i+1})$ denotes the length of the subcurve of C from p_i to p_{i+1}. This sum is trivially upper-bounded by n times the length of C.

We can choose the curve C to be an $O(1)$-approximation to the minimum-length Hamiltonian path (i.e., the traveling salesman path) for P (e.g., the simplest option would be the standard 2-approximate solution obtained from the minimum spanning tree). Since the length of the traveling salesman path is a lower bound for the problem in the path case, this would give an $O(n)$-approximate solution overall.

In Section 3 we will extend this idea to obtain an $O(n)$ approximation for general planar graphs. In the remainder of the current section we show how to improve the approximation factor for a path or matching by choosing a better curve C. The property we need is that points that are close in the plane are

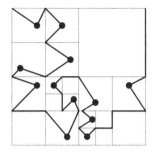

Fig. 5. The curve that is used as routing layout for embedding paths and matchings

close on the curve. The idea is to use a construction based on *shifted quadtrees*, similar to certain well known families of space-filling curves such as the *Z-order curve* or the *Hilbert curve*.

Let D_0 be the diameter of P. Without loss of generality, assume that $P \subset [0, D_0]^2$. We initially shift all the points in P by a random vector $v \in [0, D_0]^2$. Now, $P \subset [0, 2D_0]^2$.

Given a square S, the following procedure returns a simple polygonal curve, with the property that the curve starts at one corner of S, ends at another corner, and stays inside S while visiting all points of $P \cap S$.

CURVE(S):

1. if $|P \cap S| \leq 1$ then
2.   return a curve with the stated property, using at most 2 line segments
3. divide S into 4 subsquares $S_1, \ldots, S_4$
4. for $i = 1, \ldots, 4$, compute $C_i = $ CURVE(S_i)
5. return a curve with the stated property by joining $C_1, \ldots, C_4$,
    using $O(1)$ connecting line segments

Slight perturbation may be needed in line 5 to ensure that we obtain a simple curve. There is flexibility as to which corner we choose to start or end. (If we always start at the upper left corner and end at the lower right corner, the construction is similar to the Z-order curve. If we choose starting and ending corners to be adjacent in a manner similar to the Hilbert curve instead, the connecting line segments in line 5 may even be avoided; see Fig. 5. The main difference with standard space-filling curves is that we terminate the recursion as soon as we reach a square containing zero or one point.)

Lemma 2. *The length of the curve returned by* CURVE(S) *is at most* $O(D_S \sqrt{n_S})$, *where* $n_S = |P \cap S|$ *and* D_S *is the side length of* S.

Proof. Let L_i be the sum of the lengths of all line segments generated in line 2 or 5 at the i-th level of the recursion. The squares at the i-th level have side length $D_S/2^i$, and the number of squares at the i-th level is upper-bounded by both 4^i and n_S. Thus,

$$L_i \leq O(D_S/2^i) \cdot \min\{4^i, n_S\} = O(\min\{D_S 2^i, D_S n_S/2^i\}).$$

The total length of the curve is then at most

$$\sum_{i=0}^{\infty} L_i \leq \sum_{i=0}^{\lceil (1/2)\log n_S\rceil - 1} D_S 2^i + \sum_{i=\lceil (1/2)\log n_S\rceil}^{\infty} D_S n_S/2^i = O(D_S\sqrt{n_S}). \quad \square$$

We now let C be the curve returned by $\text{CURVE}(S_0)$ with $S_0 = [0, 2D_0]^2$. All the squares generated by the recursive calls are *quadtree squares*.

Define $D(p, q)$ to be the side length of the smallest quadtree square enclosing p and q. Note that $D(p, q) \geq d(p, q)/\sqrt{2}$. It is known that after random shifting, $D(p, q)$ approximates $d(p, q)$ to within a logarithmic factor in expectation (e.g., see [23, Lemma 5.1]). We include a quick proof for the sake of completeness.

Lemma 3. *For a fixed pair of points $p, q \in P$,*

$$\mathbf{E}[D(p, q)] \leq O(\log(D_0/d(p, q))) \cdot d(p, q).$$

Proof. $D(p, q) > D_0/2^i$ if and only if $\overline{pq}$ crosses a horizontal or vertical grid line in the grid formed by the quadtree squares of side length $D_0/2^i$. The probability that this happens is $O\left(\frac{d(p,q)}{D_0/2^i}\right)$. Thus,

$$\mathbf{E}[D(p, q)] \leq O\left(\sum_{i=0}^{\lceil \log(D_0/d(p,q))\rceil} \frac{d(p,q)}{D_0/2^i} \cdot D_0/2^i\right) \leq O(\log(D_0/d(p,q))) \cdot d(p,q).$$

$$\square$$

Lemma 4. *For a fixed pair of points $p, q \in P$,*

$$\mathbf{E}[d_C(p, q)] \leq O(\sqrt{n}\log(D_0/d(p, q))) \cdot d(p, q).$$

Proof. The portion of the curve C from p to q lies inside a quadtree square with side length $D(p, q)$. Thus, by Lemma 2, $d_C(p, q)$ is at most $O(D(p, q)\sqrt{n})$. The conclusion then follows from Lemma 3. $\square$

Theorem 2. *For a path G with fixed vertex locations, there is a polynomial-time randomized algorithm which computes a planar embedding of expected length at most $O(\sqrt{n}\log n) \cdot L$ where $L = \sum_{(p,q)\in E} d(p, q)$.*

Proof. By (1) and linearity of expectation, we obtain an embedding of expected length at most $(1+\epsilon)\sum_{(p,q)\in E} \mathbf{E}[d_C(p, q)]$. By Lemma 4, this quantity is at most

$$\sum_{(p,q)\in E} O(\sqrt{n}\log(D_0/d(p, q))) \cdot d(p, q) \leq O(\sqrt{n}\log(nD_0/L)) \cdot L$$

because the logarithm function is concave. The theorem follows since $L \geq \Omega(D_0)$ for the case of a path G. $\square$

Since L is a lower bound on the optimal cost, we obtain an $O(\sqrt{n} \log n)$-approximation algorithm for the case of a path G. For the case of a matching, we obtain the same result with just slightly more effort:

Theorem 3. *For a matching G with fixed vertex locations, there is a polynomial-time randomized algorithm which computes a planar embedding of expected length at most $O(\sqrt{n} \log n) \cdot L$ where $L = \sum_{(p,q) \in E} d(p, q)$.*

Proof. We can use essentially the same algorithm. The cost of the solution is still bounded by (1) and the same analysis goes through, except that $L \geq \Omega(D_0)$ may no longer be true.

Observe that if there is a vertical line that separates the line segments $\{\overline{pq} : (p, q) \in E\}$ into two nonempty parts, then we can just recursively compute a planar embedding on both sides, since each embedding can be shrunk to lie within the minimum (axis-aligned) bounding box of its points. We may thus assume that no such vertical separating line exists, which implies that L is at least the width of the bounding box of P. Similarly, we may assume that no horizontal separating line exists, which implies that L is at least the height of the bounding box of P. These two assumptions imply $L \geq \Omega(D_0)$ and we may proceed as before. ☐

Remarks. To obtain a time bound not sensitive to the bit complexity of the input, we can adopt a variant of the method where we compress long chains of degree-1 nodes in the tree (called the *compressed quadtree*), to ensure that the number of recursive calls is $O(n)$.

On the other hand, if input coordinate values are $O(\log n)$ bits long, we can derandomize the algorithm in polynomial time by trying all possible shifts.

The upper bound relative to L in Theorems 2 and 3 is tight up to a logarithmic factor: Liebling et al. [4] provide examples (due to Peter Shor) with points in the unit square for which any shortest non-crossing matching has length $\Omega(n\sqrt{n})$, proving a lower bound of $\Omega(\sqrt{n}) \cdot L$.

3   Embedding General Planar Graphs

In this section we give an $O(n)$-approximation algorithm for constructing a planar embedding of a planar graph G at fixed vertex locations P.

The construction is based on the algorithm by Pach and Wenger [1] for finding a planar polygonal embedding of a graph with fixed vertex locations and with $O(n)$ bends per edge. Pach and Wenger draw the edges of the graph by tracing around a tree of n edges drawn in the plane. Each edge of the graph is drawn as a curve that walks around the tree a constant number of times, which gives the bound of $O(n)$ bends per edge. For their tree Pach and Wenger use a star with a leaf at each vertex.

In our case we want to bound the length of each edge, which can be done by bounding the length of the tree. We cannot use a star; instead, we will use a tree that is a subset of the (non-planar) drawing of G where each edge is drawn as a straight line segment. This ensures that the tree has total length at most $L = \sum_{(p,q) \in E} d(p,q)$. Because of connectivity issues, we will actually use a set of disjoint trees:

Lemma 5. *Given a graph G, and fixed vertex locations P, we can construct in $O(n^2)$ time an embedded forest F with $O(n)$ vertices and total length at most L, such that for every edge (p,q) in G, p and q are in the same tree of F.*

Proof. We construct the forest F iteratively by adding edges of the graph one by one. For each edge $(p,q) \in E$ we will add some subsegments of the line segment pq to F. The forest will be a subset of an arrangement of $O(n)$ lines. The arrangement can be constructed in $O(n^2)$ time [24]. Consider edge (p,q). If p and q are already in the same tree of F, we are done. Otherwise consider the line segment pq. It crosses at most n segments of F, and these crossing points subdivide it into $p = p_1, p_2, \ldots, p_k = q$ with $k \leq n$. We treat these segments one by one in order. Consider segment $p_i p_{i+1}$. If p_i and p_{i+1} are already in the same tree of F, we are done. Otherwise we add segment $p_i p_{i+1}$ to F. Fig. 6 illustrates this idea. We use a union-find data structure to test if points are in the same tree of F. By construction, the length of F is bounded by L. Furthermore, we only add a segment when we join two trees of F, and this can happen at most n times. Thus F has $O(n)$ vertices and is a subset of an arrangement of n lines.   □

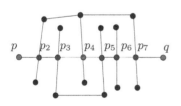

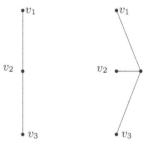

Fig. 6. Constructing the forest F in Lemma 5. Segment pq crosses multiple components of F. Segments p_3p_4, p_4p_5, and p_6p_7 are not added to F.

Fig. 7. Perturbing the tree to change v_2 from an internal vertex to a leaf.

Theorem 4. *Given a planar graph G with n vertices and fixed vertex locations P, there is an $O(n^2)$-time approximation algorithm to construct a planar embedding of G on P with total length $O(n) \cdot L$ where $L = \sum_{(p,q) \in E} d(p,q)$.*

Proof. Use Lemma 5 to construct a forest F, that will serve as the basis for our edge routing. Because we do not want paths to travel through intermediate vertices, we perturb the trees in F slightly so that each vertex of G is a leaf of the tree that contains it. See Fig. 7. This can be done while keeping the trees disjoint and of total length $O(L)$.

Consider a single tree T of the forest F, together with the induced graph G_T on the vertices of G that lie in T. We will follow the approach of Pach and Wenger and draw the edges of G_T as paths hugging the tree T. Because every edge of G lies in some G_T, and the trees are disjoint (as objects in the plane), it suffices to describe the solution for a single tree T. To simplify notation, we will assume for the remainder of the proof that we have a single tree T and $G = G_T$.

We now follow Pach and Wenger's solution, the main difference being that we have a more general tree than their star. We outline their solution and remark on the modifications required for our situation.

Pach and Wenger's solution is based on a Hamiltonian cycle that they construct by adding vertices and edges to the graph. Specifically, they subdivide each edge of the graph by at most two new vertices and add some edges between vertices to obtain a planar graph with a Hamiltonian cycle [1, Lemma 5]. (Note that the new edges do not appear in the final drawing.) The Hamiltonian cycle C partitions edges of the planar graph relative to some (arbitrary) planar embedding into the edges *inside* C and the edges *outside* C. They first draw the edges of the Hamiltonian cycle C and then draw the inside and outside edges.

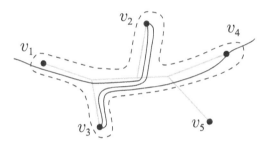

Fig. 8. Drawing the graph G around the tree T (drawn in gray) whose leaves are the graph vertices. The portion of the Hamiltonian cycle C from v_1 to v_4 is drawn as a solid curve. The dashed curve Λ_4 surrounds T_4 and is split by C into two paths between v_4 and v_1, one inside C and one outside C.

To draw the edges of C they use an approach similar to the weaving technique described in Lemma 1. Renumber vertices so they appear in the order $v_1, v_2, \ldots, v_n$ along the Hamiltonian cycle. Some of these are new vertices that were added to create the Hamiltonian cycle. Pach and Wenger assign arbitrary locations to the new vertices, but we locate them very close to the tree T, adding them as leaves of T and keeping the length of T in $O(L)$. We will use v_i to refer to the vertex of G, the corresponding point in the plane, and the corresponding leaf of T. Edge (v_{i-1}, v_i) of C will be drawn around a subtree T_i of T. We define T_i more carefully for our situation: T_i is the connected subtree of T induced on leaves $v_1, v_2, \ldots, v_i$. With these modifications, the rest of Pach and Wenger's solution applies unaltered.

As they draw C they add (multiple copies of) auxiliary paths Λ_i from v_i to v_1, one inside and one outside C. See Fig. 8. Then each edge (v_i, v_j) of G inside [outside] C is routed using the paths inside [outside] C from v_i to v_1 and from v_1 to v_j. For further details please refer to their paper [1]. The end result is a planar drawing of G on vertex locations P. Every original edge e of G has been subdivided by at most two new vertices, and each of the resulting three edges has been drawn as two paths in the tree. The total length of the drawing of e is therefore bounded by 6 times the length of T, and thus in $O(L)$.

Pach and Wenger's algorithm takes $O(n^2)$ time so our overall running time is $O(n^2)$ as well.   □

4   Conclusion and Open Problems

The problem of drawing a planar graph at fixed vertex locations while minimizing the total edge length seems to be very difficult although we are not aware of any hardness of approximation results. In fact, for the case of a path, even an NP-hardness result is lacking. Our algorithms achieve approximation factors of $O(n)$ for general graphs and $O(\sqrt{n} \log n)$ for paths and matchings. Besides the obvious question of improving these approximation factors (or proving hardness), we suggest looking at: (1) the problem of drawing a graph at fixed vertex locations with thick edges; and (2) looking at the case where some vertex locations are not fixed, which is related to drawing Steiner trees with fixed topology [18].

Acknowledgments. This work was done in the Algorithms Problem Session at Waterloo and we thank the other participants for good discussions. We learned about Sam Loyd's disjoint paths puzzle (which is not original to him) from Marcus Schaefer who has studied the history of such planarity puzzles.

References

1. Pach, J., Wenger, R.: Embedding planar graphs at fixed vertex locations. Graphs and Combinatorics 17(4), 717–728 (2001)
2. Sam Loyd, J.: Sam Loyd's Cyclopedia of 5000 Puzzles Tricks and Conundrums. Lamb Publishing Company (1914)
3. Angelini, P., Di Battista, G., Frati, F., Jelínek, V., Kratochvíl, J., Patrignani, M., Rutter, I.: Testing planarity of partially embedded graphs. In: Proceedings of the 21st Annual ACM-SIAM Symposium on Discrete Algorithms (SODA), pp. 202–221 (2010)
4. Liebling, T.M., Margot, F., Müller, D., Prodon, A., Stauffer, L.: Disjoint paths in the plane. ORSA Journal on Computing 7(1), 84–88 (1995)
5. Bastert, O., Fekete, S.P.: Geometric wire routing. Technical Report 332, Angewandte Mathematik und Informatik, Universität zu Köln (1996) (in German)
6. Papadopoulou, E.: k-pairs non-crossing shortest paths in a simple polygon. In: Nagamochi, H., Suri, S., Igarashi, Y., Miyano, S., Asano, T. (eds.) ISAAC 1996. LNCS, vol. 1178, pp. 305–314. Springer, Heidelberg (1996)

7. Erickson, J., Nayyeri, A.: Shortest non-crossing walks in the plane. In: Proceedings of the 22nd Annual ACM-SIAM Symposium on Discrete Algorithms (SODA), pp. 297–308 (2011)
8. Polishchuk, V., Mitchell, J.S.: Touring convex bodies – a conic programming solution. In: Proceedings of the 17th Canadian Conference on Computational Geometry, pp. 290–293 (2005)
9. Dror, M., Efrat, A., Lubiw, A., Mitchell, J.S.B.: Touring a sequence of polygons. In: Proceedings of the 35th Annual ACM Symposium on Theory of Computing (STOC), pp. 473–482 (2003)
10. Patrignani, M.: On extending a partial straight-line drawing. International Journal of Foundations of Computer Science 17(5), 1061–1069 (2006)
11. Bespamyatnikh, S.: Computing homotopic shortest paths in the plane. Journal of Algorithms 49(2), 284–303 (2003)
12. Efrat, A., Kobourov, S.G., Lubiw, A.: Computing homotopic shortest paths efficiently. Computational Geometry 35(3), 162–172 (2006)
13. Duncan, C.A., Efrat, A., Kobourov, S.G., Wenk, C.: Drawing with fat edges. International Journal of Foundations of Computer Science 17(5), 1143–1164 (2006)
14. Mitchell, J.S., Polishchuk, V.: Thick non-crossing paths and minimum-cost flows in polygonal domains. In: Proceedings of the 23rd Annual Symposium on Computational Geometry (SoCG), pp. 56–65 (2007)
15. Cabello, S.: Planar embeddability of the vertices of a graph using a fixed point set is NP-hard. Journal of Graph Algorithms and Applications 10(2), 353–363 (2006)
16. Dujmović, V., Evans, W.S., Lazard, S., Lenhart, W., Liotta, G., Rappaport, D., Wismath, S.K.: On point-sets that support planar graphs. Computational Geometry 46(1), 29–50 (2013)
17. Katz, B., Krug, M., Rutter, I., Wolff, A.: Manhattan-geodesic embedding of planar graphs. In: Eppstein, D., Gansner, E.R. (eds.) GD 2009. LNCS, vol. 5849, pp. 207–218. Springer, Heidelberg (2010)
18. Hwang, F., Weng, J.: The shortest network under a given topology. Journal of Algorithms 13(3), 468–488 (1992)
19. Winter, P.: Euclidean Steiner minimal trees with obstacles and Steiner visibility graphs. Discrete Applied Mathematics 47(2), 187–206 (1993)
20. Fink, M., Haunert, J.-H., Mchedlidze, T., Spoerhase, J., Wolff, A.: Drawing graphs with vertices at specified positions and crossings at large angles. In: Rahman, M. S., Nakano, S.-i. (eds.) WALCOM 2012. LNCS, vol. 7157, pp. 186–197. Springer, Heidelberg (2012)
21. Hurtado, F., Tóth, C.: Plane geometric graph augmentation: A generic perspective. In: Pach, J. (ed.) Thirty Essays on Geometric Graph Theory, pp. 327–354. Springer, New York (2013)
22. Takahashi, J., Suzuki, H., Nishizeki, T.: Shortest noncrossing paths in plane graphs. Algorithmica 16(3), 339–357 (1996)
23. Kamousi, P., Chan, T.M., Suri, S.: Stochastic minimum spanning trees in Euclidean spaces. In: Proceedings of the 27th Annual ACM Symposium on Computational Geometry (SoCG), pp. 65–74 (2011)
24. Edelsbrunner, H., O'Rourke, J., Seidel, R.: Constructing arrangements of lines and hyperplanes with applications. SIAM Journal on Computing 15(2), 341–363 (1986)

Stub Bundling and Confluent Spirals
for Geographic Networks[*]

Arlind Nocaj and Ulrik Brandes

Department of Computer & Information Science, University of Konstanz

Abstract. Edge bundling is a technique to reduce clutter by routing parts of several edges along a shared path. In particular, it is used for visualization of geographic networks where vertices have fixed coordinates. Two main drawbacks of the common approach of bundling the interior of edges are that (i) tangents at endpoints deviate from the line connecting the two endpoints in an uncontrolled way and (ii) there is ambiguity as to which pairs of vertices are actually connected. Both severely reduce the interpretability of geographic network visualizations.

We therefore propose methods that bundle edges at their ends rather than their interior. This way, tangents at vertices point in the general direction of all neighbors of edges in the bundle, and ambiguity is avoided altogether. For undirected graphs our approach yields curves with no more than one turning point. For directed graphs we introduce a new drawing style, confluent spiral drawings, in which the direction of edges can be inferred from monotonically increasing curvature along each spiral segment.

1 Introduction

We are interested in visualizing geographic networks given as a graph with fixed vertex coordinates and possibly other attributes. Although, for substantive reasons, there is often a relationship between the graph's adjacency structure and the spatial arrangement of its vertices, straight-line drawings are generally cluttered with areas of high edge-density and small-angle crossings.

A technique to reduce such clutter is edge bundling. Generalizing the idea of edge concentrators [17], edge bundles have been introduced in the context of hierarchically clustered graphs [11]. Sets of related edges are routed so that they meet, run concurrently, and then separate again, where edges are considered related if their projections on the cluster tree share a subpath. Note that the nodes of the cluster tree directly yield shared edge control points. Different bundling strategies have been introduced in force-directed layout of general graphs [12,24,19] and layered layout of directed acyclic graphs [23].

For graphs with given vertex coordinates, relatedness of edges is usually defined in terms of similarity of their straight-line realizations. Examples include similarities obtained from grid approximations [6,16,15], visibility graphs [22], or clusters in the four-dimensional space of pairs of vertex coordinates [9]. In the extreme, bundling techniques operate on the pixel level [25,8,14,27,13].

[*] Research supported in part by DFG under grant GRK 1024. We are grateful to Sabine Cornelsen for valuable comments and suggestions on earlier drafts of this paper.

S. Wismath and A. Wolff (Eds.): GD 2013, LNCS 8242, pp. 388–399, 2013.

Because of the shared inner segments, it cannot be inferred from the drawing which subgraph of a bipartite clique a bundle actually represents, i.e., we cannot know whether the drawing is faithful [18]. Moreover, having edges meet requires that they deviate from the line through their vertices in a way that has no substantive meaning.

Both these problems can be avoided by bundling edges at their ends rather than in their interior. This idea has indeed been introduced in the context of geographic networks [4,20] and also forms the basis of flow maps [21,5,26] which can be seen as drawings of in- or out-stars.

We present novel such methods for drawing geographic networks with edges bundled at their ends. For undirected graphs, we refine the approach of Peng et al. [20]. Our main contribution is an approach for directed graphs based on a new drawing style for in-/out-stars called *confluent spiral drawings*. Confluent spirals consist of smooth drawings of each bundle in which edge directions are represented by increasing curvature so that no ambiguity is created in a combined drawing of all, say, in-stars of a directed graph.

The remainder of this paper is divided into three sections. In Sect. 2, we outline how edges are assigned to bundles. Our approaches for undirected and directed graphs are then described in Sects. 3 and 4 with a short discussion in Sect. 5.

2   Stub Bundling

We consider geographic networks consisting of a graph $G = (V, E)$ with fixed vertex coordinates $p = (p_v)_{v \in V}$ where $p_v = \binom{x_v}{y_v} \in \mathbb{R}^2$. Coordinates might be defined extrinsically by, say, geographic locations, or derived from, say, a precomputed layout.

Our goal is to route the edges in such a way that readability of the network is improved over the corresponding straight-line drawing. The means in this work are bundled edges, curved routing, and color gradients.

A common objective of edge bundling is to reduce the total length of edges drawn. Since multiplicity along shared paths is ignored, bundling of interior segments of edges is attempted. For geographic networks, however, a more substantively relevant criterion is to be able to read off the general direction in which adjacent vertices are located. To represent this more accurately, and in addition to provide a faithful representation of adjacency, we bundle edges only at their ends, i.e., only with edges that share an endpoint. This type of bundling is referred to here as *stub bundling* to distinguish it from the bundling of edge interiors, or *interior bundling*.

The first step is to find a partition of the edges around each vertex into bundles (see fig. 1). To preserve their general direction, we use the angles between consecutive straight-line edges as our partition criterion. Each bundle is a set of half-edges incident on the same vertex and with similar direction.

For given angles α, γ, an (α, γ)-bundling is a coarsest partition such that

- the angle between any two half-edges in a bundle is at most α, and
- the angle between two consecutive edges in a bundle is at most γ.

Such bundlings are obtained easily by iteratively splitting adjacency lists at maximum angles (between consecutive edges). For each vertex $v \in V$, we start with a bundle containing all incident half-edges. The bundle is split at all occurrences of the maximum angle between consecutive edges in this bundle; in case of equiangular half-edges,

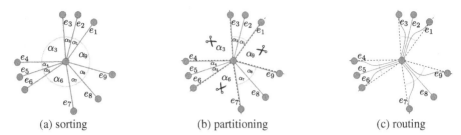

| (a) sorting | (b) partitioning | (c) routing |

Fig. 1. Stub bundling: a cyclic sequence of edges is (a) split at maximum angles ($\alpha_3, \alpha_6, \alpha_9$) until (b) the angle range of each bundle is below a given threshold; then, (c) the first segment of each half-edge in a bundle is routed with the same tangent

where the consecutive angles are equal inside a bundle we split symmetrically into two (for bundles with even cardinality) or three (odd cardinality) smaller bundles. In the latter case the resulting bundles may have different (but symmetrical) cardinalities. This process is iterated until we obtain an (α, γ)-bundling. Note that, in contrast to the counterclockwise greedy splitting of [20], we do not accidentally split at small angles and we maintain a higher degree of symmetry.

The entire bundling step is thus carried out in time $\mathcal{O}(m \log \Delta)$, where m is the number of edges and Δ the maximum degree of a vertex, by sorting adjacency lists and splitting them hierarchically. Clearly, other bundling strategies, e.g. also based on distances rather than just directions, may be more appropriate for specific applications. It remains to show how to route the edges beyond the constraint that half-edges in the same bundle share an initial path.

3   Undirected Graphs

After bundling as described in the previous section, we need to decide on two things: in which direction to route the stub of a bundle, and how to connect the two extremal segments of each edge.

To ensure that edges can be followed easily, we allow only one turning point in the routing of an edge. More precisely, we draw each half-edge as a cubic Bézier curve (without turning point) to gain more control over the curve shape. Bézier curves are especially convenient because their tangents at endpoints can be prescribed so that edges in a bundle start in parallel and the two half edges of an edge can be linked smoothly.

Stub directions are determined from the straight-line segment connecting a vertex with the centroid of all neighbors in a bundle. This incorporates not only their directions but also their distances. For the present purpose this is considered a good approximation to the general direction of all edges in a bundle, but more general nodal templates for outgoing edges could be used [3].

We now describe in detail how to choose the control points of the Bézier curves. See fig. 2 for illustration. Let $e = \{v, w\} \in E$ be an edge and let $\Gamma(e, v)$ and $\Gamma(e, w)$ be

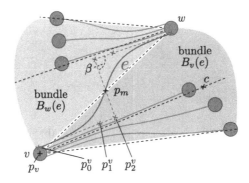

Fig. 2. Control points for routing undirected edge e

the two Bézier curves representing the half-edges of e. Further, let $B_v(e)$ and $B_w(e)$ be the bundles around v and w containing e. Let

$$p_m = \frac{1}{2}(p_v + p_w) + \left(\frac{|B_v(e)|}{|B_v(e)| + |B_w(e)|} - \frac{1}{2} \right) \cdot t_{\text{shift}} \cdot (p_w - p_v)$$

be the weighted midpoint on the segment between p_v and p_w and $t_{\text{shift}} \in [0, 1]$. More intuitively p_m is closer to p_v if $B_w(e)$ contains more edges then $B_v(e)$. This has the effect that larger bundles have longer parallel parts. Let c be the centroid of the end vertices of the edges bundled in $B_v(e)$ (excluding v). In order to define the control points of $\Gamma(e, v)$ we consider the *baseline* going through p_v and the centroid c. We first compute temporary control points, which lie on the baseline. Later these points will be shifted left and right to obtain a parallel routing.

We choose a branching angle β which is the same for every edge. This angle determines how long the edge will stay with the bundle until it branches off to enter the other bundle, and thus the smoothness of edges. Denote by p_2^v the point on the baseline such that the angle $\sphericalangle(p_v, p_2^v, p_m)$ is β. Another intermediate control point p_1^v is chosen on the segment between p_v and p_2^v with a smoothing parameter $t \in (0, 1)$, i.e., $p_1^v = p_v + t \cdot (p_2^v - p_v)$.

So far we have determined temporary control points p_v, p_1^v, p_2^v, and $\frac{1}{2}(p_2^v + p_2^w)$ for $\Gamma(e, v)$ and symmetrically for $\Gamma(e, w)$. These are refined to avoid overlap without introducing many crossings. Ordering edges around each vertex is a special case of the more general metro-line crossing minimization problem [2] but we find the simple heuristic of ordering stubs in a bundle $B_v(e)$ according to the opposite control point p_2^w to work sufficiently well. Control points are shifted left and right according to this ordering. Determining control points and ordering stubs in the same bundle does not increase the asymptotic running time of $\mathcal{O}(m \log \Delta)$ already caused by bundling.

Stub bundling is motivated by faithfulness and the substantive interest in directions at the ends of edges. Therefore, non-uniform rendering of edges can be used to highlight bundles and reduce the visual dominance of the less important interior of an edge by fading out colors toward the middle of an edge. Note the emphasis this creates in fig. 3 without eliminating the possibility to trace individual edges. In addition to the alternate bundling strategy, non-overlapping stub routing distinguishes our approach from that

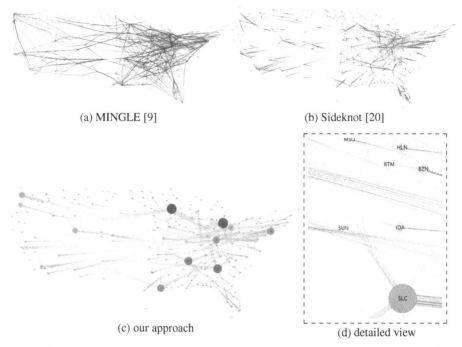

(a) MINGLE [9]   (b) Sideknot [20]

(c) our approach   (d) detailed view

Fig. 3. US airlines graph: In our approach main edge directions and their strengths are visible from an overview perspective (c) but single edges can still be traced in a detailed view (d). Unlike the approach of (b), ours uses parallel routing of edges to facilitate the display of additional data attributes by varying width or color.

of [20] and facilitates the use of different widths and colors for edges in the same bundle to convey additional attribute information such as volume, frequency, time, and so on.

The total bundling process, edge partitioning and control point computation, of the US airlines graph took 0.07 seconds on an Intel Core i7-2600K CPU@3.40GHz with a single core (impl. in Java 6).

4   Directed Graphs: Confluent Spiral Drawings

To visualize directed geographic networks we break the symmetry of the previous approach as arrows and color gradients do not seem to work well for displaying orientation of stub-bundled edges. Depending on the meaning of edge orientations, we bundle only incoming or outgoing stubs. The problem of drawing a directed graph is thus reduced to the problem of drawing one in- or out-star per vertex.

We introduce a new drawing convention for such star-configurations. It is a variation of spiral trees [26] which have been introduced for flow maps [21], but based on confluent logarithmic spirals for smoother appearance and easier inference of orientation.

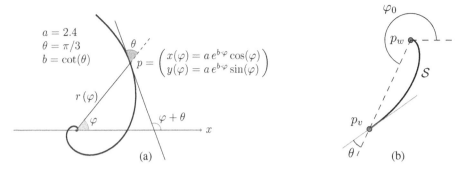

Fig. 4. (a) Logarithmic spiral with angle $\theta = \pi/3$, centered at origin, going through a point p. (b) Spiral used to represent a directed edge (v, w). The constantly increasing curvature on the path from p_v towards p_w unambiguously indicates the orientation.

4.1 Logarithmic Spirals

A *logarithmic* (or *equiangular*) *spiral* is a curve which winds around a center, or vortex, and approaches it with an exponentially increasing curvature. The increase in curvature is determined by a constant θ. In effect, all rays out of the vortex are at angle θ to the tangents of intersection points with the spiral.

Formally, a logarithmic spiral in Euclidean space can be defined in polar coordinates (r, φ) relative to its vortex by $r(\varphi) = a \cdot e^{b \cdot \varphi}$, where $b = \cot \theta$ and $a \in \mathbb{R} \setminus \{0\}$ are fixed, and $\varphi \in \mathbb{R}$. Figure 4(a) shows an example.

We use a sequence of spiral segments to represent an edge between two vertices, and the vortex of each spiral corresponds to a target vertex. We define a spiral segment S from a start point $p_v \in \mathbb{R}^2$ to an end point $p_w \in \mathbb{R}^2$ with tangent angle $\theta \in (-\frac{\pi}{2}, 0) \cup (0, \frac{\pi}{2})$ as

$$S(t) = \begin{pmatrix} x(t) \\ y(t) \end{pmatrix} = p_w + |p_v - p_w| e^{-|b| \cdot t} \begin{pmatrix} \cos\left(\frac{b}{|b|}(\varphi_0 + t)\right) \\ \sin\left(\frac{b}{|b|}(\varphi_0 + t)\right) \end{pmatrix}, t \in [0, \infty),$$

where $b = \cot \theta$ and $\varphi_0 = \angle\left(\overrightarrow{p_w p_v}, \binom{1}{0}\right)$ is counterclockwise around p_w.

Note that $S(0) = p_v$. Although the curve has finite length, it never reaches p_w. Practically this is not a problem, since vertex w is represented by a graphical element with non-zero dimensions such as a disc. The spiral S goes clockwise around p_w if $\theta < 0$ and counterclockwise if $\theta > 0$. Figure 4(b) shows how a logarithmic spiral can be used to represent a directed edge from v to w.

Fig. 5. Two drawings of the same graph. The absence of edge (w_1, w_2) is apparent because the curve from p_{w_1} to p_{w_2} is not smooth.

Confluent Spirals. The term *confluent* was introduced in Dickerson et al. [7] for a drawing style that allows to draw larger classes of graphs in a planar way. We here use it more loosely, not requiring planarity. We say a drawing is a *confluent spiral drawing*, if each edge $e = (v, w) \in E$ is represented as follows:

- There is a continuously differentiable curve from p_v to p_w consisting of logarithmic-spiral segments.
- The logarithmic spiral of the last segment has vortex $\left(\begin{smallmatrix} x_w \\ y_w \end{smallmatrix}\right)$ and the segment starts at p where
  - either $p = \left(\begin{smallmatrix} x_v \\ y_v \end{smallmatrix}\right)$ or
  - p lies in the interior of another edge $(v, w') \in E$ with $w' \neq w$.

Furthermore, we do require that the curves representing outgoing edges of the same vertex do not intersect but in a shared prefix. Figure 5 shows a small graph and a corresponding drawing with confluent spirals.

4.2   Determining Confluent Spiral Trees

In this section, we introduce an algorithm to compute a confluent spiral drawing by computing a confluent spiral tree for each vertex. Later, we extend this algorithm to handle obstacles by adding further constraints to it.

The main difference compared to the spiral trees suggested by Buchin et al. [5] is that we want to have confluent drawings, which means that the intersection angle between two spirals is zero. This means that following a path from the root to some other vertex one never has to make a sharp turn. Due to this property it is not possible to apply the method of Buchin et al. [5]. Later the same authors [26] use a spiral tree as a basis to generate flow maps by minimizing a complex cost function to smooth the curves.

In contrast to the previous approach, we require the vortex of spirals not to be on the source but on the target vertex of an edge, which directly results in smooth curves, see fig. 6 for an example.

As a first step we apply the edge partitioning, as described in section 2. The result is for each vertex $v \in V$ a set of bundles containing outgoing edges of v. Each bundle is handled separately. Let B be a bundle with outgoing edges of v. For every edge $e = (v, w) \in B$ we determine a logarithmic spiral S that is centered at p_w and either starts at p_v or branches out of another spiral in a confluent way such that $\sum_{e \in B} \text{length}(S_e)$ is minimal. Note that although a spiral never reaches the vortex its length is finite:

$$\text{length}(S) = \int_0^\infty |S'(t)|\, dt = ||p_v - p_w|| \frac{\sqrt{1 + b^2}}{|b|}.$$

(a)   (b)   (c)

Fig. 6. (a) directed graph, (b) spiral tree approach of Buchin et al. [5]: spiral vortices at source, postprocessing required for smoothness, (c) our confluent spiral tree approach: spiral vortices at targets, smoothness inherent in confluent design

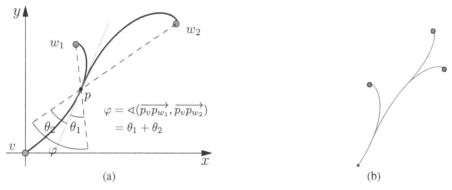

Fig. 7. (a) Construction of a confluent spiral segment with vertex w_2 branching off the parent spiral for (v, w_1) at p. The branching spiral is defined by p, p_{w_2} and the angle θ_2 which is determined by the parent spiral and the angle φ at p. (b) When a parent spiral is changed due to vertex movement, local adaptation by the other spirals is immediate. (NB: in the electronic version, this figure can be animated)

If a spiral $\mathcal{S}$ branches off another spiral $\mathcal{T}$ we refer to the latter as the *parent* spiral. The following heuristic is used to decide on the tree structure. Intuitively, we want edges leading to local targets to branch out earlier in the tree. Thus, consider the edges of a bundle B ordered by the distance of their targets from p_v. Other meaningful orderings, e.g., from data attributes, could be used too here, resulting in different trees.

We start with the first edge $(v, w) \in B$ as the *trunk*. This edge is represented by a logarithmic spiral with a predefined *trunk angle* $0 < \theta_0 < \pm\pi/6$. The upper bound $\pi/6$ ensures that the spiral approaches its vertex more directly, without orbiting around it.

Spiral segments for the other edges are allowed to branch off any already existing spiral segment $\mathcal{T}$ subject to two constraints on the new spiral $\mathcal{S}$:

- $\mathcal{S}$ must have an angle $\theta \in [\theta_{\min}, \theta_{\max}]$ (typically $\theta \in (0, \pm\pi/6]$).
- The tangent of $\mathcal{S}$ at p must have the same slope and direction as that of $\mathcal{T}$.

The branching point p on the parent spiral $\mathcal{T}$ is determined by trying k candidate points $(p_i = \mathcal{T}(i\frac{\pi}{k}), i \in 0, ..., k-1)$ on $\mathcal{T}$ and choosing the point that satisfies all constraints and results in the shortest spiral length. Note that $\mathcal{T}$ and p completely determine $\mathcal{S}$ as illustrated in Figure 7(a). If we cannot find a spiral satisfying the constraints, we postpone the current edge temporarily. Our experience so far is that edges need to be postponed rarely so that the overall runtime is in $\mathcal{O}(\sum_{B \in \mathcal{B}} k \cdot |B|^2) \subset \mathcal{O}(k\,n\,\Delta^2)$, where $\mathcal{B}$ is the set of all stub bundles and $n = |V|$.

4.3 Avoiding Obstacles and Crossings

Avoiding edge crossings is very important to reduce visual complexity and improve readability. Furthermore, it is very important that edges not connected to a vertex have a certain distance to that vertex. This can be modelled by placing obstacles on the vertex positions. We extend our framework to deal with obstacles and crossings by adding them as constraints during the search of a parent spiral. We use an R-tree [10] as spatial index and add the vertices with their shape as obstacles into it. The spirals of the already finished edges are approximated by s line segments and stored in the index too. For a possible branching point p we query the spatial index with s segments of the corresponding spiral to check whether they intersect with obstacles or other edges in the index. The creation of the R-tree index needs $\mathcal{O}(n \log n)$ time on average while maintaining and querying it takes $\mathcal{O}(s \, \log(n + s \, \Delta))$.

The intuitive interpretation of this method is that, if there is an obstacle for a desired branching point we will branch out in an earlier or later phase of the parent spiral to miss that obstacle. Although the resulting spirals will be longer, the readability will be improved. See fig. 8 for an illustration.

4.4   Edge Ordering and Parallel Routing

At this stage we have a confluent spiral tree for each bundle B. To reflect the data, in this case the different number of edges in the bundle, we route them in parallel until they branch to their targets.

Intuitively, walking along the outer contour of our tree gives us the required edge ordering. We determine this ordering by sorting the edges according to the branching point and branching side when traversing the underlying tree from the root vertex. With this ordering we then compute an offset curve to the approximated spiral, which is very similar to polygon offsetting. The offset will determine the thickness of the edge, which in turn can be used to represent, e.g., an edge attribute. See fig. 10 and fig. 9 for an example. Note that after applying an offset the result is not a true logarithmic spiral anymore. In practice this is not a problem as logarithmic spirals are eventually only approximated with cubic splines anyway [1].

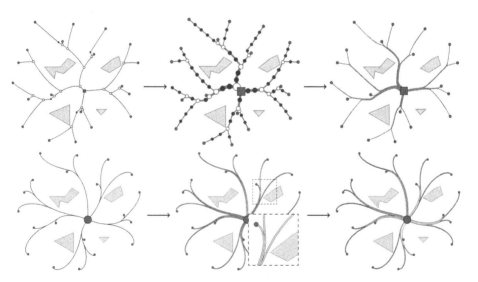

Fig. 8. Avoiding obstacles: Approach of Verbeek et al. [26] (top) and ours (bottom); spirals branch out *smoothely* from trunk at appropriate point to miss the obstacles. Parallel edge routing allows to map an edge attribute to the color; here node distance to the root is mapped (bottom-right).

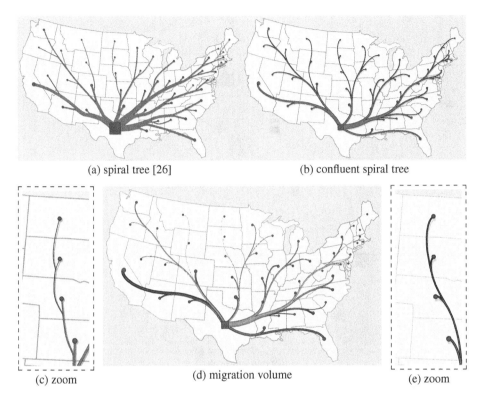

(a) spiral tree [26]   (b) confluent spiral tree

(c) zoom   (d) migration volume   (e) zoom

Fig. 9. Flow map of migration from Texas (1995-2000). The smooth linkage of confluent spirals (b) eliminates turns and thus not only yields more pleasing drawings but also facilitates the display of edge attributes (d). Note that edge direction can be inferred locally from every segment (e).

5 Discussion

We have presented drawing styles for the routing of undirected and directed edges in geographic networks using edge bundling at ends rather than interiors, and logarithmic-spiral trees that yield confluent drawings. Their main benefits are

- faithfulness (unambiguous representation of edges)
- stubs point in general direction of destinations
- edge widths and colors are still available for data attributes
- confluent flow maps of in- or out-stars

While initial feedback indicates that confluent spiral drawings are visually appealing, controlled user studies will have to show that they are effective.

As future work we plan to explore other, more data-driven, approaches to partition stubs into bundles, and we would like to prove guarantees on the tree structure of spiral segments and on avoidance of obstacles. For now, our application-oriented implementation is based largely on heuristics, but does layout networks with several thousands of edges essentially at interactive speed.

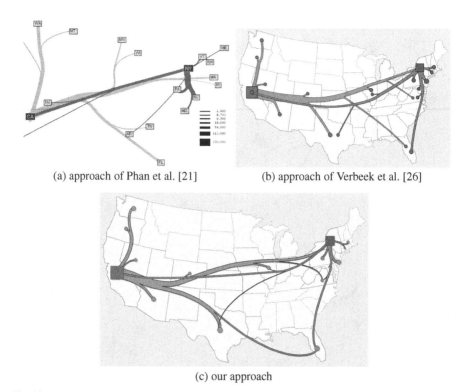

(a) approach of Phan et al. [21]   (b) approach of Verbeek et al. [26]

(c) our approach

Fig. 10. Flow map of migration to California and New York (1995-2000, top 10 states of origin). Line widths indicate migration volume and are to scale across figures.

References

1. Baumgarten, C., Farin, G.: Approximation of logarithmic spirals. Computer Aided Geometric Design 14(6), 515–532 (1997)
2. Benkert, M., Nöllenburg, M., Uno, T., Wolff, A.: Minimizing intra-edge crossings in wiring diagrams and public transportation maps. In: Kaufmann, M., Wagner, D. (eds.) GD 2006. LNCS, vol. 4372, pp. 270–281. Springer, Heidelberg (2007)
3. Brandes, U., Shubina, G., Tamassia, R.: Improving angular resolution in visualizations of geographic networks. In: de Leeuw, W.C., van Liere, R. (eds.) VisSym 2000, pp. 23–32. Springer (2000)
4. Brandes, U., Wagner, D.: Using graph layout to visualize train interconnection data. Journal of Graph Algorithms and Applications 4(3), 135–155 (2000)
5. Buchin, K., Speckmann, B., Verbeek, K.: Angle-restricted steiner arborescences for flow map layout. In: Asano, T., Nakano, S.-I., Okamoto, Y., Watanabe, O. (eds.) ISAAC 2011. LNCS, vol. 7074, pp. 250–259. Springer, Heidelberg (2011)
6. Cui, W., Zhou, H., Qu, H., Wong, P.C., Li, X.: Geometry-based edge clustering for graph visualization. IEEE Trans. on Visualization and Computer Graphics 14(6), 1277–1284 (2008)
7. Dickerson, M., Eppstein, D., Goodrich, M.T., Meng, J.Y.: Confluent drawings: Visualizing non-planar diagrams in a planar way. Journal of Graph Algorithms and Applications 9(1), 31–52 (2005)

8. Ersoy, O., Hurter, C., Paulovich, F.V., Cantareiro, G., Telea, A.: Skeleton-based edge bundling for graph visualization. IEEE Trans. on Visualization and Computer Graphics 17(12), 2364–2373 (2011)

9. Gansner, E.R., Hu, Y., North, S.C., Scheidegger, C.E.: Multilevel agglomerative edge bundling for visualizing large graphs. In: Proc. of the IEEE Pacific Visualization Symposium 2011, pp. 187–194. IEEE (2011)

10. Guttman, A.: R-trees: a dynamic index structure for spatial searching. In: SIGMOD 1984, pp. 47–57. ACM Press (1984)

11. Holten, D.: Hierarchical edge bundles: Visualization of adjacency relations in hierarchical data. IEEE Trans. on Visualization and Computer Graphics 12, 741–748 (2006)

12. Holten, D., van Wijk, J.J.: Force-directed edge bundling for graph visualization. Computer Graphics Forum 28(3), 983–990 (2009)

13. Hurter, C., Ersoy, O., Telea, A.: Smooth bundling of large streaming and sequence graphs. In: Proc. of the IEEE Pacific Visualization Symposium. IEEE (to appear, 2013)

14. Hurter, C., Ersoy, O., Telea, A.: Graph bundling by kernel density estimation. Computer Graphics Forum 31(3pt. 1), 865–874 (2012)

15. Lambert, A., Bourqui, R., Auber, D.: 3d edge bundling for geographical data visualization. IEEE Trans. on Visualization and Computer Graphics, 329–335 (2010)

16. Lambert, A., Bourqui, R., Auber, D.: Winding roads: Routing edges into bundles. Computer Graphics Forum 29(3), 853–862 (2010)

17. Newberry, F.J.: Edge concentration: A method for clustering directed graphs. SIGSOFT Software Engineering Notes 14(7), 76–85 (1989)

18. Nguyen, Q.H., Eades, P., Hong, S.: On the faithfulness of graph visualizations. In: Proc. of the IEEE Pacific Visualization Symposium. IEEE (to appear, 2013)

19. Nguyen, Q., Hong, S.-H., Eades, P.: TGI-EB: A new framework for edge bundling integrating topology, geometry and importance. In: Speckmann, B. (ed.) GD 2011. LNCS, vol. 7034, pp. 123–135. Springer, Heidelberg (2011)

20. Peng, D., Lu, N., Chen, W., Peng, Q.: Sideknot: Revealing relation patterns for graph visualization. In: Proc. of the IEEE Pacific Visualization Symposium 2012, pp. 65–72. IEEE (2012)

21. Phan, D., Xiao, L., Yeh, R., Hanrahan, P., Winograd, T.: Flow map layout. In: Proc. of IEEE Symposium of Information Visualization 2005, p. 29. IEEE (2005)

22. Pupyrev, S., Nachmanson, L., Bereg, S., Holroyd, A.E.: Edge routing with ordered bundles. In: Speckmann, B. (ed.) GD 2011. LNCS, vol. 7034, pp. 136–147. Springer, Heidelberg (2011)

23. Pupyrev, S., Nachmanson, L., Kaufmann, M.: Improving layered graph layouts with edge bundling. In: Brandes, U., Cornelsen, S. (eds.) GD 2010. LNCS, vol. 6502, pp. 329–340. Springer, Heidelberg (2011)

24. Selassie, D., Heller, B., Heer, J.: Divided edge bundling for directional network data. IEEE Trans. on Visualization and Computer Graphics 17 (2011)

25. Telea, A., Ersoy, O.: Image-based edge bundles: Simplified visualization of large graphs. Computer Graphics Forum 29(3), 843–852 (2010)

26. Verbeek, K., Buchin, K., Speckmann, B.: Flow map layout via spiral trees. IEEE Trans. on Visualization and Computer Graphics 17(12), 2536–2544 (2011)

27. Zinsmaier, M., Brandes, U., Deussen, O., Strobelt, H.: Interactive level-of-detail rendering of large graphs. IEEE Trans. on Visualization and Computer Graphics 18(12), 2486–2495 (2012)

On Orthogonally Convex Drawings
of Plane Graphs
(Extended Abstract)

Yi-Jun Chang and Hsu-Chun Yen*

Dept. of Electrical Engineering, National Taiwan University
Taipei, Taiwan 106, Republic of China

Abstract. We investigate the bend minimization problem with respect to a new drawing style called *orthogonally convex drawing*, which is orthogonal drawing with an additional requirement that each inner face is drawn as an *orthogonally convex polygon*. For the class of bi-connected plane graphs of maximum degree 3, we give a necessary and sufficient condition for the existence of a no-bend orthogonally convex drawing, which in turn, enables a linear time algorithm to check and construct such a drawing if one exists. We also develop a flow network formulation for bend-minimization in orthogonally convex drawings, yielding a polynomial time solution for the problem. An interesting application of our orthogonally convex drawing is to characterize internally triangulated plane graphs that admit floorplans using only orthogonally convex modules subject to certain boundary constraints.

Keywords: Bend minimization, floorplan, orthogonally convex drawing.

1. Introduction

An *orthogonal drawing* of a plane graph is a planar drawing such that each edge is composed of a sequence of horizontal and vertical line segments with no crossings. A classic optimization problem in orthogonal drawing is to minimize the number of *bends*, namely, the *bend-minimization* problem. The problem is NP-complete in the most general setting, i.e., for planar graphs of maximum degree 4 [5]. Subclasses of graphs with bend-minimization of orthogonal drawing tractable include planar graphs of maximum degree 3, series-parallel graphs, and graphs with fixed embeddings [3,12], etc.

Most of the orthogonal drawing algorithms reported in the literature can be roughly divided into two categories, one uses flow or matching to model the problem (e.g., [2,3,12]), whereas the other tackles the problem in a more graph-theoretic way by taking advantage of structure properties of graphs (e.g., [9,10,8]). The former usually solves a more general problem, but requires higher

---

* Corresponding author (e-mail: yen@cc.ee.ntu.edu.tw). Research supported in part by National Science Council of Taiwan under Grants NSC-100-2221-E-002-132-MY3 and NSC-100-2923-E-002-001-MY3.

S. Wismath and A. Wolff (Eds.): GD 2013, LNCS 8242, pp. 400–411, 2013.

time complexity. On the contrary, algorithms in the latter focus on specific kinds of graphs, resulting in linear time complexity in many cases.

In this paper, we introduce a new type of orthogonal drawing called *orthogonally convex drawing*, which requires that each inner face be an *orthogonally convex polygon*. A polygon is *orthogonally convex* if for any horizontal or vertical line, if two points on the line are inside a polygonal region, then the entire line segment between these two points is also inside the polygonal region. The study of this new drawing style is motivated by an attempt to learn more about the geometric aspect of orthogonal drawing, which, in the dual setting, is closely related to *rectangular dual* and *rectilinear dual* which are well-studied in floor-planning and contact graph representations [6,11,14]. Note that if we consider standard convexity instead of orthogonal convexity in the setting of no-bend orthogonal drawing, the problem becomes the "inner rectangular drawing" studied in [7]. There are several recent results on rectilinear duals in cartographic applications, see, e.g., [1]. For other perspectives of orthogonal drawing, the reader is referred to [4] for a survey chapter.

Our contributions include the following:

1. A new drawing style called *orthogonally convex drawing* is introduced, and a necessary and sufficient condition, along with a linear time testing algorithm, is given for a bi-connected plane 3-graph (i.e., of maximum degree 3) to admit a no-bend orthogonally convex drawing.
2. A flow network formulation is devised for the bend-minimization problem of orthogonally convex drawing.
3. By combining the above no-bend orthogonally convex drawing algorithm and the flow network formulation, a polynomial time algorithm (in $O(n^{1.5} \log^3 n)$ time) for constructing a bend-optimal orthogonally convex drawing is presented.
4. We apply our analysis of no-bend orthogonally convex drawing to characterizing internally triangulated graphs that admit floorplans using only orthogonally convex modules that can be embedded into a rectilinear region with its boundary order-equivalent to a given orthogonally convex polygon.

2. Preliminaries

Given a graph $G = (V, E)$, we write $\Delta(G)$ to denote the maximum degree of G. Graph G is called a d-graph if $\Delta(G) \leq d$. A *path* P of G is a sequence of vertices $(v_1, v_2, ..., v_n)$ such that $\forall 1 \leq i \leq n, v_i \in V$ and $\forall 1 \leq i \leq n - 1, (v_i, v_{i+1}) \in E$. We write $V(P)$ to denote the set of vertices $\{v_1, ..., v_n\}$, and $E(P)$ to denote the edge set $\{(v_i, v_{i+1}) | 1 \leq i < n\}$ of P. Given two paths P' and P, we write $P' \subseteq P$ if P' is a subsequence of P, and $P' \subset P$ if $P' \subseteq P$ and $P' \neq P$. P is called a *cycle* if $v_1 = v_n$. Unless stated otherwise, paths and cycles are assumed to be *simple* throughout this paper, in the sense that there are no repeated vertices other than the starting and ending vertices. A drawing of a planar graph divides the plane into a set of connected regions, called *faces*. A *contour* of a face F is the cycle formed by vertices and edges along the boundary of F. A cycle that is the boundary, ie., contour, of a face is called a *facial cycle*. The contour of the outer face is denoted as C_O. If G is bi-connected, contours of all the faces are simple cycles.

In our subsequent discussion, we adopt some of the notations and definitions used in [9,10]. A cycle C divides a plane graph G into two regions. The one that is inside (resp., outside) cycle C is called the *interior region* (resp., *outer region*) of C. We use $G(C)$ to denote the subgraph of G that contains exactly C and vertices and edges residing in its interior region. An edge $e = (u, v)$ in the outer region of C is called a *leg* of C if at least one of the two vertices u and v belongs to C. C is *k-legged* if C contains exactly k vertices that are incident to some legs of C. These k vertices are called *legged-vertices* of C. If $\Delta(G) \leq 3$, every legged-vertex v of C is incident to exactly one leg e of C. Note that 3-legged cycles coincide with the so-called *complex triangles* in the dual setting, which play a crucial role in the study of rectilinear duals [11,14].

We call a face or a cycle *inner* if it is not the outer one. If an inner face or inner cycle intersects with the outer one, then we call it *boundary face* or *boundary cycle*. A *contour path* P of a cycle C is a path on C such that P includes exactly two legged-vertices x and y of C, and x and y are the two endpoints of P. Therefore, each k-legged cycle has exactly k contour paths. If a contour path intersects with (i.e., shares some edges with) the outer cycle, we call it *boundary contour path*. In fact, each boundary contour path is a subpath of C_O. Each contour path P of C is incident to exactly one face, denoted as $F_{C,P}$, in the outer region of C. As an illustrating example, consider Figure 1. F_0 is the outer face of G. Con-

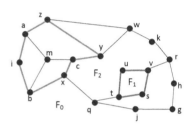

Fig. 1. Illustration of some terms about cycles and paths

sider two cycles $C_1 = (s, t, u, v, s)$ and $C_2 = (x, b, i, a, z, y, c, x)$ (both drawn in bold line). C_1 is a non-boundary 2-legged cycle, of which two legged-vertices are t and v, and two legs are (t, q) and (v, r). C_1 is also a facial cycle, which is the contour of F_1. C_2 is a boundary 3-legged cycle, of which three legged-vertices are x, y, and z. $P_1 = (t, u, v)$ is a contour path of C_1. $P_2 = (z, a, i, b, x)$ is the boundary contour path of C_2. We have $F_{C_1,P_1} = F_2$ and $F_{C_2,P_2} = F_0$.

Let $D(G)$ be an orthogonal drawing of plane graph G with outer cycle C_O. Given a cycle C, we use $D(C)$ (or equivalently $D(F)$ if C is the contour of a face F) to denote the drawing of C in $D(G)$. $D(C)$ is always a simple polygon as long as C is simple. We call $D(G)$ an *orthogonally convex drawing* of G if $D(F)$ is an *orthogonally convex polygon* for every face F other than the outer one. We use $bc(D(G))$ to denote the bend count, i.e., the total number of bends, of $D(G)$.

In an orthogonal drawing $D(G)$, $ang_C(v)$ denotes the interior angle of v in polygon $D(C)$. We called v a *convex corner*, *non-corner*, and *concave corner* of

C if $ang_C(v)$ is $90°$, $180°$, and $270°$, respectively. A *corner* in the drawing $D(G)$ is either a bend on some edge, or a vertex v of G such that $ang_C(v) \neq 180°$ for some C. If v is a non-corner of C, v is on a side of polygon $D(C)$.

From Section 3 to Section 5, graphs under the name G are assumed to be bi-connected with $\triangle(G) \leq 3$, and may have multi-edges.

3. No-Bend Orthogonally Convex Drawing

Among existing results concerning orthogonal drawings, Rahman et al. [10] gave a necessary and sufficient condition for a bi-connected plane 3-graph to admit a no-bend orthogonal drawing, and they devised an algorithm to test the condition, and subsequently construct the drawing if one exists.

Theorem 1 ([10]). *A bi-connected plane 3-graph G has a no-bend orthogonal drawing iff G satisfies the following three conditions:*

1. *There are four or more 2-vertices (i.e., vertices of degree 2) of G on $C_O(G)$.*
2. *Every 2-legged cycle contains at least two 2-vertices.*
3. *Every 3-legged cycle contains at least one 2-vertex.*

Theorem 1 clearly holds even when G has multi-edges, as such graphs do not have no-bend orthogonal drawings. The no-bend orthogonal drawing algorithm in [10] performs the following steps recursively: (1) reducing the original graph G into a structurally simpler graph G^* by collapsing the so-called "*maximal bad cycles*", (2) drawing G^* in a rectangular fashion, and (3) plugging in the orthogonal drawings of those maximal bad cycles to the rectangular drawing[1] of G^* to yield a no-bend orthogonal drawing of G.

Bad cycles in Step (1) are cycles that are 2-legged or 3-legged if the four designated corner vertices in C_O are considered as legged-vertices. Intuitively, bad cycles are cycles that violate the conditions under which a graph admits a rectangular drawing. For instance, consider the graph in Figure 1. If $\{h, i, j, k\}$ are the 4 designated vertices, then $(w, z, a, i, b, x, c, y, w)$ (a 3-legged cycle as i is considered a legged-vertex) is a bad cycle, whereas $(r, v, u, t, q, j, g, h, r)$ (a 4-legged cycle including legged-vertices h and j) is not a bad cycle. *Maximal bad cycles* are bad cycles that are not contained in $G(C)$ for another bad cycle C. Step (2) involves computing the rectangular drawing of an input graph with four designated corner vertices on $C_O(G)$. It is known that such a graph with four designated vertices admits a rectangular drawing if and only if every 2-legged cycle contains at least two designated vertices, and every 3-legged cycle contains at least one designated vertex [13]. As shown in [10], the G^* (with each of the maximal bad cycles contracted to a single vertex) always meets the condition for the existence of a rectangular drawing. The reader is referred to [10] for more.

Our goal in this section is to give a similar necessary and sufficient condition for graphs to have no-bend orthogonally convex drawings.

---

[1] A *rectangular drawing* of a graph is a no-bend orthogonal drawing such that each interior face is a rectangle and the boundary of the outer face also forms a rectangle.

Lemma 1. *Consider a no-bend orthogonally convex drawing $D(G)$ of a graph G. For every 2-legged cycle C with legged-vertices x and y and a contour path P of C, the number of convex corners of $D(C)$ in $V(P) \setminus \{x, y\}$ (i.e., the set of vertices along path P excluding x and y) must be at least 1 more than that of concave corners, if either (1) C is a boundary cycle and P is its boundary contour path, or (2) C is non-boundary and P is any of its contour paths.*

We are now in a position to give one of our main results.

Theorem 2. *A bi-connected plane 3-graph G admits a no-bend orthogonally convex drawing if and only if the three conditions in Theorem 1 and the following two additional conditions hold: (1) every non-boundary 2-legged cycle contains at least one 2-vertex on each of its contour paths, and (2) every boundary 2-legged cycle contains at least one 2-vertex on its boundary contour path.*

The necessity of Theorem 2 follows from Lemma 1. A modification to the no-bend orthogonal drawing algorithm described above yields a constructive proof of the sufficiency of Theorem 2. Based on an implementation described in [10], we have the following result.

Theorem 3. *There is a linear time algorithm to construct a no-bend orthogonally convex drawing $D(G)$ if G admits one.*

4.   An Alternative Condition

An alternative necessary and sufficient condition is given in this section to characterize bi-connected 3-plane graphs admitting no-bend orthogonally convex drawings, facilitating a min-cost flow formulation for the bend-minimization problem. As Theorem 2 indicates, contour paths along (boundary or non-boundary) 2-legged cycles play a vital role in orthogonally convex drawing. Due to possible overlaps of 2-legged cycles and complex intersections between contour paths, it becomes difficult to capture the amount of convex/concave corners along contour paths in a min-cost flow formulation. To ease this problem, we identify two types of cycles, namely, *proper* and *improper* cycles, which are later used to characterize the presence of orthogonally convex drawings.

Let G^c denote the graph resulting from contracting every 2-vertex of G. Since we require G to be of maximum degree 3, G^c must be 3-regular. A 2-legged cycle of G is called *improper* if its two legs correspond to the same edge in G^c; otherwise, it is called *proper*. See Figure 2 for instance. Due to 3-regularness of G^c and the fact that the two legs of an improper cycle C are the same edge e in G^c, there remains nothing outside $G(C)$ except the leg e. Therefore, improper cycles must be boundary cycles, or conversely, all non-boundary 2-legged cycles are proper. It is also easy to observe that a 2-legged cycle C of G with two leg-vertices x and y is improper iff for the non-boundary contour path P of C, the boundary of $F_{C,P}$ intersects C_O of G in exactly 1 path. Again consider Figure 2. Note that F_1 and F_2 correspond to the $F_{C,P}$ of the 2-legged cycles drawn as bold lines in the left and right figures, respectively. The boundary of F_1 intersects C_O

in exactly one path (x, y), whereas the boundary of F_2 intersects C_O in two paths (z, a) and (y, b, c).

Definition 1. *A path P of G is called* critical *if there is a proper 2-legged cycle C such that: (1) P is a contour path of C, (2) if C is a boundary 2-legged cycle, P is the boundary contour path of C, and (3) P does not edge-intersect with any proper 2-legged cycle other than C that is contained in $G(C)$.*

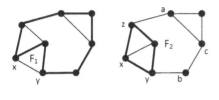

Fig. 2. Proper and improper 2-legged cycles. Left: An improper 2-legged cycle (drawn as a bold line) with leg-vertices x, y. Right: A proper 2-legged cycle (drawn as a bold line) with leg-vertices y, z.

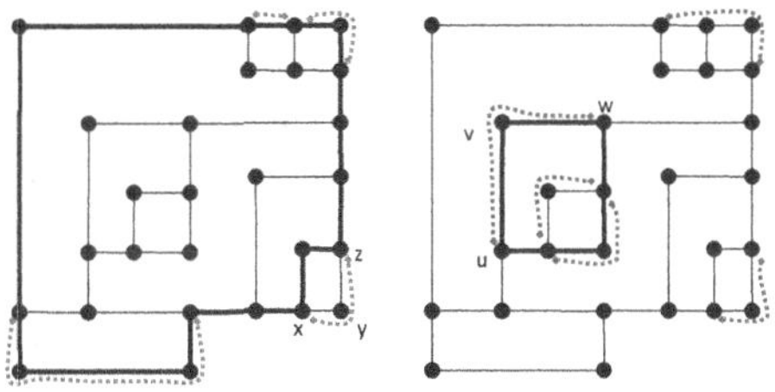

Fig. 3. Critical paths and S_G in a plane graph. Left: Paths in S_G. Right: Critical Paths

To proceed further, we require the following two lemmas.

Lemma 2. *For any bi-connected plane 3-graph G, the critical paths of G are edge-disjoint.*

Lemma 3. *Let P be a path satisfying (1) and (2) in Definition 1. If P is not critical, there must be a critical path P' such that $P' \subset P$.*

Note that the requirement of properness of 2-legged cycles in Definition 1 is essential in the sense that Lemma 2, which is crucial in the subsequent context, is not true if we remove that requirement. Given a path P with endpoints x and y, we write $P_{(x \frown y)}$ to denote the "open" version of P, i.e., excluding x and y. That is, $P_{(x \frown y)}$ consists of $V(P) \setminus \{x, y\}$ and $E(P)$.

Instead of basing on contour paths as in Theorem 2, our new characterization for no-bend orthogonally convex drawing is based upon two kinds of paths defined over proper and improper 2-legged cycles, namely, critical paths defined above for proper 2-legged cycles and a set of paths called S_G associated with improper 2-legged cycles in graph G. S_G is defined to be the set of all paths $C_O \setminus P_{(x \frown y)}$ for every boundary contour path P of an improper 2-legged cycle with two legged-vertices x, y. Note that internal vertices in paths of S_G must have degree 2, and paths in S_G must be in C_O, and hence $P \in S_G$ iff P is a boundary contour path of a facial cycle C that has only one boundary contour path. The following fact summarize the observation.

Fact 1. *Let P be a path of G with two end-vertices x and y. The following three statements are equivalent: (1) P is in S_G; (2) P is the boundary contour path of a facial cycle C that intersects C_O of G in exactly one path; (3) P is $C_O \backslash P'_{(x \frown y)}$, for some boundary contour path P' of an improper 2-legged cycle with two legged-vertices x and y.*

To have better grasp of critical paths and S_G, consider Figure 3 in which a no-bend orthogonally convex drawing of a plane graph G is shown. In the left figure, the four dotted paths are those in S_G, which are edge-disjoint. Let C be the 2-legged cycle drawn as a bold line, and P be its boundary contour path. We have $C_O \backslash P_{(x \frown z)} = (x, y, z)$. In the right figure, the five dotted paths are critical paths, which are edge-disjoint. Let C be the 2-legged cycle drawn as a bold line. We have (1) the path (u, v, w) is one of its contour paths, (2) C is a non-boundary 2-legged cycle, and (3) P does not edge-intersect with any proper 2-legged cycle other than C that is contained in $G(C)$. A path in S_G is either contained in exactly one critical path or intersects with no critical path. The reader is encouraged to verify that the graph satisfies the conditions stated in Theorem 2 and Theorem 4, and the orthogonally convex drawing satisfies the conditions in Lemma 4.

The following theorem enables us to characterize no-bend orthogonally convex drawings in terms of critical paths and S_G.

Theorem 4. *Suppose a bi-connected plane 3-graph G has a no-bend orthogonal drawing. G has a no-bend orthogonally convex drawing iff the following conditions are satisfied:*

1. *Every critical path of G contains at least one 2-vertex.*
2. *$C_O \backslash P$ contains at least one 2-vertex for every $P \in S_G$.*

We note that both S_G and the set of all critical paths can be found in linear time. The algorithm is basically a contour edge-traversal of each face with a mechanism of detecting repeated adjacent faces.

5.   Flow Formulation for Bend-Minimization

In this section, we tailor the planar min-cost flow formulation originally designed for orthogonal drawing [12] to coping with orthogonal convexity. To make our subsequent discussion clear, we use *arc* and *node* instead of edge and vertex, respectively, in describing a flow network. A *min-cost flow network* is a directed multi-graph $N = (W, A)$ associated with four functions: *lower bounds* $\lambda : A \rightarrow \mathbb{Z}_{\geq 0}$, *capacities* $\mu : A \rightarrow \mathbb{Z}_{\geq 0} \cup \{\infty\}$, *costs* $c : A \rightarrow \mathbb{Z}_{\geq 0}$, *demands* $b : W \rightarrow \mathbb{Z}$. A map $f : A \rightarrow \mathbb{Z}_{\geq 0}$ is a *flow* if the following constraints are met for each node v and arc a:

$$b(v) + \sum_{(u,v) \in A} f(u, v) - \sum_{(v,u) \in A} f(v, u) = 0, \quad \lambda(a) \leq f(a) \leq \mu(a)$$

The cost of a flow f is $c(f) = \sum_{a \in A} f(a) \times c(a)$. The flow network $N_G = (W_G, A_G)$ associated with a bi-connected plane 3-graph G is

- $W_G = W_V \cup W_F$, where W_V and W_F are the vertex set and face set (including the outer face) of G, respectively, Furthermore, $\forall u_v \in W_V$, $b(u_v) = 2$ if $deg_G(v) = 3$; $b(u_v) = 0$ if $deg_G(v) = 2$. $\forall u_F \in W_F$, $b(u_F) = -4$ if F is an inner face; $b(u_F) = 4$ if F is the outer face.
- $A_G = A_V \cup A_F$, where
  - $A_V = \{(u_v, u_F), (u_F, u_v)|deg(v) = 2\} \cup \{(u_v, u_F)|deg(v) = 3\}$, where $v \in V(G), F \in face(G), v$ incident to F. $\forall a \in A_V$, $\lambda(a) = 0$, $\mu(a) = 1$, and $c(a) = 0$.
  - $A_F = \{(u_F, u_H)|F, H \in face(G)$, and F adjacent to $H\}$ is a multi-set of arcs between faces, and the number of (u_F, u_H) in A_F equals the number of shared edges e in contours of F and H. We use $(u_F, u_H)_e$ to indicate the specific arc that corresponds to the shared edges e. $\forall a \in A_F$, $\lambda(a) = 0$, $\mu(a) = \infty$, and $c(a) = 1$.

Although our definition of N_G is slightly different from the original one given in [12], the validity of N_G is apparent as the following explains. Every flow f in N_G corresponds to an orthogonal drawing $D(G)$, and vice versa, such that

- $f(u_v, u_F) - f(u_F, u_v) = -1, 0, 1$ means v is a concave corner, non-corner, convex corner in $D(F)$, respectively,
- $f(u_F, u_H)_e$ is the number of bends on e that are concave corners in $D(F)$ and convex corners in $D(H)$, and
- the total number of bends in $D(G)$ equals $c(f)$.

Fact 2. *Let S_1 (resp., S_2) be any subset of edges (resp., vertices) along the contour of a face F. For any $e \in S_1$, we write F_e to denote the face incident to e other than F. For a flow f in N_G and its corresponding orthogonal drawing D, we must have $\sum_{e \in S_1}[f(u_{F_e}, u_F)_e - f(u_F, u_{F_e})_e] + \sum_{v \in S_2}[f(u_v, u_F) - f(u_F, u_v)]$ equals the difference between the numbers of convex corners and concave corners in S_1 and S_2 of $D(F)$.*

Lemma 4. *A bi-connected plane 3-graph G admits a no-bend orthogonally convex drawing iff there is a no-bend orthogonal drawing (not necessarily orthogonally convex) such that (1) for every critical path P along a contour path of 2-legged cycle C, $\#_{cc}(P_{(x \frown y)}) > \#_{cv}(P_{(x \frown y)})$ in $F_{C,P}$, and (2) for every P in S_G, $\#_{cc}(P_{(x \frown y)}) \leq 3 + \#_{cv}(P_{(x \frown y)})$ in the outer face, where P has endpoints x and y, and $\#_{cv}(P_{(x \frown y)})$ and $\#_{cc}(P_{(x \frown y)})$ represent the numbers of convex and concave corners, respectively, of $P_{(x \frown y)}$.*

In what follows, we show how to construct a flow network N'_G from N_G in such a way that a flow of N'_G corresponds to an orthogonal drawing meeting the conditions stated in Lemma 4. Initially we set $N'_G = N_G$.

- $\forall P \in S_G$ with endpoints x, y and outer face F', add a new node u_P to $W(N'_G)$, and two arcs $(u_{F'}, u_P)$, $(u_P, u_{F'})$ to $A(N'_G)$. We set $b(u_P) = 0$, $\lambda(u_{F'}, u_P) = \lambda(u_P, u_{F'}) = 0$, $\mu(u_{F'}, u_P) = 3$, $\mu(u_P, u_{F'}) = \infty$, and $c(u_{F'}, u_P) = c(u_P, u_{F'}) = 0$. We redirect all the arcs in the current $A(N'_G)$ of the following forms: $(u_{F'}, u_v)$, $(u_v, u_{F'})$, $(u_{F'}, u_F)_e$, $(u_F, u_{F'})_e$ for all

$v \in V(P) \setminus \{x, y\}$, $F \in S_{P,F'}$, $e \in E(P)$ by replacing $u_{F'}$ with u_P. Due to Fact 2, Statement 2 of Lemma 4 holds.

– $\forall$ critical path P with endpoints x, y, C the 2-legged cycle for which P is its contour path, and S the set of faces in $G(C)$ that border P, add a new node u_P to $W(N_G')$, and a new arc $(u_{F_C,P}, u_P)$ to $A(N_G')$. We set $b(u_P) = 0$, $\lambda(u_{F_C,P}, u_P) = 1$, $\mu(u_{F_C,P}, u_P) = \infty$, and $c(u_{F_C,P}, u_P) = 0$. We redirect all the arcs in the current $A(N_G')$ of the following forms: $(u_{F_C,P}, u_{P'})$, $(u_{P'}, u_{F_C,P})$, $(u_{F_C,P}, u_v)$, $(u_v, u_{F_C,P})$, $(u_{F_C,P}, u_F)_e$, $(u_F, u_{F_C,P})_e$ for all $P' \in S_G$ such that $P' \subseteq P$, $v \in V(P) \setminus \{x, y\}$, $F \in S$, $e \in E(P)$ by replacing $u_{F_C,P}$ with u_P. Due to Fact 2, Statement 1 of Lemma 4 holds.

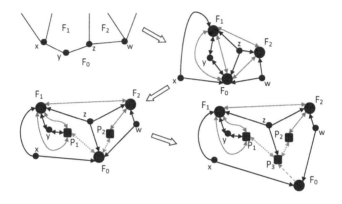

Fig. 4. Illustration of the construction of N_G': F_0 is the outer face, $P_1 = (x, y, z)$ and $P_2 = (z, w)$ are two paths in S_G, $P_3 = (x, y, z, w)$ is a critical path

For an illustrating example, consider Figure 4 in which the up-left picture is a portion of a graph G with N_G depicted in the up-right. The down-left one illustrates the result of adding two additional nodes representing P_1 and P_2 (the newly added arcs are drawn in dotted line). The down-right one illustrates the result of adding an additional node representing critical path P_3 (the newly added arc is drawn in dashed line).

The validity of the above construction follows from critical paths being mutually edge-disjoint (Lemma 2), and every path in S_G is either a subpath of a critical path or intersects with no critical paths. Note that the number of newly added arcs and nodes is linear in $n = V(G)$, and the maximum possible value of the minimum cost is also $O(n)$. Following an $O(n^{1.5} \log^3 n)$ time algorithm in [2], we have

Theorem 5. *For any bi-connected plane 3-graph G, we can construct a bend-minimized orthogonally convex drawing in $O(n^{1.5} \log^3 n)$ time.*

6. An Application to Floor Planning

In this section, we show an application of orthogonally convex drawing to floor planning. A plane graph is *internally triangulated* if all the inner faces are triangles. For any internally triangulated plane graph $G = (V, E)$, a *rectilinear dual* is a partition of a simple orthogonal polygon (denoted as R) into $|V(G)|$ simple orthogonal regions, one for each vertex, such that two region have a side-contact iff their corresponding vertices adjacent to each other.

Two polygons are said to be *order-equivalent* if they admit the same circular order (in counter-clockwise orientation) of angles. For instance, the following two figures ⌐ ⌐ are order-equivalent. Let Q be an orthogonal polygon, we use Q-*floorplan* to denote a rectilinear dual whose boundary (the R in the definition of rectilinear dual) is order-equivalent to Q. A floorplan is called orthogonally convex if all the boundaries of $|V(G)|$ simple orthogonal regions are orthogonally convex polygons. In this section, graphs under the name G_{dual} are assumed to be simple, connected, internally triangulated plane graph.

Lemma 5. *For any simple, connected, internally triangulated plane graph G_{dual}, there is a unique bi-connected 3-regular plane multi-graph G_{primal} such that G_{dual} is the weak dual[2] of G_{primal}, and the following properties are hold: (1) G_{primal} does not have any non-boundary 2-legged cycle, and (2) internal faces (which are orthogonal polygons) of an orthogonal drawing of G_{primal} form a rectilinear dual of G_{dual}.*

We remark that although G_{dual} is required to be simple, G_{primal} may still have multi-edges. Since G_{primal} is bi-connected and $\triangle(G_{primal}) \leq 3$, the results in the previous sections can be applied.

Let $C_O = (v_1, v_2, \ldots, v_s, v_1)$ be the boundary cycle of G_{dual}, which need not be a simple cycle. Then, a triangulated plane multi-graph G' is constructed by adding a new vertex t in the outer face of G_{dual}, and then triangulate the outer face by adding edge (v_i, t) for $1 \leq i \leq s$. Take the dual of G' yields G_{primal}. See Figure 5 for an illustration.

Given an orthogonally convex polygon Q, our goal is to characterize graphs that admit orthogonally convex Q-floorplans, and subsequently realize such floorplans. We use numSide(P) to denote the number of sides of polygon P with non-corner vertices neglected.

Lemma 6. *Let G be a bi-connected plane 3-graph (may have multi-edges) with k boundary critical paths. We have $\min\{\text{numSide}(D(C_O)) \mid D$ is an orthogonally convex drawing of $G\} = \max\{4, 2k - 4\}$. Further, for any orthogonally convex polygon Q of numSide(Q) $\geq \max\{4, 2k - 4\}$, there is an orthogonally convex drawing $D(G)$ such that $D(C_O)$ is order-equivalent to Q.*

The concept of critical paths turns out to be pretty clean in the dual setting. We use T_G to denote the block-cutvertex tree of G. We will see in Lemma 7

---

[2] The weak dual of a plane graph is the subgraph of the dual graph excluding the vertex (and edges) corresponding to the unbounded (i.e., outer) face.

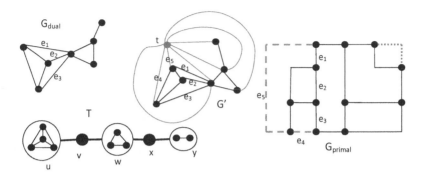

Fig. 5. The construction of G_{primal}

that leaves in $T_{G_{dual}}$ actually have one-to-one correspondence to critical paths in G_{primal}. Let (v, u) be an edge in $E(T_{G_{dual}})$ such that v is a cut-vertex. Now u must be a block. Let $V_{v,u}$ be the vertex set of the component in $G_{dual} \setminus \{v\}$ that contains some vertices in block u, and $F_{v,u}$ denote the corresponding face set in G_{primal}. Since G_{dual} is internally triangulated, the edges in $E(G_{dual})$ that link v to vertices in $V_{v,u}$ must be located consecutively in the circular list of edges incident to v that describes the combinatorial embedding of G_{dual}. We denote the edge set as $E_{v,u}$. According to the definition of duality of plane graphs and the algorithm for constructing G_{primal} from G_{dual}, these edges form a path in G_{primal}. We write $C_{v,u}$ to denote the cycle that is the boundary of union of faces in $F_{v,u}$. For instance, in Figure 5 the set $E_{v,u}$ is $\{e_1, e_2, e_3\}$, which forms the non-boundary contour path with respect to $C_{v,u} = (e_1, e_2, e_3, e_4, e_5)$.

Lemma 7. {*Boundary contour path of $C_{v,u}$ | u is a leaf of $T_{G_{dual}}, (v, u) \in E(T_{G_{dual}})$*} *is the set of boundary critical paths in G_{primal}.*

In Figure 5, the boundary contour paths of $C_{v,u}$ and $C_{x,y}$ are the paths drawn in dashed and dotted lines, respectively. These two paths are the boundary critical paths of G_{primal}. Following Lemmas 5, 6, 7 and Theorem 3, we have

Theorem 6. *For any internally triangulated graph G_{dual} and orthogonally convex polygon Q, let k be the number of leaves in the block-cutvertex tree of G_{dual}. G_{dual} admits an orthogonally convex Q-floorplan iff* numSide$(Q) \geq \max\{4, 2k - 4\}$. *The floorplan can be constructed in linear time.*

7.   Conclusion

We studied a new drawing style called orthogonally convex drawing from both combinatorial and algorithmic viewpoints. It would be interesting to see whether results/techniques developed in our work could be extended to other types of convex versions of contact graph representations or floorplans.

References

1. Alam, Md. J., Biedl, T., Felsner, S., Kaufmann, M., Kobourov, S. G., Ueckert, T.: Computing Cartograms with Optimal Complexity. In: Symposium on Computational Geometry (SoCG 2012), pp. 21–30 (2012)
2. Cornelsen, S., Karrenbauer, A.: Accelerated Bend Minimization. In: Speckmann, B. (ed.) GD 2011. LNCS, vol. 7034, pp. 111–122. Springer, Heidelberg (2011)
3. Di Battista, G., Liotta, G., Vargiu, F.: Spirality and Optimal Orthogonal Drawings. SIAM Journal on Computing 27(6), 1764–1811 (1998)
4. Duncan, C.A., Goodrich, M.T.: Planar Orthogonal and Polyline Drawing Algorithms. In: Tamassia, R. (ed.) Handbook of Graph Drawing and Visualization, ch. 7. CRC Press (2013)
5. Garg, A., Tamassia, R.: On the Computational Complexity of Upward and Rectilinear Planarity Testing. SIAM Journal on Computing 31(2), 601–625 (2001)
6. Kozminski, K., Kinnen, E.: Rectangular Dual of Planar Graphs. Networks 15, 145–157 (1985)
7. Miura, K., Haga, H., Nishizeki, T.: Inner Rectangular Drawings of Plane Graphs. International Journal of Computational Geometry and Applications 16(2-3), 249–270 (2006)
8. Rahman, M. S., Egi, N., Nishizeki, T.: No-bend Orthogonal Drawings of Series-Parallel Graphs. In: Healy, P., Nikolov, N.S. (eds.) GD 2005. LNCS, vol. 3843, pp. 409–420. Springer, Heidelberg (2006)
9. Rahman, M.S., Nakano, S., Nishizeki, T.: A Linear Algorithm for Bend-Optimal Orthogonal Drawings of Triconnected Cubic Plane Graphs. Journal of Graph Algorithms and Applications 3, 31–62 (1999)
10. Rahman, M.S., Nishizeki, T.: Orthogonal Drawings of Plane Graphs Without Bends. Journal of Graph Algorithms and Applications 7, 335–362 (2003)
11. Sun, Y., Sarrafzadeh, M.: Floorplanning by Graph Dualization: L-shaped Modules. Algorithmica 10, 429–456 (1993)
12. Tamassia, R.: On Embedding a Graph in the Grid with the Minimum Number of Bends. SIAM Journal on Computing 16, 421–444 (1987)
13. Thomassen, C.: Plane Representations of Graphs. In: Bondy, J.A., Murty, U.S.R. (eds.) Progress in Graph Theory, pp. 43–69. Academic Press, Canada (1984)
14. Yeap, K., Sarrafzadeh, M.: Floor-Planning by Graph Dualization: 2-Concave Rectilinear Modules. SIAM Journal on Computing 22, 500–526 (1993)

Planar and Plane Slope Number of Partial 2-Trees

William Lenhart[1], Giuseppe Liotta[2],
Debajyoti Mondal[3], and Rahnuma Islam Nishat[4]

[1]Department of Computer Science, Williams College, MA, USA
[2]Dipartimento di Ingegneria Elettronica e dell'Informazione,
Universita' degli Studi di Perugia, Italy
[3]Department of Computer Science, University of Manitoba, Canada
[4]Department of Computer Science, University of Victoria, Canada
lenhart@cs.williams.edu, liotta@diei.unipg.it,
jyoti@cs.umanitoba.ca, rnishat@uvic.ca

Abstract. We prove tight bounds (up to a small multiplicative or additive constant) for the plane and the planar slope numbers of partial 2-trees of bounded degree. As a byproduct of our techniques, we answer a long standing question by Garg and Tamassia about the angular resolution of the planar straight-line drawings of series-parallel graphs of bounded degree.

1 Introduction

A *drawing* of a graph G in $\mathbb{R}^2$ maps each vertex of G to a point and each edge of G to a Jordan arc such that an edge does not contain a vertex other than its endpoints, no edge crosses itself, edges do not meet tangentially, and edges sharing a common end-vertex do not cross each other. A *planar graph* is a graph that admits a *planar drawing*, i.e. a drawing such that no two edges intersect except at their common end-points. A *plane graph* is a planar graph together with a combinatorial embedding, i.e. a prescribed set of faces including a prescribed outer face. A *plane drawing* of a plane graph G is a planar drawing that realizes the combinatorial embedding of G.

The *slope number* of a straight-line drawing Γ of a planar graph G is the number of distinct slopes of the edges of Γ. Every plane (planar) graph admits a plane (planar) straight-line drawing [1], i.e. a drawing where the edges are mapped to straight line segments. The *planar slope number* of G is the smallest slope number over all planar straight-line drawings of G. If G is a plane graph, the *plane slope number* of G is the smallest slope number over all plane straight-line drawings of G.

The problem of computing drawings of planar graphs with maximum degree four, using only horizontal and vertical slopes, has long been studied in graph drawing through the research on orthogonal and rectilinear graph drawing (see, e.g., [1]). In a seminal paper, Dujmović et al. [2] extend this study to non-orthogonal slopes, and give tight upper and lower bounds (expressed as functions of the number n of vertices) on the plane slope numbers of several graph families including plane 3-trees and plane 3-connected graphs. They also ask whether the plane slope number of a plane graph of maximum degree Δ can be bounded by a function $f(\Delta)$. Keszegh et al. [7] answer the question affirmatively proving that, for a suitable constant c, the plane slope number of a plane

S. Wismath and A. Wolff (Eds.): GD 2013, LNCS 8242, pp. 412–423, 2013.

graph of bounded degree Δ is at most $O(c^\Delta)$. In the same paper, Keszegh et al. establish a $3\Delta - 6$ lower bound for the plane slope number of the plane graphs of maximum degree at most Δ, which motivates additional research on reducing the gap between upper and lower bound. The question is studied by Jelínek et al. [5] who prove that the plane slope number of plane partial 3-trees is $O(\Delta^5)$. Also Kant, Dujmović et al., Mondal et al. independently show that the plane slope number of cubic 3-connected plane graphs is six [2,6,9], whereas the slope number (i.e., when the drawings may contain edge crossings) of cubic graphs is four [10].

In this paper we prove tight bounds (up to a small multiplicative or additive constant) for the plane and the planar slope numbers of planar 2-trees of bounded degree. Our results extend previous papers concerning the planar and plane slope numbers of proper subfamilies of the partial 2-trees. Namely, Jelínek et al. [5] prove that the planar slope number of series-parallel graphs with maximum degree three is at most three. Knauer et al. [8] show that the plane slope number of outerplane graphs with maximum degree $\Delta \geq 4$ is at most $\Delta - 1$ and that $\Delta - 1$ slopes are sometimes necessary. As a byproduct of our techniques, we answer a long standing open problem by Garg and Tamassia [3], who ask whether $\Omega(\frac{1}{\Delta^2})$ is a tight lower bound on the *angular resolution* of series-parallel graphs of degree Δ (i.e. they ask whether these graphs admit planar straight-line drawings where minimum angle between any two consecutive edges is $\Omega(\frac{1}{\Delta^2})$). More precisely, our results can be listed as follows.

- We prove that the planar slope number of a partial 2-tree of maximum degree Δ is at most 2Δ and there exist partial 2-trees whose planar slope number is at least Δ if Δ is odd and at least $\Delta + 1$ if Δ is even (Section 3).
- We prove that the plane slope number of a plane partial 2-tree of maximum degree Δ is at most 3Δ and there exist plane 2-trees whose plane slope number is at least $3\Delta - 3$ if Δ is even and at least $3\Delta - 4$ if Δ is odd (Section 4).
- We show that a partial 2-tree G of maximum degree Δ admits a planar straight-line drawing with angular resolution $\frac{\pi}{2\Delta}$. If G is a plane graph, a plane straight-line drawing of G exists whose angular resolution is $\frac{\pi}{3\Delta}$ (Section 5). The previously best known bound was $\frac{1}{48\pi\Delta^2}$, established by varying the input embedding [3].

2 Decomposition Trees and Universal Slope Sets

In this section we recall some known concepts. Throughout the paper "drawing" means "planar straight-line drawing"; "plane drawing" means "plane straight-line drawing".

SPQ Trees and Block-cut Vertex Trees. Let G be a 2-connected graph. A *separation pair* is a pair of vertices whose removal disconnects G. A *split pair* of G is either a separation pair or a pair of adjacent vertices. A *split component* of a split pair $\{u, v\}$ is either an edge (u, v) or a maximal subgraph G_{uv} of G such that $\{u, v\}$ is not a split pair of G_{uv}. We call vertices u and v the *poles* of G_{uv}. Note that a split component of G need no t be 2-connected.

A *2-connected series-parallel graph* is recursively defined as follows. A simple cycle with three edges is a 2-connected series-parallel graph. The graph obtained by replacing an edge of a 2-connected series-parallel graph with a path is a 2-connected series-parallel graph. The graph obtained by adding an edge between the vertices of a non-adjacent separation pair $\{u, v\}$ of a 2-connected series-parallel graph is a 2-connected series-parallel graph. Let G be a 2-connected series-parallel graph. An SPQ-tree T of G is a rooted tree describing a recursive decomposition of G into its split components. The nodes of T are of three types: S, P, or Q. Each node μ has an associated graph called the *skeleton of* μ and denoted by $skeleton(\mu)$. Starting from a split pair $\{s, t\}$ of G, T recursively describes the split components of G as follows. The root of T is a P-node corresponding to G; its skeleton is defined as in the "Parallel case" below.

- *Base case:* The split component H is an edge. Then H corresponds to a Q-node of T whose skeleton is this edge. The Q-nodes are the leaves of T.
- *Series case:* The split component H is a 1-connected graph with split components $H_1, \ldots H_k$ ($k \geq 2$) and cut vertices $c_i = H_i \cap H_{i+1}$. Then H corresponds to an S-node μ of T. The graph $skeleton(\mu)$ is a chain $e_1, \ldots, e_k$ of edges such that $e_i = (c_{i-1}, c_i)$, where $c_0 = s$ and $c_k = t$. The children of μ are the roots of the SPQ-trees of $H_1, \ldots, H_k$.
- *Parallel case:* Otherwise, the split component H is 2-connected and its split components are $H_1, \ldots, H_k (k \geq 2)$. Then H has $skeleton(\mu)$ consisting of a set of parallel edges $e_1, \ldots, e_k$ between s and t, one for each H_i. The children of μ are the roots of the SPQ-trees of $H_1, \ldots, H_k$.

Figure 1(a) and (b) show a 2-connected series-parallel graph and its SPQ-tree, which is uniquely determined by the choice of the initial split pair. Note that no P-node (S-node) has a P-node (S-node) as a child. Let T be an SPQ-tree of a 2-connected series-parallel graph G and let μ be a node of T. The *pertinent graph* of μ is the subgraph of G whose SPQ-tree is the subtree of T rooted at μ, as shown in Figure 1(c). The *virtual edge* of μ is an edge in the skeleton of the parent of μ that represents the pertinent graph of μ. Hence for every internal (i.e., non-Q) node μ in T, each edge in $skeleton(\mu)$ is a virtual edge of some child of μ.

If μ is P-node, then we associate with μ another graph $frame(\mu)$, called the *frame of* μ, which is formed by replacing each edge e in $skeleton(\mu)$ with the skeleton of the child node whose virtual edge is e, as shown in Figure 1(d). Every vertex in a frame corresponds to a unique vertex of G. Given a vertex v of G, the *first frame of* v is the frame that is closest to the root of T and contains v. For any split pair $\{u, v\}$ in a 2-connected series-parallel graph G with n vertices, an SPQ-tree having $\{u, v\}$ as reference pair can be computed in $O(n)$ time [4].

A graph G is a *partial 2-tree* (or, has tree-width at most 2) if and only if each 2-connected component of G is either series-parallel or consists of a single edge. Let G be a 1-connected graph. The *block-cut vertex tree of* G, denoted by BC-tree, is a graph with vertex set $B \cup C$ such that B consists of one vertex for each *block* (maximal 2-connected subgraph) of G and C consists of one vertex for each cut vertex of G. There is an edge from $b \in B$ to $c \in C$ in the BC-tree if and only if the vertex of G represented by c belongs to the block represented by b.

Universal Slope Sets and Free Wedges. Let G be a graph with vertex set V. For a vertex $v \in V$, we denote the degree of v by $\delta(v)$. Hence the *maximum degree of* G is

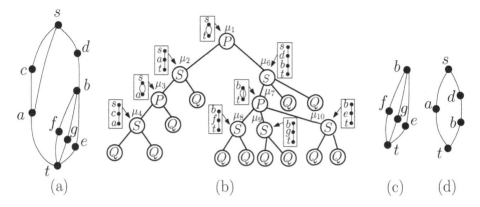

Fig. 1. (a) A 2-connected series-parallel graph G. (b) A SPQ-tree of G, where the internal nodes are labeled with $\mu_1, \ldots, \mu_{10}$. The skeleton of $\mu_i, 1 \leq i \leq 10$, is drawn in the box associated with μ_i. (c) The pertinent graph of μ_7. (d) The frame graph of μ_1.

$\Delta = \max_{v \in V} \delta(v)$. The *excess of* v is $\epsilon(v) = \Delta - \delta(v)$. If H is a subgraph of G, $\delta_H(v)$ and $\epsilon_H(v)$ are the restrictions of $\delta(v)$ and $\epsilon(v)$ to H.

A *set S of slopes is universal* for a family of planar graphs $\mathcal{G}$ if every graph $G \in \mathcal{G}$ admits a planar straight-line drawing such that the slope of every edge in the drawing belongs to S. We consider universal slope sets defined as follows.

Definition 1. *Given a positive integer $k \geq 2$, let $\alpha = \frac{\pi}{2k}$. Define S_k to be the set of slopes $i \cdot \alpha$, for $0 \leq i \leq 2k - 1$.*

We prove the upper bounds on plane and planar slope numbers showing that there is a value of k depending on Δ such that S_k is universal for the partial 2-trees. In our constructions, we guarantee that some wedge shaped regions of the plane can be used for recursive drawing. In particular, for any $r > 0$, point $p \in \mathbb{R}^2$, and angle ϕ, a ϕ-*wedge at p of radius r* is a sector of angular measure ϕ in the disk of radius r centered at p. For convenience, we will often omit reference to r, since any suitably small value of r suffices for our purposes. Let v be a vertex in some planar straight-line drawing Γ. A wedge with its apex at v in Γ is a *free wedge at v* if the wedge intersects the drawing Γ only at v. The angular measure of our free wedges will depend on the degree and excess of the corresponding vertices.

3 Slope Number of Partial 2-trees

In this section we present upper and lower bounds on the planar slope number of partial 2-trees of maximum degree Δ. We start by studying 2-connected series-parallel graphs (Section 3.1) and then we extend the study to all partial 2-trees (Section 3.2).

3.1 2-connected Series-parallel Graphs

In this section we show that S_Δ is a universal slope set for the family of 2-connected series-parallel graphs with maximum degree Δ.

Lemma 1. *Let $\mathcal{G}$ be the family of 2-connected series-parallel graphs having maximum degree at most Δ. Then S_Δ is universal for $\mathcal{G}$.*

Proof. The argument is based on a construction that recursively computes a drawing of G; the proof is by induction on the number of P-nodes in an SPQ-tree of G. Let T be an SPQ-tree of G having split pair $\{s, t\}$ associated with its root and let $m \in S_\Delta$. Since G is 2-connected, T must have at least one P-node, e.g., the root of T. We show that G admits a drawing Γ using only slopes from S_Δ that satisfies the following properties.

(1) Graph G is drawn within a triangle $\triangle abc$ having $\angle bac = (\delta(s) - 0.5)\alpha$ and $\angle abc = (\delta(t) - 0.5)\alpha$ (i.e. every edge is either in the interior or on the boundary of $\triangle abc$).
(2) Vertices s and t are located at a and b, respectively.
(3) Segment $\overline{ab}$ has slope m.
(4) The edges incident to s are drawn using consecutive slopes of S_Δ, as are the edges incident to t.
(5) At each vertex $v \notin \{s, t\}$ in the drawing of G, there is a free $\epsilon(v)\alpha$-wedge at v contained in $\triangle abc$.

Let a and b be two distinct points on a line of slope $m \in S_\Delta$, and let c be the point of intersection of the lines through a and b having slopes $m + (\delta(s) - 0.5)\alpha$ and $m - (\delta(t) - 0.5)\alpha$, respectively.

Base Case: Assume that T has a single P-node, which must be the root of T since G is 2-connected. The frame of the P-node consists of a set of paths of length at least 1 (and at most one path of length 1). We draw s and t on a and b, respectively. We draw one of these paths between s and t along the segment ab with slope m (if there is a path of length one, then we draw that path along ab; otherwise, any of the paths can be used). The remaining paths are drawn inside $\triangle abc$ using slopes $m + i\alpha, i = 1, \ldots, \delta(s) - 1$ at a and slopes $m - i\alpha, i = 1, \ldots, \delta(t) - 1$ at b for the edges incident to s and t, respectively; we use slope m for all other edges. See Figure 2 (a) for an example.

Let Γ be the computed drawing; by construction, Γ is crossing-free and it only uses slopes from S_Δ. Also, the paths from s to t lie within $\triangle abc$ and for each vertex $v \notin \{s, t\}$ there is an empty wedge of angle at least $\epsilon(v)\alpha$ with its apex at v that is completely contained within $\triangle abc$; this wedge is in the "c side" of the path containing v, i.e., in the half-plane defined by the line through a, b and containing c. Hence, Γ satisfies all invariant Properties (1)-(5).

Induction Step: Suppose now that any 2-connected series-parallel graph having at most j P-nodes in some SPQ-tree admits a drawing that only uses slopes from S_Δ and that satisfies Properties (1)-(5). Let G be a 2-connected series-parallel graph having $j + 1$ P-nodes in some SPQ-tree T. As above, the root of T is a P-node and its frame consists of a set of paths $\Pi_1, \ldots, \Pi_k$ of length at least 1. We will draw them in a fashion similar to the base case but with one important difference: we do not use consecutive slopes for the edges of the paths incident to s and t, but we leave room for the (recursive) drawings of the pertinent graphs associated with each virtual edge incident to s or t.

To do this, for each $i = 1, \ldots, k$, let e_i be the virtual edge incident to s in Π_i and let μ_{e_i} be the node of T corresponding to the virtual edge e_i (note that μ_{e_i} is either a P- or Q-node of T). Further, let $\delta_{e_i}(s)$ be the degree of s in the pertinent graph of μ_{e_i}. Then e_1 is drawn using slope m, and for each $i > 1$, e_i is drawn using

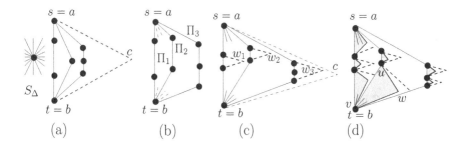

Fig. 2. (a) An example drawing for Base case, where $\delta(s) = \delta(t) = 3$ and $\Delta = 4$. (b) Illustration for Inductive step, where $\delta_{e_1}(s) = 3$, $\delta_{e_2}(s) = 1$, $\delta_{e_1}(t) = 2$, and $\delta_{e_2}(t) = 3$. (c) Illustration for $u_i v_i w_i$ in dashed line, and (d) recursive construction.

slope $slope(e_{i-1}) + \delta_{e_{i-1}}(s)\alpha$. The edges of Π_i incident to t are positioned similarly, beginning with slope m but decreasing the slopes as we move from one path to the next. See Figure 2 (b) for an example.

For every subpath $\Pi_i \setminus \{s,t\}$ with at least two vertices ($1 \leq i \leq k$), we draw the subpath using a sufficiently small line segment as follows. Let u_i, v_i be the endvertices of that subpath $\Pi_i \setminus \{s,t\}$. Let w_i be the point of intersection of the half-lines from u_i and v_i having slopes $slope(u_i v_i) + (\Delta - 0.5)\alpha$ and $slope(u_i v_i) - (\Delta - 0.5)\alpha$, respectively. We draw the paths such that $\triangle u_i v_i w_i$ lies in the region bounded by Π_i and Π_{i+1}. If $i + 1 < k$, then $u_i v_i w_i$ lies within $\triangle abc$. An example is illustrated in Figure 2(c) in dashed line.

Now that the frame of the root of T has been drawn, let $e = uv$ be a (drawn) virtual edge of a path Π_i of the frame and let μ_e be its corresponding P- or Q-node in T. Let w be the point of intersection of the lines from u and v having slopes $slope(e) + (\delta_e(u) - 0.5)\alpha$ and $slope(e) - (\delta_e(v) - 0.5)\alpha$, respectively. We recursively draw the pertinent graph of μ_e within $\triangle uvw$. If $\{u,v\} \cap \{s,t\} = \phi$, then $\triangle uvw$ is contained in $\triangle u_i v_i w_i$, which by construction does not intersect any part of the already drawn edges. If e is incident to s or t (i.e. either $u = s$ or $v = t$), then $\triangle uvw$ does not intersect the edge e' of Π_{i+1} incident to s or t, because by construction, the slope of e' is $slope(e) + \delta_e(s)\alpha$. Finally, observe that $\triangle uvw$ does not intersect any other triangle that contains the drawing of a pertinent graph associated with a node of T which is not in the subtree rooted at μ_e. See Figure 2(d) for an illustration.

The observations above, together with the fact that the drawing of the frame graph of the root is crossing-free and that it satisfies Properties (1)-(4), imply that G admits a drawing that only uses slopes from S_Δ satisfying Properties (1)-(4). To see that Property (5) is also satisfied, note that each path Π_i is drawn as a convex (or linear, for $i = 1$) chain. Thus at each vertex $v \notin \{s,t\}$ has an angle of at least π between its two consecutive (virtual) edges in its first frame. The drawing of G uses two consecutive sets of slopes at v, since the pertinent graph for the two nodes corresponding to each of those two virtual edges is drawn using consecutive slopes. This leaves a free wedge of angular measure $\pi - \delta(v)\alpha = 2\Delta\alpha - \delta(v)\alpha = \Delta\alpha + \epsilon(v)\alpha > \epsilon(v)\alpha$ at v, establishing Property (5). □

3.2 Partial 2-trees

In this section we extend the result of Lemma 1 to partial 2-trees of maximum degree Δ by proving that S_Δ is universal for these graphs. We shall focus on connected partial 2-trees, since every connected components can be drawn independently of the others.

Lemma 2. *Let $\mathcal{G}$ be the family of partial 2-trees having maximum degree at most Δ. Then S_Δ is universal for $\mathcal{G}$.*

Proof. We assume that G is connected and has at least one edge. Let T be the block-cutvertex tree of G (see Figures 3(a)–(b)). We build the desired drawing of G by drawing subgraphs of G corresponding to subtrees of T, starting with the leaves of T. The drawing of G produced will have the following properties:

(a) For some split pair $\{s, t\}$ of G, G is drawn inside a $\delta(s)\alpha$-wedge with apex s.
(b) s and t are located on a line of slope $m \in S_\Delta$.
(c) The edges incident to s are drawn using consecutive slopes of S_Δ, as are the edges incident to t.
(d) The wedge of Property (a) is bounded by rays from s in directions $m - 0.5\alpha$ and $m + (\delta(s) - 0.5)\alpha$.

If G is 2-connected, then Lemma 1 establishes the existence of a drawing with the desired properties: Property (b) follows obviously from Property (1) of Lemma 1, Property (c) from Property (3) of Lemma 1, and properties (a) and (d) from properties $(1), (2)$ and (4) of Lemma 1.

Otherwise, let s be a cut vertex of G and choose s as the root of T. The parent of each block vertex B of T is a cut vertex s_B of G. For each such block B, choose a vertex t_B in B such that $\{s_B, t_B\}$ is a split pair for B. Each leaf of T is a block vertex B and by Lemma 1, it can be drawn with split pair $\{s_B, t_B\}$ inside a wedge with apex s_B and angular measure $\delta_B(s_B)\alpha$ so that properties (a)–(d) hold.

Assume now that x is a vertex of T for which all subgraphs of T have been drawn respecting properties (a)–(d). Then x represents either a block of G or a cut vertex of G. If x represents a cut vertex v of G, then let $T_1, \ldots, T_k$ represent the subtrees of x. Let G_v be the subgraph of G corresponding to the subtree of T with root v. For each of the trees T_i, v is at the apex of the $\delta_{T_i}(v)\alpha$-wedge in which the subgraph G_i of G corresponding to T_i has been drawn. These wedges can all be rotated about v so that they use consecutive slopes in S_Δ, as shown in Figure 3(c). Thus the subgraph of G corresponding to the union of $T_1, \ldots, T_k$ has been drawn in a wedge of angle $\delta_{G_1}(v)\alpha + \ldots + \delta_{G_k}(v)\alpha = \delta_{G_v}(v)\alpha$ with apex v such that properties (a)–(d) are satisfied. Note that by Property (a), the drawing of G_v can be made small enough to lie completely inside the free wedge and so does not intersect any other portion of the drawing.

Suppose now that x represents a block B of G. The children of B in T represent the cut vertices of G that belong to B; let v be one of the child cut vertices of B. Draw B in the manner described by Lemma 1. We consider two cases: $v \neq t_B$ and $v = t_B$ (note that s_B is the parent cut vertex of B, which is handled by the previous case).

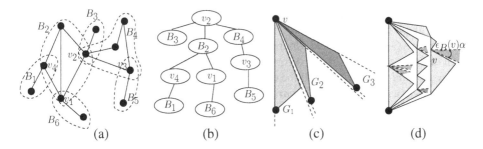

Fig. 3. (a) A 1-connected graph, and (b) corresponding block-cut vertex tree. (c)–(d) Illustration for the proof of Lemma 2.

Assume first that $v \neq t_B$. Then by Lemma 1, there is a free $\epsilon_B(v)\alpha$-wedge with apex v. Now $\epsilon_B(v)\alpha = (\Delta - \delta_B(v))\alpha > (\delta_G(v) - \delta_B(v))\alpha = \delta_{G_v}(v)\alpha$ and so G_v can be drawn completely inside the free $\epsilon_B(v)\alpha$-wedge with apex v as shown in Figure 3(d).

If $v = t_B$, then the drawing of G_v already produced can be rotated so that all of the edges in B and of G_v adjacent to t_B use consecutive (clockwise from the line containing s_B and t_B) directions in S_Δ as shown in Figure 3(d). It is an easy observation that the addition of the subgraphs G_v for each cut vertex of B into the free wedge preserves properties (a)–(d) of the drawing of B (and of the drawing of G). □

Lemma 3. *For any $\Delta > 3$, there exists a 2-connected series-parallel graph G of maximum degree Δ whose planar slope number is at least $\Delta + 1$ if Δ is even and at least Δ if Δ is odd.*

Proof. Consider the graph G obtained from $K_{2,\Delta-1}$ by adding the edge (u, v) connecting the two vertices u and v of degree $\Delta - 1$. Thus G has maximum degree Δ. Now consider any drawing of G. At least half of the remaining $\Delta - 1$ vertices are on one side of the line determined by the segment representing uv in the drawing. Each of these $\lfloor \Delta/2 \rfloor$ vertices forms a triangle with uv and these triangles are nested. Thus no two of the $2\lfloor \Delta/2 \rfloor + 1$ edges in this portion of the drawing of G have the same slope. □

The following theorem is an immediate consequence of Lemmas 2, and 3.

Theorem 1. *Let G be a partial 2-tree having maximum degree Δ and let $psl(G)$ denote the planar slope number of G. Then $psl(G) \leq 2\Delta$. Also, for every even $\Delta > 3$ there exists a partial 2-tree G such that $psl(G) \geq \Delta + 1$ and for every odd $\Delta \geq 3$ there exists a partial 2-tree G such that $psl(G) \geq \Delta$.*

4 Plane Slope Number of Partial 2-trees

In this section we show that the plane slope number of 2-connected series-parallel graphs, i.e., when the output drawings respect the input embeddings, is at least $3\Delta - 4$ and at most 3Δ. In fact, we show that $S_{1.5\Delta}$ is universal for the family of 2-connected series-parallel graphs with fixed embeddings.

We introduce some additional notation. For an embedded planar 2-connected series-parallel graph G with poles s and t, we call the edge (s, t) (if exists) the *central edge* of G. Since G is 2-connected, the root of its SPQ-tree is a P-node μ. Observe that each of the edges in $skeleton(\mu)$ corresponds to either the edge (s, t) or a *2-connected series-parallel subgraph of G*. If the edge (s, t) exists, then we categorize each of those subgraphs as a *left or right series-parallel subgraph* of G depending on whether it lies to the left or right of the edge (s, t), while walking from s to t. By $G_1^-, \ldots, G_l^-$, (respectively, $G_1^+, \ldots, G_r^+$) we denote the left (respectively, right) series-parallel subgraphs of G. If (s, t) does not exist, then we assume that all the series-parallel subgraphs are right series-parallel, i.e., $G_1^+, \ldots, G_r^+$. Furthermore, we assume that the subgraphs are ordered as follows: $G_l^-, \ldots, G_1^-, (s, t), G_1^+, \ldots, G_r^+$, reflecting their left to right ordering in the embedding.

Lemma 4. *Let $\mathcal{G}$ be the family of plane 2-connected series-parallel graphs of maximum degree at most Δ. Then $S_{1.5\Delta}$ is universal for $\mathcal{G}$.*

Sketch of Proof: Similar to Lemma 1 we employ an induction on the number of P-nodes in an SPQ-tree T of G. Since the proof follows a similar argument, for reasons of space we sketch here the main idea of the proof. Let $\{s, t\}$ be the split pair associated with the root of T, and let $\overline{ab}$ be a straight line segment with slope m, where $m \in S_{1.5\Delta}$. We show that G admits a drawing Γ using only slopes from $S_{1.5\Delta}$ such that the following properties hold.

(1) Graph G is drawn within a convex quadrilateral $\square adbc$ having $\angle dac = \delta(s)\alpha$ and $\angle dbc = \delta(t)\alpha$ (i.e. every edge is in the interior of $\square adbc$).
(2) Vertices s and t are located at a and b, respectively.
(3) Segment $\overline{ab}$ has slope m.
(4) The edges incident to s are drawn using consecutive slopes of $S_{1.5\Delta}$, as are the edges incident to t.
(5) At each vertex $v \notin \{s, t\}$ in the drawing of G, there are two free $\epsilon(v)\alpha$-wedges at v contained in $\square adbc$, one in the region between the subgraph G_i^+ (or G_i^-) containing v and the previous series-parallel subgraph of G in the ordering, and one in the region between G_i^+ (or G_i^-) and the next series-parallel subgraph of G in the ordering.

Base Case: Assume that T has a single P-node, which must be the root of T since G is 2-connected. We draw s and t on a and b, respectively. The frame of the P-node consists of a set of paths of length at least one (and at most one path of length one). If the central edge exists, then we draw that edge along $\overline{ab}$ (otherwise, we draw the leftmost path along $\overline{ab}$). We then draw the paths $\Pi_i^+, i = 1, \ldots, r$ corresponding to each G_i^+ between a and b using consecutive slopes at s and t, as in the proof of Lemma 1. All the remaining paths $\Pi_i^-, i = 1, \ldots, l$ corresponding to G_i^- are drawn symmetrically to the left of segment $\overline{ab}$, as shown in Figure 4(a). While drawing the paths, we maintain the input embedding. To construct the quadrilateral $\square adbc$, let d be the intersection of the line through a having slope 0.5α plus the slope of the edge of Π_r^+ incident to s with the line through b having slope -0.5α plus the slope of the edge of Π_r^+ incident to t. Similarly, let c be the intersection of the line through a having slope -0.5α plus the

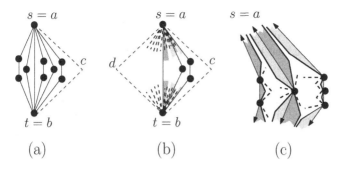

Fig. 4. (a) Base case. (b) Drawing frame, where the light-gray (respectively, dark-gray) regions correspond to the right (respectively, left) series-parallel subgraphs of the corresponding pertinent graph. (c) Recursive construction.

slope of the edge of Π_l^- incident to s with the line through b having slope 0.5α plus the slope of the edge of Π_l^- incident to t. This convex quadrilateral has angle $\delta(s)\alpha$ at a and $\delta(t)\alpha$ at b. See Figure 4 (a) for an example.

It is now straightforward to observe that the resulting drawing is planar and satisfies properties (1)-(4). As for Property (5), consider $v \notin \{s,t\}$ in one of the frame paths and its neighbors u and w on that path. Each of the two angles $\angle uvw$ is either non-acute or at least $\pi - (\delta(u) - 1)\alpha - (\delta(w) - 1)\alpha \geq (\Delta + 2)\alpha \geq \epsilon(v)\alpha$, and so the required empty wedges exist at v.

Induction Step: In a way similar to the proof of Lemma 1, we first draw the frame of the root of the SPQ-tree and then define disjoint convex quadrilaterals for each of the virtual edges. Finally, we recursively compute the drawings of the pertinent graphs inside the corresponding quadrilaterals. The idea is illustrated in Figures 4(b)-(c). □

By Property (5) of Lemma 4, for every vertex $v \notin \{s,t\}$ in G, there is a free wedge on each side of the drawing of the pertinent subgraph G_i^+ (G_i^-) of G containing v. Thus the arguments used in the proof of Lemma 2 can be directly applied in the fixed embedding case to establish the following lemma.

Lemma 5. *Let $\mathcal{G}$ be the family of plane partial 2-trees of maximum degree at most Δ. Then $S_{1.5\Delta}$ is universal for $\mathcal{G}$.*

We observe that the $\Delta - 1$ lower bound proved by Knauer et al. [8] for the outerplane slope number implies a lower bound for plane slope number of plane partial 2-trees, because outerplane graphs are plane partial 2-trees. The next lemma shows a better lower bound for partial 2-trees.

Lemma 6. *For every $\Delta \geq 2$, there exists a plane 2-connected series-parallel graph G of maximum degree Δ whose plane slope number is at least $3\Delta - 3$ if Δ is even and at least $3\Delta - 4$ if Δ is odd.*

Proof. Suppose first that $\Delta \geq 2$ is an even number and consider a plane partial 2-tree G defined as follows. The external face of G is a 3-cycle with vertices a, b, c. In its interior there are $\Delta/2 - 1$ paths of length two connecting each pair from $\{a, b, c\}$.

The external face of every plane drawing Γ of G is a triangle $\triangle abc$ that contains the paths of length two in its interior. From elementary geometry, no two edges of Γ can have common slope, and hence the graph, which has $3\Delta - 3$ edges, has plane slope number at least $3\Delta - 3$. Suppose now that Δ is odd and let $\Delta' = \Delta + 1$. Construct a graph G' of maximum degree Δ' as described above. Now remove one of those paths of length 2 connecting a and b and one connecting c and b from G'. This new graph, G, has maximum degree $\Delta = \Delta' - 1$ and it requires a different slope for each edge. Since we deleted four edges from G', G has $(3\Delta' - 3) - 4 = 3\Delta - 4$ edges. Thus $3\Delta - 4$ slopes are required, and the result is established. □

Lemmas 4 and 6 imply the following.

Theorem 2. *Let G be a plane partial 2-tree having maximum degree Δ and let $epsl(G)$ denote the plane slope number of G. Then $epsl(G) \leq 3\Delta$. Also, for every even $\Delta > 3$ there exists a plane partial 2-tree G such that $epsl(G) \geq 3\Delta - 3$ and for every odd $\Delta \geq 3$ there exists a plane partial 2-trees G such that $epsl(G) \geq 3\Delta - 4$.*

5   Angular Resolution

The *angular resolution* of a planar straight-line drawing is the minimum angle between any two edges incident to a common vertex. The angular resolution of a planar graph G is the maximum angular resolution over all possible drawings of G.

Malitz and Papakostas [11] show that the angular resolution of a planar graph of maximum degree Δ is $\Omega(\frac{1}{7^\Delta})$. Garg and Tamassia [3] show that there exist planar 3-trees of maximum degree Δ that require angular resolution $O(\sqrt{\frac{\log \Delta}{\Delta^3}})$. They also show that for a subfamily of the partial 2-trees of bounded degree, namely the series-parallel graphs of maximum degree at most Δ, the angular resolution is at least $\frac{1}{48\pi\Delta^2}$. Their drawing technique does not apply to plane graphs since it may vary a given combinatorial embedding. Garg and Tamassia leave as open the problem about whether $\Omega(\frac{1}{\Delta^2})$ is a tight lower bound for the angular resolution of the series-parallel graphs.

An implication of the drawing techniques of the previous sections of this paper is that the angular resolution of partial 2-trees (and thus also of series-parallel graphs) is in fact $\Omega(\frac{1}{\Delta})$. Namely, the constructions of Lemmas 2 and 4 either use the universal set S_Δ or the universal set $S_{1.5\Delta}$ which consist of equally spaced slopes; therefore, the minimum angle between any two edges sharing a common end-vertex is either $\frac{\pi}{2\Delta}$ in the variable embedding setting or it is $\frac{\pi}{3\Delta}$ in the fixed embedding setting. Therefore:

Theorem 3. *A partial 2-tree of maximum degree Δ admits a planar straight-line drawing with angular resolution $\frac{\pi}{2\Delta}$. A plane partial 2-tree of maximum degree Δ admits a planar straight-line drawing with angular resolution $\frac{\pi}{2\Delta}$.*

6   Open Problems

An interesting research direction is to study the trade-off between the slope number and the area requirement of planar graphs. Similar studies have been carried out for the angular resolution and the area requirements of planar graphs having maximum degree at most Δ (see, e.g., [3]).

Another fascinating open problem is to close the gap between upper and lower bounds on the planar/plane slope number of planar/plane graphs of bounded degree (see [7]). This would be interesting even restricted to partial 3-trees (see [5]).

Acknowledgments. Part of this research took place during the 12th INRIA-McGill-Victoria Workshop on Computational Geometry, held in February 2–8, 2013, at the Bellairs Research Institute of McGill University. We thank the organizers and the participants for the opportunity and the useful discussions. Special thanks go to Zahed Rahmati for early thoughtful insights on the problems of this paper.

References

1. Di Battista, G., Eades, P., Tamassia, R., Tollis, I.G.: Graph Drawing. Prentice- Hall, Englewood Cliffs (1999)
2. Dujmović, V., Eppstein, D., Suderman, M., Wood, D.R.: Drawings of planar graphs with few slopes and segments. Computational Geometry 38(3), 194–212 (2007)
3. Garg, A., Tamassia, R.: Planar drawings and angular resolution: Algorithms and bounds. In: van Leeuwen, J. (ed.) ESA 1994. LNCS, vol. 855, pp. 12–23. Springer, Heidelberg (1994)
4. Gutwenger, C., Mutzel, P.: A linear time implementation of SPQR-trees. In: Marks, J. (ed.) GD 2000. LNCS, vol. 1984, pp. 77–90. Springer, Heidelberg (2001)
5. Jelínek, V., Jelínková, E., Kratochvíl, J., Lidický, B., Tesar, M., Vyskocil, T.: The planar slope number of planar partial 3-trees of bounded degree. Graphs and Combinatorics 29(4), 981–1005 (2013)
6. Kant, G.: Hexagonal grid drawings. In: Mayr, E.W. (ed.) WG 1992. LNCS, vol. 657, pp. 263–276. Springer, Heidelberg (1993)
7. Keszegh, B., Pach, J., Pálvölgyi, D.: Drawing planar graphs of bounded degree with few slopes. SIAM J. Discrete Math. 27(2), 1171–1183 (2013)
8. Knauer, K., Micek, P., Walczak, B.: Outerplanar graph drawings with few slopes. In: Gudmundsson, J., Mestre, J., Viglas, T. (eds.) COCOON 2012. LNCS, vol. 7434, pp. 323–334. Springer, Heidelberg (2012)
9. Mondal, D., Nishat, R.I., Biswas, S., Rahman, M.S.: Minimum-segment convex drawings of 3-connected cubic plane graphs. Journal of Combinatorial Optimization 25(3), 460–480 (2013)
10. Mukkamala, P., Pálvölgyi, D.: Drawing cubic graphs with the four basic slopes. In: Speckmann, B. (ed.) GD 2011. LNCS, vol. 7034, pp. 254–265. Springer, Heidelberg (2011)
11. Papakostas, A., Tollis, I.G.: Improved algorithms and bounds for orthogonal drawings. In: Tamassia, R., Tollis, I.G. (eds.) GD 1994. LNCS, vol. 894, pp. 40–51. Springer, Heidelberg (1995)

Slanted Orthogonal Drawings*

Michael A. Bekos[1], Michael Kaufmann[1], Robert Krug[1],
Stefan Näher[2], and Vincenzo Roselli[3]

[1] Institute for Informatics, University of Tübingen, Germany
{bekos,mk,krug}@informatik.uni-tuebingen.de
[2] Institute for Computer Science, University of Trier, Germany
naeher@uni-trier.de
[3] Engineering Department, Roma Tre University, Italy
roselli@dia.uniroma3.it

Abstract. We introduce a new model that we call *slanted orthogonal graph drawing*. While in traditional orthogonal drawings each edge is made of axis-aligned line-segments, in slanted orthogonal drawings intermediate diagonal segments on the edges are also permitted, which allows for: (a) smoothening the bends of the produced drawing (as they are replaced by pairs of "half-bends"), and, (b) emphasizing the crossings of the drawing (as they always appear at the intersection of two diagonal segments). We present an approach to compute bend-optimal slanted orthogonal representations, an efficient heuristic to compute close-to-optimal drawings in terms of the total number of bends using quadratic area, and a corresponding LP formulation, when insisting on bend optimality. On the negative side, we show that bend-optimal slanted orthogonal drawings may require exponential area.

1 Introduction

In this paper, we introduce and study a new model in the context of non-planar orthogonal graph drawing: Given a graph G of max-degree 4, determine a drawing Γ of G in which (a) each vertex occupies a point on the integer grid and has four available ports, as in the ordinary orthogonal model, (b) each edge is drawn as a sequence of horizontal, vertical and diagonal segments, (c) a diagonal segment is never incident to a vertex (due to port constraints mentioned above), (d) crossings always involve diagonal segments, and, (e) the minimum of the angles formed by two consecutive segments of any edge always is 135°. We refer to Γ as the *slanted orthogonal drawing* of G, or, shortly, *slog drawing*. Figs.1(a) and 1(b) indicate what we might expect from the new model: crossings on the diagonals are more visible than in the traditional model and the use of area seems to be more effective.

* Part of the research was conducted in the framework of ESF project 10-EuroGIGA-OP-003 GraDR "Graph Drawings and Representations". The work of M.A. Bekos is implemented within the framework of the Action "Supporting Postdoctoral Researchers" of the Operational Program "Education and Lifelong Learning" (Action's Beneficiary: General Secretariat for Research and Technology), and is co-financed by the European Social Fund (ESF) and the Greek State.

S. Wismath and A. Wolff (Eds.): GD 2013, LNCS 8242, pp. 424–435, 2013.

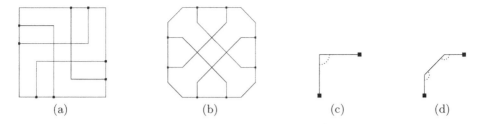

Fig. 1. (a)-(b) Traditional orthogonal and slanted orthogonal drawings of the same graph, assuming fixed ports. (c)-(d) Replacing a 90° bend by two half-bends of 135°.

Orthogonal graph drawing dates back to VLSI layouts and floor-planning applications [6,9,10]. In an *orthogonal drawing* of a graph of max-degree 4, each edge is drawn as a sequence of alternating horizontal and vertical line segments. Typical optimization functions include minimizing the used area [9], the total number of bends [4,8] or the maximum number of bends per edge [1].

Slog drawings are of improved readability and more aesthetic appeal than orthogonal drawings, since bends, which negatively affect the quality of orthogonal drawings are replaced by half-bends that have a smoother shape (see Figs.1(c) and 1(d)). In addition, slog drawings reveal crossings and help distinguishing them from vertices, since crossings are defined by diagonal segments, while vertices are incident to rectilinear segments. Our model resembles an octilinear model as it is heavily used for example in the drawing of Metro Maps [7] but it is closer to the traditional orthogonal style. In particular angles of 45° do not occur at all. So, the complexity results for the octilinear models do not apply.

For minimizing the total number of bends in orthogonal graph drawing Tamassia laid important foundations by the *topology-shape-metrics* (*TSM*) approach in [8], that works in three phases. In the first *planarization* phase a "planar" embedding is computed for a given (non)planar graph by replacing edge crossings by dummy vertices (referred to as *crossing or c-vertices*). The output is called *planar representation*. In the next *orthogonalization* phase, angles and bends of the drawing are computed, producing an *orthogonal representation*. In the third *compaction* phase the coordinates for vertices and edges are computed. The core is a min-cost flow algorithm to minimize the number of bends in the second phase [2]. We will adopt the TSM approach for our model. Note that the general problem of determining a planar embedding with the minimum number of bends is NP-hard [5], which is also the case for slog drawings. Therefore we assume in the following that a planar representation of the input graph is given.

While constructing the slog drawing we observe the following requirements: (a) all non-dummy vertices (referred to as *real or r-vertices*) use orthogonal ports and, (b) all c-vertices use diagonal ports. This ensures that the computed drawing will be a valid slog drawing that corresponds to the initial planar representation. Edges connecting real (crossing) vertices are referred to as rr-edges (cc-edges), and edges between r- and c-vertices as rc-edges.

This paper is structured as follows: In Section 2 we present an approach to compute bend-optimal slog representations. Afterwards, we present a heuristic to compute close-to-optimal slog drawings, that require polynomial drawing area, based on a given slog representation. To compute the optimal drawing, we give a formulation as a linear program in Section 4. In Section 5 we show that the optimal drawing may require exponential area. We conclude in Section 6.

2   A Flow-Based Approach

In this section, we present a modification of an algorithm by Tamassia [8], which we briefly describe in Section 2.1. Section 2.2 explains our modification and in Section 2.3, we present properties of a bend-minimal slog representation.

2.1   Preliminaries

A central notion to the algorithm of Tamassia [8] is the *orthogonal representation*, that captures the "shape" of the drawing without the exact geometry. An orthogonal representation of a plane graph $G = (V, E)$ is an assignment of four labels to each edge $(u, v) \in E$; two for each direction. Label $\alpha(u, v) \cdot 90°$ corresponds to the angle at vertex u formed by (u, v) and its counterclockwise next incident edge. Label $\beta(u, v)$ corresponds to the number of left turns along (u, v), when traversing it from u to v. Clearly, $1 \leq \alpha(u, v) \leq 4$ and $\beta(u, v) \geq 0$. The sum of angles around a vertex equals to $360°$, so for each vertex $u \in V$, $\sum_{(u,v)\in N(u)} \alpha(u, v) = 4$, where $N(u)$ denotes the neighbors of u. Similarly, since the sum of the angles formed at the vertices and at the bends of a bounded face f equals to $180°(p(f) - 2)$, where $p(f)$ denotes the number of such angles, it follows that $\sum_{(u,v)\in E(f)} \alpha(u, v) + \beta(v, u) - \beta(u, v) = 2a(f) - 4$, where $a(f)$ denotes the number of vertex angles in f, and, $E(f)$ the directed arcs of f in its counterclockwise traversal. If f is unbounded, the sum is increased by 8.

In the flow network one can think of each unit of flow as a $90°$ angle. The vertices (vertex-nodes; sources) supply 4 units of flow, and each face (face-nodes; sinks) f demands $2a(f) - 4$ units of flow (plus 8 if f is unbounded). To maintain the properties described above each edge from a vertex-node to a face-node in the flow network has a capacity of 4 and a minimum flow of 1, while an edge between adjacent faces has infinite capacity, no lower bound but each unit of flow through it costs one unit. The total cost is actually the number of bends along the corresponding edge. Hence, the min-cost flow solution corresponds to a representation with the minimum number of bends.

2.2   Modifying the Flow Network

We now modify the algorithm of Tamassia, to obtain a slog representation of a planarized graph G with minimum number of half-bends. Recall that G contains two types of vertices, namely real and crossing vertices. Real (crossing) vertices use orthogonal (diagonal) ports. Observe that a pair of half-bends on an rr- or

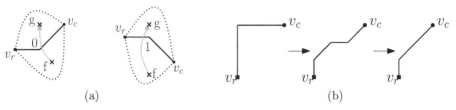

(a)   (b)

Fig. 2. (a) Two configurations corresponding to zero or one unit of flow over an rc-edge (with f and g being the two adjacent faces) (b) The right rotation of a c-vertex v_c can save a half-bend.

cc-edge of a slog drawing corresponds to a bend of an orthogonal drawing. But a rc-edge changes from an orthogonal port (incident to the r-vertex) to a diagonal port (incident to the c-vertex), requiring at least one half-bend. Consider an rc-edge (v_r, v_c) incident to faces f and g (see Fig.2(a)) and assume that the port of the real vertex v_r is fixed. Depending on the rotation of the crossing vertex v_c (clockwise or counterclockwise) we obtain two different representations with the same number of bends. To model this "free-of-cost" choice, we introduce an edge into the flow network connecting f and g with unit capacity and zero cost. For consistency we assume that, if in the solution of the min-cost flow problem there is no flow over (f, g), then there exists a left turn from the real to the crossing vertex; otherwise a right turn, as illustrated in Fig.2(a).

2.3   Properties of Optimal Slanted Orthogonal Representations

In this section we give properties of optimal slog representations. We prove that, for a planarized graph G, the computation of a slog representation with minimum number of half-bends respecting the embedding of G is always feasible. Then, we give an upper bound on the number of half-bends in optimal slog representations.

Theorem 1. *For a planarized graph G with max-degree 4, we can efficiently compute a slog representation with minimum number of half-bends respecting the embedding of G.*

Proof. We use a reduction to Tamassia's network flow algorithm. In particular, since the original flow network computes a (bend-minimal) orthogonal representation for the input plane graph, we will also obtain a slog representation with our modification. We now prove that this representation is also bend-minimal.

Assume that we are given an orthogonal representation F. We can uniquely convert F into a slog representation $S(F)$ by turning all crossing vertices counterclockwise by $45°$. More precisely, the last segment of every rc-edge before the crossing vertex will become a left half-bend. Furthermore, every orthogonal bend is converted into two half-bends, bending in the same direction as the orthogonal bend (see Fig.1(c) and 1(d)). Note that the left half-bends at the crossings might neutralize with one of the half-bends originating from an orthogonal bend, if the orthogonal bend is turning to the right (see Fig.2(b)). In this case, only the second one of the right half-bends remains. Note that this is the only possible saving operation. Therefore, since the number of rc-edges is fixed from the given embedding, a slog representation with minimum number of half-bends should minimize the difference between the number of orthogonal bends of F and the number of first right-bends on rc-edges. However, this is exactly what is done by our min-cost flow network formulation, as the objective is the minimization of the total number of bends in F without the first right-bends on rc-edges.   □

This constructive approach can be reversed so that for each slog representation S, we can get a unique orthogonal representation $F(S)$. Clearly, $F(S(F)) = F$ and $S(F(S)) = S$. Note that this is true only for bend-minimal representations; otherwise we might have staircases of bends which is impossible by min-cost flow computations. From the construction, we can also derive the following.

Corollary 1. *Let $S(F)$ be a slog representation, and F a corresponding orthogonal representation. Let b_S, rb_S and rc_S be the number of half-bends, the number of first right-bends on rc-edges and the number of rc-edges in $S(F)$. Let also b_F be the number of orthogonal bends in F. Then, $b_S = 2 \cdot (b_F - rb_S) + rc_S$.*

The following theorem gives an upper bound for the number of half-bends in optimal slog representations.

Theorem 2. *The number of half-bends of a bend-minimal slog representation is at least twice the number of bends of its related bend-minimal orthogonal representation.*

Proof. Bends of a bend-minimal orthogonal representation correspond to pairs of half-bends on cc and rr edges of a bend-minimal slog representation. In this case, the claim holds with equality. For rc-edges we need a different argument. Let $\mathcal{C}$ be a maximal component spanned by cc-edges. Then, all edges with exactly one endpoint in $\mathcal{C}$ are rc-edges, which can be split into independent cycles around components of crossings. Let C be such a cycle of length k. Clearly, there should be k first half-bends on the rc-edges in the slanted representation. In the corresponding orthogonal representation, the second and third bend of each rc-edge correspond to pairs of half-bends on the same edge in the slog representation. Similarly, in the orthogonal representation the first orthogonal left-bend of each rc-edge corresponds to the second and third left half-bend of the same edge in the slog representation. The only bends that have not been paired (i.e., have no correspondence) are the first right-bends on rc-edges. We claim that in any bend-minimal orthogonal representation, there exist at most $\frac{k}{2}$ first right bends on the edges of C. Assume to the contrary, that in a bend-minimal orthogonal representation, there exist $r > \frac{k}{2}$ first right-bends on the edges of C. If we send the flow along C in reverse direction, we decrease the number of right-bends by r and increase the number of left bends by $k - r$. So, the total number of bends decreases, which shows that the input orthogonal representation was not minimal. From the claim, it follows that the number of first right-bends in the orthogonal representation is at most half of the number of first half-bends of C (in the corresponding slog representation), which concludes the proof since all other half-bends come in pairs and have their correspondences.   □

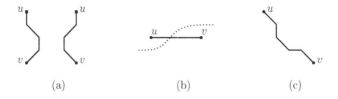

Fig. 3. (a) Spoon gadget for rc-edges. (b) Moving everything above the dotted cut up transforms the orthogonal input to a slanted drawing (c) containing 4 half-bends.

3 A Heuristic to Compute a Close to Optimal Drawing

In this section we present a heuristic which, given an optimal slog representation, computes an actual drawing, which is close to optimal, with respect to the total number of bends, and requires quadratic area. This is quite reasonable, since as we will see in Section 5, insisting in optimal slog drawing may result in exponential area. The basic steps of our approach are given in Algorithm 1. In the following, we describe them in detail.

Algorithm 1. Spoon Based Algorithm

Input: A slanted orthogonal representation F of a given planarized graph G
Output: A slanted orthogonal drawing $\Gamma_s(G)$

S1: Compute an orthogonal drawing $\Gamma_o(G)$ based on F
S2: Replace each orthogonal bend by 2 half-bends   {see Figs.1(c) and 1(d)}
S3: Fix ports on rc-edges by inserting the spoon gadget   {see Fig.3(a)}
S4: Apply cuts to fix ports on cc-edges   {see Figs.3(b) and 3(c)}
S5: Optimize number of rc half-bends   {see Fig.4(a)}
S6: Optimize number of cc half-bends   {see Fig.4(b)}
S7: Compact drawing

In step 1 of Algorithm 1 we compute a bend-minimal orthogonal drawing $\Gamma_o(G)$ from the slog representation. We use the original algorithm of Tamassia [8], ignoring the flow on the additional edges and the rotation of the crossing vertices. In step 2 of Algorithm 1, we replace all orthogonal bends with pairs of half-bends. In step 3 of Algorithm 1, we connect r-vertices with c-vertices by replacing the segment incident to the c-vertex by a gadget called *spoon*, due to its shape (see Fig.3(a)). It allows switching between orthogonal and diagonal ports on an edge. Note that the slog representation specifies how each c-vertex is rotated, thereby defining the configuration it uses.

In step 4 of Algorithm 1, we employ *cuts* (i.e., is a standard technique to perform layout stretching in orthogonal graph drawing; see [3]) to fix the ports of cc-edges, which still use orthogonal ports. To apply a horizontal (vertical) cut, we have to ensure that each edge crossed by the cut has at least one horizontal

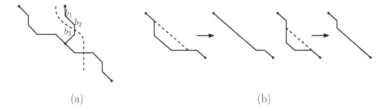

(a)   (b)

Fig. 4. (a) There is always a cut like the dashed line enabling us to move everything on its left side to the left to save half-bends b_1 and b_2. (b) Saving bends on cc-edges.

(vertical) segment. This trivially holds before the introduction of the spoons, as $\Gamma_o(G)$ is an orthogonal drawing. It also holds afterwards since a spoon replacing a horizontal (vertical) segment has two horizontal (vertical) segments. To fix a horizontal cc-edge (u,v) with u being to the left of v in the drawing, we use a "horizontal cut" which from left to right and up to vertex u either (a) lies exactly above u, then crosses edge (u,v) and stays exactly below v, or, (b) lies exactly below u, then crosses edge (u,v) and stays exactly above v (see Fig.3(b)). Our choice depends on the slog representation that specifies the rotation of each c-vertex. The result is depicted in Fig.3(c). Observe that the edge has now a horizontal and a vertical segment. Hence, we can fix all remaining cc-edges. To cope with cc-edges with bends we apply the same technique only to the first and last segments of the edge.

The resulting drawing has 2 additional half-bends on rc-edges (the spoon gadget adds 3 half-bends; one is required) and 4 additional half-bends on cc-edges (none is required), with respect to the ones suggested by the representation. By applying cuts again, we can save 2 half-bends for each rc-edge (see Fig.4(a)), by eliminating the diagonal segment of the spoon gadget (step 5 of Algorithm 1). The rectilinear segments of the edge are not affected, to be able to apply future cuts.

It is always possible to remove two of the half-bends on cc-edges (step 6 of Algorithm 1) by a local modification as depicted in Fig.4(b). If the horizontal part of a cc-edge is longer than the vertical one, a shortcut as in the left part of Fig.4(b) can be applied. If the horizontal and the vertical segments of the cc-edge have the same length all four half-bends can be saved.

After applying this operations, the drawing will contain zero additional half-bends on rr-edges and at most two additional half-bends on cc-edges, with respect to the input representation. Note that to apply our technique we need to scale up the initial drawing by a factor of 4 at the beginning, to provide enough space for additional half-bends. In subsequent steps, we increase the drawing area by cuts. However, we can reduce it by contracting along horizontal and vertical cuts at the end (step 7 of Algorithm 1). After the compaction, each horizontal and vertical grid line will be occupied by at least a half-bend, an edge or a vertex, and since all of those are linear in number, the required area of the final slanted drawing is $O(n^2)$. The following theorem summarizes our approach.

Theorem 3. *Given a slog representation of a planarized graph G with max-degree 4, we can efficiently compute a slog drawing requiring $O(n^2)$ area with (i) optimal number of half-bends on rr edges and rc edges without bends and (ii) at most two additional half-bends on cc edges and rc edges with bends.*

4   A Linear Program to Compute Optimal Drawings

This section describes how to model a given slog representation $\mathcal{S}$ of a plane graph G as a Linear Program (LP). Based on $\mathcal{S}$, we modify G and obtain a graph G', that is a subdivision of G and has at most one half-bend on each edge.

Let $\langle b_1, \ldots, b_k \rangle$, $k \geq 2$, be the half-bends of edge (u, w) of G in $\mathcal{S}$, appearing in this order along (u, w) from u to w. Say that u is an r-vertex. We add a new c-vertex v and replace (u, w) by edges (u, v) and (v, w). The first half-bend b_1 is assigned to (u, v) and $\langle b_2, \ldots, b_k \rangle$ to (v, w). If u is a c-vertex, then v would have been an r-vertex. Observe that the type of v and its ports are defined by the slope of the segments incident to b_1 in $\mathcal{S}$. By repeating this procedure, we obtain G', that is a subdivision of G having at most one half-bend on each edge.

Each face f of G has a corresponding face f' in G' such that: (i) the vertices of G' incident to f' are the same as those incident to f in G, plus the ones from the subdivision; and (ii) the sequence of slopes assigned to the segments bounding f' is the same as that of the segments bounding f. So a drawing Γ' of G' realizing the slog representation is also a drawing of G realizing $\mathcal{S}$.

For each vertex v of G' we define two variables x_v and y_v, representing its coordinates on the plane. For each edge (a, b) of G', we define a pair of constraints similar to those in Table 1, depending on the type of vertices of a and b. The table provides an example for each type, the other configurations are analogous.

Table 1. Examples of constraints of the linear program for (a) rr-edges, (b) cc-edges and (c) rc-edges, assuming the y-axis points downwards.

$$
\begin{cases} y_a = y_b \\ x_b - x_a \geq 1 \end{cases}
\qquad
\begin{cases} y_a - y_b = x_b - x_a \\ y_a - y_b \geq 1 \end{cases}
\qquad
\begin{cases} x_b - x_a \geq y_a - y_b + 1 \\ y_a \geq y_b + 1 \end{cases}
$$

(a)   (b)   (c)

We indirectly minimize the area of the produced drawing by minimizing the total edge length in the objective function. The slopes of the segments allow us to express the Euclidean length of each edge.

Despite the fact that every experiment we made on random and crafted graphs led to a feasible solution, we could not prove the feasibility of the LP. Nevertheless we believe that our LP always admits a feasible solution.

4.1   Addressing Planarity Issues in the LP

The LP models the shape of the edges and the relative positions of all nodes connected by an edge. Since there are no constraints for non-connected nodes the resulting drawing could be non-planar. An example is given in Fig.5 where the relative position of nodes r_3 and c_2 is not defined by the LP. To solve this problem, we cannot apply the approach used in the original Tamassia algorithm (cutting all faces into rectangles), since our faces are, in general, not rectilinear.

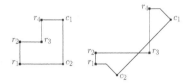

Fig. 5. A configuration that could result in a non-planar solution

In slog drawings we distinguish different corner types. There are *vertex-corners* (or simply vertices) and *bend-corners* (or simply bends). A corner is either *convex* with respect to a face, if the inner angle is $\leq 135°$, or *non-convex* otherwise. An angle of 180° at a vertex is not a corner, since it will be aligned with its neighbors by construction. This gives four possible corners: (non-)/convex vertex(or bend). To ensure planarity, we use *split-edges* and the notion of *almost-convex* faces.

Definition 1. *A* split-edge *is an edge that connects either (a) a non-convex vertex-corner v with a new vertex that subdivides a side parallel to one of the edges incident to v, or, (b) two new vertices that subdivide two parallel edges, when one of them is incident to a non-convex bend-corner (see Figs.6(a) and 6(b)).*

Definition 2. *A face f is* almost-convex *if it does not contain any non-convex vertex-corners and no split-edge exists that separates f into two non-convex faces.*

First we make all faces almost-convex. To remove non-convex vertex-corners we introduce a new split-edge; see Fig.6(a). If there is no parallel side to one of the segments of a vertex-corner we use a special structure, which we call *nose-gadget* (due to its shape); see Fig.6(c). The dashed line in the figure represents the split-edge we can apply once the gadget (dotted line) is added. The two vertices that are added on the diagonals are c-vertices, while the third one is an r-vertex. By applying the split edge the non-convex vertex corner is removed.

After the previous step no face contains non-convex vertex-corners. If a face contains non-convex bends and is not almost-convex, we search for a split-edge that creates two non-convex faces (as in Fig.6(b)). We apply such split-edges until all faces are almost-convex. Observe that all additional edges can be expressed by using the original set of constraints from the LP. To prove that we can always make all faces almost-convex, we give the following lemmas.

Lemma 1. *If a face contains two segments s_1 and s_2 defining a non-convex bend-corner, then it contains a segment s_3 parallel to either s_1 or s_2.*

Sketch of Proof. Assume to the contrary that there is no segment parallel to s_1 and s_2. Let b be the non-convex bend, and, r and c its adjacent corners;

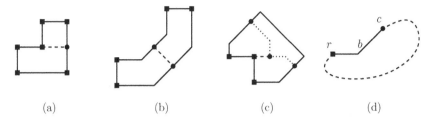

Fig. 6. Split-edges are dashed lines on (a) a vertex and (b) a bend. (c) If no split-edge is possible the nose gadget (dotted line) is used. Circles are additional nodes.(d) This face can not be completed without using at least one segment parallel to (r, b) or (b, c).

see Fig.6(d). Then, r and c cannot be connected by a polygonal chain of edges without using one of the two slopes defined by (r, b) and (b, c).

Lemma 2. *An almost-convex face f has at most two consecutive non-convex bend-corners.*

Proof. Assume that f has three consecutive non-convex corners c_1, c_2, c_3. By Lemma 1, there exists a parallel segment to one of those defining c_2. Then, there exists a split-edge separating f into two non-convex faces, one containing c_1 and one containing c_3, which is a contradiction.   □

Lemma 3. *An almost-convex face f has at most two non-convex bend-corners.*

Sketch of Proof. Assuming that f contains at least three non-convex corners, one can lead to a contradiction the fact that f is almost-convex.

Lemma 4. *An almost-convex face f is always drawn planar.*

Sketch of Proof. From Lemma 3, it follows that f has at most two non-convex bend-corners. If f contains no or a single non-convex bend-corner, then it is easy to see that f is always drawn planar. If f contains exactly two non-convex bend-corners, then by Lemma 2 f cannot have more than two consecutive non-convex bend-corners. Hence, there exist in total four cases that one has to consider with respect to the shape of f. In all of them f is drawn planar.

With these lemmas we have shown how to subdivide all faces of a graph $G = (V, E)$ to obtain a graph $G' = (V', E')$ that only contains almost-convex faces. For G' we can compute a planar drawing using our linear program, giving a planar drawing for G. We ensured the planarity of the drawing by adding N new vertices and edges, where N is $O(|E|)$, since there is at most one split-edge for each vertex and for each non-convex bend. All additional edges can be modeled by the original set of constraints. Experiments on random and crafted graphs seem to confirm that our linear program always has a feasible solution.

5   Area Bounds

Slog drawings have aesthetic appeal and improve the readability of non-planar graphs, when compared to traditional orthogonal drawings. However, in this section we show that such drawings might require increased drawing area. Note that most of the orthogonal drawing algorithms require $O(n) \times O(n)$ area. The situation is different if we want to generate slog drawings of optimal number of bends. In particular, we show that the area penalty can be exponential.

Theorem 4. *There exists a graph G whose slanted orthogonal drawing $\Gamma(G)$ of minimum number of bends requires exponential area, assuming that a planarized version $\sigma(G)$ of the resulting drawing is given.*

Proof. The planarized version $\sigma(G)$ of G is given in Fig.7(a) and consists of $n+1$ layers $L_0, L_1, \ldots, L_n$. Layer L_0 is the square grid graph on 9 vertices. Each layer L_i, $i = 1, 2, \ldots, n$, is a cycle on 20 vertices. Consecutive layers L_{i-1} and L_i, $i = 1, 2, \ldots, n$, are connected by 12 edges which define 12 crossings. Hence, G consists of $20n + 9$ vertices and $32n + 13$ edges that define $12n$ crossings.

A slog drawing $\Gamma(G)$ of G with minimum number of bends derived from $\sigma(G)$ ideally introduces (a) no bends on crossing-free edges of $\sigma(G)$, and, (b) two half-bends in total for each rc-edge. Now observe that at each layer there exist four vertices, that have two ports pointing to the next layer (gray-colored in Fig.7(a)). This together with requirements (a) and (b) suggests that the vertices of each layer L_i should reside along the edges of a rectangle, say R_i, such that the vertices of L_i whose ports point to the next layer coincide with the corners of R_i, $i = 0, 1, 2, \ldots, n$ (with the only exception of the "innermost" vertex of L_0; in Fig.7(b), R_i is identified with cycle L_i). Hence, the routing of the edges that connect consecutive layers should be done as illustrated in Fig.7(b). Since L_0 is always drawable in a 3×3 box meeting all requirements mentioned above, and, $\sigma(G)$ is highly symmetric, we can assume that each R_i is a square of side length w_i, $i = 0, 1, 2, \ldots, n$. Then, it is not difficult to see that $w_0 = 3$ and $w_{i+1} = 2w_i + 8$, $i = 1, 2, \ldots, n$. This implies that the area of $\Gamma(G)$ is exponential in the number of layers of G and therefore exponential in the number of vertices of G (recall that G has $n + 1$ layers and $20n + 9$ vertices).   □

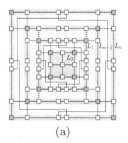   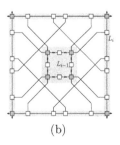

(a)   (b)

Fig. 7. (a) A planarized version $\sigma(G)$ of a graph G. (b) Edges involved in crossings in $\sigma(G)$ contribute two half-bends.

6   Conclusion and Open Problems

We introduced a new model for drawing graphs of max-degree four, in which orthogonal bends are replaced by pairs of "slanted" bends and crossings occur on diagonal segments only. The main advantage of this model is that, even in drawings of large graphs (where vertices might not be clearly visible), it is immediately clear which pair of edges induce a crossing and where such a crossing is located in the drawing. We presented an algorithm to construct slog drawings with almost-optimal number of bends and quadratic area, for general max-degree four graphs. By a modification of Tamassia's min-cost flow approach, we showed that a bend-optimal representation of the graph can efficiently be computed in polynomial time and we presented an LP-approach to compute a corresponding drawing. A natural problem is whether every max-degree four graph admits such a drawing. Our experiments on randomly generated and crafted inputs led us to believe that it is possible, although we could not prove it.

References

1. Biedl, T.C., Kant, G.: A better heuristic for orthogonal graph drawings. In: van Leeuwen, J. (ed.) ESA 1994. LNCS, vol. 855, pp. 24–35. Springer, Heidelberg (1994)
2. Cornelsen, S., Karrenbauer, A.: Accelerated bend minimization. Journal of Graph Algorithms Applications 16(3), 635–650 (2012)
3. Fößmeier, U., Heß, C., Kaufmann, M.: On improving orthogonal drawings: The 4M-algorithm. In: Whitesides, S.H. (ed.) GD 1998. LNCS, vol. 1547, pp. 125–137. Springer, Heidelberg (1999)
4. Fößmeier, U., Kaufmann, M.: Drawing high degree graphs with low bend numbers. In: Brandenburg, F.J. (ed.) GD 1995. LNCS, vol. 1027, pp. 254–266. Springer, Heidelberg (1996)
5. Garg, A., Tamassia, R.: On the computational complexity of upward and rectilinear planarity testing. SIAM Journal of Computing 31(2), 601–625 (2001)
6. Leiserson, C.E.: Area-efficient graph layouts (for VLSI). In: 21st Symposium on Foundations of Computer Science, vol. 1547, pp. 270–281. IEEE (1980)
7. Nöllenburg, M., Wolff, A.: Drawing and labeling high-quality metro maps by mixed-integer programming. IEEE Trans. Vis. Comput. Graph. 17(5), 626–641 (2011)
8. Tamassia, R.: On embedding a graph in the grid with the minimum number of bends. SIAM Journal of Computing 16(3), 421–444 (1987)
9. Tamassia, R., Tollis, I.G.: Planar grid embedding in linear time. IEEE Transactions on Circuits and Systems 36(9), 1230–1234 (1989)
10. Valiant, L.G.: Universality considerations in VLSI circuits. IEEE Transaction on Computers 30(2), 135–140 (1981)

Drawing Arrangement Graphs in Small Grids, or How to Play Planarity

David Eppstein[*]

Department of Computer Science, University of California, Irvine, USA

Abstract. We describe a linear-time algorithm that finds a planar drawing of every graph of a simple line or pseudoline arrangement within a grid of area $O(n^{7/6})$. No known input causes our algorithm to use area $\Omega(n^{1+\epsilon})$ for any $\epsilon > 0$; finding such an input would represent significant progress on the famous k-set problem from discrete geometry. Drawing line arrangement graphs is the main task in the *Planarity* puzzle.

1 Introduction

Planarity (http://planarity.net/) is a puzzle developed by John Tantalo and Mary Radcliffe in which the user moves the vertices of a planar graph, starting from a tangled circular layout (Figure 1), into a position where its edges (drawn as line segments) do not cross. The game is played in a sequence of levels of increasing difficulty. To construct the graph for the ith level, the game applet chooses $\ell = i+3$ random lines in general position in the plane. It creates a vertex for each of the $\ell(\ell - 1)/2$ crossings of two lines, and an

Fig. 1. Initial state of Planarity

edge for each of the $\ell(\ell - 2)$ consecutive pairs of crossings on the same line.

One strategy for solving Planarity would be to reconstruct a set of lines forming the given graph (Figure 2, left). However, this is tedious to do by hand, and has high computational complexity: testing whether an arrangement of curves is combinatorially equivalent to a line arrangement is NP-hard [1], from which it follows that recognizing line arrangement graphs is also NP-hard [2]. More precisely, both problems are complete for the existential theory of the reals [3]. And although drawings constructed in this way accurately convey the underlying construction of the graph, they have low angular resolution (at most π/ℓ) and close vertex spacing, making them hard to read and hard to place by hand. In practice these puzzles may be solved more easily by an incremental strategy

[*] This research was supported in part by NSF grants 0830403 and 1217322 and by the Office of Naval Research under grant N00014-08-1-1015.

S. Wismath and A. Wolff (Eds.): GD 2013, LNCS 8242, pp. 436–447, 2013.

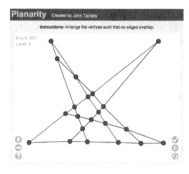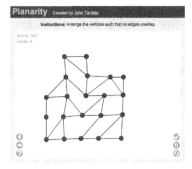

Fig. 2. Two manually constructed solutions to the puzzle from Figure 1. Left: a set of lines with this graph as its arrangement. Right: an (approximate) grid layout.

that maintains a planar embedding of a subgraph of the input, starting from a single short cycle (such as a triangle or quadrilateral), and that at each step extends the embedding by a single face, bounded by a short path connecting two vertices on the boundary of the previous embedding. When using this strategy to solve a Planarity puzzle, the embedding may be kept tidy by placing each vertex into an approximate grid (Figure 2, right). Curiously, the grid drawings found by this incremental grid-placement heuristic appear to have near-linear area; in contrast, there exist planar graphs such as the *nested triangles graph* that cannot be drawn planarly in a grid of less than $\Theta(n^2)$ area [4,5].

In this paper we explain this empirical finding of small grid area by developing an efficient algorithm for constructing compact grid drawings of the arrangement graphs arising in Planarity. Because recognizing line arrangement graphs is NP-hard, we identify a larger family of planar graphs (the graphs of simple pseudoline arrangements) that may be recognized and decomposed into pseudolines in linear time. We show that every n-vertex simple pseudoline arrangement graph may be drawn in linear time in a grid of size $\kappa_{\max}(O(\sqrt{n})) \times O(\sqrt{n})$; here $\kappa_{\max}(\ell)$ is the maximum complexity of a k-level of a pseudoline arrangement with ℓ pseudolines [6–8], a topological variant of the famous k-set problem from discrete geometry (see Section 3 for a formal definition). The best proven upper bounds of $O(\ell^{4/3})$ on the complexity of k-levels [7–9] imply that the grid in which our algorithm draws these graphs has size $O(n^{2/3}) \times O(\sqrt{n})$ and area $O(n^{7/6})$. However, all known lower bounds on k-level complexity are of the form $\Omega(\ell^{1+o(1)})$ [6,10], suggesting that our algorithm is likely to perform even better in practice than our worst-case bound. If we could find a constant $\epsilon > 0$ and a family of inputs that would cause our algorithm to use area $\Omega(\ell^{1+\epsilon})$, such a result would represent significant progress on the k-set problem.

We also investigate the construction of *universal point sets* for arrangement graphs, sets of points that can be used as the vertices for a straight-line planar drawing of every n-vertex arrangement graph. Our construction directly provides a universal point set consisting of $O(n^{7/6})$ grid points; we show how to sparsify this structure, leading to the construction of a universal set of $O(n \log n)$ points within a grid whose dimensions are again $O(n^{2/3}) \times O(\sqrt{n})$.

Finally, we formalize and justify an algorithm for manual solution of these puzzles that greedily finds short cycles and adds them as faces to a partial planar embedding. Although this algorithm may fail for general planar graphs, we show that for arrangement graphs it always finds a planar embedding that is combinatorially equivalent to the original arrangement.

2 Preliminaries

Following Shor [1], we define a *pseudoline* to be the image of a line under a homeomorphism of the Euclidean plane. Pseudolines include lines, non-self-crossing polygonal chains starting and ending in infinite rays, and the graphs of continuous real functions. Two pseudolines *cross* at a point x if a neighborhood of x is homeomorphic to a neighborhood of the crossing point of two lines, with the homeomorphism taking the pseudolines to the lines. An *arrangement* of pseudolines is a finite set of pseudolines, the intersection of every two of which is a single crossing point. An arrangement is *simple* if

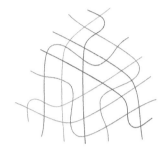

Fig. 3. A simple pseudoline arrangement that cannot be transformed into a line arrangement. Redrawn from Figure 5.3.2 of [11], who attribute this arrangement to Ringel.

all pairs of pseudolines have distinct crossing points. A *pseudoline arrangement graph* is a planar graph whose vertices are the crossings in a simple pseudoline arrangement, and whose edges connect consecutive crossings on a pseudoline.

Most of the ideas in the following result are from Bose et al. [2], but we elaborate on that paper to show that linear time recognition of arrangement graphs is possible. (See [12] for a more complicated linear time algorithm that recognizes the dual graphs of a wider class of arrangement graphs, the graphs of *weak* pseudoline arrangements in which pairs of pseudolines need not cross)

Lemma 1. *If we are given as input a graph G, then in linear time we can determine whether it is a pseudoline arrangement graph, determine its (unique) embedding as an arrangement graph, and find a pseudoline arrangement for which it is the arrangement graph.*

Proof. Let G^* be formed from a pseudoline arrangement graph G by adding a new vertex v_∞ adjacent to all vertices in G of degree less than four. As Bose et al. [2] show, G^* is 3-connected and planar, and its unique planar embedding is compatible with the embedding of G as an arrangement graph. For convenience we include two edges in G^* from v_∞ to each degree two vertex in G, so that, in G^*, all vertices except v_∞ have degree four. With this modification, the pseudolines of the arrangement for G are represented in G^* by paths starting and ending at v_∞ that, at each other vertex, connect two opposite edges in the embedding.

For any given graph G of maximum degree four we may, in linear time, add a new vertex v_∞, test planarity of the augmented graph G^*, and embed G^*

in the plane. The edge partition of G^* into paths through opposite edges at each degree four vertex may be found in linear time by connected component analysis. By labeling each edge with the identity of its path, we may verify that this partition does not include cycles disjoint from v_∞ and that no path crosses itself. We additionally check that G has $\ell(\ell-1)/2$ vertices, where ℓ is the number of paths. Finally, by listing the pairs of paths passing through each vertex and bucket sorting this list, we may verify in linear time that no two paths cross more than once. If G passes all of these checks, its decomposition into paths gives a valid pseudoline arrangement, which may be constructed by viewing the embedding of G^* as being on a sphere, puncturing the sphere at point v_∞, and homeomorphically mapping the punctured sphere to the plane.   □

3   Small Grids

To describe our grid drawing algorithm for pseudoline arrangement graphs, we need to introduce the concept of a *wiring diagram*. A wiring diagram is a particular kind of pseudoline arrangement, in which the ℓ pseudolines largely lie on ℓ horizontal lines (with coordinates $y = 1$, $y = 2$, ..., $y = \ell$). The pseudolines on two adjacent tracks may cross each other, swapping which track they lie on, near points with coordinates $x = 1$, $x = 2$, ..., $x = \ell(\ell - 1)/2$; each crossing is formed by removing two short segments of track and replacing them by two crossing line segments between the tracks. It is convenient to require different crossings to have different x coordinates, following Goodman [13], although some later sources omit this requirement. Figure 4 depicts an example. Wiring diagrams already provide reasonably nice grid drawings of arrangement graphs [14], but are unsuitable for our purposes, for two reasons: they draw the edges connecting pairs of adjacent crossings as polygonal chains with two bends, and for some arrangements, even allowing crossings to share x-coordinates, drawing the wiring diagram of ℓ lines in a grid may require width $\Omega(\ell^2)$ (Figure 5), much larger than our bounds. Instead, we will use these diagrams as a tool for constructing a different and more compact straight-line drawing.

For an arrangement of non-vertical lines in general position, an equivalent wiring diagram may be constructed by a *plane sweep* algorithm [15], which simulates the left-to-right motion of a vertical line across the arrangement. At most

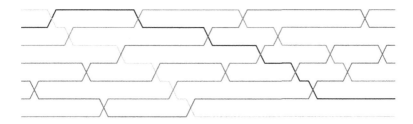

Fig. 4. A wiring diagram formed by a plane sweep of the arrangement from Figure 2

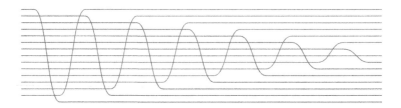

Fig. 5. Cocktail shaker sort corresponds to an arrangement of ℓ pseudolines for which drawing the wiring diagram in a grid requires width $\Omega(\ell^2)$

points in the sweep, the intersection points of the arrangement lines with the sweep line maintain a fixed top-to-bottom order with each other, with their positions in this order reflected in the assignment of the corresponding pseudolines to tracks. When the sweep line crosses a vertex of the arrangement, two intersection points swap positions in the top-to-bottom order, corresponding to a crossing in the wiring diagram. The left-to-right order of crossings in the wiring diagram is thus exactly the sorted order of the crossing points of the arrangement, as sorted by their x coordinates. The wiring diagram in Figure 4 was constructed in this way from the approximate line arrangement depicted in Figure 1.

Every simple pseudoline arrangement, also, has an equivalent wiring diagram, that may be constructed in time linear in its number of crossings. The proof of this fact uses *topological sweeping*, a variant of plane sweeping originally developed to speed up sweeping of straight line arrangements by relaxing the strict left-to-right ordering of the crossing points [16], that can also be extended to apply to pseudoline arrangements [17]. The steps of the topological sweeping algorithm require only determining the relative ordering of crossings along each of the input pseudolines, something that may easily be determined from our path decomposition of a pseudoline arrangement graph by precomputing the position of each crossing on each of the two pseudolines it belongs to.

We define the ith *level* $L_{\mathcal{D}}(i)$ in a wiring diagram $\mathcal{D}$ to be the set of crossings that occur between tracks i and $i + 1$. A crossing belongs to $L_{\mathcal{D}}(i)$ if and only if $i - 1$ lines pass between it and the bottom face of the arrangement (the face below all of the tracks in the wiring diagram); therefore, once this bottom face is determined, the levels are fixed by this choice regardless of how the crossings are ordered to form a wiring diagram. If we define the size $|\mathcal{D}|$ of a diagram to be its number of pseudolines, and the level complexity $\kappa(\mathcal{D})$ to be $\max_i |L_{\mathcal{D}}(i)|$, then it is a longstanding open problem in discrete geometry (a variant of the k-set problem) to determine the maximum level complexity of an arrangement of ℓ pseudolines, $\kappa_{\max}(\ell) = \max_{|\mathcal{D}|=\ell} \kappa(\mathcal{D})$. (Often this problem is stated in terms of the middle level of an arrangement, rather than the maximum-complexity level, but this variation makes no difference to the asymptotic behavior of the level complexity.) The known bounds on this quantity are $\kappa_{\max}(\ell) = O(\ell^{4/3})$ [7–9], and $\kappa_{\max}(\ell) = \Omega(\ell\, c^{\sqrt{\log \ell}})$ for some constant $c > 1$ [6,10], where the last bound is $O(n^{1+\epsilon})$ for all constants $\epsilon > 0$.

Theorem 1. *Let G be a pseudoline arrangement graph with n vertices, determined by $\ell = \Theta(\sqrt{n})$ pseudolines. Then in time $O(n)$ we may construct a planar straight-line drawing of G, in a grid of size $(\ell-1) \times \kappa_{\max}(\ell) = O(n^{1/2}) \times O(n^{2/3})$.*

Proof. We find a decomposition of G into pseudoline paths, by the algorithm of Lemma 1, and use topological sweeping to convert this decomposition into a wiring diagram. We place each vertex v of G at the coordinates (i, j), where i is the position of v within its level of the wiring diagram and j is the number of tracks below its level of the wiring diagram.

With this layout, every edge of G either connects consecutive vertices within the same level as each other, or it connects vertices on two consecutive levels. In the latter case, each edge between two consecutive levels corresponds to a horizontal segment of the wiring diagram that lies on the track between the two levels; the left-to-right ordering of these horizontal segments is the same as the left-to-right ordering of both the lower endpoints and the upper endpoints of these edges. Because of this consistent ordering of endpoints, no two edges between the same two consecutive levels can cross. There can also not be any crossings between edges that do not both lie in the same level or connect the same two consecutive levels. Therefore, the drawing we have constructed is planar. By construction, it has the dimensions given in the theorem.   □

Figure 6 depicts the output of our algorithm, using the wiring diagram of Figure 4, for the graph of Figure 1. The arrangement has six levels, with at most five vertices per level, giving a 6×5 grid. Although not as compact as the manually-found 5×5 grid of Figure 2, it is much smaller than standard grid drawings that do not take advantage of the arrangement structure of this graph. A more careful placement of vertices within each row would improve the angular resolution and edge length of the drawing but we have omitted this step in order to make the construction more clear.

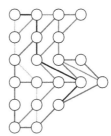

Fig. 6. Output of the drawing algorithm of Theorem 1, based on the wiring diagram of Figure 4

4   Universal Point Sets

A *universal point set* for the n-vertex graphs in a class $\mathcal{C}$ of graphs is a set U_n of points in the plane such that every n-vertex graph in C can be drawn with its vertices in U_n and with its edges drawn as non-crossing straight line segments [18]. Grids of $O(n) \times O(n)$ points form universal sets of quadratic size for the planar graphs [19, 20], and despite very recent improvements the best upper bound known remains quadratic [21]. A rectangular grid that is universal must have $\Omega(n^2)$ points [4, 5]; the best known lower bounds for universal point sets that are not required to be grids are only linear [18].

Subquadratic bounds are known on the size of universal point sets for subclasses of the planar graphs including the outerplanar graphs [22], simply-nested planar graphs [21,23], planar 3-trees [24], and graphs of bounded pathwidth [21]; however, these results do not apply to arrangement graphs. The grid drawing technique of Theorem 1 immediately provides a universal point set for arrangement graphs of size $O(n^{7/6})$; in this section we significantly improve this bound, while only increasing the area of our drawings by a constant factor.

Following Bannister et al. [21], define a sequence of positive integers ξ_i for $i = 1, 2, 3, \dots$ by the equation $\xi_i = i \oplus (i - 1)$ where $\oplus$ denotes the bitwise binary exclusive or operation. The sequence of these values begins

$$1, 3, 1, 7, 1, 3, 1, 15, 1, 3, 1, 7, 1, 3, 1, \dots.$$

Lemma 2 (Bannister et al. [21]). *Let the finite sequence $\alpha_1, \alpha_2, \dots \alpha_k$ have sum s. Then there is a subsequence $\beta_1, \beta_2, \dots \beta_k$ of the first s terms of ξ such that, for all i, $\alpha_i \le \beta_i$. The sum of the first s terms of ξ is between $s \log_2 s - 2s$ and $s \log_2 s + s$.*

Recall that the grid drawing technique of Theorem 1 produces a drawing in which the vertices are organized into $\ell - 1$ rows of at most $\kappa_{\max}(\ell) = O(\ell^{4/3})$ vertices per row, where $\ell = O(\sqrt{n})$ is the number of lines in the underlying n-vertex arrangement. In this drawing, suppose that there are n_i vertices on the ith row of the drawing, and define a sequence $\alpha_i = \lceil n_i/\ell \rceil$.

Lemma 3. $\sum \alpha_i \le 3(\ell - 1)/2.$

Proof. We may partition the n_i vertices in the ith row n_i into $\lfloor n_i/\ell \rfloor$ groups of exactly ℓ vertices, together with at most one smaller group; then α_i is the number of groups. The contribution to $\sum \alpha_i$ from the groups of exactly ℓ vertices is at most $n/\ell = (\ell - 1)/2$. There is at most one smaller group per row so the contribution from the smaller groups is at most $\ell - 1$. Thus the total value of the sum is at most $3(\ell - 1)/2$. ☐

Theorem 2. *There is a universal point set of $O(n \log n)$ points for the n-vertex arrangement graphs, forming a subset of a grid of dimensions $O(\ell) \times \kappa_{\max}(\ell)$.*

Proof. Let $s = 3(\ell - 1)/2$. We form our universal point set as a subset of an $s \times \kappa_{\max}(\ell)$ grid; the area of the grid from which the points are drawn is exactly $3/2$ times the area of the $(\ell - 1)$-row grid drawing technique of Theorem 1. In the ith row of this grid, we include in our universal point set $\min(\ell \xi_i, \kappa_{\max}(\ell))$ of the grid vertices in that row. It does not matter for our construction exactly which points of the row are chosen to make this number of points.

By Lemma 2, there is a subsequence β_i of the first s rows of sequence ξ, such that the β is termwise greater than or equal to α. This subsequence corresponds to a subsequence $(r_1, r_2, \dots r_{\ell-1})$ of the rows of our universal point set, such that row r_i has at least $\min(\ell \beta_i, \kappa_{\max}(ell)) \ge n_i$ points in it. Mapping the ith row of the drawing of Theorem 1 to row r_i of this point set will not create any crossings, because the mapping is monotonic within each row and because all edges of the drawing connect pairs of vertices that are either in the same row or in rows that are consecutive in the selected subsequence.

The number of points in the point set is $O(\ell s \log s)$ where $s = O(\ell)$. Therefore, this number of points is $O(\ell^2 \log \ell) = O(n \log n)$. ☐

5   Greedy Embedding Algorithm

The algorithm of Lemma 1 uses as a subroutine a linear-time planarity testing algorithm. Although such algorithms may be efficiently implemented on computers, they are not really suitable for hand solution of Planarity puzzles. Instead, it is more effective in practice to build up a planar embedding one face at a time, by repeatedly finding a short cycle in the input graph and attaching it to the previously constructed partial embedding. Here "short" means as short as can be found; it is not possible to limit attention to cycles of length three, four, or any fixed bound. For instance in Figure 3 the central triangle is separated from the rest of the graph by faces with five sides, and by modifying this example it is possible to separate part of an arrangement graph from the rest of the graph by faces with arbitrarily many sides. Thus, this hand-solution heuristic may be formalized by the following steps.

1. Choose an arbitrary starting vertex v.
2. Find a cycle C_1 of minimum possible length containing v.
3. Embed C_1 as a simple cycle in the plane.
4. While some of the edges of the input graph have not yet been embedded:
   (a) Let C_i be the cycle bounding the current partial embedding. Define an *attachment vertex* of C_i to be a vertex that is incident with edges not already part of the current embedding.
   (b) Choose two attachment vertices u and v, and a path P_i in C_i from u to v, such that there are no attachment vertices interior to P_i.
   (c) Find a shortest path S_i from u to v, using only edges that are not already part of the current partial embedding.
   (d) If necessary, adjust the positions of the embedded vertices (without changing the combinatorial structure of the embedding) so that S_i may be drawn with straight line edges.
   (e) Add S_i to the embedding, outside C_i, so that the new face between P_i and S_i does not contain C_i. After this change, the new bounding cycle C_{i+1} of the partial embedding is formed from C_i by replacing P_i by S_i.

When it is successful, this algorithm decomposes the input graph into the cycle C_1 and a sequence of edge-disjoint paths S_1, S_2, etc. Such a decomposition is known as an *open ear decomposition* [25].

This greedy ear decomposition algorithm does not always work for arbitrary planar graphs: even the initial cycle that is found by the algorithm may not be a face of an embedding of the given graph, causing the algorithm to make incorrect assumptions about the structure of the embedding. However (ignoring the possible difficulty of performing step d) the algorithm does always correctly embed the arrangement graphs used by Planarity. These graphs may have multiple embeddings; to distinguish among them, define the *canonical embedding* of an arrangement graph to be the one given by the arrangement from which it was constructed. By Lemma 1, the canonical embedding is unique. As we prove below, the cycles of an arrangement graph that the algorithm assumes to be faces really are faces of the canonical embedding.

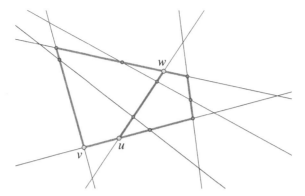

Fig. 7. Illustration for the proof of Lemma 4. Every non-facial cycle C through vertex v (blue and green edges) is crossed by at least one line $\ell = uw$ (red edges), forming a theta-graph. All the vertices on the red path of the theta are matched by an equal number of vertices on each of the other two paths, caused by crossings with the same lines, and the other two paths have additional vertices at their bends, so the red-blue cycle is shorter than the blue-green cycle.

Lemma 4. *Let v be an arbitrary vertex of arrangement graph G, and C be a shortest cycle containing v. Then C is a face of the canonical embedding of G.*

Proof. Let C be an arbitrary simple cycle through v. Then if C is not a face of the arrangement forming G, there is a line ℓ that crosses it; let u and w be two vertices on the boundary of C connected through the interior of C by ℓ (Figure 7). Then C together with the path along ℓ from u to w form a theta-graph, a graph with two degree three vertices (u and w) connected by three paths. Every vertex of ℓ between u and w is caused by a crossing of ℓ with another line that also must cross the other two paths of the theta-graph; in addition, each of these two paths must bend at least once at a vertex that does not correspond to a line that crosses ℓ. Therefore, the path through ℓ is strictly shorter than the other two paths in the theta-graph. Replacing one of the two paths of C from u to w by the path through ℓ produces a shorter cycle that still contains v. Since an arbitrary cycle C that is not a face can be replaced by a shorter cycle through v, it follows that every shortest cycle through v is a face.   □

Lemma 5. *Let D be a drawing of a subset of the faces of the canonical embedding of an arrangement graph G whose union is a topological disk, let u and v be two attachment vertices on the boundary of D with no attachment vertices interior to the boundary path P from u to v, and let S be a shortest path from u to v using only edges not already part of D. Then the cycle formed by the union of P and S is a face of the canonical embedding of G.*

Proof. Assume for a contradiction that $P \cup S$ is not a face; then as in the proof of Lemma 4, this cycle must be crossed by a line ℓ, a path L of which forms a

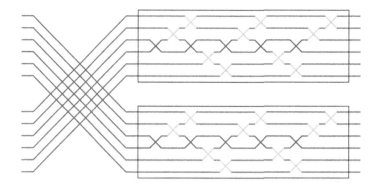

Fig. 8. Two stacked arrangements of $\ell/2$ pseudolines, each with high level complexity, cause our algorithm to create wide drawings no matter how it chooses a wiring diagram

theta-graph together with $P \cup S$. Additionally, because P is assumed to be part of a drawing of a subset of the faces of G, it cannot be crossed by ℓ, for any crossing would cause it to have an attachment vertex between u and v. Therefore, the two degree-three vertices of the theta-graph both belong to S. By the same reasoning as in the proof of Lemma 4, L must be shorter than the other two paths of the theta-graph, so replacing the path that is entirely within S by L would produce a shorter path from u to v, contradicting the construction of S as a shortest path. This contradiction shows that $P \cup S$ must be a face.   □

Theorem 3. *When the greedy ear decomposition embedding algorithm described above is applied to an arrangement graph G, it correctly constructs the canonical embedding of G.*

Proof. We prove by induction on the number of steps of the algorithm that after each step the partial embedding consists of faces of the canonical embedding whose union is a disk. Lemma 4 shows as a base case that the induction hypothesis is true after the first step. In each subsequent step, the ability to find two attachment vertices follows from the fact that arrangement graphs are 2-vertex-connected, which in turn follows from the fact that they can be augmented by a single vertex to be 3-vertex-connected [2]. Lemma 5 shows that, if the induction hypothesis is true after i steps then it remains true after $i + 1$ steps.   □

6   Conclusions

We have found a grid drawing algorithm for pseudoline arrangement graphs that uses area within a small factor of linear, much smaller than the known quadratic grid area lower bounds for arbitrary planar graphs. We have also shown that these graphs have near-linear universal point sets within a constant factor of the same area, and that a simple greedy embedding heuristic suitable for hand solution of Planarity puzzles is guaranteed to find a correct embedding.

The precise area used by our grid drawing algorithm depends on the worst-case behavior of the function $\kappa(\mathcal{D})$ counting the number of crossings in a k-level of an arrangement; closing the gap between the upper and lower bounds for this function remains an important and difficult open problem in combinatorial geometry. However, closing this gap is not the only possible method for improving our drawing algorithm.

A tempting avenue for improvement is to observe that a single pseudoline arrangement may be represented by many different wiring diagrams; therefore, we can select the wiring diagram $\mathcal{D}$ that represent the same pseudoline arrangement and that minimizes $\kappa(\mathcal{D})$. However, this would not improve our worst case width by more than a constant factor. For, if the input forms a pseudoline arrangement constructed by stacking two arrangements of $\ell/2$ lines with maximal k-level complexity, one above the other (Figure 8), then one of these two instances will survive intact in any wiring diagram for the arrangement, forcing our algorithm to produce a drawing with width at least $\kappa_{\max}(\ell/2)$. Further improvements in our algorithm will likely come by finding an alternative layout that avoids the complexity of k-levels, by proving that k-levels are small in the average case if not the worst case, or by reducing the known combinatorial bounds on k-levels.

It is also tempting to consider other drawing styles for arrangement graphs, such as orthogonal drawing (in which each edge is an axis-aligned polyline). Because arrangement graphs contain triangles, such a drawing may be forced to have at least one bend per edge. However, the need either to align neighboring vertices on adjacent rows of the drawing, or to provide space between rows for parallel edge tracks, may cause these drawings to be significantly larger than the straight-line drawings we study, so we have not found an area bound for orthogonal drawing that is as tight as our bound for straight-line drawing.

References

[1] Shor, P.W.: Stretchability of pseudolines is NP-hard. In: Gritzmann, P., Sturm-fels, B. (eds.) Applied Geometry and Discrete Mathematics: The Victor Klee Festschrift. DIMACS Series in Discrete Mathematics and Theoretical Computer Science, vol. 4, pp. 531–554. Amer. Math. Soc., Providence (1991)

[2] Bose, P., Everett, H., Wismath, S.: Properties of arrangement graphs. International Journal of Computational Geometry & Applications 13(6), 447–462 (2003)

[3] Schaefer, M.: Complexity of some geometric and topological problems. In: Eppstein, D., Gansner, E.R. (eds.) GD 2009. LNCS, vol. 5849, pp. 334–344. Springer, Heidelberg (2010)

[4] Dolev, D., Leighton, T., Trickey, H.: Planar embedding of planar graphs. Advances in Computing Research 2, 147–161 (1984)

[5] Valiant, L.G.: Universality considerations in VLSI circuits. IEEE Transactions on Computers C-30(2), 135–140 (1981)

[6] Klawe, M., Paterson, M., Pippenger, N.: Inversions with $n^{1+\Omega(1/\sqrt{\log n})}$ transpositions at the median (September 21, 1982) (unpublished manuscript)

[7] Tamaki, H., Tokuyama, T.: A characterization of planar graphs by pseudo-line arrangements. Algorithmica 35(3), 269–285 (2003)

[8] Sharir, M., Smorodinsky, S.: Extremal configurations and levels in pseudoline arrangements. In: Dehne, F., Sack, J.-R., Smid, M. (eds.) WADS 2003. LNCS, vol. 2748, pp. 127–139. Springer, Heidelberg (2003)

[9] Dey, T.L.: Improved bounds for planar k-sets and related problems. Discrete and Computational Geometry 19(3), 373–382 (1998)

[10] Tóth, G.: Point sets with many k-sets. Discrete and Computational Geometry 26(2), 187–194 (2001)

[11] Goodman, J.E.: Pseudoline arrangements. In: Goodman, J.E., O'Rourke, J. (eds.) Handbook of Discrete and Computational Geometry, pp. 83–109. CRC Press (1997)

[12] Eppstein, D.: Algorithms for drawing media. In: Pach, J. (ed.) GD 2004. LNCS, vol. 3383, pp. 173–183. Springer, Heidelberg (2005)

[13] Goodman, J.E.: Proof of a conjecture of Burr, Grünbaum, and Sloane. Discrete Mathematics 32(1), 27–35 (1980)

[14] Muthukrishnan, S., Sahinalp, S.C., Paterson, M.S.: Grid layout of switching and sorting networks. US Patent 6185220 (2001)

[15] Bentley, J.L., Ottmann, T.A.: Algorithms for reporting and counting geometric intersections. IEEE Transactions on Computers C-28(9), 643–647 (1979)

[16] Edelsbrunner, H., Guibas, L.J.: Topologically sweeping an arrangement. Journal of Computer and System Sciences 38(1), 165–194 (1989); Corrigendum, JCSS 42(2), 249–251 (1991)

[17] Snoeyink, J., Hershberger, J.: Sweeping arrangements of curves. In: Proc. 5th ACM Symp. on Computational Geometry (SCG 1989), pp. 354–363 (1989)

[18] Chrobak, M., Karloff, H.: A lower bound on the size of universal sets for planar graphs. SIGACT News 20, 83–86 (1989)

[19] de Fraysseix, H., Pach, J., Pollack, R.: Small sets supporting Fáry embeddings of planar graphs. In: Proc. 20th ACM Symp. Theory of Computing (STOC 1988), pp. 426–433 (1988)

[20] Schnyder, W.: Embedding planar graphs on the grid. In: Proc. 1st ACM/SIAM Symp. Discrete Algorithms (SODA 1990), pp. 138–148 (1990)

[21] Bannister, M.J., Cheng, Z., Devanny, W.E., Eppstein, D.: Superpatterns and universal point sets. In: Wismath, S., Wolff, A. (eds.) GD 2013. LNCS, vol. 8242, pp. 208–219. Springer, Heidelberg (2013)

[22] Gritzmann, P., Mohar, B., Pach, J., Pollack, R.: Embedding a planar triangulation with vertices at specified positions. Amer. Math. Monthly 98(2), 165–166 (1991)

[23] Angelini, P., Di Battista, G., Kaufmann, M., Mchedlidze, T., Roselli, V., Squarcella, C.: Small point sets for simply-nested planar graphs. In: van Kreveld, M., Speckmann, B. (eds.) GD 2011. LNCS, vol. 7034, pp. 75–85. Springer, Heidelberg (2011)

[24] Fulek, R., Tóth, C.D.: Universal point sets for planar three-trees. In: Dehne, F., Solis-Oba, R., Sack, J.-R. (eds.) WADS 2013. LNCS, vol. 8037, pp. 341–352. Springer, Heidelberg (2013)

[25] Khuller, S.: Ear decompositions. SIGACT News 20(1), 128 (1989)

Incremental Grid-Like Layout
Using Soft and Hard Constraints

Steve Kieffer, Tim Dwyer, Kim Marriott, and Michael Wybrow

Caulfield School of Information Technology,
Monash University, Caulfield, Victoria 3145, Australia,
National ICT Australia, Victoria Laboratory,
{Steve.Kieffer,Tim.Dwyer,Kim.Marriott,Michael.Wybrow}@monash.edu

Abstract. We explore various techniques to incorporate grid-like layout conventions into a force-directed, constraint-based graph layout framework. In doing so we are able to provide high-quality layout—with predominantly axis-aligned edges—that is more flexible than previous grid-like layout methods and which can capture layout conventions in notations such as SBGN (Systems Biology Graphical Notation). Furthermore, the layout is easily able to respect user-defined constraints and adapt to interaction in online systems and diagram editors such as Dunnart.

Keywords: constraint-based layout, grid layout, interaction, diagram editors.

1 Introduction

Force-directed layout remains the most popular approach to automatic layout of undirected graphs. By and large these methods untangle the graph to show underlying structure and symmetries with a layout style that is organic in appearance [4]. Constrained graph layout methods extend force-directed layout to take into account user-specified constraints on node positions such as alignment, hierarchical containment and non-overlap [5]. These methods have proven a good basis for semi-automated graph layout in tools such as Dunnart [7] that allow the user to interactively guide the layout by moving nodes or adding constraints.

However, when undirected graphs (and other kinds of diagrams) are drawn by hand it is common for a more grid-like layout style to be used. Grid-based layout is widely used by graphic designers and it is common in hand-drawn biological networks and metro-map layouts. Previous research has shown that grid-based layouts are more memorable than unaligned placements [14]. Virtually all diagram creation tools provide some kind of snap-to-grid feature.

In this paper we investigate how to modify constrained force-directed graph layout methods [5] to create more orthogonal and grid-like layouts with a particular focus on interactive applications such as Dunnart. In Figure 1 we show undirected graphs arranged with our various layout approaches compared with traditional force-directed layout.

S. Wismath and A. Wolff (Eds.): GD 2013, LNCS 8242, pp. 448–459, 2013.

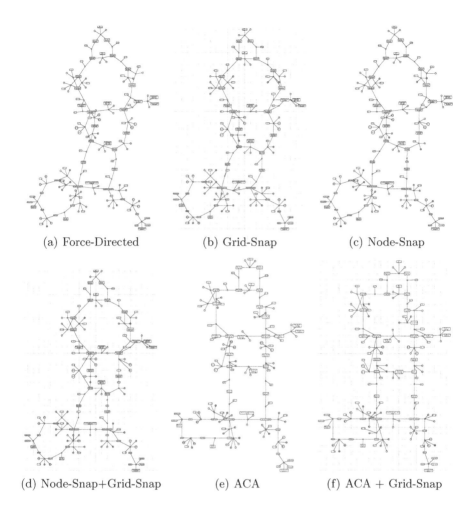

(a) Force-Directed   (b) Grid-Snap   (c) Node-Snap

(d) Node-Snap+Grid-Snap   (e) ACA   (f) ACA + Grid-Snap

Fig. 1. Different combinations of our automatic layout techniques for grid-like layout compared with standard force-directed layout. The layout is for an SBGN (Systems Biology Graphical Notation) diagram of the Glycolysis-Gluconeogenesis pathway obtained from MetaCrop [15]. In SBGN diagrams, *process nodes* represent individual chemical reactions which typically form links in long metabolic pathways, and are often connected to several degree-1 nodes representing "currency molecules" like ATP and ADP, while precisely two of their neighbours are degree-2 nodes representing principal metabolites. It is conventional that the edges connecting main chemicals and process nodes be axis-aligned in long chains, but not the leaf edges. We achieve this by tailoring the cost functions discussed in Sect. 4.

Before proceeding, it is worth defining what we mean by a *grid-like layout*. It is commonly used to mean some combination of the following properties:

1. nodes are positioned at points on a fairly coarse grid;
2. edges are simple horizontal or vertical lines or in some cases 45° diagonals;
3. nodes of the same kind are horizontally or vertically aligned;
4. edges are orthogonal, i.e., any bends are 90°.

and thus is different from the notion of a *grid layout*, which is simply property (1). In this paper we are primarily interested in producing layouts with properties (1) and (2), though our methods could also achieve (3). We do not consider edges with orthogonal bends, though this could be an extension or achieved through a routing post-process (a simple example of this is provided in [10]).

The standard approach to extending force-directed methods to handle new aesthetic criteria is to add extra "forces" which push nodes in order to satisfy particular aesthetics. One of the most commonly used functions is *stress* [9]. Our first contribution (Sect. 3) is to develop penalty terms that can be added to the stress function to reward placement on points in a grid (Property 1) and to reward horizontal or vertical node alignment and/or horizontal or vertical edges (Property 2 or 3). We call these the *Grid-Snap* and *Node-Snap* methods respectively.

However, additional terms can make the goal function rich in local minima that impede convergence to a more aesthetically pleasing global minimum. Also, such "soft" constraints cannot guarantee satisfaction and so layouts in which nodes are *nearly-but-not-quite aligned* can occur. For this reason we investigate a second approach based on constrained graph layout in which *hard* alignment constraints are automatically added to the layout so as to ensure horizontal or vertical node alignment and thus horizontal or vertical edges (Property 2 or 3). This *adaptive constrained alignment (ACA)* method (Sect. 4) is the most innovative contribution of our paper.

In Sect. 6 we provide an empirical investigation of the speed of these approaches and the quality of layout with respect to various features encoding what we feel are the aesthetic criteria important in grid-like network layout.

While the above approaches can be used in once-off network layout, our original motivation was for interactive-layout applications. In Sect. 5 we discuss an interaction model based on the above for the use of grid-like layout in interactive semi-automatic layout tools such as Dunnart.

Related Work: Our research is related to proposals for automatic grid-like layout of biological networks [1,13,11]. These arrange biological networks with grid coordinates for nodes in addition to various layout constraints. In particular Barsky *et al.* [1] consider alignment constraints between biologically similar nodes and Kojima *et al.* [11] perform layout subject to rectangular containers around functionally significant groups of nodes (e.g., metabolites inside the nucleus of a cell). In general they use fairly straight-forward simulated annealing or simple incremental local-search strategies. Such methods work to a degree but are slow and may never reach a particularly aesthetically appealing minimum.

Another application where grid-like layout is an important aesthetic is automatic metro-map layout. Stott *et al.* [18] use a simple local-search ("hill-climbing") technique to obtain layout on grid points subject to a number of constraints, such as octilinear edge orientation. Wang and Chi [20] seek similar layout aesthetics but using continuous non-linear optimization subject to octilinearity and planarity constraints. This work, like ours, is based on a quasi-Newton optimization method, but it is very specific to metro-map layout and it is not at all clear how these techniques could be adapted to general-purpose interactive diagramming applications. Metro-map layout algorithms such as [16] are designed to run for many hours before finding a solution.

Another family of algorithms that compute grid-like layout are so-called *orthogonal* graph drawing methods. There have been some efforts to make these incremental, for example Brandes *et al.* [2] can produce an orthogonal drawing of a graph that respects the topology for a given set of initial node positions. Being based on the "Kandinski" orthogonal layout pipeline, extending such a method with user-defined constraints such as alignment or hierarchical containment would require non-trivial engineering of each stage in the pipeline. There is also a body of theoretical work considering the computability and geometric properties of layout with grid-constraints for various classes of graphs, e.g. [3]. Though interesting in its own right, such work is usually not intended for practical application, which is the primary concern of this paper.

There are several examples of the application of soft-constraints to layout. Sugiyama and Misue [19] augment the standard force-model with "magnetic" edge-alignment forces. Ryall *et al.* [17] explored the use of various force-based constraints in the context of an interactive diagramming editor. It is the limitations of such soft constraints (discussed below) which prompt the development of the techniques described in Sect. 4.

2   Aesthetic Criteria

Throughout this paper we assume that we have a graph $G = (V, E, w, h)$ consisting of a set of nodes V, a set of edges $E \subseteq V \times V$ and w_v, h_v are the width and height of node $v \in V$. We wish to find a straight-line 2D drawing for G. This is specified by a pair (x, y) where (x_v, y_v) is the centre point of each $v \in V$.

We quantify grid-like layout quality through the following metrics. In subsequent sections we use these to develop soft and hard constraints that directly or indirectly aim to optimise them. We use these metrics in our evaluation Sect. 6.

Embedding quality We measure this using the *P-stress* function [8], a variant of *stress* [9] that does not penalise unconnected nodes being more than their desired distance apart. It measures the separation between each pair of nodes $u, v \in V$ in the drawing and their *ideal distance* d_{uv} proportional to the graph theoretic path between them:

$$\sum_{u<v\in V} w_{uv} \left((d_{uv} - d(u,v))^+ \right)^2 + \sum_{(u,v)\in E} wp \left((d(u,v) - d_L)^+ \right)^2$$

where $d(u,v)$ is the Euclidean distance between u and v, $(z)^+ = z$ if $z \geq 0$ otherwise 0, d_L is an ideal edge length, $wp = \frac{1}{d_L}$, and $w_{uv} = \frac{1}{d_{uv}^2}$.

Edge crossings The number of edge crossings in the drawing.

Edge/node overlap The number of edges intersecting a node box. With straight-line edges this also penalises coincident edges.[1]

Angular resolution Edges incident on the same node have a uniform angular separation. Stott *et al.* [18] give a useful formulation:

$$\sum_{v \in V} \sum_{\{e_1, e_2\} \in E} |2\pi / degree(v) - \theta(e_1, e_2)|$$

Edge obliqueness We prefer horizontal or vertical edges and then—with weaker preference—edges at a 45° orientation. Our precise metric is $M \left| \tan^{-1} \frac{y_u - y_v}{x_u - x_v} \right|$ where $M(\theta)$ is an "M-shaped function" over $[0, \pi/2]$ that highly penalizes edges which are almost but not quite axis-aligned and gives a lower penalty for edges midway between horizontal and vertical.[2] Other functions like those of [18,11] could be used instead.

Grid placement Average of distances of nodes from their closest grid point.

3   Soft-Constraint Approaches

In this section we describe two new terms that can be combined with the *P-stress* function to achieve more grid-like layout: *NS-stress* for "node-snap stress" and *GS-stress* for "grid-snap stress." An additional term *EN-sep* gives good separation between nodes and edges. Layout is then achieved by minimizing

$$\text{P-stress} + k_{ns} \cdot \text{NS-stress} + k_{gs} \cdot \text{GS-stress} + k_{en} \cdot \text{EN-sep}$$

where $k_{ns,gs,en}$ control the "strength" of the various components. These extra terms, as defined below, tend to make nodes lie on top of one another. It is essential to avoid this by solving subject to node-overlap prevention constraints, as described in [6]. To obtain an initial "untangled" layout we run with $k_{ns} = k_{gs} = k_{en} = 0$ and without non-overlap constraints (Fig. 1(a)), and then run again with the extra terms and constraints to perform "grid beautification".

Minimization of the *NS-stress* term favours horizontal or vertical alignment of pairs of connected nodes (Figs. 1(c) and 5). Specifically, taking σ as the distance at which nodes should snap into alignment with one another, we define:

$$\text{NS-stress} = \sum_{(u,v) \in E} q_\sigma(x_u - x_v) + q_\sigma(y_u - y_v) \quad \text{where } q_\sigma(z) = \begin{cases} z^2/\sigma^2 & |z| \leq \sigma \\ 0 & \text{otherwise.} \end{cases}$$

---

[1] Node/node overlaps are also undesirable. We avoid them completely by using hard non-overlap constraints [6] in all our tests and examples.

[2] Note that $[0, \pi/2]$ is the range of $|tan^{-1}|$. The "M" function is zero at 0 and $\pi/2$, a small value $p \geq 0$ at $\pi/4$, a large value $P > 0$ at δ and $\pi/2 - \delta$ for some small $\delta > 0$, and linear in-between.

We originally tried several other penalty functions which turned out not to have good convergence. In particular any smooth function with local maxima at $\pm\sigma$ must be concave-down somewhere over the interval $[-\sigma, \sigma]$, and while differentiability may seem intuitively desirable for quadratic optimization it is in fact trumped by downward concavity, which plays havoc with standard step-size calculations on which our gradient-projection algorithm is based. Thus, obvious choices like an inverted quartic $(1 + (z^2 - \sigma^2)^2)^{-1}$ or a sum of inverted quadratics $(1 + (z + \sigma)^2)^{-1} + (1 + (z - \sigma)^2)^{-1}$ proved unsuitable in place of $q_\sigma(z)$. We review the step size, gradient, and Hessian formulae for our snap-stress functions in [10].

We designed our *GS-stress* function likewise to make the lines of a virtual grid exert a similar attractive force on nodes once within the snap distance σ:

$$\text{GS-stress} = \sum_{u \in V} q_\sigma(x_u - a_u) + q_\sigma(y_u - b_u)$$

where (a_u, b_u) is the closest grid point to (x_u, y_u) (with ties broken by favouring the point closer to the origin), see Fig. 1(b). The grid is defined to be the set of all points $(n\tau, m\tau)$, where n and m are integers, and τ is the "grid size". With *GS-stress* active it is important to set some other parameters proportional to τ. First, we take $\sigma = \tau/2$. Next, we modify the non-overlap constraints to allow no more than one node centre to be in the vicinity of any one grid point by increasing the minimum separation distance allowed between adjacent nodes to τ. Finally, we found that setting the ideal edge length equal to τ for initial force-directed layout, before activating *GS-stress*, helped to put nodes in positions compatible with the grid.

Our third term *EN-sep* is also a quadratic function based on $q_\sigma(z)$ that separates nodes and nearby axis-aligned edges to avoid node/edge overlaps and coincident edges:

$$\text{EN-sep} = \sum_{e \in E_V \cup E_H} \sum_{u \in V} q_\sigma\left((\sigma - d(u, e))^+\right),$$

where E_V and E_H are the sets of vertically and horizontally aligned edges, respectively, and the distance $d(u, e)$ between a node u and an edge e is defined as the length of the normal from u to e if that exists, or $+\infty$ if it does not. Here again we took $\sigma = \tau/2$.

In our experiments we refer to various combinations of these terms and constraints:
Node-Snap: *NS-stress*, *EN-sep*, non-overlap constraints, $k_{gs} = 0$
Grid-Snap: *GS-stress*, *EN-sep*, ideal edge lengths equal to grid size, non-overlap, constraints with separations tailored to grid size, $k_{ns} = 0$.
Node-Snap+Grid-Snap: achieves extra alignment by adding *NS-stress* to the above **Grid-Snap** recipe (i.e. $k_{ns} \neq 0$)

---

4   Adaptive Constrained Alignment

Another way to customize constrained force-directed layout is by adding *hard* constraints, and in this section we describe how to make force-directed layouts more grid-like simply by adding alignment and separation constraints (Fig. 1(e)).

The algorithm, which we call *Adaptive Constrained Alignment or ACA*, is a greedy algorithm which repeatedly chooses an edge in G and aligns it horizontally or vertically (see *adapt_const_align* procedure of Figure 2). It *adapts* to user specified constraints by not adding alignments that violate these. The algorithm halts when the heuristic can no longer apply alignments without creating edge overlaps. Since each edge is aligned at most once, there are at most $|E|$ iterations.

We tried the algorithm with three different heuristics for choosing potential alignments, which we discuss below.

Node overlaps and edge/node overlaps can be prevented with hard non-overlap constraints and the *EN-sep* soft constraint discussed in Section 3, applied either before or after the ACA process. However, coincident edges can be accidentally created and then enforced as we apply alignments if we do not take care to maintain the orthogonal ordering of nodes. If for example two edges (u, v) and (v, w) sharing a common endpoint v are both horizontally aligned, then we must maintain either the ordering $x_u < x_v < x_w$ or the opposite ordering $x_w < x_v < x_u$.

Therefore we define the notion of a *separated alignment*, written $\mathsf{SA}(u, v, D)$ where $u, v \in V$ and $D \in \{\mathbb{N}, \mathbb{S}, \mathbb{W}, \mathbb{E}\}$ is a compass direction. Applying a separated alignment means applying two constraints to the force-directed layout— one alignment and one separation—as follows:

$$\mathsf{SA}(u, v, \mathbb{N}) \equiv x_u = x_v \text{ and } y_v + \beta(u, v) \leq y_u, \quad \mathsf{SA}(u, v, \mathbb{S}) \equiv \mathsf{SA}(v, u, \mathbb{N}),$$
$$\mathsf{SA}(u, v, \mathbb{W}) \equiv y_u = y_v \text{ and } x_v + \alpha(u, v) \leq x_u, \quad \mathsf{SA}(u, v, \mathbb{E}) \equiv \mathsf{SA}(v, u, \mathbb{W}),$$

where $\alpha(u, v) = (w_u + w_v)/2$ and $\beta(u, v) = (h_u + h_v)/2$. (Thus for example $\mathsf{SA}(u, v, \mathbb{N})$ can be read as, "the ray from u through v points north," where we think of v as lying north of u when its y-coordinate is smaller.)

```
proc adapt_const_align(G, C, H)              proc chooseSA(G, C, x, y, K)
   (x, y) ← cfdl(G, C)                          S ← NULL
   SA ← H(G, C, x, y)                           cost ← ∞
   while SA ! = NULL                            for each (u, v) ∈ E and dir. D
      C.append(SA)                                 if not creates_coincidence(C, x, y, u, v, D)
      (x, y) ← cfdl(G, C)                             if K(u, v, D) < cost
      SA ← H(G, C, x, y)                                 S ← SA(u, v, D)
   return (x, y, C)                                      cost ← K(u, v, D)
                                                return S
```

Fig. 2. Adaptive constrained alignment algorithm. G is the given graph, C the set of user-defined constraints, H the alignment choice heuristic, and *cfdl* the constrained force-directed layout procedure.

Alignment Choice Heuristics. We describe two kinds of alignment choice heuristics: *generic*, which can be applied to any graph, and *convention-based*, which are intended for use with layouts that conform to special conventions, e.g. SBGN diagrams [12]. Our heuristics are designed according to two principles: (1) try to retain the overall shape of the initial force-directed layout; (2) do not obscure the graph structure by creating undesirable overlaps and differ only in the choice of a *cost function* K which is plugged into the procedure `chooseSA` in Figure 2. This relies on procedure `creates_coincidence` which implements the edge coincidence test described by Theorem 1. Among separated alignments which would not lead to an edge coincidence, `chooseSA` selects one of lowest cost. Cost functions may return a special value of ∞ to mark an alignment as never to be chosen.

The `creates_coincidence` procedure works by maintaining a $|V|$-by-$|V|$ array of flags which indicate for each pair of nodes u, v whether they are aligned in either dimension and whether there is an edge between them. The cost of initializing the array is $\mathcal{O}(|V|^2 + |E| + |C|)$, but this is done only once in ACA. Each time a new alignment constraint is added the flags are updated in $\mathcal{O}(|V|)$ time, due to transitivity of the alignment relation. Checking whether a proposed separated alignment would create an edge coincidence also takes $\mathcal{O}(|V|)$ time, and works according to Theorem 1. (Proof is provided in [10].) Note that the validity of Theorem 1 relies on the fact that we apply separated alignments $SA(u, v, D)$ only when (u, v) is an edge in the graph.

Theorem 1. *Let G be a graph with separated alignments. Let u, v be nodes in G which are not yet constrained to one another. Then the separated alignment $SA(u, v, \mathbb{E})$ creates an edge coincidence in G if and only if there is a node w which is horizontally aligned with either u or v and satisfies either of the following two conditions: (i) $(u, w) \in E$ while $x_u < x_w$ or $x_v < x_w$; or (ii) $(w, v) \in E$ while $x_w < x_v$ or $x_w < x_u$. The case of vertical alignments is similar.*

We tried various cost functions, which addressed the aesthetic criteria of Section 2 in different ways. We began with a *basic cost*, which was either an estimate $K_{dS}(u, v, D)$ of the change in the stress function after applying the proposed alignment $SA(u, v, D)$, or else the negation of the obliqueness of the edge, $K_{ob}(u, v, D) = -obliqueness((u, v))$, as measured by the function of Section 2. In this way we could choose to address the aesthetic criteria of *embedding quality* or *edge obliqueness*, and we found that the results were similar. Both rules favour placing the first alignments on edges which are almost axis-aligned, and this satisfies our first principle of being guided as much as possible by the shape of the initial force-directed layout. See for example Figure 1.

On top of this basic cost we considered *angular resolution* of degree-2 nodes by adding a large but finite cost that would postpone certain alignments until after others had been attempted; namely, we added a fixed cost of 1000 (ten times larger than average values of K_{dS} and K_{ob}) for any alignment that would make a degree-2 node into a "bend point," i.e., would make one of its edges horizontally aligned while the other was vertically aligned. This allows long chains of degree-2 nodes to form straight lines, and cycles of degree-2 nodes to form perfect rectangles. For SBGN diagrams we used a modified rule based on *non-leaf degree*, or number of neighbouring nodes which are not leaves (Figs. 1(e) and (f)).

Respecting User-Defined Constraints. Layout constraints can easily wind up in conflict with one another if not chosen carefully. In Dunnart such conflicts are detected during the projection operation described in [5], an *active set*

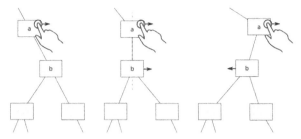

Fig. 3. Interacting with Node-Snap. The user is dragging node a steadily to the right. When the horizontal distance between a and b is less than the average width of these two nodes, the *NS-stress* function causes b to align with a. As the user continues dragging, the now aligned node b will follow until either a quick jerk of node a breaks the alignment, or else edges attached to b pull it back to the left, overcoming its attraction to a. To the user, the impression is that the alignment persisted until it was "torn" by the underlying forces in the system.

method which iteratively determines the most violated constraint c and satisfies it by minimal disturbance of the node positions. When it is impossible to satisfy c without violating one of the constraints that is already in the active set, c is simply marked *unsatisfiable*, and the operation carries on without it.

For ACA it is important that user-defined constraints are never marked un-satisfiable in deference to an alignment imposed by the process; therefore we term the former *definite* constraints and the latter *tentative* constraints. We employ a modified projection operation which always chooses to mark one or more tentative constraint as unsatisfiable if they are involved in a conflict.

For conflicts involving more than one tentative constraint, we use Lagrange multipliers to choose which one to reject. These are computed as a part of the projection process. Since alignment constraints are equalities (not inequalities) the sign of their Lagrange multiplier does not matter, and a constraint whose Lagrange multiplier is maximal in absolute value is one whose rejection should permit the greatest decrease in the stress function. Therefore we choose this one.

ACA does not snap nodes to grid-points: if desired this can be achieved once ACA has added the alignment constraints by activating Grid-Snap.

5   Interaction

One benefit of the approaches described above is that they are immediately applicable for use in interactive tools where the underlying graph, the prerequisite constraint system, or ideal positions for nodes can all change dynamically. We implemented Node-Snap, Grid-Snap and Adaptive Constrained Alignment for interactive use in the Dunnart diagram editor.[3] In Dunnart, automatic layout

---

[3] A video demonstrating interactive use of the approaches described in this paper is available at http://www.dunnart.org.

runs continuously in a background worker thread, allowing the layout to adapt immediately to user-specified changes to positions or constraints.

For example, Figure 3 illustrates user interaction with Node-Snap. As the user drags a node around the canvas, it may snap into alignment with an adjacent node. Slowly dragging a node aligned with other nodes will move them together and keep them in alignment, while quickly dragging a node will instead cause it to be torn from any alignments.

When we tried Node-Snap interactively in Dunnart we found that nodes tended to stick together in clumps if the σ parameter of *NS-stress* was larger than their average size in either dimension. We solved this problem by replacing the snap-stress term by

$$\sum_{(u,v)\in E} q_{\alpha(u,v)}(x_u - x_v) + q_{\beta(u,v)}(y_u - y_v)$$

where α, β are as defined in Sect. 4.

In Dunnart, a dragged object is always pinned to the mouse cursor. In the case of Grid-Snap, the dragged node is unpinned and will immediately snap to a grid point on mouse-up. Other nodes, however, will snap-to or tear-away from grid points in response to changing dynamics in the layout system. During dragging we also turn off non-overlap constraints and reapply them on mouse-up. This prevents nodes being unexpectedly pushed out of place as a result of the expanded non-overlap region (Sect. 3). Additionally, since *GS-stress* holds nodes in place, we allow the user to quickly drag a node to temporarily overcome the grid forces and allow the layout to untangle with standard force-directed layout. Once it converges we automatically reapply *GS-stress*.

6   Evaluation

To evaluate the various techniques we applied each to 252 graphs from the "AT&T Graphs" corpus (`ftp://ftp.research.att.com/dist/drawdag/ug.gz`) with between 10 and 244 nodes. We excluded graphs with fewer than 10 nodes and two outlier graphs: one with 1103 nodes and one with 0 edges. We recorded running times of each stage in the automated batch process and the various aesthetic metrics described in Sect. 2, using a MacBook Pro with a 2.3GHz Intel Core I7 CPU. Details of collected data etc. are given in [10].

We found that ACA was the slowest, often taking up to 10 times as long as the other methods, on average around 5 seconds for graphs with around 100 nodes, while the other approaches took around a second. ACA was also sensitive to the density of edges. Of the soft constraint approaches, Grid-Snap (being very local) added very little time over the unconstrained force-directed approach.

The Edge Obliqueness results are shown in Fig. 4 as this is arguably the metric that is most indicative of grid-like layout. Another desirable property of grid-like layout, as noted in Sect. 4 is that longer paths in the graph also be aligned. ACA does a good job of aligning such paths, as is visible in Fig. 1 and 5.

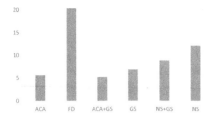

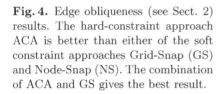

Fig. 4. Edge obliqueness (see Sect. 2) results. The hard-constraint approach ACA is better than either of the soft constraint approaches Grid-Snap (GS) and Node-Snap (NS). The combination of ACA and GS gives the best result.

Fig. 5. Layout of a SBGN diagram of Calvin Cycle pathway shows how ACA (right) gives a more pleasing rectangular layout than Node-Snap (left).

7   Conclusion

We explored incorporating grid-like layout conventions into constraint-based graph layout. We give two *soft* approaches (Node-Snap, Grid-Snap) based on adding terms to the goal function, and an adaptive constraint based approach (ACA) in which *hard* alignment constraints are added greedily. ACA is slower but gives more grid-like layout and so is preferred for medium sized graphs.

We have also discussed how the approaches can be integrated into interactive diagramming tools like Dunnart. Here ACA and Grid-Snap provide good initial layouts, while Node-Snap helps the user create further alignments by hand.

Future work is to improve the speed of ACA by adding more than one alignment constraint at a time and also to use Lagrange multipliers (LM) to improve the adaptivity of ACA. One idea is to reject any alignment whose LM exceeds a threshold on each iteration of ACA. Then, running ACA continuously during interaction would allow us to achieve the behaviour illustrated in Fig. 3 through hard rather than soft constraints. Another issue with all the techniques described is the many fiddly parameters, weights and thresholds. We intend to further investigate principled ways to automatically set these.

References

1. Barsky, A., Gardy, J.L., Hancock, R.E., Munzner, T.: Cerebral: a cytoscape plugin for layout of and interaction with biological networks using subcellular localization annotation. Bioinformatics 23(8), 1040–1042 (2007)
2. Brandes, U., Eiglsperger, M., Kaufmann, M., Wagner, D.: Sketch-driven orthogonal graph drawing. In: Goodrich, M.T., Kobourov, S.G. (eds.) GD 2002. LNCS, vol. 2528, pp. 1–11. Springer, Heidelberg (2002)
3. Chrobak, M., Payne, T.H.: A linear-time algorithm for drawing a planar graph on a grid. Information Processing Letters 54(4), 241–246 (1995)
4. Di Battista, G., Eades, P., Tamassia, R., Tollis, I.G.: Graph Drawing: Algorithms for the Visualization of Graphs. Prentice-Hall, Inc. (1999)

5. Dwyer, T., Koren, Y., Marriott, K.: IPSep-CoLa: An incremental procedure for separation constraint layout of graphs. IEEE Transactions on Visualization and Computer Graphics 12(5), 821–828 (2006)
6. Dwyer, T., Marriott, K., Stuckey, P.J.: Fast node overlap removal. In: Healy, P., Nikolov, N.S. (eds.) GD 2005. LNCS, vol. 3843, pp. 153–164. Springer, Heidelberg (2006)
7. Dwyer, T., Marriott, K., Wybrow, M.: Dunnart: A constraint-based network diagram authoring tool. In: Tollis, I.G., Patrignani, M. (eds.) GD 2008. LNCS, vol. 5417, pp. 420–431. Springer, Heidelberg (2009)
8. Dwyer, T., Marriott, K., Wybrow, M.: Topology preserving constrained graph layout. In: Tollis, I.G., Patrignani, M. (eds.) GD 2008. LNCS, vol. 5417, pp. 230–241. Springer, Heidelberg (2009)
9. Gansner, E.R., Koren, Y., North, S.C.: Graph drawing by stress majorization. In: Pach, J. (ed.) GD 2004. LNCS, vol. 3383, pp. 239–250. Springer, Heidelberg (2005)
10. Kieffer, S., Dwyer, T., Marriott, K., Wybrow, M.: Incremental grid-like layout using soft and hard constraints. Tech. Rep. 2013/275, Monash University (2013), http://www.csse.monash.edu.au/publications/2013/tr-2013-275-full.pdf
11. Kojima, K., Nagasaki, M., Jeong, E., Kato, M., Miyano, S.: An efficient grid layout algorithm for biological networks utilizing various biological attributes. BMC Bioinformatics 8(1), 76 (2007)
12. Le Novère, N., et al.: The Systems Biology Graphical Notation. Nature Biotechnology 27, 735–741 (2009)
13. Li, W., Kurata, H.: A grid layout algorithm for automatic drawing of biochemical networks. Bioinformatics 21(9), 2036–2042 (2005)
14. Marriott, K., Purchase, H., Wybrow, M., Goncu, C.: Memorability of visual features in network diagrams. IEEE Transactions on Visualization and Computer Graphics 18(12), 2477–2485 (2012)
15. MetaCrop, http://metacrop.ipk-gatersleben.de
16. Nöllenburg, M., Wolff, A.: Drawing and labeling high-quality metro maps by mixed-integer programming. IEEE Transactions on Visualization and Computer Graphics 17(5), 626–641 (2011)
17. Ryall, K., Marks, J., Shieber, S.: An interactive constraint-based system for drawing graphs. In: Robertson, G.G., Schmandt, C. (eds.) Proceedings of the 10th Annual ACM Symposium on User Interface Software and Technology, pp. 97–104. ACM Press (1997)
18. Stott, J., Rodgers, P., Martinez-Ovando, J.C., Walker, S.G.: Automatic metro map layout using multicriteria optimization. IEEE Transactions on Visualization and Computer Graphics 17(1), 101–114 (2011)
19. Sugiyama, K., Misue, K.: Graph drawing by the magnetic spring model. Journal of Visual Languages and Computing 6(3), 217–231 (1995)
20. Wang, Y.S., Chi, M.T.: Focus+context metro maps. IEEE Transactions on Visualization and Computer Graphics 17(12), 2528–2535 (2011)

Using ILP/SAT to Determine Pathwidth, Visibility Representations, and other Grid-Based Graph Drawings[*]

Therese Biedl[1], Thomas Bläsius[2], Benjamin Niedermann[2], Martin Nöllenburg[2], Roman Prutkin[2], and Ignaz Rutter[2]

[1] David R. Cheriton School of Computer Science, University of Waterloo, Canada
[2] Institute of Theoretical Informatics, Karlsruhe Institute of Technology, Germany

Abstract. We present a simple and versatile formulation of grid-based graph representation problems as an integer linear program (ILP) and a corresponding SAT instance. In a grid-based representation vertices and edges correspond to axis-parallel boxes on an underlying integer grid; boxes can be further constrained in their shapes and interactions by additional problem-specific constraints. We describe a general d-dimensional model for grid representation problems. This model can be used to solve a variety of NP-hard graph problems, including pathwidth, bandwidth, optimum st-orientation, area-minimal (bar-k) visibility representation, boxicity-k graphs and others. We implemented SAT-models for all of the above problems and evaluated them on the Rome graphs collection. The experiments show that our model successfully solves NP-hard problems within few minutes on small to medium-size Rome graphs.

1 Introduction

Integer linear programming (ILP) and Boolean satisfiability testing (SAT) are indispensable and widely used tools in solving many hard combinatorial optimization and decision problems in practical applications [3,9]. In graph drawing, especially for planar graphs, these methods are not frequently applied. A few notable exceptions are crossing minimization [8, 10, 19, 22], orthogonal graph drawing with vertex and edge labels [4] and metro-map layout [26]. Recent work by Chimani et al. [11] uses SAT formulations for testing upward planarity. All these approaches have in common that they exploit problem-specific properties to derive small and efficiently solvable models, but they do not generalize to larger classes of problems.

In this paper we propose a generic ILP model that is flexible enough to capture a large variety of different grid-based graph layout problems, both polynomially-solvable and NP-complete. We demonstrate this broad applicability by adapting the base model to six different NP-complete example problems: pathwidth, bandwidth, optimum st-orientation, minimum area bar- and bar k-visibility representation, and boxicity-k testing. For minimum-area visibility representations and boxicity this is, to the best of our knowledge, the first implementation of an exact solution method. Of course this flexibility comes at the cost of losing some of the efficiency of more specific approaches.

---

[*] T. Biedl is supported by NSERC, M. Nöllenburg is supported by the Concept for the Future of KIT under grant YIG 10-209.

S. Wismath and A. Wolff (Eds.): GD 2013, LNCS 8242, pp. 460–471, 2013.

Our goal, however, is not to achieve maximal performance for a specific problem, but to provide an easy-to-adapt solution method for a larger class of problems, which allows quick and simple prototyping for instances that are not too large. Our ILP models can be translated into equivalent SAT formulations, which exhibit better performance in the implementation than the ILP models themselves. We illustrate the usefulness of our approach by an experimental evaluation that applies our generic model to the above six NP-complete problems using the well-known Rome graphs [1] as a benchmark set. Our evaluation shows that, depending on the problem, our model can solve small to medium-size instances (sizes varying from about 25 vertices and edges for bar-1 visibility testing up to more than 250 vertices and edges, i.e., all Rome graphs, for optimum st-orientation) to optimality within a few minutes. In Section 2, we introduce generic grid-based graph representations and formulate an ILP model for d-dimensional integer grids. We show how this model can be adapted to six concrete one-, two- and d-dimensional grid-based layout problems in Sections 3 and 4. In Section 5 we evaluate our implementations and report experimental results. The implementation is available from illwww.iti.kit.edu/gdsat. Omitted proofs are in the full version [2].

2   Generic Model for Grid-Based Graph Representations

In this section we explain how to express d-dimensional boxes in a d-dimensional integer grid as constraints of an ILP or a SAT instance. In the subsequent sections we use these boxes as basic elements for representing vertices and edges in problem-specific ILP and SAT models. Observe that we can restrict ourselves to boxes in integer grids.

Lemma 1. *Any set I of n boxes in $\mathbb{R}^d$ can be transformed into another set I' of n closed boxes on the integer grid $\{1,\ldots,n\}^d$ such that two boxes intersect in I if and only if they intersect in I'.*

2.1   Integer Linear Programming Model

We will describe our model in the general case for a d-dimensional integer grid, where $d \geq 1$. Let $\mathcal{R}^d = [1, U_1] \times \ldots \times [1, U_d]$ be a bounded d-dimensional integer grid, where $[A, B]$ denotes the set of integers $\{A, A+1, \ldots, B-1, B\}$. In a *grid-based graph representation* vertices and/or edges are represented as d-dimensional boxes in $\mathcal{R}^d$. A *grid box* R in $\mathcal{R}^d$ is a subset $[s_1, t_1] \times \ldots \times [s_d, t_d]$ of $\mathcal{R}^d$, where $1 \leq s_k \leq t_k \leq U_k$ for all $1 \leq k \leq d$. In the following we describe a set of ILP constraints that together create a non-empty box for some object v. We denote this ILP as $\mathcal{B}(d)$.

We first extend $\mathcal{R}^d$ by a margin of dummy points to $\bar{\mathcal{R}}^d = [0, U_1+1] \times \ldots \times [0, U_d+1]$. We use three sets of binary variables:

$$x_{\mathbf{i}}(v) \in \{0, 1\} \qquad \forall \mathbf{i} \in \bar{\mathcal{R}}^d \tag{1}$$

$$b_i^k(v) \in \{0, 1\} \qquad \forall 1 \leq k \leq d \text{ and } 1 \leq i \leq U_k \tag{2}$$

$$e_i^k(v) \in \{0, 1\} \qquad \forall 1 \leq k \leq d \text{ and } 1 \leq i \leq U_k \tag{3}$$

Variables $x_{\mathbf{i}}(v)$ indicate whether grid point $\mathbf{i}$ belongs to the box representing v ($x_{\mathbf{i}}(v) = 1$) or not ($x_{\mathbf{i}}(v) = 0$). Variables $b_i^k(v)$ and $e_i^k(v)$ indicate whether the box of v may start

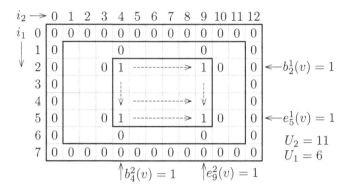

Fig. 1. Example of a 2-dimensional 8×13 grid $\bar{\mathcal{R}}^d$ with a 4×6 grid box and the corresponding variable assignments

or end at position i in dimension k. We use $\mathbf{i}[k]$ to denote the k-th coordinate of grid point $\mathbf{i} \in \mathcal{R}^d$ and $\mathbf{1_k} = (0, \dots, 0, 1, 0, \dots, 0)$ to denote the k-th d-dimensional unit vector. If $d = 1$ we will drop the dimension index of the variables to simplify the notation. The following constraints model a box in $\mathcal{R}^d$ (see Fig. 1 for an example):

$$x_{\mathbf{i}}(v) = 0 \qquad\qquad \forall \mathbf{i} \in \bar{\mathcal{R}}^d \setminus \mathcal{R}^d \qquad (4)$$

$$\sum_{\mathbf{i} \in \bar{\mathcal{R}}^d} x_{\mathbf{i}}(v) \geq 1 \qquad\qquad\qquad\qquad (5)$$

$$\sum_{i \in [1, U_k]} b_i^k(v) = 1 \qquad\qquad \forall 1 \leq k \leq d \qquad (6)$$

$$\sum_{i \in [1, U_k]} e_i^k(v) = 1 \qquad\qquad \forall 1 \leq k \leq d \qquad (7)$$

$$x_{\mathbf{i}-\mathbf{1_k}}(v) + b_{\mathbf{i}[k]}^k(v) \geq x_{\mathbf{i}}(v) \qquad \forall \mathbf{i} \in \mathcal{R}^d \text{ and } 1 \leq k \leq d \quad (8)$$

$$x_{\mathbf{i}}(v) \leq x_{\mathbf{i}+\mathbf{1_k}}(v) + e_{\mathbf{i}[k]}^k(v) \qquad \forall \mathbf{i} \in \mathcal{R}^d \text{ and } 1 \leq k \leq d \quad (9)$$

Constraint (4) creates a margin of zeroes around $\mathcal{R}^d$. Constraint (5) ensures that the shape representing v is non-empty, and constraints (6) and (7) provide exactly one start and end position in each dimension. Finally, due to constraints (8) and (9) each grid point inside the specified bounds belongs to v and all other points don't.

Lemma 2. *The ILP $\mathcal{B}(d)$ defined by constraints (1)–(9) correctly models all non-empty grid boxes in $\mathcal{R}^d$.*

Our example ILP models in Sections 3 and 4 extend ILP $\mathcal{B}(d)$ by constraints controlling additional properties of vertex and edge boxes. For instance, boxes can be easily constrained to be single points, to be horizontal or vertical line segments, to intersect if and only if they are incident or adjacent in G, to meet in endpoints etc. The definition of an objective function for the ILP depends on the specific problem at hand and will be discussed in the problem sections. In the full version [2] we explain how to translate the ILP $\mathcal{B}(d)$ into an equivalent SAT formulation with better practical performance.

3   One-Dimensional Problems

In the following, let $G = (V, E)$ be an undirected graph with $|V| = n$ and $|E| = m$. One-dimensional grid-based graph representations can be used to model vertices as intersecting intervals (one-dimensional boxes) or as disjoint points that induce a certain vertex order. We present ILP models for three such problems.

3.1   Pathwidth

The pathwidth of a graph G is a well-known graph parameter with many equivalent definitions. We use the definition via the smallest clique size of an interval supergraph. More precisely, a graph is an *interval graph* if it can be represented as intersection graph of 1-dimensional intervals. A graph G has *pathwidth* $pw(G) \leq p$ if there exists an interval graph H that contains G as a subgraph and for which all cliques have size at most $p+1$. It is NP-hard to compute the pathwidth of an arbitrary graph and even hard to approximate it [5]. There are fixed-parameter algorithms for computing the pathwidth, e.g. [6], however, we are not aware of any implementations of these algorithms. The only available implementations are exponential-time algorithms, e.g., in sage[1].

Problem 1 (Pathwidth). Given a graph $G = (V, E)$, determine the pathwidth of G, i.e., the smallest integer p so that $pw(G) \leq p$.

There is an interesting connection between pathwidth and planar graph drawings of small height. Any planar graph that has a planar drawing of height h has pathwidth at most h [18]. Also, pathwidth is a crucial ingredient in testing in polynomial time whether a graph has a planar drawing of height h [15].

We create a one-dimensional grid representation of G, in which every vertex is an interval and every edge forces the two vertex intervals to intersect. The objective is to minimize the maximum number of intervals that intersect in any given point. We use the ILP $\mathcal{B}(1)$ for a grid $\mathcal{R} = [1, n]$, which already assigns a non-empty interval to each vertex $v \in V$. We add binary variables for the edges of G, a variable $p \in \mathbb{N}$ representing the pathwidth of G, and a set of additional constraints as follows.

$$x_i(e) \in \{0, 1\} \qquad\qquad\qquad \forall i \in \mathcal{R} \text{ and } e \in E \qquad (10)$$

$$\sum_{i \in \mathcal{R}} x_i(e) \geq 1 \qquad\qquad\qquad \forall e \in E \qquad (11)$$

$$x_i(uv) \leq x_i(u) \qquad x_i(uv) \leq x_i(v) \qquad \forall uv \in E \qquad (12)$$

$$\sum_{v \in V} x_i(v) \leq p + 1 \qquad\qquad\qquad \forall i \in \mathcal{R} \qquad (13)$$

Our objective function is to minimize the value of p subject to the above constraints.

It is easy to see that every edge must be represented by some grid point (constraint (11)), and can only use those grid points, where the two end vertices intersect (constraint (12)). Hence the intervals of vertices define some interval graph H that is

---
[1] www.sagemath.org

a supergraph of G. Constraint (13) enforces that at most $p + 1$ intervals meet in any point, which by Helly's property means that H has clique-size at most $p + 1$. So G has pathwidth at most p. By minimizing p we obtain the desired result. In our implementation we translate the ILP into a SAT instance. We test satisfiability for fixed values of p, starting with $p = 1$ and increasing it incrementally until a solution is found.

Theorem 1. *There exists an ILP/SAT formulation with $O(n(n + m))$ variables and $O(n(n + m))$ constraints / $O(n^3 + n\binom{n}{p+2})$ clauses of maximum size n that has a solution of value $\leq p$ if and only if G has pathwidth $\leq p$.*

With some easy modifications, the above ILP can be used for testing whether a graph is a (proper) interval graph. Section 4.2 shows that boxicity-d graphs, the d-dimensional generalization of interval graphs, can also be recognized by our ILP.

3.2 Bandwidth

The bandwidth of a graph G with n vertices is another classic graph parameter, which is NP-hard to compute [12]; due to the practical importance of the problem there are also a few approaches to find exact solutions to the bandwidth minimization problem. For example, [14] and [25] use the branch-and-bound technique combined with various heuristics. We present a solution that can be easily described using our general framework. However, regarding the running time, it cannot be expected to compete with techniques specially tuned for solving the bandwidth minimization problem.

Let $f : V \rightarrow \{1, \ldots, n\}$ be a bijection that defines a linear vertex order. The *bandwidth* of G is defined as $bw(G) = \min_f \max\{f(v) - f(u) \mid uv \in E \text{ and } f(u) < f(v)\}$, i.e., the minimum length of the longest edge in G over all possible vertex orders.

In the full version [2] we describe an ILP that assigns the vertices of G to disjoint grid points and requires for an integer k that any pair of adjacent vertices is at most k grid points apart, i.e., we can test if $bw(G) \leq k$.

Theorem 2. *There exists an ILP/SAT formulation with $O(n^2)$ variables and $O(n \cdot m)$ constraints / $O(n^3)$ clauses of maximum size n that has a solution if and only if G has bandwidth $\leq k$.*

3.3 Optimum st-Orientation

Let G be an undirected graph and let s and t be two vertices of G with $st \in E$. An *st-orientation* of G is an orientation of the edges such that s is the unique source and t is the unique sink [17]. Such an orientation can exist only if $G \cup (s, t)$ is biconnected. Computing an st-orientation can be done in linear time [7, 17], but it is NP-complete to find an st-orientation that minimizes the length of the longest path from s to t, even for planar graphs [28]. It has many applications in graph drawing [27] and beyond.

Problem 2 (Optimum st-orientation). Given a graph $G = (V, E)$ and two vertices $s, t \in V$ with $st \in E$, find an orientation of E such that s is the only source, t is the only sink, and the length of the longest directed path from s to t is minimum.

In the full version [2] we formulate an ILP using points for vertices and non-degenerate intervals for edges that computes a *height-k st-orientation* of G, i.e., an st-orientation such that the longest path has length at most k (if one exists).

Theorem 3. *There exists an ILP with $O(n(n + m))$ variables and constraints that computes an optimum st-orientation. Alternatively, there exists an ILP/SAT formulation with $O(k(n + m))$ variables and $O(k(n + m))$ constraints / $O(k^2(n + m))$ clauses of maximum size n that has a solution if and only if a height-k st-orientation of G exists.*

4   Higher-Dimensional Problems

In this section we give examples of two-dimensional visibility graph representations and a d-dimensional grid-based graph representation problem. Let again $G = (V, E)$ be an undirected graph with $|V| = n$ and $|E| = m$.

4.1   Visibility Representations

A visibility representation (also: *bar visibility representation* or *weak visibility representation*) of a graph $G = (V, E)$ maps all vertices to disjoint horizontal line segments, called *bars*, and all edges to disjoint vertical bars, such that for each edge $uv \in E$ the bar of uv has its endpoints on the bars for u and v and does not intersect any other vertex bar. Visibility representations are an important visualization concept in graph drawing, e.g., it is well known that a graph is planar if and only if it has a visibility representation [29, 30]. An interesting recent extension are bar k-visibility representations [13], which additionally allow edges to intersect at most k non-incident vertex bars. We use our ILP to compute compact visibility and bar k-visibility representations. Minimizing the area of a visibility representation is NP-hard [24] and we are not aware of any implemented exact algorithms to solve the problem for any $k \geq 0$. By Lemma 1 we know that all bars can be described with integer coordinates of size $O(m + n)$.

Problem 3 (Bar k-Visibility Representation). Given a graph G and an integer $k \geq 0$, find a bar k-visibility representation on an integer grid of size $H \times W$ (if one exists).

Bar visibility representations. Our goal is to test whether G has a visibility representation in a grid with H columns and W rows (and thus minimize H or W). We set $\mathcal{R}^2 = [1, H] \times [1, W]$ and use ILP $\mathcal{B}(2)$ to create grid boxes for all edges and vertices in G. We add one more set of binary variables for vertex-edge incidences and the following constraints.

$$x_i(e, v) \in \{0, 1\} \qquad \forall i \in \mathcal{R}^2 \, \forall e \in E \, \forall v \in e \qquad (14)$$

$$b_i^1(v) = e_i^1(v) \qquad \forall i \in [1, U_1] \, \forall v \in V \qquad (15)$$

$$b_i^2(e) = e_i^2(e) \qquad \forall i \in [1, U_2] \, \forall e \in E \qquad (16)$$

$$\sum_{v \in V} x_i(v) \leq 1 \qquad \forall i \in \mathcal{R}^2 \qquad (17)$$

$$\sum_{v \in V \setminus e} x_i(v) \leq (1 - x_i(e)) \qquad \forall i \in \mathcal{R}^2 \, \forall e \in E \qquad (18)$$

$$x_{\mathbf{i}}(e, v) \leq x_{\mathbf{i}}(e) \qquad\qquad x_{\mathbf{i}}(e, v) \leq x_{\mathbf{i}}(v) \quad \forall \mathbf{i} \in \mathcal{R}^2 \, \forall e \in E \, \forall v \in e \quad (19)$$

$$\sum_{\mathbf{i} \in \mathcal{R}^2} x_{\mathbf{i}}(e, v) \geq 1 \qquad\qquad\qquad\qquad \forall e \in E \, \forall v \in e \quad (20)$$

$$x_{\mathbf{i}}(e, v) \leq b^1_{\mathbf{i}[1]}(e) + e^1_{\mathbf{i}[1]}(e) \qquad\qquad \forall \mathbf{i} \in \mathcal{R}^2 \, \forall e \in E \, \forall v \in e \quad (21)$$

Constraints (15) and (16) ensure that all vertex boxes are horizontal bars of height 1 and all edge boxes are vertical bars of width 1. Constraint (17) forces the vertex boxes to be disjoint; edge boxes will be implicitly disjoint (for a simple graph) due to the remaining constraints. No edge is allowed to intersect a non-incident vertex due to constraint (18). Finally, we need to set the new incidence variables $x_{\mathbf{i}}(e, v)$ for an edge e and an incident vertex v so that $x_{\mathbf{i}}(e, v) = 1$ if and only if e and v share the grid point $\mathbf{i}$. Constraints (19) and (20) ensure that each incidence in G is realized in at least one grid point, but it must be one that is used by the boxes of e and v. Finally, constraint (21) requires edge e to start and end at its two intersection points with the incident vertex boxes. This constraint is optional, but yields a tighter formulation.

Since every graph with a visibility representation is planar (and vice versa) we have $m \in O(n)$. Moreover, our ILP and SAT models can also be used to test planarity of a given graph by setting $H = n$ and $W = 2n - 4$, which is sufficient due to Tamassia and Tollis [29]. This might not look interesting at first sight since planarity testing can be done in linear time [21]. However, we think that this is still useful as one can add other constraints to the ILP model, e.g., to create simultaneous planar embeddings, and use it as a subroutine for ILP formulations of applied graph drawing problems such as metro maps [26] and cartograms.

Theorem 4. *There is an ILP/SAT formulation with $O(HWn)$ variables and $O(HWn)$ constraints / $O(HWn^2)$ clauses of maximum size HW that solves Problem 3 for $k = 0$.*

Bar k-visibility representations. It is easy to extend our previous model for $k = 0$ to test bar k-visibility representations for $k \geq 1$. See [2] for a detailed description.

Theorem 5. *There exists an ILP/SAT formulation with $O(HW(n + m))$ variables and $O(HW(m^2 + n))$ constraints / $O(\binom{HW}{k+1}m + HWm^2)$ clauses of maximum size HW that solves Problem 3 for $k \geq 1$.*

4.2 Boxicity-d Graphs

A graph is said to have *boxicity* d if it can be represented as intersection graph of d-dimensional axis-aligned boxes. Testing whether a graph has boxicity d is NP-hard, even for $d = 2$ [23]. We are not aware of any implemented algorithms to determine the boxicity of a graph. By Lemma 1 we can restrict ourselves to a grid of side length n. In the full version [2], we give an ILP model for testing whether a graph has boxicity d.

Theorem 6. *There exists an ILP with $O(n^d(n + m))$ variables and $O(n^{d+2})$ constraints as well as a SAT instance with $O(n^d(n + m))$ variables and $O(n^{d+2})$ clauses of maximum size $O(n^d)$ to test whether a graph G has boxicity d.*

5   Experiments

We implemented and tested our formulation for minimizing pathwidth, bandwidth, length of longest path in an st-orientation, and width of bar-visibility and bar 1-visibility representations, as well as deciding whether a graph has boxicity 2.

We performed the experiments on a single core of an AMD Opteron 6172 processor running Linux 3.4.11. The machine is clocked at 2.1 Ghz, and has 256 GiB RAM. Our implementation (available from http://i11www.iti.kit.edu/gdsat) is written in C++ and was compiled with GCC 4.7.1 using optimization -O3. As test sample we used the *Rome graphs* dataset [1] which consists of 11533 graphs with vertex number between 10 and 100. 18% of the Rome graphs are planar. The size distribution of the Rome graphs can be found in the full version [2].

We initially used the *Gurobi* solver [20] to test the implementation of the ILP formulations, however it turned out that even for very small graphs ($n < 10$) solving a single instance can take minutes. We therefore focused on the equivalent SAT formulations gaining a significant speed-up. As SAT solver we used *MiniSat* [16] in version 2.2.0. For each of the five minimization problems we determined obvious lower and upper bounds in $O(n)$ for the respective graph parameter. Starting with the lower bound we iteratively increased the parameter to the next integer until a solution was found (or a predefined timeout was exceeded). Each iteration consists of constructing the SAT formulation and executing the SAT solver. We measured the total time spent in all iterations. For boxicity 2 we decided to consider square grids and minimize their side lengths. Thus the same iterative procedure applies to boxicity 2.

Note that for all considered problems a binary search-like procedure for the parameter value did not prove to be efficient, since the solver usually takes more time with increasing parameter value, which is mainly due to the increasing number of variables and clauses. For the one-dimensional problems we used a timeout of 300 seconds, for the two-dimensional problems of 600 seconds.

We ran the instances sorted by size $n + m$ starting with the smallest graphs. If more than 400 consecutive graphs in this order produced timeouts, we ended the experiment prematurely and evaluated only the so far obtained results. Figures 2 and 3 summarize our experimental results and show the percentage of Rome graphs solved within the given time limit, as well as scatter plots with each solved instance represented as a point depending on its graph size and the required computation time.

Pathwidth. As Fig. 2a shows, we were able to compute the pathwidth for 17.0% of all Rome graphs, from which 82% were solved within the first minute and only 3% within the last. Therefore, we expect that a significant increase of the timeout value would be necessary for a noticeable increase of the percentage of solved instances. We note that almost all small graphs ($n + m < 45$) could be solved within the given timeout, however, for larger graphs, the percentage of solved instances rapidly drops, as the red curve in Fig. 2b shows. Almost no graphs with $n + m > 70$ were solved.

Bandwidth. We were able to compute the bandwidth for 22.3% of all Rome graphs (see Fig. 2a), from which 90% were solved within the first minute and only 1.3% within the last. Similarly to the previous case, the procedure terminated successfully within 300 seconds for almost all small graphs ($n + m < 55$ in this case), while almost none of the larger graphs ($n + m > 80$) were solved; see the red curve in Fig. 2c.

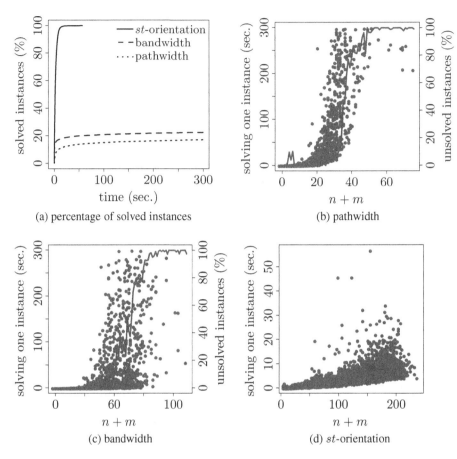

Fig. 2. Experimental results for the one-dimensional problems. (a) Percentage of solved instances. (b)–(d): Time in seconds for solving an instance (dots) and percentage of instances not solved within 300 seconds (red curves), both in relation to $n + m$

Optimum st-Orientation. Note that very few of the Rome graphs are biconnected. Therefore, to test our SAT implementation for computing the minimum number of levels in an st-orientation, we subdivided each graph into biconnected blocks and removed those with $n \leq 2$, which produced 13606 blocks in total ($3 \leq n + m \leq 230$). Then, for each such block, we randomly selected one pair of vertices s, t, $s \neq t$, connected them by an edge if it did not already exist and ran the iterative procedure. In this way, for the respective choice of s, t we were able to compute the minimum number of levels in an st-orientation for all biconnected blocks; see Fig. 2a. Moreover, no graph took longer than 57 seconds, for 97% of the graphs it took less than 10 seconds and for 68% less than 3 seconds. Even for the biggest blocks with $m + n > 200$, the procedure successfully terminated within 15 seconds in 93% of the cases; see Fig. 2d.

Bar Visibility. To compute bar-visibility representations of minimum width, we iteratively tested for each graph all widths W between 1 and $2n - 4$. We used the trivial

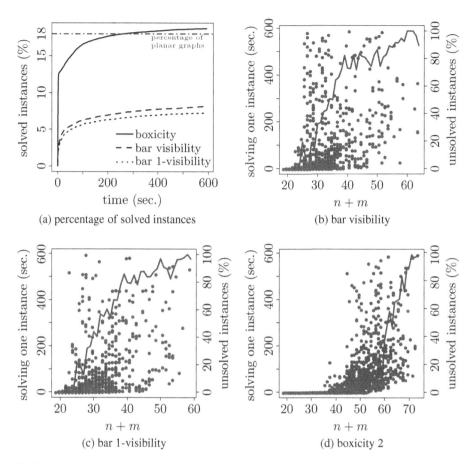

Fig. 3. Experimental results for the two-dimensional problems. (a) Percentage of solved instances. The red horizontal line shows the percentage of planar graphs over all Rome graphs. (b)–(d): Time in seconds for solving an instance (dots) and percentage of instances not solved within 600 seconds (red curves), both in relation to $n + m$.

upper bound $H = n$ for the height. We were able to compute solutions for 28.5% of all 3281 planar Rome graphs (see Fig. 3a), 69% of which were solved within the first minute and less than 0.1% within the last. We were able to solve all small instances with $n + m \leq 23$ and almost none for $n + m > 55$; see the red curve in Fig. 3b.

Bar 1-Visibility. We also ran width minimization for bar 1-visibility representations on all Rome graphs. The procedure terminated successfully within the given time for 833 graphs (7.2% of all Rome graphs), which is close to the corresponding number for bar-visibility; see Fig. 3a. For bar 1-visibility, eight graphs were solved which were not solved for bar-visibility. Interestingly, they were all planar. All but 113 graphs successfully processed in the previous experiment were also successfully processed in this one. A possible explanation for those 113 graphs is that the SAT formulation for bar

1-visibility requires more clauses. All small graphs with $n + m \leq 23$ were processed successfully. Interestingly, for none of the processed graphs the minimum width actually decreased in comparison to their minimum-width bar-visibility representation.

Boxicity-2. For testing boxicity 2, we started with a 3×3 grid for each graph and then increased height and width simultaneously after each iteration. Within the specified timeout of 600 seconds, we were able to decide whether a graph has boxicity 2 for 18.7% of all Rome graphs (see Fig. 3b), 82% of which were processed within the first minute and 0.3% within the last. All of the successfully processed graphs actually had boxicity 2. Small graphs with $n + m \leq 50$ were processed almost completely, while almost none of the graphs with $n + m > 70$ finished; see Fig. 3d.

6   Conclusion

We presented a versatile ILP formulation for determining placement of grid boxes according to problem-specific constraints. We gave six examples of how to extend this formulation for solving numerous NP-hard graph drawing and representation problems. Our experimental evaluation showed that while solving the original ILP is rather slow, the derived SAT formulations perform well for smaller graphs. While our approach is not suitable to replace faster specialized exact or heuristic algorithms, it does provide a simple-to-use tool for solving problems that can be modeled by grid-based graph representations with little implementation effort. This can be useful, e.g., for verifying counterexamples, NP-hardness gadgets, or for solving certain instances in practice.

Many other problems can easily be formulated as ILPs by assigning grid-boxes to vertices or edges. Among those are, e.g., testing whether a planar graph has a straight-line drawing of height h, whether a planar graph has a rectangular dual with integer coordinates and prescribed integral areas, whether a graph is a t-interval graph, or whether a bipartite graph can be represented as a planar bus graph. Important open problems are to reduce the complexity of our formulations and whether approximation algorithms for graph drawing can be derived from our model via fractional relaxation.

References

1. Rome graphs, www.graphdrawing.org/download/rome-graphml.tgz
2. Biedl, T., Bläsius, T., Niedermann, B., Nöllenburg, M., Prutkin, R., Rutter, I.: A versatile ILP/SAT formulation for pathwidth, optimum st-orientation, visibility representation, and other grid-based graph drawing problems. CoRR abs/1308.6778 (2013)
3. Biere, A., Heule, M., van Maaren, H., Walsh, T. (eds.): Handbook of Satisfiability. IOS Press (2009)
4. Binucci, C., Didimo, W., Liotta, G., Nonato, M.: Orthogonal drawings of graphs with vertex and edge labels. Comput. Geom. Theory Appl. 32(2), 71–114 (2005)
5. Bodlaender, H.L., Gilbert, J.R., Hafsteinsson, H., Kloks, T.: Approximating treewidth, pathwidth, frontsize, and shortest elimination tree. J. Algorithms 18(2), 238–255 (1995)
6. Bodlaender, H.L., Kloks, T.: Efficient and constructive algorithms for the pathwidth and treewidth of graphs. J. Algorithms 21(2), 358–402 (1996)
7. Brandes, U.: Eager st-ordering. In: Möhring, R.H., Raman, R. (eds.) ESA 2002. LNCS, vol. 2461, pp. 247–256. Springer, Heidelberg (2002)

8. Buchheim, C., Chimani, M., Ebner, D., Gutwenger, C., Jünger, M., Klau, G.W., Mutzel, P., Weiskircher, R.: A branch-and-cut approach to the crossing number problem. Discrete Optimization 5(2), 373–388 (2008)

9. Chen, D.S., Batson, R.G., Dang, Y.: Applied Integer Programming. Wiley (2010)

10. Chimani, M., Mutzel, P., Bomze, I.: A new approach to exact crossing minimization. In: Halperin, D., Mehlhorn, K. (eds.) ESA 2008. LNCS, vol. 5193, pp. 284–296. Springer, Heidelberg (2008)

11. Chimani, M., Zeranski, R.: Upward planarity testing via SAT. In: Didimo, W., Patrignani, M. (eds.) GD 2012. LNCS, vol. 7704, pp. 248–259. Springer, Heidelberg (2013)

12. Chinn, P.Z., Chvátalova, J., Dewdney, A.K., Gibbs, N.E.: The bandwidth problem for graphs and matrices—a survey. J. Graph Theory 6(3), 223–254 (1982)

13. Dean, A.M., Evans, W., Gethner, E., Laison, J.D., Safari, M.A., Trotter, W.T.: Bar k-visibility graphs. J. Graph Algorithms Appl. 11(1), 45–59 (2007)

14. Del Corso, G.M., Manzini, G.: Finding exact solutions to the bandwidth minimization problem. Computing 62(3), 189–203 (1999)

15. Dujmovic, V., Fellows, M.R., Kitching, M., Liotta, G., McCartin, C., Nishimura, N., Ragde, P., Rosamond, F.A., Whitesides, S., Wood, D.R.: On the parameterized complexity of layered graph drawing. Algorithmica 52(2), 267–292 (2008)

16. Eén, N., Sörensson, N.: An extensible SAT-solver. In: Giunchiglia, E., Tacchella, A. (eds.) SAT 2003. LNCS, vol. 2919, pp. 502–518. Springer, Heidelberg (2004)

17. Even, S., Tarjan, R.E.: Computing an st-numbering. Theoret. Comput. Sci. 2(3), 339–344 (1976)

18. Felsner, S., Liotta, G., Wismath, S.: Straight-line drawings on restricted integer grids in two and three dimensions. J. Graph Algorithms Appl. 7(4), 363–398 (2003)

19. Gange, G., Stuckey, P.J., Marriott, K.: Optimal k-level planarization and crossing minimization. In: Brandes, U., Cornelsen, S. (eds.) GD 2010. LNCS, vol. 6502, pp. 238–249. Springer, Heidelberg (2011)

20. Gurobi Optimization, Inc.: Gurobi optimizer reference manual (2013)

21. Hopcroft, J., Tarjan, R.: Efficient planarity testing. J. ACM 21(4), 549–568 (1974)

22. Jünger, M., Mutzel, P.: 2-layer straightline crossing minimization: Performance of exact and heuristic algorithms. J. Graph Algorithms Appl. 1(1), 1–25 (1997)

23. Kratochvíl, J.: A special planar satisfiability problem and a consequence of its NP-completeness. Discrete Appl. Math. 52(3), 233–252 (1994)

24. Lin, X., Eades, P.: Towards area requirements for drawing hierarchically planar graphs. Theoret. Comput. Sci. 292(3), 679–695 (2003)

25. Martí, R., Campos, V., Piñana, E.: A branch and bound algorithm for the matrix bandwidth minimization. Europ. J. of Operational Research 186, 513–528 (2008)

26. Nöllenburg, M., Wolff, A.: Drawing and labeling high-quality metro maps by mixed-integer programming. IEEE TVCG 17(5), 626–641 (2011)

27. Papamanthou, C., G. Tollis, I.: Applications of parameterized st-orientations. J. Graph Algorithms Appl. 14(2), 337–365 (2010)

28. Sadasivam, S., Zhang, H.: NP-completeness of st-orientations for plane graphs. Theoret. Comput. Sci. 411(7-9), 995–1003 (2010)

29. Tamassia, R., Tollis, I.: A unified approach to visibility representations of planar graphs. Discrete Comput. Geom. 1(1), 321–341 (1986)

30. Wismath, S.K.: Characterizing bar line-of-sight graphs. In: Proc. First Ann. Symp. Comput. Geom., SCG 1985, pp. 147–152. ACM, New York (1985)

Untangling Two Systems of Noncrossing Curves[*]

Jiří Matoušek[1,2], Eric Sedgwick[3], Martin Tancer[1,4], and Uli Wagner[5]

[1] Department of Applied Mathematics, Charles University, Malostranské nám. 25, 118 00 Praha 1, Czech Republic
[2] Institute of Theoretical Computer Science, ETH Zürich, 8092 Zürich, Switzerland
[3] School of CTI, DePaul University, 243 S. Wabash Ave, Chicago, IL 60604, USA
[4] Institutionen för matematik, KTH, 100 44 Stockholm, Sweden
[5] IST Austria, Am Campus 1, 3400 Klosterneuburg, Austria

Abstract. We consider two systems $(\alpha_1, \ldots, \alpha_m)$ and $(\beta_1, \ldots, \beta_n)$ of curves drawn on a compact two-dimensional surface $\mathcal{M}$ with boundary. Each α_i and each β_j is either an arc meeting the boundary of $\mathcal{M}$ at its two endpoints, or a closed curve. The α_i are pairwise disjoint except for possibly sharing endpoints, and similarly for the β_j. We want to "untangle" the β_j from the α_i by a self-homeomorphism of $\mathcal{M}$; more precisely, we seek an homeomorphism $\varphi \colon \mathcal{M} \to \mathcal{M}$ fixing the boundary of $\mathcal{M}$ pointwise such that the total number of crossings of the α_i with the $\varphi(\beta_j)$ is as small as possible. This problem is motivated by an application in the algorithmic theory of embeddings and 3-manifolds.

We prove that if $\mathcal{M}$ is planar, i.e., a sphere with $h \geq 0$ boundary components ("holes"), then $O(mn)$ crossings can be achieved (independently of h), which is asymptotically tight, as an easy lower bound shows. In general, for an arbitrary (orientable or nonorientable) surface $\mathcal{M}$ with h holes and of (orientable or nonorientable) genus $g \geq 0$, we obtain an $O((m + n)^4)$ upper bound, again independent of h and g.

Keywords: Curves on 2-manifolds, simultaneous planar drawings, Lickorish's theorem.

1 Introduction

Let $\mathcal{M}$ be a surface, by which we mean a two-dimensional compact manifold with (possibly empty) boundary $\partial\mathcal{M}$.

By the classification theorem for surfaces, if $\mathcal{M}$ is orientable, then $\mathcal{M}$ is homeomorphic to a sphere with $h \geq 0$ *holes* and $g \geq 0$ attached handles; the number g is also called the *orientable genus* of $\mathcal{M}$. If $\mathcal{M}$ is nonorientable, then it is homeomorphic to a sphere with $h \geq 0$ holes and with $g \geq 0$ *cross-caps*; in this case, the integer g is known as the *nonorientable genus* of $\mathcal{M}$. In the sequel, the word

[*] Research supported by the ERC Advanced Grant No. 267165. Moreover, the research of J.M. was also partially supported by Grant GRADR Eurogiga GIG/11/E023, the research of M.T. was supported by a Göran Gustafsson postdoctoral fellowship, and the research of U.W. was supported by the Swiss National Science Foundation (Grant SNSF-PP00P2-138948)

S. Wismath and A. Wolff (Eds.): GD 2013, LNCS 8242, pp. 472–483, 2013.

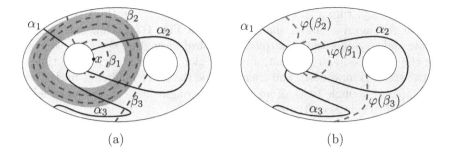

Fig. 1. Systems A and B of curves on a surface $\mathcal{M}$, with $g = 0$ and $h = 3$ (a), and a re-drawing of B via a ∂-automorphism φ reducing the number of intersections (b).

"genus" will mean orientable genus for orientable surfaces and nonorientable genus for nonorientable surfaces.

We will consider curves in $\mathcal{M}$ that are *properly embedded*, i.e., every curve is either a simple arc meeting the boundary $\partial\mathcal{M}$ exactly at its two endpoints, or a simple closed curve avoiding $\partial\mathcal{M}$. An *almost-disjoint system of curves* in $\mathcal{M}$ is a collection $A = (\alpha_1, \ldots, \alpha_m)$ of curves that are pairwise disjoint except for possibly sharing endpoints.

In this paper we consider the following problem: We are given two almost-disjoint systems $A = (\alpha_1, \ldots, \alpha_m)$ and $B = (\beta_1, \ldots, \beta_n)$ of curves in $\mathcal{M}$, where the curves of B intersect those of A possibly very many times, as in Fig. 1(a). We would like to "redraw" the curves of B in such a way that they intersect those of A as little as possible.

We consider re-drawings only in a restricted sense, namely, induced by ∂-*automorphisms* of $\mathcal{M}$, where a ∂-automorphism is an homeomorphism $\varphi \colon \mathcal{M} \to \mathcal{M}$ that fixes the boundary $\partial\mathcal{M}$ pointwise. Thus, given the α_i and the β_j, we are looking for a ∂-automorphism φ such that the number of intersections (crossings) between $\alpha_1, \ldots, \alpha_m$ and $\varphi(\beta_1), \ldots, \varphi(\beta_n)$ is as small as possible (where sharing endpoints does not count). We call this minimum number of crossings achievable through any choice of φ the *entanglement number* of the two systems A and B.

In the orientable case, let $f_{g,h}(m, n)$ denote the maximum entanglement number of any two systems $A = (\alpha_1, \ldots, \alpha_m)$ and $B = (\beta_1, \ldots, \beta_n)$ of almost-disjoint curves on an orientable surface of genus g with h holes. Analogously, we define $\hat{f}_{g,h}(m, n)$ as the maximum entanglement number of any two systems A and B of m and n curves, respectively, on a nonorientable surface of genus g with h holes. It is easy to see that f and $\hat{f}$ are nondecreasing in m and n, which we will often use in the sequel.

To give the reader some intuition about the problem, let us illustrate which re-drawings are possible with a ∂-automorphism and which are not. In the example of Fig. 1, it is clear that the two crossings of β_3 with α_3 can be avoided by sliding β_3 aside.[1] It is perhaps less obvious that the crossings of β_2 can also be

---

[1] This corresponds to an *isotopy* of the surface that fixes the boundary pointwise.

eliminated: To picture a suitable ∂-automorphism, one can think of an annular region in the interior of $\mathcal{M}$, shaded darkly in Fig. 1 (a), that surrounds the left hole and β_1 and contains most of the spiral formed by β_2. Then we cut $\mathcal{M}$ along the outer boundary of that annular region, twist the region two times (so that the spiral is unwound), and then we glue the outer boundary back. Here is an example of a single twist of an annulus; straight-line curves on the left are transformed to spirals on the right .

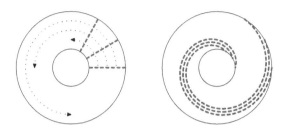

On the other hand, it is impossible to eliminate the crossings of β_1 or β_3 with α_2 by a ∂-automorphism. For example, we cannot re-route β_1 to go around the right hole and thus avoid α_2, since this re-drawing is not induced by any ∂-automorphism φ: indeed, β_1 separates the point x on the boundary of left hole from the right hole, whereas α_2 does not separate them; therefore, the curve α_2 has to intersect $\varphi(\beta_1)$ at least twice, once when it leaves the component containing x and once when it returns to this component.

A rather special case of our problem, with $m = n = 1$ and only closed curves, was already considered by Lickorish [Lic62], who showed that the intersection of a pair of simple closed curves can be simplified via Dehn twists (and thus a ∂-automorphism) so that they meet at most twice (also see Stillwell [Sti80]). The case with $m = 1$, n arbitrary, only closed curves, and $\mathcal{M}$ possibly nonorientable was proposed in 2010 as a Mathoverflow question [Huy10] by T. Huynh. In an answer A. Putman proposes an approach via the "change of coordinates principle" (see, e.g., [FM11, Sec. 1.3]), which relies on the classification of 2-dimensional surfaces—we will also use it at some points in our argument.

The results. A natural idea for bounding $f_{g,h}(m, n)$ and $\hat{f}_{g,h}(m, n)$ is to proceed by induction, employing the change of coordinates principle mentioned above. This does indeed lead to finite bounds, but the various induction schemes we have tried always led to bounds at least exponential in one of m, n. Independently of our work, Geelen, Huynh, and Richter [GHR13] also recently proved bounds of this kind; see the discussion below. Partially influenced by the results on exponentially many intersections in representations of string graphs and similar objects (see [KM91,SSŠ03]), we first suspected that an exponential behavior might be unavoidable. Then, however, we found, using a very different approach, that polynomial bounds actually do hold.

For planar $\mathcal{M}$, i.e., $g = 0$, we obtain an asymptotically tight bound:

Theorem 1. *For planar $\mathcal{M}$, we have $f_{0,h}(m, n) = O(mn)$, independent of h.*

Here and in the sequel, the constants implicit in the O-notation are absolute, independent of g and h.

A simple example providing a lower bound of $2mn$ is obtained, e.g., by replicating α_2 in Fig. 1 m times and β_1 n times.

In general, we obtain the following bounds:

Theorem 2. (i) *For the orientable case,* $f_{g,h}(m,n) = O((m+n)^4)$.
(ii) *For the nonorientable case,* $\hat{f}_{g,h}(m,n) = O((m+n)^4)$.

Both parts of Theorems 2 are derived from the planar case, Theorem 1. In the orientable case, we use the following results on genus reduction. For a more convenient notation, let us set $L = \max(m,n)$.

Proposition 1 (Orientable genus reductions).
(i) *For* $g > L$, *we have* $f_{g,h}(m,n) \leq f_{L,g+h-L}(m,n)$.
(ii) $f_{g,h}(m,n) \leq f_{0,h+1}(cg(m+g), cg(n+g))$ *for a suitable constant* $c > 0$.

Theorem 2 (i), the orientable case, follows immediately from Proposition 1 and the planar bound.

In the nonorientable case, Theorem 2 (ii) is derived in two steps. First, analogous to Proposition 1 (i), we have the following reduction:

Proposition 2 (Nonorientable genus reduction). *For* $g > 4L+2$, *we have* $\hat{f}_{g,h}(m,n) \leq \hat{f}_{g',h'}(m,n)$, *where* $g' = 4L+2-(g \bmod 2)$ *and* $h' = h+\lceil g/2 \rceil -2L-1$.

The second step is a reduction to the orientable case.

Proposition 3 (Orientability reduction). *There is a constant* c *such that* $\hat{f}_{g,h}(m,n) \leq f_{\lfloor (g-1)/2 \rfloor, h+1+(g \bmod 2)}(c(g+m), c(g+n))$.

Table 1 summarizes the proof of Theorem 2.

Motivation. We were led to the question concerning untangling curves on surfaces while working on a project on 3-manifolds and embeddings. Specifically, we are interested in an algorithm for the following problem: given a 3-manifold M with boundary, does M embed in the 3-sphere? A special case of this problem, with the boundary of M a torus, was solved in [JS03]. The problem is motivated, in turn, by the question of algorithmically testing the embeddability of a 2-dimensional simplicial complex in $\mathbb{R}^3$; see [MTW11].

In our current approach, which has not yet been completely worked out, we need just a finite bound on $f_{g,h}(m,n)$. However, we consider the problem investigated in this paper interesting in itself and contributing to a better understanding of combinatorial properties of curves on surfaces.

As mentioned above, the question studied in the present paper has also been investigated independently by Geelen, Huynh, and Richter [GHR13], with a rather different and very strong motivation stemming from the theory of graph minors, namely the question of obtaining explicit upper bounds for the graph minor algorithms of Robertson and Seymour. Phrased in the language of the present paper, Geelen et al. [GHR13, Theorem 2.1] show that $f_{g,h}(m,n)$ and $\hat{f}_{g,h}(m,n)$ are both bounded by $n3^m$.

Table 1. A summary of the proof

---

1. For a planar surface, temporarily remove the holes not incident to any α_i or β_j, and contract the remaining "active" holes, augment the resulting planar graphs to make them 3-connected. Make a simultaneous plane drawing of the resulting planar graphs G_1 and G_2 with every edge of G_1 intersecting every edge of G_2 at most $O(1)$ times. Decontract the active holes and put the remaining holes back into appropriate faces (Theorem 1; Section 2).
2. If the genus is larger than $c(m + n)$, find handles or cross-caps avoided by the α_i and β_j, temporarily remove them, untangle the α_i and β_j, and put the handles or cross-caps back (Propositions 1 (i) and 2; the proofs are omitted from this extended abstract).
3. If the surface is nonorientable, make it orientable by cutting along a suitable curve that intersects the α_i and β_j at most $O(m + n)$ times, untangle the resulting pieces of the α_i and β_j, and glue back (Proposition 3).
4. Make the surface planar by cutting along a suitable system of curves (canonical system of loops), untangle the resulting pieces of the α_i and β_j, and glue back (Proposition 1 (ii)).

---

2 Planar Surfaces

In this section we prove Theorem 1. In the proof we use the following basic fact (see, e.g., [MT01]).

Lemma 1. *If G is a maximal planar simple graph (a triangulation), then for every two planar drawings of G in S^2 there is an automorphism ψ of S^2 converting one of the drawings into the other (and preserving the labeling of the vertices and edges). Moreover, if an edge e is drawn by the same arc in both of the drawings, w.l.o.g. we may assume that ψ fixes it pointwise.*

Let us introduce the following piece of terminology. Let G be as in the lemma, and let D_G, D'_G be two planar drawings of G. We say that D_G, D'_G are *directly equivalent* if there is an orientation-preserving automorphism of S^2 mapping D_G to D'_G, and we call D_G, D'_G *mirror-equivalent* if there is an orientation-reversing automorphism of S^2 converting D_G into D'_G.

We will also rely on a result concerning simultaneous planar embeddings; see [BKR12]. Let V be a vertex set and let $G_1 = (V, E_1)$ and $G_2 = (V, E_2)$ be two planar graphs on V. A planar drawing D_{G_1} of G_1 and a planar drawing D_{G_2} of G_2 are said to form a *simultaneous embedding* of G_1 and G_2 if each vertex $v \in V$ is represented by the same point in the plane in both D_{G_1} and D_{G_2}.

We note that G_1 and G_2 may have common edges, but they are not required to be drawn in the same way in D_{G_1} and in D_{G_2}. If this requirement is added, one speaks of a *simultaneous embedding with fixed edges*. There are pairs of planar graphs known that do not admit any simultaneous embedding with fixed edges (and consequently, no simultaneous straight-line embedding). An important step in our approach is very similar to the proof of the following result.

Theorem 3 (Erten and Kobourov [EK05]). *Every two planar graphs $G_1 = (V, E_1)$ and $G_2 = (V, E_2)$ admit a simultaneous embedding in which every edge is drawn as a polygonal line with at most 3 bends.*

We will need the following result, which follows easily from the proof given in [EK05]. For the reader's convenience, instead of just pointing out the necessary modifications, we present a full proof.

Theorem 4. *Every two planar graphs $G_1 = (V, E_1)$ and $G_2 = (V, E_2)$ admit a simultaneous, piecewise linear embedding in which every two edges e_1 of G_1 and e_2 of G_2 intersect at least once and at most C times, for a suitable constant C.[2]*

In addition, if both G_1 and G_2 are maximal planar graphs, let us fix a planar drawing D'_{G_1} of G_1 and a planar drawing D'_{G_2} of G_2. The planar drawing of G_1 in the simultaneous embedding can be required to be either directly equivalent to D'_{G_1}, or mirror-equivalent to it, and similarly for the drawing of G_2 (each of the four combinations can be prescribed).

Proof. For the beginning, we assume that both graphs are Hamiltonian. Later on, we will drop this assumption.

Let $v_1, v_2, \ldots, v_n$ be the order of the vertices as they appear on (some) Hamiltonian cycle H_1 of G_1. Since the vertex set V is common for G_1 and G_2, there is a permutation $\pi \in S(n)$ such that $v_{\pi(1)}, \ldots, v_{\pi(n)}$ is the order of the vertices as they appear on some Hamiltonian cycle H_2 of G_2.

We draw the vertex v_i in the grid point $p_i = (i, \pi(i))$, $i = 1, 2, \ldots, n$. Let S be the square $[1, n] \times [1, n]$. A *bispiked* curve is an x-monotone polygonal curve with two bends such that it starts inside S; the first bend is above S, the second bend is below S and it finishes in S again.

The $n - 1$ edges $v_i v_{i+1}$, of H_1, $i = 1, 2, \ldots, n - 1$, are drawn as bispiked curves starting in p_i and finishing in p_{i+1}. In order to distinguish edges and their drawings, we denote these bispiked curves by $c(i, i + 1)$.

Similarly, we draw the edges $v_{\pi(i)} v_{\pi(i+1)}$ of H_2, $i = 1, 2, \ldots, n - 1$, as y-monotone analogs of bispiked curves, where the first bend is on the left of S and the second is on the right of S; here is an example:

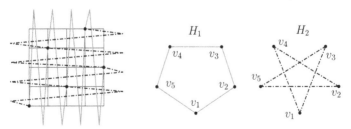

---

[2] An obvious bound from the proof is $C \leq 36$, since every edge in this embedding is drawn using at most 5 bends. By a more careful inspection, one can easily get $C \leq 25$, and a further improvement is probably possible.

We continue only with description of how to draw G_1; G_2 is drawn analogously with the grid rotated by 90 degrees.

Let D'_{G_1} be a planar drawing of G_1. Every edge from E_1 that is not contained in H_1 is drawn either inside D'_{H_1} or outside. Thus, we split $E_1 \setminus E(H_1)$ into two sets E'_1 and E''_1.

Let P_0 be the polygonal path obtained by concatenation of the curves $c(1,2)$, $c(2,3),\dots, c(n-1,n)$. Now our task is to draw the edges of $E'_1 \cup \{v_1 v_n\}$ as bispiked curves, all above P_0, and then the edges of E''_1 below P_0.

We start with E'_1 and we draw edges from it one by one, in a suitably chosen order, while keeping the following properties.

(P1) Every edge $v_i v_j$, where $i < j$, is drawn as a bispiked curve $c(i,j)$ starting in p_i and ending in p_j.

(P2) The x-coordinate of the second bend of $c(i,j)$ belongs to the interval $[j-1, j]$.

(P3) The polygonal curve P_k that we see from above after drawing the kth edge is obtained as a concatenation of some curves $c(1,i_1), c(i_1,i_2),\dots, c(i_\ell, n)$.

Here is an illustration; the square S is deformed for the purposes of the drawing:

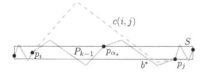

Initially, before drawing the first edge, the properties are obviously satisfied.

Let us assume that we have already drawn $k-1$ edges of E'_1, and let us focus on drawing the kth edge. Let $e = v_i v_j \in E'_1$ be an edge that is not yet drawn and such that all edges below e are already drawn, where "below e" means all edges $v_{i'} v_{j'} \in E'_1$ with $i \le i' < j' \le j$, $(i,j) \neq (i',j')$. (This choice ensures that we will draw all edges of E'_1.)

Since D'_{G_1} is a planar drawing, we know that there is no edge $v_{i'} v_{j'} \in E'_1$ with $i < i' < j < j'$ or $i' < i < j' < j$, and so the points p_i and p_j have to belong to P_{k-1}. The subpath P' of P_{k-1} between p_i and p_j is the concatenation of curves $c(i,\alpha_1), c(\alpha_1,\alpha_2),\dots, c(\alpha_s, j)$ as in the inductive assumptions. In particular, the x-coordinate of the second bend b^* of $c(\alpha_s, j)$ belongs to the interval $[j-1, j]$. We draw $c(i,j)$ as follows: The second bend of $c(i,j)$ is slightly above b^* but still below the square S. The first bend of S is sufficiently high above S (with the x-coordinate somewhere between i and $j-1$) so that the resulting bispiked curve $c(i,j)$ does not intersect P_{k-1}. The properties (P1) and (P2) are obviously satisfied by the construction. For (P3), the path P_k is obtained from P_{k-1} by replacing P' with $c(i,j)$.

After drawing the edges of E'_1, we draw $v_1 v_n$ in the same way. Then we draw the edges of E''_1 in a similar manner as those of E'_1, this time as bispiked curves below P_0. This finishes the construction for Hamiltonian graphs.

Now we describe how to adjust this construction for non-Hamiltonian graphs, in the spirit of [EK05].

First we add edges to G_1 and G_2 so that they become planar triangulations. This step does not affect the construction at all, except that we remove these edges in the final drawing.

Next, we subdivide some of the edges of G_i with *dummy* vertices. Moreover, we attach two new *extra* edges to each dummy vertex, as in the following illustration:

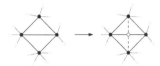

By choosing the subdivided edges suitably, one can obtain a 4-connected, and thus Hamiltonian, graph; see [EK05, Proof of Theorem 2] for details (this idea previously comes from [KW02]). An important property of this construction is that each edge of G_i is subdivided at most once.

In this way, we obtain new Hamiltonian graphs G'_1 and G'_2, for which we want to construct a simultaneous drawing as in the first part of the proof. A little catch is that G'_1 and G'_2 do not have same vertex sets, but this is easy to fix. Let d_i be the number of dummy vertices of G'_i, $i = 1, 2$, and say that $d_1 \geq d_2$. We pair the d_2 dummy vertices of G'_2 with some of the dummy vertices of G'_1. Then we iteratively add $d_1 - d_2$ new triangles to G'_2, attaching each of them to an edge of a Hamiltonian cycle. This operation keeps Hamiltonicity and introduces $d_1 - d_2$ new vertices, which can be matched with the remaining $d_1 - d_2$ dummy vertices in G'_1.

After drawing resulting graphs, we remove all extra dummy vertices and extra edges added while introducing dummy vertices. An original edge e that was subdivided by a dummy vertex is now drawn as a concatenation of two bispiked curves. Therefore, each edge is drawn with at most 5 bends.

Two edges with 5 bends each may in general have at most 36 intersections, but in our case, there can be at most 25 intersections, since the union of the two segments before and after a dummy vertex is both x-monotone and y-monotone.

Because of the bispiked drawing of all edges, it is also clear that every edge of G_1 crosses every edge of G_2 at least once.

Finally, the requirements on directly equivalent or mirror-equivalent drawings can easily be fulfilled by interchanging the role of top and bottom in the drawing of G_1 or left and right in the drawing of G_2. Theorem 4 is proved. □

Proof of Theorem 1. Let a planar surface $\mathcal{M}$ and the curves $\alpha_1, \ldots, \alpha_m, \beta_1, \ldots, \beta_n$ be given; we assume that $\mathcal{M}$ is a subset of S^2. From this we construct a set V of $O(m+n)$ vertices in S^2 and planar drawings D_{G_1} and D_{G_2} of two simple graphs $G_1 = (V, E_1)$ and $G_2 = (V, E_2)$ in S^2, as follows.

1. We put all endpoints of the α_i and of the β_j into V.

2. We choose a new vertex in the interior of each α_i and each β_j, or two distinct vertices if α_i or β_j is a loop with a single endpoint, or three vertices of α_i or β_j is a closed curve, and we add all of these vertices to V. These new vertices are all distinct and do not lie on any curves other than where they were placed.
3. If the boundary of a hole in $\mathcal{M}$ already contains a vertex introduced so far, we add more vertices so that it contains at least 3 vertices of V. This finishes the construction of V.
4. To define the edge set $E_1 = E(G_1)$ and the planar drawing D_{G_1}, we take the portions of the curves $\alpha_1, \ldots, \alpha_m$ between consecutive vertices of V as edges of E_1. Similarly, we make the arcs of the boundaries of the holes into edges in E_1; these will be called the *hole edges*. By the choice of the vertex set V above, this yields a simple plane graph.
5. Then we add new edges to E_1 so that we obtain a drawing D_{G_1} in S^2 of a maximal planar simple graph G_1 (i.e., a triangulation) on the vertex set V. While choosing these edges, we make sure that all holes containing no vertices of G lie in faces of D_{G_1} adjacent to some of the α_i. New edges drawn in the interior of a hole are also called *hole edges*.
6. We construct $G_2 = (V, E_2)$ and D_{G_2} analogously, using the curves $\beta_1, \ldots, \beta_m$. We make sure that all hole edges are common to G_1 and G_2.

After this construction, each hole of $\mathcal{M}$ contains either no vertex of V on its boundary or at least three vertices. In the former case, we speak of an *inner hole*, and in the latter case, of a *subdivided hole*. A face f of D_{G_1} or D_{G_2} is a *non-hole face* if it is not contained in a subdivided hole. An inner hole H has its *signature*, which is a pair (f_1, f_2), where f_1 is the unique non-hole face of D_{G_1} containing H, and f_2 is the unique non-hole face of D_{G_2} containing H.[3] By the construction, each f_1 appearing in a signature is adjacent to some α_i, and each f_2 is adjacent to some β_j.

In the following claim, we will consider different drawings D'_{G_1} and D'_{G_2} for G_1 and G_2. By Lemma 1, the faces of D_{G_1} are in one-to-one correspondence with the faces of D'_{G_1}. For a face f_1 of D_{G_1}, we denote the corresponding face by f'_1, and similarly for a face f_2 of D_{G_2} and f'_2.

Claim 5. *The graphs G_1 and G_2 as above have planar drawings D'_{G_1} and D'_{G_2}, respectively, that form a simultaneous embedding in which each edge of G_1 crosses each edge of G_2 at most C times, for a suitable constant C; moreover, D'_{G_1} is directly equivalent to D_{G_1}; D'_{G_2} is directly equivalent to D_{G_2}; all hole edges are drawn in the same way in D'_{G_1} and D'_{G_2}; and whenever (f_1, f_2) is a signature of an inner hole, the interior of the intersection $f'_1 \cap f'_2$ is nonempty.*

We postpone the proof of Claim 5, and we first finish the proof of Theorem 1 assuming this claim.

---

[3] Classifying inner holes according to the signature helps us to obtain a bound independent on the number of holes. Inner holes with same signature are all treated in the same way, independent of their number.

For each inner hole H with signature (f_1, f_2), we introduce a closed disk B_H in the interior of $f_1' \cap f_2'$. We require that these disks are pairwise disjoint. In sequel, we consider holes as subsets of S^2 homeomorphic to closed disks (in particular, a hole H intersects $\mathcal{M}$ in ∂H).

Claim 6. *There is an orientation-preserving automorphism φ_1 of S^2 transforming every inner hole H to B_H and D_{G_1} to D'_{G_1}.*

Proof. Using Lemma 1 again, there is an orientation-preserving automorphism ψ_1 transforming D_{G_1} into D'_{G_1} (since D_{G_1} and D'_{G_1} are directly equivalent).

Let f_1 be a face of D_{G_1}. The interior of f_1' contains images $\psi_1(H)$ of all holes H with signature $(f_1, \cdot)$, and it also contains the disks B_H for these holes. Therefore, there is a boundary- and orientation-preserving automorphism of f_1' that maps each $\psi_1(H)$ to B_H.

By composing these automorphisms on every f_1' separately, we have an orientation-preserving automorphism ψ_2 fixing D'_{G_1} and transforming each $\psi_1(H)$ to B_H. The required automorphism is $\varphi_1 = \psi_2 \psi_1$. $\qquad\square$

Claim 7. *There is an orientation-preserving automorphism φ_2 of S^2 that fixes hole edges (of subdivided holes), fixes B_H for every inner hole H, and transforms $\varphi_1(D_{G_2})$ to D'_{G_2}.*

Proof. By Lemma 1 there is an orientation-preserving automorphism ψ_3 of S^2 that fixes hole edges and transforms $\varphi_1(D_{G_2})$ to D'_{G_2}.

If an inner hole H has a signature $(\cdot, f_2)$, then both $\psi_3(B_H)$ and B_H belong to the interior of f_2'. Therefore, as in the proof of the previous claim, there is an orientation-preserving homeomorphism ψ_4 that fixes D'_{G_2} and transforms $\psi_3(B_H)$ to B_H. We can even require that $\psi_4\psi_3$ is identical on B_H. We set $\varphi_2 := \psi_4\psi_3$. $\qquad\square$

To finish the proof of Theorem 1, we set $\varphi = \varphi_1^{-1}\varphi_2\varphi_1$. We need that φ fixes the holes (inner or subdivided) and that $\alpha_1, \ldots, \alpha_m$ and $\varphi(\beta_1), \ldots, \varphi_1(\beta_m)$ have $O(mn)$ intersections. It is routine to check all the properties:

If H is a hole (inner or subdivided), then φ_2 fixes $\partial\varphi_1(H)$. Therefore, φ also restricts to a ∂-automorphism of $\mathcal{M}$.

The collections of curves $\alpha_1, \ldots, \alpha_m$ and $\varphi(\beta_1), \ldots, (\beta_m)$ have same intersection properties as the collections $\varphi_1(\alpha_1), \ldots, \varphi_1(\alpha_m)$ and $\varphi_2(\varphi_1(\beta_1)), \ldots, \varphi_2(\varphi_1(\beta_m))$. Since each α_i and each β_j was subdivided at most three times in the construction, by Claims 5, 6, and 7, these collections have at most $O(mn)$ intersections. The proof of the theorem is finished, except for Claim 5.

Proof of Claim 5. Given G_1 and G_2, we form auxiliary planar graphs $\tilde{G}_1$ and $\tilde{G}_2$ on a vertex set $\tilde{V}$ by contracting all hole edges and removing the resulting loops and multiple edges. We note that a loop cannot arise from an edge that was a part of some α_i or β_j.

Then we consider planar drawings $D_{\tilde{G}_1}$ and $D_{\tilde{G}_2}$ forming a simultaneous embedding as in Theorem 4, with each edge of $\tilde{G}_1$ crossing each edge of $\tilde{G}_2$ at least once and most a constant number of times.

Let $v_H \in \tilde{V}$ be the vertex obtained by contracting the hole edges on the boundary of a hole H. Since the drawings $D_{\tilde{G}_1}$ and $D_{\tilde{G}_2}$ are piecewise linear, in a sufficiently small neighborhood of v_H the edges are drawn as radial segments.

We would like to replace v_H by a small circle and thus turn the drawings $D_{\tilde{G}_1}$, $D_{\tilde{G}_2}$ into the required drawings D'_{G_1}, D'_{G_2}. But a potential problem is that the edges in $D_{\tilde{G}_1}$, $D_{\tilde{G}_2}$ may enter v_H in a wrong cyclic order.

We claim that the edges in $D_{\tilde{G}_1}$ entering v_H have the same cyclic ordering around v_H as the corresponding edges around the hole H in the drawing D_{G_1}. Indeed, by contracting the hole edges in the drawing D_{G_1}, we obtain a planar drawing $D^*_{\tilde{G}_1}$ of $\tilde{G}_1$ in which the cyclic order around v_H is the same as the cyclic order around H in D_{G_1} Since $\tilde{G}_1$ was obtained by edge contractions from a maximal planar graph, it is maximal as well (since an edge contraction cannot create a non-triangular face), and its drawing is unique up to an automorphism of S^2 (Lemma 1). Hence the cyclic ordering of edges around v_H in $D_{\tilde{G}_1}$ and in $D^*_{\tilde{G}_1}$ is either the same (if $D_{\tilde{G}_1}$ and $D^*_{\tilde{G}_1}$ are directly equivalent), or reverse (if $D_{\tilde{G}_1}$ and $D^*_{\tilde{G}_1}$ are mirror-equivalent). However, Theorem 4 allows us to choose the drawing $D_{\tilde{G}_1}$ so that it is directly equivalent to $D^*_{\tilde{G}_1}$, and then the cyclic orderings coincide. A similar consideration applies for the other graph G_2.

The edges of $D_{\tilde{G}_1}$ may still be placed to wrong positions among the edges in $D_{\tilde{G}_2}$, but this can be rectified at the price of at most one extra crossing for every pair of edges entering v_H, as the following picture indicates (the numbering specifies the cyclic order of the edges around H in $D_{G_1} \cup D_{G_2}$):

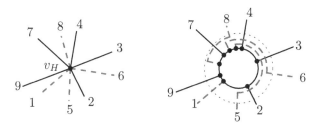

It remains to draw the edges of G_1 and G_2 that became loops or multiple edges after the contraction of the hole edges. Loops can be drawn along the circumference of the hole, and multiple edges are drawn very close to the corresponding single edge.

In this way, every edge of G_1 still has at most a constant number of intersections with every edge of G_2, and every two such edges intersect at least once unless at least one of them became a loop after the contraction. Consequently, whenever (f_1, f_2) is a signature of an inner hole, the corresponding faces f'_1 and f'_2 intersect. This finishes the proof.   □

Acknowledgement. We would like to thank the authors of [GHR13] for making a draft of their paper available to us, and, in particular, T. Huynh for an e-mail correspondence.

References

BKR12.   Bläsius, T., Kobourov, S.G., Rutter, I.: Simultaneous embedding of planar graphs (2012) (preprint), http://arxiv.org/abs/1204.5853

EK05.   Erten, C., Kobourov, S.G.: Simultaneous embedding of planar graphs with few bends. J. Graph Algorithms Appl. 9(3), 347–364 (2005) (electronic)

FM11.   Farb, B., Margalit, D.: A primer on mapping class groups. Princeton University Press, Princeton (2011)

GHR13.   Geelen, J., Huynh, T., Richter, R.B.: Explicit bounds for graph minors (May 2013) (preprint), http://arxiv.org/abs/1305.1451

Huy10.   Huynh, T.: Removing intersections of curves in surfaces. Mathoverflow (2010), http://mathoverflow.net/questions/33963/removing-intersections-of-curves-in-surfaces/33970#33970

JS03.   Jaco, W., Sedgwick, E.: Decision problems in the space of Dehn fillings. Topology 42(4), 845–906 (2003)

KM91.   Kratochvíl, J., Matoušek, J.: String graphs requiring huge representations. J. Combin. Theory Ser. B 53(1), 1–4 (1991)

KW02.   Kaufmann, M., Wiese, R.: Embedding vertices at points: few bends suffice for planar graphs. J. Graph Algorithms Appl. 6(1), 115–129 (2002); (electronic) Graph drawing and representations, Prague (1999)

Lic62.   Lickorish, W.B.R.: A representation of orientable combinatorial 3-manifolds. Ann. Math (2) 76, 531–540 (1962)

MT01.   Mohar, B., Thomassen, C.: Graphs on Surfaces. Johns Hopkins University Press, Baltimore (2001)

MTW11.   Matoušek, J., Tancer, M., Wagner, U.: Hardness of embedding simplicial complexes in $\mathbb{R}^d$. J. Eur. Math. Soc. 13(2), 259–295 (2011)

SSŠ03.   Schaefer, M., Sedgwick, E., Štefankovič, D.: Recognizing string graphs in NP. J. Comput. Syst. Sci. 67(2), 365–380 (2003)

Sti80.   Stillwell, J.: Classical Topology and Combinatorial Group Theory. Springer, New York (1980)

Drawing Permutations with Few Corners

Sergey Bereg[1], Alexander E. Holroyd[2], Lev Nachmanson[2], and Sergey Pupyrev[3,4,*]

[1] Department of Computer Science, University of Texas at Dallas, USA
[2] Microsoft Research, USA
[3] Department of Computer Science, University of Arizona, USA
[4] Institute of Mathematics and Computer Science, Ural Federal University, Russia

Abstract. A permutation may be represented by a collection of paths in the plane. We consider a natural class of such representations, which we call tangles, in which the paths consist of straight segments at 45 degree angles, and the permutation is decomposed into nearest-neighbour transpositions. We address the problem of minimizing the number of crossings together with the number of corners of the paths, focusing on classes of permutations in which both can be minimized simultaneously. We give algorithms for computing such tangles for several classes of permutations.

1 Introduction

What is a good way to visualize a permutation? In this paper we study drawings in which a permutation of interest is connected to the identity permutation via a sequence of intermediate permutations, with consecutive elements of the sequence differing by one or more non-overlapping nearest-neighbour swaps. The position of each permutation element through the sequence may then traced by a piecewise-linear path comprising segments that are vertical and $45°$ to the vertical. Our goal is to keep these paths as simple as possible and to avoid unnecessary crossings.

Such drawings have applications in various fields; for example, in channel routing for integrated circuit design [12]. Another application is the visualization of metro maps and transportation networks, where some lines (railway tracks or roads) might partially overlap [4]. A natural goal is to draw the lines along their common subpaths so that an individual line is easy to follow; minimizing the number of bends of a line and avoiding unnecessary crossings between lines are natural criteria for map readability; see Fig. 3(b) of [3]. Much recent research in the graph drawing community is devoted to edge bundling. In this setting, drawing the edges of a bundle with the minimum number of crossings and bends occurs as a subproblem [10].

Let S_n be the symmetric group of permutations $\pi = [\pi(1), \ldots, \pi(n)]$ on $\{1, \ldots, n\}$. The **identity permutation** is $[1, \ldots, n]$, and the **swap** $\sigma(i)$ transforms a permutation π into $\pi \cdot \sigma(i)$ by exchanging its ith and $(i+1)$th elements. Equivalently, $\sigma(i)$ is the transposition $(i, i+1) \in S_n$, and $\cdot$ denotes composition. Two permutations a and b of S_n are **adjacent** if b can be obtained from a by swaps $\sigma(p_1), \sigma(p_2), \ldots, \sigma(p_k)$ that are not overlapping, that is, such that $|p_i - p_j| \geq 2$ for $i \neq j$. A **tangle** is a finite sequence

---

* Research supported in part by NSF grant DEB 1053573.

S. Wismath and A. Wolff (Eds.): GD 2013, LNCS 8242, pp. 484–495, 2013.

3 6 1 4 7 2 5
3 6 1 4 7 2 5
3 6 1 4 2 7 5
3 1 6 2 4 5 7
1 3 2 6 4 5 7
1 2 3 4 6 5 7
1 2 3 4 5 6 7
1 2 3 4 5 6 7
(a)

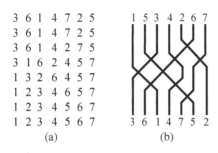

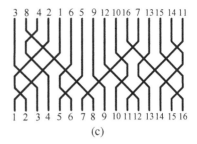

Fig. 1. (a) A tangle solving the permutation $[3, 6, 1, 4, 7, 2, 5]$. (b) A drawing of the tangle. (c) An example of a perfect tangle drawing.

of permutations in which each two consecutive permutations are adjacent. An example of a tangle is given in Fig. 1. The associated drawing is composed of polylines with vertices in $\mathbb{Z}^2$, whose segments can be vertical, or have slopes of $\pm 45°$ to the vertical. The polyline traced by element $i \in \{1, \ldots, n\}$ is called **path** i. Note that by definition all path crossings occur at right angles. We say that a tangle T **solves** the permutation π (or simply T is a tangle for π) if the tangle starts from π and ends at the identity permutation.

We are interested in tangles with informative and aesthetically pleasing drawings. Our main criterion is to keep the paths straight by using only a few turns. A **corner** of path i is a point at which it changes its direction from one of the allowed directions (vertical, $+45°$, or $-45°$) to another. A change between $+45°$ and $-45°$ is called a **double corner**. We are interested in the total number of corners of a tangle, where corners are always counted with multiplicity (so a double corner contributes 2 to the total). By convention we require that paths start and end with vertical segments. In terms of the sequence of permutations this means repeating the first and the last permutations at least once each as in Fig. 1(a).

Another natural objective is to minimize path crossings. We call a tangle for π **simple** if it has the minimum number of crossings among all tangles for π. This is equivalent to the condition that no pair of paths cross each other more than once, and this minimum number equals the *inversion number* of π. A simple tangle has no double corner since that would entail an immediate double crossing of a pair of paths.

In general, minimizing corners and minimizing crossings are conflicting goals. For example, let $n = 4k$ and $k \geq 4$ and consider the permutation

$$\pi = [2k, 3, 2, 5, 4, \ldots, 2k-1, 2k-2, 1, \quad 4k, 2k+3, 2k+2, \ldots, 4k-1, 4k-2, 2k+1].$$

It is not difficult to check that the minimum number of corners in a tangle for π is $4n-8$, while the minimum among simple tangles is $5n-20$, which is strictly greater; see Fig. 2 for the case $k = 4$. Our focus in this article is on two special classes of permutations for which corners and crossings can be minimized simultaneously. The first is relatively straightforward, while the second turns out to be much more subtle.

One may ask the following interesting question. Is there an efficient algorithm for *finding a (simple) tangle with the minimum number of corners solving a given permutation*?

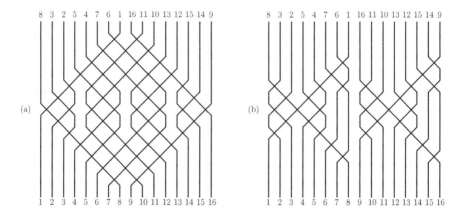

Fig. 2. (a) A tangle with 56 corners. (b) Every *simple* tangle for the same permutation has at least 60 corners.

We do not know whether there is a polynomial-time algorithm, either with or without the requirement of simplicity. Here we present polynomial-time exact algorithms for special classes of permutations.

Even the task of determining whether a given tangle has the minimum possible number of corners among tangles for its permutation does not appear to be straightforward in general (and likewise if we restrict to simple tangles). However, in certain cases, such minimality is indeed evident, and we focus on two such cases. Firstly, we call a tangle **direct** if each of its paths has at most 2 corners (equivalently, at most one non-vertical segment). Note that a direct tangle is simple. Furthermore, it clearly has the minimum number of corners among all tangles (simple or otherwise) for its permutation.

We can completely characterize permutations admitting direct tangles. We say that a permutation $\pi \in S_n$ **contains** a pattern $\mu \in S_k$ if there are integers $1 \leq i_1 < i_2 < \cdots < i_k \leq n$ such that for all $1 \leq r < s \leq k$ we have $\pi(i_r) < \pi(i_s)$ if and only if $\mu(r) < \mu(s)$; otherwise, π **avoids** the pattern (or π is μ-avoiding).

Theorem 1. *A permutation has a direct tangle if and only if it is* 321-*avoiding.*

Our proof yields a straightforward algorithm that constructs a direct tangle for a given 321-avoiding permutation.

Our second special class of tangles naturally extends the notion of a direct tangle, but turns out to have a much richer theory. A **segment** is a straight line segment of a path between two of its corners; it is an **L-segment** if it is oriented from north-east to south-west, and an **R-segment** if it is oriented from north-west to south-east. We call a tangle **perfect** if it is simple and each of its paths has at most one L-segment and at most one R-segment. Any perfect tangle has the minimum possible number of corners among all tangles solving its permutation, and indeed it has the minimum possible corners on path i for each $i = 1, \ldots, n$. To see this, note that if i has an L-segment in a perfect tangle for π then there must be an element $j > i$ with $\pi(i) > \pi(j)$, whose path crosses this L-segment. Hence, an L-segment must be present in any tangle for π. The same argument applies to R-segments. We call a permutation **perfect** if it has a perfect tangle.

Theorem 2. *There exists a polynomial-time algorithm that determines whether a given permutation is perfect and, if so, outputs a perfect tangle.*

A straightforward implementation of our algorithm takes $O(n^5)$ time, but we believe this can be reduced to $O(n^3)$, and possibly further. Our proof of Theorem 2 involves an explicit characterization of perfect permutations, but it is considerably more complicated than in the case of direct tangles. We introduce the notion of a *marking*, which is an assignment of symbols to the elements $1, \ldots, n$ indicating the directions in which their paths should be routed. We prove that a permutation is perfect if and only if it admits a marking satisfying a *balance* condition that equates numbers of elements in various categories. Finally, we show that the existence of such a marking can be decided by finding a maximum vertex-weighted matching in a certain graph with vertex set $1, \ldots, n$ constructed from the permutation.

The number of perfect permutations in S_n grows only exponentially with n (see Section 4), and is therefore $o(|S_n|)$. Nonetheless, perfect permutations are very common for small n: all permutations in S_6 are perfect, as are all but 16 in S_7, and over half in S_{13}.

Related Work. We are not aware of any other study on the number of corners in a tangle. To the best of our knowledge, the problem formulated here is new. Wang in [12] considered the same model of drawings in the field of VLSI design. However, [12] targets, in our terminology, the tangle height and the total length of the tangle paths. The heuristic suggested by Wang produces paths with many unnecessary corners.

The perfect tangle problem is related to the problem of drawing graphs in which every edge is represented by a polyline with few bends. In our setting, all the crossings occur at right angles, as in so-called RAC-drawings [6].

Decomposition of permutations into nearest-neighbour transpositions was considered in the context of permuting machines and pattern-restricted classes of permutations [1]. In our terminology, Albert et. al. [1] proved that it is possible to check in polynomial time whether for a given permutation there exists a tangle of length k (that is, consisting of k permutations), for a given k. Tangle diagrams appear in the drawings of sorting networks [8,2]. We also mention an interesting connection with change ringing (English-style church bell ringing), where similar visualizations are used [13].

2   Preliminaries

We always draw tangles oriented downwards with the sequence of permutations read from top to bottom as in Fig. 1(b). The following notation will be convenient. We write $\pi = [\ldots a \ldots b \ldots c \ldots]$ to mean that $\pi^{-1}(a) < \pi^{-1}(b) < \pi^{-1}(c)$, and $\pi = [\ldots ab \ldots]$ to mean that $\pi^{-1}(a) + 1 = \pi^{-1}(b)$. A pair of elements (a, b) is an **inversion** in a permutation $\pi \in S_n$ if $a > b$ and $\pi = [\ldots a \ldots b \ldots]$. The **inversion number** $\mathrm{inv}(\pi) \in [0, \binom{n}{2}]$ is the number of inversions of π. The following useful lemma is straightforward to prove.

Lemma 1. *In a simple tangle for permutation π, a pair (i, j) is an inversion in π if and only if some R-segment of path i intersects some L-segment of path j.*

3   Direct Tangles

Here we prove Theorem 1. We need two properties of 321-avoiding permutations.

Lemma 2. *Suppose* π, π' *are permutations with* $\mathrm{inv}(\pi') = \mathrm{inv}(\pi) - 1$ *and* $\pi' = \pi \cdot \sigma(i)$ *for some swap* $\sigma(i)$. *If* π *is 321-avoiding then so is* π'.

Proof. Let us suppose that elements i, j, k form a 321-pattern in π'. Then (i, j) and (j, k) are inversions in π'. Inversions of π' are inversions of π, hence, elements i, j, k form a 321-pattern in π. ☐

Lemma 3. *In a simple tangle solving a 321-avoiding permutation, no path has both an L-segment and an R-segment.*

Proof. Consider a simple tangle solving a 321-avoiding permutation π. Suppose path j crosses path i during j's R-segment and crosses path k during j's L-segment. By Lemma 1 we have $\pi = [\ldots k \ldots j \ldots i \ldots]$ while $i < j < k$, giving a 321-pattern, which is a contradiction. ☐

We say that a permutation $\pi \in S_n$ has a **split** at location k if $\pi(1), \ldots, \pi(k) \in \{1, \ldots, k\}$, or equivalently if $\pi(k+1), \ldots, \pi(n) \in \{k+1, \ldots, n\}$.

Theorem 1. *A permutation has a direct tangle if and only if it is 321-avoiding.*

Proof. To prove the "only if" part, suppose that tangle T solves a permutation π containing a 321-pattern. Then there are $i < j < k$ with $\pi = [\ldots k \ldots j \ldots i \ldots]$. Hence by Lemma 1, j has an L-segment and an R-segment, so T is not direct.

We prove the "if" part by induction on the inversion number of the permutation. If $\mathrm{inv}(\pi) = 0$ then π is the identity permutation, which clearly has a direct tangle. This gives us the basis of induction.

Now suppose that π is 321-avoiding and not the identity permutation, and that every 321-avoiding permutation (of every size) with inversion number less than $\mathrm{inv}(\pi)$ has a direct tangle. There exists s such that $\pi(s) > \pi(s + 1)$; fix one such. Note that $(\pi(s), \pi(s+1))$ is an inversion of π; hence, the permutation $\pi' := \pi \cdot \sigma(s)$ has $\mathrm{inv}(\pi') = \mathrm{inv}(\pi) - 1$, and is also 321-avoiding by Lemma 2. By the induction hypothesis, let T' be a direct tangle solving π'.

Perform a swap x in position s exchanging elements $\pi(s)$ and $\pi(s + 1)$, and draw it as a cross on the plane with coordinates (s, h), where $h \in \mathbb{Z}$ is the height (y-coordinate) of the cross (chosen arbitrarily). We assume that the position axis increases from left to right and the height axis increases from bottom to top. Then draw the tangle T' below the cross. This gives a tangle solving π, which is certainly simple. We show that the heights of swaps may be adjusted to make the new tangle direct. To achieve this, the L-segment and R-segment comprising the swap x must either extend existing segments in T', or must connect to vertical paths having no corners in T'. Consider two cases.

Case 1: Suppose that π' has a split at s. Then T' consists of a tangle T_1 for the permutation $[\pi'(1), \ldots, \pi'(s)]$ together with another tangle T_2 for $[\pi'(s + 1), \ldots, \pi'(n)]$; see Fig. 3. Starting with T_1 drawn below x, simultaneously shift all the swaps of T_1 upward until one of them touches x; in other words, until T_1's first swap in position

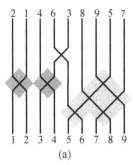

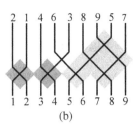

Fig. 3. Shifting two sub-tangles (T_1 is red, T_2 is blue) upward to touch the initial swap x (green) in position $s = 4$.

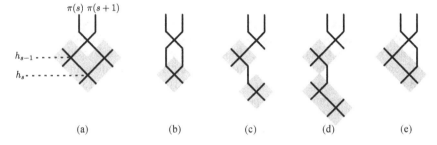

Fig. 4. (a) The tangle T' (blue) touches the swap x (green) on both sides. (b)–(e) Various impossible configurations for the proof.

$s - 1$ occurs at height $h - 1$. Or, if T_1 has no swap in position $s - 1$, no shifting is necessary. Similarly shift T_2 upward until it touches x from the right side. This results in a direct tangle.

Case 2: Suppose that π' has no split at s. Let T' be any direct tangle for π', and again shift it upward until it touches x, resulting in a tangle T for π. Write h_j for the height of the topmost swap in position j in T', or let $h_j = -\infty$ if there is none. We claim that $h_{s-1} = h_{s+1} = h_s + 1 > -\infty$, which implies in particular that $1 < s < s + 1 < n$. Thus T' has swaps in the positions immediately left and right of x, both of which touch x simultaneously in the shifting procedure as in Fig. 4(a), giving that T is direct as required. To prove the claim, first note that $h_s > -\infty$ since π' has no split at s. Therefore $\max\{h_{s-1}, h_{s+1}\} > h_s$, otherwise T would not be simple, as in Fig. 4(b). Thus, without loss of generality suppose that $h_{s-1} > h_s$ and $h_{s-1} \geq h_{s+1}$. Then $h_s = h_{s-1} - 1$, otherwise some path would have more than 2 corners in T', specifically, the path of the element that is in position s after h_{s-1}; see Fig. 4(c) or (d). Now suppose for a contradiction that $h_{s+1} < h_{s-1}$, which includes the possibility that $h_{s+1} = -\infty$, perhaps because $s + 1 = n$. Then in the new tangle T, path $\pi(s)$ contains both an L-segment and an R-segment as in Fig. 4(e), which contradicts Lemma 3.   □

The proof of Theorem 1 yields an algorithm that returns a direct tangle for $\pi \in S_n$ if one exists, and otherwise stops. The algorithm can be implemented so as to run in $O(n^2)$ time. With a suitable choice of output format, this can be improved to $O(n)$.

4 Perfect Tangles

In this section we give our characterization of perfect permutations. Given a permutation $\pi \in S_n$, we introduce the following classification scheme of elements $i \in \{1, \ldots, n\}$. The scheme reflects the possible forms of paths in a perfect tangle, although the definitions themselves are purely in terms of the permutation. We call i a **right** element if it appears in some inversion of the form (i, j), and a **left** element if it appears in some inversion (j, i). We call i **left-straight** if it is left but not right, **right-straight** if it is right but not left, and a **switchback** if it is both left and right.

In order to build a perfect tangle we use a notion of marking. A **marking** M is a function from the set $\{1, \ldots, n\}$ to strings of letters L and R. For any tangle T, we associate a corresponding marking M as follows. We trace the path i from top to bottom; as we meet an L-segment (resp. R-segment), we append an L (resp. R) to $M(i)$. Vertical segments are ignored for this purpose; hence, a vertical path with no corners is marked by an empty sequence $\emptyset$. For example, $M(3) = R$ and $M(13) = LR$ in Fig. 1(c). A marking corresponding to a perfect tangle takes only values $\emptyset$, L, R, LR, and RL. We write $M(i) = R \ldots$ to indicate that the string $M(i)$ starts with R.

Given a permutation π and a marking M, there does not necessarily exist a corresponding tangle. However, we will obtain a necessary and sufficient condition on π and M for the existence of a corresponding perfect tangle. Our strategy for proving Theorem 2 will be to find a marking satisfying this condition, and then to find a corresponding perfect tangle. We say that a marking M is a **marking for** a permutation $\pi \in S_n$ if (i) $M(i) = L$ (respectively $M(i) = R$) for all left-straight (right-straight) elements i, (ii) $M(i) \in \{LR, RL\}$ for all switchbacks, and (iii) $M(i) = \emptyset$ otherwise.

To state the necessary and sufficient condition mentioned above, we need some definitions. A quadruple (a, b, c, d) is a **rec** in permutation π if $\pi = [\ldots a \ldots b \ldots c \ldots d \ldots]$ and $\min\{a, b\} > \max\{c, d\}$. In a perfect tangle, the paths comprising a rec form a rectangle; see Fig. 5 ("rec" is an abbreviation for rectangle). Let M be a marking for $\pi \in S_n$, and let ρ be a rec (a, b, c, d) in π. We call e a **left switchback** of ρ if (i) $M(e) = RL$, (ii) $\pi = [\ldots a \ldots e \ldots b \ldots]$, and (iii) $c < e < d$ or $d < e < c$. Symmetrically, we call e a **right switchback** of ρ if $M(e) = LR$, and $\pi = [\ldots c \ldots e \ldots d \ldots]$, and $a < e < b$ or $b < e < a$. A rec (a, b, c, d) is **regular** if $a < b$ and $c < d$, otherwise it is **irregular**. A rec is called **balanced** under M if the number of its left switchbacks is equal to the number of its right switchbacks; a rec is **empty** if it has no switchbacks.

Here is our key definition. A marking M for a permutation π is called **balanced** if every regular rec of π is balanced and every irregular rec is empty under M.

Theorem 3. *A permutation is perfect if and only if it admits a balanced marking.*

The proof of Theorem 3 is technical, see full version for the complete proof [5].

Any permutation containing the pattern $[7324651]$ (for example) is not perfect since 4 must be a switchback of one of the irregular recs (7321) and (7651). It follows by [9,7]

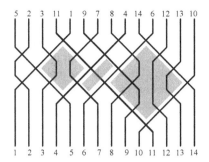

Fig. 5. A permutation with a balanced marking. Some of the recs of the permutation are: $\rho_1 =$ $(5, 11, 1, 4)$, $\rho_2 = (9, 7, 4, 6)$, $\rho_3 = (11, 14, 6, 10)$; ρ_1 and ρ_3 are regular, while ρ_2 is irregular. Left switchbacks of rec ρ_3 are 8 and 9, right switchbacks are 12 and 13. The empty irregular rec ρ_2 has neither left nor right switchbacks.

that the number of perfect permutations in S_n is at most C^n for some constant $C > 1$. Since direct tangles are perfect, it also follows from Theorem 1 that the number is at least c^n for some constant $c > 1$.

We note that Theorem 3 already yields an algorithm for determining whether a permutation is perfect in $\widetilde{O}(2^n)$ time[1] by checking all markings. In Section 5 we improve this to polynomial time.

5   Recognizing Perfect Permutations

We provide an algorithm for recognizing perfect permutations. The algorithm finds a balanced marking for a permutation, or reports that such a marking does not exist. We start with a useful lemma.

Lemma 4. *Fix a permutation. For each right (resp., left) element a there is a left-straight (right-straight) b such that the pair (a, b) (resp., (b, a)) is an inversion.*

Proof. We prove the case when a is right, the other case being symmetrical. Consider the minimal b such that (a, b) is an inversion. By definition, b is left. Suppose that it is also a right element, that is, (b, c) is an inversion for some $c < b$. It is easy to see that (a, c) is an inversion too, which contradicts to the minimality of b.   □

Recall that a marking is balanced only if (in particular) every regular rec of the permutation is balanced under the marking. We show that this is guaranteed even by balancing of recs of a restricted kind. We call a rec (a, b, c, d) of a permutation π **straight** if a, b, c, and d are straight elements of π. A marking is called **s-balanced** if every straight rec is balanced and every irregular rec is empty under the marking.

Lemma 5. *Let M be a marking of a permutation π. Then M is balanced if and only if it is s-balanced.*

---
[1] $\widetilde{O}$ hides a polynomial factor.

Proof. The "if" direction is immediate, so we turn to the converse. Let M be an s-balanced marking and $\rho = (a, b, c, d)$ be a regular rec of π. We need to prove that ρ is balanced under M. If ρ is straight then ρ is balanced by definition. Let us suppose that ρ is not straight. Then some $u \in \{a, b, c, d\}$ is not a straight element. Our goal is to show that it is possible to find a new rec ρ' in which u is replaced with a straight element so that the sets of left and right switchbacks of ρ and ρ' coincide. By symmetry, we need only consider the cases $u = a$ and $u = b$.

Case $u = a$: Let us suppose that a is not straight. By Lemma 4, there exists a right straight e such that (e, a) is an inversion. Let us denote $\rho' = (e, b, c, d)$ and show that ρ' has the same switchbacks as ρ. Let k be a left switchback of ρ; then $M(k) = RL$, and $\pi = [\ldots e \ldots a \ldots k \ldots b \ldots]$, and $c < k < d$. By definition k is a left switchback of ρ'. Let k be a left switchback of ρ'. If $\pi = [\ldots e \ldots k \ldots a \ldots b \ldots]$ then the irregular rec (e, a, c, d) has a left switchback, which is impossible. Therefore, $\pi = [\ldots e \ldots a \ldots k \ldots b \ldots]$ and k is a left-switchback of ρ.

Let us suppose that k is a right switchback of ρ, so $a < k < b$. If $k < e$ then k is a right switchback of the irregular (e, a, c, d); hence, $e < k < b$ and k is a right switchback of ρ'. On the other hand, if k is a right switchback of ρ' then $a < e < k < b$, which means that k is a right switchback of ρ.

Case $u = b$: Let us suppose that b is not straight. By Lemma 4, there exists a right straight e such that (e, b) is an inversion. Let us denote $\rho' = (a, e, c, d)$ and show that ρ' has the same switchbacks as ρ. Let k be a left switchback of ρ. We have $\pi = [\ldots a \ldots k \ldots b \ldots]$. Since k is not a left switchback of the irregular rec (e, b, c, d), we have $\pi = [\ldots a \ldots k \ldots e \ldots]$. Therefore, k is a left switchback of ρ'.

Let k be a right switchback of ρ. Then $a < k < b < e$, proving that k is a right switchback of ρ'. Let k be a right switchback of ρ'. If $b < k$ then k is a right switchback of (e, b, c, d), which is impossible. Then $k < b$ and k is a right switchback of ρ. □

We can restrict the set of recs guaranteeing the balancing of a permutation even further. We call a pair a, b of elements **right** (resp. **left**) **minimal** if a and b are right (left) straight elements of π, and $a < b$, and there is no right (left) straight element c such that $\pi = [\ldots a \ldots c \ldots b \ldots]$. We call rec $\rho = (a, b, c, d)$ **minimal** in π if a, b is a right minimal pair and c, d is a left minimal pair; see Fig. 6(a). We call a marking for a permutation **ms-balanced** if every minimal regular rec is balanced and every irregular rec is empty under the marking.

Lemma 6. *Let M be a marking of a permutation π. Then M is s-balanced if and only if it is ms-balanced.*

Before giving the proof, we introduce some further notation. Let $\rho = (a, b, c, d)$ be an arbitrary, possibly irregular, rec in π. Let us denote by ρ_ℓ (resp. ρ_r) the set of switchbacks i that can under some marking be left (resp., right) switchbacks of ρ. Formally, $i \in \rho_\ell$ if and only if $\pi = [\ldots a \ldots i \ldots b \ldots c \ldots d \ldots]$ and either $c < i < d$ or $d < i < c$. (And ρ_r is defined symmetrically.) For a rec ρ and marking M let ρ_ℓ^M (ρ_r^M) be the set of left (respectively, right) switchbacks of ρ under M. Of course, $\rho_\ell^M \subseteq \rho_\ell$ and $\rho_r^M \subseteq \rho_r$. It is easy to see from the definition that for two different minimal recs ρ and ρ' we have $\rho_\ell \cap \rho'_\ell = \emptyset$ and $\rho_r \cap \rho'_r = \emptyset$.

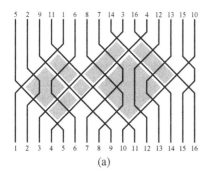

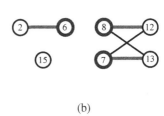

(a)   (b)

Fig. 6. (a) A perfect tangle for a permutation with 7 minimal straight recs (shown red). (b) The graph constructed in *Step 3* of our algorithm. Here, $\mathfrak{I}_\ell = \emptyset$, $\mathfrak{I}_r = \{6\}$, $\mathfrak{R}_\ell = \{6, 7, 8, 12, 13, 15\}$, and $\mathfrak{R}_r = \{2, 7, 8\}$. The vertices of the set $F = \{6, 7, 8\}$ are shown blue. The red edges are the computed maximum matching.

Proof (Lemma 6). It suffices to prove that if M is ms-balanced then it is s-balanced. Consider a straight rec $\rho = (a, b, c, d)$. Let $a = r_1, \ldots, r_p = b$ be a sequence of right straights in which each consecutive pair r_i, r_{i+1} is right minimal. Define left straights $c = \ell_1, \ldots, \ell_q = d$ similarly. Let D be the set of all recs of the form $(r_i, r_{i+1}, \ell_j, \ell_{j+1})$ for $1 \leq i < p$ and $1 \leq j < q$. Notice that all recs of D are minimal. By definition of rec switchbacks, we have $\rho_\ell^M = \bigcup_{u \in D} u_\ell^M$ and $\rho_r^M = \bigcup_{u \in D} u_r^M$. Since every rec $u \in D$ is balanced and for every pair $u, v \in D$ of different recs $u_\ell^M \cap v_\ell^M = u_r^M \cap v_r^M = \emptyset$, we have $|\rho_\ell^M| = |\rho_r^M|$; that is, ρ is balanced under M. □

Let us show how to construct an ms-balanced marking. For a permutation π, let $\mathfrak{I}_\ell = \bigcup\{\rho_\ell : \rho \text{ is an irregular rec in } \pi\}$ and $\mathfrak{R}_\ell = \bigcup\{\rho_\ell : \rho \text{ is a regular rec in } \pi\}$, and define $\mathfrak{I}_r$, $\mathfrak{R}_r$ similarly. Our algorithm is based on finding a maximum vertex-weighted matching, which can be done in polynomial time [11].

The algorithm inputs a permutation π and computes an ms-balanced marking M for π or determines that such a marking does not exist. Initially, $M(i)$ is undefined for every $i \in \{1, \ldots, n\}$. The algorithm has the following steps.

Step 1: For every element $1 \leq i \leq n$ that is neither left nor right, set $M(i) = \emptyset$. For every left straight i set $M(i) = L$. For every right straight i set $M(i) = R$.

Step 2: If $\mathfrak{I}_\ell \cap \mathfrak{I}_r \neq \emptyset$ then report that π is not perfect and stop. Otherwise, for every switchback $i \in \mathfrak{I}_\ell$ set $M(i) = LR$; for every switchback $i \in \mathfrak{I}_r$ set $M(i) = RL$.

Step 3.1: Build a directed graph $G = (V, E)$ with $V = \mathfrak{R}_\ell \cup \mathfrak{R}_r$ and $E = \bigcup\{(\rho_\ell \setminus \mathfrak{I}_\ell) \times (\rho_r \setminus \mathfrak{I}_r) : \rho \text{ is a minimal rec in } \pi\}$.

Step 3.2: Create a set $F \leftarrow (\mathfrak{R}_\ell \cap \mathfrak{R}_r) \cup (\mathfrak{I}_\ell \cap \mathfrak{R}_r) \cup (\mathfrak{I}_r \cap \mathfrak{R}_\ell)$. Create weights w for vertices of G: if $i \in F$ then set $w(i) = 1$, otherwise set $w(i) = 0$.

Step 4: Compute a maximum vertex-weighted matching U on G (viewed as an *unoriented* graph, ignoring the directions of edges) using weights w. If the total weight of U is less than $|F|$ then report that π is not perfect and stop.

Step 5.1: Assign marking based on the matching: for every edge $(i, j) \in U$ set $M(i) = RL$ provided $M(i)$ has not already been assigned, and $M(j) = LR$ provided $M(j)$ has not already been assigned.

Step 5.2: For every switchback $1 \leq i \leq n$ with still undefined marking, if $i \in \mathfrak{R}_\ell$ then set $M(i) = LR$, if $i \in \mathfrak{R}_r$ then $M(i) = RL$, otherwise choose $M(i)$ to be LR or RL arbitrarily. Note that any $i \in \mathfrak{R}_\ell \cap \mathfrak{R}_r$ was already assigned because of *Steps 3.2 and 4*.

Let us prove the correctness of the algorithm.

Lemma 7. *If the algorithm produces a marking then the marking is ms-balanced.*

Proof. Let M be a marking produced by the algorithm for a permutation π. It is easy to see that $M(i)$ is defined for all $1 \leq i \leq n$ (in *Step 1* for straights and in *Step 2* and *Step 5* for switchbacks). By construction, M is a marking for π.

Let us show that M is ms-balanced. Consider an irregular rec ρ of π, and suppose that $i \in \rho_\ell$. Since $\rho_\ell \subseteq \mathfrak{I}_\ell$, in *Step 2* we assign $M(i) = LR$, that is, $i \notin \rho_\ell^M$. Therefore, ρ does not have left switchbacks under M. Similarly, ρ does not have right switchbacks under M. Therefore, ρ is empty.

Consider a regular minimal straight rec ρ in π. Suppose that $i \in \rho_\ell^M$. Then $M(i) = RL$ and $i \in \rho_\ell \subseteq \mathfrak{R}_\ell$. If $i \in \mathfrak{I}_r$ then $i \in \mathfrak{R}_\ell \cap \mathfrak{I}_r \subseteq F$; hence i is incident to an edge in U. Since no directed edge of the form (k, i) is included in G in *Step 3.1*, there exists $(i, k) \in U$ for some k. On the other hand, if $i \notin \mathfrak{I}_r$ then string RL was not assigned to $M(i)$ in *Step 5.2*, nor in *Step 2*. Thus, it was assigned in *Step 5.1*, and again $(i, k) \in U$ for some k. By definition of E we have $k \in \rho_r$, because k cannot appear in ρ'_r for any other minimal $\rho' \neq \rho$. The algorithm sets $M(k) = LR$ at *Step 5.1*; it could not have previously set $M(k) = LR$ at *Step 2* because $k \notin \mathfrak{I}_r$ by the definition of E. Thus $k \in \rho_r^M$.

By symmetry, an identical argument to the above shows that if $k \in \rho_r^M$ then $i \in \rho_\ell^M$ for some i satisfying $(i, k) \in U$. Since U is a matching, we thus have a bijection between elements of ρ_ℓ^M and ρ_r^M. Therefore, ρ is balanced under M.   □

Lemma 8. *Let π be a perfect permutation. The algorithm produces a marking for π.*

Proof. Since π is perfect, there is a balanced marking M for π. Since M is balanced, all irregular recs are empty under M; hence, the algorithm does not stop in *Step 2*. To prove the claim, we will create a matching in the graph G with total weight $|F|$.

Let ρ be a minimal rec in π. Since ρ is balanced under M, we have $|\rho_\ell^M| = |\rho_r^M|$. Hence, let W_ρ be an arbitrary matching connecting vertices of $|\rho_\ell^M|$ with vertices of $|\rho_r^M|$. Of course, $|W_\rho| = |\rho_\ell^M|$. Let $W = \bigcup \{W_\rho : \rho$ is a minimal rec in $\pi\}$. We show that every element of set F is incident to an edge of W.

Suppose $i \in \mathfrak{R}_\ell \cap \mathfrak{R}_r$. Since i is a switchback in π, we have $M(i) = RL$ or $M(i) = LR$. In the first case $i \in \rho_\ell^M$ and in the second case $i \in \rho_r^M$ for some minimal rec ρ. Then i is incident to an edge from W_ρ.

Suppose $i \in F \setminus \{\mathfrak{R}_\ell \cap \mathfrak{R}_r\}$. Without loss of generality, let $i \in \mathfrak{I}_\ell \cap \mathfrak{R}_r$. Since M is balanced, every irregular rec has no switchbacks and hence $M(i) = LR$. Thus, $i \in \rho_r^M$ for some minimal rec ρ, and i is incident to an edge of W_ρ.

Therefore, every vertex of F is incident to an edge of the matching W, which means that the total weight of W is $|F|$.   □

Theorem 2 follows directly from Lemmas 7 and 8 and Theorem 3. A straightforward implementation of the algorithm finding a perfect tangle takes $O(n^5)$ time.

6   Conclusion

In this paper we gave algorithms for producing optimal tangles in the special cases of direct and perfect tangles, and for recognizing permutations for which this is possible. Many questions remain open. What is the complexity of determining the tangle with minimum corners for a given permutation? What is the complexity if the tangle is required to be simple? What is the asymptotic behavior of the maximum over permutations $\pi \in S_n$ of the minimum number of corners among simple tangles solving π?

Acknowledgments: We thank Omer Angel, Franz Brandenburg, David Eppstein, Martin Fink, Michael Kaufmann, Peter Winkler, and Alexander Wolff for fruitful discussions about variants of the problem.

References

1. Albert, M.H., Aldred, R.E.L., Atkinson, M., van Ditmarsch, H.P., Handley, C.C., Holton, D.A., McCaughan, D.J.: Compositions of pattern restricted sets of permutations. Australian J. Combinatorics 37, 43–56 (2007)
2. Angel, O., Holroyd, A.E., Romik, D., Virag, B.: Random sorting networks. Advances in Mathematics 215(2), 839–868 (2007)
3. Argyriou, E.N., Bekos, M.A., Kaufmann, M., Symvonis, A.: On metro-line crossing minimization. Journal of Graph Algorithms and Applications 14(1), 75–96 (2010)
4. Benkert, M., Nöllenburg, M., Uno, T., Wolff, A.: Minimizing intra-edge crossings in wiring diagrams and public transportation maps. In: Kaufmann, M., Wagner, D. (eds.) GD 2006. LNCS, vol. 4372, pp. 270–281. Springer, Heidelberg (2007)
5. Bereg, S., Holroyd, A.E., Nachmanson, L., Pupyrev, S.: Drawing permutations with few corners. ArXiv e-print abs/1306.4048 (2013)
6. Didimo, W., Eades, P., Liotta, G.: Drawing graphs with right angle crossings. Theoretical Computer Science 412(39), 5156–5166 (2011)
7. Klazar, M.: The Füredi-Hajnal conjecture implies the Stanley-Wilf conjecture. In: Krob, D., Mikhalev, A., Mikhalev, A. (eds.) Formal Power Series and Algebraic Combinatorics, pp. 250–255. Springer, Heidelberg (2000)
8. Knuth, D.: The art of computer programming. Addison-Wesley (1973)
9. Marcus, A., Tardos, G.: Excluded permutation matrices and the Stanley-Wilf conjecture. Journal of Combinatorial Theory, Series A 107(1), 153–160 (2004)
10. Pupyrev, S., Nachmanson, L., Bereg, S., Holroyd, A.E.: Edge routing with ordered bundles. In: van Kreveld, M.J., Speckmann, B. (eds.) GD 2011. LNCS, vol. 7034, pp. 136–147. Springer, Heidelberg (2012)
11. Spencer, T.H., Mayr, E.W.: Node weighted matching. In: Paredaens, J. (ed.) ICALP 1984. LNCS, vol. 172, pp. 454–464. Springer, Heidelberg (1984)
12. Wang, D.C.: Novel routing schemes for IC layout, part I: Two-layer channel routing. In: 28th ACM/IEEE Design Automation Conference, pp. 49–53 (1991)
13. White, A.T.: Ringing the changes. Mathematical Proceedings of the Cambridge Philosophical Society 94, 203–215 (1983)

Dynamic Traceroute Visualization at Multiple Abstraction Levels*

Massimo Candela, Marco Di Bartolomeo,
Giuseppe Di Battista, and Claudio Squarcella

Dipartimento di Ingegneria, Università Roma Tre, Italy
{candela,dibartolomeo,gdb,squarcel}@dia.uniroma3.it

Abstract. We present a system, called TPLAY, for the visualization of the traceroutes performed by the Internet probes deployed by active measurement projects. These traceroutes are continuously executed towards selected Internet targets. TPLAY allows to look at traceroutes at different abstraction levels and to animate the evolution of traceroutes during a selected time interval. The system has been extensively tested on traceroutes performed by RIPE Atlas [22] Internet probes.

1 Introduction

The *traceroute* command is one of the most popular computer network diagnostic tools. It can be used on computers connected to the Internet to compute the path (route) towards a given IP address, also called *traceroute path*. It is probably the simplest tool to gain some knowledge on the Internet topology. Because of its simplicity and effectiveness, it attracted the interest of several researchers that developed services for visualizing the Internet paths discovered by executing one or more traceroute commands.

Broadly speaking, there are two groups of traceroute visualization systems: tools developed for local technical debugging purposes and tools that aim at reconstructing and displaying large portions of the Internet topology. Several tools of the first group visualize a single traceroute on a map, showing the geo-location of the traversed routers. A few examples follow. Xtraceroute [10] is a graphical version of the traceroute program. It displays individual routes on an interactive rotating globe as a series of yellow lines between sites, shown as small spheres of different colors. GTrace [20] and Visual-Route [30] are traceroute and network diagnostic tools that provide a 2D geographical visualization of paths. The latter also features more abstract representations taking into account other information, e.g. the round-trip time between intermediate hops. In the second group there are several tools (see e.g. [18,5]) that merge the paths generated by multiple traceroutes into directed graphs and show them in some type of drawing.

In recent years the visualization of Internet measurements has seen a growing interest. This is mainly due to the existence of several projects that deploy *probes* in the Internet. Probes are systems that perform traceroutes and other measurements (e.g. ping,

---

* Partially supported by the ESF project 10-EuroGIGA-OP-003 GraDR "Graph Drawings and Representations" and by the European Community's Seventh Framework Programme (FP7/2007-2013) grant no. 317647 (Leone). We thank RIPE NCC for collaborating to the development of the graph animation framework used in this work.

S. Wismath and A. Wolff (Eds.): GD 2013, LNCS 8242, pp. 496–507, 2013.

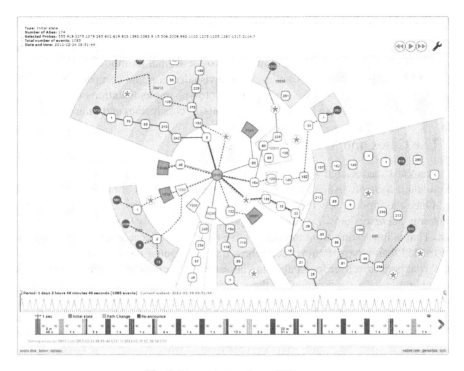

Fig. 1. The main interface of TPLAY

HTTP queries) towards selected targets. They produce a huge amount of data that is difficult to explore, especially when dealing with the network topology. Some examples follow. SamKnows [7] is a broadband measurement service for consumers. MisuraInternet [4] is an Italian project that measures the quality of broadband access. BISmark [28] is a platform for measuring the performance of ISPs. RIPE Atlas [22], CAIDA Ark [2], and M-Lab [3] continuously perform large scale measurements towards several targets.

In this paper we present a system for traceroute visualization called TPLAY, designed for supporting Internet Service Providers (ISPs) and Internet Authorities in the management and maintenance of the network. The requirements were gathered interacting with several ISPs, within the Leone FP7 EC Project, and with the RIPE Network Coordination Center (RIPE NCC). The system works as follows. The user selects a set S of probes of a certain Internet measurement project (all the experiments in this paper have been conducted using RIPE Atlas [22] probes), a target IP address τ, and a time interval $\mathcal{T}$, and obtains a visualization of how the traceroutes issued by the probes in S reach τ during $\mathcal{T}$. TPLAY can be used to study several properties of traceroute paths. These include assessing the reachability of τ over time, discovering the ISPs that provide connectivity to reach it, monitoring the length of traceroute paths as a performance indicator, and inferring how routing policies affect the paths of different probes in S.

A snapshot of TPLAY is in Fig. 1. The routing graph is presented with a radial drawing. The geometric distance between τ and any object reflects the topological distance of that object in the network. Also, since traceroutes tend to give too many details, the system allows to look at the network at different abstraction levels. Finally, the evolution of traceroute paths over time is presented by means of geometric animation.

The paper is organized as follows. In Section 2 we detail the use cases, describe the adopted visualization metaphor, and introduce some formal terminology. In Section 3 we detail the algorithms used to compute the visualization and compare them to the state-of-the-art. In Section 4 we describe the prototype implementation of our tool and the technical challenges we faced. Section 5 contains conclusions and future directions.

2   Use Cases and Visualization Metaphor

The main tasks associated with our system are detailed below. Two of them deal with *Autonomous Systems* (ASes), i.e. entities representing Internet administrative authorities. Once the input is specified as detailed in Section 1, the user is interested in the following. *Security*: knowing what ASes provide connectivity to reach the target over time. That is interesting from the perspective of security, because some ASes may be less trusted than others. *Policy*: seeing how traffic is routed inside a specific AS over time. That helps discover load balancing issues or differences in the routing applied to different probes. *Distance*: knowing the number of hops traversed by each probe over time. Longer paths are indeed potentially responsible for instability and inefficiency. *Dynamics*: seeing how the routing changes at a specific time instant, based on external key indicators. For example, the user may want to check if the routing has changed after a noticeable drop in the round-trip delay experienced when reaching the target.

We discarded solutions based on geographic representations for many reasons. First of all, the fact that a router belongs to a certain ISP or AS is the main piece of information for our purposes, whereas geography is only a secondary feature that further characterizes the nodes in the network. Also, the geo-location data associated with IP addresses is often wrong or missing, and anycast addresses (i.e., those assigned to more than one physical device) can not be mapped to a single location. Finally, the use of landmarks on geographical maps would require special care to avoid geometric cluttering. Motivated by the above, we focused on a topological representation of the data.

The visualization metaphor we adopted is presented below together with supporting motivations. Graphs are represented with *radial layered drawings*, where vertices are placed on concentric circles and targets are in the center. This style of drawing is notably effective for visualizing sparse hierarchical graphs (see, e.g., [31]); in Section 3 we show that our application domain meets such requirement. The probes originating the traceroutes are in the periphery of the drawing. This approach is effective in displaying topological distances. Moreover, radial drawings have their center as the only focus point, which avoids giving probes additional importance due to a privileged geometric position. Finally, the drawing looks like an abstract geography and hence borrows the typical user experience deriving from cartography and geographical visualization.

The need of visualizing the network at different abstraction levels is met by partitioning the set of routers into *clusters*. In our setting, clusters are in correspondence with ASes. The user can modify the representation by interacting with any cluster to either contract or expand it. A *contraction* causes all the routers in the cluster to be merged into a single object representing the cluster, while an *expansion* does the opposite. Collapsing all clusters leads to a high-level, uncluttered view of the graph. On the other hand, the user can expand all the clusters to see all the traversed routers. In general, the user can arbitrarily expand any subset of clusters to examine them in detail.

Paths for reaching the target from the probes change over time. A natural way to show the evolution of traceroutes at different time instants is to present an animation of the drawing. More precisely, for each instant in a given time interval we show a different drawing, corresponding to the traceroutes that are available at that instant. We animate the change from a drawing to a successive one by means of a geometric morph.

Since the visualization is highly interactive and the graph changes over time, preserving the mental map is of paramount importance. Indeed, the user can both animate the drawing in a specific time interval and expand/contract individual clusters. We require that the same drawing is visualized for any two sequences of cluster expansions/contractions that produce the same graph. Also, the graph should be animated smoothly, even at the expense of traversing drawings that are not aesthetically optimal.

Traceroute paths cannot simply be merged and displayed in an aggregate fashion, since each of them has its own informative value and can change over time. For this reason, we represent paths adopting a metro-line metaphor [24] and draw them using different colors. Further, paths that never change in the selected time interval should be easily distinguished. In this context we adopt the method described in [14]. Paths that do not change are partitioned into sets such that each of them determines a tree on the graph. Each tree is depicted with dashed lines and a distinctive color. This has the effect of reducing the number of lines in the drawing, while preserving the routing information for each probe. Paths that change are instead represented by solid lines.

The objects to be visualized are formally defined as follows. Consider a time interval $\mathcal{T}$ and a set of probes $\mathcal{S}$. During $\mathcal{T}$ each probe periodically issues a traceroute towards a target IP address τ. A *traceroute* from probe $\sigma \in \mathcal{S}$ produces a simple directed path on the Internet from σ to τ. If such a path is available in Internet at time $t \in \mathcal{T}$, then it is *valid* at time t. Each vertex of a traceroute originated from $\sigma \in \mathcal{S}$ is either a router or a computer. Vertices are identified as follows: (1) σ has a unique identifier selected by the RIPE NCC; (2) vertices with a *public* IP address are identified by it; (3) vertices with a *private* IP address are identified by a pair composed of their address and the identifier of σ; (4) the remaining vertices are labeled with a "*" (i.e. an unknown IP address). For the sake of simplicity, consecutive vertices labeled with "*" are merged into one. A vertex labeled with "*" is identified by the identifiers of its neighbors in the traceroute.

A digraph G_t is defined at each instant $t \in \mathcal{T}$ as the union of all the paths valid at t produced by the traceroutes issued by the probes of $\mathcal{S}$. A digraph $G_{\mathcal{T}}$ is defined as the union of all graphs G_t. Each vertex of $G_{\mathcal{T}}$ is assigned to a *cluster* as follows. (1) Each probe is assigned to the cluster that corresponds to the AS where it is hosted. (2) Each vertex identified by a public IP address [6] is assigned to a cluster that corresponds to the AS announcing that address on the Internet. This information is extracted from the RIPEstat [23] database and may occasionally be missing. (3) Each vertex v that is not assigned to a cluster after the previous steps is managed as follows. Consider all traceroute paths containing v. For traceroute p let $\mu(\nu)$ be the cluster assigned to the nearest predecessor (successor) of v with an assigned cluster. If $\mu = \nu$ then μ is added to the set of candidate clusters for v. If such set has exactly one cluster, v is assigned to it. If there is more than one candidate, an inconsistency is detected and the procedure terminates prematurely. (4) Each remaining vertex is assigned to a corresponding *fictitious* cluster. We define V_μ as the set of vertices assigned to cluster μ.

For any $t \in \mathcal{T}$ G_t can be visualized at different abstraction levels. Namely, the user can select a set $\mathcal{E}$ of clusters that are fully visualized and each cluster that is in the complement $\bar{\mathcal{E}}$ of $\mathcal{E}$ is contracted into one vertex. More formally, given the pair $G_t, \mathcal{E}$ the visualized graph $G_{t,\mathcal{E}}(V, E)$ is defined as follows. V is the union of the V_μ for all clusters $\mu \in \mathcal{E}$, plus one vertex for each cluster in $\bar{\mathcal{E}}$. E contains the following edges. Consider edge (u, v) of G_t and clusters μ and ν, with $u \in \mu$ and $v \in \nu$. If $\mu \neq \nu$, $\mu \in \mathcal{E}$, and $\nu \in \mathcal{E}$, then add edge (u, v). If both μ and ν are in $\bar{\mathcal{E}}$ then add edge (μ, ν). If $\mu \in \mathcal{E}$ ($\mu \in \bar{\mathcal{E}}$) and $\nu \in \bar{\mathcal{E}}$ ($\nu \in \mathcal{E}$) then add edge (u, ν) $((\mu, v))$. We define $G_{\mu,t}$ as the subgraph of G_t induced by V_μ. Analogously, we define $G_{\mu,\mathcal{T}}$ as the subgraph of $G_\mathcal{T}$ induced by V_μ. We define $G_{\mathcal{T},\mathcal{E}}$ as the union of the $G_{t,\mathcal{E}}$ for each $t \in \mathcal{T}$.

Fig. 1 shows an overview of our prototype implementation. Let $t \in \mathcal{T}$ be the time instant selected by the user. Graph $G_{t,\mathcal{E}}$ is represented by a radial drawing centered in τ. All vertices and clusters that appear in at least one traceroute in $\mathcal{T}$ are in the drawing, including those that are not traversed by any traceroute at time t. Probes in $\mathcal{S}$ are represented as blue circles and labeled with their identifier. Vertices are represented as white rounded rectangles and labeled with the last byte of their IP address, or with a "*". Clusters are represented as annular sectors and labeled with their AS number. Note that vertices assigned to expanded clusters are enclosed in their sectors, while sectors of contracted clusters are empty. The light red cluster contains τ. Clusters containing probes in $\mathcal{S}$ are light blue. The remaining clusters are light yellow. Fictitious clusters are not displayed. Each path from a probe $\sigma \in \mathcal{S}$ to τ is represented by a colored curve from σ to τ passing through all intermediate vertices. Paths are either solid or dashed, depending on whether they change or not during the time interval $\mathcal{T}$. Concentric circles in the background represent the increasing topological distance of vertices.

Fig. 2 contains various details on how the interaction with the visualization works. A graph with static paths and no expanded clusters is presented in Fig. 2(a). It is related to a target τ, a set of probes $\mathcal{S}$, and a small time interval $\mathcal{T}'$. Note that some vertices are not enclosed in any cluster: they belong to fictitious clusters. A graph for τ, $\mathcal{S}$ and $\mathcal{T}''$ ($|\mathcal{T}''| > |\mathcal{T}'|$) is presented in Fig. 2(b). Some dynamic paths are visible. The same graph is presented in Fig. 2(c) with one expanded cluster. Note how the ordering of clusters and vertices on the radial layers is preserved. Fig. 2(d) shows the same expanded graph at a different time instant. The intermediate vertices of two paths are different.

Fig. 2 also helps us explain how the tasks detailed at the beginning of the section can be accomplished. The *Security* task is satisfied in Fig. 2(a): we can see how ASes 1200 and 20965 provide connectivity to reach the target. The *Policy* and *Distance* tasks are addressed in Fig. 2(c), where the length and structure of the paths from each of the three probes 619, 602, 265 is clearly visible. The *Dynamics* task is solved in Figg. 2(c)-(d), where we can see how the paths change for probes 619 and 602 after a routing event.

The user interaction plays a major role in our metaphor. The reader can visit [8] for an example video of the interaction with TPLAY.

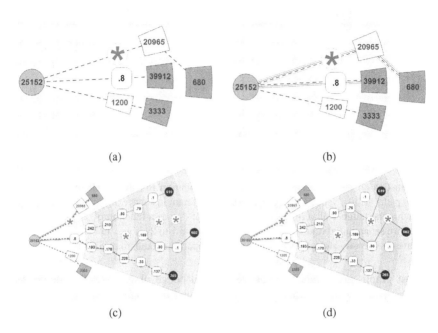

(a)   (b)

(c)   (d)

Fig. 2. Details of the interactive features of our visualization. (a) A graph $G_{\mathcal{T}'}$ relative to a target τ, a set of probes $\mathcal{S}$, and a time interval $\mathcal{T}'$. All paths in $G_{\mathcal{T}'}$ are static and all clusters contracted. (b) A graph $G_{\mathcal{T}''}$ relative to τ, $\mathcal{S}$, and $\mathcal{T}''$ ($|\mathcal{T}''| > |\mathcal{T}'|$). Some paths are dynamic and all clusters are contracted. (c) $G_{\mathcal{T}''}$ with an expanded cluster. (d) $G_{\mathcal{T}''}$ at a different time instant.

3   The Algorithms

We started our analysis by computing several statistics on the RIPE Atlas data set that we used to test the system. It consists of traceroutes executed in one month (July 2012) by 200 probes. Fig. 3 presents the main results of our analysis. In Fig. 3(a) we plot a cumulative distribution function of the length of traceroute paths. That gives us a rough indication on the maximum distance between a probe in $\mathcal{S}$ and τ. The plot shows that traceroutes with more than 15 vertices are rare, confirming the suitability of the radial metaphor. In Fig. 3(b) we plot the number of vertices and the density ($|E|/|V|$) of $G_{\mathcal{T}}$ as a function of $\mathcal{T}$. It turns out that $G_{\mathcal{T}}$ is quite sparse for time intervals that are compatible with the application domain. In particular, the density ranges between 1.2 and 1.5 for time intervals within 24 hours. The number of vertices is in the range of 2000.

As a second step, we performed experiments using spring embedders and hierarchical drawing algorithms. Layouts produced by spring embedders [29] are unsuitable for our metaphor, because the topological distance between vertices is not always represented and because they produce drawings with not enough regularity. Also, they tend to introduce crossings that are avoidable, for the expected density of the data set. For hierarchical drawing, we experimented both basic algorithms [29] and variations that allow to represent clustered graphs [25,26]. The experiments put in evidence that crossing-reduction heuristics like those in [25,26] are quite effective. However, in our

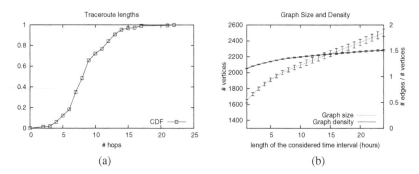

Fig. 3. Statistics on the data set. (a) Cumulative distribution function (CDF) of the length of traceroute paths at July, 1st 2012 at 00:00. CDFs at different instants exhibit similar features. (b) Plot showing the number of vertices and the density of $G_\mathcal{T}$ as a function of $\mathcal{T}$. For each day in the month we set an initial time at 00:00 and grow $\mathcal{T}$ from 1 to 24 hours. For each value of $\mathcal{T}$ we plot the average density and number of vertices. We report the standard deviation with error bars.

case most graphs are planar or quasi-planar and hence planarity-based methods are more attractive. Finally, we discarded upward planar drawings [29]. The main reason is that they tend to use vertical space to resolve crossings, which may result in large geometric distances between vertices that are topologically close.

A very high level and informal description of our algorithmic framework is the following. We pre-compute a hierarchical drawing Γ_0 of $G_\mathcal{T}$ that integrates all the traceroutes in $\mathcal{T}$. In that drawing all clusters are expanded. The layout is computed in such a way to have few crossings involving connections between clusters. The quality of the layout inside the clusters is considered with lower priority. Moreover, the quality of the drawing of edges that are part of many traceroutes in $\mathcal{T}$ is privileged among the edges of $G_\mathcal{T}$. The drawing computed for each cluster is stored and reused in any drawing where that cluster is expanded. The hierarchical drawing is mapped to a radial drawing with a suitable coordinate transformation. Changes in the drawing due to an expansion or contraction of a cluster or a change in traceroutes are visualized with an animation. At any instant $t \in \mathcal{T}$ only the traceroutes that are valid in t are displayed.

For our purposes an interesting reference is [11] that constructs radial drawings adapting techniques of the Sugiyama Framework, but, unfortunately, it does not deal with clusters. The algorithm in [16], which extends the one described in [15], inspired part of our work. However, it proposes a clustered planarity testing algorithm, while we rather need an algorithm for clustered graph planarization, and [16] is not easily extensible for this purpose (neither is the algorithm in [12] that is not suitable for hierarchical drawings). For these reasons we devised a new algorithm to produce clustered hierarchical drawings, as a planarization-oriented variation of [16]. In [21] an algorithm is proposed for the expansion/contraction of clusters of hierarchical drawings, building on [27]. Unfortunately it uses local layering for vertices, while global layering [25,26] is more suitable for our needs because it produces more compact drawings. Indeed a very common use case of TPLAY is to expand all clusters along one or more traceroutes. Local layering would visualize far from τ also vertices in unrelated paths because of the increased need for vertical space of their layers. For this reason we devised a new algorithm for expanding/contracting clusters that is based on global layering. Differently from [21] it is not a local update scheme, i.e. it computes a new drawing for the whole graph at each interaction. The lower time efficiency is negligible because the graphs commonly handled by TPLAY are small. Finally, mental map preservation during expansion/contraction of clusters is addressed by a geometric morph, implemented as an animation of objects from their initial position to their final position (see, e.g., [14]).

What follows gives more details on our the algorithmic framework. In a preprocessing step several information are computed on $G_\mathcal{T}$ that will be used for actual drawings. Given any $G_{\mu,\mathcal{T}}$, a vertex is a *source* (*sink*) of $G_{\mu,\mathcal{T}}$ if it is the last (first) vertex of $G_{\mu,\mathcal{T}}$ encountered in some traceroute path. Each graph $G_{\mu,\mathcal{T}}$ is augmented with extra vertices and edges so that all the longest paths from a source to a sink have the same length. The added vertices are called *fictitious vertices* of μ and ensure that, given an edge $(u,v) \in G_\mathcal{T}$, $u \in \mu$, $v \in \nu$, $\mu \neq \nu$, clusters μ and ν do not share a layer in any drawing of $G_{t,\mathcal{E}}$. Moreover, they force edges that leave a cluster by spanning several layers to be routed inside that cluster. A μ-*drawing* is pre-computed for each $G_{\mu,\mathcal{T}}$. It consists of 1. assigning vertices to layers so that all edges are between consecutive layers and 2. computing a total order for the vertices of each layer. A partial order $\prec$ is computed for clusters, such that for any two clusters μ and ν with $\mu \prec \nu$, the vertices of μ appear to the left of the vertices of ν for any drawing Γ where μ and ν share one or more layers. This helps preserve the mental map during expansions/contractions. The preprocessing step requires to compute a drawing Γ_0 of $G_\mathcal{T}$ with all clusters expanded. Γ_0 gives the information needed to compute a μ-drawing for each cluster and a partial order $\prec$ for clusters. The algorithm to compute Γ_0 is similar to that in [16], where a PQ-tree [13] is used to order vertices along the layers of the drawing. Our PQ-tree is initialized with a spanning tree of $G_\mathcal{T}$ and incrementally updated with the remaining edges that induce ordering constraints. An edge is added only if it does not produce a crossing (i.e. the PQ-tree does not return the null tree). A rejected edge will produce crossings in Γ_0. Edges are added with priority given by their aesthetic importance: namely, they are weighted by the number of traceroutes that traverse them in $\mathcal{T}$. As an implementation detail, we actually compute a total order for clusters to represent a partial order $\prec$. Such an order is produced by a DFS visit of the spanning tree of $G_\mathcal{T}$. The tree has an embedding induced by the layer orders produced by the PQ-tree algorithm, and children of a vertex are visited in clockwise order. Intuitively, we preserve the geometric left-to-right order for clusters from Γ_0, and reuse it to produce a drawing of any $G_{t,\mathcal{E}}$.

The computation of the drawing $\Gamma_{\mathcal{T},\mathcal{E}}$ of $G_{\mathcal{T},\mathcal{E}}$ is detailed below. Before that, note that once $\Gamma_{\mathcal{T},\mathcal{E}}$ is computed, we display, for any $t \in \mathcal{T}$ all the vertices of $G_{\mathcal{T},\mathcal{E}}$ but only the edges of $G_{t,\mathcal{E}}$. This is done to preserve the mental map of the user, using $\Gamma_{\mathcal{T},\mathcal{E}}$ as a "framework" that "hosts" the drawings of each instant. First, a layering of $G_{\mathcal{T},\mathcal{E}}$ is computed such that for each vertex the distance from τ is minimized. Also, dummy vertices, called *fictitious vertices* of $G_{\mathcal{T},\mathcal{E}}$, are added so that each edge spans two consecutive layers. Vertices are horizontally ordered on each layer such that: 1. $\prec$ is enforced; 2. for each cluster μ of $\mathcal{E}$, the orders on the layers of its μ-drawing are enforced; 3. the fictitious vertices of $G_{\mathcal{T},\mathcal{E}}$ are placed in such a way to have few crossings. In particular, they must not be interleaved with the vertices of any cluster, that is, the vertices of each cluster must be consecutive on every layer. For this reason, each fictitious vertex is assigned to a new fictitious cluster, which is inserted in the partial order $\prec$ in an intermediate position between the endpoints of the edge it belongs to. Finally, the ordered layers are used to assign geometric coordinates to vertices. The width of each cluster μ is computed as follows. Consider the layer containing the largest number of vertices assigned to μ. The cluster is assigned a width proportional to this number. Vertices of μ are assigned horizontal coordinates such that they can be enclosed by a rectangle with height proportional to the number of layers assigned to the vertices of μ and width equal to the width of μ. We avoid intersection between enclosing rectangles by means of an auxiliary directed acyclic graph where vertices are clusters of $G_{\mathcal{T},\mathcal{E}}$ and edges are selected from $\prec$ depending on which pairs of clusters share a layer in the current layering of $G_{\mathcal{T},\mathcal{E}}$. Edges are weighted based on the widths of the clusters they are incident to. The total width of the drawing is given by the longest path in this graph. The above is applied recursively to compute the horizontal spacing among all clusters. The vertical coordinate of a vertex is equal to the one assigned to its layer, which is proportional to the index of that layer in the total order of layers.

Going back to the state-of-the-art, concerning restrictions $R1$, $R2$ and $R3$, described in [16], that a planar clustered hierarchical drawing must obey, drawings produced by our algorithm satisfy $R1$ and $R2$, while we consider $R3$ too restrictive for our application. $R1$ is satisfied in the preprocessing step by merging, for each cluster, all sources into one vertex. The PQ-tree is initialized with a spanning tree that contains all these new vertices, which has the effect to keep the vertices of each cluster consecutive on any layer. $R2$, as shown in [16], is automatically satisfied for the initial drawing Γ_0, and is satisfied for any drawing of $G_{t,\mathcal{E}}$ by exploiting the partial order $\prec$.

To obtain a radial drawing, the geometric coordinates of vertices so computed are transformed as follows. Each vertex is placed on the perimeter of a circle centered in an arbitrary fixed point and having radius equal to the vertical coordinate of the vertex. Then the horizontal coordinate of the vertex is mapped to a circular coordinate on the perimeter of that circle. The perimeters of clusters are mapped with a similar radial transformation. An edge (u, v) is drawn either as a straight segment or a curved arc, depending on the angle it must sweep to connect vertices u and v. Note that in our setting each edge connects only vertices in two consecutive layers, hence a curved edge can be drawn only in the space between these layers.

4   Implementation and Technical Challenges

The implementation of TPLAY is split into three main blocks: 1. a visualization frontend; 2. a layout engine; and 3. a data back-end.

The *visualization front-end* is a Web application. It allows the user to specify input parameters and to visualize and animate interactive graphs. The main interface is presented in Fig. 1 and additional images are provided at [8]. It is composed of four main elements: the controller, the graph panel, the info panel, and the timeline panel. We detail their functionalities below.

The *controller* is a sliding panel located in the upper right corner. It allows the user to input queries composed of a target τ, a time interval $\mathcal{T}$, and a set of probes $\mathcal{S}$. Once the visualization is ready, the controller can be used to animate the graph with the traceroute paths available during $\mathcal{T}$. The play/pause/stop, repeat-last, step-back, and step-forward buttons allow for a fine-grained management of the graph animation.

The *graph panel* displays the interactive graph, initially centered and fitted to the window. The user can pan and zoom it with the mouse. The animation of the graph consists of a sequence of morphing steps. Each step transform the graph by applying the effects of an event involving one or more traceroute paths. Given a probe $\sigma \in \mathcal{S}$, an event can consist of: (a) the availability of a new traceroute path from σ to τ; (b) a change in the sequence of vertices in the traceroute path from σ to τ; (c) a disconnection resulting in an empty traceroute path. Events of type (a) are rendered with a gradual introduction of new paths in the graph. Events of type (b) are rendered with a geometric morph of curves representing the involved paths. Events of type (c) are rendered with a blinking effect after which paths disappear. We introduce a delay between each pair of consecutive animation steps. The delay is proportional to the logarithm of the elapsed time between the corresponding routing events. This gives an approximate perception of elapsed time, while limiting the overhead on the total animation time. The elements of the graph are interactive and show additional information on request. Hovering a vertex with the mouse for a few seconds highlights all the paths passing through it. Hovering a path for a few seconds highlights the path and all its vertices.

The *info panel* is in the upper part of the window. It shows all the available information about any selected network component represented in the graph. It also displays a textual description for the latest event that caused an update of the visualized graph.

The *timeline panel* is in the lower part of the window and contains two timelines that allow to accurately navigate the traceroute information in $\mathcal{T}$. The first, called *control timeline*, provides a fast overview of the trend in the number of events over time. The second, called *selection timeline*, shows individual events ordered in time and is designed for fine-grained analysis. Each block in the selection timeline contains a sequence of events happening at the same time, represented with colored rectangles. Different colors are used for different types of events. The elapsed time between any two consecutive blocks is reported in the area between them. Both timelines feature a red cursor that points at the current time instant and is continuously updated during the animation. The user can drag the cursors, changing the current instant and updating the graph accordingly. The selection timeline can only show a limited number of events due to its constrained area. In case there are more events, the animation causes involved events to be smoothly translated into the visible part of selection timeline. The user can scroll horizontally to reveal hidden events. Further, the user can limit the animation to a particularly interesting period within $\mathcal{T}$ by dragging the two green sliders at the top of the control timeline. The sliders on the selection timeline are updated accordingly.

The implementation of the front-end required to focus on some algorithmic details. The arrangement of paths in a metro-line fashion is implemented as follows. First of all, an arbitrary total ordering is computed on the set of visualized paths. For each edge without bends in the graph, the paths that traverse it are drawn as parallel segments connecting the two endpoints of the edge. The ordering of such segments reflects the total ordering of paths to promote consistency between edges. In case the edge contains bends, the drawing is computed in two steps. First, we split the bended edge in a sequence of intermediate edges $e^1, \ldots, e^n$ and compute the path segments for each of them. Second, for each pair of consecutive intermediate edges (e^i, e^{i+1}) and for each path that traverses it, we call (u, v) and (w, z) the two segments computed respectively for e^i and e^{i+1}. If there is an intersection point p between (u, v) and (w, z), we rewrite the two segments as (u, p) and (p, z). Otherwise, we add a connection (v, w) between (u, v) and (w, z). Path colors are computed with the algorithm described in [19] to ensure that they are distinguishable from each other.

The front-end is written in JavaScript and HTML. It is based on the BGPlay.js framework [1] that we developed in collaboration with the RIPE NCC. The objective of the framework is to simplify the implementation of web-based tools for the representation of evolving data described in terms of graph components. The framework consists of a solid implementation of graph domain objects and a set of modules. Modules provide functionalities and representation of data and can be used to compose ad-hoc tools. We use Scalable Vector Graphics for the representation and animation of the graph.

The visualization always starts with an overview of the traceroutes. Hence, the *layout engine* is invoked to produce a drawing of $G_{T, \emptyset}$. When the user expands/contracts a cluster (a cluster is added/removed from $\mathcal{E}$) the layout engine is invoked again on $G_{T, \mathcal{E}}$. In the implementation of the radial drawing we artificially increase the radius of each layer by an additional offset, such that vertices on dense layers are not overlapped. For the sake of simplicity, curved segments are uniformly sampled and drawn as polylines.

The layout engine is implemented in Java. We initially designed it to be implemented as part of the visualization front-end, but later moved to a back-end implementation in order to make use of already existing libraries. In particular we adopted a PQ-tree implementation [17] and Apache Commons Graph [9] for general graph models and algorithms. We optimized the output of the layout engine after the initial layout, so that only graph elements with new drawing coordinates are included.

The *data back-end* is mainly responsible of retrieving and preprocessing traceroute data. The result is then used by the front-end to animate traceroute events and by the layout engine to compute the drawings.

5   Conclusions and Open Problems

We presented a metaphor for the visualization of traceroute measurements towards specific targets on the Internet. It consists of a radial drawing of a clustered graph where vertices are routers or computers and clusters are administrative authorities that control them. Our metaphor allows the user to interact with the visualization, both exploring the content of individual clusters and animating the graph to see how traceroute paths change over a time interval of interest.

In the future we will take into account the *DNS resolution* of selected targets in the visualization. That means that some targets may be represented by more than one vertex, giving rise to an anycast behavior of the target, depending on the policies implemented at the DNS level. We will also explore the possibility to process streams of incoming data, adding or removing elements in the visualization incrementally.

References

1. BGPlayJS, http://www.dia.uniroma3.it/ compunet/www/ view/tool.php?id=bgplayjs
2. CAIDA Ark, http://www.caida.org/projects/ark/

3. Measurement Lab, http://www.measurementlab.net/
4. MisuraInternet, https://www.misurainternet.it/
5. Monitor Scout Traceroute, http://tools.monitorscout.com/traceroute/
6. RFC 1918, address allocation for private internets,
   http://www.ietf.org/rfc/rfc1918.txt
7. SamKnows, http://www.samknows.com/broadband/
8. TPlay, http://www.dia.uniroma3.it/~compunet/projects/tplay
9. Apache Software Foundation. Apache Commons Graph,
   http://commons.apache.org
10. Augustsson, B.: Xtraceroute,
    http://www.dtek.chalmers.se/~d3august/xt/index.html
11. Bachmaier, C.: A radial adaptation of the sugiyama framework for visualizing hierarchical information. IEEE Trans. on Visualization and Computer Graphics 13(3), 583–594 (2007)
12. Di Battista, G., Didimo, W., Marcandalli, A.: Planarization of clustered graphs. In: Mutzel, P., Jünger, M., Leipert, S. (eds.) GD 2001. LNCS, vol. 2265, pp. 60–74. Springer, Heidelberg (2002)
13. Booth, K.S., Lueker, G.S.: Testing for the consecutive ones property, interval graphs, and graph planarity using pq-tree algorithms. JCSS 13(3), 335–379 (1976)
14. Colitti, L., Di Battista, G., Mariani, F., Patrignani, M., Pizzonia, M.: BGPlay: A System for Visualizing the Interdomain Routing Evolution. In: Liotta, G. (ed.) GD 2003. LNCS, vol. 2912, pp. 295–306. Springer, Heidelberg (2004)
15. Di Battista, G., Nardelli, E.: Hierarchies and planarity theory. IEEE Transactions on Systems, Man and Cybernetics 18(6), 1035–1046 (1988)
16. Forster, M., Bachmaier, C.: Clustered level planarity. In: Van Emde Boas, P., Pokorný, J., Bieliková, M., Štuller, J. (eds.) SOFSEM 2004. LNCS, vol. 2932, pp. 218–228. Springer, Heidelberg (2004)
17. Harris, J.: A graphical Java implementation of PQ-Trees, http://www.jharris.ca
18. Hokstad, V.: Traceviz: Visualizing traceroute output with graphviz,
    http://www.hokstad.com
19. Kistner, G.: Generating visually distinct colors,
    http://phrogz.net/css/distinct-colors.html
20. Periakaruppan, R., Nemeth, E.: Gtrace - a graphical traceroute tool. In: Proc. 13th USENIX Conference on System Administration, pp. 69–78. USENIX Association (1999)
21. Raitner, M.: Visual navigation of compound graphs. In: Pach, J. (ed.) GD 2004. LNCS, vol. 3383, pp. 403–413. Springer, Heidelberg (2005)
22. RIPE NCC. RIPE Atlas, http://atlas.ripe.net/
23. RIPE NCC. RIPEstat, https://stat.ripe.net/
24. Roberts, M.J.: Underground Maps Unravelled - Explorations in Information Design (2012)
25. Sander, G.: Layout of compound directed graphs. Technical report, FB Informatik, Universitat Des Saarlandes (1996)
26. Sander, G.: Graph layout for applications in compiler construction. Theoretical Computer Science 217(2), 175–214 (1999)
27. Sugiyama, K., Misue, K.: Visualization of structural information: automatic drawing of compound digraphs. IEEE Trans. on Systems, Man and Cybernetics 21(4), 876–892 (1991)
28. Sundaresan, S., de Donato, W., Feamster, N., Teixeira, R., Crawford, S., Pescapè, A.: Broadband internet performance: A view from the gateway. In: Proc. SIGCOMM (2011)
29. Di Battista, G., Eades, P., Tamassia, R., Tollis, I.G.: Graph Drawing: Algorithms for the Visualization of Graphs. Prentice Hall (1998)
30. Visualware. VisualRoute, http://www.visualroute.com/
31. Yee, K.-P., Fisher, D., Dhamija, R., Hearst, M.: Animated exploration of dynamic graphs with radial layout. In: Proc. INFOVIS 2001. IEEE Computer Society (2001)

Graph Drawing Contest Report

Christian A. Duncan[1], Carsten Gutwenger[2], Lev Nachmanson[3], and Georg Sander[4]

[1] Quinnipiac University, USA
Christian.Duncan@quinnipiac.edu
[2] Technische Universität Dortmund, Germany
carsten.gutwenger@tu-dortmund.de
[3] Microsoft, USA
levnach@microsoft.com
[4] IBM, Germany
georg.sander@de.ibm.com

Abstract. This report describes the 20th Annual Graph Drawing Contest, held in conjunction with the 2013 Graph Drawing Symposium in Bordeaux (Talence), France. The purpose of the contest is to monitor and challenge the current state of graph-drawing technology.

1 Introduction

This year, the Graph Drawing Contest was divided into an *offline contest* and an *online challenge*. The offline contest had three categories: two dealt with creating and visualizing a graph from a given data set and one was a review network. The data sets for the offline contest were published months in advance, and contestants could solve and submit their results before the conference started. The submitted drawings were evaluated according to aesthetic appearance and how well the data was visually represented. For the visualization itself, typical drawings, interactive tools, animations, or other innovative ideas were allowed.

The online challenge took place during the conference in a format similar to a typical programming contest. Teams were presented with a collection of challenge graphs and had approximately one hour to submit their highest scoring drawings. This year's topic was to minimize the area for orthogonal grid layouts, where we allowed crossings (the number of crossings was not judged, only the area counted).

Overall, we received 12 submissions: 3 submissions for the offline contest and 9 submissions for the online challenge.

2 Creative

For the two categories in this topic, the task was to create a meaningful graph from data found on a specific website and visualize it in a suitable way. Any kind of visualization was allowed. We proposed pictures, map-like drawings, animations, and interactive tools, but any other innovative idea was also welcome. Submissions were to include the graph itself as well as the visualization.

S. Wismath and A. Wolff (Eds.): GD 2013, LNCS 8242, pp. 508–513, 2013.

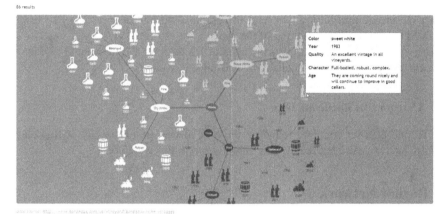

(a) de Jong, Pazienza

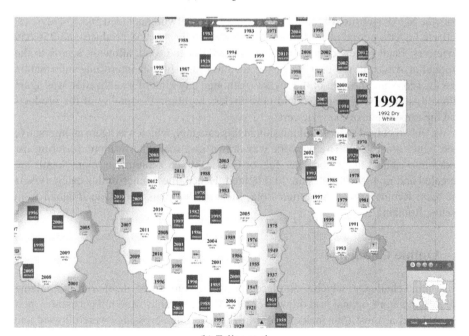

(b) Zelina et al.

Fig. 1. Creative, Category A (Bordeaux Wines)

2.1 Category A: Bordeaux Wines

The first data set could be found on the Bordeaux Wines website[1].

We received two submissions in this category, both included links to an interactive web site for visualizing the data. Fig. 1(a) shows the submission by Jos de Jong and Giovanni Pazienza; their interactive web page provides filtering and shows a dynamic layout, which is smoothly adjusted when nodes appear or disappear due to changes in the filter. The submission by Remus Zelina et al. (see Fig. 1(b)) is also an interactive web page providing filtering, but here the graph is visualized in a map-like fashion obtained by a preceding clustering of the data.

The winner in this category was the team of Jos de Jong and Giovanni Pazienza for their clear and easy-to-use visualization of Bordeaux wines.

2.2 Category B: Bordeaux City

The second data set could be found on the official website of the city of Bordeaux[2]. Unfortunately, we did not receive any submissions in this category.

3   Review Network

In this category the task was to visualize a review network that had been obtained from Amazon reviews on fine foods. The data set for the review network could be obtained from the SNAP website[3] of Stanford University. The network included 568,454 reviews, collected over a period of more than 10 years, including 74,258 products and 256,059 reviewers. Each review contained the product ID of the food, allowing access to the product at Amazon, the user ID of the reviewer, and further interesting information like the score and the date of the review. Although any kind of visualization was again allowed, a good submission should nevertheless highlight the quality of the products and the importance of the reviewers.

We only received a single submission in this category, which was again an interactive web page; see Fig. 2. The network was drawn in a map-like fashion clustering the nodes into islands like Chocolate, Tea, or Coffee. When zooming in, more details are revealed and important nodes are highlighted. Hence, the winner in this category was the team Remus Zelina, Sebastian Bota, Siebren Houtman, and Radu Balaban from Meurs, Romania.

4   Online Challenge

The online challenge, which took place during the conference, dealt with minimizing the area in an orthogonal grid drawing. The challenge graphs were not necessarily planar and had at most four incident edges per node. Edge crossings were allowed and

---

[1] http://www.bordeaux.com/us/vineyard/bordeaux-wine-vintages
[2] http://www.bordeaux.fr/
[3] http://snap.stanford.edu/data/web-FineFoods.html

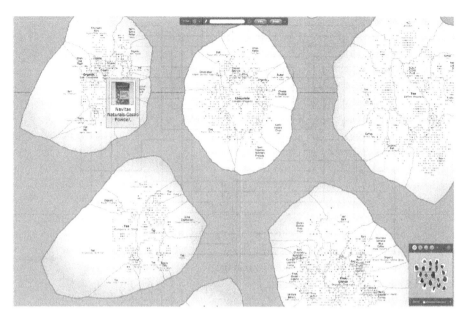

Fig. 2. Review Network

their number did not affect the score of a layout. Since typical drawing systems first try to minimize the number of crossings, which might result in long edges increasing the required area, we were in particular interested in the effect of allowing crossings on the quality of layouts when trying to reduce the area.

The task was to place nodes, edge bends, and crossings on integer coordinates so that the edge routing is orthogonal and the layout contains no overlaps. At the start of the one-hour on-site competition, the contestants were given six graphs with an initial legal layout with a large area. The goal was to rearrange the layout to reduce the area, defined as the number of grid points in the smallest rectangle enclosing the layout. Only the area was judged; other aesthetic criteria, such as the number of crossings or edge bends, were ignored.

The contestants could choose to participate in one of two categories: *automatic* and *manual*. To determine the winner in each category, the scores of each graph, determined by dividing the area of the best submission in this category by the area of the current submission and then taking the square root, were summed up. If no legal drawing of a graph was submitted (or a drawing worse than the initial solution), the score of the initial solution was used.

In the automatic category, contestants received six graphs ranging in size from 59 nodes / 87 edges to 3393 nodes / 4080 edges and were allowed to use their own sophisticated software tools with specialized algorithms. Manually fine-tuning the automatically obtained solutions was allowed. However, no team participated in this category, hence we had no winner in the automatic category this year.

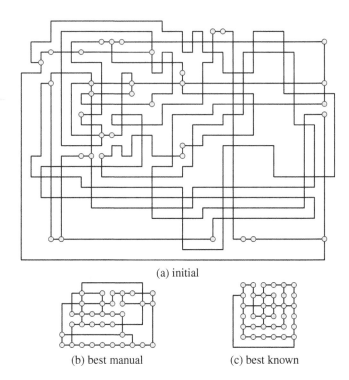

(a) initial

(b) best manual   (c) best known

Fig. 3. Challenge graph with 35 nodes and 44 edges: (a) initial layout (area: 768), (b) best manual result obtained by the team of Spisla and Gronemann (area: 70), and (c) best known solution (area 49)

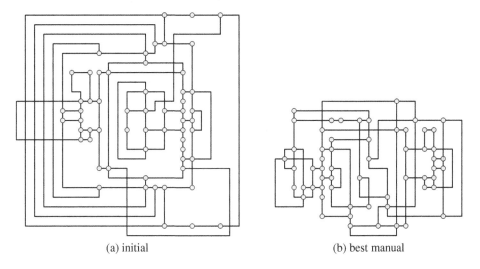

(a) initial   (b) best manual

Fig. 4. Challenge graph with 59 nodes and 87 edges: (a) initial layout (area: 600) and (b) best manual result by the team of Bläsius and Rutter (area: 315)

The 9 manual teams solved the problems by hand using IBM's *Simple Graph Editing Tool* provided by the committee. They received six graphs ranging in size from 19 nodes / 26 edges to 150 nodes / 188 edges. Three of the larger input graphs were also in the automatic category; unfortunately we could not compare automatic solutions with manual solutions this time. With a score of 5.485, the winner in the manual category was the team of Thomas Bläsius and Ignaz Rutter from Karlsruhe Institute of Technology, who found the best results for two of the six contest graphs.

Fig. 3 shows a challenge graph with 35 nodes and 44 edges and a very bad initial layout; the best manually obtained result improved the area from 768 to 70, and the best solution we know for this graph has an area of 49. Fig. 4 shows a larger challenge graph with 59 nodes and 87 edges and a better initial layout; here the best submitted solution improved the area from 600 to 315.

Acknowledgments. The contest committee would like to thank the generous sponsors of the symposium and all the contestants for their participation. Further details including winning drawings and challenge graphs can be found at the contest website:

http://www.graphdrawing.de/contest2013/results.html

3D Graph Printing in GLuskap⋆

Joel Bennett and Stephen Wismath

University of Lethbridge, Canada

Abstract. The GLuskap software package for creating and editing graphs in 3D has been extended to include 3D printing of graphs by exporting the graph to a common file format capable of being printed on most commercially available 3D printers.

The GLuskap software package [1] allows for the creation and editing of graphs in three dimensions (3D). A new plugin for GLuskap allows graphs to be exported as an STL (stereolithography) file. The STL file can then be sent to a 3D printer to create a physical representation of the graph; for example, the graph K_7 drawn orthogonally with bends is shown in Fig. 1. Printed physical models of 2D layouts can also be constructed.

Fig. 1. A graph in GLuskap and a physical 3D print of that graph

The Export to STL Plugin

Most graph drawing software, including GLuskap, rely on OpenGL, Qt, or DirectX to ultimately display a graph on some device. Although such graphics-based models are very effective for interactive use, they do not produce 3D models appropriate for printing. The only option for 3D graph printing previously, was to completely start from scratch using non-trivial 3D modeling software tools such as Blender, 3DStudioMax, AutoCAD, etc.

⋆ Research supported by N.S.E.R.C.

S. Wismath and A. Wolff (Eds.): GD 2013, LNCS 8242, pp. 514–515, 2013.
© Springer International Publishing Switzerland 2013

GLuskap's *Graph to STL* plugin generates a set of geometries that expresses the currently loaded graph. Vertices are represented by tesselated icosahedrons and edges are represented by 12-sided cylinders. The plugin parses the graph, creates the geometry, and saves it out to the specified directory as a 3D model in STL format. Options are available for overriding the vertex and edge sizes of the graph to assist in creating a printable model. The model in STL format may be directly input into printer software or imported into modeling software for post-processing. Since this conversion feature is written in python as a plugin [2] for GLuskap, users with different needs can easily tailor the code as required without recompiling the source code.

Printing Considerations
Considerations must be taken when attempting to print graphs. Depending on the size of the desired 3D print, vertex and edge sizes may need to be scaled to ensure that the graph is sufficiently rigid to be held together during printing and handling.

Printers that use an additive process and build the model up layer by layer without having a supporting substrate may require the use of supports to ensure a rigid print. Supports can later be removed or left in place to reduce the chance that later handling will break the physical print. The ideal orientation when printing a graph depends on the nature of the graph and the type of 3D printer being used. For most graphs, the printer software is used to rotate the model of the graph so that an appropriate side of the graph is aligned with the printing platform.

Graph Drawing
Although there are many theoretical results on 3D graph drawing, the visualization effectiveness of such drawings has certainly been questioned. Previously, the display of 3D drawings was hindered by its reliance on projection onto a 2D display device, at best with active or passive shutter glasses for stereoscopic viewing. The advent of inexpensive, commercial 3D printers removes a major constraint – physical models of graphs can now be produced.

References

1. Dyck, B., Johnson, G., Hanlon, S., Smith, A., Wismath, S.: GLuskap 3.0 manual (2007), http://people.uleth.ca/~wismath/gluskap/
2. GLuskap source code and plugins, http://github.com/ulethHCI/GLuskap/

Optical Graph Recognition on a Mobile Device

Christopher Auer, Christian Bachmaier, Franz J. Brandenburg,
Andreas Gleißner, and Josef Reislhuber

University of Passau, 94030 Passau, Germany
{auerc,bachmaier,brandenb,gleissner,reislhuber}@fim.uni-passau.de

In [1] we proposed Optical Graph Recognition (OGR) as the reversal of graph drawing. A drawing transforms the topological structure of a graph into a graphical representation. Primarily, it maps vertices to points and displays them by icons, and it maps edges to Jordan curves connecting the endpoints. In reverse, OGR transforms the digital image of a drawn graph into its topological structure. The recognition process is divided into the four phases preprocessing, segmentation, topology recognition, and postprocessing. In the preprocessing phase OGR detects which pixels of an image are part of the graph (graph pixels) and which pixels are not. The segmentation phase recognizes the vertices of the graph and classifies the graph pixels as vertex and edge pixels. The topology recognition phase first recognizes edge sections. Edge crossings divide edges into edge sections, i. e., the regions between crossings and vertices. The edge sections are merged into edges in the most probable way based on direction vectors. The postprocessing phase concludes OGR with tasks like converting the recognized graph into different file formats, like GraphML or adding coordinates to the vertices and edges, such that the recognized graph resembles the input graph.

Our OGR Java implementation OGR$^\mathrm{up}$ is able to recognize drawings of undirected graphs with the following properties: Vertices are drawn as filled objects such as circles, edges are drawn as contiguous curves of a width significantly smaller than the diameter of the vertices, and they should exactly end at the vertices. Our desktop version of OGR$^\mathrm{up}$ is of limited use, because it needs a camera as a second device to take a picture of the graph.

In contrast, the new Android version of OGR$^\mathrm{up}$ needs only a single device. The picture of the graph, e. g., drawn on a whiteboard, is directly taken with the camera of the mobile device at hand. The part of the image that contains the graph can be selected via touch gestures, as seen in Fig. 1. Finally, the graph is recognized and used for further processing. It can be shared, visualized and edited on the mobile device, e. g., as proposed by Da Lozzo et al. [2].

For the Android version, we had to re-implement parts of the graph recognition algorithm and we developed a GUI that fits the capabilities of a mobile device. Whereas the computation time is acceptable in the desktop version of OGR$^\mathrm{up}$ ($\approx$ 10 seconds for high resolution images), the computation time becomes unacceptably long in the Android version due to hardware limitations. The established digital image processing library OpenCV [3] helped to improve the computation time of OGR$^\mathrm{up}$. To circumvent further performance issues, and to make OGR$^\mathrm{up}$ available for different mobile operating system, like iOS or Windows Phone, we plan a web service implementation of OGR$^\mathrm{up}$.

S. Wismath and A. Wolff (Eds.): GD 2013, LNCS 8242, pp. 516–517, 2013.
© Springer International Publishing Switzerland 2013

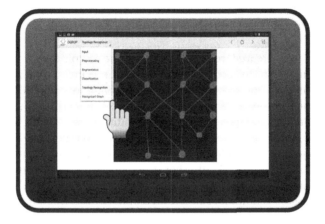

Fig. 1. Two screenshots of OGR^{up} on an Android tablet

References

1. Auer, C., Bachmaier, C., Brandenburg, F.J., Gleißner, A., Reislhuber, J.: Optical graph recognition. J. Graph Alg. App. 17(4), 541–565 (2013)
2. Da Lozzo, G., Di Battista, G., Ingrassia, F.: Drawing graphs on a smartphone. J. Graph Alg. App. 16(1), 109–126 (2012)
3. Itseez: OpenCV, http://www.opencv.org/

Browser-Based Graph Visualization
of Dynamic Data with VisGraph

Jos de Jong and Giovanni E. Pazienza

Almende B.V., Westerstraat 50, 3016 DJ, Rotterdam, The Netherlands
{jos,giovanni}@almende.org

Abstract. VisGraph is an open-source JavaScript library for the real-time graph visualization of dynamic data in a browser. Its characteristics – such as the high portability and the compatibility with multiple data formats – make it suitable for a broad range of research applications.

1   Overview of VisGraph

VisGraph has been created to fill a niche in the graph visualization market for research purposes: the need of a software for graph visualization that is comprehensive enough for a standard non-professional user and does not require any complex installation. Also, VisGraph has two key features: high portability (because it is browser independent) and possibility to work with dynamic data (for instance, coming from a sensor network or from a server). Their combination results in a key feature of VisGraph: a real-time graph representing dynamic data can be shared with a multitude of other users who work on different platforms, including smartphones and tablets. This unique characteristic of VisGraph make it particularly useful in a number of research applications ranging from real-time social network analysis to logistics.

It is crucial to emphasise that VisGraph is a lightweight browser-based library and hence it is *not* competing with desktop/server-based graph visualization softwares, such as Graphviz [1], Gephi [2], or WiGis [3], just to name a few. Other web-based tools (like D3.js [4], Springy [5], Dracula Graph Library [6], or Arbor [7]) have served as inspiration for VisGraph which, as a result, is more customizible and easier to use than its predecessors. Figure 1 shows a qualitative comparison of VisGraph with the aforementioned graph visualization softwares.

VisGraph supports multiple data formats which are all eventually translated into the native JSON format, which consists of two arrays (one for the nodes and the other for the edges) containing information regarding the topology and style of the graph; in the current version, Google DataTable and the DOT language are fully supported. As for the graph layout, VisGraph uses a force-directed algorithm where nodes try to keep a minimum distance from each other. The user can interact with the graphs by dragging the nodes and customise them by changing the colour and size of the elements etc; graphs of up to a few hundreds of nodes can be smoothly rendered by using HTML5 Canvas elements. VisGraph is part of a JavaScript library called vis.js [8], which includes other visualization

S. Wismath and A. Wolff (Eds.): GD 2013, LNCS 8242, pp. 518–519, 2013.

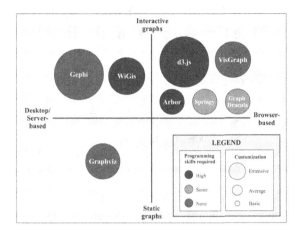

Fig. 1. Qualitative comparison of VisGraph with other graph visualization softwares

tools as well as some components to manipulate dynamic data, which VisGraph uses. The whole vis.js library (including VisGraph) requires no installation and it can be easily included in a web page by adding a single line of code.

Several additional features of VisGraph are under development, such as: i) the option to export the graphs to the formats supported by VisGraph; ii) the addition of clustering functionalities and further layout mechanisms (which will be particularly useful when graphs are visualized on portable devices); iii) the possibility for the user to edit the graph on the fly by creating and removing attributes without modifying the data set.

References

1. Ellson, J., Gansner, E.R., Koutsofios, L., North, S.C., Woodhull, G.: Graphviz - open source graph drawing tools. In: Mutzel, P., Jünger, M., Leipert, S. (eds.) GD 2001. LNCS, vol. 2265, pp. 483–484. Springer, Heidelberg (2002)
2. Bastian, M., Heymann, S., Jacomy, M.: Gephi: An open source software for exploring and manipulating networks. In: International AAAI Conference on Weblogs and Social Media, vol. 2. AAAI Press, Menlo Park (2009)
3. Gretarsson, B., Bostandjiev, S., O'Donovan, J., Höllerer, T.: WiGis: A framework for scalable web-based interactive graph visualizations. In: Eppstein, D., Gansner, E.R. (eds.) GD 2009. LNCS, vol. 5849, pp. 119–134. Springer, Heidelberg (2010)
4. Data-driven documents (2013), `http://d3js.org`
5. Springy.js: a force directed graph layout algorithm in java script (2013), `http://getspringy.com`
6. Dracula graph library (2013), `http://www.graphdracula.net`
7. Arbor.js: a graph visualisation library using web workers and query (2013), `http://arborjs.org`
8. de Jong, J.: The vis.js library (2013), `http://visjs.org/`

Exact and Fixed-Parameter Algorithms for Metro-Line Crossing Minimization Problems[*]

Yoshio Okamoto[1], Yuichi Tatsu[2], and Yushi Uno[2]

[1] The University of Electro-Communications,
1-5-1 Chofugaoka, Chofu, Tokyo 182-8585, Japan
okamotoy@uec.ac.jp
[2] Osaka Prefecture University, 1-1 Gakuen-cho, Naka-ku, Sakai 599-8531, Japan
sr301023@edu.osakafu-u.ac.jp, uno@mi.s.osakafu-u.ac.jp

The *metro-line crossing minimization problem* (MLCM), proposed by Benkert et al. [3], is to draw multiple lines on a fixed drawing of an underlying graph that models stations and rail tracks, so that the number of crossings of lines is minimum. The input of MLCM is defined by an underlying graph G and a line set $\mathcal{L}$ on G. This paper studies a variation of MLCM, called MLCM-P, with the restriction that line terminals have to be drawn at a verge of a station (*periphery condition*), which is known to be NP-hard even when its underlying graphs are paths [1]. In this paper we focus on such cases of MLCM-P, which we call MLCM-P_PATH, and present the following three results.

I. Planarity in MLCM-P_PATH. Our first result is a linear-time algorithm for deciding whether a given instance of MLCM-P_PATH has a layout without crossings. This is in a contrast with the NP-hardness of MLCM-P_PATH. Recently, Fink and Pupyrev [4] independently gave an algorithm for the same problem when a given graph is not necessarily a path but any planar graph. However, its running time is $O(|E(G)\|\mathcal{L}|^2)$.

We now describe the outline of our algorithm. First, we reduce our problem to the following PLANARITY INSIDE CIRCLE (PIC) problem in linear time. As the second step, we reduce PIC to the usual planarity testing. In the PIC problem, we are given a graph $H = (V, E)$ and a bijection $\delta: V \to \{1, 2, \ldots, |V|\}$. Then, we want to place the vertices of V on a single line in the order defined by δ and draw the edges in E and a circle passing through $\delta^{-1}(1)$ and containing all the other vertices, so that all the edges in E

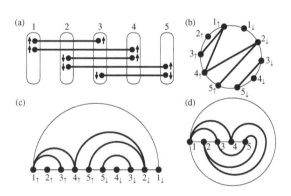

Fig. 1. (a) A layout of MLCM-P without crossings, (b) a circular drawing, (c) a transformation into a half-circle drawing, (d) a drawing inside a circle

are drawn within the circle without crossings. If such a drawing exists, the output is yes; otherwise, the output is no. The reduction at the first step is illustrated in Fig. 1. At the

---

[*] For more details, refer to the full version [6].

S. Wismath and A. Wolff (Eds.): GD 2013, LNCS 8242, pp. 520–521, 2013.
© Springer International Publishing Switzerland 2013

second step, to force the drawing to reside in a circle, we use K_4, which is a minimal non-outerplanar graph.

II. An Exact Exponential Algorithm for MLCM-P_PATH. The second result is an $O^*(2^{|\mathcal{L}|})$-time exact algorithm for MLCM-P_PATH.[1] A naive approach is to compute the number of crossings for all possible layouts of lines in $\mathcal{L}$ by using an algorithm of Bekos et al. [2], and then to output an optimal one among them. Since the number of different assignments of two ends (to top or bottom) of a line is four, this approach yields an algorithm running in $O^*(4^{|\mathcal{L}|})$ time. To improve the running time, we look at all possible assignments of left ends of lines and greedily determine the assignments of right ends for each assignment of left ends so that the number of crossings is minimum. A lemma guarantees the correctness.

III. Fixed-Parameter Tractability for MLCM-P_PATH. The third result is an $O^*(2^k)$-time exact algorithm for MLCM-P_PATH, where k is the multiplicity of lines, that is, the maximum number of lines on an edge of an underlying graph. Our result partially answers a question by Nöllenburg [5]. A recent paper by Fink and Pupyrev [4] independently gave an $O^*((k!)^2)$-time algorithm for the same problem with the same parameter. Thus, our algorithm is superior in its running time.

A naive approach performs dynamic programming along the path, and at each vertex of the path we look at all possible pairs of permutations of lines. Since the multiplicity is k, this leads to an $O^*((k!)^2)$-time algorithm. To improve the running time, we exploit several non-trivial properties of optimal solutions. For example, the algorithm by Bekos et al. [2] implies that among the optimal layouts there exists one that satisfies the following condition: If two lines $l = [i, j]$ and $l' = [i', j']$ ($j < j'$) cross, then the crossing occurs on edge $(j - 1, j)$ of the path. Such properties reduce the number of permutations we should look at, and yield an $O^*(2^k)$-time algorithm.

References

1. Argyriou, E., Bekos, M.A., Kaufmann, M., Symvonis, A.: Two polynomial time algorithms for the metro-line crossing minimization problem. In: Tollis, I.G., Patrignani, M. (eds.) GD 2008. LNCS, vol. 5417, pp. 336–347. Springer, Heidelberg (2009)
2. Bekos, M.A., Kaufmann, M., Potika, K., Symvonis, A.: Line crossing minimization on metro maps. In: Hong, S.-H., Nishizeki, T., Quan, W. (eds.) GD 2007. LNCS, vol. 4875, pp. 231–242. Springer, Heidelberg (2008)
3. Benkert, M., Nöllenburg, M., Uno, T., Wolff, A.: Minimizing intra-edge crossings in wiring diagrams and public transportation maps. In: Kaufmann, M., Wagner, D. (eds.) GD 2006. LNCS, vol. 4372, pp. 270–281. Springer, Heidelberg (2007)
4. Fink, M., Pupyrev, S.: Metro-line crossing minimization: hardness, approximations, and tractable cases. In: Wismath, S., Wolff, A. (eds.) GD 2013. LNCS, vol. 8242, pp. 328–339. Springer, Heidelberg (2013)
5. Nöllenburg, M.: An improved algorithm for the metro-line crossing minimization problem. In: Eppstein, D., Gansner, E.R. (eds.) GD 2009. LNCS, vol. 5849, pp. 381–392. Springer, Heidelberg (2010)
6. Okamoto, Y., Tatsu, Y., Uno, Y.: Exact and fixed-parameter algorithms for metro-line crossing minimization problems. arXiv:1306.3538 (cs.DS) (2013)

---

[1] The O^*-notation ignores a polynomial factor, commonly used in the exponential-time algorithm literature.

Convex-Arc Drawings of Pseudolines*

David Eppstein[1], Mereke van Garderen[2],
Bettina Speckmann[2], and Torsten Ueckerdt[3]

[1] Computer Science Dept., University of California, Irvine, USA
eppstein@uci.edu
[2] Dept. of Mathematics and Computer Science, TU Eindhoven, the Netherlands
m.v.garderen@student.tue.nl, speckman@win.tue.nl
[3] Dept. of Mathematics, KIT, Germany
torsten.ueckerdt@kit.edu

Introduction. A *pseudoline* is formed from a line by stretching the plane without tearing: it is the image of a line under a homeomorphism of the plane [13]. In *arrangements* of pseudolines, pairs of pseudolines intersect at most once and cross at their intersections. Pseudoline arrangements can be used to model sorting networks [1], tilings of convex polygons by rhombi [4], and graphs that have distance-preserving embeddings into hypercubes [6]. They are also closely related to oriented matroids [11]. We consider here the visualization of arrangements using well-shaped curves.

Primarily, we study *weak outerplanar* pseudoline arrangements. An arrangement is *weak* if it does not necessarily have a crossing for every pair of pseudolines [12], and *outerplanar* if every crossing is part of an unbounded face of the arrangement. We show that these arrangements can be drawn with all curves convex, either as polygonal chains with at most two bends per pseudoline or as semicircles above a line. Arbitrary pseudolines can also be drawn as convex curves, but may require linearly many bends.

Related Work. Several results related to the visualization of pseudoline arrangements are known. In *wiring diagrams*, pseudolines are drawn on parallel horizontal lines, with crossings on short line segments that connect pairs of horizontal lines [10]. The graphs of arrangements have drawings in small grids [8] and the dual graphs of weak arrangements have drawings in which each bounded face is centrally symmetric [5]. The pseudoline arrangements in which each pseudoline is a translated quadrant can be used to visualize *learning spaces* representing the states of a human learner [7]. Researchers in graph drawing have also studied force-directed methods for schematizing systems of curves representing metro maps by replacing each curve by a spline; these curves are not necessarily pseudolines, but they typically have few crossings [9].

Results. Below we state our results for outerplanar and arbitrary arrangements.

Theorem 1. *Every weak outerplanar pseudoline arrangement may be represented by a set of chords of a circle.*

Corollary 1. *Every weak outerplanar pseudoline arrangement may be represented by lines in the hyperbolic plane, or by semicircles with endpoints on a common line.*

---

* Parts of this work originated at Dagstuhl seminar 13151, *Drawing Graphs and Maps with Curves*. D.E. was supported in part by the National Science Foundation under grants 0830403 and 1217322, and by the Office of Naval Research under MURI grant N00014-08-1-1015.

S. Wismath and A. Wolff (Eds.): GD 2013, LNCS 8242, pp. 522–523, 2013.
© Springer International Publishing Switzerland 2013

This result complements the fact that a weak arrangement with no 3-clique can always be represented by hyperbolic lines, regardless of outerplanarity [2].

Corollary 2. *Every weak outerplanar pseudoline arrangement may be represented by convex polygonal chains with only two bends.*

Theorem 2. *Every n-element pseudoline arrangement can be drawn with convex polylines, each of complexity at most n.*

For smooth curves composed of multiple circular arcs and straight line segments, Bekos et al. [3] defined the *curve complexity* to be the maximum number of arcs and segments in a single curve. By replacing each bend of the above result by a small circular arc, one obtains a smooth convex representation of the arrangement with curve complexity $O(n)$. We can show that these bounds are optimal.

Theorem 3. *There exist simple arrangements of n pseudolines that, when represented by polygonal chains require some pseudolines to have $\Omega(n)$ bends.*

Theorem 4. *There exist simple arrangements of n pseudolines that, when represented by smooth piecewise-circular curves require some curves to have $\Omega(n)$ arcs.*

References

1. Angel, O., Holroyd, A.E., Romik, D., Virág, B.: Random sorting networks. Advances in Mathematics 215(2), 839–868 (2007)
2. Bandelt, H.-J., Chepoi, V., Eppstein, D.: Combinatorics and geometry of finite and infinite squaregraphs. SIAM Journal of Discrete Mathematics 24(4), 1399–1440 (2010)
3. Bekos, M.A., Kaufmann, M., Kobourov, S.G., Symvonis, A.: Smooth orthogonal layouts. In: Didimo, W., Patrignani, M. (eds.) GD 2012. LNCS, vol. 7704, pp. 150–161. Springer, Heidelberg (2013)
4. da Silva, I.P.F.: On fillings of 2N-gons with rhombi. Disc. Math. 111(1-3), 137–144 (1993)
5. Eppstein, D.: Algorithms for drawing media. In: Pach, J. (ed.) GD 2004. LNCS, vol. 3383, pp. 173–183. Springer, Heidelberg (2005)
6. Eppstein, D.: Cubic partial cubes from simplicial arrangements. Electronic Journal of Combinatorics 13(1), 79 (2006)
7. Eppstein, D.: Upright-quad drawing of st-planar learning spaces. Journal of Graph Algorithms and Applications 12(1), 51–72 (2008)
8. Eppstein, D.: Drawing arrangement graphs in small grids, or how to play Planarity. In: Wismath, S., Wolff, A. (eds.) GD 2013. LNCS, vol. 8242, pp. 436–447. Springer, Heidelberg (2013)
9. Fink, M., Haverkort, H., Nöllenburg, M., Roberts, M., Schuhmann, J., Wolff, A.: Drawing metro maps using bézier curves. In: Didimo, W., Patrignani, M. (eds.) GD 2012. LNCS, vol. 7704, pp. 463–474. Springer, Heidelberg (2013)
10. Goodman, J.E.: Proof of a conjecture of Burr, Grünbaum, and Sloane. Discrete Mathematics 32(1), 27–35 (1980)
11. Goodman, J.E.: Pseudoline arrangements. In: Handbook of Discrete and Computational Geometry, page 97 (2010)
12. Goodman, J.E., Pollack, R.: Allowable sequences and order types in discrete and computational geometry. In: Pach, J. (ed.) New Trends in Discrete and Computational Geometry. Algorithms and Combinatorics, vol. 10, pp. 103–134. Springer, Heidelberg (1993)
13. Shor, P.W.: Stretchability of pseudolines is NP-hard. In: Gritzmann, P., Sturmfels, B. (eds.) Applied Geometry and Discrete Mathematics: The Victor Klee Festschrift. DIMACS Series in Discrete Mathematics and Theoretical Computer Science, vol. 4, pp. 531–554. Amer. Math. Soc., Providence (1991)

The Density of Classes of 1-Planar Graphs*

Christopher Auer, Christian Bachmaier, Franz J. Brandenburg,
Andreas Gleißner, Kathrin Hanauer, and Daniel Neuwirth

University of Passau, 94030 Passau, Germany
{auerc,bachmaier,brandenb,gleissner,hanauer,neuwirth}@fim.uni-passau.de

The *density* of a graph $G = (V, E)$ is the number of edges $|E|$ as a function of the number of vertices $n = |V|$. It is an important graph parameter, and is often used to exclude a graph from a particular class. We survey the density of relevant subclasses of 1-planar graphs and establish some new and improved bounds. A graph is *1-planar* if it can be drawn in the plane such that each edge is crossed at most once.

We consider simple and connected graphs. A graph $G \in \mathcal{G}$ is *maximal* for a particular class of graphs $\mathcal{G}$ if the addition of any edge e implies $G + e \notin \mathcal{G}$. Let $M(\mathcal{G}, n)$ and $m(\mathcal{G}, n)$ denote the *maximum* and *minimum numbers* of edges of a maximal n-vertex graph in $\mathcal{G}$. Graphs $G \in \mathcal{G}$ with density $M(\mathcal{G}, |G|)$ $(m(\mathcal{G}, |G|))$ are the *densest* (*sparsest maximal*) graphs of $\mathcal{G}$. Thus $M(\mathcal{G}, n)$ is an upper and $m(\mathcal{G}, n)$ a lower bound. It is well-known that $M(\mathcal{G}, n)$ and $m(\mathcal{G}, n)$ coincide for planar, bipartite planar, and outerplanar graphs with $3n - 6$, $2n - 4$, and $2n - 3$, respectively. For 1-planar graphs the upper and lower bounds diverge. $M(\mathcal{G}, n) = 4n - 8$ was proved first of all by Bodendiek et al. [3] and was rediscovered several times. Surprisingly, there are much sparser maximal 1-planar graphs that are even sparser than maximum planar graphs. In [4] it was proved that $\frac{28}{13}n - \mathcal{O}(1) \leq m(\mathcal{G}, n) \leq \frac{45}{17}n - \mathcal{O}(1)$.

We consider the density of maximal graphs of subclasses of 1-planar graphs, with emphasis on sparse graphs. Our focus is on 3-connected [1], bipartite [7,8], and outer 1-planar [2] graphs. An *outer 1-planar* graph is drawn with all vertices in the outer face. Moreover, we restrict the drawings by fixed *rotation systems*, which specify the cyclic ordering of the edges at each vertex, and then may allow crossings of incident edges, which are generally excluded for 1-planarity.

Theorem 1. *For the classes of graphs $\mathcal{G}$ from Table 1, the stated upper bound on $M(\mathcal{G}, n)$ on the density is tight. The minimum density $m(\mathcal{G}, n)$ ranges between the functions in column "lower example" and "lower bound m".*

3-connected. For 3-connected 1-planar graphs $\mathcal{G}$ the upper bound is obvious. The lower bound $m(\mathcal{G}, n) = \frac{10}{3}n + \frac{20}{3}$ is tight. It improves the example of $3.625n + \mathcal{O}(1)$ from [6] and disproves their conjecture of $3.6n + \mathcal{O}(1)$.

A graph $G \in \mathcal{G}$ consists of non-planar K_4s and a planar remainder, which is triangulated such that two adjacent triangles imply a K_4. The removal of all pairs of crossing edges from G leaves a planar graph with t triangles and q quadrangles and the relation $t \leq q$, which together with Euler's formula yields the

---

* Research partially supported by the German Science Foundation, DFG, Grant Br-835/18-1

S. Wismath and A. Wolff (Eds.): GD 2013, LNCS 8242, pp. 524–525, 2013.

Table 1. Upper and lower bounds on the number of edges in maximal graphs

	upper bound M	lower example	lower bound m
2-connected	$4n - 8$ [3]	$\frac{45}{17}n - \frac{84}{17}$ [4]	$\frac{28}{13}n - \frac{10}{3}$ [4]
3-connected	$4n - 8$ [3]	$\frac{10}{3}n - \frac{20}{3}$	$\frac{10}{3}n - \frac{20}{3}$
straight-line	$4n - 9$ [5]	$\frac{8}{3}n - \frac{11}{3}$	$\frac{28}{13}n - \frac{10}{3}$ [4]
fixed rotation, 2-connected	$4n - 8$ [3]	$\frac{7}{3}n - 3$ [4]	$\frac{21}{10}n - \frac{10}{3}$ [4]
fixed rotation, intersect incident	$4n - 8$ [3]	$\frac{3}{2}n + 1$	$\frac{5}{4}n$
bipartite	$3n - 8$ [7]	$n - 1$	$n - 1$
bipartite, 2-connected	$3n - 8$ [7]	$2n - 4$	n
outer 1-planar	$\frac{5}{2}n - 4$	$\frac{11}{5}n - \frac{18}{5}$	$\frac{11}{5}n - \frac{18}{5}$

bound for $m(\mathcal{G}, n)$. The bound is achieved by a recursive construction of planar K_4s surrounded by non-planar K_4s surrounded by planar K_4s.

Outer 1-planar. A maximal outer 1-planar graph G is composed of planar K_3s and non-planar K_4s [2], such that two K_3s are not adjacent. Removing the pairs of crossing edges from the K_4s results in an outerplanar graph whose dual is a tree with vertices of degree 3 and 4. Each vertex of degree 3 adds one vertex and two edges, and each vertex of degree 4 adds two vertices and five edges to the density of G. Maximizing the degree-4 vertices yields $M(\mathcal{G}, n) = \frac{5}{2}n - 4$ and minimizing yields $m(\mathcal{G}, n) = \frac{11}{5}n - \frac{18}{5}$. Both bounds are tight.

References

1. Alam, M.J., Brandenburg, F.J., Kobourov, S.G.: Straight-line drawings of 3-connected 1-planar graphs. In: Wismath, S., Wolff, A. (eds.) GD 2013. LNCS, vol. 8242, pp. 83–94. Springer, Heidelberg (2013)
2. Auer, C., Bachmaier, C., Brandenburg, F.J., Gleißner, A., Hanauer, K., Neuwirth, D., Reislhuber, J.: Recognizing outer 1-planar graphs in linear time. In: Wismath, S., Wolff, A. (eds.) GD 2013. LNCS, vol. 8242, pp. 107–118. Springer, Heidelberg (2013)
3. Bodendiek, R., Schumacher, H., Wagner, K.: Über 1-optimale Graphen. Mathematische Nachrichten 117, 323–339 (1984)
4. Brandenburg, F.J., Eppstein, D., Gleißner, A., Goodrich, M.T., Hanauer, K., Reislhuber, J.: On the density of maximal 1-planar graphs. In: Didimo, W., Patrignani, M. (eds.) GD 2012. LNCS, vol. 7704, pp. 327–338. Springer, Heidelberg (2013)
5. Didimo, W.: Density of straight-line 1-planar graph drawings. Inform. Process. Lett. 113(7), 236–240 (2013)
6. Hudák, D., Madaras, T., Suzuki, Y.: On properties of maximal 1-planar graphs. Discuss. Math. Graph Theory 32(4), 737–747 (2012)
7. Karpov, D.V.: Upper bound on the number of edges of an almost planar bipartite graph. Tech. Rep. arXiv:1307.1013v1 (math.CO), Computing Research Repository (CoRR) (July 2013)
8. Xu, C.: Algorithmen für die Dichte von Graphen. Master's thesis, University of Passau (2013)

BGPlay3D: Exploiting the Ribbon Representation to Show the Evolution of Interdomain Routing

Patrizio Angelini, Lorenzo Antonetti Clarucci, Massimo Candela,
Maurizio Patrignani, Massimo Rimondini, and Roberto Sepe

Roma Tre University, Italy

Representing flows in networks is a challenging task: different flows usually traverse the same edges; flows may split and join again along their routes; and some flows may go so far as to traverse the same nodes and edges several times. Traditional 2D visualizations exploit drawing techniques like parallel multicolor curves to address these challenges and tell apart the different flows, but they tend to exhibit readability problems, for example when lots of flows traverse a single edge.

Fig. 1. A ribbon representation of a flow in a small network

We investigate a novel 2.5D visualization metaphor, conceived to clearly distinguish flows in networks. Instead of drawing flows as parallel curves on the plane, we use the third dimension to stack the flows one above the other. Namely, in a *ribbon representation* each node of the network is represented by a pin, and each flow is represented by a ribbon (a long stripe) winding around the pins corresponding to the nodes traversed by the flow [1]. Pins are distributed on a flat surface, yielding a 2.5D representation rather than a pure 3D one.

Fig. 2. A ribbon representation of a cycling flow

The potentialities of the ribbon representation with respect to a traditional 2D visualization are manifold: the order in which ribbons are stacked, as well as their thickness, may correspond to extra information associated with the ribbons themselves. Even self-intersecting flows are clearly representable (see Fig. 2). Furthermore, colors or shades of grey are not strictly needed anymore to tell the flows apart and can be therefore used with different semantics. Finally, when the thickness of the ribbon allows it, labels may be accommodated in the ribbons themselves.

On the other hand, we expect the ribbon representation to have some intrinsic limitations: navigation in a 3D scene may be less intuitive for the user than panning and zooming a 2D representation. Also, when the network is big, occlusions among pins and ribbons tend to hamper a clear perception of the 3D environment.

In order to explore both the potentialities and the limitations of using the ribbon representation to depict flows in a network, compared with a traditional 2D drawing and interface, we realized a tool, called BGPlay3D [7], which extends with a 3D interface the functionalities of the well-known BGPlay tool [6]. We took strong advantage of the modularity of the BGPlay.js JavaScript open-source framework, which offers the basis for several network visualization tools [4,3,5], including the latest release of BGPlay: starting from the latter, we replaced the existing GraphView module, responsi-

S. Wismath and A. Wolff (Eds.): GD 2013, LNCS 8242, pp. 526–527, 2013.

ble of managing events in the canvas area of the interface, with a GraphView3D module that exploits the three.js lightweight cross-browser JavaScript API [8] to create and display animated 3D computer graphics on a Web browser (three.js, in turn, relies on the WebGL JavaScript API for rendering interactive 3D graphics [9]).

The interface of BGPlay3D, like that of BGPlay, is vertically split into three panels: an information panel on the top, the main visualization window in the middle, and a control timeline at the bottom (see Fig. 3). In the visualization window the activity graph evolves based on the current event. Namely, new flows appear, old flows disappear, and some flow changes its route. Like in the 2D interface, appearances are handled by flashing the new ribbons. Flows that disappear are represented by flashing the corresponding ribbons and then letting them drop on the floor. Changes in the route of a flow have a special representation in 3D: rather than morphing the old route into the new one as it happened in the 2D interface, we let the old ribbon disappear towards the destination just as if it was a cut film, while the new ribbon originates from the source and moves towards the destination along the new route. A prototype of BGPlay3D, offering a promising alternative to BGPlay, is available online [2].

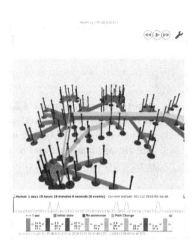

Fig. 3. The interface of BGPlay3D

Several further directions of research remain open. For example, BGPlay3D uses the same algorithm as BGPlay for laying out the pins on the plane and no specialized algorithm has been developed to take into account the 3D nature of the interface. Also, the position of the camera is chosen by the user, and no attempt has been made to offer a point of view which is more convenient for understanding the current events.

References

1. Antonetti Clarucci, L.: La ribbon representation: un nuovo modello di visualizzazione di flussi e sue applicazioni. Master's thesis, Roma Tre University, Rome, Italy (2013) (in Italian)
2. BGPlay3D, http://bgplayjs.com/?section=bgplay3d (accessed 2013)
3. BGPlay.js, http://bgplayjs.com/ (accessed 2013)
4. Candela, M.: Adaptive and responsive web-oriented visualization of evolving data: the interdomain routing case. Master's thesis, Roma Tre University, Rome, Italy (2012)
5. Candela, M., Di Bartolomeo, M., Di Battista, G., Squarcella, C.: Dynamic traceroute visualization at multiple abstraction levels. In: Wismath, S., Wolff, A. (eds.) GD 2013. LNCS, vol. 8242, pp. 496–507. Springer, Heidelberg (2013)
6. Colitti, L., Di Battista, G., Mariani, F., Patrignani, M., Pizzonia, M.: Visualizing interdomain routing with BGPlay. Journal of Graph Algorithms and Applications 9(1), 117–148 (2005)
7. Sepe, R.: Visualizzazione tridimensionale del routing interdominio. Master's thesis, Roma Tre University, Rome, Italy (2013) (in Italian)
8. three.js, http://threejs.org/ (accessed 2013)
9. WebGL (Web Graphics Library), http://www.khronos.org/webgl/ (accessed 2013)

Ravenbrook Chart: A New Library for Graph Layout and Visualisation

Nick Levine and Nick Barnes

Ravenbrook Limited, PO Box 205, Cambridge, CB2 1AN, United Kingdom
chart@ravenbrook.com

Abstract. *Ravenbrook Chart* is a newly-available library which implements layout and interactive display of large, complex graphs. It provides spring-embedded, layered and circular layouts. It has been deployed in mature applications and a free web service uses it to visualise graphs. It is now available free of charge as a permissively licensed library.

Between 2001 and 2013, the authors developed two desktop applications whose requirements included interactive graph layout and display. We assessed several commercial graphing products, and indeed used two such libraries in production versions at different times. This approach was abandoned for commercial and technical reasons: no third-party tool met all our requirements for reliability, performance, and flexible interactivity. In 2009 work began on a new layout and display library, *Ravenbrook Chart*. Its implementation was led by customer requirements and it is now used for visualisation in both applications. More recent uses of this library include *Chart Server* and *Chart Desktop* both of which, as described here, are now available for use free of charge.

Chart implements three layout types: spring-embedded (see Fig. 1), hierarchical, and circular. The time complexities of the algorithms are $O(v \log v + e \log e)$ for the circular layout, $O((v + e)^{1.5})$ for hierarchical, and $O(v^2 \log v)$ for spring-embedded, where v and e are the numbers of nodes and edges respectively.

Chart's visualisation system supports multiple graphs each of which can be shown on any number of displays. Displays can be "aligned" with each other, so that they automatically pan and zoom in tandem. The library supports standard controls over node and edge appearance, UTF-8 strings throughout, a wide range of callbacks, dynamic context-sensitive menus, incremental changes to the graph, retrieval and setting of node and edge-bend locations (per display), and object hiding. It is designed for high data thoughput, full thread-safety, and sophisticated exception handling with logging, restarts, and detailed backtraces.

Chart was originally created for two custom desktop tools. The *Critical Network Analysis Tool* is a U.S. Government off-the-shelf application for processing, visualising and analysing large social networks. It includes a large suite of analytical tools based on social network theory, including many numerical and visualisation tools independent of graph drawing. *WorldView* is used to construct, analyse, and compare "cognitive maps", identifying and contrasting core beliefs of individuals and groups based on texts such as interviews and speeches.

S. Wismath and A. Wolff (Eds.): GD 2013, LNCS 8242, pp. 528–529, 2013.

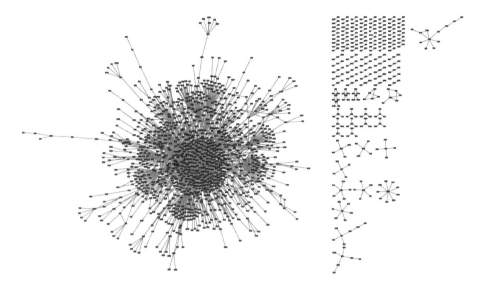

Fig. 1. Spring-embedded layout of sample social network: 1513 nodes, 3359 edges. The component on the left has 1149 nodes and 2959 edges. The two most-connected nodes have degrees d of 342 and 131; 37 nodes have $d > 10$.

Chart server (http://chart.ravenbrook.com) is a free web service for converting graph specifications into images in either PNG or PDF format. Its straightforward API supports GET and POST requests, all three layouts, and a rich control over node and edge appearance.

Chart Desktop is a freely-available version of the Chart library, distributed as a Windows DLL. Much of the Chart functionality is not exposed through the API in the initial Chart Desktop release; the focus has been on careful, comprehensive documentation and testing. The library is accompanied by a C header file, a high-level Python interface, and examples. It is available free of charge under the very permissive "BSD 2-Clause" license which in particular allows deployment in closed-source commercial applications, and it can be downloaded from: http://chart.ravenbrook.com/downloads.

The development of Chart has always been driven by end-user requirements, using a highly responsive evolutionary delivery process. We now seek new applications, users and directions for the library. Potential improvements include planar and 3D spring-embedded layouts, improved graphic quality, and further work on performance. The Chart Desktop API will gradually be enriched to support more of Chart's functionality, possibly supported by crowd-funding. Priorities for all this work will be set by users.

We are very pleased to offer up our work for public use. *Chart Desktop* differs from other available components in being a library rather than an application, in being flexibly licensed, and in being developed by a small company which can be highly responsive to users' needs.

Small Grid Embeddings of Prismatoids and the Platonic Solids*

Alexander Igamberdiev, Finn Nielsen, and André Schulz

Institut für Mathematische Logik und Grundlagenforschung,
Universität Münster, Germany
{alex.igamberdiev,andre.schulz}@uni-muenster.de

1 Embedding Prismatoids on the Grid

The question if every polyhedral graph can be embedded as a convex polyhedron on a polynomially sized 3d grid is one of the main open problems in lower dimensional polytope theory. Currently, the best known algorithm requires a grid of size $O(2^{7.21n})$, for n being the number of the vertices [2]. We show that *prismatoids* (polytopes, whose graphs are coming from triangulated polygonal annuli) can be embedded as convex polyhedra on a grid of size $O(n^4) \times O(n^3) \times 1$.

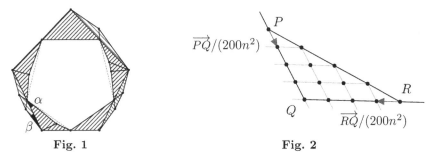

| Fig. 1 | Fig. 2 |

We call the triangles sharing an edge with the "bottom face" *red* and all other triangles *blue*. As a first step we merge consecutive red (blue) triangles to obtain a "reduced" graph of the prismatoid with $2k$ vertices, for some $k \leq n/2$. We then embed the reduced graph as a polyhedron by drawing the bottom face as a convex k-gon in the xy-plane. The top face is realized as a k-gon inscribed in the bottom face, lifted to the $z = 1$ plane. This needs a grid of size $O(n^2) \times O(n) \times 1$ [1].

We continue by reverting the initial contractions geometrically. To do so, we have to substitute edges by polygonal (convex) chains (Fig. 1). To carry out this construction we scale the whole embedding by a factor of $200n^2$. This implies that for every triangle PQR the vectors $\overrightarrow{PQ}/(200n^2)$ and $\overrightarrow{RQ}/(200n^2)$ are integral (Fig. 2). Thus, they define a basis for a grid contained in $\mathbf{Z}^2$ that is just large enough to add the missing chains back.

Note that we have to guarantee (for convexity) as additional constraint that for every edge separating a red and a blue triangle, the incident angle in the red

---

* This work was funded by the German Research Foundation (DFG) under grant SCHU 2458/2-1.

triangle is smaller than the incident angle in the blue triangle ($\beta < \alpha$, Fig. 2). This can be achieved by carefully choosing the slope of the first segment that defines the angle inside the blue triangle.

2 Grid Embeddings of the Platonic Solids

We want to realize the Platonic solids as (combinatorially equivalent) convex polyhedra on a small grid. The tetrahedron, the cube and the octahedron have all a natural realization on the $1 \times 1 \times 1$, resp., on the $1 \times 1 \times 2$ grid. A first nontrivial case is the icosahedron. Its graph can be embedded as a polyhedron using the coordinates $\{(0, \pm2, \pm1), ((\pm2, \pm1, 0), (\pm1, 0, \pm2)\}$. It is possible to reduce the grid size a bit further on one axis. Embedding the dodecahedron is more challenging. The best known realization so far requires a $8 \times 6 \times 4$ grid and is due to Francisco Santos [2]. We did a computer search to find smaller realizations. We had to make some assumptions to search efficiently. First, we fixed the grid size. Then we assumed that one of the faces lies in the xy-plane.

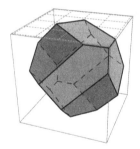

The coordinates of the vertices are:

$$\pm\{(2,2,2),(2,2,1),(2,1,2),(1,2,2),$$
$$(2,-1,0),(2,0,-1),(-1,2,0),$$
$$(0,2,-1),(0,-1,2),(-1,0,2)\}$$

Fig. 3. The smallest embedding of the dodecahedral graph as a convex polyhedron

We tried next to locate the remaining vertices face by face. The combinatorial structure of the dodecahedral graph helped us to limit the search space. In particular, most vertices are restricted to lie on a plane determined by previous choices. We were able to compute a realization on the $4 \times 4 \times 4$ grid as shown in Fig. 3. This result was found within seconds. We scanned through 4.367 choices for the first face. Our experiments have proven that this is the smallest realization in terms of minimizing the largest axis of the grid. Francisco Santos found another solution that reduces one of the axis at the expanse of the others, yielding a $3 \times 11 \times 11$ grid embedding.

References

1. Andrews, G.E.: A lower bound for the volume of strictly convex bodies with many boundary lattice points. Trans. Amer. Math. Soc. 99, 272–277 (1961)
2. Mor, A.R., Rote, G., Schulz, A.: Small grid embeddings of 3-polytopes. Discrete & Computational Geometry 45(1), 65–87 (2011)

The Graph Landscape – a Visualization of Graph Properties

Peter Eades and Karsten Klein *

School of Information Technologies, The University of Sydney, Australia
{peter.eades,karsten.klein}@sydney.edu.au

Motivation

In Graph Drawing and related fields, a large number of graph sets has been collected over years, comprising both real world graphs from application areas as well as generated graphs with specific properties. Each experimental paper makes use of some of those sets or adds a new set of graphs to the existing ones. Recently, a project was started to collect those graphs in a comprehensive database [1] to make them publicly available over a common and easy to use interface for further benchmark experiments and for analysis.

For evaluations of experimental data it is of interest to know the characteristics of the benchmark set, e.g. described in terms of a set of graph properties. Such characterics show that the set covers a specific range of properties, how the graphs are distributed over this range, and how the set differs from or overlaps with other sets. While the graph archive provides graphs plus several graph properties, it does not allow to analyze the distribution of properties within benchmark sets, or to compare those properties among different sets. We propose a web-based service that provides a visualization of graph properties, the *Graph Landscape*, and thus allows a visual analysis of benchmark sets.

Use-cases involve analyzing the properties of a local graph set in the context of one or more well-established benchmark sets. This can in particular be helpful for new sets to describe the difference and overlap in the graph characteristics compared to existing sets. As there is a variety of graph properties, but graph sets are usually generated by specifying a range for only a few of those properties, a feedback on the characteristics for the generated set and the distribution of property values will often be helpful. It might also help to improve interpretation of experimental results and to detect the shortcomings of graph sets.

Implementation

The project implementation consists of two main parts, the computational analysis of graphs and the visualization of the results for interactive analysis. The Graph Landscape visualization is implemented as a web-based service. The web-interface allows the user to select visualization views, and graphs as well as graph properties to populate the views for analysis, see Fig.1.

---

* K. Klein was partly supported by ARC grant H2814 A4421, Tom Sawyer Software, and NewtonGreen Software.

S. Wismath and A. Wolff (Eds.): GD 2013, LNCS 8242, pp. 532–533, 2013.

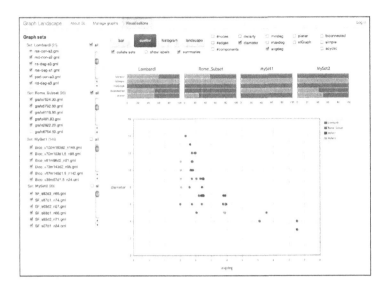

Fig. 1. User interface of the system during graph set comparison. Basic properties of the graph sets are presented in a side-by-side comparison, while user-selected properties are shown in the current main view.

Properties for graph sets available from the graph archive are precomputed, and can be extended by further analysis of other graph sources, and by uploading graphs to our server. Computation of the graph properties on the server is done by an extensible analysis engine, currently implemented as a stand-alone software using the OGDF library [2]. The handling of properties in the system is kept flexible to allow easy addition of new properties, only a few basic properties as e.g. size or connectivity are always mapped to small standard views to allow a quick overview on the graph set characteristics.

We aim at transparently connecting to the graph archive to provide up-to-date information and to allow users to retrieve or store graphs using our interface.

Remarks. We see our project as an extension to the graph archive project, with the goal to provide an easily accessible visualization that can be used as a first step in the design or analysis of benchmark sets. Our web service is still under development, but we think it should be presented to the community at this stage to allow valuable feedback and suggestions. We will have a prototype available at the Graph Drawing conference to demonstrate the system and have users try it. The feedback will then be used to guide further development decisions.

References

1. The Open Graph Archive,
   https://kandinsky.informatik.uni-tuebingen.de/forschung/graphdb/
2. The Open Graph Drawing Framework, http://www.ogdf.net

Application of Graph Layout Algorithms for the Visualization of Biological Networks in 3D

Tim Angus[1], Tom Freeman[1], and Karsten Klein[2]

[1] The Roslin Institute, University of Edinburgh, Midlothian Scotland, UK
{tim.angus,tfreeman}@roslin.ed.ac.uk
[2] School of Information Technologies, The University of Sydney, Australia
karsten.klein@sydney.edu.au

The visualization and analysis of biological systems and data as networks has become a hallmark of modern biology. Relationships between biological entities; individuals, proteins, genes, RNAs etc., can all be better understood at one level or another when modelled as networks. As the size of these data has grown, so has the need for better tools and algorithms to deal with the complex issue of network visualization and analysis. We describe application and evaluation of a state-of-the-art graph layout method for use within biological workflows.

BioLayout *Express*^{3D} is a powerful tool specifically designed for visualization, clustering, exploration, and analysis of very large networks in 2D and 3D space derived primarily from biological data [1]. In particular, its development has been driven by the need to analyse gene expression data, which typically consists of 10's of thousands of rows of quantitative gene expression measurements. First, the tool calculates a correlation matrix and then builds relationship networks, where nodes represent genes and edges expression similarities above a given r threshold. The resulting graphs can be very large e.g. 20-30,000 nodes, 5 million edges and possess a high degree of local structure with modules of co-expressed genes forming distinct cliques of high connectivity within the networks. BioLayout has for a long time used a modified CPU/GPU parallelised version of the Fruchterman-Reingold (FR) algorithm for graph layout, and visualization of the graphs in 3D offers distinct advantages when viewing such complex graph structures. MCL clustering is used to divide the graph into coexpression clusters for further analysis. Whilst the existing FR implementation is capable and in many ways adequate at laying out these types of graph, the results for other graphs derived from biological data are less satisfactory, in particular DNA assembly graphs, which are inherently different in structure. The overlapping nature of DNA fragments when joined based on read-similarity form 'chain graphs'. Layout using the FR algorithm places nodes efficiently on a local scale, but a lack of global awareness results in a knot-like graph structure (Figure 1A) inhibiting the efficient visualisation of the overall assembly.

The development of scalable approaches for high-quality layout computation is ongoing research in the graph drawing community. While successful results like multilevel methods are already established methods in graph drawing, they have as yet been little applied in the field of biology. In order to investigate their potential to improve the visual analysis of biological networks as described above, we integrated the fast multipole multilevel method (FMMM) in

S. Wismath and A. Wolff (Eds.): GD 2013, LNCS 8242, pp. 534–535, 2013.

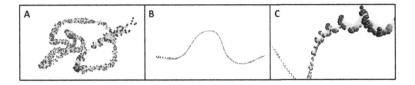

Fig. 1. Graph-based visualization of RNA-seq data for the gene BUB1. Nodes (2,886) represent individual 100 bp DNA reads and edges (132,764) the overlap between them. Graphs have appearance of a linear 'chain' reflecting the overlapping DNA sequences. (A) layout using our standard FR algorithm (B) layout using our 3D implementation of the FMMM algorithm (C) a close up of the graph shows the 'corkscrew' layout of the individual nodes.

BioLayout *Express*^{3D} such that the method can be used as part of the usual biological workflow. FMMM is available in OGDF [2], but this implementation cannot directly be used by Java tools and is also restricted to 2D drawings. We converted this implementation to Java, and extended the approach to compute 3D layouts.

We first tested this implementation of FMMM on a variety of graph types, like classic example graphs used by the graph drawing community, and were very pleased with the 3D layouts; the structure of wrapped networks being immediately obvious when visualised in 3D. After this testing, we applied it to graphs derived from biological data. In the case of graphs derived from correlation matrices the FMMM layout is quite different with major topological structures being teased apart, but the advantages of this visualisation are at this stage uncertain. The real gain comes with the DNA assembly graphs where when laid out by FR twist and turn to become knotted structures (Fig. 1A). However, when the FMMM algorithm is used the graphs are relatively linear (Fig. 1B), showing an interesting 'corkscrew' appearance at the node level (Fig. 1C). In this way the structure of the overall assembly is far more apparent and useful in cases of splice variation or other graphs structures of interest.

In summary, we are delighted in the results of this implementation, which will be available soon in the next version of BioLayout (www.biolayout.org). We would like to further improve the qualitative performance for particular graphs, and investigate and discuss the potential of other methods for use with large biological networks in 3D. We believe that our tool will also prove useful for graph drawing researchers to visualise outputs of other scalable layout methods, in particular in a 3D setting.

References

1. Theocharidis, A., van Dongen, S., Enright, A., Freeman, T.C.: Network visualisation and analysis of gene expression data using Biolayout *Express*^{3D}. Nature Protocols 4(10) (2009)
2. The Open Graph Drawing Framework, http://www.ogdf.net

Plane Cubic Graphs and the Air-Pressure Method*

Stefan Felsner and Linda Kleist

Institut für Mathematik, Technische Universität Berlin, Germany
{felsner,kleist}@math.tu-berlin.de

Abstract. Thomassen [1] proved that plane cubic graphs are area-universal, i.e., for a plane cubic graph G with prescribed face areas there exists a straight-line (re-)drawing G' that realizes these areas. Thomassen uses induction and proves the existence of a degenerate drawing where distinct vertices may be placed at the same position. We show that plane cubic graphs are area-universal using the air-pressure method. In [2,3], a similar method has been applied in the context of area-universality of rectangular layouts. With the poster, we give the idea of how the method can be adapted for other classes of plane graphs, in particular for plane cubic graphs.

The Air-Pressure Method

Let G be a plane graph and $a : F' \to \mathbb{R}_+$ a function that prescribes an area for each bounded face. Assume that the outer face of G is a k-gon and fix a convex k-gon Ω whose area is equal to the sum of the prescribed areas. We consider drawings of G such that the outer vertices of G are represented by the corners of Ω.

Let D be a drawing of G. For a face f_i let $m(i)$ be the area of f_i in D and let $a(i)$ be the prescribed area. The quotient $p(i) := \frac{a(i)}{m(i)}$ of these two values will be interpreted as the *pressure* in the face. Face f_i is pushing against an incident vertex v with a force $f(v, i)$ that depends on the pressure, this force is defined as $f(v, i) := p(i) \cdot n_s \|s\|$, where s is the segment connecting the two neighbors of v incident to f_i and n_s is the normal vector of s pointing out of f_i at v. The effective force acting on vertex v is $f(v) := \sum_{i:v \in f_i} f(v, i)$. The intuition is that shifting vertices in the direction of the effective force yields a better balance of pressure in the faces and hence a better approximation of the prescribed areas.

- A *face f_i is in balance* if $p(i) = 1$, i.e., the prescribed area is realized in the drawing.
- A *vertex v is in balance* if $f(v) = 0$, i.e., the effective force acting on v is neutral.
- A drawing is a *deadlock* if all vertices are in balance although there are faces f_i with $p(i) \neq 1$, (Fig. 1 shows an example).
- The *entropy of a drawing* is $E := \sum_i -a(i) \log p(i)$.
  Fact: For all drawings $E \geq 0$, and $E = 0$ if and only if all faces are in balance.

An *improving shift* is a shift of an unbalanced vertex v into a new position $v + d$ such that (a) the new drawing is planar and nondegenerate, and (b) the entropy increases with the shift.

---

* Partially supported by ESF EuroGIGA project GraDR.

S. Wismath and A. Wolff (Eds.): GD 2013, LNCS 8242, pp. 536–537, 2013.

Suppose that whenever there is an unbalanced vertex v there exists a d such that $v \to v + d$ is an improving shift. Then, given any initial drawing D_0, there is a sequence of improving shifts that yields a converging sequence of drawings $(D_i)_i$ such that in the limit D all vertices are in balance. The drawing D is either a deadlock or a drawing realizing the prescribed areas.

Plane Cubic Graphs

In the case of plane cubic graphs we can show that iterated shifting yields a limit drawing D realizing the prescribed areas. The result is proved with the following two claims:

1. All faces are in balance if and only if all vertices are in balance, i.e., there is no deadlock.
2. If there is an unbalanced vertex v, then there is a d such that $v \to v + d$ is an improving shift.

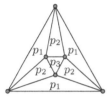

Fig. 1. Let x and 1 be the side length of the inner and outer regular triangle. If pressures satisfy $p_1 = (1 - x) p_2 + x p_3$, then this is a deadlock.

Fig. 2. This cubic graph is area-universal, it admits no deadlock.

Conclusion

We reprove area-universality for plane cubic graphs by using the air-pressure paradigm. The air-pressure method seems promising for showing area-universality for further classes of plane graphs. Time will tell how far this method can takes us.

References

1. Thomassen, C.: Plane cubic graphs with prescribed face areas. Combinatorics, Probability & Computing 1(371-381), 2–10 (1992)
2. Felsner, S.: Exploiting air-pressure to map floorplans on point sets. In: Wismath, S., Wolff, A. (eds.) GD 2013. LNCS, vol. 8242, pp. 196–207. Springer, Heidelberg (2013)
3. Izumi, T., Takahashi, A., Kajitani, Y.: Air-pressure model and fast algorithms for zero-wasted-area layout of general floorplan. IEICE Transactions on Fundamentals of Electronics, Communications and Computer Sciences 81(5), 857–865 (1998)

Author Index

Foreword

The Web plays an important role in every aspect of contemporary societies and of everyday life, i.e., in business, education, entertainment, health, and other critical activities. Web engineering, as a sub-discipline of software engineering, seeks to improve software development for this pervasive, ever-evolving platform, and strives to develop and uncover novel and cost-effective processes, models, methods, and methodologies to support rich, user-friendly, and accessible interactions between people, software, and things.

This volume collects the research articles, late-breaking results, tool demonstrations, posters, tutorials, and keynote speeches presented at the 14th International Conference on Web Engineering (ICWE 2014), held in Toulouse, France, during July 1–4, 2014.

ICWE is the flagship conference for the Web engineering community. Previous editions of ICWE took place at Aalborg, Denmark (2013), Berlin, Germany (2012), Paphos, Cyprus (2011), Vienna, Austria (2010), San Sebastian, Spain (2009), Yorktown Heights, NY, USA (2008), Como, Italy (2007), Palo Alto, CA, USA (2006), Sydney, Australia (2005), Munich, Germany (2004), Oviedo, Spain (2003), Santa Fé, Argentina (2002), and Cáceres, Spain (2001). The 2014 edition of ICWE was centered around the theme of "Engineering the Web for Users, Developers and the Crowd," hereby highlighting the importance of all the different people that, somehow, participate in the development process of interactive Web applications and, ultimately, becomes the actors and the main users of the best practices and results of the research performed in the domain of Web engineering.

ICWE 2014 featured six research tracks, namely: Cross-Media and Mobile Web Applications, HCI and the Web, Modeling and Engineering Web Applications, Quality Aspects of Web Applications, Social Web Applications and Web Applications Composition and Mashups that aimed to focus expertise and create a strong identity for the Web engineering community.

ICWE 2014 was endorsed by the International World Wide Web Conferences Steering Committee (IW3C2), the International Society for Web Engineering (ISWE), the Special Interests Groups on the Web (SIG Web) and Human-Computer Interaction (SIGCHI) of the Association for Computing Machinery (ACM), who provided the in-cooperation agreement to the conference.

ICWE 2014 attracted 100 submissions distributed over the six research tracks. Each paper was assessed by at least three members of an international panel of experts. The Program Committee accepted 20 contributions as full research papers and 13 late-breaking result papers. Additionally, ICWE 2014 welcomed 15 contributions in the form of posters and/or demonstrations, and four contributions to the PhD symposium where young research in the field of Web engineering could benefit from the advice and guidance of experts in the field.

Continuing with a healthy tradition of the ICWE conference series, three tutorials on cutting-edge topics on the field of Web engineering were presented, covering the following topics: Interaction Flow Modeling Language (IFML), Mashups and Web of Things. Moreover, three workshops were selected to be co-located at ICWE 2014.

This high-quality program would not have been possible without the help of many people that assisted the Organizing and Program Committees. We would like to thanks Marc Najork (Google) and Ricardo Baeza-Yates (Yahoo research), our keynote speakers, who accepted to give an inspiring speech at ICWE 2014, of which a written record is included in these proceedings. Many thanks to the Steering Committee liaisons Daniel Schwabe and Marco Brambilla for their advice and moral support to the organization of ICWE 2014 in Toulouse. Our sincere thanks also go out to the local organizer David Navarre, whose support was essential in hosting this conference at the University Toulouse Capitole as well as to Marlène Giamporcaro and Marie-Anne Laplaine, who oversaw all the logistic operations. We also thank Michael Krug and Martin Gaedke for the logistics required for hosting the conference website. Moreover, we address our final thanks to all the authors who submitted their scientific work to ICWE 2014, and especially to the presenters who took the time to come to Toulouse and discuss their work with their peers.

May 2014   Sven Casteleyn
                                                       Gustavo Rossi
                                                     Marco Winckler

Organization

Technical Committee

General Chair

Marco Winckler   ICS-IRIT, Université Paul Sabatier, France

Program Chairs

Sven Casteleyn   Universitat Jaume I, Castellón, Spain
Gustavo Rossi   UNLP, Argentina

Track Chairs

Cross-Media and Mobile Web Applications

Niels Olof Bouvin   Aarhus University, Denmark
In-young Ko   Korea Advanced Institute of Science and Technology, South Korea

HCI and the Web

Jose Antonio Gallud   Universidad de Castilla La Mancha, Spain
Fabio Paternò   C.N.R.-ISTI, Italy

Modeling and Engineering Web Applications

Marco Brambilla   Politecnico di Milano, Italy
Manuel Wimmer   Vienna University of Technology, Austria

Quality Aspects of Web Applications

Silvia Abrahão   Universidad Politecnica de Valencia, Spain
Filomena Ferrucci   Università di Salerno, Italy

Social Web Applications

Maria Bielikova   Slovak University of Technology in Bratislava, Slovakia
Flavius Frasincar   Erasmus University Rotterdam, The Netherlands

Web Applications Composition and Mashups

Cesare Pautasso	University of Lugano, Switzerland
Takehiro Tokuda	Tokyo Institute of Technology, Japan

Tutorials Chairs

Luis Olsina	Universidad National de la Pampa, Argentina
Oscar Pastor	Universidad Politecnica de Valencia, Spain

Workshops Chair

Santiago Meliá	University of Alicante, Spain

Demos AND Posters

Jordi Cabot	Inria/École des Mines de Nantes, France
Michael Nebeling	ETH, Switzerland

PHD Symposium Chairs

Cinzia Cappiello	Politecnico di Milano, Italy
Martin Gaedke	Technische Universität Chemnitz, Germany

Program Committee

Cross-Media and Mobile Web Applications

Wei Chen	Agricultural Information Institute, Chinese Academy of Agricultural Sciences, China
Antonella De Angeli	University of Manchester, UK
Volker Gruhn	Universität Duisburg-Essen, Germany
Célia Martinie	ICS-IRIT, Université Paul Sabatier, France
George Pallis	University of Cyprus, Cyprus
Fabio Paternò	ISTI-CNR, Pisa, Italy
Benjamin Satzger	Microsoft, USA
Quan Z. Sheng	University of Adelaide, Australia
Beat Signer	Vrije Universiteit Brussel, Belgium
Giovanni Toffetti Carughi	IBM Research Haifa, Israel
William Van Woensel	Dalhousie University, Canada
Marco Winckler	ICS-IRIT, Université Paul Sabatier, France

HCI and the Web

Julio Abascal	University of the Basque Country, Spain
Simone Barbosa	Pontificia Universidade Catolica do Rio de Janeiro, Brazil
Giorgio Brajnik	University of Udine, Italy
Carlos Duarte	University of Lisbon, Portugal
Cristina Gena	University of Turin, Italy
Luis Leiva	Universitat Politècnica de València, Spain
Maria Lozano	University of Castilla-la Mancha, Spain
Maristella Matera	Politecnico di Milano, Italy
Michael Nebeling	ETH Zurich, Switzerland
Victor Penichet	University of Castilla-La Mancha, Spain
Carmen Santoro	CNR-ISTI, Italy
Markel Vigo	University of Manchester, UK
Marco Winckler	ICS-IRIT, Université Paul Sabatier, France

Modeling and Engineering Web Applications

Luciano Baresi	Politecnico di Milano, Italy
Devis Bianchini	University of Brescia, Italy
Hubert Baumeister	Technical University of Denmark, Denmark
Alessandro Bozzon	Politecnico di Milano, Italy
Jordi Cabot	IInria École des Mines de Nantes, Italy
Richard Chbeir	LE2I-CNRS, France
Florian Daniel	University of Trento, Italy
Oscar Diaz	University of the Basque Country, Spain
Schahram Dustdar	Vienna University of Technology, Austria
Jutta Eckstein	IT communication, Germany
Marina Egea	Atos Research & Innovation Department, Spain
Flavius Frasincar	Erasmus University Rotterdam, The Netherlands
Piero Fraternali	Politecnico di Milano, Italy
Irene Garrigós	University of Alicante, Spain
Michael Grossniklaus	University of Konstanz, Germany
Guy-Vincent Jourdan	University of Ottawa, Canada
Gerti Kappel	Vienna University of Technology, Austria
Alexander Knapp	Universität Augsburg, Germany
Frank Leymann	University of Stuttgart, Germany
Maristella Matera	Politecnico di Milano, Italy
Santiago Melia	University of Alicante, Spain
Oscar Pastor	Universidad Politecnica de Valencia, Spain
Vicente Pelechano	Universidad Politecnica de Valencia, Spain
Alfonso Pierantonio	University of L'Aquila, Italy
Werner Retschitzegger	Johannes Kepler University of Linz, Austria

Fernando Sánchez   Universidad de Extremadura, Spain
Daniel Schwabe   PUC Rio, Brazil
Antonio Vallecillo   University of Málaga, Spain
Agustin Yague   Universidad Politecnica de Madrid, Spain
Gefei Zhang   arvato systems, Germany
Jürgen Ziegler   University of Duisburg-Essen, Germany

Quality Aspects of Web Applications

Joao Araujo   Universidade Nova de Lisboa, Portugal
Rami Bahsoon   University of Birmingham, UK
Michela Bertolotto   University College Dublin, Ireland
Davide Bolchini   Indiana University, USA
Giorgio Brajnik   University of Udine, Italy
Cinzia Cappiello   Politecnico di Milano, Italy
Schahram Dustdar   TU Wien, Austria
Carmine Gravino   University of Salerno, Italy
Emilio Insfran   Universitat Politècnica de València
                              (DSIC-UPV), Spain
Tahar Kechadi   University College Dublin, Ireland
Nora Koch   Ludwig Maximilians University of Munich,
                              Germany
Grace Lewis   Carnegie Mellon Software Engineering
                              Institute, USA
Maristella Matera   Politecnico di Milano, Italy
Emilia Mendes   Blekinge Institute of Technology, Sweden
Ali Mesbah   University of British Columbia, Canada
Luis Olsina   GIDIS_Web, Engineering School, UNLPam,
                              Argentina
Federica Sarro   University College, London, UK
Giovanni Toffetti Carughi   University of Lugano, Switzerland
Giuliana Vitiello   University of Salerno, Italy
Michael Weiss   Carleton University, Canada
Coral Calero   Universidad de Castilla-La Mancha, Spain
Arie van Deursen   Delft University of Technology,
                              The Netherlands
Vahid Garousi   University of Calgary, Canada
Jean Vanderdonckt   Université Catholique de Louvain, Belgium
Cristina Cachero   Universidad de Alicante, Spain

Social Web Applications

Witold Abramowicz   Poznan University of Economics, Poland
Ioannis Anagnostopoulos   University of Thessaly, Greece
Marco Brambilla   Politecnico di Milano, Italy

Richard Chbeir   Le2i - CNRS, France
Alexandra Cristea   University of Warwick, UK
Oscar Diaz   University of the Basque Country, Spain
Stefan Dietze   L3S Research Center, Germany
Roberto De Virgilio   Università di Roma Tre, Italy
Vania Dimitrova   University of Leeds, UK
Martin Gaedke   Chemnitz University of Technology, Germany
Geert-Jan Houben   Delft University of Technology,
                               The Netherlands
Zakaria Maamar   Zayed University, UAE
Jose Palazzo Moreira
  de Oliveira   UFRGS, Brazil
Jan Paralic   Technical University in Kosice, Slovakia
Oscar Pastor   Valencia University of Technology, Spain
Davide Rossi   University of Bologna, Italy
Daniel Schwabe   PUC Rio, Brazil
Markus Strohmaier   University of Koblenz-Landau, Germany
Julita Vassileva   University of Saskatchewan, Canada
Erik Wilde   UC Berkeley, USA
Guandong Xu   University of Technology Sydney, Australia
Jaroslav Zendulka   Brno University of Technology, Czech Republic

Web Applications Composition and Mashups

Saeed Aghaee   University of Lugano, Switzerland
Christoph Bussler   MercedSystems, Inc., USA
Florian Daniel   University of Trento, Italy
Oscar Diaz   University of the Basque Country, Spain
Hao Han   Kanagawa University, Japan
Gregor Hohpe   Google, Inc.
Geert-Jan Houben   Delft University of Technology,
                               The Netherlands
Peep Küngas   University of Tartu, Estonia
Maristella Matera   Politecnico di Milano, Italy
Moira Norrie   ETH Zurich, Switzerland
Tomas Vitvar   Czech Technical University of Prague,
                               Czech Repuclic
Eric Wohlstadter   University of British Columbia, Canada
Christian Zirpins   Karlsruhe Institute of Technology, Germany

Additional Reviewers

Saba Alimadadi   Michele Bianchi
Cristóbal Arellano   Hugo Brunelière
Marcos Baez   Dimoklis Despotakis

Milan Dojchinovski
Martin Fleck
Ujwal Gadiraju
Florian Geigl
Ujwal Gadiraju
Javier Luis Canovas Izquierdo
Oliver Kopp
Philip Langer
Fangfang Li
Xin Li
Jacek Mayszko
Esteban Robles Luna

Juan Carlos Preciado
Peter Purgathofer
Monica Sebillo
Simon Steyskal
Victoria Torres
Pedro Valderas
Karolina Vukojevic-Haupt
Jozef Wagner
Sebastian Wagner
Simon Walk

Local Organizing Committee

Local Chairs

David Navarre   ICS-IRIT, University of Toulouse Capitole,
                              France
Célia Martinie   ICS-IRIT, Paul Sabatier University, France

Operations

Marlène Giamporcaro   INP-Toulouse, France
Marie-Anne Laplaine   INP-Toulouse, France
Nadine Ortega   University of Toulouse 1, France

ICWE Steering Committee Liaisons

Marco Brambilla   Politecnico di Milano, Italy
Daniel Schwabe   PUC-Rio, Brazil

Acknowledgments

The conference chairs and conference organizers would like to thank our sponsors:

Sponsors

Institute of Research in Informatics of Toulouse (IRIT)
interaction-design.org
University of Toulouse Capitole (Toulouse I)
Paul Sabatier University (Toulouse III)
Institut Nationale Polytechnique de Toulouse (INP)

Scientific Sponsors

ACM In-Cooperation with Special Interests Groups SIGCHI and SIGWEB

International Society for Web Engineering (ISWE)

International World Wide Web Conferences Steering Committee (IW3C2)

Table of Contents

Research Papers

Late Breaking Results

Demos/Posters

PhD Symposium

Keynotes

Tutorials

Workshop

A Platform for Web Augmentation Requirements Specification

Diego Firmenich[1,3], Sergio Firmenich[1,2], José Matías Rivero[1,2], and Leandro Antonelli[1]

[1] LIFIA, Facultad de Informática, Universidad Nacional de La Plata
[2] CONICET, Argentina
{sergio.firmenich,mrivero,lanto}@lifia.info.unlp.edu.ar
[3] Facultad de Ingeniería, Universidad Nacional de la Patagonia San Juan Bosco
dfirmenich@tw.unp.edu.ar

Abstract. Web augmentation has emerged as a technique for customizing Web applications beyond the personalization mechanisms natively included in them. This technique usually manipulates existing Web sites on the client-side via scripts (commonly referred as *userscripts*) that can change its presentation and behavior. Large communities have surfaced around this technique and two main roles have been established. On the one hand there are *userscripters*, users with programming skills who create new scripts and share them with the community. On the other hand, there are *users* who download and install in their own Web Browsers some of those scripts that satisfy their customization requirements, adding features that the applications do not support out-of-the-box. It means that Web augmentation requirements are not formally specified and they are decided according to particular *userscripters* needs. In this paper we propose CrowdMock, a platform for managing requirements and scripts. The platform allows *users* to perform two activities: (i) specify their own scripts requirements by augmenting Web sites with high-fidelity mockups and (ii) upload these *requirements* into an online repository. Then, the platform allows the whole community (*users* and *userscripters)* to collaborate improving the definition of the augmentation requirements and building a concrete script that implements them. Two main tools have been developed and evaluated in this context. A client-side plugin called MockPlug used for augmenting Web sites with UI prototype widgets and UserRequirements, a repository enabling sharing and managing the requirements.

1 Introduction

Nowadays, the Web is a really complex platform used for a great variety of goals. This advanced use of the Web is possible because of the evolution of its supporting technology and the continuous and massive demands from users that enforces and accelerates its rapid evolution. The Web has not evolved only in terms of technical possibilities but also in terms of supporting an increasing number and types of users too. The same crowd of users has evolved in parallel with the Web itself and the user expectations in terms of what the applications provide to them are very demanding nowadays. Even more, users have found ways to satisfy their own requirements when

S. Casteleyn, G. Rossi, and M. Winckler (Eds.): ICWE 2014, LNCS 8541, pp. 1–20, 2014.
© Springer International Publishing Switzerland 2014

they are not taken into account by Web applications developers. It is not casual that several Web augmentation [5] tools have gone emerging in this context. These tools pursue the goal of modifying existing Web sites with specific content or functionality that were not originally supported. The crowd of users continuously builds a lot of such as tools (distributed within the community as browsers plugins, scripts, etc.). It is clear that users are interested in using the Web in the way they prefer and they have evolved to know how to achieve it. Even in the academy there is a dizzying trend in the area of end-user programming [11]. The reason is that (1) part of the crowd of users may deal with different kinds of development tools, and (2) part of the crowd of users want to utilize these tools in order to satisfy a particular need/requirement.

Although several tools such as Platypus[1] allow users to perform the adaptation of Web pages directly by following the idea of "what you see is what you get", these kinds of tools have some limitations regarding what can be expressed and done. Thus, some Web augmentation requirements could not be addressed with them, and these have to be tackled with other kind of script, which require of advance programming skills to be developed. In this sense, we think that prototyping could be also a solution in the context of Web augmentation requirements, and augmenting Web pages with the prototypes is really a straightway for telling others what a user wants.

In addition, users are more familiarized in dealing with software products, which is partially proven by new agile development methodologies like Scrum [19] that center the development in tackling user needs by prioritizing requirements considering its business value. The main purpose of such processes is providing valuable functionality as early as possible giving a more active role to the stakeholders to assess that the implemented application is what they need and expect. It is usual to ask users for defining a requirement by using one of several techniques such as mockups [20][6] or User Stories [4], which are completely understandable by users. Since agile methodologies try to include stakeholders (including users) as much as possible in the development process, their processes usually are compatible with User-Centered Design (UCD) approaches. Among the most common requirements artifacts used in UCD, low-fidelity prototypes (informally known as mockups) are the most used [9]. Mockups are more suitable than textual requirements artifacts because they are a shared language between development team (including developers and analysts) and users [14], avoiding domain-related jargon that may lead to faults in delivered software artifacts originated by requirements misunderstandings. In addition, mockups have been proposed as a successful tool to capture and register fluid requirements [17] – those that are usually expressed orally or informally and are an implicit (and usually lost) part of the elicitation process. However, they are rarely used in an isolated way, being usually combined with User Stories [4] or Use Cases [13][8]. Finally, UI Mockups have the advantage of being simple since stakeholders can build these by themselves using popular and emergent tools like Balsamiq[2] or Pencil[3].

---

[1] https://addons.mozilla.org/es/firefox/addon/platypus/
[2] http://balsamiq.com
[3] http://pencil.evolus.vn/

Most of the development approaches regarding using mockup are centered on applications in which the number and type of end-users are predictable. However, this is not true for massive Web applications, since the crowds that use them are not only less predictable in quantity and type, but also are changing and evolving all the time. In order to satisfy as many users as possible, the application should be customized for each user. Although the mechanism of adaptation and personalization of Web applications, the so-called *adaptive Web* [2], have also evolved (for instance, complex user models [7] have been proposed and assessed), it is really hard to give adaptation mechanisms that contemplate the requirements of all the users.

In this context, two trends with two main goals have emerged. On the one hand, well-known mash-ups [23] approaches have surfaced for integrating existing Web content and services. On the other hand, a technique called Web augmentation [5] has emerged in order to allow users to customize both content and functionalities of Web pages modifying their DOMs on the client-side. Both approaches have communities that serve as a proof of how important is for the users to improve their Web experiences. An example of the former is Yahoo Pipes [22], a mash-up tool broadly utilized for combing services over the Web that has a big active community [3]. Regarding Web augmentation, the userscripts.org [21] community represents an emblematic example. This community has created thousand of scripts that run over the client for modifying the visited Web sites. Some scripts have been installed for end users over one hundred thousand times.

In both communities we observed two different user groups: (1) the crowd of users (at least part of them) want to personalize the Web applications they use and, (2) the part of this crowd can develop software artifacts for such purpose. Our main concern is to facilitate how these communities currently work. In the design of Web applications from scratch, different stakeholders participate on the definition of the functionality. This can be done since part of these reduced and well-known set of stakeholders directly *express* their requirements. There is an explicit delegation. On the contrary, most of the open communities that intend to introduce custom enhancements to existing Web sites work usually in the opposite way: a user from the crowd with programming skills develops an artifact with some features that may be useful to other users. If the artifact is good enough and responds to a common requirement, then, it will possibly be installed for many other users of the crowd.

In some cases both user and userscripters work together by asking new requirements and implementing them (correspondently) in two ways: (1) asking for new scripts in the community forums or (2) asking scripts improvements in the script's discussion Web page. In this work we aim to create new communication mechanisms for improve both the process and how resultant scripts match user requirements. The main motivation behind the approach presented here relies in the fact that, besides informal forums, the community has no clear ways to communicate to those users with development skills in the crowd, which are the *augmentation requirements* desired.

In this paper we propose to rely in mechanisms used on agile methodologies, such as User Stories or mockups, to empower users with tools for specifying their enhancements requirements over existing Web applications. The novelty of the approach is the usage of Web augmentation techniques: with our approach, a user can augment a Web page with those widgets that are relevant for describing his requirement. Then, our approach allows sharing the augmentation requirement so other *users* and *userscripters* may reproduce the same augmentation in their own Web browsers in order to understand that user's needs in a more detailed and formal way.

The paper is organized as follows. Section 2 introduces our approach. Section 3 presents the tools, and section 4 shows the results of an evaluation made with end-users. The state of the art is tackled in section 5. Finally, section 6 gives some concluding remarks and further works.

2   The Approach

We present a novel approach for UI prototyping based on Web augmentation. In this work, Web augmentation is used as the technique for defining requirements while Web augmentation communities are taken as the source of both requirements and scripts for these requirements.

Our goal is to improve the communication between *users* and *userscripters*. The essence of the approach can be described through the following example: let's consider that a user is navigating a common video streaming Web site like YouTube. When he searches for videos, he realizes that additional information in the results screen will be helpful for him in order to decide which video to watch. For instance, he would like to see the amount of likes/dislikes and also further information about video's audio quality, resolution, etc. Consider that the user knows how to install and run GreaseMonkey scripts to augment Web sites, but he does not how to develop one of such scripts. If there were no scripts satisfying his concrete expectations, he would like to have a way to communicate to GreaseMonkey scripters his requirements. This communication should be as concrete and clear as possible. Currently, userscripting communities like userscripts.org rely on textual messages interchanges as the solely artifacts to let users and userscripters collaborate. The presented approach proposes to help the end-users and userscripters collaboration through:

- Linking users requirements with *userscripters* who can implement the requirements.
- Using a concrete language for defining those requirements. The language must be understood by both developers and users and must be formal enough to describe the requirement in terms of UI components and features.
- Managing and evolving the requirements in a well-defined process.

Our approach follows the general schema shown in Figure 1. The main goal is that any user can specify the visual components related to a particular requirement by augmenting the target Web page with mockups.

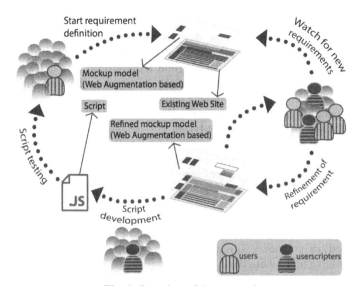

Fig. 1. Overview of the approach

Once the requirement is concretely defined (which is expressed through a model that will be introduced later); it can be uploaded into an online repository, which makes it accessible by the community to *reproduce* and enhance this. In this context, to reproduce means that any other user can augment the target Web page in the same way the user that specified the requirement did, in order to watch in detail which the required UI alterations are.

The community (i.e. users and userscripters) may colaborate in the refinement of a requirement in the repository. If a user is not capable to define the requirement with the required level of detail, another user who wants to colaborate (for example, having the same requirement) can enrich the definition. Every change is validated by the requirement's owner. When a stable version of the requirements is met, a userscripter may create the script that implements it, and then upload it into the repository. At this point, users can install and use the script, contributing with its testing.

Therefore, our approach empowers users with:

- Mechanisms for defining their requirements over existing Web pages.
- Mechanisms for managing and publishing their requirements.
- Mechanisms for proposing and bringing solutions to users, who can evaluate these solutions and eventually start the cycle again.

The main purpose of our approach, called *CrowdMock*, relies on allowing the *crowd of users* to *plug* high-fidelity *mockups* into existing Web pages. Thus, the approach involves two main contributions. On the one hand, we designed a metamodel (the *MockPlug metamodel*) for describing how high-fidelity mockups are woven into existing Web sites via Web augmentation. On the other hand, to support our approach technologically we developed a client-side tool for weaving augmentation models (called *MockPlug*) and a server-side platform for managing them collaboratively (called *UserRequirements*). MockPlug is a tool designed for allowing end-users to *plug* mockups on existing Web pages to specify their augmentation requirements and them in a concrete, formal model. With MockPlug, users may create, update, and reproduce requirements. UserRequirements is a platform deployed as a Web application[4] and also integrated with MockPlug, for allowing end-users to share their augmentation requirements and asking for scripts that implement them. This also contemplates the evolution of the requirement and the evaluation of a particular solution allowing collaboration, traceability, and validation.

Web augmentation operates over existing Web sites, usually by manipulating their DOM[5]. If we want to augment a DOM with new widgets, we have to specify how those new widgets are woven into the existing DOM. With this in mind, we defined the MockPlug metamodel, which specifies, which kinds of augmentation (widgets and their properties) are possible within our approach and also how these are materialized for a particular Web page's DOM. Within MockPlug metamodel, which is depicted in Figure 2, both the name and the URL of the augmented Web site compose a requirement. The specification of new widgets relevant for the requirement (Widgets) may be defined in the model. Widgets can be simple (SimpleWidgets, atomic and self-represented) or composite (CompositeWidgets, acting as a container of another set of Widgets), but all of them have several characteristics in common:

- Every Widget belonging to a MockPlug model has both a type and properties.
- A Widget is prepared to respond to several events, such as *mouseover* or *click*.
- A Widget can react to an event with predefined operations.

A Widget has information related to how and where it was inserted into the DOM tree. Note that, each widget has an insertion strategy associated (*float, leftElement, rightElement, afterElement, beforeElement, etc.*).

It is worth mentioning the importance of the property *url* in a MockPlug model, since this is the property that defines which Web site (o set of Web sites when using a regular expression) the model is augmenting.

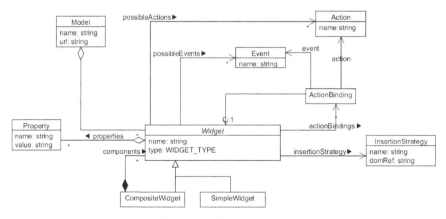

Fig. 2. MockPlug metamodel

---

[4] See a live demo of our approach on: http://www.userrequirements.org
[5] http://www.w3.org/DOM/

3   Tools for Web Augmentation Requirements Management

In this section we introduce the technical details behind the aforementioned MockPlug and UserRequirements implementation.

3.1   MockPlug

We considered two issues in order to empower users with mechanisms for defining their own requirements, (1) the supporting tool for such purpose has to be easy to use and (2) specifying a new requirement should not require too much effort and high technical knowledge. We developed MockPlug with this in mind. MockPlug is a Firefox extension that allows end-users to specify requirements by manipulating high-fidelity mockups over their Web sites that they want to augment. For each MockPlug requirement, a User Story is also defined.

From MockPlug's point of view, a requirement represents a set of modifications made over an existing Web site expressed in terms of the aforementioned metamodel. Thus, every requirement expressed in our approach has a MockPlug model associated, and the changes are expressed by adding widgets. From the user's point of view, these have a visual style like hand-drawn mockups similar to mockup tools like Pencil or Balsamiq. However, they are represented at runtime as ordinary DOM elements. In order to refine his requirements visually, a user may drag&drop widgets over the existing Web page, like it is shown in Figure 3. This figure depicts how the user can add a new button over the existing IMDB Web site. On the top of Figure 4, the button already added over the existing Web page is shown and the widget editor is shown at the bottom.

3.1.1   Widgets

The tool considers three kinds of widgets: *predefined* widgets, *packetized* widgets and *collected* widgets, which are described below.

Predefined Widgets
We have defined an initial set of predefined widgets (Figure 5), which are grouped in categories as Table 1 describes. Different properties may be set for each kind of predefined widget. For instance, the widget *Button* has the properties *name*, *title* and *label;* while a menu list has the *name* and a *collection of options*. Figure 5 shows how the predefined widgets are listed in the MockPlug Panel.

Pocketized Widgets
In order to allow users to easily specify integration requirements we created a functionality called *Pocket*. The Pocket (depicted in Figure 6) allows users to collect existing UI components as widgets. Its name comes from its provided functionality, which consists in allowing collecting and storing any part of an existing Web page and then place it as a *new* widget in other Web applications or pages. As an example, Figure 6 shows three already collected widgets (one collected from YouTube and two

collected from IMDB) that could be added in any other Web application in order to define some requirement of integration between these applications. The user must drag&drop the widget that had collected in the same way it did with predefined widgets, having the same tool aid and insertion strategies to apply. Pocketized widgets can be added as static *screenshots* of the original DOM or as real DOM elements. However, so far the tool does not guarantee the correct behavior defined with JavaScript routines taken from the original Web page from where the packetized widget was collected.

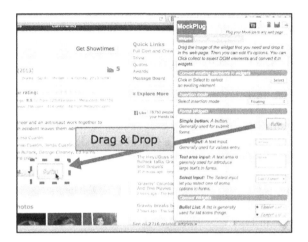

Fig. 3. MockPlug main panel

Fig. 4. Widget added and widget editor

Table 1. Predefined widgets

Category	Goal	Examples
Form Widgets	Specify requirements that involve some kind of information gathering.	Button, text input, text area, select menu
Content Widgets	Specify requirements focused on non-editable contents	Lists, images, text, anchors
Annotation Widgets	Empower users with mechanisms for describing textually, some expected behavior, presentation or content	Post-its notes, bubble comments

The user may add both predefined and pocket widgets in the DOM by using an insertion strategy. This function provides an interactive highlighting of existing elements in the DOM in order to help the user.

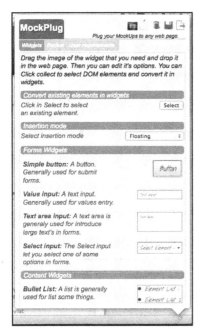

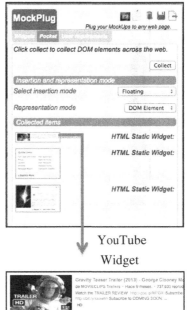

YouTube
Widget

Fig. 5. Predefined Widgets panel   **Fig. 6.** Pocket panel

Collected Widgets

Although the Pocket allows users to specify some requirements of integration and also to reuse existing UI components, sometimes it could be useful to convert an existing DOM element into widgets in order to refine it. This can be accomplished using the *collected widgets* functionality provided by the tool. A *collected widget* is a widget that has been created or imported from an existing DOM element in order to manipulate it. For example, let's assume that the user wishes to indicate that a new column should be displayed in an existing table. Then, the user can convert the existing table into a new widget with the final goal of editing its contents and columns, as it is shown in Figure 7. In this case, the user collected the table containing the 250 top movies from IMDB.com.

The ability of converting existing DOM elements into widgets, also gives the possibility of removing useless items from the target web site. Since the DOM element is converted into a widget, it could be also moved to another part of the Web page. Moreover, converting previously existing DOM elements into widgets makes possible to use them as a target or reference in actions define for other widgets. For example, the user might want to add a button to show/hide a particular element of the website on which the requirement is made.

It is worth noting that collecting a DOM element with the goal of creating a collected widget is not the same that putting a widget into the Pocket. While the first one associates the underlying DOM element as a widget in the MockPlug requirements model, the second ones can be used as a sort of *clipboard* for moving UI pieces among Web applications.

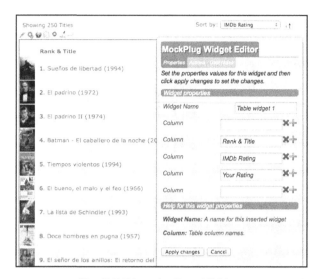

Fig. 7. Edition of collected Widget

3.1.2   Widgets Management

The widgets added with MockPlug are clearly distinguishable from the rest of the elements found in the website. However, MockPlug makes it possible to copy styles from existing DOM elements to a Widget in order to emulate an existing page style. In the example from Figure 8, the user has added a button (predefined widget) called "Button text", which is actually a DOM element (as any widget) and which with the user can interact. Different operations commonly found in mockup tools can be applied through the widget contextual toolbar (shown when the widget is selected), such as cloning, removing, moving, annotating, property editing, among others.

3.1.3   Code Generation

User requirements are abstracted with the MockPlug metamodel and their instances may be processed in order to generate code through MockPlug code generators as in well-known Model-Driven approaches [10]. The code generation capabilities in MockPlug make it possible to generate a first code stub implementing basic and repetitive features of the augmentation requirement, thus reducing the workload for the *userscripter* who wants to write a script to satisfy it. The code generator has been included by default with MockPlug and currently it is focused on obtaining GreaseMonkey scripts.

3.2   Requirements Repository

In addition to MockPlug, we have implemented UserRequirements, a repository with social features with two main purposes: (1) allowing users to collaborate in the definition and the evolution of augmentation requirements and (2) enabling *userscripters* to describe and reproduce the requirements in order to develop the corresponding script. A requirement in our repository is defined by two components in our repository:

- **A User Story (US)**, which highlights the core concerns of the requirement from a particular user role pursuing some goal in the business [4].
- **A MockPlug model**, which is the UI prototype for answering to the US. A MockPlug model is associated with only one US, but one US can be referenced in multiple MockPlug models.

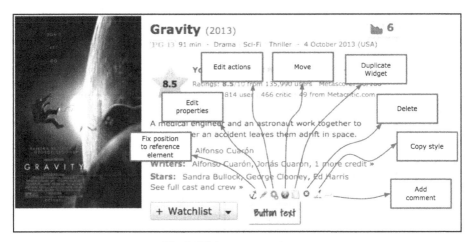

Fig. 8. Widget contextual menu

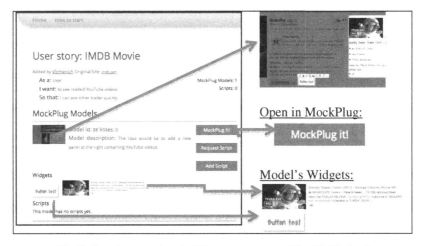

Fig. 9. Requirement view in UR   **Fig. 10.** UI details

Figure 9 and 10 shows a screenshot of how a requirement is described in the repository. It includes information about who created the requirement, over which Web site it was defined, its US, and a screenshot of how the target Web site was augmented with new widgets by using MockPlug.

It is really important to allow developers to reproduce and watch the resulting *augmented mockup* in action. This functionality is provided by the button labeled *MockPlug it* (see Figure 10). When the user clicks this button, a new tab is opened in order to load the target Web site and augment it with the MockPlug model built interactively using the MockPlug tool. Others users can collaborate by evolving the definition of that requirement using the tool.

The integration between the social repository and MockPlug tool is essential for our approach. We show how the user can add his current requirement in the repository on the left of Figure 11. To add a new requirement, the user has to define the User Story plus a general description of it. Finally, the requirement (formed by the US and the MockPlug model) is uploaded to the repository by clicking the *Save* button.

 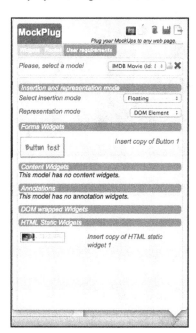

Fig. 11. Integration between MockPlug and UserRequirements

We depict how a user may choose from the menu between his already existing models in the repository on the right in the same figure. When a model is selected, all the widgets involved in that model are listed at the bottom from the same panel. The user also may choose to *open* the MockPlug model, whose action will render it in a new tab. The difference between *pre-open* (just for listing its widgets) and open a model (which actually renders it) was meant on purpose, because by pre-opening a model the user can reuse widgets defined on the model that is being defined.

For conciseness reasons we can not explain it deeply, but it is important to know that UserRequirements make it possible to "branch" and "vote" the requirements models. In this way the community may reach a personalization level of the requirement that fulfills the expectations of the majority while someone other person of the community may create a new branch of one model in order to specify further requirements.

4   Evaluation

This section describes a preliminary evaluation of the CrowdMock approach and its supporting tools. We have performed a controlled experiment focused on assessing the effectiveness achieved by experienced users when specifying requirements on existing Web applications. The experiment compared the process of specifying requirements using the proposed approach against using existing approaches. Our hypothesis was:

> *Defining augmentation requirements with CrowdMock produces better specifications than using traditional user-oriented methods (such as user stories, mockups, etc.), and consequently, CrowdMock improves the understandability of the requirements.*

4.1   Protocol and Participants

The experiment was organized into two phases: (1) defining a requirement using well-known techniques (e.g. user stories, mockups, etc...) and (2) defining requirements using MockPlug. The requirements had to be defined over well-known existing Web applications: IMDB[6] and YouTube[7]; and they had to be based on one of the following features. The features for IMDB were:

- R1.1: Filter the list of 250 best movies
- R1.2: Add some information to the 250 best movies list.
- R1.3: Change the layout and/or content of a movie's page based on the following issues:
    - o   Move elements of the page.
    - o   Remove elements from the page
    - o   Add new widgets into the page (button, input box, menu, etc.)
    - o   Add contents related to the movie from another IMDB page.

The features for YouTube:

- R2.1: Add information in the search results about the videos.
- R2.2: Idem to R1.3 focusing on a YouTube video's Web page.

Participants had to choose 2 features in each phase. Thus, each participant specified 4 requirements. Since 8 subjects participated in the experiment, a total of 32 requirements were specified. Half of them (16 requirements) had to be defined with the participants' preferred requirements artifacts and elicitation techniques, and the other 16 requirements had to be defined using MockPlug. All the 32 requirements were implemented by means of GreaseMonkey scripts, which were developed by a specific team at LIFIA.

---

[6] Internet Movie Data Base - http://imdb.com; last accessed 4-Feb-2014.
[7] YouTube – http://youtube.com; last accessed 4-Feb-2014.

The whole experiment was organized as follows:

1. First phase:
   a. A first face-to-face meeting with participants was organized for presenting the experiment and explaining the tasks. Before this meeting, participants had been introduced in the use of US as well as in user interface mockup building techniques and tools. In this meeting the definition of two requirements choosing one specific technique and one or more requirements artifacts was exemplified.
   b. Participants had 3 days to write two requirements. Then, they had to send their specifications (including the requirements artifacts) and an activity diary where they registered all their work and the time spent on it.
   c. When all the requirements were collected, developers had 2 days to create the 16 GreaseMonkey scripts. Both 2 scripts implementing the requirements were sent to every participant.
   d. Participants had 2 days for installing, using and evaluating both scripts and filling in two questionnaires:
      i. Questionnaire A, oriented to measure and describe the difficulty perceived during the specification of the requirements.
      ii. Questionnaire B: oriented to measure and describe how well the script satisfied his expectations.
2. Second phase:
   a. Another face-to-face meeting was organized for presenting both the new tasks to be accomplished and the tools involved in this phase:
      i. The UserRequirements (UR) application (i.e., the requirements repository) was introduced. Participants were asked to create a user account on the platform in order to start specifying requirements through it.
      ii. MockPlug was also explained and participants learned about how to upload the requirement from MockPlug to UR.
      iii. The second evaluation task was presented. It consisted in defining two requirements using MockPlug and UR. Participants had to choose only requirements not chosen in the first phase.
   b. Participants had 3 days to define the requirements and then upload them to the repository. Then, they had to send separately a document including the activity diary.
   c. With every requirement uploaded to UR by participants, the developers received a notification. Again, developers had 2 days to create the 16 GreaseMonkey scripts. When all the requirements were uploaded, instead of sending the scripts personally, developers uploaded the script for each requirement into UR and participants received the corresponding notification.
   d. Participants had 2 days for downloading and using both scripts. After that, they were asked to filling in the same two questionnaires (A and B). Also, they were asked to fill in another SUS [1] questionnaire for evaluating the usability of our tools.

Following a within subjects design, a total of 8 participants were involved in this evaluation. All participants were professionals on computer science from a post-graduate course on Web Engineering; 4 female and 4 male and aged between 24 and 60. It is important to mention that the 8 participants were instructed (before the experiment) in the installation and use of GreaseMonkey scripts.

4.2   Results

This section describes the results of the experiment. First, we describe the requirements specified by every participant in Table 2. Afterwards, we present the analysis.

4.2.1   First Phase

Table 2 shows depicted with "X" the requirements selected in the first phase for each participant. The average time for defining each requirement was 92.5 minutes (SD = 62.12 minutes). The techniques used were mockups (traditional ones), User Stories and some more textual description.

The difficulty perceived when specifying each requirement ranged from "normal" to "very easy". Most of the requirement specifications (11) were considered as a "normal" task in terms of difficulty (68.75%) in a scale from 1 to 5 (where 1="very easy", 2="easy", 3="normal", 4="difficult", 5="very difficult"). This task was considered "easy" for 4 requirements (25%), and "very easy" only for 1 (6.25%).

Table 2. Requirements selected by each participant (X = First Phase, O = Second phase)

	R1.1	R1.2	R1.3	R2.1	R2.2
Participant 1		X	X	O	O
Participant 2		O	O	X	X
Participant 3		O	O	X	X
Participant 4		X	X	O	O
Participant 5		X	O	O	X
Participant 6		X	O	O	X
Participant 7	O	X		X	O
Participant 8		X	O	X	O

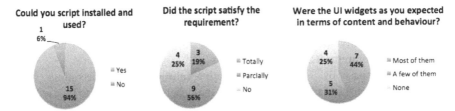

Fig. 12. Phase 1 results

After receiving all the augmentation requirements specifications, we implemented them and sent the scripts to every participant. They had to download, install and use these scripts in order to evaluate them. The results are shown in Figure 12.

4.2.2   Second Phase Results

Requirements chosen for the second phase of the experiment are depicted with "O" in Table 2. As commented previously, participants had to use MockPlug and UserRequirements in this phase. Defining a new requirement took in average 86.56 minutes (SD = 66.35). However, the difficulty perceived in the specification of each requirement had a wide range. The difficulty of the specification task was considerer "normal" (25%) in 4 requirements. Another 25% were considered "difficult" while "very difficult" was used in one requirement (6.25%). Other 37.5% (6 requirements) were qualified as "easy". Finally, 1 requirement specification was considered "very easy".

We believe the reasons were: (1) people were not experienced in the use of Web augmentation tools and (2) some incompatibilities between participants' Firefox extension version and our plug-in. This was finally confirmed by the SUS questionnaire. MockPlug scored 74.9, which is an improvable result regarding the usability of the tool.

Despite of these contingencies, we observed that participants were more satisfied regarding the implementation of their requirements when using MockPlug. It would indicate that the scripts in this phase were closer to user expectations than those developed for the first one. This information is depicted in Figure 13.

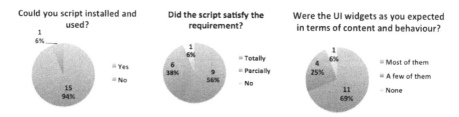

Fig. 13. Phase 2 (CrowdMock) results

4.2.3   Discussion

The results of the controlled experiment showed that when users freely *choose* the specification technique only 18.75% of the requirements were totally satisfied and 56.25% were partially specified. The remaining 25% of the scripts did not satisfy absolutely user requirements.

In the second phase, by contrast, 56.25% scripts fulfilled user's expectations, while 37.5% satisfied them partially. Because one of the scripts was not able to be executed, just one requirement (6.25%) was not satisfied.

Our tools were not well known by the participants, who reported several usability issues that are reflected in the difficulty they perceived during the task of specifying a requirement with MockPlug. In spite of these problems, our approach allowed to participants to be more satisfied. We think that this difference is due to: (1) aid provided by mockup for focusing on specific elements (UI widgets) which resulted in more concrete and clearer specifications for scripters, (2) the possibility of manipulating existing UI components easily, and (3) the possibility of defining behavior for each widget without programming knowledge.

Another additional benefit of using our approach was related to the development process of the scripts. By using MockPlug and MockPlug models, some code could be generated automatically, at least what is related to the fact of creating new widgets and get the existing DOM elements from the target Web page that were manipulated.

5   Related Works

In this section we compare our approach with two kinds of works. First, we analyse other mechanisms and tools to specify UI prototypes. On the other hand, we review some approaches to elicit requirements in the context large crowd of users.

Using UI mockups as a requirement elicitation technique is a growing trend that can be appreciated just observing the amount of different Web and desktop-based prototyping tools like Balsamiq and Mockingbird[8] that appeared during the last years. Statistical studies have proven that mockups use can improve the whole development process in comparison to use other requirements artefacts [15]. Mockups use has been introduced in different and heterogeneous approaches and methodologies. On the one hand, they have been included in traditional, code-based Agile settings as an essential requirement artifact [20] [6]. On the other hand, they have been used as formal specifications in different Model-Driven Development approaches like MockupDD [16], ranging from interaction requirements to data intensive Web Applications modelling. In this work we propose to specify augmentation changes using (among other techniques) mockup-style widgets. Also, MockPlug combines mockup-style augmentation techniques with well known annotations capabilities of common mockup tooling, that also can be used as formal specifications in the future as in [14].

There are new works on collaboration in requirements elicitation. Some authors have used social networks to support requirements elicitation in large-scale systems with many stakeholders [18]. They have used collaborative filtering techniques in order to obtain the most important and significant requirements from a huge number of stakeholders. A more formal and process-centered approach is proposed in [12]. However, we consider that none of the approaches provide a mechanism to collaboratively elicit, describe and validate requirements by using prototype definition for large-scale systems with a large number of stakeholders. The aforementioned works are limited to eliciting narrative requirements. We propose to define requirements by specifying high-fidelity prototypes that are later managed collaboratively from a common repository.

---

[8] http://gomockingbird.com

The ability of creating, validating and evolving prototypes collaboratively just by dragging and dropping widgets over existing Web sites, may allow stakeholders to express their needs in more accurate way, beyond textual descriptions [1]. In the particular context of Web augmentation communities, by improving communication between users and userscripters, we ensure that user requirements are correctly understood before future development of the concrete scripts. In this way, our integrated platform aims to reduce users' efforts when defining their requirements and upload them to the repository where, at the same time, it enables whole the community to improve them iteratively.

In the context of Web augmentation for annotations, there are also many tools to share user's annotations – for instance, Appotate[9] or Diigo[10]. These kinds of tools allow users to share annotations across the web, although these are very useful to share and highlight ideas, annotations are not enough to define complex augmentation requirements.

By using traditional mockups tools users can create their own prototypes, but usually have to start from scratch and widgets are not linked to the Web site where the requirement surfaces. This simple fact would demand much effort from scratch (in order to give some context to the augmentation requirement). Besides that, Web augmentation requirements could be extremely related to existing UI components, and desirably, the prototype should show how these existing components interact with the new widgets. Our approach enables users to quickly define high-fidelity prototypes related to the augmentation requirement, manipulating the same Web page that is going to be transformed by a script in real time. Our MockPlug metamodel makes it possible to share and reproduce these prototypes, but additionally, it also makes possible to process a prototype in order to generate a first script template based on those manipulated widgets.

6   Conclusions and Future Work

It is very complex to gather requirements from a crowd of users for widely used Web applications. In this work we have presented a methodology and its implementation for gathering requirements in the context of Web augmentation communities. Our approach is inspired in agile methodologies were users have an active role. We found that using high-fidelity mockups is very useful for specifying Web augmentation requirements using the running Web page as a foundation for specifying them graphically. With *CrowdMock*, users may define their own requirements and share them with the community, who can reproduce, edit and evolve them collaboratively. CrowdMock and its tools have been evaluated obtaining some benefits in comparison with traditional approaches. By using the proposed high-fidelity mockups that run over existing Web pages, we agilize the definition process since it is not necessary to construct an abstract conceptualization from existing UI. Users know what they want from Web applications and we try to give them the tools and mechanisms for naturally expressing their requirements.

---

[9] http://appotate.com
[10] https://www.diigo.com

We described the details of an evaluation of CrowdMock approach, which showed some positive results in Web augmentation requirements gathering and also some usability issues in the MockPlug tool that we are currently addressing. Integration between MockPlug and UserRequirements is also being improved. Additionally, we have an ongoing work on the server-side application (UserRequirements), which is being enhanced to support better collaboration mechanisms.

Another interesting further work path includes testing support and better code generation. We are planning to automate test for userscripts. Code generation is currently made without considering good *userscripters* practices and patterns; thus, the generated code may be not be easy to understand or manually adapt by userscripters. Thus, we are working on improvements our MockPlug code generators API in order to allow *userscripters* to develop their own code generators, which are a particular type of MockPlug plug-ins.

Although we are focused on Web augmentation requirements, our approach could also be useful for projects intended to build software products from scratch instead of augment existing ones. Thus, we are planning to improve the expressiveness of our metamodel, extending the tool with new kind of widgets and finally improve the server-side repository with features related to general developer support.

References

1. Brooke, J.: SUS: A 'quick and dirty' usability scale. In: Usability Evaluation in Industry, pp. 189–194. Taylor and Francis, London (1996)
2. Brusilovsky, P., Kobsa, A., Nejdl, W.: The Adaptive Web (2008)
3. Cameron Jones, M., Churchill, E.: Conversations in Developer Communities: a Preliminary Analysis of the Yahoo! Pipes Community. In: Proceedings of the Fourth International Conference on Communities and Technologies, pp. 195–204. ACM (2009)
4. Cohn, M.: User stories applied: for agile software development, p. 268. Addison-Wesley (2004)
5. Díaz, O.: Understanding Web augmentation. In: Grossniklaus, M., Wimmer, M. (eds.) ICWE Workshops 2012. LNCS, vol. 7703, pp. 79–80. Springer, Heidelberg (2012)
6. Ferreira, F., Noble, J., Biddle, R.: Agile Development Iterations and UI Design. In: Proceedings of AGILE 2007 Conference, pp. 50–58 (2007)
7. Gauch, S., Speretta, M., Chandramouli, A., Micarelli, A.: User Profiles for Personalized Information Access. In: Brusilovsky, P., Kobsa, A., Nejdl, W. (eds.) Adaptive Web 2007. LNCS, vol. 4321, pp. 54–89. Springer, Heidelberg (2007)
8. Homrighausen, A., Six, H., Winter, M.: Round-Trip Prototyping Based on Integrated Functional and User Interface Requirements Specifications. Requir. Eng. 7(1), 34–45 (2002)
9. Hussain, Z., Slany, W., Holzinger, A.: Current state of agile user-centered design: A survey. In: Holzinger, A., Miesenberger, K. (eds.) USAB 2009. LNCS, vol. 5889, pp. 416–427. Springer, Heidelberg (2009)
10. Kelly, S., Tolvanen, J.P.: Domain-Specific Modeling: Enabling Full Code Generation. Wiley-IEEE Computer Society (2008)

11. Ko, A., Abraham, R., Beckwith, L., Blcakwell, A., Burnett, M., Erwig, M., Scaffidi, C., Lawrance, J., Lieberman, H., Myers, B., Rosson, M., Rothermel, G., Shaw, M., Wiedenbeck, S.: The State of the Art in End-User Software Engineering. ACM Computing Surveys, 1–44 (2011)
12. Konaté, J., El Kader Sahraoui, A., Kolfschoten, G.: Collaborative Requirements Elicitation: A Process-Centred Approach. Group Decision and Negotiation Journal (2013)
13. Kulak, D., Guiney, E.: Use cases: requirements in context. Addison-Wesley (2004)
14. Mukasa, K., Kaindl, H.: An Integration of Requirements and User Interface Specifications. In: Proceedings of 6th IEEE International Requirements Engineering Conference, pp. 327–328 (2008)
15. Ricca, F., Scanniello, G., Torchiano, M., Reggio, G., Astesiano, E.: On the effectiveness of screen mockups in requirements engineering. In: Proceedings of ACM-IEEE Int. Symp. Empir. Softw. Eng. Meas. ACM Press, New York (2010)
16. Rivero, J.M., Rossi, G.: MockupDD: Facilitating Agile Support for Model-Driven Web Engineering. In: Sheng, Q.Z., Kjeldskov, J. (eds.) ICWE Workshops 2013. LNCS, vol. 8295, pp. 325–329. Springer, Heidelberg (2013)
17. Schneider, K.: Generating fast feedback in requirements elicitation. In: Proceedings of the 13th International Working Conference on Requirements Engineering: Foundation for Software Quality, pp. 160–174 (2007)
18. Lim, S.L., Finkelstein, A.: StakeRare: Using Social Networks and Collaborative Filtering for Large-Scale Requirements Elicitation. IEEE Transactions on Software Engineering 38(3), 707–735
19. Sutherland, J., Schwaber, K.: The Scrum Papers: Nuts, Bolts, and Origins of an Agile Process, http://assets.scrumfoundation.com/downloads/2/scrumpapers.pdf?1285932052 (accessed: February 16, 2014)
20. Ton, H.: A Strategy for Balancing Business Value and Story Size. In: Proceedings of AGILE 2007 Conference, pp. 279–284 (2007)
21. UserScripts, http://userscripts.org (accessed: February 16, 2014)
22. Yahoo Pipes, http://pipes.yahoo.com/pipes/ (accessed: February 16, 2014)
23. Yu, J., Benatallah, B., Casati, F., Florian, D.: Understanding mashup development. IEEE Internet Computing 12(5), 44–52 (2008)

An Empirical Study on Categorizing User Input Parameters for User Inputs Reuse

Shaohua Wang[1], Ying Zou[2], Bipin Upadhyaya[2], Iman Keivanloo[2], and Joanna Ng[3]

[1] School of Computing, Queen's University, Kingston, Ontario, Canada
`shaohua@cs.queensu.ca`
[2] Electrical and Computer Engineering, Queen's University, Kingston, Canada
`{ying.zou,bipin.upadhyaya,iman.keivanloo}@queensu.ca`
[3] CAS Research, IBM Canada Software Laboratory, Markham, Ontario, Canada
`jwng@ca.ibm.com`

Abstract. End-users often have to enter the same information to various services (*e.g.*, websites and mobile applications) repetitively. To save end-users from typing redundant information, it becomes more convenient for an end-user if the previous inputs of the end-user can be pre-filled to applications based on end-user's contexts. The existing pre-filling approaches have poor accuracy of pre-filling information, and only provide limited support of reusing user inputs within one application and propagating the inputs across different applications. The existing approaches do not distinguish parameters, however different user input parameters can have very varied natures. Some parameters should be pre-filled and some should not. In this paper, we propose an ontology model to express the common parameters and the relations among them and an approach using the ontology model to address the shortcomings of the existing pre-filling techniques. The basis of our approach is to categorize the input parameters based on their characteristics. We propose categories for user inputs parameters to explore the types of parameters suitable for pre-filling. Our empirical study shows that the proposed categories successfully cover all the parameters in a representative corpus. The proposed approach achieves an average precision of 75% and an average recall of 45% on the category identification for parameters. Compared with a baseline approach, our approach can improve the existing pre-filling approach, *i.e.*, 19% improvement on precision on average.

Keywords: User Input Parameters Categories, User Inputs Reuse, Auto-filling, Ontology, Web Forms.

1 Introduction

Web is becoming an essential part of our daily life. Various on-line services (*e.g.*, web services and mobile applications) allow end-users to conduct different web tasks, such as on-line shopping and holiday trip planning. These services require end-users to provide data to interact with them and some of the data

S. Casteleyn, G. Rossi, and M. Winckler (Eds.): ICWE 2014, LNCS 8541, pp. 21–39, 2014.

provided by end-users are usually repetitive. For example the AVIS[1], a car rental website, and the Homes [2], a real estate website, require end-users to enter their personal information such as *First Name* and *Last Name* illustrated in Figure 1 and Figure 2. It could be a cumbersome and annoying process for an end-user, especially a smartphone end-user, to fill the same information into web forms with multiple input fields repeatedly. Therefore information pre-filling becomes critical to save end-users from this cumbersome process. Rukzio et al. [1] found that end-users are four times faster on smartphones when they just have to correct pre-filled form entries compared to entering the information from scratch. A web form within web or mobile applications usually consists of a set of input fields assigned with a label, for example the input field *Contact Phone Number* in Figure 1. The label of this input field is "Contact Phone Number", the type of this input field is *text field*. Throughout this paper, we consider an input field as a user input parameter and an end-user input (*i.e.*, a piece of information) as the value which can be filled into an input field.

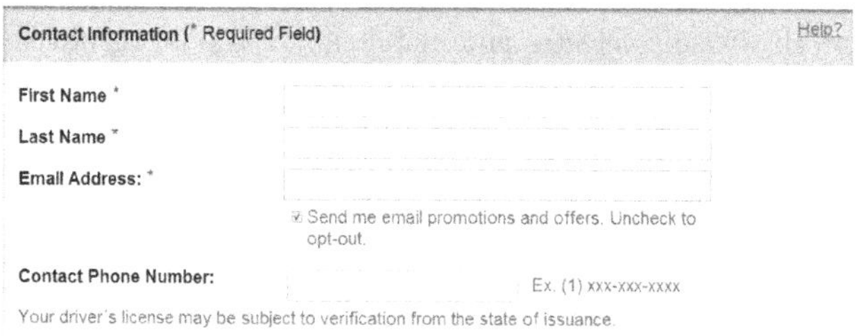

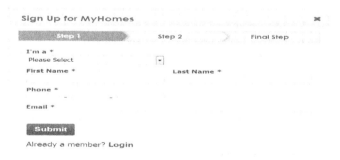

Fig. 1. A sample screen shot of a web form requiring user's personal information to reserve a car

Fig. 2. A sample screen shot of a web form requiring user's personal information to sign up the website

Recently, several industrial tools and academic approaches have been developed to pre-fill user inputs into web forms automatically. Web browsers usually

1 `http://www.avis.ca/car-rental/avisHome/home.ac`
2 `http://www.homes.com/`

provide web form Auto-filling tools, such as Firefox form auto-filling Addon [2] and Chrome form auto-filling tool [3] to help users fill in forms. In general, the auto-filling tools record the values entered by a user for specific input fields pre-configured by the tools in a given form. Then, they identify the input fields which are identical or similar to the pre-configured input fields by string matching, and fill the entered values into the identified inputs fields when the users visit websites. Since such approaches are error-prone, they allow end-users to modify the values. Recently, a few academic studies such as [4][5][6][7] proposed several approaches to help end-users. Hartman et al. [4] propose a context-aware approach using user's contextual information to fill in web forms. Toda et al. [5] propose a probabilistic approach using the information extracted from data-rich text as the input source to fill values to web forms. Wang et al. [6] propose an approach using the relations between similar input parameters to fill values to web forms. Araujo et al. [7] extract the semantic of concepts behind the web form fields and use the relations of concepts to fill in web forms. However, all these tools and approaches suffer from two main drawbacks:

- **Poor accuracy of pre-filling values to input parameters of web forms or applications**. The pre-filled information can be out of date and out of context. The user often has to modify the pre-filled values passively. With the rapid growth of the population of mobile application users, it is even more frustrating for mobile application users to modify the pre-filled values due to the restrictions in screen size and typing. Accurate pre-filling values becomes a critical step to enhance the user experience.
- **Limited support of reusing user inputs within an application and propagating them across different applications.** The existing methods can only reuse a few types of user inputs within an application or across different applications. For example an end-user is planning a holiday trip from Toronto to Miami, he or she could conduct two tasks of searching for cheap flight tickets and cheap transportation from airport to hotel. The end-user needs browse different websites to find the most appropriate solution for him or her. During the process of conducting these two tasks, the similar or identical input parameters from different websites or within a website, such as departure and return cities, should be linked together, and reuse user inputs among them. In this scenario, for example, Chrome form auto-filling tool [3] cannot help the end-users, since it can only reuse basic personal information such as credit card information and addresses.

The existing tools and approaches treat all the user input parameters equally, and the matching mechanism of existing approaches is only based on the string matching or semantic similarity calculation of field names. If a match is identified, the existing approaches pre-fill a value to an input parameter. However different user input parameters can have very varied natures. For example, the coupon number is only valid for a single purchase, therefore may not be suitable for pre-filling. The input fields for departure and destination cities from a flight booking website should not be pre-filled with previous user inputs without knowing user's context, the end-user's name can be pre-filled, since the end-user's name does not change in most cases.

It is crucial to improve the accuracy of automatic value pre-filling by understanding the characteristics of different user input parameters. In this paper, we propose an ontology model for expressing common parameters and approach based on the proposed ontology model to automatically identify a category for an input parameter. We conducted two empirical studies. The first empirical study was served as an initial empirical investigation to explore the characteristics of user input parameters. The first empirical study was conducted on 30 popular websites and 30 Android mobile applications from Google Play Android Market [3]. Based on the results of our first study, we identify and propose four categories for input parameters from web and mobile applications. The proposed categories offer a new view for understanding input parameters and provide tool developers guidelines to identify pre-fillable input parameters and the necessary conditions for pre-filling. To the best of our knowledge, we are the first to propose categories for input parameters to explore the characteristics of user input parameters. We conducted the second empirical study to verify the effectiveness of proposed categories and approach for category identification. The second study was conducted on 50 websites from three domains and 100 mobile applications from five different categories in Google Play Android Market.

The major contributions of our paper are listed as follows:

- We propose four categories of user input parameters to capture the nature of different user input parameters through an empirical investigation. For pre-filling, each category of parameters should be collected, analyzed and reused differently. The results of our empirical study show that our categories are effective to cover all the input parameters in a representative corpus.
- We propose an ontology model to express the common parameters and the relations among them. Moreover, we use the proposed ontology model to carry the category information for input parameters. We propose a WordNet-based approach that automatically updates the core ontology for unseen parameters from new applications. The results of our empirical study show that our approach obtains a precision of 88% and a recall of 64% on updating the ontology to include the unseen parameters on average.
- We propose an ontology-based approach to identify a category for input parameters automatically. On average, our approach for category identification can achieve a precision of 90.5% and a recall of 72.5% on the identification of a category for parameters on average.
- We test the effectiveness of our proposed categories on improving the existing approaches. We build a baseline approach which does not distinguish the different characteristics of input parameters, and incorporate our proposed categories with the baseline approach to form a category-enabled approach. We compare two approaches through an empirical experiment. The results show that our approach can improve the baseline approach significantly, *i.e.*, on average 19% in terms of precision.

The rest of the paper is organized as follows. Section2 describes the user interfaces of web and mobile applications. Section 3 presents our proposed approach

---

[3] https://play.google.com/store?hl=en

for categorizing input parameters. Section 4 introduces the empirical studies. Section 5 discusses the threats to validity. Section 6 summarizes the related literature. Finally, section 7 concludes the paper and outlines some avenues for future work.

2   Web and Mobile Application User Interface

In this paper, we study the user input parameters from the user interfaces of web applications and mobile applications.

Web pages are mainly built with HTML and Cascading Style Sheets (CSS), and can be converted into an HTML DOM tree [4]. A Web form is defined by an HTML FORM tag <form> and the closing tag </form>. A web form usually consists of a set of input elements, such as the text fields in Figure 2, to capture user information. An input element is defined by an HTML INPUT tag <input> specifying an input field where the user can enter data. The input element can contain several attributes, such as *name* specifying the name of an <input> element, *type* defining the type <input> to display (*e.g.*, displayed as a button or checkbox) and *value* stating the value of an <input> element. An <input> element can be associated with a human-readable label, such as First Name. This information can be accessible by parsing HTML source code.

There are three types of mobile applications:

- *Native Apps:* The native apps are developed specifically for one platform such as Android[5], and can access all the device features such as camera.
- *Web Apps:* The web apps are developed using standard Web technologies such as Javascript. They are really websites having *look and feel* like native apps. They are accessed through a web browser on mobile devices.
- *Hybrid Apps:* The hybrid apps are developed by embedding HTML5 apps inside a thin native container.

An Android application is encapsulated as an Android application package file (APK) [6]. An APK file can be decoded into a nearly original form which contains resources such as source code (*e.g.*, Java) and user interface layout templates in XML files that define a user interface page or a user interface component if it is not a HTML5 application. In this study, we only study the native mobile applications because the user interface templates in XML files can be obtained by decoding APK files.

3   Our Proposed Approach for Categorizing User Input Parameters

In this section, we first present an ontology model for expressing common user input parameters and their relations. Second, we propose an automatic approach for updating the ontology. Third, we propose an ontology-based approach for categorizing input parameters.

---

[4] http://www.w3schools.com/htmldom/dom_nodes.asp
[5] http://www.android.com/
[6] http://en.wikipedia.org/wiki/APK_(file_format)

3.1   An Ontology Definition Model

In this study, we build an ontology to capture the common user input parameters and the five relations among parameters. Figure 3 illustrates the main components of ontology definition model and their relations.

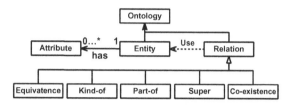

Fig. 3. Components of ontology definition model

Fig. 4. An example of 4 user inputs parameters: City, Zip Code, Home and Cell

The components are listed as follows:

- *Entity:* is an input parameter from a website or mobile application, or a concept description of a group of resources with similar characteristics.
- *Attribute:* is a property that a parameter can have. There are four attributes:
  - *Category.* The category defines the category information of a parameter.
  - *Label.* The category stores the *Label* (*i.e.*, a human-readable description) of an input field from websites or mobile applications.
  - *Coding Information.* The category stores the HTML coding information (*i.e.*, websites) or XML templates (*i.e.*, mobile applications).
  - *Concepts.* The category stores the concepts related to the parameter.
- *Relation:* defines various ways that entities can be related to one another. The five relations are listed as follows:
  - *Equivalence.* Two entities have the same concept such as Zip Code and Postal Code.
  - *Kind-of.* One entity is a kind-of another one. For example, Home (*i.e.*, home phone) is a kind of Telephone Details illustrated in Figure 4.
  - *Part-of.* One entity is a part-of another one, such as Zip Code and Address Details illustrated in Figure 4.
  - *Super.* One entity is a super of another. The super relation is the inverse of the Kind-of relation. For example, Telephone Details is a super of Home (*i.e.*, home phone) illustrated in Figure 4.

- *Co-existence.* Two entities are both a part-of an entity and they are not in the relation of equivalence, such as City and Zip Code illustrated in Figure 4.

Usually the user input parameters are terminal concepts [8], such as *Phone Number* and *Price*, which have no sub-concepts. During the process of ontology creation, we use the UI structure and semantic meanings of the parameters to identify the relations. If two input elements have the same parent node in an HTML DOM tree, their relation is co-existence, and the relation between the two input parameters and their parent HTML DOM node is part-of. For example, the user input parameters *City* and *Zip Code* co-exist and they have a part-of relation with Address Details in Figure 4. Figure 5 shows the visualization of the ontology that is built based on the example in Figure 4.

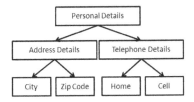

Fig. 5. An example ontology of personal details containing four user inputs parameters: City, Zip Code, Home and Cell

3.2   Our Approach of Updating Ontology

Once the initial ontology based on the proposed ontology model (Section 3.1) is established, an automatic approach for updating the ontology is required to add a new parameter into the ontology. We use WordNet [9], a large lexical database for English and containing semantic relations between words, to identify the relations between the new parameter and the existing ones in the ontology. The following relations of words defined in WordNet are used to identify our 5 relations:

1. If two words have the same synsets, they have a same semantic meaning, we convert it to Equivalence relation.
2. **Hypernym** shows a kind-of relation. For example, car is a hypernym of vehicle. We convert it to a kind-of relation.
3. **Meronym** represents a part-whole relation. We convert it to a part-of relation.
4. **Hyponym** defines that a word is a super name of another. We convert it to super relation. Hyponym is the inverse of hypernym meaning that a concept is a super name of another. For example, vehicle is a hyponym of car.
5. We use the part-of relation to identify the co-existence relation. If a word with another word both have a part-of relation with a same word, this word and the other word have co-existence relation.

3.3   Our Approach for Category Identification

It is important for a pre-filling approach to know the category of a user input parameter automatically. In this section, we introduce our ontology-based approach for identifying a category of an input parameter in details. Our approach uses the proposed ontology definition model proposed in Section 3.1. Our approach uses two strategies which are listed as follows:

- *Concept-based Strategy.* This strategy relies on an ontology of user input parameters to identify a category for a user input parameter automatically. If two input parameters have the same concepts, these two input parameters belong to the same category.
- *UI-based Strategy.* This strategy relies on the design of the UI layouts. We consider two user input parameters are the nearest neighbours to each other if their input fields are contained in the same parent UI component. Figure 4 shows an example of 4 user input parameters: *City, Zip Code, Home and Cell.* The *City and Zip Code* are the nearest neighbours to each other since they have the same parent UI node which has a label *Address Details.* We assume that if all the neighbours of a user input have been categorized and belong to the same category, there exists a high chance that they belong to the same category.

Given an ontology having a set of parameters, which is denoted as $O = < P_1^O, \ldots, P_n^O >$, where n is the number of parameters, and an input parameter P, our approach uses the following steps to identify a category for P:

- Step 1. We use WordNet synsets to retrieve synonyms of the concepts of the parameter P to expand the concept pool (*i.e.*, a bag of words) of P and the concept pool of each P_i^O, where $1 \leq i \leq n$.
- Step 2. We identify the P_j^O (where $1 \leq j \leq n$) whose concept pool has the concepts which are the synonyms of (or identical to) any concepts in the concept pool of P (*i.e.*, having the same semantic meaning).
- Step 3. If multiple parameters in the ontology are identified for P in Step 2, we choose the parameter sharing the most common concepts with the concept pool of P as the identical parameter to P. We assign the category of the chosen parameter to P
- Step 4. If no parameter is identified in Step 2, we apply the UI-strategy on the given ontology O and input parameter P. We identify the neighbors of P from its UI and repeat Step 1-3 to identify a category for every neighbor. If any neighbor of P cannot be categorized (*i.e.*, no parameters in the ontology have the same concepts as the neighbor does) or the neighbors of P have different categories, we cannot categorize P. If all the neighbors of P belong to a same category, we assign this category to input parameter P.

4   Empirical Study

In this study, we conduct two studies on different datasets. The first study is designed to study the different characteristics of parameters and propose categories for input parameters. The second empirical study is designed to evaluate the effectiveness of the proposed categories and the approach for categorizing input parameters.

4.1   First Empirical Study

We conduct an initial study to understand the nature of user input parameters and categorize the parameters. Understanding the different characteristics of parameters of different categories can help analyze, process and pre-fill the parameters differently to improve the accuracy of pre-filling. In this section, we introduce the study setup, the analysis method and the findings of our empirical investigation.

4.1.1   Data Collection and Processing

The study is conducted on 30 websites (*i.e.*, 15 shopping websites and 15 travel planning websites) and 30 mobile applications (*i.e.*, 15 shopping apps and 15 travel apps). We collect the user input parameters from web and mobile applications in different ways.

```
<input type="text" name="resForm.firstName.value" value
id="firstName" class="txtSize pickupInput" size="30"
maxlength="12">
```

Fig. 6. A sample screen shot of source code of an input field

Collecting input parameters from websites: First, we manually visit the website to bypass the login step. Second, we use *Inspect Element*[7] of Google Chrome[8] to collect the following information of a user input parameter:

- **Label:** The value of the label (*i.e.*, the descriptive text shown to users).
- **Attributes:** The values of the attributes of the input element such as id and name. For example, Figure 6 shows the source code of the input field First Name in Figure 1.

Collecting input parameters from mobile applications: Instead of running each mobile application, we use Android-apktool [9], a tool for reverse engineering Android APK files, to decode resources in APKs to readable format. We build a tool to extract the information of a form from UI layout XML files if exist, then collect the information related to a user input parameter in the same way as we collect from websites.

Data cleaning: The extracted information need to be cleaned for further processing. For example, the extracted information for the input parameter *First Name* in Figure 1 is {First Name, resForm.firstName.value, FirstName, text} need to be cleaned. We remove the duplicated phrases and programming expressions. For the given example above, the output after cleaning is {First Name, res, Form, value, text}.

---

[7] https://developers.google.com/chrome-developer-tools/docs/elements
[8] https://www.google.com/intl/en/chrome/browser/
[9] https://code.google.com/p/android-apktool/

Data processing: We manually process the information of parameters as follows: First, we identify the concepts from the information of a user input parameter. A concept is a semantic notion or a keyword for describing a subject (*e.g.,* "taxi" and "city"). Second, the parameters having the same concepts or same semantical meanings are merged and considered as one parameter which we save all the unique concepts for. For example, two input parameters from two different flight searching websites, one with the label *Departure* and the other one with the label *Leave* should be merged and considered as one parameter having two concept words *Departure* and *Leave.*

4.1.2 Results of First Empirical Study

We collected 76 and 32 unique user input parameters from websites and mobile applications respectively. Due to the fact that the user input parameters are repeated among different applications, we observe that the number of unique parameters identified decreases with the increase of the number of applications studied. The 76 parameters extracted from the websites (*i.e.,* shopping and travel) contain all the 32 parameters from mobile applications from the same domains. This is due to the fact that mobile applications usually are the simplified versions of corresponding websites.

After studied the user input parameters, we found that input parameters can be categorized into four categories. The four categories are listed as follows:

– *Volatile Parameters:* This type of parameters have no change of state. They can be further categorized into two sub-categories:

  • One-Time Parameters: The values of this type of parameters are valid for one interaction, such as coupon number. This type of parameters should not be used for pre-filling at all.

  • Service-Specific Parameters: The values of this type of parameters can only be used for a specific service (*e.g.,* a website or a mobile application). For example, Member ID of the BESTBUY Reward Zone[10] in Canada can only be used for "Sign In" function or "Account Set Up" function of reward service. When end-users receive a membership card, they need enter the Member ID to set up an account in BESTBUY Reward Zone. This type of parameters can be pre-filled, however the parameters cannot be reused by different services.

– *Short-time Parameters:* The values of this type of parameters change with high frequency. For example during the course of conducting an on-line clothes shopping, an end-user may use the different colors as the search criteria to find the best suitable clothes from different services for the user, during a short period of time, the value of the color can be pre-filled. This type of parameters can be pre-filled and reused by different services, however it is extremely hard to pre-fill this type of parameters unless some conditions are met. For example, in the above example of searching clothes, the value of the color should be pre-filled only if the end-user has not switched to a new task.

---

[10] https://www.bestbuyrewardzone.ca/

- *Persistent Parameters:* The values of this type of parameters do not change over a long period of time. For example, the gender of the end-user and permanent home address. This type of parameters are suitable for pre-filling and being reused across different services.
- *Context-aware Parameters:* There exist some user input parameters which are context-dependent, such as the user's location and current local time. This type of parameters can be pre-filled based on user's contextual information. The value of a context-aware parameter can be obtained from two types of data sources:
  - *Direct Data Sources.* The context data is directly available and accessible with little computation, such as entries in Google calender, time from mobile phone clock, "my" To-do lists from task manager and a friend's preferences on Facebook.
  - *Analyzed Data Sources.* With the accumulation of user history, additional contextual data can be mined from history data such as user behaviors. For example, a user always books a round-trip business class flight ticket from Toronto to Los Angeles for business trips and a round-trip economic class flight ticket from Toronto to Vancouver for personal trips.

In this paper, we consider only the user input parameters whose values can be obtained directly from available sources.

4.2   Second Empirical Study

The goals of the second empirical study are to: 1) examine the representativeness of the proposed categories of parameters; 2) validate the effectiveness of the approach of updating ontology; 3) test the effectiveness of the proposed approach for category identification; 4) validate the effectiveness of our proposed categories on improving existing approaches of pre-filling values to web forms.

4.2.1   Data Collection and Processing

We conduct our second empirical study on 45 websites from three different domains: Finance, Healthcare and Sports (*i.e.*, 15 websites from each domain) and 100 Android mobile Applications from five different categories: Cards& Casino, Personalization, Photography, Social and Weather in Google Play Market (*i.e.*, 20 applications from each studied category). The studied websites and mobile applications are different from the ones studied in our first empirical study. We use the same approach as discussed in section 4.1.1 to collect user input parameters from web and mobile applications. In total, there are 146 and 68 unique user input parameters from web and mobile applications respectively.

4.2.2   Research Questions

We presents four research questions. For each research question, we discuss the motivation of the research question, analysis approach and the corresponding findings.

RQ1. Are the proposed categories of parameters sufficient to cover the different types of user input parameters?

Motivation. The existing tools and approaches do not distinguish the characteristics of user input parameters. The parameters are processed equally. Therefore the lack of knowledge on the user input parameters negatively impacts the accuracy of pre-filling approaches, and is the main reason for limited support of reusing parameters within one application or across multiple applications. Categorizing the users input parameters can help us in understanding the different characteristics of parameters. Each category of parameters has its unique characteristics which need to be fully understood.

In Section 4.1, we identified 4 categories for user input parameters via a manual empirical study. In this research question, we validate the proposed categories to see if they can explain the further unseen user input parameters collected in the empirical study.

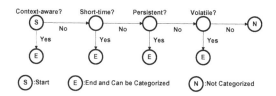

Fig. 7. The decision tree for the judgment process

Analysis Approach. To answer this question, we study the user input parameters collected from the 45 websites and 100 mobile apps. For each user input parameter, we manually process it to see whether a user input parameter belongs to any category or not. The judgment process is based on a decision tree as shown in Figure 7. When we process a parameter, we first decide whether it is a context-aware parameter or not. If no, we further decide whether the parameter can be categorized as a short-time parameter or not. If no, then we decide whether the parameter is a persistent parameter or not. If no, finally we decide whether the parameter is a volatile parameter or not. If it is still a no, the parameter cannot be categorized by using our proposed categories of input parameters.

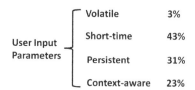

Fig. 8. The percentage of each category

Results. Figure 8 shows that all of the unseen user inputs parameters can be categorized into our proposed categories and there is no "not Categorized". More specifically, 43% of the parameters are short-time and only 3% of the parameters are volatile parameters. The results suggest that our categories are enough to describe all the unique user input parameters in our empirical study.

RQ2. Is the proposed approach for updating ontology effective?

Motivation. Usually the ontologies are created manually by web filling tool builders or contributors from the communities of knowledge sharing. It is helpful to have an automatic approach to update and expand the ontologies. We test the effectiveness of the WordNet-based approach of updating ontology proposed in Section 3.2.

Analysis Approach. We conduct two experiments to answer this question. In the first experiment, our approach for updating ontology uses the input parameters from one domain of applications to update the ontology to include the parameters from other domains of applications. In the second experiment, our approach for updating ontology uses the parameters from one domain of applications to update the ontology to include the parameters from the same domain of applications.

To conduct *experiment 1*, we construct an ontology[11], denoted as $O^{Initial}$, using the 76 user inputs parameters from the first empirical study (Section 4.1). Second, we use the ontology $O^{Initial}$ as the existing ontology to include the parameters of 146 parameters from web applications and 68 parameters from mobile applications (Section 4.2.1). Third, we compute the precision and the recall of the approach using Equation (1) and Equation (2) respectively. The precision metric measures the fraction of retrieved user input parameters that are placed in the correct place (*i.e.*, having the right relations with other parameters), while the recall value measures the fraction of correct user input parameters that are retrieved.

$$precision = \frac{|\{correct\ Parameters\} \bigcap \{retrieved\ Parameters\}|}{|\{retrieved\ Parameters\}|} \quad (1)$$

$$recall = \frac{|\{correct\ Paras\} \bigcap \{retrieved\ Paras\}|}{|\{correct\ Paras\}|} \quad (2)$$

The goal of *experiment 2* is to see whether we can obtain a better result if we build a domain dependent ontology for each domain in our empirical study. To conduct *experiment 2*, we conduct the following steps on the 45 websites:

1. randomly select 5 of 15 Finance websites and construct the domain specific ontology for the user input parameters in Finance using the user input parameters from the selected websites.
2. use the domain dependent ontology of parameters in Finance as an existing ontology and perform a manual updating on the existing ontology to include the user input parameters from the rest of 10 Finance websites.

---

[11] https://dl.dropboxusercontent.com/u/42298856/icwe2014.html

3. apply the automatic ontology updating approach on the parameters from the rest of 10 Finance websites to verify its performance.

We replicate the previous steps on 15 Health Care websites and 15 Sports websites. We conduct our analysis on mobile applications in the same way as we do on websites, the only difference is that we randomly select 5 mobile applications from each domain. We compute the precision and the recall of the approach using Equation (1) and Equation (2) respectively.

Table 1. Results of our approach for updating ontology using domain-specific ontology

Type	Domain	Precision(%)	Recall(%)
website	Finance	83	66
website	Health Care	78	69
website	Sports	95	65
mobile	Cards&Casino	92	74
mobile	Personalization	87	48
mobile	Photography	90	62
mobile	Social	85	42
mobile	Weather	97	85

Results. In the experiment 1, our approach for updating existing ontology obtains a precision of 74% and a recall of 42% on websites, and a precision of 82% and a recall of 30% on mobile applications. We further investigate our results to gain further insights regarding the low recalls. We found that 35 of 146 website user input parameters and 25 of 68 mobile applications user input parameters can be found in the ontology constructed using the input parameters collected during the initial empirical investigation. The low recalls indicate that it is hard to use one general ontology as the existing ontology to include the parameters from other domains automatically, although the precisions are considerably high.

In experiment 2, Table 1 shows that our approach for updating ontology obtain a precision of 85% and a recall of 67% on websites, and a precision of 90% and a recall of 62% on mobile applications. The results of our approach in **experiment 1** can be improved by using the domain-specific ontology as the existing ontology. On average, the approach can be improved by 11% in terms of precision and 25% in terms of recall on websites, and 8% in terms of precision and 32% in terms of recall on mobile applications.

RQ3. Is our approach of category identification effective?

Motivation. After showing the representativeness of our categories for input parameters, we propose an ontology-based approach to identify a category for an input parameter. In this section, we investigate the effectiveness of our approach for identifying a category for an input parameter.

Analysis Approach. To assess the performance of the proposed approach, we proceed as follows:

1. We use the domain-specific ontologies constructed in RQ2 as the training ontologies.
2. We categorize the parameters in ontologies by assigning a category to each parameter.
3. We locate the nearest neighbours for the input parameters which are not in the training ontologies.
4. We apply our approach on the parameters in Step 3

We compute the precision and the recall of the approaches using Equation (1) and Equation (2) respectively. The precision metric measures the fraction of retrieved user input parameter that are categorized correctly, while the recall value measures the fraction of correct user input parameters that are retrieved.

Table 2. Results of our approach of category identification

Type	Domain	Precision(%)	Recall(%)
website	Finance	90	79
website	Health Care	92	82
website	Sports	89	70
mobile	Cards&Casino	90	78
mobile	Personalization	91	58
mobile	Photography	95	62
mobile	Social	77	55
mobile	Weather	100	88

Results. Table 2 shows the precision and recall values of our approach to identify a category for input parameters. On average, our approach can achieve a precision of 90% and a recall of 77% on websites, and a precision of 91% and a recall of 68% on mobile applications. Our approach works well on the input parameters from mobile applications in the domain of weather, because usually the weather mobile apps relatively have fewer number of input parameters and very similar functionalities.

RQ4. Can the proposed categories improve the performance of pre-filling?

Motivation. The existing pre-filling approaches (*e.g.*, [7][14]) treat all the input parameters equally. They do not take into consideration the characteristics of input parameters. These approaches are usually based on the string matching or semantic similarity calculation between the user inputs and the labels of input parameters (*e.g.*, input fields). Essentially they fill a value to an input field as long as a match between a user input and an input field is identified, however the value required by the input parameter may not be suitable for pre-filling due to curtain conditions as mentioned in Section 4.1.2. In this research question, we evaluate the effectiveness of our categories on improving the existing techniques of reusing user inputs.

Analysis Approach. To answer this question, we build a baseline approach which adopts the approach in [7]. The baseline approach [7] pre-fills values to input parameters based on the semantic similarity between user inputs and textual information (*e.g.*, words mined from labels and the attributes of HTML DOM elements defining the input parameters in the user interface) of input parameters. The baseline approach calculates the similarity between user inputs and labels of input parameters using WordNet [9]. The stopword removal, word stemming and non-English words removal are conducted on the textual information of input parameters to identify meaningful words. Then, WordNet is used to expand each word with synsets (*i.e.*, a set of words or collation synonyms) to retrieve synonyms of the found terms.

We build our approach by enriching the baseline approach with the proposed categories for input parameters (Baseline + Categories). We evaluate the improvement of applying our categories with the baseline approach. Our approach use the ontology-based approach proposed in Section 3.3 to identify a category for an input parameter to see whether the input parameter is suitable for pre-filling or not. If the input parameter is suitable for being pre-filled, our approach pre-fills an input parameter. We compute precision using Equation (1) and recall using Equation (2) to measure the improvement. The precision metric measures the fraction of retrieved user input parameters that are pre-filled correctly, while the recall value measures the fraction of correct user input parameters that are retrieved.

To pre-fill input parameters using the user previous inputs, we collect user inputs through our input collector [6] which modifies the Sahi[12] tool to track the user's Web activities. The first author of this paper used the tool to track his inputs on web forms from three domains: Finance, Health and Sports. We apply both the baseline approach and our approach on the 45 websites.

Table 3. Results of the evaluations of our categories on improving the baseline approach of filling input parameters

Domain	Baseline		Baseline+Categories	
	Precision(%)	Recall(%)	Precision(%)	Recall(%)
Finance	50	34	64	36
Health	41	22	67	30
Sports	36	16	53	20

Results. Table 3 shows that our approach incorporating categories of input parameters can improve the baseline approach on average 19% in terms of precision. We further inspect the results and found that our approach can reduce the number of wrong filled values compared with the baseline approach. Some of

---

[12] http://sahi.co.in/

the wrong filled values should not be pre-filled to the input parameters in the fist place.

5   Threats to Validity

This section discusses the threats to validity of our study following the guidelines for case study research [10].

Construct validity threats concern the relation between theory and observation. In this study, the construct validity threats are mainly from the human judgment in categorizing the parameters and ontology construction. We set guidelines before we conduct manual study and we paid attention not to violate any guidelines to avoid the big fluctuation of results with the change of the experiment conductor.

Reliability validity threats concern the possibility of replicating this study. We attempt to provide all the necessary details to replicate our study. The websites and mobile apps we used are publicly accessible[13].

6   Related Work

In this section, we summarize the related work on form auto-filling approaches.

Several industrial tools have been developed to help users fill in the Web forms. Web browsing softwares provide Web form Auto-filling tools (*e.g.*, Firefox form auto-filling addon [2] and Chrome form auto-filling tool [3]) to help users fill in forms. RoboForm [11] is specialized in password management and provides form auto-filling function. Lastpass [12] is an on-line password manager and form filler. 1Password [13] is a password manager integrating directly into web browsers to automatically log the user into websites and fill in forms. These three tools store user's information in central space, and automatically fills in the fields with the saved credentials once the user revisit the page.

Some studies (*e.g.*,Winckler et al. [14] Bownik et al. [15] Wang et al. [16]) explore the use of semantic web for developing data binding schemas. The data binding schemas are essential techniques helping connect user interface elements with data objects of applications. This technology needs an ontology to perform the data integration. Instead of focusing on custom ontology, some binding schemas rely on the emergence of open standard data types, such as Microformats [17] and Micodata [18]. Winckler et al. [14] explore the effectiveness of the data schemas and the interaction techniques supporting the data exchange between personal information and web forms. Wang et al. [6] propose an intelligent framework to identify similar user input fields among different web applications by clustering them into different semantic groups. Araujo et. al [7] propose a concept-based approach, using the semantic of concepts behind the web form fields, for automatically filling out web forms. Some studies (*e.g.*, [19]) require apriori ([20]) tagging of websites, or a manually crafted list that includes the labels or names of input element to describe a semantic concept. These approaches

---

[13] https://dl.dropboxusercontent.com/u/42298856/icwe2014.html

can only be applicable to a specific domain or need explicit advice from the user. Hartman and Muhlhauser [4] present a novel mapping process for matching contextual data with UI element. Their method can deal with dynamic contextual information like calendar entries.

All the above approaches do not identify the variety of input parameters and treat input parameters equally by attempting to fill them into input parameters. They do not take into consideration the meaning of input parameters. In this study, we study the nature of input parameters and try to understand them from the end-user's point of view, not just by the similarity between the user inputs and the input parameters. Our study shows that taking into consideration that characteristics of input parameters via the identified categories significantly improves the performance of pre-filling.

7   Conclusion and Future Work

Reusing user inputs efficiently and accurately is critical to save end-users from repetitive data entry tasks. In this paper, we study the distinct characteristics of user input parameters through an empirical investigation. We propose four categories, *Volatile, Short-time, Persistent and Context-aware*, for input parameters. The proposed categories help pre-filling tool builders understand that which type of parameters should be pre-filled and which ones should not.

In this paper, we propose an ontology model to express the common parameters and the relations among them. In addition, we propose a WordNet-based approach to update the ontology to include the unseen parameters automatically. We also propose an approach to categorize user input parameters automatically. Our approach for category identification uses the proposed ontology model to carry the category information.

Through an empirical study, the results show that our proposed categories are effective to explain the unseen parameters. Our approach for updating ontology obtains a precision of 88% and a recall of 64% on updating the ontology to include unseen parameters on average. On average, our approach of category identification achieves a precision of 90.5% and a recall of 72.5% on the identification of a category for parameters on average. Moreover, the empirical results show that our categories can improve the precision of a baseline approach which does not distinguish different characteristics of parameters by 19%. Our proposed categories of input parameters can be the guidelines for pre-filling tool builders and our ontologies can be consumed by existing tools.

In the future, we plan to expand our definition of categories for input parameters by considering the scenarios of pre-filling web forms by multiple end-users. We plan to recruit end-users to conduct user case study on our approach.

Acknowledgments. We would like to thank all the IBM researchers at IBM Toronto Laboratory CAS research for their valuable feedback to this research. This research is partially supported by IBM Canada Centers for Advance Studies.

References

1. Rukzio, E., Noda, C., De Luca, A., Hamard, J., Coskun, F.: Automatic form filling on mobile devices. Pervasive Mobile Computing 4(2), 161–181 (2008)
2. Mozilla Firefox Add-on Autofill Forms, `https://addons.mozilla.org/en-US/firefox/addon/autofill-forms/?src=ss` (last accessed on February 4, 2014)
3. Google Chrome Autofill forms,
   `https://support.google.com/chrome/answer/142893?hl=en`
   (last accesed on February 4, 2014)
4. Hartman, M., Muhlhauser, M.: Context-Aware Form Filling for Web Applications. In: IEEE International Conference on Semantic Computing, ICSC 2009, pp. 221-228 (2009)
5. Toda, G., Cortez, E., Silva, A., Moura, E.: A Probabilistic Approach for Automatically Filling Form-Based Web Interfaces. In: The 37th International Conference on Very Large Data Base, Seattle, Washington, August 29-September 3 (2011)
6. Wang, S., Zou, Y., Upadhyaya, B., Ng, J.: An Intelligent Framework for Auto-filling Web Forms from Different Web Applications. In: 1st International Workshop on Personalized Web Tasking, Co-located with IEEE 20th ICWS, Santa Clara Marriott, California, USA, June 27 (2013)
7. Araujo, S., Gao, Q., Leonardi, E., Houben, G.-J.: Carbon: domain-independent automatic web form filling. In: Benatallah, B., Casati, F., Kappel, G., Rossi, G. (eds.) ICWE 2010. LNCS, vol. 6189, pp. 292–306. Springer, Heidelberg (2010)
8. Xiao, H., Zou, Y., Tang, R., Ng, J., Nigul, L.: An Automatic Approach for Ontology-Driven Service Composition. In: Proc. IEEE International Conference on Service-Oriented Computing and Applications, Taipei, Taiwan, December 14-15 (2009)
9. WordNet, `http://wordnet.princeton.edu/` (last accessed on March 25, 2013)
10. Yin, R.K.: Case Study Research: Design and Methods, 3rd edn. SAGE Publications (2002)
11. RoboForm, `http://www.roboform.com/` (last accessed on March 25, 2013)
12. LastPass, `http://www.lastpass.com/` (last accessed on March 25, 2013)
13. 1Password, `https://agilebits.com/` (last accessed on March 25, 2013)
14. Winckler, M., Gaits, V., Vo, D., Firmenich, S., Rossi, G.: An Approach and Tool Support for Assisting Users to Fill-in Web Forms with Personal Information. In: Proceedings of the 29th ACM International Conference on Design of Communication, SIGDOC 2011, October 3-5, pp. 195–202 (2011)
15. Bownik, L., Gorka, W., Piasecki, A.: Assisted Form Filling. Engineering the Computer Science and IT, vol. 4. InTech (October 2009) ISBN 978-953-307-012-4
16. Wang, Y., Peng, T., Zuo, W., Li, R.: Automatic Filling Forms of Deep Web Entries Based on Ontology. In: Proceedings of the 2009, International Conference on Web Information Systems and Mining (WISM 2009), Washington, DC, USA, pp. 376–380 (2009)
17. Khare, R.: Microformats: The Next (Small) Thing on the Semantic Web? IEEE Internet Computing 10(1), 68–75 (2006)
18. Hickson, I.: HTML Microdata, `http://www.w3.org/TR/microdata` (last accessed on March 25, 2013)
19. Stylos, J., Myers, B.A., Faulring, A.: Citrine: providing intelligent copy-and-paste. In: Proceedings of UIST, pp. 185–188 (2004)
20. Apriori, `http://en.wikipedia.org/wiki/Apriori_algorithm` (last accessed on March 25, 2013)

Analysis and Evaluation of Web Application Performance Enhancement Techniques

Igor Jugo[1], Dragutin Kermek[2], and Ana Meštrović[1]

[1] Department of Informatics, University of Rijeka,
Radmile Matejčić 2, 51000 Rijeka, Croatia
{ijugo,amestrovic}@inf.uniri.hr
http://www.inf.uniri.hr
[2] Faculty of Organization and Informatics, University of Zagreb,
Pavlinska 2, 42000 Varazdin, Croatia
dkermek@foi.hr
http://www.foi.unizg.hr

Abstract. Performance is one of the key factors of web application success. Nowadays, users expect constant availability and immediate response following their actions. To meet those expectations, many new performance enhancement techniques have been created. We have identified almost twenty such techniques with various levels of implementation complexity. Each technique enhances one or more tiers of the application. Our goal was to measure the efficiency and effectiveness of such techniques when applied to finished products (we used three popular open source applications). We argue that it is possible to significantly enhance the performance of web applications by using even a small set of performance enhancement techniques. In this paper we analyse these techniques, describe our approach to testing and measuring their performance and present our results. Finally, we calculate the overall efficiency of each technique using weights given to each of the measured performance indicators, including the technique implementation time.

Keywords: Web application, performance, enhancement, techniques.

1 Introduction

Web applications (WAs) have become ubiquitous allowing anyone, even with only basic IT knowledge, to start an online business using a free and open sourced WA or a commercial one. There are three basic factors that have led to great importance of their performance today. First, the spread of broadband Internet connections has changed visitors expectations and tolerance to waiting for the application to respond, lowering the expected time to 1 or 2 seconds. The second factor is the increasing workload generated by the constantly growing number of Internet users. The third factor is the new usage paradigm (user content creation = write-intensive applications) that has been put forth by Web 2.0. All this has increased the pressure on performance of web applications. Our research had the following objectives: a) make a systematic overview of various techniques

S. Casteleyn, G. Rossi, and M. Winckler (Eds.): ICWE 2014, LNCS 8541, pp. 40–56, 2014.

for enhancing WA performance, b) experimentally test, measure and evaluate the effectiveness of each technique, and c) measure the effect of these techniques on WA quality characteristics. The hypotheses we set were: a) it is possible to significantly increase application performance on the same hardware basis, independently of its category (type), by using even a small subset of performance enhancement techniques and b) implementation of performance enhancement techniques has a positive effect on the quality characteristics of such applications (efficiency, availability, reliability). The first hypothesis will be confirmed by achieving a 30% or more increase in throughput, while keeping the 90% of response times under 2 seconds and the average CPU usage under 70%. The second hypothesis will be confirmed by achieving a 10% or more decrease in average CPU usage, while still enhancing the throughput for at least 30%. In order to confirm our hypotheses experimentally, we selected three well known open source applications of different categories. The applications that we selected were: a) Joomla content management system (marked APP1 in this paper), b) PhpBB online community system (APP2) and c) OsCommerce e-commerce system (APP3). We have selected these applications because they have been under development for a long period of time; they are the solution of choice for many successful web companies and are used by millions of users every day.

The paper is organized into six sections. Section Two describes the theoretical foundation of this research and situates the work in the area. Section Three presents some motivations for performance enhancement of web applications, possible approaches and an overview of our analysis of performance enhancement techniques. In Section Four we present our experiment and discuss the results. In Section Five we calculate the effectiveness of each technique which suggest implications for the problem of performance enhancement technique selection in relation to the type and workload of WA whose performance we are trying to enhance. Finally, Section Six draws conclusions and suggests further work.

2   Background

There are three basic approaches to enhancing performance of WAs. While caching and prefetching are well known from other areas of computing, response time and size minimization is a relatively new approach developed by some of the most visited applications on the World Wide Web (Web) today. In this section we will point out some of the most important work within these approaches. Caching is one of the oldest methods of performance enhancement used in various information systems. In the area of web application development, caching has been analyzed in [18], [14], [5], [2] and [24], while various caching strategies have been analyzed in [25]. Another expanding area of performance enhancement is prefetching, i.e. preparing and/or sending data that is most likely to be requested after the last request. Domenech analyzed the performance of various prefetching algorithms in [8] and created prefetching algorithms with better prediction results in [7]. One of the crucial elements of prefetching is the prediction on which content will be requested next. Prediction is usually based on the analysis and modeling of user behavior as described [9], or by extracting usage patterns from web server logs, which has been done in [16] and in [11]. Adding a time component to better determine the optimal time for prefetching web objects was suggested in [15]. The latest approach is response time and size minimization. This approach is based on the experiences and methods developed by professionals that constructed some of the most popular applications on the Internet today, such as like Yahoo.com [26] and Flickr.com [10]. As said earlier Web 2.0 and AJAX have caused a paradigm shift in WA usage, which also brought on a change in the nature of WA workload. This was analyzed in [21],[20] and in [19]. Throughout 2009 and 2010, authors have studied the possibilities for the overall application performance enhancement [22], [6]. In this paper we analyze techniques based on all the mentioned approaches and measure their effect on the performance of WAs.

3   Performance of Web Applications

The quality, and with that, the performance of WAs is primarily defined by the quality of its architecture. Badly planned elements of application architecture can limit or completely block (by becoming a bottleneck) the expected or needed levels of performance. With the new focus on the importance of application performance, production teams have begun to implement performance risk management in all phases of its lifecycle. According to the Aberdeen group research [1], companies that have adopted performance management in the form of the "Pressures, Actions, Capabilities, Enablers" (PACE) model have seen a 106% increase in availability, 11.4 times better response times and 85% problem solving rate before the problem becomes obvious to their users. However, there are already many WAs used worldwide, and with the rise in the number of users, their performance has to be enhanced to meet the desired service levels. The motivation for performance enhancement usually comes from three main directions: a) the desire to increase the number of users or profit, b) expected and planned peak workload periods and c) performance enhancement based on a business plan, or inspired by a real or expected (projected) performance problem with a goal of ensuring availability and efficiency under increased workload. Performance enhancement is always a tradeoff between the number of opposing factors like hardware and working hours investment limits, desired response times and throughput values, etc. In our research we had to balance between two limiting factors (response time and CPU usage) while trying to maximize a third one (throughput).

These factors can be visualized as given in Figure 1. The X axis shows the response time, with a maximum value of eight seconds, which was considered by some authors to be the limit of tolerable waiting time a few years ago. We believe that today, this limit for an average user is much lower so we have set the average response time for 90% of requests to 2 seconds. The Z axis displays the average CPU usage during load tests. The usage limit here has been set to 70% according to some best practices in the hardware industry (constant usage over 70%

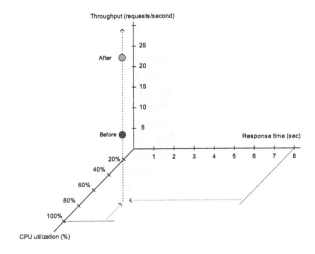

Fig. 1. Performance enhancement area

significantly raises malfunction probability and shortens the Mean Time Between Failure (MTBF), along with the Total Cost of Ownership (TCO)). In this way, we have set the margins of an area within which we can try to achieve the highest possible throughput (Y axis) by implementing various performance enhancement techniques.

Performance enhancement requires a complete team of experts from all fields of WA development - from front-end (HTML, CSS, JavaScript) to back-end (database design, SQL queries, DBMS configuration and administration) in order to plan and implement performance enhancement techniques, as well as verify changes through benchmarking. Such teams must also check the performance of application architecture at each level, as some techniques may have hidden effects that can propagate throughout the application and even cause a decrease in the performance at another level. Some techniques require additional services that must be installed on the server. This increases the complexity of application architecture and increases CPU and memory usage (if additional services are running on the same server), which has to be taken into account when doing post-implementation load testing. When trying to enhance the performance of a WA, a starting approach can be general systems theory that considers the application to be a black box with a large number of requests per second as input, and a large number of responses per second as output. When we start to decompose the black box into subsystems, we identify the ones that spend most of the time needed to generate responses. If one subsystem takes 80% of this time, then our choice is to try to enhance the performance of that subsystem, instead of another subsystem that spends just 5% of the time. After enhancing performance of one subsystem, another becomes the primary bottleneck. It is an iterative process of defining quantitative parameters that define the acceptable behavior of the system, benchmarking, identifying of bottlenecks, modifying, and benchmarking again. Performance measuring is a process of developing measurable indicators that can be systematically tracked for supervising advances in service level goal fulfillment. With WAs, there is usually a discrepancy between the required or expected, and real performance results. Performance levels can address quality, quantity, time or service price. There are two basic approaches to performance measuring: benchmarking and profiling. While benchmarking answers the question how well the application works, profiling answers the question why does the application have such performance. In this research we have used both approaches to measure the overall performance, as well as to profile the response time structure of individual requests. There are dozens of parameters that can be measured while benchmarking a production or a staging version of an application, which are selected based on identified bottleneck or performance enhancement goals. It is usually not possible to test the application in production environment, so we use test versions, or test part(s) of a cluster, and use an artificial workload. Workload modeling is a very important part of the performance benchmarking process. To achieve the correct measurements, the artificial workload must mirror the sample of the real workload, i.e., live users of the application. This can be achieved using different data sources: real time user observations, planned interactions or log file analysis. Test workload implementation must be as realistic as possible; otherwise, incorrect data may be collected and wrong decisions could be made.

Peformance Enhancement Techniques. Over the years of Web development and research, many different performance enhancement techniques (some more, some less complex to implement) have been proposed and used in practice. As the nature and complexity of the content offered online have changed [23], so have the techniques for performance enhancement. During our preliminary research of this subject we have identified many different techniques aimed at enhancing the performance of multi-tier web application architecture. Some of them are now obsolete; some are effective only with highly distributed systems; others are effective only at most extreme workloads where even 1kB makes a difference; etc. In our research we have analyzed those techniques that are currently most widely used and can be implemented on most WAs in use today - 19 techniques in total (as shown in Table 1). These techniques have been analyzed in the following manner: we defined the objective of each technique and the way in which it tries to achieve its objective, described the implementation process and gave an estimate of duration. Selected techniques have been sorted by the tier of WA architecture that they affect. As expected, the lowest number of techniques affect the business logic tier of application architecture. Although the business logic of an application can cause the bottleneck effect, it is by far the most expensive and complex problem to resolve which cannot be done by any enhancement technique. If the business logic of the application appears to be the causing the bottleneck, that is an indication of bad architecture decisions or implementation. When functioning normally, code execution takes from 5-15% of total response

Table 1. List of analyzed techniques

No code changes required	Code changes required
T1 Caching objects using HTTP expires	T2 Reducing number of HTTP requests
Compressing textual content	Positioning of HTML content objects
JavaScript minification and obfuscation	Reducing number of DNS requests
Reducing size of graphical objects	Managing redirects
RDBMS conguration tuning	Delayed content download
T3 Caching of interpreted PHP scripts	Enhancing performance of core logic
Standalone PHP application server	T4 Caching query results (Memcache)
Choice of web server	Query and database optimization
T5 Web server conguration tuning	Using stored procedures
T6 Caching pages using a proxy server	

(round trip) time and as such it is not a particularly fruitful subsystem for performance enhancement. According to [26], most of the response time is spent on content delivery (up to 80%) and query execution (database lag), so most of the analyzed techniques try to enhance the performance of these subsystems.

All the techniques displayed in Table 1 have been analyzed in detail. Due to the limited scope of this paper, we omit a detailed analysis which can be found in [13]. The analysis consists of 6 properties: 1) Tier - of the WA architecture affected by the technique, 2) Idea - basic explanation of how the technique increases performance, 3) Implementation - how is the technique implemented, 4) Time - how long it took us (on average if not stated differently) to implement this technique in the three applications used in this research. Implementation time for these techniques will vary with respect to the size and complexity of the application (e.g. amount of code, number of servers), 5) Expected results - performance characteristics the technique affects and 6) Verification - how is the performance gain measured. For the experimental part of our research we have selected six techniques with various degrees of complexity and implementation time. Those techniques have been selected based on the assumption that they will have the highest impact on the overall performance of WAs that have been used in this research. A list of analyzed techniques, as well as those that have been selected for experimental testing, can be seen in Table 1. Techniques selected for the experimental part of our research are labeled with T1-T6.

4   Performance Testing and Analysis

In the experimental part of our research we used three open-source WAs of different types/categories: portal (Joomla), community (PhpBB), and e-commerce (OsCommerce) to implement and test the effectiveness of performance enhancement techniques.

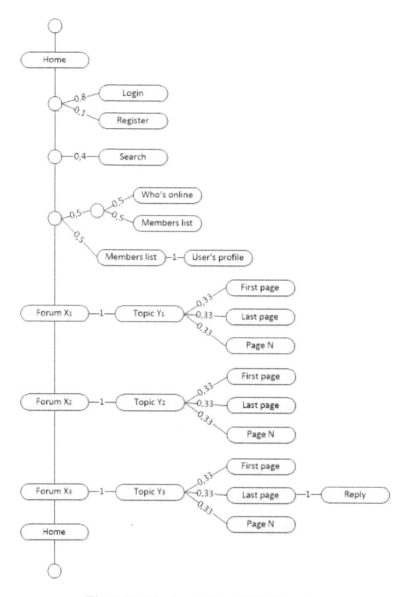

Fig. 2. Workload model for APP2 (Forum)

4.1 Preparation Phase

The first step of the research was to develop a realistic workload model (for all three applications) that will be used in all performance tests. Test workloads were modeled using various approaches: server log analysis (requires shell scripting to extract usage patterns), planned interactions (set by probability or importance, e.g. product browsing in an e-commerce application has the highest probability) and real user auditing. More about these approaches can be found in [17]. Figure 2 displays the workload model developed for load testing APP2. All actions (HTTP requests) on the baseline will be executed each time the test is performed, while others have a probability indicated on the connecting lines. The test can be described in the following way: all users start at the home page, most of them log on to the system, while some register and some (10%) will remain as guests. About 40% of users will use the search option. Visitors then look at one of the three popular pages (members list, who's online and a users profile). Then they look at three forums (two of them are most popular while the third one is randomly selected) and look at one topic inside those forums. When accessing the topic page some users start at the first post, others go directly to the last page and some select one of the pages of the topic. A third of the visitors will post a message to the forum. Finally, they will go back to the home page (check that there are no more new posts) and leave the page. Workload models for other applications were developed in a similar fashion (ie. most popular articles on the portal, featured products in the webshop). To implement workload models and perform load tests the JMeter [3] tool was used. With complex test models we simulated many requests from the individual user which were randomized by timers (simulating pauses for reading, thinking and form submissions) and random branching (between sets of actions). To ensure randomization of URL requests, test data sources (TXT files) were prepared from the real WA databases. These TXT files are constructed by selecting a subset of primary key identifiers from the real application database and written in a tab delimited format. JMeter then reads one line from the TXT file for test run and adds identifiers to the generated HTTP requests (links). In this way we can simulate the behavior of real users (reading about products, commenting

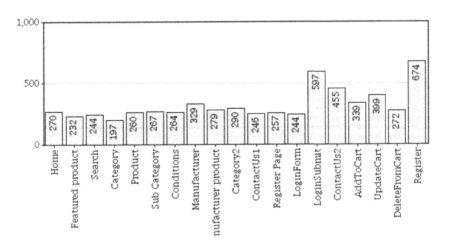

Fig. 3. JMeter aggregate graph

products, adding products to cart, making purchases, etc.). After the workload models were designed and implemented we tested them and verified that they executed correctly (reading identifiers from TXT files, adding them to HTTP requests, all response codes 200, database inserts correct, etc.). After the workload models were developed and implemented as load tests in JMeter we had to find a way to log data about hardware utilization during the tests. We decided to use Linux command line tool **vmstat** which proved to provide enough data on a fine enough time scale with a small enough footprint on server CPU usage. The next step was preparing data visualizations for the analysis phase. The first visualization was done by JMeter. After performing a test, JMeter displays data about each performed request, such as response time and server response, and calculates throughput and average response times. JMeter then offers to generate various visualizations based on this data. In relation to our first measuring constraint (90% of all requests under 2 seconds) we chose to generate an aggregate graph (shown in Figure 3) that displays average response times for each HTTP request (link) defined in the workload model. We have also used **Httpload** [12] and **ab** [4] for fast tests and to verify the results recorded by JMeter. The second visualization was done after the collected hardware utilization data was processed using an **awk** script (to remove unnecessary columns and calculate column averages) and handed over to a **gnuplot** script we developed for this research. Figure 4 demonstrates the output of our gnuplot script. Duration of the test is shown on the X axis in each chart. This visualization

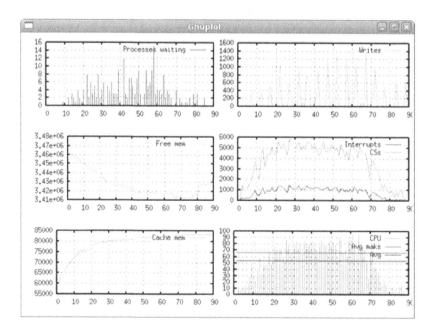

Fig. 4. Hardware utilization visualization (vmstat >awk >gnuplot)

consists of six charts representing various hardware utilization metrics received from vmstat. Some charts contain two data series. The top most graph on the left hand side displays the number of processes waiting during the test (lower number of processes waiting means lower response time), middle-left displays the amount of free memory, and bottom-left the amount of memory used for cache (more data in the cache means lower response time and higher throughput). The top right hand side graph displays the number of writes on the disc (the more files are cached, the less time is spent fetching data from the hard drive, resulting in lower response time, higher throughput and lower CPU usage). The middle-right graph contains two data sets: the number of interrupts and the number of context switches (the higher these values are, the higher CPU usage is, which results in higher response time and lower throughput). The bottom-right graph displays CPU usage during the test and 2 lines: one is the 70% limit set as one of the two main constraints, and the other is the calculated average of CPU usage during the test. The calculated average for the duration of the test had to be less than or equal to 70%. We increased the load on the test server by increasing the number of simulated users in JMeter test plans, until we came as close as possible to one or both limiting factors. Then we performed the three official tests (following the procedure described the following section). All tests were performed in an isolated LAN consisting of a load generator PC, a 100Mbit passive switch and the testbed PC. The testbed PC had the following hardware configuration: Asus IntelG43 Chipset, CPU: Intel Core2Duo 2.5Ghz, 1MB Cache, RAM 4GB DDR2, HDD WD 250GB SATA, OS Debian Linux. Network bandwidth was not a bottleneck in any of the preparation tests so we did not measure network usage.

4.2   Testing Phase

Before starting load tests we analyzed each application (file structure, number of files included per request, number of graphical objects, measured individual request response time). An important measure was the number of SQL queries performed per request. While the first two applications performed 28 and 27 queries, the third one did 83, which was a probable bottleneck. The testing phase consisted of 3 stages. First, we measured the initial performance of all three applications (after installation, using a test data set). Second, we implemented one of the selected performance enhancement techniques and repeated the measurement to determine the effect of the technique on performance (we performed this test 3 times). Then the application was restored to its initial state. This process was repeated for each technique. Third, we implemented all of the selected techniques to all three WAs and made the final performance measurement to determine final performance enhancement achieved using these techniques. In total we performed over 200 load tests. When doing performance testing, ensuring equal conditions in order to achieve the correctness of the acquired data is obligatory and was integrated in our testing procedure:

```
For APP1, APP2, APP3
  For Technique T1 - Tn
    Implement technique Tn
    Run multiple tests until as close as possible to two
    limiting factors // increment number of simulated users
    Reset()
    Restart server (clear memory, cache, etc.)
    Run a warm-up test (prime caches)
    Repeat 3 times
      Run test and gather data (JMeter, vmstat)
      Reset()
    Remove implementation of technique Tn

Reset()
  Reset database to initial state (remove inserts from the
  previous test)
  Reset server logs
```

We have performed each test three times for each implemented technique in order to get a more robust measurement and reduce the possibility of errors. Average values have been calculated from these measurements and are displayed in Table 2. Before starting our measurements, we set the throughput as the key performance indicator. Our goal was to enhance it using various performance enhancement techniques. **The limiting factors were: a) average response time for the 90% of requests had to be less than 2 seconds and b) average CPU utilization had to be kept below 70%.** First, we present the results obtained from the initial testing, then the results for each technique, and, lastly, the final testing results. These results display only the values of the aforementioned key performance characteristics. During each benchmark we collected data for 10 performance indicators which will be taken into consideration later. Second, we display the overall results that confirm our hypotheses. Third, we take into consideration all the data obtained from benchmarking and add weights to each performance indicator in order to define the effectiveness of each technique.

4.3   Results

The summary of our measurements is presented in Table 2. Let us first explain the structure of the Table. The markings in the first column are: B (for Beginning) - this row displays data about the initial performance of un-altered applications (after installation); T1-T6 (for Technique) - these rows display data about performance after each of the techniques was implemented individually; F (for Final) displays data about the performance of applications after all six techniques were implemented and finally C (for Comparison) displays data about the performance of applications under the same workload used in the B row. For each application there are three columns: The Response time (in miliseconds) is given

Table 2. Overall load testing results

-	APP1 Portal		-	APP2 Forum		-	APP3 Webshop	
	RT (ms)	T (R/s)	U (%)	RT (ms)	T (R/s)	U (%)	RT (ms)	T (R/s)	U (%)
B	834	8,3	69	791	15,2	66	794	24	66
T1	1025	8,3	68	661	**16,8**	67	794	24	66
T2	816	8,4	67	789	16,4	67	857	**27**	67
T3	793	**14,4**	68	815	**32,3**	69	1218	**35**	67
T4	858	**13,5**	68	958	**16,8**	69	993	25	67
T5	823	8,6	67	527	15,8	66	726	25	69
T6	958	8,4	69	954	17	70	1670	25	66
F	614	24,9	68	671	38,5	68	1740	35	64
C	91	8,5	21	58	16	23	442	24,1	48

LEGEND: RT= Response time, T = Throughput, U = Utilization, ms = miliseconds, R/s = Requests per second, CPU = average percent of CPU utilisation (measured by vmstat).

in the first column, the Throughput (in requests/second) in the second, and the CPU utilization (average, in %) in the third column. Although some techniques appear to increase the average response time and do not increase the overall throughput, they have a positive effect on other performance indicators that were measured and will be discussed later on. Furthermore, the first technique reduces individual response time which is not visible here, but can be observed in individual response profiling using YSlow (a Firefox profiling add-on). It can be seen that PHP script caching using APC (T3) clearly has the biggest impact on the performance of all tested applications. Storing SQL query data in memory cache using Memcache (T4) follows closely being the second most effective with two of three applications. We can also see that different techniques have different effects on each application. The comparison row (C) demonstrates the performance of applications under the same beginning workload, which was the highest possible before the implementation of all performance enhancement techniques. This data shows how the application would perform under normal daily workload, after its performance has been enhanced using the 6 mentioned techniques. In this row we can see that the throughput value has decreased significantly. This is caused by the fact that JMeter uses the number of simulated users/second (which is one of the arguments used when starting the load test) to calculate throughput. To make the comparison we had to use the load that was used to make the initial performance measurement for the WAs in their original (un-enhanced) state. Therefore, the number of simulated users/second was much smaller than it was for the final versions of WAs with all 6 techniques implemented. The result of final load testing also demonstrates that combining all of the techniques has a cumulative effect, because the maximum throughput for APP1 and APP2 is higher than the contribution of any individual technique. This was expected based on the effect of performance enhancement techniques

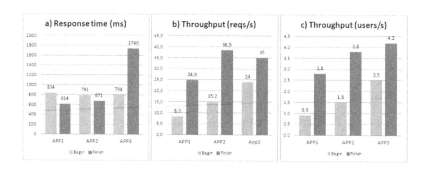

Fig. 5. Final performance enhancement results

on other performance indicators we observed (which will be described in the following section). This was not the case with APP3, whose maximum throughput was limited by database lag (due to the large number of database queries per request (an architectural problem)). To visualize the performance improvements we chose three key indicators - response time, throughput measured by number of requests/second (displayed in Column "T" in Table 2) and throughput measured by the number of users/second (not displayed in Table 2, but measured along with other indicators listed in Table 3). The differences between the values of these indicators in the first(B) and final (F) measurements are given in Figure 5 (a), (b) and c) respectively. The average response time for APP1 and APP2 was reduced even with a much higher workload while the throughput was increased by almost three times. For APP3, our hypothesis was confirmed (we achieved a 30% increase in throughput) but due to the very large number of database queries performed for each request, the increase in throughput was almost 2 times smaller than with APP1 and APP2. The increase in response time is caused by the same problem but within the limit of 2 seconds average for 90% of requests.

5   Calculating Technique Effectiveness

Although the results obtained show that it is possible to significantly enhance the performance of various types of WAs by using just a small subset of performance enhancement techniques, we were interested in defining an overall "quality indicator" of used techniques for each application type, and checking whether they appear in the same order with all WAs. This would mean that there is a uniform top list of techniques that can be implemented on any web application. To precisely determine the efficiency and effectiveness of a technique, we took all the recorded indicators into consideration. The full list of indicators whose values have been recorded during testing is displayed in Table 3. The most important indicator (after the three previously mentioned) is the time needed to implement the technique, which we recorded for each technique. Each indicator is given a

Table 3. List of indicators with appointed weights

No	Name	Weight	Proportional	Acquired from
1	Response time	5	Inversely	JMeter
2	Throughput (requests/sec)	5	Directly	JMeter
3	Throughput (users/sec)	5	Directly	JMeter
4	Utilization (cpu)	2	Inversely	vmstat
5	Processes waiting	4	Inversely	vmstat
6	RAM usage	2	Directly	vmstat
7	Cache mem. usage	3	Directly	vmstat
8	Disc usage (writes)	3	Inversely	vmstat
9	Context switches	4	Inversely	vmstat
10	Implementation time	5	Inversely	Measured / Estimated

weight (range 1-5), marking its importance in the overall performance gains. The weights were given according to our perception of each indicators importance in the overall performance enhancement. The column titled "Proportional" indicates whether the measured indicator is directly or inversely proportional to the technique effectiveness. Directly proportional means the higher the measured value, the better, while inversely proportional means the lower the value, the better, e.g. an increase of requests per second is directly proportional while the implementation time is inversely proportional. We measured the time needed for one developer to implement the technique (change server configuration, change application configuration, change code of the application) in each application. To determine the "quality indicator" of a technique we used the following procedure and equations.

```
For each WA (APP1, APP2, APP3)
 For each technique (T1-T6)
    For each of 10 performance indicators
        Calculate the effect E of indicator n using the value of
  indicator Vn in relation to the indicators maximum value change
  (Cnmax) and minimum indicators value change (Cnmin). Depending on
  whether the indicator is inversely proportional or directly
  proportional formulas (1) or (2) are used (respectively)
```

$$E_n = C_{n_{max}} - V_n / C_{n_{max}} - C_{n_{min}} . \tag{1}$$

$$E_n = V_n - C_{n_{min}} / C_{n_{max}} - C_{n_{min}} . \tag{2}$$

```
Calculate indicator weight using (3)
```

$$W_i = W_n / W_{max} . \tag{3}$$

Sum up to get the technique efficiency using (4)

$$\sum T_i = \sum_{i=1}^{10} W_i * E_i \, . \tag{4}$$

In this way, we calculated the top list of performance enhancement techniques with respect to type/category of the WA. The results are displayed in Table 4. Technique 1 (Reducing number of HTTP requests) was not implemented on APP 3 (marked "N/A" in Table 4.) because there were not enough graphical objects that could have been merged into a single larger one. It is clear that the order (ranking) of techniques is different for each type of application. A few important conclusions can be made from these calculations and will be used as problem guidelines in our future work. First, we don't have a framework that defines which techniques to use for each type of WA. It is clear that the subset of techniques to be used for performance enhancement must be tailored to the specific application. Secondly, there are a number of factors that influence the decisions about the techniques to be used such as: the goals of performance enhancement (what aspect of performance are we trying to enhance), the type of content the WA delivers (e.g. text, graphic objects, large files, video, etc.) and the specific workload. In our future work we will repeat these measurements on a larger number of (various types of) WAs and try to develop and verify such a framework for identifying a subset of techniques that yields the best results based on these factors.

Table 4. Overall ranking of performance enhancement techniques effectiveness

No	Technque	APP1	APP2	APP3
1	Reducing number of HTTP requests	5	3	N/A
2	Caching objects using HTTP Expires	3	1	3
3	Caching of interpreted PHP scripts (APC)	2	2	1
4	Caching query results (memcache)	1	6	4
5	Web server conguration tuning	6	5	4
6	Caching objects using proxy server (Squid)	4	6	2

6 Conclusion

Static websites are rapidly being replaced by web applications, and the ever increasing number of Internet users demands more and more functionality while expecting lower and lower response time. Web 2.0 has brought about a paradigm shift which changed the structure of workload, moving it from read-intensive to write-intensive. Therefore, the performance of WAs has become one of the focal points of interest of both scientists and professionals in this field. The goal of performance enhancement has to be set before any of the techniques are implemented or tests performed. This goal depends on the problem perceived in the performance of an application and can be aimed at any aspect of its performance (e.g. minimizing CPU or memory usage). In this research, our goal was to maximize throughput and lower response time of different finished systems (web applications) on the same hardware basis. Performance measurement itself is a complex process that requires careful monitoring of the real workload, identification of bottlenecks, planning and modelling test workloads, identifying key characteristics, goals, technical knowledge on all elements of the content creation and delivery system, etc. We have proved that it is possible, in a controlled environment at least, to significantly enhance the performance of WAs using just a small set of performance enhancement techniques with a total implementation time ranging from 10 to 50 working hours for applications running on one multiple-role (e.g. web, proxy, application) server. We found that the results of each technique vary from application to application and that further research is needed to develop a generalised framework that would take into consideration all the factors mentioned above (goals, content type, system architecture, etc.) and suggest what techniques would be best suitable for a selected application.

References

1. Aberdeen Group: Application Performance Management,
   `http://www.aberdeen.com/Aberdeen-Library/5807/`
   `RA-application-performance-management.aspx`
2. Amza, C., Soundararajan, G., Cecchet, E.: Transparent Caching with strong consistency in dynamic content web sites. ICS Boston (2005)
3. Apache JMeter, `http://jakarta.apache.org/jmeter`
4. Apache Benchmark Tool, `http://httpd.apache.org/docs/2.0/programs/ab.html`
5. Bahn, H.: Web cache management based on the expected cost of web objects. Information and Software Technology 47, 609–621 (2005)
6. Bogardi-Meszoly, A., Levendovszky, T.: A novel algorithm for performance prediction of web-based software system. Performance Evaluation 68, 45–57 (2011)
7. Domenech, J., Pont, A., Sahuquillo, J., Gil, J.A.: A user-focused evaluation of web prefetching algorithms. Journal of Computer Communications 30, 2213–2224 (2007)
8. Domenech, J., Pont, A., Sahuquillo, J., Gil, J.A.: Web prefetching performance metrics: a survey. Performance Evaluation 63, 988–1004 (2006)
9. Georgakis, H.: User behavior modeling and content based speculative web page prefetching. Data and Knowledge Engineering 59, 770–788 (2006)
10. Henderson, C.: Building Scalable Web Sites. OReilly, Sebastopol (2006)
11. Huang, Y., Hsu, J.: Mining web logs to improve hit ratios of prefetching and caching. Knowledge-Based Systems 21, 149–169 (2008)
12. Http Load Tool, `http://www.acme.com/software/httpload/`
13. Jugo, I.: Analysis and evaluation of techniques for web application performance enhancement, Master of Science Thesis, in Croatian (2010)
14. Khayari, R.: Design and evaluation of web proxies by leveraging self- similarity of web traffic. Computer Networks 50, 1952–1973 (2006)
15. Lam, K., Ngan, C.: Temporal prefetching of dynamic web pages. Information Systems 31, 149–169 (2006)

16. Liu, H., Keelj, V.: Combined mining of Web server logs and web contents for classifying user navigation patterns and predicting users future requests. Data and Knowledge Engineering 61, 304–330 (2007)
17. Meier, J.D., Farre, C., Banside, P., Barber, S., Rea, D.: Performance Testing Guidance for Web Applications. Microsoft Press, Redmond (2007)
18. Na, Y.J., Leem, C.S., Ko, I.S.: ACASH: an adaptive web caching method based on the heterogenity of web object and reference characteristics. Information Sciences 176, 1695–1711 (2006)
19. Nagpurkar, P., et al.: Workload characterization of selected J2EE-based Web 2.0 applications. In: 4th International Symposium on Workload Characterization, pp. 109–118. IEEE Press, Seattle (2008)
20. Ohara, M., Nagpurkar, P., Ueda, Y., Ishizaki, K.: The Data-centricity of Web 2.0 Workloads and its impact on server performance. In: IEEE International Symposium on Performance Analysis of Systems and Software, pp. 133–142. IEEE Press, Bostin (2009)
21. Pea-Ortiz, R., Sahuquillo, J., Pont, A., Gil, J.A.: Dweb model: Representing Web 2.0 dynamism. Computer Communications 32, 1118–1128 (2009)
22. Ravi, J., Yu, Z., Shi, W.: A survey on dynamic Web content generation and delivery techniques. Network and Computer Applications 32, 943–960 (2009)
23. Sadre, R., Haverkort, B.R.: Changes in the web from 2000 to 2007. In: De Turck, F., Kellerer, W., Kormentzas, G. (eds.) DSOM 2008. LNCS, vol. 5273, pp. 136–148. Springer, Heidelberg (2008)
24. Sajeev, G., Sebastian, M.: Analyzing the Long Range Dependence and Object Popularity in Evaluating the Performance of Web Caching. Information Technology and Web Engineering 4(3), 25–37 (2009)
25. Sivasubramanian, S., Pierre, G., van Steen, M., Alonso, G.: Analysis of Caching and Replication Strategies for Web Applications. Internet Computing 11(1), 60–66 (2007)
26. Souders, S.: High Performance Web Sites. O'Reilly, Sebastopol (2007)

CRAWL·E: Distributed Skill Endorsements in Expert Finding

Sebastian Heil, Stefan Wild, and Martin Gaedke

Technische Universität Chemnitz, Germany
`firstname.lastname@informatik.tu-chemnitz.de`

Abstract. Finding suitable workers for specific functions largely relies on human assessment. In web-scale environments this assessment exceeds human capability. Thus we introduced the CRAWL approach for Adaptive Case Management (ACM) in previous work. For finding experts in distributed social networks, CRAWL leverages various Web technologies. It supports knowledge workers in handling collaborative, emergent and unpredictable types of work. To recommend eligible workers, CRAWL utilizes Linked Open Data, enriched WebID-based user profiles and information gathered from ACM case descriptions. By matching case requirements against profiles, it retrieves a ranked list of contributors. Yet it only takes statements people made about themselves into account. We propose the CRAWL·E approach to exploit the knowledge of people *about* people available within social networks. We demonstrate the recommendation process for by prototypical implementation using a WebID-based distributed social network.

Keywords: Endorsements, Expert Finding, Linked Data, ACM, WebID.

1 Introduction

Knowledge work constitutes an ever increasing share of today's work. The nature of this type of work is collaborative, emergent, unpredictable and goal-oriented. It relies on knowledge and experience [17]. Traditional process-oriented Business Process Management (BPM) is not well applicable to areas with a high degree of knowledge work [21]. Addressing this issue, non-workflow approaches [5], in particular Adaptive Case Management (ACM), gain more relevance [8].

ACM systems assist knowledge workers. They provide the infrastructure to handle dynamic processes in a goal-oriented way. Traditional BPM solutions feature a-priori processes modeling. Contrary to them, ACM systems enable adaptivity to unpredictable conditions. Uniting modeling and execution phases contributes accomplishing this adaptivity. A case represents an instance of an unpredictable process and aggregates all relevant data. For adapting it to emergent processes, case owners can add ad-hoc goals. There are cases where persons currently involved cannot achieve all goals. In these cases it is necessary to identify suitable experts based on the skills and experience required for that particular part of work.

S. Casteleyn, G. Rossi, and M. Winckler (Eds.): ICWE 2014, LNCS 8541, pp. 57–75, 2014.

The nature of knowledge work often implies cross-enterprise collaboration. It necessitates access to information about the persons involved, e.g., CV and contact data. It is unlikely that all potential collaborators use the same social network platform for storing personal information. Cross-platform relationships are hard to follow. Such "walled gardens" [1] would complicate expert finding. Distributed social networks are well-suited for the knowledge work domain. Companies or knowledge workers can host their own profiles. The profile can include work experience and skill information. Interlinking these distributed profiles establishes a social network. Such social network could overcome the data silo characteristic of walled gardens. This would enable crawling the network to identify experts.

Finding suitable workers for specific functions largely relies on human assessment. Assessors have to make their decisions depending on the requirements at hand. This decision making requires knowledge of potential contributors and their experience. The selection complexity increases with the amount of eligible contributors and work requirements. Human assignment does not scale well, especially not with web-scale processes [12]. Often work is assigned to workers who are not the most suitable experts available. This can cause mediocre outcomes and longer times to completion. Dealing with this problem requires software support for finding and addressing knowledge workers to contribute to cases.

In [15] we introduced CRAWL, an approach for Collaborative Adaptive Case Management. It leverages various Web technologies to automatically identify experts for contributing to an ACM case. CRAWL recommends a set of eligible workers. It uses Linked Open Data, enriched WebID-based user profiles and information gathered from project or case descriptions. We created a vocabulary to express the skills available and the skills required. It extends user profiles in WebID-based distributed social networks and case descriptions. CRAWL's semantic recommendation method retrieves a ranked list of suitable contributors whose worker profiles match the case requirements.

Problem. The skill information about a person is limited to the expressive power and will of this particular person. As a consequence, CRAWL only takes statements people made about themselves into account. Such statements, however, might be unspecific, exaggerated or even wrong. This affects the expert finding process and makes the assessor's task more difficult and time-consuming.

There are three possible kinds of statements about skills, as shown in Figure 1. The most basic form are *skill self-claims*, statements by someone claiming that he himself has a certain skill. *Skill assignments*, on the other hand, are statements by someone claiming that someone else has a certain skill. Statement claimed by someone for himself and confirmed by someone else are *skill affirmations*. We refer to these three kinds of skill statements together as *skill endorsements*.

With knowledge work increasingly becoming an important and widespread part of work [9] and ACM evolving as an approach addressing this type of work, we are convinced that enabling knowledge workers to find the right collaborators to contribute to multi-disciplinary cases impacts the performance of future enterprises [5]. The value add by skill endorsements will trigger a demand to incorporate them into distributed worker profiles and expert finding algorithms.

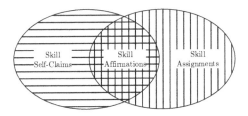

Fig. 1. Three kinds of skill endorsements

Overall Objective. To exploit the knowledge of people *about* people available within social networks, we aim at integrating skill assignments and skill affirmations in addition to skill self-claims into the distributed expert finding process.

Contributions. To contribute to the overall objective, we must achieve the following objectives to fully use skill endorsements in distributed social networks:

1. To enable expert finding in distributed social networks
2. To increase credibility of skill self-claims
3. To allow assigning skills the endorsee did not consider
4. To prevent unwanted skill endorsements
5. To express skill endorsements in distributed user profiles
6. To incorporate skill endorsements in distributed expert finding
7. To facilitate a differentiated consideration of skill endorsements

The paper is organized as follows: Section 2 illustrates the necessary objectives in order to achieve the overall objective. The background is provided in Section 3. We present the CRAWL·E approach in Section 4. Section 5 evaluates the approach. We discuss work related to ours in Section 6 and conclude the paper in Section 7.

2   Objectives of Distributed Expert Finding

This section describes the objectives in greater details. To illustrate the need for achieving each objective, we use different personae. All of them are knowledge-workers and members of a distributed social network. They have a different character and pursue different goals. Figure 2 shows the corresponding social graph. Black solid arrows indicate knows-relationships, blue dotted arrows symbolize endorsements. The personae are characterized in the following:

Alice wants to record her skills. She likes to include all skills from her current job, past jobs and education. Alice intends to record them in a way others can easily access them. She does not want to spend too much effort in achieving this goal.

Bob is a co-worker of Alice. He knows Alice very well because he worked together with her in many projects. Bob trusts Alice and Alice trusts him.

Casey is a case owner who wants to find and recruit the best persons for a job.

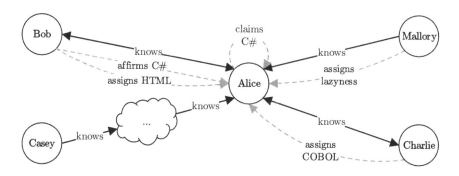

Fig. 2. Social Graph with Endorsements

Charlie is another co-worker of Alice. Compared to Bob, he is not that close to Alice. Charlie worked together with Alice in only one project long time ago.

Mallory is a bad guy. He dislikes Alice and wants to damage Alice's reputation.

Having described the personae that are used throughout this paper, we continue with outlining the objectives.

Objective 1: To enable expert finding in distributed social networks.
Interoperability, compatibility, and portability of skill information is no issue in conventional centralized social networks like LinkedIn. Expert finding benefits from the (virtually) monolithic data layer of such social networks. By contrast, this does not exist in distributed social networks which are formed by interlinked, distributed profiles[1]. Therefore, skill endorsements in profiles are also inherently distributed over the network. This requires to ensure discovery, comparability and description of skills across organization boundaries. Each skill endorsement needs to be described properly to facilitate comprehension and avoid misunderstandings. To allow adding further skill information, both skill set and skill descriptions need to be accessible, extensible and linkable. So, persons can easily refer to descriptions of the skill endorsements they made. When Alice claims she has a skill, this endorsement must be associated to the person making the claim i.e., to her. This is necessary because information a person produces, belongs to her. To persist skill endorsements associated to the person stating them, they need to be stored and connected with the person's identity. This enables persons and machines to detect skill endorsements, provided that relevant data is accessible in an easy-to-process manner. To achieve this objective, we need to deliver 1) an extensible description for each skill, 2) a way to attach all kinds of skill endorsements to persons, 3) a place to store each person's skill endorsements, and 4) a procedure to find experts in distributed social networks based on skills.

Objective 2: To increase credibility of skill self-claims. Skill self-claims are the most basic form of skill endorsements. Persons can use them to declare

---

[1] Distributed profiles are documents, which are accessible from different URLs and hosted on different servers, referencing each other. They describe persons.

that they have a certain skill set. So, Alice can claim she has a specific skill like C# programming. To increase the credibility of self-claimed skills, other persons should be enabled to affirm them. Skill affirmations allow persons to testify that someone they know has a specific skill. This affirmation may be based on past collaboration where certain skills were involved and demonstrated. For example, Bob can affirm Alice's C# skill because he worked with her on a project requiring this particular skill. In endorsing someone's skill, the endorser uses his own reputation to give more weight to the endorsee's claimed skill. This contributes to increasing the credibility of claimed skills. Achieving this objective requires delivering a procedure that allows persons to affirm self-claimed skills.

Objective 3: To allow assigning skills the endorsee did not consider. In many respects skill assignments are similar to skill affirmations. A skill assignment suggests a person, like Alice, to claim a skill she has not considered so far. As an example, Bob knows Alice very well. So, he might assign Alice the HTML skill she did not think of. While skill affirmations rely on prior skill self-claims, skill assignments do not. For achieving this objective, we need to deliver a procedure that allows persons to assign skills that have not been self-claimed beforehand.

Objective 4: To prevent unwanted skill endorsements. Centralized social networks can easily incorporate the concept of skill affirmations and skill assignments. They form a single point of truth. The skill endorsements are part of the database of the networking platform. Unless integrity of the data stock has been violated, it is impossible for Mallory to claim negative skills upon Alice. That is, the endorsee needs to self-claim skills beforehand or confirm an assignment.

Adopting this policy to distributed social networks without a central data base is more complicated. First, we must avoid maliciously negative affirmations and assignments. Otherwise, Mallory could affirm negative skills to damage the Alice's reputation by publicly claiming Alice has an "incompetence" skill. Second, persons might be found by expert finding systems due to outdated affirmations of skills they deliberately removed from their profiles. For example, an engineer who has been working for arms industry but now decided against this branch removes corresponding skills from his profile. Distributed expert finding should not consider outdated skill endorsements. Therefore, we need to strive for an agreement between the endorser and the endorsee. As a side effect, this would also contribute to increasing the credibility of skill claims.

Objective 5: To express skill endorsements in distributed user profiles. When Alice claims she has a certain skill set, this information must be recognizable by all authorized members of the distributed social network. The same holds true, when Bob endorses a skill of Alice or when Charlie makes a skill assignment. All three kinds of skill endorsements differ in who is claiming which skill for whom. Thus, each skill endorsement consists of three basic elements: endorser, skill and endorsee. So, a vocabulary able to express such triples in a unified and linkable manner would allow covering all kinds of skill endorsements. Associating and storing skill endorsements with the person claiming them, as suggested in Objective 1, requires delivering a vocabulary for specifying skill endorsements.

Objective 6: To incorporate skill endorsements in distributed expert finding. Achieving Objective 1 fulfills the basic requirements to incorporate skill endorsements in distributed expert finding. The expert finder needs to compare all skill endorsements associated to a candidate with the skills required for a task. For determining a person's suitability for a case, distributed expert finding must consider all kinds of skill endorsements. This assists Casey in deciding about assigning a task to Alice, Bob etc.. To achieve this objective, we need to deliver 1) a method to compare skill endorsements with case requirements and 2) a ranked list of experts fitting to the case requirements.

Objective 7: To facilitate a differentiated consideration of skill endorsements. Case owners benefit from an extensive knowledge about a candidate's suitability for a case. Taking all kinds of skill endorsements into account would enable Casey to gain a rich picture of each candidate's capabilities. Depending on the quantity and quality of a personal social network, the number of skill endorsements differs from person to person. For example, Alice's many social connections also entail many skill affirmations and assignments. The number of skill endorsements could be one criterion for Casey. She knows, however, that this would discriminate persons who have fewer or less diligent social connections.

Distributed expert finding could address such issues by statically weighting each kind of skill endorsement differently. This would, however, reduce adaptability of expert finding and favor persons who share similar characteristics. To preserve customizability, distributed expert finding has to enable adaptably factoring in all kinds of skill endorsements. So, Casey could weight skill self-claims more than skill affirmations or assignments. This is in line with Objectives 2 to 4.

3   Expert Finding with CRAWL

In this section we describe how CRAWL [15] assists expert finding in distributed social networks. The scenario shown in Figure 3 demonstrates our approach. Casey works as a second-level-support worker for a software development company. A key customer reports a bug in a software product developed by the company. Casey is responsible for the handling of this support case. She uses an ACM system to assist her work. As she investigates the problem, she defines several goals and asks experts from the third-level-support department to contribute. At some point during the analysis of the bug, a detailed profiling is required to rule out concurrency issues. However, there is no expert on this topic available. To assist Casey in finding a person with the required expertise, CRAWL facilitates the following workflow (cf. numbers in Figure 3):

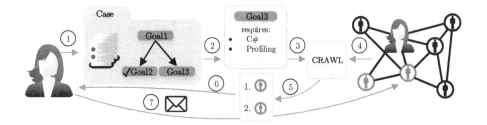

Fig. 3. CRAWL Overview

1. Casey adds a corresponding goal to the case.
2. Casey defines requirements (e.g., C# and Profiling).
3. Casey starts CRAWL.
4. CRAWL traverses Casey's social graph.
5. CRAWL generates a list of eligible workers.
6. Casey selects the most suitable candidates.
7. Casey asks them for contribution to the goal.

Finding suitable workers requires a traversal of the requestor's social graph. This graph is established by `foaf:knows` connections in WebID profiles. WebID profiles are essential artifacts of the WebID identification approach. They contain an identity owner's personal data described in a machine-readable way using Linked Data. For this, WebID relies on several RDF-vocabularies such as FOAF. With WebID, users are enabled to globally authenticate themselves, connect to each other, manage their profile data at a self-defined place and specify customized views [23]. Users can rely on WebID identity providers for creating new WebID identities and managing their WebID profile data [24].

The traversal algorithm is implemented as a depth-limited breadth-first search. It dequeues a WebID URI identifying a person, retrieves the corresponding WebID profile, calculates the rating R, marks the WebID URI as visited and adds all unvisited WebID URIs referenced via `foaf:knows` and their depth value to the queue. The initial queue consists of the WebID URIs of the persons already involved in the case. A maximum depth is used due to the exponentially rising number of nodes in a social graph with increasing depth [13]. CRAWL allows for additional limits like the number of suitable candidates rated above a certain threshold. Following the rating, the WebID profile graph of the candidates is added to a triplestore. A statement containing the calculated rating is asserted into the graph. The final ordered list of rated candidates results from executing the SPARQL query shown in Listing 1.1 on the triplestore.

Sequential traversal of WebID profiles and rating calculation have a huge impact on performance due to the distributed nature of profiles. We addressed this issue by concurrency and caching of user profiles and skill descriptions [15].

```
1  SELECT ?candidate ?rating
2  WHERE { ?candidate a foaf:Person .
3      ?candidate vsrcm:rating ?rating.
4      FILTER( ?rating > ?minRating )}
5  ORDER BY DESC( ?rating )
```

Listing 1.1. SPARQL query for candidates

Having retrieved and rated a subset of the social graph, CRAWL presents a
list of recommended candidates and contact information to the person initiating
the search. This step allows for later extension to enable applying constraint criteria, e.g., filter candidates from a specific company or within the same country.
CRAWL demonstrates the basic concept of expert finding in distributed social
networks leveraging knowledge from profiles, case descriptions and Linked Open
Data (LOD) [15]. Therefore it addresses Objective 1. Yet, it does not consider
knowledge of people about people such as skill assignments and skill affirmations.

4   CRAWL·E: Extending CRAWL with Endorsements

In order to addresses all objectives from Section 1, we propose CRAWL·E which
extends CRAWL with endorsements. The first part of this section introduces a
vocabulary to express skill endorsements. Part two explains the expert finding
algorithm and the integration of endorsements in the candidate rating.

4.1   Integrating Skill Endorsements in Distributed Profiles

In [15] we introduced a vocabulary to add skill self-claims to WebID profiles.
Linked Data provides CRAWL with a large knowledge base for concepts describing skills. CRAWL references this data to describe existing experience for persons
and experience required to achieve a case goal or contribute to it. In a WebID
profile, the RDF property `vsrcm:experiencedIn` connects a `foaf:Person` with
a URI which represents this person's experience in *something*. For referring to
the actual skills URIs are used to reference concepts which are available as dbpedia[2] resources. With dbpedia being a central element of the linked open data
cloud, this intends to increase the degree of reusability and extensibility of skill
data.

To express endorsements, we reuse this vocabulary as seen in Listing 1.2. The
important aspect to note is the distributed nature of profiles. An endorser has
no write access to foreign endorsees' profiles.

As there is no specific platform or protocol defined for adding statements to
WebID profiles, skill assignments have to be expressed in the endorser's own profile. Leveraging the RDF data model and FOAF vocabulary, CRAWL·E enables
persons to add skill assignments to their WebID profiles. These skill assignments

---

[2] http://dbpedia.org/

```
1   @prefix endorser: <http://company.org/>.
2   @prefix endorsee: <http://minisoft.ru/>.
3
4   <endorser:bob> a foaf:PersonalProfileDocument;
5       foaf:primaryTopic <endorser:bob#me>;
6       foaf:title "Bob Endorser's WebID profile".
7
8   <endorser:bob#me> a foaf:Person;
9       foaf:name "Bob Endorser";
10      foaf:knows <endorsee:alice#me>;
11      cert:key [a cert:RSAPublicKey ;
12               cert:exponent 65537  ;
13               cert:modulus "1234..."^^xsd:hexBinary].
14
15  <endorsee:alice#me> vsrcm:experiencedIn <dbp:Linux>,
16                                          <dbp:Mysql>.
```

Listing 1.2. Skill assignment in endorser's WebID profile

Fig. 4. Skill definition in Sociddea

reference the WebID URI of the endorsee who is connected to the endorser via foaf:knows.

Supporting users in specifying their expertise and case requirements, we exemplarily extended the user interfaces of Sociddea and VSRCM [15] to allow specifying skills using regular English words. We use prefix search of dbpedia lookup service to match user input against dbpedia resources. A list of skills is updated live as the user is typing. This is illustrated in Figure 4.

4.2   Extending Distributed Expert Finding to Leverage Skill Endorsements

This section describes how CRAWL·E incorporates endorsements in candidate rating. Figure 5 shows the traversal, rating and candidate recommendation which form steps 4 and 5 in Figure 3. The required skills s_{r0}, s_{r1}, s_{r2} of Goal3 and Casey's social graph are the input. In this example, Casey knows B and C. B and C know D, C knows E. CRAWL·E has already rated B with $R(B) = 15$, C with $R(C) = 0$ and D with $R(D) = 10$. To get the rating of E, the similarities between required skills and existing skills are calculated using linked open data. For a proof-of-concept, we use a prototypical rating function adapted from [16].

According to Objective 5, a skill endorsement is a triple (p_1, p_2, s) of endorser p_1, endorsee p_2 and skill s. A self-claimed skill can be represented as (p, p, s): by an endorsement with identical endorser and endorsee. A candidate c is described by E, the set of all endorsements regarding c in c's social graph as in Equation (1).

$$E = \{(p_1, p_2, s) | p_2 = c\} \tag{1}$$

S_S is the set of self-claimed skills by the candidate, S_O is the set of skills endorsed (assigned) by others and S_B is the set of skills claimed by the candidate and affirmed by others (both) (also cf. to Figure 1).

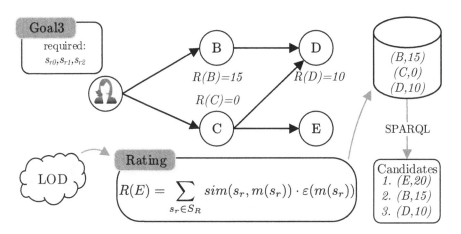

Fig. 5. Traversal, rating and recommendation

$$S_S = \{s|\exists(c,c,s) \in E\} \tag{2a}$$
$$S_O = \{s|\exists(p_1,c,s) \in E, p_1 \neq c\} \tag{2b}$$
$$S_B = S_S \cap S_O \tag{2c}$$
$$S = \{s|\exists(p_1,c,s) \in E\} = S_S \cup S_O \tag{2d}$$

CRAWL·E compares the set of required skills S_R to the set of skills S of each candidate as defined in Equation (2d). Both skill sets are represented by sets of dbpedia URIs. The similarity $sim(s_1, s_2)$ between two skills distinguishes different concept matches:

1. Exact Concept Match - URIs are identical ($s_1 = s_2$)
2. Same Concept As Match - URIs connected via `owl:sameAs` (s_1 `owl:sameAs` s_2)
3. Related Concept Match - URIs connected via `dbprop:paradigm`, `dcterms:subject`, `skos:narrower` etc.

This forms the similarity function in Equation (3).

$$sim(s_1, s_2) = \begin{cases} \sigma_1 & \text{if Exact Concept Match} \\ \sigma_2 & \text{if Same Concept As Match} \\ \sigma_3 & \text{if Related Concept Match} \\ 0 & \text{else} \end{cases} \tag{3}$$

These concept match types can easily be extended to facilitate an adapted rating. The basic idea is that each type yields a different similarity rating. For the moment, we use these values: $\sigma_1 = 10$ for 1), $\sigma_2 = 9$ for 2) and $\sigma_3 = 5$ for 3).

Per candidate, for each combination of a required skill $s_r \in S_R$ and a candidate skill $s_c \in S$, the similarity $sim(s_r, s_c)$ is computed according to Equation (3). As seen in Equation (5), to calculate the candidate rating R in CRAWL, only the skill with maximum similarity per required skill (from function $m : S_R \to S$ in eq. (4)) is considered. [15]

$$m(s_r) = s_c \Leftrightarrow sim(s_r, s_c) = \max_{s \in S} sim(s_r, s) \tag{4}$$

$$R_{\text{CRAWL}}(c) = \sum_{s_r \in S_R} sim(s_r, m(s_r)) \tag{5}$$

In CRAWL·E, an affirmed skill is given higher influence compared to a self-claimed skill. To accomodate this influence, we introduce the endorsement factor ε as defined in Equation (7). Our updated CRAWL·E rating function is shown in Equation (6).

$$R_{\text{CRAWL·E}}(c) = \sum_{s_r \in S_R} sim(s_r, m(s_r)) \cdot \varepsilon(m(s_r)) \tag{6}$$

Let c be a candidate with the set of endorsements E. The set of all endorsements of the candidate skill s_c is defined by $E_{s_c} = \{(p_1, p_2, s) \in E | s = s_c\} \subseteq E$. With this, we define the endorsement factor using the skill sets defined in 2

$$\varepsilon(s_c) = \begin{cases} 1 & \text{if } s_c \in S_S \setminus S_B \\ \alpha\sqrt{|E_{s_c}|} & \text{if } s_c \in S_B \\ \beta\sqrt{|E_{s_c}|} & \text{if } s_c \in S_O \setminus S_B \end{cases} \tag{7}$$

This factor distinguishes between the three types of skills: Self-claimed-only skills - from $S_S \setminus S_B$, skills that have been claimed by the candidate and endorsed by others - from S_B - and skills that have only been endorsed by others but are not stated in the candidate's profile - from $S_O \setminus S_B$. It yields 1 for a candidate skill without endorsements, i.e., no additional influence is given to self-acclaimed skills. Parameters α and β allow for adaption, currently we use $\alpha = 1.5$ and $\beta = 2$. To ignore unilateral skill assignments one can set $\beta \overset{!}{=} 0$.

The endorsement factor ε increases with the number of endorsements. However, the higher the number of endorsements, the slower the factor increases. This is to avoid overrating candidates with very high endorsement counts. Other function types such as a mirrored $1/x$ function are possible, too. We decided in favor of the square root function type, because it does not converge against a limit as there is no theoretical foundation to reason the limit. When the rating is finished, recommended candidates can be listed as in Figure 6.

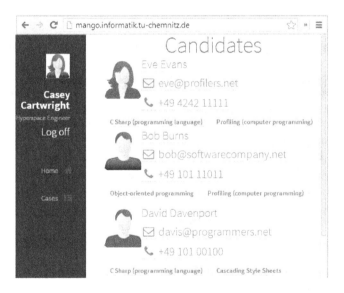

Fig. 6. Candidate recommendation in VSRCM

5   Evaluation

This section discusses the evaluation of our approach. We claim that CRAWL·E achieved the overall objective stated in Section 1 by considering all three kinds of skill endorsements for expert finding in distributed social networks. To prove this claim, we first outline the evaluation setup and then discuss our findings.

5.1   Evaluation Setup

To evaluate the extent to which CRAWL·E's objectives have been achieved, we chose the objective-based evaluation method. Event though a field experiment or a case study would allow for a profound review also, CRAWL·E's results highly depend on the underlying distributed social network. There are various characteristics to be considered, including total and average amount of social connections and of each kind of skill endorsements, richness of user profile data, and level of networking. While a prototypical implementation of CRAWL·E is publicly available, the adoption[3] by users has not yet reached a certain level. Such adoption is, however, required for conducting a field experiment and gaining both extensive and reliable evaluation results. Yet, we are convinced that enabling knowledge workers to find the right collaborators impacts the performance of future enterprises [5]. Thus, the value add by skill endorsements will soon trigger a demand to incorporate them into distributed worker profiles and expert finding.

An objective-based study is the most prevalent approach in program evaluation and is applicable to be performed internally by program developers [20]. As Stufflebeam describes, the "objectives-based approach is especially applicable in assessing tightly focused projects that have clear, supportable objectives" and that "such studies can be strengthened by judging project objectives against the intended beneficiaries' assessed needs, searching for side effects, and studying the process as well as the outcomes". For devising CRAWL we already defined objectives in Section 2. They are well-suited to be reused as evaluation criteria in this objective-based study. We incorporate the information collected during development and application of CRAWL and CRAWL·E, cf. Heil et al. in [15], to determine how well each operational objective was achieved. After outlining the evaluation setup, we discuss the findings for each objective in the following.

5.2   Discussion of Findings

To enable expert finding in distributed social networks, we follow the idea of finding suitable experts to invite. We think that searching for experts by utilizing personal social graphs is more beneficial compared to the open call approach discussed in Crowdsourcing research. Case owners intending to delegate tasks know their social network. So, they know whether a candidate fits the task description. CRAWL·E allows describing skills associated persons and requirements associated to cases. Rather than building our own database to manage skills and skill descriptions as typical for centralized social networks, our approach relies on the collective knowledge of dbpedia. In CRAWL·E, linking to a dbpedia resource refers either to a skill or to a requirement. Unlike related work, we do not want to restrict the number of options for skills and requirements. The availability of a description for a referred concept is not just a requirement in our approach, but also a good practice in general. dbpedia is a central part of the Linked

---

[3] With regard to the quantity of skill self-claims, skill affirmations and skill assignments per user and in total, and the use of the skill endorsement vocabulary in general.

Open Data cloud. So, resources are machine-readable through RDF and highly connected to each other. This allows for classification and association. The set of resources as well as each resource as such is extensible and maintained by a large community. When attaching skills to persons, we benefit from skills that are both referenced and identified by URIs. Similar to a skill URI pointing to a skill description, we use a WebID URI to refer to a person. Contrary to creating separate resources for storing each person's skill endorsements, we embedded them in machine-readable WebID profiles. As all skill endorsements a person, like Bob, issues belong to him, they also remain in his profile. With all concepts described using RDF, discovery and query become possible through interlinkage with URIs and retrieval using SPARQL. Thus, we delivered all four results necessary for achieving this objective.

To increase credibility of skill self-claims, we introduced the concept of skill affirmations. With a skill self-claim, a person describes that he possesses this skill. The credibility of skills self-claimed by a person, however, depends on the level of trust in this person. A skill self-claim which has been endorsed by someone else gains credibility depending on the trust in the endorsing person. In CRAWL·E, we assume that the more persons have endorsed a person for a particular skill, the more likely is this person to *really* possess the skill. Thus, the higher endorsed a certain skill self-claim is, the more influence it has on the candidate ranking.

To allow assigning skills the endorsee did not consider, our approach enables endorsers to claim skills a person has, without the person having to self-claim those skills beforehand. That is, it is not required for an endorsee to state such skills in their own WebID profiles. This is useful for instance to provide a more complete skill profile that includes information which the described person did not think of. Skill assignments available in distributed social networks can be exploited for various purposes including requests for adding assigned skills to own user profiles. While the increased expressiveness by skill assignments allowed achieving this objective, it comes at the cost of loosing control about what is being endorsed. We addressed this issue as explained in the following paragraph.

To prevent unwanted skill endorsements, we imposed the requirement that endorser and endorsee are bilaterally connected. This indicates that both persons *deliberately* know each other and, hence, accept each other's opinion. As an example, Charlie assigned a certain skill, like COBOL programming, to Alice some time ago. She followed Charlie's suggestion and claims this skill herself. So, Charlie's assignment became an affirmation. Alice was also endorsed by other persons for this skill, i.e., Bob or even Mallory. Today, she is not that interested in this topic anymore. Therefore, Alice does not want to be found for this skill. She removes her self-claimed COBOL skill. Thus, all affirmations become assignments. By excluding Mallory from her social network, all his affirmations and assignments are going to be ignored in CRAWL·E. That is, affirming or assigning a skill necessitates that both the endorsee and the endorser know each other. In addition to this approach, we enable avoiding malicious and outdated skill endorsements by

1) simply removing corresponding skill self-claims the endorsee made in his profile and 2) appropriate consideration within distributed expert finding.

To express skill endorsements in distributed user profiles, we had to find an alternative to what is known from centralized social networks. There, Bob can endorse Alice for a skill she did not think of and the platform triggers a notification to Alice. The endorsed skill will not be added to her profile unless she approves of it. This is not possible with distributed social networks. Alice' and Bob's WebID profiles are documents typically hosted on different servers, and identified and retrievable by URIs. There is no platform to trigger a notification to Alice let alone to allow Bob writing to her profile. Objectives 2 and 3 are inherently adverse to Objective 4 because there is no standard means of asking approval. We therefore delivered an RDF vocabulary to specify skill endorsements using only one RDF triple per skill definition, no matter if it is a skill self-claim, affirmation or assignment. Associating endorser, endorsee and endorsed skill, and storing the triple within the endorser's WebID profile allows for expressing all three kinds of skill endorsements. Due to RDF's flexible and extensible nature, skill endorsements can be attached to the endorser (for self-claims) or to one of his social connections expressed via `foaf:knows` (for affirmations and assignments).

To incorporate skill endorsements in distributed expert finding, CRAWL·E queries personal social networks for skill information, retrieves and processes the skill endorsements, and compares them with the case requirements before showing them a ranked list of suited candidates. Our approach hereby involves how we have addressed Objective 5 for query and retrieval, and Objectives 2 to 4 for comparing and ranking. As manual assessment by crawling personal social networks is time-consuming, our approach assists case owners in their recruitment tasks, but also leaves the final assessment decision to them. While CRAWL only considers self-claimed skills for the candidate recommendation, CRAWL·E also takes skill affirmations and assignments into account. However, all skill endorsements made by persons not knowing each other are ignored in this process. For comparing case requirements with all three kinds of skill endorsements, our approach computes the similarity between both concepts. Although CRAWL·E differentiates exact from similar or related concepts via different weights, finding precise and profound weights requires further empirical evaluation.

To facilitate a differentiated consideration of skill endorsements, we developed CRAWL·E in a parametrized way which allows to choose whether to include foreign endorsed skills or to operate on confirmationary endorsements only. By employing our approach, case owners can adjust the influence of unilaterally endorsed statements, i.e., which value a statement is given that potentially many others claim upon a person but which this person does not claim himself. To reduce the effect of unusually many skill endorsements per person through a large amount of diligent social connections, we introduced a function for only partially factoring in large accumulations of skill affirmations and assignments. Considering Equation (7), CRAWL·E facilitates fine-tuning and even excluding the impact skill affirmations and skill assignments have on distributed expert finding. Users can align the rating with their individual needs and preferences. Finding a more exact function type and evidenced-based default values for the parameters requires a larger empirical evaluation to be conduct in future work.

6   Related Work

Expert Finding has long been a research interest. For example, Becerra-Fernandez provides an overview on Web-based expert finding approaches in [3]. Like [2], many of them are based on information retrieval techniques. To achieve expert finding, they analyze document topics and connect them to the document authors. Perugini et al. surveys expert finding systems with a focus on social context including communication and blogs [19]. Particularly for the research domain, there are several approaches, e.g., by Xu et al. [25] or by Uddin et al. [22].

The approach described by Xu et al. in [25] is similar to CRAWL·E in that it unites social graphs with skill relationship semantics. To achieve this, network analysis on interlinked concept (expertise) and research (social) layer is employed. While this approach also considers hierarchical and correlation relationships in the expertise layer, tacit knowledge is used. CRAWL·E uses explicit knowledge from profiles and Linked Open Data, whereas [25, 2] extract information from unstructured text sources, [25] supported by WordNet. This works well in a specific domain like research, because characteristics of the domain can be exploited. For instance, citations and co-authorship can be analysed from publications [25].

By contrast, CRAWL·E is a generic approach not limited to a specific domain. None of the above approaches works in distributed social networks, nor do they explicitly consider endorsements.

In [6], Bozzon et al. present an expert finding method based on user's activities in centralized social networks like Facebook, Twitter etc. It analyzes social resources directly related (e.g. tweets, likes) or indirectly related (e.g. posts of liked pages) to persons. This approach employs text analysis: entity recognition for skills is performed on the resources, they are identified with Wikipedia URIs. CRAWL·E by contrast targets distributed social networks, uses explicit expertise information, supports skill endorsements and leverages linked data.

In spite of their benefits, skill endorsements have not yet gained much attention in research. Platforms like LinkedIn[4] and ResearchGate[5] have successfully included them. Within the first six month, more than 2 billion endorsements were created on LinkedIn allowing for interesting analysis [14]. Donston-Miller states in [10] that endorsements provide a streamlined version of a resume and can reduce the risk of hiring new personnel. Also quality aspects should be considered in addition to mere quantity (endorsement count) measures. Doyle also mentions in [11] the problem of unwanted endorsements and argues that getting the "right" endorsements is important. While Berk suspects that LinkedIn is using endorsement data in its secret search algorithm [4], there is not much public

---

[4] http://www.linkedin.com/
[5] http://www.researchgate.net/

information available about skill endorsements in expert finding. Even if current platforms are internally implementing this, the major difference to CRAWL·E is the central nature of these social networks.

Pérez-Rosés et al. presents in [18] an approach which combines social graphs with skill endorsements. It uses an undirected social graph and a directed endorsement graph per skill. Skill relationships like correlation or implied skills are considered. The PageRank algorithm is applied to a deduced graph. Its deduction matrix is similar to the similarity matrix in CRAWL·E. However, the definition of the deduction matrix is an open problem whereas we get the values of the matrix leveraging Linked Open Data. Friendship-like bilateral relationships between social network members are assumed, while the foaf:knows semantics employed in CRAWL·E allows for unilateral relationships. Unlike CRAWL·E, this work focuses on social graph analysis, lacks a complete expert finding workflow, and does not support distributed social networks.

Our approach is an application of the *social routing principle* [12] to the ACM domain. Unlike task delegation through an open call known from Crowdsourcing research [7], we follow the idea of inviting suitable experts to contribute to a case by utilizing social graphs. The *conceptual routing table* described by Dustdar and Gaedke is formed by foaf:knows statements and contact info in WebID profiles.

7   Conclusions and Future Work

In this work, we presented the CRAWL·E approach leveraging distributed endorsements for finding eligible workers to contribute to ACM cases. It comprises a vocabulary for skill endorsements in WebID profiles, a method for traversing distributed social networks based on foaf:knows relationships and an adaptable rating function for WebID profiles. We demonstrated CRAWL by implementation based on the WebID identity provider and management platform Sociddea and the case management system VSRCM.

Our future research interest will be to consider not only endorsement quantity, but also quality. If a renowned expert endorses someone else for his very own field, his endorsement should be given more weight compared to endorsements of less renowned persons. This needs considering the endorsements of the endorsers in addition to the endorsements of the candidate to rate. Empirical data and machine learning can be used to provide adapted parameters. Providing Distributed Expert Finding as a Service is desirable to enable easy integration in other systems. For this, an endpoint structure and protocol must be defined.

References

[1] Appelquist, D., et al.: A Standards-based, Open and Privacy-aware Social Web: W3C Incubator Group Report. Tech. rep., W3C (2010)
[2] Balog, K., Azzopardi, L., de Rijke, M.: Formal models for expert finding in enterprise corpora. In: Proceedings of the 29th Annual International ACM SIGIR Conference, pp. 43–50. ACM, New York (2006)

[3] Becerra-Fernandez, I.: Searching for experts on the Web: A review of contemporary expertise locator systems. ACM TOIT 6(4), 333–355 (2006)

[4] Berk, R.A.: Linkedin Triology: Part 3. Top 20 Sources for Connections and How to Add Recommendations. The Journal of Faculty Development 28(2), 1–13 (2014)

[5] Bider, I., Johannesson, P., Perjons, E.: Do workflow-based systems satisfy the demands of the agile enterprise of the future? In: La Rosa, M., Soffer, P. (eds.) BPM Workshops 2012. LNBIP, vol. 132, pp. 59–64. Springer, Heidelberg (2013)

[6] Bozzon, A.,et al.: Choosing the Right Crowd: Expert Finding in Social Networks Categories and Subject Descriptors. In: Proceedings of the 16th International Conference on Extending Database Technology, New York, NY, USA, pp. 637–348 (2013)

[7] Brabham, D.C.: Crowdsourcing as a Model for Problem Solving: An Introduction and Cases. Convergence: The International Journal of Research into New Media Technologies 14(1), 75–90 (2008)

[8] Clair, C.L., Miers, D.: The Forrester Wave^{TM}: Dynamic Case Management, Q1 2011. Tech. rep., Forrester Research (2011)

[9] Davenport, T.H.: Rethinking knowledge work: A strategic approach. McKinsey Quarterly (2011)

[10] Donston-Miller, D.: What LinkedIn Endorsements Mean To You (2012), http://www.informationweek.com/infrastructure/networking/what-linkedin-endorsements-mean-to-you/d/d-id/1106795

[11] Doyle, A.: How To Use LinkedIn Endorsements (2012), http://jobsearch.about.com/od/linkedin/qt/linkedin-endorsements.htm

[12] Dustdar, S., Gaedke, M.: The social routing principle. IEEE Internet Computing 15(4), 80–83 (2011)

[13] Goel, S., Muhamad, R., Watts, D.: Social search in "small-world" experiments. In: Proceedings of the 18th International Conference on World Wide Web, WWW 2009, pp. 701–710. ACM, New York (2009)

[14] Gupta, S.: Geographic trends in skills using LinkedIn's Endorsement feature (2013), http://engineering.linkedin.com/endorsements/geographic-trends-skills-using-linkedins-endorsement-feature

[15] Heil, S., et al.: Collaborative Adaptive Case Management with Linked Data. To appear in WWW 2014 Companion: Proceedings of the 23rd International Conference on World Wide Web Companion, Seoul, Korea (2014)

[16] Lv, H., Zhu, B.: Skill ontology-based semantic model and its matching algorithm. In: CAIDCD 2006, pp. 1–4. IEEE (2006)

[17] Mundbrod, N., Kolb, J., Reichert, M.: Towards a system support of collaborative knowledge work. In: La Rosa, M., Soffer, P. (eds.) BPM Workshops 2012. LNBIP, vol. 132, pp. 31–42. Springer, Heidelberg (2013)

[18] Pérez-Rosés, H., Sebé, F., Ribó, J.M.: Endorsement Deduction and Ranking in Social Networks. In: 7th GraphMasters Workshop, Lleida, Spain (2013)

[19] Perugini, S., Goncalves, M.A., Fox, E.A.: A connection-centric survey of recommender systems research. Journal of Intelligent Information Systems 23(2), 107–143 (2004)

[20] Stufflebeam, D.: Evaluation Models. New Directions for Evaluation 2001(89), 7–98 (2001)

[21] Swenson, K.D.: Position: BPMN Is Incompatible with ACM. In: La Rosa, M., Soffer, P. (eds.) BPM Workshops 2012. LNBIP, vol. 132, pp. 55–58. Springer, Heidelberg (2013)

[22] Uddin, M.N., Duong, T.H., Oh, K.-j., Jo, G.-S.: An ontology based model for experts search and ranking. In: Nguyen, N.T., Kim, C.-G., Janiak, A. (eds.) ACIIDS 2011, Part II. LNCS, vol. 6592, pp. 150–160. Springer, Heidelberg (2011)

[23] Wild, S., Chudnovskyy, O., Heil, S., Gaedke, M.: Customized Views on Profiles in WebID-Based Distributed Social Networks. In: Daniel, F., Dolog, P., Li, Q. (eds.) ICWE 2013. LNCS, vol. 7977, pp. 498–501. Springer, Heidelberg (2013)

[24] Wild, S., Chudnovskyy, O., Heil, S., Gaedke, M.: Protecting User Profile Data in WebID-Based Social Networks Through Fine-Grained Filtering. In: Sheng, Q.Z., Kjeldskov, J. (eds.) ICWE Workshops 2013. LNCS, vol. 8295, pp. 269–280. Springer, Heidelberg (2013)

[25] Xu, Y., et al.: Combining social network and semantic concept analysis for personalized academic researcher recommendation. Decision Support Systems 54(1), 564–573 (2012)

Cross Publishing 2.0: Letting Users Define Their Sharing Practices on Top of YQL

Jon Iturrioz, Iker Azpeitia, and Oscar Díaz

ONEKIN Group, University of the Basque Country, San Sebastián, Spain
{jon.iturrioz,iker.azpeitia,oscar.diaz}@ehu.es

Abstract. One of Web2.0 hallmarks is the empowerment of users in the transit from consumers to producers. So far, the focus has been on content: text, video or pictures on the Web has increasingly a layman's origin. This paper looks at another Web functionality, cross publishing, whereby items in one website might also impact on sister websites. The *Like* and *ShareThis* buttons are forerunners of this tendency whereby websites strive to influence and be influenced by the actions of their users in the websphere (e.g. clicking on *Like* in site A impacts a different site B, i.e. *Facebook*). This brings cross publishing into the users' hands but in a "canned" way, i.e. the 'what' (i.e. the resource) and the 'whom' (the addressee website) is set by the hosting website. However, this built-in focus does not preclude the need for a 'do-it-yourself' approach where users themselves are empowered to define their cross publishing strategies. The goal is to turn cross publishing into a crosscut, i.e. an ubiquitous, website-agnostic, do-it-yourself service. This vision is confronted with two main challenges: website application programming interface (API) heterogeneity and finding appropriate metaphors that shield users from the technical complexities while evoking familiar mental models. This work introduces *Trygger*, a plugin for *Firefox* that permits to define cross publishing rules on top of the *Yahoo Query Language* (*YQL*) console. We capitalize on *YQL* to hide API complexity, and envision cross publishing as triggers upon the *YQL*'s virtual database. Using *SQL*-like syntax, *Trygger* permits *YQL* users to specify custom cross publishing strategies.

Keywords: Data sharing, YQL, triggers, cross publishing, web service.

1 Introduction

People interaction on the Web has drastically evolved in the last few years. From business-to-consumer (B2C) interactions, the Web is now a major means for direct person-to-person interaction through social-media websites (hereafter referred to as "websites"). *Facebook*, *WordPress*, *Wikipedia* showcase this new crop of media where content is provided for and by end users. As the new content providers, end users are also amenable to define their own cross publishing strategies. Traditionally, cross publishing means allowing content items in one website to also appear in sister websites where the sharing strategies are set by

S. Casteleyn, G. Rossi, and M. Winckler (Eds.): ICWE 2014, LNCS 8541, pp. 76–92, 2014.

the web masters. However, when the source of the content is the user (like in social-media websites), cross publishing should also be made 2.0, i.e. amenable to be defined by the end user. A first step is the *Like* and *ShareThis* buttons which let users *enact* cross publishing whereby the website hosting the button (e.g. pictures in *Flickr*) also impacts the content of a sister website (e.g. *Flickr* provides sharing for *Facebook*, *Twitter* and *Tumblr*). The enactment is up to the user. However, the 'what' (i.e. the resource) and 'to whom' (the addressee website) are still built-in.

Current mechanisms for cross publishing (e.g. the *Like* button) might fall short in some scenarios. First, minority websites will be rarely offered *"ShareThis"* links within main players' websites like *Twitter* or *Youtube*. Second, sharing might be conducted in a routinary way (e.g. every time I upload a presentation in *Slideshare*, communicate it to my followers in *Twitter*). Rather than forcing users to click the *Twitter* icon when in *Slideshare*, this situation can be automatized through a "sharing rule". Third, the rationales behind sharing are not explicit but kept in the user's mind. Sharing might be dictated by some resource characterization that could be captured through a "sharing rule" (e.g. create a task in my *ToDo* account when I tag a bookmark in *Delicious* with "toread").

In a previous approach [10], we provide the infrastructure for web masters to define cross publishing. However, as the previous scenarios highlight, it is difficult for a webmaster (e.g. the *Flickr*'s one) to foresee the different settings in which its resources will need to be shared and talked about (using tweets, posts, articles, etc.). This calls for cross publishing to become a crosscut website-agnostic end-user service. Web2.0 demonstrates that end users are ready [7] and willing to adapt their Web experience, if only the tools become available that make it sufficiently easy to do so (e.g. [6]). Cross publishing should not be an exception.

This work aims at ascertaining to which extend this vision is feasible both technically and humanly. From a technical perspective, we develop **Trygger**, a *Firefox* plugin, that allows to define sharing rules as services over websites. The human factor is being considered by conceiving websites as database tables where sharing is realized as database-like triggers over these tables. By taping on a familiar representation (the relational model), our hope is to easy adoption. As an example, consider *Twitter* and *Facebook* as hosting a table of "tweets" and "wall posts", respectively. Users can be interested in publishing into their *Facebook* wall, tweets from their colleagues that contain the hashtag *#ICWE2014*. The *Trygger* expression will look something like: ***ON** INSERT a new tweet LIKE "%#ICWE2014%" **DO** INSERT a message (tweet permalink) INTO my wall*. This is the vision *Trygger* strives to accomplish. Compared with related work, the distinctive aspects of this approach include:

- Cross publishing is conceived as a crosscut service. We regard cross publishing as an idiosyncratic, traversal activity to be conducted in the cloud but managed locally. Users can define their own sharing rules, not being limited to those hardwired into the websites.

– Cross publishing is specified as triggers (Section 3 & 4). A familiar *SQL*-like syntax is offered from expressing sharing rules. We heavily rely on *Yahoo Query Language (YQL)* [13]. *YQL* provides a database vision of websites' APIs. *Trygger* extends this vision by providing a trigger-like mechanism for sharing over disparate websites.

– Cross publishing is implemented via a reactive architecture (Section 5). We provide a loosely coupled architecture for distributed trigger management. The implementation relies on existing standards and protocols, and it is entirely *HTTP*-based.

Trygger is fully realized as a *Firefox* plugin. This plugin, the working examples and a brief installation guide can be downloaded from *http://www.onekin.org /trygger/*

2   A Brief on YQL

A tenant of current Web development is the release of data through accessible APIs. The problem is that APIs are as heterogeneous as the applications they support. As a practitioner of the Programmable Web puts it "the main difficulty with APIs in general is that they require the developer to create and remember a number of cryptic URLs via which certain data can be retrieved from the API"[1]. *Yahoo Query Language (YQL)* tries to address this by converting API calls into SQL-like commands, which are somewhat more human readable. Figure 1 (1) illustrates *YQL*'s *SELECT* to access the *Slideshare* API but now conceived as the table *slideshare.slideshows*. To realize this metaphor, *YQL* offers both a service and a language.

YQL Language. So far, *YQL* offers *SELECT, INSERT* and *DELETE* statements that, behind the curtains, invoke the corresponding API methods. *YQL* alleviates programmers from details concerning parameter passing, credential handling or implicit iterations (for joins). The mapping from the *YQL* syntax to API requests is achieved through the so-called **"Open Data Tables"** (ODT). ODTs hold all the intricacies of the underlying APIs. It is worth noticing that this sample query (Figure 1 (1)) actually conducts three API requests, one for each user ('cheilmann', 'ydn', 'jonathantrevor'). *YQL* conceals these interactions, enacts the three requests, combines the outputs, and filters out those results based on the query parameters. In this way, *YQL* off-loads processing that programmers would normally do on the client/server side to the *YQL* engine.

YQL Service. This is realized through the *YQL* console[2]. The console permits to preview the query, and obtain the REST query counterpart. This query that can then be bookmarked or included in the programs. This interface contains the following main sections (see Figure 1): (1) the *YQL* statement section is

---

[1] http://blog.programmableweb.com/2012/02/08/
  like-yql-try-it-with-the-programmableweb-api/
[2] http://developer.yahoo.com/yql/console/

Fig. 1. *YQL* console

where *YQL* queries are edited, (2) the results section displays the output of the query, once the source Web service was enacted, (3) the REST query section provides the URL for *YQL* queries, (4) the queries section gives access to queries previously entered, and (5) the data tables section lists all the APIs that can be accessed using *YQL*.

The bottom line is that SQL-like syntax becomes the means to simplify Web development based on open APIs. Open APIs are the enablers of cross publishing. Therefore, and akin to *YQL*'s teachings, we aim at making cross publishing scripts look like the development of DB scripts. We build upon the similitude of SQL's triggers and cross publishing as for keeping in sync data repositories, let these be database tables or website silos, respectively. Details follow.

3 Cross Publishing Scripts: The Notion of *Trygger*

Supporting cross publishing as a service implies the monitoring of a website account so that actions on this account ripple to other accounts. This can be described as event-condition-action rules. Rules permit to enact small pieces of code ('DO') when some update operation is conducted ('ON') provided some condition is met ('WHEN'). An example follows:

> *ON a new tweet is uploaded in account ('Twitter', 'oscar')*
> *WHEN new tweet.status LIKE "%ICWE2014%"*
> *DO create a wall post in account ('Facebook', 'iker')*

Akin to the *YQL* metaphor, we use *SQL*-like syntax for triggers to specify cross publishing rules (hereafter referred to as "**try**ggers"). The whole idea is to tap into the existing ODT repository. Indeed, most of the ODT tables used throughout the paper were provided by the *YQL* community. Even so, *YQL* permits ODTs external to the system to be operated upon through the *USE* clause. This clause just provides the URL that holds the ODT definition. From then on, no difference exists in handling ODTs kept outside *YQL* boundaries. Similarly, the *ENV* clause offers a way to access hidden data (see section 5). Figure 2 outlines the *trygger* syntax. Next paragraphs address each of the *trygger* clauses.

```
{ENV <access_key>;}*
{USE <url> AS <odtTable>;}*
CREATE TRYGGER <tryggerName>
AFTER SELECT CHANGES ON <odtTable>
WHEN <yqlFilter>
BEGIN
{ [XMLNAMESPACES (<url> AS <alias>)] <yqlStatement> | <jsCode> }
END
```

Fig. 2. *Trygger*'s syntax. *Tryggers* are based on *YQL*'s ODT tables.

The Event. In *SQL*, events are risen by insertions, deletions or updates upon database tables. Since *YQL*'s ODT handle both insertions and deletions, it could be possible to define *trygger* events upon *YQL* operations. Notice however, that *YQL* is just another client built on top of somewhere else's APIs. These APIs can be accessed by other clients, and hence, "the tables" can be changed by applications other than *YQL*. Monitoring only *YQL* operations would make these other changes go unnoticed. Therefore, the only way to be aware of insertions is by change monitoring. Hence, the *trygger* event is *"after_select_changes"* which happens when two consecutive pollings incrementally differ. No "statement" *tryggers* are supported. Only "for each row", i.e. each new tuple causes the enactment of the *trygger*. The event payload refers to this new tuple, kept in the system variable *NEW*.

The Condition. In *SQL*, conditions check the event payload, i.e. the new/old tuple being inserted/deleted. Likewise, *tryggers* can check conditions in the new "ODT tuples" being inserted. Each new tuple causes the enactment of the trigger only if the conditions are satisfied.

The Action. *SQL* actions stand for procedures that are stored in the database. Procedures can be atomic (i.e. a single statement) or full-fledged programs (described using e.g. PL/SQL or Java). Back to *Trygger*, actions can be either atomic or composed. Atomic actions comprise a single statement, specifically, any "insert/update/delete" statement defined upon an ODT table. Action parameters can refer to either constants or *NEW*. However, single-statement actions fall short in numerous scenarios, namely: when payload parameters need some processing before being consumed; when data kept in other ODT tables is to be retrieved; when more than one ODT table needs to be changed[3]. This leads to composed actions, i.e. programs. Akin to *YQL*, the programming language is JavaScript. *Trygger*'s JavaScript permits both *YQL* actions and access to the

---

[3] Action statements are enacted as independent entities. So far, *Trygger* does not support transaction-like mechanisms.

```
A)  CREATE TRYGGER Twitter2Facebook
    AFTER SELECT CHANGES ON twitter.search
    WHEN q in ("from:iturrioz", "from:oscaronekin") AND text LIKE "%#ICWE2014%"
    BEGIN
    INSERT INTO facebook.setStatus (uid, status, access_token)
    VALUES ("689274514", NEW.results.text.*, "254|2.AQY00514|FbS4U_w")
    END

B)  CREATE TRYGGER Arxiv2Instapaper
    AFTER SELECT CHANGES ON arxiv.search
    WHEN search_query = "ti:YQL" AND journal_ref LIKE "% SPRINGER %"
    BEGIN
    XMLNAMESPACES ("http://www.w3.org/2005/Atom" AS atom)
    INSERT INTO instapaper.unread (username, password, title, selection, url)
    VALUES ("oscaronekin@gmail.com","12pass34", NEW.atom:entry.atom:title.*,
                NEW.atom:entry.atom:summary.*, NEW.atom:entry.atom:id.*);
    END
```

Fig. 3. A) *Twitter2Facebook trygger*. *NEW* refers to the ODT tuple being inserted since the last time the *twitter.search* table was pulled. B) *Arxiv2Instapaper trygger*.

system variable NEW. Next section illustrates this syntax throughout different examples[4].

4   Trygger at Work

Cross Publishing from *Twitter* to *Facebook*. *Facebook* launched *SelectiveTwitter*[5] whereby tweets ending in the hashtag *#fb* are directly propagated to the user's *Facebook* status. This application can be conceptually described as a cross publishing rule: "**on** introducing a tweet that contains *#fb*, **do** update my *Facebook* wall". *Facebook* developers were forced to provide a generic hashtag (i.e. *#fb*) to accommodate no matter the user. However, users can certainly be interested in monitoring *domain-specific* hashtags. For instance, when in a conference (identified through a hashtag, e.g. *#ICWE2014*), users might be interested in tracking tweets referring to *#ICWE2014* from the *Twitter* accounts of some attendees. Unlike the *#fb* case, this approach does not force the user's colleagues to introduce the *#fb* hashtag but a tag that might already exist. It is also domain-specific as for attaching distinct sharing behaviour to different hashtags. Figure 3-A provides a *trygger* that supports domain-specific hashtag tracking. We resort to two ODT tables, "*twitter.search*" and "*facebook.setStatus*" that support selections and insertions on *Twitter* and *Facebook*, respectively.

This rule is triggered if the delta of two consecutive selects on *twitter.search* is not empty. This ODT table includes two columns: "*q*" and "*text*" that keep the username of the *Twitter* account and the tweet message, respectively.

---

[4] Used ODT's can be consulted at **http://github.com/yql/yql-tables** or in the YQL console.

[5] **http://www.facebook.com/selectivetwitter**. To date, this enhancement obtains 3.6 out of 5 review summary based on 2,666 reviews.

The *trygger* conditions checks first whether *"q"* holds either *"from:iturrioz"* or *"from:oscaronekin"*, and second, if the *"text"* contains the string *"#ICWE2014"*. If met, the *trygger*'s action results in a new tuple being inserted into the *facebook.setStatus* table. The newly created tuple (i.e. *uid, status, access_token*) is obtained from the *NEW* variable and the credentials for the *Facebook* account.

Cross Publishing from Arxiv to Instapaper. *Arxiv*[6] is an online archive for electronic preprints of scientific papers. *Instapaper*[7] is a neat tool to save web pages for later reading. In this example, the *trygger* will monitor new preprints in *Arxiv* published in any *"Springer"* journal that contains *"YQL"* in the title. On detecting such preprints, the *trygger* will create a new entry in the *Instapaper* account of *"oscaronekin"*. This example involves two ODT tables: *arxiv.search* and *instapaper.unread* (see Figure 3-B). The former includes two columns, i.e. *search_query* and *journal_ref* that hold the title and the journal of the manuscript respectively.

On adding a new preprint in *arxiv.search*, the *trygger* checks whether the search expression is *"ti:YQL"* and the manuscript venue contains *"Springer"*. The interesting part is the *trygger*'s action. The action constructs a tuple out of NEW. Since "NEW.atom" holds an XML document, its content can be obtained using E4X dot [8]. To avoid clumsy expressions, an *XMLNAMESPACES* declaration is introduced (so does *SQL*).

5   Trygger Architecture

YQL console is a tester of third parties' APIs. This implies to use credentials (passwords, API keys,...) on the statements. *Tryggers* are composed of *YQL* statements, hence the credentials would be exposed to the *Trygger* service. This could prevent users from creating *tryggers*. However, *Trygger* does not analyze or decompose the statements in a *trygger*, the event part and the action part of a *trygger* are used as black boxes. Even though, users are able to hide credentials using the storage capabilities offered by *YQL*[9]. *YQL* allows to set up key values for use within ODTs and to storage keys outside of the *YQL* statements including them on environment files. In these way, users are in control of their credentials and manage them directly on the *YQL* console (see Figure 4). As an example, let's analyze the action part of the *trygger* on Figure 3-A. The credential (i.e. *access_token*) is explicitly shown (*254/2.AQY00514/FbS4U_w*). The *SET* keyword on the *YQL* language binds a key with a value. So, the *SET access_token = "254/2.AQY00514/FbS4U_w" ON facebook.setStatus* instruction establish that the *254/2.AQY00514/FbS4U_w* value is assigned to the *access_token* field each

---

[6] http://arxiv.org/

[7] http://www.instapaper.com/

[8] ECMAScript for XML (E4X) is a programming language extension that adds native XML support to ECMAScript.
http://en.wikipedia.org/wiki/ECMAScript_for_XML

[9] See http://developer.yahoo.com/yql/guide/yql-setting-key-values.html for more information.

time the *facebook.setStatus* table is invoked. The *SET* instruction should be hidden in an environment file through the *YQL* console. The response is a public key (e.g *store://kgnRBearlKjAI4rBdRntdf*) to use the hidden data in the environment file. Therefore the action part on our working trygger is simplified as:

> *ENV 'store://kgnRBearlKjAI4rBdRntdf';*
> *INSERT INTO facebook.setStatus (uid, status)*
> *VALUES ("689274514", NEW.results.text.*);*

This grounds *Trygger* to be supported using a client-side architecture, i.e. as a browser extension. Since *Trygger*'s first audience is the *YQL* community, the *Trygger* extension should be tightly integrated with the *YQL* console (not to be confused with the *YQL* system). That is, users should define *tryggers* using the same console they are accustomed to setting *YQL* requests: i.e. the *YQL* website. This implies the *Trygger* extension to be *locally* supported as an HTML wrapper upon the *YQL* website. To this end, we superimpose on the *YQL* console, GUI elements (e.g. buttons and panels) for *trygger* management. Interacting with these GUI elements will enact the services of server-side components that reside on the cloud, completely detached from the *YQL* system.

Database triggers are kept in the catalogue of the Database Management System (DBMS). *YQL* mimics the role of the DBMS as the container of ODT tables. This seems to suggest for *tryggers* to be held within the boundaries of *YQL*. However, some differences between *YQL* and DBMS advice a closer look. DBMSs are the only custodian of their data. Hence, DBMSs also act as event detectors since any table operations must go through the DBMS. By contrast, *YQL* is just another client built on top of somewhere else's data. This data can be updated by agents other than *YQL*. Therefore, the event detector can not stop at the ODT table but go down to the website. These observations unfold two divergent requirements. On one hand, we aim at a seamless integration with *YQL* so that users feel a unique experience. On the other hand, we require separate *trygger* management to detect events on monitored sites.

5.1   Deployment

The main components on the Trygger system complementing the *YQL* system are these (see Figure 4):

1. **The *Trygger* console.** It is a Web wrapper built on top of the *YQL* console. Implementation wise, this is achieved using *Greasemonkey*[10], a weaver that permits locally executed scripts to modify web pages on the fly. The wrapper provides local management for *tryggers*, i.e. creation, verification, deletion, enabling and disabling. Although the tryggers are send to the *Trygger* service a local copy is storage for rapid accessing and backing up.

---

[10] http://www.greasespot.net/

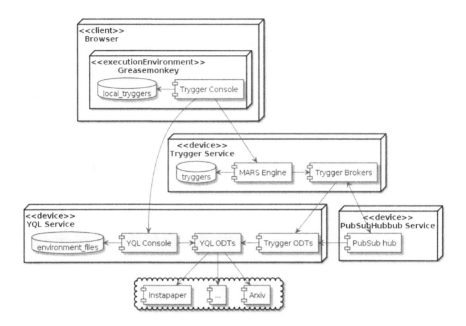

Fig. 4. Deployment diagram. In yellow color *Trygger* components, in green color *YQL* components and in pink color websites on the Internet

2. **The *Mars* Engine.** *Tryggers* are realized as Event Condition Action (ECA) rules. The *MARS* framework running into the *Trygger* service's machine is used as the rule engine [2]. Its functions include (see Figure 5): ECA rule registration (ECA engine), event occurrence signaling (Event Engine) and rule firing (i.e. rule's action enactment) (Action Engine). *MARS* provides the general framework for supporting ECA rules in a distributed environment. However, both events and actions are domain specific. Their realization and definition need to be externalized from *MARS* into so-called brokers.

3. **The *Trygger* Brokers**. The brokers are deployed in the *Trygger* service's machine. Two brokers are introduced (see Figure 5). The **Event Broker** monitors and generates events. So far, only one type of event is handled: changes in ODT table. Hence, the way to detect a change is by periodically polling the underlying website. For this purpose, we use the *PubSubHubbub* protocol [8]. On receiving the signal from the **PubSub hub**, the Event Broker generates an event occurrence for each NEW change. On event signaling, the *MARS* engine triggers the associated rules. Rule triggering is conducted through the **Action Broker**.

4. **The PubSub hub.** The *PubSubHubbub* service manages subscriptions to RSS data sources. The subscriber (i.e. the Event Broker) provides the publisher's feed URL (i.e. the *YQL* REST call to a ODT table) to be monitored as well as a callback URL to be called when the feed is updated. The hub

periodically polls the feed for new updates. On the feed being updated, the hub is noticed. This in turn propagates the feed to the subscriber callback endpoint: the Event Broker. There are some online *PubSubHubbub* services available on the Internet, for example, *Google PubSubHubbub Hub*[11] or *Superfeedr*[12]. However, we opt for deploy a dedicated service on the *Google App Engine* since the source code is freely downloadable[13].

5. **The *Trygger* ODTs**. They act as mediators from *MARS* to *YQL* (i.e. the **JSinterpreter ODT** shown in the Figure 5) and from *YQL* to PubSub hub (i.e. the **RSS-izator ODT** shown in the Figure 5)[14]. The former is a JavaScript snippet that knows how to process the code of the the rule's action part. The *Trygger* Action Broker requests this service providing as parameters the action code and the event occurrence. As for the RSS-izator, it resolves the format mismatch between the RSS required by the PubSub hub and the XML returned by *YQL*. The *trygger* ODTs reside into the *YQL*'s repository.

These components are already available in the cloud except the *Greasemonkey* wrapper. The wrapper needs to be locally installed as a *Firefox* plugin available at *http://www.onekin.org/trygger*. Next subsections outline how these components interact to support the main trygger life cycle stages: edition, verification, creation and enactment.

5.2   Interaction

The *trygger* definition and execution follows four steps:

***Trygger*Edition**. *Tryggers* are edited through the *YQL* console (see Figure 6). This console (1) detects the specification of a *trygger* in the *YQL* statement input box, and if this is the case, (2) provides a hyperlink for creating the *trygger* which is stored locally. So-stored *tryggers* can latter by managed through the new *TRYGGER ALIASES* tab (3). This tab displays a list of the locally kept *tryggers* together with icons for deletion, enabling/disabling and reading the log (4).

***Trygger* Verification**. Once edited, the *trygger* can be verified by clicking the "test" button (5). So far, *Trygger* verification comprises: (1) *Trygger* syntax, (2) checking for the triggering ODT to hold a *<select>* tag (this is needed for monitoring).

***Trygger* Creation**. Once the *trygger* is satisfactory verified, the "Create Trygger Alias" link is activated. Besides making the *trygger* locally available, *trygger* creation involves the generation of a *MARS* rule [3]. The so-generated rule is

---

[11] https://pubsubhubbub.appspot.com/
[12] http://pubsubhubbub.superfeedr.com/
[13] http://onekinpush.appspot.com/
[14] These ODT tables are available for inspection at
https://github.com/yql/yql-tables/blob/master/trygger/JSinterpreter.xml
and .../*RSS-izator.xml*

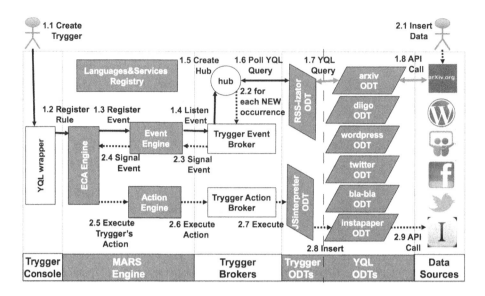

Fig. 5. Interaction diagram among *Trygger* components: *trygger* creation (continuous line) and *trygger* enactment (dotted line) propagating the NEW event occurrence

next registered in MARS (see Figure 5, continuous line). As a result, *MARS* requests the *Event Engine* to register the rule's (1.3) *who* sends the event specification to this listener (1.4). This listener is realized by creating a hub in a *PubSubHubbub* server (1.5). This server periodically polls the triggering website by issuing the *YQL SELECT* query that is obtained from the rule's event (1.6).

***Trygger* Enactment.** The hub detects an incremental change at the next polling time in terms of feeds (2.1 in Figure 5, dotted line). Next, it sends to the *Trygger Event Broker* the updated feed (2.2). The broker creates an event occurrence from the feed, and signals this happening to the *Event Engine* (2.3). The event engine forwards the occurrence to the *ECA Engine* (2.4). The *ECA Engine* retrieves those rules that match the type of new event occurrence, checks their condition and if met, fires the rules in parallel. Rule firing, i.e. enacting rule's action, is conducted through the *Action Engine* (2.5). The *Action Engine* forwards the petition to the domain-specific broker that knows how to handle the statements of the rule's action, i.e. the *Trygger Action Broker* (2.6) which is realized as the *JSinterpreter* ODT (2.7). The *JSinterpreter* processes the *trygger*'s action which can in turn, contain insertion or selection on other ODT tables.

6   Validation

ISO-9126 [5] provides a framework to evaluate quality in use. This section provides a preliminary evaluation of *Trygger* along the ISO-9126's quality-in-use dimensions. We evaluate usability along three aspects:

Fig. 6. *YQL* console wrapped for *trygger* management

- Effectiveness, which is the capability of the software product to enable users to achieve specified goals with accuracy and completeness. Indicators of effectiveness include quality of solution and error rates. The *"quality of solution"* is used as the primary indicator of effectiveness, i.e. a measure of the outcome of the user's interaction with the system.
- Productivity, which is the relation between the capability of the software product to enable users to expend appropriate amounts of resources in relation to the effectiveness. Indicators of efficiency include task completion time and learning time. In this study, we use *"task completion time"* as the primary indicator of productivity.
- Satisfaction, which is the users' comfort with and positive attitudes toward the use of the system. Users' satisfaction can be measured by attitude rating scales such as SUMI [11]. In this study, we use *"preference"* as the primary indicator of satisfaction.

6.1   Research Method

Setting. In order to eliminate differences in the perception of *Trygger* due to hardware or bandwidth differences, the study was conducted in a Faculty laboratory. All participants used computers with the same features and a clean installation of Firefox 12.0.

Subjects. The experiment was conducted among 12 graduate students applying in a Master in Web Engineering. They satisfactorily passed a 10 hour course in Web Programmable issues, where they worked with *YQL*. They had accounts in *Twitter* and *Facebook*, and a 92% access social networks on a daily basis while 41% tweet with a weekly frequency. They know about *SlideShare, Arxiv* and *Instapaper* and they were familiarized with the notion of database triggers. Six students knew the existence of *Greasemonkey* but only two had previously installed Greasemonkey scripts.

Procedure. Before starting, a 30' talk with some practical examples of the *Trygger* syntax and the main functions of the *Trygger* console were given.

Initially, the participants installed the *Greasemonkey* extension and the *Trygger* plug-in. Once the framework was deployed, subjects were faced with the creation of the *tryggers* whose outputs correspond to those in Figure 3. In order to measure productivity, participants had to annotate the start time and the finishing time. Finally, the subjects were directed to a *GoogleDocs* questionnaire to gather their opinion about *Trygger*.

Instrument. An online questionnaire served to gather users' experience using *Trygger*. It consisted of four parts, one to collect the participants' background and one for each of the quality-in-use dimensions. In order to evaluate effectiveness, the questionnaire contained the proposed tasks so that participants could indicate if they had performed them and the main problems encountered, while productivity was measured using the minutes taken in such tasks. Satisfaction was measured using 10 questions, using a 5-point Likert scale (1=completely disagree, 5=completely agree).

6.2   Results

Effectiveness. 10 students completed the first *trygger* without additional help. Two had problems with the *Trygger* grammar. This makes us think that the basic *Trygger* syntax is intuitive enough. However, only 6 students completed the second *trygger*. All the participants managed to write the main structure of the *trygger*, but they had problems in dealing with namespaces and correctly identifying the path expressions that obtained the values of the action out of the event occurrence (e.g. *NEW.atom:entry.atom.title.**).

Productivity. Installation of the *Greasemonkey* and the *Trygger* plugs-in took on average 15'. We appreciated a considerable dispersion on the time involved in *trygger* specification. The first *trygger* involved 4' on average while the second took 12' on average. Rationales rest on the same issued previously mentioned: the difficulty in copying with namespaces and Path expressions.

Satisfaction. We evaluated satisfaction for two concerns: the *Trygger* console, and the *Trygger* syntax. According to the results (see Figure 7), the *Trygger* console was in general considered quite supportive. One exception was error messages. Users did not find useful the messages generated by *Trygger* (see next section). As for the *Trygger* syntax, users found easy to equate the *Trygger* syntax to the syntax of triggers in SQL. This "semantic anchoring" might explain the perception of *tryggers* being easy to specify. However, subjects were very critical with the specification of the *trygger*'s action. This is mainly due to the action taken its parameters from NEW. This variable is populated out of an ODT query. This query is based on an ODT table, but the query results are dynamically constructed, i.e. the shape of the output cannot be inferred from the shape of the input ODT. Therefore, users need first to enact the query and look at the output before specifying the *trygger*'s action. This certainly is a main stumbling block.

7 Discussion

Trygger rises different non-functional issues as for usability, extensibility, scalability and reliability. This section discusses these concerns.

Usability (i.e. the ability to easily use *Trygger*). From the aforementioned evaluation, two weaknesses can be identified: the difficulty in describing the *trygger*'s action (the last question of the questionnaire) and the limited information provided by the *Trygger* engine (question 4 of the questionnaire). The former advices from the introduction of enhanced interfaces that traces the *trygger's* enactment, basically, event signaling and action enactment. Debuggers are needed that show the trace of both operations so that users will notice the existence of a mismatch between the returned event occurrence, and the path expressions used to extract the values for the action. This moves us to the second problem: the limitations of the error messages in *Trygger*. So far, the *Trygger* engine does not have its own error messages. Rather, it propagates *YQL* message to the console. But *YQL* is about simple *YQL* statements whereas a *trygger* can be regarded as a compound: a query (the *trygger*'s event) + an update (the *trygger*'s action). This means that a perfectly valid query and a perfectly valid update might deliver a wrong *trygger*. A Trygger-specific error module is needed.

Extensibility (i.e. ability to define *tryggers* no matter the social website). *Trygger* relies upon ODT tables. Yahoo manages an ODT repository which is open to the community[15]. At the time of this writing, above 11054 APIs are documented at *http://www.programmableweb.com/*. Out of this number, 194 have been made *YQL* accessible by the community. We tap into the *YQL* community to keep the ODT repository updated with new comers.

Scalability. *Trygger* processing is highly parallelized. We rely on MARS architecture for this purpose. Besides communication, the polling frequency of the hub also introduces some latency in the process. However, we envision *Trygger* not to be used for time-critical scenarios but rather Web2.0 scenarios where a 5-hour latency between the rising of the event and the enactment of the action is acceptable. Finally, *Trygger* is an agent acting on third-party APIs. Some APIs limit the number of calls per day. Since all calls are made on behalf of *Trygger*, this could be a problem for popular sites. Notice however that popular APIs (e.g. *Twitter*) have a reasonable allowance threshold (around three calls per minute[16]).

Reliability. Reliability is manifold. First, response time. This is not an issue since our scenarios are not time critical. Second, recoverability. We rely on *MARS* for providing storage of rules. In addition, the *Trygger* console keeps a copy of *tryggers* as well. This permits to re-create the rule in *MARS* from the copy kept locally in the *Trygger* installation. Third, safety, specifically about user credentials. We illustrate this scenario for the *Twitter2Facebook trygger* (see Figure 3-

---

[15] http://www.datatables.org/
[16] https://dev.twitter.com/docs/rate-limiting

Trygger console	Results
There were no errors during the instalation and execution of *Trygger*	4,3
The interface is well integrated with a YQL console	3,5
I understand the enabled operations at each time	3,6
The error messages of *Trygger* are valuable	2,5
The managament of *tryggers* is easy	4,0
I find *Trygger* useful to keep synchronized my sites	3,3
I found the demo interesting enough to share with my colleagues	2,9
Trygger specification	**Results**
The syntax and semantic of the trygger is clear	3,7
Creation of the trygger is intuitive	3,3
The definition of the event part is well defined	3,8
The definition of the action part is well defined	2,6

Fig. 7. Questionnaire to assess *Trygger* usability. Likert scale is used from Strongly agree (5) to Strongly disagree (1).

A) where *Facebook* credentials were required. Credentials are captured as ODT columns (*i.e. access_token*). This is commonly realized through *OAuth*[17], an open standard for authorization that allows users to share their private resources through API access without having to hand out their credentials, typically username and password.

8   Related Work

Traditionally, cross publishing address two main challenges. First, crossing the border among different media (e.g., print, Web, and TV) [17][16]. Cross media can be defined as any content (news, music, text, and images) published in multiple media. Multiple media indicates that the same content is delivered to end users in more than one medium. The second challenge is to leverage hyperlink-like mechanisms to expand besides HTML pages to other media (i.e. hypermedia). For instance, in [14] the iServer framework is introduced as a generic link management and extensible integration platform that push the boundaries of hyperlinks to paper documents. This work introduces a third challenge akin to the Web 2.0: empowering end-users for defining their own cross publishing strategies. Cross publishing 2.0 admits different variations based on the expressiveness of the strategy (i.e. the criteria for selection, and the associated reactions). *Trygger* provides a push approach to tracking ODT-described resources, and proactively responds by enacting an *YQL*-powered action. Next paragraphs introduce other solutions.

In [12], the focus is on tracking tweets. Rather than using hashtags, the selection criteria to determine tweets of interest is described through an *Resource Description Framework* (*RDF*) query. This implies that tweets need first to be

---

[17] http://oauth.net/

automatically annotated along a pre-set list of ontologies. The *RDF* query is then addressed along the so-annotated tweets. The reaction is limited to the publications of the tweets meeting the *RDF* query. The architecture also relies on *PuSH* hubs. On the other hand, *ReactiveTags* [9] tracks tagging sites' annotations with specific tags (the so-called reactive tags). Unlike Mendes et al., *ReactiveTags* tracks multiple target sites where the selection and impact criteria are defined semantically in terms of Semantically-Interlinked Online Communities [4] Items. The data is tracked on the target through the API mechanism.

The *Social RSS* [1] *Facebook*'s app links *RSS* sources to a user's *Facebook* account. The app is a feed reader that monitors *RSS* sources and next, publishes each new feed in the user's wall. Reactions are limited to publishing feeds in *Facebook* walls.

Another reactive system for *RSS* tracking is *"Atomate It"* [15]. Both the criteria and the reactions are similar to those of *Trygger*. The difference stems from the architecture. *"Atomate It"* distinguishes two stages. First, *RSS* feeds are loaded. Second, condition-action rules (rather than ECA rules) are run over the previously stored feeds. The tool is aimed at end users so graphical assistance is provided. The downside is expressivity. Complex rules can not be defined, and extending either the *RSS* providers or the action list involves considerable programming. Similarly, *Ifttt*[18] permits users create cross publishing rules using a very intuitive interface, offering a place where users can comment and share their *Ifttt* rules. However, simplicity comes at the expense of customizability. *Ifttt* fixes the way the information is transformed between the two involved services (a.k.a. channels) while also canning the websites that can act as channels.

Which approach is better? Design choices are subject to tradeoffs between factors that will value some attributes while penalizing others. This in turn is influenced by the target audience and the target websites. The work by Mendes et al. focuses on *Twitter* and RDF as main technological platforms. *"Atomate It"* and *Ifttt* are opened to a larger though fixed set of websites. Their efforts favour easy production by trading expressiveness for learnability. *Trygger* explores a different scenario by taping into an existing community (*YQL*). While other approaches depart from raw open APIs (or their RSS counterparts), *Trygger* sits at a higher abstraction layer by starting from ODT tables. Not only does ODT simplifies the development of *Trygger*, it also provides a community that, on expanding *YQL*, is also extending the scope of *Trygger*. In addition, *Trygger* expressiveness attempts to find a compromise by using SQL-like syntax while leaving a backdoor for permitting the use of JavaScript to express more complex reactions.

9   Conclusion

The increasing number of resources kept in the Web together with the growing-up of digital natives make us hypothesize a need for sophisticated cross publishing capabilities. This paper advocates for an ubiquitous, platform-agnostic, do-it-yourself approach to cross publishing. This vision is borne out in *Trygger*, a

---

[18] http://ifttt.com

Firefox plugin that works on top of the *YQL* console. Capitalizing on *YQL*, *Trygger* permits to express cross publishing as *SQL*-like triggers. Next follow-on includes to come up with more elaborate Graphical User Interfaces that open *Trygger* to a wider, less technical, audience.

References

1. Social RSS homepage (2013), http://apps.facebook.com/social-rss/
2. Alferes, J., Amador, R., Behrends, E., Fritzen, O., May, W., Schenk, F.: Pre-standardization of the language. Technical report i5-d10. Technical report (2008), http://www.rewerse.net/deliverables/m48/i5-d10.pdf
3. Alferes, J., Amador, R., May, W.: A general language for evolution and reactivity in the semantic web. In: Fages, F., Soliman, S. (eds.) PPSWR 2005. LNCS, vol. 3703, pp. 101–115. Springer, Heidelberg (2005)
4. Bojars, U., Breslin, J.G.: SIOC core ontology specification (2007), http://rdfs.org/sioc/spec/
5. Davis, I., Vitiello Jr., E.: ISO 9241-11. Ergonomic requirements for office work with visual displays terminals(vdts) (1998)
6. Díaz, O., Arellano, C., Azanza, M.: A language for end-user web augmentation: Caring for producers and consumers alike. ACM Transactions on the Web (TWEB) 7(2), 9 (2013)
7. Fischer, G.: End-user development and meta-design: Foundations for cultures of participation. In: Pipek, V., Rosson, M.B., de Ruyter, B., Wulf, V. (eds.) IS-EUD 2009. LNCS, vol. 5435, pp. 3–14. Springer, Heidelberg (2009)
8. Fitzpatrick, B., Slatkin, B., Atkins, M.: Pubsubhubbub homepage, http://pubsubhubbub.googlecode.com/
9. Iturrioz, J., Díaz, O., Azpeitia, I.: Reactive tags: Associating behaviour to pre-scriptive tags. In: Proceedings of the 22nd ACM Conference on Hypertext and Hypermedia, HT 2011. ACM (2011)
10. Iturrioz, J., Díaz, O., Azpeitia, I.: Generalizing the "like" button: empowering web-sites with monitoring capabilities. In: ACM (ed.) 29th Symposium On Applied Computing (volume to be published, 2014)
11. Kirakowski, J., Corbett, M.: SUMI: The software usability measurement inventory. Journal of Educational Technology 24(3), 210–212 (1993)
12. Mendes, P.N., Passant, A., Kapanipathi, P., Sheth, A.P.: Linked open social signals. In: Proceedings of the 2010 IEEE/WIC/ACM International Conference on Web Intelligence and Intelligent Agent Technology, Washington, DC, USA, vol. 01, pp. 224–231 (2010)
13. Yahoo! Developer Network. Yahoo query language (YQL) guide (2011), http://developer.yahoo.com/yql/guide/yql_set.pdf
14. Norrie, M.C., Signer, B.: Information server for highly-connected cross-media pub-lishing. Information Systems 30(7), 526–542 (2003)
15. VanKleek, M., Moore, B., Karger, D.R., André, P., Schraefel, M.C.: Atomate it! end-user context-sensitive automation using heterogeneous information sources on the web. In: Proceedings of the 19th International Conference on World Wide Web, New York, NY, USA, pp. 951–960 (2010)
16. Veglis, A.: Comparison of alternative channels in cross media publishing. Publishing Research Quarterly 24(2), 111–123 (2008)
17. Veglis, A.A.: Modeling cross media publishing. In: The Third International Con-ference on Internet and Web Applications and Services, pp. 267–272 (2008)

Ensuring Web Interface Quality through Usability-Based Split Testing

Maximilian Speicher[1,2], Andreas Both[2], and Martin Gaedke[1]

[1] Chemnitz University of Technology, 09111 Chemnitz, Germany
`maximilian.speicher@s2013.tu-chemnitz.de`,
`martin.gaedke@informatik.tu-chemnitz.de`
[2] R&D, Unister GmbH, 04109 Leipzig, Germany
`{maximilian.speicher,andreas.both}@unister.de`

Abstract. Usability is a crucial quality aspect of web applications, as it guarantees customer satisfaction and loyalty. Yet, effective approaches to usability evaluation are only applied at very slow iteration cycles in today's industry. In contrast, conversion-based split testing seems more attractive to e-commerce companies due to its more efficient and easy-to-deploy nature. We introduce *Usability-based Split Testing* as an alternative to the above approaches for ensuring web interface quality, along with a corresponding tool called *WaPPU*. By design, our novel method yields better effectiveness than using conversions at higher efficiency than traditional evaluation methods. To achieve this, we build upon the concept of split testing but leverage user interactions for deriving quantitative metrics of usability. From these interactions, we can also learn models for predicting usability in the absence of explicit user feedback. We have applied our approach in a split test of a real-world search engine interface. Results show that we are able to effectively detect even subtle differences in usability. Moreover, WaPPU can learn usability models of reasonable prediction quality, from which we also derived interaction-based heuristics that can be instantly applied to search engine results pages.

Keywords: Usability, Metrics, Heuristics, Interaction Tracking, Search Engines, Interfaces, Context-Awareness.

1 Introduction

In e-commerce, the usability of a web interface is a crucial factor for ensuring customer satisfaction and loyalty [18]. In fact, Sauro [18] states that "[p]erceptions of usability explain around 1/3 of the changes in customer loyalty." Yet, when it comes to interface evaluation, there is too much emphasis on so-called *conversions* in today's industry. A conversion is, e.g., a submitted registration form or a completed checkout process. While such metrics can be tracked very precisely, they lack information about the actual usability of the involved interface. For example, a checkout process might be completed accidentally due to wrongly labeled buttons. Nielsen [17] even states that a greater number of conversions can

S. Casteleyn, G. Rossi, and M. Winckler (Eds.): ICWE 2014, LNCS 8541, pp. 93–110, 2014.

be *contradictory* to good usability. In the following, we illustrate this challenge by introducing a typical example scenario.

Scenario. A large e-commerce company runs several successful travel search engines. For continuous optimization, about 10 split tests are carried out per live website and week. That is, slightly different versions of the same interface are deployed online. Then, the one gaining the most conversions is chosen after a predefined test period. The main stakeholder, who studied business administration and founded the company, prefers the usage of *Google Analytics*[1] or similar tools due to their precise and easy-to-understand metrics. Yet, the split testing division would like to gain deeper insights into users' behavior since they know that conversions do not represent usability. Thus, they regularly request more elaborate usability evaluations, such as expert inspections for assessing the interfaces. The stakeholder, however, approves these only for novel websites or major redesigns of an existing one. To him, such methods—although he knows they are highly effective[2]—appear to be overly costly and time-consuming. *Conversion-based split testing* seems more attractive from the company's point of view and is the prime method applied for optimization.

Requirements. The situation just described is a common shortcoming in today's e-commerce industry, which is working at increasingly fast iteration cycles. This leads to many interfaces having a suboptimal usability and potentially deterring novel customers. Thus, we formulate three requirements for a novel usability testing approach that would be feasible for everyday use in industry and support a short time-to-market:

(R1) Effectiveness A novel approach must be more effective than conversion-based split testing w.r.t. determining the usability of an interface.
(R2) Efficiency A novel approach must ensure that evaluations are carried out with minimal effort for both, developers and users. Particularly, deployment and integration must be easier compared to established methods such as expert inspections or controlled lab studies.
(R3) Precision A novel approach must deliver precise yet easy-to-understand metrics to be able to compete with conversion-based split testing. That is, it must be possible to make statements like "Interface A has a usability of 99% and Interface B has a usability of 42%. Thus, 'A' should be preferred."

A solution to the above is to derive usability directly from interactions of real users, such as proposed by [21]. However, they conclude that user intention and even small deviations in the low-level structures of similar webpages influence interactions considerably. This makes it difficult to train an adequate *usability model M* that predicts usability U from interactions $\boldsymbol{I}$ only: $M(\boldsymbol{I}) = U$. Still, the described approach yields great potential.

We consider the pragmatic definition of usability as presented in [20], which is based on ISO 9241-11. Using a corresponding instrument specifically designed

---

[1] http://www.google.com/analytics/ (2014-02-01).
[2] For example, [16] state that only five evaluators can find up to 90% of the usability problems in an interface.

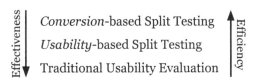

Fig. 1. Web Interface Usability Evaluation: the competing approaches (rough overview)

for correlation with client-side interactions [20], we propose a general approach to *Usability-based Split Testing* rather than considering conversions. To achieve this, we provide *WaPPU*—a tool that caters for (a) user interaction tracking, (b) collecting usability judgments from real users, (c) training usability models and (d) correlation of the obtained data. By design, the concept of Usability-based Split Testing enables developers to *ensure the quality of a web application* w.r.t. its interface usability at higher effectiveness than conversion-based split testing and higher efficiency than traditional approaches to usability evaluation (Fig. 1).

Making use of WaPPU (*"Was* that *Page Pleasant* to *Use?"*) we performed a usability-based split test of a real-world search engine results page (SERP). We paid specific attention to user intention and differences in low-level page structure to overcome the problems pointed out in [21]. From the study results, we derived interaction-based usability models and quantitative heuristic rules for SERPs. These can be instantly applied to user interactions collected on a SERP for a reasonable approximation of usability at very high efficiency.

In the following section, we give an overview of related work and describe the initial user study motivating our novel approach. Subsequently, we explain the concept of Usability-based Split Testing (Sec. 3) and the corresponding tool WaPPU (Sec. 4). The evaluation involving two web interfaces of a real-world search engine are presented in Section 5, followed by our findings (Sec. 6). Current limitations of our approach and potential future work are discussed in Section 7 before giving concluding remarks in Section 8.

2   Related Work

Our research is related to a wide variety of existing work. In particular, we are going to refer to *automatic* and *metrics-based* approaches to usability evaluation that are partly based on *user interaction analysis*. We also present an earlier study on the feasibility of quantitative interaction-based usability evaluation.

2.1   Automatic Approaches to Usability Evaluation

User Interaction Analysis. Atterer et al. [1] present a tool for client-side user interaction tracking. After having collected information about cursor behavior, keyboard strokes or dwell time, one can use these events to visualize a users interactions on a webpage. From these, the authors aim to infer implicit interactions, such as hesitation before filling in an input field [1]. This is a useful tool for facilitating more automatic usability tests and provides developers with valuable information. *m-pathy*[3] is a commercial tool for qualitative user behavior inspection that follows the concept described by [1]. The tool features additional metrics that are, however, in analogy to conversion-based split testing, e.g., the number of checkout processes and similar.

Web Usability Probe [2] is a more sophisticated tool also allowing for automatic remote usability testing. It is possible to define optimal logs for given tasks, which are then compared to the client-side logs actually produced by the user. De Vasconcelos and Baldochi Jr. [4] follow a similar approach that compares users' interactions against pre-defined patterns.

In contrast to our novel approach, all of the above methods—although as well aiming at usability improvement—have different focuses. None derives *quantitative* statements about usability from the observed interactions, which would enable direct comparison of interfaces. Rather, interpretation of the delivered qualitative information is largely up to a developer or dedicated usability evaluator.

Navalpakkam and Churchill [12] investigate the possibility to infer the user experience of a web interface from mouse movements. In a user study, they find that certain features of interaction (e.g., hovers, arrival time at an element) can be used to predict reading experience and user distraction with reasonable accuracy. Yet, they investigate only these specific aspects. Particularly, the authors do not focus on providing interaction-based measures of usability or user experience for quantitative comparison of interfaces.

Website Checking. Tools such as *AChecker* [6] and *NAUTICUS* [3] aim at automatic checking of websites according to certain criteria and heuristics. While the first specifically focuses on web accessibility, the second tool also takes into account usability improvements for visually impaired users. Both tools are particularly able to automatically suggest improvements regarding the investigated interfaces. Yet, they only consider static criteria concerned with structure and content of a website rather than actual users' interactions.

A/B Testing. *AttrakDiff*[4] is a tool that enables A/B testing of e-commerce products for user experience optimization. That is, based on a dedicated instrument, the hedonic as well as pragmatic quality of the products are compared [11]. While this may seem very similar to our proposed approach, it has to be noted that the aim of AttrakDiff is different from Usability-based Split Testing. Particularly, the tool leverages questionnaire-based remote asynchronous evaluation rather than focusing on user interactions. Also, qualitative, two-dimensional statements about user experience are derived, which has to be clearly distinguished from usability and quantitative metrics thereof.

---

[3] http://www.m-pathy.com/cms/ (2014-02-24).

[4] http://attrakdiff.de/ (2014-02-24).

2.2   Metrics-Based Approaches to Usability Evaluation

Contrary to the above approaches, Nebeling et al. [14] take a step into the direction of providing quantitative metrics for webpage evaluation. Their tool analyzes a set of spatial and layout aspects, such as *small text ratio* or *media–content ratio*. These metrics are static (i.e., purely based on the structure of the HTML document) and specifically aimed at large-screen contexts. In contrast, our goal is to provide *usability-in-use* metrics based on users' dynamic interactions with the webpage.

W3Touch [15] is a metrics-based approach to adaptation of webpages for touch devices. This means certain metrics of a website, e.g., *average zoom*, are determined from user interactions on a per-component basis. Components with values above a certain threshold are then assumed to be "critical" and adapted according to rules defined by the developer. This is a very promising approach that is, however, specifically aimed at touch devices. Moreover, the webpage metrics that identify potentially critical parts of a webpage are not transferred into more precise statements about usability.

2.3   Motivating Study

In the following, we address earlier work of the authors of this paper [21] that motivates the concept of Usability-based Split Testing.

In [21], we have tried to solve the already described conflict between traditional usability evaluations and conversion-based split testing by learning usability models from user interactions. For this, we have collected user interaction data on four similarly structured online news articles. Study participants had to answer a specific question about their assigned article. However, only two of the articles contained an appropriate answer. Once they found the answer or were absolutely sure the article did not contain it, participants had to indicate they finished their task and rate the web interface of the article based on the novel INUIT usability instrument [20]. INUIT contains seven items designed for meaningful correlation with client-side interactions (e.g., , "cursor speed positively correlates with confusion") from which an overall usability score can be derived.

All articles featured a single text column with a larger image on top and a sidebar which contained additional information and/or hyperlinks. Two articles featured a short text (~1 page) while the remaining two featured a longer text (≥2 pages). Moreover, two of the articles featured images within or close to the text column. Still, the high-level structures of the articles were similar. The lower-level differences were chosen by purpose to provoke differences in user behavior and usability judgments. Also, by providing an answer to the participant's task in only two of the four articles, we have simulated different user intentions according to [8]. That is, participants who could find an answer acted like *fact finders* while the remaining participants behaved like *information gatherers*.

We used a dedicated jQuery plug-in to track a set of well-defined interactions (e.g., clicks, hovers, scrolling distance etc.). These interactions were recorded separately for: (1) the whole page; (2) a manually annotated area of interest, i.e., the article text; (3) all text elements; (4) all media elements; (5) text elements within the area of interest; and (6) media elements within the area of interest.

To give a representative example, the interaction feature $hoverTime_{text}$ describes the aggregated time the user spent hovering text elements anywhere on the page. Furthermore, all interaction feature values were normalized using appropriate page-specific measures. For example, the page dwell time was divided by the length of the article text to ensure comparability across the four webpages.

The main hypothesis investigated in the study was *whether it is possible to learn a common usability model from similarly structured webpages.* Such a model should be able to also predict the usability of a webpage which did not contribute training data, as long as its structure is similar to a certain degree. However, when we correlated the collected interactions and usability judgments, we found huge differences between the four news articles. In fact, there was no interaction feature that showed a considerable correlation with usability for all four investigated webpages.

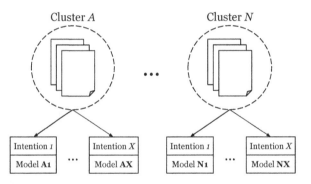

Fig. 2. Concept of a general framework for providing interaction-based usability models

This result indicates that user behavior depends on low-level webpage structure and intention more strongly than assumed, i.e., *interactions* are a function of *usability*, *structure* and *intention.*

Thus, we concluded that a general framework for interaction-based usability evaluation requires additional preprocessing steps: (1) structure-based clustering of webpages; (2) determining user intention, e.g., following the approach proposed by [8]; and (3) providing a common usability model per cluster and intention.

That is, for X types of user intention, a corresponding framework would have to provide X usability models per webpage cluster (Fig. 2). For example, assume a cluster contains all blogs using the same WordPress template. Then one would have to train different models for users *just browsing around* and users looking for a *certain piece of information* since these two intentions cause considerable differences in behavior. In the remainder of this paper, we address how to derive appropriate usability models and heuristics w.r.t. the requirements just described.

3   Usability-Based Split Testing

We propose *Usability-based Split Testing*, which is a feasible trade-off between effectiveness and efficiency, as an alternative to established approaches (Fig. 1). That is, we aim at significantly better predictions of usability than can be done using conversions. Besides, we want to be more efficient than established methods of usability evaluation. To achieve this, we have designed a two-step process:

1. Track user behavior on the interfaces of a split test—i.e., the *interfaces-under-test*—and apply the resulting interaction metrics to **heuristic rules** for usability evaluation, e.g., "a higher cursor speed indicates more confusion". Test whether the difference between the interfaces is *significant*.
2. If the result is not significant, more specific information is required. Thus, add a *usability questionnaire* to one of the interfaces. From the answers and the tracked interactions, learn more specific **usability models** that can be applied to the remaining interfaces for predictions of usability.

For realizing these steps and meeting requirements **(R1)–(R3)**, as described in Section 1, our novel approach follows a set of well-defined principles that will be introduced in the following.

3.1   Component-Based Interaction Tracking

The major goal of our approach is to overcome the problems regarding interactions and low-level page structure, as described in Section 2.3. During the study motivating this paper, interaction feature values were calculated on a very fine-grained basis. Particularly, interactions on any text or media element were considered for analysis, no matter how tiny or unimportant. This means that removing some minor elements from a webpage—such as text snippets in a sidebar—would already impact the values of interaction features. Also, only normalized absolute values were considered, rather than paying attention to relative distributions of interactions across the webpage.

To address this issue of interactions being highly dependent on low-level structure, we follow a *component-based approach*. For this, an interface-under-test, i.e., a single webpage, is divided into higher-level components such as the whole navigation bar rather than considering individual links. This approach partly follows the concepts of areas of interest [7] and critical components [15]. The rest of the webpage is treated as a separate, remaining component while the lower-level structure within a component is considered a *black box*. It is also possible to apply this to components in the context of other approaches, e.g., the WebComposition approach [5]. Since we intend to track interactions on a per-component basis, in this way small changes to the lower-level structure—e.g., removing minor text snippets—do not have an impact on feature values. *Usability models learned from such component-based interactions can then be applied to different webpages as long as the large-scale structure remains the same.*

3.2   Interaction-Based Heuristic Rules for Usability Evaluation

Interactions tracked in the context of a usability-based split test can be easily applied to pre-defined heuristic rules. To give just one example, assume a rule stating that *higher cursor speed positively correlates with user confusion*. Then, if the users of one interface-under-test produce significantly lower cursor speeds than users of another, this is a clear indicator of less confusion. By design, this variant of our approach is as efficient as conversion-based split testing *(R2)*. That is, it can be very quickly deployed on online webpages and does not bother the user with requests for explicit feedback. Moreover, the collected interaction-based metrics are precise and easily interpretable using the given heuristic rules *(R3)*.

A drawback of this variant is the fact that the rules used need to be determined in a different setting of the same context (i.e., similar high-level structure, similar user intention) first. That is, a dedicated training phase is required, e.g., a controlled user study during which explicit usability judgments are correlated with interactions. Since the applied heuristic rules originate from a different setting, they cannot be a perfect measure of usability for the interfaces-under-test. Rather, they can only give reasonable approximations, but still provide more insights into users' actual behavior than conversions *(R1)*. However, if this approach fails to deliver significant results, one needs to obtain more precise information for predicting usability by leveraging corresponding models.

3.3   Leveraging Usability Models

The second variant of our Usability-based Split Testing approach uses models for predicting usability. For this, one interface-under-test is chosen to deliver training data. That is, users of the interface are presented with a questionnaire asking for explicit, quantitative judgments of usability. This questionnaire is based on INUIT [20], an instrument describing usability by a set of only seven items: *informativeness, understandability, confusion, distraction, readability, information density* and *accessibility*. In this way, the number of questions a user has to answer is kept to a minimum [20]. The items have also been specifically designed for meaningful correlation with client-side user behavior [20]. Together with the collected interactions, explicit judgments are then used for training appropriate models based on existing machine learning classifiers. Since the interfaces-under-test all feature the same high-level structure—in accordance with *component-based interaction tracking*—these models can be applied to the interactions of the remaining interfaces for predictions of usability.

This variant of our approach cannot reach the same efficiency as conversion-based split testing since parts of the users are faced with requests for explicit feedback. Yet, by design, it is more efficient than traditional methods such as remote asynchronous user studies *(R2)*. Given the minimum of two interfaces-under-test, only 50% of our users are presented with questionnaires, compared to 100% of the participants in a controlled user study. Moreover, our approach can be easily applied to online web interfaces. It does not require a cumbersome study set-up since we rely on interactions and judgments of real users only.

In comparison to conversions or heuristic rules, models provide considerably more precise insights into users' behavior and its connection to usability *(R1)*. Also, questionnaires and models deliver an easily interpretable set of quantitative metrics in terms different usability aspects (*informativeness*, *understandability* etc.) for comparing interface performance *(R3)*.

4   The WaPPU Tool

To provide a ready-to-use framework for Usability-based Split Testing, we have designed a novel context-aware tool called *WaPPU.*The tool caters for the whole process from interaction tracking to deriving correlations and learning usability models. Based on the principles of Usability-based Split Testing, we have implemented WaPPU in terms of a central split testing service. This service has been realized using *node.js*[5]. Split testing projects are created in the WaPPU dashboard (Fig. 3), which provides the developer with ready-to-use JavaScript snippets that simply have to be included in the different interfaces-under-test. The only other thing required for deployment of the split test is a client-side jQuery plug-in for component-based interaction tracking. The overall architecture of WaPPU can be seen in Figure 3. The current implementation supports at most two interfaces per split test, i.e., only A/B testing is supported.

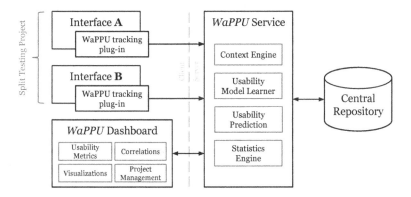

Fig. 3. Architecture of WaPPU

Interaction Tracking. Our tool tracks a total of 27 well-defined user interaction features, of which the most expressive ones are shown in Table 1. They have been derived from existing research [7,12,15,19] as well as video analyses of real search engine users. The features are tracked for each component defined by the developer, except for features annotated with an asterisk, which cannot be applied to individual components. Moreover, each feature is tracked for the

---

[5] http://nodejs.org/ (2014-02-21).

Table 1. Selection of interaction features tracked by WaPPU (* = whole page feature only). The complete list can be found in our online appendix [22].

label	description	source
charsTyped	# characters typed	
cursorMoveTime	time the mouse cursor spends moving	[19]
cursorSpeed	cursorTrail divided by cursorMoveTime	[7,19]
cursorSpeedX	cursor speed in X direction	[7]
cursorStops	# cursor stops	[7]
cursorTrail	length of cursor trail	[7,19]
hovers	# hovers	[19]
hoverTime	total time spent hovering the component	[19]
pageDwellTime*	time elapsed between loading and leaving the page	[7]
scrollDirChanges*	# changes in scrolling direction	[15]
scrollMaxY*	maximum scrolling distance from top	[7]
scrollPixelAmount*	total amount of scrolling (in pixels)	[7]
textSelections	# text selections	
textSelectionLength	total length of all text selections	

whole web interface, which gives an additional implicit component. This gives us the chance to derive the relative distribution of features across the page, e.g., "25% of the total cursor trail lie in the navigation component". If a developer defines x components in their web interface and specifies that all features shall be considered, WaPPU tracks a total of $20(x + 1) + 7$ features during the split test (7 features are applied to the whole page instead of components).

Usability Judgments. WaPPU offers the option to automatically show a questionnaire when users leave an interface-under-test, in case they have agreed to contribute training data. This questionnaire contains the seven usability items of INUIT [20], each formulated as a question and to be answered based on a 3-point Likert scale. Since the value of an item is thus either -1, 0 or $+1$, we get an overall usability value that lies between -7 and $+7$. These values are what we refer to as *quantitative measures/metrics of (aspects of) usability* in the remainder of this paper. The questionnaire can be shown on either none, one or all of the interfaces-under-test in a split test. If no interface features a questionnaire, the functionality of WaPPU is reduced to collecting interactions only, i.e., for use with *usability heuristics* (cf. Sec. 3.2).

If it is featured on one interface, WaPPU automatically learns seven models—one per usability item—based on the users' answers. These models are associated with the corresponding split testing project and stored in WaPPU's central repository (Fig. 3). They are automatically applied to the remaining interfaces for *model-based usability prediction* (cf. Sec. 3.3). The current implementation of

WaPPU uses the updateable version of the Naïve Bayes classifier[6] provided by the WEKA API [10].

Finally, in case all interfaces feature the questionnaire, the developer receives the most precise data possible. This case requires no models and is particularly useful for *remote asynchronous user studies* from which one can derive heuristic rules for usability evaluation (cf. Sec. 3.2). It is not intended for evaluation of online interfaces since the amount of questionnaires shown to real users should be minimized.

Context-Awareness. The context of a user is automatically determined by WaPPU and all collected interactions and usability judgments are annotated accordingly. In this way, it is possible to integrate context into a usability model since different contexts trigger different user behaviors. Currently, we consider two aspects that to a high degree influence a user's interactions: *ad blockers* and *screen size*. That is, the context determined by our tool is a tuple $(adBlock, screenSize)$ with $adBlock \in \{true, false\}$ and $screenSize \in \{small, standard, HD, fullHD\}$. For this, we refer to the most common screen sizes and define: $small < 1024 \times 768 \leq standard < 1360 \times 768 \leq HD < 1920 \times 1080 \leq fullHD$.[7]

Small-screen and touch devices are not supported in the current version of WaPPU. They are detected using the MobileESP library[8] and corresponding data are ignored.

5   Evaluation

We have engaged the novel concept of Usability-based Split Testing for evaluating the interface of a real-world web search, which is currently a closed beta version and developed by the R&D department of *Unister GmbH*. For this, we have re-designed the standard search engine results page (SERP) and put both the old and redesigned versions of the interface into an A/B test using WaPPU. According to component-based interaction tracking as one principle of Usability-based Split Testing (cf. Sec. 3.1), we have defined two high-level components within the SERPs: the container element containing all search results (#serpResults) and the container element containing the search box (#searchForm). From this split test, we have obtained (a) an evaluation of the two SERPs, (b) corresponding usability models and (c) a set of interaction-based heuristics for general use with SERPs of the same high-level structure. Results suggest that our approach can effectively detect differences in interface usability and train models of reasonable quality despite a rather limited set of data.

The following describes the research method for evaluating the approach, the concrete test scenario, and presents the evaluation results. Datasets for reproducing results and detailed figures can be found in our online appendix [22].

---

[6] http://weka.sourceforge.net/doc.dev/weka/classifiers/bayes/
NaiveBayesUpdateable.html (2013-10-07).
[7] Cf. http://en.wikipedia.org/wiki/Display_resolution (2014-02-12).
[8] http://blog.mobileesp.com/ (2014-02-12).

Fig. 4. Search result from the original SERP (left) vs. search result from the novel SERP redesigned by three usability experts (right)

5.1 Method

The evaluation was carried out as a remote asynchronous user study whose workflow oriented at [13]. Participants were recruited via internal mailing lists of the cooperating company. Since user intention considerably affects interactions (cf. Sec. 2.3), we intended to minimze fluctuations in this respect. Thus, we definined a *semi-structured* task to simulate that all participants act according to a common intention, i.e., "Find a birthday present for a good friend that does not cost more than 50 Euros." We assumed that the vast majority of users would not immediately have an adequate present in mind and thus behave like *information gatherers* [8]. Additionally, in order to reduce context to different screen sizes only, participants were instructed to disable any ad blockers.

Each participant was randomly presented with one of the two SERP interfaces for completing their task. Before leaving a results page, they had to rate its usability using the INUIT questionnaire displayed by WaPPU. That is, we used a configuration of WaPPU which triggers a questionnaire in both interfaces in the A/B test. Since a user might trigger several searches and thus view several different SERPs, they potentially had to answer the questionnaire more than one time during the study. This means that one study participant could produce several datasets, each containing interactions and usability judgments. Answering one questionnaire per results page is necessary since different searches lead to different results, which influences usability items such as *informativeness* and *information density*. Participants were instructed to click a "finish study" button, once they found an adequate present. Clicking the button redirected to a final questionnaire asking for demographic information.

5.2 Interface Redesign

The web search's standard SERP interface was redesigned by three experts in order to increase its usability. The redesign was carried out according to established guidelines as well as the experts' own experiences with interface design and usability evaluations. One of the experts was a graphic designer and one was an interaction designer holding an according Master's degree, both with several years of experience in industry. The third expert was a PhD student with focus on human-computer interaction.

Some representative points concerning the redesign were: (a) better visual separation of results, (b) more whitespace, (c) giving text more space in terms of a greater line height, (d) aligning text and image elements more consistently,

(e) removing unnecessary comment and social media buttons and (f) reducing the amount of advertisements. Although the changes were rather subtle, we particularly assumed less *confusion* and *distraction* as well as better *readability* and *information density*. A visual comparison of exemplary search results from the two interfaces can be found in Fig. 4.

5.3   Results

We recruited 81 unique participants who contributed 198 individual datasets, i.e., they triggered 198 searches/SERPs. 17 of the participants were familiar with the investigated web search (i.e., they had used it before); 37 answered the final questionnaire (23 male). In general, participants stated they privately surf the internet for 2–3 hours per day, mostly for social networking (N=23) and reading news (N=22). The mostly used search engine is Google (N=35), in general several times a day (N=34) and usually for knowledge (N=31) and product search (N=21). On average, participants were 31.08 years old (σ=5.28).

During the study, we registered two different contexts: *HD* (N=46) and *full HD* (N=35). One participant was excluded from the analysis, because they delivered invalid data. For our evaluation, we additonally distinguish between users who were *not familiar* and users who were *familiar* with the web search since they produced considerably different results.

Interface Evaluation. First, we have a look at the usability evaluations of the two SERP interfaces w.r.t. the questionnaires filled out by the participants. That is, we investigate whether our approach was able to detect the difference in usability originating from the experts' interface redesign. For this, we have carried out four analyses regarding the two contexts *HD* and *full HD* as well as familiarity with the investigated web search.

The largest amount of datasets (89) was produced by users with HD screens who were not familiar with the web search. Participants also not familiar with the interface, but using full HD screens contributed 52 datasets. This makes a total of 141 datasets from novel users. In contrast, participants who were familiar with the web search contributed 47 datasets (30 HD, 17 full HD).

As is apparent from Table 2, the largest group of users (*HD/not familiar*) found the redesigned SERP to be significantly[9] better regarding the aggregated usability (i.e., the sum of all individual items; μ=2.45, σ=2.46). Moreover, the new interface performed significantly better concerning *distraction* (μ=0.62, σ=0.62) and *information density* (μ=0.43, σ=0.67). This matches our assumptions based on the experts' redesign. In general, the new SERP performs better regarding all usability items, although not always statistically significant.

In contrast, analysis of HD screen users familiar with the web search did not show a significant overall difference between the two SERPs. Yet, they judged the old interface to be significantly better concerning the individual item *confusion* (μ=0.69, σ=0.48). On average, it was also judged to be less *distracting*

---

[9] All tests of significance were carried out using the *Mann–Whitney U test* (α=0.05), which is particularly suitable for non-normally distributed independent samples.

Table 2. Evaluations by participants not familiar with the web search who used an HD screen (A = old interface, B = new interface)

usability item	A (N=47)		B (N=42)		significance
	μ	σ	μ	σ	
informativeness	-0.17	0.84	-0.02	0.84	—
understandability	0.34	0.70	0.45	0.67	—
confusion	0.30	0.78	0.38	0.70	—
distraction	0.36	0.74	0.62	0.62	p<0.05, W=798.5
readability	0.45	0.65	0.52	0.71	—
information density	0.04	0.69	0.43	0.67	p<0.01, W=692
accessibility	0.06	0.67	0.07	0.75	—
usability	1.38	2.96	2.45	2.46	p<0.05, W=782

(μ=0.85, σ=0.38) and have better *readability* (μ=0.62, σ=0.51) and *accessibility* (μ=0.38, σ=0.65). This finding is contrary to our assumptions. Rather, it indicates that users get accustomed to suboptimal interfaces and seem to be confused by changes even if they yield better usability from a more objective point of view.

Concerning the context *full HD/not familiar*, our analysis shows no significant differences between the two interfaces. However, results suggest that the usability of the old interface is better on average. Contrary, the redesigned SERP on average indicates better performance regarding *information density* (μ=0.19, σ=0.83) and *confusion* (μ=0.06, σ=0.85). Finally, full HD users who were familiar with the web search saw the biggest difference between the two interfaces. They judged the new SERP to be significantly better concerning *distraction* (μ=0.80, σ=0.42), *readability* (μ=0.60, σ=0.52), *information density* (μ=0.10, σ=0.57), *accessibility* (μ=0.50, σ=0.71) and aggregated usability (μ=2.60, σ=2.41). However, this context contained the smallest number of datasets (17) and therefore cannot be considered to be representative.

Usability Models. Based on the most representative dataset (Tab. 2) we have trained and tested Random Forest[10] classifiers for predicting usability across interfaces. This is in analogy to WaPPU's functionality of providing the questionnaire only in one interface and guessing the usability of the second interface from automatically learned models. Particularly, we intend to investigate whether component-based interaction tracking (cf. Sec. 3.1) is feasible for predicting the usability of a different webpage that did not contribute training data. For this, we take interaction data and usability judgments from the old SERP and train models from these—one for each INUIT usability item. The interaction data from the redesigned SERP are used as the test set for these models.

In a first step, we have selected the most expressive interaction features for each model. This has been done using *Correlation-based Feature Subset Selection*

---

[10] http://weka.sourceforge.net/doc.dev/weka/classifiers/trees/
RandomForest.html (2014-03-23).

[9] in combination with best-first search. That is, we have selected "[s]ubsets of features that are highly correlated with the class [to be predicted] while having low intercorrelation"[11]. Both functions are provided by the WEKA API [10]. Subsequently, we have trained the models based on the selected features and used our test set to evaluate them.

In general, the quality of the trained models was reasonably good. We obtained the most precise predictions for the item *distraction* (F-measure = 0.518), which was also one of the significant items for the considered context. In contrast, the item *readability* yielded the least precise predictions (F-measure = 0.296). The amount of training and test data was rather small for the investigated context (47 and 42 data sets, respectively). Thus, we assume better prediction quality with a larger amount of real-world users since correlations would then become more homogeneous, as has been observed in [19]. The precise results of the model evaluation can be found in our online appendix [22].

6 Key Findings of the User Study

The results from the largest and most representative group of participants *HD/not familiar* (Tab. 2) confirm that our approach is able to effectively detect differences in the usability of two versions of the same interface. We moreover found that users get used to interfaces and thus become less receptive to adjustments, even if these aim at better usability. What remains to be investigated are the differences between judgments from HD and full HD users. In particular, it requires deeper insights into users' actual behavior to understand why users familiar with the investigated web search produced very contradictory evaluations when differentiating between screen resolutions.

The usability models trained from our data underpin that the component-based tracking approach can reduce variations in users' interactions, which are caused by differences in lower-level structure. This was a major problem during the motivating study (cf. Sec. 2.3). Results suggest that WaPPU is able to predict interface usability based on adequate models with reasonable effectiveness.

Based on the feature selection process for learning usability models and Pearson's correlations r, we have additionally derived heuristic rules for SERPs, which are summarized in Figure 5. Regarding the dataset these rules are based on, their validity is theoretically restricted to *HD screen* users. Still, they can be applied to any SERP—as long as it is of similar structure and has the same components defined as the SERPs investigated in our evaluation—since many of the included features (e.g., *page dwell time*) do not strongly depend on screen resolution. To give just one example, a developer could monitor interactions on two SERPs. If *page dwell time* and *maximum scrolling distance* are significantly lower on one SERP, this is a clear signal for better *information density*. Yet, results must be interpreted with caution, as we have only investigated the user type *information gatherer* in our study. If it is not possible to obtain significant results

---

[11] http://weka.sourceforge.net/doc.dev/weka/attributeSelection/
CfsSubsetEval.html (2014-02-20).

- Better **informativeness** is indicated by
  - a lower absolute *cursor speed* on the search box (r=-0.21);
  - a higher relative amount of *hovers* on the search results (r=0.40).
- Better **understandability** is indicated by
  - a lower absolute *cursor speed* on the search box (r=-0.46);
  - a higher relative amount of *hovers* on the search results (r=0.24).
- Less **confusion** is indicated by
  - a lower relative *cursor speed (X axis)* on the search box (r=-0.49);
  - a lower absolute *maximum scrolling distance from top* (r=-0.44);
  - a lower absolute *amount of scrolling (in pixels)* (r=-0.33).
- Less **distraction** is indicated by
  - a lower absolute amount of *cursor stops* (r=-0.26);
  - a smaller absolute *length of the cursor trail* (r=-0.25).
- Better **readability** is indicated by
  - a lower absolute *page dwell time* (r=-0.21);
  - a smaller absolute amount of *text selections* (r=-0.27);
  - a smaller absolute *length of text selections* (r=-0.39).
- Better **information density** is indicated by
  - a lower absolute *page dwell time* on the search box (r=-0.11);
  - a lower absolute *maximum scrolling distance from top* (r=-0.27).
- Better **accessibility** is indicated by
  - a lower absolute amount of *characters typed* into the search box (r=-0.27);
  - a lower absolute amount of *changes in scrolling direction* (r=-0.31).

Fig. 5. Heuristic rules for usability evaluation of SERPs, as derived from our user study

from the heuristics, one must switch to a more effective method—e.g., leveraging specifically trained models, as described earlier.

7   Limitations and Future Work

We are aware of the fact that usability is a hard-to-grasp concept that is difficult to measure in an objective manner—if possible at all. However, our approach is able to yield reasonable approximations of usability in a quantitative and easy-to-understand form. This is particularly valuable in today's IT industry with its short time-to-market. If existing conversion-based analyses are augmented with Usability-based Split Testing, it will be possible to detect major shortcomings in web interfaces without having to carry out costly and/or time-consuming evaluations (yet, our approach only detects differences between interfaces-under-test and does not directly drive adequate changes for better usability). If results delivered by our method are not significant, it is still possible to apply such evaluations, which are more effective yet less efficient. However, we intend to minimize the need for the latter.

As has been pointed out earlier (cf. Sec. 2.3), a user's intention has considerable impact on their behavior. However, we have not yet considered intention as a factor in the design of our Usability-based Split Testing tool WaPPU. Rather, we modeled the user study for our evaluation in such a way that all users had to behave the same. Currently, we are investigating how intention can be derived from the interaction features tracked by WaPPU. According to [8], features such as the *page dwell time* can indicate user behavior. In future versions of WaPPU, we intend to add an extra question to the questionnaire asking for the user's intention. In this way, we can train an additional model for determining intention before applying adequate usability models or heuristics.

The current version of WaPPU is restricted to processing mouse and keyboard input. Yet, small-screen touch devices are gaining more and more popularity. Therefore, a major part of our future work will be to transfer Usability-based Split Testing into the context of touch devices. It will be particularly interesting to investigate how the different set of interaction features (e.g., missing *cursor trail*, new *zooming interaction*) affects usability prediction quality. First steps into this direction have already been taken by Nebeling et al. [15].

Finally, we intend to integrate our approach into the WebComposition process model [5] for enabling continuous evaluation of evolving widget-based interfaces.

8 Conclusions

This paper has presented *Usability-based Split Testing*—a novel method for *ensuring web inferface quality* based on quantitative metrics and user interactions. We have also introduced a corresponding A/B testing tool called *WaPPU*. Our approach intends to determine the usability of an interface more effectively than conversion-based methods while being more efficient than traditional approaches like expert inspections or controlled lab studies. To realize this, our method determines usability based on users' interactions. That is, we track interactions and apply them to either pre-defined heuristic rules or models trained with the help of users who answered an additional questionnaire. In this way, we obtain quantitative approximations of usability for empirically comparing web interfaces.

In a user study with 81 participants, we have applied our approach to the standard version and a redesigned version of a real-world SERP. Results show that our tool is able to detect the predicted differences in usability at a statistically significant level. Moreover, we were able to train usability models with reasonable prediction quality. Additionally, a set of key usability heuristics for SERPs could be derived based on user interactions. The study findings underpin the feasibility of the proposed approach.

Future work includes transferring the approach into the context of touch devices. Moreover, future versions of WaPPU shall be able to determine a user's intention before selecting appropriate usability heuristics and models.

Acknowledgments. We thank Robert Frankenstein, Björn Freiberg, Viet Nguyen, Thomas Stangneth and Katja Zatinschikow for helping with design and implementation of the user study. This work has been supported by the ESF and the Free State of Saxony.

References

1. Atterer, R., Wnuk, M., Schmidt, A.: Knowing the Users Every Move – User Activity Tracking for Website Usability Evaluation and Implicit Interaction. In: Proc. WWW (2006)
2. Carta, T., Paternò, F., de Santana, V.F.: Web Usability Probe: A Tool for Supporting Remote Usability Evaluation of Web Sites. In: Campos, P., Graham, N., Jorge, J., Nunes, N., Palanque, P., Winckler, M. (eds.) INTERACT 2011, Part IV. LNCS, vol. 6949, pp. 349–357. Springer, Heidelberg (2011)
3. Correani, F., Leporini, B., Patern, F.: Automatic Inspection-based Support for Obtaining Usable Web Sites for Vision-Impaired Users. UAIS 5(1) (2006)
4. de Vasconcelos, L.G., Baldochi Jr., L.A.: Towards an Automatic Evaluation of Web Applications. In: Proc. SAC (2012)
5. Gaedke, M., Gräf, G.: Development and Evolution of Web-Applications using the WebComposition Process Model. In: WWW9-WebE Workshop, Amsterdam (2000)
6. Gay, G.R., Li, C.Q.: AChecker: Open, Interactive, Customizable, Web Accessibility Checking. In: Proc. W4A (2010)
7. Guo, Q., Agichtein, E.: Beyond Dwell Time: Estimating Document Relevance from Cursor Movements and other Post-click Searcher Behavior. In: Proc. WWW (2012)
8. Gutschmidt, A.: Classification of User Tasks by the User Behavior. PhD thesis, University of Rostock (2012)
9. Hall, M.A.: Correlation-based Feature Subset Selection for Machine Learning. PhD thesis, University of Waikato (1998)
10. Hall, M., Frank, E., Holmes, G., Pfahringer, B., Reutemann, P., Witten, I.H.: The WEKA Data Mining Software: An Update. SIGKDD Explor. Newsl. 11(1) (2009)
11. Hassenzahl, M.: Hedonic, emotional and experiential perspectives on product quality. In: Ghaoui, C. (ed.) Encyclopedia of Human Computer Interaction, pp. 266–272. IGI Global (2006)
12. Navalpakkam, V., Churchill, E.F.: Mouse Tracking: Measuring and Predicting Users' Experience of Web-based Content. In: Proc. CHI (2012)
13. Nebeling, M., Speicher, M., Norrie, M.C.: CrowdStudy: General Toolkit for Crowdsourced Evaluation of Web Interfaces. In: Proc. EICS (2013)
14. Nebeling, M., Matulic, F., Norrie, M.C.: Metrics for the Evaluation of News Site Content Layout in Large-Screen Contexts. In: Proc. CHI (2011)
15. Nebeling, M., Speicher, M., Norrie, M.C.: W3Touch: Metrics-based Web Page Adaptation for Touch. In: Proc. CHI (2013)
16. Nielsen, J., Molich, R.: Heuristic Evaluation of User Interfaces. In: Proc. CHI (1990)
17. Nielsen, J.: Putting A/B Testing in Its Place,
    http://www.nngroup.com/articles/putting-ab-testing-in-its-place/
18. Sauro, J.: Does Better Usability Increase Customer Loyalty?
    http://www.measuringusability.com/usability-loyalty.php
19. Speicher, M., Both, A., Gaedke, M.: TellMyRelevance! Predicting the Relevance of Web Search Results from Cursor Interactions. In: Proc. CIKM (2013)
20. Speicher, M., Both, A., Gaedke, M.: Towards Metric-based Usability Evaluation of Online Web Interfaces. In: Mensch & Computer Workshopband (2013)
21. Speicher, M., Both, A., Gaedke, M.: Was that Webpage Pleasant to Use? Predicting Usability Quantitatively from Interactions. In: Sheng, Q.Z., Kjeldskov, J. (eds.) ICWE Workshops 2013. LNCS, vol. 8295, pp. 335–339. Springer, Heidelberg (2013)
22. WaPPU Online Appendix, http://vsr.informatik.tu-chemnitz.de/demo/WaPPU

Evaluating Mobileapp Usability:
A Holistic Quality Approach

Luis Olsina[1], Lucas Santos[1], and Philip Lew[2]

[1] GIDIS, Web Engineering School at Universidad Nacional de La Pampa, Argentina
[2] School of Computer Science and Engineering, Beihang University, China
olsinal@ing.unlpam.edu.ar, santos.ls@live.com,
philiplew@gmail.com

Abstract. As newer-generation smartphones enhance functionalities, interactions and services become more complex, leading to usability issues that are increasingly critical and challenging. Also mobile apps have several particular features that pose challenges evaluating their usability using current quality models, usability views, and their relations with target and context entities. With respect to the current literature, usability, actual usability, and user experience are poorly related to target entities (e.g. system and system in use) and context entities, to quality views (e.g. external quality and quality in use), in addition to measurement and evaluation building blocks. In this paper, we propose a holistic quality approach for evaluating usability and user experience of mobile apps. Practical use of our strategy is demonstrated through evaluation for the Facebook mobile app from the system usability viewpoint. Ultimately, a usability evaluation strategy should help designers to understand usability problems effectively and produce better design solutions so we analyze in the context of the framework's applicability toward this goal.

Keywords: Mobile app, Quality model, Usability, User Experience, Evaluation.

1 Introduction

Nowadays, for mobile apps, more robust network infrastructures and smarter mobile devices have led to increased functionality, integration and interactivity thereby warranting special attention in understanding their differences from apps on other platforms from the usability and user experience (UX) point of view because user requirements, expectations, and behavior can be somewhat different with the mobile platform. For instance, the quality design of Operability from a system viewpoint has a much different and greater influence for mobileapp Usability and UX due to the size of the screen and context of the user. Attributes such as button size, placement, color visibility, and widget usage have, for example, a much greater impact on task completion rates and task error rates [1, 5, 16] than for desktop platforms.

Nielsen *et al.* [16] indicate in recent mobile phone studies that usability varies by device category, which is mainly differentiated by screen size such as regular cellphones with small screen; smartphones with midsize screen and full A-Z keypad;

S. Casteleyn, G. Rossi, and M. Winckler (Eds.): ICWE 2014, LNCS 8541, pp. 111–129, 2014.
© Springer International Publishing Switzerland 2014

and full-screen smartphones with a nearly device-sized touch screen. Authors state that regular cellphones "*offer horrible usability, enabling only minimal interaction with websites*" (i.e. mobile webapps); and conclude "*unsurprisingly, the bigger the screen, the better the user experience when accessing websites*". This is supported by authors across several user testing studies from 2009 to 2012, in which the average success rate metric (which measures the percentage of users who were able to accomplish the proposed mobileapp tasks) rated for each mobile device category 44%, 55% and 74% respectively.

Despite these findings, the reader can ask him/herself what do "horrible usability" and "better UX" mean? What is the relationship between Usability and UX? Are they synonym concepts? Evaluating the success rate of users completing tasks correctly (as a performance indicator of effectiveness) is directly related to UX? If users are highly effective in completing tasks but they are unsatisfied due to perceived low app usefulness, then does UX score still high? Does UX depend on app Usability only or also from other characteristics such as Functional and Information Quality, Security, Reliability, and Efficiency? Is UX a quality characteristic of the system (e.g. a mobile app) or of an app in use? And, what about Usability?

Looking for the answers to these questions, we examined the current literature and found that Usability, Actual Usability (in-use) and UX are poorly linked to target entities (e.g. system and system in use) and context entities (e.g. device, environment, user, etc.), in addition to quality views (e.g. external quality and quality in use) and their quality models. Regarding quality models, ISO 25010 [9] outlines a flexible model with product/system quality –also known as internal and external quality (EQ)- and system-in-use quality –also referred to as quality in use (QinU). System quality consists of those characteristics and attributes that can be evaluated with the app in execution state both in testing and in operative stages; while system-in-use quality consists of characteristics and attributes as evaluated by end users when actually executing app tasks in a real context of use. ISO 25010 also delineates a relationship between the two quality views whereby system quality 'influences' system-in-use quality and system-in-use quality is determined by ('depends on') system quality. Usability is a system quality characteristic, while Effectiveness, Efficiency and Satisfaction are QinU characteristics. However, Actual Usability and UX, as experienced by the end user are missing concepts in the quoted standard.

From the QinU viewpoint, Hassenzahl [7] characterizes a user's goals into pragmatic, do goals and hedonic, be goals and categorizes system-in-use quality to be perceived in two dimensions, pragmatic and hedonic. Pragmatic quality refers to the system's perceived ability to support the achievement of tasks and focuses on the system's actual usability in completing tasks that are the 'do-goals' of the user. Hedonic quality refers to the system's perceived ability to support the user's achievement of 'be-goals', such as being happy, or satisfied with a focus on self.

Based on ISO 25010 among other works, such as [2, 7], we have developed 2Q2U (*Quality, Quality in use, actual Usability* and *User experience*) v2.0 [17], which ties together all of these quality concepts by relating system quality characteristics with Actual Usability and UX. Using the 2Q2U quality framework and a tailored strategy, evaluators can instantiate the quality characteristics to evaluate and conduct a systematic evaluation using the 'depends' and 'influences' relationships [14]. Besides in [12], we have addressed relevant features of mobile apps with regard to Usability and UX in the light of 2Q2U v2.0 quality models, but a global scheme which links main relationships among mobile target and context entities, quality views, characteristics and measurable properties were left for future endeavors.

Therefore, the major contributions of this research are: i) Represent relevant Usability and UX features of mobile apps with regard to system, system-in-use and context entities; ii) Analyze Usability and UX relationships, as well characteristics and attributes for mobile apps in the light of a conceptual framework and evaluation strategy; and iii) Illustrate an evaluation study for Facebook mobile app from the system usability viewpoint, showing the potential positive impact in designing quality interfaces. Lastly, we hope most of the above raised issues will be answered after reading this work.

Following this introduction, Section 2 describes a global scheme which links main relationships among mobile target and context entities, quality views, characteristics and measurable properties, measurement, and evaluation building blocks, with a focus on Usability and UX. Section 3, outlines our conceptual framework and evaluation approach which give support to the above building blocks. Section 4 demonstrates the practical use of our quality framework and evaluation strategy through the Facebook's mobileapp usability case study. Section 5 describes related work and, finally, Section 6 draws our main conclusions and outlines future work.

2   Featuring Mobileapp Usability and UX

For mobile phones, Usability and UX become crucial because users interact with apps –both native mobile apps and mobile webapps- in different contexts using devices with reduced display real estate. In particular the user's activity at the time of usage, location, and daytime, amongst other influencing factors such as user profile and network performance have an actual impact on the quality of the user's experience.

This section examines several features relevant for understanding and evaluating Usability and UX for mobile apps. To do this, Fig. 1 depicts the main building blocks which link some relationships among: i) entity categories, quality views/characteristics, and measurable properties (green/orange boxes); ii) measurement (light-blue box); and iii) evaluation (pink box). Next, we examine particularly i) features which allow better understanding non-functional requirements to further specify Usability and UX attributes for interface-, task-, and perception-based evaluations. First, we give a summary followed by deeper discussion.

The 'entity' label in the upper box represents the potential target entity category to be evaluated. It is defined as the "*object category that is to be characterized by measuring its attributes*", while an attribute is "*a measurable physical or abstract property of an entity category*" [18]. There are two instances of (super) categories for target entities that are of interest for evaluating Usability and UX, viz. product/system and system in use. In turn, an entity category can aggregate sub-entities, e.g., a mobile app is composed of basic and advanced GUI objects from the interface standpoint, as shown in Fig. 1. Moreover, lower level entities can be identified for GUI objects like button, menu, widget, etc. Another label in the upper box is 'context' which is defined as "*a special kind of entity representing the state of the situation of a target entity category, which is relevant for a particular information need*". So, system in use is characterized by a context-in-use entity (upper-right orange box) which in turn can aggregate environment, user and task contextual sub-entities.

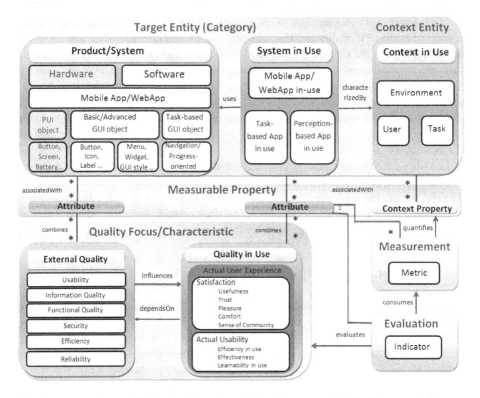

Fig. 1. Global scheme which links the main relationships among mobile Target and Context Entities, Quality Focuses, Measurable Properties, Measurement, and Evaluation building blocks. Note that the two lower-level Product/System sub-entity categories are just addressed for the Usability characteristic, which deals with user interface-oriented evaluation issues –PUI/GUI stands for Physical/Graphical User Interface.

On the other hand, an entity (category) and their sub-entities cannot be measured and evaluated directly but only by means of the associated measurable properties, i.e. attributes and context properties accordingly (see Fig. 1). Quality models can be the focus for different entities, and usually specify product/system or system-in-use quality requirements regarding main characteristics that can be further subdivided into sub-characteristics, which combine attributes. Product/system quality requirements are modeled by the EQ focus (view), which includes higher-level characteristics such as Usability, Security, Functional Quality, etc. Instead, system-in-use quality requirements are modeled by the QinU view, which include higher-level characteristics such as Actual UX, Satisfaction and Actual Usability.

Lastly, looking at the entity building block relationships, we see the 'uses' and 'characterized by' relations. Also between EQ and QinU views, we observe that system quality 'influences' system-in-use quality or system-in-use quality 'depends on' system quality. Note that for instantiated EQ and QinU models these relationships can be explored in light of concrete entity attributes by performing evaluations. For instance, using an evaluation strategy we can explore relationships between system quality and system-in-use quality attributes that may contribute to usage improvements. Regarding the above global scheme, in the next three sub-sections we closely examine the features of mobileapp Usability, UX, and context.

2.1   Featuring Mobileapp Usability

Usability is a characteristic for a system from the EQ viewpoint. It is one out of eight EQ characteristics in 2Q2U v2.0 (see [17] for quality models details). We define Usability as the "*degree to which the product or system has attributes that enable it to be understood, learned, operated, error protected, attractive and accessible to the user, when used under specified conditions*". (Note this definition is very close to that in ISO 9126-1 [11] rather than to [9], as we discuss in related work).

Examining the first part of the above definition, products are entities at early phases of a software life cycle (e.g., textual or graphical documents, etc.); while systems are executable software products (e.g. a mobile app in a testing or operative stage), which could include hardware and software together. Examining the second part of the above definition, we observe that the system (particularly, interface-related objects of the app) has attributes that enable the user to interact considering certain factors. These are the Usability sub-characteristics which can be evaluated through Understandability, Learnability, Operability, User error protection, UI aesthetics and Accessibility. Table 1 shows the Usability sub-characteristics and attributes definitions used in the Facebook evaluation study, in Section 4.

Recalling that characteristics and sub-characteristics combine attributes which are associated to entities (see Fig. 1) some typical mobileapp sub-entities that should be considered for Usability design and evaluation are entry fields and widgets, menus, carousels, breadcrumb path, amongst others. Entity sub-categories specific for Usability evaluation can be physical and graphical user interface (PUI/GUI) objects [6], and task-based GUI objects. A possible categorization for GUI objects can be basic or advanced objects (similar to that described in [8]).

All definitions for sub-characteristics and attributes in Table 1 include to a great extent the referred target sub-entity. For instance, the Visibility (1.3.2 coded) sub-characteristic is defined as "*degree to which the application enables ease of operation through controls and text which can be seen and discerned by the user in order to take appropriate actions*", and one combined attribute viz. Brightness difference appropriateness (1.3.2.1.1) is defined as "*degree to which the foreground color of the GUI object (e.g. text, control, etc.) compared to the background color provide appropriate brightness difference*". Actually, many attributes can determine whether or not the application is easily visible to the user. Depending on the context, different text colors and backgrounds can have a positive or negative impact. Remember that mobile users want to glance quickly and understand and operate almost immediately and there may be glare on their screen if they are outdoors. Also this means that appropriate usage of control/text colors (and size) can greatly impact the user's speed of comprehension and therefore, operational effectiveness and efficiency.

In addition to Usability, characteristics to evaluate other mobileapp EQ aspects are Security, Functional and Information Quality, etc. in which the target sub-entities should be defined accordingly, at least to the system lower layers.

Table 1. Definition of EQ/Usability sub-characteristics and attributes –in *italic*

Characteristic/*Attribute*	2Q2U v2.0 Definition
1 Usability	Degree to which the product or system has attributes that enable it to be understood, learned, operated, error protected, attractive and accessible to the user, when used under specified conditions.
1.1 Understandability (synonym Appropriateness Recognizability)	Degree to which users can recognize whether a product or system is appropriate for their needs. <u>Note</u>: Same ISO 25010 definition.
1.1.1 Familiarity	Degree to which the user understand what the application, system's functions or tasks are about, and their functionality almost instantly, mainly from initial impressions
1.1.1.1 Global organization scheme understandability	Degree to which the application scheme or layout is consistent and adheres to either de facto or industry standard to enable users to instantly understand its function and content.
1.1.1.2 Control icon ease to be recognized	Degree to which the representation of the control icon follows or adheres to an international standard or agreed convention.
1.1.1.2.1 Main control icon ease to be recognized	Degree to which the representation of the main controls icons follows or adheres to an international standard or agreed convention.
1.1.1.2.2 Contextual control icon ease to be recognized	Degree to which the representation of the contextual controls icons follows or adheres to an international standard or agreed convention.
1.1.1.3 Foreign language support	Degree to which the application functions, controls and content has multi-language support enabling user to change his/her language of preference.
1.2 Learnability	Degree to which the product or system enables users to learn its app.
1.2.1 Feedback Suitability	Degree to which mechanisms and information regarding the success, failure or awareness of actions is provided to users to help them interact with the application.
1.2.1.1 Current location feedback appropriateness	Degree to which users are made aware of where they are at the current location by an appropriate mechanism.
1.2.1.2 Alert notification feedback appropriateness	Degree to which users are made aware of new triggered alerts that they are involved by an appropriate mechanism.
1.2.1.3 Error message appropriateness	Degree to which meaningful error messages are provided upon invalid operation so that users know what they did wrong, what information was missing, or what other options are available.
1.2.2 Helpfulness	Degree to which the software product provides help that is easy to find, comprehensive and effective when users need assistance
1.2.2.1 Context-sensitive help appropriateness	Degree to which the application provides context sensitive help depending on the user profile and goal, and current interaction.
1.2.2.2 First-time visitor help appropriateness	Degree to which the application provides an appropriate mechanism (e.g. a guided tour, etc) to help beginner users to understand the main tasks that they can do.
1.3 Operability	Degree to which a product or system has attributes that make it easy to operate and control. <u>Note</u>: Same ISO 25010 definition
1.3.1 Data Entry Ease	Degree to which mechanisms are provided which make entering data as easy and as accurate as possible.

Table 1. (*Continued*)

1.3.1.1 Defaults	Degree to which the application provides support for default data.
1.3.1.2 Mandatory entry	Degree to which the application provides support for mandatory data entry.
1.3.1.3 Widget entry appropriateness	Degree to which the application provides the appropriate type of entry mechanism in order to reduce the effort required.
1.3.2 Visibility (synonym Optical Legibility)	Degree to which the application enables ease of operation through controls and text that can be seen and discerned by the user in order to take appropriate actions.
1.3.2.1 Color visibility appropriateness	Degree to which the main GUI object (e.g. text, control, etc.) color compared to the background color provide sufficient contrast and ultimately appropriate visibility.
1.3.2.1.1 Brightness difference appropriateness	Degree to which the foreground color of the GUI object (e.g. text, control, etc.) compared to the background color provide appropriate brightness difference.
1.3.2.1.2 Color difference appropriateness	Degree to which the foreground text or control color compared to the background color provide appropriate color difference.
1.3.2.2 GUI object size appropriateness	Degree to which the size of GUI objects (e.g. text, buttons, and controls in general) are appropriate in order to enable users to easily identify and operate them.
1.3.2.2.1 Control (widget) size appropriateness	Degree to which the size of GUI controls are appropriate in order to enable users to easily identify and operate them.
1.3.2.2.2 Text size appropriateness	Degree to which text sizes and font types are appropriate to enable users to easily determine and understand their meaning.
1.3.3 Consistency	Degree to which users can operate the task controls and actions in a consistent and coherent way even in different contexts and platforms.
1.3.3.1 Permanence of controls	Degree to which main and contextual controls are consistently available for users in all appropriate screens or pages.
1.3.3.1.1 Permanence of main controls	Degree to which main controls are consistently available for users in all appropriate screens or pages.
1.3.3.1.2 Permanence of contextual controls	Degree to which contextual controls are consistently available for users in all appropriate screens or pages.
1.3.3.2 Stability of controls	Degree to which main controls are in the same location (placement) and order in all appropriate screens.
1.4 User Error Protection	Degree to which a product or system protects and prevents users against making errors and provides support to error tolerance.
1.4.1 Error Management	Degree to which users can avoid and recover from errors easily.
1.4.1.1 Error prevention	Degree to which mechanisms are provided to prevent mistakes.
1.4.1.2 Error recovery	Degree to which the application provides support for error recovery.
1.5 UI Aesthetics (synonym Attractiveness)	Degree to which the UI enables pleasing and satisfying interaction for the user. <u>Note</u>: Same ISO 25010 definition.
1.5.1 UI Style Uniformity	Degree to which the UI provides consistency in style and meaning.
1.5.1.1 Text color style uniformity	Degree to which text colors are used consistently throughout the UI with the same meaning and purpose.
1.5.1.2 Aesthetic harmony	Degree to which the UI shows and maintains an aesthetic harmony regarding the usage and combination of colors, texts, images, controls and layouts throughout the whole application.

Table 2. Definition of QinU (sub-)characteristics absent in [9] or were rephrased in 2Q2U v2.0

Characteristic/Sub-characteristic Definition	ISO 25010 QinU Definition
Actual User Experience: Degree to which a system in use enable specified users to meet their needs to achieve specific goals with satisfaction, actual usability, and freedom from risk in specified contexts of use	Note: Absent characteristic in ISO 25010, but similar definition to QinU in this standard
Actual Usability (synonym Usability in use):Degree to which specified users can achieve specified goals with effectiveness, efficiency, learnability in use, and without communicability breakdowns in specified contexts of use	Note: Absent characteristic, but similar concept (i.e. *usability in use*) was in the ISO 25010 draft, and in [2]
Effectiveness: Degree to which specified users can achieve specified goals with accuracy and completeness in specified contexts of use	*Effectiveness*: Accuracy and completeness with which users achieve specified goals
Efficiency (in use): Degree to which specified users expend appropriate amounts of resources in relation to the effectiveness achieved in specified contexts of use	*Efficiency:* Resources expended in relation to the accuracy and completeness with which users achieve goals
Learnability (in use): Degree to which specified users can learn efficiently and effectively while achieving specified goals in specified contexts of use	Note: Absent characteristic
Sense of Community: Degree to which a user is satisfied when meeting, collaborating and communicating with other users with similar interest and needs	Note: Absent characteristic

As we depict in the next sub-section, UX is a broader concept that depends not only on Usability but also on other system characteristics such as Functional and Information Quality, Security, Reliability, Efficiency, and contexts of use as well.

2.2   Featuring Mobileapp UX

Fig. 1 shows UX as the higher-level characteristic for QinU evaluations. The QinU view characterizes the impact that the system in use (e.g. a mobile app) has on actual users in real contexts of use, i.e., while users perform application tasks in a real environment. Actual UX is defined in Table 2, as "*degree to which a system in use enable specified users to meet their needs to achieve specific goals with satisfaction, actual usability, and freedom from risk in specified contexts of use*".

UX is determined by the satisfaction of the user's be goals (hedonic), and do goals (pragmatic) as noted by Hassenzahl [7]. Moreover, do-goals relate to the user being able to accomplish what they want with Effectiveness and Efficiency (i.e. Actual Usability or Usability in use), while be-goals relate to the user's satisfaction. Satisfaction in [9] includes those subjective, perception-oriented sub-characteristics including Usefulness, Trust, Pleasure, and Comfort -also Sense of Community in [17].

Ultimately, Usability deals with the specification and evaluation of interface-based sub-characteristics and attributes of a system, while Actual Usability deals with the specification and evaluation of task-based sub-characteristics and attributes of an app in use, and Satisfaction with perception-based sub-characteristics and attributes. Recalling that sub-characteristics combine attributes which are associated to entities, we have considered in Fig. 1 two typical app-in-use target sub-entities, namely: Task-based and Perception-based App in-use. The Task-based App-in-use sub-entity can be evaluated using Effectiveness, Efficiency and Learnability in-use attributes. In [14], for the JIRA webapp in-use, we evaluated the "Entering a new defect" task performed by 50 beginner tester users, in which for example Effectiveness combined three attributes such as Sub-task correctness, Sub-task completeness and Task successfulness. (Note that Task successfulness attribute was measured in a similar way that the Average success rate used by Nielsen *et al.* [16]). On the other hand, the Perception-based App-in-use sub-entity can be measured and evaluated using Satisfaction sub-characteristics and attributes that can be included in questionnaire items as in [13], or evaluated through other methods such as observation.

As a final remark, mobileapp selected tasks should be evaluated with respect to real users performing real tasks. This issue includes several key design concerns that have significant impact on the Effectiveness and Efficiency of the final user. For example, *task workflows* need to be designed with the most common tasks in mind that would be suited to mobile usage. Because of the context of use of a mobile user, and the mobile user's limited attention span, the choice of tasks, task workflow and length are extremely important for this limited task set. If task workflows are not designed to be short, there is a higher probability of user error and a lower rate of completion –see Effectiveness definition in Table 2. Workflows therefore need to be compressed by combining several steps into one through careful task definition, evaluation and analysis. Reduced workflows, in turn, reduce task times and increase Efficiency (see definition in Table 2) while, at the same time, reducing error rates and error rate reduction is extremely critical for users with short attention spans. If you are driving and executing a task and get an error, do you continue trying?

2.3   Featuring Mobileapp Context

As mentioned above, the Context entity (category) is a special kind of entity representing the state of the situation of a target entity to be assessed, which is relevant for a particular measurement and evaluation (M&E) information need. Context for a given QinU M&E project is particularly important –i.e. a must-, as instantiation of QinU requirements must be done consistently and in the same context so that evaluations and improvements can be accurately assessed and compared. But also context is important regarding the EQ view, as we describe in Section 4. (Note that in order to reduce clutter in Fig 1, we did not draw an upper-left orange box for product/system context). For instance, system in use is characterized by a context-in-use entity, which in turn can aggregate environment, user and task sub-entities, while a product/system context (for idem target entity), can be characterized by sub-entities such as device (hardware), software, etc.

As commented in [12], the context of a mobileapp user is much different than a traditional desktop webapp user not only due to the size of the screen but also to other situations that influence the user's environment and therefore its behavior. In particular the user's activity at the time of usage, location, amongst other influencing factors such as user profile have actual impact on the quality of the user's experience.

Some a few of these factors considering context sub-entities and properties include: i) *Activity*: What users are doing at the time of usage have a significant influence on the user's attention span; e.g. if they are driving, then they have a very short attention span, maybe 1 second, versus if they are in the middle of a conversation, perhaps they have an attention span of 3 seconds; ii) *Day/time of day*: The day and time can impact what a user is doing, and the level of natural light. Unlike desktop apps which are typically accessed indoors, the usage of mobile apps is particularly sensitive to this contextual factor influencing visibility; ii) *Location*: The location of the user influences many elements; e.g. indoors, outdoors, in a car or in an elevator, all of which can also be related to the user activity; iii) *Network performance*: Obviously the speed at which an app uploads and downloads data is going to have a great impact because of the decreased attention span; iv) *User profile*: The increasing complexity of software combined with an aging user demographic has an interesting effect on the usability of mobile apps. For aging users, usually their close range vision capability has diminished along with their dexterity. On the other hand, apps have become complex, and therefore function and content simplicity and understandability are also critical and influenced by the particular user group. Not only are there more aging users, there are also more younger users as children these days begin using computing devices as toddlers; v) *Device*: The size and type of the device and its physical features influence what the user can see (or not see) as well as the placement and number of controls and widgets in reduced real-estate displays.

This shortage of resources and particular contexts of use all impact on the UX. Lastly, context properties are not part of the EQ or QinU models, but should be recorded accordingly for characterizing the situation of the target entities at hand.

3   Conceptual Framework and Evaluation Approach

3.1   M&E Conceptual Framework

At this point, it is worth mentioning that the main building blocks depicted in Fig. 1 are grounded in a M&E conceptual framework. We have built –as part of evaluation strategies- the C-INCAMI (*Contextual-Information Need, Concept model, Attribute, Metric* and *Indicator*) conceptual framework [18], which is structured in six components, namely: (a) Measurement and Evaluation Project; (b) Non-functional Requirements; (c) Context; (d) Measurement; (e) Evaluation; and (f) Analysis and Recommendation. Each component contains key terms and relationships. Fig 2 shows, for illustration purpose, the (b), (c), and (d) components whose colors match those green/orange/light-blue boxes of Fig. 1.

In fact, the different labels in Fig 1 are mostly instances of the concepts, properties, and relationships included in the C-INCAMI conceptual framework. For instance "System" and "System in Use" in Fig. 1 are two instances of the *Entity Category* term; specifically, each string is the value of the *name* field in Fig. 2. Since an entity category can have *sub-entity* categories, "Basic/Advanced GUI object", "Menu", etc. are instances of sub-entity categories. *Entity* term represents a concrete object; for example, "Facebook mobile app" is the entity *name* that *belongs to* the "System" category regarding the EQ *focus*.

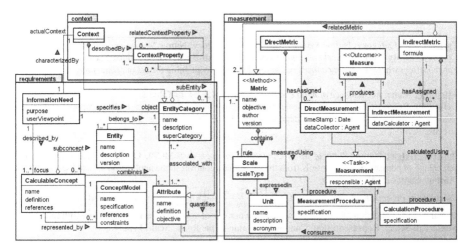

Fig. 2. C-INCAMI Nonfunctional Requirements, Context, and Measurement components

Therefore, the *requirements* component specifies the *Information Need* for a M&E project, i.e., the *purpose* (e.g. "understand", "improve") and the *user Viewpoint* (e.g. "final user", "developer"). In turn, it *focuses* on a *Calculable Concept* (i.e. characteristics whose *names* are for example "External Quality", "User Experience", etc.) and *specifies* the *Entity Category* to be evaluated. On the other hand, a calculable concept and its *sub-concepts* (e.g. "Usability") can be *represented by* a *Concept Model* (e.g. an "EQ model") where the leaves of an instantiated quality model are *Attributes* which are *associated with* a target entity. Table 1 specifies the requirements tree for "Usability", which contains the *names* and *definitions* for the selected sub-characteristics and attributes used in the Facebook mobileapp evaluation.

The *context* component (in Fig. 2) shows explicitly that *Context* is a special kind of *Entity Category*. Context represents the state of the situation of a target entity, which is relevant for a particular information need. To describe the context sub-entities (e.g. "Environment", "Device", etc.) *Context Properties* are used, which are also attributes. Additionally, attributes –as measurable properties- can be *quantified* by metrics and interpreted by indicators.

Metric is a key term in the *measurement* component in Fig 2 (see also Fig. 1). This component allows specifying *Direct* and *Indirect Metrics* used by *Direct* and *Indirect Measurement* tasks which produce *Measures*. A metric is "*the defined measurement or calculation procedure and the scale*". So a metric represents the *how*, that is to say, the method that should be assigned to the steps of a measurement task (the *what*). Lastly, a *Measure* is the number or category assigned to an attribute by making a measurement upon a concrete entity. In order to illustrate the added value of a well-defined measurement component, Table 3 shows the derived template for indirect and direct metric specifications to the "Permanence of main controls" attribute. The "Operability" sub-characteristic combines this attribute that is coded 1.3.3.1.1 in Table 1. Additionally, the screenshot in Fig. 3.b shows the concrete sub-entity named "Main controls bar" that can be further measured.

Table 3. Indirect and direct metric specifications to the *Permanence of main controls* attribute

<u>Target Entity Category:</u> **Name**: *System*; **Sub-Entity Category: Name**: *Smartphone mobile app*;
<u>Concrete Entity:</u> **Name**: *Facebook app*; **Version**: 3.8; **Sub-Entity Description:** Set of *Screens* of the *Facebook app* where the *Main controls bar* is (or should be) containing the set of *Main controls (Buttons)*

<u>Attribute:</u> **Name**: *Permanence of main controls*; **Code**: *1.3.3.1.1* in Table 1
Definition: Degree to which main controls are consistently available for users in all appropriate screens or pages; **Objective**: To determine the degree to which the main controls are present in all appropriate screens.

<u>Indirect Metric:</u> **Name**: *Ratio of Main Controls Permanence* (%MCP); **Objective**: To determine the percentage of permanence for controls from the set of main controls in the application selected screens; **Author**: Santos L.; **Version**: 1.0;

Calculation Procedure: Formula: $\%MCP = \left[\frac{\sum_{i=1}^{m} \sum_{j=1}^{n} MCPLij}{(m*n)} \right] * 100$; for i=1 to m and j=1 to n, where m is the number of application main controls and n is the number of application selected screens; with m, n > 0
Numerical Scale: Representation: Continuous; **Value Type**: Real; **Scale Type**: Ratio;
Unit Name: Percentage; **Acronym**: %
Related Metrics: Main control permanence level (MCPL)

<u>Related</u> **Direct Metric: Name**: *Main Control Permanence Level* (MCPL); **Objective:** To determine the permanence level of a selected control in a given application screen; **Author**: Santos L.; **Version**: 1.0;
Measurement Procedure: Type: Objective; **Specification**: The expert inspects the main controls bar in a given screen in order to determine whether the button is available or not, using the 0 or 1 allowed values. Where 0 means the main button is absent in the screen, and 1 means the main button is present in the screen;
Numerical Scale: Representation: Discrete; **Value Type**: Integer; **Scale Type**: Absolute;
Unit: Name Control

3.2   Evaluation Approach and Strategies

This sub-section gives a summary of our generic evaluation approach, which is made up of a *quality modeling framework* and *M&E strategies*, where a concrete strategy should be selected for purposefully instantiating quality models, processes, and performing evaluations for a concrete project information need. Particularly, the generic evaluation approach relies on two pillars, namely: i) a *quality modeling framework* –where 2Q2U v2.0 is a subset [17], which includes the EQ and QinU views and the 'depends' and 'influences' relationships between them; and ii) *M&E strategies*, which in turn are based on three principles viz. a *M&E conceptual framework* (as introduced in the previous sub-section), *process view specifications*, and *method specifications*.

So far, we have developed two integrated strategies which include these three principles, namely: GOCAME (*Goal-Oriented Context-Aware Measurement and Evaluation*) [17, 18], and SIQinU (*Strategy for understanding and Improving Quality in Use*) [14]. GOCAME is a multi-purpose strategy that follows a goal-oriented and context-based approach in defining and performing M&E projects. GOCAME is a multi-purpose strategy because it can be used to evaluate (e.g. "understand", "improve", etc.) quality not only for product, system and system-in-use entities but also for others such as resource, by using their instantiated quality models and tailored process accordingly. However, GOCAME does not incorporate the QinU/EQ/QinU relationships and improvement cycles as in SIQinU. Rather it can be used to understand the current or future situation, as an evaluation snapshot, of concrete entities. On the other hand, SIQinU is a specific-purpose strategy, which has specific processes, methods and change procedures that are not specified in GOCAME. Ultimately, given the target information need and objective, we can select the specific strategy and its tailored processes and methods in order to fulfill that specific goal.

For example, GOCAME has a well-defined M&E process specification, which is composed of six generic activities, namely: (A1) *Define Non-functional Requirements*; (A2) *Design the Measurement*; (A3) *Implement the Measurement*; (A4) *Design the Evaluation*; (A5) *Implement the Evaluation*; and (A6) *Analyze and Recommend*. Each activity can be accordingly tailored for a specific quality focus regarding the information need, e.g. if the focus is on EQ then A1 is named *Define Non-functional Requirements for EQ*, and so on. Instead, if the focus is on QinU then A1 is named *Define Non-functional Requirements for QinU*, and so forth. Note that in our process specifications each activity is not atomic, so it should be decomposed into tasks.

Lastly, the strategies' activities are supported by different method specifications. Since the M&E strategies rely on the quality modeling framework which is made up of quality models, inspection of characteristics and attributes is the basic method category. Attributes are supported by metric and elementary indicator method specifications, while quality models are calculated using different indicator aggregation methods such as LSP (*Logic Scoring of Preference*) [4], which is a weighted multi-criteria aggregation method. However, user testing and inquiry method categories can be used –mainly for QinU- meanwhile attributes of Efficiency, Effectiveness, Learnability in use and Satisfaction can be derived from task usage log files, questionnaire items, etc., as we did in [13, 14]. For planning and performing changes traditional methods and techniques such as refactoring, re-structuring, re-parameterization, among others can be used as well. The next section demonstrates a practical use of our quality framework and GOCAME strategy through excerpts of our Facebook's mobileapp Usability evaluation study.

4   Usability Evaluation for the Facebook Mobile App

The abovementioned A1 activity named *Define Non-functional Requirements for EQ* has a specific goal or problem as input and a nonfunctional specification document as output. A1 consists of: *Establish EQ Information Need* (A1.1), *Select an EQ Model* (A1.2), and *Specify (System) Context* (A1.3) sub-activities [17].

Considering A1.1, the *purpose* of the information need is to "understand" the current EQ satisfaction level achieved, particularly by evaluating the "Usability" strengths and weaknesses from the "final user" *viewpoint*. "Facebook mobile app" is the concrete entity whose sub-entities for the Usability *focus* are related to basic-, advanced- and task-based GUI objects (recall Fig. 1). For example, in the Fig. 3.b screenshot the "Main controls bar" contains a set of main controls or "Buttons".

For the given focus, the A1.2 sub-activity allows selecting from a repository the sub-characteristics and attributes to be included. Table 1 documents the resulting requirements tree, which includes "Understandability" (1.1), "Learnability" (1.2), "Operability" (1.3), "User error protection" (1.4), and "Aesthetics" (1.5) sub-characteristics. For example, "Operability" includes in turn sub-characteristics such as "Visibility" (1.3.2), "Consistency" (1.3.3), etc., which *combine* attributes *associated* to the entities. Particularly, "Color visibility appropriateness" (1.3.2.1) combines two attributes associated to "Color/Text" objects (see Fig.3.a); while "Permanence of controls" (1.3.3.1) combines two attributes associated to "Main controls bar" and "Contextual control" objects (Fig.3.a and b). For instance, the "Permanence of main controls" attribute (1.3.3.1.1) is defined in Table 1 as "*degree to which main controls are consistently available for users in all appropriate screens or pages*". Note that in the main controls bar of the three shown screenshots all buttons are not always available, so the measured value will be produced in the A3 (*Implement the EQ Measurement*) activity, using the appropriate metric from Table 3.

Fig. 3. Three Facebook screenshots: a) Contextual control and Color/Text objects are highlighted; b) The Main controls bar, and the chat Button, which is not available in the other two screenshots; and c) A typical date widget used when create event task is performed

Lastly, the A1.3 sub-activity deals with the selection of *context properties* (and further values) like "mobile device type" (e.g. tablet, mobile phone); "mobilephone generation" (e.g. regular cellphone, mid-sized smartphone, full-sized smartphone [16]); "mobilephone device brand-model"; "target mobileapp type" (e.g. native mobile app, mobile webapp), among many others, recalling that *context* is a special kind of entity related to the entity category to be evaluated, as mentioned in sub-section 2.3.

Once the information need, EQ requirements, and context specifications were yielded, the next A2 activity, *Design the EQ Measurement*, consists of selecting the meaningful metrics from a repository to quantify the 23 measurable attributes. One direct or indirect metric should be assigned per each attribute of the requirements tree respectively. For example, the "Ratio of Main Controls Permanence" (%MCP) *indirect metric* whose *objective* is "*to determine the percentage of permanence for controls from the set of main controls in the application selected screens*" was chosen to quantify the 1.3.3.1.1 attribute, as shown in Table 3. While an indirect metric has a *calculation procedure* for its *formula* specification, a direct metric has a *measurement procedure*. %MCP includes a related direct metric where the measurement procedure indicates "*the expert inspects the main controls bar in a given screen in order to determine whether the button is available or not, using the 0 or 1 allowed values. Where 0 means the main button is absent in the screen, and 1 means the main button is present in the screen*". In summary, 48 metrics were designed for this study taking into account direct, indirect and related metrics for the latter.

The A3 activity produces the *measures* for all metrics at given moments in time as well the linked data to concrete object references and parameters. Data collection for metrics on the Facebook app (ver.3.8 for Android) were performed from Dec 26-28, 2013. The measure for %MCP gave 54.9% of permanence of main controls, regarding that 5 main buttons should be placed in the main controls bar in 35 appropriate screens (out of 38 app screens). Looking at Fig 3, we can observe for example that the "chat" button is absent in screens a) and c), so if the end user wants to trigger this action on those screens, he/she needs to perform more clicks than needed to initiate the task.

Table 4. Excerpt of *Usability* sub-characteristics and attributes from Table 1. Only *Operability* sub-characteristics and attributes are fully shown with Elementary Indicator values (2nd column); the 3rd column shows Partial/Global Indicator values, which are all in % scale unit.

1 Usability		60.5
1.1 Understandability		76.1
1.2 Learnability		59.7
1.3 Operability		80.7
1.3.1 Data Entry Ease		90
1.3.1.1 Defaults	100	
1.3.1.2 Mandatory entry	50	
1.3.1.3 Widget appropriateness	100	
1.3.2 Visibility (synonym **Optical Legibility**)		81.5
1.3.2.1 Color visibility appropriateness		100
1.3.2.1.1 Brightness difference appropriateness	100	
1.3.2.1.2 Color difference appropriateness	100	
1.3.2.2 GUI object size appropriateness		63
1.3.2.2.1 Control (widget) size appropriateness	100	
1.3.2.2.2 Text size appropriateness	42.1	
1.3.3 Consistency		75.5
1.3.3.1 Permanence of controls		57.3
1.3.3.1.1 Permanence of main controls	54.9	
1.3.3.1.2 Permanence of contextual controls	67.4	
1.3.3.2 Stability of controls	95.5	
1.4 User Error Protection		8.4
1.5 UI Aesthetics		80.8

Once metrics were selected for quantifying all attributes, then A4 can be performed, which deals with designing the EQ evaluation. For space reasons, we did not describe the *evaluation* component in sub-section 3.1, but a key concept is *Indicator*, as shown in Fig. 1. While an *elementary indicator* evaluates the satisfaction level met for an elementary requirement, i.e., an attribute of the requirements tree, a *partial/global indicator* evaluates the satisfaction level achieved for partial (sub-characteristic) and global (characteristic) requirements represented in the quality model. Therefore, a new *scale* transformation and *decision criteria* (in terms of *acceptability levels* and ranges) are defined. In this study, we used three acceptability ranges in a percentage scale: a value within 60-80 (a marginal –yellow- range) indicates a need for improvement actions; a value within 0-60 (an unsatisfactory –red- range) means change actions must take place with high priority; and a score within 80-100 indicates a satisfactory level – green- for the analyzed attribute or characteristic.

Details of elementary and global evaluation, as well as the LSP model used in this study to calculate indicators (A5 activity) can be referred elsewhere [18]. Table 4 shows the elementary and partial indicators' values for "Operability", and only partial and global indicators' values for the other sub-characteristics in addition to the acceptability levels achieved.

Finally, GOCAME projects record all data, metadata and information coming from metrics and indicators as well as the quality model and context specifications and values. The *Analyze and Recommend* (A6) activity produces a recommendation document, which can facilitate planning actions for further improvement.

Based on indicator results shown in the 3rd column of Table 4, we can observe that the Usability characteristic in the Facebook app reached a marginal acceptability level (60.5%), which means a need for improvement actions. Taking into account its sub-characteristics viz. Understandability (1.1), Learnability (1.2) and User Error Protection (1.4) reached a marginal and unsatisfactory acceptability levels respectively. Therefore some of their elementary indicators are performing weakly and surely need recommendations for attribute changes.

On the other hand, Operability (1.3) met the satisfactory level of 80.7%. However, this does not imply that there are no weakly performing attributes for Operability. While Color Visibility Appropriateness (1.3.2.1) scored 100 in its two attributes, Permanence of Controls (1.3.3.1) scored in its two attributes, 54.9 and 67.4 respectively (see 2nd column), so a recommendation for further improvement can be made. For understanding the reasons and planning change actions, the metric specification and the measured values are central in GOCAME for these endeavors. %MCP metric allowed (in A3) to store per each main button its availability in each corresponding app screen in which must stay. So evaluators can easily understand, for instance, where the "chat" button is absent, and so for each button of the main controls bar. Also, the tailored strategy may help designers to understand and act on (system) usability problems effectively to produce better design solutions as well.

Finally, as the reader can surmise the metric design specification helps not only planning the change action, but also gauging (predicting) the improvement gain once the action is performed. Ultimately, if we add a new activity to the six described GOCAME activities (as we did it in fact previously with SIQinU), e.g. (A7) *Plan and Perform Improvement Actions*, then the A3, A5 and A6 can be fully reused for re-evaluation and analysis of the improvement gain with regard to the previous app version. Note that changes should be made on the app entity not on the app in use.

5   Related Work and Discussion

As commented previously, in the state-of-the-art literature, Usability, Actual Usability and UX features are very often poorly linked to target entities (e.g. system and system in use) and context entities (e.g. device, environment, user, etc.), in addition to EQ and QinU views and their relationships. Bevan [3] states that international standards for Usability should be more widely used because one of their main purposes is to impose consistency, compatibility, and safety. Usability has also been integrated into standards for software quality and evaluation; e.g. ISO 25010 (which supersedes to ISO 9126-1 [11]) provides a comprehensive structure for the role of Usability as part of system quality as well a broader concept of QinU increasing the business relevance of usability in many situations. Besides, author indicates that referring to the terminology from the field of software quality, it can be said that UX is more related to the concept of QinU, whereas Usability more to EQ.

From our viewpoint, one of the strengths in ISO 25010 is not only the quality models and included characteristics but also the two quality views and relationship whereby system quality 'influences' system-in-use quality and system-in-use quality 'depends on' system quality. However, some weaknesses we point out are: UX is still an absent characteristic in ISO; there exists a dual Usability definition which blurs system Usability with QinU meanings (i.e. Usability is defined in [9] as "*degree to which a product or system can be used by specified users to achieve specified goals with effectiveness, efficiency and satisfaction in a specified context of use*", while QinU as "*degree to which a product or system can be used by specific users to meet their needs to achieve specific goals with effectiveness, efficiency, freedom from risk and satisfaction in specific contexts of use*") so, we consider the Usability definition given in [11] (adapted in Table 1) is closer to the intended aim; and, the Context coverage characteristic included in the [9] QinU model, which can be represented independently of quality models, as shown in previous sections and in [17]. Therefore, in order to bridge this gap, we have developed the 2Q2U v2.0 quality modeling framework, considering also contributions such as [2, 7], amongst others.

On the other hand, in the Apple [1] and Google [5] design and user interface guidelines, the relationship between mobileapp entities with Usability and UX concepts is not definitively explicit in models, nor is it represented in the Usability works in [16], nor in other quality-related research such as [15, 19]. For instance, Nielsen *et al.* list out in [16] many features and checklists of mobile apps in that would be desired or needed in certain contexts of use but do not use quality views and modeling approaches. Therefore, the capability for consistent application using a conceptual framework and strategies to systematically apply concepts and evaluate and improve a mobile app is rather limited. (Recall the raised issues in Section 1).

Lastly, a holistic approach similar to ours for evaluating the Usability of mobilephones in an analytical way is documented in [8], which is based on a multi-level, hierarchical model of Usability factors. These factors related to views and entities are collectively measured to give a single score with the use of checklists. Moreover, the conceptual framework and strategy involves a hierarchical model of usability factors, four sets of checklists, a quantification method, and an evaluation process. The conceptual framework for usability indicators [6] is based on the ISO 15939 [10] measurement model, which its terms are structured in a glossary. Conversely, we developed an ontology [18] for the C-INCAMI M&E components (recall Fig. 2) where [10] was one of the used sources. Consequently, from components we derive metric an indicator metadata in templates (as in Table 3) that allows consistency and repetitively among projects and analysis of data. As added value, a well-designed metric helps not only to yield measures but also to plan change actions on the product/system attribute or capability. Finally, the process in the [8] strategy is poorly specified compared to that in GOCAME or SIQinU [14] strategies.

6   Conclusions

As the contributions mentioned in the Introduction Section, firstly, we have characterized and represented relevant Usability and UX features of mobile apps with regard to system, system-in-use and context entities. Secondly, we have analyzed Usability, Actual Usability and UX relationships regarding also EQ and QinU views, as well as specific Usability sub-characteristics and attributes for mobile apps in the light of a holistic evaluation approach. This evaluation approach is made up of a *quality modeling framework* (where 2Q2U is a subset) and *M&E strategies*, which in turn are based on three principles namely: a *M&E conceptual framework* (i.e. the C-INCAMI conceptual framework which is rooted in ontologies), *process view specifications*, and *method specifications*. So given the target information need, we can select the specific strategy and its tailored processes and methods in order to fulfill that specific purpose aimed at performing evaluations, analysis and recommendations. To this, we illustrated an evaluation study for the Facebook mobile app from the system Usability viewpoint, using the GOCAME strategy.

Of course, the Facebook study was made on the basis of a proof of concept as a typical social network app. But if we could have had control of the source code obviously GOCAME can be tailored to support change actions based on recommendations for improvement on weak performing indicators and re-evaluation of the new app version. An app with design features that jeopardize Effectiveness, Efficiency, Safety or Satisfaction (i.e. the do and be UX goals) can potentiate risks that it will not meet its business objectives. Evaluating these high level non-functional requirements such as Satisfaction, Actual Usability may feed back into detailed Usability, Functional and Information Quality, Security, etc. attributes and design requirements to maximize the quality of the user's experience and to minimize the likelihood of adverse consequences. Hence, our holistic evaluation approach can give support by means of specific strategies –as SIQinU- to the QinU/EQ/QinU improvement cycles. Ongoing research focuses on further utilizing our evaluation approach for QinU/EQ/QinU cycles for improving the design of mobile apps.

References

1. Apple iOS Human Interface Guidelines (2014), `http://developer.apple.com/library/ios#documentation/UserExperience/Conceptual/MobileHIG/Introduction/Introduction.html` (retrieved by January)
2. Bevan, N.: Extending Quality in Use to provide a Framework for Usability Measurement. In: Kurosu, M. (ed.) HCD 2009. LNCS, vol. 5619, pp. 13–22. Springer, Heidelberg (2009)

3. Bevan, N.: International Standards for Usability Should Be More Widely Used. Journal of Usability Studies 4(3), 106–113 (2009)
4. Dujmovic, J.: Continuous Preference Logic for System Evaluation. IEEE Transactions on Fuzzy Systems 15(6), 1082–1099 (2007)
5. Google User Interface Guidelines (2014), `http://developer.android.com/guide/practices/ui_guidelines/index.html` (retrieved by January)
6. Ham, D.-H., Heo, J., Fossick, P., Wong, W., Park, S., Song, C., Bradley, M.: Model-based Approaches to Quantifying the Usability of Mobile Phones. In: Jacko, J.A. (ed.) HCI 2007. LNCS, vol. 4551, pp. 288–297. Springer, Heidelberg (2007)
7. Hassenzahl, M.: User Experience: Towards an experiential perspective on product quality. In: 20th Int'l Conference of the Assoc. Francophone d'IHM, vol. 339, pp. 11–15 (2008)
8. Heo, J., Ham, D.-H., Park, S., Song, C., Chul, W.: A framework for evaluating the usability of mobile phones based on multi-level, hierarchical model of usability factors. Interacting with Computers 21(4), 263–275 (2009)
9. ISO/IEC 25010: Systems and software engineering - Systems and software Quality Requirements and Evaluation (SQuaRE) - System and software quality models (2011)
10. ISO/IEC 15939: Software Engineering - Software Measurement Process (2002)
11. ISO/IEC 9126-1: Software Engineering - Product Quality - Part 1: Quality Model (2001)
12. Lew, P., Olsina, L.: Relating User Experience with MobileApp Quality Evaluation and Design. In: Sheng, Q.Z., Kjeldskov, J. (eds.) ICWE Workshops 2013. LNCS, vol. 8295, pp. 253–268. Springer, Heidelberg (2013)
13. Lew, P., Qanber, A.M., Rafique, I., Wang, X., Olsina, L.: Using Web Quality Models and Questionnaires for Web Applications Evaluation. In: IEEE Proc., QUATIC, pp. 20–29 (2012)
14. Lew, P., Olsina, L., Becker, P., Zhang, L.: An Integrated Strategy to Systematically Understand and Manage Quality in Use for Web Applications. Requirements Engineering Journal 17(4), 299–330 (2012)
15. Nayebi, F., Desharnais, J.-M., Abran, A.: The state of the art of mobile application usability evaluation. In: 25th IEEE Canadian Conference on Electrical Computer Engineering, pp. 1–4 (2012)
16. Nielsen, J., Budiu, R.: Mobile Usability. New Riders, Berkeley (2012)
17. Olsina, L., Lew, P., Dieser, A., Rivera, B.: Updating Quality Models for Evaluating New Generation Web Applications. In: Abrahão, S., Cachero, C., Cappiello, C., Matera, M. (eds.) Journal of Web Engineering, Special Issue: Quality in New Generation Web Applications, vol. 11(3), pp. 209–246. Rinton Press, USA (2012)
18. Olsina, L., Papa, F., Molina, H.: How to Measure and Evaluate Web Applications in a Consistent Way. In: Rossi, G., Pastor, O., Schwabe, D., Olsina, L. (eds.) Web Engineering: Modeling and Implementing Web Applications. HCIS, pp. 385–420. Springer (2008)
19. Sohn, T., Li, K.A., Griswold, W.G., Holland, J.: A diary study of mobile information needs. In: ACM, Conference CHI 2008, Florence, Italy, pp. 433–442 (2008)

Finding Implicit Features in Consumer Reviews for Sentiment Analysis

Kim Schouten and Flavius Frasincar

Erasmus University Rotterdam
P.O. Box 1738, NL-3000 DR
Rotterdam, The Netherlands
{schouten,frasincar}@ese.eur.nl

Abstract. With the explosion of e-commerce shopping, customer reviews on the Web have become essential in the decision making process for consumers. Much of the research in this field focuses on explicit feature extraction and sentiment extraction. However, implicit feature extraction is a relatively new research field. Whereas previous works focused on finding the correct implicit feature in a sentence, given the fact that one is known to be present, this research aims at finding the right implicit feature without this pre-knowledge. Potential implicit features are assigned a score based on their co-occurrence frequencies with the words of a sentence, with the highest-scoring one being assigned to that sentence. To distinguish between sentences that have an implicit feature and the ones that do not, a threshold parameter is introduced, filtering out potential features whose score is too low. Using restaurant reviews and product reviews, the threshold-based approach improves the F_1-measure by 3.6 and 8.7 percentage points, respectively.

1 Introduction

With the explosion of online shopping at e-commerce companies like Amazon (US), Bol (NL), Alibaba (CN), etc., the use of consumer product reviews has become instrumental in the decision making process of consumers. In fact, potential consumers trust reviews from other consumers more than information on the vendor's website [1]. As a result, the number of reviews for a single product can be quite high, especially for a popular product. When a consumer is interested in the overall sentiment of a product, (s)he must first read through many of the reviews to come to a conclusion. Since reading through these reviews is a tedious process, this may hinder decision making. Therefore an efficient way of displaying the overall sentiment of a product based on costumer reviews is desirable.

Much of the current research in the analysis of product reviews is concerned with classifying the overall sentiment for a certain product. To better describe the overall sentiment of a product, it is useful to look at the sentiment per product aspect, from now on referred to as a feature. Sentiment classification per feature can be difficult as a customer review does not have a standard structure and may

S. Casteleyn, G. Rossi, and M. Winckler (Eds.): ICWE 2014, LNCS 8541, pp. 130–144, 2014.

include spelling errors and synonyms for product features. Although a consumer might explicitly mention a feature for a product, many of the important features are mentioned implicitly as well. For example:

"The battery of this phone is quite good."
"The phone lasts all day."

In the first sentence, the battery is explicitly mentioned and the second one refers to the battery lasting all day. Notice that while in the second sentence the battery is not explicitly mentioned, we can infer that the comment is about the battery. This inference is based on the other words in the sentence that direct the reader towards the actual feature being described. This mapping from words in the sentence to the implied feature must be shared between writer and reader of a text in order for the reader to understand what the writer meant to imply. Because of this, it is usually a small group of well-known, coarse-grained features that is used implicitly. Examples include generic features like price, size, weight, etc., or very important product-specific features like the already mentioned battery, sound quality, ease of use, etc. Since it is this class of features that is often implied, it is important to include them in any sentiment analysis application, as they represent key features for consumers.

This research presents a method to both determine whether an implicit feature is present in a sentence, and if so, which one it is. After describing some of the related work that inspired this research, the method will be presented. Then, the two data sets that are used in the experiments are discussed, followed by the evaluation of the proposed method. This will lead to the conclusions and suggestions for future work in the last section.

2 Related Work

While many methods have been proposed to find features for the task of aspect-level sentiment analysis, most of them focus on explicit features only. This is logical, given that the vast majority of the features in consumer reviews is mentioned explicitly. However, as discussed in the previous section, it is often the important features that are mentioned implicitly. Alas, only few works focus on this task. One of the first to address the problem of detecting implicit features is [8]. An interesting solution is presented in the form of semantic association analysis based on Pointwise Mutual Information. However, since no quantitative results are given, it is impossible to know how well this method performs.

In [5], a method based on co-occurrence Association Rule Mining is proposed. It is making use of the co-occurrence counts between opinion words and explicit features. The latter can be extracted from labeled data, or can be provided by an existing method that finds explicit features. Association rule mining is used to create a mapping from the opinion words to possible features. The opinion word then functions as the antecedent and the feature as the consequent in the rules that are found. When an opinion word is encountered without a linked feature, the list of rules is checked to see which feature is most likely implied by that opinion word. On a custom set of Chinese mobile phone reviews, this method is reported to yield an F_1-measure of 74%.

Similar to [5], the same idea of association rule mining is used in [9]. With association rule mining being used to find a set of basic rules, three possible ways of extending the set of rules are investigated: adding substring rules, adding dependency rules, and adding constrained topic model rules. Especially the latter turned out to be a successful way of improving the results. By constraining the topic model (e.g., Latent Dirichlet Allocation [2] in this case), to include one of the feature words and build the topic around that word, meaningful clusters are generated. Thus, a different way of finding co-occurrences between features and other words in the text is used, and it is reported that this complements the association rule mining method. The best reported result is an F_1-measure of 75.51% on a Chinese data set of mobile phone reviews.

Instead of using annotated explicit features, [10] uses the idea of double propagation [7] to find a set of explicit words and a set of opinion words. An advantage is that the found explicit features are already linked to appropriate opinion words. Then a co-occurrence matrix is created, not between only opinion words and explicit features, but between the words in the sentences and the found explicit features. In this way, the right implicit feature is chosen, not based on just the opinion words in the sentence, but based on all words in the sentence. The opinion words in the sentence are used to constrain the number of possible features from which the right one must be chosen: only features that have co-occurred with the encountered opinion word before, are eligible to be chosen.

In the previously introduced method, for each eligible explicit feature, a score is computed that represents the average conditional probability of a feature being implied, given the set of words in the sentence. The feature with the highest score is chosen as the implicit feature for this sentence. This method is reported to yield an F_1-measure of 0.80 and 0.79 on a Chinese corpus of mobile phone reviews, and a Chinese collection of clothes reviews, respectively. Like [9], it uses all words to find implicit features instead of only opinion words as in [5], and, apart from a small seed set of opinion words, it operates completely unsupervised.

However, there are several drawbacks that are apparent, both in [5], [9], and in [10]. The first problem is that only features that have been found as explicit features somewhere in the corpus can be chosen as implicit features. This assumes that the same features are present in reviews, both explicitly and implicitly. However, as we have discussed before, well-known or important features are implied more often than features that are less important or less described. Furthermore, by counting the co-occurrence frequencies between a feature that is mentioned explicitly and the words in the sentence, it is assumed that when the feature is used implicitly, the same sentential context is present. We argue, however, that this is not necessarily the case. For example, when saying that 'this phone is too expensive', the word 'expensive' prevents the word 'price' from being used. Either one uses the word 'expensive', or one uses the word 'price'. Because of that, there is no real co-occurrence between 'expensive' and 'price', even though the first definitely points to the latter as its implicit feature.

3   Method

In this section the issues discussed in the previous section are addressed and an algorithm is presented that improves upon previous work in the given, more realistic, scenario. This scenario entails the following:

- Sentences can have both explicit and implicit features;
- Sentences can have zero or more implicit features;
- Implicit features do not have to appear explicitly as well;
- The sentential context of explicit features does not have to be the same as the sentential context for implicit features.

The algorithm first scans the training data and constructs a list F of all unique implicit features, a list O of all unique lemmas (i.e., the syntactic root form of a word) and their frequencies, and a matrix C to store all co-occurrences between annotated implicit features and the words in a sentence. Hence, matrix C has dimensions $|F|$ x $|O|$.

When F, O, and C have been constructed, processing the test data goes as follows. For each potential implicit feature f_i, a score is computed that is the sum of the co-occurrence of each word in the sentence divided by the frequency of that word:

$$score_{f_i} = \frac{1}{v} \sum_{j=1}^{v} \frac{c_{i,j}}{o_j}, \tag{1}$$

where v is the number of words, f_i is the ith feature in F for which the *score* is computed, j represents the jth word in the sentence, $c_{i,j}$ is the co-occurrence frequency of feature i and lemma j in C, and o_j is the frequency of lemma o in O. Subsequently, for each sentence the highest scoring feature is chosen.

However, since there are many sentences without any implicit feature, a threshold is added, such that the highest scoring feature must exceed the threshold in order to be chosen. If the computed score does not exceed the threshold, the considered implicit feature is not assigned to that sentence. The pseudocode for the whole process is shown in Alg. 1, where the training process is shown (i.e., constructing co-occurrence matrix C and lists O and F), and in Alg. 2, where the processing of new sentences using the trained algorithm is shown.

The optimal threshold is computed based on the training data only, and consists of a simple linear search. A range of values is manually defined, all of them which are then tested consequently. The values ranged from 0 to 1, with a step size of 0.001. The best performing threshold is then used when evaluating on the test data. Since there is only one parameter to train and the range of possible values is rather limited, more advanced machine learning techniques were not deemed necessary to arrive at a good threshold value.

A limitation of this method is the fact that it will choose at most one implicit feature for each sentence. Both of our data sets, as can be seen in the next section, contain sentences that have more than one implicit feature. In these cases, chances are higher that the chosen implicit feature is in the golden standard,

---

Algorithm 1. Training the algorithm with annotated data

Initialize list of unique word lemmas with frequencies O
Initialize list of unique implicit features F
Initialize co-occurrence matrix C
for sentence $s \in$ training data **do**
    for word $w \in s$ **do**
        if $\neg(w \in O)$ **then**
            add w to O
        end if
        $O(w) = O(w) + 1$
    end for
    for implicit feature $f \in s$ **do**
        if $\neg(f \in F)$ **then**
            add f to F
        end if
        for word $w \in s$ **do**
            if $\neg((w, f) \in C)$ **then**
                add (w, f) to C
            end if
            $C(w, f) = C(w, f) + 1$
        end for
    end for
    Determine optimal threshold.
end for

Algorithm 2. Executing the algorithm to process new sentences

---

for sentence $s \in$ test data **do**
    $currentBestFeature = empty$
    $scoreOfCurrentBestFeature = 0$
    for feature $f \in F$ **do**
        $score = 0$
        for word $w \in s$ **do**
            $score = score + C(w, f)/O(w)$
        end for
        if $score > scoreOfCurrentBestFeature$ **then**
            $currentBestFeature = f$
            $scoreOfCurrentBestFeature = score$
        end if
    end for
    if $scoreOfCurrentBestFeature > threshold$ **then**
        Assign $currentBestFeature$ to s as its implicit feature
    end if
end for

but all features beyond the first will be missed by the algorithm. Another limitation is the obvious need for labeled data. Since this method is trained, not on explicit features, which can be determined by some other method, but on annotated implicit features, a sufficient amount of annotated data is required for our method to work properly.

4 Data Analysis

This section presents an overview of the two data sets that are used to train and evaluate the proposed method and its variants. The first data set is a collection of product reviews [6], where both explicit and implicit features are labeled. The second data set consists of restaurant reviews [4], where explicit aspects are labeled, as well as implicit aspect categories. Each sentence can have zero or more of these coarse-grained aspect categories. The restaurant set features five different aspect categories: 'food', 'service', 'ambience', 'price', and 'anecdotes/miscellaneous'. Since these aspects are implied by the sentence instead of being referred to explicitly, they function as implicit features as well. However, since there are only five options to choose from, it is much easier to obtain good performance on the restaurant set compared to the product set, where there are many different implicit features. Because of this, results for both data sets are not directly comparable. Even so, it is interesting to see how the proposed method performs on different data.

4.1 Product Reviews

The collection of product reviews are extracted from amazon.com, covering five different products: Apex AD2600 Progressive-scan DVD player, Canon G3, Creative Labs Nomad Jukebox Zen Xtra 40GB, Nikon Coolpix 4300, and Nokia 6610. Because the primary purpose of this data set is to perform aspect-level sentiment analysis, it is the case that features are only labeled as a feature when an opinion is expressed about that feature in the same sentence. In the example below, both sentences have a feature 'camera', but only in the second sentence is 'camera' labeled as a feature since only in the second sentence it is associated with a sentiment word.

> "I took a picture with my phone's *camera*."
> "The *camera* on this phone takes *great* pictures."

Because the product data set contains a lot of different, but sometimes similar, features, a manual clustering step has been performed. This makes the set of features more uniform and reduces unnecessary differences between similar features. It also removes some misspellings that were present in the data set. In total, the number of unique implicit features is reduced from 47 to 25.

As can be seen in Fig. 1, there are not many sentences with an implicit feature. This only stresses the need for a good selection criterion to distinguish the ones with an implicit feature from the ones that do not have one. There is also a small number of sentences (0.2%) that have two implicit features. Since the algorithm will only choose zero or one implicit feature for each sentence, this can potentially impact performance in a negative way. The second implicit feature will always be missed, leading to a lower recall. This is however slightly mitigated by the fact that it is easier to pick a correct feature, as it is checked against both annotated features in the sentence.

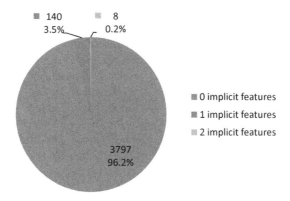

Fig. 1. Distribution of sentences in the product review data set, according to the number of implicit features they contain

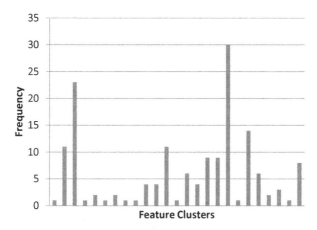

Fig. 2. Frequencies for all 25 unique feature clusters in the product review data set

In Fig. 2, the frequency distribution of the set of implicit features is given. Frequency is measured as the number of sentences a certain implicit feature appears in. As can be seen, there are quite a few implicit features which appear in only a couple of sentences. Fifteen out of the 25 feature appear in less than 5 sentences, with eight features occurring in only one sentence. This makes it extremely difficult to learn a classifier that is able to find these features. In case of the features that appear only once, it is completely impossible to devise a classifier, since they cannot both appear in the test and in the training set.

4.2   Restaurant Reviews

Compared to the product reviews, the restaurant review data set has clearly different statistical characteristics, as shown in Fig. 3. Where the product review set has only a few sentences that contain an implicit feature, in the restaurant set, all of them have an aspect category, which we will regard as an implicit feature in this research. The much bigger size, together with the already mentioned fact that there are only five different implicit features in this data set, makes for a much easier task. To measure the influence of the threshold parameter, the fifth category of 'anecdotes/miscellaneous' is removed from the data set. Since this category does not really describe a concrete implicit feature, removing it leaves us with sentences that do not have any implicit feature, allowing the performance of the threshold to be assessed on this data as well.

Compared to the product reviews data set, the frequency distribution of the implicit features in the restaurant reviews set, shown in Fig. 4 is more balanced. Every features has at least a couple of hundred sentences in which it is appearing. The one outlier is the 'food' category, which appears twice as much as the second largest feature which is 'service'. Still, the difference between the feature that appears the most ('food') and the one that appears the least ('price') is only a factor of three, whereas for the product features, this would be much higher (i.e., around 30).

5   Evaluation

All evaluations are performed using 10-fold cross-evaluation. Each tenth of the data set is used to evaluate an instance of the algorithm that is trained on the other 90% of the data. Both the co-occurrence frequencies and the threshold parameter are determined based on the training data only. When evaluating the algorithm's output, the following definitions are used:

- *truePositives* are the features that have been correctly identified by the algorithm;
- *falsePositives* are those features that have been annotated by the algorithm, that are not present in the golden standard;
- *falseNegatives* are those features that are present in the golden standard, but that have not been annotated by the algorithm;
- *trueNegatives* are features that are not present in the golden standard, and are correctly not annotated by the algorithm.

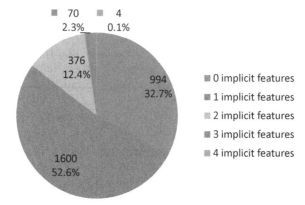

Fig. 3. Distribution of sentences in the restaurant review data set, according to the number of implicit features they contain

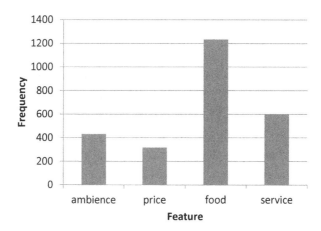

Fig. 4. Frequencies for all 4 unique features in the restaurant review data set

When evaluating, a feature always has to be the same one as in the golden feature to count as a true positive. Simply stating that there is some implicit feature in a sentence, which might be true, is not enough. In order to count as a true positive, it has to be the right implicit feature. From this follows that, given a sentence with only one annotated implicit feature and one golden implicit feature, when the algorithm correctly identifies that a sentence contains an implicit feature, but it chooses the wrong one, the wrongly assigned feature will count as a false positive and the annotated one will count as a false negative. As such, both precision and recall will be lower. In general the algorithm can make three kinds of mistakes:

- State that a sentence contains an implicit feature, while actually it does not: precision will be lower;
- State that a sentence does not contain an implicit feature, while actually it does: recall will be lower;
- Correctly stating that a sentence contains an implicit feature, but picking the wrong one: both precision and recall will be lower.

Because of the ten-fold cross-validation, the reported scores are computed on the sum of the ten confusion matrices (i.e., derived from the ten folds). For example, precision would be computed as:

$$precision = \frac{\sum_{fold=1}^{10} truePositives_{fold}}{\sum_{fold=1}^{10} truePositives_{fold} + falsePositives_{fold}}. \tag{2}$$

Recall is computed in a similar way, leaving the F_1-measure, being the harmonic mean of precision and recall, to be computed as usual. In the end, each sentence will be processed exactly once, but will be used nine times as training instance.

The proposed algorithm is tested both with and without the proposed threshold, to assess the benefit of training such a threshold. Furthermore, both versions are evaluated using a Part-of-Speech filter. The latter is used to filter out words in the co-occurrence matrix that may not be useful to find implicit features. Besides evaluating using all words (i.e., including stopwords), both algorithms are evaluated using an exhaustive combination of four word groups, namely nouns, verbs, adjectives, and adverbs.

Since the algorithm without a threshold will generally choose some implicit feature for every sentence, any trained threshold is expected to surpass that score. To provide more insight in this problem, a maximum score is also provided. This maximum score is computed by filtering out all sentences without any implicit feature and then letting the algorithm simply pick the most appropriate feature. This situation reflects a perfect threshold that is always able to make the distinction between the presence or absence of an implicit feature. Obviously, in reality, the trained threshold does not come close to this ideal performance, but including this ideal line allows the separation of errors due to threshold problems from errors due to not picking the right feature. The latter is an intrinsic problem of the algorithm, not of the threshold. With this in mind, one can see that the gap between the ideal line and the bars represents errors that can be attributed to the threshold, while the gap between 100% performance and the ideal line represents errors that can be attributed to the method of using co-occurrence frequencies to find the right feature.

The results on the product review data set are presented in Fig. 5, whereas the results on the restaurant review data set are presented in Fig. 6. In each graph there are two grouped bars for each Part-of-Speech filter, where the first bar shows the performance without a threshold and the second bar the performance with the trained threshold. The line above the bars represents the ideal, or maximum possible, performance with respect to the threshold, as discussed above. There are 16 different Part-of-Speech filters shown in both graphs. The first `all`, simply means that all words, including stopwords, are used in the co-occurrence matrix. The other fifteen filters only allow words of the types that are mentioned, where NN stands for nouns, VB stands for verbs, JJ stands for adjectives, and RB stands for adverbs.

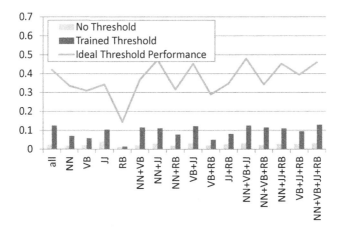

Fig. 5. The performance on the product review data set in F_1-measure for the various PoS-filters

For the product set, it is beneficial to keep as many words as possible, something that is probably caused by the small size of the data set. However, removing stopwords results in a slightly higher performance: the NN+VB+JJ+RB filter scores highest. Looking at the four individual categories, it is clear that adjectives are

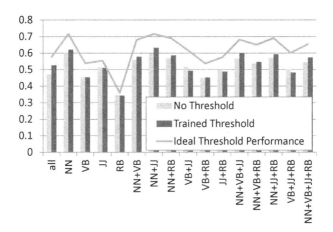

Fig. 6. The performance on the restaurant review data set in F_1-measure for the various PoS-filters

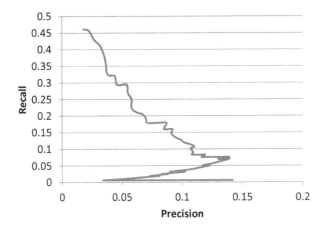

Fig. 7. The precision-recall trade-off on the product review data set, when manipulating the threshold variable (using the NN+VB+JJ+RB filter)

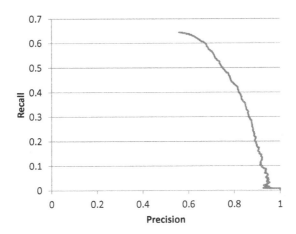

Fig. 8. The precision-recall trade-off on the restaurant review data set, when manipulating the threshold variable (using the NN+JJ filter)

most important to find implicit features. For the restaurant set, the situation is a bit different. Here, nouns are the most important word group, followed by adjectives. Because of its larger size, it is possible to remove verbs and adverbs without any detrimental effects. Hence, the NN+JJ filter yields the best performance.

Another observation we can draw from comparing Fig. 5 and Fig. 6 is that the restaurant set is in general much easier for the algorithm to process. Not only are the ideal performances higher on the restaurant set, also the gap between the ideal and the realized performance is smaller. The most likely reason for this

Table 1. Comparison of results with Zhang & Zhu [10], with and without the proposed threshold. Reported scores are F_1-measures for the best scoring Part-of-Speech filter. Differences between scores are expressed in percentage points (pp.), the arithmetic difference between two percentages.

product review data set

method	no threshold	trained threshold	difference
Zhang & Zhu	1.2% (all)	1.4% (NN+VB+JJ+RB)	+0.2 pp.
proposed method	4.2% (JJ)	12.9% (NN+VB+JJ+RB)	+8.7 pp.
difference	+3 pp.	+11.5 pp.	

restaurant review data set

method	no threshold	trained threshold	difference
Zhang & Zhu	31.5% (all)	32.4% (all)	+0.9 pp.
proposed method	59.7% (NN+JJ)	63.3% (NN+JJ)	+3.6 pp.
difference	+28.2 pp.	31.1 pp.	

difference is the fact that in the restaurant set there are roughly 2000 sentences that contain at least one of the four possible implicit features, whereas in the product set, there are 140 sentences that contain at least one of 25 possible implicit features. Not only does this render the task of picking the right feature more difficult, it also increases the complexity of judging whether a sentence contains one of these features.

The fact that the vast majority of the product set has no implicit feature at all makes the utilization of a threshold all the more important. This is in contrast to the restaurant set, where two-thirds of the sentences have an implicit feature. Again, this is shown clearly in Fig 5 and Fig 6: the relative improvement of the threshold is much higher for the product data than the restaurant data.

In Fig. 7 and Fig. 8, the precision-recall trade-off is shown for the best scoring Part-of-Speech filter. The restaurant set yields a well-defined curve, which is to be expected due to the large quantity of available data. Note that as in all other graphs, two tasks are being evaluated: determine whether or not there is an implicit feature in a sentence, and if so, determine which one it is. This is the reason that, even with a threshold of zero, the recall will not be 100%: while it does state that every sentence will have an implicit feature, it still has to pick the right one in order to avoid a lower recall (and precision for that matter).

A comparison with the method of Zhang & Zhu [10] is given in Table 1. To increase comparability, both methods are tested with all sixteen possible Part-of-Speech filters (only the best one is reported). To be fair, the original method is also tested with a threshold added, using the same procedure as for the proposed method, even though this contributes little to its performance.

Interestingly, the effect of the threshold is bigger on the product set compared to the restaurant set. This might point to the fact that training this parameter can partly mitigate the negative effect of having a small data set. Consequently, when the data set is larger, the algorithm on its own already performs quite well, leaving less room for improvement by other methods, like adding a threshold.

6   Conclusion

Based on the diagnosed shortcomings in previous work, we proposed a method that directly maps between implicit features and words in a sentence. While the method effectively becomes a supervised one, it is not flawed in its assumptions as previous work, and performance is reported to increase on the two used data sets. Furthermore, a more realistic scenario is implemented wherein the proposed method not only has to determine the right implicit feature, but also whether one is actually present or not.

The proposed algorithm shows a clear improvement with respect to an existing algorithm on the two data sets considered, as it is better in distinguishing between sentences that have an implicit feature and the ones that do not. Both for product reviews and restaurant reviews, the same general improvement is observed when implementing this threshold, even though the actual performance differs much between the two data sets.

Analysis of the performance of the algorithm in relation to the characteristics of the two data sets clearly shows that having less data, but more unique implicit features to detect severely decreases performance. While the proposed algorithm is much better in dealing with this lack of data, the results for that particular data set are still too low to be useful in practice. On the set of restaurant reviews, being of adequate size and having only four unique implicit features, the proposed algorithm yields promising results. Adding a threshold further boosts the performance by another 3 percentage points, which is highly desirable for this kind of user generated content.

A primary suggestion for future work is to learn a threshold for each individual implicit feature, instead of one general threshold that applies to all implicit features. We hypothesize that because some features are used more often in an implicit way than others, and the sentential context differs from feature to feature as well, it makes sense to learn a different threshold for each unique implicit feature.

Also interesting could be to adjust the algorithm to be able to choose more than one implicit feature. Especially on the restaurant set, where about 14% of the sentences have more than one implicit feature, performance could be improved. Possible ways of doing this include choosing all features whose score exceeds the threshold, or employ a classifier that determines how many implicit features are likely to be present. The latter could also be investigated as a possible alternative for the threshold.

Last, a move from word based methods, like this one, toward concept-based methods, as advocated in [3], would be interesting as well. For example, cases like:

"This phone doesn't fit in my pocket."

is very hard to process based on words alone. It is probably feasible to determine that the implicit feature here is 'size', if enough training data is at hand, but determining that this sentence represents a negative sentiment, since mobile phones are *supposed* to fit in ones pocket, seems extremely hard for word-based methods. While concept level methods are still in their infancy, they might be up to this challenge, since common sense knowledge, world knowledge, and domain knowledge are integrated in such an approach.

Acknowledgments. The authors are partially supported by the Dutch national program COMMIT. We also would like to thank Andrew Hagens, Gino Mangnoesing, Lotte Snoek, and Arno de Wolf for their contributions to this research.

References

1. Bickart, B., Schindler, R.M.: Internet Forums as Influential Sources of Consumer Information. Journal of Interactive Marketing 15, 31–40 (2001)
2. Blei, D.M., Ng, A.Y., Jordan, M.I.: Latent Dirichlet Allocation. Journal of Machine Learning Research 3, 993–1022 (2003)
3. Cambria, E., Schuller, B., Xia, Y., Havasi, C.: New Avenues in Opinion Mining and Sentiment Analysis. IEEE Intelligent Systems 28, 15–21 (2013)
4. Ganu, G., Elhadad, N., Marian, A.: Beyond the Stars: Improving Rating Predictions using Review Content. In: Proceedings of the 12th International Workshop on the Web and Databases, WebDB 2009 (2009)
5. Hai, Z., Chang, K., Kim, J.: Implicit Feature Identification via Co-occurrence Association Rule Mining. In: Gelbukh, A.F. (ed.) CICLing 2011, Part I. LNCS, vol. 6608, pp. 393–404. Springer, Heidelberg (2011)
6. Hu, M., Liu, B.: Mining and Summarizing Customer Reviews. In: Proceedings of 10th ACM SIGKDD International Conference on Knowledge Discovery and Data Mining (KDD 2004), pp. 168–177. ACM (2004)
7. Qiu, G., Liu, B., Bu, J., Chen, C.: Opinion Word Expansion and Target Extraction through Double Propagation. Computational Linguistics 37, 9–27 (2011)
8. Su, Q., Xu, X., Guo, H., Guo, Z., Wu, X., Zhang, X., Swen, B., Su, Z.: Hidden Sentiment Association in Chinese Web Opinion Mining. In: Proceedings of the 17th International Conference on World Wide Web (WWW 2008), pp. 959–968. ACM (2008)
9. Wang, W., Xu, H., Wan, W.: Implicit Feature Identification via Hybrid Association Rule Mining. Expert Systems with Applications 40, 3518–3531 (2013)
10. Zhang, Y., Zhu, W.: Extracting Implicit Features in Online Customer Reviews for Opinion Mining. In: Proceedings of the 22nd International Conference on World Wide Web Companion (WWW 2013 Companion), pp. 103–104. International World Wide Web Conferences Steering Committee (2013)

From Choreographed to Hybrid User Interface Mashups: A Generic Transformation Approach

Alexey Tschudnowsky[1], Stefan Pietschmann[2], Matthias Niederhausen[2], Michael Hertel[1], and Martin Gaedke[1]

[1] Technische Universität Chemnitz, Germany
{alexey.tschudnowsky,michael.hertel,gaedke}@informatik.tu-chemnitz.de
[2] T-Systems MMS, Germany
{stefan.pietschmann,matthias.niederhausen}@tu-dresden.de

Abstract. Inter-widget communication (IWC) becomes an increasingly important topic in the field of user interface mashups. Recent research has focused on so-called choreographed IWC approaches that enable self-organization of the aggregated components based on their messaging capabilities. Though a manual configuration of communication paths is not required anymore, such solutions bear several problems related to awareness and control of the emerging message flow. This paper presents a systematic approach to tackle these problems in the context of *hybrid* user interface mashups. We show how users can be made aware of the emerged IWC configuration and how they can adjust it to their needs. A reference architecture for development of hybrid mashup platforms, is derived and one implementation based on the publish-subscribe choreography model is given. We report on the results of a first user study and outline directions for the future research.

Keywords: inter-widget communication, user interface mashup, widgets, end-user development.

1 Introduction

User interface (UI) mashups have become a popular approach for end-user development. Based on autonomous but cooperative visual components called widgets, they promise to significantly lower the barrier for Web application development [2,19,10]. The development process of UI mashups usually implies three steps: First, finding appropriate widgets for composition; Second, placement and configuration of widgets on a common canvas; and finally, configuration of the cooperative behaviour by means of inter-widget communication (IWC). IWC hereby refers to the process of exchanging data between widgets and can be used to synchronize internal states of the aggregated components.

The IWC behaviour of a mashup can be defined either explicitly by a mashup designer (*orchestrated mashups*), emerge from the capabilities of the integrated components (*choreographed mashups*) or be defined by a combination of both (*hybrid mashups*) [22]. While orchestrated approaches aim at providing flexibility during mashup development by enabling designers to define the desired data

S. Casteleyn, G. Rossi, and M. Winckler (Eds.): ICWE 2014, LNCS 8541, pp. 145–162, 2014.

flow manually, choreographed and hybrid solutions focus on keeping the development process lean and fast. In choreographed solutions, widgets "decide" autonomously on how to communicate and with whom. Hybrid mashups behave as choreographed ones, but provide additional means to restrict the emerging communication. EDYRA [19], DashMash [3] or EzWeb [13] are some approaches that exemplify orchestrated mashups. Using the "wiring" metaphor, they enable mashup designers to connect "inputs" and "outputs" of components and, thus, specify data flow in a mashup. The target group of such platforms are skilled users and hobby developers, who are experienced with the concepts of operations, input/output parameters and data structures. OMELETTE [4], ROLE [10] and Open Application [8] projects follow the choreographed approach. Widgets communicate without a prior configuration - using the publish-subscribe messaging pattern each widget decides autonomously on which messages to send and which messages to receive. Chrooma+ [12] is an example of a hybrid platform. While widgets publish and subscribe for messages on their own, a mashup designer is still able to "isolate" one or more widgets from their environment. The target group of choreographed and hybrid platforms are end-users, who have little to no programming skills but are experts in their corresponding business domains.

Though choreographed and hybrid mashups are considered to be more "end-user-friendly"[8], they also pose some challenges with regard to awareness and control of what is happening in a mashup [21,8]. The major awareness problem caused by implicitly defined IWC is that users do not know which pairs of widgets *could* and which actually *do* communicate. Users have to learn the data and control flows as they use and explore the mashup. While in general this may merely frustrate users, such "exploratory" interaction can also accidentally affect live data, causing undesired side effects. The major control problem is that, being defined implicitly and not as first-level concepts, communication paths cannot be blocked, modified or added directly by end-users. Possible reasons for intervention are, e.g., untrusted widgets or unexpected or undesired state synchronisations. A detailed analysis of these and other problems related to awareness and control in UI mashups can be found in our prior work in [5].

This paper presents an approach for systematic development of hybrid mashup platforms with IWC awareness and control in mind. The goal is to support end-user development of UI mashups by combining advantages of choreographed and orchestrated mashup platforms - self-emerging IWC with flexible visualization and tailoring facilities. In summary, the contributions of this paper are as follows:

- *A generic IWC model including corresponding visualization and tailoring mechanisms.* The model and mechanisms are used to communicate IWC behaviour to end-users and facilitate its configuration.
- *A reference architecture for hybrid mashup platforms.* The architecture enables systematic development of hybrid mashup platforms - both from scratch and by extension of existing choreographed ones.
- *Evaluation of the proposed concepts with end-users.* Experiments with 27 end-users assessed the efficiency and usability of the proposed mechanisms.

The rest of the paper is structured as follows. Section 2 gives a background on choreographed UI mashups and presents the steps required for systematic development of hybrid mashup platforms. Section 3 demonstrates one implementation of the proposed architecture for the publish-subscribe choreography strategy. An evaluation of the proposed awareness and control facilities is given in section 4. Finally, section 6 concludes the paper and derives directions for future research.

2   From Choreographed to Hybrid UI Mashups

The section presents a systematic approach to build hybrid mashup platforms. As the working principle of hybrid mashups is close to choreographed ones, the idea is to leverage existing choreographed platforms and to extend them with the missing visualization and tailoring functionality. The approach can also be used to build hybrid platforms from scratch.

In the following, different types of choreographed mashups with regard to utilized IWC models are presented. Afterwards, a unified communication model for visualization and tailoring of data flow is described. Based on the model, several awareness and control mechanisms are proposed. Finally, a reference architecture to support these mechanisms is given.

2.1   Choreographed UI Mashups

Choreographed UI mashups do not require mashup designers to specify data flow in a mashup. Instead, communication emerges in a self-organizing fashion depending on the messaging capabilities of widgets. Technically, different IWC strategies exist to enable "self-organization" of autonomous but cooperative components: message passing, publish-subscribe, remote procedure calls (RPC) and shared memory[24]. *Message passing* considers widgets as senders and recipients of structured application-specific data. Delivery can take place in uni-, multi- or broadcast fashion. *RPC* solutions enable widgets to offer operations, which can be invoked by others in synchronous or asynchronous way. Discovery of available operations happens either at widget design-time or at run-time, e.g., by means of centralized widget registries. In *publish-subscribe* systems, widgets emit and receive messages on different channels, also called "topics". The decision, which channels to publish or subscribe on, is met by widgets autonomously without intervention of a mashup designer. Finally, *shared memory* solutions enable widgets to read and to write to a common data space and, thus, autonomously exchange data among each other. Independently of the concrete communication model, widgets can communicate either directly or by means of the platform middleware. Platform-mediated communication enables loose coupling of aggregated components as well as additional services such as traffic monitoring and management. As to the authors' knowledge, all of the current UI mashups make use of platform middleware to implement IWC. The presented strategies offer techniques for implementation of self-organizing widget compositions. User awareness and control are out of their scope and thus require additional engineering on the side of the mashup development/execution platform.

2.2 Communication Model

The goal of the unified communication model presented here is to provide a common data structure for visualization and control mechanisms. The assumption for the definition of the model is that – from an end-users' point of view – widgets communicate in pairs by means of unidirectional message transfers. In terms of a concrete choreography model, messages have different semantics, e. g., invocation of a remote procedure or publication/subscription to some topic. Regardless of the model, the user-perceived result is that one widget receives data from another one. These considerations build a basis for the unified communication model described by the following data structure:

The unified communication model M is a graph $G = (V, E)$ with

- $V = \{v | v = (id, s)\}$ set of vertices with identifier id and state $s \in \{ENABLED, ISOLATED\}$. Each vertex corresponds to exactly one widget in a mashup.
- $E = \{e | e = (v_1, v_2, s, t)\}$ set of edges corresponding to possible communication paths between widgets corresponding to $v_1, v_2 \in V$ with state $s \in \{ENABLED, BLOCKED\}$ and with label t.

For all of the presented choreography models it is possible to define an algorithm which yields a unified communication model M. Section 3.2 presents one possible algorithm for publish-subscribe-based choreography models.

A data flow restricted by the unified model M takes place as follows:

- A widget corresponding to the vertex v is allowed to emit or receive messages only if the state s of the vertex v is $ENABLED$.
- A message m from a widget corresponding to v_1 is allowed to be delivered to a widget corresponding to v_2 only if $\exists e \in E : e = (v_1, v_2, ENABLED, t)$.
- The data flow takes place according to the utilized choreography strategy if none of the above restrictions apply.

2.3 Visualization and Tailoring Facilities

The following visualization and tailoring facilities are proposed to make mashup designers aware of the data flow in a UI mashup and to enable them to adjust it:

- States $s \in \{ENABLED, ISOLATED\}$ of vertices are visualized using borders of different color and type around the corresponding widgets.
- Potential communication paths $e = (v_1, v_2, s, t) \in E$ are visualized using arrows between widgets corresponding to v_1 and v_2. The arrow style indicates the state of the communication path $s \in \{ENABLED, BLOCKED\}$. Annotation t is displayed above the corresponding arrow to provide additional information on the communication path.

- Flashing icons on widget borders show which widgets corresponding to v_1 and v_2 are currently communicating along a communication path $e = (v_1, v_2, s, t) \in E$.
- For every vertex v, visualization of its state $s \in \{ENABLED, ISOLATED\}$ and in-/outgoing edges $e = (v, *) \in E$ can be turned on or off to avoid cognitive overload in case of strong connectivity.
- For every vertex v, its state $s \in \{ENABLED, ISOLATED\}$ can be toggled using corresponding user interface controls.
- For each edge $e = (v_1, v_2)$, its state $s \in \{ENABLED, BLOCKED\}$ can be toggled by clicking on the corresponding arrows.

2.4 Reference Architecture

The reference architecture (cf. Figure 1) acts as a blueprint for the development of awareness- and control-enabled hybrid mashup platforms. Some of the platform components (such as the *Awareness and Control Module* and the *Communication Model*) are independent of a chosen choreography model, whereas others (*Widget and Mashup Descriptors, Model Importer, Model Exporter* and *Message Broker*) are choreography-model-specific.

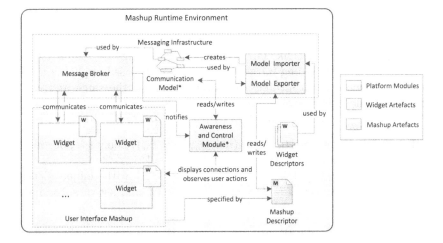

Fig. 1. Reference architecture for implementation of hybrid UI mashups. * - implementation can be shared between platforms with different choreography strategies.

The components of the reference architecture provide resources and services required for UI mashup development. *Widget Descriptors* describe IWC capabilities and default configuration parameters of installed widgets. Mashup layout, aggregated widget instances, user preferences and IWC configuration are specified by the *Mashup Descriptor*. All artefacts are dependent on the selected choreography model, but can be used by the *Model Importer* to derive a unified *Communication Model*. The *Awareness and Control Module* displays the model in the composition canvas according to the rules described in 2.3. It is also responsible for updating the *Communication Model* upon changes triggered by users, e. g., changing state of an edge upon clicks on the corresponding arrow. The *Message Broker* is the platform's communication middleware. Its goal is, first, to provide messaging functionality according to the rules of the underlying choreography approach and, second, to assure that restrictions of the unified *Communication Model* such as isolated widgets or blocked communication paths are respected. The *Awareness and Control Module* gets notified about activities within the *Message Broker* and displays activated or blocked communication paths to mashup users.

In the following, we present one example implementation of the reference architecture. It reuses an existing publish-subscribe-based UI mashup platform and extends it towards the missing awareness and control functionality. We selected a platform based on the publish-subscribe strategy because of the wide use of this approach in current choreography platforms.

3 Hybrid Mashups Based on Publish-Subscribe Choreography Model

Publish-subscribe-based IWC is a part of many choreographed UI mashup platforms [4,12,10]. Frameworks like OpenAjaxHub[1] or AmplifyJS[2] simplify the integration of the corresponding infrastructure into Web-based applications. The underlying communication strategy is, however, always the same and can be formalized as follows:

Let $m = (W, T)$ be a widget mashup with

- Reference ontology $T = \{t_n : t_n = (name_n, TYPE_n)\}$ being a set of concepts and associated message types $TYPE_n = \{value_n\}$
- Widget set $W = \{w_j : w_j = (id_j, PUB_j, SUB_j)\}$ with unique identifier id_j, set of publications $PUB_j = \{p_{jl} : p_{jl} \in T\}$ and set of subscriptions $SUB_j = \{s_{jk} : s_{jk} \in T\}$

Let $a = (wsender, t, data)$ with $wsender \in W, t \in wsender.PUB, data \in t.TYPE$ be a message emitted by a widget *wsender*. The message a is delivered to all widgets $w_i \in W : t \in w_i.SUB$.

The following implementation of a publish-subscribe-based hybrid mashup platform is based on two open-source projects - Apache Wookie[3] and Apache Rave[4]. Apache Wookie is a widget container for hosting W3C widgets[5], which are

---

[1] http://www.openajax.org/member/wiki/OpenAjax_Hub_2.0_Specification
[2] http://amplifyjs.com
[3] http://wookie.apache.org
[4] http://rave.apache.org
[5] http://www.w3.org/TR/widgets

stand-alone Web applications packaged for distribution on the Web. Apache Rave is a widget mashup environment, which enables aggregation of both W3C and OpenSocial[6] widgets on one canvas and provides a publish-subscribe messaging infrastructure for communication between the widgets. In the following, those parts of the projects are described in detail, which were extended towards the reference architecture.

3.1   Widget Descriptors

According to the W3C specification, widget metadata contains only basic information on the packaged application such as title, description, author, license etc. Currently there is no standard way to describe widget communication capabilities. However, this is essential for the proposed awareness and control mechanisms (cf. section 2.2). Therefore, an extension of the W3C metadata file is proposed to describe publications and subscriptions of widgets together with used topics. Listing 1.1 shows the proposed extension for a fictitious *Contacts* widget, which maintains a list of structured contact entries. An incoming phone number (topic *http://example.org/phoneNumber*) causes the widget to filter the list towards a contact with the passed number. If a contact is selected by user, its complete data is published on the *http://example.org/contact* topic. The schema of the involved messages is given in the *oa:topics* section. Both Apache Wookie and Rave were extended to support the extended widget metadata.

Listing 1.1. Extension of the W3C metadata file

```
<feature name="http://www.openajax.org/hub"
xmlns:oa="http://www.openajax.org/hub">

<!-- Declaration of topics -->
<oa:topics>
  <oa:topic oa:name="http://example.org/contact">
    <oa:schema oa:schemaType="JSON">
    <![CDATA[
    {"description":"A person",
      "type":"object",
      "properties":{
        "name":{"type":"string"},
        "age" :{
          "type":"integer",
          "maximum":125 }
    }}]]>
    </oa:schema>
  </oa:topic>

  <oa:topic oa:name="http://example.org/phoneNumber">
```

---
[6] https://developers.google.com/gadgets/docs/dev_guide?csw=1

```
  <oa:schema oa:schemaType="XML" oa:simpleType="xs:string"/>
  </oa:topic>

  <oa:topic ...>
  </oa:topics>

  <!-- Declaration of publications -->
  <oa:publications>

  <oa:publication oa:topicRef="http://example.org/contact"/>
  <oa:publication ...>
  </oa:publications>

  <!-- Declaration of subscribtions -->
  <oa:subscriptions>
  <oa:subscription oa:topicRef="http://example.org/phoneNumber"/>

  <oa:subscription ...>
  </oa:subscriptions>

</feature>
```

3.2   Model Importer

The module has been added to Apache Rave and is responsible for construction of the *Communication Model* based on *Widget Descriptors* and *Mashup Descriptors*. The construction of the model for publish-subscribe-based UI mashups is specified by the Algorithm 1.

Algorithm 1. Creating a Unified Communication Model for Publish-Subscribe-based UI Mashups

---

Input   : Publish-subscribe-based UI mashup $m = (W, T)$ as defined above
Output: Communication model $G = (V, E)$ as defined in section 2.2
1. Set $V = \emptyset$, $E = \emptyset$
2. For each widget $w_i \in W$, create a new vertex v_i and set $V = V \cup v_i$
3. For each pair of widgets (w_i, w_j) and for each $p_{ik} \in w_i.PUB : p_{ik} \in w_j.SUB$, create a new edge $e_{ik} = (v_i, v_j, ENABLED, p_{ik})$ and set $E = E \cup e_{ik}$
4. Return $G = (V, E)$

The resulting communication model reflects potential message flows in a publish-subscribe-based UI mashup. By default, all communication paths and widgets are in the $ENABLED$ state.

3.3   Awareness and Control Module

The *Awareness and Control Module* weaves the *Communication Model* into the composition canvas and updates it, based on user actions such as widget isolations or path blockades. It also highlights communicating partners according to notifications from the *Message Broker*. Visualization of the model takes place as proposed in Section 2.3 and is implemented using the jsPlumb[7] JavaScript drawing library (cf. Figure 2).

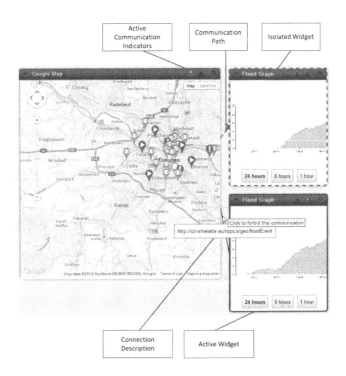

Fig. 2. Awareness and Control Mechanisms integrated into Apache Rave

3.4   Message Broker

The *Message Broker* is an existing component in the Apache Rave platform and is implemented using the OpenAjaxHub framework. It is responsible for routing messages between widgets according to the publish-subscribe strategy as specified above. The component has been extended to take the *Communication Model* into account while routing messages. The routing algorithm was refined as follows: a message $a = (wsender, t, data)$ from widget $wsender$ is allowed to be delivered to $\forall w_i \in W : t \in w_i.SUB$ if and only if $v_{wsender}.s = ENABLED \wedge v_{w_i}.s = ENABLED \wedge e.s = ENABLED : e = (v_{wsender}, v_{w_i})$.

---

[7] http://jsplumbtoolkit.com/

3.5 Model Exporter

The *Model Exporter* component enables serialization of the current communication model and its integration into the platform-specific *Mashup Descriptor*. Apache Rave makes use of the OMDL[8] format to describe the configuration of UI mashups. However, the current specification doesn't provide any means to include IWC configuration into the mashup specification. Thus, we propose to extend OMDL documents with missing information on the *Communication Model*. An example of an extended *Mashup Descriptor* in OMDL format is given in Listing 1.2. The goal of the described mashup is to provide aggregated information on emergency incidents during natural disasters such as in case of flood. The mashup aggregates four widgets and defines IWC restrictions for the current composition. One of the widgets (flood graph) is completely isolated. Such configuration is useful to fix view of a widget and to avoid refreshes caused by changes others. Communication between the map widget and the contacts widget is forbidden only for the topic *http://example.org/phoneNumber*. It results in the behavior, that no marker selection in the map will refresh the contacts widget, while activities in other widgets may do.

Listing 1.2. OMDL mashup description and proposed extension

```
<workspace xmlns="http://omdl.org/"
    xmlns:oa="http://www.openajax.org/hub">
  <identifier>http://example.org/mashup/379</identifier>
  <title>Dresden Flood</title>
  <description></description>
  <creator>Alexey</creator>
  <date>2013-06-18T14:39:58+0200</date>
  <layout>THREE COLUMNS</layout>

  <app id="http://example.org/incidentMap-1">
    <link href="http://example.org/s/incidentsMap.wgt"
    type="application/widget" rel="source"/>
    <position>LEFT TOP</position>
  </app>

  <app id="http://example.org/contacts-1">
    <link href="http://example.org/s/contacts.wgt"
    type="application/widget" rel="source"/>
    <position>LEFT MIDDLE</position>
  </app>

  <app id="http://example.org/floodGraph-1">
    <link href="http://example.org/s/floodGraph.wgt"
    type="application/widget" rel="source"/>
```

---

[8] http://omdl.org

```
  <position>RIGHT TOP</position>
</app>

<app id="http://example.org/floodGraph-2">
  <link href="http://example.org/s/floodGraph.wgt"
  type="application/widget" rel="source"/>
  <position>RIGHT MIDDLE</position>
</app>

<!-- Proposed extension specifying mashup IWC behaviour -->

<!-- Isolate a widget completely -->
<oa:pubsub-restriction
  oa:source="*"
  oa:target="http://example.org/floodGraph-1"
  oa:topicRef="*"/>
<oa:pubsub-restriction
  oa:source="http://example.org/floodGraph-1"
  oa:target="*"
  oa:topicRef="*"/>

<!-- Forbid two widgets to communicate over the specified topic -->
<oa:pubsub-restriction
  oa:source="http://example.org/incidentMap-1"
  oa:target="http://example.org/contacts-1"
  oa:topicRef="http://example.org/phoneNumber"/>

</workspace>
```

This section presented an implementation of the reference architecture for the publish-subscribe choreography strategy. Application of the approach to other strategies requires to implement strategy-specific *Model Importer*, *Model Exporter* and *Message Broker* components. The *Awareness and Control Module* and *Communication Model*, however, can be reused across different platforms with minor adaptations.

4   Evaluation

To evaluate the presented approach in practice, we explored the following three hypotheses:

1. *Users solve tasks that require IWC faster if awareness and control mechanisms are used.* The intent was to check if the proposed mechanisms increase efficiency of end-users while working with UI mashups. The hypothesis was considered to be approved if the time needed to complete a given task with activated awareness and control facilities was lower than without them.

2. *Users find out easier, which widgets are connected, if IWC is visualized.* The intent was to check if the awareness mechanisms help users to spot, understand, and make use of the connections between widgets. The hypothesis was considered to be approved if average assessment on the ease of finding connections between widgets was higher with awareness facilities than without. The hypothesis was checked with two different types of widgets being employed: one group had mostly static content, while the other included animated changes. The rationale was that animated changes in widgets themselves might be sufficient on their own to identify communication partners, thereby making additional awareness mechanisms unnecessary.

3. *Users find the proposed control mechanisms easy to use.* The intent was to check if users felt comfortable with the proposed tailoring facilities. The approval condition was that more than 60% of participants agree or strongly agree that control facilities are easy to use.

To test the introduced hypotheses we applied the laboratory experiment methodology. Overall, 27 participants took part in the user study. Almost 90% of participants had no programming skills but were experts in the domain of marketing and telecommunication. The majority of users (74%) had an understanding of the term "widget", which was mostly related to mobile devices and the "Windows Vista Sidebar". Only 4 out of 27 users had ever configured a portal UI (mostly intranet portal) on their own, e. g., by repositioning of widgets and changing the colour scheme.

For each of the 27 participants, the evaluation procedure took about one hour and involved the following steps: Before the task execution, participants filled in a pre-evaluation questionnaire to judge their skill levels. Based on the results, they were evenly distributed over test and control groups. After that, users were given an introduction on the widget mashup platform, its purpose, concepts (mashups, widgets, etc.) and core functionalities. Following the introduction, users had the chance to explore and try out different aspects of the portal as they liked. After then, participants were asked to complete three tasks targeting different aspects of the system. The completion time was measured for each group and for each task. In a standardized post-questionnaire, users could express their subjective opinions on the introduced facilities. The study used a two-tailed t test, with a significance level of 95

Awareness Facilities. In the first experiment, users were asked to play a game on a dedicated mashup, called "Easter Egg Hunt" (cf. Figure 3, left). The goal of the game was to find the Easter egg in each of the nine visible widgets. Only one egg was visible at a time. Upon clicking it, the egg would disappear and appear in another widget. Thus, users quickly had to find the corresponding widget and the egg inside. For the test group, visualization of communicating widget pairs was enabled, so that outgoing and incoming data were displayed in the widget header. Thereby, the data flow could be perceived by users and, ideally, they could deduce where to look next. The control group accomplished the task without IWC visualization. The experiment was conducted in two different setups:

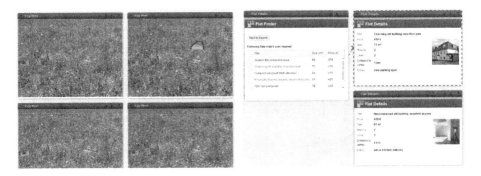

Fig. 3. Evaluation mashups for testing end-user efficiency with awareness (left) and control (right) facilities

for animated content (the egg would "fall" into place) and static content (the egg would just appear). The time required for completion of the task was measured.

Control Facilities. In the second experiment, participants had to solve a comparison task using two widgets (cf. Figure 3, right). One widget gave an overview of flat offers in a selected city and the other one showed details of the offer. The task was to find the cheapest flat in one city and then a comparable one in another city. While the control group used the default setting with one "overview" and one "detail" widget, the test group was provided with two "detail" widgets, one of which they could isolate from communication to simplify the comparison. Once isolation was enabled for a detail widget, it would "freeze", so that new details could be loaded in the second widget and easily be compared to the first one. The time required for completion of the task was measured.

4.1 Results

Time measurements and evaluation of the post-questionnaire results yielded the following findings:

Hypotheses 1: End-User Efficiency. The results indicate a possible advantage of IWC visualization for widgets with mostly static content, i. e., whenever changes due to IWC are rather subtle as opposed to animations (cf. Figure 4, left). However, this difference is not statistically significant. When changes were animated by widgets, the average task completion times in test and control groups were roughly the same. Statistically, the test group was 18% slower than the control group (95% confidence), revealing a possible distraction of users due to the visualization.

This can be partly attributed to the overlapping of indicator flashes for fast users. The indicators were still flashing from their last interaction, when data was published by a new widget. This confused several users, who expected updated data, i. e., the egg, to appear in those widgets. The hypothesis for the awareness mechanisms is, thus, considered to be *not approved*.

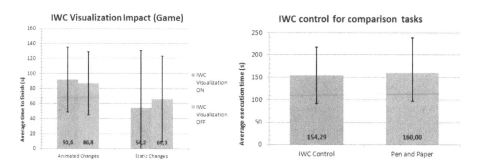

Fig. 4. Impact of awareness (left) and control (right) facilities on end-user efficiency

The control mechanisms for IWC, namely the possibility to isolate widgets, gave users a slight benefit when solving the flat comparison task (cf. Figure 4, right). As the time advantage is not significant, this hypothesis is also considered to be *not approved*.

Hypotheses 2: Usability of the Awareness Mechanisms. According to user ratings (cf. Figure 5, left), IWC visualization does not help users in subtly suggesting "where to look next" and thus understanding which widget are communicating. For users in the test group, it made no difference whether changes in widgets were animated or not. In contrast, users from the control group obviously found it easier to "follow the egg" if the changes were animated, since those were easier to spot in their peripheral field of vision. Thus, for mostly static widgets IWC visualization seems to compensate the missing IWC indicator and to facilitate the recognition of communicating parties.

In the light of the above results, the hypothesis is considered to be *approved* for widgets with mostly static content and *not approved* for widgets with animated changes. Based on this fact, we can derive a guideline for widget developers to animate changes triggered by IWC in order to improve usability of the future mashups.

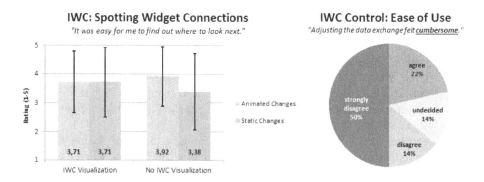

Fig. 5. Usability of the proposed awareness (left) and control (right) facilities

Hypotheses 3: Usability of the Control Mechanisms. IWC control mechanisms got very positive response. 64% of users found them easy to use (cf. Figure 5, right). The controls for this feature, namely the integration with the widget menu, were also rated positively. The hypothesis is considered to be *approved*.

In the post-questionnaire, the vast majority of the participants described the visualization and control facilities as helpful and easy to use. Users recommended making the IWC visualization optional and as such less subtle. One suggestion was to enable/disable visualization per widget by clicking the indicators directly. Furthermore, they suggested using a more noticeable colour scheme for indicating communication partners. It was proposed to investigate if a distinction between incoming/outgoing data is necessary to be visualized, or rather one indicator for taking part in data exchange is enough, e.g., a cogwheel. Finally, some participants suggested IWC controls, e.g., "Isolation"/"Pinning", to be accessible from the widget header bar.

5   Related Work

The need for appropriate awareness and control facilities in choreographed user interface mashups has been recognized by a number of research projects and initiatives [8,21]. As for authors' knowledge, the only hybrid mashup platform proposed at the moment is Chrooma+ [12], whereas only spare configuration of widget IWC behaviour is possible. In [21], Wilson proposes several ideas to tackle awareness and control challenges in UI mashups. Our work continues the research on this field and proposes actual solutions as well as corresponding architectural components.

The problem of visualizing and controlling interactions between autonomous entities is also tackled in other research areas. For example, in the field of self-organizing multi-agent systems, much research has been performed on the visualization of agent interactions. Typical strategies are, e.g., either to draw a graph of interactions in a whole system [1,6] or to highlight relationships between single agents and their environment.[14,17,15]. In [20], the authors propose to draw a causality graph to visualize message exchange in a multi-agent system. Though awareness of relationships in a multi-agent system can be increased with their approach, an a priori simulation and an event log are needed for the system to work. For mashups that employ cost-causing widgets (e.g., with telecommunication functionality), an a priori simulation might be undesirable.

An approach similar to the presented one is given in [16]. The authors propose a set of tools with different perspectives for monitoring and debugging of multi-agent systems. Relationships between agents and possible interactions are shown in the so-called "society tool". A requirement for the tool is that agents expose their partial knowledge about the outer world and communication capabilities to the tool, which then visualizes them in a graph fashion. Several controlling tools enable modification of agent states and configuration of their reactions on incoming messages. Though fine-grained exploration and adaptation of the system is possible, the target group of the approach are skilled developers.

In the field of natural programming environments, a common practice is to use natural language and question-answering games to explore a system. The WhyLine tool [11] applies natural language to enable unskilled developers to debug their algorithms. Using menus and pictograms of objects involved into an algorithm, developers can construct "why did" and "why did not" questions in order to explore system behaviour. A user study revealed that the participants were more efficient with this system than with traditional debugging tools. The approach is applied for explanation of static information and recorded event log. However, it doesn't foresee any means for visualization of active data transfer as it is required for IWC scenarios.

Similar research on control facilities in the context of loosely coupled communicating components can be found among visual programming tools. Lego-Mindstorms products[9] based on LabVIEW [10] enable unskilled developers to design their algorithms in a graphical way. Different boxes representing robot functionalities like move, rotate or stop can be connected with each other and controlled using loops or branches. The tool applies the "wiring" metaphor to connect inputs and outputs of the components, which makes the approach flexible and extensible. Many mashup platforms have adopted the "wiring" technique to enable their users to define data flow in compositions [9,23] and have shown its suitability in the context of end-user development [7]. However, the direct adoption of the technique to choreographed UI mashups is impossible due to the self-organizing nature of the aggregated widgets. The awareness and control facilities in this paper are inspired by the "wiring" approach and apply it to enable visualization and tailoring of the unified communication model.

6   Conclusions and Outlook

Missing understanding of inter-widget dependencies and lack of IWC control facilities can significantly impact usability and user experience within choreographed UI mashups. This paper presented an approach to systematically develop hybrid mashups with integrated IWC awareness and control mechanisms. The resulting solutions differ from the current state of the art in that they both enable self-emerging ("automatic" from end-users' point of view) IWC and keep users in control of how widgets communicate at the same time. The proposed reference architecture can be used as a guidance to build UI mashup platforms either from scratch or by extension of existing choreographed ones. One implementation of the reference architecture has been demonstrated in the context of an existing UI mashup platform based on publish-subscribe IWC strategy. The implementation is easily portable to other communication models such as RPC or shared memory.

A user study with 27 participants confirmed usability of the proposed control mechanisms and helped to discover shortcomings in the awareness ones.

---

[9] http://education.lego.com/en-us/preschool-and-school/
secondary/mindstorms-education-ev3

[10] http://www.ni.com/labview

The drawn consequence for the future work is therefore to explore alternative non-ambiguous visualization techniques (e. g., flashing arrows instead of blinking icons) and to make the control mechanisms more prominent (e. g., by making the isolation icons accessible from the composition canvas). Finally, explanation of data being transferred between widgets has not been tackled sufficiently so far. The challenge here is to present the technical data (such as message syntax and semantic) in end-user-friendly way. Use of dedicated widget annotations or semantically enriched messages (as proposed in [18]) should be explored in the future.

Online Demonstration. A demonstration of the proposed awareness and control facilities integrated into Apache Rave and Apache Wookie projects is available at `http://vsr.cs.tu-chemnitz.de/demo/hybrid-ui-mashups`.

Acknowledgement. This work was supported by the European Commission (project OMELETTE, contract 257635). The authors thank Vadim Chepegin and TIE Kinetix b.v. for their contributions to the user study.

References

1. Bellifemine, F., Caire, G., Poggi, A., Rimassa, G.: JADE: A White Paper. EXP in Search of Innovation 3(3), 6–19 (2003)
2. Cappiello, C., Daniel, F., Matera, M., Picozzi, M., Weiss, M.: Enabling end user development through mashups: Requirements, abstractions and innovation toolkits. In: Piccinno, A. (ed.) IS-EUD 2011. LNCS, vol. 6654, pp. 9–24. Springer, Heidelberg (2011)
3. Cappiello, C., Matera, M., Picozzi, M., Sprega, G., Barbagallo, D., Francalanci, C.: DashMash: A mashup environment for end user development. In: Auer, S., Díaz, O., Papadopoulos, G.A. (eds.) ICWE 2011. LNCS, vol. 6757, pp. 152–166. Springer, Heidelberg (2011)
4. Chudnovskyy, O., Nestler, T., Gaedke, M., Daniel, F., Ignacio, J.: End-User-Oriented Telco Mashups: The OMELETTE Approach. In: WWW 2012 Companion Volume, pp. 235–238 (2012)
5. Chudnovskyy, O., Pietschmann, S., Niederhausen, M., Chepegin, V., Griffiths, D., Gaedke, M.: Awareness and control for inter-widget communication: Challenges and solutions. In: Daniel, F., Dolog, P., Li, Q. (eds.) ICWE 2013. LNCS, vol. 7977, pp. 114–122. Springer, Heidelberg (2013)
6. Collis, J.C., Ndumu, D.T., Nwana, H.S., Lee, L.C.: The zeus agent building toolkit. BT Technology Journal 16(3), 60–68 (1998)
7. Imran, M., Soi, S., Kling, F., Daniel, F., Casati, F., Marchese, M.: On the systematic development of domain-specific mashup tools for end users. In: Brambilla, M., Tokuda, T., Tolksdorf, R. (eds.) ICWE 2012. LNCS, vol. 7387, pp. 291–298. Springer, Heidelberg (2012)
8. Isaksson, E., Palmer, M.: Usability and inter-widget communication in PLEs. In: Proceedings of the 3rd Workshop on Mashup Personal Learning Environments (2010)
9. JackBe. Presto Wires, `http://www.jackbe.com/products/wires.php`

10. Kirschenmann, U., Scheffel, M., Friedrich, M., Niemann, K., Wolpers, M.: Demands of modern pLEs and the ROLE approach. In: Wolpers, M., Kirschner, P.A., Scheffel, M., Lindstaedt, S., Dimitrova, V. (eds.) EC-TEL 2010. LNCS, vol. 6383, pp. 167–182. Springer, Heidelberg (2010)

11. Ko, A.J., Myers, B.A.: Designing the whyline: a debugging interface for asking questions about program behavior. In: Proceedings of the SIGCHI Conf. on Human Factors in Computing Systems, vol. 6, pp. 151–158 (2004)

12. Krug, M., Wiedemann, F., Gaedke, M.: Enhancing media enrichment by semantic extraction. In: Proceedings of the 23nd International Conference on World Wide Web Companion, WWW 2014 Companion (to appear, 2014)

13. Lizcano, D., Soriano, J., Reyes, M., Hierro, J.J.: Ezweb/fast: Reporting on a successful mashup-based solution for developing and deploying composite applications in the upcoming ubiquitous soa. In: Proceedings of the 2nd Intl. Conf. on Mobile Ubiquitous Computing Systems, Services and Technologies, pp. 488–495. IEEE (September 2008)

14. Luke, S., Cioffi-Revilla, C., Panait, L., Sullivan, K., Balan, G.: Mason: A multiagent simulation environment. Simulation 81(7), 517–527 (2005)

15. Minar, N., Burkhart, R., Langton, C.: The swarm simulation system: A toolkit for building multi-agent simulations. Technical report (1996)

16. Ndumu, D.T., Nwana, H.S., Lee, L.C., Collis, J.C.: Visualising and debugging distributed multi-agent systems. In: Proceedings of the Third Annual Conference on Autonomous Agents, AGENTS 1999, pp. 326–333. ACM, New York (1999)

17. North, M.J., Howe, T.R., Collier, N.T., Vos, J.R.: The Repast Simphony runtime system. In: Proceedings of the Agent 2005 Conference on Generative Social Processes, Models, and Mechanisms, ANL/DIS-06-1, co-sponsored by Argonne National Laboratory and The University of Chicago (2005)

18. Radeck, C., Blichmann, G., Meißner, K.: CapView – functionality-aware visual mashup development for non-programmers. In: Daniel, F., Dolog, P., Li, Q. (eds.) ICWE 2013. LNCS, vol. 7977, pp. 140–155. Springer, Heidelberg (2013)

19. Rümpel, A., Radeck, C., Blichmann, G., Lorz, A., Meißner, K.: Towards do-it-yourself development of composite web applications. In: Proceedings of International Conference on Internet Technologies & Society 2011, pp. 330–332 (2011)

20. Vigueras, G., Botia, J.A.: Tracking causality by visualization of multi-agent interactions using causality graphs. In: Dastani, M., El Fallah Seghrouchni, A., Ricci, A., Winikoff, M. (eds.) ProMAS 2007. LNCS (LNAI), vol. 4908, pp. 190–204. Springer, Heidelberg (2008)

21. Wilson, S.: Design challenges for user-interface mashups user control and usability in inter-widget communications (2012)

22. Wilson, S., Daniel, F., Jugel, U., Soi, S.: Orchestrated user interface mashups using W3C widgets. In: Harth, A., Koch, N. (eds.) ICWE 2011. LNCS, vol. 7059, pp. 49–61. Springer, Heidelberg (2012)

23. Yahoo! Yahoo! Pipes, http://pipes.yahoo.com/

24. Zuzak, I., Ivankovic, M., Budiselic, I.: A classification framework for web browser cross-context communication. CoRR, abs/1108.4770 (2011)

Identifying Patterns in Eyetracking Scanpaths in Terms of Visual Elements of Web Pages

Sukru Eraslan[1,2], Yeliz Yesilada[1], and Simon Harper[2]

[1] Middle East Technical University, Northern Cyprus Campus, Guzelyurt, Mersin 10, Turkey
{seraslan,yyeliz}@metu.edu.tr
[2] University of Manchester, School of Computer Science, United Kingdom
sukru.eraslan@postgrad.manchester.ac.uk,
simon.harper@manchester.ac.uk

Abstract. Web pages are typically decorated with different kinds of visual elements that help sighted people complete their tasks. Unfortunately, this is not the case for people accessing web pages in constraint environments such as visually disabled or small screen device users. In our previous work, we show that tracking the eye movements of sighted users provide good understanding of how people use these visual elements. We also show that people's experience in constraint environments can be improved by reengineering web pages by using these visual elements. However, in order to reengineer web pages based on eyetracking, we first need to aggregate, analyse and understand how a group of people's eyetracking data can be combined to create a common scanpath (namely, eye movement sequence) in terms of visual elements. This paper presents an algorithm that aims to achieve this. This algorithm was developed iteratively and experimentally evaluated with an eyetracking study. This study shows that the proposed algorithm is able to identify patterns in eyetracking scanpaths and it is fairly scalable. This study also shows that this algorithm can be improved by considering different techniques for pre-processing the data, by addressing the drawbacks of using the hierarchical structure and by taking into account the underlying cognitive processes.

Keywords: eyetracking, scanpaths, commonality, transcoding, reengineering.

1 Introduction

Web pages mainly consist of different kinds of visual elements, such as menu, logo and hyperlinks. These visual elements help sighted people complete their tasks, but unfortunately small screen device users and disabled users cannot benefit from these elements. When people access web pages with small screen devices, they typically experience many difficulties [1]. For example, on small screen devices, only some parts of web pages are accessible or the complete web page is available with very small text size. Hence, they may need to scroll or zoom a lot which can be annoying. Moreover, they may need more time and effort to find their targets. Similarly, web experience can be challenging for visually disabled users who typically use screen readers to access the web [2]. Since screen readers follow the source code of web pages, visually disabled users have to listen to unnecessary clutter to get to the main content [3].

S. Casteleyn, G. Rossi, and M. Winckler (Eds.): ICWE 2014, LNCS 8541, pp. 163–180, 2014.

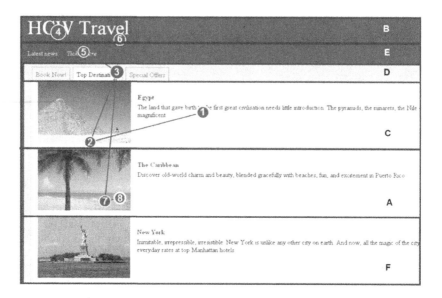

Fig. 1. A scanpath on a segmented web page

In our previous work, we show that reengineering web pages by using the visual elements can improve the user experience in constraint environments [4]. However, identifying visual elements and their role is the key for such reengineering process. To automatically process a web page and identify these elements, in our previous work we have extended and improved the Vision Based Page Segmentation (VIPS) algorithm [5,6]. This extended algorithm automatically discovers visual elements and relates them to the underlying source code. It allows direct access to these visual elements via XPath. However, this algorithm does not provide any information on how these visual elements are used. In our previous work, we also show that tracking the eye movements of sighted users provide good understanding of how they are used [2]. Eyes make quick movements which are called saccades. Between saccades, eyes make fixations where they become relatively stationary. Both fixations and saccades create scanpaths which are eye movement sequences [7]. Fig. 1 shows how a web page is segmented and illustrates a scanpath on a segmented web page. The circles represent fixations where the larger circles represent longer fixations. The numbers in the circles show the sequence. Also, the lines between circles are saccades.

In order to be able to use eyetracking data for reengineering web pages, this paper presents an algorithm called *"eMine scanpath algorithm"*[1]. This algorithm analyses and aggregates a group of people's eyetracking data to create a common scanpath in terms of visual elements of web pages (Section 3). Web pages are first automatically segmented into visual elements with the extended and improved version of the VIPS algorithm [6,5]. Eyetracking data is then exported and related to these visual elements. This creates individual scanpaths of users in terms of visual elements. These individual

---

[1] http://emine.ncc.metu.edu.tr/

scanpaths are then used by eMine scanpath algorithm to create a common scanpath. eMine scanpath algorithm was iteratively developed with the existing eyetracking data and our preliminary evaluation of this algorithm with the existing data was promising [8]. But in order to experientially evaluate validity and scalability of this algorithm, we conducted a new eyetracking study with 40 participants (Section 4). This study illustrates that eMine scanpath algorithm is able to identify a common scanpath in terms of visual elements of web pages and it is fairly scalable (Section 5 and Section 6). It has also revealed some weaknesses which can be improved in the future (Section 7).

2 Related Work

Eyetracking scanpaths have been analysed with different methods for different purposes. These methods typically use string representations of scanpaths which are generated using the sequence of Areas of Interest (AoIs) [9]. For example, the string representation of the scanpath in Fig. 1 is generated as CCDBEBAA. Different ways can be used to generate these AoIs such as using a grid layout directly [9] or the fixations' distribution over web pages [10]. However, these existing approaches typically treat a web page as an image to identify these AoIs which means these scanpaths cannot be used to process web pages. In order to address this, our previous work automatically segments a web page and each segment becomes an AoI [5,6]. This allows relating AoIs with the underlying source code which is important for being able to process web pages by using the eyetracking data.

The Levenshtein Distance (String-Edit) algorithm has commonly been used to analyse scanpaths [11,9]. This algorithm calculates the dissimilarity between the string representations of two scanpaths by transforming one to another with a minimum number of operations (insertion, deletion and substitution). For example, the dissimilarity between XYCZ and XYSZ is calculated as 1 (one) by the String-Edit algorithm because the substitution C with S is sufficient to transform one to another. Although the String-edit algorithm can be used to categorise scanpaths [12] and investigate differences between the behaviours of people on web pages [11], the algorithm itself is not able to identify a common scanpath for multiple scanpaths.

Transition Matrix is one of the methods which use multiple scanpaths to create a matrix [12]. This matrix allows identifying the possible next and previous AoI of the particular AoI. However, when this method is considered for identifying a common scanpath, some considerable problems arise, such as What is the start and end point of the common scanpath? Which probabilities should be considered?

To address these problems, some other methods can be considered. For example, the Shortest Common Supersequence method has been mentioned in literature to identify a common scanpath for multiple people but it has considerable weaknesses [13]. For example, it identifies XABCDEZ as a common scanpath for the individual scanpaths XAT, XBZ, XCZ, XDZ and XEZ. As can be easily recognised, the common scanpath is not supported by the individual scanpaths, for instance, the common scanpath has E which is included by only one individual scanpath (XEZ). Furthermore, the common scanpath is quite longer compared to the individual scanpaths.

Some methods, such as T-Pattern [14] and eyePatterns's Discover Patterns [12], have been proposed to detect subpatterns in eyetracking scanpaths. However, eyePatterns's Discover Patterns method [12] is not tolerant of extra items in scanpaths. For instance, XYZ can be detected as a subpattern for XYZ and WXYZ but it cannot be detected for XYZ and WXUYZ because of the extra item U. This shows that this method is reductionist which means it is likely to produce unacceptable short scanpaths.

The Multiple Sequence Alignment method was proposed to identify a common scanpath but this method was not validated [15]. Moreover, the Dotplots-based algorithm was proposed to identify a common scanpath for multiple people [16]. This algorithm creates a hierarchical structure by combining a pair of scanpaths with the Dotplots algorithm. The individual scanpaths are located at leafs whereas the common scanpath is located at the root. Some statistical methods have been applied to address the reductionist approach of the Dotplots algorithm [16].

We are interested in common patterns in eyetracking data instead of individual patterns to be able to reengineer web pages. However, as can be seen above, there is not much research in identifying common scanpaths and the existing ones are likely to produce unacceptable short common scanpaths. In this paper, we present our eMine scanpath algorithm to address the limitations of these existing approaches, especially the problem of being reductionist.

3   eMine Scanpath Algorithm

Algorithm 1 shows our proposed eMine scanpath algorithm [8] which takes a list of scanpaths and returns a scanpath which is common in all the given scanpaths. If there is only one scanpath, it returns that one as the common scanpath, if there is more than one scanpath, then it tries to find the most similar two scanpaths in the list by using the String-edit algorithm [11]. It then removes these two scanpaths from the list of scanpaths and introduces their common scanpath produced by the Longest Common Subsequence method [17] to the list of scanpaths. This continues until there is only one scanpath.

Algorithm 1. Find common scanpath

Input: Scanpath List
Output: Scanpath
 1: **if** the size of Scanpath List is equal to 1 **then**
 2:     **return** the scanpath in Scanpath List
 3: **end if**
 4: **while** the size of Scanpath List is not equal to 1 **do**
 5:     Find the two most similar scanpaths in Scanpath List with the String-edit algorithm
 6:     Find the common scanpath by using the Longest Common Subsequence method
 7:     Remove the similar scanpaths from the Scanpath List
 8:     Add the common scanpath to the Scanpath List
 9: **end while**
10: **return** the scanpath in Scanpath List

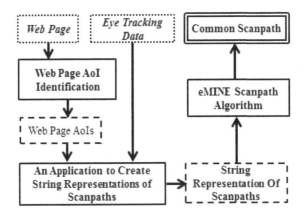

Fig. 2. System architecture where '...' shows the input parts, '_ _ _' represents intermediate parts, '_' illustrates the functional parts and '=' is used for the output part

3.1 System Architecture and Implementation

eMine scanpath algorithm was integrated with the extended and improved version of the VIPS algorithm [6,5]. Fig. 2 illustrates the system architecture which consists of the following parts: two input parts (web page and eyetracking data), three functional parts (web page AoI identification, an application to create string representations of scanpaths, eMine scanpath algorithm), two intermediate parts which are created as an output of one functional part and used as an input for another functional part (web page AoIs, string representations of scanpaths) and one output part (common scanpath). The functional parts are explained below.

Web Page AoI Identification. A web page is used as an input for the web page AoI identification part. This part creates AoIs automatically by using the extended and improved version of the VIPS algorithm [6,5]. Even though, the extended VIPS was used, it would be easily replaced by an alternative method of AoI identification approach. These AoIs represent visual elements of web pages.

An Application to Create String Representations of Scanpaths. The automatically generated web page AoIs and eyetracking data, provided by eyetracking software, are then used by an application to create string representations of scanpaths.

eMine Scanpath Algorithm. Once the string representations are created, our scanpath algorithm is applied to them to produce a common scanpath in terms of AoIs.

eMine scanpath algorithm[2] was implemented on the Accessibility Tools Framework (ACTF)[3] which is an open-source Eclipse project.

---

[2] http://emine.ncc.metu.edu.tr/software.html
[3] http://www.eclipse.org/actf/

4   An Eyetracking Study

In order to experimentally evaluate validity and scalability of eMine scanpath algorithm, we conducted an eyetracking study. This study aims to investigate the following two research questions:

1. *Validity:* The aim is to investigate whether or not eMine scanpath algorithm can successfully identify common scanpaths in terms of visual elements of web pages. Thus, we ask *"Can eMine algorithm identify common scanpaths in terms of visual elements of web pages?"*.
2. *Scalability:* We would like to investigate whether or not eMine scanpath algorithm works well for different numbers of participants on different web pages. Hence, the research question here is *"How does the number of individual scanpaths affect common scanpaths?"*.

4.1   Equipment

Participants sat in front of a 17" monitor with a built-in TOBII T60 eye tracker with screen resolution 1280 x 1024. The web pages were on a HP ELiteBook 8530p laptop and these web pages were shown to the participants using the eye tracker's screen. Tobii Studio eye gaze analysis software was used to record the data. Eyetracking data was also stored on that laptop, too. The collected eyetracking data were analysed on a 17" monitor with the screen resolution 1280 x 1024.

4.2   Materials

Six web pages were randomly selected from a group of pages that were used in our previous study. That study focused on evaluating the extended and improved version of the VIPS algorithm and to have continuity in our studies we used same set of pages [6,5]. These web pages were categorised based on their complexity, which were low, medium and high [6,5,18]. Two web pages were chosen randomly from each level of complexity for our study. These pages with their complexity levels are as follow: Apple (Low), Babylon (Low), AVG (Medium), Yahoo (Medium), Godaddy (High) and BBC (High). Since the 5th segmentation granularity level was found as the most successful level with approximately 74% user satisfaction, we decided to use the 5th level for our experiments [6,5]. The segmented web pages can be seen in Fig. 3, 4, 5, 6, 7 and 8.

4.3   Procedure

This eyetracking study consists of the following three parts.

Introduction: The participants read the information sheet and signed the consent form. Next, they filled in the short questionnaire which was for the purpose of collecting basic demographic information of participants, which are gender, age groups and education level. The participants were also asked to rank their web page usage for the six web pages with 1 (Daily), 2 (Weekly), 3 (Monthly), 4 (Less than once a month) or 5 (Never).

Main Part: The participants sat in front of the eye tracker which calibrated to their gaze. They then viewed all of the six web pages twice, one view for searching (maximum 120 seconds) and one view for browsing in a random order. For browsing tasks, the participants were given 30 seconds as used in other studies [19]. The searching and browsing tasks are shown in Table 1. The researcher was responsible to check if the participants complete the tasks successfully and take notes if necessary.

Conclusion: At the end, the participants were asked to redraw three web pages from three different complexity levels.

4.4 User Tasks

User tasks are categorised into two groups for this study: searching and browsing. In the literature, many studies were conducted to categorise user tasks on the web [20]. G. Marchionini Search Activities Model is one of the most popular models in this field [20]. It consists of three groups which are lookup, learn and investigate [20]. Our searching category is related to fact finding which is associated with the lookup group whereas our browsing category is related to serendipitous browsing which is associated with the investigation group. The tasks which are defined for the six web pages are listed in Table 1.

We designed the system to ensure that half of the participants complete searching tasks firstly and then complete browsing tasks. Other half completed browsing task firstly and then completed searching tasks. The reason is to prevent familiarity effects on eye movements which can be caused by the user tasks.

4.5 Participants

The majority of the participants comprised students, along with some academic and administrative staff at Middle East Technical University Northern Cyprus Campus and the University of Manchester. Twenty male and twenty female volunteers participated. One male participant changed his body position during the study, so the eye tracker could not record his eye movements. Another male participant had no successful eye calibration. Unfortunately, these two participants were excluded from the study. Therefore, the eyetracking data of 18 males and 20 females were used to evaluate eMine scanpath algorithm.

All of the participants use the web daily. Most of the participants (18 participants) are aged between 18 and 24 years old, then 25-34 group (14 participants) and 35-54 group (6 participants). Moreover, 14 participants completed their high/secondary schools, 6 participants have a bachelor's degree, 9 participants have a master's degree and 9 participants completed their doctorate degrees.

5   Results

In this section, we present the major findings of this study in terms of the two research questions presented in Section 4.

Table 1. Tasks used in the eyetracking study

Apple	
Browsing	1. Can you scan the web page if you find something interesting for you?
Searching	1. Can you locate a link which allows watching the TV ads relating to iPad mini?
	2. Can you locate a link labelled iPad on the main menu?
Babylon	
Browsing	1. Can you scan the web page if you find something interesting for you?
Searching	1. Can you locate a link you can download the free version of Babylon?
	2. Can you find and read the names of other products of Babylon?
Yahoo	
Browsing	1. Can you scan the web page if you find something interesting for you?
Searching	1. Can you read the titles of the main headlines which have smaller images?
	2. Can you read the first item under News title?
AVG	
Browsing	1. Can you scan the web page if you find something interesting for you?
Searching	1. Can you locate a link which you can download a free trial of AVG Internet Security 2013?
	2. Can you locate a link which allows you to download AVG Antivirus FREE 2013?
GoDaddy	
Browsing	1. Can you scan the web page if you find something interesting for you?
Searching	1. Can you find a telephone number for technical support and read it?
	2. Can you locate a text box where you can search a new domain?
BBC	
Browsing	1. Can you scan the web page if you find something interesting for you?
Searching	1. Can you read the first item of Sport News?
	2. Can you locate the table that shows market data under Business title?

5.1 Validity

"Can eMine scanpath algorithm identify common scanpaths in terms of visual elements of web pages?"

The participants were asked to complete some searching tasks on web pages, therefore we are expecting to see that the common scanpath supports those tasks. We used eMine scanpath algorithm to identify a common scanpath for each of the six web pages. Some participants could not complete the searching tasks successfully and/or had calibration problems. These participants were defined as unsuccessful participants and excluded from the study. The success rates in completing searching tasks are as follow: Apple: 81.58 %, Babylon: 94.74 %, AVG: 94.74 %, Yahoo: 84.21 %, Godaddy: 73.68 % and BBC: 100 %. These values are calculated by dividing the number of the successful participants by the total number of the participants on the page.

Table 2 shows the common scanpaths and the abstracted common scanpaths produced by eMine scanpath algorithm for the web pages where 'P' represents the number of successful participants. In order to have abstracted common scanpaths, their string representations are simplified by abstracting consecutive repetitions [21,22]. For instance, MMPPQRSS becomes MPQRS.

Table 2. The common scanpaths produced by eMine scanpath algorithm for the web pages

Page Name	P	Common Scanpath	Abstracted Common Scanpath
Apple	31	EEB	EB
Babylon	36	MMPPQRSS	MPQRS
AVG	36	GGGGGGGGGGGGGGGGGGGIIIIIIIII	GI
Yahoo	32	IIIIIIIIIIIIIIIIIIIIIIIIIIIIIIIIIIIII	I
Godaddy	28	OOOOMMMMMM	OM
BBC	38	RNNNNN	RN

On the Apple web page, 31 out of 38 participants were successful. On this page, the participants were asked to locate a link which allows watching the TV ads relating to iPad mini and then locate a main menu item iPad. EB is identified as a common scanpath for these participants. Since E is associated with the first part and B is related to the second part of the searching task, this common scanpath completely supports the searching task. Fig. 3 shows this common scanpath on the Apple web page.

Fig. 3. Common scanpath on the Apple web page

On the Babylon web page, only 2 participants out of 38 were not successful. On this page, the participants were requested to locate a link which allows downloading a free version of Babylon and then read the names of other products of Babylon. The common scanpath for the 36 participants was identified as MPQRS shown in Fig. 4. M is related with a free version of Babylon whereas P, Q, R and S are associated with four other products of Babylon. Therefore, the common scanpath thoroughly supports the searching task.

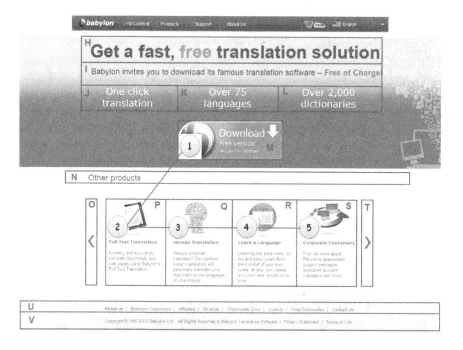

Fig. 4. Common scanpath on the Babylon web page

Similar to the Babylon web page, only 2 participants were unsuccessful on the AVG web page. The searching task here was locating a link which allows downloading a free trial of AVG Internet Security 2013 and then locating a link which allows downloading AVG Antivirus FREE 2013. The common scanpath was produced as GI where G has a link to download a free trial of AVG Internet Security 2013 and I contains a link to download AVG Antivirus FREE 2013. Therefore, the common scanpath, shown in Fig. 5, entirely supports the searching task.

For the Yahoo web page, 6 participants could not be successful. The participants required to read the titles of the main headlines which have smaller images and then read the first item under News title. Since only I is produced as a common scanpath on this web page and I contains both parts of the task, the common scanpath nicely supports the searching task, too. Fig. 6 shows this common scanpath.

Since 28 out of 38 participants were successful, 10 participants were excluded for the Godaddy web page. The successful participants read the telephone number for technical support and then located a text box where they can search for a new domain. eMine scanpath algorithm produced OM as a common scanpath shown in Fig. 7. Since M

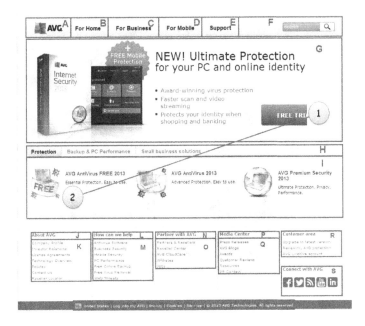

Fig. 5. Common scanpath on the AVG web page

Fig. 6. Common scanpath on the Yahoo web page

Fig. 7. Common scanpath on the Godaddy web page

Fig. 8. Common scanpath on the BBC web page

contains the text box and there is no AoI in the scanpath which is related with the telephone number, the common scanpath partially supports the searching task on the Godaddy web page.

On the BBC web page, all participants completed the searching task successfully. The participants were asked to read the first item of the sports news and then locate a table which shows the market data. Therefore, the participants needed to locate R and then N. As the common scanpath RN is produced, it supports the searching task very well. Fig. 8 illustrates this common scanpath on the BBC web page.

To sum up, the common scanpaths on the Apple, Babylon, AVG, Yahoo and BBC web pages completely support the searching tasks whereas the common scanpath on the Godaddy web page partially supports the searching task.

5.2 Scalability

"How does the number of individual scanpaths affect common scanpaths?"

In order to test whether or not eMine scanpath algorithm works well with different numbers of individual scanpaths, we tested the algorithm with different numbers of individual participants. The participants were selected randomly from all of the successful participants. Table 3 illustrates the common scanpaths in terms of AoIs on the different web pages for 10, 20, 30 and 30+ participants while browsing and searching.

Table 3. The common scanpaths on the different web pages for 10, 20, 30 and 30+ participants while browsing and searching where '-' means that there was no sufficient number of successful participants and '——' means that no common scanpath was detected

Task	Page Name	P=10	P=20	P=30	P=30+
Browsing	Apple	IF	F	F	F
	Babylon	MS	M	M	M
	AVG	GIG	G	G	G
	Yahoo	IJI	I	I	I
	Godaddy	O	O	O	O
	BBC	LP	LP	P	——
Searching	Apple	EB	EB	EB	EB
	Babylon	MPQRS	MPQRS	MPQRS	MPQRS
	AVG	IGI	GI	GI	GI
	Yahoo	I	I	I	I
	Godaddy	OM	OM	-	-
	BBC	LPRN	RN	RN	RN

In order to see how the common scanpaths are affected when the number of participants increases, we calculated the similarities between the scanpaths which were produced for 10, 20, 30 and 30+ participants. To calculate the similarity between two common scanpaths the String-edit distance between two common scanpaths is divided by the length of the longer common scanpath to have a normalised score [23]. The purpose of a normalised score is to prevent any inconsistencies in similarities caused by different lengths [23,24]. Finally, the normalised score is subtracted from 1 [23]. For example, the common scanpath for 10 participants is LPRN and the common scanpath for 20 participants is RN on the BBC web page for the searching task. The String-edit distance is calculated as 2 between two scanpaths. After that, since the length of the longer scanpath (LPRN) is equal to 4, this distance is divided by 4. As a result, the normalised score is equal to 0.5. To calculate the similarity 0.5 is subtracted from 1, so the similarity between the two common scanpaths is equal to 0.5 (50 %). Table 4 shows these similarities between the common scanpaths for the searching task on the BBC web page whereas Table 5 illustrates the similarities between the common scanpaths for the browsing task on the Yahoo web page as examples.

Table 4. The similarities between the common scanpaths on the BBC web page for 10, 20, 30 and 30+ participants while searching

BBC Searching	P = 10	P = 20	P = 30	P = 30+
P = 10	—	50	50	50
P = 20	50	—	100	100
P = 30	50	100	—	100
P = 30+	50	100	100	—

Table 5. The similarities between the common scanpaths on the Yahoo web page for 10, 20, 30 and 30+ participants while browsing

Yahoo Searching	P = 10	P = 20	P = 30	P = 30+
P = 10	—	33.3	33.3	33.3
P = 20	33.3	—	100	100
P = 30	33.3	100	—	100
P = 30+	33.3	100	100	—

For both the browsing and searching tasks, we calculated the average similarity between the common scanpaths on each web page. To calculate these average similarities we divided the sum of the similarities between the scanpaths for 10, 20, 30 and 30+ participants by the total number of the similarities. In addition, we calculated the average similarity for both the browsing and searching tasks. Since each web page typically has four scanpaths (for 10, 20, 30 and 30+ participants), we determined their weights based on the number of scanpaths. All of the pages' weights are set to 4, except the Godaddy page because of the searching task. The Godaddy page has one common scanpath for 10 participants and one common scanpath for 20 participants, therefore its weight is set to 2. When the average is calculated, we multiplied the value with its weight to find the weighted value. After that, we found the sum of the weighted value and divided it by the sum of the weights. It was found that the average similarity for searching tasks (92.42%) is higher than the average similarity for the browsing task (69.44 %).

Table 6. The average of the similarities between the common scanpaths on each web page for 10, 20, 30 and 30+ participants

Page Name	Task	Average Similarity for Each Page
Apple	Browsing	75
Babylon	Browsing	75
AVG	Browsing	66.65
Yahoo	Browsing	66.65
Godaddy	Browsing	100
BBC	Browsing	33.33
Average Similarity for the 6 Pages	Browsing	69.44
Apple	Searching	100
Babylon	Searching	100
AVG	Searching	83.3
Yahoo	Searching	100
Godaddy	Searching	100
BBC	Searching	75
Average Similarity for the 6 Pages	Searching	92.42

6   Discussion

The eMine scanpath algorithm was experimentally evaluated with an eyetracking study and this study illustrates that the algorithm is able to successfully identify common scanpaths in terms of visual elements of web pages and it is fairly scalable.

The searching tasks completed by the participants on the given pages were used to validate eMine scanpath algorithm. We expected that the common scanpaths should support these searching tasks. For instance, on the Babylon web page, the participants were asked to locate the link which allows downloading the free version of Babylon (related to AoI M) and then read the names of other products of Babylon (related to AoIs P, Q, R and S). Therefore, we expected that the common scanpath on the Babylon web page should involve at least MPQRS for the searching tasks.

The results in Section 5.1 show that the common scanpaths produced by eMine scanpath algorithm completely support these tasks, except the common scanpath on the Godaddy page. On that page, the participants were asked to read a telephone number for technical support and locate the text box where they can search for a new domain. The common scanpath involves the AoI for the text box but does not include the AoI for the telephone number. Thus, it partially supports the searching task. There may be various reasons: (1) The participants might make a very few fixations on that AoI (2) Some participants might find the telephone number directly whereas some of them looked at many AoIs to find the telephone number. Therefore, it would be good to pre-process eyetracking data in depth to investigate the individual differences and their reasons.

Some other methods could also be used to validate eMine scanpath algorithm. One might consider calculating the similarities between the individual scanpaths and the common scanpath. Besides, the AoIs appeared in all individual scanpaths might be detected and then one part of the validation process could be done by using these AoIs.

The scalability of eMine scanpath algorithm was tested by using the different numbers of individual scanpaths as mentioned in Section 5.2. As expected, we can see that the algorithm is more scalable with the searching tasks because the participants were asked to complete some specific searching tasks. The average similarity is equal to 92.42 % between the common scanpaths which were produced with the different number of scanpaths for the searching tasks. However, the average similarity is equal to 69.44 % for the browsing tasks. Based on these values we can suggest that our algorithm is fairly scalable, especially in searching tasks.

There are some differences between scanpaths, such as producing LPRN for 10 participants and RN for 30+ participants on the BBC page. It is caused by using the hierarchical structure. As mentioned in Section 3, eMine scanpath algorithm uses a hierarchical structure while identifying common scanpaths. It selects the two most similar scanpaths from the list and finds their longest common subsequence. It is iteratively repeated until a single scanpath left. Because of the hierarchical structure, some information in intermediate levels can be lost because of combining two scanpaths.

Assume that there are three sequences: S1: GATACCAT S2: CTAAAGTC and S3: GCTATTGCG [17]. S1 and S2 can be aligned firstly and then S1'= - - A - A - - A - - - can obtained [17]. Following this, S1' and S3 can be aligned and then S3'= - - - A - - - - - - - - can be obtained [17]. This example clearly illustrates that the hierarchical structure can make the method reductionist. Here, all of the three scanpaths have G and T in different locations but G and T do not exist at the end. This may cause some differences in common scanpaths. Because of this reason, eMine scanpath algorithm was not able to identify any common scanpath on the BBC page for the browsing task. When a number of individual scanpaths is increased, the different most similar scanpath pairs can be generated and this may affect common scanpaths. Although eMINE scanpath algorithm has some drawbacks because of the hierarchical structure, it still partly addresses the reductionist problem of the other existing approaches (See Section 2).

To address the drawbacks of using the hierarchical structure a constraint might be created to prevent losing the AoIs appeared in all individual scanpaths in intermediate levels. Alternatively, some statistical approaches can be used to sort these AoIs and then create a common scanpath for multiple people.

7   Concluding Remarks and Future Work

This paper presents an algorithm and its evaluation that identifies common scanpaths in terms of visual elements of web pages. These visual elements are first automatically generated with the extended and improved version of the VIPS algorithm [6,5]. Eyetracking data is then related to these visual elements and individual scanpaths are created in terms of these visual elements. This algorithm then uses these individual scanpaths and generates a common scanpath in terms of these visual elements. This common scanpath can be used for reengineering web pages to improve the user experience in constraint environments.

To our knowledge, there is no work on correlating scanpaths with visual elements of web pages and the underlying source code, and this work is novel from that perspective [6,5]. This paper also shows how the validity and scalability of eMine scanpath algorithm was demonstrated with an eyetracking study. The results clearly show that this algorithm is able to identify common scanpaths in terms of visual elements of web pages and it is fairly stable. This algorithm aims to address the reductionist problem that the other existing work has, but the results show that there is still room for improvement.

The eyetracking study also suggests some directions for future work. It indicates that the individual differences can affect the identification of patterns in eyetracking scanpaths. Thus, eyetracking data should be pre-processed to investigate the individual differences and their reasons. Since an eye tracker collects a large amount of data, pre-processing is also required to eliminate noisy data. It is important because noisy data are likely to decrease the commonality in scanpaths. Another benefit of pre-processing is to identify outliers which are potential to decrease the commonality, too.

Finally, as with the existing scanpath methods, eMine scanpath algorithm also tends to ignore the complexities of the underlying cognitive processes. However, when people follow a path to complete their tasks on web pages, there may be some reasons that affect their decisions. Underlying cognitive processes can be taken into account while identifying common scanpaths.

Acknowledgments. The project is supported by the Scientific and Technological Research Council of Turkey (TÜBİTAK) with the grant number 109E251. As such the authors would like to thank to TÜBİTAK for their continued support. We would also like to thank our participants for their time and effort.

References

1. W3C WAI Research and Development Working Group (RDWG): Research Report on Mobile Web Accessibility. In: Harper, S., Thiessen, P., Yesilada, Y., (eds.): W3C WAI Symposium on Mobile Web Accessibility. W3C WAI Research and Development Working Group (RDWG) Notes. First public working draft edn. W3C Web Accessibility Initiative (WAI) (December 2012)

2. Yesilada, Y., Jay, C., Stevens, R., Harper, S.: Validating the Use and Role of Visual Elements of Web Pages in Navigation with an Eye-tracking Study. In: The 17th international Conference on World Wide Web, WWW 2008, pp. 11–20. ACM, New York (2008)

3. Yesilada, Y., Stevens, R., Harper, S., Goble, C.: Evaluating DANTE: Semantic Transcoding for Visually Disabled Users. ACM Trans. Comput.-Hum. Interact. 14(3), 14 (2007)

4. Brown, A., Jay, C., Harper, S.: Audio access to calendars. In: W4A 2010, pp. 1–10. ACM, New York (2010)

5. Akpınar, M.E., Yesilada, Y.: Heuristic Role Detection of Visual Elements of Web Pages. In: Daniel, F., Dolog, P., Li, Q. (eds.) ICWE 2013. LNCS, vol. 7977, pp. 123–131. Springer, Heidelberg (2013)

6. Akpınar, M.E., Yesilada, Y.: Vision Based Page Segmentation Algorithm: Extended and Perceived Success. In: Sheng, Q.Z., Kjeldskov, J. (eds.) ICWE Workshops 2013. LNCS, vol. 8295, pp. 238–252. Springer, Heidelberg (2013)

7. Poole, A., Ball, L.J.: Eye tracking in human-computer interaction and usability research: Current status and future Prospects. In: Ghaoui, C. (ed.) Encyclopedia of Human-Computer Interaction. Idea Group, Inc., Pennsylvania (2005)

8. Yesilada, Y., Harper, S., Eraslan, S.: Experiential transcoding: An EyeTracking approach. In: W4A 2013, p. 30. ACM (2013)

9. Takeuchi, H., Habuchi, Y.: A quantitative method for analyzing scan path data obtained by eye tracker. In: CIDM 2007, April 1-5, pp. 283–286 (2007)

10. Santella, A., DeCarlo, D.: Robust clustering of eye movement recordings for quantification of visual interest. In: ETRA 2004, pp. 27–34. ACM, New York (2004)

11. Josephson, S., Holmes, M.E.: Visual attention to repeated internet images: testing the scanpath theory on the world wide web. In: ETRA 2002, pp. 43–49. ACM, NY (2002)

12. West, J.M., Haake, A.R., Rozanski, E.P., Karn, K.S.: EyePatterns: software for identifying patterns and similarities across fixation sequences. In: ETRA 2006, pp. 149–154. ACM, New York (2006)

13. Räihä, K.-J.: Some applications of string algorithms in human-computer interaction. In: Elomaa, T., Mannila, H., Orponen, P. (eds.) Ukkonen Festschrift 2010. LNCS, vol. 6060, pp. 196–209. Springer, Heidelberg (2010)

14. Mast, M., Burmeister, M.: Exposing repetitive scanning in eye movement sequences with t-pattern detection. In: IADIS IHCI 2011, Rome, Italy, pp. 137–145 (2011)

15. Hembrooke, H., Feusner, M., Gay, G.: Averaging scan patterns and what they can tell us. In: ETRA 2006, p. 41. ACM, New York (2006)

16. Goldberg, J.H., Helfman, J.I.: Scanpath clustering and aggregation. In: ETRA 2010, pp. 227–234. ACM, New York (2010)

17. Chiang, C.H.: A Genetic Algorithm for the Longest Common Subsequence of Multiple Sequences. Master's thesis, National Sun Yat-sen University (2009)

18. Michailidou, E.: ViCRAM: Visual Complexity Rankings and Accessibility Metrics. PhD thesis, University of Manchester (2010)

19. Jay, C., Brown, A.: User Review Document: Results of Initial Sighted and Visually Disabled User Investigations. Technical report, University of Manchester (2008)

20. Marchionini, G.: Exploratory Search: From Finding to Understanding. Commun. ACM 49(4), 41–46 (2006)

21. Brandt, S.A., Stark, L.W.: Spontaneous Eye Movements During Visual Imagery Reflect the Content of the Visual Scene. J. Cognitive Neuroscience 9(1), 27–38 (1997)

22. Jarodzka, H., Holmqvist, K., Nyström, M.: A Vector-based, Multidimensional Scanpath Similarity Measure. In: ETRA 2010, pp. 211–218. ACM, New York (2010)

23. Foulsham, T., Underwood, G.: What can Saliency Models Predict about Eye Movements? Spatial and Sequential Aspects of Fixations during Encoding and Recognition. Journal of Vision 8(2), 1–17 (2008)

24. Cristino, F., Mathot, S., Theeuwes, J., Gilchrist, I.D.: Scanmatch: a novel method for comparing fixation sequences. Behavior Research Methods 42(3), 692–700 (2010)

Identifying Root Causes of Web Performance Degradation Using Changepoint Analysis

Jürgen Cito[1], Dritan Suljoti[2], Philipp Leitner[1], and Schahram Dustdar[3]

[1] s.e.a.l. – Software Evolution & Architecture Lab, University of Zurich, Switzerland
{cito,leitner}@ifi.uzh.ch
[2] Catchpoint Systems, Inc., New York, USA
drit@catchpoint.com
[3] Distributed Systems Group, Vienna University of Technology, Austria
dustdar@dsg.tuwien.ac.at

Abstract. The large scale of the Internet has offered unique economic opportunities, that in turn introduce overwhelming challenges for development and operations to provide reliable and fast services in order to meet the high demands on the performance of online services. In this paper, we investigate how performance engineers can identify three different classes of externally-visible performance problems (global delays, partial delays, periodic delays) from concrete traces. We develop a simulation model based on a taxonomy of root causes in server performance degradation. Within an experimental setup, we obtain results through synthetic monitoring of a target Web service, and observe changes in Web performance over time through exploratory visual analysis and changepoint detection. Finally, we interpret our findings and discuss various challenges and pitfalls.

1 Introduction

The large scale of the Internet has offered unique economic opportunities by enabling the ability to reach a tremendous, global user base for businesses and individuals alike. The great success and opportunities also open up overwhelming challenges due to the drastic growth and increasing complexity of the Internet in the last decade. The main challenge for development and operations is to provide reliable and fast service, despite of fast growth in both traffic and frequency of requests. When it comes to speed, Internet users have high demands on the performance of online services. Research has shown that nowadays 47% of online consumers expect load times of *two seconds or less* [7, 14, 21]. With the growth of the Internet and its user base, the underlying infrastructure has drastically transformed from single server systems to heterogeneous, distributed systems. Thus, the end performance depends on diverse factors in different levels of server systems, networks and infrastructure, which makes providing a satisfying end-user experience and QoS (Quality of Service) a challenge for large scale Internet applications. Generally, providing a consistent QoS requires continually collecting data on Web performance on the Web service provider side, in order to observe and track changes in desired metrics, e.g., service response time. Reasons to observe

S. Casteleyn, G. Rossi, and M. Winckler (Eds.): ICWE 2014, LNCS 8541, pp. 181–199, 2014.
© Springer International Publishing Switzerland 2014

these changes are different in their nature, and range from detecting anomalies, identifying patterns, ensuring service reliability, measuring performance changes after new software releases, or discovering performance degradation.

Early detection and resolution of root causes of performance degradations can be achieved through monitoring of various components of the system. Monitoring can be classified as either *active* or *passive* monitoring, and, orthogonally, as *external* or *internal*. In active monitoring, monitoring agents are actively trying to connect to the target system in order to collect performance data, whether the system is accessed by real end-users or not. Passive monitoring, on the other hand, only collects measurements if the system is actively used. Internal and external monitoring differentiate in whether the measurements are obtained in systems within the organization's data center or through end-to-end monitoring over the network outside the data center. This has ramifications in terms of what the level of detail of monitoring data that is available. In our experiments, we make use of active, external monitoring, which provides a way of capturing an end user perspective and enables the detection of issues before they affect real users [17].

Whatever the reason to observe changes may be, the measurements are only useful when we know how to properly analyze them and turn our data into informed decisions. The main contribution of this paper is a model for understanding performance data via analyzing how common underlying root causes of Web performance issues manifest themselves in data gathered through *external, active* monitoring. We introduce a taxonomy of root causes in server performance degradations, which serves as the basis for our experiments. Furthermore, we describe the methods and steps we take to obtain our results and explain how the results will be examined and discussed. Following this, we will outline the design of the simulations that will be conducted, as well as the physical experimental setup enabling the simulations. We conclude by providing interpretation of the simulation based results, explaining how we can derive certain conclusions based on exploratory visual analysis and statistical changepoint analysis.

2   Root Causes of Server Performance Degradation

In general, if we consider performance and computation power of a system, we must consider resources that enable computation. These are usually hardware resources, such as processors, memory, disk I/O, and network bandwidth. We also need to consider the ways and methods these resources are allocated and utilized. The demand on resources of a computer system increases as the workload for the application of interest increases. When the demand of the application is greater than the resources that can be supplied by the underlying system, the system has hit its resource constraints. This means the maximum workload of the application has been reached and, typically, the time taken for each request to the application will increase. In case of extreme oversaturation, the system stops reacting entirely. For Web applications and Web services, this translates into poor response times or (temporary) unavailability. A delay in performance as observed through active monitoring can be defined as a negative change in response time at a certain point in time t. This means we look at two observations of response times x_t and x_{t+1} where $\lambda = |x_t - x_{t+1}| > c$ and c denotes a certain threshold accounting for possible volatility. In the following, this simplified notion of performance delays λ over a threshold c will be used in the description of the elements of the taxonomy.

The underlying causes of performance degradations in Web application and Web service backends are diverse. They differ significantly in the way they manifest themselves in performance data gathered through active monitoring. We propose a simple taxonomy of root causes in Web performance degradation. First, we divide a possible root cause in three main categories, which can be determined through external monitoring: *global delay*, *partial delay*, and *periodic delay*. Further classifications and proper assignment to the main categories in the taxonomy have been derived together with domain experts in the area of Web performance monitoring and optimization. In the following, we provide a brief explanation of the general causes of performance delays in computer systems. We then classify the main categories of our taxonomy by grouping the root causes by the distinguishable effect they have. The taxonomy is depicted in Figure 1, and explained in more detail the following. Note that this taxonomy does by no means raise the claim of completeness. It is rather an attempt to give an overview of common pitfalls that cause slowness in performance.

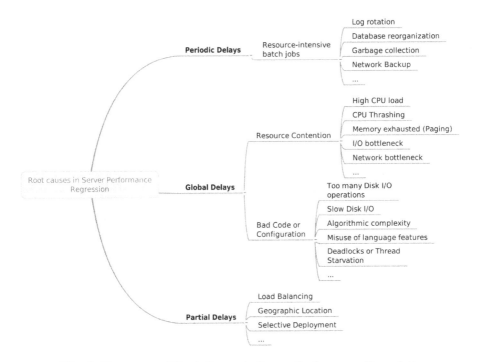

Fig. 1. Taxonomy of Root Causes in Server Performance Degradation

2.1   Global Delay

A global delay means that a change in a new deployment or release of the backend system introduced a significant difference in response time λ, which is higher than a defined threshold c on *all* incoming requests. We distinguish between global delays that are caused through resource contention and code or configuration issues. Global delays caused by resource contention include, for instance, delays due to a bottleneck in disk I/O. In this case, the flow of data from the disk to the application (e.g., via a database query) is contended. This means the query takes longer to perform and return data which causes an overall performance delay. Global delays caused by problems in the application code can for instance be caused by the induction of logical units or algorithms with high computational complexity in the backend service. Alternatively, such global delays may be caused by overzealous synchronization in the application code, which may even lead to temporary service outages when a deadlock situation cannot be immediately resolved by the underlying operating system or virtual machine.

2.2   Periodic Delay

Periodic delays are global delays that are not continuous, but happen, e.g., a few times a day. A periodic delay is mostly not induced through a new deployment or release, but rather through a background process causing the system to use an increased amount of resources. One practical example of such a background job is log rotation. Log rotation is an automated process in server systems administration, where log files that exhibit certain characteristics are archived. The process is usually configured to run as a periodic cronjob to be fully automated. Log rotating often includes archiving and transferring large text files, and, hence, can oversaturate the backend system for a short period of time.

2.3   Partial Delay

Partial delays are global delays that occur only for a subset of all requests, e.g., for a subset of all customers of the Web service. This situation can occur if the application employs a form of redundancy, e.g., load balancing or content distribution networks. In such scenarios, any problems that lead to global delays can potentially be inflicting only one or a subset of all backend servers, hence delaying only those requests that happen to be handled by one of the inflicted backends. Partial delays are interesting, as they are hard to detect (especially if the number of inflicted backends is small).

3   Identifying Root Causes of Performance Degradation

After introducing externally visible classes of root causes of performance degradation (global, partial, periodic), we want to identify characteristics in performance data associated with each class. Furthermore, we want to present statistical methods that are well suited for identifying such changes. Our approach is to generate realistic performance traces for each class through a testbed Web service, which we have modified in order to be able to inject specific *performance degradation scenarios* associated with each class. We collect data through active monitoring as described in Section 3.1. The specific scenarios we used are described and defined in Section 3.2. Afterwards, we apply the methods described in Section 4.1 to the traces we generated, in order to be able to make general statements about how these methods are able to detect root causes of performance degradation.

3.1   Simulation Design

We consider a simulation model of a simple Web service environment for our experiments. The model consists of the following components: *Synthetic Agent Nodes*, *Scenario Generation Component*, and *Dynamic Web Service Component*. Synthetic agents send out HTTP GET requests every n minutes from m agents and collect response times. In the simulation, we sample every 1 minute resulting in 1 new observation of our system every minute. Each observation is stored in a database with the corresponding timestamp.

The simulation design and its communication channels are depicted in Figure 2. We consider a system with this architecture where requests incoming from the synthetic nodes are governed by a stochastic process $\{Y(t), t \in T\}$, with T being an index set representing time.

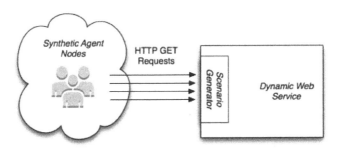

Fig. 2. Simulation Design

Synthetic Agent Nodes. We gather data from the *Dynamic Web Service Component* via active, periodic monitoring through *Synthetic Agent Nodes*. A synthetic monitoring agent acts as a client in order to measure availability and performance metrics, such as response time. Every synthetic agent is able to perform active measurements or synthetic tests. An active measurement is a request to a target URL, where subsequently all performance metrics that are available through the response are obtained. When configuring a set of synthetic tests, we can configure the following parameters: (1) URL of the Web service that should be tested; (2) sampling interval, e.g., every n minutes; (3) test duration, i.e., how many sample requests to issue in each run of the test agent.

Scenario Generation Component. As only the main classifications of root causes can be identified through synthetic monitoring, changes following the primary notions of *global delays*, *partial delays*, and *periodic delays* (see Section 2) are injected in the simulation. We achieve this by introducing a *Scenario Generation Component* into our model. It functions as an intermediary between the request sent by the synthetic agent nodes and the Web service. Instead of manually injecting faults into our test Web server, we define a set of scenarios that, subsequently, reflect the desired scenario, i.e., the faults over time in our system. The scenarios need to reflect performance degradation and performance volatility within a certain system, i.e., a single Web server. The Scenario Generation Component also needs to take into account possible geographic distribution of the agents, as well as load balancing mechanisms. In the following, we introduce a formal model for defining scenarios that reflects these notions that can be used to formally describe scenarios.

We consider a given set of parameters to compose a complete scenario within our simulation model. Within a scenario, we need to be able to specify how performance metrics (i.e., response times) develop over time, as well as synthetic agents that are actively probing our target system.

- A development $D \in \mathcal{D}$ maps from a certain point in $t \in T$ of the stochastic process $\{Y(t), t \in T\}$ (driving the requests of the synthetic agent nodes) to an independent random variable $X_i \in \mathcal{X}$, where $\mathcal{X}$ being the set of possible random variables (Equation 1).

$$D : T \mapsto \mathcal{X} \tag{1}$$

  where $X_i \in \mathcal{X}$ and $\forall \, X_i \sim \mathrm{U}(a, b)$

- On the top-level, we define a scenario $\mathcal{S}$ that describes how the target system that is observed by each synthetic agent $\mathcal{A} = \{a_1, a_2, ..., a_n\}$ develops over time. Each agent observes a development in performance $D_i \in \mathcal{D}$, with $\mathcal{D}$ being the set of all possible developments (Equation 2).

$$\mathcal{S} : \mathcal{A} \mapsto \mathcal{D} \tag{2}$$

This formalization allows us to express any performance changes (either positive or negative) as a classification of performance developments over time attributed to specific synthetic agents. More accurately, it models a performance metric as a uniformly distributed random variable of a system at any given time point $t \in T$. Specifying an assignment for every point in time is a tedious and unnecessary exercise. In order to define scenarios in a more efficient and convenient way, we introduce the following notation for developments $D \in \mathcal{D}$:

- Simple Developments: $[X_0, t_1, X_1, ..., t_n, X_n]$ defines a *simple development* as a sequence of independent random variables X_i and points in time t_i, further defined in Equation 3.

$$[X_0, t_1, X_1, ..., t_n, X_n](t) = \begin{cases} X_0 & 0 \le t < t_1 \\ X_1 & t_1 \le t < t_2 \\ \vdots & \\ X_n & t_n \le t \end{cases} \tag{3}$$

This allows us to easily define developments X_i in terms of time spans $t_{i+1} - t_i \ge 0$ within the total time of observation. The last development defined through X_n remains until the observation of the system terminates.

– Periodic Developments: A *periodic development* is essentially a simple development, which occurs in a periodic interval p. It is preceded by a "normal phase" up until time point n. The "periodic phase" lasts for $p - n$ time units until the development returns to the "normal phase". A periodic development $[X_0, n, X_1, p]^*$ is defined in Equation 4.

$$[X_0, n, X_1, p]^*(t) = \begin{cases} X_1 & \text{for } kp + n \le t < (k+1)p \\ X_0 & \text{otherwise} \end{cases} \tag{4}$$

where $k \ge 0$.

Figure 3 depicts how a periodic development can be seen over time with given parameters X_0 as the "normal phase" random variable, X_1 as the "periodic phase" random variable, n to define the time span for a "normal phase", p as the periodic interval and $(p - n)$ as the time span for the "periodic phase".

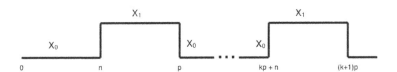

Fig. 3. Depiction of the periodic development scheme defined in Equation 4

These two defined functions allow us to conveniently define scenarios which adhere to our notions of global, partial and periodic delays. We can now define changes in performance at certain points in time in a declarative way, as well as define changes in periodic time intervals that result in changes in response times for a specific amount of time.

Dynamic Web Service Component. The *Dynamic Web Service Component* works together with the *Scenario Generation Component* to achieve the reflection of performance issues for the *Synthetic Agent Nodes*. In order to do so, it offers an endpoint that simulates delays over parameters passed from a specific scenario script. This means that the declared scenarios are executed within the Web service component and thus simulate workload through the parameters given in the scenarios.

3.2   Simulation Scenarios

Here, we formally define the parameters that actually form a certain scenario, and give an example of a real-life situation that would lead to such a performance behavior. The aim of each scenario is to properly represent one of global, partial or periodic delay.

Global Delay. A global delay is the introduction of a significant difference in response time on all incoming requests, i.e., it affects all users accessing the resource in question.

Example Scenario Use Case. A new feature needs to be implemented for a new release. A junior developer in charge of the new feature introduces a new (slow) database query, causing significantly higher overall response times. The slow query is not caught in QA (Quality Assurance) and the new release is deployed to all users.

Scenario Parameters. The parameter for this delay is given in Equation 5. Further, for every index i, we define the initial response time range as in Equation 6, as well as the range for the global change over all agents in Equation 7.

$$\mathcal{S}_G = \{a_i \mapsto [X_{a_i,0}, 420, X_{a,1}] \mid a_i \in \{a_1, a_2, a_3, a_4, a_5\}\} \tag{5}$$

$$X_{a_1,0} \sim U(90, 115), X_{a_2,0} \sim U(100, 130), X_{a_3,0} \sim U(110, 140),$$
$$X_{a_4,0} \sim U(95, 110), X_{a_5,0} \sim U(100, 110) \tag{6}$$

$$X_{a,1} \sim U(150, 175) \tag{7}$$

Partial Delay. A partial delay scenario consists of requests that, at some point in time, cause a delay on a subset of the incoming requests.

Example Scenario Use Case. A Web application sits behind a load balancer handling 5 servers. One of the servers encounters unexpected hardware issues, which result in higher response times. The balancer uses "Round Robin" as its load balancing algorithm [22]. 20% of all users perceive the application with higher response times.

Scenario Parameters. The parameter for this delay is defined in Equation 8. For every index i we define the initial response time range as in Equation 9, as well as the range for the partial change for agent a_5 in Equation 10.

$$\mathcal{S}_P = \{a_i \mapsto [X_{a_i,0}, \infty] | a_i \in \{a_1, a_2, a_3, a_4, a_5\}, a_5 \mapsto [X_{a_5,0}, 360, X_{a_5,1}]\} \tag{8}$$

$$X_{a_1,0} \sim U(115, 125), X_{a_2,0} \sim U(115, 120), X_{a_3,0} \sim U(120, 145),$$
$$X_{a_4,0} \sim U(105, 115), X_{a_5,0} \sim U(110, 120) \tag{9}$$

$$X_{a_5,1} \sim U(140, 165) \tag{10}$$

Periodic Delay. A periodic delay takes place when, due to a (background) process, resource contention occurs and, subsequently, higher usage of hardware resources leads to higher response times for a certain amount of time. This scenario addresses those processes that are (usually) planned ahead and are executed within a specific interval.

Example Scenario Use Case. Log files of an application make up a large amount of the server's disk space. The system administrator creates a cron job to process older log files and move them over the network. The process induces heavy load on CPU (processing) and I/O (moving over network), which result into temporarily higher response times. The cron job is configured to take place in periodic intervals to ensure the server has enough disk space.

Scenario Parameters. The parameter for this periodic delay is defined in Equation 11. The response time ranges are defined as in Equation 12.

$$\mathcal{S}_{PD} = \{a_1 \mapsto [X_0, 45, X_1, 65]^*\} \tag{11}$$

$$X_0 \sim U(95, 115), X_1 \sim U(160, 175) \tag{12}$$

4 Experiments

We now describe the methods of analysis and execution of the experiments introduced in Section 3, as well as the concrete results we achieved.

4.1 Methods of Analysis

The specified scenarios are executed within a physical testbed (described in Section 4.2) and results are analyzed and interpreted. The following sections outline the methods of analysis that are applied to the results.

Exploratory Data Analysis. Our first analysis approach is to examine the time series of monitored response times over time visually over graphs in order to explore and gain further understanding on the effect specific underlying causes have on the resulting data. We also determine what kind of statistical attributes are well suited to identify key characteristics of the data sample and for humans to properly observe the performance change. For this initial analysis we plot the raw time series data[1] as line charts. This allows for visual exploratory examination, which we further complement with proper interpretations of the data displayed in the visualization that correlates with induced changes in the server backend.

Statistical Analysis. After manually inspecting the time series over visual charts, we evaluate the observation of performance changes by the means of statistical changepoint analysis. Specifically, we evaluate algorithms that are employed in the R [20] package "changepoint". This approach makes sense, as we are not interested in the detection of spikes or other phenomena that can be considered as outliers in the statistical sense, but rather in the detection of fundamental shifts in our data that reflect a longer standing performance degradation. We also want to keep false positives low, and determine whether a change is actually a change that implies a root cause that requires some kind of action. Thus, we also need to determine the magnitude of the change. For this, we recall the simple view on delays we introduced in our taxonomy: $\lambda = |x_t - x_{t+1}| > c$. We adapt this model of change as follows. Instead of comparing two consecutive data points x_t and x_{t+1}, we compare the changes in the mean at the time point where changepoint have been detected by the algorithm. In other words, we compute the difference between the mean of the distribution before the detected changepoint occurred, $\mu_{<\tau}$, and the mean of the distribution after the detected changepoint occured, $\mu_{>\tau}$, where τ denotes the changepoint. This difference is then denoted as λ, and can be defined as $\lambda = |\mu_{<\tau} - \mu_{>\tau}| > c$ via replacing two variables.

The threshold c is challenging to determine optimally. When c is set up too high, legitimate changes in performance that were caused by problems may not be detected and the system is in risk of performance degradation. When c is defined unnecessarily sensitive, the monitoring system is prone to false positives. The value of the threshold depends on the application and must be either set by a domain expert or be determined by statistical learning methods through analysis of past data and patterns. Sometimes it is useful to compare new metrics in relation to old metrics, this is a very simple way of statistical learning through past data. In the conducted experiments, we set the threshold as $c = \mu_{<\tau} \cdot 0.4$. This means that if the new metric after the changepoint $\mu_{>\tau}$ is 40% above or below the old metric $\mu_{<\tau}$, the change is considered a *real change* as opposed to a *false positive*. If we want to consider positive changes as well, the calculation of the threshold must be extended in a minor way to not yield into false negatives due to a baseline that is too high: $c = \min(\mu_{>\tau}, \mu_{<\tau}) \cdot 0.4$.

Note that, in the case of this paper, the threshold 40% of the mean was chosen after discussions with a domain expert, as it is seen as an empirically estimated

---

[1] Raw in this context means that we will not smooth the data by any means and will not apply statistical models in any way.

baseline. In practice, a proper threshold depends on the type of application, SLAs, and other factors and is usually determined by own empirical studies.

4.2 Testbed Setup

The experiments based on the described simulation model were executed in a local testbed consisting of 5 *Synthetic Agent Nodes* and 1 *Dynamic Web Service Component*. The agents operate in the local network as well. Each of the 5 nodes sends out a request every 5 minutes that is handled by a scheduler that uniformly distributes the requests over time. This results into 1 request/minute that is being send out by an agent that records the data. The physical node setup consists of Intel Pentium 4, 3.4 GHz x 2 (Dual Core), 1.5 GB RAM on Windows and is running a Catchpoint agent node instance for monitoring. The Web service running on the target server has been implemented in the Ruby programming language and runs over the HTTP server and reverse proxy nginx and Unicorn.The physical web server setup is the same as the synthetic agent node setup, but is running on Ubuntu 12.04 (64 bit). During simulation, a random number generator is applied to generate artificial user behavior as specified in the distributions, which can be represented efficiently with common random numbers [12]. However, as with deterministic load tests, replaying user behavior data may not always result into the same server response. Even with the server state being the same, server actions may behave nondeterministically. To adhere the production of independent random variables that are uniformly distributed we use MT19937 (Mersenne twister) [18] as a random number generator.

4.3 Results and Interpretation

We now discuss the results of our scenario data generation, and the results of applying the statistical methods described in Section 4.1. At first, we display a raw plot of the resulting time series data without any filters and interpret its meaning. Further, we apply a moving average smoothing filter (with a window size $w = 5$) to each resulting time series and conduct a changepoint analysis.

Global Delay. The global delay forms the basis of our assumptions on how performance changes can be perceived on server backends. Both, the partial and periodic delay, are essentially variations in the variables time, interval and location of a global delay. Hence, the findings and interpretations of this section on global delays are the foundation for every further analysis and discussion.

Exploratory Visual Analysis. In Figure 4c, we see the results for the global delay scenario. We can clearly see the fundamental change in performance right after around 400 minutes of testing. The data before this significant change does seem volatile. There are a large amount of spikes occurring, though most of them seem to be outliers that might be caused in the network layer. None of the spikes sustain for a longer period of time, in fact between the interval around 150 and 250 there seem to be no heavy spikes at all. The mean seems stable around 115ms in response time and there is no steady increase over time that might suggest higher load variations. Thus, we can conclude that the significant performance change has occurred due to a new global release deployment of the application.

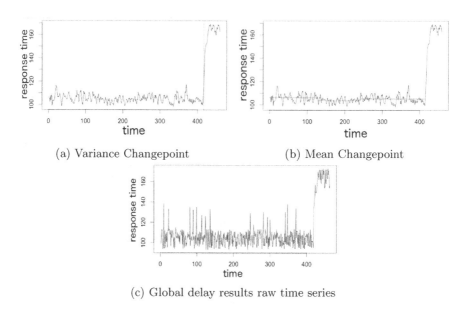

(a) Variance Changepoint   (b) Mean Changepoint

(c) Global delay results raw time series

Fig. 4. Global Delay results

Statistical Changepoint Analysis. We apply changepoint analysis to the smoothed time series with a moving average window size of 5. In Figure 4a, we can immediately see how the smoothing affected the chart, compared to the chart with the raw data in Figure 4c: The spikes, i.e., the random noise, have been canceled out to a certain extent, making the main signal stronger and easier to identify. This makes it easier for our statistical analysis to focus on our signal and detect the proper underlying distributions. While this is definitely a pleasant effect of every smoothing technique, we also need to keep in mind that every model that we apply contains its own assumptions and own errors that need to be considered. What can further be seen in Figure 4a is the changepoint in the variance, denoted by a vertical red line at the changepoint location. Table 1 contains the numerical results of the changepoint in variance analysis and indicates the estimation for the changepoint τ_1 at 422. This number coincides approximately with our own estimation we concluded in the previous section.

Next, we look at Figure 4b, where the change in the mean is indicated by horizontal lines depicting the mean value for each segment that has been detected. This method has detected more changepoints than the previous analysis, which can be not clearly seen in Figure 4b due to the very small change in the mean. The estimated numerical values for the changepoints are listed in Table 2.

The table also lists the mean values, as well as the calculated threshold $c = \mu_{<\tau} \cdot 0.4$. The last column also states whether or not a detected changepoint in the mean is an *actual changepoint* (CP) as defined by the notion of *significant change* where $\lambda = |\mu_{<\tau} - \mu_{>\tau}| > c$, or a *false positive* (FP). Only one of the estimated changepoints has been identified as CP when considering the threshold c. This shows that detecting a fundamental change is a difficult undertaking, especially considering non-parametric statistical analysis, as in our case. Post-processing and analysis of the estimated changepoints and its according mean values is important to avoid false positives.

Table 1. Variance CP for $\mathcal{S}_G$

τ	$\sigma^2_{<\tau}$	$\sigma^2_{>\tau}$
422	10.695	67.731

Table 2. Mean CP for $\mathcal{S}_G$

τ	$\mu_{<\tau}$	$\mu_{>\tau}$	λ	c	CP/FP
127	106.18	103.7	2.48	42.47	FP
241	103.7	105.32	1.62	41.48	FP
366	105.32	110.85	5.53	42.12	FP
374	110.8	103.62	7.23	44.32	FP
421	103.62	150.65	47.03	41.44	**CP**
427	150.65	165.62	14.97	60.62	FP

Partial Delay. Partial delays are global delays that only occur on a certain subset of requests and, therefore, need different techniques to properly examine the time series data and diagnose a performance degradation. Experiments on simulating partial delays have found that detection of a changepoint in partial delays, or even the visual detection of performance change, is not at all trivial.

Exploratory Visual Analysis. As before, we plot the time series data and look for changes in our performance. In Figures 5a and 5b, we see rather stable time series charts, relatively volatile (spiky), due to the higher amount of conducted tests and random network noise. Around the time points 460-500 and 1600-1900, we see a slight shift, but nothing alarming that would be considered a significant change. From this first visual analysis, we would probably conclude that the system is running stable enough to not be alerted. However, that aggregation hides a significant performance change. The convenience, or sometimes necessity, of computing summary statistics and grouping data to infer information this time concealed important facts about the underlying system. In order to detect this performance change, we have to look at additional charts and metrics.

In Figure 5c, we plot the same aggregation as in Figures 5a and 5b, but also plot the 90th percentile of the data to come to a more profound conclusion: There actually has been a performance change that is now very clear due to our new plot of the 90th percentile. While this can mean that percentiles also show temporary spikes or noise, it also means that if we see a significant and persisting shift in these percentiles, but not in our average or median, that a subset of our data, i.e., a subset of our users, indeed has experienced issues in performance and we need to act upon this information. Another way of detecting issues of this kind is to plot all data points in a scatterplot. This allows us to have an overview of what is going on and to identify anomalies and patterns more quickly. As we can see in Figure 5d, a majority of data points is still gathered around the lower response time mean. But we can also see clearly that there has been a movement around the 400 time point mark that sustains over the whole course of the observation, building its own anomaly pattern.

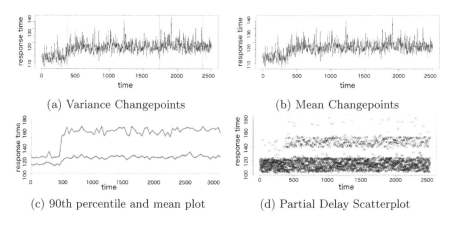

(a) Variance Changepoints

(b) Mean Changepoints

(c) 90th percentile and mean plot

(d) Partial Delay Scatterplot

Fig. 5. Partial Delay results

Statistical Changepoint Analysis. Analyzing both the changepoints in the variance in Table 3 and Figure 5a, as well as the changepoints in the mean in Table 4 and Figure 5b yields no surprise following our initial exploratory visual analysis. The changes that have been identified are not significant enough to be detected through the mean and the variance respectively. Although we need to point out that both analyses actually detected the actual significant changepoint around the time point 374-376, but are disregarded as false positives by our post-processing step of checking the threshold. Thus, our post-process actually resulted into a *false negative*. This is usually a sign that an indicator (in our case the threshold c) needs to be adjusted or rethought completely. However, before this kind of decision can be made, more information has to gathered on how this indicator has performed in general (i.e., more empirical evidence on false negatives, ratio between false positives and false negatives, etc.).

In order for our regular statistical analysis process, as applied previously, to properly work we need a further pre-processing step. Neither mean nor variance can detect the performance change, therefore, we need to consider a different metric. In our exploratory analysis, we concluded that the 90th percentile was able to detect the change. Thus, we need to apply our changepoint analysis to percentiles in order to detect a significant shift for partial delays.

Table 3. Variance CP for $\mathcal{S}_P$

τ	$\sigma^2_{<\tau}$	$\sigma^2_{>\tau}$
374	12.88	24.25
1731	24.25	12.43
1911	12.43	6.63

Table 4. Mean CP for $\mathcal{S}_P$

τ	$\mu_{<\tau}$	$\mu_{>\tau}$	λ	c	CP/FP
376	114.74	121.92	7.18	45.88	FP
1710	121.92	118.91	3.01	48.76	FP
1756	118.91	124.1	5.19	47.56	FP
2035	124.1	122.27	1.83	49.64	FP

Periodic Delay. Periodic delays are either global or partial delays that occur in specific intervals, persist for a certain amount of time, and then performance goes back to its initial state.

Exploratory Visual Analysis. For this experiment we recorded enough values to result into three periods that indicate a performance degradation for a certain amount of time before returning back the system returns to its normal operation. The intuition we have on periodic delays can be clearly observed in Figure 6. Between the phases, we have usual performance operation with usual volatility, and in between we see fundamental shifts that persist for approximately 20 minutes before another shift occurs. Seeing periodic delays is fairly simple, as long as we are looking at a large enough scale. If we would have observed the time series within the periodic phase, before the second shift to the normal state occurred, we might have concluded it to be a simple global delay. Thus, looking at time series at a larger scale might help to identify periodic delays that would have otherwise been disregarded as short-term trends or spikes.

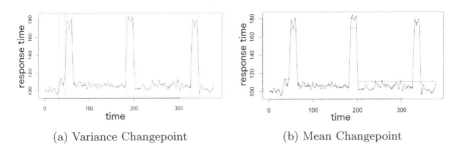

(a) Variance Changepoint   (b) Mean Changepoint

Fig. 6. Periodic Delay results

Statistical Changepoint Analysis. While the exploratory visual analysis in periodic delays was straight forward, the changepoint analysis brings some interesting insights. Figure 6a and Table 5 show the analysis for the changepoints in the variance, which mostly coincide with our visual assessment. Merely the first detected changepoint in the variance is a false negative.

The changepoint in the mean yields more interesting results, as seen in Table 6 Contrary to the results in the other experiments the number of false positives is very low at only one. When looking at the result in Figure 6b, we can observe another interesting phenomena we have not seen before, a false negative. The third period was not detected as a changepoint in the mean by the algorithm, although it was detected by the changepoint in the variance method. This means, in order to avoid false negatives, we should apply both analyses for changepoint in the mean and variance.

Table 5. Variance CP for $\mathcal{S}_{PD}$

τ	$\sigma^2_{<\tau}$	$\sigma^2_{>\tau}$
32	5.21	20.22
46	20.22	156.25
62	156.25	11.73
181	11.73	233.96
199	233.96	16.3
327	16.3	180.3
343	180.3	25.16

Table 6. Mean CP for $\mathcal{S}_{PD}$

τ	$\mu_{<\tau}$	$\mu_{>\tau}$	λ	c	CP/FP
33	100.47	113.27	12.8	40.18	FP
47	113.27	168.11	54.84	45.31	**CP**
63	168.11	106.18	61.93	42.47	**CP**
180	106.18	169.35	63.17	42.46	**CP**
199	169.35	110.52	58.83	44.20	**CP**

4.4 Threats to Validity

In this research, we employ a simplified simulation model in order to show how significant performance changes can be detected in continuously observed data of a constructed target system. The creation and design of a simulation model comes with inherent risk concerning the validity of the results when applied to real-life systems. Most importantly, in our simulation design, we do not actively inject the notion of web traffic workloads (as has been, for instance, taken in consideration in [16]), but rather simulate the variability of workloads (and therefore response times) in web services through specific scenarios parameters. Further, the notions of network traffic and network volatility in wide area networks (WAN) are completely omitted to limit the interference of network noise in the recorded data. Of course, noise is still present even in LANs, but is limited to very few factors within the data center, as opposed to the possible factors in WANs. This also means that there is, for instance, no increased response time due to DNS resolution in our experiments.

5   Related Work

Analyzing performance data observed through continuous monitoring in distributed systems and web services has produced a large body of research dealing with statistical modeling of underlying systems, anomaly detection and other traditional methods in statistics to address problems in performance monitoring and analysis. For instance, Pinpoint [8] collects end-to-end traces of requests in large, dynamic systems and performs statistical analysis over a large number of requests to identify components that are likely to fail within the system. It uses a hierarchical statistical clustering method, using the arithmetic mean to calculate the difference between components by employing the Jaccard similarity coefficient [13]. [11] makes use of statistical modeling to derive signatures of systems state in order to enable identification and quantification of recurrent performance problems through automated clustering and similarity-based comparisons of the signatures. [1] proposes an approach which identifies root causes of latency in communication paths for distributed systems using statistical techniques. An analysis framework to observe and differentiate systemic and anomalous variations in response times through time series analysis was developed in [9]. A host of research from Borzemski *et al.* highlights the use of statistical methods in web performance monitoring and analysis [2–6]. The research work spans from statistical approaches to predict Web performance to empirical studies to assess Web quality by employing statistical modeling. [19] suggests an approach to automated detection of performance degradations using control charts. The lower and upper control limits for the control charts are determined through load testing that establish a baseline for performance testing. The baselines are scaled employing a linear regression model to minimize the effect of load differences. [10] proposes an automated anomaly detection and performance modeling approach in large scale enterprise applications. A framework is proposed that integrates a degradation based transaction model reflecting resource consumption and an application performance signature that provides a model of runtime behavior. After software releases, the application signatures are compared in order to detect changes in performance. [17] conducted an experimental study comparing different monitoring tools on their ability to detect performance anomalies through correlation analysis among application parameters. In the past, we have also applied time series analysis to the prediction of SLA violations in Web service compositions [15].

6 Conclusion and Future Work

A taxonomy of root causes in performance degradations in web systems has been introduced, which was further used to construct scenarios to simulate issues in web performance within an experimental setup. In a series of simulations, we measured how performance metrics develop over time and presented the results. Furthermore, we provided analysis and interpretation of the results. Following the work presented in this paper, there are possible improvements and further work we were not able to address sufficiently. Performance data gathered through external synthetic monitoring only allows for a black box view of the system and is often not sufficient for in-depth root cause analysis of performance issues. Combining data from external monitoring and internal monitoring in order to automate or assist in root cause analysis and correlation of issues is a possible approach that should be considered. The simulation and analysis in this paper is limited to performance issues on the server. Further work might include extending the taxonomy of root causes and simulation scenarios to also represent frontend performance issues.

Acknowledgements. The research leading to these results has received partial funding from the European Community's Seventh Framework Programme (FP7/2007-2013) under grant agreement 610802 (CloudWave). Numerical simulation was carried out during a research visit of the first author at Catchpoint Systems, Inc., New York, USA. The authors would like to thank Prof. Gernot Salzer for his comments on earlier versions of this work.

References

1. Aguilera, M.K., Mogul, J.C., Wiener, J.L., Reynolds, P., Muthitacharoen, A.: Performance debugging for distributed systems of black boxes. ACM SIGOPS Operating Systems Review 37, 74–89 (2003)
2. Borzemski, L.: The experimental design for data mining to discover web performance issues in a wide area network. Cybernetics and Systems 41(1), 31–45 (2010)
3. Borzemski, L., Drwal, M.: Time series forecasting of web performance data monitored by MWING multiagent distributed system. In: Pan, J.-S., Chen, S.-M., Nguyen, N.T. (eds.) ICCCI 2010, Part I. LNCS, vol. 6421, pp. 20–29. Springer, Heidelberg (2010)
4. Borzemski, L., Kamińska-Chuchmała, A.: Knowledge discovery about web performance with geostatistical turning bands method. In: König, A., Dengel, A., Hinkelmann, K., Kise, K., Howlett, R.J., Jain, L.C. (eds.) KES 2011, Part II. LNCS, vol. 6882, pp. 581–590. Springer, Heidelberg (2011)
5. Borzemski, L., Kaminska-Chuchmala, A.: Knowledge engineering relating to spatial web performance forecasting with sequential gaussian simulation method. In: KES, pp. 1439–1448 (2012)
6. Borzemski, L., Kliber, M., Nowak, Z.: Using data mining algorithms in web performance prediction. Cybernetics and Systems 40(2), 176–187 (2009)
7. Bouch, A., Kuchinsky, A., Bhatti, N.: Quality is in the eye of the beholder: meeting users' requirements for internet quality of service. In: Proceedings of the SIGCHI Conference on Human Factors in Computing Systems, pp. 297–304. ACM (2000)
8. Chen, M.Y., Kiciman, E., Fratkin, E., Fox, A., Brewer, E.: Pinpoint: Problem determination in large, dynamic internet services. In: Proceedings of International Conference on Dependable Systems and Networks, DSN 2002, pp. 595–604. IEEE (2002)
9. Chen, Y., Mahajan, R., Sridharan, B., Zhang, Z.-L.: A provider-side view of web search response time. SIGCOMM Comput. Commun. Rev. 43(4), 243–254 (2013)
10. Cherkasova, L., Ozonat, K., Mi, N., Symons, J., Smirni, E.: Automated anomaly detection and performance modeling of enterprise applications. ACM Transactions on Computer Systems (TOCS) 27(3), 6 (2009)
11. Cohen, I., Zhang, S., Goldszmidt, M., Symons, J., Kelly, T., Fox, A.: Capturing, indexing, clustering, and retrieving system history. ACM SIGOPS Operating Systems Review 39, 105–118 (2005)
12. Heikes, R.G., Montgomery, D.C., Rardin, R.L.: Using common random numbers in simulation experiments - an approach to statistical analysis. Simulation 27(3), 81–85 (1976)
13. Jaccard, P.: The distribution of the flora in the alpine zone. 1. New Phytologist 11(2), 37–50 (1912)
14. King, A.: Speed up your site: Web site optimization. New Riders, Indianapolis (2003)

15. Leitner, P., Ferner, J., Hummer, W., Dustdar, S.: Data-Driven Automated Prediction of Service Level Agreement Violations in Service Compositions. Distributed and Parallel Databases 31(3), 447–470 (2013)
16. Liu, Z., Niclausse, N., Jalpa-Villanueva, C., Barbier, S.: Traffic Model and Performance Evaluation of Web Servers. Technical Report RR-3840, INRIA (December 1999)
17. Magalhaes, J.P., Silva, L.M.: Anomaly detection techniques for web-based applications: An experimental study. In: 2012 11th IEEE International Symposium on Network Computing and Applications (NCA), pp. 181–190. IEEE (2012)
18. Matsumoto, M., Nishimura, T.: Mersenne twister: A 623-dimensionally equidistributed uniform pseudo-random number generator. ACM Transactions on Modeling and Computer Simulation (TOMACS) 8(1), 3–30 (1998)
19. Nguyen, T.H., Adams, B., Jiang, Z.M., Hassan, A.E., Nasser, M., Flora, P.: Automated detection of performance regressions using statistical process control techniques. In: Proceedings of the Third Joint WOSP/SIPEW International Conference on Performance Engineering, pp. 299–310. ACM (2012)
20. R Core Team. R: A Language and Environment for Statistical Computing. R Foundation for Statistical Computing, Vienna, Austria, (2013)
21. Forrester research. Ecommerce web site performance today: An updated look at consumer reaction to a poor online shopping experience (August 2009)
22. Shirazi, B.A., Kavi, K.M., Hurson, A.R. (eds.): Scheduling and Load Balancing in Parallel and Distributed Systems. IEEE Computer Society Press, Los Alamitos (1995)

Indexing Rich Internet Applications
Using Components-Based Crawling

Ali Moosavi[1], Salman Hooshmand[1], Sara Baghbanzadeh[1],
Guy-Vincent Jourdan[1,2], Gregor V. Bochmann[1,2], and Iosif Viorel Onut[3,4]

[1] EECS - University of Ottawa
[2] Fellow of IBM Canada CAS Research, Canada
[3] Research and Development, IBM® Security AppScan® Enterprise
[4] IBM Canada Software Lab., Canada
{smousav2,shooshma,sbaghban}@uottawa.ca,
{gvj,bochmann}@eecs.uottawa.ca,
vioonut@ca.ibm.com

Abstract. Automatic crawling of Rich Internet Applications (RIAs) is a challenge because client-side code modifies the client dynamically, fetching server-side data asynchronously. Most existing solutions model RIAs as state machines with DOMs as states and JavaScript events execution as transitions. This approach fails when used with "real-life", complex RIAs, because the size of the produced model is much too large to be practical. In this paper, we propose a new method to crawl AJAX-based RIAs in an efficient manner by detecting "components", which are areas of the DOM that are independent from each other, and by crawling each component separately. This leads to a dramatic reduction of the required state space for the model, without loss of content coverage. Our method does not require prior knowledge of the RIA nor predefined definition of components. Instead, we infer the components by observing the behavior of the RIA during crawling. Our experimental results show that our method can index quickly and completely industrial RIAs that are simply out of reach for traditional methods.

Keywords: Rich Internet Applications, Web Crawling, Web Application Modeling.

1 Introduction

In the past decade, modern web technologies such as AJAX, Flash, Silverlight, etc. have given emergence to a new class of more responsive and interactive web applications commonly referred to as Rich Internet Applications (RIAs). RIAs make Web-applications more interactive and efficient by introducing client-side computation and updates, and asynchronous communications with the server [1]. Crawling RIAs is more challenging than crawling traditional web applications because some core characteristics of traditional web applications are violated by RIAs. Client states no longer correspond to unique URLs as modern web technologies enable the ability to change the client state and even populate it with

S. Casteleyn, G. Rossi, and M. Winckler (Eds.): ICWE 2014, LNCS 8541, pp. 200–217, 2014.

new data without changing the URL, up to the point that it is possible to have complete complex web applications with a single URL. Moreover, JavaScript events (from here on, called *"events"*) can take the place of hyperlinks; and unlike hyperlinks, the crawler cannot predict the outcome of an event before executing it. In that sense, the behavior and user interface of a RIA is more similar to an event-driven software like a desktop software GUI than to a traditional web application.

The problem of crawling AJAX-based RIAs has been a focus of research in the past few years. In such RIAs, executing events can change the Document Object Model (DOM), hence leading the RIA to a new client state, and can possibly leaned to message exchanges with the server, changing the server state as well. A common approach is to model a RIA as a finite state machine (FSM). In the FSM, DOMs are represented as states and event executions are represented as transitions. Events can lead from one DOM-state[1] to another.

One simplifying assumption that is usually made is that server states are in sync with client states. Therefore by covering all client states (i.e. DOM-states), the crawler has also covered all server states. Based on this assumption, by executing each event from each DOM-state once and building a complete FSM model from the RIA, the crawler can assume that the RIA has been entirely covered and modeled. In order to stay in sync with server, however, the crawler cannot jump to arbitrary DOM-states at will (e.g. by saving DOM-state in advance and restoring it when desired); instead, it must follow a sequence of events. If the desired DOM-state is not reachable from the current DOM-state using a chain of transitions (called a "transfer sequence"), the crawler needs to issue a "reset" (reloading the URL of the initial page) to go to the initial DOM-state and follow a valid transfer sequence from there. Resets are usually modelled as special transitions from all DOM-states to the initial DOM-state. In the beginning, the only known DOM-state is the initial DOM-state and all its events are unexecuted yet. By executing an unexecuted event, the crawler discovers its destination, which might be a known DOM-state or a new one. The event execution can then be modelled as a transition between its source and destination DOM-states. A state machine can be represented as a directed graph. The problem of crawling a RIA is therefore that of exploring an unknown graph. At any given time, the crawler needs to execute an unexecuted event, or use the known portion of the graph to traverse to another DOM-state to execute one, until all events in the graph have been executed, at which point the graph is fully uncovered and crawling is done. Based on this model, different exploration strategies (such as depth-first search, Greedy and Model-Based strategies, see Section 2) have been suggested. Comparing different exploration strategies can be done by comparing the number of events and resets executed during the crawl. We call this the *"exploration cost"*.

---

[1] Related works in the literature commonly refer to DOM-states simply as "states". In this work we differentiate between "DOM-states" and what we will call "component-states".

One major challenge in this field is a state space explosion: the model being built grows exponentially in the size of the RIAs being crawled. This state space explosion not only leads to production of a large model that is difficult to analyze, but also makes the crawlers unable to finish in a reasonable time. Because of this excessive running times, all DOM-based methods that have been published so far are essentially unsuitable for real-life scenarios. These methods cannot be used in the industrial world. Current studies use different notions of DOM equivalence to map several DOMs to one state. These approaches usually involve applying reduction and normalization functions on the DOM. While these approaches have been used and tested successfully on experimental RIAs, they fail to provide satisfactory equivalency criteria when faced with real-world large-scale RIAs.

Most of the time, this state space explosion is caused by having the same data being displayed in different combinations, leading to large sets of new DOMs and large state space for a small set of functionalities. In a typical RIA, it is common to encounter a new DOM-state which is simply a different combination of already-known data (Figure 1). We call this situation *new DOM-state without new data*. Such DOM-states should be ideally regarded as already known. Today's complex RIA interfaces consist of many interactive parts that are independent from one another, and the Cartesian product of different content that each part can show easily leads to an exponential blow-up of the number of DOM-states. In the following, we call these independent parts *components*, and each of their values *component-state*. A fairly intuitive example is widget-based RIAs, in which various combination of contents that each widget can show creates a very large number of different DOM-states. Not all these DOM-states are of interest to the crawler. A content indexing crawler, for instance, needs to retrieve the content once and finish in a timely fashion. These "rehashed" DOM-states only prolong crawling while providing no new data. Figure 1 provides an example. This issue is not just limited to widgets, but is present in any independent part in RIAs down to every single popup or list item. Typical everyday websites such as Facebook, Gmail and Yahoo, and any typical RIA mail client, enterprise portal or CMS contain dozens if not hundreds of independent parts. Different combinations of these independent parts lead crawlers through a seemingly endless string of new DOM-states with no new data. A human user, on the other hand, is not confused by this issue since she views these components as separate entities, and in fact would be surprised if the behavior of one of these parts turns out to be dependent on another.

We observe that one major drawback inherent to all these methods is that they model client states of RIA at the DOM level. We propose a novel method to crawl RIAs efficiently by modeling in terms of states of individual sub-trees of the DOM that are deemed independent, which we call *components*. Our method detects independent components of RIA automatically, using the result of diffs between DOMs. By modeling at the component level rather than at the DOM level, the crawler is able to crawl complex RIAs completely (and in fact quickly) while covering all the content. The resulting end-model is smaller and therefore easier for humans to understand and for machines to analyze, while providing

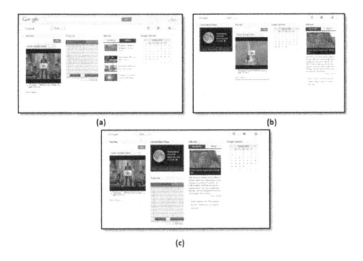

Fig. 1. Example of a new DOM-state with no new data. The DOM in **(c)** is only a combination of data already present in **(b)** and **(a)**, but will have a new DOM-state in the existing methods.

more information about the RIAs being modeled. As we will show in our experimental results, the method presented here is suitable to crawl and index real-life, complex RIAs, without loss of content in our experiments, where previously published methods failed to do the same thing even on much simplified version of the same RIAs.

The remainder of this paper is organized as follows. In Section 2, we provide a review of related work. In Section 3 we present the general overview of our solution. We first describe the model that the crawler builds, and we then describe how the crawler builds this model and makes use of it during the crawl. Experimental results and comparisons are presented in Section 4, and we conclude in Section 5.

2   Related Works

Crawling RIAs using a state transition model has been extensively studied. Duda et al. use a breadth-first search approach to explore RIA, assuming the ability to cache and restore client states at will [2, 3]. In [2], they point to the state space explosion problem caused by independent parts as an unresolved challenge. Amalfitano et al. use manual user-sessions to build a state machine [4]. In a follow-up work, they automate their tool by using depth-first exploration [5]. Peng et al. propose using a greedy algorithm as exploration strategy that outperforms depth-first and breadth-first search exploration significantly [6]. We use here the same greedy approach as exploration strategy. A different approach, called "model-based crawling" [7], focuses on finding the clients states as soon as possible by assuming some particular behavior from the RIAs [8,9]. The model used in [9] accumulates statical information about the result of previous event executions to infer what event to execute next. All the these method suffer from the problem of accumulating new DOM states that do no contain new data. In [10], an approach similar to model-based crawling is used, but for sites that have a known structure.

DynaRIA [11] provides a tool for tracing AJAX-Based application executions. It generates abstract views on the structure and run-time behavior of the application. The generated crawling model has been used for accessibility testing [12] or for generating test sequences [13]. It also also been used for modelling native android apps [14, 15] and native iOS apps [16].

All the above mentioned works use DOM-level state machines and use different DOM equivalence criteria to guide crawling: in [17], an edit distance between DOM-states is used. Methods based on DOM manipulations are used in [7–9,18]. These various DOM equivalence criteria do help but ultimately fail to address completely the state space explosion problem. To alleviate this problem, in [17] it is proposed to explore only new events that appear on a DOM after an event execution.This limits the crawler's ability to reach complete coverage, and does not prevent exploring redundant data when different event execution paths lead to the same structure (e.g. a widget frame) but in different DOMs. *FeedEx* [19] extends [6] by selecting states and events to be explored based on probability to discover new states and increase coverage. They include four factors to prioritize the events : code coverage, navigational diversity, page structural diversity and test model size. Surveys of RIAs crawling can be found in [20, 21].

In the context of detecting independent parts, static widget detection methods such as [22] and [23], and detection of underlying source dataset [24] have been developed. In [23], the use of patterns for detecting widgets based on static JavaScript code analysis and interaction between widget parts is proposed. However, these methods designed only to detect widgets or source datasets, which are a small subset of independent entities in RIAs.

3   Component-Based Crawling

3.1   Overview of Our Solution

Our solution is to model RIAs at a "finer" level, using subtrees of the DOM (called "components") instead of modeling in terms of DOMs. By building a state-machine at the component level, we get a better understanding of how the RIA behaves, which helps addressing the aforementioned state explosion problem [25]. The crawler can use this model along with its exploration strategy. Our prototype implementation uses the greedy algorithm presented in [26] as the exploration strategy. In this section we present a brief general overview of the concept of components, before providing more details in the following Sections.

In a typical real-life RIA such as the one depicted in Figure 2, a given "page" (DOM-state) contains a collection of independent entities. We call these entities "components". They are subtrees of the current DOM. Examples of components include menu bars, draggable windows in Twitter, as well as each individual tweet, chat windows in Gmail, as well as the frame around each chat window, the notifications drop-down and mouse-over balloons in Facebook, etc. Users normally expect to be able to interact with each of these components independently, without paying attention to the state of the other components on the page. Classical crawling methods do not consider these components, and consequently generate every possible combination of these components states while building the model. Our aim is to detect these components to crawl each of them separately. The assumption of independency between components enables our method to "divide and conquer" the RIA to overcome state space explosion without loss of coverage. We expect this assumption to hold true in almost all real-life RIAs as it follows human user intuition. We did not encounter any counterexamples in our investigation of real-life RIAs. If, however, there are components on a particular RIA that affect each other, the crawler might lose coverage of some of the content of the RIA since it does not analyze the interactions between the components. In our experiments, this situation did not occur.

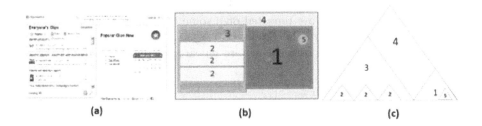

Fig. 2. (a) A webpage, (b) components on the page the way a human user sees them as entities of the page, and (c) the way the crawler sees them as subtrees of the DOM

The input of the crawler after each event execution is the DOM tree. Since components appear as subtrees in the DOM tree, we partition the DOM into multiple subtrees that are deemed independent of each other. Each of these subtrees correspond to a particular state of a component (a *component-state*). We model the RIAs as a combination of independent component states instead of assigning a DOM-state to the entire DOM. The idea of components and their associated component-states completely replace use of DOM-states in our method. Each component has a set of possible component-states, and a component-state of a particular component is only compared to other component-states of its own.

As explained, in our model, at any given time, the RIA is in a set of component-states, since it consists of different components each in its own component-state. It is worth mentioning that the DOM is partitioned into components in a collectively exhaustive and mutually exclusive manner, meaning that each XML-node on the DOM tree belongs to one and only one component. Modeling RIAs at the component level as several benefits. The most obvious one is that it reduces the state space by avoiding modeling separately every combination of component states, including the many instances in which the combination is new but each component state has been seen before (as depicted previously in Figure 2). Moreover, this fine-grained view of RIA helps the crawler map the effect of event executions more precisely, resulting in a simpler model of the RIA with fewer states and transitions. As a result, the crawler will traverse the RIA more efficiently by taking fewer steps when aiming to revisit a particular structure (such as a text or event) in the RIA that is not present in the current DOM. The resulting model of the RIA will also be more easily understandable by humans because it has fewer states and transitions and the effect of each event execution on the DOM is defined more clearly.

To illustrate the potential gain of our methods, imagine that the current DOM is made of k independent components $C_1, C_2, \ldots, C_k$. Assume that each component C_i has $\bar{C}_i$ components states. Using the traditional, DOM-state based method, this will lead to $\prod_{i=1}^{k} \bar{C}_i$ DOM-states. If in addition the components can be moved freely on the page, this number will be repeated $k!$ times, leading to $k! \prod_{i=1}^{k} \bar{C}_i$ DOM-states. This already intractable number will increase even more of some components are repeated or if some components can be removed from the DOM. Using our method yields only $\sum_{i=1}^{k} \bar{C}_i$ component-states for the same RIAs, even when the components are repeated or removed from the page.

3.2 Model Description

In our model, we partition each DOM into independent components. Each component has its own component-state so the current DOM corresponds to a set of component-states in the state machine. Because JavaScript events are attached to XML nodes, each event resides in one of the component-states present in the DOM. We call it the *"owner component-state"* of the event[2].

3.2.1 Multistate Machine

An event is represented as a transition that starts from its owner component-state. Since the execution of the event can affect multiple components, the corresponding transition can end in multiple component-states. Therefore, our model is a multi-state-machine. Figure 3 illustrates how an event execution is modeled in our method versus other methods. The destination component-states of a transition correspond to component-states that were not present in the DOM, and appeared as a result of the execution of the event.

Our model is a multistate-machine, defined as a tuple $M = (A, I, \Sigma, \partial)$ where A is the set of component-states, I is the set of initial component-states (those that are present in the DOM when the URL is loaded), Σ is the set of events, and $\partial : A \times \Sigma \to 2^A$ is a function that defines the set of valid transitions. Similarly to

---

[2] For events that are not attached to an XML-node on the DOM, such as timer events, a special global always-present component is defined as their owner component.

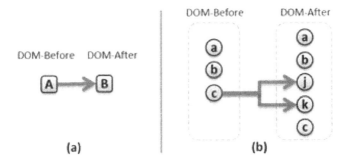

Fig. 3. An event execution modeled with **(a)** DOM-states, and **(b)** component-states. Rectangles in **(b)** represent DOM-states and are not used in the actual model.

the classical state-transition model, ∂ is a partial function, since not all events are available on all component states. Unlike the state-transition model, we have a set of initial states, and executing an event can modify any number of component-states. Our multistate-machine is resilient to shuffling components around in a DOM, and does not store information about exact location of the component-states in a DOM.

3.2.2   Components Definition

We have mentioned that components must be "independent" from one another. By this, we mean that the outcome of execution of an event only depends on the component-state of its owner component. In other words, the behavior of an events in a component is independent from the other components in the DOM and their individual component states. As an example, the border around a widget that has minimize/close buttons is independent of the widget itself, since it minimizes or closes regardless of the widget that it is displaying. Therefore, the widget border and the widget itself can be considered separate independent components. On the other hand, the next/previous buttons around a picture frame are dependent on that picture frame, since their outcome depends on the picture currently being shown. So the next/previous buttons should be put in the same component as the picture frame. Note that event executions outcome can affect any number of components and this does not violate the constraint of independency.

Two notions are important in our definition: first, we need to specify how we define a component, then how we capture the various component states that component might have.

Component are identified by an XPath, which specifies the root of the sub-tree that contains this component. In order to find a particular component in the DOM, one should start from the document root and follow the component's associated XPath. The element reached is the root of the component i.e. the component is the subtree under that element. Note that an XPath can potentially map to several nodes, therefore several instances of a component can be present in a DOM at the same time. Since the XPath serves as an identifier for a component, we need the XPath to be consistent throughout the RIA, i.e. it should be able to point to the intended subtree across different DOMs of the RIA. However, some attributes commonly used in XPath are too volatile to be consistent throughout the RIA. Hence, we only use the "id" and "class" attributes for each node in the XPath, and omit other predicates such as the position predicate.

Here is how we build an XPath for an element e: we consider the path p from the root of the DOM to e, and for each HTML element in p, we include the tag name of the element, the *id* attribute if it has one, and the *class* attribute if it has one.

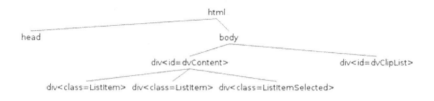

Fig. 4. Part of a shopping website's DOM

Figure 4 is an example of DOM for a shopping website. Individual list items in the product list are instances of a component "product list item". In this example, there are two instances of this component, which is identified by the XPath */html/body/div[@id= "dvContent"]/div[@class= "ListItem"]*. But the selected item in the list yields a different XPath since it is assigned a different class attribute, *[@class= "ListItemSelected"]*.

Each component, identified by its XPath which points at the root of the component, can have a series of component states that are going to be uncovered throughout the crawl. These states are simply the content of the subtree found under the XPath. For efficiency reasons, we are not recording the entire subtree for each component state. Instead, we derive a unique identifier for the component state by hashing the content of the corresponding subtree[3]. In practice, the crawler keeps information on each component-state of each component in a data structure for use by the greedy algorithm. A simplified version of the data structure, called *stateDictionary*, is depicted in Table 1. In general, on any given DOM, some components are present in the DOM and some are not. Using the "Component Location" column, the crawler can find out which components are present on the DOM, then use the component's content to compute its ID.

---

[3] In reality, we first prune nested sub-components as explained later, and we also perform some transformations on the subtree to detect equivalents sub-components. See [27] for more details.

Using the "Component-State ID" column it can look up additional info on that component-state (event/transition destinations, unexecuted events, etc.), or discover that it is a new component-state.

Table 1. The StateDictionary

#	Component Location (XPATH)	Component State-ID (Hash)	Info
1	/	@J#F@)J#403rn0f29r3m19	....
2	/html/body/div[@id="dvClipList"]	*&^$@J$$P@@$#$#_!$_*$_*	...
		GPDFJD}{PL"!{#R$$)%$*!_$#!!	...
3	/html/body/div[@id="dvContent"]	VMLCVCPQ!#$!_()_IKEF)_I)	...
		{:$%@)(@#*GRJPGFD{#@)(	...
		?"$#%*%@$)(!#HI!_D}{|||#R!#!	...

Since components are identified by their XPath, it is possible to have nested components, with the root of one component (identified by its XPath) failing inside the subtree of another component. If a component contains other components, these components should be removed from the containing component when defining that component's states. The pseudo-algorithm below explains how to list all the known component states that are present in the current DOM. Refer to [27] for more details.

Algorithm 1. Pseudo-code to find current component-states

> **procedure** FINDCURRENTSTATES
>> **for all** *xpath* in *stateDictionary* **do**
>>> *ComponentInstances* := go through the *xpath* and give the subtree
>>> **for all** *Instance* in *ComponentInstances* **do**
>>>> **for all** *sub-path* under the current *xpath* **do**
>>>>> go through the *sub-path* and prune the subtrees
>>>> **end for**
>>>> *stateID* := ReadContentsAndComputeStateID(*instance*)
>>>> Add the *stateID* to *SetOfCurrentState*
>>> **end for**
>> **end for**
>> **return** *SetOfCurrentStates*
> **end procedure**

Note that the only location information for component states is an XPath to the root of the component state. Therefore, while our model is able to break a DOM into component-states, it is not possible to reconstruct an exact DOM using the multi-state-machine. While the resulting model of a RIA can be used to infer an execution trace to any given component state (thus any content of the RIA), it cannot be used to infer an execution trace to any given DOM.

3.3   Crawling Algorithm

As explained above, a crawling algorithm relies on an exploration strategy that tells it which events to execute next. Several exploration strategies can be used with our model. We explain our algorithm independently from the chosen strategy (which is "greedy" in our prototype).

Generally, using the "Component Location" list in the stateDictionary, the crawler can discover new component-states during the crawl and populate the "Component-State ID" lists. Our proposed component discovery algorithm populates the Component Location and Component-State ID lists incrementally during the crawl as it observes the behavior of the RIA. The algorithm is based on comparing the DOM tree snapshots before and after each event execution. Every time an event is executed by the crawler, the subtree of the DOM that has changed as a result of the event execution is considered a component.

The way we compare the DOM trees to obtain the changed subtree is defined as follows: suppose the DOM-tree before the event execution is T_{before} and the DOM tree after the event execution is T_{after}. We traverse T_{before} using breadth-first search. For each node x in T_{before}, we compute the path from root to x, and find the node y in T_{after} that has the same path. If x and y are different, or have different number of children, x is considered the root of a component; its XPath is added to the stateDictionary if not already existing, and the search is discontinued in the subtree of x . If several such nodes y exist in T_{after}, their deepest common ancestor is used as the root of the component.

Initially, the stateDictionary contains only one component with XPath "/". Additional components are discovered and added to the stateDictionary as the crawling proceeds. The algorithm is summarized in the pseudo-code below. One important practical point to note is that the discovery of a new component can lead to a modification of previously known component states, if the new component is nested inside these component states. As explained before, the new component must be pruned from the containing component states, so their component state ID must be recomputed. It is not practical to save all component states DOM to be able to recompute their ID when this occurs. Instead, in our prototype we mark these component states as invalid and visit them again later during the crawl.

4   Experimental Results

4.1   Test Cases

In order to evaluate the efficiency of our method, we have run some crawling experiments with a number of experimental and real RIAs. We split these results in two categories. In the first category, we have seven simple RIAs[4]. Two of these are test application that we have built ourselves for testing purpose, while the five other ones are real, but simple (or simplified) RIAs. We have also run our

---

[4] http://ssrg.eecs.uottawa.ca/testbeds.html

Algorithm 2. Pseudo-code of proposed crawling algorithm

1: **procedure** COMPONENTBASEDCRAWL
2:     **for** as long as crawling goes **do**
3:         *event* := select next event to be executed based on the exploration strategy
4:         execute (*event*)
5:         *delta* := diff (*dom*$_{before}$, *dom*$_{after}$)
6:         *xpath* := getXpath (*delta*)
7:         **if** *stateDictionary* does not contain *xpath* **then**
8:             add *xpath* to *stateDictionary*
9:         **end if**
10:        *resultingStates* := FindCurrentStates(*delta*)
11:        **for all** *state* in *resultingState* **do**
12:            **if** *stateDictionary* does not contain *state* **then**
13:                add *state* to *stateDictionary*
14:            **end if**
15:            *event.destinations* := *resultingStates*
16:        **end for**
17:    **end for**
18:    **return** *stateDictionary*
19: **end procedure**

---

test on two real "complex" RIAs: IBM *Rational Team Concert* (RTC[5]), an agile application life-cycle management web-based application, and *MODX*[6], an open source content management system. For these two test cases, the complexity of the web site made it impossible for us to crawl with classical method for comparison, so we report the results separately in Section 4.3. Note that the number of test cases is not as large as we would like, but we are faced with the limitation of the tools we use to execute the crawl on RIAs[7]. We provide the characteristics of the model built for each of these nine RAIs in table 2. Note that these are the numbers for our component-based model, which is much smaller than the classical DOM-based model (see [27] for more details).

Table 2. Applications tested, along with their number of states and transitions in component-based model

Name	# States	# Trans.	Type	Name	# States	# Trans.	Type
Bebop	119	774	Simple	TestRIA	67	191	Test
Elfinder	152	3,239	Simple	Altoro	87	536	Test
Periodic	365	2,019	Simple	RTC	432	3,667	Complex
Clipmarks	31	377	Simple	MODX	1,291	7,868	Complex
DynaTable	24	49	Simple				

---

[5] https://jazz.net/products/rational-team-concert
[6] http://modx.com
[7] We stress that the work in question is not related to the strategy described here, but to the limitation of the available tools.

We have implemented all the mentioned crawling strategies in a prototype of IBM® Security AppScan® Enterprise[8]. Each strategy is implemented as a separate class in the same code base, so they use the same DOM equivalence mechanism [28], the same event identification mechanism [29], and the same embedded browser. For this reason, in Section 4.2 all strategies find the same model for each application. We crawl each application with a given strategy ten times and present the average of these crawls. In each crawl, the events of each state are randomly shuffled before they are passed to the strategy. The aim here is to eliminate influences that may be caused by exploring the events of a state in a given order since some strategies may explore the events on a given state sequentially.

4.2   Results on Simple RIAs

This first set of test case were simple enough to allow crawling with the traditional method. We report here comparisons with the greedy exploration [6] and the probability strategy [8], which are known to be to two most efficient strategy for building an exhaustive model [9]. This gives us complete knowledge of the model, allowing us to see whether our optimized strategy provides 100% coverage.

4.2.1   Complete Exploration Cost

Our first set of results are about the "total exploration costs", that is, the cost of finishing the crawl, expressed in terms of number of events executed. Most results are detailed in Figure 5. As can be seen, our component-based crawling method consistently outperforms both probability method and the greedy method by a very wide margin. The difference is more dramatic in RIAs that have a complex behavior, though even for the smaller ones, TestRIA and Altoro Mutual (not shown), the cost of component-based crawling is about 30% of the cost of the other methods. The best example among our test cases is Bebop, which contains very few data items shown on the page, but can sort and filter and expand/collapse those items in different manners. Even in an instance of the RIA with only 3 items, component-based crawling is 200 times more efficient than the other methods. This difference in performance quickly gets even bigger in an instance of the RIA with more items, as shown in Section 4.2.4.

4.2.2   Time Measurement

Since component-based crawling requires a fair amount of computation at each step, we also measured time in similar experiments to ensure this processing overhead does not degrade the overall performance. As can be seen on Table 3, even in terms of absolute time component-based crawling significantly outperforms the two other methods.

---

[8] Details are available at http://ssrg.eecs.uottawa.ca/docs/prototype.pdf Since our crawler is built on top of the architecture of a commercial product, we are not currently able to provide open-source implementations of the strategies.

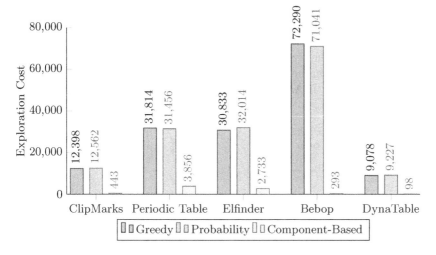

Fig. 5. Comparison of exploration costs of finishing crawl for different methods

Table 3. Time of finishing crawl for different methods (hh:mm:ss)

	TestRIA	Altor-Mutual	ClipMarks	Periodic Table	Elfinder	Bebop	DynaTable
Greedy	00 : 00 : 18	00 : 00 : 34	00 : 03 : 38	01 : 13 : 08	00 : 51 : 22	01 : 25 : 11	00 : 05 : 35
Probability	00 : 00 : 11	00 : 00 : 20	00 : 02 : 50	01 : 09 : 42	00 : 49 : 00	01 : 17 : 32	00 : 04 : 51
Component-Based	00 : 00 : 06	00 : 00 : 04	00 : 00 : 13	00 : 01 : 21	00 : 08 : 21	00 : 00 : 29	00 : 00 : 06

4.2.3   Coverage

Unlike previous DOM-based methods, component-based crawling does not guaranty complete coverage. This is because the method is based on discovering automatically independent components, and if the method wrongly identifies as "components" sections of the DOM that are not independent from each other, some coverage might be lost. It is difficult to know in general the amount of coverage that can be lost, but in the case of our seven test cases, our method systematically reached 100% coverage. No information was lost despite the dramatic decrease in crawling time.

4.2.4   Scalability

Some of the test RIAs that we used had to be significantly "trimmed" before they could be crawled by the traditional methods. One example is Clipmark, which displays a number of items on the page. Although it had initially 40 items, we had to reduce it to 3 in order to finish the crawl with the traditional methods! When using component-based crawling, on all of our test beds we were easily able to finish the crawl on the original data, and we could increase the number of items beyond that without problem. We show the data for two examples, Clipmarks on Figure 6 and Bebop on Figure 7. As can clearly been seen on both examples, while the crawling time increases linearly with the number of items in the page when using component-based crawling, it grows exponentially with the DOM-based, greedy method and soon becomes intractable. The results are the same with probability, and will *necessarily* be similar for *any* DOM-based crawling method, since the size of the end-model itself increases exponentially with the number of items in the page. This shows that component-based crawling is able to crawl and index RIAs that are simply out of reach to any DOM-based strategy.

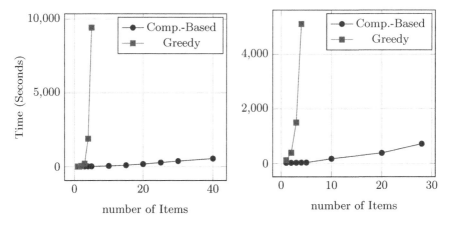

Fig. 6. Scalability with Clipmarks   **Fig. 7.** Scalability with BeBop

4.3 Results on Complex RIAs

When crawling large and complex RIAs, comparison with DOM-based methods do not tell much, since these methods are essentially unable to crawl them. In addition, for many of these RIAs (e.g. Facebook, Gmail etc.), the amount of data available is very large. "Finishing" the crawl is often not a realistic proposition. Instead, the question becomes how efficient the crawl can be as it progresses overtime. In order to measure this, we focus on the question of crawling for indexing, where we argue that a fair definition of "efficient" is the ability of the crawler to keep finding new content. In our experiment, we have measured how much the textual information accumulated increases overtime. An efficient method will provide a steady increasing amount of information, while an inefficient method might stop providing any new information for long period of times (basically re-fetching known data many times). We have measured how "efficient" component-based crawling is, by counting how many "lines" of text (excluding html tags) are accumulated overtime. As can been seen from the Figures 8 and 9, for both of our examples, the method provides a nice steadily increasing line, showing that the method is efficient at fetching new information overtime. An inefficient method would have plateaued, during which the crawl is not adding any new data.

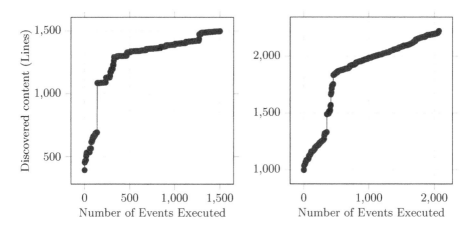

Fig. 8. Progress overtime with RTC   **Fig. 9.** Progress overtime with MODX

5   Conclusions and Future Work

This work addresses one of the most prevalent problems in the context of crawling AJAX-based RIAs: state space explosion. The presented method detects independent components and models the RIA at component level rather than DOM level, resulting in exponential reduction of the overall crawling complexity, with minimal or no loss of data coverage. The method captures the effect of event execution more precisely, resulting in a simpler model with fewer states and transitions. The produced model can point to any desired data with an event execution trace from the initial state, but cannot necessarily produce a path to lead to any valid DOM-state. The methods has been implemented using a greedy exploration strategy and DOM diff as automatic component discovery algorithm. Our experimental results verify the significant performance gain of the method while covering equal content as DOM-based methods.

This work can be improved in several areas. In particular, our future work include devising better algorithms for component discovery. One important improvement can be done on detection and handling of violations: currently, we have no effective way of recovering from a situation where components that have been assumed to be independent turn out not to be. We simply ignore theses violation. Although that did not impact negatively our experimental results, we cannot be sure that this won't be the case on every RIAs. A naive approach for detecting violations and adapting the strategy accordingly is not particularly difficult, but it would be too costly to be practical.

Acknowledgments. This work is partially supported by the IBM Center for Advanced Studies, and the Natural Sciences and Engineering Research Council of Canada (NSERC). The views expressed in this article are the sole responsibility of the authors and do not necessarily reflect those of IBM.

Trademarks: IBM and AppScan are trademarks or registered trademarks of International Business Machines Corp., registered in many jurisdictions worldwide. Other product and service names might be trademarks of IBM or other companies. A current list of IBM trademarks is available on the Web at *Copyright and trademark information* at www.ibm.com/legal/copytrade.shtml. Java and all Java-based trademarks and logos are trademarks or registered trademarks of Oracle and/or its affiliates.

References

1. Fraternali, P., Rossi, G., Sánchez-Figueroa, F.: Rich internet applications. IEEE Internet Computing 14(3), 9–12 (2010)
2. Duda, C., Frey, G., Kossmann, D., Zhou, C.: Ajaxsearch: crawling, indexing and searching web 2.0 applications. Proceedings of the VLDB Endowment 1(2), 1440–1443 (2008)
3. Duda, C., Frey, G., Kossmann, D., Matter, R., Zhou, C.: Ajax crawl: making ajax applications searchable. In: ICDE 2009, pp. 78–89. IEEE (2009)
4. Amalfitano, D., Fasolino, A.R., Tramontana, P.: Reverse engineering finite state machines from rich internet applications. In: Proceedings of WCRE, pp. 69–73. IEEE (2008)
5. Amalfitano, D., Fasolino, A.R., Tramontana, P.: Rich internet application testing using execution trace data. In: Proceedings of ICSTW, pp. 274–283. IEEE (2010)
6. Peng, Z., He, N., Jiang, C., Li, Z., Xu, L., Li, Y., Ren, Y.: Graph-based ajax crawl: Mining data from rich internet applications. In: Proceedings of ICCSEE, vol. 3, pp. 590–594 (March 2012)
7. Dincturk, M.E., Jourdan, G.V., Bochmann, G.v., Onut, I.V.: A model-based approach for crawling rich internet applications. ACM Transactions on the WEB (to appear, 2014)
8. Choudhary, S., Dincturk, M.E., Mirtaheri, S.M., Jourdan, G.-V., Bochmann, G.v., Onut, I.V.: Model-based rich internet applications crawling:menu and probability models. Journal of Web Engineering 13(3) (to appear, 2014)
9. Choudhary, S., Dincturk, M.E., Mirtaheri, S.M., Jourdan, G.-V., Bochmann, G.v., Onut, I.V.: Building rich internet applications models: Example of a better strategy. In: Daniel, F., Dolog, P., Li, Q. (eds.) ICWE 2013. LNCS, vol. 7977, pp. 291–305. Springer, Heidelberg (2013)
10. Faheem, M., Senellart, P.: Intelligent and adaptive crawling of web applications for web archiving. In: Daniel, F., Dolog, P., Li, Q. (eds.) ICWE 2013. LNCS, vol. 7977, pp. 306–322. Springer, Heidelberg (2013)
11. Amalfitano, D., Fasolino, A.R., Polcaro, A., Tramontana, P.: The dynaria tool for the comprehension of ajax web applications by dynamic analysis. In: Innovations in Systems and Software Engineering, pp. 1–17 (2013)
12. Doush, I.A., Alkhateeb, F., Maghayreh, E.A., Al-Betar, M.A.: The design of ria accessibility evaluation tool. Advances in Engineering Software 57, 1–7 (2013)
13. Mesbah, A., van Deursen, A.: Invariant-based automatic testing of ajax user interfaces. In: ICSE, pp. 210–220 (May 2009)
14. Amalfitano, D., Fasolino, A.R., Tramontana, P.: A gui crawling-based technique for android mobile application testing. In: Proceedings of ICSTW, pp. 252–261. IEEE Computer Society, Washington, DC (2011)

15. Amalfitano, D., Fasolino, A.R., Tramontana, P., De Carmine, S., Memon, A.M.: Using gui ripping for automated testing of android applications. In: Proceedings of ASE, pp. 258–261. ACM, New York (2012)

16. Erfani, M., Mesbah, A.: Reverse engineering ios mobile applications. In: Proceedings of WCRE (2012)

17. Mesbah, A., Bozdag, E., van Deursen, A.: Crawling ajax by inferring user interface state changes. In: Proceedings of ICWE, pp. 122–134. IEEE (2008)

18. Ayoub, K., Aly, H., Walsh, J.: Dom based page uniqueness identification, canada patent ca2706743a1 (2010)

19. Milani Fard, A., Mesbah, A.: Feedback-directed exploration of web applications to derive test models. In: Proceedings of ISSRE, 10 pages. IEEE Computer Society (2013)

20. Choudhary, S., Dincturk, M.E., Mirtaheri, S.M., Moosavi, A., Bochmann, G.v., Jourdan, G.-V., Onut, I.-V.: Crawling rich internet applications: the state of the art. In: CASCON, pp. 146–160 (2012)

21. Mirtaheri, S.M., Dinçtürk, M.E., Hooshmand, S., Bochmann, G.v., Jourdan, G.-V., Onut, I.V.: A brief history of web crawlers. In: Proceedings of CASCON, pp. 40–54. IBM Corp. (2013)

22. Bezemer, C.P., Mesbah, A., van Deursen, A.: Automated security testing of web widget interactions. In: Proceedings of ESEC/FSE, pp. 81–90. ACM (2009)

23. Chen, A.Q.: Widget identification and modification for web 2.0 access technologies (wimwat). ACM SIGACCESS Accessibility and Computing (96), 11–18 (2010)

24. Crescenzi, V., Mecca, G., Paolo, Merialdo, et al.: Roadrunner: Towards automatic data extraction from large web sites. In: VLDB, vol. 1, pp. 109–118 (2001)

25. Harel, D.: Statecharts: A visual formalism for complex systems. Science of Computer Programming 8(3), 231–274 (1987)

26. Peng, Z., He, N., Jiang, C., Li, Z., Xu, L., Li, Y., Ren, Y.: Graph-based ajax crawl: Mining data from rich internet applications. In: Proceedings of ICCSEE, vol. 3, pp. 590–594. IEEE (2012)

27. Moosavi, A.: Component-based crawling of complex rich internet applications. Master's thesis, EECS - University of Ottawa (2014), http://ssrg.site.uottawa.ca/docs/Ali-Moosavi-Thesis.pdf

28. Benjamin, K., Bochmann, G.v., Jourdan, G.-V., Onut, I.-V.: Some modeling challenges when testing rich internet applications for security. In: Proceedings of ICSTW, pp. 403–409. IEEE Computer Society, Washington, DC (2010)

29. Choudhary, S., Dincturk, M.E., Bochmann, G.v., Jourdan, G.-V., Onut, I.V., Ionescu, P.: Solving some modeling challenges when testing rich internet applications for security. In: Proceedings of ICST, pp. 850–857 (2012)

Pattern-Based Specification
of Crowdsourcing Applications

Alessandro Bozzon[1], Marco Brambilla[2], Stefano Ceri[2],
Andrea Mauri[2], and Riccardo Volonterio[2]

[1] Software and Computer Technologies Department, Delft University of Technology,
Postbus 5 2600 AA, Delft, The Netherlands
a.bozzon@tudelft.nl
[2] Dipartimento di Elettronica, Informazione e Bioingegneria (DEIB)
Politecnico di Milano, Piazza Leonardo da Vinci, 32, 20133 Milano, Italy
{name.surmame}@polimi.it

Abstract. In many crowd-based applications, the interaction with performers is decomposed in several tasks that, collectively, produce the desired results. Tasks interactions give rise to arbitrarily complex workflows. In this paper we propose methods and tools for designing crowd-based workflows as interacting tasks. We describe the modelling concepts that are useful in such framework, including typical workflow patterns, whose function is to decompose a cognitively complex task into simple interacting tasks so that the complex task is co-operatively solved.

We then discuss how workflows and patterns are managed by Crowd-Searcher, a system for designing, deploying and monitoring applications on top of crowd-based systems, including social networks and crowdsourcing platforms. Tasks performed by humans consist of simple operations which apply to homogeneous objects; the complexity of aggregating and interpreting task results is embodied within the framework. We show our approach at work on a validation scenario and we report quantitative findings, which highlight the effect of workflow design on the final results.

1 Introduction

Crowd-based applications are becoming more and more widespread; their common aspect is that they deal with solving a problem by involving a vast set of performers, who are typically extracted from a wide population (the "crowd"). In many cases, the problem is expressed in the form of simple questions, and the performers provide a set of answers; a software system is in charge of organising a crowd-based computation – typically by distributing questions, collecting responses and feedbacks, and organising them as a well-structured result of the original problem.

Crowdsourcing systems, such as Amazon Mechanical Turk (AMT), are natural environments for deploying such applications, since they support the assignment to humans of simple and repeated tasks, such as translation, proofing,

S. Casteleyn, G. Rossi, and M. Winckler (Eds.): ICWE 2014, LNCS 8541, pp. 218–235, 2014.
© Springer International Publishing Switzerland 2014

content tagging and items classification, by combining human contribution and automatic analysis of results [1]. But a recent trend (emerging, e.g., during the CrowdDB Workshop[1]), is to use many other kinds of platforms for engaging crowds, such as proprietary community-building systems (e.g., FourSquare or Yelp) or general-purpose social networks (e.g., Facebook or Twitter). In the various platforms, crowds take part to social computations both for monetary rewards and for non-monetary motivations, such as public recognition, fun, or genuine will of sharing knowledge.

In previous work, we presented CrowdSearcher [2, 3], offering a conceptual framework, a specification paradigm and a reactive execution control environment for designing, deploying, and monitoring applications on top of crowd-based systems, including social networks and crowdsourcing platforms. In Crowdsearcher, we advocate a top-down approach to application design which is independent on the particular crowd-based system. We adopt an abstract model of crowdsourcing activities in terms of elementary task types (such as: labelling, liking, sorting, classifying, grouping) performed upon a data set, and then we define a crowdsourcing task as an arbitrary composition of these task types; this model does not introduce limitations, as arbitrary crowdsourcing tasks can always be defined by aggregating several operation types or by decomposing the tasks into smaller granularity tasks, each one of the suitable elementary type. In general, an application cannot be submitted to the crowd in its initial formulation; transformations are required to organise and simplify the initial problem, by structuring it into a *workflow of crowd-based tasks* that can be effectively performed by individuals, and can be submitted and executed, possibly in parallel. Several works [4, 5] have analysed typical *crowdsourcing patterns*, i.e. typical cooperative schemes used for organising crowd-based applications.

The goal of this paper is to present a systematic approach to the design and deployment of crowd-based applications as arbitrarily complex workflows of elementary tasks, which emphasises the use of crowdsourcing patterns. While our previous work was addressing the design and deployment of a single task, in this paper we model and deploy applications consisting of arbitrarily complex task interactions, organised as a workflow; we use either *data streams* or *data batches* for data exchange between tasks, and illustrate that tasks can be controlled through *tight coupling* or *loose coupling*. We also show that our model supports the known crowd management patterns, and in particular we use our model as a unifying framework for a systematic classification of patterns.

The paper is structured as follows. Section 2 presents related work; Section 3 introduces the task and workflow models and design processes. Section 4 details a set of relevant crowdsourcing patterns. Section 5 illustrates how workflow specifications are embodied within the execution control structures of Crowdsearcher, and finally Section 6.3 discusses several experiments, showing how differences in workflow design lead to different application results.

---

[1] http://dbweb.enst.fr/events/dbcrowd2013/

2   Related Work

Many crowdsourcing startups[2] and systems [6] have been proposed in the last years. Crowd programming approaches rely on imperative programming models to specify the interaction with crowdsourcing services (e.g., see *Turkit* [7], *RABJ* [8], *Jabberwocky* [9]). Several programmatic methods for human computation have been proposed [7][8][9][10], but they do not support yet the complexity required by real-world, enterprise–scale applications, especially in terms of designing and controlling complex flows of crowd activities.

Due to its flexibility and extensibility, our approach covers the expressive power exhibited by any of the cited systems, and provides fine grained targeting to desired application behaviour, performer profiles, and adaptive control over the executions.

Several works studied how to involve humans in the creation and execution of workflows, and how to codify common into modular and reusable patterns. Process-centric workflow languages [11] define business artefacts, their transformations, and interdependencies trough tasks and their dependencies. Scientists and practitioners put a lot of effort in defining a rich set of control-driven workflow patterns.[3] However, this class of process specification languages: focus mainly on control flow, often abstracting away data almost entirely; disregard the functional and non-functional properties of the involved resources; do not specify intra- and inter-task execution and performer controls; and provide no explicit modelling primitives for data processing operations.

In contrast, data- driven workflows have recently become very popular, typically in domains where database are central to processes [12][13], and data consistency and soundness is a strong requirement. Data-driven workflows treat data as first-class citizens, emphasise the role of control intra- and inter-task control, and ultimately served as an inspiration for our work.

Very few works studied workflow-driven approaches for crowd work. Crowd-Lang [5] is notable exception, which supports process-driven workflow design and execution of tasks involving human activities, and provides an executable model-based programming language for crowd and machines. The language, however, focuses on the modelling of coordination mechanisms and group decision processes, and it is oblivious to the design and specification of task-specific aspects.

Several works tried to codify patterns for crowdsourcing. At task level, a great wealth of approaches has been proposed for the problems of output agreement [14], and performer control [15]. At workflow level, less variety can be witnessed, but a set of very consolidated patterns emerge [7][16][4][17][18]. In Section 4 we will provide an extensive review of the most adopted pattern in crowdsourcing, classifying them in the light of the workflow model of Section 3.

---

[2] E.g., CrowdFlower, Microtask, uTest.
[3] http://workflowpatterns.com/

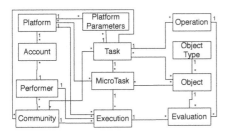

Fig. 1. Metamodel of task properties

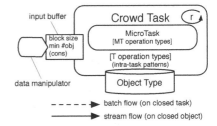

Fig. 2. Task notation

3   Models and Design of Crowd-Based Workflows

Although humans are capable of solving complex tasks by using their full cognitive capacity, the approaches used in crowd-based computations prefer to decompose complex tasks into simpler tasks and then elaborate their results [16]. Following this approach, we restrict crowdsourcing tasks be to simple operations which apply to homogeneous objects; operations are simple actions (e.g. labelling, liking, sorting, classifying, grouping, adding), while objects have an arbitrary schema and are assumed to be either available to the application or to be produced as effect of application execution.

3.1   Task Model

Tasks of a crowd-based application are described in terms of an abstract model, that was initially presented in [2], and represented in Fig. 1. We assume that each **task** receives as input a list of **objects** (e.g., photos, texts, but also arbitrarily complex objects, all conforming to the same **object type**) and asks performers to do one or more **operations** upon them, which belong to a predefined set of abstract **operation types**. Examples of operation types are *Like*, for assigning a preference to an item; or *Classify*, for assigning each item to one or more classes. The full list of currently supported operation types is reported in [2]. Task management requires specific sets of objects to be assembled into a unit of execution, called **micro-task**, that is associated with a given **performer**. Each micro-task can be invited or executed on different **platforms** and/or **communities**. The relation with platform is specified through a series of **platform parameters**, specific for each platform, that are needed in order retrieve the answers of the performers (e.g., the HIT identifier on AMT). A **performer** may be registered on several platforms (with different accounts) and can be part of several communities. Micro-task **execution** contains some statistics (e.g., start and end timestamps). The **evaluation** contains the answer of the performer for each object, whose schema depends on the operation type.

For example, a *like* evaluation is a counter that registers how many performers like the object, while a *classify* evaluation contains the category selected by the performers for that object.

3.2 Workflow Model

A **crowdsourcing workflow** is defined as a control structure involving two or more interacting tasks performed by humans. Tasks have an input buffer that collects incoming data objects, described by two parameters: 1) The **task size**, i.e. the minimum number of objects (m) that allow starting a task execution; 2) The **block size**, i.e. the number of objects (n) consumed by each executions.

Clearly, $n \leq m$, but in certain cases at least m objects must be present in the buffer before starting an execution; in fact n can vary between 1 and the whole buffer, when a task execution consumes all the items currently in the buffer. Task execution can cause **object removal**, when objects are removed from the buffer, or **object conservation**, when objects are left in the buffer, and in such case the term **new items** denotes those items loaded in the buffer since the last execution.

Tasks communicate with each other with **data flows**, produced by extracting objects from existing data sources or by other tasks, as streams or batches. **Data streams** occur when objects are communicated between tasks one by one, typically in response to events which identify the completion of object's computations. **Data batches** occur when all the objects are communicated together from one task to another, typically in response to events related to the closing of task's computations.

Flows can be constrained based on a condition associated with the arrow representing the flow. The condition applies to properties of the produced objects and allows transferring only the instances that satisfy the condition. Prior to task execution, a **data manipulator** may be used to compose the objects in input to a task, possibly by merging or joining incoming data flows.

We can represent tasks within workflows as described in Fig. 2, where each task is equipped with an input buffer and an optional data manipulator, and may receive data streams or data batches from other tasks. Each task consists of micro-tasks which perform given operations upon objects of a given object type; the parameter r indicates the number of executions that are performed for each micro-tasks, when statically defined (default value is 1). Execution of tasks can be performed according to intra-task patterns, as described in Section 4.

3.3 Workflow Design

Workflow design consists of designing tasks interaction; specifically, it consists of defining the workflow schema as a directed graph whose nodes are tasks and whose edges describe dataflows between tasks, distinguishing streams and batches. In addition, the coupling between tasks working on the same object type can be defined as loose or tight.

Loose coupling is recommended when two tasks act independently upon the objects (e.g. in sequence); although it is possible that the result of one task may have side effects on the other task, such side effects normally occur as an exception and affect only a subset of the objects. Loosely coupled tasks have independent control marts and monitoring rules (as described in Section 3.4).

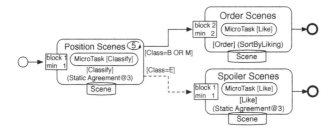

Fig. 3. Example of crowd flow

Tight coupling is recommended when the tasks intertwine operations upon the same objects, whose evolution occurs as combined effect of the tasks' evolution; tightly coupled tasks share the same control mart and monitoring rules.

Figure 3 shows a simple workflow example in the domain of movie scenes annotation. The *Position Scenes* tasks asks performers to say whether a scene appears at the beginning, middle or end of the film; it is a classification task, one scene at a time, with 5 repetitions and acceptance of results based on an agreement threshold of 3. Scenes in the ending part of the movies are transmitted to the *Spoiler Scenes* task, which asks performers whether the scene is a spoiler or not;[4] scenes at the beginning or in the middle of the movie are transmitted to the *Order Scenes* task, which asks performers to order them according to the movie script; each micro-task orders just two scenes, by asking the performer to select the one that comes first. The global order is then reconstructed. Given that all scenes are communicated within the three tasks, they are considered as tightly coupled.

3.4 Task Design

Crowdsourcing tasks are targeted to a single object type and are used in order to perform simple operations which either apply to a single object (such as **like**, **tag**, or **classify**) or require comparison between objects (such **choice** or **score**); more complex tasks perform operations inspired by database languages, such as **select**, **join**, **sort**, **rank**, or **group by**.

Task design consists of the following phases: 1) **Operations design** – deciding how a task is assembled as a set of operation types; 2) **Object and performer design** – defining the set of objects and performers for the task; 3) **Strategy design** – Defining how a task is split into micro-tasks, and how micro-tasks are assigned to subsets of objects and performers; 4) **Control Design** – Defining the rules that enable the run-time control of objects, tasks, and performers.

For monitoring task execution, a data structure called **control mart** was introduced in [3]; Control consists of four aspects:

- **Object control** is concerned with deciding when and how responses should be generated for each object.

---

[4] A *spoiler* is a scene that gives information about the movie's plot and as such should not be used in its advertisement.

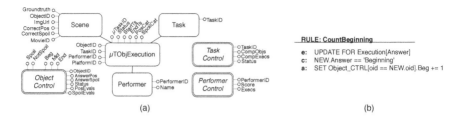

Fig. 4. (a) Example of control mart for the tasks of Fig. 3; (b) Example of control rule that updates the number of responses in the *Position of Scenes* task

- **Performer control** is concerned with deciding how performers should be dynamically selected or rejected, on the basis of their performance.
- **Task control** is concerned with completing a task or re-planning task execution.

The control of objects, performers and tasks is performed by **active rules**, expressed according to the *event-condition-action* (ECA) paradigm. Each rule is triggered by **events** (e) generated upon changes in the control mart or periodically; the rule's **condition** (c) is a predicate that must be satisfied on order for the action to be executed; the rule's **actions** (a) change the content of the control mart. Rules properties (e.g., termination) can be been proven in the context of a well-organised computational framework [3].

Figure 4(a) shows a sample control mart for the three tasks in the example scenario, which we assume to be tightly connected, thus using the same data mart. The control mart stores all the required information for controlling the task's evolution and is automatically defined from the task specifications. Figure 4(b) reports a simple control rule that updates the number of responses with value "Beginning" after receiving an answer.

This rule has the following behaviour: every time a performer perform a new evaluation on a specific object (UPDATE event on μTObjExecution), if the selected answer is "Beginning" (the condition part of the rule), then it increases the counter of the "Beginning" category for that object (Object_CTRL[oid == New.oid] selected the correct object, then the correct property can be acessed with the dot notation). For a deeper description of the rule grammar and structure see our previous work [3].

4   Crowdsourcing Patterns

Several patterns for crowd-based operations are defined in the literature. We review them in light of the workflow model of Section 3. We distinguish them in three classes and we implement them in Crowdsearcher (see Section 5):

- **Intra-Task Patterns.** They are typically used for executing a complex task by means of a collection of operations which are cognitively simpler than the original task. Although these patterns do not appear explicitly in the workflow, they are an essential ingredient of crowd-based computations.
- **Workflow Patterns.** They are used for solving a problem by involving different tasks, which require a different cognitive approach; results of the different tasks, once collected and elaborated, solve the original problem.
- **Auxiliary Patterns.** They are typically performed before or after both intra-task and workflow patterns in order either to simplify their operations or to improve theirs results.

4.1   Intra-Task Patterns

Intra-task patterns apply to complex operations, whose result is obtained by composing the results of simpler operations. They focus on problems related to the planning, assignment, and aggregation of micro tasks; they also include quality and performer control aspects. Figure 5 describes the typical set of design dimensions involved in the specification of a task. When the operation applies to a large number of objects and as such cannot be mapped to a single pattern instantiation, it is customary to put in place a *splitting strategy*, in order to distribute the work, followed by an *aggregation strategy*, to put together results. This is the case in many data-driven tasks stemming from traditional relational data processing which are next reviewed.

Consensus Patterns. The most commonly used intra-task patterns aim at producing responses by replicating the operations which apply to each object, collecting multiple assessments from human workers, and then returning the answer which is more likely to be correct. These patterns are referred to as *consensus* or *agreement patterns*. Typical consensus patterns are: *a*) **StaticAgreement** [3]: accepts a response when it is supported by a given number of performers. For instance, in a tag operation we consider as valid responses all the tags that have been added by at least 5 performers. *b*) **MajorityVoting** [19]: accepts a response only if a given number of performers produce the same response, given a fixed number of total executions. *c*) **ExpectationMaximisation** [20]: adaptively alternates between estimating correct answers from task parameters (e.g. complexity), and estimating task parameters from the estimated answers, eventually converging to maximum-likelihood answer values.

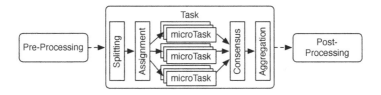

Fig. 5. Building blocks of an Intra-Task Pattern

Join Patterns. Crowd join patterns, studied in [14], are used to build an equality relationship between matching objects in the context of crowdsourcing tasks. We identify: *a*) **SimpleJoin** consists in defining microtasks performing a simple classification operation, where each execution contains a single pair of items to be joined, together with the join predicate question, and two buttons (Yes, No) for responding whether the predicate evaluates to true or false; *b*) **One-ToManyJoin** is a simple variant that includes in the same microtask one left object and several right candidates to be joined; *c*) **ManyToManyJoin** includes in the same microtask several candidate pairs to be joined;

Sort Patterns. Sort patterns determine the total ordering of a set of input objects. The list includes: *a*) **SortByGrouping** [14] orders a large set of objects by aggregating the results of the ordering of several small subsets of them. *b*) **SortByScoring** [14] asks performers to rate each item in the dataset according to a numerical scale. *c*) **SortByLiking** [3] is a variant that simply asks the performer to select/like the items they prefer. The mean (or sum) of the scores achieved by each image is used to order the dataset. *d*) **SortByPairElection** [3] asks workers to perform a pairwise comparison of two items and indicate which one they like most. Then ranking algorithms calculate their ordering. *e*) **SortByTournament** [18], presents to performers a tournament-like structure of sort tasks; each tournament elect its champions that progress to the next level, eventually converging to a final order.

Grouping Patterns. Grouping patterns are used in order to classify or clustered several objects according to their properties. We distinguish:
a) **GroupingByPredefinedClasses**[21] occurs when workers are provided with a set of known classes. *b*) **GroupingByPreference** [22] occurs when groups are formed by performers, for instance by asking workers to select the items they prefer the most, and then clustering inputs according to ranges of preferences.

Performer Control Patterns. Quality control of performers consists in deciding how to engage qualified workers for a given task and how to detect malicious or poorly performing workers. The most established patterns for performer control include: *a*) **QualificationQuestion** [23], at the beginning of a microtask, for assessing the workers expertise and deciding whether to accept his contribution or not. *b*) **GoldStandard**, [3] for both training and assessing worker's quality through a initial subtask whose answers are known (they belong to the so-called *gold truth*. *c*) **MajorityComparison**, [3] for assessing performers' quality against responses of the majority of other performers, when no gold truth is available.

4.2 Auxiliary Intra-Task Patterns

The above tasks can be assisted by auxiliary operations, performed before or after their executions, as shown in Figure 5. *Pre-processing steps* are in charge of assembling, re-shaping, or filtering the input data so to ease or optimise the main task. *Post-processing steps* is typically devoted to the refinement or transformation of the task outputs into their final form.

Examples of auxiliary patterns are: *a)* **PruningPattern** [14], consisting of applying simple preconditions on input data in order to reduce the number of evaluations to be performed. For instance, in a join task between sets of actors (where we want to identify the same person in two sets), classifying items by gender, so as to compared only pairs of the same gender. *b)* **TieBreakPattern** [14], used when a sorting task produces uncertain rankings (e.g. because of ties in the evaluated item scores); the post-processing includes an additional step that asks for an explicit comparison of the uncertainly ordered items.

4.3   Workflow Patterns

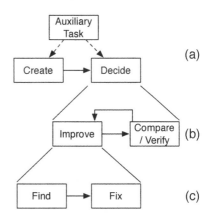

Very often, a single type of task does not suffice to attain the desired crowd business logic. For instance, with open-ended multimedia content creation and/or modification, it is difficult to assess the quality of a given answer, or to aggregate the output of several executions. A **Workflow Pattern** is a workflow of heterogeneous crowdsourcing tasks with co-ordinated goals. Several workflow patterns defined in the literature are next reviewed; they are comparatively shown in Figure 6:

Fig. 6. Template for complex task patterns

a) **Create/Decide** [16], shown in Figure 6(a), is a two-staged pattern where first workers create various options for new content, then a second group of workers vote for the best option. Note that the *create* step can include any type of basic task. This pattern can have several variants: for instance, with a stream data flow, the vote is typically restricted to the solutions which are produced faster, while with a batch data flow the second task operates on all the generated content, in order to pick the best option overall. *b)* **Improve/Compare** [7], shown in Figure 6(b), iterates on the decide step to progressively improve the result. In this pattern, a first pool of workers creates a first version of a content; upon this version, a second pool of workers creates an improved version, which is then compared, in a third task, to decide which one is the best (the original or the improved one). The improvement/compare cycle can be repeated until the improved solution is deemed as final. *c)* **Find/Fix/Verify** [4], shown in Figure 6(c), further decomposes the *improve* step, by splitting the task of finding potential improvements from the task of actually implementing them.

4.4   Auxiliary Workflow Patterns

Auxiliary tasks can be designed to support the creation and/or the decision tasks. They include: *a)* **AnswerBySuggestion** [17]: given a create operations as input, the provided solution can be achieved by asking suggestions from the crowd as follows. During each execution, a worker can choose one of two actions: it can either stop and submit the most likely answer, or it can create another job and receive another response to the task from another performer. The auxiliary suggestion task produces content that can be used by the original worker to complete or improve her answer. *b)* **ReviewSpotcheck** strengthens the decision step by means of a two-staged review process: an additional quality check is performed after the corrections and suggestions provided by the performers of the decision step. The revision step can be performed by the same performer of the decision step or by a different performer.

5   Workflow Execution

Starting from the declarative specification described in Sections 3 and 4, an automatic process generates task descriptors and their relations. Single tasks and their internal strategies and patterns are transformed into executable specification; we support all the intra-task patterns described in Section 4, through model transformations that generate the control marts and control rules for each task [3]. Task interactions are implemented differently depending on whether interacting tasks are tightly coupled or loosely coupled.

Tightly coupled tasks share the control mart structure (and the respective data instances), thus coordination is implemented directly on data. Each task posts its own results and control values in the mart. Dependencies between tasks are transformed into rules that trigger the creation of new micro-tasks and their executions, upon production of new results by events of object or task closure.

Loosely coupled tasks have independent control marts, hence their interaction is more complex. Each task produces in output events such as `ClosedTask`, `ClosedObject`, `ClosedMicrotask`, `ClosedExecution`. We rely on an event based, publish-subscribe mechanism, which allows tasks to be notified by other tasks about some happening. Loosely coupled tasks do not rely on a shared data space, therefore events carry with them all the relevant associated pieces of information (e.g., a `ClosedObject` event carries the information about that object; a `ClosedTask` event carries the information about the closed objects of the task).

The workflow structure dictates how tasks subscribe to events of other tasks. Once a task is notified by an incoming event, the corresponding data is incorporated in its control mart by a-priori application of the data manipulation program, specified in the data manipulator stage of the task. Then, reactive processing takes place within the control mart of the task.

Modularity allows executability through model transformations which are separately applied to each task specification. Automatically generated rules and mart structures can be manually refined or enriched when non-standard behaviour is needed.

This approach is supported by CrowdSearcher, a platform for crowd management written in JavaScript. CrowdSearcher runs on *Node.js*, a full-fledged event-based system, which fits the need of our rule-based approach. Each control rule is translated into scripts; triggering is modelled through internal platform

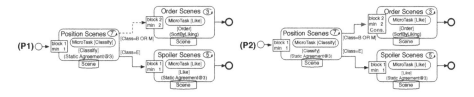

Fig. 7. Flow variants for the Positioning scenario

events. Precedence between rules is implicitly obtained by defining the scripts in the proper order. CrowdSearcher offers a cloud-based environment to transparently interface with social networks and crowdsourcing platforms, according to the task model described in Section 3.1. It features an online configuration interface where designers build complex crowdsourcing applications through a wizard–driven, step by step approach. A built-in *Task Execution Framework* (TEF) provides support for the creation of custom task user interfaces, to be deployed as stand-alone application, or embedded within third-party platforms such as Amazon Mechanical Turk. Specific modules are devoted to the invitation, identification, and management of performers, thus offering support for a broad range of expert selection paradigms, from pure pull approaches of open marketplaces, to pre-assigned execution to selected performers. Alternatives for the implementation of operations on crowd-based systems are discussed in [2].

6   Experiments

We demonstrate various pattern-based workflow scenarios, defined using our model and method and deployed by using Crowdsearcher as design framework and Amazon Mechanical Turk as execution platform. We consider several scenes taken from popular movies, and we enrich them with crowd-sourced information regarding their position in the movie, whether the scene is a spoiler, and the presence of given actors in each scene. In the experiments reported here we considered the movie "The Lord of the Rings: the Fellowship of the Ring". We extracted 20 scenes and we created a groundtruth dataset regarding temporal positioning and actors playing in the scenes. We compare cost and quality of executions for different workflow configurations.

Table 1. *Scenario 1 (Positioning)*: number of evaluated objects, microtask executions, elapsed execution time, performers, and executions per performer (for each task and for each scenario configuration)

	Position Scenes (payed $0.01)				Order Scene (payed $0.01)					TOTAL		
	#Obj	#Exe	Time	#Perf	#Exe/Perf	#Obj	#Exe	Time	#Perf	#Exe/Perf	Time	Cost	#Perf
P1	20	147	123	16	9.19	17	252	157	14	18.00	342	3.99$	26
P2	20	152	182	12	12.67	17	230	318	17	13.53	349	3.82$	26

Table 2. *Scenario 2 (Actor)*: number of evaluated objects, microtask executions, elapsed execution time, performers, and executions per performer (for each task and for each scenario configuration)

	Find Actors (payed $0.03)					Validate Actors (payed $0.02)					TOTAL		
	#Obj	#Exe	Time	#Perf	#Exe/Perf	#Obj	#Exe	Time	#Perf	#Exe/Perf	Time	Cost	#Perf
A1	20	100	120	18	5.56	–	–	–	–	–	120	3.00$	18
A2	20	100	128	10	10.00	–	–	–	–	–	128	3.00$	10
A3	20	100	123	14	7.15	20	21	154	10	2.10	159	3.42$	20
A4	20	100	132	10	10.00	41	19	157	9	2.10	164	3.38$	16
A5	20	100	126	13	7.69	69	60	242	17	3.53	257	4.20$	24
A6	66	336	778	56	6.00	311	201	821	50	4.02	855	14.10$	84

6.1 Scenario 1: Scene Positioning

The first scenario deals with extracting information about the temporal position of scenes in the movie and whether they can be considered as as spoilers. Two variants of the scenario have been tested, as shown in Figure 7: the task *Position Scenes* classifies each scene as belonging to the beginning, middle or ending part of the movie. If the scene belongs to the final part, we ask the crowd if it is a spoiler (*Spoile Scenes* task); otherwise, we ask the crowd to order it with respect to the other scenes in the same class (*Order Scenes* task).

Tasks have been configured according to the following patterns:

- *Position Scene*: task and microtask types are both set as *Classify*, using a *StaticAgreement* pattern with threshold 3. Having 3 classes, a maximum number of 7 executions grants that one class will get at least 3 selections. Each microtask evaluates 1 scene.
- *Order Scene*: task type is *Order*, while microtask type is set as *Like*. Each microtask comprises two scenes of the same class. Using a *SortByLiking* pattern, we ask performers to select (Like) which scene comes first in the movie script. A rank aggregation pattern calculates the resulting total order upon task completion.
- *Spoiler Scene*: Task and microtask type both set as *Like*. A *StaticAgreement* pattern with threshold 3 (2 classes, maximum 5 executions) defines the consensus requirements. Each microtask evaluates 1 scene.

We experiment with two workflow configurations. The first (**P1**) defines a *batch* data flow between the *Position Scene* and *Order Scene* tasks, while the second configuration (**P2**) defines the same flow as *stream*. In both variants, the data flow between *Position Scene* and *Spoiler Scenes* is defined as *stream*.

The **P2** configuration features a dynamical task planning strategy for the the *Order Scenes* task, where the construction of the scene pairs to be compared in is performed every time a new object is made available by the *Position Scenes* task. A conservation policy in the *Order Scenes* data manipulator ensures that all the new scenes are combined with the one previously received.

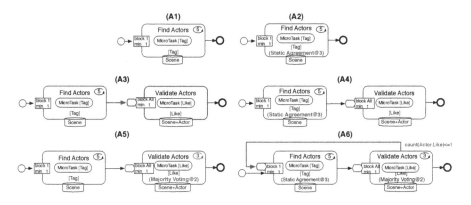

Fig. 8. Flow variants for the Actor scenario

6.2   Scenario 2: Actors

In the second scenario, we model a **create/decide** workflow pattern by asking the crowd to identify the actors that take part in the movie scenes; in *Find Actors*, performers indicate actors, in *Validate Actor* they confirm them. Tasks are designed as follows:

- *Find Actors*: Task and microtask types are set as *Tag*. Each microtask evaluates one scene; each scene is evaluated five times. Depending on the configuration, either no consensus pattern (**A1, A3, A5**) or a *StaticAgreement* pattern with threshold three (**A2, A4, A6**) is employed.
- *Validate Actors*: the task is preceded by a data manipulator function that transform the input Scene object and associated tags into a set of tuples (*Scene, Actor*), which compose an object list subject to evaluation. In all configurations, microtasks are triggered if at least one object is available in the buffer. Note that each generated microtask features a different number of objects, according to the number of actors tagged in the corresponding scene. Configurations **A5** and **A6** features an additional *MajorityVoting* pattern to establish the final actor validation.

We tested this scenario with five workflow configurations, shown in Figure 8, and designed as follows:

- Configuration **A1** performs 5 executions and for each scene collects all the actors tagged at least once;
- Configuration **A2** performs 5 executions and for each scene collects all the actors tagged at least three times (StaticAgreement@3);
- Configuration **A3** adds the validation task to **A1**; the validation asks one performer to accept or reject the list of actors selected in the previous step;
- Configuration **A4** adds a validation task to **A3**, performed as in **A3**;
- Configuration **A5** is similar to **A3**, but the validation task is performed 3 times and a MajorityVoting@2 is applied for deciding whether to accept or not the object;

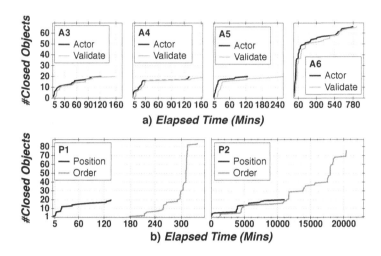

Fig. 9. Temporal distributions of closed objects

- Configuration **A6** extends **A5** by adding a StaticAgreement@3 on FindActors a feedback stream flow, originating from the *Validate Actors* task and directed to the *Find Actors* task, which notifies the latter about actors that were wrongly tagged in a scene (i.e., for which agreement on acceptance was not reached). Misjudged scenes are then re-planned for evaluation; for each scene, the whole process is configured to repeat until validation succeeds, or at most 4 re-evaluations are performed.

6.3   Results

We tested the performance of the described scenarios in a set of experiments performed on Amazon Mechanical Turk during the last week of September 2013. Table 1 and Table 2 summarise the experiment statistics for the two scenarios, 1700 HITS for a total cost of 39$.

Streaming Vs. Batch (Scenario 1: Positioning). In the first scenario we tested the impact on the application performance of the adoption of a stream data flow in a crowd workflow.

Time. Figure 9(b) shows the temporal distribution of closed objects for the **P1** and **P2** configurations. As expected, a stream flow (**P2**) allows for almost synchronous activation of the subsequent task in the flow, while batch scenario (**P1**) shows a strict sequential triggering of the second task. However, the overall duration of the workflow is not significantly affected by the change. While the first task of the flow behaves similarly in the two configurations, the second task runs significantly quicker in the batch flow, thus recovering the delay due to the sequential execution.

Quality. Table 3a shows the precision of the classification results of task *Position Scenes* (note that for this first part the two configurations are exactly the same, it makes no sense to compare the two results). Table 3b shows a measure of the quality of the obtained orders of scenes, i.e., Spearman's rank correlation coefficient of the resulting ranks from the *Order Scenes* task against the real order of scenes. Both tables show that the attained quality was not significantly influenced by the different task activation modes.

In summary, we didn't notice a different behaviour due to streaming. One possible reason is that in the batch configuration the entire set of assignments is posted at once on AMT, thus becoming more prominent in terms of number of available executions (and thus being preferred by performers, as widely studied [1]), while in a stream execution a small number of assignments is posted on AMT at every closing event of objects from the previous tasks.

Intra-Task Consensus Vs. Workflow Decision (Scenario 2: Actors). The second scenario aimed at verifying the impact that different intra-task and workflow patterns produced on the quality, execution time, and cost. We focused in particular on different validation techniques.

Time. Figure 9(a) and (c) shows the temporal distribution of closed object for configurations **A3-A6**. Configurations **A1** and **A2** are not reported because they are composed of one single task and thus their temporal distribution is not comparable. The temporal behaviour of the first and second tasks in the flow are rather similar (in the sense that the second one immediately follows the other). Validation is more delayed in **A5** due to the MajorityVoting pattern that postpones object close events. Configuration **A6** (Figure 9(c)) is significantly slower due to the feedback loop, which also generates a much higher cost of the campaign, as reported in Table 1. Indeed, due to the feedback, many tasks are executed several times before converging to validated results.

Quality. Table 4 reports the precision, recall and F-Score figures of the six configurations. The adoption of increasingly refined validation-based solutions (configurations **A3-A4-A5**) provides better results with respect to the baseline configuration **A1**, and also to the intra-task agreement based solution **A2**; validations do not have a negative impact in terms of execution times and costs. On the other hand, the complexity of of case **A6**, with the introduction of feedback, proved counter-productive, because the validation logic harmed the performance, both in monetary (much higher cost) and qualitative (lower results quality) senses, bringing as well overhead in terms of execution time. Notice

Table 3. Scenario 1 (Positioning), configuration P1 and P2: a) Precision of the *Position Scenes* classification task; b) Spearman's rank correlation coefficient of the resulting ranks from the *Order Scenes* task against the real order of scenes

	(a)		
Config.	P Beg.	P Mid.	P End
P1	0.50	1	0.11
P2	0.50	0.80	0.33

	(b) Spearman Beg.	Spearman Mid.
P1	0.500	0.543
P2	0.900	0.517

Table 4. Scenario 2 (Actor): Precision, Recall, and F-score of the 6 configurations

	A1	A2	A3	A4	A5	A6
Precision	0.79	1	0.92	0.99	0.95	0.89
Recall	0.98	0.87	0.97	0.90	1	0.96
F-Score	0.85	0.91	0.93	0.93	0.97	0.90

that the configuration **A3** reaches the highest precision score. That's because the StaticAgreement strategy ensures that all the selected actors really appear in the image, while using the crowd for the validation part can add some errors (for instance some actors recognized in the *Find Actor* can be discarded in the *Validate Actors*). However note that the other configurations (**A3 - A5**) reach an higher recall and F-score value, meaning an overall better quality of the final result.

In summary, the above tests show an advantage of concentrating design efforts in defining better workflows, instead of just optimising intra-task validation mechanisms (based e.g. on majority or agreement), although overly complex configurations should be avoided.

7   Conclusions

We present a comprehensive approach to the modeling, design, and pattern-based specification of crowd-based workflows. We discuss how crowd-based tasks communicate by means of stream-based or batch data flows, and we define the option between loose and tight coupling. We also discuss known patterns that are used to create crowd-based computations either within a task or between tasks and we show how the workflow model is translated into executable specifications which are based upon control data, reactive rules, and event-based notifications.

A set of experiments demonstrate the viability of the approach and show how the different choices in workfllow design may impact on the cost, time and quality of crowd-based activities.

References

[1] Law, E., von Ahn, L.: Human Computation. Synthesis Lectures on Artificial Intelligence and Machine Learning. Morgan & Claypool Publishers (2011)

[2] Bozzon, A., Brambilla, M., Ceri, S.: Answering search queries with crowdsearcher. In: 21st Int.l Conf. on World Wide Web, WWW 2012, pp. 1009–1018. ACM (2012)

[3] Bozzon, A., Brambilla, M., Ceri, S., Mauri, A.: Reactive crowdsourcing. In: 22nd World Wide Web Conf., WWW 2013, pp. 153–164 (2013)

[4] Bernstein, M.S., Little, G., Miller, R.C., Hartmann, B., Ackerman, M.S., Karger, D.R., Crowell, D., Panovich, K.: Soylent: a word processor with a crowd inside. In: Proceedings of the 23nd Annual ACM Symposium on User Interface Software and Technology, UIST 2010, pp. 313–322. ACM, New York (2010)

[5] Minder, P., Bernstein, A.: How to translate a book within an hour: towards general purpose programmable human computers with crowdlang. In: WebScience 2012, Evanston, IL, USA, pp. 209–212. ACM (June 2012)

[6] Doan, A., Ramakrishnan, R., Halevy, A.Y.: Crowdsourcing systems on the worldwide web. Commun. ACM 54(4), 86–96 (2011)

[7] Little, G., Chilton, L.B., Goldman, M., Miller, R.C.: Turkit: tools for iterative tasks on mechanical turk. In: HCOMP 2009, pp. 29–30. ACM (2009)

[8] Kochhar, S., Mazzocchi, S., Paritosh, P.: The anatomy of a large-scale human computation engine. In: HCOMP 2010, pp. 10–17. ACM (2010)

[9] Ahmad, S., Battle, A., Malkani, Z., Kamvar, S.: The jabberwocky programming environment for structured social computing. In: UIST 2011, pp. 53–64. ACM (2011)

[10] Marcus, A., Wu, E., Madden, S., Miller, R.C.: Crowdsourced databases: Query processing with people. In: CIDR 2011, pp. 211–214 (January 2011), www.cidrdb.org

[11] (OMG), O.M.G.: Business process model and notation (bpmn) version 2.0. Technical report (January 2011)

[12] Wang, J., Kumar, A.: A framework for document-driven workflow systems. In: van der Aalst, W.M.P., Benatallah, B., Casati, F., Curbera, F. (eds.) BPM 2005. LNCS, vol. 3649, pp. 285–301. Springer, Heidelberg (2005)

[13] Nigam, A., Caswell, N.: Business artifacts: An approach to operational specification. IBM Systems Journal 42(3), 428–445 (2003)

[14] Marcus, A., Wu, E., Karger, D., Madden, S., Miller, R.: Human-powered sorts and joins. Proc. VLDB Endow. 5(1), 13–24 (2011)

[15] Kazai, G., Kamps, J., Milic-Frayling, N.: An analysis of human factors and label accuracy in crowdsourcing relevance judgments. Inf. Retr. 16(2), 138–178 (2013)

[16] Little, G., Chilton, L.B., Goldman, M., Miller, R.C.: Exploring iterative and parallel human computation processes. In: Proceedings of the ACM SIGKDD Workshop on Human Computation, HCOMP 2010, pp. 68–76. ACM, New York (2010)

[17] Lin, C.H., Mausam, Weld, D.S.: Crowdsourcing control: Moving beyond multiple choice. In: UAI, pp. 491–500 (2012)

[18] Venetis, P., Garcia-Molina, H., Huang, K., Polyzotis, N.: Max algorithms in crowdsourcing environments. In: WWW 2012, pp. 989–998. ACM, New York (2012)

[19] Nowak, S., Rüger, S.: How reliable are annotations via crowdsourcing: a study about inter-annotator agreement for multi-label image annotation. In: Proceedings of the International Conference on Multimedia Information Retrieval, MIR 2010, pp. 557–566. ACM, New York (2010)

[20] Dempster, A.P., Laird, N.M., Rubin, D.B.: Maximum likelihood from incomplete data via the em algorithm. Journal of the Royal Statistical Society 39(1), 1–38 (1977)

[21] Davidson, S.B., Khanna, S., Milo, T., Roy, S.: Using the crowd for top-k and group-by queries. In: Proceedings of the 16th International Conference on Database Theory, ICDT 2013, pp. 225–236. ACM, New York (2013)

[22] Adomavicius, G., Tuzhilin, A.: Toward the next generation of recommender systems: a survey of the state-of-the-art and possible extensions. IEEE Transactions on Knowledge and Data Engineering 17(6), 734–749 (2005)

[23] Alonso, O., Rose, D.E., Stewart, B.: Crowdsourcing for relevance evaluation. SIGIR Forum 42(2), 9–15 (2008)

SmartComposition: A Component-Based Approach for Creating Multi-screen Mashups

Michael Krug, Fabian Wiedemann, and Martin Gaedke

Technische Universität Chemnitz, Germany
{firstname.lastname}@informatik.tu-chemnitz.de

Abstract. The spread and usage of mobile devices, such as smartphones or tablets, increases continuously. While most of the applications developed for these devices can only be used on the device itself, mobile devices also offer a way to create a new kind of applications: multi-screen applications. These applications run distributedly on multiple screens, like a PC, tablet, smartphone or TV. The composition of all these screens creates a new user experience for single as well as for several users. While creating mashups is a common way for designing end user interfaces, they fail in supporting multiple screens. This paper presents a component-based approach for developing multi-screen mashups, named SmartComposition. The SmartComposition approach extends the OMELETTE reference architecture to deal with multiple screens. Furthermore, we enhance the OMDL for describing multi-screen mashups platform independently. We draw up several scenarios that illustrate the opportunities of multi-screen mashups. From these scenarios we derive requirements SmartComposition needs to comply with. A huge challenge we face is the synchronization between the screens. SmartComposition solves this through real-time communication via WebSockets or Peer-to-Peer communication. We present a first prototype and evaluate our approach by developing two different multi-screen mashups. Finally, next research steps are discussed and challenges for further research are defined.

Keywords: Mobile, distributed user interface, distributed displays, multi-screen applications, web applications, mashup, widgets.

1 Introduction

Internet-enabled devices are not anymore limited to computers and laptops. Devices, like smartphones, tablets or TVs, also offer users access to the Internet [8]. The number of mobile devices sold in 2013 exceeds the numbers of computers and laptops by factor three [10]. These new devices as well as the new capabilities of the Internet enable applications that can be used in several areas, such as entertainment, communication or productivity. While most of these applications can only be used on the device itself, mobile devices offer a way to create a new kind of applications: multi-screen applications. Multi-screen applications run distributedly on different devices, like a PC, smartphone, tablet or TV. The

S. Casteleyn, G. Rossi, and M. Winckler (Eds.): ICWE 2014, LNCS 8541, pp. 236–253, 2014.

composition of these multiple screens creates a new user experience. Multi-screen applications are suitable for single as well as several users.

A particular kind of multi-screen applications are multi-screen mashups. These multi-screen mashups are also a special type of user interface mashups (UI mashups), the importance of which has increased significantly within the last years [1]. A mashup consists of several widgets that offer a limited functionality. By combining and aggregating these widgets more complex tasks can be solved. While each widget can work by itself, it has to be extended to support the communication with other widgets inside the mashup. Examples for these UI mashups are platforms like iGoogle or Yahoo! Pipes. The compositional way UI mashups are built eases the development of new rich web applications. Current approaches regarding UI mashups focus on applications that run on a single screen. They offer the opportunity to easily create end user interfaces [3] on different devices but not across these devices. Although the EU FP7 project OMELETTE has focused on telco service mashups, it proposed a reference architecture for UI mashups in [6,18]. This architecture facilitates the deployment of UI mashups to desktop as well as mobile devices, but cannot distribute the UI across several devices. Another outcome of this EU project was the Open Markup Description Language (OMDL) [18] which supports the process of designing, evolving and deploying UI mashups, but is also limited to single-screen mashups. Thus, we extend the OMELETTE reference architecture and the OMDL to deal with multi-screen mashups.

Developing web applications is a methodical process our approach needs to deal with. Therefore, aspects of software and web engineering have to be considered. The WebComposition approach [4] describes a way to design and develop web applications based on reusable components. These components can be exchanged or reused within other web applications. The purpose of this paper is to propose an approach to support end users in creating multi-screen mashups. Therefore, we want to extend the UI mashup approach to support multiple screens. We present an architecture that supports and eases the development process of multi-screen mashups.

The rest of the paper is organized as follows: In Section 2 we describe three scenarios to illustrate the use cases we focus on. From these scenarios we derive challenges that a solution has to deal with in Section 3. The SmartComposition approach is proposed in Section 4. We describe our extension of the OMELETTE reference architecture and the OMDL. A first prototype which follows the SmartComposition approach is demonstrated in Section 5. In Section 6 we evaluate our approach. We discuss related work in Section 7. Finally, we provide a conclusion and give an outlook towards future work in Section 8.

2   Scenarios

In this section we present three scenarios that illustrate where multi-screen applications can create a benefit for users. Using these scenarios we want to show what users could expect from such applications and what challenges exist that need to be solved.

Scenario 1. The first scenario focuses on media enrichment using multiple screens. While consuming a documentary on television, the user, called Amy, wants to see additional content about the currently watched show. This additional content could be a related article on Wikipedia, images from online photo services, like Google or Flickr, or where to buy items that were shown in the video. For easing the reading or enabling interaction with the additional content, it will be displayed on Amys smartphone or tablet, while she is still watching the documentary on her television. Figure 1 illustrates this scenario. While watching a documentary about the United Nations (cf. 1) the location of the UN headquarters in New York is shown on a map on Amys smartphone (cf. 2) as soon as the video reaches the scene where it is presented. In the same way other kinds of information, which are directly related to the current scene, are provided. Amy can easily decide what types of additional content she wants to receive. For deleting this information two use cases are applicable. First, the information about the UN headquarters' location will be removed from Amy's smartphone, if the documentary does not talk anymore about it. Second, by clicking on the information Amy can prevent the automatic removing of the information.

Fig. 1. Mockup illustrating media enrichment in Scenario 1

Scenario 2. The second scenario addresses the enhancement of a presentation by distributing additional information to multiple secondary screens. As an example, a professor is giving a talk in front of many students. While showing slides of his presentation via a projector, the students receive further information on their mobile devices. Thus, the students can read and see details about the current topic without messing up the slides with a lot of text. Furthermore, the students can use the additional information on their laptops as a script. The professor can keep his slides clean, while the students get a synced script related to the current lecture. While the professor is authenticated as a lecturer, he can broadcast slides and information. This functionality is denied to the students who do not have this privilege.

Scenario 3. Our third scenario aims at collaborative work on multiple screens. Therefore, we want to reuse a scenario described by Husmann et al. in [9]. In this scenario "two users, Alex and Bill, are planning a mountain biking trip at home in their living room". While Alex is planning the mountain biking route on her tablet, Bill is searching for railway connections on his smartphone. Both are using the television in their living room to share the results of their tasks on a larger display. When Alex chooses a start point for their mountain biking trip, this start point will be pushed to Bill's smartphone, where it is automatically inserted in the input field for the destination of his railway connection query. Bill can choose his favorite railway connections to be displayed on the television. Alex' chosen mountain biking route will be displayed as a map on the television next to the railway connection information.

3   Analysis

We use the presented scenarios in Section 2 to derive objectives our approach needs to deal with for creating multi-screen mashups. First, we begin with some basic findings. In all scenarios there is always a common context shared between all participating screens. This context has to be maintained. We identified two possible screen constellations. In the first one a primary screen publishes information to all secondary screens (cf. Scenarios 1 and 2). The secondary screens are only consuming and displaying information. An interaction with the information is possible, but without reflection to the original publisher. In the second constellation all screens are in an equal role (cf. Scenario 3). All of the participants are publishing and consuming information. A more interactive and collaborative usage of the application is possible. This setup requires a more complicated synchronization mechanism. Furthermore, there are also two scenario types regarding the number of users: single-user (cf. Scenario 1) and multi-user (cf. Scenarios 2 and 3). Both will require a different application design.

Based on the scenarios in Section 2, we now state several objectives that our approach should fulfill.

Simplicity of Mashup Creation. We targeted several scenarios in Section 2 where multi-screen mashups can provide a suitable solution. Thus, our approach should enable end users to easily create these multi-screen mashups for their specific tasks. This includes the reuse of developed widgets as well as the easy configuration of a set of devices which interact as a multi-screen. Inspired by the WebComposition approach from Gaedke et al. in [4,5] we apprehend widgets as loosely-coupled components, which were assembled in a mashup to a new application. Furthermore, the end users should not fiddle with designing UI mashups for mobile use only. The process of creating a multi-screen mashup is independent of the targeted device which can be a mobile as well as a PC or laptop.

Support for Multi-screen Usage. The focus of our proposal is the development of multi-screen mashups. Therefore, our approach should provide functionality to develop applications that can be used with more than one screen or device. Thus, there is a need for unique identification and data synchronization. The types of the participating screens should be detected and used to deliver an according user interface. While there can be a lot of devices which use the system, they should not be connected to any other available screen. This objective implies that devices can be grouped to so called workspaces and cannot communicate with devices in different workspaces. As we described in Scenarios 1 to 3 the support for multiple screen is the central point of our approach.

Support for Real-Time Communication and Synchronization. To propagate information without noticeable delay between server and client a component for real-time communication and synchronization should be available. We define the communication as real-time, when the latency of the transmission of a message from one widget to another one is 50 milliseconds or less [15]. That ensures no noticeable delay when distributing information across one screen as well as multiple screens. Furthermore, the components on one screen must be able to communicate among themselves.

Collaboration Support. The applications to develop should not only be used by a single user. We want to support the development of multi-screen multi-user mashups. Thus, our approach should provide the basis for collaborative work, such as user identification and data synchronization. This is highly related to the real-time synchronization requirement. Scenario 3 illustrates this objective best, while Scenario 2 also partially requires some support in collaboration.

Authentication. Nowadays it is important to provide a personalized user experience as well as social interaction. Thus it is required to offer the users a way to authenticate themselves. Using protocols like WebID even an authentication without requiring a password is possible. A user and role based identity management would provide a flexible way to handle most scenarios. The usage of multiple different devices creates a challenge to provide user-friendly login mechanisms for different use cases.

After we specified our objectives, we propose our own approach within the next section.

4 The SmartComposition Approach

To fulfill the requirements gathered from the analysis we propose a component-based approach to develop multi-screen mashups, called SmartComposition. We extend the OMELETTE reference architecture [18] for fitting the requirements in Section 3. In the following paragraphs we describe the SmartComposition architecture, which is depicted in Figure 2. We highlight our significant changes related to the OMELETTE reference architecture with underlined text. Our proposal consists of four parts: SmartComposition Runtime Environment, Multi-Screen Workspace, Information Store, SmartScreen. These parts interact like described in the following.

The SmartComposition Runtime Environment runs a multiple instances of a mashup. Thus, it handles multiple Multi-Screen Workspaces which are separated and secured against other existing Multi-Screen Workspaces. To enable the security between the Multi-Screen Workspaces we introduce the Session-Handler.

The Session-Handler knows all running workspaces and instantiates the Multi-Screen Workspaces. Each Multi-Screen Workspace contains a Workspace Manager.

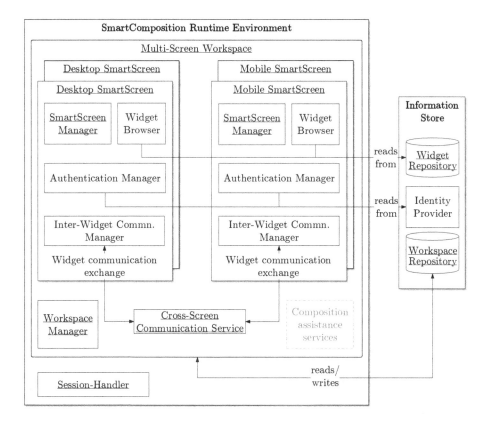

Fig. 2. Architecture of the SmartComposition approach

The Workspace Manager handles the workspace of a single user as well as several users who use the mashup in a collaborative way. Furthermore, the Workspace Manager is responsible for initialization of the SmartScreens and for the state management of the whole workspace. The illustrated component composition assistance services [16] in Figure 2 are not covered by our approach yet. Therefore, a workspace is defined as a set of one or more SmartScreens, which share the same information space and can communicate among themselves. Thus, the simplest form of a workspace consists of a single SmartScreen. A workspace is identified by a workspace ID and described by its workspace configuration. The configuration contains information about the SmartScreens which are connected, the widgets on each screen and additional properties. We use the OMDL as the basis for our configuration notation. Therefore, we extend the OMDL to support multi-screen aspects. How we did this and what changes we made is described later in this section.

The inter-widget communication is a highly important topic when developing mashups [1]. Our approach does not improve the inter-widget communication by itself but extends it for multi-screen mashups. On a single SmartScreen we use the publish-subscribe pattern for transmitting messages from one widget to one or several other widgets. For supporting inter-widget communication across multiple screens we extend the inter-widget communication component to publish the message to other connected SmartScreens (cf. Figure 3). That is, if a widget publishes a message on a certain topic, the inter-widget communication component publishes this message to all widgets on the same SmartScreen that are subscribed on the same topic. Furthermore, the message is transmitted to a Cross-Screen Communication Service where the message is sent to all connected SmartScreens within the same session. The assignment of SmartScreens and the workspace session they belong to is done by the Session-Handler. This ensures that a SmartScreen can only communicate with other SmartScreens of the same workspace. On each SmartScreen the inter-widget communication component publishes the message also to the subscribed widgets on their SmartScreen. The Cross-Screen Communication Service can work in two modes: broadcast mode and multi-layered publish-subscribe mode. In broadcast mode the Cross-Screen Communication Service sends every incoming message of one SmartScreen to all other SmartScreens of the same workspace. However, in the multi-layered publish-subscribe mode the inter-widget communication component of each SmartScreen subscribes to topics at the Cross-Screen Communication Service. The topics to subscribe depend on the topics that the widgets subscribe to on their SmartScreen. After publishing a message from a widget to the inter-widget communication component of its SmartScreen the message will be published to all subscribed widgets. Furthermore, the message will be published to the Cross-Screen Communication Service. There, the message will be published to inter-widget communication component of the SmartScreens, if they are subscribed to the topic. Each inter-widget communication component then publishes the message to the subscribed widgets.

As defined before, a workspace consists of SmartScreens. A SmartScreen is an abstract representation of a web browser window. It provides the runtime environment for all client-side components. Each SmartScreen has its own identifier. This identifier has to be at least unique in the workspace. This is important for the communication between the screens. The SmartScreen Manger within the SmartScreen initializes the widgets, handles the state management and provides access to device-specific properties. It detects the type and resolution of the device and uses this information to adapt the presentation. Thus, it is responsible for composition of the user interface.

The information store is the part which is dealing with all required and available persistent data. It consists of three components: the widget repository, the identity provider and the workspace repository.

Described in the OMELETTE reference architecture in [18] the widget repository is part of the SmartComposition Runtime Environment, but there are also meta data about the widgets available in the information store. Thus, we decided

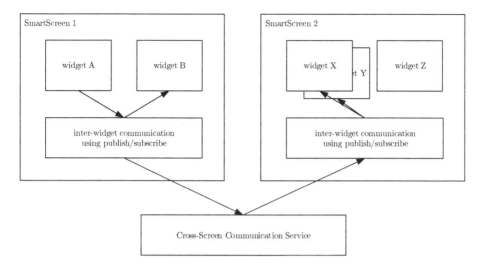

Fig. 3. Cross-screen message flow of inter-widget communication

to combine both parts and move the widget repository to the information store. There it is responsible for delivering the executables as well as providing meta data about the widgets. The end user can search for a specific widget based on the name or id of the widget. Furthermore, the end user can discover widgets based on their meta data. That is, a widget can be selected by the messages it can consume or the way it presents the proposed information, such as a map widget which shows given coordinates on a map excerpt. The widget repository is also responsible for delivering the executables to the SmartComposition Runtime Environment. In case of W3C widgets [19] the executables contains HTML-, CSS- and JavaScript-files which can be delivered to the SmartScreen without much effort.

The identity provider offers information about the current user or users. It contains endpoints for verifying the user's identity. While our approach enables using different identity management concepts, we propose the usage of WebID [17]. When using WebID the identity provider gives access to the users' WebID profiles. The supplied information about the authenticated user can also be used for accessing resources from the widget repository as well as the workspace repository [7]. Closely related to the identity provider is the Authentication Manager within the SmartScreens.

Therefore, the Authentication Manager requests the users WebID certificate. The WebID URI is extracted from the Subject Alternative Name of the WebID certificate. Then, the Authentication Manager requests the WebID profile from the Identity Provider and verifies that the public key of the users WebID certificate is equal to the public key in the corresponding WebID profile. This is the authentication part within the WebID flow described in [17]. Afterwards, based on the users WebID, authorization could be granted. Based on the authenticated user access to restricted resources can be permitted or denied, such as the users workspace or some special widgets, which are limited to specific users. To ease the access to a users roles, the Authentication Manager offers an interface to request the users WebID profile for personal information, like name, birthdate and more. Furthermore, the users roles can be requested by other widgets or core components of the SmartComposition Runtime Environment. However, the users relationships can be queried without requesting the users WebID profile again.

The workspace repository stores and offers predefined or user-specific Multi-Screen Workspaces. Predefined workspaces will be used if there is no user-specific workspace existing. Thus, our approach enables the developer to define the default appearance of a single SmartScreen. Defining a default workspace with more than one SmartScreen is not useful, because the developer cannot predict with how many devices the user will use the application. However, it is possible to define multiple different default workspaces regarding to context specific information, such as the users role, device or geo-location. Within the default workspace a set of widgets and their position on the SmartScreen is defined.

When a user or several users are using the multi-screen web application, it could be necessary to save the current Multi-Screen Workspace. The users workspace will be saved by storing the OMDL description of the workspace in the workspace repository. While a Multi-Screen Workspace can contain multiple SmartScreens, it is at least necessary that one device wants to restore its SmartScreen as part of the workspace. Therefore, the device needs to save at least the workspace identifier within its client-side storage. The workspace repository offers an interface where the OMDL description corresponding to a given identifier is responded. From this OMDL description the device can extract its workspace configuration and can so restore the widgets on this SmartScreen. When another device of this workspace wants to restore its SmartScreen, it also sends the workspace identifier and retrieves the OMDL description of the workspace where it can extract its SmartScreen. On restoring the second or another SmartScreen the communication channel between these SmartScreen will be established by the Cross-Screen Communication Service.

4.1   Extension of OMDL

We use the OMDL for describing our Multi-Screen Workspaces and for storing them in the workspace repository. While the OMDL defines three levels for describing mashups [18], we just use the Physical Level for the SmartComposition approach. However, the defined Physical Level needs to be extended to support multiple screens, because the OMDL has no support for distributed mashups. The OMDL uses XML documents to describe mashups, their layout and the apps they contain. The root element of an OMDL document is the workspace. In the SmartComposition approach we apply the OMDL workspace to theMulti-Screen Workspace described above.

While the OMDL does not support multiple screens yet, we extend the vocabulary of the OMDL by adding several elements. The first new element is the SmartScreen which is added within the workspace element. The SmartScreen element is used to define a SmartScreen within a Multi-Screen Workspace. Thus, there can be multiple SmartScreen elements. The SmartScreen element has an attribute id, which is unique and identifies the SmartScreen. Furthermore, the SmartScreen element has a child element named type, which defines the type of the device the SmartScreen runs on. For dealing with collaborative multi-user workspaces we define a user element within the SmartScreen element, which refers to the authenticated users WebID URI.

```
1   <workspace xmlns="http://omdl.org/">
2   <identifier>http://example.org/workspacerepo/ws1</identifier>
3   <title>Media</title>
4   <date>2012-07-03T14:23+37:00</date>
5   <SmartScreen id="phone1">
6     <type>Smartphone</type>
7     <user>http://example.org/people/alice.rdf#me</user>
8     <grid><height>4</height><width>2</width></grid>
9   </SmartScreen>
10  <SmartScreen id="pc1">
11    <type>PC</type>
12    <user>http://example.org/people/bob.rdf#me</user>
13    <grid><height>6</height><width>10</width></grid>
14  </SmartScreen>
15  <app id="http://example.org/workspacerepo/ws-of-alice/1">
16    <SmartScreen>phone1</SmartScreen>
17    <type>MAP</type>
18    <link rel="source" href="http://example.org/repo/w1"
19      type="application/widget"/>
20    <position><x>0</x><y>0</y></position>
21    <size><height>2</height><width>2</width></size>
22  </app>
23  <app id="http://example.org/workspacerepo/ws-of-alice/2">
24    <SmartScreen>pc1</SmartScreen>
25    <type>LIST</type>
26    <link rel="source" href="http://example.org/repo/w2"
27      type="application/widget"/>
28    <position><x>0</x><y>0</y></position>
29    <size><height>5</height><width>2</width></size>
30  </app>
31  </workspace>
```

Listing 1.1. Example of a workspace description in OMDL

Each SmartScreen has an individual fine grained grid to arrange the widgets. To store the position of the widgets the SmartScreen element has a child element named grid. This element has two child elements height and width which represent the number of rows and columns the grid has. Another element we adapted is the app element. In OMDL this element describes a part of the mashup. In the

SmartComposition approach we define an app as a widget. While the OMDL has no multi-screen support yet, we also need to extend the app element to define on which SmartScreen the widget is located. Therefore, we introduce a SmartScreen element within the app element. This SmartScreen element contains the id of the containing SmartScreen. Furthermore, we assume that the type element of the app can be extended by several types which are required for a specific multi-screen web application. However, the alignment of apps in OMDL does not fit our approach, which focuses on a finer grained positioning. Therefore, we extended the position element by two new elements which are x and y. These elements represent the position of the widget regarding to a fine grained grid on the SmartScreen, where x means the distance to the left side and y means the distance to the top. Furthermore, we added a new element to the app element named size. This element has two child elements height and width, which mean the size of the widget regarding to fine grained grid on the SmartScreen. An example of a SmartComposition workspace described in OMDL can be seen in Listing 1.1.

5   Prototype

To validate the SmartComposition approach we have implemented a first prototype, which is based on another work we recently presented at the ICWE 2013 [12,14] and WWW 2014 [13]. The prototype demonstrates the Scenarios 1 and 2 we described in this paper (cf. Figure 4). All the client-side components we describe are implemented in JavaScript running as a web application in a web browser.

To achieve a multi-screen experience we implemented the following components. We created a SmartScreen class. This class works as described in our approach and is the host for the widgets. For our user interface components we implemented a basic widget class which provides the interface and basic functionality. This basic class is used to derive various new widget types using prototype-based inheritance. We have implemented widgets that can display a map, images from different web sources, excerpts from Wikipedia, text, translations or tweets. The widgets can be added and removed on runtime. The user can arrange them in a grid-based layout by using drag-and-drop.

To handle the information exchange between the widgets we implemented a component as part of the inter-widget communication. This component provides the publish-subscribe pattern and thus offers loosely coupled communication. Once a widget is added to the SmartScreen it can subscribe to one or more topics on which events are published.

The workspace configuration of the SmartScreens is currently stored on the client-side using the HTML5 feature LocalStorage. Using the proposed workspace identifier the users customized arrangement can be restored. A workspace repository is not yet implemented.

To make the prototype work on multiple screens we extended the inter-widget communication component with a synchronization mechanism. This mechanism uses WebSockets to propagate the events which were published on one

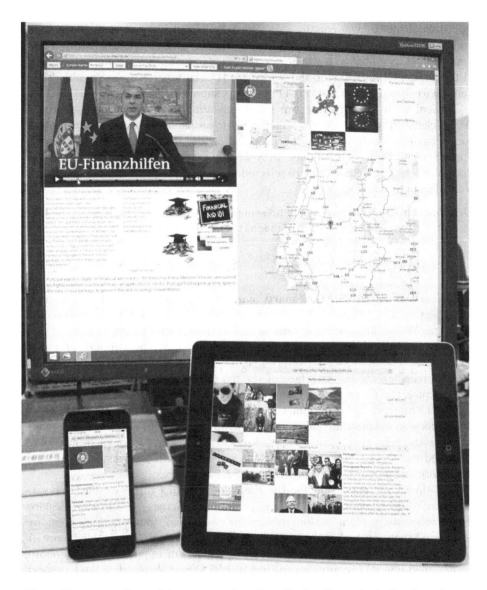

Fig. 4. Prototype of a multi-screen mashup described in Scenario 1. On the primary screen (top) a german news cast is played. Meta data from this video are published to all widgets within the workspace. Additional information are displayed on the different screens, such as Google Maps on the primary screen, Flickr images on the iPad (bottom right), and Twitter feed on the iPhone (bottom left).

SmartScreen to the other connected screens. This extension also handles the re-publishing of received events to the widgets. Thus, the whole communication is synchronized with all connected SmartScreens. This approach assures that each client behaves equally regardless of where the event was published. To make that work we also had to implement a server-side component. Once a new SmartScreen is added to a workspace the RTC components registers the screen at the Session-Handler using the workspace ID, the SmartScreens own identifier and device specific information. This data is supplied to the other participants.

To deal with the Cross-Screen Communication Service we implemented a WebSockets server that provides methods for propagating information to other connected clients. It handles the propagation of the events it receives to all SmartScreens that are combined in a workspace. Using the workspace identifier the related SmartScreens are selected. This component is implemented as a Node.JS application.

Demonstration. The prototype presented in this paper is available for testing at: http://vsr.informatik.tu-chemnitz.de/demo/chrooma/icwe14/

6   Evaluation

In this section we evaluate the SmartComposition approach. Based on the analysis in Section 3, we examine how the approach assists in developing multi-screen mashups. We outline our expectations and explain actual findings.

A big challenge our approach has to deal with is the support of multiple screens. Therefore, we introduced the concept of workspaces where several SmartScreens create the interface of the mashup. The introduced parts of our architecture, the Cross-Screen Communication Service and the Session-Handler, enable the inter-widget communication across multiple screen (cf. Figure 2). This two parts also ensure that only screens within the same workspace can communicate with each other. Thus, the workspaces are isolated from each other. We extended the OMDL to support multiple screens within a workspace and used it to store and restore the configuration of a workspace.

Real-time communication for the SmartScreens has to be enabled for synchronization and communication of multiple SmartScreens within a workspace. While web systems often use long-polling for nearly real-time communication, we considered that new technologies, such as WebRTC or WebSockets, which are introduced with HTML5, might better correspond to our requirements. Therefore, we designed our RTC component using WebRTC and WebSockets and support long-polling as a fallback mechanism for browsers, which do not support these new technologies. We evaluate the performance of the RTC component in our prototype by measuring the duration of transporting an information from one widget to another. A special widget was developed, which publishes a timestamp to a second widget. The second one re-publishes the timestamp to its source. After receiving the timestamp on the first widget the difference between the current timestamp and the received one is calculated and then halved. We examined different constellations. The first one was where both widgets were

Table 1. Transport duration of a message from one widget to another in milliseconds

	Average	Standard Deviation
Same SmartScreen	5,121	0,654
Same PC	4,100	0,921
PC - iPad	38,735	26,842
PC - PC (WiFi)	26,668	19,036
PC - PC (Wired)	3,494	1,254

on the same SmartScreen. Another one was where the widgets were on different SmartScreens which run on the same PC. The third constellation was one SmartScreen on a PC and the other on a WiFi-connected iPad. The fourth one was two SmartScreens on two PCs which are connected via WiFi. Two PCs are connected via wire in the last constellation. The statistical values of our evaluation can be seen in Table 1. As we proposed in Section 3 the real-time communication is satisfied as the transport of a message from one widget to another is less than 50 ms. Our findings show that the largest latency occurs at the communication between a widget on the PC and a widget on the iPad with 38,735 ms. We can argue the difference in the average communication between the same SmartScreen and two SmartScreens on the same PC with the multi-threading architecture of the PC. That is, with two SmartScreens in two browser windows enables more parallelism in execution, because each browser window can be run in a different thread. The high standard deviation in the communication via WiFi (PC-iPad and PC-PC) can be ascribed to some side effect caused by the WiFi. To ensure that the higher latency when using WiFi does not is caused by constant term we give the relation of the IP-latency measured by ping in Table 2.

Table 2. Ping latency between two PCs in milliseconds

	Average	Standard Deviation
PC - PC (WiFi)	3,800	1,279
PC - PC (Wired)	1,025	0,798

We evaluate the simplicity of creation by developing two mashups for Scenario 1 and Scenario 2. In Scenario 1 we developed some new widgets for displaying a video and processing the meta data, such as the given subtitles. Other widgets could be reused in both scenarios, like a Wikipedia-widget which is displaying an article from Wikipedia or a GoogleMaps-widget which shows a map excerpt. Since the separation of the SmartScreens into their workspaces is an essential part of our approach, the end user has not to deal with the configuration of the workspaces. A mashup created based on our approach runs on PCs, laptops, smartphones, tablets as well as on web-enabled TVs.

For supporting authentication we propose a two tier approach The Authentication Manager within the SmartComposition Runtime Environment deals with the process of authentication and offering interfaces for accessing roles and permissions. The Identity Provider in the Information Store gives access to the users' profile and personal data, such as name, birth date or e-mail address. The usage of WebID offers the opportunity to let users easily login with their different devices. They can use a WebID certificate on each device which refers to the same WebID profile of the user.

Performing collaborative tasks is facilitated by multi-user workspaces and real-time communication. For collaborative work on a multi-screen web application we extended the OMDL to define a specific user for each SmartScreen. Thus, a device, which is authenticated by the users WebID, receives the corresponding workspace and can load the SmartScreen associated to the device. This enables several users to use the same workspace. Utilizing the real-time communication functionality of the RTC component synchronization of different users SmartScreens can be achieved.

7 Related Work

Our approach is closely related to work in three research domains: distributing and migrating Web UIs, data mashups, and classic user interface mashups (UI mashups). The DireWolf approach proposed in [11] enables the distribution and migration of Web widgets between multiple devices. While our approach focuses on developing multi-screen mashups from scratch, the DireWolf approach extends the functionality of single-device mashups to run across multiple devices. The framework is involved in every layer of a widget-based Web Application: the widget itself, the client browser, the backend service and the data storage. DireWolf manages communication between widgets on one device as well as between widgets on multiple devices. It also extends the functionality of common widget spaces with a shared application state. While our approach uses HTML5 features, such as WebSockets, for communication the DireWolf approach uses the XMPP protocol and its publish-subscribe extensions. That is, the DireWolf approach uses an additional layer within the communication architecture, which increases the effort for maintaining and exchanging parts of this communication layer. Furthermore, the usage of XMPP can increase the incompatibility when using mobile devices inside restricted GSM networks.

MultiMasher is a visual tool to create multi-device mashups from existing web content [9]. It aims at designing mashups without the need for modeling or programming. The MultiMasher runs in the browser, connects to the MultiMasher server and provides the user with a toolbar. After the user has connected his devices he can load any web site, which he can then mashup. UI elements can be selected visually and distributed (move or copy) to other connected devices. The mashups can be saved, loaded and reused as basis for other mashups. MultiMasher uses event forwarding to propagate changes to the connected clients and only displays the previously selected UI elements on each device. MultiMasher focuses on the creation of mashups and not on the creation of new applications.

The EU FP7 Project OMELETTE is closely related to our work, because it proposes a reference architecture for designing platforms for UI mashups [18]. They focus on end users with less or no programming skills to create their own mashups for fulfilling a certain goal. The outcome of this project was a prototype based on Apache Rave and Apache Wooky [2]. After a user log in, she can restore recent workspaces or create a new one by adding widgets to her workspace. The focus of the OMELETTE project was on the automatic composition of these mashups. Therefore, they invented some great strategies for suggesting widgets to add or for establishing the inter-widget communication between two widgets based on the user's behavior. While the OMELETTE reference architecture is a well-proven approach, it lacks in supporting UI mashups which run distributed across multiple screens. Thus, we extend the proposed reference architecture to support workspaces which include multiple screens. We also showed that concepts of inter-widget communication designed for a workspace on one screen works also for a workspace across multiple screens.

Mashup tools, like Yahoo! Pipes, offer end users developing data mashups. End user in this context means people with no or less development skills. While our approach focuses on user interface mashups, Yahoo! pipes is focusing on mixing popular data feeds to create data mashups via a visual editor. Therefore, it lets users transform, aggregate, transform and filter one or more data sources, such as RSS/Atom feeds or XML sources, and output this as a RSS feed. Thus, it enables end users to develop a kind of business logic for processing data [20]. However, Yahoo! Pipes does not offer a multi-screen environment where the created mashups can be executed.

8   Lessons Learned and Outlook

The SmartComposition approach proposed in this paper enables end users to easily create multi-screen mashups. We extended the OMELETTE reference architecture to deal with mashups across multiple screens. The Cross-Screen Communication Service extends classic inter-widget communication to work on a trans-screen level, while separating different workspaces from each other via the Session-Handler. Furthermore, our approach handles the challenge of real-time communication. The component-based architecture enables adding required and removing obsolete components. Thus, SmartComposition can be adjusted for different multi-screen web application scenarios. Both, the workspace repository as well as the widget repository offer a generalized access to widgets and workspaces.

Our future work will focus on integrating commonly used widget formats, such as W3C-widgets or Opera widgets. We also plan to evolve the cross-screen inter-widget communication by considering constraints given by several roles. That is, in a multi-user scenario, e.g., Scenario 3, one user is permitted to update a specific widget on a shared screen, while another user just has read-only access to the presented information. Another aspect we want to focus on in future work is the authorization of widgets to use the users' external services, such as Facebook or Google Drive.

As described in Section 4 the composition assistance services, such as automatic composer and workspace pattern recommender, are not yet a part of our approach for multi-screen mashups. We plan to adapt the approaches proposed by Roy Chowdhury in [16] for multi-screen mashups. Different use cases in this field of research are possible: The user gets a recommendation to add an additional screen, like a smartphone or a laptop, when she is using the mashup to accomplish her desired goal in a better way. When a user starts a new workspace, the widgets will be deployed to all available screens. This can be done by learning from the behavior of the users or by designing a default workspace.

Acknowledgment. This work was supported by the Sächsische Aufbaubank within the European Social Fund in the Free State of Saxony, Germany (Project Crossmediale Mehrwertdienste für die digitale Mediendistribution).

References

1. Chudnovskyy, O., Fischer, C., Gaedke, M., Pietschmann, S.: Inter-widget communication by demonstration in user interface mashups. In: Daniel, F., Dolog, P., Li, Q. (eds.) ICWE 2013. LNCS, vol. 7977, pp. 502–505. Springer, Heidelberg (2013)
2. Chudnovskyy, O., Nestler, T., Gaedke, M., Daniel, F., Fernández-Villamor, J.I., Chepegin, V., Fornas, J.A., Wilson, S., Kögler, C., Chang, H.: End-user-oriented telco mashups: the omelette approach. In: Proceedings of the 21st International Conference Companion on World Wide Web, pp. 235–238. ACM (2012)
3. Chudnovskyy, O., Pietschmann, S., Niederhausen, M., Chepegin, V., Griffiths, D., Gaedke, M.: Awareness and control for inter-widget communication: Challenges and solutions. In: Daniel, F., Dolog, P., Li, Q. (eds.) ICWE 2013. LNCS, vol. 7977, pp. 114–122. Springer, Heidelberg (2013)
4. Gaedke, M., Rehse, J.: Supporting compositional reuse in component-based web engineering. In: Proceedings of the 2000 ACM Symposium on Applied Computing, vol. 2, pp. 927–933. ACM (2000)
5. Gaedke, M., Turowski, K.: Specification of components based on the web-composition component model. In: Managing Information Technology in a Global Economy, p. 411 (2001)
6. Gebhardt, H., Gaedke, M., Daniel, F., Soi, S., Casati, F., Iglesias, C., Wilson, S.: From mashups to telco mashups: A survey. IEEE Internet Computing 16(3) (2012)
7. Hollenbach, J., Presbrey, J., Berners-Lee, T.: Using rdf metadata to enable access control on the social semantic web. In: Proceedings of the Workshop on Collaborative Construction, Management and Linking of Structured Knowledge (CK 2009), vol. 514 (2009)
8. Horizont: Report TV-Marketing, Ausgabe 17, p. 40 (April 2012), http://www.horizont.net/report
9. Husmann, M., Nebeling, M., Norrie, M.C.: MultiMasher: A visual tool for multi-device mashups. In: Sheng, Q.Z., Kjeldskov, J. (eds.) ICWE Workshops 2013. LNCS, vol. 8295, pp. 27–38. Springer, Heidelberg (2013)
10. IDC Coroporate USA: Tablet Shipments Forecast to Top Total PC Shipments in the Fourth Quarter of 2013 and Annually by 2015, According to IDC (2013), http://www.idc.com/getdoc.jsp?containerId=prUS24314413

11. Kovachev, D., Renzel, D., Nicolaescu, P., Klamma, R.: DireWolf - distributing and migrating user interfaces for widget-based web applications. In: Daniel, F., Dolog, P., Li, Q. (eds.) ICWE 2013. LNCS, vol. 7977, pp. 99–113. Springer, Heidelberg (2013)

12. Krug, M., Wiedemann, F., Gaedke, M.: Media enrichment on distributed displays by selective information presentation: A first prototype. In: Sheng, Q.Z., Kjeldskov, J. (eds.) ICWE Workshops 2013. LNCS, vol. 8295, pp. 51–53. Springer, Heidelberg (2013)

13. Krug, M., Wiedemann, F., Gaedke, M.: Enhancing Media Enrichment by Semantic Extraction. In: Proceedings of the Companion Publication of the 23rd International Conference on World Wide Web Companion, pp. 111–114. International World Wide Web Conferences Steering Committee (2014)

14. Oehme, P., Krug, M., Wiedemann, F., Gaedke, M.: The chrooma+ approach to enrich video content using html5. In: Proceedings of the 22nd International Conference on World Wide Web Companion, pp. 479–480. International World Wide Web Conferences Steering Committee (2013)

15. Pantel, L., Wolf, L.C.: On the impact of delay on real-time multiplayer games. In: Proceedings of the 12th International Workshop on Network and Operating Systems Support for Digital Audio and Video, pp. 23–29. ACM (2002)

16. Roy Chowdhury, S., Chudnovskyy, O., Niederhausen, M., Pietschmann, S., Sharples, P., Daniel, F., Gaedke, M.: Complementary assistance mechanisms for end user mashup composition. In: Proceedings of the 22nd International Conference on World Wide Web Companion, pp. 269–272. International World Wide Web Conferences Steering Committee (2013)

17. Sporny, M., Inkster, T., Story, H.: WebID 1.0: Web Identication and Discovery (2011), http://www.w3.org/2005/Incubator/webid/spec/

18. University of Trento: D2.2 - Initial Specification of Mashup Description Language and Telco Mashup Architecture. Tech. rep., University of Trento (2011)

19. W3C: Packaged Web Apps (Widgets) - Packaging and XML Configuration, 2nd edn. (2012), http://www.w3.org/TR/2012/REC-widgets-20121127/

20. Yu, J., Benatallah, B., Casati, F., Daniel, F.: Understanding mashup development. IEEE Internet Computing 12(5), 44–52 (2008)

SSUP – A URL-Based Method
to Entity-Page Discovery

Edimar Manica[1,2], Renata Galante[2], and Carina F. Dorneles[3]

[1] Campus Avançado Ibirubá - IFRS - Ibirubá, RS, Brazil
edimar.manica@ibiruba.ifrs.edu.br
[2] PPGC - INF - UFRGS - Porto Alegre, RS, Brazil
galante@inf.ufrgs.br
[3] INE/CTC - UFSC - Florianópolis, SC, Brazil
dorneles@inf.ufsc.br

Abstract. Entity-pages are Web pages that publish data representing one only instance of a certain conceptual entity. In this paper we propose **SSUP**, a new method to entity-page discovery. Specifically, given a sample entity-page from a Web site (e.g., `Jolyon Palmer` entity-page from `GP2` Web site) we aim to find all same type entity-pages (driver entity-pages) from this Web site. We propose two structural URL similarity metrics and a set of algorithms to combine URL features with HTML features in order to improve the quality results and minimize the number of downloaded pages and processing time. We evaluate our method in real world Web sites and compare it with two baselines to demonstrate the effectiveness of our method.

Keywords: entity-pages, structural similarity, URL features, HTML features.

1 Introduction

The Web contains an increasing number of Web sites that can be considered a repository of pages with valuable information about real world entities. These pages are called *entity-pages* (or object-pages). Weninger et al. [8] define an entity-page as a Web page that describes a specific entity. For example, a Web page that describes a GP2 driver or a city council member.

Making use of the data presented in the entity-pages is an opportunity to create knowledge useful to several real applications (such as: comparative shopping, vertical search, named entity recognition and query suggestion). For instance, when we type the query "`players of real madrid`" on Google[1], it shows a list with all the players of Real Madrid Football Club including the attributes: `name`, `position` and `image`. If we click in one player of this list, the original query is replaced by the player name. The data for this kind of suggestion can be extracted from Wikipedia[2]. However, the Wikipedia does not contain all entities.

---

[1] http://www.google.com/
[2] http://en.wikipedia.org/

S. Casteleyn, G. Rossi, and M. Winckler (Eds.): ICWE 2014, LNCS 8541, pp. 254–271, 2014.
© Springer International Publishing Switzerland 2014

For example, the council members of Natal (a Brazilian state capital) are not presented in Wikipedia, but there is an official Web site with an entity-page for each council member describing the attributes: `name`, `political party`, `image`, `phone`, among others. As in general the entity-pages in the same Web site share a common template, HTML patterns can be learned to extract their data.

One important problem in this context is how to discover the entity-pages of the same type in one Web site. This is not a trivial task due to the variety of Web sites and structures. Some sites contain a page with a link to all entity-pages, while others divide these links in pages organized in a hierarchical manner and others divide these links in discrete pages (pagination). Moreover, some entity-pages have more details about an entity while others have less influencing in the HTML structure of the entity-pages. On the other hand, how the data in the entity-pages can change and new entity-pages can become available, it is necessary to re-execute the method to entity-page discovery periodically. Then, the number of downloaded pages and the processing time are important metrics to choose an effective method.

We found some methods to entity-page discovery in the literature. Methods based on HTML-tables [7] or human-compiled encyclopedias [6] are very restrictive. The method proposed by Weninger et al. [8] is based on the HTML and visual features, but in order to analyze the visual information is necessary to render the page in a browser loading all images, css and javascripts, then increasing the processing time. On the other hand, the method proposed by Blanco et al. [2] uses only the HTML features, then collect the information required is faster, but being based on only one kind of feature makes the method very sensitive to the parameter configuration.

Our goal is to find the set of entity-pages of the same type given a sample entity-page minimizing the number of downloaded pages and the processing time. For example, given the entity-page about `Jolyon Palmer` on the `GP2` Web site, we intent to find the set of driver entity-pages on the `GP2` Web site. The main contributions of this paper can be summarized as follows: (i) two structural URL similarity metrics ($weaksim_{url}$ and $strongsim_{url}$); (ii) **SSUP**: a new method to entity-page discovery that combines URL features with HTML features; (iii) show through experiments that combining URL features with HTML features improves the quality results of entity-page discovery and decreases the number of downloaded pages and the processing time; (iv) create datasets from real World Web sites for experiments.

The rest of this paper is organized as follows. Section 2 discusses the related work. Section 3 specifies the key concepts at the basis of our method. Section 4 presents the two structural URL similarity metrics proposed and the **SSUP** method. Section 5 presents a set of experiments we have conducted to evaluate our method on real Web sites comparing it with two baselines. Section 6 concludes the paper and presents future directions.

2   Related Work

Yu et al. [9] uses an SVM-based method to classify Web pages according to features from the content and URL of a Web page. Blanco et al. [3] presents a method for structurally clustering Web pages using only the URLs of the Web pages and simple content features. Although both methods analyze the URL of the Web page, their assumptions are different from **SSUP**, since they have as input all Web pages of the Web site in order to classify or cluster them while **SSUP** wants to download the minimum number of Web pages as possible to discover the entity-pages. **SSUP** represents URL in a way very similar to Blanco et al. [3], but **SSUP** treats this representation using the principle of TF-IDF [1] while Blanco et al. [3] uses the principle of minimum description length [4].

Kaptein et al. [6] present a Wikipedia-based method for entity-page discovery searching the Wiki-page for a link to the entity's home page. Lerman et al. [7] present a method that relies on the common structure of many Web sites, which present information as a list or a table, with a link in each entry pointing to a detail page containing additional information about that item. **SSUP** differs from these methods because it depends neither on the specific HTML markup nor on a human-compiled encyclopedia. He et al. [5] focus on deep Web, then the entity-pages are found through HTML form submissions while **SSUP** focuses on surface Web, then the entity-pages are found through browsing the Web site by its hyperlinks.

Indesit [2] and GPP (Growing Parallel Paths) [8] aim to find the set of entity-pages of the same type given a sample entity-page. Indesit models Web pages in terms of the DOM-structure of the HTML of the Web page and measure the structural similarity between two Web pages with respect to this feature. GPP combines both DOM-structure of the HTML and visual information. SSUP combines both DOM-structure of the HTML and URL information that is more robust that consider only DOM-structure of HTML and less costly that consider visual information. We choose Indesit and GPP as our baselines in the experiments because they are the most similar methods to **SSUP** and they depend neither on the specific HTML markup nor on a human-compiled encyclopedia. It is important to note that we reuse the definitions of Indesit related to the HTML of the Web pages because our contribution is how to use the URL features and to combine them with HTML features. The use of HTML features to entity-discovery seems to be consolidated since it was proposed in Indesit and reused by GPP.

3   Definitions

Here, we specify the key concepts at the basis of our method. First, we define the basic structures of a Web site (Section 3.1). Finally, we define the basic structures of a Web page (Section 3.2).

3.1   Web Site

A Web site is a directed graph whose nodes correspond to the pages of the Web that exist within the same domain and the edges correspond to the hyperlinks. According to our intuition, a part of a Web site is designed to allow the user to navigate in it from homepage until the entity-pages through a logical hierarchy of topics. The pages that are part of this logical hierarchy are classified as entity-pages or index-pages. Entity-pages were defined in Section 1. Index-pages are pages with links to the entity-pages or with links to other index-pages, whose role is to group entity-pages/index-pages in order to allow the user to navigate through a logical hierarchy of topics.

Definition 1. *(Index page) Let p be a Web page in the Web site s, EP be the set of all entity-pages in s, sEP be a subset of EP, IP be the set of all index-pages in s, sIP be a subset of IP. p is an index-page if it contains at least one link for each entity-page in sEP (or index-page in sIP) and its functional role in s is group sEP (or sIP).*

Each index-page and entity-page has a degree of topic specificity that represents the number of attributes that the page restricts. Entity-pages have the highest degree of topic specificity, since always entity-pages restrict more attributes that their index-pages, i.e., they represent the more specific topic in the logical hierarchy.

Definition 2. *(Degree of topic specificity) Let an entity e with two attributes a_1 and a_2. Let $P_x[w]$ ($P_y[z]$) be an index-page with links to entity-pages that describe instances where the value of a_1 is x (y) and the value of a_2 is w (z). Let P_x be an index-page with links to index-pages with links to entity-pages that describe instances where the value of a_1 is x, then we say that $P_x[w]$ and $P_y[z]$ has the same degree of topic specificity (because they restrict the same number of attributes) and $P_x[w]$ has higher degree of topic specificity that P_x (because $P_x[w]$ restricts more attributes that P_x).*

Example 1. In Figure 1, `Page A` has the lowest degree of topic specificity restricting the value of the attribute `type` to `car`. The pages B, C and D, besides restricting the `type` attribute value, restricts the value of the `manufacture year` attribute, then they have a degree of topic specificity higher than `Page A`. The pages E, F, G, H, I and J have the highest degree of topic specificity because they are entity-pages.

The logical hierarchy of topics is a tree, named entity-tree, whose root is the index-page with the lowest degree of topic specificity, the leaves are the entity-pages and non-leaves are index-pages.

Definition 3. *(Entity-tree) Let Et be a tree where the node set and the edge set are subsets of that of a given Web Site. Et is an entity-tree if it satisfies the following properties: (1) all leaf nodes are entity-pages of the same type; (2) all leaf nodes are at the same level; (3) all non-leaf nodes are index-pages; (4) the*

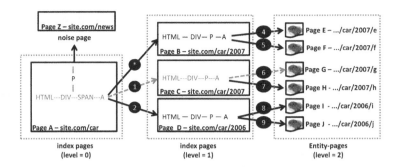

Fig. 1. Entity-tree example. The `Page Z` is highlighed because it is not part of the entity-tree.

root is the index-page with links to all index-pages of the next level (level+1); (5) all pages at a same level have the same degree of topic specificity; (6) the higher the level of a page the higher its degree of topic specificity; (7) all index-pages at a same level have links to their child nodes in a same link DOM path; and (8) all nodes at the same level of a same entity-tree share a similar URL and HTML structure.

Example 2. The illustration in Figure 1 describes an entity-tree with height two. The pages that are not part of the logical hierarchy of topics (e.g. `Page Z`) do not belong to the entity-tree and are referred as `noise pages`.

A Web Site can have more than one entity-tree to the same entity-type with a different topic structure. For example, a Web Site with software project entity-pages can be designed in such way that user may navigate through a logical hierarchy of licenses or programming languages. For each type of entity-pages available in a Web site must be at least one entity-tree.

In real Web sites usually occurs a situation that violates the property 5: "*all pages at a same level have the same degree of topic specificity*". This situation, called *pagination*, occurs when an index-page is pointed by an index-page of the same degree of topic specificity instead of an index-page of the immediately preceding degree of topic specificity (more general topic). In this case, we have a pagination page and need a pagination operation.

Definition 4. *(Pagination page) Let $p1$ be an index-page, $pp1$ be the parent node of $p1$, the Web page $p2$ is a pagination page if $p1$ and $p2$ have the same degree of topic specificity, $p1$ points to $p2$ and there is not an index-page of the same degree of topic specificity of $pp1$ pointing to $p2$.*

Definition 5. *(Pagination operation) Let $p1$ be an index-page, $pp1$ be the parent node of $p1$, and $p2$ be a pagination page pointed by $p1$, then pagination operation states that we need to remove the edge from $p1$ to $p2$ and add an edge from $pp1$ to $p2$.*

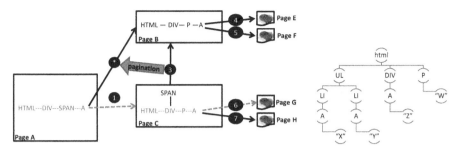

Fig. 2. Example of pagination operation

Fig. 3. Example of the DOM tree of a Web page

Example 3. The illustration in Figure 2 shows the pagination operation that generates the valid entity-tree presented in Figure 1. Originally the Web site had the edge 3 and did not have the edge *, violating the property 5. Then, after pagination operation, the edge 3 was removed and the edge * was added.

3.2   Web Page

We consider that the structure of a Web page is defined by its URL and HTML features. The URL feature is defined through a url-schema. A url-schema is the set of terms of the URL.

Definition 6. *(url-schema) A url-schema of a Web page p, denoted $\upsilon(p)$, is the set of terms of the URL of the p.*

In order to understand our concept of URL term is necessary to know that we see a URL as a sequence of substrings (called *tokens*) split by "/", "?" or "&" characters. Each token is a set of substrings (*sub-tokens*) split by non-alphanumeric characters, changing from letter to digit and vice versa. Then, a URL term is a sub-token associated with the position of the token that contains it. We do not consider the position of each sub-token in a token, since URLs of the Web pages with same entity-type can have different number of sub-tokens.

Definition 7. *(URL Term) Let T be a sequence of tokens of a URL u $(T_1, T_2, ..., T_n)$, where T_i occurs in u before T_{i+1}. Each token T_i is a set of sub-tokens $(S_i[1], S_i[2], ..., S_i[n])$, where $S_i[j]$ is the jth sub-token of the token T_i. Each sub-token $S_i[j]$ associated with i is a URL term. An additional URL term is the size of T.*

Example 4. Figure 4 shows an example of a URL, describing all tokens, the sub-tokens of the token T_3 and all URL terms. The additional URL term is highlighted.

The HTML feature is defined through a html-schema. An html-schema is the set of link DOM paths of the HTML.

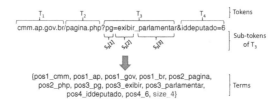

Fig. 4. Example of tokens, sub-tokens and URL terms

Definition 8. *(*html-schema*) A html-schema of a Web page p, denoted $\Delta(p)$, is the set of link DOM paths in p.*

In order to understand our concept of link DOM path is necessary to know that we see the HTML of Web pages as a Document Object Model (DOM) tree. Then, the link DOM path is a path from the root node to an anchor node.

Definition 9. *(*Link DOM Path*) Let p be a Web page, a link DOM path is a path through the DOM-tree of p that starts from the root and terminates into an anchor node.*

Example 5. The html-schema of the page presented in Figure 3 is $(HTML - UL - LI - A, HTML - DIV - A)$. The path $HTML - P$ does not belong to the html-schema because it does not terminate into an anchor node.

4   SSUP – A URL-Based Method to Entity-Page Discovery

The **SSUP** (**S**tructurally **S**imilar **U**RLs and **P**ages) is a method for entity-page discovery. Specifically, **SSUP** aims to find the set of same type entity-pages in a Web site given a sample entity-page.

We present an overview of our method using the Figure 1. **SSUP** starts with an entity-page as sample page (**Page G**) and find its index-page (**Page C**). Then, a new instance of our problem is recursively triggered using now the index-page found as sample page (finding **Page A** as index-page of the **Page C**). This process is performed until to find the root of the entity-tree (**Page A** in the case). Then, **SSUP** obtains a *sample page path* from root until the given entity-page (**Page A** -> **Page C** -> **Page G**). We call *sample page* each page from the sample page path. We call *sample link DOM path* each link DOM path that points to a sample page. The **Page C** is the sample page of the level 1 and "HTML-DIV-SPAN-A" is its sample link DOM path.

After, **SSUP** transverses the sample page path from root until the leaf (which is the sample entity-page) catching same level pages. To perform this, in each level x of the entity-tree, **SSUP** catches all pages pointed by the pages of that level and analyzes the structural URL and HTML similarity between these pages and the sample page in the level $x+1$ in order to discard pages that do not belong to the entity-tree. For example, consider Figure 1, analyzing the level 0, **SSUP** catches all pages pointed by `Page A`, including pages `C`, `D` and `Z`. Remember that `Page B` is actually pointed by `Page C`. **SSUP** analyzes the structural URL and HTML similarity between each page with the `Page C` (sample page of level 1) in order to prune noise pages (`Page Z`). By this time, we have pages `C` and `D` in the level 1. Then, **SSUP** needs to discover if these pages contain pagination pages. To perform this, **SSUP** catches all pages pointed by pages `C` and `D` that are structural URL and HTML similar to them (finding the `Page B` as a pagination page of the `Page C`). The process is performed to the next level catching all pages pointed by the pages `B`, `C` and `D`, analyzing the structural URL and HTML similarity between each page with `Page G` (sample page of level 2) and so on.

In the next subsections, we present the similarity metrics used to determine the structural URL and HTML similarity between Web pages (Section 4.1). Finally, we propose four algorithms (Section 4.2). The first algorithm finds the index-page of a given entity/index-page. The second algorithm catches same level pages. The third algorithm catches pagination pages. The last algorithm combines the previous algorithms to perform entity-page discovery.

4.1   Structural Similarity Metrics

In this subsection we present two metrics based on the Web page URL (Weak URL Similarity and Strong URL Similarity) and reuse a metric based on the Web page HTML (sim_{html}) proposed by Blanco et al. [2] in order to measure the structural similarity between two Web pages. The similarity metrics Weak URL Similarity and HTML Similarity are used to determine if a page is a pagination of other. The similarity metrics Strong URL Similarity and HTML Similarity are used to determine if two pages are in the same level in the entity-tree.

Weak URL Similarity is a simple similarity metric that compares two Web pages based on the terms that belong to their `url-schema`. This metric gives the same importance to each term. So, if two URLs share the term "`www`" or the term "`driver`" has the same impact in the result.

Definition 10. *(Weak URL Similarity) Let $v(p_1)$ be the url-schema of the Web page p_1 and $v(p_2)$ be the url-schema of the Web page p_2, the weak URL similarity between p_1 and p_2 is defined as:*

$$weaksim_{url}(p_1, p_2) = \frac{|v(p_1) \cap v(p_2)|}{|v(p_1) \cup v(p_2)|} \tag{1}$$

The Strong URL Similarity also compares two Web pages based on the terms that belong to their `url-schema`. However, it is more robust than Weak URL similarity since it assigns a different weight to each URL term according to its importance to distinguish same level pages from non-same level pages. In this metric, the URL terms are assumed to be all mutually independent. The URL of the sample page is represented as vector of URL term weights in a n-dimensional space, in which n is the total number of URL terms.

Definition 11. *(Strong URL Similarity) Let $v(sp)$ be the url-schema of the sample page sp, $v(p)$ be the url-schema of the Web page p, which we desire to compare with sp, and ip be the index-page of sp, the strong URL similarity between sp and p is defined as:*

$$strongsim_{url}(ip, sp, p) = \frac{\sum_{t \in v(sp) \cap v(p)} W_t(ip, sp)}{\sum_{t \in v(sp)} W_t(ip, sp)} \qquad (2)$$

The weight of a URL term used in the Strong URL Similarity is based on the observation of Weninger et al. [8] "lists usually contain items which are similar in type or in content" and our observation "entity-pages of the same type usually share a common URL structure". In our context, we have a sample page and we want to know how are other web pages similar to sample page based on their URLs. The idea is that terms that occur in many URLs, from links of the index-pages, in the same link DOM path that the sample page, and occur in few link DOM paths from the index-pages are more important because they have a high discriminating power. For example, in Table 1, considering L1 as sample page and $P1$ as its index-page, the term "pos2_car" should have the highest weight because it occurs in five URLs that are in the same link DOM path of the sample page ("HTML-UL-LI-A") and does not occur in other link DOM paths.

Definition 12. *(URL Term Weight) Let ip be the index-page of the sample page sp. Let P be a set of link DOM paths $(P_1, P_2, ..., P_n)$ from ip. Let $P_i = (P_i[1], P_i[2], ..., P_i[n])$, where $P_i[j]$ is the url-schema of the jth URL in the link DOM path P_i. Let P_{sp} be the P_i that contains $v(sp)$, and t a URL term from $v(sp)$, the weight of t from sp in ip is defined as:*

$$W_t(ip, sp) = TF_{t,P_{sp}} \times IDF_{t,P} \qquad (3)$$

where, $TF_{t,P_{sp}}$ is the number of url-schemas in P_{sp} that contains the term t and $IDF_{t,P}$ is defined as:

$$IDF_{t,P}(ip) = log(\frac{|P|}{|P_i \in P : t \in P_i|})$$

Example 6. The Table 1 presents an example of same level index-pages with their links and the link DOM path that points to each link. For each link, we show its strong URL similarity with the sample page L1. The highest strong URL similarity is with itself (1.00), followed by pages L2 and L3 that share with $L1$ the sub-token "car" in the second token and the sub-token "ford" in the third token. These sub-tokens occur only in the link DOM path of the sample page. The strong URL similarity of the **Page L7** is 0 because all sub-tokens shared between this page and the sample page (e.g., "www") occurs in all link DOM paths and all occurrences are in the first token. Pages L0 and L8 do not have ss_u because they are in a link DOM path different from sample page.

HTML Similarity is a simple similarity metric that compares two Web pages based on the link DOM paths that belong to their **html-schema**. This metric gives the same importance to each link DOM Path.

Table 1. An example of same level index-pages with their link DOM paths and the pages pointed by each link DOM path (column `Links`). The column ss_u shows the strong URL similarity between each page in the column `Links` with the sample page L1.

Same Level Index-pages	Link DOM Path	Links	ss_u
	HTML-DIV-A	L0 = www.website.com/privacy_policy.htm	-
P1		L1 = www.website.com/car/ford/ba-falcon-rx-8	1.00
		L2 = www.website.com/car/ford/svt-mustang-cobra-coupe	0.68
		L3 = www.website.com/car/ford/mustang-mach-1	0.68
P2	HTML-UL-LI-A	L4 = www.website.com/car/ferrari/360-challenge-stradale	0.53
P3		L5 = www.website.com/car/audi/a4-cabriolet-3.0	0.53
P4		L6 = www.website.com/test_drive/rx-8.htm	0.21
		L7 = www.website.com/news.htm	0.00
P5			
	HTML-DIV-A	L8 =www.website.com/wallpapers.htm	-

Definition 13. *(HTML Similarity) Let $\Delta(p_1)$ be the html-schema of the Web page p_1 and $\Delta(p_2)$ be the html-schema of the Web page p_2, the HTML similarity between p_1 and p_2 is defined as:*

$$sim_{html}(p_1, p_2) = \frac{|\Delta(p_1) \cap \Delta(p_2)|}{|\Delta(p_1) \cup \Delta(p_2)|} \tag{4}$$

4.2   Algorithms

In this subsection we present the algorithms that compose the method **SSUP**. The Algorithm 1 aims to find the index-page of a given page. The intuition behind the algorithm is that the index-page of a given page sp is the page delivering the largest number of more structurally similar URLs to sp. The solution presented by the algorithm collects the pages pointed by the given page (line 4). Then, the pages that do not have a link to the given page are removed (line 5). Finally, the algorithm returns the page with the max candidate index-page weight (line 6).

The candidate index-page weight accumulates the Strong URL Similarity between its links and a given page. It is considered only the links in the same link DOM path that the given page. The Algorithm 1 uses this weight to find the index-page of a given page, choosing the candidate index-page with the highest weight.

Algorithm 1. Top Index-Page algorithm

1: INPUT: a sample page sp;
2: OUTPUT: the top index-page;
3: **begin** *top_index_page(sp)*
4: $P = \text{GetLinks}(sp)$;
5: remove p_i from P where p_i does not have a link to sp;
6: **return** p_i from P with max $W_{ci}(p_i, sp)$; //Equation 5
7: **end**

Algorithm 2. Catching Same Level Pages algorithm

1: INPUT: a set of index-pages of the level x (IP), the sample page of the level $x + 1$
   (sp), the sample link DOM path of sp (sdp);
2: OUTPUT: the set of pages of the level $x + 1$ pointed by IP;
3: **begin** *catching_same_level_pages(IP, sp, sdp)*
4: add sp to rs;
5: **for each** ip_i **IN** IP **do**
6:     add GetLinksInPath(ip_i, sdp) to L;
7: **end for**
8: create a list of groups $G = (G_1, G_2, ..., G_n)$, where each group G_i contains each
   link $l_i \in L$ (except sp) with the same $strongsim_{url}$ with sp;
9: sort G by $strongsim_{url}$ decreasing order;
10: add all links from G_1 to rs;
11: $minsim_{html} = \min_{l_i \in G_1} sim_{html}(sp, l_i)$;
12: remove G_1 from G
13: **for each** G_i **IN** G **do**
14:     **if** $\max_{l_i \in G_i} sim_{html}(sp, l_i) >= minsim_{html}$ **then**
15:         add all links from G_i to rs;
16:     **else**
17:         break;
18:     **end if**
19: **end for**
20: **return** rs
21: **end**

Definition 14. *(Candidate index-page weight) Let p be a candidate index-page for the Web page sp, L be a set of pages pointed by p through the same link DOM path that sp, the candidate index-page weight of p for sp is defined as:*

$$W_{ci}(p, sp) = \sum_{l \in L} strongsim_{url}(p, sp, l) \tag{5}$$

The goal of the Algorithm 2 is given a set of index-pages of level x and the sample page of the level $x + 1$, finds all pages of level $x + 1$ delivered by the index-pages. The intuition behind the algorithm is that same level pages have structurally similar URLs and HTMLs. The solution proposed is that same level pages have larger $strongsim_{url}$ and sim_{html} than non-same level pages.

The Figure 5 presents an example of the execution of the Algorithm 2, considering the pages P1, P2, P3, P4 and P5 as index-pages of level x (called just index-pages); L1 as the sample page of level $x + 1$ (called just sample page) and "HTML-UL-LI-A" as the sample link DOM path of the sample page, i.e., the path in the index-page that points to sample page. The Table 1 describes these pages. The goal is to return all pages of level $x + 1$. In the line 4, the sample page is added to the result set. In lines 5-9, the Web pages pointed by the index-pages through the sample link DOM path, except sample page, are collected and grouped by their strong URL similarity with the sample page. The

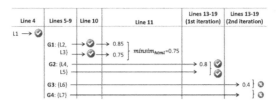

Fig. 5. An execution of the Algorithm 2. Input: IP=(P1, P2, P3, P4, P5), SP=L1, sdp=HTML-UL-LI-A (described in Table 1). Ok icon means that the Web page was added to the result set. Not ok icon means that the Web page was discarded.

groups are sorted by their strong URL similarity decreasing order. Four groups are created: G1, G2, G3, G4. The group G1 (G4) has the Web pages with largest (lowest) strong URL similarity value with the sample page. In line 10, all Web pages from G1 (L2 and L3) are added to result set. In line 11, we compute the min HTML similarity between the sample page and the pages from G1, called $minsim_{html}$. In this example, the $minsim_{html}$ is the HTML similarity between sample page and L3 (0.75). In line 12, omitted in the figure, the G1 is removed from the list of groups. The first iteration of the "*for*" (lines 13-19) analyzes the group G2, since it has the largest strong URL similarity after G1 has been removed. It is computed the HTML similarity between the first Web page of the G2 (L4) and the sample page. As the HTML similarity value is greater than $minsim_{html}$ all Web pages from G2 are added to result set, without compute the HTML similarity for other Web pages of G2. The second iteration of the "*for*" (lines 13-19) analyzes the group G3. Since the min HTML similarity between sample page and the pages of G3 is less than $minsim_{html}$, so all pages of the group (L6) are discarded and all remain groups (G4) are discarded too. This follows the idea "the same level pages have larger $strongsim_{url}$ and sim_{html} than non-same level pages". Finally, the result set (with pages L1, L2, L3, L4 and L5) is returned in line 20.

The Algorithm 2 fails in catch pagination pages because these pages are not pointed by index-pages of the preceding level, but by pages of the same level. So, the Algorithm 3 aims to catch these pages. The intuition behind this algorithm is that pages delivered by index-pages of the level x with HTML similarity to pages of the level x larger than to pages of the level $x+1$ are strong candidate to be pagination pages. The solution presented in the algorithm receives as input: (i) a set of index-pages of the level x; (ii) a sample page of the level x (sp_x); and (iii) a sample page of the level $x+1$ (sp_{x+1}). Then, in lines 4-6, the pages pointed by the index-pages are collected. These pages are grouped according to their Weak URL Similarity with sp_{x+1} (line 7). After, one page of each group has its HTML similarity computed with sp_x and with sp_{x+1}, and if the first score is greater or equal to the second score, then all pages of the group are added to the result set (lines 8-13). Finally, the result set is returned (line 14).

It is important to note that this step catches noise pages. But these pages are pruned when a next level is analyzed because noise pages do not point to pages with structurally similar URL and HTML to the sample page.

Algorithm 3. Pagination algorithm

1: INPUT: the sample page of the level x (sp_x), the sample page of the level $x + 1$
   (sp_{x+1}), index-pages of the level x (IP);
2: OUTPUT: a set of pagination pages of the level x pointed by the IP;
3: **begin** $pagination(sp_x, sp_{x+1}, IP)$
4: **for each** ip_i **IN** IP **do**
5:     add GetLinks(ip_i) to L;
6: **end for**
7: create a list of groups $G = (G_1, G_2, ..., G_n)$, where each group G_i contains each link
   $l_i \in L$ with the same $weaksim_{url}(sp_{x+1}, l_i)$ and has the same link DOM path;
8: **for each** G_i **IN** G **do**
9:     $l_0 = $ first link from G_i
10:     **if** $sim_{html}(sp_x, l_0) >= sim_{html}(sp_{x+1}, l_0)$ **then**
11:         add all links from G_i to rs;
12:     **end if**
13: **end for**
14: **return** rs
15: **end**

The Algorithm 4 shows the overall SSUP algorithm. In lines 4-12 the sample page path is created by calling recursively the function top_index_page. Note that we do not have an approach to discovery the root node, instead we have a parameter that estimates the height of the entity-tree. If this parameter is less than the real entity-tree height then some entity-pages will not be discovered. If this parameter is greater than the real entity-tree height then some unnecessary pages will be downloaded, but the quality results is not affected since the noise pages are pruned when we analyze the URL and HTML structural similarity. However, we add an additional stopping criterion (lines 7-9): when an index-page discovered for a level x was already discovered for a level y where $y > x$. In lines 13-22, we transverse the entity-tree from root to leaves catching same level entity-pages through the functions $catching\_same\_level\_pages$ and $pagination$. It is important to note in lines 19-21 that entity-pages do not have pagination.

5   Experiments

In this section, we describe the experiments we performed in order to evaluate the effectiveness of **SSUP**. We compared **SSUP** with Indesit [2] and GPP [8]. We chose to compare our method with these baselines among the methods that focus on surface Web because they depend neither on the specific HTML markup [7] nor on a human-compiled encyclopedia [6].

In order to analyze the results of experiments, we evaluated them considering the following measures: (i) recall, (ii) precision; (iii) F1-measure; (iv) number of downloaded pages; (v) processing time and (vi) T-test. T-test is a statistical hypothesis test, and recall, precision and F1-measure are well known quality measure metrics in IR community [1]. We show just F1-measure results in terms

Algorithm 4. SSUP algorithm

1: INPUT: one sample entity-page sp, the entity-tree height h;
2: OUTPUT: the set of entity-pages of the same type that sp contained in the Web site;

3: **begin** $ssup(sp, r)$
4: add sp to $SPP[h]$; //SPP = Sample Page Path
5: **for** $i{=}0$ **until** $h - 1$ **do**
6:     $tmp = top\_index\_page(SPP[h - i])$;
7:     **if** SPP contains tmp **then**
8:         break;
9:     **end if**
10:     $aux = h - i - 1$;
11:     add tmp to $SSP[aux]$;
12: **end for**
13: **for** $i{=}aux$ **until** $h - 1$ **do**
14:     **if** $i = aux$ **then**
15:         add $SPP[i]$ to IP;
16:     **end if**
17:     $sdp = $ the link DOM path in $SPP[i]$ that contains the link to $SPP[i + 1]$;
18:     add $catching\_same\_level\_pages(IP, SPP[i + 1], sdp)$ to IP;
19:     **if** $i < h$ - 2 **then**
20:         add $pagination(SPP[i + 1], SPP[i + 2], IP)$ to IP;
21:     **end if**
22: **end for**
23: **return** IP
24: **end**

of quality because recall and precision are combined in F1-measure. In the efficiency aspect we evaluated the total number of downloaded pages and the processing time. We previously downloaded the pages of each Web site used in order to avoid the network interference in the processing time.

5.1   Setup

The experiments were performed on a Intel Core 2 Quad 2.66GHz running Ubuntu 9.10, with 8GB of main memory and 5TB of disk space. **SSUP**, `Indesit` and GPP were implemented in Java. We created 28 datasets, each one from a specific Web site, considering a specific entity-type. We created two groups with these datasets: (i) `multiple type group` with 10 datasets of different entity-types (except `association` that repeats once), described in Table 2; and (ii) `council type group` with 18 datasets of council member entity-type. In the last group, we analyzed the official Web sites of the council of the 26 Brazilian state capitals. However, we excluded: 4 Web sites because they do not have an entity-page for each council member; 3 Web sites because the council member entity-pages are internal frames of the index-page; and 1 Web site because it does not allow crawling its pages.

Table 2. Datasets of the multiple type group

ID	Web site	Entity-type	NEP
Fifa	www.fifa.com	Association	209
GP2	www.gp2series.com	Driver	32
Guitar	www.guitaretab.com	Group	32318
MIT EECS	www.eecs.mit.edu	People	1042
Olympic-A	www.olympic.org	Association	204
Olympic-S	www.olympic.org	Sport	56
Pgfoundry	pgfoundry.org	Software Project	382
Senado	www.senado.gov.br	Senator	121
Stanford EE	engineering.stanford.edu	Staff	473
Supercar	www.supercarsite.net	Car	512

Label: NEP: number of entity-pages

Indesit has two parameters: (i) T-value - the html-schema similarity threshold between two Web pages varying from 0 to 1; (ii) Tag features - defines if the Indesit considers as a link DOM path only tags; tags and attribute names; or tags, attribute names and attribute values. We tested all combinations between T-values (0.1; 0.2; ...; 0.9) and tag features. The best parameter configuration related to quality of results in the multiple type group was T-value= 0.5 and Tag feature = tags and attribute names; and in the council type group was T-value= 0.8 and Tag feature = tags and attribute names. SSUP has one parameter: H-value - the estimate of the entity-tree height. We tested the H-values: 1, 2, ..., 5. The best H-value related to quality of results was 2 and 1 in the multiple type group and council type group, respectively. GPP has one parameter: K-value - the number of iterations. We tested only one iteration because the processing time of GPP is too high compared with Indesit and SSUP. We ran all methods with each configuration three times, each one with a different sample entity-page in order to verify that the behavior of the method does not depend on specific features of the sample entity-page. The three sample entity-pages were chosen randomly from the first page displayed (when there is pagination).

5.2   Comparison

The goal of these experiments is to compare the methods **SSUP**, Indesit and GPP in terms of quality and efficiency. For **SSUP** and Indesit we show the results with the best parameter configuration in terms of quality results in each individual dataset (individual configuration) and with the best parameter configuration in terms of quality results considering all datasets of the group analyzed (general configuration). We run GPP with only one iteration (so, individual and general configuration are the same one) and only on the multiple type group because its processing time is too high.

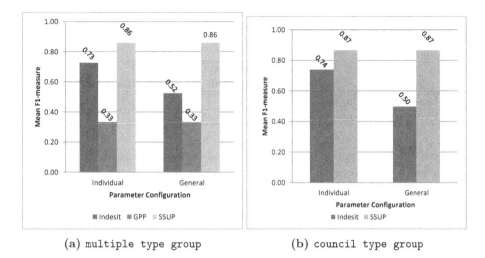

(a) multiple type group   (b) council type group

Fig. 6. Mean F1-measure comparison

Figure 6 shows a comparison between the F1-measure of **SSUP**, Indesit and GPP on the `multiple type group` (Figure 6a) and between **SSUP** and Indesit on the `council type group` (Figure 6b). SSUP outperformed Indesit and GPP. There is a statistical significant difference (p-value < 0.05) between the F1-measure values of the methods (except between Indesit and SSUP with individual configuration on the `multiple type group` where p-value was 0.50126). Considering the general configuration, **SSUP** had better quality results than Indesit because it learns the threshold of html-schema similarity in each dataset from the groups of Web pages with high structurally URL similarity with the sample entity-page and that are in the same link DOM path that the sample entity-page. On the other hand, in Indesit the user should define a threshold of html-schema similarity between entity-pages and this threshold varies from dataset to dataset. Considering the individual configuration (the smallest difference in terms of F1-measure between **SSUP** and Indesit), **SSUP** only outperformed Indesit in datasets that contain an intersection of entity-pages and non-entity pages related to the html-schema similarity. In these cases, in Indesit if we increase the T-value the precision increases, but the recall decreases. If we decrease the T-value the recall increases, but the precision decreases because the method catches erroneous pages. On the other hand, **SSUP** uses the structurally URL similarity as an additional feature to discriminate entity-pages from non entity-pages. GPP presented the worst values of F1-measure because, in many cases, one iteration was not enough to find most entity-pages. However, even on the `Olympic - S` dataset where GPP found all entity-pages (recall $= 1$) many erroneous pages were returned decreasing the precision to 0.32 because the host city entity-pages have similar HTML and visual features to sport entity-pages.

The main SSUP fail cases occurred when: (i) the entity-pages do not contain a link to their index-page affecting the recall of SSUP; (ii) some entity-pages have different link DOM paths affecting the recall of SSUP; (iii) the URL of the

Table 3. Average of downloaded pages and processing time on the `multiple type group`

	Indesit Individual	Indesit General	GPP	SSUP Individual	SSUP General
Downloaded Pages	1680	980	212	778	781
Processing Time	854	779	2984	46	47

entity-pages and non entity-pages are composed of the domain plus "/" plus one token affecting the precision.

The average of downloaded pages on the `multiple type group` for each method is presented in Table 3. Considering individual configuration, the mean number of downloaded pages of `Indesit` was 1680 while **SSUP** was 778 (53.69% less than `Indesit` and p-value < 0.05). This occurred because **SSUP** filters out some pages just by their URLs, without needing to download them while `Indesit` needs to download a Web page to evaluate it. Considering general configuration, the mean number of downloaded pages of `Indesit` was 980 while SSUP was 781 (20.31% less than `Indesit`). However, this difference is not statistically significant (p-value > 0.05). This occurred because `Indesit` with general configuration had a low recall in many datasets, consecutively it downloaded fewer pages. GPP presented the lowest number of downloaded pages (72.86% less than **SSUP** with general configuration and p-value < 0.05). The main reason of this big difference is the low recall in most datasets. However, even in the `Olympic - S` dataset where all entity-pages were found, the number of downloaded pages was lower. This occurred because we used only one iteration, but if we increase the number of iterations to increase the recall, more pages are downloaded.

The average of processing time for **SSUP**, `Indesit` and GPP on the `multiple type group` is presented in Table 3. **SSUP** had the shortest processing time (with p-value < 0.05 in all cases). This behavior can be explained because **SSUP** filters out some pages just by their URLs, without needing to analyze the HTML of them while `Indesit` needs to analyze the HTML of all downloaded pages. Moreover, after computing the Sample Page Path, **SSUP** analyzes in each Web page only the sample link DOM path while `Indesit` analyzes all link DOM paths in all Web pages downloaded. GPP had the worst values of processing time, even having downloaded less Web pages, because it explores visual information and to perform this it is necessary to render the Web page in a browser loading all images, css and javascripts, which greatly increases the processing time.

The experiments showed that **SSUP** outperforms `Indesit` related to the quality results when we define one unique parameter configuration for all datasets. When we use the best parameter configuration for each individual dataset, **SSUP** improves the quality results when the dataset has an intersection between entity-pages and non-entity pages related to the html-schema similarity. In other cases, **SSUP** does not improve the quality results, but it reduces the number of downloaded pages and processing time. Furthermore, the **SSUP** parameter - H - can be defined by a user visually navigating through the Web site while the `Indesit` parameters (T and tag feature) require a processing of the entity-pages. Then, we can conclude that combining URL and HTML features improves the quality results of entity-page discovery and decreases the number of downloaded pages and the processing time.

6   Conclusions

In this paper we propose a new method to entity-page discovery, called **SSUP**. Specifically, given a sample entity page of a Web site, we want to find the set of same type entity-pages of that Web site. The main contribution of **SSUP** is to combine URL and HTML features to entity-page discovery. We first define the basic structures of a Web site and a Web page according to our method. Second, we propose two structural URL similarity metrics and reuse a HTML similarity metric. Thirdly, we develop a set of algorithms that compose the method. We demonstrated the effectiveness of **SSUP** by comparing it with `Indesit` and GPP using real datasets.

The experiments have showed that combining URL and HTML features improves the results in terms of quality, number of downloaded pages and processing time. Entity-pages dynamically generated by populating fixed pages templates with content from a back-end DBMS have the better results. However, we are working on incorporating other features besides URL and HTML to achieve good results in people entity-pages created by different users, since these pages usually do not contain a link to its index-page affecting the recall of SSUP. We are also creating a parallelized version of SSUP.

References

1. Baeza-Yates, R., Ribeiro-Neto, B.: Modern Information Retrieval: The Concepts and Technology Behind Search. Addison Wesley Professional (2011)
2. Blanco, L., Crescenzi, V., Merialdo, P.: Efficiently locating collections of web pages to wrap. In: WEBIST, pp. 247–254. INSTICC Press (2005)
3. Blanco, L., Dalvi, N.N., Machanavajjhala, A.: Highly efficient algorithms for structural clustering of large websites. In: WWW, pp. 437–446. ACM (2011)
4. Grünwald, P.D.: The Minimum Description Length Principle (Adaptive Computation and Machine Learning). The MIT Press (2007)
5. He, Y., Xin, D., Ganti, V., Rajaraman, S., Shah, N.: Crawling deep web entity pages. In: WSDM, pp. 355–364. ACM (2013)
6. Kaptein, R., Serdyukov, P., de Vries, A.P., Kamps, J.: Entity ranking using wikipedia as a pivot. In: CIKM, pp. 69–78. ACM (2010)
7. Lerman, K., Getoor, L., Minton, S., Knoblock, C.A.: Using the structure of web sites for automatic segmentation of tables. In: SIGMOD Conf., pp. 119–130. ACM (2004)
8. Weninger, T., Johnston, T.J., Han, J.: The parallel path framework for entity discovery on the web. ACM Trans. Web 7(3), 16:1–16:29 (2013)
9. Yu, H., Han, J., Chang, K.C.C.: Pebl: Web page classification without negative examples. IEEE Trans. on Knowl. and Data Eng. 16(1), 70–81 (2004)

StreamMyRelevance!

Prediction of Result Relevance from Real-Time Interactions and Its Application to Hotel Search

Maximilian Speicher[1,2], Sebastian Nuck[2,3],
Andreas Both[2], and Martin Gaedke[1]

[1] Chemnitz University of Technology, 09111 Chemnitz, Germany
[2] R&D, Unister GmbH, 04109 Leipzig, Germany
[3] Leipzig University of Applied Sciences, 04277 Leipzig, Germany
maximilian.speicher@s2013.tu-chemnitz.de,
martin.gaedke@informatik.tu-chemnitz.de,
andreas.both@unister.de, sebnuck@gmail.com

Abstract. The prime aspect of quality for search-driven web applications is to provide users with the best possible results for a given query. Thus, it is necessary to predict the relevance of results *a priori*. Current solutions mostly engage clicks on results for respective predictions, but research has shown that it is highly beneficial to also consider additional features of user interaction. Nowadays, such interactions are produced in steadily growing amounts by internet users. Processing these amounts calls for streaming-based approaches and incrementally updateable relevance models. We present *StreamMyRelevance!*—a novel streaming-based system for ensuring quality of ranking in search engines. Our approach provides a complete pipeline from collecting interactions in real-time to processing them incrementally on the server side. We conducted a large-scale evaluation with real-world data from the hotel search domain. Results show that our system yields predictions as good as those of competing state-of-the-art systems, but by design of the underlying framework at higher efficiency, robustness, and scalability.

Keywords: Streaming, Real-Time, Interaction Tracking, Learning to Rank, Relevance Prediction.

1   Introduction

Nowadays, search engines are among the most important and most popular web applications. They are essential for supporting users with finding specific pieces of information on the web. Thus, their *prime aspect of quality* is to ensure that relevant results are displayed where they receive the highest attention. In other words, the ranking of results is a major quality aspect in the context of the search application as a whole. This makes it necessary to estimate the relevance of results *a priori*. Common methods for obtaining such estimates are generative *click models* (e.g., [3,4,15]). Based on certain assumptions about user behavior, these models predict the relevance of a certain result taking into account the

S. Casteleyn, G. Rossi, and M. Winckler (Eds.): ICWE 2014, LNCS 8541, pp. 272–289, 2014.

number of clicks it has received for a given query. However, click data are not a perfect indicator concerning relevance since users might return to the search engine results page (SERP) after having clicked a useless result. Additionally, search engines more and more try to answer queries directly on the SERP, e.g., as Google do with their *Knowledge Graph*[1]. Thus, additional information that complement click data should be taken into account for predicting relevance, e.g., in terms of dwell times on landing pages [9] or other client-side user behavior (e.g., [9,13,18]). Previous research has shown the value of such page-level interactions [11,13,20]. Also, generative [12] as well as discriminative [20] approaches to relevance prediction exist that engage user behavior other than clicks only.

With a growing amount of users, it is possible for search engine providers to collect enormous amounts of client-side data. This is particularly the case if we consider interactions other than clicks. Along with the increasing quantities of tracking data, a short time-to-market becomes more and more important. That is, providers need to quickly analyze collected information and feed potential findings back into their products to ensure user satisfaction. This calls for the use of novel systems for data stream mining, such as *Storm*[2], which are currently gaining popularity in industry and research. These systems can help to cope with the seemingly endless streams of data produced by today's internet users. Yet, none of the approaches for relevance prediction mentioned above leverages data stream mining to process collected information.

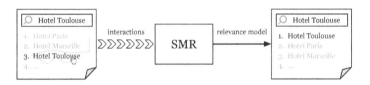

Fig. 1. The intention behind StreamMyRelevance!—from collecting a stream of user interactions to reordering search results based on relevance models

We present *StreamMyRelevance!* (SMR), which is a novel streaming-based system for ensuring ranking quality in search engines. Our system caters for the whole process from tracking interactions to learning incremental *relevance models*, i.e., models that predict the relevance of a search result (for a given query) based on certain features of user interaction. The latter can be used to directly feed predictions back into the ranking process of the search engine, e.g., as a weighted factor in a learning-to-rank function (cf. Fig. 1). SMR is based on Storm and leverages tracking and data processing functionalities provided by *TellMyRelevance!* (TMR)—a pipeline that has proven its effectiveness in predicting search result relevance [20]. Yet, TMR is a batch-oriented approach that does not provide means for incrementally learning relevance models on

---

[1] http://www.google.com/insidesearch/features/search/knowledge.html (2013-09-06).

[2] http://www.storm-project.net/ (2013-12-30).

a streaming basis. Thus, SMR wraps the borrowed functionalities into a new system that is able to handle real-time streams. Our system has three main advantages over existing approaches, i.e., (1) considering interactions other than clicks for predicting relevance, (2) collecting and processing these interactions as a stream and (3) providing incremental relevance models that do not require re-processing of previously processed data. Based on this, the main hypothesis investigated in this paper is as follows: *SMR is able to achieve the same relevance prediction quality as TMR at better efficiency, robustness and scalability.*

We have evaluated SMR in terms of its feasibility and quality of relevance predictions. For this, large amounts of real-world data from two *hotel booking portals* were available. A comparison to TMR has been performed, which due to its batch-oriented design has look-ahead capabilities and thus more information available [20]. Still, our results show that SMR's prediction quality is not significantly worse compared to TMR. Moreover, our system in parts compares favorably with predictions of the Bayesian Browsing Model (BBM) [16], a *state-of-the-art* generative click model successfully applied in industry. Furthermore, reviews of efficiency, robustness and scalability show that SMR compares favorably with the competing approaches in these respects.

In the following section, we describe important concepts our work is based on, before giving an overview of related work. Section 3 explains the design and architecture of SMR, followed by an evaluation of effectiveness, efficiency, robustness and scalability of SMR and competing approaches in Section 4. Limitations and potential future work are addressed in Section 5, before giving concluding remarks in Section 6.

2   Background and Related Work

The following gives background information on the underlying concepts of Storm [17], which are important for understanding the architecture of SMR.

The logic of a Storm application is represented as a graph consisting of **spouts** and **bolts** that are connected by *streams*, i.e., unbounded sequences of data tuples. This concept is called a **topology**. On the one hand, spouts act as sources of streams by reading from external data sources (e.g., a DB) and emitting tuples into the topology. On the other hand, bolts are the core processing units of a topology. They receive tuples, process the contained data and emit results as a new stream. Spouts and bolts can have multiple outgoing streams, which provides the possibility of separating tuples within bolts and emitting them using different streams.

The direct competitor to Storm is Yahoo!'s S4[3]. It as well provides distributed stream computing functionality, but its underlying concepts and configuration are more complex[4]. As described in [22], benchmarks have shown that S4 is almost 10 times slower than Storm.

This research is related to a variety of existing work in the fields of *relevance prediction* and *data stream mining*. An overview will be given in the following.

---

[3] http://incubator.apache.org/s4/ (2013-09-28).
[4] http://demeter.inf.ed.ac.uk/cross/docs/s4vStorm.pdf (2014-01-06).

Concerning the relevance of search results, it is necessary to rely on human relevance judgments—i.e., asking the user to explicitly rate the relevance of a result—for the best possible predictions. However, since such data are usually not available in large numbers, different solutions are required. Joachims [15] proposes to use *clickthrough data* instead of human relevance judgments. Based on the *cascade hypothesis* [4,16], i.e., the user examines results top-down and neglects results below the first click, it is possible to infer relative relevances. That is, the clicked result is more relevant than the non-clicked results at higher positions. Using such relative relevances, Joachims engages clickthrough data as training data for learning retrieval functions with a support vector machine approach [15]. In contrast to the above, models like the *Dependent Click Model* [8] assume that more than one result can receive clicks. That is, results below a clicked position might be examined and thus also clicked if they are relevant.

The *Dynamic Bayesian Network Click Model* (DBN) described in [3] generalizes the *Cascade Model* [4] by aiming at relevance predictions that are not influenced by position bias. To achieve this, the authors (besides the perceived relevance of a search result) also consider users' satisfaction with the website linked by the clicked result.

Generally, click models are based on the *examination hypothesis*, which states that only relevant search results that have been examined are clicked [16]. Yet, not all of these models follow the cascade hypothesis. All of the above described are *generative* click models that try to provide an alternative to explicit human judgments by *predicting the relevance of search results* based on click logs. The main differences to SMR are that we aim at predicting relevance using a *discriminative* approach also taking into account *interactions other than clicks*. Moreover, the above click models are not designed for efficient processing of *massive data streams* or *incremental updates*.

The *Bayesian Browsing Model* (BBM) [16] is based on the *User Browsing Model* (UBM) [7], which assumes that the probability of examination depends on the position of the last click and the distance to the current result [16]. Contrary to UBM, BBM aims at scalability to petabyte-scale data and incremental updates. The authors compute "relevance posterior[s] in closed form after a single pass over the log data" [16]. This enables incremental learning of the click model while making iterations unnecessary. Still, contrary to SMR, BBM is again a *generative* model that does not leverage the advantages of additional interaction data.

Concerning user interactions other than clicks, in [11], Huang has found that these are a valuable source of information for relevance prediction. Following, Huang et al. [13] investigate the correlations between human relevance judgments and mouse features such as hover time and unclicked hovers, among others. They find positive correlationsand conclude that these can be used for inferring search result relevance. Also, part of our system is based on a scalable approach for collecting client-side interactions described by the authors [13].

In [9], Guo and Agichtein present their *Post-Click Behavior Model*. They incorporate interactions like cursor or scrolling speed on a landing page into determining its relevance, i.e., interactions that happen *post-click*. This is also partly related to DBN [3], where the relevance of the landing page is modeled separately from the perceived relevance of the result. While this approach is promising for inferring the actual usefulness of a landing page, it would be difficult to realize since search engines would need access to landing page interactions through, e.g., a browser plug-in or tracking scripts.

Making use of scrolling and hover interactions, Huang et al. [12] extend the Dynamic Bayesian Network Click Model described earlier to leverage information beyond click logs. Their results show that this improves the performance in terms of predicting future clicks compared to the baseline model. While this *generative* approach involves interactions other than clicks, in contrast to SMR, it does not specifically aim at incremental learning or efficient processing of massive data streams.

TMR is a system described by Speicher et al. [20]. Parts of SMR are based on this work, particularly in terms of client-side interaction tracking, preprocessing of raw data and computation of interaction features. Like SMR, TMR is a *discriminative* approach to relevance prediction, but in contrast is a batch-oriented system. In particular, its relevance models are not trained incrementally, i.e., *all* data have to be re-processed before obtaining an updated model.

3   SMR: Streaming Interaction Data for Learning Relevance Models

The following Section describes SMR, which is organized as a streaming-based process. Its aim is to enable processing of big data streams while leveraging the advantages of user interaction data for the prediction of search result relevance. This supports more optimal ranking of results, which is a major quality aspect of search-driven web applications.

The system comprises four main components as illustrated in Fig. 2: The **Client-Side Interaction Tracking** component in terms of a jQuery plug-in; The **Preprocessor** for reading and preprocessing streams of tracking data and

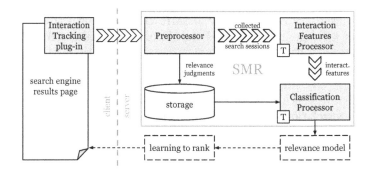

Fig. 2. The main components and process flow of SMR (Streams are visualized by sequences of chevrons; Storm topologies are annotated using a "T")

relevance judgments; The **Interaction Features Processor** for calculating interaction features from tracking data; The **Classification Processor** for incrementally training a relevance model using the previously computed features and collected relevance judgments.

Our Storm-based system has been specifically designed with an incremental approach in mind. The four steps above can be regarded as a sequence of independent processes. That is, the results of each step as well as the resulting relevance models are persisted (temporarily). As a result, in case of a crash within the system, SMR can resume its work at the step prior to the incident without starting over from the very beginning.

3.1 Client-Side Interaction Tracking

For client-side interaction tracking, SMR builds upon a "minimally invasive jQuery plug-in" [20] that is provided by TMR. This plug-in tracks `mouseenter`, `mousepause`, `mousestart`, `mouseleave` and `click` events that happen within the bounds of a search result on a SERP [20]. Each mouse event is extended with the search query, a user ID and the ID of the corresponding result [20]. The resulting data packets are then sent to a specified key-value store at suitable intervals (Fig. 2) [20]. For integration, the developer has to specify jQuery selectors for (a) the HTML container element holding all results, (b) a single search result, (c) an element within a result holding the result ID and (d) links to landing pages.

The second function provided by the plug-in is intended for recording human relevance judgments, often also referred to as *conversions*, which are crucial for learning relevance models. It is realized as a JavaScript method that can be called from anywhere, e.g., upon clicking an upvote button next to a search result [20]. This method has to be provided with the value of the judgment (e.g., -1 for a downvote and $+1$ for an upvote) as well as the corresponding search query, session ID and user ID by the developer.

3.2 Preprocessor

After having been recorded using the above jQuery plug-in, all interaction data is received by SMR as a *stream of individual events* for preprocessing (Fig. 2). Additionally, information about a corresponding search session[5] is transferred when a user enters a SERP. These contain an anonymous user ID, the current search query and the ordered list of all results, among others [20]. Every event received by SMR is subsequently associated with its respective search session. This concept is referred to as a *collected search session*.

It is logically not possible to process events from search sessions that have not ended yet. Thus, all events are passed on in the SMR pipeline on a per-search session basis. Since it is unreliable to fire client-side `unload` events on

---

[5] For our purposes, a search session starts when entering and ends when leaving a SERP. For example, a reload triggers a new session, even for the same user and query.

a SERP, this is realized using a configurable time-out on the server side. For example, if no events related to a given session have been received for 2 minutes, it is considered finished and the collected search session is passed on for interaction feature computation (Fig. 2).

Moreover, the preprocessing component receives human relevance judgments that are required for learning actual models. These judgments are checked for validity, i.e., whether a corresponding search session exists during which the judgment happened. The latter is not the case if a judgment is triggered by a user who did not perform a search beforehand, e.g., because they received a link to a result from a friend. Relevance judgments are persisted at this point for later use by the Classification Processor (Fig. 2). Finally, for later filtering purposes, each valid judgment is associated with the list of queries triggered by the corresponding user ID.

3.3 Interaction Features Processor

The Interaction Features Processor is realized as a separate topology within our Storm-based system (Fig. 2). It receives *collected search sessions* from the preprocessor that are emitted as a stream by a dedicated spout. To ensure that all interaction events associated with a search session are ordered logically, invalid sequences of events are filtered out. This prevents the computation of faulty interaction feature values. An invalid sequence would be, e.g., if a `mouseleave` happens before a `mouseenter` event on the same search result. Typical causes for such a case can be faulty time stamps or latency while transferring data from client to server. Since at the moment we specifically focus on *mouse* interactions, search sessions that have been recorded on touch devices are eliminated as well.

Subsequently, the values of the actual interaction features are calculated per query–result pair. For example, the value of the *arrival time* is determined by subtracting the time stamp of the first `mouseenter` event on a result from the time stamp of the page load (which is available as meta information about the associated search session). The features we are considering are:

(i) ARRIVAL TIME, (ii) CLICKS (not leading to a landing page), (iii) CLICK-THROUGHS (leading to a landing page), (iv) CURSOR MOVEMENT TIME, (v) CURSOR SPEED (cursor trail divided by cursor movement time), (vi) CURSOR TRAIL, (vii) HOVERS, (viii) HOVER TIME, (ix) MAXIMUM HOVER TIME, (x) POSITION and (xi) UNCLICKED HOVERS (hovers during which no clickthrough happened).

These are in accordance with [20]. Features are moreover averaged over the number of hovers, if possible. This applies to *clicks, clickthroughs, cursor movement time, cursor trail, hover time* and *unclicked hovers* [20]. Finally, the computed values are persisted, which is important for later normalization purposes and actual use of SMR's relevance models (see below). In case feature values are already present for a query–result pair, they are automatically updated by adding the new values and taking the average over all values.

Within this topology, emitting a stream of collected search sessions is realized using a *spout*. Contrary, checking event sequence validity, the actual computation of feature values and updating values of already existing query–result pairs are realized through *bolts*.

The raw search sessions and associated events are not necessarily lost after they have been used for computing interaction features. Rather, SMR provides the option to persist all processed data. In this way, it is possible to batch-wise train a new model from parts of old data (e.g., after removing outdated information) before continuing to incrementally update this new model using real-time interactions and judgments.

3.4 Classification Processor

The Classification Processor is as well realized as a separate topology within our system (Fig. 2). It receives the previously calculated *interaction features* (one set per query–result pair) in terms of a stream that is emitted into the Storm cluster by a dedicated spout. Using the lists of queries associated to judgments during preprocessing, we filter out sets of interaction feature values that are not associated with a user who triggered at least one relevance judgment. This helps to ensure a good quality of our training data.

Moreover, relevance models provided by SMR highly depend on the layout of a SERP [20]. Thus, normalization of feature values is necessary to guarantee comparability between models related to different SERP layouts [20]. This happens in terms of dividing feature values by the maximum value of the respective feature across all results for the given query. Since interaction feature values arrive as a stream, maximum values change over time and have to be constantly updated. Hence, they become more precise the longer the system runs. This is a major difference compared to TMR, which—due to its batch-oriented nature—has look-ahead capabilities and knows exact maximum values from the start.

In the next step, we derive the normalized relevance rel_N for a query–result pair using the human relevance judgments that have been persisted in the preprocessing step. For this, all relevance judgments *judg* corresponding to the query–result pair (q,r) are summed up before dividing them by the sum of all judgments for the given query [20]:

$$rel_N(q, r) = \frac{\sum\limits_{u \in U} \mathrm{judg}(u, q, r)}{\sum\limits_{s \in R} \sum\limits_{u \in U} \mathrm{judg}(u, q, s)} ,$$

with U the set of users who triggered a judgment and R the set of possible results for the query q. Normalizing judgments is important since otherwise, a result X that was among the results of 20 queries and received 10 positive judgments (rel_N=0.5) would be considered more relevant than a result Y that was among the results of only 5 queries and received 5 positive judgments (rel_N=1).

Having available interaction feature values and normalized relevance of a query–result pair, it is possible to use them as a training instance for SMR's relevance model. For this, the query–result pair is transformed into an instance that can be interpreted by the WEKA API [10]. The interaction features are labeled as attributes while "relevance" is labeled as the target attribute on which we train the model. At the moment, SMR has two built-in classifiers available that are provided by the WEKA API and trained in parallel. That is, a Hoeffding Tree, which is specifically aimed at incremental learning and is suitable for very large datasets [6], and an updateable version of Naïve Bayes[6], which also works for smaller datasets. The current states of the relevance models are serialized and persisted after each incremental update. These models are ready-to-use and can be instantly engaged for obtaining relevance predictions and feeding them back into a SERP for results optimization (Fig. 2). Moreover, all training instances are persisted to a file to enable manual inspections using, e.g., the WEKA GUI.

Within this topology, emitting a stream of interaction feature values is realized using a *spout*. Contrary, filtering and normalization tasks as well as incrementally training the relevance models are realized as *bolts*.

The incrementally trained relevance models are serialized and persisted after every update. This makes it possible to manually review the quality of the current model and interrupt or stop training if the model is reasonably stable, which helps to prevent overfitting. Moreover, SMR does not require to directly feed predictions by the incremental relevance model back into the ranking process of the underlying search engine. Rather, as just described, search engine owners are given the option to review the model before usage to ensure ranking quality.

3.5   Making Use of Relevance Models

SMR only caters for learning and providing relevance models. This means that the actual usage of a model is up to the search engine owner. A relevance model RM takes a vector of interaction feature values I for a given query–result pair (q,r) and returns a corresponding relevance prediction $\widehat{rel}$, i.e., $RM(q, r, I) = \widehat{rel}(q, r)$.

This can, e.g., be integrated into a scheduled process of updating search result ranking according to a learning-to-rank function that contains $\widehat{rel}$ as a parameter (Fig. 2). The interaction feature values used for prediction could be those recorded by SMR and persisted by the Interaction Features Processor.

4   Evaluation

To show SMR's capability of coping with realistic workloads, we have performed a large-scale log analysis of real-world user interactions. The anonymous data used were collected on two large hotel booking portals. We used the *number of conversions* (i.e., when a hotel has been actually booked by users) as relevance judgments for training our models. This stands in contrast to commonly used click models, where clicks are the prime indicators of relevance. First, we compare SMR to its analogous batch-wise approach TMR (cf. [20]) in terms of the

---

[6] http://weka.sourceforge.net/doc.dev/weka/classifiers/bayes/
NaiveBayesUpdateable.html (2013-10-07).

prediction quality of the two systems. Second, we provide BBM (as a state-of-the art generative click model aiming at stream processing; cf. [16]) with the same set of raw interaction logs and compare its quality of relevance prediction against that of SMR. Third, we check SMR against a version of itself that considers click-throughs only (SMR_{click}) as well as an analogous version of TMR, i.e., TMR_{click}. Results indicate that SMR is able to provide reasonably good relevance predictions that are not significantly different from those of TMR and might compare favorably to those of BBM—although the difference is not significant. Moreover, our system is superior to corresponding discriminative approaches that do not consider interactions other than clickthroughs. Subsequently, we have a look at the efficiency, robustness and scalability of the evaluated approaches. Results show that SMR can easily cope with realistic workloads in a manner that is robust to external influences. This is especially important in real-world settings with big data streams.

For detailed figures and descriptive statistics, see http://vsr.informatik. tu-chemnitz.de/demo/SMR. Also, we provide training data and serialized models for reproducing this evaluation using WEKA (cf. [10]).

4.1 Effectiveness

Method. Approximately 32 GB of raw tracking data were collected by SMR's interaction tracking facilities in May 2013 on two large hotel booking portals. Of these, $\sim$10 GB of interaction logs were chosen for evaluation, which correspond to $\sim$3.8 million search sessions over a period of 10 days. Based on these, we computed interaction features for a total of 86,915 query–result pairs. Because the collected data contained critical information about the cooperating company's business model, it was a requirement that all data was saved to a key-value store controlled by the company. In particular, we are not allowed to publish the concrete conversion–to–search session (CTS) ratio. Yet, it can be stated that this ratio is very low, i.e., $\#conversions \ll \#search\ sessions$.

We divided the chosen raw interaction data into *10 distinct datasets DS0–DS9* ($\sim$0.7–1.5 GB each) that were intended for training relevance models and corresponded to one day each. Since SMR cannot—due to its streaming-based nature—use fixed maximum values for interaction feature normalization (cf. Section 3.4), it produces different feature values for the same tracking data compared to TMR. Thus, processing the above raw datasets with both systems yields a total of 20 datasets containing interaction features and relevances (i.e., normalized conversions) of the extracted query–result pairs: DS^0_{TMR}–DS^9_{TMR} from TMR and DS^0_{SMR}–DS^9_{SMR} from SMR. For this, we considered only search sessions that were produced by users who triggered at least one conversion (in terms of booking a hotel). Conversions are treated as relevance judgments in analogy to [20], i.e., a greater number of conversions implies higher relevance and vice versa. For evaluating SMR, we *simulated a stream* of search sessions based on the logs containing raw interaction data.

In analogy to [20], we observed a very low ratio of booked hotels to search sessions. In addition with a high query diversity this leads to more than 99% of the query–result pairs having a relevance of either 0.0 or 1.0. Therefore, in this evaluation, we treat relevance prediction as a binary classification problem with two classes: "bad" (relevance < 0.5) and "good" (relevance ≥ 0.5). With more than 90% of the query–result pairs having a *bad* relevance and less than 10% having a *good* relevance, these classes are rather unbalanced. Thus, we use the *Matthews Correlation Coefficient* (MCC) for evaluations of model quality, which is suitable for cases with unbalanced classes [1].

Relevance models as provided by SMR and TMR are highly sensitive to layout specifics of the corresponding SERPs [20]. Yet, since the two hotel booking portals feature the exact same layout template, it is valid to use combined data from both portals for training the same model(s).

The Storm cluster used for evaluation was based on *Amazon EC2*[7]. It comprised four computing instances. An additional machine was used for logging purposes and hosting the database used. All computers in the Storm cluster were instances of type `m1.large`, featuring two CPUs and 7.5 GB RAM[8].

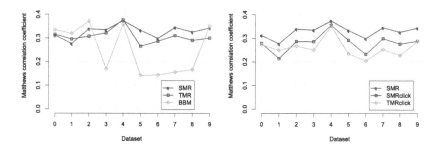

Fig. 3. MCC values for *DS0–DS9* (threshold = 0.5)

SMR vs. TMR. Based on the datasets described above, we trained a total of 20 Naïve Bayes classifiers (10 per system), as provided by TMR and SMR through the WEKA API. Thereby, our system used the updateable version of the classifier for incremental learning. The Naïve Bayes classifier was chosen because the amount of data available for evaluation was too small to train reasonably good Hoeffding Tree classifiers [6]. All classifiers learned have been evaluated using *10-fold cross validation*, from which we obtained corresponding MCC values. As can be seen in Fig. 3, the difference between SMR and TMR is not significant across the 10 datasets. This result has been validated using a Wilcoxon rank sum test, with $p > 0.05$ ($\alpha = 0.05$, $W = 75$, 95.67% conf. int. = [-0.047, 0.004]). It implies that statistically, SMR yields the same prediction quality as TMR, even though it has less information available; particularly in terms of feature normalization and missing look-ahead capabilities. While Fig. 3 shows only MCC values at a

---

[7] http://aws.amazon.com/ec2 (2013-09-30).

[8] http://aws.amazon.com/en/ec2/instance-types/#instance-details (2013-10-05).

threshold of 0.5, our result is underpinned by the exemplary receiver operating characteristic (ROC) curves depicted in Fig. 4, where SMR does not dominate TMR or vice versa. This is similar for the remaining nine datasets.

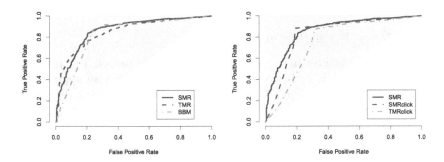

Fig. 4. ROC values for *DS7*

SMR vs. BBM. Additionally, we have compared SMR's prediction quality to that of a state-of-the-art generative click model designed for very large amounts of data and incremental learning. For this, we have used an existing re-implementation of BBM—as described in [16]—and provided it with the exact same raw interaction logs. Fig. 3 shows that BBM yields slightly better predictions for four out of ten datasets (*DS0–DS2, DS9*) at a threshold of 0.5 while SMR has a better prediction quality for the remaining six datasets. For this, predictions of BBM have been compared to the normalized relevances computed by SMR based on the available conversions. The difference between the two approaches is not significant according to a Wilcoxon rank sum test (α=0.05, W=64.5, p>0.05, 95.67% conf. int. = [-0.177, 0.021]). Still, our result indicates that SMR has the potential to provide relevance predictions that compare favorably to BBM. Particularly, Fig. 4 suggests that predictions of BBM can be partly dominated by SMR's predictions for certain datasets. We expect SMR's prediction quality to increase with amounts of data larger than used in this evaluation. Thus, we hypothesize that our system can predict relevance at least as good as BBM, whose predictions are being successfully used in industry.

SMR vs. SMR$_{click}$ vs. TMR$_{click}$. To investigate the influence of the additional user interactions, we have performed a comparison of SMR to versions of itself and TMR that consider clickthroughs only, named SMR$_{click}$ and TMR$_{click}$. Results show that SMR outperforms the click-only approaches across all 10 datasets (Fig. 3) based on *10-fold cross-validation*. Moreover, the MCC differences between SMR and SMR$_{click}$/TMR$_{click}$ are significant, as has been shown by two Wilcoxon rank sum tests (SMR$_{click}$: α=0.05, W=84.5, p<0.05, 95.67% conf. int. = [-0.075, -0.020]; TMR$_{click}$: α=0.05, W=90, p<0.01, 95.67% conf. int. = [-0.101, -0.044]). Our results are further supported by the ROC curves shown in Fig. 4, where SMR (area under ROC = 0.861) performs better than both SMR$_{click}$ (area under ROC = 0.834) and TMR$_{click}$ (area under ROC = 0.759).

These findings underpin that adding interaction data other than clicks yields considerable improvements for discriminative approaches, as has also been outlined in [11,13]. This is true even if clickthroughs show a correlation with relevance that is notably higher than those of the additional attributes (e.g., $r=0.34$ for DS^2_{TMR}).

4.2   Efficiency, Scalability and Robustness

Efficiency and Scalability. SMR is a feasible approach for processing web-scale interaction data. In contrast, TMR uses a batch-wise approach and non-incremental classifiers [20]. This means that all training data (in terms of query–result pairs, i.e., interaction features and relevances) already put into a model have to be re-processed for an update, which yields a time-complexity of $O(q) + O(s)$ with s = #search sessions in new log, q = #previously processed query–result pairs. Assume we receive one log with raw interaction data per day and want a daily model update. Then the amount of data that needs to be re-processed grows linearly. At some point, processing these data would take longer than 24 hours unless we add more/faster hardware to the system, which is, however, not a feasible approach in the long-term. Particularly, re-processing previously processed query–result pairs involves numerous slow database requests. To give just one concrete example from our evaluation, TMR needs ~5 hours for processing a single 1.5 GB log on a dual-core machine with a 2.3 GHz *Intel Core i5* CPU and 4 GB RAM. Since this corresponds to one day, processing the logs for two days would already take ~10 hours etc. This means that after *five days*, we exceed a processing time of 24 hours, which makes it impossible to provide a daily model update unless we use a better machine than the given one.

In contrast, SMR does not need to re-process logs from previous days since data is processed on a per–search session basis and models are learned incrementally. Thus, a model update considers only one search session at a time and the time-complexity of the update depends on the complexity of the classifier used. For example, "constant time per example [i.e., a query–result pair in our case]" [6] if using a Hoeffding Tree. SMR needs ~2 hours for processing all search sessions in a 1.5 GB log using the cluster described in Section 4.1. For this, the search sessions have been put into the system at the highest possible frequency. The log used corresponds to one day of real-world traffic from two hotel booking portals. This means that—using simple interpolation—SMR would be able to cope with approximately *12 times the load* based on the relatively simple cluster set-up used.

Finally, BBM has been specifically designed for incremental updates and web-scalability. As described in [16], 0.25 PB of data were processed using the generative click model. The authors state that it was possible to compute relevances for 1.15 billion query–result pairs in three hours on a MapReduce [5] cluster. BBM's time-complexity for updating a relevance model is $O(s)$.

Due to the differences in system architecture—TMR runs on a single node while the other two approaches require a cluster—the above is not an absolute, hardware-independent comparison of performance. Rather, it describes *relative performances* between the three systems. An overall, relative comparison of efficiency and scalability of the compared approaches is shown in Table 1.

Robustness. Being based on Storm, SMR is a highly robust system by design. In particular, it features guaranteed message passing[9] and high fault-tolerance[10] if one or more nodes die due to external reasons—which happened numerous times during our evaluation. In such a case, SMR continued processing the current interaction data from the step prior to the incident.

In [16], Liu et al. do not explicitly address the robustness of their approach. Rather, BBM has been designed for use as a MapReduce job on a *Hadoop* cluster. That is, differences in robustness between SMR and BBM originate from corresponding differences between Storm and Hadoop. Particularly, Hadoop has disadvantages when it comes to guaranteed message processing or when supervising/master nodes are killed.

Finally, TMR is the least robust of the compared approaches. In case the processing of a batch of data is stopped due to external reasons (e.g., a memory overflow), all data need to be re-processed. In particular, this means that already computed values of interaction features are useless since contributions of already processed data can not be subtracted out before starting over an iteration. Therefore, careful evaluation and set-up of the required hardware are necessary before using TMR to minimize the risk of costly and time-consuming errors.

4.3 Discussion and Summary

In this evaluation, we have shown that SMR does not perform significantly less effective than TMR, even though it relies on lower-quality information for training its relevance models. Moreover, SMR is more efficient, robust and scalable compared to its batch-wise predecessor. The difference of SMR's predictions to those of the generative state-of-the-art click model BBM were not significant as well. Yet, our results indicate that our discriminative approach can be advantageous over BBM for certain datasets and that it is more robust at similar efficiency and scalability. Finally, we have underpinned the value of interaction data other than clicks for relevance prediction, with clickthrough-only versions SMR_{click} and TMR_{click} performing significantly worse than SMR. However, there are some points remaining for discussion.

Discussion. *Why does SMR show the tendency to perform better than TMR, although its training data are of lower quality?* As described in Section 3.4, the maximum values for feature normalization change during the processing of a dataset due to SMR's streaming-based nature (i.e., no look-ahead is possible). This means that SMR has less information available and as a result, the training

---

[9] https://github.com/nathanmarz/storm/wiki/Guaranteeing-message-processing (2013-12-30).

[10] https://github.com/nathanmarz/storm/wiki/Fault-tolerance (2013-12-30).

data has lower quality. However, the different feature values for query–result pairs that appear early in a dataset can—purely by chance—lead to better predictions of SMR. This is especially the case because in this evaluation we were working with relatively small and closed datasets, as compared to a real-world setting. Hence, we strongly assume that in such a setting, the already non-significant difference between SMR and TMR would become even smaller.

Why does BBM make better predictions than SMR for DS2 but predicts worse for DS7? SMR computes almost the same amount of query–result pairs for the two datasets, with nearly identical means and distributions of the individual interaction features. In contrast, BBM has approximately 12% less search sessions available in *DS7* compared to *DS2*, which is due to the fact that search sessions are treated differently by BBM. Our system treats every page load event on a SERP as the beginning of a new search session. That is, if a user clicks a result and then returns to the SERP for clicking another result, SMR interprets this as two separate sessions. However, BBM handles this as a single search session with two clickthrough events. Besides containing more of these "combined" search sessions, *DS7* also features ∼12% less clickthrough events. All in all, this results in BBM having less data available for training its relevance model, which is an explanation for the lower-quality prediction compared to *DS2*. The same holds for other datasets showing similar differences, *DS2* and *DS7* are only used for representative purposes here.

Why are the MCC values relatively low (< 0.5) in general? The data collected for evaluation featured a very low CTS ratio, i.e., the amount of interaction data exceeded the available relevance judgments by far. To give just one example, the CTS ratios of both *DS0* and *DS1* lie under 1%, which is similar for the remaining datasets. This and the fact that the datasets used for evaluation were relatively small (compared to a realistic long-term scenario) leads to a rather low data quality. Yet, in an evaluation with larger amounts of data, we would expect increasing MCC values. This is, e.g., indicated in [20], where the authors work with datasets that are notably larger than 1.5 GB. Also, Huang et al. state that "adding more data can result in an order of magnitude of greater improvement in the system than making incremental improvements to the processing algorithms" [12].

How does SMR deal with click spam? Click spam is a major problem in systems where clicks are the main indicator for relevance [19]. However, in the specific setting we are focusing on in this paper, a high number of *conversions* indicates high relevance. Since conversions imply a confirmed payment, we do not have to deal with "traditional" click spam as described in [19]. Yet, in settings where no conversions are available, our discriminative approach has to rely on other indicators of relevance, such as clicks on social media buttons, for training its models. In such cases, additional measures have to be taken that prevent fraudulent behavior aiming at manipulating relevance models. Potential measures could be based on, e.g., filtering pre-defined behavior profiles, blacklists, personalized search [19] or the ranking framework described by [2].

Table 1. Overall relative comparison of the considered approaches

	effectiveness	efficiency	robustness	scalability
SMR	0	++	++	++
BBM	−	++	+	++
TMR (baseline)	*0*	*0*	*0*	*0*
SMR_{click}	− −	++	++	++
TMR_{click}	− −	0	0	0

Summary. Table 1 shows a comparison of all approaches considered in the evaluation. Since the systems—due to differences in the underlying architectures—are difficult to compare in an absolute, hardware-independent manner, we give a comparison of relative performances. Using TMR as the baseline, "0" indicates similar performance, "+"/"−" indicate a tendency and "++"/"− −" indicate a major or significant difference.

5 Limitations and Future Work

The following section discusses limitations of SMR and provides an overview of potential future work.

As described, in this paper SMR specifically aims at relevance prediction in the context of *travel search*. One specific feature of this setting is the fact that we can use hotel booking conversions as indicators of relevance. However, in a more general setting, other implicit or explicit relevance judgments are necessary. For example, one could obtain such judgments by providing optional vote up/down buttons to visitors or tracking clicks on Facebook "Like" buttons of a search result. Transferring SMR into such a more general context is our current work-in-progress.

Concerning the evaluation of our system, we had to rely on relatively small datasets compared to the real-world settings the system is intended for in the long-term. As part of our future work, we intend to evaluate SMR with larger datasets that simulate a real-world setting of a timespan considerably longer than 10 days. This will also give us the chance to investigate the performance of the Hoeffding Tree classifier, which becomes feasible only for very massive amounts of data [6].

Currently, SMR is only able to track client-side interactions on desktop PCs, i.e., mouse input. However, since the mobile market is steadily growing, an increasing number of users access search engines using their (small-screen) touch devices. This demands for also making use of touch interactions for predicting the relevance of results. Leveraging these valuable information is especially important for search engine owners and intended in future versions of SMR.

Finally, interaction features are often coupled with temporal features or their values change over time. This has to be addressed in the context of *concept drift* [21]. SMR is generally capable of handling changing data streams, as Tsymbal states that "[i]ncremental learning is more suited for the task of handling concept drift" [21]. However, the Naïve Bayes classifier used in the context of this paper would have to be replaced by an adequate concept drift–ready learner. A potential candidate is the CVFDT learner, which is based on Hoeffding trees and dismisses a subtree based on old data whenever a subtree based on recent data becomes more accurate [14].

6   Conclusions

This paper presented SMR, which is a novel approach to providing incremental models for predicting the relevance of web search results from real-time user interaction data. Our approach helps to ensure one of the *prime aspects of search engine quality*, i.e., providing users with the most relevant results for their queries. In contrast to numerous existing approaches, SMR does not require re-processing of already processed data for obtaining an up-to-date relevance model. Moreover, our system involves interaction features other than clicks and was specifically designed for coping with large amounts of data in real-time. This allows for feeding relevance predictions back into SERPs with relatively low latency.

For evaluating SMR, we have simulated a *real-world setting* with large amounts of interaction data from two *large hotel booking portals*. Comparison of our system to an analogous batch-wise approach showed that SMR is able to predict relevances that do not differ significantly, although it has less information available for training. Furthermore, we have compared the discriminative SMR approach to BBM—a generative state-of-the-art click model for incrementally processing big data streams that is successfully applied in industry. Results show that prediction quality does not differ significantly between the two systems. Still, they indicate that predictions by SMR might compare favorably to those of BBM, as it outperforms the click model for the majority of datasets. Additionally, we have considered a click-only version of SMR that was compared to the complete system. From the significantly better predictions of the latter, we conclude that interactions other than clicks yield valuable information for relevance prediction and should not be neglected.

As future work, we plan to adjust SMR to more general settings besides travel search. Moreover, it is planned to further optimize the system regarding performance and perform an evaluation with even larger amounts of real-world interaction data.

Acknowledgments. We thank Christiane Lemke and Liliya Avdiyenko for supporting us with their implementation of BBM. This work has been supported by the ESF and the Free State of Saxony.

References

1. Baldi, P., Brunak, S., Chauvin, Y., Andersen, C.A., Nielsen, H.: Assessing the accuracy of prediction algorithms for classification: an overview. Bioinformatics 16(5) (2000)

2. Bian, J., Liu, Y., Agichtein, E., Zha, H.: A Few Bad Votes Too Many? Towards Robust Ranking in Social Media. In: Proc. AIRWeb (2008)
3. Chapelle, O., Zhang, Y.: A Dynamic Bayesian Network Click Model for Web Search Ranking. In: Proc. WWW (2009)
4. Craswell, N., Zoeter, O., Tylor, M., Ramsey, B.: An Experimental Comparison of Click Position-Bias Models. In: Proc. WSDM (2008)
5. Dean, J., Ghemawat, S.: MapReduce: Simplified Data Processing on Large Clusters. CACM 51(1) (2008)
6. Domingos, P., Hulten, G.: Mining High-Speed Data Streams. In: Proc. KDD (2000)
7. Dupret, G.E., Piwowarski, B.: A User Browsing Model to Predict Search Engine Click Data from Past Observations. In: Proc. SIGIR (2008)
8. Guo, F., Liu, C., Wang, Y.M.: Efficient Multiple-Click Models in Web Search. In: Proc. WSDM (2009)
9. Guo, Q., Agichtein, E.: Beyond Dwell Time: Estimating Document Relevance from Cursor Movements and other Post-click Searcher Behavior. In: Proc. WWW (2012)
10. Hall, M., Frank, E., Holmes, G., Pfahringer, B., Reutemann, P., Witten, I.H.: The WEKA Data Mining Software: An Update. SIGKDD Explor. Newsl. 11(1) (2009)
11. Huang, J.: On the Value of Page-Level Interactions in Web Search. In: HCIR Workshop (2011)
12. Huang, J., White, R.W., Buscher, G., Wang, K.: Improving Searcher Models Using Mouse Cursor Activity. In: Proc. SIGIR (2012)
13. Huang, J., White, R.W., Dumais, S.: No Clicks, No Problem: Using Cursor Movements to Understand and Improve Search. In: Proc. CHI (2011)
14. Hulten, G., Spencer, L., Domingos, P.: Mining Time-Changing Data Streams. In: Proc. KDD (2001)
15. Joachims, T.: Optimizing Search Engines using Clickthrough Data. In: Proc. KDD (2002)
16. Liu, C., Guo, F., Faloutsos, C.: BBM: Bayesian Browsing Model from Petabyte-scale Data. In: Proc. KDD (2009)
17. Marz, N.: Storm Wiki, http://github.com/nathanmarz/storm/wiki
18. Navalpakkam, V., Churchill, E.F.: Mouse Tracking: Measuring and Predicting Users' Experience of Web-based Content. In: Proc. CHI (2012)
19. Radlinski, F.: Addressing Malicious Noise in Clickthrough Data. In: LR4IR Workshop at SIGIR (2007)
20. Speicher, M., Both, A., Gaedke, M.: TellMyRelevance! Predicting the Relevance of Web Search Results from Cursor Interactions. In: Proc. CIKM (2013)
21. Tsymbal, A.: The problem of concept drift: definitions and related work. Technical Report, Trinity College Dublin (2004)
22. Zaharia, M., Das, T., Li, H., Hunter, T., Shenker, S., Stoica, I.: Discretized streams: A fault-tolerant model for scalable stream processing. Technical Report, UC Berkeley (2012)

The Forgotten Many? A Survey of Modern Web Development Practices

Moira C. Norrie, Linda Di Geronimo, Alfonso Murolo, and Michael Nebeling

Department of Computer Science, ETH Zurich
CH-8092 Zurich, Switzerland
{norrie,lindad,amurolo,nebeling}@inf.ethz.ch

Abstract. With an estimated 21.9% of the top 10 million web sites running on WordPress, a significant proportion of the web development community consists of WordPress developers. We report on a survey that was carried out to gain a better understanding of the profile of these developers and their web development practices. The first two parts of the survey on the background and development practices were not exclusive to WordPress developers and therefore provide insight into general web developer profiles and methods, while the third part focussed on WordPress specifics such as theme development. We present the results of the survey along with a discussion of implications for web engineering research.

Keywords: web engineering, web development practices, WordPress developers.

1 Introduction

Second-generation content management systems (CMS) such as WordPress[1] and Drupal[2] are based on a crowdsourcing model where vast developer communities share themes and plugins. It is possible for endusers to create a web site without any programming effort by selecting an existing theme and adding content, even adding or customising the functionality through the user interface. At the same time, developers with programming skills and knowledge of the platform can create or edit PHP templates, CSS stylesheets and JavaScript functions to extend the functionality or create their own themes and/or plugins.

The availability of these platforms has radically changed the web development landscape with estimates that 21.9% of the top 10 million web sites are running on WordPress which has 60.3% of the CMS market share[3]. While many sites running on WordPress are personal web sites, the platform also supports everything from web sites created by professional designers for small businesses

---

[1] http://www.wordpress.com, http://www.wordpress.org
[2] http://www.drupal.org
[3] http://w3techs.com/technologies/overview/content_management/all (10.4.2014).

S. Casteleyn, G. Rossi, and M. Winckler (Eds.): ICWE 2014, LNCS 8541, pp. 290–307, 2014.

to large, complex sites created by teams of developers. WordPress has gone well beyond its origins as a blogging platform and its web sites include popular online newspapers, e.g. Metro UK[4], as well as e-commerce sites, e.g. LK Bennett[5].

Yet, WordPress and its developers have received little attention within the web engineering research community. Information gleaned from books about WordPress, online articles and forums as well as talking to personal contacts, suggests that many WordPress web sites are developed by individuals with a mix of technical and design skills. Books on developing WordPress themes such as [1] propose an *interface-driven* approach where the main steps are to develop a mockup of the interface, add client-side functionality and then migrate to the WordPress platform. This contrasts with the *model-driven* approaches [2] widely promoted within the web engineering research community.

Since WordPress developers form a significant part of the development community, we think it is important to get a better understanding of their development practices with a view towards identifying requirements and research challenges. We therefore decided to carry out a survey of web development practices which, although not exclusively limited to WordPress developers, made efforts to reach out to this community.

The results of our survey show that there is a need to support alternative methods to model-driven web engineering that are more in line with widely-used interface-driven practices and can be integrated with platforms such as WordPress. Further, since many developers seek inspiration from existing web sites and frequently reuse elements of design and implementation from other projects, a major issue is how to provide better support for reuse in all aspects of web engineering.

In Sect. 2, we discuss the background to this work including previous surveys of web development practices. Section 3 provides details of our survey and how it was carried out. The results are presented in three sections. Section 4 reports on results related to developer profiles in terms of experience, educational background and the size of team and organisation for which they work. Results on general methods and tools used in development are then presented in Sect. 5. The third part of the survey was specific to WordPress developers and we report on the results for this part in Sect. 6. Implications for web engineering research are discussed in Sect. 7, while concluding remarks are given in Sect. 8.

2   Background

The discipline of web engineering emerged in the late 1990s with calls for systematic methods for the development of web applications, e.g. [3,4]. This in turn led to the first of the ICWE series of conferences in 2001 and the appearance of the Journal of Web Engineering (JWE) in 2002. A position paper [5] in the first issue of JWE defined web engineering as "the application of systematic, disciplined and quantifiable approaches to the development, operation and maintenance of

---

[4] http://metro.co.uk
[5] http://www.lkbennett.com

web-based applications". The paper presented the characteristics of both simple and advanced web-based systems, discussing how the development of such systems differed from traditional software engineering. The authors concluded that "web engineering at this stage is a moving target since web technologies are constantly evolving, making new types of applications possible, which in turn may require innovations in how they are built, deployed and maintained."

It is certainly true that both web technologies and the kinds of web-based applications in everyday use have changed dramatically over the last decade. Further, the emergence of second-generation CMS such as WordPress which offer powerful platforms for both the development and operation of all kinds of web sites has also changed how a significant proportion of web sites are built, deployed and maintained. By offering a WordPress hosting platform[6], it is even possible for endusers to literally create and deploy a web site in a few clicks. Meanwhile, developers with technical skills and knowledge of the WordPress model can develop both plugins and themes offering rich functionality for their own use and to share with others.

Examining the research literature in web engineering over the past decade reveals less radical changes in proposals for how web sites should be developed. Model-driven approaches such as OOHDM [6], UWE [7], WebML [8] and WSDM [9] were introduced in the 1990s and early 2000s. Many of these still prevail although the modelling languages may have been extended to cater for new kinds of technologies and applications. For example, the web modelling language WebML has been extended to cater for service-enabled applications [10] and context-awareness [11]. The continued emphasis on model-driven approaches may be due to the fact that the main focus still appears to be on development within, or for, large enterprises using multi-disciplinary development teams involving programmers, database architects and graphic designers. In such settings, it might be expected that the model-driven approaches widely used in software engineering and information systems would be familiar to both programmers and database architects and hence adaptations for web engineering would be more likely to be adopted. However, it is interesting to note that in a recent paper analysing model-driven web engineering methodologies [2], they comment on the fact that model-driven web engineering approaches have still not been widely adopted and they accredit this mainly to the lack of tools.

There is little recent research literature reporting on modern web development practices, especially concerning the use of platforms such as WordPress. A number of surveys were carried out in the early 2000s in conjunction with the call for web engineering to be established as a discipline. Barry and Lang [12] reported on a study in Ireland on multimedia software development methods, which included web-based information systems. Almost a quarter (24.6%) reported that they did not use a methodology while the rest stated that they used an in-house variant, with most using what the researchers considered as outdated methods and only 6.2% using UML. Reasons given for not using methodologies were that

---

[6] http://www.wordpress.com

they were "too cumbersone", "not suited to the real world" or "long training is required".

Taylor et al. [13] carried out a study of web development activities in 25 UK organisations based on interviews. They found that few formalised techniques were used and most "web site development activities appeared to be undertaken in an ad hoc manner" with only 8 of the 25 using design techniques such as hierarchy charts, flowcharts and storyboards. They reported little or no use of established software development techniques. Around the same time, McDonald and Welland [14] carried out a study of web development practices based on in-depth interviews, in this case involving 9 UK organisations. Only 7 of the 15 interviewees claimed to have a development process in place, with only 2 of these 7 using industry standard software development processes. Although the majority of interviewees were using prototyping or user-centred design techniques, none of them mentioned involving endusers in validating the success of a project.

More recently, El Sheikh and Tarawneh [15] reported on a survey of web engineering practices in small Jordanian companies. The results of their study showed that many developers had 5 or fewer years of software experience and that the development processes were still mainly ad hoc, with little application of established web practices.

We wanted to find out how much the situation has changed over the years in terms of the profile of web developers and also the methods used. In particular, we were interested in the community of WordPress developers and whether their backgrounds, work settings and methods differ significantly from developers that use some form of web development framework rather than a CMS as the basis for their implementation.

3 Survey

The survey was designed to address both web developers and designers including those specifically developing with and for WordPress. We designed a questionnaire consisting of 31 questions distributed over three parts: *background*, *development practices* and *WordPress development*. We used a mix of 5-point Likert-scale questions for frequency-based answers or where agreement with different statements was to be expressed as well as open-ended questions.

The first part collected demographics by asking participants to provide their age, gender and country of residence and origin. We also enquired about any formal qualifications in computer science, design and web development. Other questions addressed the participant's professional background and experience. We asked for the number of years working as a professional web developer, as well as the size of both the organisation they work for and their web development team. Participants were also encouraged to share any recent projects they developed and their role and specific contributions to the projects. These questions together enabled us to determine developer profiles that we will report in the next section.

The second part concerned their development practices with the goal of finding out about particular methods and tools used by participants. This part started with a question on how much they look at existing web sites for inspiration in the beginning of a project. Participants were asked how often they start by modifying an existing web site or theme as opposed to creating a new one from scratch. This was followed by questions on the use of sketching and digital mockups as well as the modelling of data and functional requirements. Participants were also asked to list any tools used for creating mockups and for modelling. These questions enabled us to better assess current development practices and identify trends between different groups of developers.

We also included a question on the reuse of resources published by other developers in terms of design or layout (HTML, CSS, etc.) and functionality (JavaScript, PHP, etc.). The goal was to get a better understanding of how different types of developers work and whether and how they make use of existing resources and material provided by other developers.

The second part closed with a question on the use of CMS such as WordPress or Drupal as opposed to web development frameworks as the starting point for web development. Participants were also asked to list the specific CMS and frameworks that they use. The answers to these questions were used to classify developers based on the software tools they typically use as the basis for development. These classifications were then used for comparison purposes in the analysis of other results.

Finally, the third part specifically dealt with WordPress development. Only participants indicating that they were WordPress developers were asked to complete this part of the survey. We asked participants whether they mainly use an existing theme, modify an existing theme, create a child theme or create their own theme from scratch when they create a web site using WordPress. These questions tried to characterise the role of themes as one of the main concepts supporting reuse in WordPress.

The last set of questions allowed us to further profile WordPress developers and identify their specific needs and requirements. We asked how often they reuse code from previous WordPress projects and find themselves in the situation that they would like to mix parts of two or more themes. As before, we again distinguished between layout/style and functionality for mixing and matching parts. Finally, participants were asked to indicate the need for more customisation options of WordPress themes and which additional features they would like to see added in future versions of WordPress to support theme development.

Before starting the online survey, we first asked members of our research group to fill it in and provide feedback on the design of the questionnaire. This allowed us to fix minor issues in the phrasing of some questions and calibrate the time typically required to answer all questions which was around 10 minutes. For dissemination, we primarily recruited via Twitter, reaching out to members of the web design and development community as well as the WordPress community, asking them to contribute to our survey and retweet our request for participation with a link to the online questionnaire. Targeted Twitter users ranged from users who frequently post and retweet links to articles related to web design and development to organisers of WordPress Meetup groups, giving us access to a network of several thousand followers of these active Twitter users. We also used Facebook and Reddit as well as directly contacting web developers known to us personally via email. Between January and February 2014, the survey was accessed 622 times and we received 208 complete responses that we included in the following analysis.

4   Developer Profiles

The 208 participants (83% male, 17% female) were from 24 different countries, with the majority living in the USA (49), Switzerland (45), Germany (39) or the UK (22). The age groups are shown in Fig. 1a and the years of professional web development in Fig. 1b.

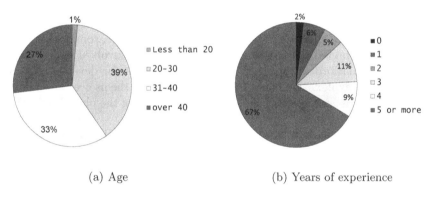

(a) Age   (b) Years of experience

Fig. 1. Age and experience of participants

It is interesting to note that we had good coverage of the different age groups and the majority of our participants (67%) had 5 or more more years of experience as a professional developer. This contrasts with the survey of El Sheikh and Tarawneh [15] where 63% had 5 or fewer years of software experience.

Since one of the aims of our survey was to compare the profiles and methods of developers using a CMS as their main development platform with those using web development frameworks such as Django[7], Ruby on Rails[8] and Bootstrap[9], we asked participants how often they use each of these approaches.

The results in Fig. 2 show that the CMS developers are more likely to stick with this approach as 39% of participants answered that they always use this approach while only 18% always use a development framework. It is important

---

[7] http://www.djangoproject.com
[8] http://rubyonrails.org
[9] http://getboostrap.com

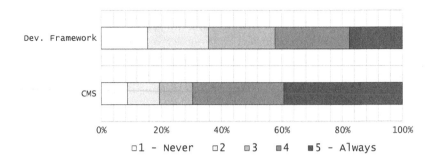

Fig. 2. Use of CMS or development framework

to note that these are not disjoint communities and 9% said that they always use both. 53% of the participants classified themselves as WordPress developers and 47% did not. Since we made efforts to target WordPress developers, it is not surprising that the majority were in this category, but we also achieved our aim to have a good mix of WordPress and non-WordPress developers. We note that some developers listed WordPress as a CMS that they use, but answered 'No' to the question asking if they are a WordPress developer. One reason for this might be that they interpreted the question as whether they are involved in developing the WordPress platform rather than whether they use it for developing applications. Another explanation could be that they classify themselves as endusers rather than developers since they create applications using the platform without actually doing any coding.

We classified the participants into three disjoint categories: those who answered 'Yes' to the question asking if they are a WordPress developer (WP), those who are not in WP but answered in the range 3-5 (sometimes to always) when asked if they use a CMS (CMS) and those who are not in WP and answered 1 or 2 (rarely or never) when asked if they use a CMS (Other). Thus developers who mostly use Drupal would be in the CMS category, while those who mainly use a web development framework and only occasionally use a CMS would be in the Other category. The sample sizes of each category are 111 (WP), 62 (CMS) and 35 (Other).

We asked participants how they would classify themselves in terms of whether they are designers, developers or both. Figure 3 reveals that, in all three categories, a significant proportion classified themselves as half-designer/half-developer (WP: 40%, CMS: 29%, Other: 34%), but there was also a significant proportion who classified themselves as 'developer' or 'mainly developer' (WP: 48%, CMS: 48%, Other: 63%). Since we were mainly targeting developer communities rather than design communities, we did not expect many participants to classify themselves as 'designer' or 'mainly designer'. Nevertheless, this shows that, rather than considering themselves as pure developers, many web development practitioners would see themselves as a mix of web developer and web designer.

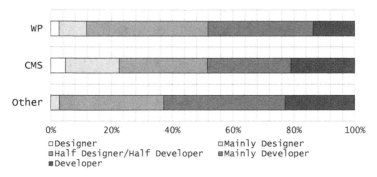

Fig. 3. Designer and/or developer

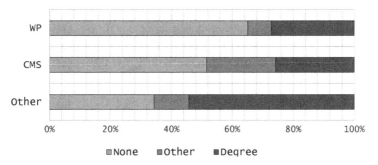

Fig. 4. Computer Science education

The educational background of the three categories is shown in Fig. 4 and 5. In all three categories, a significant proportion of participants have no formal qualification in computer science (WP: 65%, CMS: 52%, Other: 34%). Although we targeted developer communities rather than design communities, we also asked what, if any, qualification the participants have in design. As might be expected, relatively few have any formal qualification in design (WP: 31%, CMS: 25%, Other: 24%), although a significant proportion in each of the three categories classified themselves as half-designer/half-developer.

42% of all participants have no qualification in either computer science or design. In the case of participants who classified themselves as half-designer/half-developer, we also had 42% with no qualification in computer science or design. We also asked participants whether they have any kind of qualification in web development or specific web technologies. Taking this information into account, we still had 38% of participants with no formal education in computer science, design or web development.

Next, participants were asked to indicate the size of the organisation for which they work and also the size of their project team. The results shown in Fig. 6 indicate that a significant proportion of WordPress developers are either self-employed (42%) or belong to organisations with 5 or fewer employees (16%). On the other hand, it also shows that WordPress is not solely used by

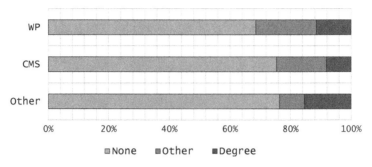

Fig. 5. Design education

individuals and small businesses since 10% of WordPress developers are working in organisations with more than 250 employees. While a significant proportion of non-CMS developers are also self-employed (14%) or in organisations with 5 or fewer employees (9%), the proportion working in organisations with more than 50 employees (51%) is far greater than for WordPress (20%).

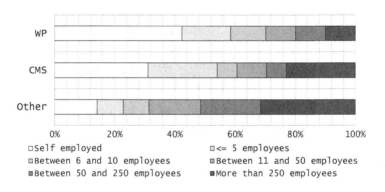

Fig. 6. Size of organisation

Previous surveys have tended to target organisations rather than individuals, and therefore have not involved developers who are self-employed. The smallest organisation involved in the survey by Taylor et al [13] had 20 employees. While the survey of El Sheikh and Tarawneh [15] targeted small companies in Jordan and 75% of companies had fewer than 10 employees, they also did not include self-employed developers.

Since many of our participants are self-employed or working in organisations with 5 or fewer members, clearly these developers either work alone or in very small teams and therefore the percentages for 'no team' and a team size of '5 or fewer' would be expected to reflect this. Still, even in larger organisations, participants often work in small teams and 75% of WordPress developers work in teams with 5 or fewer members, with only 7% working in teams with more than 10 members. In the case of the non-CMS participants (Other), team sizes still tend to be small with 51% working in teams with 5 or fewer members, but 26% of them do work in teams with more than 10 members.

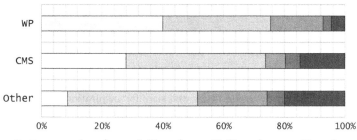

Fig. 7. Size of team

McDonald and Welland [14] estimated the average team size of web development projects in the organisations that they surveyed as 6. They argued that the small size of web development teams is one of the major differences to traditional software development teams, citing an article published in 2000 by Reifer [16] where he estimated the size of web development teams as 3-5 compared with traditional software development projects with hundreds of team members.

Interestingly, a survey of 200 Java developers carried out in 2011 by Munoz[10] reported that 40.7% worked in a team size of 1-5, 26.6% in a team size of 5-10 and 32.6% in teams larger than 10. These figures are actually not so different from the Other group where 52% work in teams of 1-5, 23% work in teams of 6-10 and 26% in teams larger than 10.

5   Methods and Tools

In this section, we report on the second part of the survey where all participants were asked questions about the methods and tools that they use in development projects. We started by asking them if they use existing web sites for inspiration at the beginning of a new project.

As shown in Fig. 8, more than 20% always look at examples of web sites for inspiration (WP: 23%, CMS: 24%, Other: 30%) and more than 50% answered 4 or 5 indicating that they often inspect examples (WP: 53%, CMS: 54%, Other: 67%). Although WordPress explicitly supports design-by-example through its notion of themes that can be easily accessed and previewed in online galleries, it is interesting to note that examples are used as much, if not more, in the Other group.

---

[10] http://www.antelink.com/blog/software-developer-survey-first-chapter.html

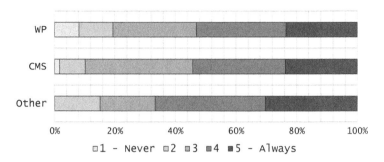

Fig. 8. Use of other web sites for inspiration

The survey by Taylor et al [13] published in 2002 also noted that examples of other web sites were often used for inspiration: "Roughly a third of those interviewed across 25 organisations studied indicated that they used other organisations' websites for design ideas in order to supplement their website design activities".

We also asked how often developers create a website or theme based on the modification of an existing web site or theme, either of their own or of another developer. As shown in Fig. 9, 61% of WordPress developers answered that they sometimes, often or always base the design of a web site or theme on an existing web site or theme. While none of the Other group answered that they always base a new design on an existing one, 50% said that they do this sometimes or often.

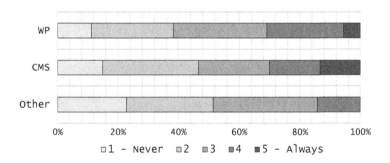

Fig. 9. Modify existing web sites or themes

Participants were asked how often they sketch mockups or create digital mockups and the results are shown in Fig. 10 and 11.

It is clear that sketching plays an important role with 43% of WordPress developers and 39% of the CMS group saying that they usually or always sketch. Sketching is used even more in the Other group with 57% stating that they usually or always sketch.

It is also common to produce digital mockups with more than 45% of the WordPress developers, 56% of the CMS group and 57% of the Other group answering 4 or 5 to indicate that they often or always use them. A range of tools

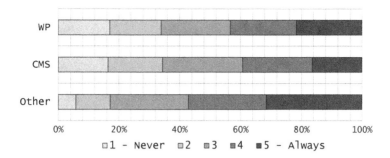

Fig. 10. Sketching mockups

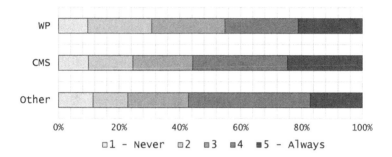

Fig. 11. Digital mockups

were listed including graphics editing tools such as Adobe Photoshop, diagram editors such as Microsoft Visio and wireframing tools such as Balsamiq[11].

Some developers wrote that they do not sketch or create digital mockups because they have a pure development role and implement the mockups produced by a graphic designer.

Since model-driven web engineering is widely promoted in the research community, we were interested in how frequently data and functional requirements are modelled. The results are shown in Fig. 12 and Fig. 13, respectively.

While only 28% of WordPress developers answered 4 or 5 to indicate that they often or always model data, 52% of Other developers answered that they often or always model data. The percentages answering that they often or always model functional requirements were also higher in the Other group (60%) compared to the WordPress developers (42%).

26% of participants answered that they never model data or functional requirements, leaving 74% who indicated that they use modelling at least some of the time. However, further analysis of the written comments provided by participants showed that the figures presented in Fig. 12 and 13 are very misleading as, in many cases, the participants had no idea what was meant by "modelling data" or "modelling functional requirements". We asked participants to list tools

---

[11] http://balsamiq.com

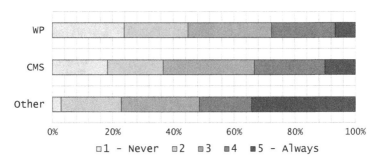

Fig. 12. Modelling data

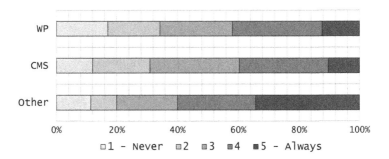

Fig. 13. Modelling functional requirements

that they use for modelling data and/or functional requirements and answers included "text documents", "spreadsheets and/or code editors", "Django to create prototypes" and "WordPress". One participant wrote "Not sure if I misunderstand this, but I usually just write requirements out—paper, text edit, google doc spreadsheets etc." Some listed project management tools and one participant even wrote something about testing and deployment. Only 11% of all participants listed an application or suite of tools that provides support for data or functional modelling. A further 5% wrote something general such as "paper and pen" or "whiteboard" that could also be considered as tools for modelling. This suggests that the number of developers actually doing some form of modelling of data or functional requirements is well below the figures reported.

This leads us to conclude that many of the participants are not even aware of software engineering practices, let alone applying them in even an informal way. This could be a consequence of the fact that a significant proportion of participants (WP 65%, CMS 52%, Other 34%) have no formal education in computer science.

6   WordPress Development Practices

The third part of the survey was only for WordPress developers as it deals specifically with the development of WordPress themes. A theme is a set of

PHP templates, CSS stylesheets and media objects that define the structure, navigation, functionality and presentation of a web site. The media objects included in a theme are generally static images used in the presentation of a web site such as the arrows used in sliders, buttons used in navigation and images that appear in the header. Endusers can select a theme from a gallery and create their own web site by simply adding content. A theme can also have a number of associated parameters to make it customisable through the general administrative interface. A professional developer will typically develop a theme to meet the requirements of a client, but they may also develop a theme for a particular class of clients such as restaurants, photographers or professional societies and make it customisable to the needs of a specific client.

The questions in this part of the survey were designed to find out more about how developers generate themes and specifically the forms of reuse that they employ or would like to have supported. Figure 14 presents an overview of the answers to a set of questions asking if and how they develop new themes for a specific project.

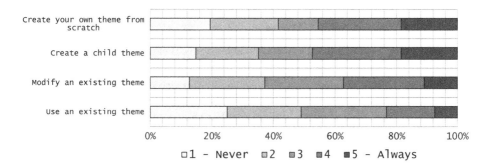

Fig. 14. Developing themes

7% of developers answered that they always use an existing theme. This means that these developers simply select a theme already provided by another developer and customise it for a client. This could involve the design of logos and other presentation features as well as the choice of layout, navigation and content.

19% of developers indicated that they always develop their own themes from scratch while 19% specified that they never do this. A developer can create a new theme based on an existing theme. This can be done by formally creating a child theme, but often developers will simply modify the PHP templates and CSS stylesheets provided. The results show that it is common for developers to build on existing themes using either of these approaches. 47% indicated that they often or always create a child theme of an existing theme for a project, while 37% answered that they often or always create a theme for a project by modifying an existing theme.

Since a theme defines an entire web site, a developer can only select a single theme as the starting point for a web site which means that they support all-or-nothing reuse. Once a theme has been selected, reuse of features from other themes can only be done by copying pieces of code and making any necessary modifications to integrate it into the theme under development.

We asked developers how often they find themselves in the situation where they would like to be able to mix parts of two or more existing themes. In the questions, we differentiated between the reuse of layout as specified by HTML and CSS and the reuse of functionality which could be either PHP code or JavaScript.

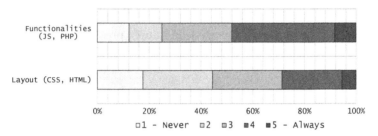

Fig. 15. Desire to be able to mix themes

Only 12% said they never find themselves in the situation where they would like to mix functionality from different themes, while 18% said they never want to mix layout. 75% answered 3-5 indicating that they sometimes, often or always find themselves wanting to mix functionality, while 56% answered 3-5 for layout.

We included a question asking participants to list what, if any, features they would like to see added to WordPress to support theme development. Most participants left this empty and the suggestions covered a range of issues from better means of managing media to easier ways of handling custom post types. One participant wrote: "I think where WordPress needs to go is to click-and-play development. Get rid of the need to code and it will take over the Internet". This comment can be interpreted as a request that it should go further in its support of enduser development.

7   Discussion

The results of our survey confirmed our impression that a significant proportion (40%) of WordPress developers work alone and act as both designer and developer. Since previous studies targeted organisations and tended to omit self-employed developers, it is impossible to say whether this is an increasing trend. However, the tendency for web developers to work in small teams as reported in earlier surveys is still the case, with 75% of WordPress developers working in teams with 1–5 members and only 26% of non-CMS developers working in teams with more than 10 members.

The fact that a significant proportion (40%) of both WordPress and non-CMS developers classified themselves as half-designer/half-developer, taken together with the fact that 41% of participants in this category have no qualification in either computer science or design, suggests that many of these developers have a mix of some design skills and some technical skills. Without a formal education in computer science and working alone or in very small organisations where there is likely to be a lack of in-house training, it may well be the case that many of these developers are not aware of modern software engineering methods, let alone using them. This would certainly be suggested by the answers that we received to our question about the tools that they use for the modelling of data and/or functional requirements. It is interesting to compare this with the results of the survey of Java developers carried out in 2011 by Munoz[12] where he reported that almost all participants had either a Bachelor or Masters degree in Computer Science.

This raises the question of whether efforts to adapt and promote software engineering methods, and specifically model-driven approaches, for web engineering are ever likely to have an impact in the web development community at large. Not only are many of these developers unaware of the underlying principles and techniques as well as the details of the methods, but many CMS developers have good reasons to employ interface-driven approaches rather than model-driven approaches. Therefore, while model-driven approaches may have their place in larger enterprises, we believe that the research community should also be exploring alternative methods that target practitioners at large.

One of the key findings of our study is how much developers build on the work of other designers and developers in their projects. This includes everything from using examples of other web sites for inspiration down to the detailed reuse of code. At the moment, there is little engineering support for reuse in CMS other than the concept of themes which support all-or-nothing reuse. Even the concept of child themes which is intended to provide a controlled way of developers building on existing themes is frequently not used and themes modified directly instead.

Within the web engineering research community, support for reuse has mainly been at the level of services. For example, WebComposition [17] allows applications to be built through hierarchical compositions of reusable application components. There has also been a lot of research in the area of web mashups to allow applications to be created through compositions of existing web sites, e.g. [18,19]. While this research is certainly relevant, the focus is purely on reuse rather on the design and development of new web sites as a whole and, as far as we know, there has been no attempt yet to adapt or integrate these methods into platforms such a WordPress. It is however important to mention that the work on mashups is also significant within the web engineering research community in its efforts to support enduser development.

Some researchers within the HCI community advocate a design-by-example approach [20,21] where the focus is very much on the reuse of the design aspects

---

[12] http://www.antelink.com/blog/software-developer-survey-first-chapter.html

of a web site. The idea is to allow users with little or no technical knowledge to develop their web site by selecting and combining elements of example web sites accessed in galleries. While the results of their studies are promising, they only deal with static elements and have not addressed the technical challenges of extracting and reusing functionality.

We believe that design-by-example is a promising paradigm worthy of detailed investigation within the web engineering research community. It is compatible with the interface-driven approaches that are currently in widespread use where mockups lead to prototypes that are gradually refined and migrated to platforms such as WordPress. With this goal in mind, we have started investigating how design-by-example could be supported in WordPress so that users could design and develop a fully functioning web site by selecting and reusing components of existing themes [22].

8   Conclusion

With a view to providing an insight into modern web practices, especially among the vast communities of WordPress developers, we have reported on the results of an online survey involving 208 participants working with CMS and/or web development frameworks. Unlike many previous surveys, we were keen to reach out to self-employed developers as well as developers within larger organisations and this we achieved.

The results point to the need for alternatives to model-driven approaches with a stronger focus on interface-driven development and enduser tools suited to the large numbers of developers with a lack of formal education in computer science and a mix of design and technical skills. Further, there is a need for methods that support the reuse of all aspects of web engineering and can be integrated into platforms such as WordPress that already have a significant proportion of the CMS market share and are continuing to grow.

Acknowledgements. We acknowledge the support of the Swiss National Science Foundation who financially supported this research under project FZFSP0_147257.

References

1. Silver, T.B., McCollin, R.: WordPress Theme Development - Beginner's Guide. Packt Publishing (2013)
2. Aragon, G., Escalona, M.-J., Lang, M., Hilera, J.R.: An Analysis of Model-Driven Web Engineering Methodologies. International Journal of Innovative Computing, Information and Control 9(1) (2013)
3. Coda, F., Ghezzi, G., Vigna, G., Garzotto, F.: Towards a Software Engineering Approach to Web Site Development. In: Proc. 9th Intl. Workshop on Software Specification and Design (1998)
4. Gellersen, H.W., Gaedke, M.: Object-Oriented Web Application Development. IEEE Internet Computing 3(1) (1999)

5. Deshpande, Y., Murugesan, S., Ginige, A., Hansen, S., Schwabe, D., Gaedke, M., White, B.: Web Engineering. Journal of Web Engineering 1(1) (2002)
6. Schwabe, D., Rossi, G.: The Object-Oriented Hypermedia Design Model. Communications of the ACM 38(8) (1995)
7. Koch, N., Kraus, A.: The Expessive Power of UML-based Web Engineering. In: Proc. 2nd Intl. Workshop on Web-Oriented Software Technology (IWWOST) (2002)
8. Ceri, S., Fraternali, P., Bongio, A.: Web Modeling Language (WebML): A Modeling Language For Designing Web Sites. Computer Networks 33(1-6), 137–157 (2000)
9. de Troyer, O., Leune, C.: WSDM: A User-Centred Design Method for Web Sites. Computer Networks and ISDN Systems 30(1-7) (1998)
10. Brambilla, M., Ceri, S., Fraternali, P., Acerbis, R., Bongio, A.: Model-Driven Design of Service-Enabled Web Applications. In: Proc. SIGMOD Industrial (2005)
11. Ceri, S., Daniel, F., Matera, M., Facca, F.M.: Model-Driven Development of Context-Aware Web Applications. TOIT 7(1) (2007)
12. Barry, C., Lang, M.: A Survey of Multimedia and Web Development Techniques and Methodology Usage. IEEE Multimedia (April–June 2001)
13. Taylor, M.J., McWilliam, J., Forsyth, H., Wade, S.: Methodologies and Web Site Development: A Survey of Practice. Information and Software Technology (44) (2002)
14. McDonald, A., Welland, R.: Web Engineering in Practice. In: Proc. 4th WWW Workshop on Web Engineering (2001)
15. Sheikh, A.E., Tarawneh, H.: A Survey of Web Engineering Practice in Small Jordanian Web Development Firms. In: Proc. 6th Joint Meeting on European Software Engineering Conference and ACM SIGSOFT Symposium on Foundations of Software Engineering (ESEC/FSE) (2007)
16. Reifer, D.: Web Development: Estimating Quick-to-Market Software. IEEE Software (November–December 2000)
17. Gellersen, H.-W., Wicke, R., Gaedke, M.: WebComposition: An Object-Oriented Support System for the Web Engineering Lifecycle. Computer Networks 29(8-13) (1997)
18. Daniel, F., Casati, F., Benatallah, B., Shan, M.C.: Hosted Universal Composition: Models, Languages and Infrastructure in mashArt. In: Laender, A.H.F., Castano, S., Dayal, U., Casati, F., de Oliveira, J.P.M. (eds.) ER 2009. LNCS, vol. 5829, pp. 428–443. Springer, Heidelberg (2009)
19. Cappiello, C., Matera, M., Picozzi, M., Sprega, G., Barbagallo, D., Francalanci, C.: DashMash: A Mashup Environment for End User Development. In: Auer, S., Díaz, O., Papadopoulos, G.A. (eds.) ICWE 2011. LNCS, vol. 6757, pp. 152–166. Springer, Heidelberg (2011)
20. Hartmann, B., Wu, L., Collins, K., Klemmer, S.R.: Programming by a Sample: Rapidly Creating Web Applications with d.mix. In: Proc. 20th ACM User Interface Software and Technology Symposium (UIST) (2007)
21. Lee, B., Srivastava, S., Kumar, R., Brafman, R., Klemmer, S.: Designing with Interactive Example Galleries. In: Proc. Conf. on Human Factors in Computings Systems (CHI) (2010)
22. Norrie, M.C., Di Geronimo, L., Murolo, A., Nebeling, M.: X-Themes: Supporting Design-by-Example. In: Casteleyn, S., Rossi, G., Winckler, M. (eds.) ICWE 2014. LNCS, vol. 8541, Springer, Heidelberg (2014)

Using Path-Dependent Types to Build Type Safe JavaScript Foreign Function Interfaces

Julien Richard-Foy, Olivier Barais, and Jean-Marc Jézéquel

IRISA, Université de Rennes, France

Abstract. The popularity of statically typed programming languages compiling to JavaScript shows that there exists a fringe of the programmer population interested in leveraging the benefits of static typing to write Web applications. To be of any use, these languages need to statically expose the Web browser dynamically typed native API, which seems to be a contradiction in terms. Indeed, we observe that existing statically typed languages compiling to JavaScript expose the browser API in ways that either are not type safe, or when they are, typically over constrain the programmers. This article presents new ways to encode the challenging parts of the Web browser API in static type systems such that both type safety and expressive power are preserved. Our first encoding relies on type parameters and can be implemented in most mainstream languages but drags phantom types up to the usage sites. The second encoding does not suffer from this inconvenience but requires the support of dependent types in the language.

1 Introduction

We recently observed the emergence of several statically typed programming languages compiling to JavaScript (*e.g.* Java/GWT [10], Dart [9], TypeScript [8], Kotlin[1], Opa[2], SharpKit[3], Haxe [2], Scala [7], Idris [1], Elm [6]). Though dynamic typing has its merits and supporters, the mere existence of these statically typed languages shows that there is also a community interested in benefiting from static typing features (allowing *e.g.* better refactoring support in IDE, earlier error detection, etc.) to write Web applications. Nevertheless, at some point developers need a way to interface with the underlying Web browser dynamically typed native API using a *foreign function interface* mechanism.

We observe that, even though these languages are statically typed, their integration of the browser API either is not type safe or over constrain the programmers. Indeed, integrating an API designed for a dynamically typed language into a statically typed language can be challenging. For instance, the `createElement` function return type depends on the value of its parameter: `createElement('div')` returns a `DivElement`, `createElement('input')` returns an `InputElement`, *etc.*

---

[1] `http://kotlin.jetbrains.org`
[2] `http://opalang.org`
[3] `http://sharpkit.net`

S. Casteleyn, G. Rossi, and M. Winckler (Eds.): ICWE 2014, LNCS 8541, pp. 308–321, 2014.

Most of the aforementionned languages expose this function by making it return an `Element`, the least upper bound of the types of all the possible returned values, thus loosing type information and requiring users to explicitly downcast the returned value to its expected, more precise type. Another way to expose this function consists in exposing several functions, each one fixing the value of the initial parameter along with its return type: `createDivElement`, `createInputElement`, *etc.* are parameterless functions returning a `DivElement` and an `InputElement`, respectively. This encoding forces to hard-code the name of the to-be created element: it cannot anymore be a parameter. In summary, the first solution is not type safe and the second solution reduces the expressive power of the API.

This paper reviews some common functions of the browser API, identifies the patterns that are difficult to encode in static type systems and shows new ways to encode them in such a way that both type safety and expressive power are preserved. We show that type parameters are sufficient to achieve this goal and that *path-dependent types* provide an even more convenient encoding of the browser API.

The remainder of the paper is organized as follows. The next section reviews the most common functions of the browser API and how they are typically integrated into statically typed languages. Section 3 shows two ways to improve their integration such that type safety and expressiveness are preserved. Section 4 validates our contribution and discusses its limits. Section 5 discusses some related works and Section 6 concludes.

2   Background

This section reviews the most commonly used browser functions and presents the different integration strategies currently used by the statically typed programming languages GWT, Dart, TypeScript, Kotlin, Opa, SharpKit, Haxe, Scala, Idris and Elm.

All these languages support a foreign function interface mechanism, allowing developers to write JavaScript expressions from their programs. Since this mechanism is generally untyped and error prone, most languages (TypeScript, Kotlin, Opa, SharpKit, Haxe, Scala and Elm) support a way to define *external* typed interfaces. Most of them also expose the browser API this way, as described in the next section.

2.1   The Browser API and Its Integration in Statically Typed Languages

The client-side part of the code of a Web application essentially reacts to user events (*e.g.* mouse clicks), triggers actions and updates the document (DOM) according to their effect. Table 1 lists the main functions supported by Web

Table 1. Web browsers main functions that are challenging to encode in a static type system

Name	Description
getElementsByTagName(name)	Find elements by their tag name
getElementById(id)	Find an element by its id attribute
createElement(name)	Create an element
target.addEventListener(name, listener)	React to events

browsers according to the Mozilla Developer Network[4] (we omit the functions that can trivially be encoded in a static type system).

To illustrate the challenges raised by these functions, we present a simple JavaScript program using them and show how it can be implemented in statically typed programming languages according to the different strategies used to encode these functions. Listing 1 shows the initial JavaScript code of the program. It defines a function slideshow that creates a slide show from an array of image URLs. The function returns an image element displaying the first image of the slide show, and each time a user clicks on it with the mouse left button the next image is displayed.

```
function slideshow(sources) {
  var img = document.createElement('img');
  var current = 0;
  img.src = sources[current];
  img.addEventListener('click', function (event) {
    if (event.button == 0) {
      current = (current + 1) % (sources.length - 1);
      img.src = sources[current];
    }
  });
  return img
}
```

Listing 1. JavaScript function creating a slide show from an array of image URLs

The most common way to encode the DOM API in statically typed languages is to follow the standard interface specifications of HTML [18] and DOM [4].

The main challenge comes from the fact that the parameter types and return types of these functions are often too general. Indeed, functions getElementsByTagName(name), getElementById(id) and createElement(name) can return values of type DivElement or InputElement or any other subtype of Element (their least upper bound). The interface of Element is more general and provides less features than its subtypes. For instance, the ImageElement type (representing images) has a src property that does not exist at the Element level. Similarly, the MouseEvent type has a button property that does not exist at the (more general) Event level, used by the function addEventListener.

---

[4] https://developer.mozilla.org/en-US/docs/DOM/DOM_Reference/Introduction

```
def slideshow(sources: Array[String]): ImageElement = {
  val img =
    document.createElement("img").asInstanceOf[ImageElement]
  var current = 0
  img.src = sources(current)
  img.addEventListener("click", event => {
    if (event.asInstanceOf[MouseEvent].button == 0) {
      current = (current + 1) % (sources.size - 1)
      img.src = sources(current)
    }
  })
  img
}
```

Listing 2. Scala implementation of `slideshow` using the standard HTML and DOM API

Listing 2 shows a Scala implementation of the `slideshow` program using an API following the standard specifications of HTML and DOM. The listing contains two type casts, needed to use the `src` property on the `img` value and the `button` property on the `event` value, respectively.

These type casts make the code more fragile and less convenient to read and write. That's why some statically typed languages attempt to provide an API preserving types as precisely as possible.

Of course, in the case of `getElementById(id)`, the `id` parameter does not give any clue on the possible type of the searched element, so it is hard to more precisely infer the return type of this function. Hence, most implementations use `Element` as their return type.

However, in the case of `getElementsByTagName(name)` and `createElement(name)`, there is exactly one possible return type for each value of the `name` parameter: e.g. `getElementsByTagName('input')` always returns a list of `InputElement` and `createElement('div')` always returns a `DivElement`. This feature makes it possible to encode these two functions by defining as many parameterless functions as there are possible tag names, where each function fixes the initial `name` parameter to be one of the possible values and exposes the corresponding specialized return type.

The case of `target.addEventListener(name, listener)` is a bit different. The `name` parameter defines the event to listen to while the `listener` parameter identifies the function to call back each time such an event occurs. Instead of being polymorphic in its return type, it is polymorphic in its `listener` parameter. Nevertheless, a similar property as above holds: there is exactly one possible type for the `listener` parameter for each value of the `name` parameter. For instance, a listener of `'click'` events is a function taking a `MouseEvent` parameter, a listener of `'keydown'` events is a function taking a `KeyboardEvent` parameter, and so on. The same pattern as above (defining a set of functions fixing the `name` parameter value) can be used to encode this function in statically typed languages.

```scala
def slideshow(sources: Array[String]): ImageElement = {
  val img = document.createImageElement()
  var current = 0
  img.src = sources(current)
  img.addClickEventListener { event =>
    if (event.button == 0) {
      current = (current + 1) % (sources.size - 1)
      img.src = sources(current)
    }
  }
  img
}
```

Listing 3. Scala implementation of `slideshow` using specialized functions

Listing 3 shows what would our `slideshow` implementation look like using such an encoding. There are two modifications compared to Listing 2: we use `document.createImageElement` instead of `document.createElement`, and we use `img.addClickEventListener` instead of `img.addEventListener`.

The `createImageElement` function takes no parameter and returns a value of type `ImageElement`, and the `addClickEventListener` function takes as parameter a function that takes a `MouseEvent` value as parameter, ruling out the need for type casts.

In the case of the `addEventListener` function we also encountered a slight variation of the encoding, consisting in defining one general function taking one parameter carrying both the information of the event name and the event listener.

```scala
img.addEventListener(ClickEventListener { event =>
  // ...
})
```

Listing 4. Implementation of `slideshow` using a general `addEventListener` function taking one parameter containing both the event name and the even listener

Listing 4 shows the relevant changes in our program if we use this encoding. The `addEventListener` function takes one parameter, a `ClickEventListener`, carrying both the name of the event and the event listener code.

Most of the studied languages expose the browser API following the standard specification, but some of them (GWT, Dart, HaXe and Elm) define a modified API getting rid of (or at least reducing) the need for downcasting, following the approaches described above.

2.2   Limitations of Existing Encoding Approaches

We distinguished three approaches to integrate the challenging parts of the browser API into statically typed languages. This section shows that each approach favours either type safety or expressive power but none provides both type safety *and* the same expressive power as the native browser API. We indeed consider that an API requiring users to do type casts is not type safe, while an API making it impossible to implement a function that can readily be implemented using the native API gives less expressive power to the programmer.

The first approach, consisting in using the least upper bound of all the possible types has the same expressive power as the native browser API, but is not type safe because it sometimes requires developers to explicitly downcast values to their expected specialized type.

The second approach, consisting in defining as many functions as there are possible return types of the encoded function, is type safe but leads to a less general API: each function fixes a parameter value of the encoded function, hence being less general. The limits of this approach are better illustrated when one tries to combine several functions. Consider for instance Listing 5 defining a JavaScript function `findAndListenTo` that both finds elements and registers an event listener when a given event occurs on them. Note that the event listener is passed both the event and the element: its type depends on both the tag name and the event name. This function cannot be implemented if the general functions `getElementsByTagName` and `addEventListener` are not available. The best that could be done would be to create one function for each combination of tag name and event name, leading to an explosion of the number of functions to implement. Thus, this approach gives less expressive power than the native browser API. Moreover, we find that defining many functions for the same task (creating a DOM element or listening to an event) clutters the API documentation: functions serving other purposes are hidden by these *same-purpose-functions*.

The third approach, consisting in combining two parameters into one parameter carrying all the required information, is type safe too, but reduces the expressive power because it forbids developers to partially apply the function by supplying only one parameter. Consider for instance Listing 6 that defines a function `observe` partially applying the `addEventListener` function[5]. Such a function cannot be implemented with this approach because the name of the event and the code of the listener cannot be decoupled. Thus, this one gives less expressive power than the native browser API.

In summary, the current integration of the browser API by statically typed languages compiling to JavaScript is either not type safe or not as expressive as the underlying JavaScript API. Indeed, we showed that our simple `slideshow` program requires type casts if the browser API is exposed according to the standard specification. We are able to get rid of type casts on this program by using modified browser APIs, but we presented two functions that we were not able to implement using these APIs, showing that they give less expressive power than the native API.

---

[5] The code of this function has been taken (and simplified) from the existing functional reactive programming libraries Rx.js [14] and Bacon.js
(http://baconjs.github.io/).

```
function findAndListenTo(tagName, eventName, listener) {
  var elements = document.getElementsByTagName(tagName);
  elements.forEach(function (element) {
    element.addEventListener(eventName, function (event) {
      listener(event, element);
    });
  });
}
```

Listing 5. Combination of use of `getElementsByTagName` and `addEventListener`

```
function observe(target, name) {
  return function (listener) {
    target.addEventListener(name, listener);
  }
}
```

Listing 6. Partial application of `addEventListener` parameters

This article aims to answer the following questions: is it possible to expose the browser API in statically typed languages in a way that both reduces the need for type casts and preserves the same expressive power? What typing mechanisms do we need to achieve this? Would it be convenient to be used by end developers?

3 Contribution

In this section we show how we can encode the challenging main functions of the DOM API in a type safe way while keeping the same expressive power.

The listings in this paper use the Scala language, though our first solution could be implemented in any language with basic type parameters support, such as Java's *generics*[6]. Our second solution is an improvement over the first one, using *path-dependent types*.

3.1 Parametric Polymorphism

In all the cases where a type `T` involved in a function depends on the value of a parameter `p` of this function (all the aforementionned functions of the DOM API are in this case), we can encode this relationship in the type system using type parameters as follows:

1. Define a parameterized class `P[U]`
2. Set the type of `p` to `P[U]`
3. Use type `U` instead of type `T`
4. Define as many values of type `P[U]` as there are possible values for `p`, each one fixing its `U` type parameter to the corresponding more precise type

---

[6] For a lack of space, we do not present them here but all Java versions of all the Scala listings (excepted those using type members) are available online at http://github.com/js-scala/js-scala/wiki/ICWE'14

```
class ElementName[E]

trait Document {
  def createElement[E](name: ElementName[E]): E
  def getElementsByTagName[E](name: ElementName[E]): Array[E]
}

val Input = new ElementName[InputElement]
val Img = new ElementName[ImageElement]
// etc. for each possible element name
```

Listing 7. Encoding of the `createElement` function using type parameters

Listing 7 shows this approach applied to the `createElement` and `getElementsByTagName` functions which return type depends on their `name` parameter value: a type `ElementName[E]` has been created, the type of the `name` parameter has been set to `ElementName[E]` instead of `String`, and the return type of the function is `E` instead of `Element` (or `Array[E]` instead of `Array[Element]`, in the case of `getElementsByTagName`). The `ElementName[E]` type encodes the relationship between the name of an element and the type of this element[7]. For instance, we created a value `Input` of type `ElementName[InputElement]`.

Listing 8 shows the encoding of the `addEventListener` function. The `EventName[E]` type represents the name of an event which type is `E`. For instance, `Click` is a value of type `EventName[MouseEvent]`: when a user adds an event listener to the `Click` event, it fixes to `MouseEvent` the type parameter `E` of the `callback` function passed to `addEventListener`.

```
class EventName[E]

trait EventTarget {
  def addEventListener[E](
        name: EventName[E], callback: E => Unit): Unit
}

val Click = new EventName[MouseEvent]
val KeyUp = new EventName[KeyboardEvent]
// etc. for each possible event name
```

Listing 8. Encoding of the `addEventListener` function using type parameters

Listing 9 illustrates the usage of such an encoding by implementing our `slideshow` program presented in the introduction. Passing the `Img` value as a parameter to the `createElement` function fixes its `E` type parameter to `ImageElement`

---

[7] The type parameter `E` is also called a *phantom type* [12] because `ElementName` values never hold a `E` value.

```
def slideshow(sources: Array[String]) {
  val img = document.createElement(Img)
  var current = 0
  img.src = sources(current)
  img.addEventListener(Click, event => {
    if (event.button == 0) {
      current = (current + 1) % (sources.length - 1)
      img.src = sources(current)
    }
  })
  img
}
```

Listing 9. Scala implementation of the `slideshow` function using generics

so the returned value has the most possible precise type and the `src` property can be used on it. Similarly, passing the `Click` value to the `addEventListener` function fixes its `E` type parameter to `MouseEvent`, so the event listener has the most possible precise type and the `button` property can be used on the `event` parameter.

It is worth noting that this code is actually exactly the same as in Listing 2 excepted that type casts are not anymore required because the browser API is exposed in a way that preserves enough type information. Our way to encode the browser API is more type safe, but is it as expressive as the native API?

Listings 10 and 11 show how the challenging functions of Section 2.2, `findAndListenTo` and `observe`, can be implemented with our encoding. They are basically a direct translation from JavaScript syntax to Scala syntax, with additional type annotations.

```
def findAndListenTo[A, B](
      tagName: ElementName[A],
      eventName: EventName[B],
      listener: (A, B) => Unit) = {
  for (element <- document.getElementsByTagName(tagName)) {
    element.addEventListener(eventName, event => {
      listener(event, element)
    })
  }
}
```

Listing 10. Combination of `getElementsByTagName` and `addEventListener` functions encoded using type parameters

In summary, our encoding is type safe and gives as much expressive power as the native API since it is possible to implement exactly the same functions as we are able to implement in plain JavaScript.

```
def observe[A](target: EventTarget, name: EventName[A]) = {
  (listener: A => Unit) => {
    target.addEventListener(name, listener)
  }
}
```

Listing 11. Partial application of **addEventListener** encoded with type parameters

However, every function taking an element name or an event name as parameter has its type signature cluttered with phantom types (extra type parameters): the **observe** function takes a phantom type parameter **A** and the **findAndListenTo** function takes two phantom type parameters, **A** and **B**. These extra type parameters are redundant with their corresponding value parameters and they make type signatures harder to read and reason about.

3.2 Path-Dependent Types

This section shows how we can remove the extra type parameters needed in the previous section by using *path-dependent types* [16]. Essentially, the idea is to model type parameters using *type members*, as suggested in [17].

Programming languages generally support two means of abstraction: parameterization and abstract members. For instance Java supports parameterization for values (method parameters) and types (*generics*), and member abstraction for values (abstract methods). Scala also supports member abstraction for types through type members [5,16]. An abstract type member of a class is an inner abstract type that can be used to qualify values. Subclasses can implement and override their methods, and similarly they can define or refine their type mem-

```
trait ElementName {
  type Element
}

trait Document {
  def createElement(name: ElementName): name.Element
  def getElementsByTagName(
      name: ElementName): Array[name.Element]
}

object Div extends ElementName {
  type Element = DivElement
}
object Input extends ElementName {
  type Element = InputElement
}
// etc. for each possible element name
```

Listing 12. Encoding of **createElement** using path-dependent types bers. A concrete subclass must provide a concrete implementation of its type members. Outside of the class, type members can be referred to using a type selection on an instance of the class: the type designator p.C refers to the C type member of the value p and expands to the C type member implementation of the singleton type of p.

```
trait EventName {
  type Event
}

object Click extends EventName { type Event = MouseEvent }

trait EventTarget {
  def addEventListener(name: EventName)
                      (callback: name.Event => Unit): Unit
}
```

Listing 13. Encoding of addEventListener using path-dependent types

Listings 12 and 13 show an encoding of createElement, getElementsByTagName and addEventListener in Scala using type members. Now, the ElementName type has no type parameter but a type member Element. The return type of the createElement function is name.Element: it refers to the Element type member of its name parameter. The Div and Input values illustrate how their corresponding element type is fixed: if one writes createElement(Input), the return type is the Element type member of the Input value, namely InputElement. The same idea applies to EventName and addEventListener: the name of the event fixes the type of the callback.

The implementation of the slideshow function with this encoding is exactly the same as with the previous approach using generics. However, functions findAndListenTo and observe can be implemented more straightforwardly, as shown by listings 14 and 15, respectively.

With this encoding, the functions using event names or element names are not anymore cluttered with phantom types, and type safety is still preserved.

```
def findAndListenTo(eltName: ElementName, evtName: EventName)
    (listener: (evtName.Event, eltName.Element) => Unit) = {
  for (element <- document.getElementsByTagName(eltName) {
    element.addEventListener(evtName) { event =>
      listener(event, element)
    }
  }
}
```

Listing 14. Combination of getElementsByTagName and addEventListener using path-dependent types

```
def observe(target: EventTarget, name: EventName) =
  (listener: (name.Event => Unit)) => {
    target.addEventListener(name)(listener)
  }
```

Listing 15. Partial application of `addEventListener` using path-dependent types

4 Validation

4.1 Implementation in js-scala

We implemented our encoding in js-scala [11], a Scala library providing composable JavaScript code generators[8]. On top of that we implemented various samples, including non trivial ones like a realtime chat application and a poll application.

We have shown in this paper that our encoding leverages types as precisely as possible (our `slideshow` program is free of type casts) while being expressive enough to implement the challenging `findAndListenTo` and `observe` functions that were impossible to implement with other approaches.

4.2 API Clarity

We mentioned in the background section that a common drawback of existing approaches to bring more type safety was the multiplication of functions having the same purpose, making the API documentation harder to read.

Our encoding preserves a one to one mapping with browser API functions whereas existing approaches often have more than 30 functions for a same purpose. For instance, the `createElement` function is mapped by 31 specialized functions in GWT and 62 in Dart, the `addEventListener` is mapped by 32 specialized functions in GWT and 49 in Dart.

4.3 Convenience for End Developers

Statically typed languages are often criticized for the verbosity of the information they add compared to dynamically typed languages [15]. In our case, what is the price to pay to get accurate type ascriptions?

The first encoding, using type parameters, can be implemented in most programming languages because it only requires a basic support of type parameters (for instance Java, Dart, TypeScript, Kotlin, HaXe, Opa, Idris and Elm can implement it). However this encoding leads to cluttered type signatures and forces functions parameterized by event or element names to also take phantom type parameters.

However, the second encoding, using type members, leads to type signatures that are not more verbose than those of the standard specifications of the HTML

---

[8] Source code is available at `http://github.com/js-scala`

and DOM APIs, so we argue that there is no price to pay. However, this encoding can only be implemented in language supporting type members or dependent types (Scala and Idris).

4.4 Limitations

Our encodings only work with cases where a polymorphic type can be fixed by a value. In our examples, the only one that is not in this case is `getElementById`. Therefore we are not able to type this function more accurately (achieving this would require to support the DOM tree itself in the type system as in [13]).

Our solution is actually slightly less expressive than the JavaScript API: indeed, the value representing the name of an event or an element is not anymore a `String`, so it cannot anymore be the result of a `String` manipulation, like *e.g.* a concatenation. Fortunately, this case is uncommon.

5   Related Works

The idea of using dependent types to type JavaScript has already been explored by Ravi Chugh *et. al.* [3]. They showed how to make a subset of JavaScript statically typed using a dependent type system. However, their solution requires complex and verbose type annotations to be written by developers.

Sebastien Doreane proposed a way to integrate JavaScript APIs in Scala [7]. His approach allows developers to seamlessly use JavaScript APIs from statically typed Scala code. However, his work does not expose types as precise as ours (*e.g.* in their encoding the return type of `createElement` is always `Element`).

TypeScript supports overloading on constant values: the type of the expression `createElement("div")` is statically resolved to `DivElement` by the constant parameter value `"div"`. This solution is type safe, as expressive and as easy to learn as the native API because its functions have a one to one mapping. However, this kind of overloading has limited applicability because overload resolution requires parameters to be constant values: indeed, the `findAndListenTo` function would be weakly typed with this approach.

6   Conclusion

Having a statically typed programming language compiling to JavaScript is not enough to leverage static typing in Web applications. The native browser API has to be exposed in a statically typed way, but this is not an easy task.

We presented two ways to encode dynamically typed browser functions in mainstream statically typed languages like Java and Scala, using type parameters or path-dependent types. Our encodings give more type safety than existing solutions while keeping the same expressive power as the native API.

We feel that parametric polymorphism and, even more, dependent types are precious type system features for languages aiming to bring static typing to Web applications.

Acknowledgements. The authors would like to thank Thomas Degueule.

References

1. Brady, E.: Idris, a general-purpose dependently typed programming language: Design and implementation. Journal of Functional Programming 23(05), 552–593 (2013)
2. Cannasse, N.: Using haxe. The Essential Guide to Open Source Flash Development, 227–244 (2008)
3. Chugh, R., Herman, D., Jhala, R.: Dependent types for javascript. SIGPLAN Not. 47(10), 587–606 (2012)
4. W3C-World Wide Web Consortium et al.: Document object model (dom) level 3 core specification. W3C recommendation (2004)
5. Cremet, V., Garillot, F., Lenglet, S., Odersky, M.: A core calculus for scala type checking. In: Královič, R., Urzyczyn, P. (eds.) MFCS 2006. LNCS, vol. 4162, pp. 1–23. Springer, Heidelberg (2006)
6. Czaplicki, E.: Elm: Concurrent frp for functional guis (2012)
7. Doeraene, S.: Scala.js: Type-Directed Interoperability with Dynamically Typed Languages. Technical report (2013)
8. Fenton, S.: Typescript for javascript programmers (2012)
9. Griffith, R.: The dart programming language for non-programmers-overview (2011)
10. Kereki, F.: Web 2.0 development with the Google web toolkit. Linux J., 2009(178) (February 2009)
11. Kossakowski, G., Amin, N., Rompf, T., Odersky, M.: JavaScript as an embedded DSL. In: Noble, J. (ed.) ECOOP 2012. LNCS, vol. 7313, pp. 409–434. Springer, Heidelberg (2012)
12. Leijen, D., Meijer, E.: Domain specific embedded compilers. ACM SIGPLAN Notices 35, 109–122 (1999)
13. Lerner, B.S., Elberty, L., Li, J., Krishnamurthi, S.: Combining Form and Function: Static Types for JQuery Programs. In: Castagna, G. (ed.) ECOOP 2013. LNCS, vol. 7920, pp. 79–103. Springer, Heidelberg (2013)
14. Liberty, J., Betts, P.: Reactive extensions for javascript. In: Programming Reactive Extensions and LINQ, pp. 111–124. Springer (2011)
15. Meijer, E., Drayton, P.: Static typing where possible, dynamic typing when needed: The end of the cold war between programming languages
16. Odersky, M., Cremet, V., Röckl, C., Zenger, M.: A nominal theory of objects with dependent types. In: Cardelli, L. (ed.) ECOOP 2003. LNCS, vol. 2743, pp. 201–224. Springer, Heidelberg (2003)
17. Odersky, M., Zenger, M.: Scalable component abstractions. ACM SIGPLAN Notices 40, 41–57 (2005)
18. Raggett, D., Le Hors, A., Jacobs, I., et al.: Html 4.01 specification. W3C Recommendation 24 (1999)

Visual vs. DOM-Based Web Locators:
An Empirical Study

Maurizio Leotta[1], Diego Clerissi[1], Filippo Ricca[1], and Paolo Tonella[2]

[1] DIBRIS, Università di Genova, Italy
[2] Fondazione Bruno Kessler, Trento, Italy
{maurizio.leotta,filippo.ricca}@unige.it,
diego.clerissi@gmail.com, tonella@fbk.eu

Abstract. Automation in Web testing has been successfully supported by DOM-based tools that allow testers to program the interactions of their test cases with the Web application under test. More recently a new generation of visual tools has been proposed where a test case interacts with the Web application by recognising the images of the widgets that can be actioned upon and by asserting the expected visual appearance of the result.

In this paper, we first discuss the inherent robustness of the locators created by following the visual and DOM-based approaches and we then compare empirically a visual and a DOM-based tool, taking into account both the cost for initial test suite development from scratch and the cost for test suite maintenance during code evolution. Since visual tools are known to be computationally demanding, we also measure the test suite execution time.

Results indicate that DOM-based locators are generally more robust than visual ones and that DOM-based test cases can be developed from scratch and evolved at lower cost. Moreover, DOM-based test cases require a lower execution time. However, depending on the specific features of the Web application under test and its expected evolution, in some cases visual locators might be the best choice (e.g., when the visual appearance is more stable than the structure).

1 Introduction

The importance of test automation in Web engineering comes from the widespread use of Web applications (Web apps) and the associated demand for code quality. Test automation is considered crucial for delivering the quality levels expected by users [14], since it can save a lot of time in testing and it helps developers to release Web apps with fewer defects [1]. The main advantage of test automation comes from fast, unattended execution of a set of tests after some changes have been made to a Web app.

Several approaches can be employed to automate functional Web testing. They can be classified using two main criteria: the first concerns how test cases are developed, while, the second concerns how test cases localize the Web elements (i.e., GUI components) to interact with, that is what kind of *locators* (i.e., objects that select the target web elements) are used. Concerning the first criterion, it is possible to use the capture-replay or the programmable approach. Concerning the second criterion, there are three main approaches, Visual (where image recognition techniques are used to *locate* GUI components and a locator consists of an image), DOM-based (where Web

S. Casteleyn, G. Rossi, and M. Winckler (Eds.): ICWE 2014, LNCS 8541, pp. 322–340, 2014.

page elements are *located* using the information contained in the Document Object Model and a locator is, for instance, an XPath expression) and Coordinates-based (where screen coordinates of the Web page elements are used to interact with the Web app under test). This categorization will be deeply analysed in the next section.

For developers and project managers it is not easy to select the most suitable automated functional web testing approach for their needs among the existing ones. For this reason, we are carrying out a long term research project aimed at empirically investigating the strengths and weaknesses of the various approaches (see also our previous work [9]).

In this work we evaluate and compare the visual and DOM-based approaches considering: the robustness of *locators*, the initial test suite development effort, the test suite evolution cost, and the test suite execution time. Our empirical assessment of the robustness of locators is quite general and tool independent, while the developers' effort for initial test suite development and the effort for test suite evolution were measured with reference to specific implementations of the two approaches. We instantiated such analysis for two specific tools, Sikuli API and Selenium WebDriver, both adopting the programmable approach but differing in the way they localize the Web elements to interact with during the execution of the test cases. Indeed, Sikuli API adopts the visual approach, thus using images representing portions of the Web pages, while Selenium WebDriver employs the DOM-based approach, thus relying on the HTML structure. We selected six open source Web apps and for each tool, we first developed a test suite per application and then we evolved them to a subsequent version. Moreover, since visual tools are known to be computational demanding, we also measured and compared the test suite execution time.

The paper is organized as follows: Sect. 2 gives some background on test case development using the visual and the programmable approaches, including examples for the two specific tools used in this work. In the same section, we describe test case repair activities. Sect. 3 describes our empirical study, reports the obtained results and discusses the pros and cons of the two considered approaches. We then present the related works (Sect. 4), followed by conclusions and future work (Sect. 5).

2   Background

There are several approaches for functional Web testing [13] and the choice among them depends on a number of factors, including the technology used by the Web app and the tools (if any) used for Web testing. Broadly speaking, there are two main criteria to classify the approaches to functional Web testing that are related to: (1) test case construction; and, (2) Web page element localisation.

For what concerns the first criterion, we can find two main approaches:

1) *Capture-Replay (C&R) Web Testing*: this approach consists of recording the actions performed by the tester on the Web app GUI and generating a script that provides such actions for automated, unattended re-execution.

2) *Programmable Web Testing*: this approach aims at unifying Web testing with traditional testing, where test cases are themselves software artefacts that developers write, with the help of specific testing frameworks. For Web apps, this means that the framework supports programming of the interaction with a Web page and its elements, so

that test cases can, for instance, automatically fill-in and submit forms or click on hyperlinks.

An automated functional test case interacts with several Web page elements such as links, buttons, and input fields, and different methods can be employed to locate them. Thus, concerning the second criterion, we can find three different cases[1]:

1) *Coordinate-Based Localisation:* first generation tools just record the screen coordinates of the Web page elements and then use this information to locate the elements during test case replay. This approach is nowadays considered obsolete, because it produces test cases that are extremely fragile.

2) *DOM-Based Localisation:* second generation tools locate the Web page elements using the information contained in the Document Object Model. For example, the tools Selenium IDE and WebDriver employ this approach and offer several different ways to locate the elements composing a Web page (e.g., ID, XPath and LinkText).

3) *Visual Localisation:* third generation tools have emerged recently. They make use of image recognition techniques to identify and control GUI components. The tool Sikuli API belongs to this category.

In our previous work [9], we compared the capture-replay approach and the programmable approach using two 2nd generation tools: Selenium IDE and Selenium WebDriver. In this work, we fixed the test case definition method (i.e., programmable) and changed the locator type, with the aim of comparing the visual approach and the DOM-based approach. Empirical results refer to two specific programmable tools: Sikuli API and Selenium WebDriver.

Let us consider a running example, consisting of a typical login web page (login.asp). The login page requires the users to enter their credentials, i.e., *username* and *password* (see Fig. 1). After having inserted the credentials and clicked on "Login", the application shows the home page (homepage.asp). If credentials are correct, the username (contained in an HTML tag with the attribute ID="uname") and the logout button are reported in the upper right corner of the home page (e.g., *John.Doe*). Otherwise, *Guest User* and login button are shown. For the sake of simplicity, the application does not report any error message in case of invalid credentials or unrecognised users.

Fig. 1. login.asp – Page and Source

2.1 Programmable Web Testing

Programmable Web testing is based on manual creation of a test script. Web test scripts can be written using ad-hoc languages and frameworks or general purpose programming languages (such as Java and Ruby) with the aid of specific libraries able to play the role of the browser. Usually, these libraries extend the programming language with user friendly APIs, providing commands to, e.g., click a button, fill a field and submit

---

[1] http://jautomate.com/2013/08/22/730/

```
public class LoginPage {                              public class HomePage {
 private final WebDriver driver;                       private final WebDriver driver;
 public LoginPage(WebDriver driver) {this.driver=driver;}  public HomePage(WebDriver driver)
 public HomePage login(String UID, String PW) {          {this.driver = driver;}
  driver.findElement(By.id("UID")).sendKeys(UID);       public String getUsername() {
  driver.findElement(By.xpath("./input[2]")).sendKeys(PW);  return
  driver.findElement(By.linkText("Login")).click();      driver.findElement(By.id("uname")).getText;
  return new HomePage(driver);                          }
 }                                                     }
}
```

Fig. 2. LoginPage and HomePage page objects in Selenium WebDriver

a form. Test scripts are completed with assertions (e.g., JUnit assertions if the language chosen is Java).

A best practice often used when developing programmable test cases is the *page object* pattern. This pattern is used to model the Web pages involved in the test process as objects, employing the same programming language used to write the test cases. In this way, the functionalities offered by a Web page become methods exposed by the corresponding page object, which can be easily called within any test case. Thus, all the details and mechanics of the Web page are encapsulated inside the page object. Adopting the *page object* pattern allows the test developer to work at a higher level of abstraction and it is used to reduce the coupling between Web pages and test cases, and the amount of duplicate code. For these reasons, the adoption of the *page object* pattern is expected to improve the test suite maintainability and evolvability [8].

DOM-Based Programmable Test Case Creation: The tool for Web app testing belonging to the DOM-based/Programmable category that we used in this work is Selenium WebDriver release 2.25.0 (in the following shortly referred to as WebDriver - http://seleniumhq.org/projects/webdriver/). WebDriver is a tool for automating Web app testing that provides a comprehensive programming interface used to control the browser. WebDriver test cases are implemented manually in a programming language (in our case Java) integrating WebDriver commands with JUnit or TestNG assertions. We chose WebDriver as the representative of this category, because: (1) it is a quite mature tool, (2) it is open-source, (3) it is one of the most widely-used open-source solutions for Web test automation (even in the industry), (4) during a previous industrial collaboration [8], we have gained a considerable experience on its usage.

As an example, we here use a simple WebDriver test case (Fig. 3 left): a successful authentication test case. It submits a valid login, using correct credentials (i.e., *username=John.Doe* and *password=123456*) and verifies that in the home page the user appears as correctly authenticated ("John.Doe" must be displayed in the top-right corner of the home page).

The first step is to create two page objects (LoginPage.java and HomePage.java) corresponding to the Web pages login.asp and homepage.asp respectively (see Fig. 2). The page object LoginPage.java offers a method to log into the application. This method takes username and password as inputs, inserts them in the corresponding input fields, clicks the Login button and returns a page object of kind HomePage.java, because after clicking the Login button the application moves to the page homepage.asp. HomePage.java contains a method that returns the username authenticated in the application or *"Guest"* when no user is authenticated. As shown in Fig. 2, the Web page elements

```
public void testLogin() { // WebDriver        public void testLogin(){ // Sikuli
  WebDriver driver = new FirefoxDriver();
  // we start from the 'login.asp' page          // we start from the 'login.asp' page
  driver.get("http://www.....com/login.asp");     CommonPage.open("http://www.....com/login.asp");
  LoginPage LP = new LoginPage(driver);           LoginPage LP = new LoginPage();
  HomePage HP = LP.login("John.Doe","123456");    HomePage HP = LP.login("John.Doe", "123456"); ⌐John.Doe
  // we are in the 'homepage.asp' page             // we are in the 'homepage.asp' page
  assertEquals("John.Doe", HP.getUsername());     assertTrue(HP.isUsernamePresent(new URL("JohnDoe.png")));
}                                               }
```

Fig. 3. TestLogin test case in Selenium WebDriver (left) and in Sikuli API (right)

are located by searching for values in the DOM (using ID and LinkText locators) or navigating it (using XPath locators). While WebDriver offers several alternative ways to locate the Web elements in a Web page, the most effective one, according to Web-Driver developers (http://docs.seleniumhq.org/docs/03_webdriver.jsp), is searching the elements by their ID values. Hence, whenever possible, we used this method. The second step is to develop the test case making use of the page objects (see Fig. 3 left). In the test method, first, a WebDriver of type FirefoxDriver is created, so that the test case can control a Firefox browser as a real user does; second, the WebDriver (i.e., the browser) opens the specified URL and creates a page object that instantiates Login-Page.java (modelling the login.asp page); third, using method login(...) offered by the page object, a new page object (HP) representing the page homepage.asp is created; finally, the test case assertion is checked.

Visual Programmable Test Case Creation: The Web app testing tool, belonging to the Visual/Programmable category, that we used in this work is Sikuli API release 1.0.2 (in the following shortly referred to as Sikuli - http://code.google.com/p/sikuli-api/). Sikuli is a visual technology able to automate and test graphical user interfaces using screen-shot images. It provides image-based GUI automation functionalities to Java program-mers. We chose Sikuli as the representative of this category mainly because: (1) it is open-source and (2) it is similar to WebDriver, thus, we can create test cases and page objects similarly to the ones produced for WebDriver. In this way, using Sikuli, we are able to make the comparison between visual and DOM-based programmable tools fair and focused as much as possible on the differences of the two approaches. In fact, in this way we can use the same programming environment: programming language (Java), IDE (Eclipse), and testing framework (JUnit). Sikuli allows software testers to write scripts based on images that define the GUI widgets to be tested and the asser-tions to be checked. This is substantially different from the way in which WebDriver performs page element localisation and assertion checking.

As an example, the Sikuli version of the testLogin test case is shown in Fig. 3 (right) while the related page objects are given in Fig. 4. The test case developed in Sikuli performs the same conceptual steps[2] as the WebDriver test case, as apparent from Fig. 3 (left) and Fig. 3 (right). The page objects (shown in Fig. 4) are instead quite different. To locate a Web page element, an instruction (based on the Sikuli Java API)

---

[2] Actually, in Sikuli there is no command to open Firefox at a specified URL as in WebDriver. We have encapsulated this functionality in a method, CommonPage.open(...), that clicks the Firefox icon on the desktop, inserts the URL into the address bar and then clicks on the "go" arrow.

```
public class LoginPage {                            public class HomePage {
 private ScreenRegion s = new DesktopScreenRegion();  private ScreenRegion s = new
 private Mouse m = new DesktopMouse();                          DesktopScreenRegion();
 private Keyboard keyboard = new DesktopKeyboard();   private Mouse m = new DesktopMouse();
 public HomePage login(String UID, String PW){       public boolean isUsernamePresent(URL uname){
  m.click(s.find(new ImageTarget(new URL("un.png"))).getCenter());   try{m.click(s.find(new
                                                        ImageTarget(uname)).getCenter());
  keyboard.type(UID);        Username: [        ]    return true;
                                                      } catch(Exception e) {return false;}
  m.click(s.find(new ImageTarget(new URL("pw.png"))).getCenter());   }
                                                    }
  keyboard.type(PW);         Password: [        ]
  m.click(s.find(new ImageTarget(new URL("log.png"))).getCenter());
  return new HomePage();     [ Login ]
 }
}
```

Fig. 4. LoginPage and HomePage page objects in Sikuli API

is used which searches for the portion of Web page that looks like the image saved for the test suite (e.g., in a png file). Thus, in Sikuli, locators are always images. In addition, some other minor differences can be noticed in the test case implementation. For instance, in the case of WebDriver it is possible to assert that an element must contain a certain text (see the last line in Fig. 3 (left)), while in Sikuli it is necessary to assert that the page shows a portion equal to an image where the desired text is displayed (see the last line in Fig. 3 (right)).

2.2   Test Case Evolution

When a Web app evolves to accommodate requirement changes, bug fixes, or functionality extensions, test cases may become broken. For example, test cases may be unable to locate some links, input fields and submission buttons and software testers will have to repair them. This is a tedious and expensive task since it has to be performed manually (automatic evolution of test suites is far from being consolidated [11]).

Depending on the kind of maintenance task that has been performed on the target Web app, a software tester has to execute a series of test case repairment activities that can be categorised, for the sake of simplicity, in two types: *logical* and *structural*.

Logical Changes involve the modification of the Web app functionality. To repair the test cases, the tester has to modify the broken test cases and the corresponding page objects and in some cases, new page objects have to be created. An example of a change request that needs a logical repairment activity is enforcing the security by means of stronger authentication and thus adding a new Web page, containing an additional question, after the login.asp page of Fig. 1. In this case, the tester has to create a new page object for the Web page providing the additional authentication question. Moreover, she has to repair both the testLogin test cases shown in Fig. 3, adding a new Java command that calls the method offered by the new page object.

Structural Changes involve the modification of the Web page layout/structure only. For example, in the Web page of Fig. 1 the string of the login button may be changed to Submit. Usually, the impact of a structural change is smaller than a logical change. To repair the test cases, often, it is sufficient to modify one or more localisation lines, i.e., lines containing locators. In the example, the tester has to modify the LoginPage.java page object (see Fig. 2 and 4) by: (1) repairing the line:

```
driver.findElement(By.linkText("Login")).click()
```

in the case of Selenium WebDriver; or, (2) changing the image that represents the Login button in the case of Sikuli.

3   Empirical Study

This section reports the design, objects, research questions, metrics, procedure, results, discussion and threats to validity of the empirical study conducted to compare visual vs. DOM-based Web testing.

3.1   Study Design

The primary *goal* of this study is to investigate the difference in terms of robustness (if any) that can be achieved by adopting visual and DOM-based locators with the purpose of understanding the strengths and the weaknesses of the two approaches. Then, after having selected two tools that respectively belong to the two considered categories, as secondary *goal* we investigated the cost/benefit trade-off of visual vs. DOM-based test cases for Web apps. In this case, the *cost/benefit focus* regards the effort required for the creation of the initial test suites from scratch, as well as the effort required for their evolution across successive releases of the software. The results of this study are interpreted according to two *perspectives*: (1) *project managers*, interested in understanding which approach could lead to potentially more robust test cases, and in data about the costs and the returns of the investment associated with both the approaches; (2) *researchers*, interested in empirical data about the impact of different approaches on Web testing. The *context* of the study is defined as follows: two *human subjects* have been involved, a PhD student (the first author of this paper) and a junior developer (the second author, a master student with 1-year industrial experience as software tester); the *software objects* are six open source Web apps. The two human subjects participating in the study are referred below using the term "developers".

3.2   Web Applications

We have selected and downloaded six open-source Web apps from *SourceForge.net*. We have included only applications that: (1) are quite recent, so that they can work without problems on the latest versions of Apache, PHP and MySQL, technologies we are familiar with (actually, since Sikuli and WebDriver implement a black-box approach, the server side technologies do not affect the results of the study); (2) are well-known and used (some of them have been downloaded more than one hundred thousand times last year); (3) have at least two major releases (we have excluded minor releases because with small differences between versions the majority of the test cases are expected to work without problems); (4) belong to different application domains; and, (5) are non-RIA – Rich Internet Applications (to make the comparison fair, since RIAs can be handled better by the visual approach, see Sect. 3.7).

Table 1 reports some information about the selected applications. We can see that all of them are quite recent (ranging from 2008 to 2013). On the contrary, they are considerably different in terms of number of source files (ranging from 46 to 840) and number of lines of code (ranging from 4 kLOC to 285 kLOC, considering only the

Table 1. Objects: Web Applications from *SourceForge.net*

	Description	Web Site	1st Release				2nd Release			
			Vers.	Date	File[a]	kLOC[b]	Vers.	Date	File[a]	kLOC[b]
MantisBT	bug tracking system	sourceforge.net/projects/mantisbt/	1.1.8	2009	492	90	1.2.0	2010	733	115
PPMA[c]	password manager	sourceforge.net/projects/ppma/	0.2	2011	93	4	0.3.5.1	2013	108	5
Claroline	learning environment	sourceforge.net/projects/claroline/	1.10.7	2011	840	277	1.11.5	2013	835	285
Address Book	address/phone book	sourceforge.net/projects/php-addressbook/	4.0	2009	46	4	8.2.5	2012	239	30
MRBS	meeting rooms manager	sourceforge.net/projects/mrbs/	1.2.6.1	2008	63	9	1.4.9	2012	128	27
Collabtive	collaboration software	sourceforge.net/projects/collabtive/	0.65	2010	148	68	1.0	2013	151	73

[a] Only PHP source files were considered   [b] PHP LOC - Comment and Blank lines are not considered

lines of code contained in the PHP source files, comments and blank lines excluded). The difference in lines of code between 1st and 2nd release (columns 7 and 11) gives a rough idea of how different the two chosen releases are.

3.3 Research Questions and Metrics

Our empirical study aims at answering the following research questions, split between considerably tool-independent (A) and significantly tool-dependent (B) questions:

RQ A.1: *Do Visual and DOM-based test suites require the same number of locators?*
The goal is to quantify and compare the number of locators required when adopting the two different approaches. This would give developers and project managers a rough idea of the inherent effort required to build the test suites by following the two approaches. Moreover, the total number of locators could influence also the maintenance effort, since the more the locators are, the more the potential locators to repair could be. The metrics used to answer the research question is the number of created locators.

RQ A.2: *What is the robustness of visual vs. DOM-based locators?*
The goal is to quantify and compare the robustness of the visual and the DOM-based locators. This would give developers and project managers an idea of the inherent robustness of the locators created by following the two approaches. The metrics used to answer this research question is the number of broken locators.

RQ B.1: *What is the initial development effort for the creation of visual vs. DOM-based test suites?*
The goal is to quantify and compare the development cost of visual and DOM-based tests. This would give developers and project managers an idea of the initial investment (tester training excluded) to be made if visual test suites are adopted, as compared to DOM-based test suites. The metrics used to answer this research question is the time (measured in minutes) the two developers spent in developing visual test suites vs. DOM-based test suites.

RQ B.2: *What is the effort involved in the evolution of visual vs. DOM-based test suites when a new release of the software is produced?*
This research question involves a software evolution scenario. For the next major release of each Web app under test, the two developers evolved the test suites so as to make them applicable to the new software release. The test case evolution effort for visual and for DOM-based test suites was measured as the time (measured in minutes) spent to update the test suites, until they were all working on the new release.

RQ B.3: *What is the execution time required by visual vs. DOM-based test suites?*
Image processing algorithms are known to be quite computation-intensive [2] and execution time is often reported as one of the weaknesses of visual testing. We want to quantitatively measure the execution time difference (in seconds) between visual and DOM-based test execution tools.

It should be noticed that the findings for research questions A.x are mostly influenced by the approaches adopted, independently of the tools that implement them, since the number of (DOM-based/visual) locators and the number of broken locators depend mostly on the test cases (and on the tested Web app), not on the tools. On the other hand, the metrics for research questions B.x (effort and execution time) are influenced by the specific tools adopted in the empirical evaluation.

3.4   Experimental Procedure

The experiment has been performed as follows:

– Six open-source Web apps have been selected from *SourceForge.net* as explained in Section 3.2.

– For each selected application, two equivalent test suites (written for Sikuli and WebDriver) have been built by the two developers, working in pair-programming and adopting a systematic approach consisting of three steps: (1) the main functionalities of the target Web app are identified from the available documentation; (2) each discovered functionality is covered with at least one test case (developers have assigned a meaningful name to each test case, so as to keep the mapping between test cases and functionalities); (3) each test case is implemented with Sikuli and WebDriver. For both approaches, we considered the following best practices: (1) we used the page object pattern and (2) we preferred, for WebDriver, the ID locators when possible (i.e., when HTML tags are provided with IDs), otherwise Name, LinkText, CSS and XPath locators were used following this order. Overall, for the WebDriver case we created: 82 ID, 99 Name, 65 LinkText, 64 CSS and 177 XPath locators. For each test suite, we measured the number of produced locators (to answer RQ A.1) and the development effort for the implementation as clock time (to answer RQ B.1). Each pair of test suites (i.e., visual and DOM-based) are equivalent because the included test cases test exactly the same functionalities, using the same sequences of actions (e.g., locating the same web page elements) and the same input data. The WebDriver test suites had been created by the same two developers about a year ago during a previous case study [9], while the Sikuli test suites have been created more recently, for the present study, which potentially gives a slight advantage to Sikuli (see *Threats to Validity* section).

– Each test suite has been executed against the second release of the Web app (see Table 1). First, we recorded the failed test cases (we highlight that no real bugs have been detected in the considered applications; all the failures are due to broken locators and minimally to modifications to the test cases logic) and then, in a second phase, we repaired them. We measured the number of broken locators (to answer RQ A.2) and the repair effort as clock time (to answer RQ B.2). Finally, to answer RQ B.3 we executed 10 times (to average over any random fluctuation of the execution time) each of the 12 repaired test suites (both WebDriver and Sikuli test cases) and recorded the execution times. We executed the test suites on a machine hosting an Intel Core i5 dual-core processor (2.5 GHz) and 8 GB RAM, with no other computation or communication load, in order to avoid CPU or memory saturation. To avoid as much as possible network delays we installed the web server hosting the applications on a machine belonging to the same LAN.

– The results obtained for each test suite have been compared to answer our research questions. On the results, we conducted both a quantitative analysis and a qualitative analysis, completed with a final discussion where we report our lessons learnt. The test suites source code can be found at: http://softeng.disi.unige.it/2014-Visual-DOM.php

3.5   Quantitative Results

This section reports the quantitative results of the empirical study, while the reasons and implications of the results are further analysed in Section 3.6.

RQ A.1. Table 2 shows the data to answer the A.x research questions by reporting, for each application, the number of visual and DOM-based broken/total locators. The number of locators required to build the test suites varies from 81 to 158 when adopting the visual approach and from 42 to 126 when adopting the DOM-based one. For all the six applications the number of required locators is higher when adopting the visual approach. Considering the data in aggregate form, we created 45% more Visual locators than DOM locators (706 visual vs. 487 DOM-based locators). To summarise, with respect to the research question RQ A.1, the visual approach has always required to create a higher number of locators.

RQ A.2. From the data shown in Table 2, we can see that in only two test suites out of six visual locators result more robust than the DOM-based ones (i.e., Mantis and Collabtive), while in the remaining four cases the DOM-based locators are more robust. Overall, 312 visual locators out of 706 result broken, while only 162 DOM-based locators out of 487 have been repaired (i.e., 93% more broken locators in the case of the visual approach). To summarise, with respect to RQ A.2, the result is not clear-cut. Generally, DOM-based locators are more robust but in certain cases (i.e., depending on the kind of modifications among the considered releases), visual locators proved to be the most robust (e.g., in Collabtive only 4 broken visual locators vs. 36 DOM-based ones).

RQ B.1. Table 3 reports some data about the developed test suites. For each application, it reports: the number of test cases and page objects in the test suites built for the newer release of each application (c. 1- 2); the time required for the initial development of the test suites (c. 3- 4); the percentage difference between the initial development time required by WebDriver vs. Sikuli (c. 5); the p-value of the Wilcoxon paired test used

Table 2. Visual vs. DOM-Based Locators Robustness

	Visual		DOM-Based	
	Broken Locators	Total Locators	Broken Locators	Total Locators
MantisBT	15	127	29	106
PPMA	78	81	24	42
Claroline	56	158	30	126
Address Book	103	122	14	54
MRBS	56	83	29	51
Collabtive	4	135	36	108

Table 3. Test Suites Development

	Number of		Time (Minutes)			Code (Java LOC)					
	Test Cases	Page Objects	Sikuli API	Web Driver	p-value	Sikuli API			WebDriver		
						Test	PO	Total	Test	PO	Total
MantisBT	41	30	498	383	-23%	< 0.01	1645	1291	2936	1577	1054	2631
PPMA	23	6	229	98	-57%	< 0.01	958	589	1547	867	346	1213
Claroline	40	22	381	239	-37%	< 0.01	1613	1267	2880	1564	1043	2607
Address Book	28	7	283	153	-46%	< 0.01	1080	686	1766	1078	394	1472
MRBS	24	8	266	133	-50%	< 0.01	1051	601	1652	949	372	1321
Collabtive	40	8	493	383	-22%	< 0.01	1585	961	2546	1565	650	2215

to assess whether the development time difference is statistically significant (c. 6) and finally the test suites size (measured in Lines Of Code (LOC), comment and blank lines have not been not considered), split between page objects and test cases, for the newer release of each application (c. 7-12). The development of the Sikuli test suites required from 229 to 498 minutes, while the WebDriver suites required from 98 to 383 minutes. In all the six cases, the development of the WebDriver test suites required less time than the Sikuli test suites (from 22% to 57%). This is related with the lower number of locators required when adopting the DOM-based approach (see RQ A.1). According to the Wilcoxon paired test (see c. 6 of Table 3), the difference in test suite development time between Sikuli and WebDriver is statistically significant (at $\alpha = 0.05$) for all test suites. For what concerns the size of the test suites (Table 3, c. 9, *Total*), we can notice that in all the cases, the majority of the code is devoted to the test case logics, while only a small part is devoted to the implementation of the page objects. Moreover, the number of LOCs composing the test cases is very similar for both Sikuli and WebDriver, while it is always smaller in WebDriver for what concerns the page objects (often, in the Sikuli page objects, two lines are needed to locate and perform an action on a web element while in WebDriver just one is sufficient, see for example Fig. 2 and 4). To summarise, with respect to the research question RQ B.1 we can say that for all the six considered applications, the effort involved in the development of the Sikuli test suites is higher than the one required by WebDriver.

RQ B.2. Table 4 shows data about the test suites repairing process. In detail, the table reports, for each application, the time required to repair the test suites (Sikuli and Web-Driver), and the number of repaired test cases over the total number of test cases. The WebDriver repair time is compared to the Sikuli repair time by computing the percentage difference between the two and by running the Wilcoxon paired test, to check for statistical significance of the difference. Sikuli test suites required from 7 to 126 minutes to be repaired, while WebDriver test suites required from 46 to 95 minutes. The results are associated with the robustness of the two kinds of locators (RQ A.2) employed by the two tools and thus follow the same trend: in four cases out of six, repairing of the

Table 4. Test Suites Maintenance

	Sikuli API		WebDriver			
	Time Minutes	Test Repaired	Time Minutes		p-value	Test Repaired
MantisBT	76	37 / 41	95	+ 25%	0.04	32 / 41
PPMA	112	20 / 23	55	- 51%	< 0.01	17 / 23
Claroline	71	21 / 40	46	- 35%	0.30	20 / 40
Address Book	126	28 / 28	54	- 57%	< 0.01	28 / 28
MRBS	108	21 / 24	72	- 33%	0.02	23 / 24
Collabtive	7	4 / 40	79	+ 1029%	< 0.01	23 / 40

Table 5. Test Suites Execution

	Number of Test Cases	Sikuli API			WebDriver			
		Mean	σ		Mean		σ	
		Seconds	Absolute	Relative	Seconds	p-value	Absolute	Relative
MantisBT	41	2774	60	2,2%	1567	- 43%	< 0.01	70	4,5%
PPMA	23	1654	12	0,7%	924	- 44%	< 0.01	35	3,8%
Claroline	40	2414	34	1,4%	1679	- 30%	< 0.01	99	5,9%
Address Book	28	1591	19	1,2%	977	- 39%	< 0.01	106	10,9%
MRBS	24	1595	19	1,2%	837	- 48%	< 0.01	54	6,5%
Collabtive	40	2542	72	2,8%	1741	- 31%	< 0.01	59	3,4%

WebDriver test suites required less time (from 33% to 57% less) than Sikuli. In one case (i.e., MantisBT), the WebDriver test suite required slightly more time (25% more) to be repaired than the corresponding Sikuli test suite. In another case (i.e., Collabtive), WebDriver required a huge amount of time for test suite repairment with respect to the time required by Sikuli (about 10x more time with WebDriver where, as seen before, we have about 10x more locators to repair). According to the Wilcoxon paired test, the difference in test suite evolution time between Sikuli and WebDriver is statistically significant (at $\alpha = 0.05$) for all test suites (sometimes in opposite directions) except for Claroline (see Table 4). Note that, the maintenance effort is almost entirely due to repair the broken locators (i.e., structural changes) and minimally to modifications to the test cases logic (i.e., logical changes). Indeed, during maintenance, we have approximately modified only the 1% of the LOCs composing the test suites in order to address logical changes of the Web apps. Very often the modifications were exactly the same for both the approaches (i.e., Visual and DOM-based). To summarise, with respect to RQ B.2, the result is not clear-cut. For four out of six considered applications, the effort involved in the evolution of the Sikuli test suites, when a new release of the software is produced, is higher than with WebDriver, but in two cases the opposite is true.

RQ B.3. Table 5 shows data about the time required to execute the test suites. For both tools we report: the mean execution time, computed on 10 replications of each execution; the standard deviation (absolute and relative); the difference in percentage between the time required by the Sikuli test suites and the WebDriver test suites; and, the p value reported by the Wilcoxon paired test, used to compare Sikuli vs. WebDriver's execution times. Execution times range from 1591s to 2774s for Sikuli and from 837s to 1741s for WebDriver. In all the cases, the WebDriver test suites required less time to complete their execution (from -30% to -48%). According to the Wilcoxon paired test, the difference is statistically significant (at $\alpha = 0.05$) for all test suites. To summarise, with respect to the research question RQ B.3 the time required to execute the Sikuli test suites is higher than the execution time of WebDriver for all the six considered applications.

3.6 Qualitative Results

In this section, we discuss on the factors behind the results presented in the previous section, focusing more on the ones that are related to the two approaches and, for space reasons, less on the factors related to the specific tools used:

Web Elements Changing Their State. When a Web element changes its state (e.g., a check box is checked or unchecked, or an input field is emptied or filled), a visual locator must be created for each state, while with the DOM-based approach only one locator is required. This occurred in all the six Sikuli test suites and it is one of the reasons why, in all of them, we have more locators than in the WebDriver test suites (see RQ A.1 and Table 2). As a consequence, more effort both during the development (RQ B.1) and maintenance (RQ B.2) is required in the case of Sikuli test suites (more than one locator had to be created and later repaired for each Web element, RQ A.2). For instance, in MRBS, when we tested the update functionality for the information associated with a room reservation, we had to create two locators for the same check box (corresponding to the slot: Monday from 9:00 to 10:00) to verify that the new state has been saved (e.g., from booked, *checked*, to available, *unchecked*). Similarly, in Collabtive, we had to verify the changes in the check boxes used to update the permissions assigned to the system users.

Changes behind the Scene. Sometimes it could happen that the HTML code is modified without any perceptible impact on how the Web app appears. An extreme example is changing the layout of a Web app from the "deprecated" table-based structure to a div-based structure, without affecting its visual aspect in any respect. In this case, the vast majority of the DOM-based locators (in particular the navigational ones, e.g., XPath) used by DOM-based tools may be broken. On the contrary, this change is almost insignificant for visual test tools. A similar problem occurs when auto-generated ID locators are used (e.g., *id1, id2, id3, ... , idN*) by DOM-based locators. In fact, these tend to change across different releases, while leaving completely unaffected the visual appearance of the Web page (hence, no maintenance is required on the visual test suites). For example, the addition of a new link in a Web page might result in a change of all IDs of the elements following the new link [8]. Such "changes behind the scene" occurred in our empirical study and explain why, in the case of Collabtive, the Sikuli test suite has required by far a lower maintenance effort (see RQ B.2 and Table 4). In detail, across the two considered releases, a minor change has been applied to almost all the HTML pages of Collabtive: an unused div tag has been removed. This little change impacted quite strongly several of the XPath locators (XPath locators were used because IDs were not present) in the WebDriver test suite (see RQ A.2). The majority of the 36 locators (all of them are XPaths) was broken and had to be repaired (an example of repairment is from .../div[2]/... to .../div[1]/...). No change was necessary on the Sikuli visual test suite for this structural change. Overall, in Sikuli, we had only few locators broken. For this reason, there is a large difference in the maintenance effort between the two test suites. A similar change across releases occurred also in MantisBT, although it had a lower impact in this application.

Repeated Web Elements. When in a Web page there are multiple instances of the same kind of Web element (e.g., an input box), creating a visual locator requires more time than creating a DOM-based one. Let us consider a common situation, consisting of a form with multiple, repeated input fields to fill (e.g., multiple lines, each with *Name*, *Surname*, etc.), all of which have the same size, thus appearing identical. In such cases, it is not possible to create a visual locator using only an image of the Web element of our interest (e.g., the repeated *Name* input field), but we have to: (i) include also some context around (e.g., a label as shown in Fig. 4) in order to create an unambiguous locator (i.e., an image that matches only one specific portion of the Web page) or, when this is not easily feasible, (ii) locate directly a unique Web element close to the input field of interest and then move the mouse of a certain amount of pixels, in order to reach the input field. Both solutions locate the target Web element by means of another, easier to locate, element (e.g., a label). This is not straightforward and natural for the test developer (i.e., it requires more effort and time). Actually, both solutions are not quite convenient. Solution (i) requires to create large image locators, including more than one Web element (e.g., the label and the corresponding input field). On the other hand, even if it allows to create a small locator image for only one Web element (e.g., the label), Solution (ii) requires to calculate a distance in pixels (similarly to 1st generation tools), not so simple to determine. Both solutions have problems in case of variation of the relative positions of the elements in the next releases of the application. Thus, this factor has a negative effect on both the development and maintenance of Sikuli test suites. Repeated Web elements occurred in all test suites. For instance, in Claroline, a form contains a set of radio buttons used to select the course type to create. In Sikuli, localisation of these buttons requires either Solution (i) or (ii). Similarly, in AddressBook/MantisBT, when a new entry/user is inserted, a list of input fields, all with the same appearance, has to be filled. In these cases, we created the Sikuli locators as shown in Fig. 4. Recently, JAutomate (http://jautomate.com/), a commercial GUI test automation tool, provided a different solution to this problem by mixing visual locators and position indexes. When a visual locator selects more than one element, it is possible to use an index to select the desired element among the retrieved ones.

Elements with Complex Interaction. Complex Web elements, such as drop-down lists and multilevel drop-down menus, are quite common in modern Web apps. For instance, let us consider a registration form that asks for the nationality of the submitter. Typically, this is implemented using a drop-down list containing a list of countries. A DOM-based tool like WebDriver can provide a command to select directly an element from a drop-down list (only one locator is required). On the contrary, when adopting the visual approach the task is much more complex. Once could, for instance: (1) locate the drop-down list (more precisely the arrow that shows the menu) using an image locator; (2) click on it; (3) if the required list element is not shown, locate and move the scrollbar (e.g., by clicking the arrow); (4) locate the required element using another image locator; and, finally, (5) click on it. All these steps together require more LOCs (in the page objects, see RQ B.1) and locators. Actually, in this case the visual approach performs exactly the same steps that a human tester would do.

Execution Time. The execution time required by the Sikuli tool is always higher than the one required by WebDriver (see RQ B.3 and Table 5). This was expected, since executing an image recognition algorithm requires more computational resources (and thus, generally, more time) than navigating the DOM. However, surprisingly, the difference in percentage between the two approaches is not high, being only 30-48%. It is not very much considering that: (1) Sikuli is a quite experimental tool, (2) it is not focused on Web app testing and, (3) the needed manual management of the pages loading delay (through *sleep* commands) we applied is not optimal[3]. For what concerns the latter point, according to our estimates, the overhead due to the Web page loading delay is not a major penalty for Sikuli (only 20-40 seconds per test suite) as compared to the total processing time. Indeed, we carefully tuned the delays in order to find the smallest required. The standard deviation (see Table 5) is always greater in the case of WebDriver given that, sometimes, it unexpectedly and randomly stops for short periods during test suites execution (e.g., 2-3s between two test cases).

Lesson Learnt: In the following, we report some lessons learnt during the use of the two experimented approaches and tools:

Data-driven Test Cases. Often in the industrial practice [8], to improve the coverage reached by a test suite, test cases are re-executed multiple times using different values. This is very well supported by a programmable testing approach. However, benefits depend on the specific programmable approach that is adopted (e.g., visual vs. DOM-based). For instance, in WebDriver it is possible to use data from various sources, such as CSV files or databases, or even to generate them at runtime. In Sikuli it is necessary to have images of the target Web elements, so even if we can use various data sources (e.g., to fill input fields), when assertions are evaluated, images are still needed to represent the expected data (see Fig. 3). For this reason, in the visual approach it is not possible to create complete data-driven test cases (i.e., for both input and assertions). In fact, while it is indeed possible to parameterise the usage of image locators in the assertions, it is not possible to generate them from data. This happens because using a DOM-based tool there is a clear separation between the locator for a Web element (e.g., an ID value) and the content of that Web element (e.g. the displayed string), so that we can reuse the same locator with different contents (e.g., test assertion values). On the contrary, using a visual tool, the locator for a Web element and the displayed content are the same thing, thus if the content changes, the locator must be also modified. Moreover, it is important to highlight that, if necessary, parameterising the creation of DOM-based locators is usually an easy task (e.g., .//*[@id='list']/tr[**x**]/td[1] with **x**=1..n), while it is infeasible in the visual approach. In our case study, we experienced this limitation of the visual approach since we had, in each test suite, at least one test case that performs multiple, repeated operations that change only in the data values being manipulated, such as: insert/remove multiple different users, projects, addresses, or groups (depending on the considered application). In such cases we used: (1) a single parameterized locator in WebDriver, and (2) several different image locators in Sikuli (e.g., for evaluating the assertions), with the effect that, in the second case, the number of locators required is higher.

Test Case Comprehensibility. The locators used by the two approaches have often a different degree of comprehensibility. For instance, by comparing Fig. 2 with Fig. 4, it is clear that the visual locator pw.png (password) is much easier to understand than the corresponding XPath locator. In fact, the visual approach works in a manner that

---

[3] A browser needs time to open a Web page. Thus, before starting to perform actions on the page the test automation tool has to wait. WebDriver provides specific commands to deal with this problem (i.e., waiting for the web page loading). In Sikuli this is not available and testers have to insert an explicit delay (e.g., `Thread.sleep(200)`).

is closer to humans than the DOM-based approach. In our case study, we experienced this fact several times. For instance, during test suites maintenance, understanding why a locator is broken is generally easier and faster with Sikuli than with WebDriver.

Test Suites Portability. If a Sikuli test suite is executed on a different machine where the screen resolution or the font properties are different, Sikuli test cases may not work properly. We experienced this problem two times while executing the Sikuli test suites on two different computers: in one case because the default font size was different, resulting in broken image locators, and in another case because the screen resolution was lower than expected, thus more mouse scroll operations were required.

3.7 Threats to Validity

The main threats to validity that affect this study are: Construct (authors' bias), Internal and External validity threats.

Authors' Bias threat concerns the involvement of the authors in manual activities conducted during the empirical study and the influence of the authors' expectations about the empirical study on such activities. In our case, two of the authors developed the test suites and evolved them to match the next major release of each application under test. Since none of the authors was involved in the development of any of the tools assessed in the empirical study, the authors' expectations were in no particular direction for what concerns the performance of the tools. Hence, we think that the authors' involvement in some manual activities does not introduce any specific bias.

Internal Validity threats concern confounding factors that may affect a dependent variable (number of locators, number of broken locators, development, repair, and execution time of the test suites). One such factor is associated with the approach used to produce the test cases (i.e., the chosen functional coverage criterion). Moreover, the variability involved in the selection of the input data and of the locators could have played a role. To mitigate this threat, we have adopted a systematic approach and applied all known good-practices in the construction of programmable test suites. Concerning RQ B.1, learning effects may have occurred between the construction of the test suites for WebDriver and Sikuli. However, this is quite unlikely given the long time (several months) elapsed between the development of WebDriver and Sikuli test suites and the kind of locators (DOM-based vs. visual), which is quite different. Moreover, given the high level of similarity of the test code (in practice, only locators are different), learning would favour Sikuli, which eventually showed lower performance than WebDriver, so if any learning occurred, we expect that without learning the results would be just amplified, but still in the same direction.

External Validity threats are related to the generalisation of results. The selected applications are real open source Web apps belonging to different domains. This makes the context quite realistic, even though further studies with other applications are necessary to confirm or confute the obtained results. In particular, our findings could not hold for RIAs providing sophisticated user interactions, like, for instance, Google Maps or Google Docs. In fact, using a visual approach it is possible to create test cases that are very difficult (if not impossible) to realise with the DOM-based approach. For instance, it is possible to verify that in Google Docs, after clicking the "center" button, a portion of text becomes centred in the page, which is in practice impossible using just the DOM.

The results about number and robustness of locators used by the visual and DOM-based approaches (RQ A.1 and RQ A.2) are not tied to any particular tool, thus we expect they hold whatever tool is chosen in the two categories. On the other hand, the same is not completely true for RQ B.1 and RQ B.2, where the results about the development and maintenance effort are also influenced by the chosen tools, and different results could be obtained with other Web testing frameworks/tools. The problem of the generalisation of the results concerns also RQ B.3 where, for instance, employing a different image recognition algorithm could lead to different execution times.

4 Related Works

We focus our related work discussion considering studies about test suite development and evolution using visual tools; we also consider automatic repairment of test cases.

Several works show that the visual testing automation approach has been recently adopted by the industry [2,6] and governmental institutions [3]. Borjesson and Feldt in [2], evaluate two visual GUI testing tools (Sikuli and a commercial tool) on a real-world, safety-critical industrial software system with the goal of assessing their usability and applicability in an industrial setting. Results show that visual GUI testing tools are applicable to automate acceptance tests for industrial systems with both cost and potentially quality gains over state-of-practice manual testing. Differently from us, they compared two tools both employing the visual approach and did not focus specifically on Web app testing. Moreover, our goal (comparing visual vs. DOM-based locators) is completely different from theirs.

Collins et al. [6], present three testing automation strategies applied in three different industrial projects adopting the Scrum agile methodology. The functional GUI test automation tools used in these three projects were respectively: Sikuli, Selenium RC and IDE, and Fitnesse. Capocchi et al. [3], propose an approach, based on the DEVSimPy environment and employing both Selenium and Sikuli, aimed at facilitating and speeding up the testing of GUI software. They validated this approach on a real application dealing with medical software.

Chang et al. [4] present a small experiment to analyse the long-term reusability of Sikuli test cases. They selected two open-source applications (Capivara and jEdit) and built a test suite for each application (10 test cases for Capivara and 13 test cases for jEdit). Using some subsequent releases of the two selected applications, they evaluated how many test cases turned out to be broken in each release. The lesson drawn from this experiment is: as long as a GUI evolves incrementally a significant number of Sikuli test cases can still be reusable. Differently from us, the authors employed only a visual tool (Sikuli) without executing a direct comparison with other tools.

It is well-known that maintaining automated test cases is expensive and time consuming (costs are more significant for automated than for manual testing [15]), and that often test cases are discarded by software developers due to huge maintenance costs. For this reason, several researchers proposed techniques and tools for automatically repairing test cases. For instance, Mirzaaghaei et al. [12] presents TestCareAssistant (TcA), a tool that combines data-flow analysis and program differencing to automatically repair test compilation errors caused by changes in the declaration of method parameters.

Other tools for automatically repairing GUI test cases or reducing their maintenance effort have been presented in the literature [16,7,10]. Choudhary *et al.* [5] extended these proposals to Web apps, presenting a technique able to automatically suggest repairs for Web app test cases.

5 Conclusions and Future Work

We have conducted an empirical study to compare the robustness of visual vs. a DOM-based locators. For six subject applications, two equivalent test suites have been developed respectively in WebDriver and Sikuli. In addition to the robustness variable, we have also investigated: the initial test suite development effort, the test suite evolution cost, and the test suite execution time. Results indicate that DOM-based locators are generally more robust than visual ones and that DOM-based test cases can be developed from scratch at lower cost and most of the times they can be evolved at lower cost. However, on specific Web apps (MantisBT and Collabtive) visual locators were easier to repair, because the visual appearance of those applications remained stable across releases, while their structure changed a lot. DOM-based test cases required a lower execution time (due to the computational demands of image recognition algorithms used by the visual approach), although the difference was not that dramatic. Overall, the choice between DOM-based and visual locators is application-specific and depends quite strongly on the expected structural and visual evolution of the application. Other factors may also affect the testers' decision, such as the availability/unavailability of visual locators for Web elements that are important during testing and the presence of advanced, RIA functionalities which cannot be tested using DOM-based locators. Moreover, visual test cases are definitely easier to understand, which, depending on the skills of the involved testers, might also play a role in the decision.

In our future work we intend to conduct further studies to corroborate our findings. We plan to complete the empirical assessment of the Web testing approaches by considering also tools that implement capture-replay with visual Web element localisation (e.g., JAutomate). Finally, we plan to evaluate tools that combine the two approaches, such as SikuliFirefoxDriver (http://code.google.com/p/sikuli-api/wiki/SikuliWebDriver), that extends WebDriver by adding the Sikuli image search capability, combining in this way the respective strengths.

References

1. Berner, S., Weber, R., Keller, R.: Observations and lessons learned from automated testing. In: Proc. of ICSE 2005, pp. 571–579. IEEE (2005)
2. Borjesson, E., Feldt, R.: Automated system testing using visual GUI testing tools: A comparative study in industry. In: Proc. of ICST 2012, pp. 350–359 (2012)
3. Capocchi, L., Santucci, J.-F., Ville, T.: Software test automation using DEVSimPy environment. In: Proc. of SIGSIM-PADS 2013, pp. 343–348. ACM (2013)
4. Chang, T.-H., Yeh, T., Miller, R.C.: Gui testing using computer vision. In: Proc. of CHI 2010, pp. 1535–1544. ACM (2010)
5. Choudhary, S.R., Zhao, D., Versee, H., Orso, A.: Water: Web application test repair. In: Proc. of ETSE 2011, pp. 24–29. ACM (2011)

6. Collins, E., Dias-Neto, A., de Lucena, V.: Strategies for agile software testing automation: An industrial experience. In: Proc. of COMPSACW 2012, pp. 440–445. IEEE (2012)
7. Grechanik, M., Xie, Q., Fu, C.: Maintaining and evolving GUI-directed test scripts. In: Proc. of ICSE 2009, pp. 408–418. IEEE (2009)
8. Leotta, M., Clerissi, D., Ricca, F., Spadaro, C.: Improving test suites maintainability with the page object pattern: An industrial case study. In: Proc. of 6th Int. Conference on Software Testing, Verification and Validation Workshops, ICSTW 2013, pp. 108–113. IEEE (2013)
9. Leotta, M., Clerissi, D., Ricca, F., Tonella, P.: Capture-replay vs. programmable web testing: An empirical assessment during test case evolution. In: Proc. of 20th Working Conference on Reverse Engineering, WCRE 2013, pp. 272–281. IEEE (2013)
10. Memon, A.M.: Automatically repairing event sequence-based GUI test suites for regression testing. TOSEM, 18(2), 4:1–4:36 (2008)
11. Mirzaaghaei, M.: Automatic test suite evolution. In: Proc. of ESEC/FSE 2011, pp. 396–399. ACM (2011)
12. Mirzaaghaei, M., Pastore, F., Pezze, M.: Automatically repairing test cases for evolving method declarations. In: Proc. of ICSM 2010, pp. 1–5. IEEE (2010)
13. Ricca, F., Tonella, P.: Testing processes of web applications. Ann. Softw. Eng. 14(1-4), 93–114 (2002)
14. Ricca, F., Tonella, P.: Detecting anomaly and failure in web applications. IEEE MultiMedia 13(2), 44–51 (2006)
15. Skoglund, M., Runeson, P.: A case study on regression test suite maintenance in system evolution. In: Proc. of ICSM 2004, pp. 438–442. IEEE (2004)
16. Xie, Q., Grechanik, M., Fu, C.: Rest: A tool for reducing effort in script-based testing. In: Proc. of ICSM 2008, pp. 468–469. IEEE (2008)

Widget Classification
with Applications to Web Accessibility

Valentyn Melnyk[1], Vikas Ashok[1], Yury Puzis[2],
Andrii Soviak[2], Yevgen Borodin[2], and I.V. Ramakrishnan[2]

[1] Computer Science Department, Stony Brook University, Stony Brook, NY, USA
{vmelnyk,vganjiguntea}@cs.stonybrook.edu
[2] Charmtech Labs LLC, 1500 Stony Brook Rd., Stony Brook, NY, USA
{yury.puzis,and.soviak,borodin,ram}@charmtechlabs.com

Abstract. Once simple and static, many web pages have now evolved into complex web applications. Hundreds of web development libraries are providing ready-to-use dynamic widgets, which can be further customized to fit the needs of individual web application. With such wide selection of widgets and a lack of standardization, dynamic widgets have proven to be an insurmountable problem for blind users who rely on screen readers to make web pages accessible. Screen readers generally do not recognize widgets that dynamically appear on the screen; as a result, blind users either cannot benefit from the convenience of using widgets (e.g., a date picker) or get stuck on inaccessible content (e.g., alert windows). In this paper, we propose a general approach to identifying or classifying *dynamic* widgets with the purpose of "reverse engineering" web applications and improving their accessibility. To demonstrate the feasibility of the approach, we report on the experiments that show how very popular dynamic widgets such as date picker, popup menu, suggestion list, and alert window can be effectively and accurately recognized in live web applications.

Keywords: web applications, reverse engineering, widget classification, widget localization, dynamic widgets, screen reader, web accessibility, ARIA.

1 Introduction

The Web has permeated many aspects of our lives; we use it to obtain and exchange information, shop, pay bills, make travel arrangements, apply for college or employment, connect with others, participate in civic activities, etc. A 2012 report by the Internet World Stats shows that Internet usage has skyrocketed by more than 566% since 2000, to include over a third of the global population in 2012 (over 2.4 billion people) [23]. However, over this time period, the Web has evolved from text-based web pages to interactive web applications, becoming less accessible to blind people.

Many popular websites such as Blackboard, Gmail, Linked-In, Google Drive, eBay, Kayak, YouTube, etc. have turned into sophisticated web applications that utilize dynamic (appearing and disappearing) widgets such as dropdown menus, date pickers, suggestion boxes, etc. to enhance user experience by adding convenient tools.

S. Casteleyn, G. Rossi, and M. Winckler (Eds.): ICWE 2014, LNCS 8541, pp. 341–358, 2014.

According to the W3C definition [31], a *widget* (called "widget" thereafter) is defined as a "discrete user interface object with which the user can interact". Widgets can be simple objects including standard HTML controls with a single value or operation (e.g., buttons and textboxes) or they can be complex objects (e.g., trees).

Hundreds of web development libraries and toolkits (e.g., [10, 13, 28] to mention a few) are providing an ever growing number of ready-to-use web widgets that can be further customized by web developers to fit the needs of individual web sites. Unfortunately, the diversity of the libraries and a lack of standardization and enforcement of W3C specifications have proven to be an insurmountable problem for blind users.

For web browsing, blind people employ screen readers (*e.g.*, JAWS [17], Windows-Eyes [33], VoiceOver [30], Dolphin [27], Sa To Go [26], NVDA [24], etc.), which convert the Web to speech, generally ignoring layout and graphics, and reading aloud all the textual content in web pages. Screen readers enable their users to listen to and navigate web content sequentially in the order it is laid out in the HTML source code, which often does not correspond to visual layout. Screen readers provide many shortcuts to navigate among elements of a particular type, e.g., links, buttons, edit fields, etc. Although not very efficient for web browsing [3], screen readers enable visually-impaired people to browse the Web and perform online activities.

Unfortunately, screen readers do not recognize widgets that dynamically appear on the screen, so the user has no easy way to find them; at best, a dynamically appearing widget will be "navigable," meaning that it can be found and narrated by the screen reader, but without giving any indication as to what kind of widget it is. As a result, blind users either cannot benefit from the convenience of using widgets (e.g., use a date picker) or they even get stuck on inaccessible content (e.g., an HTML alert window). So, while sighted people can enjoy Rich Internet Applications (RIA), blind people either cannot access them at all (e.g., cannot use Google Docs) or have to use basic versions of the websites (Gmail and Facebook).

To make web applications more accessible, web developers have to follow Accessible Rich Internet Applications (ARIA) specifications [32]. For instance, ARIA allows developers to mark up live regions where the content may update, specify the importance of those updates, and provide simple roles such as "progress meter." Unfortunately, web developers do not follow ARIA specifications consistently, and ARIA does not have predefined roles for complex widgets such as date picker.

In this paper, we propose an ARIA-independent approach towards improving the accessibility of dynamic widgets. Specifically, the contribution of this paper is a scalable machine learning approach to identification/classification of *dynamic* widgets. To demonstrate the feasibility of the approach, we demonstrate high accuracy in classifying popular dynamic widgets such as date pickers, popup menus, suggestion lists, and HTML alert windows. Sample screenshots of these widgets are shown in Fig. 1.

The identification of *static* widgets (the ones that are already in the web page) and the design and evaluation of accessible user interfaces for widgets are beyond the scope of this paper, as these have been well explored in the literature, e.g., [7, 29].

a) Date Picker   b) Alert Box

c) Suggestion Box   d) Popup Menu

Fig. 1. Sample screen shots of the four widget types collected in the corpus: a) Date Picker, b) Alert window, c) Suggestion Box, and d) Popup menu

In the remainder of this paper, we provide some background on the complexities of dealing with dynamic widgets as well as on the specifics of using ARIA in Section 2; we review prior work relevant to widget localization and classification in Section 3; we describe the setup of the experiments in Section 4; we present and discuss the obtained results in Section 5, and, finally, we conclude by providing the broader impact view of this work and propose future directions of research in Section 6.

2   Background

In order to help understand the technology behind dynamic widgets, and thus the complexity of identifying them, we give a short overview of a variety of methods employed for making the widgets "appear" and "disappear" in web pages. We also provide background on how ARIA [32] can be used to mark up widgets in a way that would make them accessible with screen readers.

2.1   Methods for Displaying Dynamic Widgets

The most straightforward approach to displaying a new widget on the screen is to use JavaScript to insert a new sub-tree representing the widget into the HTML DOM tree. However, in order for this newly inserted widget, or any other widget that is already a part of the DOM tree to appear on the screen, the web developer must set a long list of widget's properties. Modifying any one of them can make the widget invisible or visible for the user. The following is a non-exhaustive list of "tricks" that web developers use to make the widget "appear" or "disappear" (this list was collected from HTML specification [16], numerous developer forums, and confirmed by observing the behavior of hundreds of widgets in the wild):

- Coordinates can be set to a negative value (off-screen) to hide a widget, and set to a positive value to show a widget within the webpage viewport;
- Coordinates can be set to place a widget within or outside of the viewport of a containing object, which can also be another widget;
- Dimensions of a containing object's viewport can be set in a way that the contained widget is hidden or shown on the screen;
- The width and height of a widget can be set to a value that would make it big enough to be visible or make it so small that it disappears;
- The widget's color can be set to blend in with the background or make it stand out;
- The font can be set to a human-readable size or to 0 to make the textual content of widgets appear or disappear;
- The opacity of the widget can be set to make it transparent or opaque;
- The "visibility" or "display" style of the widget can be set as desired;
- The Z-index of the widget can be set to place it behind or in front of other objects.

All these properties can be modified directly in the HTML code of the widget or indirectly, e.g., by changing the widget's "style" attribute. Further, some of these properties (e.g., "display") are inherited from ancestor DOM nodes, so adding a widget as a sub-tree to the DOM will immediately result in the application of ancestor's properties unless they are overridden.

The code and content of widgets can be in the HTML, Cascading Style Sheets (CSS), in JavaScript, or it can be delivered to the webpage via AJAX from the web server. Given the variety of ways to hide and show a widget and deliver it to the web page, the only good way to detect the appearance of dynamic widgets is by monitoring the changes in the DOM tree [12], as we describe in Section 4.1.

2.2   Screen Readers and Accessible Rich Internet Applications (ARIA)

As the blind user is "navigating" on a webpage, screen readers maintain the virtual cursor pointing at the currently narrated object. When dealing with widgets, screen-readers have three main tasks: 1) allow the user to move a virtual cursor from widget to widget, and from one atomic object on the webpage to another; 2) announce relevant information at the current (new) virtual cursor position, including: textual content of the widget, semantics or purpose of the visited widget (e.g., text box, alert window, etc.), its state (e.g., "blank", "disabled", etc.), and the results of user actions affecting the widget (e.g., typed letter, selected item, etc.), and 3) announce important changes to the webpage not localized at the current virtual cursor position.

Unfortunately, the best screen readers can only accomplish the tasks above if widgets are marked up following the Accessible Rich Internet Applications standard (ARIA) [32]. And the responsibility for making web content accessible lies entirely on web developers most of whom take the off-the-shelf ready-to-use widgets from a widget library. Alas, ARIA is not widely followed by web developers or even by screen reader developers. Furthermore, due to its complexity, it is often incorrectly implemented by web developers, which makes the accessibility problem even worse.

Making a Widget Accessible with ARIA. ARIA markup is not necessary for making standard widgets (e.g., textbox, link, listbox, button, etc.) accessible, because screen readers already have a long list of sophisticated hardcoded interaction rules that enable: a) virtual cursor's movement between widgets, b) user interaction with the widgets, and c) relevant announcements that are made to the user when the cursor arrives or leaves the widget or changes its value.

For custom widgets which model the standard widgets, web developers can use the ARIA roles, states, and properties that enable the screen reader to map the custom widget to a standard widget and activate corresponding hardcoded rules. For example, a checked checkbox implemented using JavaScript and html tag "div" can be marked up with attributes role="checkbox" and aria-checked="true". When the checkbox is unchecked, the author needs to use JavaScript to set aria-checked="false".

For custom widgets that model a combination of standard widgets web authors can use a complex combination of ARIA roles, states and properties. ARIA authoring practices can be found in [19], and a good overview of accessible widgets in [21].

For custom widgets that, due to their functional complexity or uniqueness, cannot be mapped to any standard widget (e.g., calendar), web developers need to: a) inform the screen-reader that this is a custom widget without explicit mapping to a standard widget by specifying the widget's role as "application"; b) process relevant keyboard commands, including those that would normally be handled by the screen-reader (e.g., arrow keys), and map those keys to the widget's functionality; c) if the widget consists of multiple interactive components (e.g., a grid with editable grid cells), the web author needs to assign some shortcuts to move the browser's focus between all those components (this will hint the screen-reader to follow the focus with the virtual cursor, and narrate the relevant content).

To announce changes to the webpage not localized at the current virtual cursor position the authors can use ARIA Live Regions; designated by attribute "live", it can be set to one of four politeness levels ("off", "polite", "assertive", "rude") determining the urgency (importance) of the changes in the region. The attribute "relevant" is used to set the relevance of specific types of DOM changes within the live region.

The Problems with Using ARIA. Using standard widgets or a custom implementation of standard widgets enables the screen-reader to implement a powerful, accessible interface to a very limited set of widgets. This requires no or little overhead for the web developer, but, in case of custom widgets, this often depends on the diligence in assigning roles and dynamically modifying attributes as a result of user interactions.

Using a complex combination of ARIA roles and states to implement a single widget has the potential to expand the range of functionality and types of widgets supported by the screen-readers. However, this approach requires advanced understanding of ARIA by the developer, increases cost of implementation and support, depends on advanced support of ARIA by screen-readers, which is not yet available for the full ARIA specification, and is limited by the scope of ARIA specifications that do not cover all possible types of functionality. As a result, the use of ARIA in complex widgets (e.g., date picker, etc.) to make them accessible is not widely used.

The use of the "application" role in combination with JavaScript code for processing keyboard events and controlling browser focus has the potential to enable even more complicated widgets than the other approaches. However, predictably, this requires even more effort on the part of the developers than the purely ARIA approach, and, hence, this approach is not widely deployed either.

Issues with ARIA live regions, including difficulties with determining causality of the region updates, giving developers the ability to combine technically discrete but semantically atomic updates, handling interim updates, and providing higher-level abstractions for web developers, are well summarized in [29].

Overcoming the Widget Accessibility Problems. In order to make a widget accessible, it needs to be first localized and identified. Once localized, the widget needs to be enabled for screen reader interaction which can be accomplished either by injecting ARIA into the widget components [9, 18] or by enabling the screen-reader to map the widget (or its components) to the hardcoded interaction rules that will make this widget accessible, similar to the way it is done with the standard widgets. The latter approach is both more powerful than ARIA injection (because it is not restricted by the expressively of the ARIA specifications) and is less prone to errors since it is not restricted by the ability of other screen-readers to correctly interpret ARIA.

In this paper, we focus on the classification or identification of dynamic widgets, which is the first step toward making them accessible. Customizing and evaluating user interfaces for widgets [7, 29] is not in scope of this paper. However, it is notable that the proposed method does not limit the way widgets will be made accessible.

3   Related Work

Many researchers have recognized the accessibility problems caused by dynamic content and dynamic widgets (those that dynamically appear and disappear on the screen), as well as the deficiency of screen readers in handling this problem [1, 2, 4, 6, 8, 9, 11, 14, 18, 20]. Several approaches were proposed to make dynamic content and widgets accessible, localize widgets, and classify the widgets into types.

3.1   Making Dynamic Widgets Accessible

The majority of the approaches described in the literature focused primarily on retro-fitting the existing web application by adding ARIA markup [32] to the web page, while a few attempted to provide a custom screen-reader interface.

An early approach proposed by the authors [4] enabled users to review any dynamic updates on the web page using a special layered view isolating the changes. While useful, that approach, however, did not help identify the type of changes or their importance; and, without the ability to isolate and classify the type of widgets, it is impossible to provide a usable interface that works best with a particular widget.

For example, if a slideshow widget changes, unless the screen-reader user is on it, it is a low priority event that should not interrupt what the user is listening to. On the other hand, if an alert window appears, the event has to be audibly announced, screen-reader navigation should only be available within the scope of the window, and the text of the window needs to be read out. If a date picker appears, current date has to be announced, and table navigation (left/right/up/down) has to be enabled in the calendar.

AxsJax [9] was one of the pioneering approaches that showed (using the example of Google chat) how web developers can make their web applications more accessible by injecting ARIA metadata into AJAX (Asynchronous JavaScript) responses sent from the server to the web browser. The injection would happen on the server-side and would have to be enabled by the web developers. However, if the developers were motivated to make their websites accessible, they would have ARIA from the start. So, AxsJax, does not fix the accessibility problem, but rather provides web developers with an alternative method for making their web applications accessible.

Single Structured Accessibility Stream for Web 2.0 Access Technologies (SASWAT) [18] is a project performing the AJAX injection on a proxy server. This releases the web developers of responsibility and puts it into the hands of the SASWAT supporters, who would have to provide the ARIA metadata describing the dynamic content and store it into a repository. Then, if any of the specified webpages were loaded through the proxy, the proxy would inject the ARIA metadata into the webpages, thus making them accessible. Of course, this approach breaks down if the web pages are changed by the developers and it is not scalable across websites, because any variation of a widget would require modifications of the scripts. A truly scalable approach would require automatic widget localization and classification as we discuss in the following section.

3.2   Widget Localization and Classification

In [1], authors propose an approach to widget security vulnerability detection. For this purpose, the authors locate widgets using DOM dynamic updates analyses. They track the parts of the DOM, which were changed as a result of some user action or Java-Script event. Later, they test identified widgets for inter change, i.e. when one widget is updated automatically by another widget. While there is some similarity in track-ing mutation events used in this paper, authors do not care about the types of widgets.

The approach proposed in [2] utilizes end-user-programming to enable automation and customization of web application. Based on the selected keywords, one can build a pattern to localize a widget on the web page, which could potentially be reused across websites. However, this approach only works for simple widgets (search box or a button) and requires manual effort for selected keywords for a particular widget.

Several research teams have looked into the possibility of recognizing complex dynamic widgets. For instance, [7] has explored the possibility of detecting calendar widgets; however, the approach was a special-case heuristic classification that de-tected the calendar based on user interaction, which is not scalable to other widget types.

An early attempt to classify different widgets was made in [8]. The proposed algo-rithm analyzed webpage sources (HTML, JS, and CSS) using regular expressions to find the display window (the area containing the widget) and then matching the con-tent to an ontology with predefined widget features. Unfortunately, the classification was done by constructing a hierarchy of widgets based on widget implementations in specific libraries, which means that the classification will work only for the widgets taken from these libraries. Also, the ontology has to be created manually, and the approach will not be able to detect the (dis)appearance of a dynamic widget.

A more dynamic approach to desktop widget classification is proposed in [11]. The main goal of the paper is to simplify a desktop interface and make it "visually access-ible". The system is built as an extension of the Prefab pixel-based recognition sys-tem. The system recognizes some simple widgets such as "Windows 7 steal buttons" based on the visual markers. Besides requiring compute-intensive vision based analy-sis of the screen, unfortunately, this approach is also limited by the visual distinction between widgets. For example, it will not able to distinguish a suggestion box from the popup menu because both look like popups. Neither will it be able to find a widget if it is hidden behind some other control such as a drop down menu.

In contrast, the method proposed in this paper overcomes the limitations listed above by using a more general and lightweight approach to widget classification. It employs machine learning to classify widgets automatically regardless of the library they come from. Since a variety of features can be extracted from the DOM tree (Sec-tion 4.2), the approach can classify widgets even if they are visually similar, e.g., based on trigger events. Furthermore, once the widgets are classified, this information can be used either to inject ARIA metadata into the webpage or provide this informa-tion directly to the screen reader to provide customized interaction with the widget.

The limitation of the proposed approach is that it can only identify *dynamic* wid-gets that appear and disappear in web pages. However, some of the reviewed methods [2, 8, 11] can be employed to detect *static* widgets that are already on the web page.

4   Experimental Setup

4.1   The Corpus and Widget Localization

To experiment with widgets, we collected a corpus with four types of popular widgets (suggestion list, HTML alert window, popup window, and date picker, shown in Fig. 1) with 50 examples of each widget type. To this corpus, we added an additional 50 examples of other randomly selected widget types; we refer to this generic widget type as "others". The corpus was collected from widget libraries and live websites using custom tool developed specifically for this purpose.

The data collection tool, based on the Capti Narrator (www.captivoice.com) for Mac/Windows [5], consists of a Firefox browser extension and a Java application. The browser extension listens to all DOM mutation events (updates), and communicates them over an open socket to the Java application. Java application uses the updates to construct a timeline of DOM mutations, reconstructs the DOM at any given point of time, and displays it in a separate window, in the form of a tree. Once a website with a widget has loaded, the process of collecting the data of that single widget is semi-automatic:

1. Press a button control shortcut to start "recording" all DOM mutation events;
2. Trigger opening of the widget and wait for it to open (usually instantaneous), e.g., press a button opening an alert window, focus on a textbox with a date picker, etc.;
3. Press the same button to terminate the "recording" of DOM mutations;
4. Verify that the recorded DOM mutations represent the target widget;
5. Save the resulting timeline of DOM mutations into the corpus of widgets.

The recorded timeline spans the period of "recording" and includes only the events in two sub-trees: one representing the trigger object and the other representing the widget that opened as a result of the trigger; all other events are ignored. The data collection tool was designed with the following considerations in mind.

A webpage may have many scheduled DOM mutation events, so it is very important to localize the relevant DOM mutations. While JavaScript is executed synchronously in most browsers, the exact localization of the mutation events relevant to a particular widget is an unsolved problem; multiple unrelated mutation events can happen immediately one after another. So, automated analysis of the underlying Java-Script would be required to understand the relationship between the user action (e.g., pressing a shortcut) and the system reaction (e.g., displaying a widget).

However, in practice, a heuristic approach that uses both temporal and spatial information helps minimize the risk of collecting irrelevant mutation events. Specifically, any user event such as focus change or control activation can be considered to be the potential trigger event, starting an observation period. Any subsequent DOM mutation events occurring within time t of the trigger event can be considered candidates for the widget. If more than one DOM sub-tree has updated, the collected mutation events can be further filtered by their spatial proximity to the trigger object.

4.2   Features for Widget Classification

The selection of features was inspired by the observations made during a manual inspection of the corpus. The vector representation of features was assembled from the features extracted from four different categories listed in Table 1. Most of the examined features are binary with the exception of the "Proportion of text nodes with only numbers in them" and the "Number of text nodes".

Table 1. Feature space for widget classification

Feature	Description	Binary
PRESENCE OF HTML TAGS & KEYWORDS		
P_{table}	Presence of table tag <table> in the HTML associated with widget	Yes
P_{list}	Presence of list related tags like <ul>, <li>, etc, in the HTML associated with widget	Yes
$P_{textbox}$	Presence of textboxes (e.g. <input type= "text">) in the HTML associated with the widget	Yes
P_{name}	Presence of widget name in "class" attribute of any tag in the HTML associated with the widget	Yes
P_{date}	Presence of "date" as the value of "type" attribute in any tag in the HTML associated with the widget	Yes
P_{image}	Presence of an image (<img>) in the HTML associated with the widget	Yes
CONTEXT RELATED		
T_{link}	Widget appears due to click of a button or link	Yes
T_{input}	Widget appears due to a keyboard entry in an input box	Yes
CHARACTERISTICS OF NODES IN WIDGET DOM SUBTREE		
$C_{text.num}$	Number of text(<text>) nodes	No
$C_{text.prop}$	Proportion of text (<text>) nodes containing only numbers in them	No
$C_{table.list}$	A table (<table>) or list (<ul>) is present and over 80% of its content are links	Yes
DISPLAY PROPERTIES OF WIDGET		
D_{widget}	Widget appears right below the "triggering" element	Yes

Presence of HTML Tags and Keywords (P). Analysis of the corpus revealed that, in some widgets, certain HTML elements are almost always present. For example, a list of links (<ul><li>...</li></ul>) can be found with high probability in a popup menu, a table (<table>) is likely to be present in a date picker widget to format the calendar, and an input textbox is always a part of suggestion list. In addition to HTML tags, attribute values can also be used to identify the widget. For example, we observed that, in many cases, the "class" attribute of one of the <div> or <span> HTML tags contained the name of the widget (e.g., "suggestion-box", "date-picker", etc.).

Context-Related Features (T). The local context surrounding a widget provides valuable information and important cues for identifying that widget. We refer to any HTML element that causes the appearance of a widget as a "trigger". The fact that different widgets are triggered in different ways by different HTML elements can be exploited to improve widget classification performance. For example, a suggestion box is almost always triggered when a user types something in an input text box; a menu popup appears on-screen when a user presses the corresponding button; a date picker widget appears when the user goes in focus on the textbox, etc.

Characteristics of Nodes in Widget DOM Sub-tree (C). This set of features was crafted after an extensive analysis of the DOM sub-trees corresponding to different widgets in the corpus. These features strive to leverage differences in the composition of DOM sub-trees corresponding to different widgets. For example, we observed that the DOM sub-tree of a suggestion box contains relatively higher number of <text> nodes compared to the DOM sub-tree of an alert box. Variation in composition also includes the type of content stored in the DOM nodes. For example, a list in a menu pop-up widget is likely to contain a high percentage of links, whereas a list in a suggestion box widget is likely to contain a large number of text nodes.

Display Characteristics of Widget (D). The position of a widget on the screen provides an important cue for its classification. Specifically, we are interested in the screen location of the widget relative to its triggering HTML element. This is based on our observation that different widgets exhibit different display patterns relative to their corresponding trigger. For example, in our corpus, the suggestion box widget appeared right below the triggering input textbox 100% of the time, and the menu pop widget always appeared either below or to the right of the corresponding triggering button or link. This feature is especially useful to filter out "irrelevant" dynamic mutations that can happen in the same time frame when the widget appears.

Having identified the salient features that could help distinguish different widgets, we conducted experiments to identify which combination of these features with which machine learning tools would yield the best widget classification results.

5   Experiments and Results

To conduct the experiments, we used the Weka [15] toolkit. We considered several popular machine learning classifiers and selected the following classifiers that yielded the best performances: Support Vector Machine, frequently used for benchmarking, and the J48 Decision Tree classifiers, which is a simple rule-based classifier that is appropriate for the mostly binary widget features.

As described in the previous section, we divided the features into five thematic categories according to the type of the feature. In order not to evaluate each feature separately, we combined the categories into five groups (Groups 1-5 in Table 2) in a way that would allow us to evaluate the impact of each individual category on the results. For example, Presence (P) features are absent in Group 1, Context (T) features are absent in Group 2 and so on. Group 5 has all the feature sets, and hence the performance of group 1-4 can be compared with Group 5 to assess the importance of the corresponding missing feature set.

Table 2. Feature groups used for widget classification

Group	Features
Group 1	Context (T) + Characteristics (C) + Display (D)
Group 2	Presence (P) + Characteristics (C) + Display (D)
Group 3	Presence (P) + Context (T) + Display (D)
Group 4	Presence (P) + Context (T) + Characteristics (C)
Group 5	Presence (P) + Context (T) + Characteristics (C) + Display (D)

Finally, we ran both classifiers on each of the five groups of features. We used 5-fold cross validation: 200 widget examples (40 of each type) were used for training and 50 (10 of each type) for testing, repeated 5 times with different divisions into folds. The results of the experiments are detailed in Table 3; the winning group-widget type combinations are in bold.

The absolute winners (J48 on Groups 4 and 5) have shaded background. As can be seen from the averages in Table 3, SVM and J48 classifiers yielded similar performance: SVM showed highest performance: 86% recall and precision in groups 4 and 5, while J48 won by a single percent point in both precision and recall. In all groups, J48 was performing better than SVM on average. In general, SVM yielded slightly better performance than J48 in identifying Date Pickers, but J48 was better at distinguishing Alert Boxes and Popup Menus. The absolute best performance was shown by J48 in Groups 4 and 5, yielding precision: 87% and recall: 87%, beating the SVM by 1% in both precision and recall.

Group 5 yielded the best average performance (Precision: 91%, Recall: 94%) when the generic widget type "*Others*" was excluded from the analysis, beating Group 4 (Precision: 90%, Recall: 94%) by a narrow margin of 1% in precision. In all of the feature groups, the average precision excluding "*Others*" is only slightly better than average precision with "*Others*". However, the average recall excluding "*Others*" is significantly better than the overall average recall, in all of the feature groups. These performance results demonstrate the effectiveness of our models in accurately identifying the 4 core widget types considered in our work.

Table 3. Widget classification results, Notation: P - Precision, R - Recall, J48 - Decision Tree, SVM - Support Vector Machine; the values in bold indicate the best performances per group

Widget	Classifier	Group 1		Group 2		Group 3		Group 4		Group 5	
		P	R	P	R	P	R	P	R	P	R
Suggestion box	SVM	**0.96**	**0.96**	0.81	0.88	0.76	0.68	**0.96**	**0.96**	**0.96**	**0.96**
	J48	**0.96**	**0.96**	0.80	0.96	0.94	0.58	**0.96**	**0.96**	**0.96**	**0.96**
Alert Box	SVM	0.65	0.76	0.73	0.90	0.67	0.84	0.74	0.94	0.77	0.94
	J48	0.92	0.78	0.84	0.78	0.76	0.84	**0.83**	**0.90**	0.85	0.88
Menu Popup	SVM	0.47	0.68	0.82	0.82	0.82	0.82	0.82	0.82	0.82	0.82
	J48	0.61	0.62	0.79	0.88	0.80	0.80	0.83	0.90	**0.85**	**0.90**
Date Picker	SVM	**0.98**	**1.00**	**0.98**	**1.00**	0.61	0.64	**0.98**	**1.00**	0.97	1.00
	J48	0.96	1.00	0.96	1.00	0.61	0.82	0.96	1.00	0.96	1.00
Others	SVM	0.70	0.24	0.59	0.36	0.64	0.44	0.79	0.54	0.80	0.56
	J48	0.59	0.60	0.58	0.42	0.62	0.54	**0.78**	**0.60**	0.74	0.60
Overall Avg.	SVM	0.75	0.73	0.78	0.79	0.70	0.68	0.86	0.85	0.86	0.86
	J48	0.81	0.79	0.79	0.81	0.75	0.72	**0.87**	**0.87**	**0.87**	**0.87**
Avg. excl. Others	SVM	0.77	0.85	0.83	0.90	0.72	0.75	0.88	0.93	0.88	0.93
	J48	0.87	0.84	0.85	0.91	0.78	0.76	0.90	0.94	**0.91**	**0.94**

Group 3, missing the Characteristics (C) features had the worst results both with the SVM (average P: 70%, R: 68%) and J48 (average P: 75%, R: 72%) classifiers. This shows that Characteristics (C) features were very important for classification of widgets in general. The results of both Group 1 and Group 2 are also significantly worse than Group 5, thereby demonstrating the importance of Presence (P) and Context (T) features. However, the absence of Display (D) feature (Group 4) did not cause any significant drop in performance (compared to Group 5 containing D feature). Overall, it can be inferred from Table 3 that feature sets P, T and C are critical for high performance, while the feature set D has minimal impact on the performance.

Notice in Table 3 that the accuracy in identifying Suggestion Boxes is heavily dependent on the T and C features (there is a significant drop in performance in Groups 2 and 3 compared to Group 5), while the performance is least influenced by the P and D features (Groups 1 and 4 yield the same performance as Group 5). Similarly, in case of Date Pickers, the C features were seen to contribute the most towards performance improvement compared to P, T and D features.

Table 4 lists the top discriminatory features for each widget as determined by the SVM classifier. As expected, T_{input} is the topmost predictive feature for Suggestion Box widget since all suggestion boxes are always activated when a user types something in an input textbox (often a search box). Also, observe that the feature P_{image} is also highly predictive of Suggestion Boxes, which indicates that a lot of websites provide suggestion boxes that contain image icons in addition to the suggested links or text. An example of such a suggestion box is shown in Fig. 1(c).

Table 4. Top discriminatory features as determined by the SVM classifier weights. According to [25], high positive weights indicate high predictiability of the corresponding class.

Widget	Top Discriminatory Features
Suggestion Box	T_{input}, P_{list}, P_{image}, $C_{table.list}$
Alert Box	P_{image}, T_{input}, $P_{textbox}$
Menu Popup	P_{list}, P_{name}, $C_{table.list}$
Date Picker	$C_{text.prop}$, P_{table}, T_{input}, P_{name}, $C_{text.num}$, $C_{table.list}$

It can be also seen in Table 4 that P_{name} is predictive of Menu Popup and Date Picker widgets, but not the other two widgets, thereby, highlighting a difference in the way these types of widgets are implemented with reference to CSS; compared to Alert windows and Suggestion boxes, a higher percentage of Popup Menu and Date Picker widgets have at least one node in their DOM sub-trees storing the corresponding widget name as a class attribute. Also, as expected, P_{list} is a top discriminatory feature for Menu Popup, since almost all pop-up menus contain a list of selectable items.

It can be inferred from Table 4 that overall, the features related to the DOM subtree characteristics (C) are extremely useful for widget classification.. This claim is supported by two observations: (i) Feature Group 3 yielded the lowest performance among the feature groups as previously noted in Table 3 analysis, and (ii) 3 out of 4 classes in Table 4 have at least one top discriminatory feature belonging to this category, e.g., $C_{text.num}$ points to variations in textual composition of different widgets.

Figure 2 depicts the decision tree (considering the entire dataset and all the features) produced by the J48 decision tree algorithm supported by WEKA toolkit. It can be inferred from Figure 2 that the feature $C_{text.prop}$ is the most important feature for identifying Date Picker widget type. More specifically, the proportion of text nodes with only numbers, in all Date Picker widgets in the dataset was above 0.21. Only one other widget of a different type in the dataset had the value of $C_{text.prop}$ greater than 0.21 (The label '51/1' of Date Picker leaf node in Figure 2 indicates that out of 51 data points placed in that group, 1 of them is incorrectly classified).

Similar inferences can be made from the decision tree in Figure 2 with respect to other widget types. For example, it can be observed that the context feature T_{input} is critical for correctly identifying the Suggestion Box widget type. Recall that even the SVM classifier determined T_{input} to be the most discriminating feature for identifying the Suggestion Box widget type (Table 4).

Similarly, it can be seen that P_{list} is the most important feature required for the accurate classification of Menu Popup widget type. Also observe in Figure 2 that the feature P_{name} plays an important role in identifying those Menu Popup data points that are not covered by P_{list}. These observations are in accordance with the SVM results presented in Table 4 where P_{list} and P_{name} are the top two discriminating features for Menu Popup widget type. However, no such straightforward comparisons between SVM and J48 decision tree results can be made with respect to the Alert box widget type as it can be observed in Figure 2 that the Alert Box widget type relies on different combinations of a wide variety of features for their accurate identification.

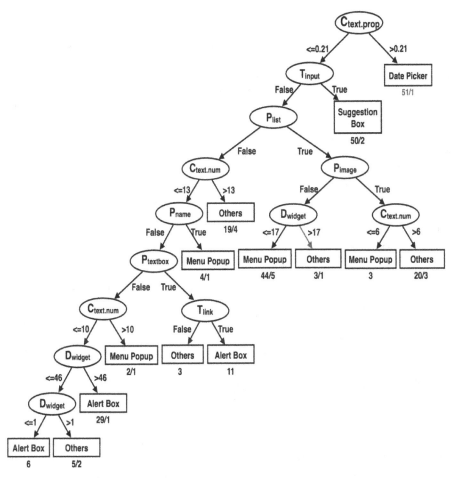

Fig. 2. Decision tree produced by the J48 algorithm on the entire dataset for feature group 5. Each leaf label indicates total classified data points followed by total incorrect classifications.

The importance of each group of features is apparent from the decision tree in Figure 2, where all features except P_{table} and P_{date} are used for classification. The absence of P_{table} in Figure 2 is a bit surprising since P_{table} was determined to be one of the top discriminatory features for Date Picker widget type. This observation adds weight to our earlier deduction that $C_{text.prop}$ is the single most important feature for identifying Date Picker widgets.

6   Conclusion and Future Work

In this paper, we have proposed and evaluated a scalable method for classification of dynamic web widgets using machine learning. The experiments on a corpus of 250 widgets showed that the decision tree learning was the most accurate machine learning technique for identifying and distinguishing among four popular types of widgets: the popup menu, the HTML alert window, suggestion box, and data picker (Fig. 1).

To date, there exists no assistive technology capable of enabling consistent and usable interaction with dynamic content and dynamic widgets. As was discussed in Sections 2 and 3, the existing solutions are very limited and are inadequate given the continuing and rapid adoption of dynamic web technologies. The ability to access the same web applications that are available to sighted people will enable people with vision impairments to enjoy the latest assistive technology, making them more productive. The approach proposed in this paper is the first step towards making widgets accessible, requiring further research and development in this direction.

While this paper handled four popular types of dynamic widgets, there are many more types of dynamic widgets that are used more rarely, but are used nonetheless. Although the proposed approach is scalable for more types of widgets, an extensive dataset has to be first assembled in order to handle more types of widgets. New approaches to dynamic widget localization have to be explored and tested. A reliable method needs to be developed for identifying static widgets that are already in the web page as soon as it loads.

In parallel, automatically identified widgets need to be made accessible to screen readers. While injecting ARIA (Section 3.1) may provide general accessibility to all screen readers supporting ARIA, it is also possible to embed the method described in this paper into a screen reader. The latter approach will allow for a more powerful user interface that is not possible given the limited expressivity of ARIA. Longitudinal user studies will have to be conducted to evaluate the usability of the user interfaces and verify the accuracy of the widget identification approach in the wild.

Finally, the approach proposed in this paper can find use in other web-based application areas. For instance, widget identification can be useful in website crawling [22], website simplification [11], and other reverse engineering of web applications.

Acknowledgements. This work was developed under a grant from the Department of Education, NIDRR grant number H133S130028. However, contents do not represent the policy of the Department of Education, and you should not assume endorsement by the Federal Government. We are also grateful to our Accessibility Consultant, Glenn Dausch, for his insightful feedback on accessibility problems with dynamic widgets.

References

[1] Bezemer, C.-P., Mesbah, A., Deursen, A.V.: Automated security testing of web widget interactions. In: Proceedings of the the the 7th Joint Meeting of the European Software Engineering Conference and the ACM SIGSOFT Symposium on The Foundations of Software Engineering, pp. 81–90. ACM, Amsterdam (2009)

[2] Bolin, M., Webber, M., Rha, P., Wilson, T., Miller, R.C.: Automation and customization of rendered web pages. In: Proceedings of the 18th Annual ACM Symposium on User Interface Software and Technology, pp. 163–172. ACM, Seattle (2005)

[3] Borodin, Y., Bigham, J.P., Dausch, G., Ramakrishnan, I.V.: More than meets the eye: a survey of screen-reader browsing strategies. In: Proceedings of the 2010 International Cross Disciplinary Conference on Web Accessibility (W4A), pp. 1–10. ACM, Raleigh (2010)

[4] Borodin, Y., Bigham, J.P., Raman, R., Ramakrishnan, I.V.: What's new?: making web page updates accessible. In: Proceedings of the 10th International ACM SIGACCESS Conference on Computers and Accessibility. ACM, Halifax (2008)

[5] Borodin, Y., Sovyak, A., Dimitriyadi, A., Puzis, Y., Melnyk, V., Ahmed, F., Dausch, G., Ramakrishnan, I.V.: Universal and ubiquitous web access with Capti. In: Proceedings of the International Cross-Disciplinary Conference on Web Accessibility, pp. 1–2. ACM, Lyon (2012)

[6] Brown, A., Jay, C., Chen, A.Q., Harper, S.: The uptake of Web 2.0 technologies, and its impact on visually disabled users. Univers. Access Inf. Soc. 11(2), 185–199 (2012)

[7] Brown, A., Jay, C., Harper, S.: Audio access to calendars. In: Proceedings of the 2010 International Cross Disciplinary Conference on Web Accessibility (W4A), pp. 1–10. ACM, Raleigh (2010)

[8] Chen, A., Harper, S., Lunn, D., Brown, A.: Widget Identification: A High-Level Approach to Accessibility. World Wide Web 16(1), 73–89 (2013)

[9] Chen, C.L., Raman, T.V.: AxsJAX: a talking translation bot using Google IM: bringing Web-2.0 applications to life. In: Proceedings of the 2008 International Cross-Disciplinary Conference on Web Accessibility (W4A). ACM, Beijing (2008)

[10] DevExpress. DevExpress Widget Library (2014), https://www.devexpress.com (cited 2014)

[11] Dixon, M., Leventhal, D., Fogarty, J.: Content and hierarchy in pixel-based methods for reverse engineering interface structure. In: Proceedings of the SIGCHI Conference on Human Factors in Computing Systems, pp. 969–978. ACM, Vancouver (2011)

[12] DOM. W3C Document Object Model (2004), http://www.w3.org/DOM/DOMTR (cited 2010)

[13] Google. Google Web Toolkit (2014), http://gwt-ext.com/demo/ (cited 2014)

[14] Hailpern, J., Guarino-Reid, L., Boardman, R., Annam, S.: Web 2.0: blind to an accessible new world. In: Proceedings of the 18th International Conference on World Wide Web, pp. 821–830. ACM, Madrid (2009)

[15] Hall, M., Frank, E., Holmes, G., Pfahringer, B., Reutemann, P., Witten, I.H.: The WEKA Data Mining Software: An Update. SIGKDD Explorations (2009)

[16] HTML5. Hyper-Text Markup Language v.5.0 (2010), http://dev.w3.org/html5/spec/ (cited 2010)

[17] JAWS. Screen reader from Freedom Scientific (2013), http://www.freedomscientific.com/products/fs/jaws-product-page.asp (cited 2013)

[18] Jay, C., Brown, A.J., Harper, S.: Internal evaluation of the SASWAT audio browser: method, results and experimental materials, The University of Manchester (2010)

[19] Joseph Scheuhammer, M.C.: WAI-ARIA 1.0 Authoring Practices (2013), http://www.w3.org/TR/wai-aria-practices/ (cited 2014)

[20] Linaje, M., Lozano-Tello, A., Perez-Toledano, M.A., Preciado, J.C., Rodriguez-Echeverria, R., Sanchez-Figueroa, F.: Providing RIA user interfaces with accessibility properties. Journal of Symbolic Computation 46(2), 207–217 (2011)

[21] Lourdes, M.: Toward an Equal Opportunity Web: Applications, Standards, and Tools that Increase Accessibility. In: Paloma, M., Belen, R., Ana, I. (eds.), pp. 18–26 (2011)

[22] Mesbah, A., Bozdag, E., Deursen, A.V.: Crawling AJAX by inferring user interface state changes. In: Proceedings of the 2008 8th International Conference on Web Engineering. IEEE Computer Society (2008)

[23] MiniwattsMarketingGroup. Internet Usage Statistics: The Internet Big Picture World Internet Users and Population Stats (2013),
http://www.internetworldstats.com/stats.htm (cited 2013)

[24] NVDA. NonVisual Desktop Access (2013), http://www.nvda-project.org/ (cited 2013)

[25] Rayson, P., Wilson, A., Leech, G.: Grammatical word class variation within the British National Corpus sampler. Language and Computers 36(1), 295–306 (2001)

[26] SaToGo, Screen reader from Serotek (2010)

[27] SuperNova. Screen Reader from Dolphin (2013),
http://www.yourdolphin.com/productdetail.asp?id=1 (cited 2013)

[28] Telerik. Telerik Widget Library, http://www.telerik.com (cited 2014)

[29] Thiessen, P., Chen, C.: Ajax live regions: chat as a case example. In: Proceedings of the 2007 International Cross-Disciplinary Conference on Web Accessibility (W4A), pp. 7–14. ACM, Banff (2007)

[30] VoiceOver, Screen reader from Apple (2010)

[31] W3C. Important Terms (2014), http://www.w3.org/TR/wai-aria/terms (cited 2014)

[32] WAI-ARIA. W3C Accessible Rich Internet Applications (2013),
http://www.w3.org/TR/wai-aria (cited 2013)

[33] Window-Eyes, Screen Reader GW Micro (2010)

(De-)Composing Web Augmenters

Sergio Firmenich[1], Irene Garrigós[2], and Manuel Wimmer[3]

[1] LIFIA, Universidad Nacional de La Plata and CONICET Argentina
sergio.firmenich@lifia.info.unlp.edu.ar
[2] WaKe Research, University of Alicante, Spain
igarrigos@dlsi.ua.es
[3] Business Informatics Group, Vienna University of Technology, Austria
wimmer@big.tuwien.ac.at

Abstract. Immersed in social and mobile Web, users are expecting personalized browsing experiences, based on their needs, goals, and preferences. This may be complex since the users' Web navigations usually imply several (related) Web applications. A very popular technique to tackle this challenge is Web augmentation. Previously, we presented an approach to orchestrate user tasks over multiple websites, creating so-called procedures. However, these procedures are not easily editable, and thus not reusable and maintainable. In this paper, we present a complementary model-based approach, which allows treating procedures as (de)composable activities for improving their maintainability and reusability. For this purpose we introduce a dedicated UML profile for Activity Diagrams (ADs) and translators from procedures to ADs as well as back-translators to execute new compositions of these procedures. By combining benefits of end-user development for creation and model-driven engineering for maintenance, our approach proposes to have the best of both worlds as is demonstrated by a case study for trip planning.

1   Introduction

The evolution of the Web is a complex and constant process. Nowadays, immersed in social and mobile Web, users are expecting a personalized browsing experience, which adapt to their needs, goals, and preferences. One of the main limitations of how to adapt the application to each user is the current use of the Web. When performing a concrete task (e.g., organizing a trip) the user normally exceeds the application's boundaries, visiting several (related) Web applications. In cases like these, the user may feel a loss of context every time she navigates from one application to another, because the new application used has no way of tracking the previous user navigation. This missing integration, and also a lack in customization, has a deep impact in the user's browsing experience.

These limitations motivated the development of mash-ups tools [21] in order to merge a set of resources that are scattered among different websites into specialized applications. One often occurring limitation is that mash-ups are used straightforward when most of the tasks users perform are volatile and do not require the creation of entirely new applications. In the same context of managing existing Web applications, another technique that has emerged is called Web augmentation [3]. Web augmentation

S. Casteleyn, G. Rossi, and M. Winckler (Eds.): ICWE 2014, LNCS 8541, pp. 359–369, 2014.
© Springer International Publishing Switzerland 2014

is the activity of navigating the Web using a "layer" over the visited websites. This layer may manipulate the original UI of existing third-party websites; in this way, users perceive an augmented website instead of the original one. Generally, these augmentations are performed on the client-side, once the content is delivered from the server. Normally, users having some kind of programming skills are the ones who develop the software artifacts that perform these augmentations. Web augmentation as a technique may be applied with different aims; from simple presentation changes to task-based Web integration mechanisms.

In [6] we presented an approach based on Web augmentation to orchestrate user tasks over multiple websites. It supports flexible processes by allowing the users to combine manual and automated tasks from a repertoire of patterns of tasks performed over the Web, creating so-called *procedures*, which are persisted in XML files. Although the tools around our previous approach allow users to record their own *procedures* by-example, and subsequently edit the details; larger editions, such as replacing several tasks with other equivalent ones or building reusable chunks, is challenging. However, this may be often required, since several large tasks such as planning a trip involve several smaller ones (book flights, hotel rooms, cars, etc.) and the requirements involved change, as well as the browsed websites. If the user wants to change a larger part of the procedure, the process order may have to be changed, additional tasks have to be intermingled, or complete procedures have to be substituted or executed in series. The importance of these aspects has been studied before in the field of Web applications [13] [17]. These are also relevant issues in the context of Web Augmentation, not only because the Web changes constantly and consequently the scripts may stop working, but also because the same script could be reused in several Web pages under the same domain [13].

Therefore, one challenging aspect for those approaches that support users tasks based on Web augmentation, is the maintenance of procedures, which has associated two dimensions: *(i)* how to reuse existing augmentation units in order to support complex scenarios (i.e., how to compose them to fulfill a larger goal), *(ii)* how to decompose subtasks and make them reusable chunks. In order to tackle these challenges, this paper extends our previous work with a modeling language based on UML Activity Diagrams (ADs) to represent the procedure's tasks involving dedicated transformations from procedures to activities. Models allow raising the abstraction level and the separation from the applications functional specification [19], which improve the reusability and maintenance of the procedures. In this way, the maintenance of existing procedures as well as the composition of new ones based on existing building blocks is supported by graphical modeling. By having the transformations from activities to procedures, we are able to execute new compositions of Web augmenters. With this approach, we combine the benefits of end-user development for creating procedures based on Web augmentation and model-driven engineering for maintaining Web augmenters to have the best of both worlds as is demonstrated by a case study for trip planning.

The remainder of this paper is as follows: Section 2 briefly summarizes our previous work on Web augmentation and introduces an example used to illustrate our approach. Section 3 elaborates on the proposed model-based approach for representing procedures. Section 4 discusses the state-of-the-art on Web augmentation, and finally, we conclude with pointers to future work in Section 5.

2   Background

Web augmentation is used for improving the user experience in several aspects. In particular, we have previously proposed an approach for supporting Web tasks by supporting users with *procedures* [6]. Procedures are programs focused on executing augmentation tasks when some user interaction is detected. These artifacts support tasks involving more than one application, and also give some mechanisms for moving information from one application to another one. In order to specify *procedures* we have previously designed a DSL based on XML that defines a procedure as a sequence of tasks. This DSL has been improved in the context of this work. The current version of the procedures metamodel is shown in Figure 1.

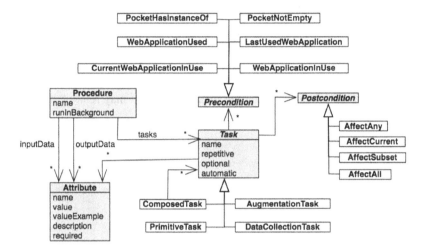

Fig. 1. The metamodel of the Web Augmentation DSL

The main concepts around the DSL are explained in the following.

- There are four types of tasks:
    - *Primitive tasks*: are based on common actions that users perform when navigating the Web (e.g., clicking an anchor).
    - *Augmentation tasks*: are tasks that allow the execution of a specific augmenter developed with our underlying framework for Web augmentation [7].
    - *DataCollection tasks*: this kind of tasks enables procedures to contemplate data collected by users. These tasks are also strongly related to *DataCollectors* and *Pocket*, two tools distributed with the framework supporting the procedures, which allow users to move information among Web applications.
    - *Composed tasks*: these tasks make possible to group other instances of tasks in order to manage them altogether. As an example, imagine the need of executing an augmenter each time that the user collects some information. In this case both tasks may be grouped in order to do repetitive the whole set.
- All tasks have three properties: *(i) repetition* property for specifying if the task may be executed more than once; *(ii) optional* property allows skipping the

execution of the task; *(iii) automatic* property is *true*, then the Web augmentation framework automatically triggers the task.

- Tasks have attributes representing information needed for the execution, e.g., if an augmenter is applied for filling in a form and it is marked as automatic, the augmenter needs to know which form fields are filled with which value.
- Tasks may have preconditions. Preconditions are used to decide if the task will be executed or not according to which information is currently available. There are two main kinds of preconditions: on the one side, *preconditions about collected data,* and on the other side, *preconditions about navigation history.*

Our approach gives support to the end-user with visual tools, deployed as Web browser plugins, for creating and executing *procedures.* Figure 2 shows the editor: a sidebar that allows users to specify tasks into the procedure while analyzing websites. The tool provides an assisted mode: users may *record* their interaction with the Web and the corresponding tasks will be added to the procedure automatically. This mode contemplates primitive tasks, augmentation tasks, and data collection tasks. Figure 3 shows how to edit a particular task. It allows users to specify the name, pre and post-conditions as well as values for both properties and attributes. If some sensitive information is saved when recording the interaction, users may remove it by editing the corresponding task.

Fig. 2. General view of the tool   **Fig. 3.** Edition of a single task

In order to give more insights to the real use of *procedures,* consider the following example (it will be used as a running example during the rest of the paper). The example responds to the following situation:

> *"Peter is going to travel to Paris for vacation. In that context, he has to buy flights from his town to there, and also book a taxi from the airport to downtown. For accomplishing these tasks, he uses different websites, e.g., expedia.com and wecab.com. In each of these two subtasks, Peter has to enter the same information. Besides booking flights and taxi, Peter is also interested in getting touristic information about Paris and nearby areas".*

In scenarios like this, users may take advantage of using Web augmentation approaches, since these may support users on moving relevant information from one application to other while using this information for executing augmenters in the visited websites. It is important to note that not only each augmenter is configurable (even replaceable by other similar ones) but also the subtasks (*book flight* or *book taxi*) may be also reordered and replaced according with the user's interest. Also the information used for performing the tasks (both primitive and augmentation tasks) may vary in distinct executions of the same procedure. It could be achieved by using conceptual tags during data collection tasks. In this way, if different users prefer, for example, different hotel's location or airlines, the procedure can be defined for consuming information through concept names such as "Hotel Location" or "Airline" instead of concrete data.

Although this is a common scenario, the order used for each subtasks may vary for different instantiations of the same scenario, when these are more complex. It may vary even more when Web augmentation is involved, because it is desirable to allow users to vary the augmentations applied in an easy way and to compose different procedures to solve larger examples. Thus, we provide a complementary extension to the end-user based development of Web augmenters, namely a model-based maintenance approach as explained in the next section.

3   (De-)Composing Procedures – A Model-Based Perspective

In order to solve the before mentioned drawbacks, we present a model-based approach, which allows treating procedures as composable activities. For this purpose we introduce transformations from procedures to activities as well as back-transformations to be able to execute new compositions of augmenters. In order to do so, we first need to be able to represent the procedures on the model level. With this goal in mind, we propose to use UML activity diagrams (ADs).

3.1   Model-Based Representation of Procedures

Representing procedures with ADs [16], in particular following the fUML execution semantics proposed by the OMG [15], requires a systematic mapping between our DSL and ADs. Here we follow existing methodologies for deriving UML profiles from DSL metamodels [18,20]. After investigating ADs for the purpose of modeling *procedures*, we identified a high overlap, although the later are, of course, more specific as the former. The following table illustrates the identified mappings between our DSL and ADs from a Web augmentation point of view, i.e., only the AD concepts are shown that are corresponding to the DSL concepts.

In addition to the mappings, to explicitly represent the specifics of Web augmenters (cf. Table 1 – column comments), we introduce a Web Augmentation profile for ADs. By using this profile, we are able to provide information preserving the transformations between the executable procedures expressed as XML files and the corresponding ADs. This property is one of the main building blocks of our approach to allow the continuous development on the front-end side (recording and testing procedures) as well as on the model side (maintaining and composing

Table 1. Mapping of Web Augmentation concepts to UML activity diagrams

Web Augmentation Procedures	UML Activity Diagrams	Comments
Procedure	Activity Diagram	*runInBackground* attribute has no direct mapping to UML, rest has direct mapping to UML
Task	Activity	*Optional, automatic, repetitive* attributes have no direct mapping to UML, rest has direct mapping to UML
PrimitiveTask	Activity	May be mapped to *Action* metaclass, but to allow for properties, *Activity* is used as metaclass
AugmentationTask	Activity	Same comment as for PrimitiveTask
DataCollectionTask	Activity	Same comment as for PrimitiveTask
ComposedTask	Activity	Activity may contain other activities by using CallBehaviorAction
Attribute	Property	*Value* and *example* attributes have no direct mapping to UML, rest has direct mapping to UML
Precondition	Constraint (LocalPreCondition)	*Precondition* subclasses have no direct mapping to UML
Postcondition	Constraint (LocalPostCondition)	*Postcondition* subclasses have no direct mapping to UML

procedures). Figure 4 shows the introduced stereotypes (for each meta-class we introduce a stereotype) as well as the extended metaclasses of UML. Please note that we only introduce tagged values in the profile for properties of the DSL that miss corresponding properties in the base UML metaclasses. By this, we ensure to reuse as much as possible the UML language and to keep the profile concise and minimal.

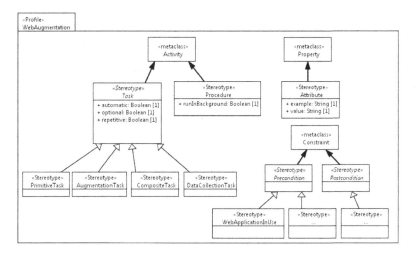

Fig. 4. Web Augmentation Profile – an Extension for UML Activity Diagrams

To summarize the syntax of the developed profile, we use the *Activity* metaclass as the main metaclass for our extension. We map the *Procedure* metaclass to the *Activity* metaclass, because ADs are UML internally represented by a root activity that contains all the elements shown within the ADs. Also the different kinds of *Task*s are mapped to the *Activity* metaclass. Because the *Activity* metaclass inherits from the

Classifier metaclass in UML, activities may contain properties. This is exactly what is required for reflecting the *Attribute* concept of *Tasks*. Besides these aspects, activities may also be nested, by using the *CallBehaviourAction* that is also able to trigger another activity from a context activity. By this, we can simulate the *CompositeTask* concept of the Web Augmentation DSL.

Moreover, we extend the metaclass *Property* with a stereotype to represent the *Attribute* metaclass of the DSL and to introduce additional attributes to allow the definition of an actual *value* and an *example value* for properties. Finally, we extend the *Constraint* metaclass of UML with specific stereotypes to reflect the specific pre- and post-condition types contemplated in our DSL.

Concerning the semantics of ADs, we consider an explicit control flow, normally a sequence of tasks, by defining *control flow links*. This is quite analogue to the sequence of tasks involved in a procedure and the information flow among these. In addition, we also exploit other control structure possibilities of ADs such as parallelization, conditions, etc. However, these constructs are not explicitly available in the current version of the Web augmentation DSL and thus, have to be compiled to a more verbose representation on the execution level. In addition to the control flow, we explicitly model the data flow, i.e., to represent the pocket and data collectors of the Web augmentation DSL, by making use of the *object flow links* supported by UML activity diagrams. By using this type of links, we are able to connect *activity parameter nodes*, i.e., parameters that are set externally before calling a certain activity as well as parameters that provide values after the execution of an activity to its environment, with so called *pins*. Activities may have input and output pins that represent input and output parameters, respectively. Pins are a powerful modeling concept in UML, e.g., by setting the multiplicity of input pins, input pins may be defined as mandatory or optional (i.e., a value is available or not for a given pin) for the execution of an activity. By linking output pins with input pins, data exchange between two activities is defined.

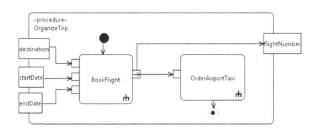

Fig. 5. Procedure OrganizeTrip in UML Activity Diagram Notation

Fig. 6. Composite Task *Book Flight* as UML Activity Diagram

Consider again our main example. If we take only one of the main subtasks, such as *book flights* (a *CompositeTask* called *BookFlight*), an activity diagram with stereotype *«procedure»* is generated (cf. Figure 5) for visualizing the execution of the sequence of *tasks* as illustrated in Figure 6. In this specific case, and for reason of conciseness, we only contemplated *AugmentationTasks*, but the given activity may also include several *PrimitiveTasks* allowing the *procedure* developer to specify specific user interactions. Again we use the *CallBehaviorActions* to call the primitive and augmentation tasks.

3.2   Transformation Chain: Procedures to Activities and Back Again

In order to allow for a transparent transition from Web Augmentation (WA) DSL expressed in XML to UML activity diagrams (ADs) and back again, we implemented a bi-directional transformation chain consisting of a set of transformations as explained in the following paragraphs. More information on the implementation may be found at our project website[1].

Model Injection/Extraction Transformations. We developed an XML 2 WA DSL transformation that parses the XML-based representations and produces models conform to an Ecore-based WA DSL metamodel. In addition, we developed a WA DSL 2 XML transformation for printing models back to executable XML code. These transformations have been implemented in Groovy[2] due to its dynamic programming features and the support by the XmlSlurper and XmlMarkupBuilder APIs.

DSL/UML Integration Transformations. We developed a WA DSL 2 UML AD transformation that produces UML models from WA DSL models and applies automatically the Web augmentation profile to the UML models. In addition, we also developed the inverse transformation that takes a profiled UML model and produces a WA DSL model. These transformations have been implemented in ATL [11] due to its support for EMF models as well as UML models and the possibility to deal with profile information within the transformations.

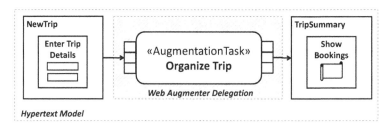

Fig. 7. Composing Web Augmentation Tasks with Hypertext Models

3.3   Composing Web Augmenter Models and Hypertext Models

One additional benefit of having Web augmenters explicitly modeled is the possibility to compose them with traditional Web design models such as supported by WebML,

---

[1] https://sites.google.com/site/decomposingwebaugmenters
[2] http://groovy.codehaus.org

OO-H [10], or UWE [12]. By this, Web augmentation techniques may be used by Web applications by delegating to pre-defined Web augmenters or Web augmenters may be developed for a specific Web application and integrated in the hypertext models of such applications. Consider the following example. Assume one would like to provide for a Web application that offers specific events the possibility to book a hotel room at an external website. Navigating to the external website with the specific information such as place and time may be provided by the hypertext model. This information may be passed by typical transport links transferring parameters to the Web augmenter activity (as it is done for standard hypertext nodes) and the Web augmenter activity may provide information of the booked hotel room back to the hypertext model again as parameters of a transport link. In Figure 7 we show such a composition of a hypertext model and a Web augmenter activity for the WebML language. We leave as subject for future work the creation of Web augmenter units for WebML based on the WebRatio inherent extension mechanism and the integration of the profile presented in this paper with the UWE profile for modeling hypertext models. We think this is an important line of future work to close the gap between traditional Web modeling and Web augmentation.

4   Related Work

Several approaches for supporting Web user tasks have been created, and different abstraction levels have been used. For example, CoScripter [1] proposes a DSL for supporting recurrent tasks, which may be parameterized in order to alter the data used in each step. The main idea of CoScripter is to automate some tasks by recording the user interactions (based on DOM events) and then the script may reproduce the same steps automatically. A similar approach, ChickenFoot [2], also proposes a DSL that raises the abstraction level of JavaScript programs in order to emulate user behaviour easily. However, although these approaches support slight changes in the task processes, considerable changes over these cannot be contemplated. These tools allow modifying end-user programs to vary the way that tasks are going to be performed, but usually, the *augmentation* effect is limited to a predefined subset of possibilities.

Although we share the philosophy behind these approaches, we think that further efforts should be made for making this kind of tools closer to the actual use of the Web, because users navigate the Web in a volatile way, and some tasks may be achieved in different ways (Web applications involved, data used, navigation) under different circumstances. In previous work we have presented our approach called *procedures*. Although this involves a composition of tasks where each task may be preconditioned and parameterized, the reuse of *parts* of procedures related to a particular subtask is not foreseen. All the mentioned approaches would improve taking into account some aspects from task modelling such as HAMSTERS [14], in which "abstract tasks" may be defined and the execution order may be more flexible.

The most related work in this context is [4], which proposed to model the user navigation using state machines in order to create the so-called webflows. This work defines a DSL, which allows users to specify the navigation flow as well as the data associated with each transition. One of the main differences to our work is the fact that [4] does not foresee the inclusion of third-party augmentations (i.e. developed by users), which again implies a limitation of augmentation effects. In our approach, this is contemplated by the execution of augmenters [7]. Finally, [9] define a UML profile for data mashups, but the integration with Web augmenters is not considered.

5   Conclusions and Future Work

Web augmentation is an emerging trend that allows users to improve their experiences while navigating the Web. Several approaches have been proposed to improve websites with different goals, from accessibility aspects over data integration to complex user task support, which is the focus of this work.

Although there are currently several works aiming to support specific navigation scenarios, user navigation is not always systematic as current approaches assume. In this way, one of the main challenges in this context is to support users even under volatile requirements. There are several other issues in the middle, such as how easy users may define their own artifacts for these approaches. The key is to find a good trade-off between the expressivity of the approach (what can be specified) and the usability of the tools (how it is specified). Reaching this point is challenging, and in this work, we aim to address a solution of maintaining procedures by using activity diagrams, where each activity represents a relevant subtask in a more general navigation scenario. Of course, the target users of the proposed modeling approach may no longer be end-users, but Web engineers may decompose, recompose, and maintain already existing Web augmenters and integrate these pieces in their developed hypertext models. The next steps imply defining mechanisms for including the transformations developed in this work in our Web augmentation tools and performing experiments with different kinds of users. Since our underlying Web augmentation framework allows tracking the user interaction, we plan to incorporate aspect orientation concepts [8] in order to further (de)compose procedures when cross-cutting concerns occur.

References

1. Bogart, C., Burnett, M., Cypher, A., Scaffidi, C.: End-user programming in the wild: a field study of CoScripter scripts. In: VL/HCC, pp. 39–46 (2008)
2. Bolin, M., Webber, M., Rha, P., Wilson, T.: C. Miller R.: Automation and customization of rendered web pages. In: UIST, pp. 163–172 (2005)
3. Díaz, O.: Understanding Web augmentation. In: Grossniklaus, M., Wimmer, M. (eds.) ICWE Workshops 2012. LNCS, vol. 7703, pp. 79–80. Springer, Heidelberg (2012)
4. Diaz, O., De Sosa, J., Trujillo, S.: Activity fragmentation in the Web: empowering users to support their own webflows. In: Hypertext, pp. 69–78 (2013)
5. Díaz, O., Arellano, C., Iturrioz, J.: Interfaces for Scripting: Making Greasemonkey Scripts Resilient to Website Upgrades. In: Benatallah, B., Casati, F., Kappel, G., Rossi, G. (eds.) ICWE 2010. LNCS, vol. 6189, pp. 233–247. Springer, Heidelberg (2010)
6. Firmenich, S., Rossi, G., Winckler, M.: A Domain Specific Language for Orchestrating User Tasks Whilst Navigation Web Sites. In: Daniel, F., Dolog, P., Li, Q. (eds.) ICWE 2013. LNCS, vol. 7977, pp. 224–232. Springer, Heidelberg (2013)

7. Firmenich, S., Winckler, M., Rossi, G., Gordillo, S.: A crowdsourced approach for concern-sensitive integration of information across the web. JWE 10(4), 289–315 (2011)
8. Garrigós, I., Wimmer, M., Mazón, J.-N.: Weaving Aspect-Orientation into Web Modeling Languages. In: Sheng, Q.Z., Kjeldskov, J. (eds.) ICWE Workshops 2013. LNCS, vol. 8295, pp. 117–132. Springer, Heidelberg (2013)
9. Gaubatz, P., Zdun, U.: UML2 Profile and Model-Driven Approach for Supporting System Integration and Adaptation of Web Data Mashups. In: Grossniklaus, M., Wimmer, M. (eds.) ICWE Workshops 2012. LNCS, vol. 7703, pp. 81–92. Springer, Heidelberg (2012)
10. Gómez, J., Cachero, C., Pastor, O.: Extending a Conceptual Modelling Approach to Web Application Design. In: Wangler, B., Bergman, L. (eds.) CAiSE 2000. LNCS, vol. 1789, pp. 79–93. Springer, Heidelberg (2000)
11. Jouault, F., Allilaire, F., Bézivin, J., Kurtev, I.: ATL: A model transformation tool. Sci. Comput. Program. 72(1-2), 31–39 (2008)
12. Koch, N., Kraus, A., Zhang, G., Baumeister, H.: UML-Based Web Engineering - An Approach Based on Standards. In: Web Engineering, pp. 157–191 (2008)
13. Li, J., Gupta, A., Arvid, J., Borretzen, B., Conradi, R.: The empirical studies on quality benefits of reusing software components. In: COMPSAC, pp. 399–402 (2007)
14. Martinie, C., Palanque, P., Winckler, M.: Structuring and composition mechanisms to address scalability issues in task models. In: Campos, P., Graham, N., Jorge, J., Nunes, N., Palanque, P., Winckler, M. (eds.) INTERACT 2011, Part III. LNCS, vol. 6948, pp. 589–609. Springer, Heidelberg (2011)
15. Object Management Group. Unified Modeling Language (UML), Superstructure, Version 2.4.1 (2011), http://www.omg.org/spec/UML/2.4.1
16. Object Management Group. Semantics of a Foundational Subset for Executable UML Models (fUML), Version 1.0 (2011), http://www.omg.org/spec/FUML/1.0
17. Rossi, G., Schwabe, D., Lyardet, F.: Abstraction and Reuse Mechanisms in Web Application Models. In: Mayr, H.C., Liddle, S.W., Thalheim, B. (eds.) ER Workshops 2000. LNCS, vol. 1921, p. 76. Springer, Heidelberg (2000)
18. Selic, B.: A Systematic Approach to Domain-Specific Language Design Using UML. In: ISORC, pp. 2–9 (2007)
19. Van Deursen, A., Visser, E., Warmer, J.: Model-driven software evolution: A research agenda. In: Workshop on Model-Driven Software Evolution (2007)
20. Wimmer, M.: A semi-automatic approach for bridging DSMLs with UML. IJWIS 5(3), 372–404 (2009)
21. Yu, J., Benatallah, B., Casati, F., Florian, D.: Understanding mashup development. IEEE Internet Computing 12(5), 44–52 (2008)

An Exploratory Study on the Relation between User Interface Complexity and the Perceived Quality

Seyyed Ehsan Salamati Taba[1], Iman Keivanloo[2], Ying Zou[2],
Joanna Ng[3], and Tinny Ng[3]

[1] School of Computing, Queen's University, Canada
[2] Department of Electrical and Computer Engineering, Queen's University, Canada
[3] IBM Toronto Lab, Markham, Ontario, Canada
taba@cs.queensu.ca, {iman.keivanloo,ying.zou}@queensu.ca,
{jwng,tinny.ng}@ca.ibm.com

Abstract. The number of mobile applications has increased drastically in the past few years. Some applications are superior to the others in terms of user-perceived quality. User-perceived quality can be defined as the user's opinion of a product. For mobile applications, it can be quantified by the number of downloads and ratings. Earlier studies suggested that user interface (UI) barriers (*i.e.,* input or output challenges) can affect the user-perceived quality of mobile applications. In this paper, we explore the relation between UI complexity and user-perceived quality in Android applications. Furthermore, we strive to provide guidelines for the proper amount of UI complexity that helps an application achieve high user-perceived quality through an empirical study on 1,292 mobile applications in 8 different categories.

1 Introduction

Mobile applications are pervasive in our society and play a vital role in our daily lives. Users can perform similar tasks both on smartphones and PCs [1] such as: checking e-mails or browsing the web. Due to the limitations of smartphones (*e.g.,* small screen size, network problems and computational power) developers should be more careful in designing their applications on smartphones than PCs. Developers' negligence in the importance of UI design is one of the major reasons for users to abandon a task on smartphone and switch to PC [2].

User-perceived quality can be defined as user's opinion of a mobile application. It can be quantified by the number of downloads and ratings in mobile stores. It is important to mention that based on our definition user-perceived quality has no relation with usability in this context. The studies conducted by Karlson et al. [2] and Kane et al. [3] demonstrate that improper use of UI elements (*e.g.,* input and output) on mobile applications increases end-user frustration. For example, the excessive use of input fields in mobile applications negatively affect user-perceived quality. Although mobile applications seem to be simple and easy

S. Casteleyn, G. Rossi, and M. Winckler (Eds.): ICWE 2014, LNCS 8541, pp. 370–379, 2014.

to develop, these studies illustrate that designing UI for mobile applications is not a trivial task.

Software metrics are widely used to derive guidelines for programmers. For example, McCabe [4] defines a complexity metric for functions, and recommends a proper implementation should hold a value below 10. Such guidelines can be exploited either during the development process for on-the-fly recommendation or during the quality assurance process. There exist several studies on the design patterns for UI development of mobile applications [5]. However, they do not provide a concrete number of appropriate UI complexity for mobile applications in order to achieve high user-perceived quality. In this paper, we focus on UI complexity and its relation with the user-perceived quality of mobile applications. Moreover, we aim to derive guidelines for UI complexity by mining the available mobile applications on Android Market. We define seven UI complexity metrics that can be calculated using static analysis. We calculate the metrics in two different granularities: i) *category*, and ii) *functionality* of 1,292 mobile applications. A category reflects the purpose of a group of mobile applications (*e.g.*, Shopping or Health) extracted from mobile stores. A functionality defines a fine-grained capability of a mobile application (*e.g.*, Payment or Sign in). We observe that there exists a relation between UI complexity and user-perceived quality of application pages (activities) belong to a similar functionality. UI complexity is dependent on the corresponding functionality. Activities with high user-perceived quality tend to be simpler in terms of UI complexity in general.

2   Background

In this section, we briefly talk about the architecture of Android applications. Android applications are written in Java programming language using Android Software Development Kit (SDK). The Android SDK compiles the code into an Android PaKage (APK) file which is an archive file with a ".apk" extension. One APK file contains all the content of an Android application.

Application components are the essential building blocks of an Android application. There are four different types of application components, including activities, services, content providers and broadcast receivers. Among those, users only interact with activities. An Android application consists of several activities. An activity is a single, focused task that the user can do. Each activity represents a single-screen user interface (UI). As a result, only one activity can be in the foreground for the users to interact with.

There are two ways to declare a UI layout for an activity: i) Declaring UI layout elements in an XML file (standard), or ii) Instantiating UI layout elements programmatically. Our premise in this work is towards the former approach since it is the recommended way by Android design guidelines [6]. Applications using the latter way are excluded from our study since our analysis and data gathering approach cannot handle them.

Every Android application has an AndroidManifest.xml (manifest) file in root directory. It contains meta-data information of an application (*e.g.*, the path to the source code of activities, permissions).

3   Study Design

3.1   Data Collection

In Android Market, there are 34 different kinds of categories from which we analyze 8 different categories. The 8 different categories are: Shopping, Health, Transportation, News, Weather, Travel, Finance and Social. Table 1 shows descriptive statistics for different categories. In total, we study 1,292 free android applications crawled in the first quarter of 2013.

Table 1. Summary of the characteristics of different categories

Category	# Applications	# Activities	# Inputs	# Outputs	# Elements
Shopping	193	2,822	12,529	25,058	68,468
Health	286	4,129	23,232	40,330	108,366
Transportation	128	1,078	5,603	7,718	22,991
News	114	1,302	4,725	7,407	23,507
Weather	244	1,608	6,713	38,659	84,739
Travel	106	1,711	7,164	15,210	38,285
Finance	103	1,167	5,989	12,899	33,818
Social	118	1,107	4,948	7,646	24,091

Extracting User-Perceived Quality. In Android Market, users can rate applications from 1 to 5 (i.e., Low to High), and write comments. The rating reflects the user-perceived quality of applications.However, Ruiz et al. [7] have shown that the rating of an application reported by Android Market is not solely a reliable quality measure. They found that 86% of the five-star applications throughout the Android Market in 2011 are applications with very few raters (less than 10 raters). Moreover, Harman et al. [8] show that the ratings have a high correlation with the download counts which is a key measure of the success for mobile applications. To overcome these challenges, we measure user-perceived quality by considering both rating and popularity factors (i.e., the number of downloads and raters) using Equation (1):

$$UPQ(A) = (\frac{1}{n} * \sum_{j=1}^{n} log(Q_j)) * Rating(A). \tag{1}$$

Where $UPQ(A)$ is the measured user-perceived quality for an application; A refers to an application; n is the total number of quality attributes (i.e., the number of downloads and raters) extracted from Android Market for A. Q_j shows a quality attribute. To normalize the value of quality attributes, we used log transform. Rating(A) is the rating score extracted for A from the Android Market.

3.2   Data Processing

Extracting APK Files. To extract the content and the needed information from APKs, we use apktool [9], a tool for reverse engineering closed, binary

Android applications. It decodes APK files almost to the original form and structure. It provides the source code of the application in an intermediate "Smali" format [10] which is an assembler for the dex format used in Android Java virtual machine implementation.

Inspecting Decoded APK Files. Given an activity, there does not exist any direct mapping between its source code and its UI page. To measure UI complexity, we need to recover this linking.

Given an application, we extract the path to the source code of activities from the manifest file. To map the activities to their corresponding XML layouts, similar to Shirazi et al.'s work [11], we parse the source code of an activity (*i.e.*, Smali file) to look for a call of the *SetContentView()* method, which includes an ID to the corresponding UI XML layout file. However, this heuristic cannot map an activity to the corresponding XML layout file if the input argument to this method is the name of the UI XML layout file. To overcome this issue, we trace both IDs and names.

Calculating Metrics. We parse the XML layout files to calculate different UI metrics that is used to quantify UI complexity. We consider two sets of metrics in different granularities (*i.e.*, application and activity levels) as shown in Table 2. For the application level metrics, we compute the UI complexity metrics for each activity, and lift the metrics up to the application level by using the *average* values for ANI, ANO, ANE and *sum* for NA. We categorize the elements as inputs and outputs as shown in Table 3. We use input and output tags listed in Table 3 since such elements are frequently used in Android applications [11].

Table 2. Proposed Application and Activity Level Metrics

	Metric Names	Description
Activity Level	NI	Number of Inputs in an activity
	NO	Number of Outputs in an activity
	NE	Number of Elements in an activity
Application Level	ANI	Average Number of Inputs in an application
	ANO	Average Number of Outputs in an application
	ANE	Average Number of Elements in an application
	NA	Average Number of Activities in an application

Table 3. Input and Output Tags

	Element Names
Inputs	Button, EditText, AutoCompleteTextView, RadioGroup, RadioButton ToggleButton, DatePicker, TimePicker, ImageButton, CheckBox, Spinner
Outputs	TextView, ListView, GridView, View, ImageView, ProgressBar, GroupView

Extracting Functionalities. We extract the functionalities of each mobile application using text mining techniques. For each activity, we extract contents, strings, labels and filenames associated to the source code of activities and their corresponding UI XML layout files. We use two different heuristics to extract the texts shown to a user from an activity: i) labels assigned to each element in the UI XML layout file, and ii) strings assigned from the source code. Finally, we use LDA [12] to automatically extract the functionalities in each category.

4 Study Results

This section presents and discusses the results of our two research questions.

RQ1: Can our measurement approach quantify UI complexity?

Motivation. Measuring the complexity of a UI is not a trivial task. As the first step, we evaluate if our UI complexity metrics and our measurement approach (*i.e.*, static analysis) can be used to quantify UI complexity. We want to answer this concern by testing whether our UI complexity metrics can testify hypotheses reported by previous different studies. A user study by Kane et al. [3] has shown that user-perceived quality of some categories of mobile applications is lower than the others. For example, users are reluctant to use smartphones for shopping purposes. As a result, we aim to find out whether we can make similar observations using our metrics and approach. If we provide evidence that our measured metrics for quantifying UI complexity can correlate with the findings of previous studies, we will conjecture that our proposed metrics can be used for studies on the UI complexity of mobile applications.

Approach. For each APK file (application), we use the approach mentioned in Section 3.2 to map the source code of activities to their corresponding UI XML layout files. Next, to quantify UI complexity within each category (see Table 2), we calculate four application level UI metrics (*i.e.*, ANI, ANO, ANE and NA). Finally, based on each metric, we observe whether the UI complexity is different between categories. We test the following null hypothesis among categories:

H_0^1: *there is no difference in UI complexity of various categories.*

We perform Kruskal Wallis test [13] using the 5% confidence level (*i.e.*, p-value < 0.05) among categories. This test assesses whether two or more samples are originated from the same distribution.

To testify the previous findings by Kane et al. [3], we classify our categories based on their study into two categories: i) applications that belong to the categories with high user-perceived quality, and ii) the ones that belong to categories with low user-perceived quality (*i.e.*, Shopping, Health, Travel, Finance, Social). Then, we investigate whether UI complexity is different among these two groups. We test the following null hypothesis for these two groups:

H_0^2: *there is no difference in the UI complexity of applications related to categories with high and low user-perceived quality.*

We perform a Wilcoxon rank sum test [13] to evaluate H_0^2, using the 5% level (*i.e.*, p-value < 0.05).

Findings. Our Approach for Quantifying UI Complexity Confirms the Findings of Previous Studies. The Kruskal Wallis test was statistically significant for each application level UI metric between different categories (Table 4)

Table 4. Kruskal-Wallis test results for application level UI metrics in different categories

Metric	p-value
ANI	0.001148
ANO	<2.2e-16
ANE	<2.2e-16
NA	4.842e-05

Table 5. Wilcoxon rank sum test results for the usage of application level UI metrics in categories with high and low user-perceived quality

Metric	p-value	ΔCliff
ANI	1.23e-11	-0.21
ANO	1.927e-10	-0.19
ANE	0.001	-0.10
NA	0.007	-0.05

meaning that there exists a significant difference in the UI complexity of various categories. Moreover, there also exists a difference between the UI complexity of applications related to categories with high and low user-perceived quality. As shown in Table 5, there exists a significant difference in UI complexity quantified by the four studied metrics that are used to quantify the applications in the categories of high and low user-perceived quality. Therefore, by quantifying UI complexity of mobile applications, we found the similar findings as the earlier user studies ([2], [3]) that UI complexity is important on user-perceived quality of mobile applications, and it varies among different categories. Therefore, our measurement approach based on static analysis can quantify UI complexity.

RQ2: Does UI complexity have an impact on the user-perceived quality of the functionalities in mobile applications?

Motivation. Mobile applications have a lot of variety even in the same category. To perform a fine-grained analysis, we cluster the activities based on their functionalities. We investigate whether there is a relation between UI complexity and the user-perceived quality among various functionalities of mobile applications. If yes, we can provide guidelines to developers of the proper number of activity level UI metrics required to have a high quality functionality.

Approach. For each application, we extract the corresponding activities and their UI XML layouts (see Section 3.2). Next, to label each activity with a fine-grained functionality, we use LDA [12] which clusters the activities (documents) based on their functionalities (*i.e.,* topics). In other words, for each activity, we extract all the strings and labels shown to the users (see Section 3.2). We apply LDA to all the activities retrieved from the existing applications in a category to extract their corresponding functionalities.

Since mobile applications perform a limited number of functionalities, the number of topics (*i.e.,* K) should be small in our research context. As we are interested in the major functionalities of applications, we empirically found that $K = 9$ is a proper number for our dataset by manual labeling and analysis of randomly selected mobile applications. We use MALLET [14] as our LDA implementation. We run the algorithm with 1000 sampling iterations, and use the parameter optimization provided by the tool to optimize α and β. In our corpus, for each category, we have n activities (extracted from the applications in the corresponding category) A = $\{a_1, ..., a_n\}$, and we name the set of our topics (*i.e.*, functionalities) F = $\{f_1, ..., f_K\}$. These functionalities are different in each category, but the number of them is the same ($K = 9$). For instance, f1 in the Shopping category is about *"Login"* and *"Sign in"* functionality. However, in the Health category, it is about *"information seeking"* functionality. LDA automatically discovers a set of functionalities (*i.e.*, F), as well as the mapping (*i.e.*, θ) between functionalities and activities. We use the notation θ_{ij} to describe the topic membership value of functionality f_i in activity a_j.

Each application (A) is consisted of several activities ($\{a_1, a_2, ..., a_n\}$), and it has a user-perceived quality calculated by Equation (1). To compute the user-perceived quality for each activity, we assign each activity the user-perceived quality obtained from the application that they belong to. All the activities from the same application acquire the same user-perceived quality. However, by applying LDA [12] each activity acquires a weight of relevance to each functionality. Therefore, the user-perceived quality for an activity can originate from two sources: i) the user-perceived quality of its corresponding application, and ii) the probability that this activity belongs to a functionality. Moreover, we use a cut-off threshold for θ (*i.e.*, 0.1) that determines if the relatedness of an activity to a functionality is important. A similar decision has been made by Chen et al. [15]. We calculate the user-perceived quality for each activity as the following:

$$AUPQ(a_j) = \theta_{ij} * UPQ(a_j), \tag{2}$$

Where AUPQ(a_j) reflects the activity level user-perceived quality for activity j (a_j); θ_{ij} is the probability that activity j (a_j) is related to functionality i (f_i); UPQ(a_j) is the user-perceived quality of the application which a_j belongs to it.

For each functionality, we sort the activities based on the user-perceived quality. Then, we break the data into four equal parts, and named the ones in the highest quartile, activities with high user-perceived quality, and the ones in the lowest quartile, activities with low user-perceived quality. Finally, we investigate whether there exists any difference in the distribution of activity level UI metrics (*i.e.*, quantifiers of UI complexity in functionality level) between activities of low and high user-perceived quality. We test the following null hypothesis for each activity level UI metric in each category for each functionality:

H_0^3: *there is no difference in UI complexity between activities with low and high user-perceived quality.*

We perform a Wilcoxon rank sum test [13] to evaluate H_0^3. To control family-wise errors, we apply Bonferroni correction which adjusts the threshold p-value by dividing the number of tests (*i.e.*, 216). There exists a statistically significant difference, if p-value is less than $0.05/216 = 2.31\text{e-}04$.

Findings. There is a significant difference between UI complexity of activities with low and high user-perceived quality. For each cell of Table 6, we report three pieces of information. Let's consider the cell related to the Shopping category for the first functionality (*i.e.*, f1) which refers to *"Login"* and *"Sign in"* functionalities, for the NI (*i.e.*, Number of Inputs) metric. In this

Table 6. Average usage of activity level UI metrics in the activities with low and high user-perceived quality for each functionality in each category. (p<0.0002/50*; p<0.0002/5°; p <0.0002+)

		f1	f2	f3	f4	f5	f6	f7	f8	f9
Shopping	NI	↗2.23*	↗2.38*	↗3.92	↗2.83*	↘3.89	↗3.22*	↗2.53*	↘3.44*	↗2.25
	NO	↗4.01*	↗3.55	↗4.77*	↗4.92*	↘8.55	↗5.40*	↗4.28*	↗6.27*	↗3.57
	NE	↗11.13*	↗9.84*	↗16.17*	↗13.32*	↘22.80	↗15.71*	↗11.83*	↗16.90*	↗10.00*
Health	NI	↗2.92*	↗2.01*	↘2.57*	↗3.25*	↗2.41*	↗2.54*	↘2.55	↗3.23	↘2.46
	NO	↘4.20*	↗3.16*	↗3.22*	↗5.24*	↗2.70*	↗3.70*	↗3.78*	↗4.66*	↘3.11*
	NE	↗13.41*	↘13.43*	↗10.35*	↗14.55*	↗8.32*	↗10.89*	↗10.86*	↗13.77*	↘8.95*
News	NI	↗2.63*	↘2.36*	↘3.25	↗2.50*	↗2.21*	↗2.45*	↗3.26*	↗2.13*	↘2.47*
	NO	↘3.70*	↘3.15*	↗3.50*	↗3.03*	↗3.00*	↗3.91*	↗3.58*	↗2.51*	↗3.72*
	NE	↘11.66*	↘10.40*	↘11.94*	↘9.69*	↗10.06*	↗12.38*	↗12.59*	↗8.63*	↗12.75*
Transportation	NI	↘3.49*	↗4.17	↗3.11*	↘1.77*	↘3.83*	↗4.39	↗3.22*	↗4.09+	↗3.78*
	NO	↗2.81*	↗5.07+	↗2.84*	↗3.35*	↘3.49*	↘4.21*	↗4.85*	↗3.53*	↘5.25°
	NE	↗10.39*	↗15.91*	↗8.96*	↗9.44*	↗12.83*	↘12.72*	↗13.34*	↗12.70*	↘12.98*
Weather	NI	↗2.08*	↘3.33	↗2.35	↗1.23*	↗2.87*	↗2.04	↗2.03*	↗3.78+	↘3.29°
	NO	↘5.35*	↘6.17*	↗5.57*	↘11.77*	↘5.48*	↗4.23*	↗1.82*	↗6.38*	↗3.79*
	NE	↘10.92*	↘16.61*	↗14.03*	↘30.97*	↗14.15*	↗8.99*	↗5.92*	↗14.65*	↗11.51*
Travel	NI	↗3.36*	↘3.81	↗2.36*	↘2.36*	↗3.51*	↗2.87*	↗3.64*	↗3.03*	↗2.41*
	NO	↗4.22*	↘5.64*	↗2.61*	↘5.66*	↗4.77*	↘4.03*	↗4.26*	↘3.57*	↘3.21*
	NE	↗12.94*	↗16.62*	↗7.78*	↘16.19*	↗14.52*	↗12.38*	↗11.94*	↗12.68*	↗10.46*
Finance	NI	↗3.38*	↘2.81*	↗3.97*	↘2.81*	↗2.40*	↗2.37*	↘4.14	↗4.02*	↘5.29
	NO	↗6.42*	↘4.59*	↗7.85*	↗6.30*	↗4.00*	↗4.01*	↗7.22*	↗5.98*	↘9.41°
	NE	↗15.90*	↘11.60*	↗18.16*	↗18.16*	↗11.24*	↗10.58*	↗18.24*	↗16.85*	↘24.77
Social	NI	↗2.77*	↘3.17*	↗2.48*	↘2.04*	↗2.96+	↗3.02*	↗3.12*	↗2.06*	↘4.07
	NO	↗4.57*	↘3.81*	↗3.86*	↗2.56*	↗3.50*	↗3.86*	↗5.38*	↗2.55*	↘6.16
	NE	↗14.45*	↘14.10*	↗13.43*	↘9.39*	↗12.41*	↗14.36*	↗16.38*	↗9.35*	↘19.34

cell, first, there is a "↗" or "↘" sign which implies whether the difference for the corresponding metric (*i.e.*, NI) between activities with low and high user-perceived quality is positive or negative. In this example, it is positive ("↗") which means that activities related to this functionality (f1) in the Shopping category with low user-perceived quality have more complexity for NI than the ones with high user-perceived quality. Moreover, we report the average usage of the corresponding metric (NI) for the activities with high user-perceived quality which is 2.23 in this example. Such average values can be used to derive software development guidelines (*e.g.*, McCabe [4]). Here, it implies the average number of the corresponding activity level UI metric for high quality activities. Finally, we report whether the difference in the usage of the corresponding metric (*i.e.*, NI) is statistically significant between low quality activities and high quality ones. In this example, the difference is statistically significant (⋆).

As it can be seen from Table 6, we can reject H_0^3, and conclude that there exists a significant difference in UI complexity between activities with low and high user-perceived quality. Furthermore, we observe that UI complexity is dependent on the corresponding functionality and category. For some functionalities higher UI complexity can result in a better user-perceived quality. However, in some cases this relation is quite different. In most cases this difference is a positive number ("↗") meaning that low quality activities tend to use more activity level UI metrics than the high quality ones. In other words, simpler activities in terms of our used activity level UI metrics may results in a better perceived quality by the users. Our guidelines can be exploited by developers to use the proper UI complexity required to have functionalities with high user-perceived quality.

5   Threats to Validity

We now discuss the threats to validity of our study following common guidelines for empirical studies [16].

Construct validity threats concern the relation between theory and observation. They are mainly due to measurement errors. Szydlowski et al. discuss the challenges for dynamic analysis of iOS applications [17]. They mention that these challenges are user interface driven. Due to such challenges, we were not able to use dynamic analysis for mobile application UI reverse engineering for a large scale study. In this study, our premise is based on the UI elements declared in XML files since it is the recommended approach by Android guidelines [6].

Threats to internal validity concern our selection of subject systems, tools, and analysis method. The accuracy of apktool impacts our results since the extracted activity and XML files are provided by this tool. Moreover, the choice of the optimal number of topics in LDA is a difficult task. However, through a manual analysis approach, we found that in all categories there exist at least 9 common functionalities.

Conclusion validity threats concern the relation between the treatment and the outcome. We paid attention not to violate assumptions of the constructed statistical models; in particular we used non-parametric tests that do not require any assumption on the underlying data distribution.

Reliability validity threats concern the possibility of replicating this study. Every result obtained through empirical studies is threatened by potential bias from data sets [18]. To mitigate these threats we tested our hypotheses over 1,292 mobile applications in 8 different categories. We chose these categories since they contain both categories with high and low user-perceived quality, and they are from different domains. Also, we attempt to provide all the necessary details to replicate our study.

Threats to external validity concern the possibility to generalize our results. We try to study several mobile applications (1,292) from different categories. Our study analyzes free (as in "no cost") mobile applications in 8 different categories of the Android Market. To find out if our results apply to other mobile stores and mobile platforms, we need to perform additional studies on those environments.

6   Conclusion

In this paper, we provided empirical evidence that UI complexity has an impact on user-perceived quality of Android applications. To quantify UI complexity, we proposed various UI metrics. Then, we performed a detailed case study using 1,292 free Android applications distributed in 8 categories, to investigate the impact of UI complexity on user-perceived quality of mobile applications. The highlights of our analysis include: i) We can quantify UI complexity based on our measurement approach (**RQ1**) and ii) There is a significant difference between UI complexity of activities with low and high user-perceived quality. Activities with high user-perceived quality tend to use less activity level UI metrics (*i.e.,* simpler) than activities with low user-perceived quality. Moreover, we derive guidelines for the proper amount of UI complexity required to have functionalities with high user-perceived quality (**RQ2**).

In future work, we plan to replicate this study on more categories existing on Android Market. Moreover, we should investigate whether our findings are consistent among other platforms (iOS and BlackBerry).

References

1. Karlson, A.K., Meyers, B.R., Jacobs, A., Johns, P., Kane, S.K.: Working overtime: Patterns of smartphone and pc usage in the day of an information worker. In: PerCom (2009)
2. Karlson, A.K., Iqbal, S.T., Meyers, B., Ramos, G., Lee, K., Tang, J.C.: Mobile taskflow in context: A screenshot study of smartphone usage. In: SIGCHI (2010)
3. Kane, S.K., Karlson, A.K., Meyers, B.R., Johns, P., Jacobs, A., Smith, G.: Exploring cross-device web use on pcs and mobile devices. In: Gross, T., Gulliksen, J., Kotzé, P., Oestreicher, L., Palanque, P., Prates, R.O., Winckler, M. (eds.) INTERACT 2009. LNCS, vol. 5726, pp. 722–735. Springer, Heidelberg (2009)
4. McCabe, T.: A complexity measure. IEEE Transactions on Software Engineering SE-2(4), 308–320 (1976)
5. Nilsson, E.G.: Design patterns for user interface for mobile applications. Advances in Engineering Software 40(12), 1318–1328 (2009)
6. Android guidelines (April 2014), http://developer.android.com/guide/developing/building/index.html
7. Mojica Ruiz, I.J.: Large-scale empirical studies of mobile apps. Master's thesis, Queen's University (2013)
8. Harman, M., Jia, Y., Zhang, Y.: App store mining and analysis: Msr for app stores. In: MSR (2012)
9. apktool, http://code.google.com/p/android-apktool/
10. smali, http://code.google.com/p/smali/
11. Sahami Shirazi, A., Henze, N., Schmidt, A., Goldberg, R., Schmidt, B., Schmauder, H.: Insights into layout patterns of mobile user interfaces by an automatic analysis of android apps. In: SIGCHI (2013)
12. Blei, D.M., Ng, A.Y., Jordan, M.I.: Latent dirichlet allocation. The J. of Machine Learning Research 3, 993–1022 (2003)
13. Sheskin, D.J.: Handbook of parametric and nonparametric statistical procedures. CRC Press (2003)
14. McCallum, A.K.: Mallet: A machine learning for language toolkit (2002), http://mallet.cs.umass.edu
15. Chen, T.-H., Thomas, S.W., Nagappan, M., Hassan, A.E.: Explaining software defects using topic models. In: MSR (2012)
16. Yin, R.K.: Case study research: Design and methods, vol. 5. Sage (2009)
17. Szydlowski, M., Egele, M., Kruegel, C., Vigna, G.: Challenges for dynamic analysis of ios applications. In: iNetSec (2012)
18. Menzies, T., Greenwald, J., Frank, A.: Data mining static code attributes to learn defect predictors. IEEE Transactions on Software Engineering 33(1), 2–13 (2007)

Beyond Responsive Design:
Adaptation to Touch and Multitouch

Michael Nebeling and Moira C. Norrie

Department of Computer Science, ETH Zurich
CH-8092 Zurich, Switzerland
{nebeling,norrie}@inf.ethz.ch

Abstract. The new generation of touch devices are often used for web browsing, but the majority of web interfaces are still not adapted for touch and multi-touch interaction. Using an example of an existing web site, we experiment with different adaptations for touch and multi-touch. The goal is to inform the design of a new class of web interfaces that could leverage gesture-based interaction to better support application-specific tasks. We also discuss how current responsive design techniques would need to be extended to cater for the proposed adaptations.

Keywords: responsive web design, adaptation to touch and multi-touch.

1 Introduction

Given the proliferation of touch devices, web applications in particular are increasingly accessed using input modalities other than mouse and keyboard. However, to date many web sites still do not provide an interface that is optimised for touch input, and multi-touch interaction is generally still limited to gestures for scrolling and zooming content as interpreted by web browsers [1,2]. This paper aims to show how web sites could be adapted and instead provide carefully designed multi-touch features that are tailored to the web interface and therefore of potential benefit when carrying out application-specific tasks.

We investigate the adaptation to touch and multi-touch as a two-layered web design problem with a new set of technical and design challenges beyond responsive design [3]. In the original proposal[1], responsive web design was conceived as a way of developing flexible web page layouts that can dynamically adapt to the viewing environment by building on fluid, proportional grids and flexible images. Similar to some of the techniques used to support the study presented in this paper, this is mainly based on using CSS3 media queries for defining breakpoints for different viewing conditions and switching styles to adapt the layout. While responsive design is nowadays used as a broader term to describe techniques for delivering optimised content across a wide range of devices, the focus is still on dealing with different screen sizes, rather than adapting to different input modalities such as touch, which is the focus of this paper. While both

---

[1] http://alistapart.com/article/responsive-web-design

S. Casteleyn, G. Rossi, and M. Winckler (Eds.): ICWE 2014, LNCS 8541, pp. 380–389, 2014.
© Springer International Publishing Switzerland 2014

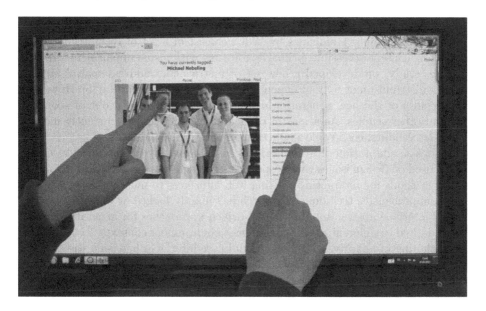

Fig. 1. One of the multi-touch versions we have designed and evaluated for a simple picture tagging application similar to Facebook, here using a two-point interaction

issues come together when designing for small, mobile touch devices, we want to separate the concerns by focusing on the adaptation to touch and multi-touch on devices where screen size is not the primary issue.

As one example, we examine alternative designs with a number of touch enhancements and different multi-touch features for *FBTouch*, a simple picture tagging application that we designed and gradually adapted based on the one provided by Facebook (Fig. 1). The goal of our work is two-fold: First, we want to identify the key issues related to touch specifically in a web context and consider various aspects of web design and the adaptation of interfaces for touch input. Second, we want to demonstrate how a new class of web interfaces with active support for multi-touch could be designed to improve task performance and the overall user experience on touch devices.

We begin by discussing related work and the background to our study in the next section. We then present the different designs created for the picture tagging application and the results of an initial user experiment. The paper discusses the implications for designing the required forms of adaptation and technical considerations for extending current responsive design techniques.

2   Background

Over the years, many different web design guidelines and best practices have been developed by practitioners and experts as well as in research. For example, several metrics to quantify usability factors such as the total word count in a page, the number of links and media as well as the spectrum of colours and font styles have been proposed [4]. More directed guidelines such as WCAG, the Web Content Accessibility Guidelines by W3C, consist of a set of recommendations on making content accessible, primarily for users with visual impairments. In our previous study, we developed a new set of metrics that address spatial factors and the distribution of content depending on the viewing condition [5]. However, the specifics of the new generation of touch devices have been out of scope of this and other studies. As a matter of fact, best practices are currently mostly driven by vendors, e.g. Apple's iOS Human Interface Guidelines.

As for the adaptation of web interfaces to different devices, previous research has focused on design issues with respect to small screens, e.g. [6], and fully automated methods for retargeting existing web interfaces to mobile phones [7]. For more comprehensive techniques, research has mainly looked at different models of user interface abstraction and model-driven approaches for generating interfaces adapted to different user, platform and environment contexts [8]. Although the authoring of adaptive and multi-modal user interfaces has been the subject of extensive research, e.g. [9], the concrete design and layout requirements have received relatively little attention. Specifically for touch, the key adaptation techniques have not been established, and studies so far have been limited to touch without considering multi-touch [2]. Despite the increasing availability of multi-touch web development frameworks such as jQMultiTouch [1], we are still far from an advanced use of multi-touch gestures in web interfaces. Rather, users currently employ simple pinch and pan gestures primarily for scrolling web pages and navigating between them, or as a workaround, and then for dealing with low-level issues such as precise selection of hyperlinks and other forms of active web content that are often inappropriately sized and placed for touch.

3   FBTouch

The initial design of our *FBTouch* example application is based on the original Facebook design. Rather than creating a new touch interface specifically designed for our study, we wanted to experiment with an existing interface with which many users are familiar, and study the effects of our adaptations. We chose picture tagging primarily because it is a common interaction technique for many Web 2.0 sites and current support in web interfaces is rather limited. As challenging and beneficial as it is for multi-touch interaction, it is also an example of a real-world task likely to be performed by users.

Common to all our FBTouch prototypes shown in Fig. 2 is the fact that users have to select from a list of names to tag people in pictures. Apart from the fact that this is the case in Facebook, we intentionally designed it in this way as it requires precise selection from multiple options which is difficult if the size and spacing of elements is too narrow for touch input. Also similar to Facebook, text search functions are provided in all interfaces to narrow down the list of names to only those that match the input. However, these are not necessarily needed because of the limited amount of scrolling required to view all name tags. The intention was not to give too much of an advantage to the mouse and keyboard

Fig. 2. Different interfaces designed and compared as part of the experiment

interface since text input on a touch screen requires users to switch between the browser and the on-screen keyboard and text input via a physical keyboard is usually faster. The alignment of the two links, "Previous" and "Next", next to each other above the picture is also similar to Facebook. At the same time, it is interesting for the purpose of our study since this alignment requires precise selection from a horizontal set of options where padding and margin are often less generous in web interfaces.

In the following, we will start by presenting the features of the standard interface and then show how it was gradually adapted, focussing mostly on what was changed between versions in terms of the tagging interaction.

The standard interface can be operated using mouse and keyboard or touch input. The tagging interaction is started by pointing with the mouse and clicking on people in the picture, or directly by touching the picture. This will then open a pop-up window next to where the input occurred, showing the search field on top and a list of names with a vertical scrollbar below. Clicking or touching the check boxes or names in the list will tag the selected person, as indicated by the checkmarks, or remove the tag if the name was already selected in a previous tagging interaction. Tagging can be cancelled by pressing ESC as well as by clicking or touching somewhere outside the pop-up window. The pop-up will be hidden after a person was successfully tagged or if the tagging interaction is cancelled. Pressing the "Previous" and "Next" links will navigate between pictures. Alternatively, the left and right arrow keys can be used to navigate to the previous or next picture. We will refer to this interface as A1 for mouse and keyboard input or A2 for touch.

Interface B uses the same tagging interaction, but in preparation for touch input, scales text slightly larger and increases the padding and spacing for active content areas such as links. In particular, the size of the "Next" link was adjusted to match the touch area of the "Previous" link. While this may not be as important for mouse input, it was considered because the touch area for the dominant action "Next" in the standard interface is significantly smaller, simply because the text link consists of only four letters. Also, the design now uses increased line height and spacing in the list of names, which required further adjustments of the list's height to have the same number of visible names. To prevent undesired default behaviour in browsers when using touch, the interface was further modified to disable text selection within touch-sensitive areas and prevent users from accidentally dragging the picture when touching it. Except for these small enhancements for touch input, the interface does not make use of any multi-touch features yet.

Interface C1 is based on the touch enhancements from the previous interface B, but in addition introduces a simple set of basic multi-touch gestures. These include swipe right or left to navigate to the previous or next picture, spread to overlay a larger version of the picture in higher quality and pinch to hide the overlay again. The tagging interaction itself then requires two-point interaction with one finger touching the picture and the other a name in the list. The order in which the touches occur determines the meaning of the interaction. Tapping the picture at different positions while touching the same name only changes the position of the tagging box. Tapping other names while touching the picture overwrites the current tag with the currently selected name. Tagged names are marked with a background colour slightly lighter than the highlight colour, and untagging can be performed by simply tapping a marked name. Interestingly, Windows 7 used on the TouchSmart with which the interfaces were developed and tested, did not allow for simultaneous touches on the picture while interacting with Windows standard controls such as the list of tags. We therefore enhanced the scrolling mechanisms of the list control to support scrolling when users touch the picture at the same time and prevent accidental tagging/untagging when scrolling occurred prior to the interaction.

The last interface C2 uses an alternative design of version C1 so that tagging now requires dragging a name from the list and dropping it on a person shown in the picture; dropping the tag outside the picture will cancel the operation. The interface supports the simple gesture set introduced in the previous version and additionally allows for performing multiple such drag-n-drop operations at a time. Not to remove names from the list via drag-n-drop, we implemented a way for touch events to be delegated to other elements that was previously not available in browsers. We used this method to drag a thumbnail of the person's photo as an intermediate representation of the original touch target. We also developed our own event capture technique so that simultaneous dragging of two or more photo tags using multi-finger or hand interaction can be supported. This new interaction is further enabled by the horizontal layout of the list now placed below the picture. Finally, the scrolling mechanisms of the previous version were adapted for horizontal scrolling not to interfere with active dragging operations, which required special event handling mechanisms.

We are aware of the fact that the design space for adaptations to support touch and gesture-based interaction is very large and that the different touch interfaces we created represent only two possible adaptations of the Facebook interface for touch devices. Nevertheless, the intention was to make only minor design modifications and then test their effects on users.

The first multi-touch tagging interaction using a two-point concept was designed as an alternative to the pop-up window from the standard versions of the interface as well as a simple way of employing two-touch to interact with two interface controls at a time. The intended meaning of this interaction was a natural mapping of pointing at somebody while calling out the name, or linking elements by holding them at the same time. However, this kind of layout that assigns rather fixed roles to hands also requires an alternative design for left-handed users who may prefer to use their left hand for the selection task.

The design modifications for the second multi-touch interface were made to allow users to use two hands independently and also to see whether users would employ multi-drag to improve their performance in the picture tagging task. The fact that this interface shows slightly more options in the list of names has two reasons. First, pilot testing showed that users are not as comfortable with the horizontal scroll layout and, second, it is also a countermeasure to make up for the fact that usually one hand hides a significant portion of the list when the tagging interaction is performed.

The swipe gestures available in both multi-touch interfaces were added to provide users with basic gestural support also known from other multi-touch applications. However, we intentionally kept the "Previous" and "Next" links in the interfaces primarily to assess the gestural support as an optional feature and make switching between them easier for users since the main navigation controls are present in every interface. Also we were interested to see how often users would make use of the gestures or the basic controls. Finally, the pinch/spread gestures for zooming only the picture rather than the entire interface were added to see whether users would appreciate an adapted zoom for the task.

The device used both for the active development as well as for user testing was an HP TouchSmart 600-1200 with 58.4 cm (23") screen diagonal and 16:9 wide-format screen at full HD resolution (1920x1080 pixels). The multi-touch features were built on top of Firefox which has included support for Windows 7 touch events since Version 4. The FBTouch prototypes were implemented using jQMultiTouch [1] and are published on the project web site[2].

---

[2] http://dev.globis.ethz.ch/fbtouch

4   User Experiment

To assess the user performance and experience of the different interfaces shown in Fig. 2, we carried out a small lab study with 13 participants on the Touch-Smart all-in-one computer also used for designing the interfaces. The majority of participants were between 25 and 35 years of age and right-handed. Participants' overall background using touch input was generally high and more than half of them (8 people) stated that they use touch devices several times a week to every day. The experiment consisted of performing the same simple picture tagging task with every interface. Rather than actually connecting to Facebook and tagging personal pictures and friends, we used a pre-defined set of pictures and names to make for equal conditions for all participants. The order was randomised and counterbalanced so that participants were not necessarily provided with gradually more features. They were therefore given a short time to get familiar with each interface before the actual experiment started.

The results of the user study reveal a number of interesting aspects concerning task performance and rated user experience for the different interfaces and input modalities that were tested. In general, one can see that participants were most efficient using mouse and keyboard, but the touch enhancements and multi-touch prototypes were well received by participants and helped them to achieve a better performance on the touch screen. Also, the relatively high user experience of the mouse and keyboard interface was only matched by the multi-touch interfaces. In the following, we provide a brief analysis.

As for task performance, Figure 3a shows the average times required by participants. A one-way repeated measures ANOVA found a significant effect of time required to complete the task using the different interfaces. The mean task completion times were best for interface A1 using mouse and keyboard. On the other hand, the same interface using touch provided the worst performance as participants took on average almost one minute longer to complete the task. As a result, the differences between interfaces A1 and A2 were significant.

Quite promising when comparing interface A2 to the adapted interfaces for touch is the fact that the relatively simple touch enhancements in interface B already contributed to almost 15% faster task completion times. Also, the second best mean time overall was only then achieved as participants used the drag-n-drop multi-touch features of interface C2, which was significantly faster compared to interface A2. The mean task completion time using interface C1 was close to B. Overall, interface C2 seemed the best version for touch and multi-touch in terms of the time required to complete the picture tagging task, but the differences to B and C1 were not significant.

Moreover, interface A1 showed significant differences compared to interfaces A2, B and C1. Even though participants were on average still 9% slower with interface C2 than using mouse and keyboard, there was no significant difference between interfaces A1 and C2. This high suitability of interface C2 for touch input is also underlined by the fact that participants completed the task almost 26% faster than with interface B on the touch screen. We can therefore say that only complementing the simple touch enhancements of interface B with the

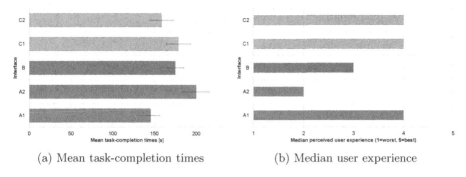

(a) Mean task-completion times   (b) Median user experience

Fig. 3. Compared to the mouse and keyboard interface A1, only interfaces C1 and C2 with task-specific adaptations for multi-touch produced similar task-completion times and user experience—the basic touch enhancements in B were not sufficient

multi-touch features of interface C2 helped participants to achieve accuracies that come a lot closer to the mouse and keyboard interface A1.

In terms of user feedback, Figure 3b illustrates that interfaces A1, C1 and C2 seemed to provide the best user experience for participants. The standard interface A2 executed on the touch display was rated by far the lowest. While interface B performed slightly better than interface C1, the user experience of the latter was on average rated considerably higher. This supports the general rule that faster execution times do not necessarily reflect in higher user experience. As a matter of fact, for nearly half of participants, interface A1 with mouse and keyboard was the fastest (6 times). Taking the touch interfaces only, interface C2 was first (5 times), closely followed by interface B (4 times). However, when looking at the best interfaces in terms of the user experience as rated by participants, we are presented with a slightly different picture where interface C1 was selected six times and interface C2 only four times.

In general, most participants felt very comfortable and fairly efficient with the touch adaptations specifically designed for the Facebook interface. When using the multi-touch interfaces C1 and C2, all participants but one favoured the swipe gestures rather than clicking the "Previous" and "Next" links for navigating between pictures. The majority of participants (10 people) also found the tagging interactions more tangible compared to the standard interfaces, with minor differences between the two-point interaction used in interface C1 and the drag-n-drop interaction in interface C2, which was however not significant. For the eight participants that effectively used the pinch/spread gestures, namely to zoom either the entire web site in interfaces A2 and B or specifically the picture in interfaces C1 and C2, it can be said that the adapted zoom to view a larger version of the picture was appreciated and rated higher. Still, due to the relatively large screen used in the study, the adapted zoom was generally not so often requested by participants and further studies on small-form factor devices should therefore aim to update these results. For other aspects, e.g. the almost vertical position of the touch display or technical limitations, such as limited precision and number of touch points that can be recognised with the TouchSmart, participants were generally neutral.

5 Observations and Implications for Design

The relatively high ratings for the user experience of the touch enhanced and multi-touch interfaces generally support our design decisions. One of the key factors that contributed to the fast times using interface A is that precise selection was not an issue with the mouse. Some participants even stated that the interface was optimally designed for mouse input because of the short distances between target elements. On the contrary, using the same interface on the touch display, participants had to concentrate on very precise selection and, for the nearly vertical setup that we used, often expected the touch to be recognised much higher than it was. Hence, we could often observe that participants developed a sort of counter technique in generally touching the screen slightly above the targets they actually wanted to hit, but this usually required some time to get used to. This was most often observed with activating the "Previous" and "Next" links or when trying to directly check the boxes associated with the listed tags rather than selecting the names. For the simple touch enhancements that we applied to the main navigation controls as well as the list of tags, the number of times participants missed the intended target elements on average were effectively reduced from 8 in the standard interface A2 using touch to a consistent .08 in all touch-enhanced interfaces B, C1 and C2—a significant improvement. As a result, the touch enhancements were sufficient to counteract some issues related to the precise selection of content.

We have already mentioned the fact that users rated the gestures for navigating between pictures relatively high. It also happened that users switched between techniques by sometimes using gestures and sometimes referring back to the "Previous" and "Next" links. In particular, participants found it faster to use the links when the tagging interaction previously occurred close to them. Web designers should therefore think of employing gestures as an alternative way of interacting with web content, not to completely replace standard means for interaction. This is especially important when users are already familiar with the traditional interface on non-touch devices and frequently switch between versions depending on the device in use, as for example in the case of Facebook.

6 Moving Forward

The study presented in this paper was driven by the current need to adapt interfaces for the emerging forms of multi-touch devices often used for web browsing. We have not only demonstrated that basic adaptations for touch can already contribute to better user experience, but also that it seems beneficial to further adapt interfaces to multi-touch interaction. In particular, we found it practical to start by addressing the low-level design issues first, such as the appropriate size and position of touch areas, and then address any issues with the existing interaction model as it is translated to a multi-touch design. We have already started to operationalise the key adaptation techniques used in this paper. Recently, we have added initial support in a multi-touch web interface toolkit, jQMultiTouch [1], and successfully used them as the basis for W3Touch [2], a metrics-based interface adaptation tool for touch. Our ongoing investigations have revealed several shortcomings of current web standards. Media queries provide a foundation for responsive design but, for the adaptations presented in this paper, they were not sufficient. One issue is that not all device aspects can be queried. For example, whether the TouchSmart was configured for touch rather than mouse input cannot be detected. Also information on the number of touch points supported by the device in use, namely two on the TouchSmart, is not available. Given that the latest proposals for CSS4 media queries cover only a few interaction media features, namely pointer and hover, this may not change in the near future[3]. In this regard, we want to critically note the remaining problem that state-of-the-art web technologies still lack common concepts and vocabulary, let alone a unified method, for the specification of multi-device web applications. In a related project, we have therefore investigated ways of enhancing existing languages with powerful context-adaptive mechanisms [10]. We would hope that similar concepts will make it to the web standards and be natively and consistently supported in future web browsers.

References

1. Nebeling, M., Norrie, M.C.: jQMultiTouch: Lightweight Toolkit and Development Framework for Multi-touch/Multi-device Web Interfaces. In: Proc. EICS (2012)
2. Nebeling, M., Speicher, M., Norrie, M.C.: W3Touch: Metrics-based Web Page Adaptation for Touch. In: Proc. CHI (2013)
3. Nebeling, M., Norrie, M.C.: Responsive Design and Development: Methods, Technologies and Current Issues. In: Daniel, F., Dolog, P., Li, Q. (eds.) ICWE 2013. LNCS, vol. 7977, pp. 510–513. Springer, Heidelberg (2013)
4. Ivory, M., Megraw, R.: Evolution of Web Site Design Patterns. ACM Trans. on Information Systems 23(4) (2005)
5. Nebeling, M., Matulic, F., Norrie, M.C.: Metrics for the Evaluation of News Site Content Layout in Large-Screen Contexts. In: Proc. CHI (2011)
6. Findlater, L., McGrenere, J.: Impact of screen size on performance, awareness, and user satisfaction with adaptive graphical user interfaces. In: Proc. CHI (2008)
7. Hattori, G., Hoashi, K., Matsumoto, K., Sugaya, F.: Robust Web Page Segmentation for Mobile Terminal Using Content-Distances and Page Layout Information. In: Proc. WWW (2007)
8. Calvary, G., Coutaz, J., Thevenin, D., Limbourg, Q., Bouillon, L., Vanderdonckt, J.: A Unifying Reference Framework for Multi- Target User Interfaces. IWC 15 (2003)
9. Paternò, F., Santoro, C., Spano, L.: MARIA: A Universal, Declarative, Multiple Abstraction-Level Language for Service-Oriented Applications in Ubiquitous Environments. TOCHI 16(4) (2009)
10. Nebeling, M., Grossniklaus, M., Leone, S., Norrie, M.C.: XCML: Providing Context-Aware Language Extensions for the Specification of Multi-Device Web Applications. WWW 15(4) (2012)

---

[3] http://dev.w3.org/csswg/mediaqueries-4

Composing JSON-Based Web APIs

Javier Luis Cánovas Izquierdo and Jordi Cabot

AtlanMod, École des Mines de Nantes – INRIA – LINA, Nantes, France
{javier.canovas,jordi.cabot}@inria.fr

Abstract. The development of Web APIs has become a discipline that companies have to master to succeed in the Web. The so-called API economy is pushing companies to provide access to their data by means of Web APIs, thus requiring web developers to study and integrate such APIs into their applications. The exchange of data with these APIs is usually performed by using JSON, a schemaless data format easy for computers to parse and use. While JSON data is easy to read, its structure is implicit, thus entailing serious problems when integrating APIs coming from different vendors. Web developers have therefore to understand the domain behind each API and study how they can be composed. We tackle this issue by presenting an approach able to both discover the domain of JSON-based Web APIs and identify composition links among them. Our approach allows developers to easily visualize what is behind APIs and how they can be composed to be used in their applications.

1 Introduction

The use and composition of different APIs is in the basis of computer programming. Software applications have largely used APIs to access different assets such as databases or middleware. In the last years, a new economy based on APIs has been emerging in the web field. To be competitive, companies are not only providing attractive websites but also useful Web APIs to access their data. Web developers have therefore to cope with the existing plethora of web APIs in order to create new web applications.

More and more web APIs use the JavaScript Object Notation (JSON) to exchange data (more than 47% of the APIs included in ProgrammableWeb[1] return JSON data). JSON is a schemaless data format easy for computers to parse and use. While JSON data is easy to read, its structure is implicit, thus entailing serious problems when integrating APIs coming from different vendors. In order to integrate external JSON-based web APIs, developers have to deeply analyze them in order to understand and manage the JSON data returned by their services. After analyzing JSON-based web APIs individually, it is still required to identify how to map the data coming from an API to call others since their implicit structure can differ.

Some approaches have appeared to make easier the understanding of JSON-based APIs, but they are still under development (e.g., RAML[2]) or are not widely used (e.g., JSON Schema[3] or Swagger[4]). Furthermore, the support for easily identifying how

---

[1] http://www.programmableweb.com
[2] http://raml.org
[3] http://json-schema.org
[4] http://swagger.wordnik.com

S. Casteleyn, G. Rossi, and M. Winckler (Eds.): ICWE 2014, LNCS 8541, pp. 390–399, 2014.

JSON-based web APIs can be composed is still limited. We believe that an approach intented to help developers to both understand and compose JSON-based web APIs would be a significant improvement.

In a previous work [1] we shown how to discover the schema which is implicit in JSON data. In this paper we build on that contribution to study how schemas coming from different JSON-based web APIs can be composed. Thus, we present an approach able to identify composition links between schemas of different APIs. This composition information plus the API schemas are used to render a graph where paths represent API compositions and are used to easily identify how to compose the APIs. For instance, we illustrate one application based on generating sequence diagrams from graph paths, where the diagram includes the API calls (and their corresponding parameters) that web developers have to perform in order to compose one or more APIs.

The paper is structured as follows. Section 2 motivates the problem. Sections 3, 4 and 5 describe our approach to discover the domain and composition links among JSON-based web APIs, respectively. Section 6 illustrates how our approach can be used to compose JSON-based web APIs and Section 7 discusses additional applications. Section 8 presents the related work and finally Section 9 concludes the paper and describes further work.

2   Using and Composing JSON-Based Web APIs

The development of web applications usually involves the composition of different web APIs. With the emergence of JSON-based APIs, web developers have to cope with the lack of documentation of these APIs and, when it exists, its non-standard format. Nowadays it is therefore usual to devote a significant amount of time to study JSON-based web APIs and to understand the implicit structure of the data they return. However, this is only the beginning since once APIs have been studied, it is required to explore how they can be composed (if possible). In this section we will show a simple example using two JSON-based web APIs we want to compose. From now on, we will refer JSON-based web APIs as APIs for the sake of conciseness.

Our example consists of a web application for tourists which includes a set of places to visit in our city and shows the routes to follow to reach them. The application includes a set of predefined places and needs to calculate the best route the user has to follow. Furthermore, the application also visualizes the bus/tram stops throughout the route, thus facilitating the route for old or handicapped people. Thus, we need two APIs to (1) calculate the best route between two points and (2) discover the bus/tram stops.

To calculate routes between points we will use the Google Maps API[5]. In particular, we will use the service to calculate the route to follow from a source point to a target one, which we will refer as *routeCalculation* service. This service receives as inputs: (1) the origin and (2) the destination of the route (expressed as addresses), and (3) whether a location sensor is available. The service returns a route to follow including the bounds and steps. Figure 1a shows an example of this service.

---

[5] https://developers.google.com/maps

```
┌─────────────────────────────────────────────┐   ┌──────────────────────────────────────┐
│          routeCalculation service           │   │         stopPosition service         │
├─────────────────────────────────────────────┤   ├──────────────────────────────────────┤
 Input data:                                       Input data:
   origin:      28 Boulevard des Belges              latitude:   47.2342334
   destination: 3 Rue de Rue de l'Arche Sèche        longitude:  -1.5385382
   sensor:      false

"routes" : [                                       [ {
 {                                                   "placeCode": "BBEL",
  "bounds" : {                                       "tag": "Bd des Belges",
   "northeast" : { "lat" : 47.23464389999999, "lng" : -1.5385382 },   "distance": "119 m",
   "southwest" : { "lat" : 47.2155321, "lng" : -1.5583926 }    "line": [ { "lineNum": "C6" },
  },                                                             { "lineNum": "70" },
  "legs" : [                                                     { "lineNum": "LU" }
   {                                                 ]
    ...                                            },
    "end_address" : "3 Rue de...",                 {
    "end_location" : { "lat" : 47.2155321, "lng" : -1.5583558 },   "placeCode": "RPAR",
    "start_address" : "28 Boulevard...",             "tag": "Rond-Point de Paris",
    "start_location" : { "lat" : 47.2342334, "lng" : -1.5385382 },   "distance": "149 m",
    "steps" : [                                      "line": [ { "lineNum": "C1" },
     {                                                          { "lineNum": "70" },
      ...                                                       { "lineNum": "LU" }
      "start_location" : { "lat" : 47.2342334, "lng" : -1.5385382 },   ]
  ...                                              }, ...
                    (a)                                            (b)
```

Fig. 1. Two API calls examples: (a) the *routeCalculation* service from the Google Maps API and (2) the *stopPosition* service from the TAN API. The input data is shown on top while the resulting JSON is listed below. For the sake of clarity, the resulting data is shown partially and strings of the TAN API have been translated into English.

As we plan to deploy our example application in Nantes, France, we will use the API provided by TAN[6], the transportation entity of the city of Nantes, to discover the bus/tram stops along the calculated route. In particular, we will use the service we call *stopPosition*, which allows knowing the set of bus/tram stops near a given location. The service receives a position determined by the latitude and longitude, and returns the nearest tram/bus stops. Figure 1b shows an example of this service.

In this example the developer must first explore these APIs and then study how they can be composed (if possible). The analysis of the inputs/outputs allows identifying the main concepts used in each API (i.e., the domain). For instance, *routeCalculation* uses addresses to specify both the origin and destination of the route. Regarding its output data, locations are represented by `lat` and `lng` to specify the latitude and longitude, respectively. On the other hand, *stopPosition* receives a location as input and returns a set of bus/tram stops. After this study, the developer may come up with a possible mapping involving the values representing locations in the output of *routeCalculation* and the input of *stopPosition*, thus enabling their composition.

As can be seen, composing JSON-based web APIs require deeply studying the involved APIs and also how to compose them, which is a time-consuming and hard task, in particular, when dealing with a number of candidate APIs. In the remainder of this paper we will show our proposal to identify the domain behind APIs as well as data mappings which can help developers to easily compose them.

---

[6] http://data.nantes.fr/donnees/detail/
info-trafic-temps-reel-de-la-tan

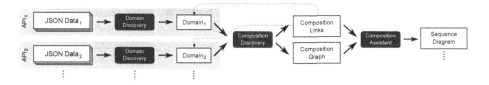

Fig. 2. Overall view of our approach. The main phases are represented with black-filled rounded boxes while input/output data is represented with white-filled boxes.

3 Our Approach

We propose an approach to study the composition of JSON-based web APIs. Our approach applies a discovery process which first analyzes the domains behind each involved API and then identifies composition links among them. The discovered information is used to render a graph in which calculations can be made to assist developers to compose APIs (e.g., sequence diagrams can be generated). Figure 2 illustrates our approach including the main two discovery phases (i.e., *Domain Discovery* and *Composition Discovery*) and facilities to realize the composition (see *Composition Assistant*).

Our approach represents domain information as class diagrams, including concepts (i.e., classes) and their relationships (i.e., attributes/associations), while composition links will be represented as relationships between concepts from different domains. We will leverage on model-driven techniques to represent both the domain and composition information as models and model references, respectively. The following sections will describe the main phases of our approach.

4 Domain Discovery in JSON-Based Web APIs

The domain of an API can be discovered by merging the domain of its services, which in turn can be discovered by analyzing the JSON data used as input/output. We devised a two-phase process to obtain the API domain represented as a model [1], which has been extended and adapted to enable the subsequent composition discovery phase (i.e., enriching the generated metadata). Next, we describe the basis of the process to facilitate the understanding of the remainder of the paper[7].

The first phase, called *Single-service discovery*, analyzes each service in order to discover its domain. Since JSON-based API services do not necessarily return JSON data conforming to the same structure, the accuracy of this phase increases when a number of JSON examples are provided. Thus, the single-service discovery phase analyzes a set of JSON examples (including inputs/outputs defined as JSON data) per API service. This phase is launched for each API service and has two execution modes: creation, which initializes the model concepts from JSON objects representing new concepts; and refinement, which refines existing model concepts with information coming from new JSON objects representing such concepts. Both execution modes are driven by a set of mapping rules transforming JSON elements into model elements[7]. As result, the single-service discovery phase returns a model representing the service domain.

---

[7] A detailed description of the process and mapping rules applied can be found in [1].

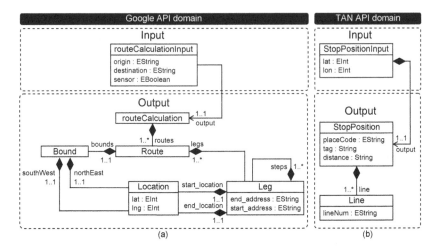

Fig. 3. Discovered domain for (a) Google API (including *routeCalculation* service) and (b) TAN API (including *stopPosition* service)

The second phase, called *Multi-service discovery*, composes the models generated by the previous phase and produces a new model representing the overall domain of the API. Similarly to what the single-service discovery phase does, several mapping rules are applied to obtain the composed model[7].

Figure 3 shows the API domains for the Google Maps and TAN APIs. For the sake of conciseness, we only show the excerpt of the model regarding the data shown in Figure 1. Note that since some JSON name/value pairs represent the same information, some concepts have been merged (e.g., the `Location` concept represents `northeast`, `southwest`, `end_location` and `start_location` JSON objects).

5 Composition Discovery in JSON-Based Web APIs

Composition links among APIs are discovered by means of matching concepts among their domains and analyzing whether they are part of the input parameters of API services. In this section we describe how to identify matching concepts and create composition links. These links can be later digested to facilitate the composition of the involved APIs, as we will explain below.

The discovery process of composition links analyzes the API domains to discover differences and similarities. However, this is not an easy task when dealing with models since the problem can be reduced to the problem of finding correspondences between two graphs (i.e., an NP-hard problem [2]). Based on our experience, we have identified a set of core rules but they can be extended by implementing other existing approaches (e.g., the ones presented in [3]):

R1 Two domain concepts *c1* and *c2* contained in different API domains are considered the same concept if *c1.name* = *c2.name*.

R2 As an API domain concept can represent several JSON objects (e.g., `Location` in Figure 3), only concept attributes/references found in every object are considered.

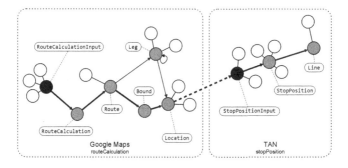

Fig. 4. Composition graph for the *routeCalculation* and *stopPosition* services

R3 Two attributes/references *a1* and *a2* are similar if *a1.name* = *a2.name* and *a1.type* = *a2.type*. Otherwise heuristics based on their name/type may be applied (e.g., the number of matching letters in their names must be higher than a given threshold).

R4 Two domain concepts *c1* and *c2* contained in different API domains are similar if they contain a number of similar attributes/references higher than a given threshold.

R5 There is a composition link between two domain concepts *c1* and *c2* contained in different API domains if they are the same (or similar) and *c2* is an input concept. The source of the composition link will be *c1* and the target will be *c2*.

The application of rules to our example will result in only one composition link from `Location` to `StopPositionInput` since R2, R3, R4 and R5 are fulfilled.

Composition links plus the API domains can be used to render a graph where nodes represent concepts/attributes and edges represent composition links or attribute composition. Figure 4 shows an example of this graph representation for our example. For the sake of clarity, nodes have been annotated with the name of the concept they represent. Gray-filled nodes represent the concepts used in each API, black nodes the concepts used as input to call an API, and white nodes the concept attributes, which are linked to the concept by an un-directed edge. Nodes are connected by directed edges, which can link nodes from the same (filled arrow) or different (dashed arrow) APIs. Nodes from the same API are linked when there is a reference between them, whereas nodes from different APIs are linked when a composition link has been detected.

6   Assisting Developers to Compose APIs

Paths in the graph can be used to assist developers in the composition of APIs. To calculate a path, developers must specify both the input information (by selecting the concepts/attributes they have available) and what they want to get (by selecting the desired concepts/attributes). Well-known graph algorithms can then be applied to calculate paths (if exist) among the selected nodes (through the directed edges). For instance, in our example we provide the attributes of the node *RouteCalculationInput* and our target node is *Line*. A possible path between these two nodes is highlighted in Figure 4, which indicates that a composition between these two APIs is possible. In particular,

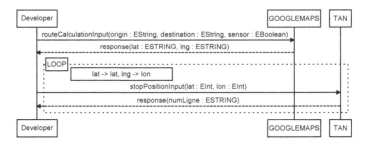

Fig. 5. Sequence diagram generated from a path between `RouteCalculationInput` and `StopPositionInput` nodes of the graph shown in Figure 4

the composition can be performed calling the *RouteCalculation* service and using the attributes of the resulting `Location` concept to call to the *stopPosition* service.

Given this graph and the API domain models, several calculations can be applied to make easier the composition of the involved APIs and the understanding of paths in the graph. For instance, a sequence diagram can illustrate the calls and parameters to realize the composition. Figure 5 shows the sequence diagram for our example. Sequence diagrams can be drawn following these rules:

– There are as many actors as APIs are traversed by the path plus the developer actor.
– The diagram includes as many synchronous calls as APIs are traversed by the path.
– A method call is included for each API crossed. The method calls is named as the first node of the sub-path traversing the API and the parameters are its attributes. The method returns the set of attributes of the ending node of the sub-path.
– If the sub-path traverses a multivalued reference, the call for such path is a loop.
– A mapping between the output/input parameters of intermediate calls may be provided as annotation following the rules explained in Section 5.

7   Additional Applications

In previous sections we used a simple example to illustrate our approach and how paths in the graph can facilitate the composition of APIs. In this section we will increase the scope and size of the graph in order to study additional applications. In particular, we will focus on a cost-aware composition mechanism and obtaining the minimal branching subgraph. We will show first the extended graph we will use and then we will describe these applications.

Figure 6 shows the composition graph obtained from three real JSON-based web APIs, namely: Google Maps, TAN and an adapted version of the Foursquare API. Foursquare[8] is a social network allowing users to share their experiences when visiting places. For the sake of conciseness, we do not present the real name of each service but we will use an identifier[9]. Thus, the Google API includes four services (i.e., G_1,

---

[8] `http://foursquare.com`

[9] The graph can be found at `https://github.com/atlanmod/json-discoverer/tree/master/fr.inria.atlanmod.json.discoverer.zoo/exampleThreeAPIs`

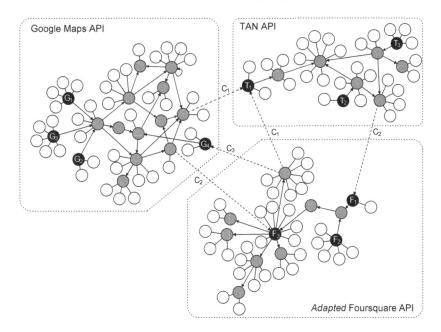

Fig. 6. Composition graph obtained from the services provided by three JSON-based APIs, namely, Google Maps, TAN and an adapted version of the Foursquare API

G_2, G_3 and G_4), the TAN API includes three services (i.e., T_1, T_2 and T_3) and the Foursquare API includes three services (i.e., F_1, F_2 and F_3).

Cost-Aware Composition. Some APIs follow a pay-per-use schema, e.g., fixed price per call, special price according to agreements, etc. To enable cost calculation, edges connecting different APIs can be annotated with the cost value. This information can then be used to obtain the best path (e.g., the cheapest path) among APIs.

Figure 6 includes annotations with the cost value in those edges connecting differente APIs. Thus, calling the TAN, adapted Foursquare and Google APIs costs C_1, C_2 and C_3 respectively. A possible scenario could be as follows. As described before, developers can compose the Google and TAN APIs by means of the services G_1 and T_1, which costs C_1. However, there exists a second option which involves composing the three APIs of the graph (i.e., the path will start with the G_1 node of the Google API, then will cross the F_3 node of the adapted Foursquare API and finally the T_1 node of the TAN API), which costs $C_2 + C_1$. Depending on the concrete values of these costs, developers can decide which one is the most suitable to their needs.

Minimal Subgraph. The graph shown in Figure 6 also allows developers to discover all the composition paths among the analyzed APIs. In order to facilitate the identification of all the API compositions, it is possible to apply traditional graph algorithms to calculate the optimum branching (such as [4]), which will provide the minimal path among every node of the graph. The developer can also prune some nodes and recalculate the graph in those cases in which a path crosses some nodes representing concepts/attributes that the developer cannot provide.

It is important to note that composition paths may not access to every API node since the API subgraph may not be a strongly connected graph. For instance, the composition of the Google API with the TAN API provides access to the latter API through the service T_1, which is connected with a limited number of nodes of the TAN subgraph. Thus, when the API subgraph is not a strongly connected graph, to be precise, composition paths should be indicated in terms of API services.

8   Related Work

The discovery of the implicit structure in JSON data is related to works focused on obtaining structured information from unstructured data such as [5]. Our approach integrates some of their ideas. Furthermore, the use of metadata in the model discoevry phase has been inspired by works such as [6, 7].

Composition link discovery applies some basic tecniques to detect matching modeling elements. Several works such as [8–11] and tools such as EMFCompare [12] could be used here to improve our discovery process.

In the field of web engineering, our approach is related to those ones focused on web services. For instance, [13] proposes an approach based on semantic web services which are analyzed to discover how they can be coreographed (i.e., composed). A similar approach to ours has been presented in [14], where a solution to integrate and query web data services is presented. The approach resorts in the web service definitions (i.e., WSDL) to define *service interface*s which are later analyzed to discover possible ways to integrate and query them. The Yahoo Query Language (YQL)[10] is also related to our approach, since they allow perfoming queries among web services with the aim of composing them. Finally, in the particular field of mashups, the works [15, 16] also address the problem of composing different web services. Regarding the data used, the main difference with these approaches is that ours is specifically adapted to deal with JSON-based web APIs, where generally there are no formal definitions of the services (as it could happen with web services by means of definition such as WSDL or semantic web services with OWL-S). However, with regard to the mechanisms to discover potential composition links, our approach can be enriched adapting their proposals.

9   Conclusion and Future Work

In this work we have presented an approach to study how JSON-based web APIs can be composed. Our approach leverages on a previous work and extends it to infer composition links among APIs. Composition information is represented as graphs, where paths represent concrete API compositions. Furthermore, these paths are used to create sequence diagrams to facilitate the understanding of the composition. Our tool has been fully implemented and is available as a free service[11].

---

[10] https://developer.yahoo.com/yql
[11] http://atlanmod.github.io/json-discoverer

As future work, we plan to explore other possible ways to facilitate API composition, such as generating the glue code among them. We would also like to study new mechanisms to detect concept similarities (e.g., using WordNet[12]) as well as conduct a quantitative evaluation of our approach to study its scalability.

References

1. Cánovas Izquierdo, J.L., Cabot, J.: Discovering Implicit Schemas in JSON Data. In: Daniel, F., Dolog, P., Li, Q. (eds.) ICWE 2013. LNCS, vol. 7977, pp. 68–83. Springer, Heidelberg (2013)
2. Lin, Y., Gray, J., Jouault, F.: DSMDiff: a differentiation tool for domain-specific models. Europ. Inf. Syst. 16(4), 349–361 (2007)
3. Kolovos, D.S., Di Ruscio, D., Pierantonio, A., Paige, R.F.: Different models for model matching: An analysis of approaches to support model differencing. In: CVSM Conf., pp. 1–6 (2009)
4. Edmonds, J.: Optimum Branchings. J. Res. Nat. Bur. Standards 71B, 233–240 (1967)
5. Nestorov, S., Abiteboul, S., Motwani, R.: Inferring structure in semistructured data. ACM SIGMOD Record 26(4), 39–43 (1997)
6. Famelis, M., Salay, R., Di Sandro, A., Chechik, M.: Transformation of Models Containing Uncertainty. In: Moreira, A., Schätz, B., Gray, J., Vallecillo, A., Clarke, P. (eds.) MODELS 2013. LNCS, vol. 8107, pp. 673–689. Springer, Heidelberg (2013)
7. Famelis, M., Salay, R., Chechik, M.: Partial models: Towards modeling and reasoning with uncertainty. In: ICSE Conf., pp. 573–583 (2012)
8. Alanen, M., Porres, I.: Difference and union of models. In: Stevens, P., Whittle, J., Booch, G. (eds.) UML 2003. LNCS, vol. 2863, pp. 2–17. Springer, Heidelberg (2003)
9. Ohst, D., Welle, M., Kelter, U.: Differences between versions of UML diagrams. In: ACM SIGSOFT Conf., pp. 227–236 (2003)
10. Selonen, P., Kettunen, M.: Metamodel-Based Inference of Inter-Model Correspondence. In: CSMR Conf., pp. 71–80 (2007)
11. Melnik, S., Garcia-molina, H., Rahm, E.: Similarity Flooding: A Versatile Graph Matching Algorithm. In: DE Conf., pp. 117–128 (2002)
12. Brun, C., Pierantonio, A.: Model Differences in the Eclipse Modeling Framework. UP-GRADE, The European Journal for the Informatics Professional 9(2), 29–34 (2008)
13. Sycara, K.P., Paolucci, M., Ankolekar, A., Srinivasan, N.: Automated discovery, interaction and composition of Semantic Web services. J. Web Sem. 1(1), 27–46 (2003)
14. Quarteroni, S., Brambilla, M., Ceri, S.: A bottom-up, knowledge-aware approach to integrating and querying web data services. TWEB 7(4), 19 (2013)
15. Daniel, F., Rodríguez, C., Chowdhury, S.R., Nezhad, H.R.M., Casati, F.: Discovery and reuse of composition knowledge for assisted mashup development. In: WWW Conf., pp. 493–494 (2012)
16. Chowdhury, S.R., Daniel, F., Casati, F.: Efficient, interactive recommendation of mashup composition knowledge. In: ICSOC Conf., pp. 374–388 (2011)

---

[12] http://wordnet.princeton.edu/

Design Criteria for Web Applications Adapted to Emotions

Giulio Mori, Fabio Paternò, and Ferdinando Furci

ISTI – CNR, Via Moruzzi, 1, 57126 Pisa, Italy
{giulio.mori,fabio.paterno,ferdinando.furci}@isti.cnr.it

Abstract. The main goal of this work is to identify a set of design criteria for Web applications taking into account the users' emotions. The results are based on the analysis of a user study with 50 participants who tested six Web interfaces, each one designed to elicit a specific emotion (hate, anxiety, boredom, fun, serenity, love). The design criteria applied to the six emotion-based Web interfaces were drawn from the results of a previous survey, which involved 57 different users, on the relationships between emotional state and Web interfaces. This initial survey asked the users to indicate the emotions most often associated with Web interaction, and then assign each emotion with some specific Web design characteristics. The resulting design criteria can form the basis for a set of emotion-related guidelines for Web application user interfaces.

Keywords: Web guidelines, emotions, affective interfaces.

1 Introduction

The important role of emotions in HCI is widely accepted [3-5]. However, little work has been dedicated to how to take them into account in Web applications, which are the most widely used applications. Thus, Web developers and designers need support on some design criteria (such as choice of user interface elements, navigation style, suitable colors, etc.) associated to emotional states. Emotions are complex and depend on individual preferences, attitudes, moods, affect dispositions, and interpersonal stances; "there is no single standard gold-method for their measurement" [10].

The literature reports different methodological approaches to classifying emotions, such as Geneva [8] or Feeltrace [9], and scales and questionnaires to measure either two primary (negative-positive) dimensions of moods [1], or hedonic and pragmatic dimensions of user-experience [7][12], but no work has focused on typical emotions during Web interaction. There are some works comparing different versions of Web pages [5] to investigate the impact of their attractiveness [2] or aesthetics, or if one Web site is better than another at eliciting emotions [4]. An analysis of existing Web sites about the hedonic elements (such as color, images, shapes and photographs) has been carried out [11] to investigate the emotional appeal, the sense of the aesthetic or positive impression resulting from the overall graphical look of a Website. However, none of these studies provide precise indications on how the various aspects of Web interfaces can elicit a specific emotion. Modeling the key subjective aspects of user

S. Casteleyn, G. Rossi, and M. Winckler (Eds.): ICWE 2014, LNCS 8541, pp. 400–409, 2014.
© Springer International Publishing Switzerland 2014

experience and how it affects the perception of the final product or emotional responses [4] are important contributions, but there is the need to better address the emotion-based Web design aspects.

In this paper we aim to investigate the impact of some Web design criteria to elicit a particular emotional state on the user, independently of the application domain. In order to better understand what design choices are most suitable for most recurrent user emotions in Web applications, we have conducted two user studies described herein. Since there were no specific indications in the literature about the effects of Web design on eliciting a specific emotion, we organized a first survey to start our research, with the goal of collecting some basic indications from a sample of 57 users. The survey aimed to better understand the most recurrent emotions during Web interaction and the related Web design features. Next, we wanted to check the effectiveness of the data gathered in the survey through a user study. We applied the criteria to six Web interfaces, each one designed to elicit a specific emotion. Fifty different users judged their emotional impact. Positive emotions are certainly important for improving the user experience, however, also understanding the Web design criteria eliciting negative emotions has its importance for Web designers in order to improve their awareness of the risks and kinds of emotions their Web sites may elicit. In particular, section 2 describes the initial survey having the goal to investigate the relationships between emotional state and Web interfaces. Section 3 reports on the user study having the goal to validate the design principles through six Web Interfaces, and finally, we draw some conclusions and provide indications for future work.

2   A Survey: Collecting Opinions about Emotional Web Design

The questionnaire was completed by 57 users in two sessions (the average completion time per user was about two hours): the first session was carried out in the presence of the authors (to provide the users with explanations, when necessary), and a second session in which the users completed the questionnaire alone. The questionnaire was composed of three parts: 1) personal information, 2) classification of emotions in Web interaction, 3) emotion-based Web design opinions. The first part aimed to collect some personal information from the users and their experiences with the Web. In the second part, users had to propose some emotions they considered relevant during Web interaction and, for each of them, they had to freely associate colours and some attributes characterizing the user activity. The third part was more oriented to the Web design, where users had to give their opinions by associating each emotion with various Web interface features (depending on the question, we showed them various graphical examples of typical elements of a Web interface).

2.1   Personal Information

The participants were 25 females and 32 males, with an average age of 38,21 years (ranging from 26 to 59). Seventeen users had a PhD, 21 users had a five-year degree, 2 users had a four-year degree, 8 users had a bachelor's degree, while 7 users had a high school diploma and 2 users had a school diploma. Users were used to surf the Internet (46 users were connected to the Web every day, 9 users navigated three times per week, 2 users used the Web one time every fifteen days). The sample considered both experienced and inexperienced users in Web development (at different levels, 38 users had implemented some Web interfaces, while 19 users had little or no knowledge on Web programming).

2.2   Classification of Emotions in Web Interaction

In this part, each user had to indicate a certain number of emotions (maximum 8) which s/he considered relevant during the Web interaction. The only constraint of their suggestions was that for each chosen emotion, they also had to indicate the opposite emotion (depending on the emotional valence, negative or positive, in their perception). The reason for this request was that we wanted to define a complete design space that given a negative emotion could allow the identification of the positive counterpart. At the beginning of the questionnaire, some users found difficulties to define emotions and needed support, however at the end, the total average number of the proposed emotions was 3.84 per user, with a total of 219 distinct proposed emotion names perceived as negative and other 219 perceived as positive emotions. Each user had to associate a value (in an ordered scale from -8 to +8) to each emotion, as a measure to indicate how negative or positive s/he considered it. Analyzing the proposed emotions of the 57 users, we decided to consider as more "significant" only the emotions with at least 10 preferences, discarding the remaining others, which had received just only 1, 2 or at maximum 3 preferences each one. In addition, after having analyzed the complete final results of the questionnaire, we noticed that some emotions (having different proposed names by the users, but similar meaning), were characterized by the same Web features. In these cases, with the consent of the users who proposed the emotions, synonyms have been considered the same emotion, and finally, we have obtained an essential basic ordered scale of 6 emotions (3 negative and 3 positive) to express the typical affective states of a user interacting with Web (see Table 1), corresponding to well distinguishable Web design characteristics. The final emotions would have been the same also first joining the synonyms and then filtering the emotions with at least 10 preferences because in addition to the "primary emotions" [13] (such as hate, love, etc.) many users proposed a lot of different "secondary emotions" [13] (such as jealousy, nostalgia, loneliness, etc.), which were proposed only by few other users.

Considering the complexity of the emotions world for human beings and the many emotion classifications existing in literature [8][9][10] (even if no one is specifically oriented to the Web interaction and design), this classification is not exhaustive. However, the goal of this work is not to provide a further emotion classification, but rather to investigate whether some clearly distinguishable design characteristics can elicit a specific emotion on the user independently of the application domain. So this small starting set composed of 6 emotions oriented to Web interaction (obtained by the users suggestions) has been the basis of our study. The goal is to understand if some different Web design features can concretely have a specific impact on the users perception and on the personal emotional state, independently of the contents. Looking at Table 1, some considerations about the meanings of the emotions in the scale proposed by the users are necessary: a) *hate* and *love* express the sense of disliking/liking or indifference/empathy for something or somebody (typical of Web social network environments and Web 2.0). English language expresses well these meanings (i.e.: I hate/love Louis Armstrong music); b) *anxiety* and *serenity* express the emotional state during critical/safe actions (i.e.: the user is booking/buying something on Web and s/he need to fill a form inserting personal or credit card data); c) *boredom* or *fun* depends on the interest of users for the Web content, and the way the contents are presented is fundamental. Table 1 shows the ordered scale of the emotions with the corresponding average values proposed by the users.

Table 1. The main 6 emotions indicated by the users. The order is determined by the average values assigned by the users on the scale of a negative or positive personal perceptions.

Average values of the main Web Emotions considered relevant by the users					
Hate	Anxiety	Boredom	Fun	Serenity	Love
-3.48	-2.54	-1.99	+2.1	+2.25	+3.6
Standard Deviation values of the main Web Emotions considered relevant by the users					
1.29	1.04	0.76	0.92	0.83	1.23

In this part of the questionnaire, we also asked the subjects to associate some attributes related to the user activities for each emotion. Users had to select from the following pairs: static or dynamic (perceived level of changes of the interfaces reflecting the changes of the personal emotional state), passive or active (perceived level of the user's involvement in doing actions), simple or complex (perceived level of how the interface can be elaborated in design). The results are summarized in Table 2. As a criteria of choice (for the results of Table 2 and for the other tables presented for this first survey), we took into considerations an attribute as *strongly characterizing an emotion*, when the total number of users' preferences for a value of a pair (or group) was at least the double (over 50%) of the other/s preferred choices; in borderline cases, conclusions could be ambiguous. The double threshold was a prudential strategy, because we noticed that when a characteristic was chosen quite unanimously by most of the users, that characteristic collected a number of preferences higher than the double (in comparison with the other choices).

Table 2. User activity attributes for each emotion

Emotions	User activities attributes
Love	dynamic, active, complex
Serenity	static, simple
Fun	dynamic, active
Boredom	static, passive, simple
Anxiety	dynamic, complex
Hate	static, active

The following part of the questionnaire asked also to the users to associate one or more colors (preferably belonging to the 16 HTML standard color palette supported by all browsers [6]) to each emotion, and then it was asked to associate freely some visual characteristics, real objects or abstract ideas. The exact tint of the 16 colors were showed to each user during the questionnaire. Table 3 summarizes the results.

Table 3. Colours and visual characteristics for each emotion

Emotions	Main Colors	Visual Characteristics
Love	red, pink	bright, transparent, indefinite, heart
Serenity	blue, aqua, white, green, lime	clear, bright, calm waters, large open spaces, open sky, nature, flat, smooth, light colors
Fun	fuchsia, red, orange, yellow, green, lime, teal, aqua	brilliant, bright, colorful, spring, light, sun, flowers, light fire
Boredom	silver, gray, black	night, dark, blurred, indefinite, hazy, fog, opaque, rain, tears, dim, empty
Anxiety	black, gray, navy, yellow	night, dark, wavy, intermittent, storm, throbbing, blurred, indefinite, fog
Hate	black	night, dark

2.3   Emotion-Based Web Design

The last part of the questionnaire asked the users suggestions regarding specific associations between each emotion and many Web design features. This part aimed to investigate the structure, the multimedia elements and type of interactive elements being the most effective to evoke a specific emotional state. The users could express their preference about any elements of the interface or Web design features, (we showed for each question some graphical samples with the goal that each proposed choice was clear for the users). In particular, first the users had to choose between the following groups related to the Web site structure and the interaction elements for data insertion: 1) few or many pages, 2) blurred or clear text, 3) short or long text, 4) presence of textbox (to insert short data) or textarea (to insert long information, e.g. requests). Table 4 summarizes the results.

Table 4. Page contents & structure for each emotion

Emotions	Pages & Content structure
Love	few pages, long text, clear text, textarea
Serenity	little content, short text, clear text, textarea
Fun	many pages, short text, clear text, textbox
Boredom	few pages, long text, textbox
Anxiety	many pages, blurred text, textarea
Hate	few pages, blurred text, textbox

We then asked the users their opinions about the emotional impact of the multimedia element style, whereby the users had to choose one option from each group: 1) presence of video, animations, images, or no multimedia, 2) small, medium or large size of the images, 3) definition of the images (blurred or clear), 4) color, black & white, or de-saturated (the color was reduced) images. Table 5 shows the results. It is not trivial at all (we cannot say if it depends on some cultural factors or something unconscious in human beings), observing that every user suggested unanimously color and clear images for every positive emotion, and blurred and no color images for two negative emotions hate and anxiety (for boredom absence of images or video was perceived by the users as a factor more emphasizing boredom).

Table 5. Multimedia elements style for each emotion

Emotions	Multimedia Elements
Love	color medium/large clear images
Serenity	color medium-clear images, videos
Fun	animations, color medium-clear images, videos
Boredom	no images, no videos
Anxiety	de-saturated small blurred images, videos
Hate	black & white medium/large blurred images, videos

The questionnaire also sought to explore the users' opinions about the navigation elements (Table 6), choosing amongst the following options: 1) links, standard buttons, graphic buttons or tabs, 2) static or dynamic effects on navigation elements.

Table 6. Navigation elements for each emotion

Emotions	Navigation Elements
Love	graphic buttons, dynamic effects
Serenity	tabs, link, graphic/standard buttons
Fun	graphic buttons, dynamic effects
Boredom	links/standard buttons, static effects
Anxiety	standard/graphic buttons, dynamic effects
Hate	standard/graphic buttons, static effects

Finally, the users gave their opinions about the interactive elements (Table 7), choosing from among the following options: 1) interactive elements with static or dynamic effects 2) textual or graphic interactive elements, 3) radio-button or pull-down single selection, 4) checkboxes, scroll or fixed multiple selection.

Table 7. Interactive elements for each emotion

Emotions	Interactive Elements
Love	dynamic, graphic, pull-down, checkboxes/scroll selection
Serenity	static, textual
Fun	dynamic, graphic
Boredom	static, textual, checkboxes
Anxiety	dynamic, graphic, pull-down menu, scroll select
Hate	checkboxes

As a confirmation of the complexity of the emotions, it is not surprising that tables contain some overlapping Web features between the six emotions. With this survey we received indications about many other aspects of Web design (such as position of the elements, font type/dimension, alignment of text, etc.). We do not report these extra data, because the results are too ambiguous and further tests are necessary.

2.4   Additional Emotion-Related Design Aspects

In addition to the analysed data, we received interesting comments of the participants suggesting us useful indications to improve the emotional effect of Web design. Even if enumeration of these comments was not possible because some users did not provide any comment, we decided to apply them to the Web interfaces design (see section 3) to verify their effects. We report below the summary of the key indications:

Hate-Love. A user can feel *hate* due to the design of the interface, if the interface is difficult to use (bad usability), the layout of the interface and/or the positioning of the widgets are confused (not easy to understand). In the worst case, the interface obstacles the interaction of the user, or/and something is not working (i.e. elements of the interface are not responding to the user input). On the contrary, a user can feel *love* due to the design of the interface, if the appealing aesthetic of the Web interface stimulates its use. Besides, the interface should be easy to use (good usability), the layout and disposition of the widgets should be disposed in a way easy to understand.

Anxiety-Serenity. A user can feel *anxiety* due to the design of the interface, if the interface emphasizes particular stress factors (i.e. a deadline, risk, or sense of losing something, etc.). In these stress conditions, the interface does not allow the user to reason comfortably (i.e. adding intermittent light effects, distortion or jerky transformations of the elements in the interface, etc.). On the contrary, a user can feel *serenity* due to the design of the interface, if the interface let the user to feel safe, (i.e. providing always feedbacks, or showing well known reassuring elements, as a logo of secure transactions, etc.). Besides, the simplicity of the interface allows the user to interact easily, reducing her/his effort and giving the time s/he needs.

Boredom-Fun. A user can feel *boredom* due to the design of the interface, if the interface provides or requires lots of information (i.e. very long texts, or many required fields in a form, etc.). Much text without images or multimedia elements, increases boredom. On the contrary, a user can feel *fun* due to the design of the interface, if unexpected elements, animations or effects surprise the user in a positive way. The animations or dynamic effects should be oriented to facilitate the interaction, otherwise they are perceived as annoying.

3   A User Study: Applying the Emotion-Based Design Principles

On the basis of the results of the survey, we developed six Web interfaces, applying the collected Web designs principles. Each Web interface had the goal to elicit one of the six emotions of the scale. Each Web interface presented the same content (except very minimal additions suggested by the users) in a different design style. Considering that most users in the survey suggested music as topic for an emotion-based Web application, we chose the Beatles' musical history as a topic for the Web six interfaces. In details, the interfaces contained: a short textual biography, a media player to listen to five famous songs, a musical video, a form where the user could buy some virtual tickets for revival musical events, and some clickable graphical covers of six famous albums. Finally, we recruited 50 different users (through a mailing list of our institute), who had not participated to the previous survey, to test the six Web interfaces through some interactive tasks, and after that, to judge (through a questionnaire) the Web design effectiveness in stimulating a specific emotion.

3.1   Description of the User Test and Discussion of the Results

We showed the six Web interfaces to the 50 users in random order. We wanted to avoid that the order could influence the emotional perception. Each user had to test each Web interface by completing three tasks, and then s/he had to fill in an online questionnaire. The questionnaire was composed of three sections, and it asked the users: a) personal information, b) judgment about the effectiveness of the interface to stimulate the proposed emotion and comments, c) suggestions about other emotions we did not consider in the classification, opinions about the utility of the adaptation of the Web design to elicit more positive emotions on the users, and some suggestions about useful application fields. Most users considered the six emotions exhaustive for the Web interaction (even if some users proposed anger or surprise as examples of additional emotions). Nearly all the participants found the adaptation of the Web design to elicit positive emotions very useful. As important applications for applying the emotional Web design, they suggested educational environments, telemedicine and online psychological platforms, games, home automation applications, or tools oriented to helping people with disabilities or the elderly. The users preferred Web design stimulating positive emotions because it improves the user experience. However, they also considered the utility of eliciting negative emotions not only to improve the awareness of Web designers, or to recreate particular thrilling atmospheres in games, but also in the telemedicine field. In this area the ability to understand the reactions of patients in a good or bad affective state is important. Moreover, it could improve children's awareness about the difference between good and bad behavior in some educational or learning tools. The users gave their judgment in a scale from 1 to 5 (where the value 1 indicated that the page was very ineffective to elicit the proposed emotion, while value 5 indicated that the page was very effective, and the value 3 represented the neutrality). Considering that the survey was "open answer" (concerning the users' proposal of Web emotions), we decided for this user test that the users could know in advance the emotional goal of the currently tested interface. We wanted to avoid that another "open-answer user test" could produce too vague results. In particular for each emotion, the users were asked to accomplish three tasks: a) find the answer to a proposed question in the biography text (the goal was obliging the user to read the text with specific style characteristics), b) click on one proposed cover of one album (the goal was obliging the user to evaluate the interaction with the elements of navigation), c) fill in a form (with proposed data corresponding to one imaginary user) to buy some virtual tickets (the goal was obliging the user to evaluate the elements of interaction). The questionnaire was completed by 50 users (average completion time per user of about one hour). The participants were 21 females and 29 males, with an average age of 38,28 years (ranging from 26 to 77). Ten users had a PhD, 21 users had a five-year degree, 4 users had a four-year degree, 9 users had a bachelor's degree, while 6 users had an high school diploma. Users were used to surf the Internet (42 users were connected to the Web every day, 5 users navigated three times per week, 2 users used the Web one time every fifteen days, and one used Web when it happens). The sample considered both experienced and inexperienced users in Web development (at different levels, 28 users had implemented some Web interfaces, while 22 users had little or no knowledge on Web programming).

The six Web interfaces were designed as follows: a) the Web interface to elicit hate was completely unusable with a confused layout, b) the Web interface to elicit anxiety showed intermittent light effects and jerky transformations, with a countdown as a pressure factor to fill in the form, c) the Web interface to elicit boredom, was neutral, without images or videos, requiring more fields to fill in the form, d) the Web interface to elicit fun, showed unpredictable animations and dynamic effects, e) the Web interface to elicit serenity, was very simple to minimize the user's effort, f) the Web interface to elicit love had an appealing graphics and it was usable. For lack of space, we have to omit further details. The average judgment of the 50 users about the effectiveness of the interfaces to elicit the proposed emotion was high: (over value 4) for hate, anxiety, boredom and serenity, while it was slightly effective (over value 3) for fun and love. The results of the user test showed that the usability, even if it is an important factor, it is not the unique aspect responsible to elicit an emotional reaction on the user (e.g. the different unusable interfaces aiming to elicit hate and anxiety, produced different emotional effects). Even if the results are encouraging, a more detailed statistical analysis is necessary. In particular it is important to understand for each emotion, if a subset ("core") of key design factors can be responsible for eliciting a specific mood. It is necessary investigate further if the results can depend on some "confounding factors" (such as age, gender, experience in development, etc.), or if the results are domain/topic-dependent or domain/topic-independent.

4   Conclusions and Future Work

The goal of this work has been to study whether the collected preliminary user indications about the relevant emotions during Web interaction and the corresponding specific design criteria actually do have an emotional impact on users. The results obtained are encouraging, even if further refinements and investigation are necessary. In particular, other user tests will be necessary to better understand the essential Web design key factors affecting emotional state. The ultimate goal of this research is the formalization of a set of Web guidelines to design interfaces that can effectively stimulate user emotions during the interaction. For the implementation of the next user tests, we are considering various sensors to detect physiological parameters of the users in order to monitor their changes and to adapt the Web design with the goal of eliciting more positive emotions on the users.

References

1. Watson, D., Clark, L.A.: Development and Validation of Brief Measures of Positive and Negative Affect: the PANAS Scales. Journal of Personality and Social Psychology 54(6), 1063–1070 (1988)
2. Hartmann, J., Sutcliffe, A., De Angeli, A.: Investigating Attractiveness in Web User Interfaces. In: Proc. CHI 2007. ACM Press (2007)
3. Hassenzahl, M.: The Think and I: Understanding the relationship between user and product. Funology Human Computer Interaction Series 3, 31–42 (2005)
4. Karlsson, M.: Expressions, Emotions, and website design. CoDesign 3(1), 75–89 (2007)
5. Kim, J., Lee, J., Choi, D.: Designing Emotionally Evocative Homepages: An Empirical Study of the Quantitative Relations Between Design Factors and Emotional Dimensions. International Journal of Human-Computer Studies 56(6), 899–940 (2003)
6. The 16 HTML color names, http://en.wikipedia.org/wiki/Web_colors
7. Voss, K.E., Spangenberg, E.R., Grohmann, B.: Measuring the hedonic and utilitarian dimensions of consumer attitude. Journal of Marketing Research 40(3), 310–320 (2003)
8. Bànziger, T., Tran, V., Scherer, K.R.: The Geneva emotion wheel. Journal, Social Science Information 44(4), 23–34 (2005)
9. Cowie, R., Douglas-Cowie, E., Savvidou, S., McMahon, E.: Feeltrace: an instrument for recording perceived emotion in real time. In: Proceedings of the ISCA Workshop on Speech and Emotion 2000, pp. 19–24 (2000)
10. Scherer, K.R.: What are emotions? And how can they be measured? Social Science Information 44(4), 695–729 (2005)
11. Cyr, D.: Emotion and Website Design, 2nd edn., ch. 40., http://www.interaction-design.org/encyclopedia/emotion_and_website_design.html
12. Diefenbach, S., Hassenzahl, M.: The Dilemma of Hedonic – Appreciated, but hard to justify. Interacting with Computers 23(5), 461–472 (2011)
13. Damasio, A.R.: Descartes' error: Emotion, reason, and the human brain. Avon Books, New York (1997)

Driving Global Team Formation in Social Networks to Obtain Diversity

Francesco Buccafurri, Gianluca Lax, Serena Nicolazzo,
Antonino Nocera, and Domenico Ursino

DIIES, University of Reggio Calabria
Via Graziella, Località Feo di Vito
89122 Reggio Calabria, Italy
{bucca,lax,s.nicolazzo,a.nocera,ursino}@unirc.it

Abstract. In this paper, we present a preliminary idea for a crowdsourcing application aimed at driving the process of global team formation to obtain diversity in the team. Indeed, it is well known that diversity is one of the key factors of collective intelligence in crowdsourcing. The idea is based on the identification of suitable nodes in social networks, which can profitably play the role of generators of diversity in the team formation process. This paper presents a first step towards the concrete definition of the above application consisting in the identification of an effective measure that can be used to select the most promising nodes w.r.t. the above feature.

1 Introduction and Description of the Idea

It is well known that diversity is one of the key factors of collective intelligence in crowdsourcing [20]. On the other hand, it is clear that this concept fully confirms the famous principle summarized as *the strength of weak ties*, stated in the field of social networks [14]. The two worlds, social networks and crowdsourcing, have a strong overlap, as social-network users form a huge crowd. But, social-network crowd includes something more than the simple Web crowd. It has a friendship-based structure, embeds contents, and is full of knowledge about people. This opens a lot of opportunities that can reinforce the power of crowdsourcing (e.g., see [16]). For instance, consider the problem of dynamic formation of globally distributed teams for enterprisers [21]. Driving team formation can result in tangible benefits for the success of the team work, as a number of features of the individuals, such as expertise, should be considered. Social networks are repositories of a large amount of information about people, in which we can find the aimed features. But this is not enough. Indeed, it is not what a crowdsourcing process, thus spontaneous and evolutionary, requires. As a matter of fact, individuals in a social networks are not monads. So, not only friendship relationships allow the autonomous flooding of the network, enabling crowd formation, but the type of ties on which the crowd formation propagates can be dramatically important for the final quality of the global team: We have to hope that the most of crossed ties are weak, to fully reach the goal of diversity. Therefore, the

S. Casteleyn, G. Rossi, and M. Winckler (Eds.): ICWE 2014, LNCS 8541, pp. 410–419, 2014.

basic principle we are stating here is that global team formation in crowdsourcing, even though mostly spontaneous, should be driven in some way by taking
advantage of social networks with the aim of maximizing diversity.

In this paper, we try to answer a simple question: How to translate the above
principle in a social-web application aimed at improving the quality of global
team formation in crowdsourcing? This is a preliminary research, as it contains
an idea on how to do this, but it is focused on a very important aspect which is
the starting point for transforming this idea into a concrete application. The idea
is that team formation should be driven by weak ties, as mentioned earlier. But,
how to find weak ties? From social network analysis we know that the concept of
weak tie is related to behavioral aspects (for example, the number of interactions
in the dyad), which are very difficult to capture in a feasible way. Actually, the
concept of weak tie is also related to structural properties, since weak ties are
typically bridges between two different communities. In particular, it is related
to centrality measures [11]. Among these, betweenness centrality has a primary
role, as it is the fraction of shortest paths between node pairs that pass through
a node, and thus it is capable of measuring the influence of this node over the
information spread through the network [1,17]. Therefore, to detect nodes that
interconnect different communities we have to find nodes that are central in
terms of betweenness centrality. In general, this search could appear difficult,
if no restriction of the domain is done. The idea we present here is based on
the consideration that some special nodes exist that exhibit explicitly a role of
connectors between two different worlds. To understand this, we have to move
towards the perspective of the social internetworking scenario (SIS) [18,3,7,6,8],
in which the scope of the action of both people and applications is not confined to
a single social network. As a matter of fact, in the current scenario characterized
by hundreds of online social networks, a single user can join more of them.
This leads to have membership overlap among social networks as expression
of different traits of users' personality (sometimes almost different identities),
also enabling, as side effect, the passage of information from one social network
to another. As a consequence, membership overlap can be viewed as a feature
that gives a specific power to users in terms of capability of connecting different
worlds. But the good news is that, differently from a generic central node in
a social network, a node interconnecting two different social networks (that we
call *i-bridge*) often exhibits some explicit information showing this feature, thus
it can be easily (automatically) identified. Indeed, often, a user shows in the
home page of her account in a social network the link to her account in another
social network. As a consequence, it is possible to define crawling strategies that
privilege these nodes, allowing their easy discovery. One of these strategies is
that called Bridge-Driven Search (BDS) presented in [9].

On the basis of this reasoning, our idea is to use BDS to find a number of
seeds for our team formation, possibly iterating the process until a suitable stage
is reached. But, the basic question is now: Are all i-bridges good for our purpose? How to compare different i-bridges in terms of capability of generating
diversity? On the basis of the above considerations, one could think that the

classic measure of betweennes centrality could be used to do this. In this paper, we show that the above hypothesis is not correct. In particular, we show that we need a measure of centrality that, differently from classic betweenness centrality, is able to take into account paths crossing distinct social networks in a different way from internal paths. Thus, we study a new measure of betweenness central-ity, called *cross betweenness centrality (CBC)*, which allows us to characterize nodes in a social internetworking scenario in terms of importance w.r.t. inter-social-network information flows and to characterize candidate nodes in terms of suitability to be actors in global team formation.

It is clear that all the above reasoning appears well-founded only if the basic (even intuitive) assertion really holds that i-bridges play the role of ties connect-ing two social networks in a weak way. Thus, in this paper, as a first contribution, we prove the above claim through an experimental validation.

The plan of this paper is as follows. In the next section, we model our reference scenario. A description of the testbed adopted for the experimental campaign is described in Section 3. In Section 4, we prove the basic claim that i-bridges play the role of ties connecting two social networks in a weak way. In Section 5, we introduce the new measure, called *cross betweenness centrality (CBC)*, needed for our application to detect good i-bridges. An experimental proof about the need and the validity of CBC is presented in Section 6. In Section 7, a review of the related literature is discussed. Finally, in Section 8, we draw our conclusions and identify how to continue our research.

2   The Multi-Social-Network Model

In this section, we model our reference scenario, which is called *social inter-networking scenario* (SIS) and takes into account the multiplicity of social net-works and their interconnections through me edges. A *SIS* is a directed graph $G = \langle N, E \rangle$, where N is the set of *nodes* representing (social network) user accounts and E is the set of *edges* (i.e., ordered pairs of nodes) representing re-lationships between user accounts. Given a node $a \in N$, we denote by $S(a)$ the social network which a belongs to. E is partitioned into two subsets E_f and E_m. E_f is said the set of *friendship edges* and E_m is the set of me *edges*. E_f is such that for each $(a, b) \in E_f$, $S(a) = S(b)$, while E_m is such that for each $(a, b) \in E_m$, $S(a) \neq S(b)$. Each node a of G is associated with the (social network) account of a user joining the social network $S(a)$. An edge $(a, b) \in E_f$ means that the user having the account b is a friend of the user having the account a (both a and b belong to the same social network). An edge $(c, c') \in E_m$ means that c and c' are accounts of the same user in two different social networks. As a consequence, c is an i-bridge and a me edge interconnects the two social networks $S(c)$ and $S(c')$. Given a SIS $G = \langle N, E \rangle$, we call *the social networks of G* the set of social networks S such that for each $Q \in S$ there exists $n \in N$ such that $S(n) = Q$. Observe that, the graph of a SIS differs from those underlying a single social network because of the presence of me edges, which connect nodes of different social networks. From now on, consider given a SIS $G = \langle N, E \rangle$.

3   Tools and Data for Our Experiments

Our experiments have been carried out on real-life data obtained by crawling on-line social networks. As for the crawling strategy, we used the well-known BFS [22], which performs a *Breadth First Search* on a local neighborhood of a seed it starts from. The crawling task was performed by means of the system $SNAKE$ [5], which is able to extract not only connections among the accounts of different users in the same social network, but also connections among the accounts of the same user in different social networks. These connections are represented by two standards encoding human relationships, namely XFN (XHTML Friends Network) and FOAF (Friend-Of-A-Friend). This way, we got a dataset consisting of five social networks, namely Flickr, LiveJournal, Google+, MySpace, and Twitter. They are compliant with the XFN or FOAF standards and have been largely analyzed in social network analysis in the past. Starting from this real-life dataset, we extracted several subgraphs for our tests. Each subgraph was obtained by randomly choosing an i-bridge node b and selecting all the nodes having a minimum distance from b less than or equal to 4 (observe that, due to the small diameter of real-life social networks, the chosen distance is significative). Figure 1 shows one of the subgraphs of the real-life dataset, in which black and white nodes stand for users belonging to Flickr and LiveJournal, respectively.

4   Are me edges Weak ties?

One of the most fascinating results of social network analysis is due to one of the fathers of this discipline and regards the concept of weak ties [14]. Although a complete notion of the strength of a tie can be given only if we consider dynamic and behavioral information, even the sole structural knowledge about a social network allows us to identify those ties that, informally speaking, connect two dense components keeping a low connection degree between them. The theory presented in [14] and confirmed by years of study on social networks, gives a strong importance to such ties, as they connect different communities, so they can be a formidable vehicle of cross contamination between them. The claim underlying our crowdsourcing application is that i-bridges are good candidates to be actors in a global team formation. But to prove this claim, the first step is to face this issue: Are me edges weak ties, in general?

In this section, we try to give an answer to the above question by conducting a suitable experiment on real-life social networks. To detect weak ties, we adopt the strategy proposed in [10]. In particular, we consider an edge e between the nodes n_1 and n_2 and we check if the removal of this edge would increase the distance between n_1 and n_2 to a value strictly more than two. If this occurs, then e can be considered as a weak tie. This experiment is carried out on the whole dataset described in Section 3 consisting of a set of $171,982$ normal edges (i.e., non-me edges) and a set M of 79 me edges. At the end of the experiment, the approach described above detected a set W of 5619 weak ties among the set E of $172,061$

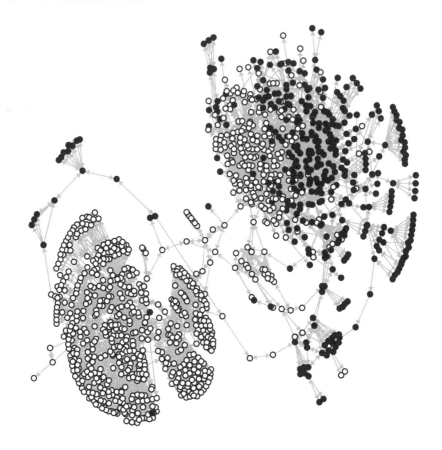

Fig. 1. Visualization of one subgraph

edges and 60 of them were also me edges. We calculated the percentage of me edges which are weak ties as $\frac{|M \cap W|}{|M|} = 0.76$, the percentage of me edges $\frac{|M|}{|E|} = 4.6 \cdot 10^{-4}$ and the percentage of weak ties $\frac{|W|}{|E|} = 3.3 \cdot 10^{-2}$. From the above results, we conclude that the probability that a me edge is a weak tie is high, whereas the probability that a generic edge is a weak tie is very low. Because $|M|$ and $|W|$ are much lesser than $|E|$, the result that $\frac{|M \cap W|}{|M|} = 0.76$ demonstrates that a strong relation between weak ties and me edges exists. Thus, the experiment concludes that the correlation between weak ties and me edges exists, so that we can sight potential powerful roles for such edges in a social internetworking scenario. It is worth remarking that the interpretation of me edges as weak ties means that, given a user u, her account in a social network S sees her account in another social network T (in case a me edge from S to T is established by u) as a weak tie. This means that u can be used as powerful disseminator of information across different communities, each belonging to a different social network. Obviously, the more the number of me edges of a user u, the higher her strength (in the

Granovetter sense [14]) in the network. Observe that the above conclusion is not in contradiction with the results given in [8], where the presence of me edges into user accounts has been proven to be assortative. Indeed, one could think that, if the friends of a user u assortatively declare me edges from the social network S to the social network T, as done by u, we obtain a very dense clique of users invalidating the result about the correspondence between me edges and weak ties. However, [8] shows that the *strict* (i.e., towards the same social network T) assortativity does not hold for the most representative real-life social networks (i.e., Facebook). In other words, the friends of u assortatively declare me edges from S to any other social network. Thus, the contradiction does not exist.

5   Measuring the Suitability of Diversity Generators: Cross Betweenness Centrality

After having proved, in Section 4, that me edges are weak ties, we know that our application is well-founded in the sense that all the i-bridges that we are able to find by using a crawling strategy as BDS [9] are good candidates to play the role of diversity generators in team formation. In this section we face a second important issue: Are all the candidates the same in terms of suitability to our application? In general, betweenness centrality (BC) [11] is used to detect weak ties (and also their *structural* strength). Unfortunately, as we will show in Section 6, the above claim cannot be applied to the case of i-bridges, in the sense that it is not able to measure their structural strength in terms of connectors of two social networks. Therefore, in this section, we introduce a new measure called *cross betweenness centrality (CBC)* to *rate* candidates in our application, overcoming the limits of BC. The need and the validity of CBC is shown in the next section. Recall that we are interested in a measure able to take into account paths crossing distinct social networks in a different way from internal paths. Even though the definition of betweenness centrality does not explicitly take into account the presence of the multiplicity of social networks, it could happen that the real-life structure of the interconnections among distinct social networks (i.e., i-bridges) is such that BC automatically favors nodes belonging to the frontier of each social network, as paths are in some way forced to cross them. Intuitively, the above claim is true if the density of the involved social networks is comparable. Otherwise, we expect that the most dense social network works as an accumulation point, biasing the centrality towards it. However, also in this case, the role of i-bridges is still crucial, so we would like not to miss it.

The definition of *cross betweenness centrality (CBC)* is the following. Let $\Omega \subseteq \mathcal{S}$. Given a node $n \in N$, we denote the *cross betweenness centrality* of n w.r.t. Ω as:

$$CBC(n, \Omega) = \begin{cases} \sum_{s,t \in N, s \neq n, t \neq n, S(s) \neq S(t), S(t) \in \Omega} \frac{\sigma_{st}(n)}{\sigma_{st}} & \text{if } \sigma_{st} > 0 \\ 0 & \text{otherwise} \end{cases}$$

where σ_{st} is the total number of the shortest paths from s to t and $\sigma_{st}(n)$ is the number of those shortest paths from s to t passing through n. In this definition,

Table 1. Results obtained for the first subgraph

Type of node	BC	CBC
Bridges	1,242,081.10	42.67
Power Users	1,543,513.59	4.43
Normal Users	3,795.34	0.01

Ω is a subset of the social networks of the SIS (see Section 2) and allows the computation of the cross betweenness centrality of a node to be limited (if this is desired) to a subset of the social networks of the SIS. In the definition of CBC, considered paths are only those (*i*) linking two nodes belonging to different social networks and (*ii*) having the target node (t) belonging to one social network in Ω (it does not matter whether the source node s belongs to a social network in Ω). In particular, we compute how many times the node n is involved in this kind of path. Interestingly, if n belongs to a fragment of a social network not connected with the rest of the SIS, then $CBC(n, \Omega) = 0$.

Observe that the following relation between cross betweenness centrality and the classical betweenness centrality can be proved: $BC(n) = CBC(n, \Omega) + CBC(n, \overline{\Omega}) + IBC(n)$, where $IBC(n) = \sum_{s,t \in N, s \neq n, t \neq n, S(s)=S(t)} \frac{\sigma_{st}(n)}{\sigma_{st}}$ and $\overline{\Omega} = \mathcal{S} \setminus \Omega$. A direct consequence of this results is that, in the trivial case of a single-social-network SIS, $BC(n) = IBC(n)$. Indeed, no inter-social-network contribution occurs.

6   Need and Validity of CBC: An Experimental Proof

In this experimental section, we show that BC is not able to capture the capability of i-bridges to be central w.r.t. cross-social-network paths. Then, we show that CBC is a measure that can be used in our application to compare different i-bridges in terms of suitability to be actors in global team formation. For this purpose, we partitioned nodes into three categories:

1. i-bridges, which are nodes with a **me** edge;
2. power users, which are non-i-bridge nodes whose degree is equal to, or higher than, the average degree of all nodes;
3. normal users, which are neither i-bridges nor power users.

Then, we computed the average values of BC and CBC for each category. The experiments presented in this section are carried out on two of the subgraphs described in Section 3. The results obtained for the first and second subgraphs are reported in Tables 1 and 2.

From the analysis of these tables, we note that there is no correlation between BC and node cathegory. Indeed, in Table 1, i-bridges and power users have comparable values, whereas normal users have a value that is about 3 magnitude orders lesser than the previous ones. By contrast, in Table 2, power users and

Table 2. Results obtained for the second subgraph

Type of node	BC	CBC
Bridges	105.01	43.33
Power Users	1,480,655.07	5.14
Normal Users	1,092,715.52	0.04

normal users have comparable values, whereas i-bridges have a value 4 magnitude orders lesser than the previous ones. Thus, it is evident that betweenness centrality is not able to correctly identify i-bridge nodes.

Consider now cross betweenness centrality. By looking at Tables 1 and 2, we observe that, for each node category, there is a great uniformity in the corresponding values. Even more interesting, i-bridges have a value of CBC always higher than power users (about one magnitude order), which, in turn, show a value much higher than normal users (about two magnitude orders). Therefore, the distinction among the three categories of nodes is evident by taking cross betweenness centrality into account. Thus, even our expectation about cross betweenness centrality is confirmed by analyzing real-life social networks.

7 Related Work

The concept of centrality, as applied to the context of human communication, was first introduced by Bavelas in 1948 [2]. He mainly focused on communication in small groups and he hypothesized a relationship between structural centrality and influence in group processes. More recently, Leavitt [15], Shaw [19] and Goldberg [13] proposed studies on speed, activity and efficiency in solving problems, on personal satisfaction and on leadership in small group settings. The concept of centrality is motivated by the idea that a person who is close to others can have access to more information, a higher status, more power, a greater prestige, or a greater influence [12] than others. Indeed, this person can facilitate or inhibit the communication of others and is, therefore, in a position to mediate their information access, power, prestige, or influence. Among all the centrality measures, betweenness centrality is one of the most popular, and its computation is the core component of a range of algorithms and applications. Both Bavelas [2] and Shaw [19] suggested that, when a person is strategically located in the middle of communication paths linking other users, she is central. A person in such a position can influence the group by holding or distorting information. By the way, the development of betweenness centrality is generally attributed to the sociologist Linton Freeman [11]. Over the past few years, betweenness centrality has become a popular strategy to measure node influence in complex networks, such as social networks. For this purpose, a lot of new metrics based on betweenness centrality have been already defined [17]. A concept strongly related to edge importance is edge classification. This task is usually performed on the basis of the kind (and, hence, of the "strength") of the relationship the edge represents. Under this assumption, an edge could be a strong or a weak tie. The concept of tie strength was introduced by Mark Granovetter in his very popular paper entitled "The Strength of Weak Ties" [14]. He identified four main features contributing to outline the strength of a tie, namely: amount of time, intimacy, intensity and reciprocal services. Finally, a first characterization of the nodes of a SIS has been proposed in [4]. However, no experimental validation has been provided therein.

8   Conclusion and Future Work

In this paper, we have presented a preliminary idea of a crowdsourcing application aimed at driving the process of global team formation to obtain diversity in the team. This paper presents a first step towards the concrete definition of the above application consisting in the identification of an effective measure that can be used to select seed nodes in the team formation. A first preliminary experimental validation has been provided showing that our idea is well-founded. The next steps are to further validate the new measure and to design the social web application in detail. We plan to do this in our future research.

Acknowledgments. This work has been partially supported by the TENACE PRIN Project (n. 20103P34XC) funded by the Italian Ministry of Education, University and Research and by the Program "Programma Operativo Nazionale Ricerca e Competitività" 2007-2013, Distretto Tecnologico CyberSecurity funded by the Italian Ministry of Education, University and Research.

References

1. Barthelemy, M.: Betweenness centrality in large complex networks. The European Physical Journal B-Condensed Matter and Complex Systems 38(2), 163–168 (2004)
2. Bavelas, A.: A Mathematical Model for Small Group Structures. Human Organization 7(3), 16–30 (1948)
3. Buccafurri, F., Foti, V., Lax, G., Nocera, A., Ursino, D.: Bridge Analysis in a Social Internetworking Scenario. Information Sciences 224, 1–18 (2013)
4. Buccafurri, F., Lax, G., Nicolazzo, S., Nocera, A., Ursino, D.: Measuring Betweenness Centrality in Social Internetworking Scenarios. In: Demey, Y.T., Panetto, H. (eds.) OTM 2013 Workshops 2013. LNCS, vol. 8186, pp. 666–673. Springer, Heidelberg (2013)
5. Buccafurri, F., Lax, G., Nocera, A., Ursino, D.: A system for extracting structural information from Social Network accounts. Technical Report. Available from the authors
6. Buccafurri, F., Lax, G., Nocera, A., Ursino, D.: Crawling Social Internetworking Systems. In: Proc. of the International Conference on Advances in Social Analysis and Mining (ASONAM 2012), Istanbul, Turkey, pp. 505–509. IEEE Computer Society (2012)
7. Buccafurri, F., Lax, G., Nocera, A., Ursino, D.: Discovering Links among Social Networks. In: Flach, P.A., De Bie, T., Cristianini, N. (eds.) ECML PKDD 2012, Part II. LNCS, vol. 7524, pp. 467–482. Springer, Heidelberg (2012)

 8. Buccafurri, F., Lax, G., Nocera, A., Ursino, D.: Internetworking assortativity in Facebook. In: Proc. of the International Conference on Social Computing and its Applications (SCA 2013), Karlsruhe, Germany, pp. 335–341. IEEE Computer Society (2013)
 9. Buccafurri, F., Lax, G., Nocera, A., Ursino, D.: Moving from social networks to social internetworking scenarios: The crawling perspective. Information Sciences 256, 126–137 (2014)
10. Easley, D., Kleinberg, J.: Networks, crowds, and markets, vol. 8. Cambridge University Press, Cambridge (2010)
11. Freeman, L.C.: Centrality in Social Networks Conceptual and Clarification. Social Networks 1(3), 215–239 (1979)
12. Friedkin, N.E.: Theoretical foundations for centrality measures. American Journal of Sociology 96(6), 1478–1504 (1991)
13. Goldberg, S.C.: Influence and leadership as a function of group structure. Journal of Abnormal and Social Psychology 51(1), 119–122 (1955)
14. Granovetter, M.: The strength of weak ties. American Journal of Sociology 78(6), 1360–1380 (1973)
15. Leavitt, H.J.: Some effects of a certain communication patterns on group performance. Journal of Abnormal and Social Psychology 46(1), 38–50 (1951)
16. Lim, S., Ncube, C.: Social networks and crowdsourcing for stakeholder analysis in system of systems projects. In: Proc. of the International Conference on System of Systems Engineering (SoSE 2013), Maui, Hawaii, USA, pp. 13–18. IEEE (2013)
17. Newman, M.: A measure of betweenness centrality based on random walks. Social Networks 27(1), 39–54 (2005)
18. Okada, Y., Masui, K., Kadobayashi, Y.: Proposal of Social Internetworking. In: Shimojo, S., Ichii, S., Ling, T.-W., Song, K.-H. (eds.) HSI 2005. LNCS, vol. 3597, pp. 114–124. Springer, Heidelberg (2005)
19. Shaw, M.E.: Group structure and the behavior of individuals in small groups. Journal of Psychology 38(1), 139–149 (1954)
20. Surowiecki, J., Silverman, M.: The Wisdom of Crowds. American Journal of Physics 75(2), 190–192 (2007)
21. Vukovic, M.: Crowdsourcing for enterprises. In: Proc. of the International Conference on Services-I 2009, Los Angeles, CA, USA, 2009, pp. 686–692. IEEE Computer Society Press (2009)
22. Ye, S., Lang, J., Wu, F.: Crawling online social graphs. In: Proc. of the International Asia-Pacific Web Conference (APWeb 2010), Busan, Korea, pp. 236–242. IEEE (2010)

Effectiveness of Incorporating Follow Relation into Searching for Twitter Users to Follow

Tomoya Noro and Takehiro Tokuda

Department of Computer Science, Tokyo Institute of Technology
Meguro, Tokyo 152-8552, Japan
{noro,tokuda}@tt.cs.titech.ac.jp

Abstract. Twitter is one of the most popular microblogging services that facilitate real-time information collection, provision, and sharing. Following influential Twitter users is one way to get valuable information related to a topic of interest efficiently. Recently many researches on this issue have been done and, in general, it is said that the follow relation is not useful for measuring user influence. In this paper, we study effectiveness of incorporating not only the tweet activity (retweet and mention) but also the follow relation into searching for good Twitter users to follow for getting information on a topic of interest. We present a method for finding Twitter users based on both the follow relation and the tweet activity, and show the follow relation could improve the performance as compared with methods based on only the tweet activity.

Keywords: Social network analysis, microblog, Twitter, search, influential users, power iteration algorithm.

1 Introduction

Currently Twitter has become a more and more important platform of information collection, provision, and sharing. If we would like to get information of a topic of interest and discuss the topic with others on a daily basis, we usually follow some users who provide valuable information on the topic [1]. For example, if we look for information about dementia, we could find that some doctors and care staff members deliver information about the topic, and some people who have family members with dementia post tweets about their daily care. We can get various information on dementia by following such users. However, it is not easy for us to find good users to follow in a massive number of users.

Many researches on this issue have been done recently. Measuring user influence on a particular topic will be one solution. They measure each user's influence based on the tweet content, the follow relation, the tweet activity such as retweet and mention, and so on, and some of them pointed out that the follow relation is not useful for measuring user influence. Cha et al. investigated

---

[1] In this paper, we do not consider temporary topics such as incidents and events (natural disaster, terrorism, FIFA World Cup, etc). If we would like to get information on such topics, we would take different actions such as keyword search.

S. Casteleyn, G. Rossi, and M. Winckler (Eds.): ICWE 2014, LNCS 8541, pp. 420–429, 2014.

characteristics of Twitter users, then concluded users who have many followers are popular but not necessarily influential, while users who are retweeted or mentioned many times have ability to post valuable tweets or ability to engage others in conversation [2]. We have also been working on this issue and showed a user search method based on the tweet activity outperforms a method based on the follow relation [6].

Here is one question. Is the follow relation really useless for finding good users to follow? Actually some users follow almost all of their followers. We can easily get many followers if we search for such users, follow them, wait for them to follow us, and remove them if they do not follow us. The follow relation built in this way is meaningless since most of the followers may not be interested in our tweets and we have little influence on them. However, in general, influential users have many followers. Users who have a small number of followers may not be good users to follow since they should have more followers if they provide a lot of valuable information. The search accuracy could be improved by dealing with both the tweet activity and the follow relation.

In this paper, we study effectiveness of incorporating not only the tweet activity but also the follow relation into searching for good Twitter users to follow for getting information on a topic of interest. We present a method for finding good users to follow based on both the follow relation among users and the tweet activity of each user. In evaluation, we compare the method with other methods without taking the follow relation into account and show incorporating both the tweet activity and the follow relation could improve the search performance.

This paper is organized as follows. In section 2, we discuss some researches on searching Twitter users to follow on a topic of interest. We present our method based on both the follow relation and the tweet activity in section 3, then show some evaluation results in section 4. Lastly we conclude this paper in section 5.

2   Related Work

Twitter provides its own services for user search and recommendation. Given some keywords, Twitter mainly shows some users whose screen names or profiles match the keywords and does not care whether they actually post tweets related to the keywords. The Twitter recommendation service shows users based on the follow relation (users who have mutual followers and/or friends), and does not consider their activity and the tweet content either.

Twittomender [3] finds users related to a particular user or query by using lists of followers, friends and terms in the user's tweets. TwitterRank [8] considers the follow relation and topical similarity to find influential users. Both of the methods are based on the follow relation and they do not consider the tweet activity such as retweet and mention.

Leavitt et al. [5] measured user influence by using ratio of being retweeted and mentioned to the number of tweets the user posted. They do not consider the follow relation. Anger et al. [1] defined user influence based on ratio of retweeted tweets and ratio of the user's followers who retweeted the user's tweets or mentioned the user. Although they consider both the tweet activity and the follow relation, they use the follow relation to observe how many followers retweeted the user's tweets or mentioned the user and do not consider who follows whom.

We presented a method for finding good users to follow for getting information about a topic of interest by using the tweet activity [6]. Although we showed that a search method based on the tweet activity outperforms a method based on the follow relation, we study effectiveness of considering both the tweet activity and the follow relation in this paper.

3   Method for Finding Good Twitter Users to Follow

3.1   Overview

Our process of finding good users to follow on a topic of interest goes as follows.

1. Given some keywords representing the topic of interest, collect tweet data and user data by using the Twitter APIs.
2. Create a user reference graph based on the follow relation and a user-tweet reference graph based on both the tweet activity and the follow relation.
3. Calculate score of each user from the two graphs and rank the users.

The data collection process in the first step goes as follows.

1. Given some keywords representing a topic of interest, get tweets matching the keywords posted in the last N days. Duplicate tweets (exactly the same tweet text posted by the same user) are removed to exclude spammers who post the same tweets repeatedly. Let this tweet set be T_0.
2. For each tweet in T_0, get ID and poster's name of the tweet and user names in the tweet text (user mention). If the tweet is a reply tweet/retweet, get ID and poster's name of the replied/retweeted tweet. Let the set of tweets and the set of users be T_{all} and U_{all} respectively.
3. Get the follow relation among users in U_{all} ($F \subseteq U_{all} \times U_{all}$).

3.2   Score Calculation

In order to define the score of each user, we assume the followings.

1. Users who post many valuable tweets about the topic are worth following.
2. Valuable tweets attract attention from many users.
3. Each user pay attention to tweets the user retweets or replies to.
4. Each user also pay attention to tweets posted by the user's friends.

The first assumption means that users who post many tweets related to the topic should be ranked higher. However, some users who post many valueless tweets such as spam tweets will also be ranked higher if we consider only this assumption. To exclude such users, we take the other assumptions into account. A user's retweeting or replying to a tweet means that the user is interested in the tweet. Each user may pay attention to a tweet posted by the user's friends to some extent even if the user did not retweet or reply to the tweet.

Based on these assumptions, we define the score of each user u as follows.

$$\text{Score}(u) = \text{TC}(u)^{w_c} \times \text{UI}(u)^{w_i} \times \text{FR}(u)^{w_f}$$

$$\text{such that} \quad w_c + w_i + w_f = 1 \wedge w_c \geq 0 \wedge w_i \geq 0 \wedge w_f \geq 0 \quad (1)$$

$\text{TC}(u)$, $\text{UI}(u)$, and $\text{FR}(u)$ are respectively "tweet count (TC) score", "user influence (UI) score", and "follow relation (FR) score" of user u ranging between 0 and 1. The TC score is based on the number of tweets each user posted, and reflects the first assumption. The FR score is based on the follow relation among users, and reflects the second and forth assumptions. The UI score is based on both the tweet activity and the follow relation, and reflects the second, third and forth assumptions.

3.3   Tweet Count Score (TC Score)

The TC score is calculated by counting not only each user's original tweets but also retweets in T_0. We count retweets as each user's own tweets [2]. The score is normalized so that the largest value should be 1.

$$\text{TC}(u) = \frac{\log(1 + |\{t|t \in T_0 \wedge t.user.id = u.id\}|)}{\max_{u' \in U_{all}} \log(1 + |\{t|t \in T_0 \wedge t.user.id = u'.id\}|)} \quad (2)$$

$t.user.id$ indicates poster's ID of tweet t and $u.id$ indicates ID of user u.

3.4   User Influence Score (UI Score)

The basic idea is as follows.

1. If user u_i retweets or replies to user u_j's tweet, u_j has an influence on u_i.
2. Users who post many tweets paid attention to by many users are influential, especially if their tweets are often paid attention to by influential users.

How much each tweet is paid attention to by others is measured according to the tweet activity (retweet and reply) and the follow relation. Based on this idea, we define not only the UI score of each user but also tweet influence score (TI score) of each tweet. The UI score is calculated using the TI score of tweets and retweets posted by the user, and the TI score is calculated using the UI score of users who pay attention to the tweet.

We create a user-tweet reference graph consisting of user nodes (U_{all}), tweet nodes (T_{all}), and directed edges each of which connects a user node and a tweet

---

[2] The retweet activity is incorporated into both the TC score and the UI score. The number of times each user retweeted is considered in the TC score while the user-tweet relation (who retweeted what) is considered in the UI score.

node. The reference graph is represented as combination of three adjacency matrices A_t, A_r, and A_s.

$$A_t(t_i, u_j) = \begin{cases} 1 & \text{if } t_i \text{ is tweeted/retweeted by } u_j \\ 0 & \text{otherwise} \end{cases} \tag{3}$$

$$A_r(u_j, t_i) = \begin{cases} 1 & \text{if } u_j \text{ retweets/replies to } t_i \\ 0 & \text{otherwise} \end{cases} \tag{4}$$

$$A_s(u_j, t_i) = \begin{cases} 1 & \text{if } u_j \text{ follows at least 1 user who tweets/retweets } t_i \\ \alpha & \text{otherwise } (0 < \alpha \leq 1) \end{cases} \tag{5}$$

t_i and u_j indicates the i-th tweet and the j-th user respectively ($1 \leq i \leq |T_{all}|$ and $1 \leq j \leq |U_{all}|$). A_t and A_r are derived from the tweet activity of each user, and A_s is derived from the follow relation among users. A_t represents what (tweet) is tweeted or retweeted by whom (user), and A_r and A_s respectively represent who retweets or replies to what and who sees what. The follow relation will be ignored if α is equal to 1.

These adjacency matrices are transformed into the following two matrices.

$$B_t(t_i, u_j) = \frac{A_t(t_i, u_j)}{\sum_k A_t(t_i, u_k)} \tag{6}$$

$$B_a(u_j, t_i) = \begin{cases} \frac{A_r(u_j, t_i)}{\sum_k A_r(u_j, t_k)}(1-d) + \frac{A_s(u_j, t_i)}{\sum_k A_s(u_j, t_k)}d & \text{if } \sum_k A_r(u_j, t_k) \neq 0 \\ \frac{A_s(u_j, t_i)}{\sum_k A_s(u_j, t_k)} & \text{otherwise} \end{cases} \tag{7}$$

d is a damping factor of $0 < d < 1$. Transformation of A_r and A_s into B_a reflects the third and forth assumptions of good users to follow described in section 3.1. Each user pay attention to tweets the user retweets or replies to, and the user also watches all tweets at a certain rate of d regardless of the user's activity of retweet and reply. Tweets posted or retweeted by the user's friends are more likely to be seen than the other tweets, and the idea is also included.

The UI score and the TI score are calculated as follows.

$$\mathbf{u} = B_t^T \mathbf{t} \qquad\qquad \mathbf{t} = B_a^T \mathbf{u} \tag{8}$$

$\mathbf{u}$ and $\mathbf{t}$ indicate a column vector of the UI score of all users and a column vector of the TI score of all tweets respectively. We can calculate the UI score and the TI score using the power iteration method. Lastly the UI score of each user is normalized so that the largest value should be 1.

$$\text{UI}(u_j) = \frac{\mathbf{u}(j)}{\max_k \mathbf{u}(k)} \tag{9}$$

3.5 Follow Relation Score (FR Score)

The FR score is calculated based on the follow relation using PageRank [7]. A user reference graph is created from the follow relation F. Adjacency matrix of the graph is represented as follows.

$$A_f(u_i, u_j) = \begin{cases} 1 & \text{if } u_i \text{ follows } u_j \text{ i.e. } (u_i, u_j) \in F \\ 0 & \text{otherwise} \end{cases} \tag{10}$$

$$B_f(u_i, u_j) = \begin{cases} \frac{A_f(u_i,u_j)}{\sum_k A_f(u_i,u_k)}(1-d) + \frac{d}{|U_{all}|} & \text{if } \sum_k A_f(u_i, u_k) \neq 0 \\ \frac{1}{|U_{all}|} & \text{otherwise} \end{cases} \tag{11}$$

$$\mathbf{f} = B_f^T \mathbf{f} \tag{12}$$

u_i and u_j indicates the i-th user and the j-th user respectively, and d is a damping factor. $\mathbf{f}$ indicates the column vector of the FR score of all users.

Unlike normalization of the TC score and the UI score, the FR score of each user is not divided by the maximum value. As described in section 1, users who have many followers are not necessarily influential, and it is said that the follow relation is not useful for measuring user influence. However, we think the follow relation could be used for excluding uninfluential users who have few (influential) followers. Instead of dividing the FR score of each user by the maximum value, we set upper limit of the FR score to the minimum score of the top-$P\%$ users and divide the score of each user by the limit.

$$\text{FR}(u_i) = \frac{\min(\mathbf{f}(i), limit)}{limit} \tag{13}$$

$limit$ indicates the minimum FR score of the top-$P\%$ users. This normalization can weaken influence of the users who have high FR score since the score of all of the top-$P\%$ users will be set to 1.

Some alternative ways for normalization may be considered. For example, some may think of normalization by setting the upper limit to a predetermined proportion to the maximum value (e.g. $limit = 0.1 \times \max_k \mathbf{f}(k)$) or determining the number of users to be capped (e.g. Top-50 users). However, the number of users to be ranked depends on topics of interest, and distribution of the FR score also varies by the topics. We think that determining the percentage of users to be capped is better from our observation.

4   Evaluation

4.1   Experimental Setup

We selected the following 7 Japanese keywords (in Japanese characters) as input query representing topics of interest: "nuclear power", "animal test", "whaling", "dementia", "digital book", "basic income" and "fair trade". We chose these topics since we expect that tweets related to the topics are posted on a daily basis (independent of season). Tweets and other data were collected 6 times on different days. For each time, we get tweets posted in the last 5 days. The average number of tweets and users we collected is shown in Table 1. "Reply" means tweets replying to tweets specified in "reply-to" attribute, and "Mention" means tweets including user names but not specifying their target tweets.

Table 1. The average number of tweets and users

| Keyword | $|T_0|$ | | | | $|T_{all}|$ | $|U_{all}|$ |
	Total	Retweet	Reply	Mention		
nuclear power	26,937.7	14,878.0	1,008.7	1,336.0	28,124.0	13,435.3
animal test	2,591.7	1,539.8	185.7	126.5	2,818.3	1,349.3
whaling	4,057.7	1,045.7	254.0	321.0	4,249.7	3,112.7
dementia	5,497.5	1,670.7	832.3	163.5	6,255.0	4,959.3
digital book	19,208.0	4,408.2	1,307.8	1,273.3	20,333.5	12,976.8
basic income	400.7	148.8	68.3	19.8	449.0	251.5
fair trade	779.2	364.8	70.7	14.7	849.8	662.2

Table 2. Value of each parameter

d in Eqs. (7) and (11)	0.15
P for FR(+lim)	5%
α in Eq. (5) for UI(+fol)	0.1
w_c and w_i in Eq. (1) for TC+UI	0.6 and 0.4
w_c, w_i and w_f in Eq. (1) for TC+UI+FR	0.6, 0.2, and 0.2

We set up the following methods for comparison.

TC+UI(+fol)+FR(+lim): Rank users based on the TC score, the UI score with the follow relation, and the FR score with upper limit.

TC+UI(-fol)+FR(+lim): Rank users based on the TC score, the UI score without the follow relation, and the FR score with upper limit (α in Eq. (5) is set to 1).

TC+UI(+fol)+FR(-lim): Rank users based on the TC score, the UI score with the follow relation, and the FR score without upper limit (normalization of the FR score is done by dividing each score by the maximum value).

TC+UI(-fol)+FR(-lim): Rank users based on the TC score, the UI score without the follow relation, and the FR score without upper limit.

TC+UI(+fol): Rank users based on the TC score and the UI score with the follow relation (w_f in Eq. (1) is set to 0).

TC+UI(-fol): Rank users based on the TC score and the UI score without the follow relation

TC: Rank users based on only the TC score.

UI(+fol): Rank users based on only the UI score with the follow relation.

UI(-fol): Rank users based on only the UI score without the follow relation

FR(-lim): Rank users based on only the FR score without upper limit

We carried out a preliminary experiment using tweet data collected on different days to determine the parameters appeared in section 3. The value of each parameter is shown in Table 2.

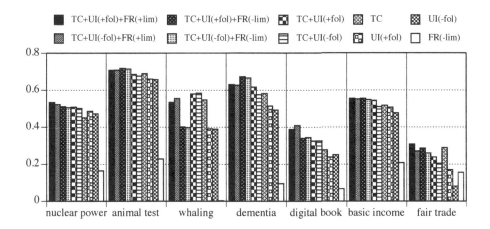

Fig. 1. Average nDCG of each method

Evaluation is done on the top 20 users ranked by each method with respect to normalized discounted cumulative gain (nDCG) [4].

$$\text{DCG}_{20} = rel_1 + \sum_{i=2}^{20} \frac{rel_i}{\log_2 i} \qquad \text{maxDCG}_{20} = 2 + \sum_{i=2}^{20} \frac{2}{\log_2 i} \qquad (14)$$

$$\text{nDCG}_{20} = \frac{\text{DCG}_{20}}{\text{maxDCG}_{20}} \qquad (15)$$

rel_i indicates relevance score between the user ranked i-th and the input keyword, which is judged on a scale of 0 to 2. Users who often post related tweets are assigned the score of 2, while users who rarely post related tweets are assigned the score of 0. Users who post a lot of unrelated tweets like advertisement and spams are also assigned the score of 0. The judgment was done by watching their tweets posted after the data collection period (for about one month) to check whether they continuously post related tweets.

4.2   Result

The result is shown in Figure 1. We can see that considering the follow relation in calculation of the UI score improves the search result on average (e.g. methods including UI(+fol) vs methods including UI(-fol)).

Except for the case of "whaling", incorporating the FR score is also effective (e.g. TC+UI(+fol)+FR(+lim) vs TC+UI(+fol)). The FR(-lim) method found no relevant user in the case of "whaling". When incidents related to whaling occur, many major news organizations will post tweets related to the topic and will get high ranking in the FR score since they have many followers. However, their interest is not always focused on the topic. On the other hand, users who usually talk about or discuss the topic do not have strong follow relation with others

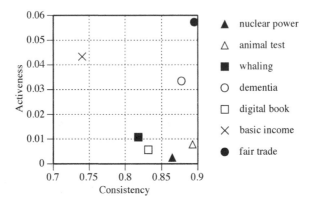

Fig. 2. Consistency and activeness of each keyword

compared to such news organizations. We think this is why the performance were not improved by incorporating the FR score.

Effectiveness of setting the upper limit for normalization of the FR score differs according to keywords (e.g. TC+UI(+fol)+FR(+lim) vs TC+UI(+fol)+FR (-lim)). In the case of "whaling", a large decline caused by incorporating the FR score is prevented. This is because influence of users who is ranked high on the FR score but do not focus on the topic (e.g. major news organizations) are reduced. Setting the upper limit worsen the search performance in the case of "dementia" since, unlike the case of "whaling", some users who usually focus on the topic also have strong follow relation with others and difference between their FR score and the score of major news organizations is not large.

If we compare methods including UI(+fol) with methods including UI(-fol) in more details, we can see that incorporating the follow relation in calculation of the UI score seems not to improve the search performance in the case of "whaling" and "digital book". To analyze this issue, we calculate two measures. One measure is "consistency", how much their tweet activity (retweet and reply) is consistent with their follow relation, and the other measure is "activeness", how many tweets posted by their friends they retweet or reply to. They are defined as follows.

$$\text{Consistency} = \frac{Consistent}{Consistent + Inconsistent} \tag{16}$$

$$\text{Activeness} = \frac{Consistent}{Consistent + Ignored} \tag{17}$$

"*Consistent*" means the number of times they retweeted or replied to their friends' tweets and retweets, and "*Inconsistent*" means the number of times they retweeted or replied to tweets and retweets posted by none of their friends. "*Ignored*" means the number of times they do not retweeted nor replied to their friends' tweets and retweets. From the result shown in Figure 2, we can see both consistency and activeness of "whaling" and "digital book" are low compared with other keywords. We guess users who are interested in the topics are likely to see tweets of non-following users and to communicate with them while they do not communicate with their friends so much. This situation would occur if they usually talk about or discuss topics of interest by using hashtags. It can also be seen in the case of "basic income" but, in this case, many of them communicate with their friends as well as other users (activeness is high in the case of "basic income"). On the other hand, both consistency and activeness of "fair trade" and "dementia" are high. In both cases, we can see effectiveness of incorporating the follow relation into calculation of the UI score. The percentage of retweets, reply tweets and mention tweets of "digital book" and "whaling" is low (36.3% and 40.0% respectively) as shown in Table 1, which would also be one factor that worsens the performance.

5   Conclusion

In this paper, we presented a method for finding good Twitter users to follow for getting information about a topic of interest based on both the tweet activity and the follow relation, and showed incorporating both of them improves the search performance on average except for the case that performance of the method based only the follow relation is extremely bad. We also measured "consistency" and "activeness", and found effectiveness of incorporating the follow relation into the UI score is high if the two scores are high.

References

1. Anger, I., Kittl, C.: Measuring influence on Twitter. In: 11th International Conference on Knowledge Management and Knowledge Technologies, vol. 31. ACM Digital Library (2011)
2. Cha, M., Haddadi, H., Benevenuto, F., Gummadi, K.P.: Measuring user influence in Twitter: The million follower fallacy. In: 4th International AAAI Conference on Weblogs and Social Media, pp. 10–17 (2010)
3. Hannon, J., Bennett, M., Smyth, B.: Recommending Twitter users to follow using content and collaborative filtering approaches. In: 4th ACM Conference on Recommender Systems, pp. 199–206 (2010)
4. Jarvelin, K., Kekalainen, J.: Cumulated gain-based evaluation of IR techniques. ACM Transactions on Information Systems 20(4), 422–446 (2002)
5. Leavitt, A., Burchard, E., Fisher, D., Gilbert, S.: The influentials: New approaches for analyzing influence on Twitter. Web Ecology Project (2009),
   http://www.webecologyproject.org/2009/09/
   analyzing-influence-on-twitter/
6. Noro, T., Ru, F., Xiao, F., Tokuda, T.: Twitter user rank using keyword search. In: Information Modelling and Knowledge Bases XXIV. Frontiers in Artificial Intelligence and Applications, vol. 251, pp. 31–48. IOS Press (2013)
7. Page, L., Brin, S., Motwani, R., Winograd, T.: The PageRank Citation Ranking: Bringing Order to the Web. Tech. rep., Stanford University (1998)
8. Weng, J., Lim, E.P., Jiang, J., He, Q.: TwitterRank: Finding topic-sensitive influential Twitterers. In: 3rd ACM International Conference on Web Search and Data Mining. pp. 261–270 (2010)

Improving the Scalability of Web Applications with Runtime Transformations

Esteban Robles Luna[1], José Matías Rivero[1,2], Matias Urbieta[1,2], and Jordi Cabot[3]

[1] LIFIA, Facultad de Informática, UNLP, La Plata, Argentina
{esteban.robles,mrivero,matias.urbieta}@lifia.info.unlp.edu.ar
[2] Also at Conicet
[3] École des Mines de Nantes / INRIA
jordi.cabot@inria.fr

Abstract. The scalability of modern Web applications has become a key aspect for any business in order to support thousands of concurrent users while reducing its computational costs. If a Web application does not scale, a few hundred users can take the application down and as a consequence cause business problems in their companies. In addition, being able to scale a Web application is not an easy task, as it involves many technical aspects such as architecture design, performance, monitoring and availability that are completely ignored by current Model Driven Web Engineering approaches. In this paper we present a model-based approach that uses runtime transformations for overcoming scalability problems in the applications derived from them. We present our approach by "scaling up" a WebML application under a stress scenario, proving that it provides a "framework" for overcoming scalability issues.

1 Introduction

Scalability is the ability of a system to handle a growing amount of work in a capable manner or its ability to be enlarged to accommodate that growth [3]. Scalability problems in the applications derived using Model Driven Web Engineering (MDWE) tools may not appear as soon as they are deployed, but rather after they has been "living" in production for some time.

In the Web engineering research area, MDWE approaches [11] have become an attractive solution for building Web applications by raising the level of abstraction and simplifying Web application development. Regardless of the approach used, the main focus is always the same: to create a Web application that satisfies functional requirements. As a consequence, little attention has been put to non-functional requirements such as scalability issues as they have been considered technological-dependent aspects [12]. Additionally, the Web applications derived from MDWE approaches pose a rigid/static architecture that cannot be easily changed hindering scalability fixes. To make things worse, diagnosing and fixing these problems becomes cumbersome and impossible to be done in the models thus forcing teams to deal with the generated code.

To overcome these problems, a recent study [7] has captured the top 20 problems that Web applications face to achieve scalability and it also shows that *"unforeseen scalability issues during development, can easily appear in a production environment"*. This list includes obstacles that affect both code and model based development

S. Casteleyn, G. Rossi, and M. Winckler (Eds.): ICWE 2014, LNCS 8541, pp. 430–439, 2014.

and includes aspects such as monitoring, logging and caching. Furthermore, though scalability is a desired aspect of a Web application, achieving it does not only require the application to run fast (within an acceptable threshold), but it also involves other technical aspects that need to be handled during the application lifecycle [7].

Identifying a scalability problem (e.g. the application runs out of DB connections) can be easy, however *diagnosing* to find the root cause of the issue and *fixing* it consumes many human resources and it has been probed to consume 50%-75% of the software development effort [2]. We argue that the current status of MDWE tools can not help on that aspect as they do not allow to decompose the model elements until their primitives and as a consequence forces teams to deal with the generated code, thus breaking the models' abstraction and, consequently, losing the productivity improvement claimed by MDWEs.

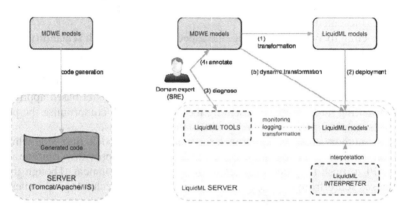

Fig. 1a. Traditional MDWE   **Fig. 1b.** LiquidML approach

In this paper we present a model-based approach called *LiquidML* that complement MDWEs to help them diagnose and fix scalability issues in running applications. Instead of deriving the code from MDWE models (Fig 1a), we (Fig 1b):

1. Transform the MDWE models into LiquidML models that can be interpreted by our interpreter.
2. Deploy those models into a LiquidML Server (final platform for running them).
3. Provide tools for domain experts such as Site Reliability Engineers (SREs) to diagnose scalability problems directly at the model level after their identification.
4. Provide scalability annotations with architectural changes so the next derivation contemplates them to produce a more scalable version of the application once the problem has been diagnosed and as development continues in the higher-level MDWE models,
5. Enable to dynamically apply a model transformation to fix the problem at runtime directly in the LiquidML servers if we want to avoid a deployment (doing steps 1 and 2 again).

The paper is structured as follows: in Section 2 we present related work and in Section 3 we present the details of our approach. In Section 4, we show how the approach is implemented and in Section 5 we present the conclusions and further work.

2   Related Work

Developing and maintaining Web applications require not only an initial construction process but also a continuous monitoring/fixing cycles that must interleave with new requirements implementation. Scalability is generally an implicit desired feature by product owners but it is the least aspect that is paid attention on during development.

In this matter, the authors in [12] clearly states *"Many MDWE approaches have been created in the last 20 years with special focus in modeling the functional aspects of Web applications. Non-functional (e.g., scalability) properties of Web applications have traditionally been a minor concern in the Web engineering community and have been seen as technology or system-related issues"*. Additionally, the author presents a theoretical proposal to deal with non-functional requirements during a model based development; however scalability is not treated as a specific concern and no implementation is presented.

Not restricting to Web Engineering, the aspects of dealing with scalability in Model-Driven Engineering (MDE) are the central topic of a recent study [8]. However, the term scalability in this area is treated not as the scalability of the derived application but to actually scale the tools and models to be able to handle relatively big applications. In this aspect, being able to scale the development of model based applications involves handling huge graph representing the models which compromise the performance of the transformations applied to obtain the applications.

In [5] the authors present an approach for transforming models into code while the application is running. To accomplish such task, the authors provide a runtime model with an API to push the code changes to the running environment. Though the approach makes sense from a conceptual point of view, many technological issues such as concurrency and well known issues in the derived language (e.g. memory issues in a Java program such as hot code replacement) are not taken into account. As a consequence, we argue that the approach, in its current state, lacks practical application.

Finally, holding models at runtime to perform runtime changes has been presented in [1]. However, the approach focuses in the representation of the actual requirements while the application is running rather than on the live models. The approach was initially applied for autonomous systems where domain experts are not able to access the system easily and thus has to run autonomously. The applications built with models @ runtime are from a different domain and seems to have less sophisticated business requirements than a Web application; as a consequence those automatic changes can be applied. In the Web engineering area, we think that those changes should be applied by SREs in a semi automatic way.

3   LiquidML

In this section we present our approach for fixing scalability issues at runtime. Though our approach can be applied to any MDWE method, we will illustrate it using WebML [4], as it is the most mature MDWE regarding to tool support.

3.1   LiquidML in a Nutshell

A LiquidML model is a composition of so-called *Flows*. A *Flow* describes a sequence of steps that need to happen within a Web request (called a *Message* in our approach).

A Message has a payload (body) and a list of properties while each step is visually identified by an icon and constitutes an *Element* of the flow. Communication between Elements happens by means of message interchanges. The way messages are moved from one Element to another one is defined by the Connections between them.

We have categorized the Elements using the categories found in [6]:

- Message source: Listens for incoming requests and generates messages from them. A concrete example is an HTTP listener that listens to incoming requests from users and creates Messages from HTTP requests.
- Processor: Processes a message by changing its payload and properties. Processors can vary from custom user logic to DB access.
- Router: Moves the message between processors depending on its type and conditions. For instance, a *ChoiceRouter* routes the message to a specific Element of its list, based on a Boolean condition.

To summarize the LiquidML concepts, a metamodel is presented in Fig. 2.

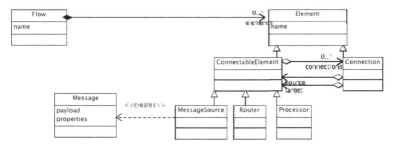

Fig. 2. LiquidML metamodel

3.2 Transforming MDWE Models to LiquidML Models

Flow models (main model of the our LiquidML models) could be manually created, however a more common scenario would be to start from higher level models (like WebML navigation models) and then transform those into our flow models to benefit from LiquidML. To provide an example of this transformation we show in Fig 3 the product details page of a WebML model. The product details page shows the basic product information and a rank that qualifies the product.

The actual transformation from the WebML model to LiquidML works iterating through the elements of the model and "unfolding" some elements that are implicit in the WebML model but that have an explicit representation in our lower level of abstraction (so that they can be configured, if needed, for scalability purposes). Fig. 4 shows the result of the transformation for the WebML example.

Fig. 3. Fragment of WebRatio model for our E-Commerce application

The Element with no incoming arrow represents the Message source listener that will receive incoming requests (implicit in WebML). The Element connected to the Message source named "Route path" is a ChoiceRouter that behaves like a choice/switch statement and it will route the message to the "Get info" processor if the request comes to a URL starting with "/product/*". This element is represented in the Page units of WebML ("Detail" unit in Fig. 3) by a property that identifies its URL. The transformation also tries to derive highly concurrent models and thus the "Get info" is another router that gets information in parallel from multiple sources. It obtains the product info from the DB: ("Get product info") and it triggers the computation of the product's rank, which involves two database (DB) queries ("Get user reputation" and "Get product reviews") and a Processor that computes the rank from this information ("Compute product rank"). The "Get product info" element is obtained from the "product detail unit" while the other elements are derived from the Groovy source code of "Compute product rank". Finally, the information gets composed ("Compose data") and used for rendering a Web page in the "Render template" processor. These elements are hidden in the WebML model as WebRatio generates an example UI; we also make these elements explicit.

Fig. 4. Product details flow

A Flow model can be executed by means of the LiquidML interpreter. In the following subsection, we provide an overview of how our model interpretation works and a brief description of the deployment process.

3.3   Model Interpretation and Deployment

As aforementioned, flows define the behavioral part of the Web application. On the contrary to all MDWE approaches, we decided to interpret rather than to derive the code of a Web application. Strong cons and pros of both approaches can be found in [9] and in many informal discussions[1,2,3]; however, we do not expect to find a definite answer to this matter but rather present the advantages we found for Web scaling in our model based approach. It is true that at a first sight, code-generation seems to be

---

[1] http://www.theenterprisearchitect.eu/archive/2010/06/28/
model-driven-development-code-generation-or-model-interpretation
[2] http://modeling-languages.com/
executable-models-vs-code-generation-vs-model-interpretation-2/
[3] http://blog.webratio.com/?p=368

the right option when aiming at scalable applications but scalability involves more than an application running fast.

As our behavioral models (Flows) are rather simple, the interpreter algorithm is quite simple too. We present a simplified version of the algorithm using a Java-based pseudocode in the next lines: the interpreter works when messages are received (event based style [6]) on Message sources (e.g. an HTTP message source) (line 1). It finds the next element (currentElement) that will handle the message (line 2 and 5) and evaluates it using the message content (line 4). An evaluation returns a Message instance that could be the same as the previous one or a new one depending on the Element intent (data transformation, routing, etc.) and it is passed to the next Element until we run out of Elements (line 3).

```
1. OnMessageReceived(MessageSource msgSource, Message message): {
2.   Element currentElement = interpreter.getNextInChain(msgSource);
3.   while (currentElement != null) {
4.     interpreter.evaluate(currentElement, message);
5.     currentElement = interpreter.getNextInChain(currentElement);
6.   }
7. }
```

Interpretation happens while engineers are building the application and when the application is run in every other deployment environment (QA, Staging, Production). Once the models satisfy the requirements, the deployment process to a specific environment occurs. The deployment is an automatic process where a copy of the models is moved to the servers where they can start receiving messages. As aforementioned, unforeseen problems may appear in a production environment; thus in the following subsection, we present two tools to help diagnosing problems while our models are running.

3.4 Diagnosing Production Problems

The LiquidML model is what the engineering team tweaks and what is being run in production by means of the LiquidML interpreter. Such interpreter resides inside the LiquidML server, which is our execution platform that provides the ability of tweaking the models at runtime to improve the overall scalability of the application.

In our approach, each Element inside a Flow can suffer of production problems and as a consequence it may need to be adjusted. In these cases our approach allows us to monitor specific Elements but only when the engineer requests so. Monitoring adds a small performance footprint on each request that must be handled with caution. To diagnose problems we provide two well-known features such as profiling and logging. However, instead of generating low-level artifacts such as Java code, we work at a high level allowing to reference elements and messages.

As shown in [7], logging is a practical tool for this purpose as we can log information about the incoming messages. In our approach, we also allow condition logging (log information if it only satisfies some conditions). This is extremely important as it helps to reduce the performance degradation [10] of the elements when thousands of messages are processed per second and because it filters messages that we know are not causing the problem. Considering the Product detail example, we can dynamically add a logger to the "Path router" Element that prints the requested resources (the incoming URL) by simply writing the following expression:

```
'Incoming traffic: ' + message.properties['actions.http.requestURI']
```

The second feature we provide is Element profiling, meaning that we allow the temporary measuring of the performance of an Element. It helps with the identification of elements that are consuming more time to complete in the Flow and orients modelers in providing quick performance solutions through the introduction of model optimizations. We can attach multiple profilers to our elements (Fig 5) and see how the average response time changes as traffic comes in. Using both logging and profiling we can better diagnose the problem and check if a runtime fix is possible or not; we will discuss this topic in the following subsections.

Fig. 5. Profiling a flow

3.5   Annotation of MDWE Models

Once the problem has been identified and diagnosed, the development team has to fix it. Our approach allows annotating the MDWE model with scalability annotations to improve the scalability properties of the applications derived from the MDWE tool. Fig. 5 shows that the computation of the rank is taking a large percent of the response time and as a consequence some optimization needs to be introduced. The development team proposes to cache the product rank that gets computed by tagging the navigation line in the WebML model of Fig. 3 with a cache strategy of the computed value. In Fig. 6 we show the tagged WebML model (the name ALC stands for "Async Lazy Compute" as shown in the following section).

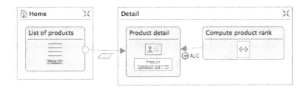

Fig. 6. Tagged WebML model

The tag applied to the WebML model generates a transformation that can be applied to the generated LiquidML model. In the following section, we present the result of applying such transformation.

3.6   Runtime Model Transformations

As mentioned in the previous subsection, some problems can be quickly fixed in our approach and do not require a build-deploy process. Those fixes tend to help satisfying non-functional requirements and thus functional requirements should not be altered. In this matter, we have created a small, non-exhaustive catalog of solutions that can be automatically applied to flows and it is based on the concepts found in [6]. Due to space constraints, we will not present the catalog but just explain the transformation applied in the example of Fig. 6.

The caching (Async Lazy Compute) is used when an Element is running slow, as a consequence we cache the results produced by the Element for a long period of time. We store when was the last time that the element was run and the result of such execution and upon incoming requests we update the cached results asynchronously. Applying this transformation to the connection between "Get info" and "Get product rank" will cache all the results coming from the subgraph starting in "Get product rank". The transformation needs some inputs to be applied which can be automatically obtained if it is specified in the WebML model:

- A key expression: an expression used for the getting and putting the results in the cache. In our case we will use message. Properties ['actions.http.productId'].
- A value expression: what we want to cache. In our case just the payload of the message: message.payload.

The application of the transformation modifies the Flow model dynamically and leaves it in the state shown in Fig 7. We have highlighted the elements and connections added by the transformation, as the remaining of the diagram is the same as shown in Fig. 4. The first element added by the transformation "Get cached value" gets the cached value for this request using the product id. In the "Is cached?" router if it is not cached we continue to the "Chain" element. Here, we compute the values as before and then we store the value in the cache ("Put cached value") and save the last update computation ("Put last update"). Finally, the "Async cache" gets asynchronously evaluated when the value has been cached before and (NOW() – lastComputedTime()) > configurable threshold. So, we recomputed the value in an asynchronous way but only when it is requested and it is old enough. The user may see an old value but after the new value has been computed it will appear for the following user.

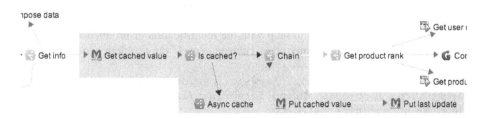

Fig. 7. Result of applying "Caching (async lazy compute)"

4   Implementation

We have developed some tools to support the construction of models using a Web based application that allows modeling each of the concepts we have mentioned. All the figures presented in this paper have been obtained from the Web tool we have built. Such tool allows handling multiple teams and each team allows the creation of multiple applications. Each application consists of multiple flows, properties and resources and it can be snapshot and deployed.

Once the application is deployed, we can move to a deployment view of the flows where we can modify the application at runtime to diagnose and fix any performance problem that may arise. In the deployment view, we can select an element and add a profiler. After some messages are received, we can start seeing the number of milliseconds it takes for a message to be processed. If necessary we can add some logging information that can help us diagnose the problem. The logging information is generated from a dynamic expression on the message and each recorded entry can be seen in the tool. Finally, if the problem can be fixed by simply applying a transformation, we can do so by selecting the Element involved, configuring the transformation (if necessary) and clicking in the apply button. The task is straightforward because the user selects a Flow Element used as a reference for the transformation and then chooses a valid transformation.

We have decided not to use the well-known Eclipse Modeling Framework (EMF)[4] for defining our language and its visual representation since, in our humble opinion, it provides a rigid structure that we want to avoid. As a consequence, our tool support required a bit of extra effort to be developed. To implement it we have used a standard Java architecture composed by Spring Framework[5] and Hibernate[6]. In addition, Cappuccino[7] was used to build the editor. On the server side, we implemented the REST API services with Jersey[8] so that the models can be deployed and share most of the Java code. We have also used Spring and Hibernate in the Server side too. We encourage the reader to visit our Web site at http://www.liquidml.com for demonstration videos.

5   Conclusions and Future Work

In this paper we have presented LiquidML, a model-based approach that complements MDWEs methodologies to help them improving the scalability of the applications they derive. To the best of our knowledge, this is the first work to propose and implement a solution to deal with this topic. Once a LiquidML model is obtained, we can monitor the application to help with the identification of production problems that can not be reproduced in any other environment. If the problem can be fixed at

---

[4] Eclipse modeling framework. http://www.eclipse.org/modeling/emf/

[5] Spring, http://projects.spring.io/spring-framework/

[6] Hibernate, http://hibernate.org/

[7] Cappuccino, http://www.cappuccino-project.org/

[8] Jersey, https://jersey.java.net/

runtime, engineers can apply a well-known solution safely and automatically, avoiding the high cost of redeploying the application.

Interpreting models poses multiple challenges and many advantages that we want to take into account as future work. From analyzing multiple empirical experiences, it is easy to abstract solutions to common patterns that appear in the application. Thus, we plan to provide these patterns that fix production problems and the rules that help their identification as first class entities. We predict that such approach cannot be fully automated as it requires knowledge about the running application (e.g. where the information is stored in the message and what are the things we want to cache) and as a consequence a semi automatic process using a rule engine seems to be a viable solution. Finally, we plan to expand the catalog of patterns that can be applied to flows, which will help SREs fixing more problems at runtime.

References

1. Blair, G., Bencomo, N., France, R.B.: Models@ run.time. Computer 42(10), 22 (2009)
2. Boehm, B.W.: Software engineering economics. Prentice-Hall, Englewood Cliffs (1981)
3. Bondi, A.: Characteristics of scalability and their impact on performance. In: Proceedings of the 2nd International Workshop on Software and Performance (WOSP 2000), pp. 195–203. ACM, New York (2000)
4. Ceri, S., Fraternali, P., Bongio, A.: Web Modeling Language (WebML): A Modeling Language for Designing Web Sites. Computer Networks and ISDN Systems 33(1-6), 137–157 (2000)
5. Sánchez Cuadrado, J., Guerra, E., de Lara, J.: *The Program Is the Model*: Enabling Transformations@run.time. In: Czarnecki, K., Hedin, G. (eds.) SLE 2012. LNCS, vol. 7745, pp. 104–123. Springer, Heidelberg (2013)
6. Hohpe, G., Woolf, B.: Enterprise Integration Patterns: Designing, Building, and Deploying Messaging Solutions, p. 735. Addison-Wesley (2012)
7. Hull, S.: 20 Obstacles to Scalability. ACM Queue 11(7), 20 (2013)
8. Kolovos, D., Rose, L., Matragkas, N., Paige, R., Guerra, E., Cuadrado, J.S., Lara, J., Ráth, I., Varró, D., Tisi, M., Cabot, J.: A research roadmap towards achieving scalability in model driven engineering. In: Proceedings of the Workshop on Scalability in Model Driven Engineering (BigMDE 2013). ACM, New York (2013)
9. Mellor, S.J., Balcer, M.: Executable UML: A Foundation for Model-Driven Architectures. Addison-Wesley Longman Publishing Co., Inc., Boston (2002)
10. Molyneaux, I.: The Art of Application Performance Testing: Help for Programmers and Quality Assurance, 1st edn. O'Reilly Media, Inc. (2009)
11. Rossi, G., Pastor, O., Schwabe, D., Olsina, L.: Web Engineering: Modelling and Implementing Web Applications. Springer (2007)
12. Toffetti, G.: Web engineering for cloud computing. In: Grossniklaus, M., Wimmer, M. (eds.) ICWE Workshops 2012. LNCS, vol. 7703, pp. 5–19. Springer, Heidelberg (2012)

Multi Matchmaking Approach for Semantic Web Services Selection Based on Fuzzy Inference

Zahira Chouiref[1,2], Karim Benouaret[3], Allel Hadjali[2], and Abdelkader Belkhir[4]

[1] Université de Bouira, 10000 Bouira, Algeria
zahira.chouiref@univ-bouira.dz
[2] LIAS, ENSMA, 86360 Chasseneuil-du-Poitou, France
{zahira.chouiref,allel.hadjali}@ensma.fr
[3] LT2C, Université Jean Monnet, 42000 Saint-Etienne, France
karim.benouaret@univ-st-etienne.fr
[4] USTHB, 16000 Algiers, Algeria
kaderbelkhir@hotmail.com

Abstract. Selecting services from those available according to user preferences plays an important role due to the exploding number of services. Current solutions for services selection focus on selecting services based either only on non functional features, on context preferences or profile preferences. This paper discusses an improvement of existing services selection approaches by considering both user context and profile. The ultimate aim is to derive maximum profit from available profile and context information of the user by inferring the most relevant preferences w.r.t his/her contextual profile. Linguistic/fuzzy preference modeling and fuzzy inference based approach are used to achieve efficiently a selection process. Some experiments are conducted to validate our approach.

Keywords: Web Services Selection, Profile, Context, Preferences, Fuzzy Logic Theory, Fuzzy Inference Rules, Contextual Profile Matching.

1 Introduction

Semantic Web services; *SWS* field plays an increasingly important role in enhancing the user interaction in the Web and enterprise search, as well as in providing a flexible solution to the problem of application integration. With the rapid worldwide deployment of offered services on Internet, SWS selection has been an active and fast growing research area. The services selection is a technique which uses functional features *(Input, Output, Precondition, Effect; IOPE)* and non functional features *(Quality of Service; QoS, etc.)* [12], to search Web services from large scale service repositories that fit best the user requirements. Development of methods which would increase research accuracy and reduce research time is one of the main challenges in SWS selection. Most of these approaches focus on satisfying the functional requirements. A service consumer copes with a difficult situation in having to make a choice from a mass of already discovered services satisfying the functional requirements. To discriminate

S. Casteleyn, G. Rossi, and M. Winckler (Eds.): ICWE 2014, LNCS 8541, pp. 440–449, 2014.
© Springer International Publishing Switzerland 2014

such discovered services, the focal point of current SWS selection is on the non-functional aspect of a service [12]. This can be done using *QoS parameters* [1], *context* [8], *preferences* [11], *profile* [5]. These approaches help to improve the service discovery, selection and composition and simplify the management process for non functional attributes of Web services. However, such approaches do not address the issue of: i) Taking into account all the information characterizing the service (offered and requested) often called contextual profile; *CP*, ii) the gradual nature of the parameters related to context/preferences in a human language, iii) deriving new relevant preferences on the basis of user *CP* information by means of fuzzy inference rules. The key concept of the approach is the user/service profile where fuzzy logic theory is used to describe information related to the profile in a faithfully way. The first objective of this paper is to propose a common profile model that can capture all information describing the user and the service. The second objective is to introduce linguistic terms to express preferences and fuzzy rules to model contextual preferences.

The remainder of this paper is structured as follows: Section 2 presents a brief background on fuzzy set theory and provides a survey on existing approaches of SWS selection. In Section 3, we set up a required and provided service model based on *CP*, then we provide an SWS selection framework based on our model. Section 4 describes the query processing in a real case study related to the field of restaurant business and the ranking mechanism as well. In Section 5, an experimental study is described to show the feasibility and effectiveness of our proposal. Finally, Section 6 concludes the paper.

2   Related Work

A key issue in service computing is selecting service providers with the best user desired quality. Recently, existing service selection approaches reviewed below are distinguished by the fact that they rely on QoS parameters [1], context information [8], user preferences [2][7][9][11], etc.

A matchmaking algorithm proposed by Adnan et al. [1] is based on tying QoS metrics of Web service with fuzzy words that are used in users request. The aim of the paper is to satisfy user's requirements and preferences regarding only QoS and not all preferences related to non-functional service parameters. In [8], authors proposed the non-functional properties that are related to local constraints which reflect the user preferences and context of the demanded service. However, in real-life systems, context information is naturally dynamic, uncertain, and incomplete, which represents an important issue when comparing the service description with user requirements. This approach, however, is based on context rather than the data of services and could not handle both exact and fuzzy requirements. They do not allow reasoning on context information to determine criteria weights automatically, also the model proposed by the authors does not consider the implicit user preferences. Web services selection based on preferences mainly consider single user's preferences. Benouaret et al [2] introduce a novel concept called collective skyline to deal with the problem of multiple users preferences to select skyline Web services. Chao et al. [4] proposed a framework, which leverages fuzzy logic to abstract and classify the underlying data of Web services as fuzzy terms and rules. The aim is to increase the efficiency of Web services discovery and allow the use of imprecise or vague terms at the level of the search query. Steffen et al [7] model Web service configurations and associated prices and preferences more compactly using utility function policies, and propose flexible and extensible framework for optimal service selection that combines declarative logic-based matching rules with optimization methods, such as linear programming. In [10], a framework of SWS discovery based on fuzzy logic and multi-phase matching is proposed in this work. The first level matchmaking is executed with service capability against the second level matching is executed with service fuzzy information. The authors do not take into account neither the vague information of the user profile (personal information, etc.) nor the vague information of the context.

The presented work links user profile, user context and user preferences and provides suitable selection method that uses inferred preferences in order to have powerful, yet scalable ranking process. These preferences are inferred from the user's contextual profile, this is done by directly applying an efficient inference method based on *an extended modus ponens*.

3   Our Model

3.1   Service Description Model

We describe a SWS (provided / required) by the following model, which not only supports service capability information (IO), but also supports *service profile*, *service context* and presents *service vague information* associated to *preferences*. For the sake of illustration, the following reference example is used.

Reference Example: Let U be a user wants to book a hotel in Australia. He sets his preferences and submits the following query Q: "return the hotels in Australia preferably {*near* **to his position**} with {*affordable* **price**} and {*at least three* **stars**} and having a **restaurant** with an {*asiatic* **cuisine**}, knowing that he has a **car** and he is accompanied by his **wife** and his **child**".

Definition 1 (Advertised Service Description Model/Required Service Description Model). An advertised Web service (A required Web service) is described by the following model respectively: $SA = \{CA, CPA, FZA\}/$ $SR = \{CR, CPR, FZR, \theta\}$, where:

- CA/CR is the advertised/required service capability information description, which contains $(NA, DA, FPA)/(NR, DR, FPR, \theta)$ where NA/NR is the name of the advertised/required service, DA/DR is the functional description of the advertised/required service and FPA/FPR is all functional parameters of the advertised/required service $(IOPE)$.

- $CPA = (CPA_1, \ldots CPA_n)$, $CPR = (CPR_1, \ldots CPR_n)$, is a set of non-functional parameters that make up the CP of SA/SR. The detail context can be explored in section 3.2.
- FZA/FZR is a set of linguistic terms which used to describe the vagueness that pervades information containing in CA/CR and CPA/CPR.
- $\theta(0<\theta<1)$ is a threshold such that if the satisfaction degree of a service w.r.t the query at hand is below this threshold, it is not retrieved.

Each I/O attribute of request a_i is characterized by a set of preferences values p_i, i.e. $(a_i,\ p_i)$. For each SA attribute, we can assign crisp constraints (e.g. MinStars) or fuzzy constraints (FZA) (e.g. CheaperPrice). The attributes are self explanatory, which indicates the pereference values which help the user to choose the service that suits its preferences. For each SR attribute, we can assign crisp preferences, or fuzzy preferences (FZR) (e.g. AffordablePrice).

3.2 Fuzzy Model to Contextual Profile

Fuzzy Contextual Profile Modeling. The information of CP can be: static such as *Profile {Personal data, etc}*, evolutionary such as *Preferences {Colour, Language, etc}* and temporary such as *Context {Devices, Localization, etc}*. These pieces of information must be captured to match demands to offers of services, in order to improve the relevance of answers during a selection process. For a given query X, we define its CP environment CPE_x as a finite set $\{(P_1, P_2, \ldots P_n)\}$ of multidimensional parameters, for instance, *{personal informations, context, preferences, etc}*. Each parameter Pi is modeled as a finite set $\{(C_1, \ldots C_m)\}$ of concepts, for instance *{demographic information, spatial context, display mode, etc}*. Each concept Ci is modeled as a finite set $\{(C'_1, \ldots C'_t)\}$ of sub concepts for instance *{gender value, family situation, country name , etc}*, and/or a finite set $\{(v_1, \ldots v_t)\}$ of attributes value, for instance *{full screen, postscript format, etc}*. Each concept (subconcept) is characterized by a set of *preference values*. The attribute value domain $dom(C_i)$ can be expressed by means of: numerical assessments, logical assessments and fuzzy linguistic assessments.

An instantiation of the CP, called *CP state*, writes:
$w = (C_1$ is $v_1 \wedge \ldots \wedge C_k$ is $v_k)$, $k \preceq m$, where each $C_i \in CPE_x$, $1 \preceq i \preceq k$ and $v_i \subseteq dom\ (C_i)$ (the symbol $\wedge$ denotes a conjunction).

Example 3. For instance, w may be (family situation is *married*, means of transport is *car*, accompanying people is *wife* and *child*) for the example above.

Definition 2 (Contextual Profile Preferences). *A contextual profile preference CPP is a fuzzy rule of the form:* **if** C_1 **is** $v_1 \wedge \ldots \wedge C_m$ **is** v_m **then** A_1 **is** $F_1 \wedge \ldots \wedge A_l$ **is** F_l*, where vi, $1 \leq i \leq m \leq n$, stands for a crisp or fuzzy value of the context or the profile parameter CP_i and F_j, $1 \leq j \leq l$ represents a fuzzy preference related to attribute A_j.*

The meaning of CPP is that in the CP state specified by the left part of the rule, the preference A_j is F_j is inferred. From the user profile, one can deduce the following preferences on the searched hotels:

Example 4. A user who has a *car*, generally prefers hotels with *parking*. This may be expressed as (CPP_1): **if** *means of transport is car* **then** *PreferencePark- ing is yes.*

User Preferences Modeling. Let us now discuss the notion of fuzzy pref- erences: for instance, *"affordable price"* and *"nearest city"* are primitive terms. A primitive term can be described thanks to fuzzy sets, allowing to obtain for a price and a given distance, the satisfaction levels defined on the interval [0, 1]. As for categorial attributes, the membership functions are modeled as follows: The membership function of *"CuisineStyle"* is modeled by: $\mu_{C\_cuisine}$={1/chinese, 0.9/japanese, 0.8/thaiwanese, 0.7/sushi, 0.7/indonesian, 0.5/vietnamese, 0.3/in- dian, 0.2/pakistani, 0.1/americain} for chinese cuisine and $\mu_{F\_cuisine}$={1/french, 0.9/belgian, 0.8/mediterranean, 0.7/italian, 0.7/dutch, 0.6/latin, 0.5/german, 0.4/british, 0.2/american} for french cuisine. The membership function of park- ing is $\mu_{parking} = \{1/yes, 0/no\}$.

4   Query Processing

4.1   Semantic Web Service Selection Framework

Let H_D be a Hotels database. The desired services should accept {Address} as inputs and return {HotelName, StarsNumber, Price, CuisineStyle} as output for the case of our *reference example*. This query Q is written as follows:
SELECT *name-Hotel* **FROM** H_D **WHERE** (Hotel.Price is *affordable* AND Hotel.Stars is *at least 3* AND Hotel.Dist is *near of city* AND Hotel.CuisineStyle is *asiatic*).

Fig. 1 gives an overview of the selection framework. The matching engine must match the list of services with the input, output specified by the user. The main steps of the framework are:

First Search Filter. The result returned by *step 1* and *2* of Fig. 1 is $\sum_Q$, the list of services that correspond to the desired *number of stars, price, asiatic cui- sine* and *distance to user*. Note that the preferences expressed on these attributes are mandatory. The system will computes the distance or similarity between all the concepts vector of the query and those of the Web services by means of suit- able measures of similarity. In presence of different types of attributes, for each attribute, an adequate similarity measure is used as the following functions: S_{p1}, S_{p2}, S_{p3} and S_{p4} represents respectively functions that compute the degrees of satisfaction of the *distance, Price, Number of Stars* and *Cuisine Style* of the SWS at hand w.r.t the fuzzy set modeling the user preference on attribute *dis- tance, Price, Stars* and *Cuisine*. For instance, $\mu_{near}(8) = 0.4$, $\mu_{affordable}(45) = 0.5$, $\mu_{stars}(4) = 1$ and $\mu_{F\_Cuisine}(italian) = 0.7$.

It is worth noticing the all the functions $(S_{p1}, S_{p2}, S_{p3}, S_{p4})$ provide degrees that belong to the same scale [0,1]. This property of commensurability allow aggregating them, in a convenient way, to obtain an overall score of an SWS.

Overall Matching *Score₁*. After calculating the different individual matching values of each attributes of a service, one way to obtain an overall matching score is to aggregate these individual matching values using a *T-norm* operator (such the *min operator*) as follows:

$Score_1 = \top(S_{p1}, S_{p2}, S_{p3}, S_{p4}) = \min(S_{p1}, S_{p2}, S_{p3}, S_{p4})$.

Now, if the *Score₁* of a service is higher than θ, then this service is added to $\sum_Q$ list of services that satisfying request's attributes.

Second Search Filter. In this phase, not only *I/O* parameters has to be considered but also contextual profile information such as accompanying people, age of child, etc. This means that service contextual profile should contribute to the development of an advanced search strategy. The steps (3, 3', 4, 5, 5', 6) of the Fig. 1 illustrate this strategy and are summarized in the following: (i) Infering a set of relevent preferences and their semantics from the fuzzy rules base B^{CPP1}, regarding the user *CP* state w, then augment the query by the inferred preferences. To achieve this, we make use of a *knowledge based model* described bellow. (ii) Calculation of the satisfaction of the result provided by the first filter w.r.t. to the inferred preferences.

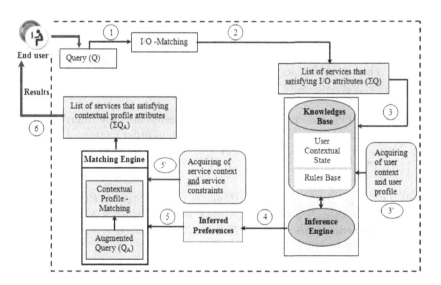

Fig. 1. Semantic Web Services Selection Framework

---

[1] Is a set of contextual profile preferences that can be built from users' experiences in the domain considered.

Knowledge Based Model. We rather propose here to specify contextual preferences by means of *fuzzy gradual rules*. The use of gradual rules for deriving preferences with respect to the user CP leads to refining the search results and returns the best ones to the user.

Inference Process: We assume available $B^{CPP} = \{CPP_1, ..., CPP_m\}$ a fuzzy rules base modeling a set of contextual profile preferences. Each rule is of the form **if** C_1 **is** $v_1 \wedge ... \wedge C_m$ **is** v_m **then** A_1 **is** $F_1 \wedge ... \wedge A_l$ **is** F_l (see Definition 2.). Now, given a user's CP state, i.e., $CP = (C_1$ **is** $v_1 \wedge ... \wedge C_k$ **is** $v_k)$, one can derive relevant preferences to the user by using a fuzzy inference schema, called the *Generalized Modus Ponens; GMP* [6]. In a simple case, from the rule: **if** C is V **then** A is F and the fact: C is V', where V, F and V' are gradual predicates modeled thanks to fuzzy sets, the *GMP* allows inferring the preference A *is* F' and the fuzzy semantics of F'. See [6] for more details.

Example 6. The preference *Parking is Yes* is inferred by applying the rule (**If** user has a *car* **then** preferenceParking is *Yes*) and the fact (user has a *car*). Note that *car* and *yes* are fuzzy sets represented by (car/1)(yes/1) respectively.

Example 7. In our booking hotels example, assume available the rules base B^{CPP} of Table 1. The preference *Hotel is Animated'* is inferred by applying the rule (**If** age is *young* **then** hotel is *animated*) and the fact (age is *about_26*) where *young* and *about_26* are fuzzy sets represented by (0,0,25,27) and (24, 26, 26, 28) and *animated* is expressed using a qualitative rating such as: *low, medium, high* and *very high* whose semantics are given by triangular membership function. The semantics of *Animated'* is calculated from the semantics of *Animated, young* and *about_26*, see [3].

Our model entirely leverages the knowledges base presented in Table 1 to derive new relevant preferences. A rule-based representation which includes different possible preferences for our reference example is proposed.

Augmented Query Process: Once the user preferences are inferred, we offer the system the possibility of augmenting the query in order to refine the selection process. Then, we have a final user query Q^A, the augmented query of Q writes: $Q^A = \{C_1 \wedge C_2 ... \wedge C_n \wedge P_1 \wedge P_2 \wedge ... P_m\}$.

Example. $Q^A = \{Price \wedge Stars \wedge Distance \wedge CuisineStyle \wedge \textbf{Parking} \wedge \textbf{Hair}$ *and Beauty Service* $\wedge$ **Sauna** $\wedge$ **Games and Activities** $\wedge$ **Kid Friendly** *Menu* $\wedge$ **Childrens Highchairs** $\wedge$ **Family** $\wedge$ **Animated'**$\}$, where the infered preferences appear in bold.

Similarity Computing: in this phase, the system computes the similarity between inferred user preferences and services' constraints by means of fuzzy semantics associated w.r.t these preferences. To accomplish this phase, the system will evaluate the similarity degree between all the inferred preferences from B^{CPP} and the services' constraints of $\sum_Q$ by using the following functions: Now,

Table 1. Knowledges Base

Knowledges Base		
Rule	**Rules Base**	**Facts Base**
R1	**If** user has car **then** preferenceParking is Yes	Car
R2	**If** accompanying people is wife **then** facilities is hair and beauty service	Married
R3	**If** accompanying people is wife **then** facilities is sauna	Wife
R4	**If** accompanying people is child **then** facilities is games and activities	Child
R5	**If** accompanying people is child **then** facilities is kid friendly menu	
R6	**If** accompanying people is child **then** facilities is childrens highchairs	
R7	**If** accompanying people is wife **then** theme is family	
R8	**If** age is young **then** hotel is animated	

assume that the initial query Q is only augmented by the inferred preferences: *Parking* and *Animated'*. To compute the satisfaction of each hotel $h \in \sum_Q$ w.r.t such preferences, we use:

S_{p5}: for each $h \in \sum_Q$, we have $\mu_{parking}(h) = 1$ if $h.parking = yes$, 0 otherwise, and S_{p6}: the fuzzy semantics of the predicate *Animated'* is computed by means of the combination/projection principle [3] (see example 7). Then, for each $h \in \sum_Q$, the degree of satisfaction is $\mu_{animated'}(h)$.

Overall Matching $Score_2$.
$Score_2 = \top(Score_1, S_{p5}, S_{p6}) = min(Score_1, S_{p5}, S_{p6})$.

Overall Score S. We use an aggregation function of $Score_1$ and $Score_2$ to compute the overall score S. Now, to give priority to the initial preferences w.r.t. to inferred preferences (IP) we make use of the following formula (where $\alpha \in]0, 1]$ is the priority of IP). Then S is given by:

$S = min(Score_1, max(Score_2, 1 - min(Score_1, \alpha)))$

Finally, the user can select the top-k answers or the answers whose score S is greater than a given threshold.

5 Experimental Evaluation

The main purpose of this evaluation is to compare the effectiveness of our proposed selection framework (referred to as IP for inference process) with the traditional frameworks that do not use the inference process (referred to as TR for traditional). We perform a case study, due to the limited availability of public services. We created a set of 100 synthetic restaurant service descriptions, and we involved different users to conduct our experiments. However, due to lake of space we only report results regarding 4 users.

Fig. 2 shows the precision of IP and TR at various ranks for 4 different users. Observe that IP has consistently better precision than TR since IP includes into the ranking process inferred preferences that are interesting for users. See also that, for user$_1$ and user$_2$ IP has an almost perfect precision, while the precision of TR is mediocre. Moreover, for user$_3$ and user$_4$ IP and TR have similar precision at rank 15 and rank 20. The reason is that the increase of the rank may increase the probability that similar services belong to the top-k list of both approaches.

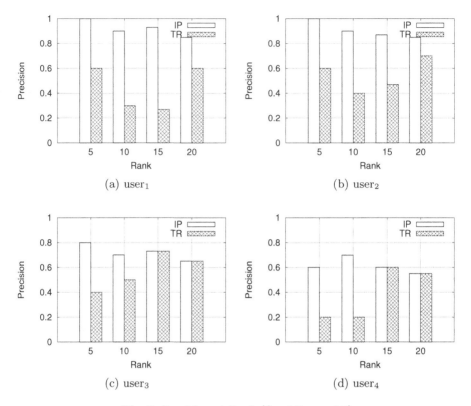

Fig. 2. Precision at Rank ($\theta = 0.5$, $\alpha = 0.5$)

6   Conclusion

This paper has proposed an SWS selection framework based on user (fuzzy) preferences. The goal of our work is to enhance accuracy of SWS search results by using multi matching level to compute similarity between advertised SWS and users request and taking into account the user preferences that may be inferred from users profile. We also showed the interest of using gradual rules for representing contextual preferences and deriving new preferences that are relevant to the user. The proposed framework makes the integration of user contextual profile and fuzzy inference rules techniques into the selection process. Some experiments are done to show the feasibility and the precision of our proposal.

References

1. Al Rabea, A.I., Al Fraihat, M.M.: A new matchmaking algorithm based on multi-level matching mechanism combined with fuzzy set. Journal of Software Engineering and Applications 5(3) (2012)
2. Benouaret, K., Benslimane, D., HadjAli, A.: Selecting skyline web services for multiple users preferences. In: ICWS, pp. 635–636 (2012)

3. Bouchon-Meunier, B., Dubois, D., Godo, L., Prade, H.: Fuzzy sets and possibility theory in approximate and plausible reasoning. In: Fuzzy Sets in Approximate Reasoning and Information Systems (1999)
4. Chao, K.M., Younas, M., Lo, C.C., Tan, T.H.: Fuzzy matchmaking for web services. In: 19th International Conference on Advanced Information Networking and Applications, AINA 2005, vol. 2, pp. 721–726. IEEE (2005)
5. Chouiref, Z., Belkhir, A., Hadjali, A.: Advanced profile similarity to enhance semantic web services matching. International Journal of Recent Contributions from Engineering, Science & IT (iJES) 1(1), 1–13 (2013)
6. Hadjali, A., Mokhtari, A., Pivert, O.: A fuzzy-rule-based approach to contextual preference queries. In: Computational Intelligence for Knowledge-Based Systems Design, pp. 532–541. Springer (2010)
7. Lamparter, S., Ankolekar, A., Studer, R., Grimm, S.: Preference-based selection of highly configurable web services. In: Proceedings of the 16th International Conference on World Wide Web, WWW 2007, pp. 1013–1022. ACM, New York (2007), http://doi.acm.org/10.1145/1242572.1242709
8. Reiff-Marganiec, S., Yu, H.Q.: An integrated approach for service selection using non-functional properties and composition context. In: Handbook of Research on Service-Oriented Systems and Non-Functional Properties: Future Directions, pp. 165–191 (2011)
9. Skoutas, D., Alrifai, M., Nejdl, W.: Re-ranking web service search results under diverse user preferences. In: VLDB, Workshop on Personalized Access, Profile Management, and Context Awareness in Databases, pp. 898–909 (2010)
10. Su, Z., Chen, H., Zhu, L., Zeng, Y.: Framework of semantic web service discovery based on fuzzy logic and multi-phase matching. Journal of Information and Computational Science 9, 203–214 (2012)
11. Wang, H., Shao, S., Zhou, X., Wan, C., Bouguettaya, A.: Web service selection with incomplete or inconsistent user preferences. In: Baresi, L., Chi, C.-H., Suzuki, J. (eds.) ICSOC-ServiceWave 2009. LNCS, vol. 5900, pp. 83–98. Springer, Heidelberg (2009)
12. Yu, H.Q., Reiff-Marganiec, S.: Non-functional property based service selection: A survey and classification of approaches (2008)

Semantic Mediation Techniques for Composite Web Applications

Carsten Radeck, Gregor Blichmann, Oliver Mroß, and Klaus Meißner

Technische Universität Dresden, Germany
{carsten.radeck,gregor.blichmann,oliver.mross,
klaus.meissner}@tu-dresden.de

Abstract. The mashup paradigm allows end users to build custom web applications by combining data-exchanging components in order to fulfill specific needs. Since such building blocks typically originate from different third party vendors, compatibility issues at component interface level are inevitable. This decreases re-usability and requires skilled users or automatisms to provide the necessary mediation to solve such issues. However, current mashup proposals are very limited in this regard.

We present techniques for data mediation that leverage semantically annotated interface descriptions to overcome a high degree of interface mismatch. We equipped the EDYRA mashup platform for end user development with automatic support for these techniques to increase the re-usability of components and to foster the long tail of user needs. In order to show the practicability of our approach, we describe the platform implementation and present benchmark results.

Keywords: mashup, semantics, data mediation, end user development.

1 Introduction

Recently, universal composition approaches like CRUISe [1] allow for platform-independent modeling of mashups and a uniform description of components spanning all application layers. Since components are typically developed by different third party providers, combining component interfaces in a meaningful way is far from trivial. Data exchanged between components may differ in various aspects leading to incompatibilities: providers use different vocabularies, schemata, units or abstraction levels when designing interface signatures. This complicates end user development (EUD) further, and connecting components in ways not anticipated becomes a cumbersome task. Semantic technologies are a potent solution to provide **data mediation**, i. e., automatic resolving of heterogeneous data structures. Although proposals in the semantic web service (SWS) domain exist, most mashup platforms neglect data mediation so far.

Within the EDYRA project, we adhere to universal composition and strive for enabling domain experts without programming skills to build and reuse composite web application (CWA). We utilize semantic annotations to refer to ontology concepts of component interfaces. Based on this, semantic data mediation techniques are applied by our platform and hidden from end users.

S. Casteleyn, G. Rossi, and M. Winckler (Eds.): ICWE 2014, LNCS 8541, pp. 450–459, 2014.

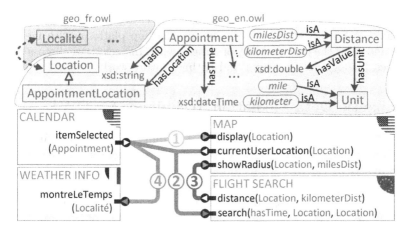

Fig. 1. Reference scenario: conference planning

To highlight arising challenges, let us consider the following use case which we use as a **reference scenario** throughout this paper.

Non-programmer Bob from the USA builds a CWA (Fig. 1) to organize a conference participation in Toulouse, France. He selects the conference, which he noted previously as an appointment, in the calendar. Then Bob wants to see the appointment location on a map. However, syntactically there is no possibility to combine the calendar and the map component's interface. While the calendar offers a data object of type `Appointment`, the map consumes a `Location`. Therefore, semantically it would be possible to take the appointment's `AppointmentLocation` and "cast" it to a general `Location` ①. Next, he wants to search for flights to Toulouse ②. In this case, he adds a flight search service and uses his current location as well as the appointment's location and time as search criteria. Besides the semantic problem of querying time and location of the appointment, it is necessary to put all three parameters together in compliance with the signature required by the flight service. Furthermore, Bob wants to visualize the distance between the airport and the conference location, which is calculated by the flight service ③. Because the flight service is located in Europe, the distance is provided in kilometers. To use the `showRadius` functionality of the map (from an American provider) this value has to be converted to miles. Finally, Bob wants to see the weather forecast for the appointment's time and location. He utilizes a French weather service which uses an ontology for annotating the interface that differs from the calendar's ④. But semantically the same concepts are described and the components can be coupled.

Current mashup proposals lack capabilities to implement this scenario. Thus, as our main contribution, we introduce data mediation techniques for CWAs and show their practicability within our EUD platform. These concepts help to combine components in more flexible ways than pure syntactic interfaces would allow, increasing re-usability and fostering the long tail of user needs.

The remaining paper is structured as follows. Sect. 2 presents mediation techniques for CWA. In Sect. 3, we describe our mashup platform with mediation support and show the practicability of our concepts. In Sect. 4, we discuss related work. Sect. 5 concludes the paper and outlines future work.

2   Semantic Data Mediation Techniques for CWA

Data mediation serves to resolve interface incompatibilities, of course within certain boundaries. In this section, incorporating results from the SWS domain like WSMO Mediators [6], we introduce a set of generic mediation techniques for CWA. We apply semantic data mediation and thereby leverage the domain knowledge defined in OWL-DL ontologies and annotated to component interfaces [1]. Essentially, annotations refer to classes, datatype and object properties or individuals in ontologies modeling the application/component domain. Since it is possible to model the same domain in various ways, we assume a certain modeling and annotation style.

In general, ontology classes can be annotated directly, e. g. `Location`, or via an OWL object property whose range the class is, if it is necessary to highlight a more specific meaning, e. g. `hasCenter`. Additionally, OWL datatype properties can be used, e. g. `hasLatitude`. However, there are circumstances where it is more appropriate to model individuals rather than subclasses. Units, currencies, and quantities, i. e., convertible concepts, may be mentioned as examples, or classes that refer to such convertibles on OWL property level (see `Distance` in Fig. 1). In this case, concrete individuals should be annotated, e. g. `milesDist`, a `Distance` individual where `hasUnit` points to `mile`.

In general, data transfer is realized through interface elements, which are *properties*, *operations* and *events* in our case. Interface elements can have one (e. g. properties) or more (e. g. operations and events) parameters, which have an identifier and a semantic type annotation. Channels combine one interface element of a source component SC with one of a target component TC. Therefore, an assignment $assi$ has to exist, that maps n parameters P_{out} of the SC bijectively to all n parameters P_{in} of the TC. A perfect match exists if $assi$ only includes mappings between parameters that are semantically identical (both refer to the identical concept). As an example, the mapping (`latitude`, `longitude`) $\rightarrow$ (`latitude`, `longitude`) is a perfect match.

Due to the usage of third-party components, a perfect match is unlikely. P_{out} and P_{in} can be *semantically compatible* if a **Semantic Connector** $SeCo$ can be defined, which is a set of channels and mediation techniques. It ensures that all parameters P_{in} of *one* interface element of a TC are connected and of the required semantic type. This may include that several channels from one or more SC can exist. In case of multiple inbound channels, the $SeCo$ takes care of an appropriate synchronization between them.

Upcast. As proposed earlier [1], the *upcast* mediation technique serves for solving different generalization levels of concepts annotated at parameters. In case of classes this means, that a more specific class is cast into a more generic one as long as they are in `subClassOf` relationship. Assume that a component outputs an `AppointmentLocation` but the target component requires a more generic `Location`. Then a upcast can be applied.

In principle, upcasts may additionally be used for OWL object properties if there is a `subPropertyOf` relation, by dealing with it as if the range would have been annotated. In case of datatype properties we presume that the underlying range stays the same, rendering upcasts simple. Upcasts are one way, i. e., only casts upwards the inheritance hierarchy are valid along the data flow.

Conversion. The *conversion* mediation technique has two main application areas. First, it resolves incompatibilities between two parameters annotated by convertible concepts, like *units, quantities* and *data types*. Please consider the example `kilometer` → `mile` from Fig. 1. The specific knowledge required can, e. g., be formalized in dedicated ontologies like the QUDT (http://qudt.org). In the latter, a base type is assigned to each unit for conversion purposes. For example, `meter` is the base type for `LengthUnit`, to which all other length units have a conversion factor to. Another use case are *scale adjustments* requiring more domain-specific transformations, e. g., mapping a five star rating to a ten point rating. Typically, these conversions require domain-specific knowledge and cannot be covered by generic algorithms or reasoners.

Second, conversion is used on class level in case of `equivalentClass` relationships, e. g. `Location` and `Localité` in Fig. 1 or `Location` and `Place`. For sake of simplicity, we pose a rather strict definition of equivalence: There have to exist `equivialentProperty` relations for all declared properties of those classes.

Semantic Split. A *semantic split* queries multiple OWL properties of an individual, which is represented as a parameter or property, and distributes them on one or more parameters of a target interface element. Fundamentally, only individuals can be "split" within the restrictions of their ontology class. This mediation technique is applicable if the following OWL constructs are annotated:

- Class: OWL data and object properties can be assigned to target parameters that reference a semantically compatible class, data or object property, e. g. connection ① in Fig. 1 (`Appointment.hasLocation` → `Location`).
- OWL object property: This case is handled as if the range class of the object property is annotated, e. g. `hasLocation` → {`hasLatitude`, `hasLongitude`}

Semantic Join. A *semantic join* creates an individual, representing a target parameter, by joining of multiple parameters of one source interface element. Assume, that a map publishes an event with parameters {`hasLatitude`, `hasLongitude`} and there is a point of interest finder that offers an operation consuming a parameter of type `Location`. Then, a semantic join is possible.

It has to be guaranteed, that the generated individual fulfills all constraints on OWL properties defined by the target class (and thus all superclasses).

Partial Substitution. Using a *partial substitution*, an OWL property of an individual represented as parameter can be updated with an individual or literal given by another parameter. With regard to Fig. 1, a partial substitution is possible between `Location` and `Event`, since the object of the OWL property `hasLocation` of an `Event` can be substituted by a `Location` individual. Partial substitutions are exclusively applicable for properties as target interface element. This is caused by the fact, that in our component model only properties expose and allow to change a partition of the component's data layer directly.

Using partial substitution increases the possibility to connect properties bidirectionally. As an example, consider that the calender has a property for its currently selected appointment and the map has a property for the currently selected location. Beside the possibility to semantically split the event to display its location, it is even feasible to connect the map to the calendar, to substitute the appointment's location with that of the map by dragging the map's marker.

Partial substitutions are not suitable for connections between events and operations for two reasons: First, it is not guaranteed that an event is correlated to an input interface element which can update the individual represented by the event. Second, events and operations in general hide the data layer.

Syntactic Join. A *syntactic join* is intended to synchronize m parameters published by n source interface elements of several *SC* and feeds them together in 1 target interface element. Several synchronization modes are supported.

- *tolerant*: The joiner waits until all sources have published at least once. Only the latest parameters are cached per source (the old value is overridden), and the cache is cleared after data transfer to the target.
- *repeating*: Here, the cache is not cleared, i. e., once all sources have sent data, each following publication causes the joiner to transfer data to the target.
- *queuing*: There is a queue per parameter. When all queues have at least one entry, the data is transferred to the target and the first element is removed.

3   Mediation-Equipped Platform for Mashups

3.1   Architecture

Our platform builds up on the CRUISe and EDYRA infrastructure we introduced earlier [1,2]. An overview is shown in Fig. 2. Universal composition is applied to create and execute presentation-oriented CWA, where components of the data, business logic and user interface (UI) layer share a generic component model. The latter characterizes components by means of several abstractions: parametrized events and operations, properties, and capabilities. The Semantic Mashup Component Description Language (SMCDL) serves as a declarative language implementing the component model. It features semantic annotations to clarify the meaning of component interfaces and capabilities [2]. Based on the component model, the declarative Mashup Composition Model (MCM) describes all aspects of a CWA, e. g. included components and event-based communication.

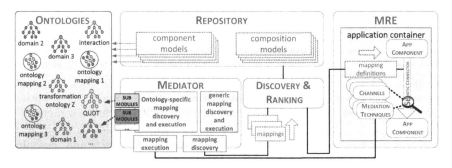

Fig. 2. Architectural overview of our mediation-equipped mashup platform

A *repository* is in charge of managing components and compositions. Furthermore it provides services for querying those artifacts.

Ontologies play an important role in our approach as they serve for annotating components and provide the schema knowledge most mediation techniques are based on. Besides domain-specific ontologies there can be (1) upper ontologies, e. g., for units, (2) ontologies defining how to transform concepts, and (3) mapping ontologies pointing out similarity relations of concepts in different knowledge representations. For the provision of mapping ontologies we assume that a state-of-the-art ontology alignment process takes place. Only confirmed mappings result in an ontology, linking concepts via predicates like `equivalentClass` and `sameAs`, e. g. `Location equivalentClass Localité`. A component can introduce new ontologies by using them for semantic annotation.

To enable recommendations, there are facilities for *discovery and ranking* of composition fragments, which represent composition knowledge, best matching user requirements and the current context. Within those modules, algorithms for recommendations utilize amongst others the semantic component annotations and ontology mappings. An essential task during discovery is the calculation of semantic connectors between components. Thereby, *mapping definitions* are derived utilizing the mediator or reused if already calculated.

A *mapping definition* specifies the mediation techniques required to align interface elements in a semantic connector. It is a data structure consisting of: an *ID*, P_{in} and P_{out} for more efficient matching and reuse, and one or more *mappings* including the composition and configuration mediation techniques, like the synchronization mode of a syntactic join.

A *mashup runtime environment (MRE)* interprets composition models in order to run mashups. With regard to data mediation, an MRE provides automatic support for the proposed mediation techniques, see Sect. 3.2 for details.

The *mediator* is responsible for mainly two tasks. First, it provides means for looking up mappings, which involve mediation techniques. To this end, the mediator takes two signatures, i. e., the URIs of concepts annotated at a source and a target interface element. In order to detect mappings e. g. between `milesDist` and `kilometerDist` in our scenario, algorithms have to inspect concepts on OWL property level, whereby we restrict the depth to one level. This task may result in multiple valid mappings, which have to be ranked further. Second, the mediator serves for the execution of mediation techniques defined in mapping definitions as requested by an MRE.

As illustrated in Fig. 2, there are generic and ontology-specific algorithms to accomplish these tasks. Generic algorithms utilize standard modeling constructs and reasoning rules of RDF/S and OWL. Mainly, relations like subsumption, property ranges and domains, sameAs and equivalentClass are leveraged.

The mediator can be extended with ontology-specific modules by implementing the required interface. They use dedicated modeling constructs and reasoning rules to derive and apply mappings and are consulted if generic algorithms cannot provide a mapping. Knowledge is encapsulated in modules for the algorithmic part and the corresponding ontologies for the terminology. Although in principle generically applicable, such modules are especially useful for conversions. There are modules per transformation ontology, responsible for interpreting and applying transformations on the payload delivered by events/properties. To identify suitable modules, each one provides a method to state if it supports the given signatures during discovery as well as execution of mappings.

Within the EDYRA project[1], we implemented a client-side thin-server MRE completely written in JavaScript. The mediator is distributed over the MRE and the SOAP-based *mediation service*. The latter is implemented in Java and uses the Jena framework for working with semantic models, including validating, reasoning and querying via SPARQL. The mapping discovery is located at the Java-based repository. We implemented the DataSemanticsMatcher as a generic algorithm that builds up on a QUDT-specific and a generic conversion module. The QUDT module supports annotated parameter types which refer to QUDT concepts. It queries the QUDT ontologies to check if both concepts are convertible, i. e., if they belong to the same unit domain. At execution time, the conversion multiplier is looked up and applied. The generic conversion module utilizes OWL constructs like equivalentClass and equivalentProperty to decide if two given classes are equal according to our definition in Sect. 2. To add new ontologies in our prototype, they have to be used for semantic annotation in SMCDL and registered manually at the repository and mediation service.

3.2   Runtime Support

Semantic connectors are implemented by associating the mapping definitions of semantic mediation techniques with a communication channel, and providing syntactic joins as built-in mediation components. The latter comply to the component model and can consequently be connected with other components. The MRE has templates for the SMCDL and the implementation of joiners and configures those as stated in the mapping definition to seamlessly instantiate and manage joiners like application components. With syntactic joins as components, single channels connect one *SC* with one *TC*.

---

[1] http://mmt.inf.tu-dresden.de/edyra (also links to our live demonstrator).

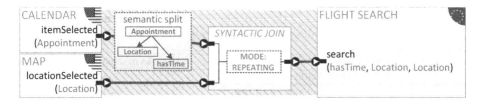

Fig. 3. Implementation of an exemplified semantic connector from the scenario

Fig. 3 shows an example (dashed area) of a semantic connector present in the mashup from our scenario. The semantic split is realized through a mapping definition which is directly attached to the channel between the calendar and a syntactic join. The latter aligns the split Location and Date as well as the Location from the map and transfers them combined to the flight search.

As described in our previous work [1], our components predominantly exchange data serialized in XML. There is a predefined *grounding* per ontology concept specifying the XML schema for individuals and literals. In case component developers utilize their own schemata or even other formats, like JSON, they have to ensure that data are transformed to the grounding, either as part of the component implementation or by transformation instructions in the SMCDL. In order to apply semantic mediation, data has to be available in RDF. Thus, we assume lifting and lowering transformations per predefined grounding.

When a SC publishes data according to the grounding, the channel delegates the execution of mediation techniques to the mediator by handing over the transported payload, the mapping definition, and in case of partial substitutions the target property's value. Per parameter, the mediator applies the lifting to get triples, which are then put in separate semantic models together with the terminological knowledge of ontologies. Using Jena in our prototype, validation and reasoning takes place automatically. Then the mediation techniques are executed as configured by the mapping definition. We utilize Jena's OWL API, e. g., to add OWL property values to individuals, and invoke conversion modules if required. Next, lowering takes place. Thereby, per target parameter, the model is queried with SPARQL, the results are serialized in the XML results format on which an XSLT transformation is applied. Finally, the mediated payload is forwarded to the TC. For simple datatypes as grounding, e. g. as input for conversions or result of a split, there is no dedicated lifting and lowering required. We programmatically map primitive data types to Literals and vice versa.

Syntactic joiners are connected to **n** inbound and one outbound channels. To achieve that, there are **n** operations in the joiner's generated SMCDL, whose parameter count and names correspond to those of the source interface element. Occurring event on an inbound channel are handled by the connected operation. The joiner extracts the parameter payload, handles it (e. g., adds it to a queue) and decides whether to fire its event according to its synchronization mode.

We conducted a benchmark where we measured the average response time of the mediation service's operations implementing the mediation techniques.

Table 1. Average response time of the mediation service in 100 runs per technique on a local server. The setup includes an Intel Core i7 2.8 GHz and 32 GB RAM.

Mediation technique	Avg. response time
Upcast (`AppointmentLocation` → `Location`)	≈15 ms
Conversion (`kilometers` → `miles`)	≈14 ms
Conversion (two equivalent classes)	≈40 ms
Semantic split (`Location` → {`hasLat`, `hasLng`})	≈23 ms
Semantic join (reversed split)	≈13 ms
Partial substitution (`Location` → `Event.hasLocation`)	≈23 ms

The results show a decent performance considering that lifting and lowering takes place in most cases. Network overhead of SOAP over HTTP may result in noticeable delays on slow connections, so that user experience may suffer in comparison to perfect match channels. That can be lowered by using WebSockets, and by integrating the mapping execution in a client-server MRE. Other crucial factors are ontology size and complexity, especially when a reasoner is attached.

4   Related Work

Proposals for semantically annotated services mostly use lifting and lowering to transfer data to the semantic layer [4,5]. While we use this, because our components do not exchange semantic data, our concept is not limited to upcasts.

Research shows that semantic web service descriptions are suited for mediation. There are different mediators in the conceptual framework of WSMO [6], which are provided as web services too and completely operate at a semantic layer. There are similar techniques involved, but due to the lack of a UI, execution performance is not that critical as for CWA. In addition, only a 1:1 communication is supported, while we can handle n:m semantic connectors.

For mashups, automated data mediation has been neglected so far [7]. Simple constructs exist that support, e. g., filtering, assignment and sorting of data, for example in Yahoo Pipes. But those rather belong to application logic than generic mediation techniques, and data semantics is not taken into consideration. Few approaches use semantically annotated component interfaces for matching at all, like [8,1]. Our previous work [1] can solve syntactical issues like different parameter naming with the help of wrappers. As a semantic issue only upcasts can be handled. Therefore, we largely extend this work.

5   Conclusion and Future Work

Since components of a CWA typically originate from different vendors, connecting them in meaningful way is challenging. Incompatibilities of signatures and exchanged data may, for instance, result from varying vocabularies or units. This complicates EUD, especially for non-programmers. In addition, connecting components in unforeseen ways becomes far from trivial. Thus, platforms for mashup EUD should feature means to automatically provide the required glue code.

We describe a set of data mediation techniques that use semantic component annotations to resolve interface mismatches. Those techniques reflect knowledge captured by ontologies, can be combined and are essential for establishing semantic connectors between components. The higher flexibility for combining components fosters re-usability and unforeseen coupling. This way, more niche requirements can be meet without the need for new components. Utilizing our implemented core platform, we show the practicability. However, due to the increased possibility of combinations, it is challenging to identify useful connectors. In addition, our approach depends on semantic annotations of OWL concepts, and we limit the depth to 1 when analyzing concepts during mapping discovery. Tough this may restrict the solution space, it lowers the algorithmic complexity.

Currently, we utilize mediation techniques for deriving and visualizing recommendations, in the CapView [2] and for synchronization of mediable components during collaboration. Future research will focus on the mediation of collections.

Acknowledgments. Carsten Radeck and Gregor Blichmann are funded by the German Federal Ministy of Economic Affairs and Energy (ref. no. 01MU13001D).

References

1. Pietschmann, S., Radeck, C., Meißner, K.: Semantics-based discovery, selection and mediation for presentation-oriented mashups. In: 5th Intl. Workshop on Web APIs and Service Mashups (Mashups), pp. 1–8. ACM (September 2011)
2. Radeck, C., Blichmann, G., Meißner, K.: Capview – functionality-aware visual mashup development for non-programmers. In: Daniel, F., Dolog, P., Li, Q. (eds.) ICWE 2013. LNCS, vol. 7977, pp. 140–155. Springer, Heidelberg (2013)
3. Shvaiko, P., Euzenat, J.: Ontology Matching: State of the Art and Future Challenges. IEEE Transactions on Knowledge and Data Engineering 25(1), 158–176 (2013)
4. Szomszor, M., Payne, T., Moreau, L.: Automated syntactic medation for web service integration. In: Proc. of the Intl. Conf. on Web Services, pp. 127–136 (September 2006)
5. Nagarajan, M., Verma, K., Sheth, A.P., Miller, J.A.: Ontology driven data mediation in web services. Intl. Journal of Web Services Research 4(4), 104–126 (2007)
6. Mocan, A., Cimpian, E., Stollberg, M., Scharffe, F., Scicluna, J.: WSMO mediators (December 2005), http://www.wsmo.org/TR/d29/
7. Di Lorenzo, G., Hacid, H., Paik, H.Y., Benatallah, B.: Data integration in mashups. SIGMOD Rec. 38(1), 59–66 (2009), http://doi.acm.org/10.1145/1558334.1558343
8. Bianchini, D., De Antonellis, V., Melchiori, M.: A recommendation system for semantic mashup design. In: Workshop on Database and Expert Systems Applications (DEXA), pp. 159–163 (September 2010)

Standard-Based Integration of W3C and GeoSpatial Services: Quality Challenges

Michela Bertolotto[1], Pasquale Di Giovanni[1,2], Monica Sebillo[2], and Giuliana Vitiello[2]

[1] School of Computer Science and Informatics
Belfield, Dublin 4, Ireland
michela.bertolotto@ucd.ie
[2] Department of Management and Information Technology (DISTRA)
Via Giovanni Paolo II, 132-84084 Fisciano, SA, Italy
{pdigiovanni,msebillo,gvitiello}@unisa.it

Abstract. In recent years, Service Oriented Computing (SOC) has become one of the leading approaches for the design and implementation of distributed solutions. The key concepts are the notion of service and the possibility to seamlessly combine several modules to offer more sophisticated functionality. Such features were soon recognized by both W3C and OGC as relevant for their purposes, although their standards are incompatible and the seamless communication and exchange of information between these types of services are not directly achievable. The current most accepted solution to address this matter is represented by the development of a wrapper that manages technical issues that arise during the translation of requests and responses between them. However, the design of such a software module presents challenges in terms of infrastructure design and Quality of Service. In this paper we describe some issues to be faced when developing a service wrapper aimed at integrating existing geospatial services into a W3C service-based infrastructure.

1   Introduction

The Service Oriented Computing (SOC) paradigm has emerged as one of the leading approaches for designing and implementing distributed applications. The key idea behind this approach is the concept of service, an autonomous software module that, combined with other services, can be used to create complex solutions.

A service exposes its functionality through its public interface whose methods can be invoked by any software system without the need, for the service client, to know any detail about the service internal structure and business logic. However, since services and their clients can be developed by different entities, the first issues to address involve describing the public interface and providing a framework for data exchange in a wide accepted way and in a technology neutral manner.

In this context, the World Wide Web Consortium (W3C) has defined a series of universally accepted standards based on the use of the Extensible Markup Language (XML) in order to guarantee their independence from a specific platform or technology. The two most important W3C standards are the Web Services Description Language (WSDL) for the description of the service interface and the SOAP

S. Casteleyn, G. Rossi, and M. Winckler (Eds.): ICWE 2014, LNCS 8541, pp. 460–469, 2014.
© Springer International Publishing Switzerland 2014

protocol for the exchange of messages [3], [13]. The flexibility and pervasiveness guaranteed by the W3C infrastructure has promoted during time the development of a growing number of service based solutions in many diverse fields [4].

The service-based approach has also become one of the preferred ways to discover, access, and manage geographic information. The ability to offer traditional Geographic Information Systems (GIS) capabilities in a distributed manner has been recognized by the geographic community as a valuable opportunity to provide new ways to use geospatial information and increase its distribution. However, despite the wide acceptance of the W3C proposals as a means to promote interoperability and platform independence, the GIS community has developed, over time, its own set of standards for the fulfilment of geospatial data oriented services. In particular, the proposals of the Open Geospatial Consortium (OGC), which represents the reference organization for "the development of international standards for geospatial interoperability" (http://www.opengeospatial.org/), have become the *de facto* standard for developing distributed geographic applications.

Unfortunately, although both W3C and OGC standards are based on XML for data exchange and HTTP as the transport protocol, some design choices make them totally incompatible. Nevertheless, a better integration between these two worlds could be of interest for both communities. In fact, the former could access and process the wide amount of geospatial data currently available only by invoking OGC services, while the latter could benefit both from additional standards, such as those for access management and security, and from the huge amount of supporting infrastructures for W3C services. Such awareness is stimulating a radical revision of such standards to remove their intrinsic incompatibilities.

The currently most accepted solution to this aim is represented by the development of a software wrapper, usually a service itself, that "translates" the requests and responses messages from the W3C services format to a format suitable for the OGC services and vice-versa, while keeping the structure of the original services unchanged [2], [5], [15]. However, the concrete development of a wrapper does not represent the only issue to solve since, independently of the actual set of adopted standards, every service-oriented solution cannot disregard fundamental nonfunctional requirements, such as the quality of the provided information, its security, the service response time. In fact, taking into account essential Quality of Service (QoS) aspects is of utmost importance to guarantee a satisfactory computation and make the SOC paradigm a feasible option for the development of complex distributed solutions.

In this paper we discuss our research in the area of geospatial service-oriented architectures. We analyze the challenges involved in the integration and reuse of heterogeneous services with focus on QoS aspects, and propose recommendations for the development of a viable solution that takes such aspects into account.

The remainder of this paper is organized as follows. In Section 2 we provide an overview of the two main W3C standards and compare them with the OGC proposals. In Section 3 we briefly describe the most important QoS attributes that directly impact on the development of service-based solutions, and discuss the QoS issues that affect the development of distributed solutions for geospatial data. In Section 4, we describe the main challenges that arise during the design of a wrapper addressed to the W3C-OGC services dialogue. Some conclusions are drawn in Section 5.

2   W3C and OGC Standards for Service Based Development

The functionality of a service is exposed through its public interface and the communication between a service and its clients is based on various messages exchange patterns. However, the actual definition of the public interface and the structure of messages are strongly related to the particular set of standards adopted. In this section we provide an overview of the main characteristics of the two major W3C standards, namely WSDL and the SOAP protocol, and compare them with the three principal OGC proposals, namely the Web Map Service (WMS), the Web Feature Service (WFS) and the Web Coverage Service (WCS) standards.

WSDL [3] is an XML based language for describing W3C services. A WSDL document separates the abstract aspects of a service description from more concrete ones, such as the binding to a certain network protocol. The typical structure of a WSDL document is made up of seven elements, namely Types, Message, Operation, Port Type, Binding, Port and Service. In particular, the former four are meant to statically define the public interface of a Web service, while the latter three are used to bind the interface to a concrete network protocol. A direct drawback of such design choices is that, in a W3C-oriented environment, the structure of the message payload has to be completely specified at design time [22].

SOAP is "a lightweight protocol for exchange of information in a decentralized, distributed environment" [13]. Three basic components characterize a typical SOAP-based message: an Envelope, a Header and a Body. The Envelope can be seen as the container of the message itself. The optional Header field can be used to carry additional information useful to guarantee some properties, such as security and reliability of exchanged messages. The Body element encompasses the real payload of the exchanged message.

When compared to the W3C choices, the design philosophy of the OGC standards is quite different, and the main dissimilarities between them concern the different approach for the public interface design, and the binding type and the binding time of operations. Two further clear differences are the basic role of the Geography Markup Language (GML) [21] and the fact that each type of OGC service is based on a separate standard explicitly designed to deal with a specific kind of data. Moreover, they also specify the functionality offered by the service interface along with the possibly needed additional data structures. A direct consequence of this choice is that, with OGC services, the actual structure of the message payload can be known only at run time, differently from what occurs in W3C environments.

As for the service specification, the most widespread and commonly used are WMS, WFS and WCS. A WMS allows clients to request georeferenced map images from one or more geospatial databases. WFS allows for accessing and manipulating geographic features. Finally, the WCS defines an interface for the exchange of geospatial information representing phenomena that can vary in space and time, known as coverages. The only functionality that is common to these three types of services is GetCapabilities, it allows a geospatial service to expose its capabilities to clients.

3   QoS Issues in Geospatial Web Services

The increasing adoption of the OGC proposals as a concrete means to access and make use of geospatial data in a distributed and vendor independent manner, has shifted the attention from data and information supply to information quality and implementation of services themselves. Thus, also for OGC services, the assessment of the most common QoS attributes is becoming fundamental to distinguish between reliable and non-reliable services. Generally speaking, QoS within the SOC paradigm represents an important and widely discussed topic, due to its basic role in various key aspects. The scientific and industrial communities have defined several main QoS categories and various attributes for each of them that contribute to the fulfillment of the desired QoS property. Moreover, a W3C working group [19] has identified a set of basic QoS requirements that have to be taken into account during the development of a Web service, namely Performance, Reliability, Scalability, Capacity, Robustness, Exception Handling, Accuracy, Integrity, Accessibility, Availability, Interoperability and Security. A complete analysis of these can be found in [20], while approaches to express and describe QoS characteristics and metrics can be found in [18].

As for the quality of geospatial Web services, basic assumptions about QoS attributes still hold. However, their evaluation must be performed according to both the specific characteristics of geospatial data and the way it is handled by OGC-compliant solutions. In addition, another relevant aspect that must be taken into account is represented by the technical differences among the various software implementations of the OGC standards and the related supporting infrastructure.

From a high level perspective, the process of obtaining knowledge from geospatial information can be viewed as a three step process, namely querying the data, assembling the retrieved subset and finally performing the effective computation [25]. The first issue along this sequence of operations is represented by the specific characteristics of geospatial data that usually is voluminous and heterogeneous, distributed among different data silos and can suffer from access restrictions due to institutional policies [11]. Such characteristics have, of course, a significant impact on the actual quality of the final information offered by geospatial services to third party users. In fact, as clearly discussed in [12], due to the common practice of combining data from multiple sources, geospatial datasets are inclined to contain errors since the various providers can make, for example, different assumptions about data structure. As defined in [12], the most important quality components for geospatial data are lineage, completeness, logical consistency, attribute accuracy and positional accuracy.

The aforementioned quality attributes are useful to assess also the quality of metadata that, due to its importance in this context, must be accurately evaluated. Indeed, a poor quality metadata determines the lack of information quality and can lead final users to formulate wrong assumptions about the received dataset.

The ISO19113 standard, instead, identifies five criteria for geospatial data quality, namely positional accuracy, temporal accuracy, logical accuracy, thematic accuracy, and completeness [17]. Finally, a recent factor that influences the quality of geospatial data is represented by the creation of user-generated geospatial content and Web 2.0. How to efficiently assess the quality of such a type of data is still an open research question. An example of filtering and composition of Web 2.0 sources can be found in [1].

As for the quality factors that mainly impact on the actual development of geospatial services dealing with significant amount of data, a first important discussion can be found in [10]. This document shows how, from a general point of view, the quality attributes proposed by W3C and mentioned in the previous section can be applied to geospatial services, except for the scalability requirement. In [9] some more specific directives and obligations for implemented services are mentioned. In particular, the three fundamental QoS criteria to respect are:

- performance: the time for sending the initial response to a discovery service request shall be maximum 3 seconds in normal situations. Normal situations represent out of   peak load periods, i.e., 90 % of the time;

- capacity: the minimum number of simultaneous requests served by a discovery service according to the performance quality of service shall be 30 per second;

- availability: the probability of a network service to be available shall be 99% of the time.

In [24] the common issues impacting on the overall QoS and concerning current proposals and implementations of OGC standards are discussed. The authors divide those issues into three levels, namely standard definition, software implementation of the standard, and software application. Among the various problems, the following are functional to the goal of the present discussion: the lack of a standardized authorization/authentication mechanism, the misuse of the standardized HTTP error codes, the version proliferation, the discrimination between mandatory and optional features, and the high level of autonomy offered by the various standard specifications.

A concrete example of QoS issues in a real software solution can be found in [26]. In the development of their prototype for real time geospatial data sharing over the Web, authors notice how the adoption of OGC standards is useful to solve problems at the syntactic level, while several issues may arise at the semantic level. System reliability represents the second important problem that is particularly accentuated when OGC services are provided by different entities. Security is another major concern. Finally, performance bottlenecks due to the transfer of redundant XML data over the network and the high cost of the parsing XML messages have a serious impact on the effective use of the proposed solution.

In [11] several OGC-compliant services implementations are tested. The results related to relevant performance parameters, show how, due to the GML verbose nature, a consistent number of bottlenecks may arise when there is the need to transfer large amount of geospatial data. Moreover, different software solutions vary in the way OGC specifications are implemented. Two direct consequences may arise from such dissimilarities, namely the reduced quality that can be perceived by final users and critical interoperability problems.

4   A Wrapper-Based Solution

In order to face effects deriving from technical and semantic differences between W3C and OGC services the currently most accepted solution is represented by a software wrapper that manages most of the technical topics that arise during the translation of requests and responses [15]. However, such a translation cannot be automated due to several issues that need to be carefully taken into account during the wrapper design to make this solution a feasible option. First of all, a wrapper is usually a service itself, then it requires a typical supporting infrastructure of service-based solutions, while its design might be influenced by the specific needs of the application under development. Indeed, two symmetrical types of wrapper can be developed, either adapting the interface of an OGC service to the technical requirements of a W3C-based infrastructure or vice-versa. Existing W3C services providing geospatial information that could be useful in an OGC-based Spatial Data Infrastructure (SDI) constitute an example of the latter case. The second cause of difficulties is represented by the number of services whose functionality has to be exposed by the intended wrapper. In fact, although the simplest solution concerns a one-to-one mapping, i.e., a wrapper adapts the interface and functionality of a single W3C / OGC service, it is also possible for it to gather functionality of different services. A typical example is constituted by a W3C service that offers, in a single WSDL document, the methods to access the data layers of either two WFSs or a WFS and a WMS. Finally, a further issue concerns the need to properly structure the WSDL document in order to distinguish among the various OGC services since the public interface and signatures of the implemented methods are rigorously standardized by the Consortium.

In the following, we discuss some challenges about the design of a one-to-one wrapper by describing a concrete example of an OGC to W3C mapping, Moreover, some basic QoS parameters are investigated that are affected when offering geospatial data coming from other OGC services and exploited through W3C standards.

As a concrete example where an OGC-to-W3C wrapper can actually promote and support a better information exchange between different entities, we illustrate its usage in the context of a research activity aimed at helping Sri Lankan farmers improve their productivity by providing them with customized and up-to-date information, such as the current selling prices of a product. Such an activity constitutes a pilot study for the Social Life Networks for the Middle of the Pyramid (SLN4MoP) project, an international collaborative research program that aims at providing real-time information to meet the daily needs of people living in developing countries [23].

The proposed system is based on a client-server architecture, although some technological constraints and the elicited needs of the involved stakeholders deeply influenced its overall design. As for the client tier, a common trait in many developing countries is the wide spread of mobile devices compared to the diffusion of traditional PCs. Such a factor led us to propose a mobile solution for the actual application with which the farmers interact. Detailed information about the implications and design challenges of our choice can be found in [6,7].

As for the back-end, the blueprint of the architecture has been organized by exploiting the principles of the SOC paradigm, which better comply with the *in progress* nature of SLN4MoP project, that is, providing its functionality as set of interacting services helped us to easily satisfy several fundamental design goals and QoS parameters. In particular, we needed both a reliable and flexible infrastructure, where new software modules can be added and can communicate with the existing ones without affecting the original design and behavior, and a reduced complexity during the access to heterogeneous and distributed data sources hiding, at the same time, the underlying different storage formats.

As for the QoS aspects, since the business processes are now decomposed into a series of interacting services, the availability, interoperability and performance parameters are of utmost importance for an efficient usage of this system. However, while availability strongly depends on the failure ratio of the underlying supporting components, interoperability and performance deserve further considerations.

To support our discussion, we consider the following real scenario. A governmental officer needs to visualize on a map the position of all local markets of a given district along with the selling prices of certain crops. The required operation corresponds to a combination of two atomic functions, namely the provision of various data units for the composition of a map, and a list of scalar values. The former is a typical functionality offered by an OGC service, the latter can be provided by a W3C service. In order to derive the expected result, it is necessary to invoke an advanced service capable to split and direct the atomic requests towards components in charge of performing them, and then combine responses deriving from them as a unique output. This capability represents a fundamental feature of the SOC paradigm: the services composition, namely the ability to compose services to obtain complex results.

One of the most common types of composition is service orchestration where the messages exchanged among services and the execution order of their interactions, is coordinated by a central controller. In order to effectively make the orchestration possible, all the involved services need to share the same Interface Description Language (IDL) and the same framework for the messages exchange. Then, the interoperability in the context of services orchestration represents a key requirement, but, as shown in our scenario, protocols based on different rules for the definition of the public interface and the message exchange system, make services orchestration not directly achievable.

The solution we have proposed is based on a wrapper addressed to a syntactic translation from OGC to W3C, which exploits existing orchestration middleware and the well-established services orchestration in W3C environments. In particular, the task performed by the proposed wrapper consists in the translation of SOAP-based messages into OGC-compliant requests and vice-versa. Such a task can be partitioned into four main steps:

1. the wrapper receives, from a W3C service a SOAP message containing a request for a specific geospatial dataset;
2. the wrapper translates the SOAP-based request into a format suitable for the underlying OGC service, and sends the query;
3. the OGC service returns the desired information;
4. the wrapper translates the received response into a SOAP-compliant format and sends it back to the requesting client.

However, a wrapper-based solution presents an important drawback, namely a serious impact on the overall composition performance. Such an aspect, in the context of our project, cannot be underestimated and requires further investigation.

Besides traditional aspects (such as, the quality of the underlying network that contributes to the achievement of a satisfactory performance level), a relevant factor for performances is represented by the specific characteristics of geospatial information (described in Section 3). As compared to the size of traditional SOAP messages, the size of geospatial data is usually several orders of magnitude larger. Since in the wrapper-based solution such data has to be packaged in the Body element of a SOAP message, it is clear that encoding, decoding and transmission of SOAP messages represent new significant issues. Such a problem has a direct impact on measurable values (like response time or throughput) directly related to the Quality of Experience of final users. Moreover, it might also influence the behavior of other aspects of the entire Service Oriented Architecture (SOA) to which the wrapper belongs, such as the transaction management protocols and the above described services orchestration. In particular, in a traditional orchestration the execution of an operation may depend on the output of a previous computation, and data complexity. The above described scenario deals with high volumes of data and long running operations, then the considerable amount of waiting time needed to process or simply transfer SOAP-encoded geographic data may cause a throughput reduction. In the worst case, a time-out error may occur that causes the entire workflow blocking. An asynchronous strategy based on appropriate SOAP message patterns (e.g., Fire and Forget) [8] represents a possible solution for all wrapper-based and time-consuming tasks.

Another aspect that adversely affects the performance and effectiveness of a wrapper is related to the supplementary delay caused by the need to query remote OGC sources. Information caching represents a feasible solution to reduce this inconvenience and improve performance and overall scalability. Some considerations about the design choices of OGC services support this option. In particular, most of OGC services are basically read-only services whose queries "access groups of features rather than individual features" [16]. Of course, traditional cache invalidation mechanisms (on demand, time limited, etc.) can be used to force the wrapper to invoke the original data source and refresh the local cache. Examples of cacheable items are the Capabilities document returned by the invocation of the GetCapabilities function, and the GML Schemas returned by the DescribeFeatureType function of a WFS.

A further service property to be taken into account when designing a wrapper, concerns its level of flexibility and reusability (a desirable property in the SOC paradigm). Such parameters are related to the granularity of a service, namely its size. In [14], the authors classify service granularity into three different categories: functionality granularity, data granularity and business value granularity. In a wrapper-based solution for service orchestration, data granularity represents the unique parameter that can be investigated during the design. Some optimizations can be done, however such parameters depend on the implementation and specific choices made for the original OGC service. A detailed discussion about this topic can be found in [16].

5   Conclusions

The goal of the research we are conducting is to define an infrastructure for the provision of heterogeneous Web services within a geographic information system. In particular, the focus of our current efforts is on the orchestration of traditional and geospatial services. The solution we have proposed is based on a wrapper that integrates W3C and OGC services in a seamlessly manner. In this paper we have discussed the QoS parameters that should be properly considered in this context. We have emphasized that, besides parameters that the literature suggests to take into account when dealing with these two standards separately, it is necessary to include some criteria that exclusively derive from the growing complexity of the integrated solution, such as supplementary delay and throughput reduction. Indeed, a wrapper-based solution implies a notable impact on the service performances and effectiveness, and then it is essential to handle those QoS parameters during the design phase in order to perform the best choices, independently from the technology used in the subsequent implementation step. In the future, we plan to complete the infrastructure proposed for SLN4MoP, and stress it by testing its performances against a large amount of data.

References

1. Barbagallo, D., Cappiello, C., Francalanci, C., Matera, M., Picozzi, M.: Informing observers: quality-driven filtering and composition of web 2.0 sources. In: EDBT/ICDT Workshops, pp. 1–8 (2012)
2. Bertolotto, M., Di Giovanni, P., Sebillo, M., Tortora, G., Vitiello, G.: The Information Technology in Support of Everyday Activities: Challenges and Opportunities of the Service Oriented Computing. Mondo Digitale (2014)
3. Christensen, E., Curbera, F., Meredith, G., Weerawarana, S.: Web Services Description Language (WSDL) 1.1. World Wide Web Consortium (2001),
   http://www.w3.org/TR/wsdl
4. Costagliola, G., Casella, G., Ferrucci, F., Polese, G., Scanniello, G.: A SCORM Thin Client Architecture for e-learning Systems Based on Web Services. Int. J. of Distance Education Technologies. 5(1), 19–36 (2007)
5. Di Giovanni, P., Bertolotto, M., Vitiello, G., Sebillo, M.: Web Services Composition and Geographic Information. In: Pourabbas, E. (ed.) Geographical Information Systems: Trends and Technologies, pp. 104–141. CRC Press (to appear, 2014)
6. Di Giovanni, P., Romano, M., Sebillo, M., Tortora, G., Vitiello, G., Ginige, T., De Silva, L., Goonethilaka, J., Wikramanayake, G., Ginige, A.: User Centered Scenario Based Approach for Developing Mobile Interfaces for Social Life Networks. In: First International Workshop on Usability and Accessibility Focused Requirements Engineering (UsARE 2012), pp. 18–24. IEEE (2012)
7. Di Giovanni, P., Romano, M., Sebillo, M., Tortora, G., Vitiello, G., Ginige, T., De Silva, L., Goonethilaka, J., Wikramanayake, G., Ginige, A.: Building Social Life Networks Through Mobile Interfaces: The Case Study of Sri Lanka Farmers. In: Spagnoletti, P. (ed.) Organizational Change and Information Systems. LNISO, vol. 2, pp. 399–408. Springer, Heidelberg (2013)
8. Erl, T.: Service-oriented architecture: concepts, technology, and design. Prentice Hall, PTR (2005)

9. European Commission: Commission Regulation 1088/2010 amending Regulation (EC) No 976/2009 as regards download services and transformation services (2010), `http:seur-lex.europa.eu/LexUriServ/LexUriServ.do?uri=CELEX:02009R0976-20101228:EN:NOT`
10. European Commission: INSPIRE Network Services Performance Guidelines (2007)
11. Giuliani, G., Dubois, A., Lacroix, P.: Testing OGC Web Feature and Coverage Service performance: Towards efficient delivery of geospatial data. J. of Spatial Information Science. 7, 1–23 (2013)
12. Goodchild, M.F., Clarke, K.C.: Data Quality in Massive Data Sets. In: Abello, J., Pardalos, P., Resende, M.G.C. (eds.) Handbook of Massive Data Sets, pp. 643–659. Kluwer Academic Publishers, The Netherlands (2002)
13. Gudgin, M., Hadley, M., Moreau, J.J., Nielsen, H.F.: SOAP Version 1.2. World Wide Web Consortium (2001), `http://www.w3.org/TR/2001/WD-soap12-20010709/`
14. Haesen, R., Snoeck, M., Lemahieu, W., Poelmans, S.: On the Definition of Service Granularity and Its Architectural Impact. In: Bellahsène, Z., Léonard, M. (eds.) CAiSE 2008. LNCS, vol. 5074, pp. 375–389. Springer, Heidelberg (2008)
15. Ioup, E., Lin, B., Sample, J., Shaw, K., Rabemanantsoa, A., Reimbold, J.: Geospatial Web Services: Bridging the Gap between OGC and Web Services. In: Sample, J.T., Shaw, K., Tu, S., Abdelguerfi, M. (eds.) Geospatial Services and Applications for the Internet, pp. 73–93. Springer, New York (2008)
16. Ioup, E., Sample, J.: Managing Granularity in Design and Implementation of Geospatial Web Services. In: Zhao, P., Di, L. (eds.) Geospatial Web Services: Advances in Information Interoperability, pp. 18–35. IGI Global, New York (2011)
17. ISO 19113: Geographic information - Quality principles. International Organization for Standardization (2002)
18. Kritikos, K., Pernici, B., Plebani, P., Cappiello, C., Comuzzi, M., Benrernou, S., Brandic, I., Kertész, A., Parkin, M., Carro, M.: A Survey on Service Quality Description. ACM Computing Surveys 46(1), Article 1 (2013)
19. Lee, K.G., Jeon, J.H., Lee, W.S., Jeong, S.-H., Park, S.-W.: QoS for Web Services: Requirements and Possible Approaches. World Wide Web Consortium (2003)
20. O'Brien, L., Bass, L., Merson, P.F.: Quality Attributes and Service-Oriented Architectures. Software Engineering Institute (2005)
21. Portele, C.: OpenGIS Geography Markup Language (GML) Encoding Standard. Open Geospatial Consortium Inc. (2007), `http://www.opengeospatial.org/standards/gml`
22. Schäffer, B.: OWS 5 SOAP/WSDL Common Engineering Report. Open Geospatial Consortium Inc. (2008), `http://www.opengeospatial.org/standards/dp`
23. Sebillo, M., Tortora, G., Vitiello, G., Di Giovanni, P., Romano, M.: A Framework for Community-Oriented Mobile Interaction Design in Emerging Regions. In: Kurosu, M. (ed.) Human-Computer Interaction, HCII 2013, Part III. LNCS, vol. 8006, pp. 342–351. Springer, Heidelberg (2013)
24. Vanmeulebrouk, B., Bulens, J., Krause, A., de Groot, H.: OGC standards in daily practice: gaps and difficulties found in their use. In: GSDI11 World Conference (2009)
25. Wei, Y., Santhana-Vannan, S.-K., Cook, R.B.: Discover, Visualize, and Deliver Geospatial Data through OGC Standards-based WebGIS System. In: 17th International Conference on Geoinformatics, pp. 1–6. IEEE (2009)
26. Zhang, C., Li, W.: The Roles of Web Feature and Web Map Services in Real-time Geospatial Data Sharing for Time-critical Applications. Cartography and Geographic Information Science 32(4), 269–283 (2005)

Tamper-Evident User Profiles
for WebID-Based Social Networks

Stefan Wild, Falko Braune, Dominik Pretzsch, Michel Rienäcker,
and Martin Gaedke

Technische Universität Chemnitz, Germany
{firstname.lastname}@informatik.tu-chemnitz.de

Abstract. Empowering people to express themselves in global communities, social networks became almost indispensable for exchanging user-generated content. User profiles are essential elements of social networks. They represent their members, but also disclose personal data to companies. W3C's WebID offers an alternative to centralized social networks that aims at providing control about personal data. WebID relies on trusting the systems that host user profiles. There is a risk that attackers exploit this trust by tampering user profile data or stealing identities. In this paper, we therefore propose the IronClad approach. It improves trustworthiness by introducing tamper-evident WebID profiles. IronClad takes protective measures to publicly discover malicious manipulation of profile data. We exemplarily implement IronClad in an existing WebID identity management platform known from previous work.

Keywords: Identity, WebID, Linked Data, Social Networks, Trust, Security, Data Integrity, Protection, Tamper Detection.

1 Introduction

Social networks have become crucial elements in modern society. They allow people to connect, communicate, and express themselves on a global scale. Joining social networks usually requires creating user profiles. Users therefore need to entrust personal information to network providers which escrow the information for them. This enables identification, eases discovery, and digitally represents the user. However, it also creates another copy of the user's digital identity. In today's ecosystem of social networks each copy needs to be maintained separately. This makes exchange and update of personal data across different domains and organizational boundaries difficult. Thus, centralized social networks like Twitter and Facebook gave rise to customer lock-in for millions of users [15].

For avoiding such walled gardens, the W3C devised WebID. As an open, universal, and decentralized identification approach, WebID allows users to be their own identity provider and establish their own personal social networks [11]. With WebID, users are enabled to manage their personal information at a self-defined place. They can also employ their WebID identities for global authentication.

WebID consists of three interrelated artifacts, which are illustrated in Figure 1: The *WebID URI* is a unique identifier referring to an agent. While an agent is

S. Casteleyn, G. Rossi, and M. Winckler (Eds.): ICWE 2014, LNCS 8541, pp. 470–479, 2014.

typically a person, it can also be a robot, group or any other entity that needs to be identified. The *WebID certificate* is a common X.509v3 client certificate. It includes a public key and a WebID URI linking to the WebID profile. The *WebID profile* is a resource containing personal information about the identity owner. Personal information are described in an extensible and machine-readable way using Linked Data. Each WebID profile also stores public keys. They are used along with the corresponding WebID certificate for an ownership-based authentication defined by the WebID protocol.

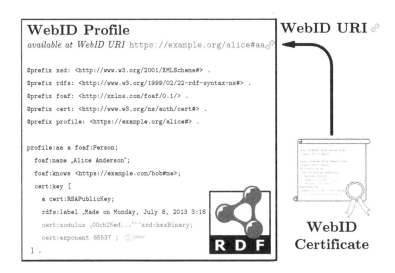

Fig. 1. Artifacts in WebID

Problem. Despite all advantages, being an identification mechanism that does not rely on authorities makes WebID vulnerable to attacks on user identities. WebID allows users to host their profiles at arbitrary locations. Yet it does neither ensure nor verify data integrity of user profiles. Experienced users know how to set up and protect a system storing their WebID profile. Inexperienced users, however, do not know this and would probably prefer using third-party managed services for hosting. Consequently, inexperienced users must trust third parties to not accessing and tampering their profile data [6]. The trustworthiness of managed services can hardly be assessed or guaranteed [5]. This shifts the problem from trusting identity providers [4] to trusting cloud storage providers. Requestors therefore depend on external means to decide whether to trust profile data. Obtaining write access to a WebID profile would enable an attacker to tamper user data stored inside [5]. Tampering user profile data, e.g., changing the e-mail address or replacing social contacts, could interfere with further transaction. Having a chance to add a public key to the identity owner's WebID profile would allow constructing a client certificate with the corresponding private key. Attackers could then use such certificate for authenticating to services as the identity owner without her knowledge and intent.

Objective. To increase trustworthiness in WebID-based social networks, we aim at providing a means for users to publicly detect tampering of profile data. For achieving this objective, we propose the novel IronClad approach for tamper-evident[1] WebID profiles. We therefore provide contributions for:

- Signing WebID profile data,
- Discovering WebID identity theft, and
- Verifying WebID profile data integrity.

Impact. Achieving the objective would allow for storing WebID profiles in potentially harmful environments, reducing the entry barrier for inexperienced users and, thus, contributing to increase the overall security and adoption of the WebID identity mechanism. Otherwise, users still risk to retrieve WebID profile data that is not in accordance with the identity owner's original intention.

The rest of the paper is organized as follows: Section 2 discusses related work. Section 3 describes and exemplarily demonstrates the IronClad approach for providing tamper-evident WebID profiles. Section 4 evaluates the approach. Section 5 concludes the paper.

2   Related Work

Literature to file systems and database systems broadly deals with the topic of ensuring data integrity. This discussion of related work focuses on the applicability of integrating features for detecting tampering attacks to Web-based systems only. To structure the discussion, we pay particular attention to interoperability, applicability and accessibility of tamper-evident features.

Centralized social networking platforms like Facebook and Google+ enable users to create views on their profile data. Such views can conceal sensitive data to external parties, e.g., groups of requestors. When targeting profile data disclosure or malicious manipulation, they are, however, inapplicable to detect internal read/write attacks without further ado, as described by Feldman et al. in [6]. Through establishing decentralized social networks, WebID distributes this problem to systems that host the profiles. Contrary to centralized social networking platforms, Web Access Control[2] (WAC) facilitates securing resources in a decoupled and decentralized way. However, WAC also focuses on access protection and not on data protection [14]. Even though protection of personal data could be accomplished through encryption, this is inappropriate in WebID. Profiles have to be, at least partially, accessible for authentication of identity owners and for queries of requestors. It would be also required to either distribute keys for decryption to an unknown number of potential requestors or establish central authorities for key management, which does not conform to WebID's idea of focusing on individuals.

---

[1] "Tamper-evident" is commonly defined as a means for making unauthorized access to a protected object easily detectable.
[2] http://www.w3.org/wiki/WebAccessControl

With regard to detect tampering of profile data, WebID shares similar disadvantages with other identity management systems like OpenID or Mozilla Persona [8]. OpenID implements only limited handling of personal attributes [7], whereas Persona is not designed for attaching profile data to an identity in a holistic way [1]. In contrast, WebID allows flexibly extending profile data, i.e., add cryptographic signatures. Such extension is not applicable to many social networking platforms and identification systems due to their centralized, closed or restricted handling of user profile data.

Public keys and signatures must also be protected from manipulation in order to provide sound proof of the identity owner's intent. This could be accomplished by a public key infrastructure (PKI) involving certificate authorities (CA) or a Web of Trust (WoT). A PKI based on CAs represents a centralized trust model that uses hierarchically organized authority chains [2]. WebID allows for adapting this model, e.g., similar to signing WebID certificates by a trusted third party instead of the identity owner himself[3]. However, we do not want to impair WebID's decentralized approach of involving and empowering individuals instead of authorities. By contrast, the WoT concept represents a flat hierarchy only relying on individuals [2]. It needs member discovery and makes updating public keys and signatures difficult due to their necessary distribution and inclusion in other data stores, e.g., user profiles.

3   Tamper-Evident WebID Profiles through IronClad

In order to detect tampering of WebID profile data, we created IronClad. It is based on the principle that only identity owners should be enabled to change their profile data in a sustainable way. For ensuring requestors that all WebID profile data is what was intended by the identity owner, IronClad incorporates three main activities: signing profile data, storing/retrieving signatures, and verifying data integrity of profiles. They are according to our key contributions. We illustrated them using BPMN in Figure 2 and describe them in the following.

3.1   Operations Supported by IronClad

Signing WebID Profile Data. By signing the WebID profile, the identity owner (cf. top of Figure 2) proves that personal data stored in his profile is sound and was not changed by another party. In order to avoid signing tampered data, the data integrity of the WebID profile needs to be checked (cf. ① in Figure 2) prior to updating relevant data (cf. ②) and creating signatures (cf. ③). Algorithm 1 specifies in pseudo code notation how IronClad creates signatures of WebID profile data. Our approach uses the RDF graph representation of a WebID profile for computing hash values independent from specific data serializations, e.g., RDF/XML or Turtle. To address different orders of RDF

---

[3] In WebID, certificates are usually self-signed as the trust does not rely on the public key, but on the WebID URI linked resource storing the public key, which we want to protect against tampering.

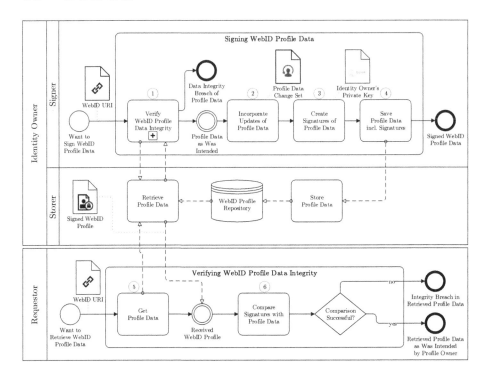

Fig. 2. Big picture of IronClad: Signing, storing, and verifying data integrity of profiles

triples and blank nodes, IronClad performs a canonicalization. We therefore utilize the One-Step Deterministic Labeling method proposed by Carroll in [3] and the methodology described by Tummarello et al. in [12].

To avoid disclosing the identity owner's private key to a third party, the signing process is split into a server and a client side part. IronClad's server side computes hash values of each *minimal self-contained graph* [12] found in the graph representation of the WebID profile. It combines all hash values to a signing request afterwards (cf. lines 4 to 8). The client side part analyzes this request and signs the content. It creates the signature through encrypting each hash value with a private key (cf. lines 9 to 12). The identity owner selects the corresponding private key beforehand. IronClad supports client side signing by a tool. It transforms a signing request into a signed response, which is then sent back to the server side.

Once received by the server side (cf. lines 13 to 16), the signed response containing the signatures is verified. Provided that the verification was successful, IronClad stores the signatures in the identity owner's WebID profile in ④. When storing (cf. middle of Figure 2), IronClad applies the method proposed by Sayers and Eshghi in [10]. This method closely links the public key of the profile with the WebID URI. Thus, it assists in detecting attacks that aim at removing profile data and signatures.

Discovering WebID Identity Theft. Following the principle of empowering individuals instead of authorities, we could not solely rely on attaching signatures to profile data[4]. IronClad creates a binding between the public key and the WebID URI. Thus, it ensures that this key cannot be changed without losing personal relationship data such as incoming social connections expressed via foaf:knows WebID URIs. Having the public key stored in the WebID URI allows detecting the same key inside the profile. This facilitates not only discovering identity theft done by malicious key manipulation, but also using the public key for signature verification. The length of the public key, e.g., 2048-bit, makes it inconvenient to store it directly inside a WebID URI. Therefore, IronClad uses the SHA-1 hash value of the public key instead.

Algorithm 1. Creating Signatures of User Profile Data

Input: WebID URI u, Private Key key
// on server side
1 get WebID profile p from u;
2 get RDF graph g from p;
3 apply canonicalization on g using [3] and [12];
4 **repeat**
5 | delete minimal self-contained graph msg from g;
6 | create hash value h of msg; // currently uses SHA-1
7 | add h to client request req;
8 **until** g *is empty*;
   // on client side to avoid private key disclosure
9 **foreach** *hash value h* in *client request req* **do**
10 | create signature sig by encrypting h with key;
11 | add sig to server response res;
12 **end**
   // on server side
13 **foreach** *signature sig* in *server response res* **do**
14 | **if** *signature sig is invalid* **then** stop
15 **end**
16 add all signatures in server response res to graph g;

Verifying WebID Profile Data Integrity. To make signed WebID profiles easily verifiable for requestors, we integrated the verification process into the WebID authentication routine. It is triggered when the WebID profile has been loaded. For verifying signed profile data (cf. bottom of Figure 2), IronClad receives the WebID profile via the WebID URI in ⑤. It tries to detect a plausible public key[5] inside the profile. Such public key has to correspond to the hash value stored in the WebID

---

[4] By gaining access to the system storing the WebID profile, an attacker could tamper identity data and manipulate signatures stored in the profile. Due to this vulnerability to attacks, an external authority would be required to provide proof of correctness.

[5] A public key with a common length, e.g., 2048 bit.

URI. Having found the public key, IronClad computes hash values of WebID profile data as mentioned in the signing process. It then compares the hash values with the hash values retrieved by decrypting the signatures using the public key, as indicated by ⑥. The data integrity of WebID profiles cannot be guaranteed in case any detection or verification step has failed. Handling failed verifications depends on the scenario and authentication target. This is not a part of the IronClad approach and, thus, needs be to addressed separately.

3.2 Implementation and Demonstration of IronClad

For showcasing the approach, we exemplarily implemented IronClad in the Sociddea WebID identity provider and management platform proposed by Wild et al. in [13]. It is important to mention here that IronClad is not limited to a specific platform. That is, it is possible to apply the approach in arbitrary WebID identity providers, management platforms or authentication methods. Thus, IronClad is line with the idea of decentralized social networks.

In Sociddea, IronClad is optionally applied when creating new WebID identities. It is furthermore used for signing and verifying profile data hosted in Sociddea's ecosystem. Figure 3 illustrates the visual representation of a tamper-evident WebID profile in four different scenarios. While the figure shows tampering by changing a personal attribute of the identity owner, data integrity breaches through adding or removing RDF triples are also detected by IronClad. In addition to visually highlighting malicious data manipulations, identity owners and service providers also benefit from tamper-evident WebID profiles during user authentication and requests for profile data.

Try It Out. For a live demonstration, screen casts, and further information about IronClad and Sociddea please visit:
http://vsr.informatik.tu-chemnitz.de/demo/sociddea/

4   Evaluation of IronClad

While the acceptance and handling of tamper-evident WebID profiles is a planned subject of a larger empirical investigation, this evaluation discusses to which extent IronClad takes the criteria into account we used for analyzing related work. From a theoretical point of view, we think that such study therefore allows for determining how well IronClad achieves the objective defined in Section 1.

WebID profiles depend on RDF for describing an identity owner's personal attributes in a machine-readable way. Appropriate RDF-based vocabularies enable describing and interlinking new contents. This facilitates extending WebID profiles by additional RDF triples. It is consequently well applicable to represent and associate signatures to personal attributes.

As the IronClad approach does not involve encrypting user profile data, it does not impair the accessibility of tamper-evident WebID profiles. Both common and tamper-evident WebID profiles share similar characteristics. While tamper-evident WebID profiles consist of additional RDF-based signatures, existing personal data

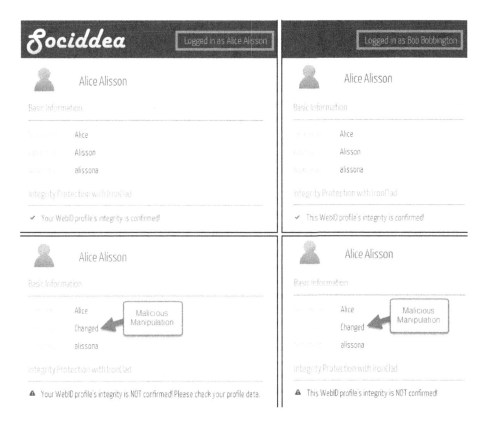

Fig. 3. Exemplary implementation of IronClad for tamper-evident user profiles with visualized results for: Identity owner Alice's view on her valid profile data (top/left), "Lastname" changed - tampering detected (bottom/left), requestor Bob's view on Alice's valid profile data (top/right), Bob's view on Alice's tampered profile (bottom/right)

remains untouched. In IronClad, signatures are loosely coupled statements about personal data. Removing all signature RDF triples would reveal the original payload of a WebID profile. The hash value included in the WebID URI, however, indicates that the corresponding WebID profile is tamper-evident. Considering an attacker would remove signatures stored in a tamper-evident WebID profile, there is still the WebID URI indicating that the WebID profile has to be data integrity protected. While there are other ways of using hash values in URIs [9], IronClad just appends them to common WebID URIs for the sake of simplicity and conformity. Such verification for tamper-evidence is part of the proposed extension of the WebID authentication sequence. Even though signatures and personal data contained in tamper-evident WebID profiles are accessible per se, view filters can assist in concealing particular RDF triples, as described by Wild et al. in [14]. For example, excluding all signatures might be beneficial for presenting profile data to users.

Through directly operating on the RDF graph representing a WebID profile, IronClad is compatible to different orders and structures of RDF triples, and diverse types of serialization. This allows for dealing with the high heterogeneity prevailing in this context. Despite all advantages, we also identified some issues restricting the interoperability and compatibility of our approach: Combining the hash value of the public key with the WebID URI creates a universal mean to detect tampering in WebID profiles. This, however, implicates that it is not possible 1) to transform common WebID URIs to tamper-evident ones and 2) to change tamper-evident WebID URIs without invalidating all incoming connections from social contacts expressing their friendship and so on. Changing a WebID URI results in creating a different and therefore new identity. This can be considered as a common shortcoming in WebID. There is a need for an approach to indicate that a new (tamper-evident) WebID URI belongs an already existing one. Creating a tamper-evident WebID profile necessitates signing the profile data worth protecting. As we wanted to avoid disclosing the private key to a third-party, we rely on a client-side signing tool at the moment. This dependency entails new requirements for system, platform and device support. We seek to resolve the dependency by a solution that is user-friendly and interoperable.

5   Conclusion and Future Work

By providing a means for users and machines to detect data integrity breaches in WebID profiles, we contributed to increase the trustworthiness of open, decentralized, and universal identification mechanisms. Taking compatibility and accessibility of user profile data into account, our approach for tamper-evident user profiles is well-applicable for new WebID identities. Focusing on empowering individuals instead of authorities, we expect that such mechanism will gain more momentum and coverage in the future. By adding security features to WebID artifacts, we enabled profile owners and requestors to detect malicious manipulation and identity theft. We integrated IronClad into the WebID authentication routine to verify the profile data integrity as a requirement for any successful identity proof. Not only does IronClad operate independently of data type and order, but also independently of the profile hosting system. We enabled verification without requiring prior knowledge except for an already known WebID identifier.

Having made an important step towards more trustworthiness in the context of WebID, we will constitute future work on that basis. For validating not solely the technical feasibility of our approach, it is necessary to conduct a more comprehensive evaluation involving how well users accept tamper-evident WebID URIs and profiles. Even though we are aware that tamper-evident WebID URIs make the management for users more difficult, e.g., compared to managing email addresses, we claim that most issues can be successfully addressed through utilizing techniques such as URI drag and drop, QR codes or WebID URIs embedded into other objects like WebID certificates. Identifying the origin of tampering attacks, facilitating key replacement and enabling signature creation with several keys are also interesting topics for future research.

References

1. Bamberg, W., et al.: Persona - Protocol Overview (2013),
   https://developer.mozilla.org/en-US/docs/
   Mozilla/Persona/Protocol_Overview (accessed February 23, 2014)
2. Caronni, G.: Walking the Web of Trust. In: Proeedings of the IEEE 9th International Workshops on Enabling Technologies: Infrastructure for Collaborative Enterprises (WET ICE 2000), pp. 153–158. IEEE (2000)
3. Carroll, J.J.: Signing RDF Graphs. In: Fensel, D., Sycara, K., Mylopoulos, J. (eds.)
   ISWC 2003. LNCS, vol. 2870, pp. 369–384. Springer, Heidelberg (2003)
4. Dhamija, R., Dusseault, L.: The Seven Flaws of Identity Management: Usability
   and security Challenges. IEEE Security & Privacy 6(2), 24–29 (2008)
5. Feldman, A.J., Blankstein, A., Freedman, M.J., Felten, E.W.: Privacy and Integrity
   are Possible in the Untrusted Cloud. Bulletin of the IEEE Computer Society Technical Committee on Data Engineering 35(4), 73–82 (2012)
6. Feldman, A.J., Blankstein, A., Freedman, M.J., Felten, E.W.: Social Networking
   with Frientegrity: Privacy and Integrity with an Untrusted Provider. In: Proceedings of the 21st USENIX Conference on Security Symposium, Security, vol. 12,
   p. 31 (2012)
7. Fitzpatrick, B., Recordon, D., Hardt, D., Hoyt, J.: OpenID Authentication 2.0 -
   Final (2007), http://openid.net/specs/openid-authentication-2_0.html (accessed February 23, 2014)
8. Hackett, M., Hawkey, K.: Security, Privacy and Usability Requirements for Feder-
   ated Identity. In: Workshop on Web 2.0 Security & Privacy (2012)
9. Sauermann, L., Cyganiak, R., Völkel, M.: Cool URIs for the Semantic Web. Tech.
   rep., Saarländische Universitäts- und Landesbibliothek (2007)
10. Sayers, C., Eshghi, K.: The case for generating URIs by hashing RDF content
    (2002)
11. Sporny, M., Inkster, T., Story, H., Harbulot, B., Bachmann-Gmür, R.: WebID 1.0:
    Web Identification and Discovery (2011),
    http://www.w3.org/2005/Incubator/webid/spec/
    (accessed February 23, 2014)
12. Tummarello, G., Morbidoni, C., Puliti, P., Piazza, F.: Signing Individual Fragments
    of an RDF Graph. In: Special Interest Tracks and Posters of the 14th International
    Conference on WWW, pp. 1020–1021. ACM (2005)
13. Wild, S., Chudnovskyy, O., Heil, S., Gaedke, M.: Customized Views on Profiles in
    WebID-Based Distributed Social Networks. In: Daniel, F., Dolog, P., Li, Q. (eds.)
    ICWE 2013. LNCS, vol. 7977, pp. 498–501. Springer, Heidelberg (2013)
14. Wild, S., Chudnovskyy, O., Heil, S., Gaedke, M.: Protecting User Profile Data
    in WebID-Based Social Networks Through Fine-Grained Filtering. In: Sheng,
    Q.Z., Kjeldskov, J. (eds.) ICWE Workshops 2013. LNCS, vol. 8295, pp. 269–280.
    Springer, Heidelberg (2013)
15. Yeung, C.M.A., Liccardi, I., Lu, K., Seneviratne, O., Berners-lee, T.: Decentraliza-
    tion: The Future of Online Social Networking. In: W3C Workshop on the Future
    of Social Networking Position Papers, vol. 2, pp. 2–7 (2009)

X-Themes: Supporting Design-by-Example

Moira C. Norrie, Michael Nebeling, Linda Di Geronimo, and Alfonso Murolo

Department of Computer Science, ETH Zurich
CH-8092 Zurich, Switzerland
{norrie,nebeling,lindad,amurolo}@inf.ethz.ch

Abstract. Design-by-example enables users with little technical knowledge to develop web sites by reusing all or parts of existing sites. In CMS such as WordPress, themes essentially offer example designs for all-or-nothing reuse. We propose an extension to the theme concept that allows web sites to be designed by reusing and combining components of different themes. In contrast to previous research advocating design-by-example, we do not restrict ourselves to static web pages, but also support the reuse of dynamic content including functionality for animations and database access. Our approach is to provide a theme generator that structures the themes that it generates in terms of reusable components which can then be reused in future themes. We present a first prototype tool, called the X-Themes Editor, developed to demonstrate the viability of the approach and investigate requirements and issues. We describe how the X-Themes Editor has been integrated into the WordPress platform as well as discussing the outcomes of these initial investigations.

Keywords: design-by-example, content management system, theme, end-user development, reuse in web design.

1 Introduction

Within the research community, model-driven approaches to web engineering have been promoted as the best way of supporting the systematic development of web sites [1]. Often methods are designed to cater for projects involving large, diverse teams including graphic designers, database architects, programmers and marketing staff.

However, an increasing number of professional as well as personal web sites are being developed by individuals using platforms such as WordPress[1] which allows users to dynamically customise crowdsourced themes. A theme defines a set of templates, stylesheets and media for an entire web site and therefore essentially supports only all-or-nothing reuse.

The idea of this paper is to show how this paradigm of design-by-example could be extended to allow users to design their web sites by reusing and combining features of different themes. For example, they might choose the colour scheme and front page layout of one theme, a slider content component from

---

[1] http://www.wordpress.org

S. Casteleyn, G. Rossi, and M. Winckler (Eds.): ICWE 2014, LNCS 8541, pp. 480–489, 2014.
© Springer International Publishing Switzerland 2014

a second theme, and the drop-down menu style from a third. As discussed in Sect. 2, this approach has been advocated by researchers in the HCI community [2,3], but typically their experiments and studies focus on the static parts of web sites and do not address the technical challenges of extracting and reusing functionality. A key factor in our research was to support the reuse of all aspects of a web site, including animations, client-side processing and database access. Further, to avoid the need for developers to change their work practices, we wanted to investigate how such an approach could be integrated into the WordPress platform.

Our approach, which is presented in Sect. 3, exploits the concept of a theme generator to produce a collection of so-called X-Themes that conform to a well-structured model designed to support reuse. The special X-Themes generated by our tool are then available to other developers who can drag-and-drop components of existing themes into their newly created themes at design time. We outline the main features of a first prototype designed to demonstrate the viability of the approach in Sect. 4 before discussing the main requirements and challenges to be addressed in future research in Sect. 5.

2   Background

Several model-driven approaches to web site development have been proposed, for example WebML [1], Hera [4] and UWE [5]. While these approaches have received acclaim in the research community, they have had limited impact in the web development community at large where many professionals work with a mix of technical knowledge and design skills and build on modern content management systems (CMS) such as WordPress or Drupal[2]. An indication of the scale of the developer community using WordPress is the estimate that 21.9% of the top 10 million web sites are implemented on WordPress[3]. The size and complexity of these web sites varies enormously, but has certainly gone well beyond the original focus on personal blogging sites and now includes online newspapers, e.g. Metro UK[4], e-commerce sites, e.g. Kuborra[5], and information platforms for communities e.g. SAP.info[6].

As highlighted by a recent survey on modern web development practices [6], many of these developers have no formal computer science training and are either self-employed or working in very small organisations. In these settings, developers typically adopt an *interface-driven* approach, starting from a mockup of the interface and first adding client-side functionality before adding server-side functionality by migrating to a CMS platform.

The WordPress platform was developed using a crowdsourcing model that allows their user community to develop and share both *themes* and *plugins*.

---

[2] http://www.drupal.org
[3] http://www.w3techs.com, accessed 10 April 2014
[4] http://metro.co.uk/
[5] http://www.kuborra.com/
[6] http://en.sap.info/

A theme is a set of PHP templates, CSS stylesheets and media objects that define the presentation, structure, functionality and types of content of a web site. To develop their own web site, a user simply has to select a theme and then start adding their own content. Themes are usually parametrised so that users can easily customise their sites and they can also extend the types of content and functionality by adding plugins.

Themes can be considered as a form of *design-by-example* [7]. While Word-Press themes only support all-or-nothing reuse, researchers have investigated approaches which would allow users to design their web sites by freely selecting and combining parts of example web sites from interactive galleries [2,3]. While these studies have demonstrated the benefits of this general approach, they did not address the technical challenges of being able to extract and combine arbitrary elements of modern web pages that are often dynamic rather than static and make heavy use of JavaScript and jQuery[7]. As a result, their methods were only able to deal with the reuse of appearance and content and not functionality.

A number of approaches for developing web applications from reusable components have been proposed. WebComposition [8] is an object-oriented support system for building web applications through hierarchical compositions of reusable application components. While some mashup editors help users to integrate information from distributed sources, others provide infrastructure for building new applications from reusable components. For example, MashArt [9] enables advanced users to create their own applications through the composition of user interface, application and data components. While our approach shares some of the goals and enables similar extraction and reuse techniques, it offers these at the theme level to base web site designs on multiple examples, which is different from mashups that are direct compositions of existing web sites. More recently, extensions to CMS such as WordPress have been proposed to allow web applications to be developed from a component model that supports composition at the data, application and interface levels [10]. The focus of this work differs in that our approach is *interface-driven* rather than *model-driven* and our component model captures concepts of popular web development platforms such as WordPress as well as the new HTML5 and CSS3 web standards together with jQuery that power the underlying themes.

Summarising, the solution proposed by many researchers has been to try and bring discipline into web engineering by requiring developers to first model different aspects of their web sites and only then generate code. But this usually requires that developers abandon popular platforms and learn new modelling skills, tools and possibly even languages. It also makes it more difficult to support rapid prototyping and allow developers to start by adapting existing web site designs developed either by themselves or other developers. Instead of trying to force developers to change their ways of working, we think it is important to instead find ways of better supporting them. This means that not only should we find ways to support the design-by-example paradigm, but this should be done using existing, popular platforms and technologies.

---

[7] http://jquery.com

3   Approach

While the WordPress platform is very flexible and powerful, the basic model behind it is relatively simple and developers are offered a very loose framework in which to work. As a result, it is possible to build advanced web sites but widely varying approaches are used to achieve similar functionality and presentation.

In the rare case that a personal or professional developer finds an existing theme that fully meets their requirements, the process of developing a web site mainly consists of setting some parameters and adding content through the WordPress dashboard. To some degree, they can even extend the functionality of the theme through this interface by selecting and adding various plugins. However, as soon as a developer is faced with the task of adapting or extending a theme, they have to start working at the level of the HTML, CSS and PHP files and learning about the core WordPress model and system operation. Developers often work on a need-to-know basis, learning only enough to solve the particular task at hand. Further, the documentation and tutorials vary a lot in terms of guidelines and solutions offered to developers. It is clear from reading tutorial-style books, e.g. [11,12] as well as online forums[8] that many developers simply copy and paste bits of CSS, HTML and PHP with the hope that it will achieve the desired effects. However, often these attempts to reuse code fail because they are inconsistent with how other parts of the site have been developed.

Fig. 1. Examples of portfolio themes from the company Elegant Themes

We want users to be able to create their own themes by selecting and combining parts of different existing themes. For example, Fig. 1 shows three portfolio themes offered by the company Elegant Themes[9]. A user might want to have the general styling of the theme on the left which includes a slider of background images and an animated drop down element in the gallery page, but want to include in the front page the slider component of the theme in the middle and have

---

[8] For example, http://www.wpbeginner.com
[9] http://www.elegantthemes.com

a blog post page with the layout shown in the theme on the right. To achieve this, we have developed a graphical theme editor, called the X-Themes Editor, which allows users to select parts of other themes that can be integrated into their own themes using simple drag-and-drop operations.

To support this kind of reuse, it is necessary that themes adhere to a well-defined component model. One way to do this would be to develop a new platform or domain specific language based on a component model but, as stated previously, this is something that we wanted to avoid. Instead, we designed our own metamodel for WordPress themes and then based the X-Themes Editor on this model. This means that any themes generated with our editor are clearly structured in terms of components that can be shared and reused. By offering a theme editor that is integrated into the WordPress platform, developers can already be offered a valuable tool for creating new themes from scratch. Themes generated using the tool are called X-Themes and the collection of X-Themes are made available for reuse in an interactive gallery.

It is important to note that a number of theme generators for WordPress already exist but many of these have serious limitations, especially when it comes to customising functionality. For example, Templatr[10] is a free web-based tool that allows users to customise static elements but they can only select from a fixed set of layouts. Some tools such as Lubith[11] allow users to customise layout via drag and drop, but they usually do not support the customisation of functionality. Another limitation of existing generators is the fact that they are not integrated into WordPress. This means that it is not possible to perform content-related tasks such as displaying the pages or latest posts and comments during the design of the theme and often there can be compatibility problems between WordPress versions. Therefore, offering a theme editor that is integrated into the WordPress platform, and enables not only presentation but also layout and functionality to be customised, is in itself a valuable contribution.

4   X-Themes Editor

Using the possibilities to customise and extend the functionality of the Word-Press dashboard, we were able to integrate the X-Themes Editor into the tool options menu offered to developers. When the X-Themes tool is selected, the standard dashboard interface is replaced by the GUI of the X-Themes Editor which, in addition to offering a menu for creating and customising elements of a theme, also provides access to an interactive gallery of previously generated X-Themes. A user can create a theme using a mix of editing operations to define new components and drag-and-drop operations to import selected components from various source themes opened in the X-Themes gallery.

Every theme created in the editor has three main components—Header, Body and Footer— that are visually separated as indicated in Fig. 2. Users can create

---

[10] http://templatr.cc

[11] http://www.lubith.com

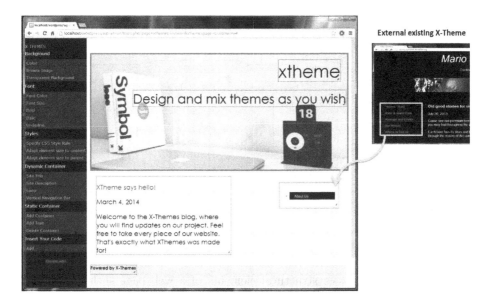

Fig. 2. Dragging a navigation menu from an existing website into the X-Themes editor

different types of containers within each of these main components and each container may itself contain other containers.

As soon as a component from an existing theme is dropped into the new theme, it will be active and any dynamic behaviour experienced. For example, if the user drag-and-drops a dropdown menu into the navigation component as illustrated in Fig. 2, they will immediately be able to try out this functionality in the theme under construction.

There are two main types of container—static and dynamic. A static container has no elements that can change dynamically. A dynamic container is one in which at least one JavaScript or PHP function appears. This could be a function to access a theme parameter or content of the database such as the site's title or application data. A container with any kind of dynamic content will be executed and show the corresponding result already at design time, even when dragged-and-dropped from a different theme.

After a drag-and-drop operation, the user can choose whether or not they want to keep the style of the source theme or have the style of the destination theme applied. For example, a user may wish to keep the information displayed in an imported database query (referred to as a Loop in WordPress) but apply a different design in the new theme. If the user decides to keep the style of the source theme, they can adjust both the format and style later using the general editing functionality of the X-Themes Editor.

When the editing of a theme is complete, the X-Themes Editor generates the set of templates for that theme together with the files defining the components of the X-Theme and associated metadata based on the X-Themes metamodel shown in Fig. 3.

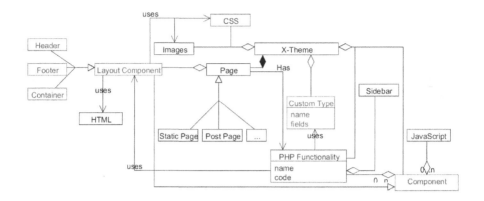

Fig. 3. X-Themes metamodel

Concepts defining basic structural elements of a web page such as header, footer and sidebar are fundamental to the WordPress model as are posts as the primary content type. To extend the model to support other types of content, custom post types can be introduced. Our metamodel introduces the concept of a component shown at the bottom right of Fig. 3. A component can include layout and style rules as well as dynamic behaviour specified as PHP logic or JavaScript. If a component is to be exported and reused, then clearly all code, including CSS rules as well as PHP logic and JavaScript, must also be exported along with the component.

To support the extraction and reuse of components, each component is described in an XML file of the form shown in Fig. 4. All the required PHP, CSS and JavaScript files are stored in a separate folder in the WordPress theme directory, and the DOM of the theme is annotated using HTML5 dataset properties to indicate reusable components based on the metamodel.

The `component_type` can be custom header, sidebar or undefined in the case of a generic component. For header and sidebar components, a `functions` element is included to specify dependencies so it is known which files have to be included in the *functions.php* file of a WordPress theme for that component to be used.

A `clientlogic` element specifies JavaScript code that needs to be exported with the component, while `serverlogic` does the same for PHP logic. Any required style files are specified in the *styles* element. The `include` element specifies the file to be included in a page of the theme to execute and show the component. Various forms of dependencies to other code files are specified in the dependency elements.

A `layout_component` is a special type of component representing the layout structure of the theme and can be a header, a footer or a container. Any of these three types of component can include containers, allowing a fully nested container structure. We distinguish layout components that contain only static content from those that include JavaScript or PHP code.

```
l version="1.0"?>                                    ependency>dropdown.js</dependency>
ture name="dropdown">                                ependency>style1.css</dependency>
omponent_type>undefined</component_type>             ependency>dropdown.code</dependency>
lientlogic>                                          ependency>dropdown.js</dependency>
  <name>dropdown.js</name>                           ependency>style1.css</dependency>
clientlogic>                                         ependency>common.css</dependency>
erverlogic>                                          ependency>demo.css</dependency>
<name>dropdown.php</name>                            ependency>icons.css</dependency>
<name>dropdown.code</name>                           ependency>dropdown.code</dependency>
serverlogic>                                         ependency>icommon.eot</dependency>
tyles>                                               ependency>icommon.svg</dependency>
<name>style1.css</name>                              ependency>icommon.ttf</dependency>
styles>                                              ependency>icommon.woff</dependency>
nclude>dropdown.php</include>                        ature>
ippath>dropdown</zippath>
```

Fig. 4. Component XML Sample

Each component is allocated a separate directory where its code, resources and XML component files are stored. This includes all PHP required to reuse the component as well as generated CSS files. In the final step of theme generation, an XML representation of the theme defining the component structure is created.

When a user opens a theme in the X-Themes gallery, reusable components are highlighted as the cursor moves over them. If the user drags a component, such as a navigation menu made in PHP/JavaScript and CSS, the X-Themes Editor accesses the metamodel information of that component and acquires references to linked resources which are then cloned for use in the editor. The component is then executed in the context of the X-Themes Editor running on the user's own WordPress installation. The ability for users to view their own content at design time is only one of the advantages of implementing the X-Themes Editor as a WordPress plugin. Another is the fact that it provides an easy means of deploying the tool to existing developer communities and hence raising awareness and chances of acceptance.

5   Discussion

While the X-Themes Editor represents an important first step in showing how design-by-example could be fully supported in CMS platforms such as Word-Press, there are a number of issues that need to be addressed in future research.

X-Themes Metamodel. The metamodel that we have worked with so far is based on the WordPress core model and therefore WordPress specific. Further, it focusses more on system and implementation features than general asbtract concepts. One of the next steps will be to generate a conceptual metamodel that could be applied to more than one CMS. Specifically, we are currently investigating how our approach could be applied in Drupal and the metamodel generalised.

Beyond the Front Page. Although the themes that we currently generate are not single page, like many WordPress themes, they very much focus on the front page which is often seen as defining the main features of a web site in terms of the header, footer, navigation and layout as well as presentation. We need to extend the X-Themes Editor to cater for web sites with more complex structures with variable page layout and content.

Transforming Themes to X-Themes. One of the main ideas behind our approach was to provide users with a graphical theme editor that would in itself be a valuable tool and hence something that developers would want to use independent of the goals of this project. In this way, we could address the cold start problem of generating a collection of X-Themes available for reuse. At the same time, we do think it important that we offer support for transforming existing WordPress themes into X-Themes. This is a topic for future research where we will investigate the extent to which this could be automated.

Interactive Galleries. The current way of searching for WordPress themes relies on descriptions and keywords provided by developers. We want to investigate alternative and complementary ways of supporting search within interactive galleries. This would include search-by-example as well as ways of automatically classifying themes based on a variety of factors.

Data-Intensive Web Sites. Data-intensive web sites require the integration of custom post types to manage application data. In previous work within our research group, a tool was developed that generates a WordPress plugin with custom post types based on an entity-relationship data model defined by a developer [13]. We have now started to investigate an alternative approach that would extend the design-by-example approach to automatically derive data schemas and custom post type definitions based on example data content.

6   Conclusion

We have shown how the support for design-by-example offered by modern CMS platforms such as WordPress could go well beyond the current all-or-nothing approach of sharing entire themes. Users should be able to selectively reuse and combine parts of existing themes, including dynamic components that define functionality. Our work therefore goes beyond previous research on design-by-example paradigms in web development [2,3] which were limited to static views on web sites.

We note however that, while the X-Themes Editor we have developed is sufficient to demonstrate the potential of the approach, as outlined in Sect. 5, there are a number of open issues that would need to be addressed in future research in order for the method to be deployed in practice. Further, the tool would need to advance beyond the prototype stage to support a full palette of editing capabilities expected of a state-of-the-art web design tool.

Acknowledgements. We acknowledge the support of the Swiss National Science Foundation who financially supported this research under project FZFSP0_147257.

References

1. Ceri, S., Fraternali, P., Bongio, A., Brambilla, M., Comai, S., Matera, M.: Designing Data-Intensive Web Applications. Morgan Kaufmann Publishers Inc. (2002)
2. Hartmann, B., Wu, L., Collins, K., Klemmer, S.R.: Programming by a Sample: Rapidly Creating Web Applications with d.mix. In: Proc. 20th ACM User Interface Software and Technology Symposium, UIST (2007)
3. Lee, B., Srivastava, S., Kumar, R., Brafman, R., Klemmer, S.: Designing with Interactive Example Galleries. In: Proc. Conf. on Human Factors in Computings Systems, CHI (2010)
4. Houben, G., Barna, P., Frasincar, F., Vdovjak, R.: Hera: Development of Semantic Web Information Systems. In: Cueva Lovelle, J.M., Rodríguez, B.M.G., Gayo, J.E.L., Ruiz, M.d.P.P., Aguilar, L.J. (eds.) ICWE 2003. LNCS, vol. 2722, pp. 529–538. Springer, Heidelberg (2003)
5. Hennicker, R., Koch, N.: A UML-based methodology for hypermedia design. In: Evans, A., Caskurlu, B., Selic, B. (eds.) UML 2000. LNCS, vol. 1939, pp. 410–424. Springer, Heidelberg (2000)
6. Norrie, M.C., Geronimo, L.D., Murolo, A., Nebeling, M.: The Forgotten Many? A Survey of Modern Web Development Practices. In: Casteleyn, S., Rossi, G., Winckler, M. (eds.) ICWE 2014. LNCS, vol. 8541, pp. 285–302. Springer, Heidelberg (2014)
7. Herring, S., Chang, C., Krantzler, J., Bailey, B.: Getting Inspired! Understanding How and Why Examples are Used in Creative Design Practice. In: Proc. Conf. on Human Factors in Computings Systems, CHI (2009)
8. Gellersen, H.W., Wicke, R., Gaedke, M.: WebComposition: An Object-Oriented Support System for the Web Engineering Lifecycle. Computer Networks 29(8-13) (1997)
9. Yu, J., Benatallah, B., Saint-Paul, R., Casati, F., Daniel, F., Matera, M.: A Framework for Rapid Integration of Presentation Components. In: Proc. 16th Intl. World Wide Web Conference, WWW (2007)
10. Leone, S., de Spindler, A., Norrie, M.C., McLeod, D.: Integrating Component-Based Web Engineering into Content Management Systems. In: Daniel, F., Dolog, P., Li, Q. (eds.) ICWE 2013. LNCS, vol. 7977, pp. 37–51. Springer, Heidelberg (2013)
11. Blakeley-Silver, T.: WordPress Theme Design. Packt Publishing (2008)
12. Casabona, J.: Building WordPress Themes from Scratch (2012)
13. Leone, S., de Spindler, A., Norrie, M.C.: A Meta-plugin for Bespoke Data Management in WordPress. In: Wang, X.S., Cruz, I., Delis, A., Huang, G. (eds.) WISE 2012. LNCS, vol. 7651, pp. 580–593. Springer, Heidelberg (2012)

A Tool for Detecting Bad Usability Smells
in an Automatic Way

Julián Grigera[1], Alejandra Garrido[1,2], and José Matías Rivero[1,2]

[1] LIFIA, Facultad de Informática, Universidad Nacional de La Plata, Argentina
[2] CONICET, Argentina
{Julian.Grigera,Garrido,MRivero}@lifia.info.unlp.edu.ar

Abstract. The refactoring technique helps developers to improve not only source code quality, but also other aspects like usability. The problems refactoring helps to solve in the specific field of web usability are considered to be issues that make common tasks complicated for end users. Finding such problems, known in the jargon as *bad smells*, is often challenging for developers, especially for those who do not have experience in usability. In an attempt to leverage this task, we introduce a tool that automatically finds bad usability smells in web applications. Since bad smells are catalogued in the literature together with their suggested refactorings, it is easier for developers to find appropriate solutions.

1 Introduction

The refactoring technique [1] has been recently brought to the usability field of web applications, allowing developers to apply usability improvements without altering the application's functionality [2]. The problems developers can solve by refactoring, called *bad usability smells* are often hard to find, so there are ways to assist this task.

Running usability tests is the most common way to find usability problems [3], but it requires supervision by usability experts, among other resources. These tests can be automated to some extent, presenting an attractive alternative to lower the costs. Some approaches in the literature automate the gathering of data from users [4, 5] but not the analysis, which still depends on usability experts. Other approaches automate part of the analysis by comparing users behavior to optimal behavior paths [6, 7], but this requires prior preparation and subjects to conduct the experiments. There are also commercial tools like CrazyEgg[1] or ClickTale[2] that offer statistical data to their customers by analyzing interaction data from real users instead of tests subjects. However, even if these tools can represent a cheaper option, the results they obtain also require analysis from usability experts.

The tool we present in this work also automates the gathering of interaction data from real users, but in addition, it preprocesses the events on the client side to report concrete usability problems, easier to interpret than mere statistics. Moreover, the tool presents the usability issues as *bad usability smells*, which are problems catalogued in the literature along with the *refactorings* that solve them. Using these catalogues, developers can find a

---

[1] http://www.crazyegg.com
[2] http://www.clicktale.com

S. Casteleyn, G. Rossi, and M. Winckler (Eds.): ICWE 2014, LNCS 8541, pp. 490–493, 2014.

concrete way to correct the detected bad smells. The *Bad Smells Finder* (as we called our tool) was developed as the first stage of a process for automatically improving usability on web applications. We explain this process in the next section.

2   The Process in a Nutshell

Our process for improving web usability is based on the refactoring technique. When applying refactoring in the context of web usability, developers first must detect bad usability smells, and then they must find refactorings to solve them, keeping the basic functionality intact. The tool helps them find bad usability smells.

The automated process for finding bad smells consists in three steps, depicted in Fig. 1. The **Threats Logger** is a client-sided script that gathers interaction events from real users. Instead of logging raw, atomic events, it processes them to generate *usability threats*, a concept we devised to represent higher-level interaction events.

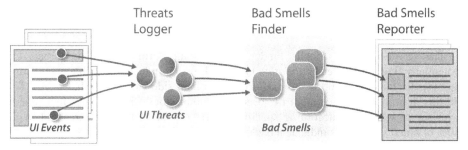

Fig. 1. Schematics of the process

The server-sided **Bad Smells Finder** receives usability threats and stores them for analysis. When a user requests a report, the Bad Smells Finder processes the threats and displays the resulting bad smells through the **Bad Smells Reporter** frontend.

3   The Tool in Action

We will show how the tool works with an example. Consider for instance a web application where users need to register before they can operate. Whenever a user fills the registration form, the threats logger captures the form submission event and evaluates what happens next:

- If no navigation follows, the threats logger considers the submission was blocked by **client-side validation**.
- If a navigation is detected, and the form is still on the destination page, the logger considers there was **server-side validation**.
- If a navigation is detected, but the form is absent in the destination page, the logger considers a **successful submission**.

The tool uses a simple algorithm to identify search forms, where validation rules do not generally apply. Combining all this information on the client-side, it creates a *Form Submission* threat and sends it to the Bad Smells Finder. The script can be set to *verbose* mode to show the threats it finds in the browser's console, as seen in Fig. 2.

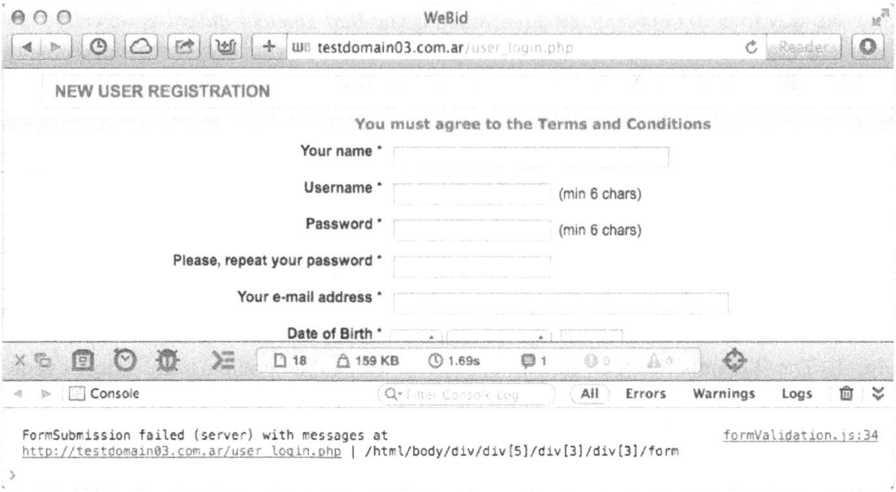

Fig. 2. Threats Logger indicating the detection of a *Failed Submission* threat

The server-side Bad Smells Finder processes all Form Submission threats to potentially find a *No Client Validation* bad smell, which indicates a problematic form that usually fails to submit without offering any client-side validation whatsoever. To do this, it compares the amount of successful submissions with the ones that failed with server validation, according specific criteria for the proportion threshold (e.g. 30% of failed validations indicate a bad smell).

The site owners may then ask for a report by accessing the tool's Reporter, where bad smells are listed with data like the URL where it happened, an XPath of the affected element, and specific extra information depending on each bad smell.

The Bad Smells Finder can detect 12 different kinds of bad usability smells, and the logics for detecting each one are diverse. Other featured bad smells are:

- **No Processing Page:** By calculating the average time of a request and watching DOM mutations, the Bad Smells Finder is able to detect that a process usually takes a long time, but users are never informed about that process taking place on the background (i.e. "loading..." widget).

- **Unnecessary Bulk Action:** Users perform actions on a list of items by first marking checkboxes, and then selecting the action – e.g. deleting emails on a webmail application. If the Bad Smells Finder detects that most of the time users apply actions one item at a time rather than many, then the Unnecessary Bulk Action is detected, implying that the UI with checkboxes and actions should be complemented with other mechanism that require less interactions.

- **Free Input For Limited Values:** A free text input is presented to the user, but the set of possible values that can be entered belong to a limited set, like countries or occupations. Two problems ensue: error-proneness, and unnecessarily time wasted in typing the whole text. The Bad Smells Finder captures all the inputs and calculates the proportion of repeated (and similar) values, in order to determine the bad smell's presence.

The rest of the bad smells are related to navigation issues (like long paths for frequently accessed pages), and misleading/misused widgets. We are currently extending the tool to detect more bad usability smells.

4 Tool Implementation and Usage

The tool has two main modules: the Threats Logger, implemented as a client-side script, and the Bad Smells Finder, a server-sided component that analyzes threats and reports bad smells.

The client-side Threats Logger (coded in JavaScript using the JQuery[3] library) captures interaction events and then processes them on the client to create (and filter) usability threats. The server-sided analyzer parses the incoming asynchronous POST requests from the client script to generate usability threats. When a report is asked, the analyzer filters all the threats to find potential bad usability smells, and then renders the bad smells report in the web frontend.

To install the tool, the site owner must include the Threats Logger script in the application's header. After completing this step, the Bad Smells Finder starts logging and reporting bad usability smells right away.

References

1. Fowler, M., Beck, K., Brant, J., Opdyke, W., Roberts, D.: Refactoring: Improving the Design of Existing Code. Object Technology Series. Addison Wesley (1999)
2. Garrido, A., Rossi, G., Distante, D.: Refactoring for Usability in Web Applications. IEEE Softw. 28, 60–67 (2011)
3. Rubin, J., Chisnell, D.: Handbook of Usability Testing: How to Plan, Design, and Conduct Effective Tests. Wiley (2008)
4. Atterer, R., Wnuk, M., Schmidt, A.: Knowing the user's every move. In: Proceedings of the 15th International Conference on World Wide Web, WWW 2006, p. 203. ACM Press, New York (2006)
5. Saadawi, G.M., Legowski, E., Medvedeva, O., Chavan, G., Crowley, R.S.: A Method for Automated Detection of Usability Problems from Client User Interface Events AMIA 2005 Symposium Proceedings, pp. 654–658 (2005)
6. Fujioka, R., Tanimoto, R., Kawai, Y., Okada, H.: Tool for detecting webpage usability problems from mouse click coordinate logs. In: Jacko, J.A. (ed.) HCI 2007. LNCS, vol. 4550, pp. 438–445. Springer, Heidelberg (2007)
7. Okada, H., Fujioka, R.: Automated Methods for Webpage Usability & Accessibility Evaluations. In: Adv. Hum. Comput. Interact., ch. 21, pp. 351–364. In-Tech Publ. (2008)

---

[3] http://jquery.com

An Extensible, Model-Driven and End-User Centric Approach for API Building

José Matías Rivero[1,2], Sebastian Heil[3], Julián Grigera[1], Esteban Robles Luna[1], and Martin Gaedke[3]

[1] LIFIA, Facultad de Informática, UNLP, La Plata, Argentina
{mrivero,julian.grigera,esteban.robles}@lifia.info.unlp.edu.ar
[2] Also at Conicet
[3] Department of Computer Science, Chemnitz University of Technology, Germany
{sebastian.heil,martin.gaedke}@informatik.tu-chemnitz.de

Abstract. The implementation of APIs in new applications is becoming a mandatory requirement due to the increasing use of cloud-based solutions, the necessity of integration with ubiquitous applications (like Facebook or Twitter) and the need to facilitate multi-platform support from scratch in the development. However, there is still no theoretically sound process for defining APIs (starting from end-user requirements) or their productive development and evolution, which represents a complex task. Moreover, high-level solutions intended to boost productivity of API development (usually based on Model-Driven Development methodologies) are often difficult to adapt to specific use cases and requirements. In this paper we propose a methodology that allows capturing requirements related to APIs using end-user-friendly artifacts. These artifacts allow quickly generating a first running version of the API with a specific architecture, which facilitates introducing refinements in it through direct coding, as is commonly accomplished in code-based Agile processes.

Keywords: API, Model-Driven Development, Agile Development, Prototyping.

1 Introduction

Over the last years, users and businesses have witnessed a trend to *move* applications and services to the cloud. Several aspects motivate this trend, most importantly cost and deployment time. In this context, developers must interact with applications and services they do not directly control, so they need APIs to facilitate this interaction. Since APIs generally centralize operations and business-logic among applications in several platforms, they are inherently complex; however, most development processes do not take this into account, particularly Agile methodologies, which do not provide clear and structure method to cope with the complexity of API design [1]. To tackle such complex requirements, we have introduced a Model-Driven Development (MDD) [2] solution called *MockAPI* [1], which is limited to providing a prototypical version of the API. In this paper we propose *ELECTRA* (standing for *Extensible modeLdriven Enduser CenTRic API*), an hybrid Agile, MDD and coding approach that

S. Casteleyn, G. Rossi, and M. Winckler (Eds.): ICWE 2014, LNCS 8541, pp. 494–497, 2014.
© Springer International Publishing Switzerland 2014

(1) uses an end-user friendly language to define API-related requirements as in MockAPI (annotated mockups), (2) allows to quickly define and generate a running API for testing integration with other software artifacts, and (3) proposes an API runtime that can be extended (hence the term *Extensible*) with custom code without breaking the model's abstraction. We chose mockups as our main requirement artifact because of their positive results in agile approaches [3], and their valuable requirements communication capabilities [4]. Using mockup and annotations as an end-user friendly language we intend to capture the complex API requirements and, at the same, time, provide a framework for quick API generation and refining.

2   The ELECTRA Approach

The ELECTRA process, depicted in Figure 1, is an adaptation of Scrum [5], the most widely used agile process in industry [6]. As in Scrum, every iteration in the ELECTRA approach starts by selecting the User Stories to be tackled (*Define Sprint Backlog* step). Then, ELECTRA mandatorily requires building mockups with essential end-user participation to concretize each of these stories (*Mockup Construction* step). After all User Stories are associated to mockups, developers use an enhanced version of the MockAPI tool [1] to tag the mockups with API-related annotations (*Mockup Tagging* step in Figure 1). These annotations are based on a simple grammar that makes them easy to understand by end-users. From the annotated mockups, an API can be derived through a code generation process (*API Generation* step). At this point, developers get a running usable API for integration testing with other software artifacts (*API testing* step). The generated API is deployed to ELECTRA's runtime environment which allows developers to refine it through direct coding (*API Refining*). The result of this process is the Final API for the current iteration (*API Increment* step).

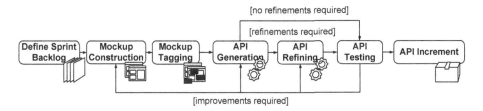

Fig. 1. The ELECTRA workflow

Any type of user interface mockup can be used with ELECTRA tooling. After mockups have been imported, different kinds of annotations can be defined by the engineers in presence and with collaboration of end-users. The three most important annotations types are *Data annotations*, which allow defining object types or business entities well-known by stakeholders, *Constraint annotations*, which enable the definition of business rules and *Action annotations*, which describe the execution of heterogeneous or complex tasks within the API. Besides its technical specifications (understood by engineers), some annotations provide an end-user friendly structured text mode to describe API requirements in natural language. While end-users require engineers help to write annotations, this text mode eases their understanding. ELECTRA tooling currently uses JavaScript as its default scripting language, but any scripting language implementing the Java Scripting API can be used instead. In Figure 2 a *Data* and *Action* annotations are shown applied over an invoice management application to specify the existence of an Invoice business object and how it must be integrated with an external API.

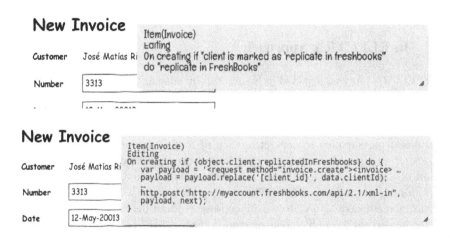

Fig. 2. Annotations in its different modes: end-user friendly (upper) and developer (lower)

3   Architecture and Code Generation

The core of the ELECTRA tool implementation (the so-called ELECTRA API runtime) defines and implements a RESTful API as a set of Endpoints, which are composed by an HTTP method, a URL regular expression and a script. When a new HTTP request is received by the runtime, it seeks for a matching Endpoint (with the same HTTP method and a matching URL regular expression). If it finds one, it executes its internal script. Endpoint scripts can access and invoke other *Scripts*, use stored *Resources* (for instance, media content as images or video) or invoke operations on *Services*, which are declared and customized by developers.

ELECTRA's code generation process consists in the generation of a set of Endpoints and Scripts, and the configuration of a default DB Service for storage purposes analyzing the annotated mockups. After triggering a code generation, developers can tune the API as much as needed just editing the required Scripts and Endpoints or changing the DB Service used. Also, they can define new Resources, Services or Endpoints manually. Scripts are generated respecting the Pipes and Filters pattern (where every Filter is formed by a Script implementing a different concern – i.e., code for a different annotation type –, thus facilitating and isolating changes in the API.

When triggering a code regeneration, edited Scripts (Filters) are not altered, thus preserving changes incorporated through direct coding.

4   Related Work

Benefits of using mockups as a primary requirement artifact in the development process have been reported through statistical studies [7]. Also, its benefits in the context of Agile and MDD processes (even API generation) have been commented [8]. In addition, in our previous work [8] we demonstrated that mockup and annotations provide a modeling framework that results more efficient than manual modeling, even considering conceptual models – which are very similar to the *Data annotations* presented in this work. However, none of these works propose a Model-Driven approach that can be extended through direct coding even in the models, as ELECTRA provides. Several MDD approaches specifically tackle RESTful APIs [9,10] (as ELECTRA), but they not provide a clear and pattern-based codebase allowing the quick introduction of detailed implementation as ELECTRA does.

References

[1] Rivero, J.M., Heil, S., Grigera, J., Gaedke, M., Rossi, G.: MockAPI: An Agile Approach Supporting API-first Web Application Development. In: Daniel, F., Dolog, P., Li, Q. (eds.) ICWE 2013. LNCS, vol. 7977, pp. 7–21. Springer, Heidelberg (2013)

[2] Kelly, S., Tolvanen, J.-P.: Domain-Specific Modeling: Enabling Full Code Generation. Wiley-IEEE Computer Society (2008)

[3] Ferreira, J., Noble, J., Biddle, R.: Agile Development Iterations and UI Design. In: Agil. 2007 Conf., pp. 50–58. IEEE Computer Society, Washington, DC (2007)

[4] Mukasa, K.S., Kaindl, H.: An Integration of Requirements and User Interface Specifications. In: 6th IEEE Int. Requir. Eng. Conf., pp. 327–328. IEEE Computer Society, Barcelona (2008)

[5] Sutherland, J., Schwaber, K.: The Scrum Papers: Nuts, Bolts, and Origins of an Agile Process

[6] VersionOne Inc., State of Agile Survey (2011)

[7] Ricca, F., Scanniello, G., Torchiano, M., Reggio, G., Astesiano, E.: On the effectiveness of screen mockups in requirements engineering. In: 2010 ACM-IEEE Int. Symp. Empir. Softw. Eng. Meas. ACM Press, New York (2010)

[8] Rivero, J.M., Grigera, J., Rossi, G., Luna, E.R., Montero, F., Gaedke, M.: Mockup-Driven Development: Providing agile support for Model-Driven Web Engineering. Inf. Softw. Technol., 1–18 (2014)

[9] Pérez, S., Durao, F., Meliá, S., Dolog, P., Díaz, O.: RESTful, Resource-Oriented Architectures: A Model-Driven Approach, in: Web Inf. In: Chen, L., Triantafillou, P., Suel, T. (eds.) WISE 2010. LNCS, vol. 6488, pp. 282–294. Springer, Heidelberg (2010)

[10] Valverde, F., Pastor, O.: Dealing with REST Services in Model-driven Web Engineering Methods. In: V Jornadas Científico-Técnicas En Serv. Web y SOA, JSWEB (2009)

Building Bridges between Diverse Identity Concepts Using WebID

Michel Rienäcker, Stefan Wild, and Martin Gaedke

Technische Universität Chemnitz, Germany
{firstname.lastname}@informatik.tu-chemnitz.de

Abstract. Single sign-on systems enable users to log into different Web services with the same credentials. Major identity providers such as Google or Facebook rely on identity concepts like OpenID or OAuth for this purpose. WebID by the W3C offers similar features, but additionally allows for storing identity data in an expressive, extensible and machine-readable way using Linked Data. Due to differences in manageable user attributes and the authentication protocols as such, the identity concepts are incompatible to each other. With more than one identity concept in use, users need to remember or keep further credentials. In this paper we therefore propose the B3IDS approach. It aims at improving the user experience and the adoption of WebID by building bridges between diverse identity concepts with WebID. We exemplarily implement B3IDS in an existing WebID identity provider and management system.

Keywords: Identity, Linked Data, Social Networks, Security, WebID.

1 Introduction

For providing a personalized experience, today's Web applications rely on user authentication. Major single sign-on systems have gradually replaced proprietary authentication solutions by consolidating the users' digital identities. So, they also addressed the password-fatigue issue [1]. Global enterprises, e.g., Google or Facebook, offer widely used authentication systems like OpenID or Facebook Connect. While this is convenient for the users, it carries the risk of analyzing user-generated content, including shared information, messages and votes. This enables tracking the user's behavior and creating tailored advertisements. To prevent this risk, the W3C created WebID, which does not only enable authentication, but also empowers people to keep control about their identity data [3].

Despite the advantages, WebID is still in development and not yet broadly accepted by common Web users. Service providers are also reluctant to integrate it as an authentication option. Reasons might be the missing involvement of username/password pairs and technological problems through requesting a certificate. An insufficient adoption by service providers reduces the chance of convincing more users from the benefits of WebID and, thus, prevents its wider use.

S. Casteleyn, G. Rossi, and M. Winckler (Eds.): ICWE 2014, LNCS 8541, pp. 498–502, 2014.

Enabling users to access their Web applications and services as usual and make use of WebID requires the availability of identity bridges. Identity bridging is predicted by Gartner as a major trend in the next year [4]. Existing approaches including Mozilla BigTent (`https://wiki.mozilla.org/Identity/BrowserID/BigTent`), CA CloudMinder Gateway [2] and ForgeRock Bridge Service [6] enable bridging between particular identity concepts such as OpenID and Mozilla Persona, but they are characterized by limitations including only providing one-way identity bridging or only enabling access to selected applications via SAML or OAuth.

In order to address the problem, we propose the B3IDS approach. It uses WebID to mediate between different identity concepts. In WebID, all identity data is based on RDF. So, it is expressive, extensible, and machine-readable. This enables representing all data required in other identity concepts in so-called WebID profiles. Users maintain exclusive control about their identity data.

To demonstrate the approach, we show how B3IDS mediates between WebID and OpenID as an example. Relying on WebID, B3IDS strives to bridge the gap between diverse identity concepts by considering 1) exchange of identity data, 2) authentication as usual, and 3) WebID creation from existing identities like an OpenID. So, B3IDS aim at giving rise to the adoption of WebID by the users.

2   Building Secure Bridges to WebID

For bridging between diverse identity concepts, B3IDS mimics two generic components: an identity provider and a relying party. The B3IDS identity provider allows users for authenticating to relying parties of other identity concepts using their WebID certificates as specified in the WebID authentication sequence. Relevant data available in a WebID profile is therefore transparently mapped to data required in the other identity concept. The B3IDS relying party enables users to create a new WebID identity based on their identity data from another identity concept like OpenID. Users can then authenticate to services with such WebID identities. As an example, Alice wants to access a WebID service, but only has a Google account. To get access, she uses B3IDS to issue an authentication request to Google. Alice authenticates herself and, thus, authorizes B3IDS to retrieve identity data from Google. The approach then creates a new WebID profile and a WebID certificate which Alice can use for logging into the WebID service.

B3IDS also incorporates logging Alice into a relying party of another identity concept like OpenID, as shown in Fig. 1. Therefore, B3IDS mimics an OpenID provider retrieving user data from Alice's WebID profile. To accomplish this, Alice passes an identifier to the OpenID relying party (cf. ① in Fig. 1) . The identifier contains the URI of the location of the identity bridge and the WebID URI linking to her WebID profile in the query part, e.g., `https://b3ids.example.org/?webid=https://webid.example.org/Alice`. The OpenID relying party redirects Alice's user agent to B3IDS using the given identifier. It then requests her identity data (cf. ②). Having received Alice's

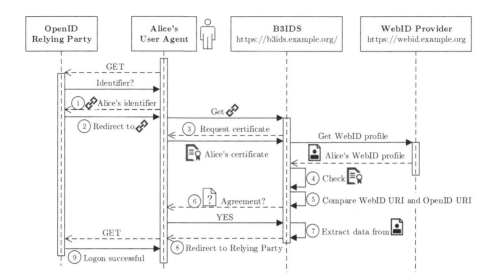

Fig. 1. Sequence Diagram of B3IDS Bridging WebID to OpenID

identifier, B3IDS asks Alice for proving her identity (cf. ③). According to the WebID authentication sequence, Alice has to legitimate herself with the right WebID certificate (cf. ④). To ban an illegitimate WebID URI, B3IDS verifies that her WebID certificate contains a URI that matches the WebID URI in the identifier (cf. ⑤). When Alice grants access to her identity data (cf. ⑥), B3IDS transfers requested identity data to the relying party (cf. ⑦) by redirecting her user agent back to it (cf. ⑧). Alice is logged in successfully (cf. ⑨).

To show B3IDS in practice, we prototypically implemented the approach in the Sociddea WebID provider known from previous work [7]. Fig. 2 presents a usage scenario of logging into the OpenID-enabled Stack Overflow using a WebID identity. The numbers in this figure correlates to the number in Fig. 1. While the implementation can only bridge between WebID and OpenID at the moment, our approach is neither restricted to this combination nor to Sociddea. In the editing box shown in Fig. 2 users can type their OpenID identifier to create a WebID identity based on their OpenID identity. To log into an OpenID relying party users have to use the following identifier: `https://vsr-demo.informatik.tu-chemnitz.de/sociddea/bridge/openid?webid=<WebID-URI>`.

Demonstration. For a live demo and further information about B3IDS and Sociddea visit: `http://vsr.informatik.tu-chemnitz.de/demo/sociddea/`

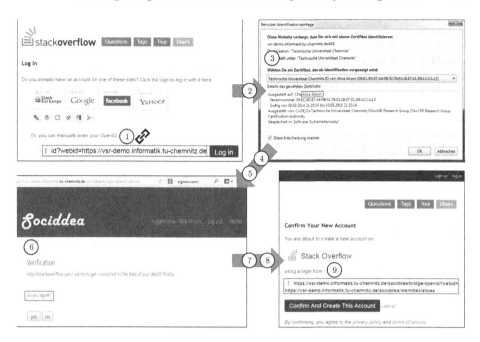

Fig. 2. Logging into Stack Overflow using a WebID identity

3   Conclusion and Future Work

By enabling users to both create WebID identities based on the personal data stored at their OpenID identity providers and use their new WebIDs to log into OpenID relying parties, B3IDS demonstrated the identity bridging concept as an example. In B3IDS, WebID users can authenticate to relying parties using a typical OpenID without impairing their normal user experience. Thus, we expect to increase the user acceptance of WebID. Future work will focus on addressing the security concerns related to B3IDS acting as a mediator between identity concepts with the need of having access to associated user data. We also plan to support further identity concepts like OpenID Connect, which obsoletes OpenID [5]. In addition to this, we will work on retrieving protected profile data taking the user's privacy and secrecy preferences into account.

References

1. Dhamija, R., Dusseault, L.: The Seven Flaws of Identity Management: Usability and security Challenges. IEEE Security & Privacy 6(2), 24–29 (2008)
2. Diodati, M.: Identity Bridges: Uniting Users and Applications Across the Hybrid Cloud (2012), https://www.gartner.com/doc/2008315/identity-bridges-uniting-users-applications (accessed March 12, 2014)
3. Inkster, T., et al.: WebID TLS - W3C Editors Draft (2014), http://www.w3.org/2005/Incubator/webid/spec/tls/ (accessed March 12, 2014)

4. Pettey, C., et al.: Gartner Identifies Six Trends That Will Drive the Evolution of Identity and Access Management and Privacy Management (2012), http://www.gartner.com/newsroom/id/1909714 (accessed March 12, 2014)
5. Sakimura, N., et al.: OpenID Connect Core 1.0 (2014), http://openid.net/specs/openid-connect-core-1_0.html (accessed March 11, 2014)
6. Simmons, H.: ForgeRock Releases Breakthrough Identity Bridge (2013), http://www.reuters.com/article/2013/07/30/ca-forgerock-idUSnBw305380a+100+BSW20130730 (accessed March 12, 2014)
7. Wild, S., Chudnovskyy, O., Heil, S., Gaedke, M.: Protecting User Profile Data in WebID-Based Social Networks Through Fine-Grained Filtering. In: Sheng, Q.Z., Kjeldskov, J. (eds.) ICWE Workshops 2013. LNCS, vol. 8295, pp. 269–280. Springer, Heidelberg (2013)

Cross-Browser Testing in Browserbite

Tõnis Saar[1], Marlon Dumas[2], Marti Kaljuve[1], and Nataliia Semenenko[2]

[1] Software Technology and Applications Competence Center, Estonia
{tonis.saar,marti.kaljuve}@stacc.ee
[2] University of Tartu, Estonia
{marlon.dumas,nataliia}@ut.ee

Abstract. Cross-browser compatibility testing aims at verifying that a web page is rendered as intended by its developers across multiple browsers and platforms. Browserbite is a tool for cross-browser testing based on comparison of screenshots with the aim of identifying differences that a user may perceive as incompatibilities. Browserbite is based on segmentation and image comparison techniques adapted from the field of computer vision. The key idea is to first extract web page regions via segmentation and then to match and compare these regions pairwise based on geometry and pixel density distribution. Additional accuracy is achieved by post-processing the output of the region comparison step via supervised machine learning techniques. In this way, compatibility checking is performed based purely on screenshots rather than relying on the Document Object Model (DOM), an alternative that often leads to missed incompatibilities. Detected incompatibilities in Browserbite are overlaid on top of screenshots in order to assist users during cross-browser testing.

Keywords: Cross-browser compatibility testing, image processing.

1 Introduction

Cross-browser (compatibility) testing aims at finding incompatibilities in the way a Web page is rendered across different combinations of a browser, a browser setting, an operating system (OS) and a hardware platform (herein called a *configuration*). The exact meaning of the term "incompatibility" varies from one testing subject to another and hence cross-browser testing has to take into account the sensitivity of the intended user(s). Incompatibilities may range from missing buttons, to misaligned text blocks, broken images or misplaced elements. In the absence of tool support for cross-browser testing, testers have to open web pages manually and check for differences. This procedure is time-consuming, monotonous and non-scalable given the growing number of configurations that need to be supported by Web applications.

Existing automated methods for cross-browser testing are generally based on an analysis of the Document Object Model (DOM) [1][2][3]. However, the fact that a Web page has very similar DOM structure and parameters across different configurations does not guarantee absence of incompatibilities, as rendering engines may display similar DOMs in rather different ways. Thus DOM-based cross-browser testing

S. Casteleyn, G. Rossi, and M. Winckler (Eds.): ICWE 2014, LNCS 8541, pp. 503–506, 2014.

techniques suffer from lower recall (high number of missed incompatibilities). Some techniques such as WebDiff [1] apply DOM-based web page segmentation in conjunction with image comparison over pairs of matching segments. But while the latter step improves recall, the DOM segmentation step may still hide incompatibilities.

In contrast to the above techniques, Browserbite employs image processing both for web page segmentation and segment comparison. Specifically, Browsebite combines an image segmentation technique based on detection of discontinuities and colour changes with an image comparison technique based on a combination of geometric features and histograms of pixel intensity distribution. These techniques are complemented by supervised machine learning, so as to take into account user sensitivity.

2   System Overview

Browserbite consists of three main components: screenshot generation, image segmentation and comparison, and classification, as shown in Fig. 1. These components are triggered sequentially when a user inserts a URL of a *web page under test* in Browserbite's interface. The URL is added to a queuing system implemented using Ruby Resque. Different Ruby workers then take specific tasks from the queue and perform the task in question, incl. generating screenshot, resizing image, comparing pair of images, or filtering potential incompatibility via a classification model.

Fig. 1. Overview of Browserbite's tool chain

In addition to a URL, the user specifies a *baseline configuration*, that is a browser-OS-platform for which the user has verified correct rendering. On average, results are displayed in 30 – 45 seconds (with initial results shown incrementally). Detected incompatibilities are highlighted on top of the baseline configuration as shown in Fig 2.

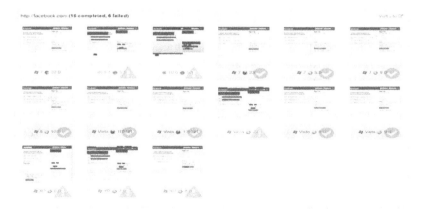

Fig. 2. Example of Browserbite report

2.1   Screenshot Capturing

A number of virtual machines are used to generate screenshots on different browsers and operating systems. Browserbite supports six OS (Windows XP, Vista, 7, 8, Apple OS X 10.6, iOS 6.0) and two browsers with default settings (Google Chrome, Firefox, IE and Safari). A screenshot is generated using the Selenium WebDriver library.

After a web page is loaded, the browser window is maximized. Then the whole window (a.k.a. viewport) is saved as an image. In case of OS X a full-page screenshot has to be composed out of fragments, in a process of scrolling, screenshotting and re-stitching. In Windows, a full-page screenshot can be taken in a single step [4].

2.2   Segmentation and Comparison

In this stage, pairs of screenshot images are compared to find significant differences. One of the images is a baseline image and the other is an image under test.

As mentioned earlier, pixel-by-pixel comparison leads to excessive false positives. Indeed, small misalignments of even one pixel may cause pixel-by-pixel comparison to immediately fail. Accordingly, Browserbite adopts a two-step comparison process. First, images are segmented into smaller so-called regions, which are then matched pairwise. Segmentation helps to prevents false alarms caused by small misalignments of web page elements. The segmented regions can represent for example buttons, forms, headings, text blocks etc. Segmentation in Browserbite is based purely on visual features (discontinuity and colour changes) and is implemented using well-known image processing techniques [5].

In a second step, segments are compared pairwise (one baseline segment versus one image-under-test segment). Pairwise comparison is performed first on geometric features (position and size) and secondly on the values of the histogram of pixel density distribution, following a well-known histogram extraction technique used in computer vision [5]. Each baseline segment is matched to the most similar segment from the image-under-test. If two matched segments have differences in feature parameters beyond a *tolerance threshold*, or if a segment in one image has no matching pair in the other, the segment(s) is/are declared *potentially incompatible*. Tolerance thresholds have been tuned experimentally based on a corpus of images (see below) in such a way as to produce a small number of false negatives (2% of missed incompatibilities, i.e. 98% recall). These thresholds however lead to a precision of 66%. To strike a better tradeoff, Browserbite relies on an additional classification stage.

2.3   Classification

The classification stage is used to classify potential incompatibilities into actual incompatibilities versus false alarms. The classifier uses the same features mentioned above, which are extracted via image comparison (full list of features is given in [6]).

For training and testing the classifier, we used the 140 most popular web pages in Estonia (from the alexa.com list). These web pages were tested manually using Browserbite without the classifier component. As a result 20 000 potential differences were found. From this set 2700 segment pairs were randomly selected. 40 people were asked to classify these 2700 potential incompatibilities into the two classes. As a result 1350 positive and negative cases were obtained.

We tested both decision trees and neural networks (using the OpenCV library) as classification techniques. Neural networks give clearly better results [6]. Plain Browserbite without neural network classifier has precision of 66% and recall 98%. The neural network classifier improves precision to 96% with a recall of 89%, illustrating the trade-offs. Despite this imperfect result, Browserbite's accuracy (F-score) is superior to that of a state-of-the-art tool (Mogotest) [6].

On the background of these trade-offs, Browserbite has been made a commercial product, available on a software-as-a-service basis at: http://www.browserbite.com. It has a growing user base (over 10 000 registered users). At present, Browserbite can produce false positive results while testing pages with dynamic regions (e.g. animations). It is planned to add dynamic region suppression. Dynamic regions are detected by taking a screenshot of a web page with an interval in-between, and comparing the two screenshots using the same technique described above.

3   Conclusion

The Browserbite development experience demonstrates the feasibility and power of cross-browser testing based on image processing. Extensions include the ability to handle Web page flows (as opposed to individual pages) and the adaptation of Browserbite to non-traditional web platforms like smart TV's, billboards and GPS devices.

References

1. Choudhary, S.R., Versee, H., Orso, A.: WEBDIFF: Automated identification of cross-browser issues in web applications. In: 2010 IEEE International Conference on Software Maintenance (ICSM), pp. 1–10 (2010)
2. Choudhary, S.R., Prasad, M.R., Orso, A.: CrossCheck: Combining Crawling and Differencing to Better Detect Cross-browser Incompatibilities in Web Applications. In: 2012 IEEE Fifth International Conference on Software Testing, Verification and Validation (ICST), pp. 171–180 (2012)
3. Mesbah, A., Prasad, M.R.: Automated cross-browser compatibility testing. In: Proceedings of the 33rd International Conference on Software Engineering, pp. 561–570 (2011)
4. Kaljuve, M.: Cross-Browser Document Capture System. Master's Thesis, University of Tartu (June 2013), http://tinyurl.com/nlze7ub
5. Shapiro, L.G., Stockman, G.C.: Computer Vision. Prentice Hall (2001)
6. Semenenko, N., Dumas, M., Saar, T.: Browserbite: Accurate Cross-Browser Testing via Machine Learning Over Image Features. In: Proceedings of the 28th International Conference on Software Maintenance (ICSM), pp. 528–531. IEEE Computer Society (2014)

DireWolf Goes Pack Hunting: A Peer-to-Peer Approach for Secure Low Latency Widget Distribution Using WebRTC

István Koren, Jens Bavendiek, and Ralf Klamma

Advanced Community Information Systems (ACIS) Group,
RWTH Aachen University,
Ahornstr. 55, 52056 Aachen, Germany
{lastname}@dbis.rwth-aachen.de
http://dbis.rwth-aachen.de

Abstract. Widget-based Web applications are outperforming mono-lithic Web applications in terms of distribution of the user interface on many devices and many standard browsers. However, latency of the re-mote inter-widget communication may be an obstacle for the uptake of Widget-based Web applications in near real-time domains like Web gaming and augmented reality. In this demo paper we show DireWolf 2.0 which is replacing the XMPP server of the DireWolf approach by a client-side relay realized by the means of WebRTC. This is not only decreas-ing the latency of the distributed interface for any application but also increasing the security by avoiding man-in-the-middle attacks on the XMPP server. This progress is enabling further uptake in Widget-based solutions in advanced Web engineering.

1 Introduction

The Web as a ubiquitous platform allows us to deal with complex tasks within the familiar environment of a Web browser. Mobile devices such as laptops, smartphones and tablets allow us to access Web resources from any possible location. This goes in hand with the shift away from fixed office settings to mobile and dynamic working environments. Yet using multiple devices in parallel is not sufficiently dealt with on the Web. Though the *Responsive Web Design (RWD)* [1] paradigm helps us to open the same Web site on devices with varying screen sizes, it only targets the UI level and does not care about data migration and synchronization issues.

To this end in earlier work we have presented the DireWolf framework [2]. DireWolf is a widget based Web application framework that allows distributing widgets over multiple devices while keeping the application state. It is based on the *Inter-Widget Communication (IWC)* capabilities of the underlying Open Source ROLE SDK[1]. However, the underlying architecture that involves mes-sages being sent over an XMPP server turned out to be an obstacle for latency-sensitive near real-time Web applications.

---

[1] http://sourceforge.net/projects/role-project/

S. Casteleyn, G. Rossi, and M. Winckler (Eds.): ICWE 2014, LNCS 8541, pp. 507–510, 2014.

To solve this problem we hereby introduce the next iteration of the DireWolf framework that uses the recent *Web Real-Time Communication (WebRTC)* draft for sending peer-to-peer messages from browser instances across multiple devices [3]. In the next sections we present our architecture that involves a refurbished message passing and show a preliminary evaluation that confirms an average decrease of message round-trip times of around 80%.

A video of our demo is available at `http://goo.gl/ZV7RJ1`.

2   Peer-to-Peer Distributed User Interfaces

DireWolf 1.0 is using a server-side `Device Manager` that maintains different devices and their respective device profiles e.g. tablet or desktop of a user; the server also stores widget arrangements i.e. which widget is present on which device. Clients initiate the migration of widgets by requesting it at the `Device Manager` which then notifies online devices. The communication between clients and the `Device Manager` as well as between the devices themselves is organized by using publish/subscribe over XMPP. The central routing leads to two major problems: First, the indirection over the server is on the expense of high latency. Second, security is threatened as the XMPP server imposes a single point of failure for man-in-the-middle attacks. Since in most cases where devices are used jointly to master a task they are in immediate vicinity, it is obvious to establish direct peer-to-peer connections for message exchange.

The recent WebRTC draft introduces the DataChannels API that allows Web applications to connect and send arbitrary data to instances running on other devices; thereby firewalls as well as NATs and proxy servers are automatically taken care of. Additionally, WebRTC connections are encrypted by default. Therefore, WebRTC fulfills two major requirements to overcome the aforementioned drawbacks of the existing DireWolf framework: Security as well as low latency through avoiding intermediary servers[2].

To leverage WebRTC we have adapted the architecture of DireWolf. Instead of sending the IWC messages through the XMPP server we introduced a *relay* that acts as a proxy server for all participating devices. We deliberately opted for the relay approach in order to avoid a full-mesh topology of the peer-to-peer network as initial tests showed a performance loss caused by resource-constrained mobile devices. The relaying device now hosts the Message Router that is responsible for setting up the proxied publish/subscribe node and the WebRTC connections to other clients. We still keep the XMPP connection in place for exchanging endpoint information and the initial connection negotiation process; all other messages are sent over the relay. We have successfully submitted an XMPP extension protocol that describes the DataChannel negotiation process over Jingle to the *XMPP Standards Foundation (XSF)* [4].

---

[2] In case all firewall-traversal techniques fail WebRTC still redirects the encrypted traffic through *Traversal Using Relays around NAT (TRUN)* servers as last resort.

3   Evaluation

We performed both a comparative technical evaluation as well as a user study to prove our conceptions. For the technical part, we measured the round-trip-times of messages sent from a widget to an instance running on a remote device. Figure 1 shows the results of the test series which included 50 runs. First we measured the round-trip times on the previous DireWolf framework that uses the XMPP server for message routing; the average delay was around 143 ms.

Fig. 1. Latency over various connections   **Fig. 2.** Evaluation Space

In contrast, test runs performed within the newly developed DireWolf 2.0 framework show a **significant reduction** of round-trip times, however results vary depending on where the computationally intensive publish/subscribe relay was hosted. On a state of the art desktop PC we reached an average round-trip time of around 26 ms. Measurements on a recent smartphone revealed an average time of around 99 ms though there were significant outliers; we assume that these peaks were caused by periodical background tasks occupying the processor.

To the usability end we performed a preliminary evaluation with 16 participants recruited from our research lab. We tested the outcome with two highly latency sensitive applications: near real-time collaborative painting as well as gaming. Both widget spaces were shown on an Android tablet and a Windows laptop side by side and executed first on DireWolf 1.0 and then on DireWolf 2.0. First, users were requested to migrate widgets from one device to another. Then, participants were asked to draw a house on the tablet; the painting was synchronized to a laptop screen via the framework. Finally, users were presented a little platform game that can be seen on Figure 2; the control widgets were placed on the touchscreen device while the interactions were shown on the laptop screen.

For the migration 60% rated DireWolf 2.0 as being "very fast" while around 40% noted it was "fast". DireWolf 1.0 scored around 20% for "very fast" and around 55% for "fast"; around 20% of the interviewed persons even considered the migration time being "slow".

For DireWolf 1.0 the overall synchronization speed was rated as "slow" by around 40% and even "very slow" by around 25%, while the proportions for DireWolf 2.0 shifted to "very fast" or "fast" by almost 100% of the respondents.

4   Conclusion and Future Work

In this paper we have presented the demo of the DireWolf framework in its newest iteration which allows distributing widget based Web applications to various devices while leveraging the recent WebRTC draft for peer-to-peer style message exchange. For that we have successfully moved the publish/subscribe functionality from the server to a dedicated client in the form of a relay node; client devices connect to the relay for getting updates on the application state.

The technical evaluation has shown that we could decrease the average latency from around 150 ms to around 25 ms with WebRTC relayed by a desktop computer. Furthermore we are now able to prevent man-in-the-middle attacks on the XMPP server for synchronization messages. A usability study verified our findings.

What remains open for future work is a comparison with other DUI or IWC solutions like in the established *Omelette* project. Technical limitations of our systems include the comparatively high initial migration time, as currently widget resources like HTML, JavaScript and image files on a new device still have to be loaded from a Web server. Employing W3C widgets might solve this problem as they are packaged in a single zip file; the system would then send the whole widget bundle over the peer-to-peer connection. Besides, responsifying the widget spaces has a high priority in order to accommodate for today's huge variation in display sizes and resolutions.

We are committed to tackle these challenges in future versions of the DireWolf framework and are dedicated to continue defining the underlying standards.

Acknowledgements. The work has received funding from the European Commission's FP7 IP Learning Layers under grant agreement no 318209.

References

1. Nebeling, M., Norrie, M.C.: Responsive Design and Development: Methods, Technologies and Current Issues. In: Daniel, F., Dolog, P., Li, Q. (eds.) ICWE 2013. LNCS, vol. 7977, pp. 510–513. Springer, Heidelberg (2013)
2. Kovachev, D., Renzel, D., Nicolaescu, P., Klamma, R.: DireWolf - Distributing and Migrating User Interfaces for Widget-Based Web Applications. In: Daniel, F., Dolog, P., Li, Q. (eds.) ICWE 2013. LNCS, vol. 7977, pp. 99–113. Springer, Heidelberg (2013)
3. Burnett, D., Bergkvist, A., Jennings, C., Narayanan, A.: WebRTC 1.0: Real-time Communication Between Browsers (2013)
4. Bavendiek, J.: XEP-0343: Use of DTLS/SCTP in Jingle ICE-UDP Version 0.1, Experimental (2014)

Easing Access for Novice Users in Multi-screen Mashups by Rule-Based Adaption

Philipp Oehme, Fabian Wiedemann, Michael Krug, and Martin Gaedke

Technische Universität Chemnitz, Germany
{firstname.lastname}@informatik.tu-chemnitz.de

Abstract. Novice users often need support to become familiar with a new mashup. The most common problem is that mashups offer a high grade of personalization, such as the user's choice which widgets she wants to use. This problem becomes more difficult in multi-screen mashups, because the user has to decide additionally on which screen the widgets should run. In our recent work we focused on creating multi-screen mashups for enriching multimedia content. That is, a user can watch a video on one screen and also can consume additional content, like a Google Maps excerpt, on another one. This paper presents an approach for rule-based adaption of multi-screen mashups to ease the access for novice users. Therefore, we analyze the users' interaction with the mashup and detect patterns. Based on these patterns we derive rules which will be applied to the mashups of novice users as well as experienced ones. Thus, widgets will be added and arranged automatically on the user's several screens when the execution of a previously generated rule is triggered.

Keywords: Mobile, distributed user interface, distributed displays, multi-screen applications, web applications, mashup, widgets, user interface adaption.

1 Introduction

Mashups allow end users without programming knowledge to easily create applications for their desired target. There, the end users combine small applications, so called widgets, to accomplish their goal. With the emerging amount of mobile devices the opportunities of mashups increase. That is, the mashup is not limited to run on one screen. Rather, multiple screens can be used to create the user interface of one mashup. The following example illustrates the opportunities of these multi-screen mashups: Alice watches a video about an animal documentary on her TV. Meanwhile, she receives additional content on her tablet. This additional content could be a map excerpt about the location of the current scene or the Wikipedia article of the animal which is presented. In this scenario several widgets are described, like a video widget, a map widget or a Wikipedia widget.

One problem for users - especially for novice ones - is that they have to decide which widget they should add. This problem gets more difficult in the

S. Casteleyn, G. Rossi, and M. Winckler (Eds.): ICWE 2014, LNCS 8541, pp. 511–514, 2014.

field of multi-screen mashups due to the choice on which screen a widget should be placed. Closely related to this topic is the automatic composition in classic single-screen mashups described in [4]. While Roy Chowdhury et al. focus on creating a useful startup configuration of a mashup, they do support neither the adaption of the mashup during runtime nor mashups across multiple screens.

In this paper, we enhance our recent approach for multi-screen mashups to ease the access for novice users to multi-screen mashups by rule-based adaption. By observing and analyzing user actions we create rules which are executed, if a defined action is triggered. For example, Alice adds a video widget to her TV and a map widget to her tablet. This interaction will be transformed into a rule. When Bob adds a video widget to his TV, the rule is triggered and a map widget is automatically added to his tablet.

The rest of this paper is organized as follows: In Section 2 we first present our previous work called SmartComposition and we extend this approach with rule-based adaption in Section 3. Finally, we provide a conclusion and give an outlook on future work in Section 4.

2   SmartComposition Overview

In recent work we described an approach for creating multi-screen mashups called SmartComposition. For illustrating this approach we developed a prototype that enriches a video with related information originating from the Web [1]. In the SmartComposition approach, we describe a workspace as the union of all SmartScreens of one user who uses a mashup. A SmartScreen is an abstract representation of the browser window. It offers the runtime environment for the widgets and also includes functionality to enable the inter widget communication. We differentiate between mobile and desktop SmartScreens that differ in size, accessibility and the available widgets. We also developed a mechanism to enable inter widget communication across multiple screens.

We extend the SmartComposition to offer rule-based adaption of multi-screen mashups for novice users.

3   SmartComposition Enhancements

Our approach enhances the existing architecture of the SmartComposition to effectively support adaption for novice users. The following is a brief description of the extensions we made.

To capture the user actions and also the goal of the users [3, 5] the interaction receiver observes every interaction of the user with the system. We consider actions as every interaction of the user with the system, such as adding new widgets to any SmartScreen as well as moving widgets across several screens or removing them. The captured actions include all required information to reproduce the action by the system, such as the device, the type of the action and other relevant options, such as the widget position. Afterwards, the captured actions are analyzed by the pattern detector to search for patterns on the different

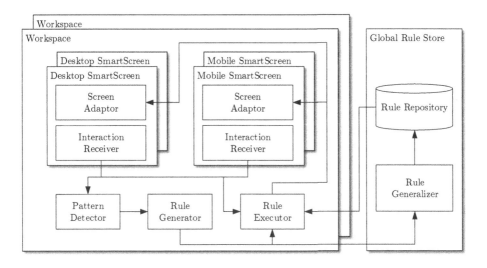

Fig. 1. Architecture of the approach

actions on all devices of a user. That is, the pattern detector examines similar actions which are performed in a specific time interval. Thereby it searches for two actions, one of them is the initiating action and the other one is the resulting action.

If the identified pattern is detected again, the rule generator creates a new rule based on the actions of the pattern. For deciding which pattern will be transformed to a rule we introduce a significance value. This significance value takes two parameters into account: the time difference between the two related actions of one pattern and the amount of occurrences of the same pattern.

The generated rules consist of the initiating action, the resulting action and further parameters to particularize the rule. While the initiating action is triggering the rule later on, the resulting action takes effect in the process of adaption. The rules are utilized to adapt the system to the user [2]. With the aid of the interaction receiver, the rule executor searches for a recurrence of an action that is part of a rule as an initiating action. If an initiating action is found, the rule executor executes the associated resulting action of the related rule. Thus, the action of adding a video widget can lead to the automatic execution of adding a map widget on one of the screens in the workspace.

The rule generalizer uses created user specific rules to derive global rules that are applicable to all users. Therefore, the rule generalizer examines the quality of a rule. This quality is associated to the user's experience in using the mashup. That is, a novice user also creates rules in her own workspace but they are ignored by the rule generalizer because of their minor quality. The generalized rules are saved in the rule repository from which rules can be retrieved from the rule executor in any workspace.

Demonstration. The prototype presented in this paper is available for testing at: http://vsr.informatik.tu-chemnitz.de/demo/chrooma/adaption/

4   Lessons Learned and Outlook

In this paper we enhanced our SmartComposition approach to adapt multi-screen mashups by a rule-based system. We observe the users' interaction with the mashup and detect patterns of these interactions. Based on these patterns the system generated rules that follow the abstract template of If-This-Then-That. These rules are executed if a condition is triggered and the multi-screen mashup will be updated. We also introduced a rule-system which differentiates two types. While rules of the first type are applicable only for one specific user, rules of the second type are applicable for all users.

Our future work will focus on a comprehensive user study to examine good thresholds for the rule generation based on the detected patterns as well as the rule generalization. This also includes the time interval within which user interactions will be grouped as one pattern and not as separated ones. We also plan to extend the evolution of generalized rules. That is, rules can be outdated, modified or merged with other rules based on their acceptance of the users.

Acknowledgment. This work was supported by the Sächsische Aufbaubank within the European Social Fund in the Free State of Saxony, Germany (Project Crossmediale Mehrwertdienste für die digitale Mediendistribution).

References

1. Krug, M., Wiedemann, F., Gaedke, M.: Enhancing Media Enrichment by Semantic Extraction. To appear in WWW 2014 Companion: Proceedings of the 23rd International Conference on World Wide Web Companion. International World Wide Web Conferences Steering Committee, Seoul (2014),
   http://vsr.informatik.tu-chemnitz.de/publications/2014/002
2. Muntean, C.H., McManis, J.: Fine grained content-based adaptation mechanism for providing high end-user quality of experience with adaptive hypermedia systems. In: Proceedings of the 15th International Conference on World Wide Web, pp. 53–62. ACM (2006)
3. Rose, D.E., Levinson, D.: Understanding user goals in web search. In: Proceedings of the 13th International Conference on World Wide Web, pp. 13–19. ACM (2004)
4. Roy Chowdhury, S., Chudnovskyy, O., Niederhausen, M., Pietschmann, S., Sharples, P., Daniel, F., Gaedke, M.: Complementary assistance mechanisms for end user mashup composition. In: Proceedings of the 22nd International Conference on World Wide Web Companion, pp. 269–272. International World Wide Web Conferences Steering Committee (2013)
5. Tsandilas, T., Schraefel, M.: Usable adaptive hypermedia systems. New Review of Hypermedia and Multimedia 10(1), 5–29 (2004)

Interactive Scalable Lectures with ASQ

Vasileios Triglianos and Cesare Pautasso

Faculty of Informatics, University of Lugano (USI), Switzerland
{name.surname}@usi.ch
http://asq.inf.usi.ch/

Abstract. Taking full advantage of the Web technology platform during in-class lectures requires a shift from the established scheme of online education delivery that utilizes the video channel to embed all types of content and gathers student feedback via multiple choice questions or textual answers. In this paper we present the design of ASQ to deliver interactive content for use in heterogeneous educational settings with a large number of students, taking advantage of the co-location of students and instructors and building upon the latest capabilities of the Web platform. ASQ is centered around interactive HTML5 presentations coupled with a versatile microformat to create and deliver various types quizzes and scalable, synchronous/asynchronous feedback mechanisms.

1 Introduction

The Web is increasingly used to deliver educational content, being the medium of choice for the popular [1] massive open online courses (MOOCs) and "flipped" (or blended or hybrid) classrooms [2]. Courses are delivered through video lectures and students are assessed either automatically, through multiple choice or text input quizzes, or via peer assessment. Student Response Systems that take advantage of the increased number of Web-enabled devices are finding their way to brick and mortar classrooms replacing traditional hardware clickers. Video as the prominent delivery format of an online lecture is not optimized for the Web medium; it is cumbersome to author and interleave with quizzes and various types of assessment; and it lacks the interactivity and the features that modern Web technologies offer. Today's Web technology can support lectures with highly interactive content such as selectable text, forms, 3D graphics, that can be reactive to a student's input and personalized [3] for different learning styles.

Also of importance are the current assessment models which have remained stale for years: formative or summative assessment is predominantly perfomed through multiple choice or free text quizzes which do not encourage experimentation and creation of original content. In terms of communication models, either between students or between students and teachers, findings suggest that synchronous communication, as a complement to asynchronous communication, can potentially enhance participation in online education [4].

Our goal is to demonstrate ASQ, a platform to create and deliver interactive lectures and gather student feedback in synchronous or asynchronous settings. ASQ makes full use of the HTML5 capabilities of modern Web browsers

S. Casteleyn, G. Rossi, and M. Winckler (Eds.): ICWE 2014, LNCS 8541, pp. 515–518, 2014.
© Springer International Publishing Switzerland 2014

and provides teachers with the ability to author, deliver and reflect upon the performance of their interactive educational content, while it gives students an additional communication channel to demonstrate their learning progress and actively participate during a traditional lecture.

2   ASQ in the Classroom

ASQ aims to promote the shift from the traditional frontal lecture paradigm (monologue) to interactive bi-directional presentations and discussions (dialogue), through the following features that will be demonstrated: **Delivering educational content:** Interactive lectures are shared with students through a simple URL pointing to the lecture slides. Retrieving the link will connect the student browsers to follow the online presentation. As the instructor navigates through the slides, the navigation events propagate to the connected students which can synchronously follow the material on their Web browsers. This can enhance the accessibility of the lecture, since the material is now displayed in front of the students as it is being explained to them.

Authoring educational content: ASQ manages different kinds of educational content: lecture slides and questions. These can be authored using all the rich multi-hyper-media (images, videos, text, interactive widgets, audio) capabilities of HTML5 compliant Web browsers. Exemplary interactions include selecting or highlighting text, keeping notes for a slide in provided placeholder, playing back audio and video, filling forms, dragging and dropping textual or iconic elements, and any sort of interaction that can be implemented in a Web page using Javascript, CSS3 and HTML5.

Interactive questions and microformat: Questions can be used for both formative and summative assessment, embedded in lectures, collected in quizzes and homework assignements or exams. Each question type is associated with a set of related statistics to be computed over the answers collected by the students. The configuration for rendering questions, assessing answers and visualizing the results of the assessment is controlled with a simple microformat, as shown in the following example:

```
<!-- a multiple choice quiz with id q-1 -->
<article class="asq-question multi-choice choose-0-n" id="q-1">
  <h3 class="stem"><img src="img/shape.png"></h3>
  <ol>
    <li class="option">This figure is a square.</li>
    <li class="option">This figure is a circle.</li>
    <li class="option">This figure is symmetric.</li>
    <li class="option">This figure has four corners.</li>
  </ol>
</article>
<!-- statistics for the quiz q-1 -->
<article class="stats" data-target-assessment="q-1"
    id="stats-q-1"></article>
```

The microformat parser searches for elements that contain the `asq-question` keyword in their `class` attribute. Once a question is identified the parser searches for its type, in this case `multi-choice`; and for configuration options, in this case students can choose as many of the available multiple choice options they want (`choose-0-n`, but also `choose-1` or `choose-1-n` are possible). Each question type features keywords that provide information about its structure. In the multiple choice example, the class `option` of the `li` elements instruct the parser to store these elements as the options of the multiple choice question. Similarly, statistics are processed with the parser searching for the `asq-stats` keyword. The `data-target-assessment` points to the associated question. Once all questions and statistics are parsed, they are stored in the database and then a markup generator is invoked, that injects necessary markup like form fields and buttons.

Innovative quiz types: Besides standard multiple choice (MC) and single-line text input (STI) quizzes, ASQ currently features two question types specifically targeting the Computer Science domain: code highlight and code input.

Classroom flow: ASQ enhances the educational material by weaving support for complex interactions. Instructors can highlight important points on the presentation and have the marking happen instantly on the students screens. Students can answer quizzes and questions embedded in the slides –individually or in teams– giving instant feedback to the instructor about their level of comprehension. ASQ supplies instructors with a continuous stream of events and statistics related to quizzes, like student progress, correct versus wrong answers, enumeration of actual solutions. Upon receiving the results an instructor may choose to discuss them with individual students, share them with the class through the projector or the students screens to present insightful statistics.

3 Architecture

ASQ follows a client server architecture (Fig. 1). In the backend the logic is implemented in express.js on top of node.js, a choice mandated by the need for efficient I/O operations. The Web server design follows a REST approach treating educational material (lectures, slides, questions, answers) as Web resources.

Bi-directional communication between instructors and students is implemented using WebSockets. This allows for real-time exchange of a fairly big volume of events that are crucial for monitoring progress, supporting complicated assessment modes (for example peer assessment) or fine-grained logging of student actions. User management in ASQ, involves user roles like professors, teaching assistants and students and user permissions defined for each course. Interactions with the website that involve regular HTTP requests use cookie-based authentication, while realtime communication uses token-based authentication and group-based WebSocket namespaces.

The database consists of MongoDB collections for question instances, question types, lectures, sessions, users, and answers. Session events and socket events are stored in a Redis key-value store for scalability and increased performance.

ASQ supports both client- and server-side rendering with Dust.js templates. Question instances are parsed and rendered on the backend, which allows for

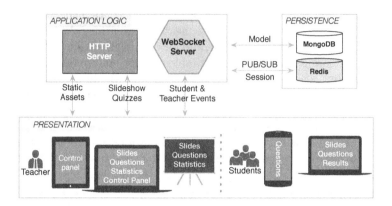

Fig. 1. ASQ Architecture

caching and saves CPU time on the clients. Dynamic elements like graphs and dialog windows are rendered client-side. Moreover, ASQ takes into account display size and user roles to optimize the rendered content. For example, in a student's smartphone with limited HTML5 features and/or a small screen, ASQ only renders the questions so that students can answer them, but does not show the statistics visualizations.

4 Conclusion

In this paper we demonstrate the main features of ASQ, a tool for delivering highly interactive educational content on the Web. Traditionally, instructors may only gather a small sample of answers from the students (who may have to speak up in front of a large audience). ASQ opens a new channel through which teachers can collect the answers of every single student attending a lecture. In the future, we will augment the question delivery system to enable the formation of groups of students to collaboratively solve problems. Finally we plan on evaluating the educational outcomes of ASQ through a summative evaluation that aims to compare a traditional classroom with an ASQ-enabled classroom.

References

1. Shah, D.: MOOCs in 2013: Breaking Down the Numbers. Teasing out trends among the unabated growth of online courses. edSurge (December 2013)
2. Bergmann, J., Sams, A.: Flip your classroom: reach every student in every class every day. International Society for Technology in Education Eugene, OR (2012)
3. Dolog, P., Henze, N., Nejdl, W., Sintek, M.: Personalization in distributed e-learning environments. In: Proceedings of WWW Alt. 2004, pp. 170–179. ACM (2004)
4. Hrastinski, S., Carlsson, S.A.: Participating in synchronous online education. Lund Studies in Informatics (6) (2007)

LiquidML: A Model Based Environment for Developing High Scalable Web Applications

Esteban Robles Luna[1], José Matías Rivero[1,2], and Matias Urbieta[1,2]

[1] LIFIA, Facultad de Informática, UNLP, La Plata, Argentina
{esteban.robles,mrivero,matias.urbieta}@lifia.info.unlp.edu.ar
[2] Also at Conicet

Abstract. The scalability of modern Web applications has become a key aspect for any business in order to support thousands of concurrent users while reducing its computational costs. However, existing model driven web engineering approaches have been focus on building Web applications that satisfy functional requirements while disregarding "technological" aspects such as scalability and performance. As a consequence, the applications derived from these approaches may not scale well and need to be adapted. In this paper we present the LiquidML environment, which allows building Web applications using a model-based approach. In contrast with existing approaches, aspects that help to improve the scalability of a Web application are modeled as first class citizens and as a consequence the applications obtained scale better than its counterparts.

Keywords: Scalability, Model driven development, Web engineering.

1 Introduction

Scalability is the ability of a system to handle a growing amount of work in a capable manner or its ability to be enlarged to accommodate that growth [2]. Scalability problems in the applications derived using Model-Driven Web Engineering (MDWE) tools may not appear as soon as they are deployed, but rather after they has been "living" in production for some time. Fixing scalability issues requires a detail diagnosis and consumes many human resources [4] and it has been shown to take 50-70% of the development time [1].

In the Web engineering research area [6], MDWE approaches [5,3,7] have become an attractive solution for building Web applications as they raise the level of abstraction and simplify the Web application development. However, according to [8], little attention has been put to non-functional requirements such as scalability issues as they have been considered a technological aspect that does not need to be modeled. In the same line, the Web applications derived from MDWE approaches pose a rigid/static architecture that cannot be easily changed thus limiting the size of the Web applications that can be built with them. In addition, diagnosing and fixing these problems in production systems (which are the ones that present overload symptoms) becomes cumbersome and impossible to be done in the models thus forcing teams to deal with the generated code losing the high level of abstraction provided by the design models.

S. Casteleyn, G. Rossi, and M. Winckler (Eds.): ICWE 2014, LNCS 8541, pp. 519–522, 2014.

In this paper we present *LiquidML*, an environment that helps building high scalable Web applications that can be manipulated at runtime. A LiquidML application can be either derived from MDWE models such as WebML, or it can be built straight inside the environment by manually creating models. The main difference with existing tools is that we do not impose a rigid architecture and those "technological" aspects are modeled; as a consequence we can keep track of their changes and alter the models at runtime in a safe way.

2   LiquidML Models

The LiquidML environment allows engineers to model a set of features that handle Web applications requests as first class citizens. A non-exhaustive list of features includes: logging and profiling, caching, service calls, parallel, asynchronous and sequence execution modes, message queuing and the deployment process. A LiquidML application is a composition of Flows and a Flow describes a sequence of steps that need to happen within a Web request (called *Message* in our approach). A Message has a payload (body) and a list of properties while each step in the Flow is visually identified by an icon and constitutes an *Element* of it. Communication between Elements happens by means of Message interchanges.

To exemplify the concepts, we present a Flow for the product's detail page of an E-commerce web application (Fig. 1). The Element with no incoming arrow represents the Message source listener that will receive incoming requests – in this case, it will receive HTTP request and will transform them in to Messages. The Element connected to the Message source named "Route path" is a ChoiceRouter, which behaves like a choice/switch statement and it will route the message to the "Get info" processor if the request comes to a URL starting with "/product/*". The "Get info" is another router that gets information in parallel from multiple sources. It obtains the product info from the DB: ("Get product info") and triggers the computation of the product's rank, which involves two database (DB) queries ("Get user reputation" and "Get product reviews") and a Processor that computes the rank from this information ("Compute product rank"). Finally, the information gets composed ("Compose data") and used for rendering a Web page in the "Render template" processor.

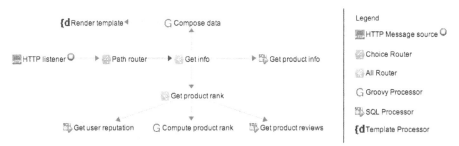

Fig. 1. Product details flow

As aforementioned, to improve the scalability of the Web application some elements suggested in [4] has been modeled as first class citizens. In Fig. 2 we show a modified version of the Product details flow of Fig 1, were some elements for caching purposes have been added to improve the overall scalability of the Web application. In LiquidML, these changes can be applied at design time by adding the elements in the diagram or by means of runtime transformations that change model elements on-the-fly while the application is running.

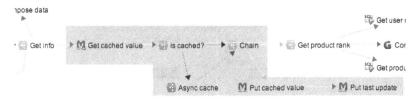

Fig. 2. Product details flow improved with caching elements

3   LiquidML Environment

The LiquidML environment is composed of 2 applications:

- LiquidML editor: it allows defining the applications, modelling Flows and to deploy them to the LiquidML servers. It also, allows developers to apply dynamic transformations to the deployed diagrams in order to deal with the scalability problems and watch for performance and logging information on-the-fly in any environments where the application is running (DEV, QA or even Production).

- LiquidML server: The server is responsible for holding the application definitions, the LiquidML interpreter and notifying the editor about how applications are running. In addition, it regularly checks if it has any pending deployments and if so, it fetches the Application and automatically deploys it.

LiquidML do not impose any rigid implementation architecture. For instance, a complete Web application can be split in 1 front-end app and 2 different service apps, which can be integrated by sending messages to their REST endpoints. In addition, each application, e.g. the front-end app, can be instantiated in multiple servers and combined by load balancers to distribute the load and improve the overall scalability of the application. As an example, in Fig. 3 we present the deployment view of a diagram, which allows us to use the diagnostic tools of LiquidML such as profiling and logging and it also supports performing runtime transformations oriented to improve the scalability of the Web application.

Both, the editor and the server have been built using open source technologies of the JEE stack. We have used Spring and Hibernate for basic service and ORM mapping, Spring MVC and Twitter bootstrap for UI and Jersey for the LiquidML API. As part of this development, we have built the CupDraw framework[1] for building Web diagram editors, which is publicly available. For the technical readers, we invite them to visit the LiquidML site http://www.liquidml.com and check the project's source code and the demonstration videos.

---

[1] https://github.com/estebanroblesluna/cupDraw

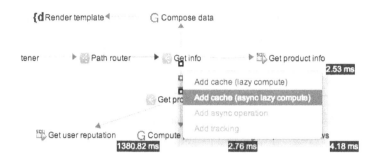

Fig. 3. Deployment view

4   Conclusions

In this demo paper, we have presented the LiquidML environment as a place where Web applications can be either derived from MDWE models or manually created from our "low level" LiquidML models. These models can be then interpreted and dynamically reconfigured at runtime. The environment is web based and it only requires a browser to be used.

References

1. Boehm, B.W.: Software engineering economics. Prentice-Hall, Englewood Cliffs (1981)
2. Bondi, A.: Characteristics of scalability and their impact on performance. In: Proceedings of the 2nd International Workshop on Software and Performance (WOSP 2000), pp. 195–203. ACM, New York (2000)
3. Ceri, S., Fraternali, P., Bongio, A.: Web Modeling Language (WebML): A Model-ing Language for Designing Web Sites. Computer Networks and ISDN Systems 33(1-6), 137–157 (2000)
4. Hull, S.: 20 Obstacles to Scalability. Queue 11(7), 20 (2013)
5. Koch, N., Knapp, A., Zhang, G., Baumeister, H.: UML-Based Web Engineering, An Approach Based On Standards. In: Web Engineering, Modelling and Implementing Web Applications, pp. 157–191. Springer, Heidelberg (2008)
6. Rossi, G., Pastor, O., Schwabe, D., Olsina, L.: Web Engineering: Modelling and Implementing Web Applications. Springer (2007)
7. Rossi, G., Schwabe, D.: Modeling and Implementing Web Applications using OOHDM. In: Web Engineering, Modelling and Implementing Web Applications, pp. 109–155. Springer, Heidelberg (2008)
8. Toffetti, G.: Web engineering for cloud computing (web engineering forecast: cloudy with a chance of opportunities). In: Proceedings of the 12th International Conference on Current Trends in Web Engineering (ICWE 2012), pp. 5–19. Springer, Heidelberg (2012)

Managing and Monitoring Elastic Cloud Applications

Demetris Trihinas, Chrystalla Sofokleous, Nicholas Loulloudes, Athanasios Foudoulis,
George Pallis, and Marios D. Dikaiakos

Department of Computer Science, University of Cyprus
{trihinas,stalosof,loulloudes.n,afoudo01,
gpallis,mdd}@cs.ucy.ac.cy

Abstract. Next generation Cloud applications present elastic features and rapidly
scale their comprised resources. Consequently, managing and monitoring Cloud
applications is becoming a challenge. This paper showcases the functionality and
novel features of: (i) c-Eclipse, a framework for describing Cloud applications
along with their elasticity requirements and deploying them on any IaaS provider;
and (ii) JCatascopia, a fully-automated, multi-layer, interoperable Cloud monitoring system. Particularly, we demonstrate how a user can manage the full lifecycle
of a three-tier web application and observe, in real-time, how an elasticity management platform automatically scales the application based on various user-defined
elasticity requirements, workloads and performance metrics.

1 Introduction

Cloud computing offers organizations the opportunity to reduce IT costs and improve
the efficiency of their services. The next generation of Cloud applications present elastic features which allow them to expand or contract their allocated resources in order
to meet their exact demands. However, managing and monitoring elastic Cloud applications is not a trivial task. For instance, organizations with large-scale distributed web
applications (e.g. online video streaming services) require a deployment comprised of
multiple virtual instances, which often have complex inter-dependencies. Despite the
fact that current elasticity management platforms such as Amazon Auto Scaling[1] and
Rackspace Auto Scale[2] can automatically and seamlessly scale large Cloud applications, these systems are proprietary and limit the application to operate only on specific
Cloud platforms. Portability imposes a level of complexity and additional effort, from
the Cloud consumer perspective, to *move* an application from one IaaS provider to another. Another downside of the aforementioned systems is that they only handle simplistic boolean requirements based on a limited number of low-level metrics (i.e. CPU,
memory usage, etc.) and only support fine-grained elasticity actions (e.g. add/remove
virtual instances). Tiramola [1] on the other hand, is an open-source alternative which
succeeds in accommodating complex elasticity requirements based on application-level
metrics but it is limited to only scale NoSQL databases.

---

[1] http://aws.amazon.com/autoscaling/
[2] http://www.rackspace.com/cloud/monitoring/

S. Casteleyn, G. Rossi, and M. Winckler (Eds.): ICWE 2014, LNCS 8541, pp. 523–527, 2014.

Furthermore, elasticity management requires the monitoring of elastic applications, a challenge that is left unaddressed by many of the existing monitoring tools (i.e. Ganglia[3], Nagios[4]). The complex nature of this task requires for the monitoring system to run in a fully automated manner, detecting configuration changes in the application topology which occur due to elasticity actions (e.g. a new VM is added) or when allocated resource-related parameters change (e.g. new disk attached to VM).

To address the aforementioned challenges which occur when managing and monitoring elastic Cloud applications, we present two powerful open-source tools:

c-Eclipse: A client-side Cloud application management framework which allows developers to describe, deploy and manage the lifecycle of their applications in a clean and graphical manner, under a unified environment. c-Eclipse is built on top of the well-established Eclipse platform. It adheres to two highly desirable Cloud application features: portability and elasticity. Current frameworks such as the ServiceMesh Agility Platform[5] limit users to describe and deploy their applications to a small number of Cloud platforms for which connectors are available and when users want to *move* their application to another Cloud, the description must be altered. Additionally, current frameworks offer limited or no elasticity support by only allowing the definition of fine-grain elasticity requirements. c-Eclipse ensures application portability by adopting the open TOSCA specification[6] for describing the provision, deployment and application re-contextualization across different Cloud platforms. In contrast to other frameworks which also adopt the TOSCA specification [2], c-Eclipse does not require for its users to have any prior knowledge of the TOSCA specification since users describe their application through an intuitive graphical interface which automatically translates the description into TOSCA. Finally, c-Eclipse facilitates the specification of complex, multi-grained elasticity requirements via the SYBL [3] directive language.

JCatascopia [4]: A fully-automated, multi-layer, interoperable Cloud monitoring system. JCatascopia addresses the aforementioned challenges by being able to run in a non-intrusive and transparent manner to any underlying virtualized infrastructure. Current monitoring systems which provide elasticity support [5] [6], require for special entities at the physical level or depend on the current hypervisor to detect topology changes. In contrast to these systems, JCatascopia uses a variation of the publish and subscribe protocol [7] to dynamically detect, at runtime, when monitoring agents have been added/removed from the overall system due to elasticity actions. This is accomplished without any human intervention, special entities or dependence on the underlying hypervisor, allowing users to even monitor applications distributed over multiple Cloud platforms. In addition, JCatascopia provides filtering capabilities to reduce the overhead for metric distribution and storage, and generates high-level application metrics dynamically at runtime by aggregating and grouping low-level metrics.

---

[3] http://ganglia.sourceforge.net/

[4] http://www.nagios.org/

[5] http://www.servicemesh.com

[6] http://docs.oasis-open.org/tosca/TOSCA/v1.0/

[7] Monitoring Agents (metric producers) subscribe to Monitoring Servers instead of the other way around, allowing for them to (dis-)appear dynamically [4].

2   Elasticity Management Platform

In this section we focus on describing the components which comprise an elasticity management platform which incorporates c-Eclipse and JCatascopia.

A developer, at first, uses the c-Eclipse **Application Description Tool** to graphically describe the application topology, software dependencies and elasticity requirements. The graphical description is translated, on the fly, into TOSCA. Then, the developer selects a Cloud provider and via the c-Eclipse **Submission Tool**, the description is submitted to the **Cloud Manager** for deployment. Subsequently, the Cloud Manager parses the portable and platform independent TOSCA description, and initiates the deployment of the Cloud application via the **Orchestrator**. The Orchestrator consists of two subcomponents. The first component is the **Cloud Orchestrator**, which is the interface that interacts with the IaaS provider to (de-)allocate the requested Cloud resources. The second component is the **App Orchestrator**, which performs the execution of application specific scripts and ensures the configuration and deployment correctness.

After successfully deploying a Cloud application, users are able, via c-Eclipse, to monitor the deployment, acquire aggregated monitoring metrics from **JCatascopia**, and configure the deployment by refining the elasticity requirements. The **Cloud Manager** (Fig. 1) constantly checks the user-defined elasticity requirements and when a violation is detected, resizing actions are issued. Specifically, the Cloud Manager requests from the **Cloud Orchestrator** to add/remove instances depending on the application demands and from the **App Orchestrator** to configure the application accordingly.

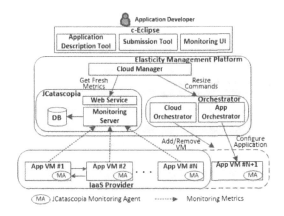

Fig. 1. Elasticity Management Platform

3   Demonstration Description

This demonstration showcases[8] the functionality of the proposed platform by managing the full lifecycle of an elastic Cloud application. Specifically, we will: (i) describe,

---

[8] Screenshots of c-Eclipse and JCatascopia can be found at:
http://linc.ucy.ac.cy/CELAR/icwe2014

via c-Eclipse, a Cloud application's topology, software dependencies and relationships between its tiers; (ii) define, via c-Eclipse, elasticity requirements for the elastic components comprising the application; (iii) select a Cloud platform and submit the generated TOSCA description to the Cloud Manager; (iv) monitor both the Cloud resources allocated for the application and its performance by utilizing JCatascopia (Fig. 2); and (v) scale the application, via the Cloud Manager, based on collected metrics and the user-defined elasticity requirements. It must be noted that both the Cloud Manager and Orchestrator are simplistic components developed only to showcase the full potential of c-Eclipse and JCatascopia. Furthermore, attendees may configure the deployment by refining the elasticity requirements. Finally, users will observe real-time graphs for each collected metric, configure monitoring parameters (i.e. sampling rate) and generate graphs by aggregating metrics originated from multiple instances.

Use Case Scenario: we consider a three-tier online video streaming service comprised of: (i) an HAProxy[9] *load balancer* which distributes client requests (i.e. download, upload video) across multiple application servers. (ii) An *application server tier*, where each instance is an Apache Tomcat[10] server containing the video streaming web service. Aggregated tier metrics such as the average *number of connections* and/or *request throughput* can be used to decide when a scaling action should occur; (iii) a Cassandra[11] NoSQL *distributed data storage backend*. Similarly, aggregated metrics such as the average *CPU utilization* and/or *query latency* can be used to scale the Cassandra ring. To stress the video service, we have developed a *workload generator* where the workload form (i.e. sinusoidal, linear), type (i.e. read-heavy, write-heavy, combination) and parameters (i.e. intensity, max execution time) are all configurable.

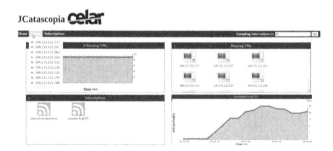

Fig. 2. Screenshot from JCatascopia while running the demo scenario

Acknowledgements. This work was partially supported by the European Commission in terms of the CELAR 317790 FP7 project (FP7-ICT-2011-8).

---

[9] http://haproxy.1wt.eu/
[10] http://tomcat.apache.org/
[11] http://cassandra.apache.org/

References

1. Tsoumakos, D., Konstantinou, I., Boumpouka, C., Sioutas, S., Koziris, N.: Automated, Elastic Resource Provisioning for NoSQL Clusters Using TIRAMOLA. In: IEEE International Symposium on Cluster Computing and the Grid, pp. 34–41 (2013)
2. Kopp, O., Binz, T., Breitenbücher, U., Leymann, F.: Winery – a modeling tool for tosca-based cloud applications. In: Basu, S., Pautasso, C., Zhang, L., Fu, X. (eds.) ICSOC 2013. LNCS, vol. 8274, pp. 700–704. Springer, Heidelberg (2013)
3. Copil, G., Moldovan, D., Truong, H.L., Dustdar, S.: SYBL: An Extensible Language for Controlling Elasticity in Cloud Applications. In: 13th IEEE/ACM International Symposium on Cluster, Cloud and Grid Computing, pp. 112–119 (2013)
4. Trihinas, D., Pallis, G., Dikaiakos, M.D.: JCatascopia: Monitoring Elastically Adaptive Applications in the Cloud. In: 14th IEEE/ACM International Symposium on Cluster, Cloud and Grid Computing (2014)
5. Clayman, S., Galis, A., Mamatas, L.: Monitoring virtual networks with lattice. In: Network Operations and Management Symposium Workshops (NOMS Wksps), pp. 239–246. IEEE/IFIP (2010)
6. de Carvalho, M.B., Granville, L.Z.: Incorporating virtualization awareness in service monitoring systems. In: IEEE Integrated Network Management, pp. 297–304 (2011)

MAPMOLTY: A Web Tool for Discovering Place Loyalty Based on Mobile Crowdsource Data

Vinicius Monterio de Lira[1], Salvatore Rinzivillo[2], Valeria Cesario Times[1],
Chiara Renso[2], and Patricia Tedesco[1]

[1] Universidade Federal de Pernambuco, Recife, Brazil
{vcml,vct,pcart}@cin.ufpe.br
[2] ISTI - CNR, Pisa, Italy
{salvatore.rinzivillo,chiara.renso}@isti.cnr.it

Abstract. Mobility crowdsourced data, like check-ins of the social networks and GPS tracks, are the digital footprints of our lifestyles. This mobility produces an impact on the places that we are visiting, characterizing them by our behavior. In this paper we concentrate on the *loyalty of places*, indicating the regularity of people in visiting a given place for performing an activity. In this demo we show a web tool called MAPMOLTY that, given a dataset of mobility crowdsourced data and a set of Points of Interests (POI), computes a number of quantitative indicators to indicate the loyalty level of each POI and displays them in a map.

Keywords: place loyality, crowdsource mobility data, activities regularity.

1 Introduction

Tracking capabilities of modern geo-based network services provide us with unprecedented opportunities of sensing both movements and activities performed by people. We can exploit these data to monitor and study traffic, animals, maritime and people [3]. Particularly, people produce crowdsourced data from which we can investigate how people use the area where they live. One interesting aspect is to analyze how regularly a person is visiting a given destination. For example, some people tend to go to their preferred restaurant for eating, while some others may tend to explore different restaurants.

We introduce the concept of *loyal user* for a place p and activity A to indicate a user who regularly visits p to perform activity A. For the sake of generality, we associate the Point of Interest's (POI) category to the activity performed in that POI, thus in the rest of the paper we refer to the category or activity as synonyms. From the loyalty measure of the user we can derive a *loyalty map* of a territory. We can discover that some areas have the tendency to be visited by loyal users, while other areas are more characterized by occasional visitors.

The purpose of this demo is to show a web tool called MAPMOLTY (MAPping MObility loyaLTY)[1] that, given a dataset of mobility crowdsourced data, like

---

[1] http://mapmolty.isti.cnr.it

S. Casteleyn, G. Rossi, and M. Winckler (Eds.): ICWE 2014, LNCS 8541, pp. 528–531, 2014.
© Springer International Publishing Switzerland 2014

tracks of individuals, and a set of Point Of Interest, computes a number of measure, called *loyalty indicators*, to summarize the loyalty level of each POI.

The analysis enabled by the tool may be useful in different scenarios. For example, an urban manager may quickly discover attractions that are visited by occasional visitors like tourists or loyal visitors like dwellers. This may be useful in supporting decision making in traffic management and / or urban planning. On the other hand, the loyalty analysis results may be used for marketing purpose by the owners of the POIs to better plan advertising to targeted individuals. The loyalty indicators are also useful for developing services for the citizen like a recommendation system suggesting new destinations according to the observed visitors' behavior.

To the best of our knowledge, this is the first web tool providing a loyalty map of Points of Interests. Other related approaches about individual regularity measures are, for example, in [1, 2, 4]. These papers focus on the individual instead of the places and they measure (using entropy or other measures) how much a user is regular in visiting specific locations. This idea is at the basis of our approach, but here we focus on geographic and activity aspects instead of concentrating on the regularity behavior of a specific user.

2   The MAPMOLTY Tool

The overview of the tool is illustrated in Figure 1. As we can see, it is organized into three components as described below.

Data Collection and Pre-processing. This module organizes the data that is fed into our system. We identify two main sources of data: list of Points of Interest (POI) and set of crowdsourced time stamped visits of individuals to these POIs. There are many POIs dataset available from several web services. To enable the analytical features of our system we require that the **POIs dataset** should provide at least the following attributes for each POI: a unique identifier *poi_id*, spatial coordinates *latitude* and *longitude*, category of the POI *poi_cat* and a name or description of the place. Many POIs collections are organized into a hierarchy,

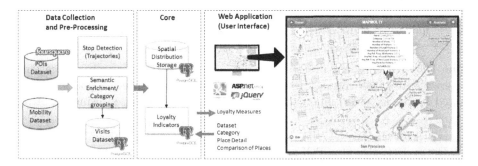

Fig. 1. The overview of MAPMOLTY with the three components

where places with similar categories are grouped under the same larger category. For example, places tagged as *Indian Restaurant* and *Fast Food*, may be associated with the super-category *Restaurant*. Since the super-category is generally informally paired with an activity, we are using the two terms as synonyms. In our experiments we used the set of POIs provided by the Foursquare API[2].

The **Mobility dataset** provides the mobility information to associate a person p to a POI *poi_id* she visited. We require that each visit is represented by a tuple $< VisitID, UserID, poi\_id, timestamp >$, where *VisitID* is the visit unique identifier, *UserID* and *poi_id* represent univocally a user and a place, and *timestamp* is the time when *UserID* arrived at *poi_id*. We call such tuple a *visit*. MAPMOLTY works with many different sources like: Location Based Mobile Social Networks (e.g. Foursquare, Jiepang, Brightkite) and GPS traces. However, a transformation may be needed to convert the mobility data into this format. For example, let us consider a sequence of time stamped GPS points for an individual (called *trajectory*). A trajectory can be transformed into a sequence of visits in a two-step process: *(1)* we detect *stops*, i.e. subsequences of points where the user stands still for a minimum amount of time; *(2)* we associate each stop to the closest POIs provided by the POIs dataset [3]. For the stop detection, we use two parameters: δ, a spatial tolerance threshold and τ, a temporal tolerance threshold. In our experiments, we used $\delta = 50m, \tau = 20min$, meaning that we detect a stop if the user stays in an area of radius $50m$ for at least 20 minutes.

MAPMOLTY uses PostGres SQL 9.3 [4] with PostGIS 2.1.1[5] as Data Base Management System (DBMS) to store the data. PostGIS provides spatial extension for the PostgreSQL database, allowing storage and query of geographical data.

Core. This module analyzes the visits dataset to derive loyalty indicators about the POIs. A visitor is *loyal* to a place when her visits to that place, for performing an activity, are regular. We measure the regularity for a visitor computing his spatial distribution over the frequency of visits to places for a specific activity. Due to lack of space we omit here the mathematical background of this computation. Starting from the spatial distributions from the visitors, MAPMOLTY computes a set of loyalty indicators for each POI: *Number of Visits*; *Number of Visitors*; *Number of Loyal Visitors*; *Number of Non Loyal Visitors*; *Average Relative Frequency of All Visitors*; *Average Relative Frequency of Loyal Visitors*; *Average Relative Frequency of Non Loyal Visitors*; *Average Visits by Visitors*. These indicators show different aspects of the loyalty and are implemented as SQL Procedures developed in PostGres SQL.

The User Interface. The user interface is implemented as a web application, where the user can interact with the map and visualize the information computed from the core component. This tool has been developed using the ASP

---

[2] https://developer.foursquare.com/
[3] More sophisticated Stop-POI association techniques can be used.
[4] http://www.postgresql.org/download/
[5] http://postgis.net/install

MVC 4 framework[6]. This technology has a Model-View-Controller architecture providing an easy separation between the data manipulation (server side) and the interaction of the user with the web application (client side). MAPMOLTY uses JQuery Mobile 1.3.2 with JQuery 1.9.1 [7] to implement the visual widgets used for the visualization for different types of web-browser devices. The web map widget also uses the JavaScript Google Maps API V3[8] for the visualization and interaction with the map.

The analytical process implemented in MAPMOLTY is structured as follows. When the analyst begins the interaction, the system proposes a list of POI datasets from which to choose the area of interest. Once the area has been selected, the user selects the super-category. The system shows a summary of the available indicators on a map. Each POI is indicated in the map by three visualization properties: marker color, circle size and circle opacity. Based on the loyalty indicators and normalization limits selected by the user, MAPMOLTY computes and plots the values of these visualization properties.

Clicking on the marker the user can visualize the detailed information about the place like the name, the super-category, category and all the indicators values. Other interesting features provided by the web interface are: visually comparison two places, filtering places by their categories and search places by their names.

One peculiarity of this web tool is that it provides a first kernel of features and it can be easily extended with new functionalities. For example, we can incorporate new analysis functions in the database like statistics functions (median, mode and so on) over the data and consequently update the web application to display this new information. Another possible extension is towards the mobile environment. Since we are using a component from JQuery Mobile, the tool can also be easily embedded into a smartphone app using a web panel. This can be particularly useful when associated to a recommendation function based on the loyalty.

Acknowledgments. This work was partially supported by EU-FP7-PEOPLE SEEK (295179), EU-FP7-FET DataSim (270833), FP7-SMARTCITIES-2013 PETRA (609042), CNPq (Brazil) grant 246263/2012-1.

References

1. Eagle, N., Pentland, A(S.): Reality mining: Sensing complex social systems. Personal Ubiquitous Comput. 10(4), 255–268 (2006)
2. Qin, S., Verkasalo, H., Mohtaschemi, M., Hartonen, T., Alava, M.: Patterns, entropy, and predictability of human mobility and life. CoRR, abs/1211.3934 (2012)
3. Renso, C., Spaccapietra, S., Zimányi, E.: Mobility Data: Modeling, Management, and Understanding. Cambridge Press (2013)
4. Song, C., Qu, Z., Blumm, N., Barabási, A.-L.: Limits of predictability in human mobility. Science 327(5968), 1018–1021 (2010)

---

[6] http://www.asp.net/mvc/mvc4
[7] http://jquerymobile.com/download/
[8] https://developers.google.com/maps/documentation/javascript/

Paving the Path to Content-Centric and Device-Agnostic Web Design

Maximilian Speicher

VSR Research Group, Chemnitz University of Technology, 09111 Chemnitz, Germany
maximilian.speicher@s2013.tu-chemnitz.de

Abstract. Content-centric and device-agnostic design are crucial parts of modern web design. They are required to cater for the rapidly growing variety of different web-enabled devices and screen resolutions. We review satire site `motherfuckingwebsite.com` as a drastic example for realizing these aspects. Additional enhancements are proposed that pave the path to up-to-date minimalistic web design. A simple example application is described to illustrate the proposed approach.

Keywords: Content-centric Web Design, Device-agnostic Web Design, Responsive Web Design, Web Interfaces.

1 Introduction

Nowadays, web developers are confronted with a growing amount of novel devices. Thus, also an increasing range of display resolutions has to be addressed (Fig. 1). When the first web-enabled smartphones became popular, it was common practice to provide separately designed versions of the same website. Yet, this approach is highly inefficient considering the range of devices and display resolutions. This calls for the application of responsive web design, i.e., a website flexibly reacts to the device it is accessed with [4]. The usual approach is to combine a fluid grid layout (cf. frameworks like *Bootstrap*[1]) with CSS3 media queries (i.e., breakpoints) to select rules based on the detected device context [4].

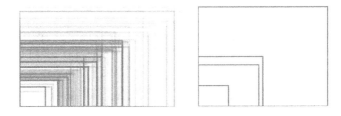

Fig. 1. Comparison of display resolutions of Android devices (left) and Apple devices (right). The graphics have been taken from [5].

---

[1] `http://getbootstrap.com/` (2014-03-17).

S. Casteleyn, G. Rossi, and M. Winckler (Eds.): ICWE 2014, LNCS 8541, pp. 532–535, 2014.

Besides, websites that are overloaded with, e.g., JavaScript libraries or extensive images, can cause problems with web-enabled mobile devices due to slow loading times. Thus, besides responsiveness, modern web design should strongly focus on delivering content while minimizing user distraction through trivialities. A promising method for realizing this is *mobile first* design [6] since the capabilities of mobile devices are still limited compared to desktop computers.

Still, many websites are developed with only three kinds of devices in mind, i.e., desktop, tablet and smartphone, while neglecting resolutions in between.[2] Moreover, cutting-edge websites—although often mobile-ready—tend to focus on a complex visual appearance that is often graphic- and animation-heavy. In this paper[3], we present a review of `motherfuckingwebsite.com` (MFW), which denounces these grievances and satirically claims that it is perfect by following the simplest possible approach to web design. That is, the website is radically content-centric and device-agnostic *without using any device- or resolution-specific breakpoints*. We find that, based on established findings from user experience design, MFW would require adjustments to three particular aspects to provide a more perfect user experience. Yet, although intended to be satire, it is a valid step into the direction of up-to-date minimalistic web design.

2   Related Work

Two well-known strategies for providing websites for different devices are *progressive enhancement* and *graceful degradation* [4]. These focus on starting from one end of the spectrum of devices and then adding more or less sophisticated variants of the website to cater for other devices [4]. The above depend on specific devices while *responsive design* is more oriented towards device-agnosticism. Still, the majority of responsive approaches depend on breakpoints for devices and/or resolutions. The *Goldilocks* approach [1] strongly focuses on text presentation and using as little breakpoints as possible. Conceptually, it is closer to MFW than other responsive approaches such as Bootstrap.

3   Review of motherfuckingwebsite.com

MFW is a website following a drastically minimalistic approach to web design (Fig. 2). Although the site is intended to be satire and highly exaggerated, we believe its concept is a valid step into the direction of content-centric and device-agnostic web design. This was underpinned by numerous positive reactions by users on different social media platforms. We provide a review of MFW w.r.t. established findings from user experience design. Particularly, the website makes the following points to underpin its initial statement that "it's [...] perfect":

---

[2] http://www.webdesignerdepot.com/2013/05/common-misconceptions-about-responsive-design/ (2014-03-17).

[3] This paper is based on an earlier blog post by the author, see http://wp.me/p4gilw-I (2014-03-17).

Fig. 2. Screenshot of `motherfuckingwebsite.com`.

1. It is lightweight and loads fast by omitting client-side scripts and graphics.
2. It is completely device-agnostic, i.e., "[the] site doesn't care if you're on an iMac or a [...] Tamagotchi". In fact, it omits any JavaScript- or CSS-based breakpoints and only makes use of the HTML viewport meta tag[4].
3. It is cross-browser compatible.
4. It is accessible for all users, particularly visually impaired ones.
5. It delivers content instead of overloading the site with trivialities.
6. It uses HTML5 tags to leverage semantics.

The above points are correct and necessary for modern web design. Yet, MFW's statement about perfectness, if not meant satirically, would not be completely correct. This is because the site's design follows a rather functional point of view while neglecting important aspects that affect user experience. In this context, we assume that a *perfect user experience*—independent of devices and resolutions—is what makes a website perfect. Particularly, MFW would need adjustments w.r.t. the following three points.

Line Width. Text lines on MFW span across the whole width of the viewport. They might exceed the optimal line length of ∼66 characters [1], particularly in large-screen contexts. More optimal than MFW would be to limit text lines to a width of about 30 characters [1]. Further optimization would include a multicolumn layout[5] and pagination for large screens [3].

Navigation. MFW omits statements about navigation. However, a website featuring larger amounts of content requires corresponding means. Optimally, developers should use a navigation bar fixed to the top of the viewport, which currently is common practice in web design (cf. *Ecosia*[6]). Following the approach of device-agnostic design, the navigation bar must react flexibly if the contained links would span more than one line on small screens.

Visual Aesthetics. Although technically flawless, MFW does not pay attention to site aesthetics, which strongly affect user satisfaction [2] and thus also user experience. Better aesthetics can be realized by using more sophisticated

---

[4] https://developer.mozilla.org/en-US/docs/Mozilla/Mobile/Viewport_meta_tag (2014-03-17).

[5] http://www.w3.org/TR/css3-multicol/ (2014-03-17).

[6] http://www.ecosia.org/what (2014-03-17).

typography and color schemes, e.g., for better contrast. This does not require the use of extensive graphics, which is a major aspect of MFW.

The above points underpin that MFW cannot be a completely *perfect* approach to web design. Yet, they also show that only little changes to the site can make for a serious approach to up-to-date minimalistic web design. Given the increasing diversity of web-enabled devices, we believe that more content-centric and device-agnostic design—as promoted by MFW—are an integral part of the future of web engineering.

Example Application. We have designed a website featuring a fixed navigation bar at the top of the viewport. Established approaches like Bootstrap engage CSS3 media queries to determine whether the standard navigation has to be adapted. Since it is cumbersome to cover all potential resolutions (Fig. 1) using media queries, we leverage a more device-agnostic approach. A script adds a second navigation bar featuring only a single dummy link to the site. This additional navigation is placed outside the viewport. On each window resize event, we compare the height of this navigation bar with the real navigation. If the latter has a greater height than the hidden navigation, we know that the contained links span more than one line and should be adjusted to form a dropdown menu. This approach is in accordance with MFW and the proposed enhancements. A corresponding demo is available at `http://www.maximilianspeicher.tk/DAD/`.

4 Conclusion

The increased proliferation of novel devices and growing number of screen resolutions makes it difficult to design websites that can react flexibly to all of them. This requires highly device-agnostic approaches to web design. Also, focusing on content more strongly supports modern, device-independent websites. We provide a review of `motherfuckingwebsite.com`, which is a minimalistic satire site drastically addressing current problems in (cross-device) web design. Additional enhancements to this website make for a valid and feasible approach to up-to-date content-centric and device-agnostic webdesign.

References

1. Armstrong, C.: The Goldilocks Approach to Responsive Design, `http://goldilocksapproach.com/article/`
2. Lavie, T., Tractinsky, N.: Assessing dimensions of perceived visual aesthetics of websites. IJHCS 60(3) (2004)
3. Nebeling, M., Matulic, F., Streit, L., Norrie, M.C.: Adaptive Layout Template for Effective Web Content Presentation in Large-Screen Contexts. In: Proc. DocEng. (2011)
4. Nebeling, M., Norrie, M.C.: Responsive Design and Development: Methods, Technologies and Current Issues. In: Proc. ICWE (Tutorials) (2013)
5. OpenSignal: Android Fragmentation Visualized, `http://opensignal.com/reports/fragmentation.php`
6. Wroblewski, L.: Mobile First. A Book Apart (2011)

Twiagle: A Tool for Engineering Applications Based on Instant Messaging over Twitter

Ángel Mora Segura, Juan de Lara, and Jesús Sánchez Cuadrado

Universidad Autónoma de Madrid, Spain
{Angel.MoraS,Juan.deLara,Jesus.Sanchez.Cuadrado}@uam.es

Abstract. Microblogging services, like Twitter, are widely used for all kind of purposes, like organizing meetings, gathering preferences among friends, or contact community managers of companies or services.

With suitable automation, tweets can be used as a dialogue mechanism between users and computer applications, and we have built a tool, named Twiagle, to construct tweet-based applications. Twiagle includes a pattern-matching language to express the interesting parts to be detected and selected from tweets, and an action language to query matched tweets, aggregate information from them or synthesize messages.

1 Introduction

Microblogging and instant messaging systems are booming nowadays, thanks in part to the proliferation of smartphones and mobile devices. Services like Twitter[1] or WhatsApp[2] are extremely used nowadays to connect with friends, or to organize social activities. These services are not only used for leisure, but most companies and brands use these services to keep in contact with clients.

In this setting, we observe a growing need to automate social activities, leveraging on popular social network platforms, like Twitter. On the one hand, users of social networks – possibly lacking any programming skills – may wish to define simple applications involving the participation of a community of users. On the other, companies may like to open their information systems to social networks platforms, but this integration effort needs to be done by hand.

We claim that social networks based on instant messaging, in particular Twitter, are suitable as front-ends for computer-based applications. We call them *tweet-based* applications, and they present many advantages in some scenarios. First, instant messaging systems are designed to support a high load of users and messages, serving as a robust front-end, difficult to achieve for companies or end users. Second, many people are familiar with Twitter, and have it already installed. Hence, they do not need to learn a new application, or even install a new one. Third, applications can leverage from Twitter's social network structure.

We foresee three main kinds of scenarios for tweet-based applications. In the first one, Twitter is used as a front-end, which then needs to be connected to an

---

[1] http://www.twitter.com
[2] http://www.whatsapp.com/

S. Casteleyn, G. Rossi, and M. Winckler (Eds.): ICWE 2014, LNCS 8541, pp. 536–539, 2014.
© Springer International Publishing Switzerland 2014

existing information system (e.g., an airport may send notifications with flight information, or with status updates via Twitter to interested users). In the second scenario, small, simple, self-contained applications can be designed by unexperienced end users (e.g., outdoors educational games based on quizzes). Finally, an important scenario is the quick construction of applications to coordinate a large amount of people upon unexpected events, like natural disasters.

These scenarios present several challenges for this technology. First, if tweets are used as simple communication mechanism with the application, the relevant information needs to be extracted from them. Second, a mechanism is needed to specify simple actions, like querying the extracted information, or synthesizing messages. Finally, a quick, easy way for constructing this kind of applications is needed, enabling their use by non-experts, but supporting also their deployment into servers, and their integration with existing information systems.

This paper presents *Twiagle*, a tool for constructing tweet-based applications, including a Domain Specific Language (DSL) for expressing patterns, and a DSLs for describing actions. Sec. 2 describes our architecture for *Tweet-based applications*, Sec. 3 describes *Twiagle* using an example, and Sec. 4 ends with the conclusions and prospects for future work.

2 Architecture

The working scheme of our solution for tweet-based applications is shown in the inset figure, where the numbers illustrate a typical interaction. Firstly (label 1), users send tweets or private messages via

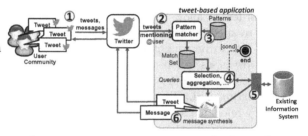

Twitter. Then, the relevant information in tweets is extracted. Our solution relies on the definition of patterns, expected to be found in tweets. Not every tweet is sought, buy only those mentioning the user associated to the application, or private messages directed to it (label 2). The patterns (label 3) are defined by social media experts, or software engineers. A typical application may include different queries, selecting the relevant concepts in matching tweets, or calculating different aggregation values from them (label 4). In addition, data can be obtained or sent to existing information systems (label 5). The data extracted from queries, or provided by the information system can be used to synthesize tweets or private messages, directed to the users (label 6). Finally, conditions can be defined to signal the end of the execution.

In order to facilitate the construction of such system, we provide a Model-Driven Engineering solution, based on two domain-specific languages (DSLs). The first DSL (called *Twittern*) helps in the definition of relevant patterns, and concepts to be found in them. The latter are sets of relevant words, or fragments, and sets of synonyms can be automatically extracted from Wordnet [??]. The second DSL, called *Twition*, is targeted to the description of the processing logic of tweet-based applications. It allows defining queries on tweets matching some pattern, using an SQL-like syntax. Queries can be used to select relevant information from tweets, or to calculate aggregated information from a set of tweets. The DSL also provides commands to synthesize private messages and tweets. Finally, it is also possible to define *data hooks*, a way to push extracted data into an existing information system, or to gather data from it. The next section describes a tool that realizes this approach.

3   Twiagle by Example

We show *Twiagle*'s capabilities through an example consisting in a simple voting among a set of users (see Fig. 1). The first step is to describe the interesting information in Tweets, using the *Twittern* DSL. A pattern is made of concepts, and in its simplest form, a concept is a set of words, which can be either defined explicitly by the designer, or can be automatically taken from a synonym set provided by Wordnet. We have also included specific Twitter concepts, like patterns to detect user names, URLs (specially pictures), and to define collections of interesting hashtags. The meta-data information present in tweets, like the originator, date or geoposition can be retrieved and does not need to be explicitly declared in patterns. Patterns also indicate if concepts have to appear in some specific order, or allow the interleaving of concepts with other words. It is also possible to specify that some concept cannot occur in a pattern, and whether concepts are to be sought ignoring upper/lower case, accents, and permitting missing vowels, as this is a usual idiom in tweets.

As a second step, our approach considers the description of actions by means of the *Twition* DSL. Twition allows issuing queries using an SQL-like syntax. They may refer to a set of matches of a pattern, as if they formed an SQL table, and the concepts in the pattern, as if they were SQL columns. Three kinds of specialized queries can be issued: Select (to select some concepts from a set of tweets matching a pattern), Adding (to perform some arithmetical operation on result sets), and Metadata queries (to obtain a result set made of some tweet metadata). Similar to data stream management systems [1] we may query using temporal windows. Currently, we support two kinds of temporal windows, one considering all data, and another one with the last tweet (@newest annotation). Once data becomes available from queries, messages can be composed and sent to a collection of users either publicly (command tweet), or in private, directed to a certain user (command message). In addition, received tweets can also be retweeted, and be categorized as favorite. Other commands include facilities to exchange data with an external source, and to signal the application end.

Each action has a name, so that actions can refer to the data they produce simply by that name. The type of data does not need to be declared, but it is inferred by simple rules. The execution model of Twition is based on data flow, relying on data dependencies, the recommended execution model for reactive, event-driven, scalable applications [3]. In this way, an action is performed as soon

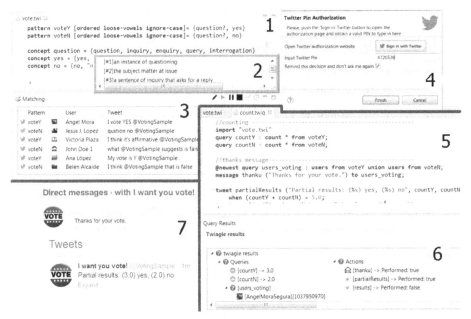

Fig. 1. (1) Defining patterns with Twittern, (2) Using Wordnet, (3) Testing with live tweets, (4) Authorizing Twiagle to use Twitter account data, (5) Defining actions with Twition, (6) Execution Debug, (7) Results shown in the Twitter console

as its data becomes available, unless it contains an explicit trigger, in which case it is executed when the data is available *and* the trigger becomes true.

Twiagle includes a console to test patterns against live tweets, as well as an execution debug, showing the results of queries and actions performed. The tool is available at `http://www.miso.es/tools/twiagle.html`.

4 Conclusions and Future Work

In this paper, we have introduced *Twiagle*, a tool to build tweet-based applications. We are currently increasing the expressiveness of *Twittern*, improving *Twition* with new primitives, taking inspiration from data-stream systems for tweet querying. We are also working on the deployment mode, and considering support for other social networks, enabling inter-platform applications.

Acknowledgements. This work has been funded by the Spanish Ministry of Economy and Competitivity with project "Go Lite" (TIN2011-24139).

References

1. Babcock, B., Babu, S., Datar, M., Motwani, R., Widom, J.: Models and issues in data stream systems. In: PODS, pp. 1–16. ACM (2002)
2. Miller, G.A.: Wordnet: A lexical database for english. CACM 38(11), 39–41 (1995)
3. Reactive manifesto, `http://www.reactivemanifesto.org`

Using Linked Data for Modeling Secure Distributed Web Applications and Services

Falko Braune, Stefan Wild, and Martin Gaedke

Technische Universität Chemnitz, Germany
{firstname.lastname}@informatik.tu-chemnitz.de

Abstract. The increasing service orientation of today's Web applications enables swift reaction on new customer needs by adjusting, extending or replacing parts of the Web application's architecture. While this allows for an agile response to change, it is inappropriate when it comes to security. Security needs to be treated as a first thought throughout the entire lifecycle of a Web application. The recently proposed WAMplus approach does not only offer an expressive, extensible and easy-to-use way to model a Web application architecture, but also puts a strong emphasis on the security. In this paper we present an exemplary implementation of WAMplus using the Sociddea WebID identity management system known from prior work. There, we show how WebID is used to identify, describe and authenticate Web applications and services while taking their protection through WAC and fine-grained data filters into account.

Keywords: Modeling, Security, Identity, Protection, Linked Data, WebID.

1 Introduction

Modern Web applications are facing changing costumer needs, short time to market, and an increasing degree of distribution. The service-oriented architecture (SOA) design pattern supports developing such Web applications by providing a set of best practices for organizing distributed capabilities. For responding to change, agile methodology fits well in this context. When it comes to security, however, it is inappropriate to apply this approach without sufficient consideration. Security needs not to be treated as an afterthought, but as a first thought throughout the entire lifecycle of a Web application [5]. While there are various tools which support developers in modeling and building Web applications, they do not holistically address the security of the entire Web application architecture [6].

In a recent work, **(author?)** proposed the WAMplus approach [8]. It does not only offer an expressive, extensible and easy-to-use way to model a Web application architecture, but also puts a strong emphasis on the security.

This work describes a prototypical implementation of the WAMplus approach into an existing WebID identity provider and management system. By exemplarily integrating the approach into Sociddea (http://www.sociddea.com/), we show its applicability, demonstrate its use in practice and strive to increase its adoption.

S. Casteleyn, G. Rossi, and M. Winckler (Eds.): ICWE 2014, LNCS 8541, pp. 540–544, 2014.

The rest of the paper is organized as follows: Sect. 2 discusses related work. Sect. 3 demonstrates the WAMplus approach. Sect. 4 concludes the paper.

2   Related Work

Model-driven Web engineering (MDWE) approaches like OOWS, UWE or WebML aim at providing means for a systematic and efficient engineering of Web applications [3]. Yet, the variety of existing domain-specific and often proprietary model description languages reduces interoperability. An integral and widely adopted solution for addressing the security topic in MDWE is missing. Dealing with the interoperability concern, semantic vocabularies for describing Web services in RDF typically consider only one type of service aspect: SAWSDL or OWL-S for SOAP-based services; SA-REST or ROSM for RESTful services [2]. OpenID or Facebook Connect are widely adopted identity management systems [1], but their limited extensibility makes them inappropriate to directly identify and attach data to Web applications and services. The Access Control Ontology (ACO) semantically specifies role-based protection of URI-identifiable resources via RDF-based access control lists. Compared to WAC, it is not yet widely used [7].

3   Utilizing WAMplus to Model Secure Web Applications

WAMplus enriches the WebComposition Architecture Model (WAM), known from prior work [4], with 1) semantic descriptions, 2) universal identification, and 3) protection through fine-grained access control for Web application and services. The semantic description of Web services allows for dynamic adaption to interface changes or feature updates. The identification facilitates interconnecting components and establishing authentication through WebID. WAMplus suggests to rely on WAC and customized views for protecting resources, also including descriptions of Web applications and services, at different granularity levels [8].

For the *description* of SOAP-based and RESTful Web services and to maintain interoperability, we use the WSDL RDF mapping (http://www.w3.org/TR/wsdl20-rdf/) combined with WebID. Being an identity concept, WebID (http://www.w3.org/2005/Incubator/webid/spec/) enables *identification* and authentication, and facilitates more detailed description. It consists of three artifacts: The *WebID URI* refers to a subject, like a Web service, and links to a *WebID profile* storing the subject's identity data and public keys using Linked Data. WebID profiles rely on RDF to semantically describe a subject's attributes. The *WebID certificate* is an X.509 certificate that includes a WebID URI identifying the subject. Matching the public key in both WebID certificate and profile enables subject authentication.

To protect resources and descriptions, the WebAccessControl (WAC) (http://www.w3.org/wiki/WebAccessControl) is a RDF-based vocabulary that defines access rules for URI-addressable resource. The WebID Profile Filter Language enables specifying fine-grained filters to protect data *within* user profiles [7].

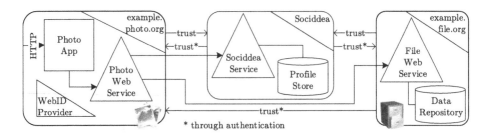

Fig. 1. WAM example

Combining these technologies facilitates a security-focused modeling of Web applications and services that also takes interoperability into account.

To demonstrate WAMplus in practice, we prototypically implemented it in the Sociddea WebID identity provider proposed by **(author?)** in [7]. Yet, WAMplus is not limited to this system. Fig. 1 illustrates an example Web application modeled with WAM. There, a file Web service contains data a photo Web service intents to use. Sociddea is used here to publish Web service descriptions as WebID profiles.

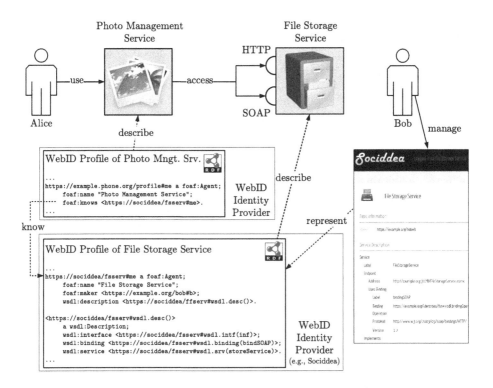

Fig. 2. Usage scenario of photo management system showing WAMplus in practice

As an example, Alice wants to use this photo service to create a new album, as illustrated by Fig. 2. Her photos are stored on the file service, which is operated by Bob. When deploying the file service, Bob also published the service description as a WebID profile using Sociddea. He uses the provided graphical user interface to create a new profile and to add the WSDL RDF description. All following steps are executed during the run-time. Requesting the description of the file service through the photo service invokes the WebID authentication process of Sociddea. In the Sociddea system, every request to a resource has to pass the WAC-type access control rules. Assuming that WAC only allows authenticated users to access profile resources, the photo service has to authenticate with its own WebID. Having access to the profile, its WSDL RDF description provides all necessary information for addressing and binding the service to interact with it. Consistent linking between the resources is established by the use of predicates like *foaf:maker* or *foaf:knows*. If the photo service encounters a problem, Alice or other services could retrieve the operator's contact data by following the WebID URI to Bob's profile being linked to in the WebID profile of the file Web service.

Demonstration. For a live demo and further information about Sociddea and WAMplus visit: `http://vsr.informatik.tu-chemnitz.de/demo/sociddea/`

4 Conclusion

Combining semantic description, universal identification, and access control with WAM's modeling capabilities, WAMplus contributes to designing and managing secure distributed Web applications and services. Our approach assists Web engineers in modeling SOA-based Web applications by creating machine-readable big pictures of their architecture, by using WebID to identify and describe Web services with WSDL+RDF, and by protecting resources with WAC and fine-grained filters. In future work we intend to apply the WAMplus approach in more scenarios to discover patterns relevant to the evolution of Web applications. There we will research the topic of dynamic service replacement and delegation.

References

1. El Maliki, T., et al.: A Survey Of User-centric Identity Management Technologies. In: The International Conference on Emerging Security Information, Systems, and Technologies, SecureWare 2007, pp. 12–17. IEEE (2007)
2. Kim, C.S., et al.: Building semantic ontologies for RESTful web services. In: CISIM, pp. 383–386 (2010)
3. Koch, N., et al.: Model-driven Web Engineering. Upgrade 9(2), 40–45 (2008)
4. Meinecke, J., et al.: Enabling Architecture Changes in Distributed Web-Applications. In: Web Conference, LA-WEB 2007, pp. 92–99. IEEE (2007)
5. Papazoglou, M.P., et al.: Service-Oriented Computing: State of the Art and Research Challenges. IEEE Computer 40(11), 38–45 (2007)

6. Saleem, M.Q., et al.: Model Driven Security Frameworks for Addressing Security Problems of Service Oriented Architecture. In: ITSim, vol. 3, pp. 1341–1346. IEEE (2010)
7. Wild, S., Chudnovskyy, O., Heil, S., Gaedke, M.: Protecting User Profile Data in WebID-Based Social Networks Through Fine-Grained Filtering. In: Sheng, Q.Z., Kjeldskov, J. (eds.) ICWE Workshops 2013. LNCS, vol. 8295, pp. 269–280. Springer, Heidelberg (2013)
8. Wild, S., Gaedke, M.: Utilizing Architecture Models for Secure Distributed Web Applications and Services. Information Technology Special Issue on Architecture of Web Applications (Accepted Journal Paper, Published Q2/2014)

WaPPU: Usability-Based A/B Testing

Maximilian Speicher[1,2], Andreas Both[2], and Martin Gaedke[1]

[1] Chemnitz University of Technology, 09111 Chemnitz, Germany
{maximilian.speicher@s2013,martin.gaedke@informatik}.tu-chemnitz.de
[2] R&D, Unister GmbH, 04109 Leipzig, Germany
{maximilian.speicher,andreas.both}@unister.de

Abstract. Popular split testing approaches to interface optimization mostly do not give insight into users' behavior. Thus, a new concept is required that leverages usability as a target metric for split tests. *WaPPU* is a tool for realizing this concept. It learns models using a minimal questionnaire and can then predict usability quantitatively based on users' interactions. The tool has been used for evaluating a real-world interface. Results underpin the effectiveness and feasibility of our approach.

Keywords: Usability, Metrics, Interaction Tracking.

1 Introduction

Usability is an utterly important factor for successful interfaces. Yet, in today's e-commerce industry, effective methods for determining usability are applied only at very slow iteration cycles (i.e., mainly before a website or redesign goes live). This is because such established usability evaluation methods are costly and time-consuming from a company's point of view. User testing and expert inspections are two of the most prominent examples (cf. [3,6]). Contrary, interfaces are mostly optimized based on *conversions* and more efficient split tests (cf. *Visual Website Optimizer*[1]) during operation. A conversion is a predefined action completed by the user, e.g., a submitted registration form. In a common split test, the interface version which generated the most conversions is considered best. However, this gives no insights into the user's actual behavior.[2] In fact, suboptimal interfaces can lead to accidentally triggered conversions, which is contrary to usability.

Based on the above, Speicher et al. [7] have pointed out the need for *usability as a target metric* in split tests. This would enable a trade-off between traditional methods and split testing. That is, a usability-based approach to split tests would provide insights into user behavior while leveraging the efficiency and affordability of split testing. As a solution, Speicher et al. [7] propose a novel concept called *Usability-based Split Testing* that must meet three requirements: **(R1)** It is more effective in measuring usability than conversions; **(R2)** Efforts for users and developers are kept to a minimum; and **(R3)** It delivers precise/easy-to-understand usability metrics that enable quantitative comparison of interfaces.

---

[1] http://visualwebsiteoptimizer.com/ (2014-03-07).
[2] http://www.nngroup.com/articles/putting-ab-testing-in-its-place/ (2014-03-16)

S. Casteleyn, G. Rossi, and M. Winckler (Eds.): ICWE 2014, LNCS 8541, pp. 545–549, 2014.
© Springer International Publishing Switzerland 2014

In this demo paper, we present *WaPPU* (*"Was that Page Pleasant to Use?"*), a tool that has been developed for realizing the above concept. Our tool enables developers to perform usability-based A/B tests and carry out metric-based comparisons of web interfaces. For this, WaPPU tracks user interactions I and, based on existing machine learning classifiers, trains models M_i for predicting aspects of usability U_i from these: $M_i(I) = U_i$. The tool offers a dedicated service for creating and monitoring A/B testing projects in real-time with minimal effort.

2   Related Work

WaPPU is related to a variety of existing research and tools concerned with usability evaluation. The following gives a representative overview.

AttrakDiff[3] is an existing A/B testing tool for optimizing e-commerce products. It measures two dimensions of user experience based on a specific instrument. Despite superficial similarities, the tool is conceptually different from WaPPU. Particularly, it does not leverage user interactions for predicting usability metrics based on models.

W3Touch [5] and *m-pathy*[4] are two tools that leverage user interactions for evaluating web interfaces. The first engages interactions to automatically adapt interfaces for touch devices. The latter collects interactions for manual interpretation by an evaluator. In analogy, tools like *WUP* [1] allow to track interactions on a website and compare them to predefined, optimal logs. Yet, all of these do not focus on usability as a quantitative target metric.

Nebeling et al. [4] derive quantitative metrics from static layout properties of a web interface, such as the amount of too small text. The specific aim of their approach is to support the adaptation of interfaces for large-screen contexts. Although they take a step into the direction of usability metrics, they do not leverage dynamic interactions. Contrary to WaPPU, their concept is based on well-grounded *a priori* assumptions rather than learning indicators of good/bad usability from actual user behavior.

3   WaPPU

Our tool is realized as a service with a central repository, which enables developers to create and monitor A/B testing projects. First, an A/B testing project is given a name and password for access to real-time analyses and visualizations. Subsequently, the developer chooses which interaction features shall be tracked for which components (defined by jQuery selectors) of their interface.

Listing 1.1. Exemplary WaPPU configuration.

```
1  WaPPU.start({
2      projectId: 42, interfaceVersion: 'A', provideQuestionnaire: true },{
3      '#nav': ['hovers', 'hoverTime'],
4      '#content': ['cursorSpeed', 'cursorTrail'] });
```

---

[3] http://attrakdiff.de/ (2014-03-07).

[4] http://www.m-pathy.com/cms/ (2014-03-07).

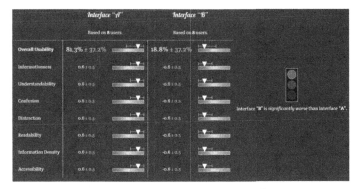

Fig. 1. The WaPPU dashboard

It is possible to choose from a range of 27 predefined interaction features including *clicks, length of the cursor trail, amount of scrolling* etc. The developer is then provided with two automatically generated JavaScript snippets to be added to the two interfaces-under-test of the A/B testing project (Listing 1.1). The above snippet defines an interface-under-test associated with the split testing project with project ID 42 and specifies that a usability questionnaire is shown to users of the interface. The *amount of hovers* and *hover time* are tracked for the component #nav. Moreover, the *length of the cursor trail* as well as the *cursor speed* are tracked for the component #content.

By default, the first interface is configured to show a dedicated usability questionnaire *(R1)* with seven items based on [8] (Fig. 1) while the second interface does not. From users' answers to this questionnaire and client-side interactions, WaPPU incrementally learns models (e.g., *Naïve Bayes*[5]) on a per-item and per-context[6] basis for the A/B testing project. These are stored in the central repository. The models are then used to predict the usability of the second interface based on user interactions only *(R1)*. In this way, the amount of questionnaires shown to users is kept to a minimum *(R2)*. Yet, WaPPU can also be configured to show the questionnaire on both (for remote asynchronous user testing) or none of the interfaces-under-test (for use with existing models/heuristics). As soon as the configuration snippets have been integrated into the interfaces *(R2)*, analyses are available via the WaPPU dashboard (Fig. 1). On the dashboard, developers are presented with users' evaluations of seven usability items for the two interfaces. Depending on whether an interface shows a questionnaire, these evaluations are either based on users' answers or on predictions by the learned models for each item. The items' value range lies between -1 and $+1$. The dashboard provides the average rating across all users of the interface along with standard deviations. From the seven individual ratings, WaPPU computes an

---

[5] http://weka.sourceforge.net/doc.dev/weka/classifiers/bayes/
NaiveBayesUpdateable.html (2014-03-16).

[6] Context is determined based on screen size and whether an ad blocker is activated. These factors influence the appearance of an interface and thus also users' interactions.

overall usability rating between 0% and 100% *(R3)*. Moreover, the tool automatically performs a *Mann–Whitney U test* to determine whether there is a statistically significant difference between the interfaces w.r.t. overall usability. The result of this test is indicated by a traffic light *(R3)*.

4   Example Study

WaPPU was used to evaluate two versions of a real-world search engine results page in a user study with 81 participants [7]. The old version of the interface had been redesigned by three usability experts. Our tool was able to detect the predicted difference in usability with statistical significance for the largest and most representative user group (screen=*HD*, new user). We were able to train reasonably good models with F-measures between 0.518 (item *distraction*) and 0.296 (item *readability*). Finally, heuristics for search engine results pages were derived based on these models. These state that, e.g., less *confusion* is indicated by a lower scrolling distance from top (Pearson's r=-0.44); and better *readability* is indicated by a lower length of text selections (r=-0.39).

5   Conclusion

We have presented *WaPPU*, a tool for realizing the novel concept of Usability-based Split Testing [7]. Our tool enables developers to set-up A/B testing projects with minimal effort and instantly provides them with quantitative usability metrics. These metrics can be derived from user interactions based on previously learned models. A user study with a real-world interface underpins the feasibility of WaPPU. Potential future work includes integration with existing approaches such as *Web-Composition* [2] for automatic optimization of widget-based interfaces.

Acknowledgments. This work has been supported by the ESF and the Free State of Saxony.

References

1. Carta, T., Paternò, F., de Santana, V.F.: Web Usability Probe: A Tool for Supporting Remote Usability Evaluation of Web Sites. In: Campos, P., Graham, N., Jorge, J., Nunes, N., Palanque, P., Winckler, M. (eds.) INTERACT 2011, Part IV. LNCS, vol. 6949, pp. 349–357. Springer, Heidelberg (2011)
2. Gaedke, M., Gräf, G.: Development and Evolution of Web-Applications using the WebComposition Process Model. In: WWW9-WebE Workshop (2000)
3. Insfran, E., Fernandez, A.: A systematic review of usability evaluation in web development. In: Hartmann, S., Zhou, X., Kirchberg, M. (eds.) WISE 2008. LNCS, vol. 5176, pp. 81–91. Springer, Heidelberg (2008)
4. Nebeling, M., Matulic, F., Norrie, M.C.: Metrics for the Evaluation of News Site Content Layout in Large-Screen Contexts. In: Proc. CHI (2011)

5. Nebeling, M., Speicher, M., Norrie, M.C.: W3Touch: Metrics-based Web Page Adaptation for Touch. In: Proc. CHI (2013)
6. Nielsen, J., Molich, R.: Heuristic Evaluation of User Interfaces. In: Proc. CHI (1990)
7. Speicher, M., Both, A., Gaedke, M.: Ensuring web interface quality through usability-based split testing. In: Casteleyn, S., Rossi, G., Winckler, M. (eds.) ICWE 2014. LNCS, vol. 8541, pp. 89–106. Springer, Heidelberg (2014)
8. Speicher, M., Both, A., Gaedke, M.: Towards Metric-based Usability Evaluation of Online Web Interfaces. In: Mensch & Computer Workshopband (2013)

Webification of Software Development: User Feedback for Developer's Modeling

Eduard Kuric and Mária Bieliková

Faculty of Informatics and Information Technologies,
Slovak University of Technology, Ilkovičova 2, 842 16 Bratislava 4, Slovakia
name.surname@stuba.sk

Abstract. In this paper we present an approach to leveraging experience from rapidly evolving field of information processing on the Web for software development. We consider a web of software artifacts (components) as an information space. Supporting any task in such environment of interconnected artifacts depends on our knowledge on user preferences and his characteristics. We envision the concept of collaborative software development to improve software quality and development efficiency by using both implicit and explicit user (software developer) feedback. It opens a space for using approaches originally devised for the Web. The core of our approach is based on our developed platform for independent code monitoring where we create a dataset of developers' implicit and explicit feedback based on monitoring developers' behavior. Employing this platform we acquire, generate and process descriptive metadata that indirectly refer source code artifacts, project documentations and developers activities via document models and user models. As an example of our concept we present an approach for estimation of student's expertise in a programming course.

Keywords: webification of software development, implicit/explicit feedback, interaction information, user modeling, monitoring user behavior.

1 Developer's Feedback in a Web of Software Artifacts

Developers often use a web of software artifacts as a giant repository of source code, which can be utilized for solving their software development tasks. Supporting any task in such environment of interconnected artifacts depends on our knowledge on user preferences and his characteristics. Relevance user feedback is typically used for user profiling during long/short-term modeling of user's interests and preferences on the Web. Relevance feedback techniques have been used to retrieve, filter and recommend a variety of items [4]. Our aim is to support software development by using both implicit and explicit user (software developer) feedback, which creates rich interconnections between software artifacts. This includes not only the shift towards the use of web-based resources in software processes, but also and more importantly it opens a space for using approaches originally devised for the Web (as a network of interconnected content) to support the software development process.

S. Casteleyn, G. Rossi, and M. Winckler (Eds.): ICWE 2014, LNCS 8541, pp. 550–553, 2014.
© Springer International Publishing Switzerland 2014

Our work is a part of a research project called PerConIK[1] (Personalized Conveying of Information and Knowledge). We cooperate with a medium size software company. We focus on support of applications development by viewing a software system as a web of information artifacts. Our aim is devising the right metrics to evaluate software artifacts and to identify particular problems and recommending corrective actions. The core of our approach is based on our developed platform for independent code monitoring [1]. We developed several agents that collect and process documentations, source code repositories, developers' activities, etc. We create within the project a dataset of developers' implicit and explicit feedback based on monitoring behavior of developers for the purpose of estimating developers' expertise. For example, we exploit implicit feedback such as searching relevant information on the Web and searching in source code during performing tasks, writing and correcting source code in development environment, and explicit feedback such as peer review (review feedback attached to source code).

To the present, we have focused mainly on modeling developer's expertise in software house environment. It is based on investigation of software artifacts which the developer creates and the way how the artifacts were created. In other words, we take into account the developer's source code contributions, their complexity and how the contributions were created to a software artifact (e.g. copy/paste actions from external resources, such as a web browser); the developer's know-how persistence about a software artifact; and technological know-how - the level of how the developer knows the used libraries, i.e., broadly/effectively. All on daily basis of software development.

In a software company estimation of developers' expertise allows, for example, on the one hand, managers and team leaders to look for specialists with desired abilities, form working teams or compare candidates for certain positions, on the other hand, developers can locate an expert in a particular library or a part of a software system (someone who knows a component or an application interface) [2,5]. It can be also used to support so-called "search-driven development". When a developer reuses a software artifact from an external source he has to trust the work of an external developer who is unknown to him. If a target developer would easily see that a developer with a good level of expertise has participated in writing the software artifact, then the target developer will be more likely to think about reusing.

On the contrary of a software company, where software is created by professionals, in academic environment, students learn how to design and develop software. Moreover, a student produces significantly less data (implicit/explicit feedback). Our goal is to provide a tool that allows a teacher to evaluate student's knowledge and skills (expertise) based on monitoring student's behavior during developing tasks and adaptation of instruments developed for the general approach. As an example of our approach we present an approach for estimation of student's expertise in a programming course. It allows, for example, teacher to adapt and modify his teaching practices.

---

[1] PerConIK: http://perconik.fiit.stuba.sk/

2   Estimation of Student's Programming Expertise

In our approach modeling student's expertise is based on estimation of a degree of student's expertise of a concept in comparing with other investigated students. In other words, if students solve a task focused on acquiring skills of particular concept, e.g., *priority queue*, then by analyzing their resultant source code, interaction data and by using appropriate software metrics, we can estimate levels of students' expertise of the concept. By using the particular students' expertise estimations of concepts we are able to estimate a degree of student's expertise for the whole course and compare the estimations among the investigated students based on the same evaluation criteria.

We experimented with data gathered during bachelor course on *Data structures and algorithms*. During seminars the students solve programming tasks. Each week is focused on training and acquiring skills of a concept such as *stack*, *binary tree, hash table, etc.* The students solve the tasks in a learning system *Peoplia*[2]. Students can select to solve a simpler or more complex task focused on acquiring skills of a concept. Students get points for their successful solutions. In autumn semester 2013/14, 251 students enrolled in the course.

When a student submits a solution of a task to *Peoplia*, its correctness and efficiency (time complexity) is evaluated. The solution is accepted if it is correct and efficiency tests are successful. The student has unlimited number of submission attempts and the solutions are checked by a plagiarism detection system. Estimation of a degree of student's expertise of a concept c based on a student's correct solution l for a programming task t is calculated as follows:

$$Exp_c(s,t,l) = CX(t) * EF(s,l) * \frac{1}{log_2(1 + CT(s,t))}, \tag{1}$$

where $CX(t)$ is complexity of the task t estimated based on a combination of Logical Source Lines of Code (SLOC-L) and McCabe VG complexity metrics. For calculation of SLOC-L we have adopted the definition from the CodeCount[3]. $EF(s,l)$ returns 1.5 if the student's s submitted solution l is effective, otherwise 1. A solution is effective if its execution time is less than or is equal to a median value of all execution times of submitted correct solutions for t (it is based on preliminary experiments). $CT(s,t)$ is a number of submitted solutions by the student s for the task t (the last solution was accepted by Peoplia). The estimation of a degree of student's expertise of the course is calculated as $\sum_i Exp_{c_i}(s,t_i,l_i)$.

We estimated expertise for all students and compared our results to results achieved on exam. 78 out of 251 students were not allowed to take the final exam because they did not achieve the qualification criteria. Both values (estimated expertise and points of the final exam) were normalized into the interval $[0,100]$. To each student a pair (X,Y) is assigned, where $X \in [0,100]$ is a number of points of the final exam and $Y \in [0,100]$ is the estimated student's expertise. Subsequently, the values in each pair were mapped as follows: $A - [92,100]$, $B - [83,91]$, $C - [74,82]$, $D - [65,73]$, $E - [56,64]$, and $FX - [0,55]$.

---

[2] Peoplia: http://www.peoplia.org/

[3] USC CodeCount: http://sunset.usc.edu/research/CODECOUNT/

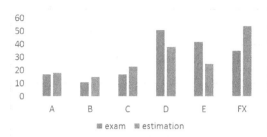

Fig. 1. Comparison of how automatic estimation of students' expertise correlates with exam results

The result of our experiment is illustrated in Figure 1. The number of concordant pairs equals 143 and the number of discordant pairs equals 30. The total number of pairs equals 173. We calculated precision as 0.83.

The idea of "webification" of software development which is based on viewing software repositories as webs is not new. Already Knuth in 1984 presented the idea that "a program is best thought of as a web"[3]. The novel aspect lies in considering not only software artifacts but also users (developers) together with their explicit and implicit feedback, which brings a new view on software and software process metrics. It helps developers be more efficient and can enrich them with the experience and knowledge of their colleagues while managers or senior developers can get advantage of improved planning and decision support via aggregation of statistical data for individual developers.

Acknowledgement. This work was partially supported by the Scientific Grant Agency of the Slovak Republic, grant No. VG1/0675/11 and it is the partial result of the Research & Development Operational Programme for the project PerConIK, ITMS 26240220039, co-funded by the ERDF.

References

1. Bieliková, M., Polášek, I., Barla, M., Kuric, E., Rástočný, K., Tvarožek, J., Lacko, P.: Platform independent software development monitoring: Design of an architecture. In: Geffert, V., Preneel, B., Rovan, B., Štuller, J., Tjoa, A.M. (eds.) SOFSEM 2014. LNCS, vol. 8327, pp. 126–137. Springer, Heidelberg (2014)
2. Fritz, T., Ou, J., Murphy, G.C., Murphy-Hill, E.: A degree-of-knowledge model to capture source code familiarity. In: Proc. of the 32nd Int. Conf. on Softw. Eng., vol. 1, pp. 385–394. ACM, USA (2010)
3. Knuth, D.E.: Literate programming. Comput. J. 27(2), 97–111 (1984)
4. Manning, C.D., Raghavan, P., Schütze, H.: Introduction to Information Retrieval. Cambridge University Press, USA (2008)
5. Minto, S., Murphy, G.C.: Recommending emergent teams. In: Proc. of the 4th Int. Workshop on Mining Softw. Repositories. IEEE Computer Society, USA (2007)

Comparing Methods of Trend Assessment

Radek Malinský and Ivan Jelínek

Department of Computer Science and Engineering, Faculty of Electrical Engineering
Czech Technical University in Prague,
Karlovo náměstí 13, 121 35 Prague, Czech republic
{malinrad,jelinek}@fel.cvut.cz
http://webing.felk.cvut.cz

Abstract. This paper deals with a comparison of selected webometric methods for the evaluation of Internet trends. Each of the selected methods uses a different methodology to the trend assessment: frequency, polarity, source quality. It can be assumed that a combination of individual methods can provide much more accurate results with respect to the desired area of interest. This will lead to improve the quality of search engines on the principle of webometrics and thereby the reduction of irrelevant web search results. The introductory part of the paper explains a concept and basic functional background for all selected webometric methods.

Keywords: Webometrics, Web Mention Analysis, Sentiment Analysis, Social Network Analysis, Trend Assessment.

1 Introduction

With the growing popularity of social networking and blogging, there have been arising a large number of comments on various topics from many different types of users on the web. Some of these comments may be totally unimportant to the other Internet users. On the contrary, other comments might be very important, and do not only for an ordinary user, but also for some commercial companies that want to know a public opinion on price, quality and other factors of their products.

However, web content diversity, variety of technologies and website structure differences, all of these make the web a network of heterogeneous data, where things are difficult to find for common internet users.

Web search engines are the easiest way to find specific information in such diversified network for ordinary users. The search engines are based on complex algorithms that allow search structured but also unstructured data sets and return the most relevant results in a correlation to user-entered query. Webometric methods are often used as supportive assessment methods for search engines algorithms; Web Mention Web Analysis, Sentiment Analysis and Social Network Analysis are among the most widely used webometric methods.

Each of the selected methods uses a different methodology to the trend assessment: frequency, polarity, source quality. It can be assumed that a combination

S. Casteleyn, G. Rossi, and M. Winckler (Eds.): ICWE 2014, LNCS 8541, pp. 554–557, 2014.

of individual methods can provide much more accurate results with respect to the desired area of interest. This will lead to improve the quality of search engines on the principle of webometrics and thereby the reduction of irrelevant web search results.

2   Selected Methods of Trend Assessment

Web Mention Analysis, Sentiment Analysis and Social Network Analysis are among frequently used methods for searching and evaluating of web pages [7]. The selected methods were primarily chosen for their diversity and applicability in various areas of web and social engineering; blogs and social networks are an ideal data source for social science research because it contains vast amount of information from many different users.

Web Mention Analysis [7] is used for the evaluation of the "web impact" of documents or ideas by counting how often they are mentioned online. This idea essentially originates due to a study of an academic research. The researches wanted to know the place and the context which their works occurred in. This approach is applied in a commercial search service Google Scholar [1], which covers not only academic works but also journal articles, patents, etc. Another example of the use of Web Mention Analysis is an identification of how often and in which countries is some product (e.g. camera, book) mentioned online [2].

Sentiment Analysis or Opinion Mining [3], [4] enables us to automatically detect opinions from structured but also unstructured data. The research in this field originated from the demand of commercial companies, who wanted to know public opinion on price, quality and other features of their products. The classification of sentences or whole documents is very often based on the identification of sentiments of individual words or phrases [5]. Several approaches for the purpose have been explored and basically divided into three categories [8]: full-text machine learning, lexicon-based methods and linguistic analysis.

Social Network Analysis is the mapping and measuring of relationships on the web [2], [6]. A centrality is a part of the social network analysis, which is very often used in the web link analysis [6]. The centrality is a single node feature, which explains the node position in a network. It is for example used to determine the most active collaborator in the collaboration scientific networks [2]. There are several methods to measure the centrality of the nodes in a network [6].

3   Methodology of Study

The methods selected for the evaluation of the trends were compared over the data from film industry. Blog posts published in 2012-2013 served as the data source for this research. The five best rated movies which premiered in 2012 were chosen as trends for the assessment. All movies were selected according to IMDb[1]

---

[1] IMDb (Internet Movie Database) - an online database of information related to films, http://www.imdb.com

ranking, which is based on the site visitors rating. The rating is performed by selecting a numeric value from 1 to 10; with 10 being the best.

Because it is very difficult to find a correlation among the methods, the output of each assessment is represented as a list of movies rated from best (1) to worst (5). Table 1 shows the evaluation of trends for each compared method. The names of movies were for all comparing methods used as exact phrase searched with quotes on both sides, e. g. "The Avengers".

Table 1. Comparing Methods of Trend Assessment

Movie	IMDb	SA	WMA	SNA	SNA+SA	Rank
Django Unchained	2	3	5	4	3	4 (12)
Life of Pi	4	5	4	5	5	5 (14)
The Avengers	3	2	3	2	4	2 (7)
The Dark Knight Rises	1	1	1	1	1	1 (3)
The Hobbit: An Unexpected Journey	5	4	2	3	2	3 (9)

For Sentiment Analysis (SA), the surroundings of each searched expression had been recognized and a list of sentences for the trend had been created. All sentences were processed using Lexicon-Based method with SentiWordNet as a lexicon of words. Web Mention Analysis (WMA) is based on counting how often searched words were mentioned in the corpus of blog posts. For Social Network Analysis (SNA), the Degree Centrality was used to determine the most read blog posts and thereby to identify the trend assessment. The combination of methods (SNA+SA) represents the evaluation of trends by using Sentiment Analysis for only the most important blogs that were selected in the previous step through Social Network Analysis.

Values in the column "Rank" were determined by the sum of (SA)+(WMA)+ (SNA); result of the sum is given in parentheses. Trend assessment is represented by the first number; lower sum means better ranking. The combination of methods (SNA+SA) had not been included into the sum because its output does not reflect all the data, but only the selected most important blogs are evaluated. The final ranking in the "Rank" column in comparison with ranking in the "IMDb" column, it represents the rating difference between "common users" and "film fans from IMDb".

The result shows that movie *The Dark Knight Rises* was rated as the best by all methods; it is also correlated with ranking from IMDb. On the contrary, *Django Unchained*, very well ranked on IMDb, it did not gain too much popularity on blogs (Rank is 4). This may be caused primarily by the value of WMA, which shows that this movie was at least mentioned on the web in comparison with other movies. Another important influence of the overall assessment by WMA is observed on the *The Hobbit: An Unexpected Journey*. This movie was the worst rated of the selected set of movies from IMDb. This negative evaluation is also reflected by SA, which shows the high number of negative words. However, the low value of WMA proves that the movie was high interested.

4   Conclusion

We have selected three webometric methods, which are often used as supportive search engines assessment algorithms. Each of the selected methods was used to analyze five trends (movie titles) over a set of blog posts published in 2012-2013. The output of the analysis is by popularity ordered ranking of trends (movies).

The output of each method represents a different view on the evaluation of trends: Web Mention Analysis - emphasizes the frequency of blog posts that mention the trend; Sentiment Analysis - defines the output based on the positive / negative feedback from bloggers; Social Network Analysis – defines the output by quality of blogs that mention the trend. The combination of individual methods can provide much more accurate results with respect to the desired area of interest. In our case the ranking defined by the all three methods in comparison with ranking from IMDb represents the rating difference between "common users" and "film fans from IMDb".

The subject of future work is especially in the finding a correlation among the methods. This means to define criteria for quality assessment of found information, and "distance" among each trend. On this basis, rules for evaluation of semantic content in relation to user's queries can be designed.

Acknowledgments. This research has been supported by the Grant Agency of the Czech Technical University in Prague, grant No. SGS12/149/OHK3/2T/13.

References

1. Gehanno, J. F., Rollin, L., Darmoni, S.: Is the coverage of Google Scholar enough to be used alone for systematic reviews. BMC medical informatics and decision making. Vol 13, No. 1 (2013)
2. Han, S. K., Shin, D., Jung, J. Y., Park, J.: Exploring the relationship between keywords and feed elements in blog post search. World Wide Web. 12:381–398 (2009)
3. Jagtap, V. S., Pawar, K.: Analysis of different approaches to Sentence-Level Sentiment Classification. In International Journal of Scientific Engineering and Technology. Vol. 2, No. 3, 164-170 (2013)
4. Liu, B.: Sentiment analysis and opinion mining. In Synthesis Lectures on Human Language Technologies. Vol. 5, No. 1, 1-167 (2012)
5. Montejo-Ráez, A., Martínez-Cámara, E., Martín-Valdivia, M. T., Urena-López, L. A.: Ranked WordNet Graph for Sentiment Polarity Classification in Twitter. In Computer Speech & Language. Vol. 41, No. 11, 373-381 (2013)
6. Pak, A., Paroubek, P.: Twitter as a corpus for sentiment analysis and opinion mining. Proceedings of the Seventh conference on International Language Resources and Evaluation (LREC'10), Valletta, Malta (2010)
7. Thelwall, M.: Introduction to webometrics: Quantitative web research for the social sciences. San Rafael, CA: Morgan & Claypool (2009)
8. Thelwall, M., Buckley, K., Paltoglou, G.: Sentiment in twitter events. Journal of the American Society for Information Science and Technology. 62:406–418 (2011)

Exploiting Different Bioinformatics Resources for Enhancing Content Recommendations

Abdullah Almuhaimeed* and Maria Fasli

University of Essex
{ansalm,mfasli}@essex.ac.uk

Abstract. To assist the user in his/her quest for information it may be possible to draw and combine information from multiple resources in order to provide more accurate answers/recommendations. Resources can be structured (such as ontologies and taxonomies) or unstructured (corpora). The purpose of this work is to explore how better recommendations can be provided to users by mining and exploiting semantic relations, hidden associations, and overlapping information between various concepts in multiple bioinformatics resources such as ontologies, websites, and corpora. The work also utilizes users' interests to enhance the provided recommendations. A number of techniques will be explored and developed, including ontology mapping, reasoning with multiple resources, and constructing adaptive user profiles.

Keywords: Semantic Techniques, Recommendations & Bioinformatics.

1 Introduction

Given the recent advances in the field of Bioinformatics, a lot more information has become available online. However, searching for such information may not necessarily be easy as resources remain unconnected and current search engine and recommendation systems are not able to combine information that may exist in different resources in order to better understand the user request, enrich it and then use it in order to extract more accurate information to satisfy the users' needs. For instance, Middleton et al. [5] developed a recommender approach which provides recommendations on online academic papers. It uses a single source (i.e. ontology) to enrich a user profile and draws recommendations based on this enrichment. This approach does not take into account the availability of multiple sources of information (ontologies, taxonomies, etc.) to enrich the user profile, and as a result this may decrease the accuracy of the provided recommendations. The motivation of this work is to bridge this gap by using multiple sources to provide the user with more accurate and rich recommendations. We also aim to develop techniques that infer semantic relations and hidden associations from different bioinformatics resources which we subsequently exploit to enhance the precision of the provided recommendations, while we also make use of user profiles to further tailor the recommendations provided.

---

* PhD student, School of Computer Science & Electronic Engineering.

S. Casteleyn, G. Rossi, and M. Winckler (Eds.): ICWE 2014, LNCS 8541, pp. 558–561, 2014.

2   Background

Ding et al. [1] purports that ontologies are a fundamental concept in the Semantic Web, used to represent researcher perspectives about a domain in a conceptualised manner. A number of formal languages have been developed to represent ontologies such as Resource Description Framework (RDF), Resource Description Framework Schema (RDFs) [6], and Web Ontology Language (OWL) [3]. Reasoning with multiple resources is an increasing need due to the large amount of resources contained on the Web. There are several challenges that need to be overcome in performing reasoning through different resources. These resources may have different structures and may not necessarily be compatible, which may lead to inconsistencies during the reasoning process [2]. In addition to exploiting semantic relations between bioinformatics resources, user profiles represent an important source of information providing recommendations to users. User profiles can be constructed based on different methods. The data used to construct user profiles can be classified into two types: static and dynamic data. Static which does not change very frequently, such as names. Dynamic information represents user preferences that can be collected explicitly or implicitly [4].

3   The Recommender Approach

This research contributes to developing semantic-based methods for identifying relations and hidden associations extracted from bioinformatics resources (e.g. ontologies such as Protein Ontology (PO)[1], Gene Ontology (GO)[2], Open Directory Project (ODP)[3] and Bioinformatics Links Directory (BLD)[4] and corpora such as Wikipedia). By studying such resources, we have concluded that there is implicit information that can be extracted through semantic analysis and our central hypothesis is that this can be used in providing better recommendations to users. In addition, we want to tailor-make the recommendations to user needs based on their profiles which will be automatically constructed by collecting user preferences and interests implicitly. We aim to demonstrate our methods through providing recommendations to bioinformaticians on the most relevant content (i.e. articles) from the bioinformatics corpus BMC[5].

Our work consists of two branches: (i) developing methods for extracting semantic information from multiple resources (such as ontologies, taxonomies, Wikipedia) and reasoning with this information to obtain new relations; (ii) constructing an ontological user profile based on information extracted implicitly from the user surfed sessions as well as interaction with the system. The information from (i) and (ii) is then combined to enrich the user query and provide more accurate recommendations to users.

---

[1] http://pir.georgetown.edu/pro/
[2] http://geneontology.org/
[3] http://www.dmoz.org/
[4] http://bioinformatics.ca/links_directory/
[5] http://code.google.com/p/bmc-bioinformatics-processed-corpus/

We first aim to develop a reasoning method to exploit overlapping
information between different bioinformatics resources such PO[1],
GO[2], etc., and to extract semantic
relations and hidden associations between different classes. This method
uses SPARQL[6] queries to extract information, and provides them to the
reasoner which combines them with
semantic rules to infer new relations
that may exist among resources. As
a result, a semantic network is created which represents the extracted
semantic relations and hidden associations from the intersection between
different resources. This includes new
identified relations, not found in the

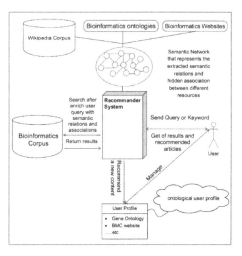

Fig. 1. Recommender System Structure

original resources. Users' profiles will then be boosted by adding the relevant
information from this network.

We suppose that such enrichment contributes to enhancing the precision and
accuracy of the returned results and recommendations. There are several challenges that need to be overcome in this branch of work in that the resources'
structure may contain inconsistencies, resources may get updated or further developed and hence the system needs to take this into account, and exploiting
semantic relations to enhance recommendations may not be straightforward. The
contribution of our approach will be in its ability to handle different resources
with various structures such as ontologies and corpora and employing extracted
semantic relations, such as siblings relations, between multiple resources to
enhance the accuracy of the articles' recommendations.

Among the requirements to reach our goal of providing a user with recommendations that are drawn from multiple resources is constructing an adaptive
ontological profile based on a bioinformatics ontology (i.e. ODP[3] bioinformatics
branch). This profile will be equipped with mechanisms to perform the main
tasks (i.e. add, update, and delete). All of the aforementioned tasks will be
handled automatically. An adaptive user profile needs to be created and examined in a manner that reflects the user's interests. This profile should be able
to accommodate the frequent changes of the user preferences, the enrichment
with valuable information that is gained from the semantic network and providing fully automated solutions. This approach will be fully automated and
tailor recommendations to each user individually based on his/her preferred
topics.

---

[6] http://www.w3.org/TR/rdf-sparql-query/query

4 Bioinformatics Recommender Pipeline

For assessing our recommendation approach, a system pipeline (Figure 1) was put together as follows: a user is interested in identifying articles of interest and will enter a bioinformatics concept (query) to be searched in the BMC[5] corpus. We use the Lucene[7] search engine to index and retrieve data from the corpus. A method has been created to automatically collect user's interests from surfed sessions and we subsequently calculate their similarity with the ODP[3] ontology to construct ontological user profiles. The semantic network will enrich the user's query with valuable information that is acquired from multiple resources. If the same class has been found in more than one resource, our approach can exploit any associated extra information but avoids duplication and repetition. The results (i.e. articles) are re-ranked and returned based on the highest degree of similarity to the user's preferences and the enrichment of the query that is provided by the semantic network. In this pipeline, we target the recommender system, semantic network and user profile. In addition, each user is presented with a hyperlink which contains some recommendations related to the entered query as well as the top five interests stored in the user profile.

5 Conclusion

This research aims to provide a set of new methods to improve recommendations. This includes a personalised recommender service, a mechanism for reasoning through multiple resources and extracting semantic relations and hidden associations. This service should include modelling and learning automatic adaptive user profiles, a method for representing semantic relations and hidden associations, and a method for filtering user interests. Finally, we have constructed the main blocks of our recommender approach to assess to what extent our assumption assists in enhancing the precision of the provided recommendations. We are currently preparing to run an experiment with the aid of bioinformaticians.

References

1. Ding, L., Kolari, P., Ding, Z., Avancha, S.: Using ontologies in the semantic web: A survey. In: Ontologies, pp. 79–113. Springer (2007)
2. Huang, Z., Van Harmelen, F.: Using semantic distances for reasoning with inconsistent ontologies. In: Sheth, A.P., Staab, S., Dean, M., Paolucci, M., Maynard, D., Finin, T., Thirunarayan, K. (eds.) ISWC 2008. LNCS, vol. 5318, pp. 178–194. Springer, Heidelberg (2008)
3. McGuinness, D.L., Van Harmelen, F., et al.: OWL web ontology language overview. W3C Recommendation 10(2004-03) (2004)
4. Mezghani, M., Zayani, C.A., Amous, I., Gargouri, F.: A user profile modelling using social annotations: a survey. In: WWW 2012, pp. 969–976 (2012)
5. Middleton, S.E., Shadbolt, N.R., De Roure, D.C.: Ontological user profiling in recommender systems. ACM Transactions on Information Systems 22(1), 54–88 (2004)
6. Miller, E.: An introduction to the resource description framework. Bulletin of the American Society for Information Science and Technology (1998)

---

[7] https://lucene.apache.org/core/

Methodologies for the Development of Crowd and Social-Based Applications*

Andrea Mauri

Dipartimento di Elettronica, Informazione e Bioingegneria (DEIB)
Politecnico di Milano. Piazza Leonardo da Vinci, 32. 20133 Milano, Italy
andrea.mauri@polimi.it

Abstract. Even though search systems are very efficient in retrieving world-wide information, they cannot capture some peculiar aspects of user needs, such as subjective opinions, or information that require local or domain specific expertise. In these scenarios the knowledge of an expert or a friend's advice can be more useful than any information retrieved by a search system. This way of exploiting human knowledge for information seeking and computational task is called Crowdsourcing. The main objective of this work is to develop methodologies for the creation of applications based on Crowdsourcing and social interaction. The outcome will be a framework based on model-driven approach that will allow end user to develop their own application with a fraction of the effort required by the traditional approaches. It will guarantee a strong control of the execution of the crowdsourcing task by mean of a declarative specification of objectives and quality measures. A prototype will be developed that will allow the creation and execution of task on various platforms. Validation of the approach will consist of quantitative and qualitative analysis of results and performance of the system upon some sample scenarios, where real users from social networks and crowdsourcing platforms will be involved.

1 Introduction

Crowd-based applications are becoming more and more widespread [4]; their common aspect is that they deal with solving a problem by involving a vast set of performers, who are typically extracted from a wide population (the "crowd"). In a typical crowdsourcing scenario is presente a *requester*, who is the one who want to solve a problem with a crowdsourcing campaign (a set of one or more tasks created to fulfill an objective), a set of *responders*, which are the people who perform some tasks and provide answers, and finally the *system* that organize the tasks and collect the results.

The main objective of the requester is to solve his problem while making the best use of responder's availability and reliability so as to get the best possible result for his campaign. For instance he wants to maximize the quality while minimizing the cost and time.

---

* This research is developed under the supervision of Professor Marco Brambilla.

S. Casteleyn, G. Rossi, and M. Winckler (Eds.): ICWE 2014, LNCS 8541, pp. 562–566, 2014.
© Springer International Publishing Switzerland 2014

Crowdsourcing is applied to different fields and within various communities, including information retrieval, databases [5], artificial intelligence and social sciences.

Moreover different platforms can be used to perform this activity, ranging from the classical crowdsourcing marketplaces (e.g Amazon Mechanical Turk), to question answering systems (e.g Quora[1]) and generic social networks (e.g Facebook).

Large crowds may take part to human computation for a variety of motivations, which include non-monetary ones, such as public recognition, fun, or the genuine wish of contributing their knowledge to a social process.

In this situation different issues arise: each type of scenario is characterized by a peculiar set of needs and requirements, that need to be mapped to the particular platform,that usually is not flexible, as it do not support a high-level, fine-tuned control upon posting and retracting tasks.

For instance if the requester wants to post and control a crowdsourcing task on Amazon Mechanical Turk he has to code the implementation with imperative and low-level programming language or using a framework like Turkit [10]. If he wants to exploit the relations between people, he may want to use a social network as crowdsourcing platform. In this case the requester has to directly use the API provided by the social network.

The objective of this work is to develop methodologies for creating applications that leverage the knowledge of the crowd or social communities. The approach developed should be platform agnostic and allow the requester to create his application without having strong technical knowledge.

Thus the research questions that lead this work are: (1)What are the main features of a crowdsourcing campaign ?(2)How can I abstract all these characteristics in a agnostic metamodel? (3) How can this model be used to facilitate the development of a crowd or social based application maximizing its performance?

2   The Approach

In this work I define a top-down approach to application design that adopts an abstract model of crowdsourcing activities in terms of elementary task operations (such as: labeling, liking, sorting, classifying, grouping) performed upon a data set. Starting from operations, strategies are defined for task splitting, replication, and assignment to performers. I also define the data structures which are needed for controlling the planning, execution, and reactive control of crowd-based applications.

The whole process follow a model driven approach, in which the data structures needed for the execution of the application are automatically generated starting from the abstract description of the crowdsourcing task.

Figure 1 shows an example of such process (taken from [3]). On the top there is the model, while on the bottom there are the instances. The image shows how the model-driven process generates the structure needed for the execution and control. The **metamodel**(1) describes which operation needs to be performed

---

[1] www.quora.com

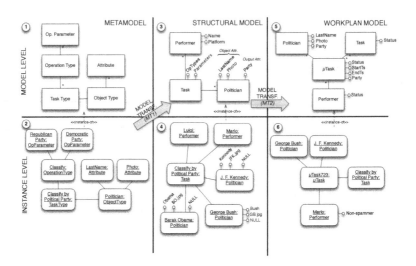

Fig. 1. The model transformations that generate the needed data structure starting from the abstract model

and the schema of the data to be used. The first model transformation generate the **structural model**(3), that contains all the information needed for the execution of the task. Then the second transformation generate the structure needed for the control (the **workplan model**(5)) . This model is composed by the **control mart**, that is analogous to data marts used for data warehousing [6] as its central entity represents the facts, surrounded by three dimension tables, and the **aggregate tables** (Performer, Object and Task), that contain aggregate information.

Control is performed on top of this last model by mean of declarative rules. A rule is composed by an event, condition and action. An event is triggered when a change in the control mart occurs, a condition is a predicate that must be satisfied in order to execute the action, finally, an action is a modification of the control mart and aggregate tables.

Together to the conceptual approach, a concrete prototype was developed[2].

Early results were published in [3].

3   Related Work

Most current approaches rely on an imperative programming models to specify the interaction with crowdsourcing services. For instance Turkit [10] offers a scripting language for programming iterative tasks on Amazon Mechanical Turk. RABJ [8] offers very simple built-in control logic, while complex controls are externalized to client applications written in HTML and Javascript. Another example is Jabberwocky [1], a framework that transparently manages several crowdsourcing platforms and can be procedurally programmed.

---

[2] http://crowdsearcher.search-computing.org/

Other works propose approaches for human computation which are based on high-level abstractions, sometimes of declarative nature. For example in [13] the authors propose a language that interleaves human-computable functions, standard relational operators and algorithmic computation in a declarative way. Qurk [11] exploits a relational data model, SQL to express queries, and a UDF-like approach to specify human tasks. CrowdDB [5] also adopts a declarative approach by extending SQL both as a language for modeling data and to ask queries; human tasks are modeled as crowd operators in query plan, from which it is possible to semi-automatically derive task execution interfaces. Similarly, the DeCo [14] system allows SQL queries to be executed on a crowd-enriched data-source. Finally CrowdLang [12] supports workflow design and execution of tasks involving human and machine activities, it incorporates explicit abstractions for group decision processes and human computation tasks. None of the these works face the problem of specifying the control associated with the execution of human tasks, leaving its management to opaque optimization strategies. Moreover all these systems and researches either deal with a single specific problem (and they focus on a single deployment platform) or they focus on a single aspect of the crowdsourcing campaign.

4   Conclusions and Future Work

The model and the process described in this paper are the starting point of my PhD work. Future steps will consist in extending the conceptual framework in the following way:

- **Crowdsourcing workflow:** In the current model the crowdsourcing campaign is composed by a single task, while complex problems need to be solved by coordinating different tasks. An example of this kind of problem is the creation and modification of content, where is difficult to evaluate and aggregate the outcome of a given task. Previous works face the problem of designing and executing workflow of tasks focusing on a specific domain [2][7] or develop a programmatic approach based on a single execution platform [9]. I aim to start from the existing model of the task extending the existing concepts in order to support task coordination.
- **Higher level of abstraction:** The requester needs to directly configure the task to be performed and has to program himself the control rules (if not provided by the system). Raising the level of abstraction of the model allows to automatically generate the correct strategies and control rules needed to solve the problem.
- **Optimization problem:** There are several possible task configurations suited to solve a particular problem. In this case it's interesting to study the problem of finding the *optimal* configuration. To achieve this, specific metrics are needed in order to evaluate the effectiveness of a selected approach.

In parallel the development of the prototype will be carried on in order to support the evolution of the conceptual model.

References

[1] Ahmad, S., Battle, A., Malkani, Z., Kamvar, S.: The jabberwocky programming environment for structured social computing. In: UIST 2011, pp. 53–64. ACM (2011)

[2] Bernstein, M.S., Little, G., Miller, R.C., Hartmann, B., Ackerman, M.S., Karger, D.R., Crowell, D., Panovich, K.: Soylent: a word processor with a crowd inside. In: Proceedings of the 23rd Annual ACM Symposium on User Interface Software and Technology, UIST 2010, pp. 313–322. ACM, New York (2010)

[3] Bozzon, A., Brambilla, M., Ceri, S., Mauri, A.: Reactive crowdsourcing. In: 22nd World Wide Web Conf., WWW 2013, pp. 153–164 (2013)

[4] Doan, A., Ramakrishnan, R., Halevy, A.Y.: Crowdsourcing systems on the worldwide web. Commun. ACM 54(4), 86–96 (2011)

[5] Franklin, M.J., Kossmann, D., Kraska, T., Ramesh, S., Xin, R.: Crowddb: answering queries with crowdsourcing. In: ACM SIGMOD 2011, pp. 61–72. ACM (2011)

[6] Inmon, W.H.: Building the Data Warehouse. John Wiley & Sons, Inc., New York (1992)

[7] Kittur, A., Smus, B., Khamkar, S., Kraut, R.E.: Crowdforge: Crowdsourcing complex work. In: Proceedings of the 24th Annual ACM Symposium on User Interface Software and Technology, UIST 2011, pp. 43–52. ACM, New York (2011)

[8] Kochhar, S., Mazzocchi, S., Paritosh, P.: The anatomy of a large-scale human computation engine. In: HCOMP 2010, pp. 10–17. ACM (2010)

[9] Kulkarni, A., Can, M., Hartmann, B.: Collaboratively crowdsourcing workflows with turkomatic. In: Proceedings of the ACM 2012 Conference on Computer Supported Cooperative Work, CSCW 2012, pp. 1003–1012. ACM, New York (2012)

[10] Little, G., Chilton, L.B., Goldman, M., Miller, R.C.: Turkit: tools for iterative tasks on mechanical turk. In: HCOMP 2009, pp. 29–30. ACM (2009)

[11] Marcus, A., Wu, E., Madden, S., Miller, R.C.: Crowdsourced databases: Query processing with people. In: CIDR 2011, pp. 211–214 (January 2011), www.cidrdb.org

[12] Minder, P., Bernstein, A.: How to translate a book within an hour: towards general purpose programmable human computers with crowdlang. In: WebScience 2012, Evanston, IL, USA, pp. 209–212. ACM (January 2012)

[13] Parameswaran, A.G., Polyzotis, N.: Answering queries using humans, algorithms and databases. In: CIDR 2011, Asilomar, CA, USA, pp. 160–166 (January 2011)

[14] Park, H., Pang, R., Parameswaran, A.G., Garcia-Molina, H., Polyzotis, N., Widom, J.: Deco: A system for declarative crowdsourcing. PVLDB 5(12), 1990–1993 (2012)

Using Semantic Techniques to Improve Service Composition by End Users

Giuseppe Desolda

Dipartimento di Informatica, Università degli Studi di Bari Aldo Moro
via Orabona, 4, 70125 - Bari, Italy
giuseppe.desolda@uniba.it

Abstract. My PhD research focuses on supporting non-technical end users to flexibly integrate, into personal interactive workspaces, heterogeneous services available in the Web, in order to satisfy their own personal needs in various contexts of daily life. End users should be enabled to shape their application at use time, and also supported on identifying data and services of interests. The latter requirement addresses an issue about improving the quality of the retrieved services with respect to the end user's goal. As described in this paper, a direction that I am exploring refers to the use of linked open data as a new source of data to be exploited, in order to better fulfill the end user's desires.

Keywords: Composition Paradigms, End-User Development, Linked Data.

1 Introduction

Nowadays we are facing the increasing amount of data and services available on Internet. This situation, together with the opportunities offered by Web 2.0, stimulates researchers to investigate new ways for effectively allowing laypeople, i.e., end users without expertise in programming (often called end users in the rest of this paper), to access, manipulate, and combine different kinds of resources in order to generate personalized contents and applications. Mashups have been very much investigated in the last years; they are applications assembled by end users through the integration of heterogeneous resources (APIs, databases, spreadsheets, HTML pages, etc.), in order to solve situational needs. Unfortunately, the proposed mashup platforms are not very end-user oriented [1]. Some mashup tools provide graphical user interfaces for combining services, but the adopted visual composition languages are not suitable for end users, who have difficulties in understanding the integration logic (e.g. data flow, parameter coupling) [2]. Moreover, platforms are usually general purpose and not adequate to the needs of specific application domains and specific end users.

The research I'm interested for my PhD aims at empowering end users to create personal interactive workspaces by combining End User Development (EUD) principles with mashup methodologies and techniques. In this way, services available on the Web can be composed to create new contents and applications by using intuitive and easy-to-use composition mechanisms and platforms. The first couple of years of my research, I investigated recent proposals of mashup compositions and identified

S. Casteleyn, G. Rossi, and M. Winckler (Eds.): ICWE 2014, LNCS 8541, pp. 567–570, 2014.

models and techniques derived from the lessons learned on End-User Development; they were used in developing a novel methodology and the prototype of a platform for service and data composition to satisfy end users' needs [3]. A key point is that the platform has to offer mechanisms to be customized to a specific domain in which it has to be used, so that, capitalizing on the knowledge of the people working in that domain, it can offer a composition process that is adequate to such people. Thus, the developed platform is based on a meta-design approach and a novel stratification into different design layers. Meta-design permits the involvement of different stakeholders in the design: the first phase (the meta-design phase) is performed by professional developers and consists of designing software environments that allow some stakeholders (including domain experts) to create templates, basic elements, and software environments appropriate for end users in the specific application domain, making possible the domain customization; in the second phase, using such environments, end users are able to compose their Personal Information Space (PIS), in which they integrate and manipulate services of interest. Thus, as explained in details in [3], there are different design layers: at the top layer, professional developers create software environments, services descriptors and visual templates for other stakeholders by using an integrated development environment; at the middle layer, domain experts and professional developers perform participatory design to customize the platform by selecting visual templates and registering and composing services; at the bottom layer, end users can create, use and update their PIS.

2   Using Linked Open Data

User studies recently performed indicated new requirements, such as adding new services, changing their visualization, composing different services. In [4], prototypes developed to satisfy such requirements are presented. Moreover, a need for mashup data quality emerged. From the end users' point of view, data quality also refers to the amount of information provided by the retrieved services and its appropriateness to their needs [5]. Actually, some people using the platform prototype observed that few information were retrieved and with few details, compared to what they expected. This because end users are very diverse and have a great variety of needs. Often, not only a single service does not provide the required information, but even the mashup of more services is not capable to completely satisfy end users, who are increasingly more demanding.

I present here an approach that exploits linked data as a new source to be combined with services: it overcomes the above problem of data quality by adding semantic annotations to the services. Previous proposals tried to annotate services on the basis of manual or semi-automatic approaches [6]. The novelty of this approach is that it exploits automatic annotation of services, as described after the following scenario.

Scenario. Tony, the main persona of this scenario, is an organizer of entertainment events. He is looking for a musician for a local event he is organizing. He has added to his workspace *Grooveshark*, a service that, receiving in input the name of a musical artist, retrieves her albums and songs. Tony wants to know details like age, band, birthplace, etc. To find these details, Tony looks in the platform for some services to be combined with Grooveshark, but he is very disappointed because he does not find anything. How can linked data be useful to solve Tony's problem? The use of linked data enable Tony to query a knowledge base and have support in finding relevant data to be composed with other services.

The term linked data refers to best practices for publishing and linking structured data on the Web. Because they have been adopted by an increasing number of data providers, a global data space containing billions of assertions - the Web of Data – has been created. Linked data will be implemented in our platform by building a meta-layer over the registered services. In this layer, all relevant properties (input, output, etc.) of registered services are annotated by using classes' name of the ontologies adopted in the knowledge base of reference. This approach faces some critical issues: 1) identification of significant service properties to annotate; 2) how to label these properties, manually or automatically; 3) how to use these annotations to 'connect' services with linked data. For 1), being the considered mashups a composition of services oriented to information retrieval, the important properties are input query, output attributes and topic of service. For 2), a solution to automatically annotate services is the following. For each service offered by the platform, the platform administrator produces some significant query examples, which are used by an annotation engine to query these services and to collect many instances. Afterwards, for each service the automatic annotation phase starts: for each instance of each service attribute, the annotation engine queries a knowledge base, e.g. DBpedia, and obtains a set of classes related to the specific attribute instance. An algorithm establishes, for each attribute, the most important classes. The annotation engine annotates each attribute using the relevant classes. The same procedure is applied to the input attribute. This flow is summarized in this pseudo-code.

Annotation algorithm
Input: Set T of Triples $t=(s,A,Q)$, s is a service, A is a set of attributes a for s, and Q is a set of query on s
1:   **for** each $t \in T$ **do**
2:     **create** I as empty set of instances results for queries
3:     **for** each $q \in Q$ **do**
4:       **query** s by using q and collect instances results into I
5:     **end for**
6:     **for** each attribute a of s **do**
7:       **create** L as empty set of label for a
8:       **for** each $i \in I$ **do**
9:         **query** DBpedia by using value of i respect to a and obtain a set C of classes
10:         **put** values of C into L
11:       **end for**
12:       **annotate** a by choosing most important class of L
13:     **end for**
14:   **end for**

Starting from the annotation of service attributes, an algorithm establishes the service topic. The proposed approach will be empowered by investigating recent algorithms and techniques proposed in ontology matching research area [7] and methods of semantic annotation of Web service based on DBPedia [8]. The result is a meta-level to each service that describes attributes in term of classes of an ontology. For example, in the case of our previous scenario, the artist_name attribute of Grooveshark is annotated with the Musical_artist class contained in DBpedia ontology. The most critical point is the third problem: how the system can exploit these annotations?

To cover the lack of information of previous scenario, the platform offers Tony the opportunity to use DBpedia as new data source. When Tony, starting from Grooveshark, opens DBpedia source in the mashup platform to add details, the system retrieves the classes used to annotate the artist_name attribute (Musical_artist in the previous example) and accesses to the DBpedia ontology to retrieve all possible links of the Musical_artist class with other classes or properties like age, instrument, style, etc. Tony can choose some of these links so that he can view in his future searches new artist's details contained in DBpedia.

This approach allows end users to deal linked data as new rich knowledge base connected with each service through an automatically built meta-layer transparent to the end user. However, this solution presents some open questions: how many classes have to be used to annotate services? How the system should choose the correct abstraction level of the class respect to the ontology? If the links of thighs to show to the users are too many, how the system should visualize all these information?

Acknowledgments. This work is partially supported by Italian Ministry of University and Research, grant VINCENTE, and by Ministry of Economic Development (MISE), grant LOGIN.

References

1. Casati, F.: How End-User Development Will Save Composition Technologies from Their Continuing Failures. In: Costabile, M.F., Dittrich, Y., Fischer, G., Piccinno, A. (eds.) IS-EUD 2011. LNCS, vol. 6654, pp. 4–6. Springer, Heidelberg (2011)
2. Namoun, A., Nestler, T., De Angeli, A.: Conceptual and Usability Issues in the Composable Web of Software Services. In: Daniel, F., Facca, F.M. (eds.) ICWE 2010. LNCS, vol. 6385, pp. 396–407. Springer, Heidelberg (2010)
3. Ardito, C., Costabile, M.F., Desolda, G., Lanzilotti, R., Matera, M., Piccinno, A., Picozzi, M.: User-driven visual composition of service-based interactive spaces. Journal of Visual Languages & Computing (in print)
4. Ardito, C., Costabile, M.F., Desolda, G., Lanzillotti, R., Matera, M., Picozzi, M.: Visual Composition of Data Sources by End-Users. In: Proc. of AVI 2014 (in print, 2014)
5. Picozzi, M., Rodolfi, M., Cappiello, C., Matera, M.: Quality-Based Recommendations for Mashup Composition. In: Daniel, F., Facca, F.M. (eds.) ICWE 2010. LNCS, vol. 6385, pp. 360–371. Springer, Heidelberg (2010)
6. Zeshan, F.: Semantic Web Service Composition Approaches: Overview and Limitations. Int. Journ. of New Computer Architectures and their Applications 3(1) (2011)
7. Shvaiko, P., Euzenat, J.: Ontology Matching: State of the Art and Future Challenges. IEEE Transactions on Knowledge and Data Engineering 25(1), 158–176 (2013)
8. Zhen, Z., Shizhan, C., Zhiyong, F.: Semantic Annotation for Web Services Based on DBpedia. In: IEEE Int. Symp. on Service Oriented System Eng., pp. 280–285 (2013)

Social Search

Marc Najork

Microsoft Research
1065 La Avenida, Mountain View, CA 94043, USA

Abstract. "Social Search" refers to two aspects of the integration of web search with social networks: how queries to a search engine may surface (socially) relevant content from social networks, and how signals from social networks may influence the (personalized) ranking of search results. The first part of the talk surveys the integration of Bing with Facebook, Twitter, Quora, Foursquare, LinkedIn, Klout, and other social platforms. The second part focuses on two technical details of this integration: a measure for quantifying the "affinity" between two users of a social network and an efficient algorithm for computing that measure, and a method for efficiently surfacing pages "liked" by your friends from a document-sharded index. The final part discusses limitations of social search, such as skewed demographics and weak homophily.

Keywords: Web search, social networks, social search.

1   Introduction

In this talk, I discuss the integration of web search with social networks. "Social Search" refers to two distinct aspects of this integration: how a user's queries to a search engine may surface content from social networks (possibly authored by the searcher's connections on that network), and how signals from social networks (say, the fact that a friend "liked" a web page) may influence the personalized ranking of algorithmic search results.

The talk is divided into three parts. In the first part, I survey the integration of Microsoft's Bing Search engine with various social platforms, including Facebook, Twitter, Quora, Foursquare, LinkedIn, Klout, and others. Bing surfaces relevant content from multiple social networks, whether it is public or authored by the searcher's connections on each network. It also promotes algorithmic search results that were endorsed by the user's friends, or that are trending social media platforms.

Bing pays particular attention to "people search", queries meant to retrieve relevant information about a person. Celebrity search is a well-studied problem, and retrieval precision is high, but this is less true for non-celebrity people search, due to the ambiguity of common names. Bing uses separation in social networks to surface individuals in the searcher's extended social circle. Furthermore, it allows registered users to claim content related to them, thereby making it possible to cluster people search results by individual. Finally, it shows summaries

S. Casteleyn, G. Rossi, and M. Winckler (Eds.): ICWE 2014, LNCS 8541, pp. 571–572, 2014.

of LinkedIn profiles directly on the results page. For celebrity searches, it will similarly show their presence in social media prominently on the results page together with related postings.

The second portion of the talk focuses on two technical details of Social Search. First, I describe an "affinity" measure [3] for quantifying how robustly connected two nodes in a graph (or two users of a social networks) are, or more precisely, what fraction of the graph's edges can be deleted before the nodes become disconnected. The affinity measure can be efficiently estimated by a randomized, sketch-based algorithm. The off-line phase of that algorithm computes a fixed-size sketch for each node of the graph, capturing a representative of its connected component at various levels of edge deletion. The online phase consists of retrieving the sketches of two nodes and performing a pointwise comparison on them to compute the affinity. The space complexity of the algorithm is $O(n)$, the time complexity of the off-line phase is $O(\alpha(n))$ (the complexity of union-find with path compression), while the time complexity of the online phase is $O(1)$.

Second, I discuss an approach for efficiently retrieving web pages "liked" by a user's friends. While seemingly trivial, it is challenging to integrate this functionality into a document-sharded distributed search index [2]. In such a setting, queries are distributed from a front-end to many index servers (each holding a part of the index), and results are sent back. Because of network constraints, both query and result transfers should be small; in particular, it is neither feasible to send the full set of the searcher's friends down the distribution tree, nor to send the full set of results up the aggregation tree. Moreover, social graphs can be very large and change continuously, making it impractical to maintain a copy of the graph on each index server.

The final part of the talk confesses to some of the limitations of Social Search; namely, that many social networks have skewed demographics [4] (in terms of gender, race, age, education and income), making it dangerous to generalize trends in networks to the overall population; and that while a user's actions on social networks *is* predictive of their proclivities [1], it is not clear that these preferences transfer to their "virtual" friends. Finally, it is challenging to "separate the wheat from the chafe" – to identify salient posts in a sea of the mundane.

References

1. Kosinski, M., Stillwell, D., Graepel, T.: Private traits and attributes are predictable from digital records of human behavior. Proceedings of the National Academy of Sciences, 5802–5805 (2013)
2. Najork, M.A., Panigrahy, R., Shenoy, R.K.: Considering document endorsements when processing queries. US Patent App. 13/218,450 (filed 2011)
3. Panigrahy, R., Najork, M., Xie, Y.: How user behavior is related to social affinity. In: 5th ACM International Conference on Web Search and Data Mining, pp. 713–722. ACM, New York (2012)
4. Rainie, L., Brenner, J., Purcell, K.: Photos and videos as social currency online. Pew Research Center (2012)

Wisdom of Crowds or Wisdom of a Few?

Ricardo Baeza-Yates

Yahoo Labs Barcelona
Barcelona, Spain

Abstract. In this keynote we focus on the concept of wisdom of crowds in the context of the Web, particularly through social media and web search usage. As expected from Zipf's principle of least effort, the wisdom is heterogeneous and biased to active people, which represent at the end the wisdom of a few. We also explore the impact on the wisdom of crowds of dimensions such as bias, privacy, scalability, and spam. We also cover related concepts such as the long tail of the special interests of people, or the digital desert, web content that nobody sees.

Summary

The Web continues to grow and evolve very fast, changing our daily lives. This activity represents the collaborative work of the millions of institutions and people that contribute content to the Web as well as more than two billion people that use it. In this ocean of hyperlinked data there is explicit and implicit information and knowledge. But how is the Web? What are the activities of people? What is the impact of these activities? Web data mining is the main approach to answer these questions. Web data comes in three main flavors: content (text, images, etc.), structure (hyperlinks) and usage (navigation, queries, etc.), implying different techniques such as text, graph or log mining. Each case reflects the wisdom of some group of people that can be used to make the Web better.

The wisdom of crowds [9] at work in the Web is best seen in social media as well as in social networks. It is also implicit in the usage of search engines [1] and other popular web applications. The wisdom behind web users is shaped by different complex factors such as the heterogeneity of user activity [10] and hence a heavy long tail [6]; different types of bias [3] that create problems such as the bubble effect [7]; privacy breaches coming from data [5]; too much data that endangers minorities [2]; or web spam in all possible ways [8].

The diversity of user activity implies that an elite of users represent most of the wisdom and that we should really talk about the wisdom of a few [4]. This diversity also generates a digital desert, web content that no one ever sees.

References

1. Baeza-Yates, R., Ribeiro-Neto, B.: Modern Information Retrieval: The Concepts and Technology Behind Search, 2nd edn. Addison-Wesley (January 2011)
2. Baeza-Yates, R., Maarek, Y.: Usage Data in Web Search: Benefits and Limitations. In: Ailamaki, A., Bowers, S. (eds.) SSDBM 2012. LNCS, vol. 7338, pp. 495–506. Springer, Heidelberg (2012)

S. Casteleyn, G. Rossi, and M. Winckler (Eds.): ICWE 2014, LNCS 8541, pp. 573–574, 2014.
© Springer International Publishing Switzerland 2014

3. Baeza-Yates, R.: Big Data or Right Data? In: AMW 2013, Puebla, Mexico (May 2013)
4. Baeza-Yates, R., Saez-Trumper, D.: Wisdom of the Crowd or Wisdom of a Few? An Analysis of Users' Content Generation (submitted, 2014)
5. Barbaro, M., Zeller. Jr., T.: A face is exposed for AOL searcher no. 4417749. The New York Times (August 9, 2006)
6. Goel, S., Broder, A., Gabrilovich, E., Pang, B.: Anatomy of the long tail: ordinary people with extraordinary tastes. In: Proceedings of the Third ACM International Conference on Web Search and Data Mining, WSDM 2010, New York, NY, USA, pp. 201–210 (2010)
7. Pariser, E.: The Filter Bubble: What the Internet Is Hiding from You. Penguin Press (2011)
8. Spirin, N., Han, J.: Survey on web spam detection: principles and algorithms. ACM SIGKDD Explorations Newsletter Archive 13(2), 50–64 (2011)
9. Surowiecki, J.: The Wisdom of Crowds: Why the Many Are Smarter Than the Few and How Collective Wisdom Shapes Business, Economies, Societies and Nations. Random House (2004)
10. Zipf, G.K.: Human behavior and the principle of least effort. Addison-Wesley Press (1949)

IFML: Building the Front-End of Web and Mobile Applications with OMG's Interaction Flow Modeling Language

Marco Brambilla

Politecnico di Milano. Dipartimento di Elettronica, Informazione e Bioingegneria
Piazza L. Da Vinci 32. I-20133 Milan, Italy
marco.brambilla@polimi.it

Abstract. Front-end design of Web applications is a complex and multidisciplinary task, where many perspectives intersect. A new standard modeling language called IFML (Interaction Flow Modeling Language) addresses this problem in a platform-independent way. IFML grants executability of models and binding to other aspects of system and enterprise design through integration with widespread software modeling languages such as UML, BPMN, SysML, SoaML and the whole MDA suite.

1 The UI Modeling Problem

In the last twenty years, capabilities such as form-based interaction, information browsing, link navigation, multimedia content fruition, and interface personalization have become mainstream and are implemented on top of a variety of technologies and platforms. However, no PIM-level design approach has emerged as a standard in the industry so far. Thus, front-end development continues to be a costly and inefficient process, where manual coding is predominant, reuse is low, and cross-platform portability is limited.

A possible solution to this problem is the Interaction Flow Modeling Language (IFML) [2], a visual notation for platform-independent design of software front end. IFML has been adopted as a standard by the Object Management Group (OMG) and features direct involvement of influential industrial players, seamless integration with widespread modeling languages such as UML, BPMN, SysML and the whole MDA suite, and availability of both open-source editors and industrial-strength implementations supporting end-to-end development.

2 The Interaction Flow Modeling Language (IFML)

The main contributions of IFML is the integration of best practices in the fields of model-driven development, software engineering, and Web engineering. IFML spawns mainly from the WebML [3] and the industrial experience obtained by the tool WebRatio [1]. IFML features user events and system events as first-class citizens, orthogonalization of business logic and interaction logic, and explicit and formal description of interoperability procedures and notations.

S. Casteleyn, G. Rossi, and M. Winckler (Eds.): ICWE 2014, LNCS 8541, pp. 575–576, 2014.

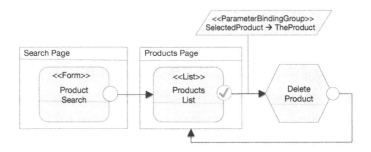

Fig. 1. IFML model example: search and listing of products, with deletion of an item

IFML supports the specification of: (1) *view structure*, i.e., visibility and reachability of view containers; (2) *view content*, , i.e., view components contained within view containers; (3) *user events and system events*; (4) *event transitions*, i.e., the effect of events on the user interface; (5) *parameter binding*, i.e., input-output dependencies between view elements; and (6) *binding to the business logic*, through references to application logic models and data models.

Figure 1 shows a simple example of IFML model where the user can search for products, gets the list of matching items and then can select one; the selection causes the selected product to be deleted and then leads back to the list.

Core Competencies. Besides learning the basics of the IFML syntax, the core competencies of IFML designers shall comprise: multi-perspective modeling, pattern-based design, UX-based design, and real-world industrial experience.

References

1. Acerbis, R., Bongio, A., Brambilla, M., Tisi, M., Ceri, S., Tosetti, E.: Developing ebusiness solutions with a model driven approach: The case of acer emea. In: Baresi, L., Fraternali, P., Houben, G.-J. (eds.) ICWE 2007. LNCS, vol. 4607, pp. 539–544. Springer, Heidelberg (2007)
2. Brambilla, M., Fraternali, P., et al.: The Interaction Flow Modeling Language (IFML), version 1.0 (2014), `http://www.ifml.org`
3. Ceri, S., Brambilla, M., Fraternali, P.: The History of WebML Lessons Learned from 10 Years of Model-Driven Development of Web Applications. In: Borgida, A.T., Chaudhri, V.K., Giorgini, P., Yu, E.S. (eds.) Mylopoulos Festschrift. LNCS, vol. 5600, pp. 273–292. Springer, Heidelberg (2009)

Mashups: A Journey from Concepts and Models to the Quality of Applications

Cinzia Cappiello[1], Florian Daniel[2], and Maristella Matera[1]

[1] Politecnico di Milano
Via Ponzio 34/5, I-20133 Milano, Italy
{cinzia.cappiello,maristella.matera}@polimi.it
[2] University of Trento,
Via Sommarive 9, I-38123, Povo (TN), Italy
daniel@disi.unit.it

Abstract. This tutorial aims to provide insight into the constantly evolving mashup ecosystem. It presents core *definitions*, overviews a representative set of *mashup components* (the resources to be integrated) and *mashup models* (how resources are integrated), illustrates *composition paradigms* supported by state-of-the-art mashup tools, and discusses the *quality* of the resulting mashups from a user perspective. The goal of the tutorial is to introduce the topic and show its applicability, benefits and limitations.

1   Context and Motivation

The term "mashup" is widely used today. There are people developing "mobile mashups," others doing research on "Web mashups," and others again selling tools for "data mashups." Yet, when it comes to a concrete discussion of the topic, it is not uncommon to discover that the parties involved in the discussion actually have very different interpretations of what mashups are and what they are not. Typical discussion points are whether a mashup must have a user interface (UI) or not to be called a "mashup", whether it must be built by using Web-accessible resources only or not, whether it must be developed with client-side technologies (e.g., JavaScript) only, and the like. That is, even after several years that the term has been around and used, there is still no common agreement on its actual meaning and implications.

Interestingly, however, in the meantime mashing up data, functionalities and user interface widgets sourced from the Web has inexorably percolated into Web Engineering as a tacitly accepted development practice. Today, it is unimaginable to develop modern Web applications without some form of reuse and integration of value-adding, third-party content or services, a task that is greatly facilitated by technologies like Web services [3], the RESTful architectural style [2], Open Data, XML, JSON, W3C widgets, and many more.

But which are the conceptual underpinnings of this practice? What does it exactly mean to "mash up" resources that can be accessed via the Web? Which are the paradigms adopted for the composition of mashups? What kinds of tools

S. Casteleyn, G. Rossi, and M. Winckler (Eds.): ICWE 2014, LNCS 8541, pp. 577–578, 2014.
© Springer International Publishing Switzerland 2014

exist that support this activity? And, eventually, what does it mean to develop "good" mashups? Working with students, discussing with colleagues, reading publications on the topic, we have seen that these questions are still open to many. We also identified a lack of suitable study material.

2   Learning Objectives

In light of these considerations, the learning objectives of this tutorial are:

- To obtain a basic understanding of the *core mashup aspects and concepts*, such as their contexts of use and target users, the most important definitions of mashups depending on the considered point of view (e.g., Web mashups, enterprise mashups, process mashups, telco mashups, mobile mashups, etc.), their intrinsic complexity and benefits.
- To get insight into the *most representative component technologies* used by mashups. Components are the basic elements of a mashup, and the comprehension of their characteristics and capability is fundamental for the understanding of what a mashup is and how it can be developed.
- To understand the *conceptual underpinning of mashups*, which developers must master and that can help them focus on the relevant issues when integrating components into mashups, as well as reference architectural patterns that can be instantiated. Both concepts and architectures can be analyzed independently of the particular technologies or sources used for an actual implementation.
- To gain insight into how mashup models and development practices can materialize into dedicated *mashup tools and composition paradigms* for assisted mashup development.
- To get insight into *quality models* for both mashup components and mashups, to understand how such models can guide the initial choice of components and the successive selection of composition patterns, and how they can also augment composition paradigms through the generation of quality-based recommendations.

The *target audience* are researchers, practitioners, advanced students who want to learn more about mashup development from a perspective that especially privileges abstractions and models, not only implementation aspects.

The tutorial is based on the authors' latest publication on mashups [1] and is complemented with an *online resource* providing additional material, slides and links for further study: http://www.floriandaniel.it/mashupsbook.

References

1. Daniel, F., Matera, M.: Mashups: Concepts, Models and Architectures. Springer (2014)
2. Fielding, R.: Architectural Styles and the Design of Network-based Software Architectures. Ph.D. Dissertation, University of California, Irvine (2007)
3. Papazoglou, M.P.: Web Services - Principles and Technology. Prentice Hall (2008)

Web of Things: Concepts, Technologies and Applications for Connecting Physical Objects to the Web

Iker Larizgoitia, Dominique Guinard, and Vlad Trifa

EVRYTHNG LIMITED, 4th Floor, 45 - 49 Leather Lane,
London EC1N 7TJ, United Kingdom
{iker,dom,vlad}@evrythng.com

Abstract. Inter-communicating devices and integrated Web-based services will open exciting opportunities. The idea of a world where everything is connected is becoming a reality, as hardware and software evolves to provide the necessary infrastructure to connect any physical device and objects to the Web. The goal of this tutorial is to present an overview on the advances in the Web of Things initiative. It will cover the key ideas behind the concept, how it works, infrastructure, as well as present the attendees with ideas and frameworks on how to build applications on top of it. The tutorial will have a practical approach and is open to both academia and industry attendees with interest in technologies and applications to interconnect physical objects to the Web. Knowledge of the architecture of the Web (e.g SOA, REST) and basic understanding of Web technologies (e.g. URIs, HTML, Javascript) are encouraged but not necessary to attend this tutorial.

1 Introduction

Inter-communicating devices and integrated Web-based services will open exciting opportunities. Smart appliances in smarter buildings will be linked in grids that can be more tightly monitored and regulated in real time. New communications protocols and software applications of fundamental importance are showing the way. Over the last several years, we have witnessed two major trends in the world of embedded devices:

- Hardware is becoming smaller, cheaper and more powerful, so that many devices will soon have communication and computation capabilities. Objects will be able to connect, interact and cooperate with other objects in their surrounding environment and with control centres—a vision generally dubbed the Internet of Things (IoT).
- The software industry is moving towards service-oriented integration technologies; especially in the business software domain, complex applications based on combined and collaborative services have been appearing. The Internet of Services (IoS) vision projects such integration on a large scale: services will reside in different layers of an enterprise; for example, in different operational units, IT networks or even run directly on devices and machines within a company.

S. Casteleyn, G. Rossi, and M. Winckler (Eds.): ICWE 2014, LNCS 8541, pp. 579–580, 2014.
© Springer International Publishing Switzerland 2014

As both of these trends are not domain-specific but common to multiple industries, we are facing a trend where the service-based information systems blur the border between the physical and virtual worlds, providing a fertile ground for a new breed of real-world aware applications. To facilitate these connections, research and industry have come up with a number of low-power network protocols such as Zigbee and Bluetooth, BLE and IPv6, in a version optimized for resource-constrained devices called 6lowpan.

Although they are increasingly part of our world, embedded devices still form multiple, small and incompatible islands at the application layer; developing applications to take advantage of them remains a very challenging task that requires expert knowledge of each platform. To ease the task, recent research initiatives have tried to provide uniform interfaces that create a loosely coupled ecosystem of services for smart things; these initiatives are often referred to as: "Web of Things".

2   Web of Things

The Web of Things is an evolution of the Internet of Things where the primary concern has been how to connect objects together at the network layer: similarly to the way the Internet addressed the lower-level connectivity of computers (layers 3-4 of the OSI model), the Internet of Things is primarily focusing on using various technologies such as RFID, Zigbee, Bluetooth or 6LoWPAN.

On the other hand, just like what the Web is to the Internet, the Web of Things regroups research and industrial initiatives looking into building an application layer for physical objects to foster their reusability and integration into innovative 3rd party applications. The envisioned approach is to reuse the already well-accepted and ubiquitous Web standards such as URI, HTTP, HTML5, REST, Web feeds, Javascript etc. Although these technologies were initially created for desktop computers, the fast increase of capabilities of embedded devices makes this possible already today.

3   Tutorial Objectives

In this tutorial we'll review a number of embedded devices and their integration to the Web, reviewing the Web of Things best practices on the way. We'll then dig into Web of Things platforms both on the commercial side, where we will present the EVRYTHNG platform and on the research side, where latest advances in object integration carried out in the COMPOSE platform will be presented. Based on these two platforms, use cases and applications will be showcased to demonstrate how to integrate objects, devices and applications following the Web of Things concepts.

Distributed User Interfaces and Multimodal Interaction

María D. Lozano[1], Jose A. Gallud[1], Víctor M.R. Penichet[1], Ricardo Tesoriero[1],
Jean Vanderdonckt[2], Habib Fardoun[3], and Abdulfattah S. Mashat[3]

[1] Computing Systems Department, University of Castilla-La Mancha, Spain
{maria.lozano,victor.penichet,jose.gallud,
ricardo.tesoriero}@uclm.es
[2] Université Catholique de Louvain, Belgium
jean.vanderdonckt@uclouvain.be
[3] King AbdulAziz University, Saudi Arabia
{hfardoun,asmashat}@kau.edu.sa

1   Objetives

This Workshop is the fourth in a series on Distributed User Interfaces. On this occasion, the workshop is focused on Distributed User Interfaces and Multimodal interaction. The main goal is to join together people working in extending the Web and other user interfaces to allow multiple modes of interaction such as GUI, TUI, Speech, Vision, Pen, Gestures, Haptic interfaces, etc. Multimodal interaction poses the challenge of transforming the way we interact with applications, creating new paradigms for developers and end-users. In a multi-device environment, we can find coupled displays, multi-touch devices, interactive table-tops, tablets, tangible user interfaces, eWatchs, etc., and this diversity of devices offers new possibilities and makes multimodal interaction even more challenging. Through active group discussion, participants will have the chance to share their knowledge and experience to advance in this field.

2   Theme and Topics

The main theme is to discuss and analyze how multimodal interaction can be applied in software applications based on Distributed User Interfaces (DUI). Moreover, interacting with the system through different means provides a richer user experience that it is worth exploring. Some of the topics to tackle are the following:

- Multimodal Interaction within Distributed User Interfaces
- Multimodal Web technologies
- Gesture-based interaction
- Multimodal Web Applications
- Multimodal mobile application
- Tangible user interfaces

S. Casteleyn, G. Rossi, and M. Winckler (Eds.): ICWE 2014, LNCS 8541, pp. 581–582, 2014.

3   Workshop Format

This one-day workshop will include a mix of paper presentations and breakout session activities according to the following scheme. With the goal of promoting debate among the workshop attendees, we will set the role of the "commentator" before the workshop, in such a way that every participant will be assigned the task of commenting the key points of other participant's paper, promoting this way the discussion and participation of everybody.

During the workshop, we will divide it into two clearly differentiated parts. During the morning, participants will present their papers followed by short questions and discussion promoted by the "commentator". Then, breakout sessions will be held during the afternoon. We will set working groups of 4-5 people to discuss about the topics of the workshop. This way we will promote active participation among all attendees to end up the workshop with useful conclusions and final remarks agreed by the attendees.

4   Expected Outcomes

We would like the workshop to be a forum to promote collaborations, generate ideas and synergies. We expect to find answers to the questions that would surely arise during the workshop and define the key guidelines that designers might consider when addressing the design of applications based on Distributed User Interfaces and the different ways of interacting within them. Besides, we also plan to organize a Special Issue in an International Journal with extended versions of a selection of papers, which will help to disseminate the workshop outcomes.

References

1. Calvary, G., Coutaz, J., Thevenin, D., Limbourg, Q., Bouillon, L., Vanderdonckt, J.: A Unifying Reference Framework for Multi-Target UI. Interacting with Computers 15(3)
2. Vandervelpen, C., Vanderhulst, G., Luyten, K., Coninx, K.: Light-weight Distributed Web Interfaces: Preparing the Web for Heterogeneous Environments. In: Lowe, D.G., Gaedke, M. (eds.) ICWE 2005. LNCS, vol. 3579, pp. 197–202. Springer, Heidelberg (2005)
3. Lozano, M.D., Gallud, J.A., Tesoriero, R., Penichet, V.M.R. (eds.): Distributed User Interfaces: Usability and Collaboration. Human–ComputerInteraction Series. Springer-Verlag London (2013) ISBN 978-1-4471-5498-3
4. Gallud, J.A., Tesoriero, R., Penichet, V.M.R. (eds.): Distributed User Interfaces. Designing Interfaces for the Distributed Ecosystem. Human-Computer Interaction Series. Springer-Verlag London (2011) ISBN: 978-1-4471-2270-8
5. Pedro, G., Villanueva, R., Tesoriero, J.A.: Gallud. Distributing web components in a display ecosystem using Proxywork. In: Proceedings of the 27th International BCS Human Computer Interaction Conference, BCS-HCI 2013, Article No. 28. British Computer Society, Swinton (2013)

Author Index